W9-ADC-906

Auto Fundamentals

How and why of the design, construction, and operation of automobiles. Applicable to all makes and models.

Martin W. Stockel
Automotive Writer

Martin T. Stockel
Automotive Writer

Chris Johanson
ASE Certified Master Automobile Technician

Publisher
The Goodheart-Willcox Company, Inc.
Tinley Park, Illinois

Copyright 2000

by

THE GOODHEART-WILLCOX COMPANY, INC.

Previous Editions Copyright 1996, 1990, 1985, 1982, 1978, 1974, 1969, 1963

All rights reserved. No part of this book may be reproduced, stored in a retrieval system, or transmitted in any form or by any means, electronic, mechanical, photocopying, recording, or otherwise, without the prior written permission of The Goodheart-Willcox Company, Inc. Manufactured in the United States of America.

Library of Congress Catalog Card Number 98-53655
International Standard Book Number 1-56637-577-0

2 3 4 5 6 7 8 9 10 00 03 02 01 00

Important Safety Notice

Proper service and repair is important to the safe, reliable operation of motor vehicles. Procedures recommended and described in this book are effective methods of performing service operations. Some require the use of tools specially designed for this purpose and should be used as recommended. Note that this book also contains various safety procedures and cautions, which should be carefully followed to minimize the risk of personal injury or the possibility that improper service methods may damage the engine or render the vehicle unsafe. It is also important to understand that these notices and cautions are not exhaustive. Those performing a given service procedure or using a particular tool must first satisfy themselves that neither their safety nor engine or vehicle safety will be jeopardized by the service method selected.

This book contains the most complete and accurate information that could be obtained from various authoritative sources at the time of publication. Goodheart-Willcox cannot assume any responsibility for any changes, errors, or omissions.

Library of Congress Cataloging-in-Publication Data

Stockel, Martin W.
Auto fundamentals : how and why of the design, construction, and operation of automobiles : applicable to all makes of and models / by Martin W. Stockel, Martin T. Stockel, Chris Johanson.
p. cm.
Includes index
ISBN 1-56637-577-0
1. Automobiles—Design and construction. I. Stockel, Martin T. II. Johanson, Chris. III. Title.
TL240.S83 1999
629.2'31—dc21 98-53655
CIP

Components used in the cover photograph were provided by Jack Loehman, owner of Elmhurst Auto Parts.

Introduction

CIP

Auto Fundamentals provides you with a thorough understanding of the design, construction, and operation of automotive systems. It contains information on the latest developments in the field of automotive technology. The material in this text is easy to understand and is applicable to all vehicles.

This text explains each system by starting with basic theory, then parts are added until the system is complete. By following this procedure, the function of each system is explained and its relationship to the complete vehicle is made clear.

A build-it-yourself approach is used in a number of areas. You will "build" on paper many of the systems being described. This provides a thorough understanding of the basic principles that are necessary to learning the how and why of automotive technology. Areas involving math, physics, chemistry, electricity, magnetism, and hydraulics are covered in the text where they apply.

Many of the illustrations were drawn especially for **Auto Fundamentals.** Important areas are featured in these illustrations and many are exaggerated to place emphasis on the parts being discussed. Color is used throughout to add emphasis and is coordinated with the specific needs of each illustration.

Each chapter of the textbook begins with Learning Objectives that provide focus for the chapter. Technical terms are printed in ***bold italic type*** and are defined when first used. Warnings are provided where a danger to life or limb may exist in the system being discussed. Each chapter also includes a Summary, a "Know These Terms" section, and Review Questions. A Dictionary of Automotive Terms is included in the back of this text.

A Workbook for **Auto Fundamentals** is also available. It is a convenient study guide and shop activity guide directly related to this textbook.

Martin W. Stockel
Martin T. Stockel
Chris Johanson

Contents

Chapter 18 Engine Clutches 323

Chapter 19 Manual Transmissions and Transaxles 335

Chapter 20 Four-Wheel Drive 355

Chapter 21 Automatic Transmissions and Transaxles 367

Chapter 22 Axles and Drives 395

Chapter 23 Brakes 419

Chapter 29 ASE Certification 555

Chapter 30 Career Opportunities 561

Conversion Tables 569

Dictionary of Automotive Terms 577

Acknowledgments 597

Index 599

Although most modern engines may seem complex, each one uses the same theories of operation. (Lexus)

1

Building an Engine

After studying this chapter, you will be able to:

- Identify the basic parts of an engine.
- Explain engine operating principles.
- Describe the function of the major parts of an engine.
- Describe the four-stroke cycle sequence.

This chapter covers the basic design and operation of an engine. This chapter is very important. To properly understand the theory and operation of the internal combustion engine, it is essential that you completely understand all the material presented in this chapter. Studying this chapter will prepare you for the chapters on engine design and construction later in this book.

What Is an Engine?

An engine is a group of related parts assembled in a specific order. In operation, it is designed to convert the energy given off by burning fuel into a useful form.

There are many parts in a modern engine, each essential to the engine's operation. For the time being, however, we may think of an engine as a device that allows us to pour fuel into one end and get power from the other, **Figure 1-1.**

Internal Combustion Engine

Before beginning to construct an engine, it would be helpful to know what the term internal combustion engine means. ***Internal combustion*** means burning within. ***Engine*** refers to the device in which the fuel is burned. ***Fuel*** indicates what is being burned. The engine converts the heat of burning fuel into useful energy.

Building an Engine on Paper

An excellent way to learn about engines is to build one on paper. Pretend that the engine has not yet been invented. You will be the inventor. You will solve the problems involved, step by step.

What to Use for Fuel

If you are going to convert fuel into useful energy, you will need a fuel that will ignite (burn) easily. It should burn cleanly, be reasonably inexpensive, and produce sufficient power. It must be available in quantity. It should also be safe to use and easy to transport. How about dynamite? No, it is expensive and burns violently. In short, it would blow up the engine, **Figure 1-2.** Kerosene? No, it is too hard to ignite and does not

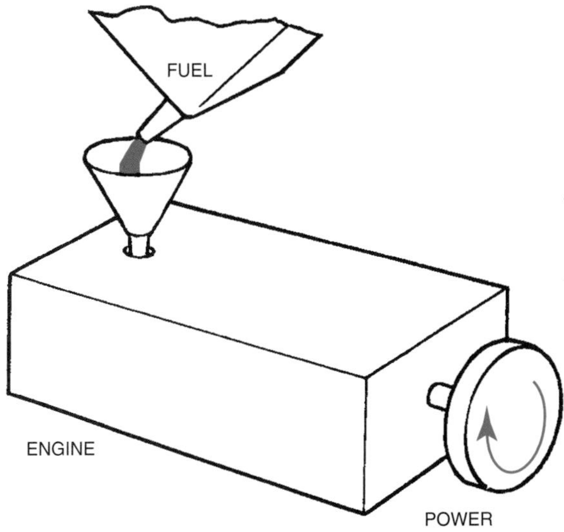

Figure 1-1. An engine converts fuel into useful energy.

Figure 1-2. Dynamite will explode. Because of this, it is not useful as a fuel.

burn cleanly. Gasoline? Now you have found a fuel that will serve your purpose.

Gasoline

Gasoline is a product obtained by refining crude oil (petroleum) obtained from wells drilled into the earth. Basically, the crude oil is treated in various ways to produce gasoline, **Figure 1-3**. Gasoline is only one of the many items produced from crude oil. Gasoline used in engines is a complex mixture of basic fuels and special additives. See **Figure 1-4**. Since gasoline is a mixture of carbon and hydrogen atoms, it is termed a ***hydrocarbon***.

Gasoline is usually available in three grades: regular, the lowest grade, plus or extra, the intermediate grade, and premium, the highest grade. All grades are assigned an ***octane rating.*** The octane rating indicates how well the gasoline will resist detonation (too rapid burning) in the cylinders. Regular gasoline usually has an octane rating of 87, plus has an octane rating of about 89, and premium has an octane rating of 92 to 94.

All gasoline sold today is ***unleaded***. Unleaded gas contains no ***tetraethyl lead***, which was used to raise octane. Unleaded gas must be used in late-model cars as tetraethyl lead will quickly destroy parts that help to reduce exhaust emissions (catalytic converter).

Many factors enter into the quality of any gasoline. It must pass exhaustive tests, both in the laboratory and in actual use. Basically, gasoline must burn cleanly, ignite readily, and resist freezing or boiling. It should contain a minimum number of harmful ingredients and prevent detonation.

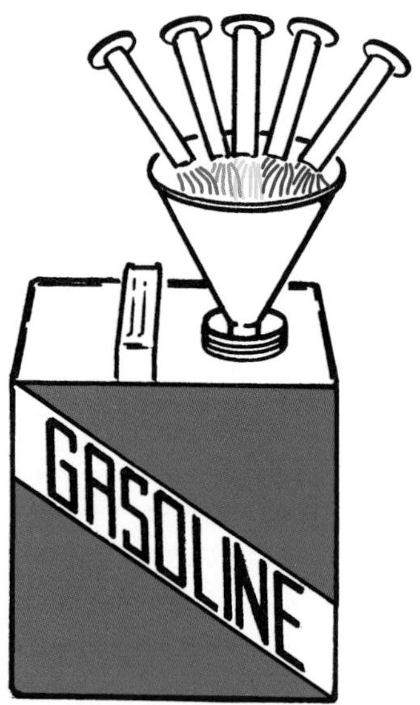

Figure 1-4. Gasoline is a complex mixture of refined crude oil and other chemicals that allow it to provide maximum performance.

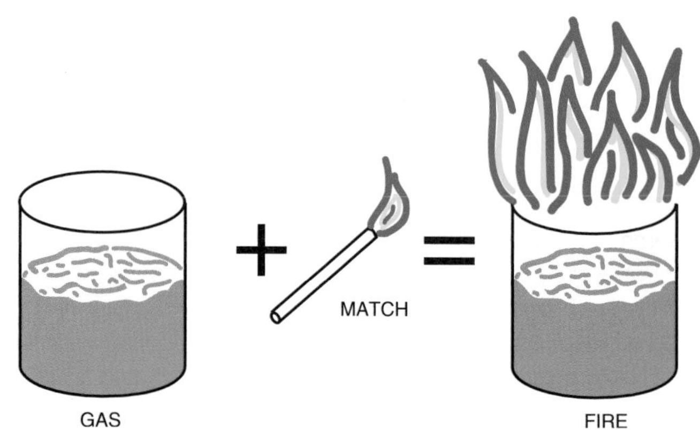

Figure 1-5. Gasoline burns. However, it will not burn outside the presence of oxygen.

Warning: The following steps, which show how gasoline is prepared for use in the engine, are for purposes of illustration only and should not be tried by the student. Gasoline is dangerous. Treat it with care and respect.

Preparing the Fuel

As you know, gasoline burns readily. However, to get the most power from this fuel and, in fact, to get it to power an engine, special treatment is required.

If you were to place a small amount of gasoline in a jar and drop a match in it, it would burn, **Figure 1-5.** Such burning is fine to produce heat, but it will not give us the explosive burning effect we need to operate an engine. We must figure out a way to get the gas to burn fast enough to produce an explosive force?

If you examine **Figure 1-5,** you will notice that the gasoline is burning on the top side. Why is it not burning along the sides and bottom of the container? In order to burn, gasoline must combine with oxygen in the air. The sides and bottom of the gasoline in the jar are not exposed to air, so they do not burn.

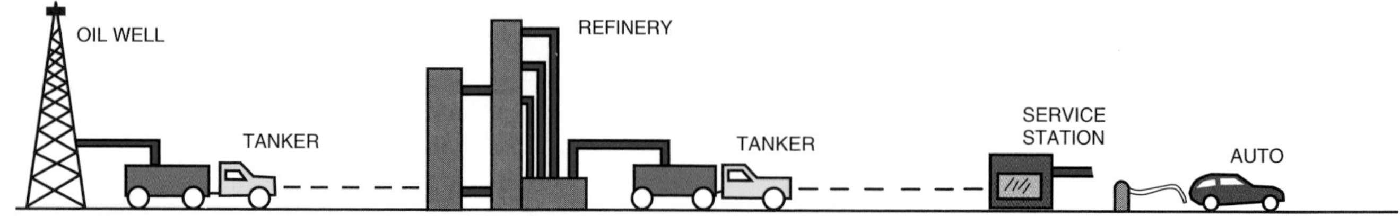

Figure 1-3. Before reaching your engine, crude oil requires much refining.

Breaking up Gasoline

For purposes of illustration, imagine that a gasoline particle is square, **Figure 1-6A.** It will burn on all sides. However, it still will not burn quickly enough for use in an engine.

To make the gasoline burn more rapidly, we can break it into smaller particles. Notice that as you divide it into smaller particles, you expose more area to the air, **Figure 1-6B.** If you ignite it now, it will burn much faster.

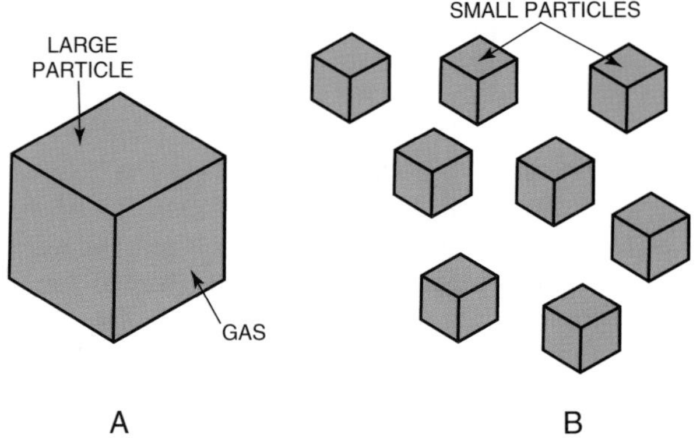

Figure 1-6. Breaking gasoline into smaller particles exposes more surface to the air, increasing the burning rate.

The Basic Force of an Engine

If you break up gasoline into very tiny particles, burning is fierce, **Figure 1-7.** Rapid burning produces a tremendous amount of heat, which in turn causes a rapid and powerful expansion. The burning gasoline gives off ***energy*** in the form of ***heat.*** You now have the necessary basic force with which to do work. The next step is to harness this force and make it work for you.

Trapping the Explosion

If you were to spray a mixture of gasoline and air into a sturdy metal container, place a lid over the top of the container, and then light the mixture, the resulting explosion would blow the lid high into the air. See **Figure 1-8.**

This is an example of using gasoline to do work. In this case, the work is blowing the lid into the air. Obviously, a flying lid will not push a car, but the flying lid does suggest a way to convert the energy of burning fuel into useful motion.

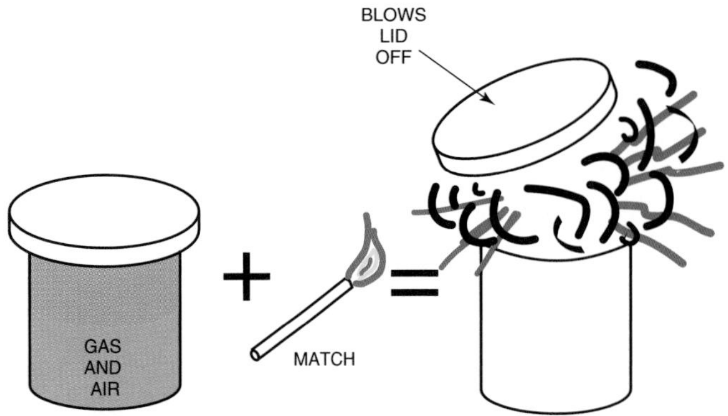

Figure 1-8. If air and fuel are mixed in a closed container and lit, it will blow the lid off of the container.

A Simple Engine

Make the same setup again, but this time use a rod to hook the lid to a shaft that is shaped like the one shown in **Figure 1-9.** Support each end of the shaft in bearings. Then, place a wheel on one end of the crankshaft, **Figure 1-10.**

Now, when the mixture explodes and the lid is blown into the air, the shaft will be given a sharp upward push, causing

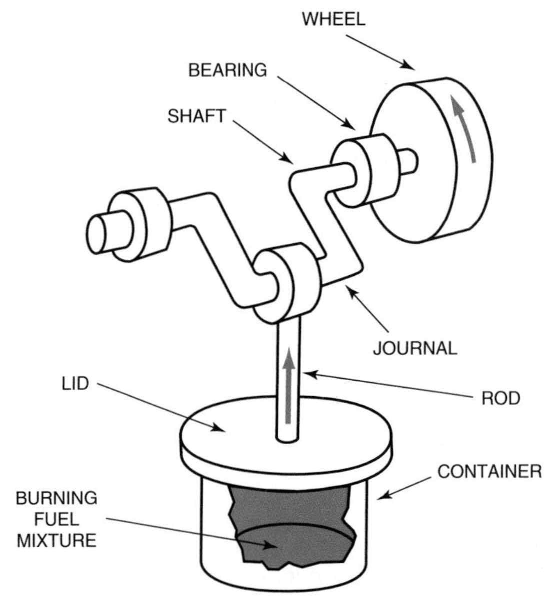

Figure 1-9. A closed container containing air and fuel can be connected to a shaft to create a simple engine.

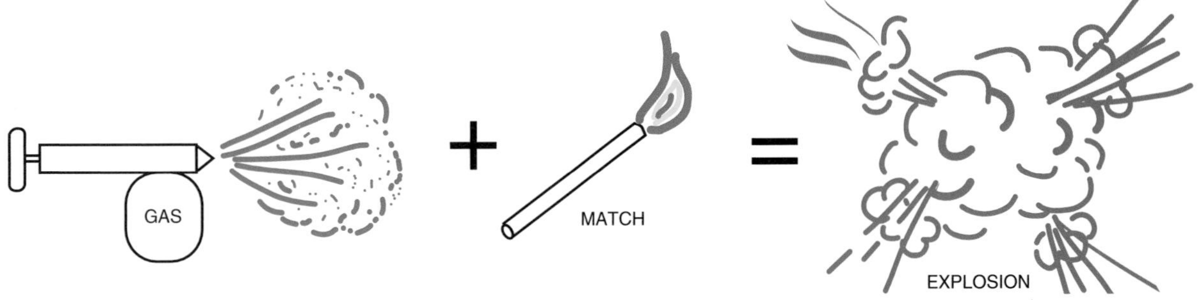

Figure 1-7. When gas particles are small enough they will explode.

the wheel to spin. You have built a very simple engine. Although this engine is not practical, it is pointing the way.

What Is Wrong With It?

What is wrong with our simple engine? Many things. Let us discuss them one at a time. The lid will fly up during the explosion. As the wheel spins, the lid will be forced down again. The lid can come down in any position, but if the engine is to work, it must come down over the container.

Instead of putting the lid over the container, cut it so that it just slips inside. Make the container longer so that the lid can push the shaft to the top of its travel and still not fly out of the container, **Figure 1-10.**

If you were to bolt the container and the shaft bearings so they could not change position, you would have an engine that would spin the wheel every time you fired a fuel mixture.

In order to cause the wheel to spin in the proper direction, you would have to fire the mixture with the crank in a position similar to the one shown in **Figure 1-10A.** If the crank is in the position shown in **Figure 1-10B,** the lid could not fly up without pushing the crank bearing up or the container down. If it were fired with the crank in the position shown in **Figure 1-10C,** the wheel would spin backwards.

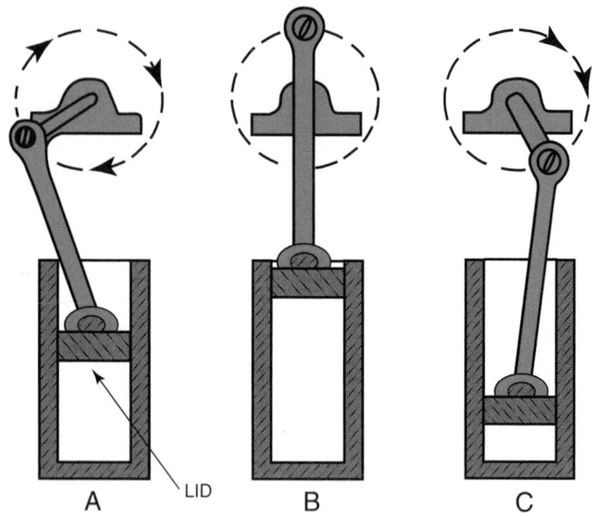

Figure 1-10. Placing the lid inside the container forms a simple engine.

The mixture must be fired when the shaft is in the proper position. By studying **Figure 1-11**, you can see that the crankshaft changes the ***reciprocating motion*** (up and down) of the lid into ***rotary motion*** (round and round).

Name the Parts

At this time, the parts you have developed should be named. By doing so, you will learn what to call the parts in a gasoline engine that serve the same general purpose.

The container will be called the ***block.*** The hole in the container, or block, will be called the ***cylinder.*** The lid will be termed the ***piston.*** The shaft, with a section bent in the shape of a crank, will be called the ***crankshaft.*** The rod that connects the crankshaft to the piston will be called the ***connecting rod.*** The bearings that support the crankshaft will be called ***main bearings.*** The connecting rod has an upper bearing and lower bearing. The lower bearing (the one on the crankshaft) is called a ***connecting rod bearing***. The wheel will be called the ***flywheel***. Refer back to **Figure 1-10** and see if you can substitute the correct names for those listed.

Fastening the Parts

Since you are going to fasten the parts so the main bearing and cylinder cannot move, it would be wise to invert your engine and place the cylinder on top. Just why this is done will soon be obvious.

Make the block heavy to give it strength to withstand the fuel explosions. Bring it down to support the main bearings, **Figure 1-11A**. By bringing the block out and down around the main bearings, you now have a strong unit. In **Figure 1-11B,** the lower block end forms a case around the crankshaft. It allows you to have two main bearings. This lower block end is called the ***crankcase.***

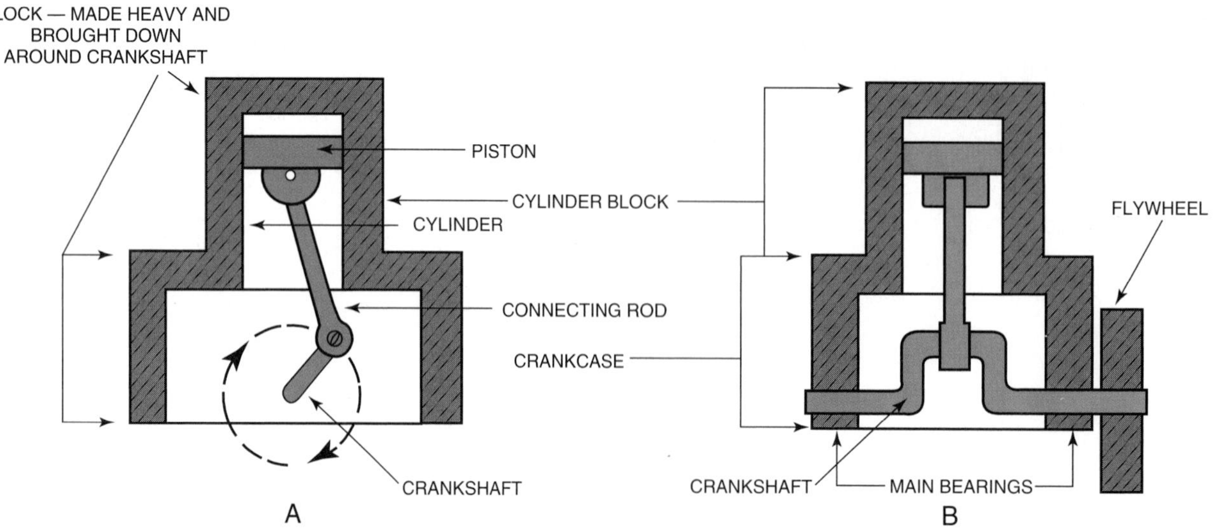

Figure 1-11. Front, side views showing how an engine is inverted, the block strengthened and brought down and around the crankshaft, forming the crankcase.

Lengthen the Piston

Now, make the piston longer. This stops it from tipping sideways in the cylinder. In order to avoid a piston that is too heavy, make it hollow. See **Figures 1-12** and **1-13.**

If the piston is to travel straight up and down and the connecting rod is to swing back and forth in order to follow the crank, it is obvious that the connecting rod must be able to move at the point where it is fastened to the piston, **Figure 1-14.**

Now, drill a hole through the upper end of the connecting rod. Also drill a hole through the piston. Line up the two holes and pass a pin through them. This pin is called a ***piston pin,*** or ***wrist pin.*** It is secured in various ways (more on this later). See **Figure 1-15.**

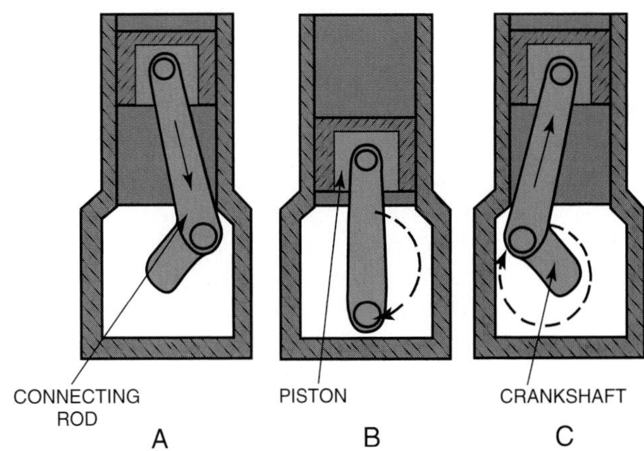

Figure 1-14. To take advantage of the full piston action, the connecting rod must swing. Note position in A compared to C.

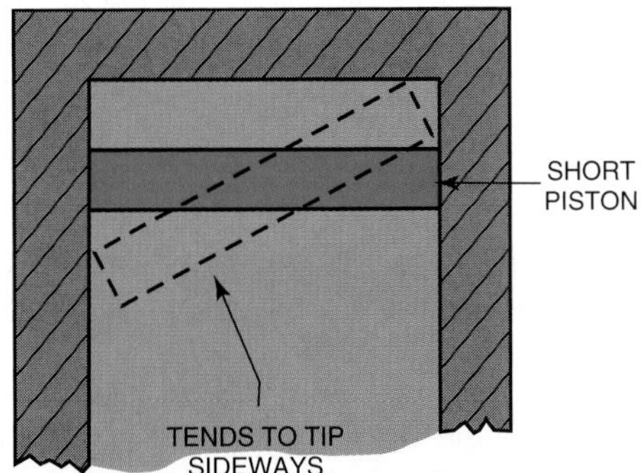

Figure 1-12. Effect of a short piston. A short piston will tip from side to side.

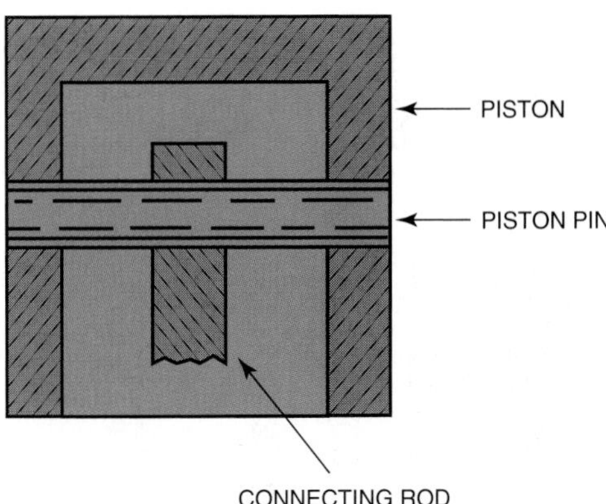

Figure 1-15. A metal pin is used to secure the connecting rod to the piston.

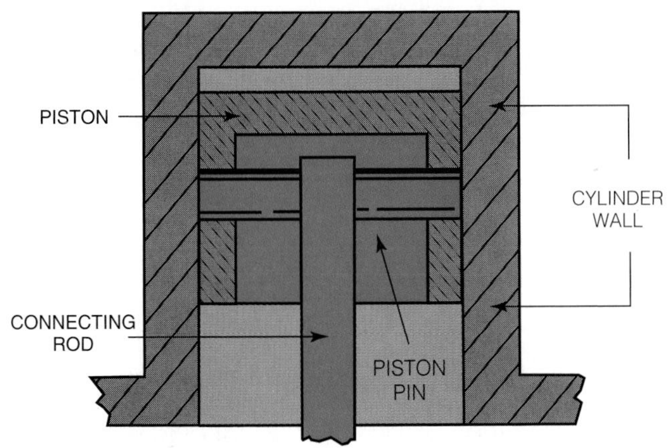

Figure 1-13. All automotive pistons are lengthened to avoid tipping.

Getting Fuel into the Engine

You have probably noticed that no provision has been made to get fuel into the upper cylinder area of your assembled engine. Your next step is to develop a system to admit fuel and to exhaust (blow out) the fuel after it is burned.

One opening in the upper cylinder area (called the ***combustion chamber***) will not be adequate. You cannot admit fuel and exhaust the burned gases from the same opening. You need *two* openings.

Removable Cylinder Head

Redesign your cylinder block and make the top removable. You will call this removable top a ***cylinder head.*** It will be fastened in place with bolts or studs and nuts, **Figure 1-16.**

Fuel Intake and Exhaust Passages

Making your cylinder head removable has not yet solved all of the problems. You cannot take the head off and put it on each time the engine fires. Now, make the head of much thicker metal and make two holes, or passages, like those shown in **Figure 1-17.** This will give you one passage to take in the fuel mixture and another passage to exhaust it. These passages are called ***valve ports***, **Figure 1-18.**

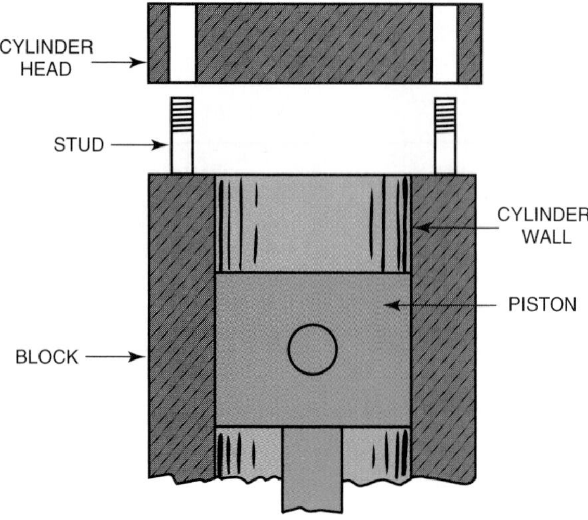

Figure 1-16. Cylinder head. Top of the cylinder has been removed.

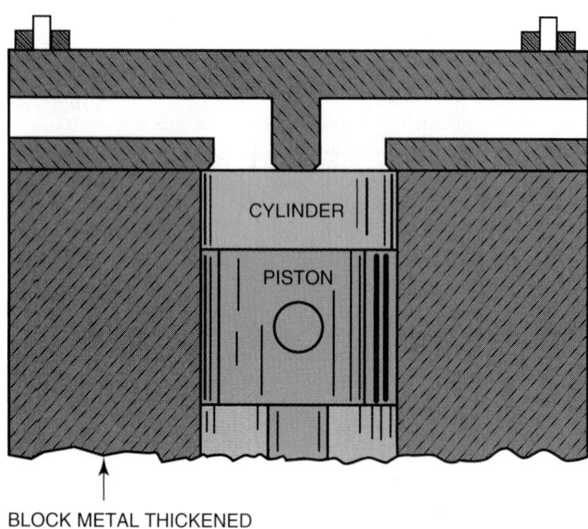

Figure 1-17. The cylinder head metal is thickened and passages have been added. The passages and additional head metal will permit the installation of additional parts.

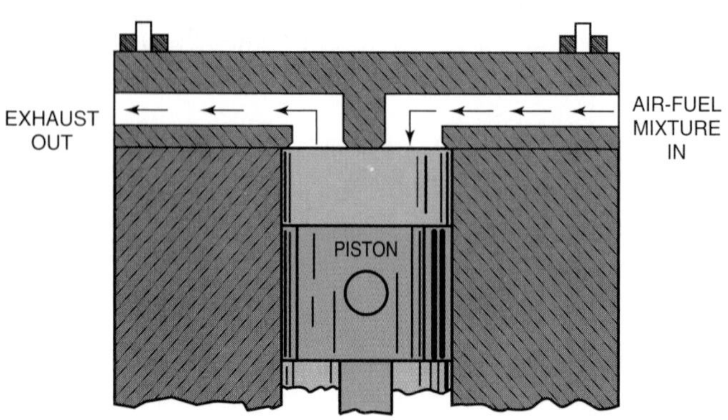

Figure 1-18. The valve ports in the cylinder head allows the air-fuel mixture to enter and exhaust gas to leave the cylinder. The space above the piston and below the valves is called the combustion chamber.

Valves

The next logical step is to provide a device to open and close the ports. If the ports are left open and fuel explodes in the combustion chamber, the forces from the explosion will blow out through the openings and fail to push the piston down.

This port control device, or ***valve***, will have to be arranged so that it can be opened and closed when desired. Place a valve in each opening. This may be done as shown in **Figure 1-19.** When the valve sits on its ***valve seat***, it seals the opening. The valves may be held in position by a hole bored in the head at A in **Figure 1-19**. This hole is called a ***valve guide*** because it guides the ***valve stem*** up and down in a straight line.

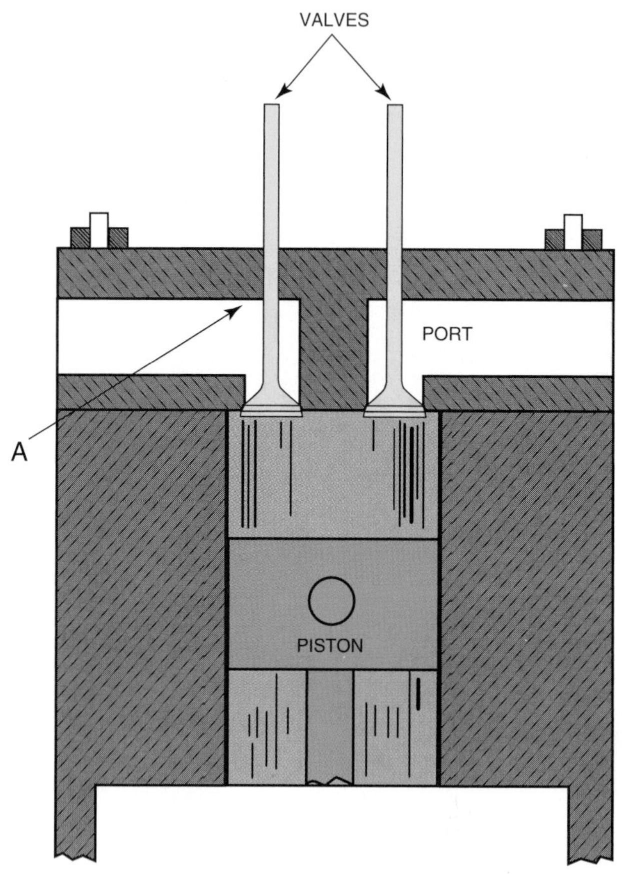

Figure 1-19. Valves are installed in the ports to seal the combustion chamber.

You will also install a ***spring***, ***spring washer,*** and ***keeper***. The spring is necessary to keep the valve tightly closed against its seat. When the valve is opened, the spring will close it again. Arrange your spring, spring washer, and keepers as shown in **Figure 1-20.** The valves now provide a satisfactory method of opening and closing the ports.

Four-Stroke Cycle

At this point you have an intake valve to allow fuel to enter the cylinder and an exhaust valve to let the burned fuel out. The next problem is how to get fresh fuel into the cylinder and how to get burned fuel out of the cylinder.

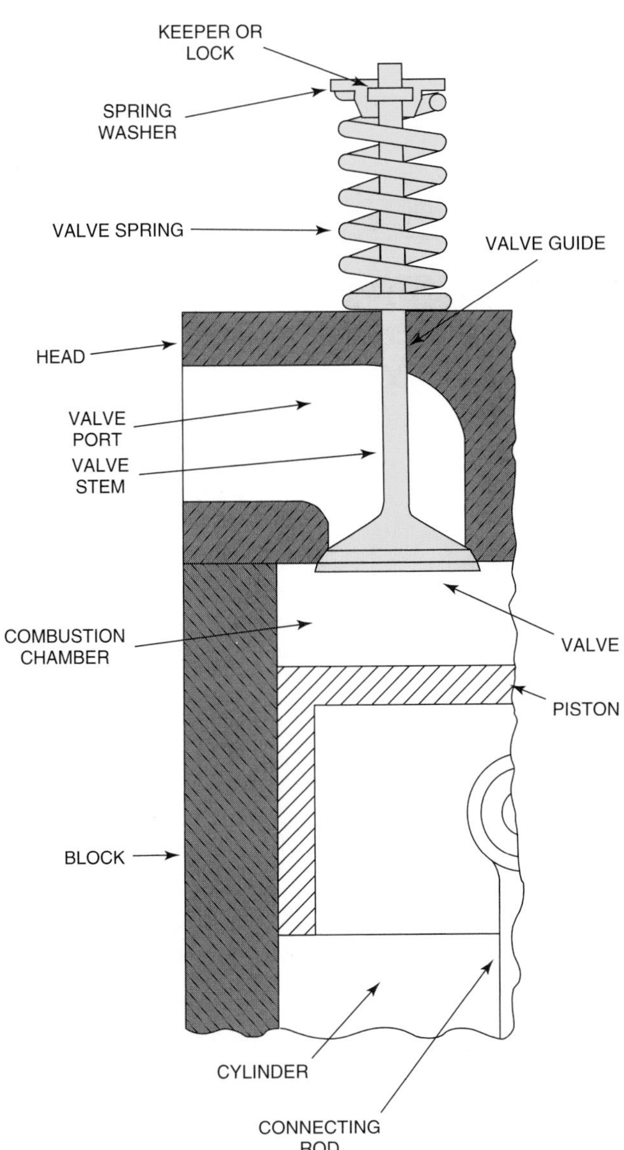

Figure 1-20. Installation of valve, spring, washer, and other valve assembly parts. This assembly closes the valves.

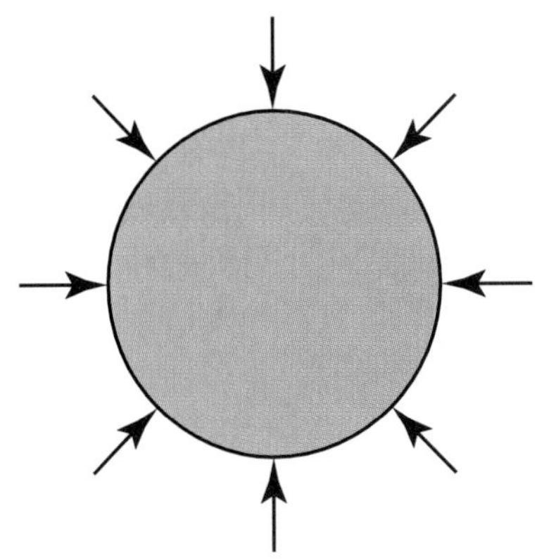

Figure 1-21. Air presses on all things with a pressure of 14.7 lb. per square inch at sea level. Air pressure varies with altitude.

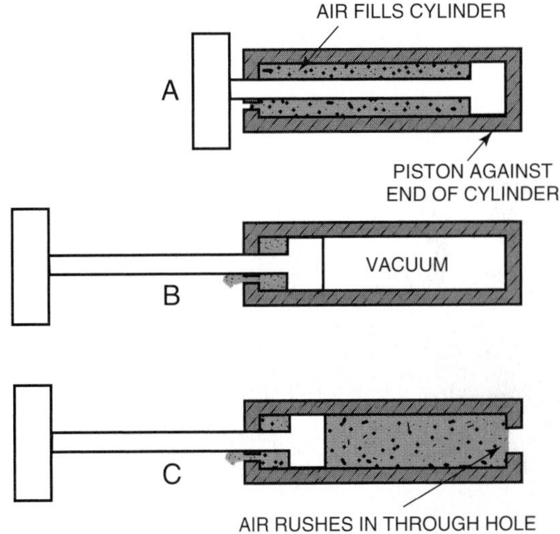

Figure 1-22. An engine cylinder is also a simple vacuum pump.

Vacuum Is the Clue

The air we live in presses on all things, **Figure 1-21.** This pressure is approximately 14.7 lb. per square inch (103 kPa) at sea level.

When we draw all the air out of a container, we form a ***vacuum.*** A vacuum is unnatural, and atmospheric pressure will do all it can to get into the low pressure area. If there is the slightest leak in the container, air will seep in until the pressure is the same on the inside of the container as it is on the outside. A vacuum, then, is any area in which the air pressure is lower than atmospheric pressure.

You Already Have a Vacuum Pump

Figure 1-22 illustrates a simple vacuum pump. This pump consists of a cylinder into which a snug fitting piston is placed. In **Figure 1-22A**, the piston is against the end of the cylinder. Obviously, there is no air between these two surfaces.

When you pull the piston through the cylinder, as in **Figure 1-22B**, you will have a large area between the face of the piston and the end of the cylinder. If there was no air between them before, and if the piston fits snugly against the cylinder, there still will be no air between them. This gives you a large area in which there is no air, or in other words, a vacuum.

If you were to drill a hole in the closed end of the cylinder, **Figure 1-22C**, air would immediately rush in and fill the cylinder. Any material, such as fuel, that happened to be in the surrounding air would also be drawn into the cylinder.

Where Is Your Vacuum Pump?

If the cylinder and piston in **Figure 1-22** form a vacuum, the cylinder and piston in your engine will do the same. If the piston is at the top of the cylinder (with both valves closed) and you turn the crankshaft, the piston will be drawn down into the cylinder. This will form a strong vacuum in the cylinder. If you now open the intake valve, the air will rush into the cylinder.

Stroke No. 1—The Intake Stroke

The first stroke in your engine is called the ***intake stroke.*** Instead of opening the intake valve after you have drawn the piston down, you will find it better to open the intake valve as the piston starts down. This allows the air to draw fuel in the entire time the piston is moving down. If you wait until the piston is down before opening the valve, the piston will be moving up before the cylinder can be filled with air, **Figure 1-23.**

Remember, the intake stroke starts with the piston at the top of the cylinder (intake valve open and exhaust valve closed) and stops with the piston at the bottom of its travel. This requires one-half turn of the crankshaft.

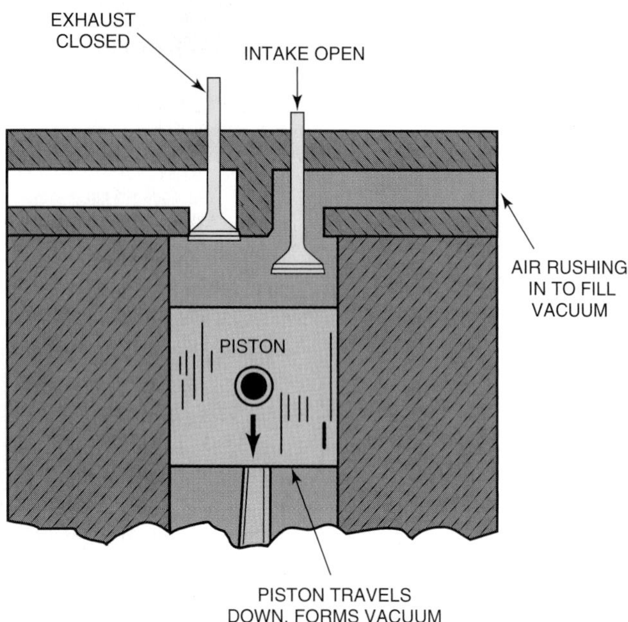

Figure 1-23. Engine cylinder on its intake stroke. As the piston travels down, it creates a vacuum, drawing in air and fuel.

Stroke No. 2—The Compression Stroke

You have discovered that the smaller the particles of gasoline mixed in the air, the more powerful the explosion.

As the crankshaft continues to move, the piston is forced up through the cylinder. If you keep both valves closed, the fuel mixture will be squeezed, or compressed, as the piston reaches the top. This is called the ***compression stroke,*** **Figure 1-24.** It, too, requires a half turn of the crankshaft.

The compression stroke serves to break up the fuel into even smaller particles. This happens due to the sudden swirling and churning of the mixture as it is compressed.

When the air-fuel mixture is subjected to a sudden sharp compression force, its temperature rises. This increase in temperature makes the mixture easier to ignite and causes it to explode with greater force. As the piston reaches the top of its travel on the compression stroke, it has returned to the proper position to be pushed back down by the explosion.

Remember, the compression stroke starts with the piston at the bottom of the cylinder (both valves closed) and stops with the piston at the top of the cylinder. This requires an additional half turn of the crankshaft.

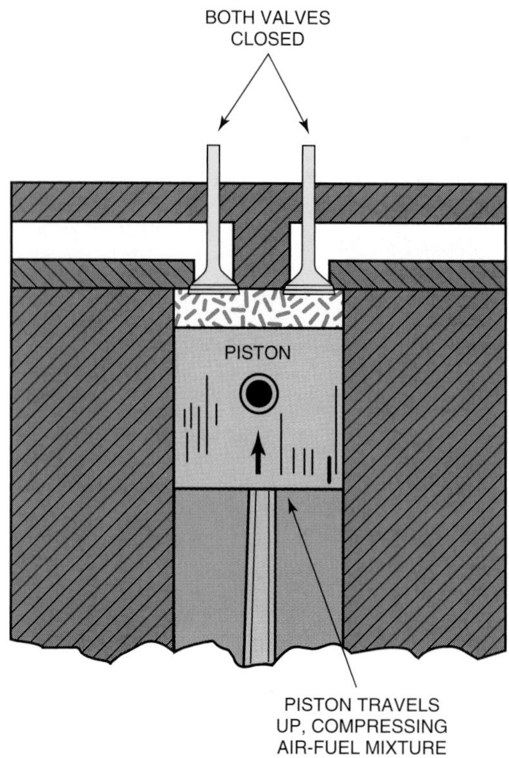

Figure 1-24. Engine cylinder compression stroke. As the piston approaches the top of the cylinder, it compresses the air and fuel.

Compression Ratio

The amount your engine will compress the fuel mixture depends on how small a space the mixture is squeezed into. In **Figure 1-25A,** the piston has traveled down 6" (152.4 mm) from the top of the cylinder. This is the intake stroke. In **Figure 1-25B,** the piston, which is on the compression stroke, has traveled up to within 1" (25.4 mm) of the cylinder top. It is obvious that 6" (152.4 mm) of cylinder volume has been squeezed into 1" (25.4 mm) of cylinder volume. This gives you a ratio of 6 to 1. This is called the ***compression ratio.***

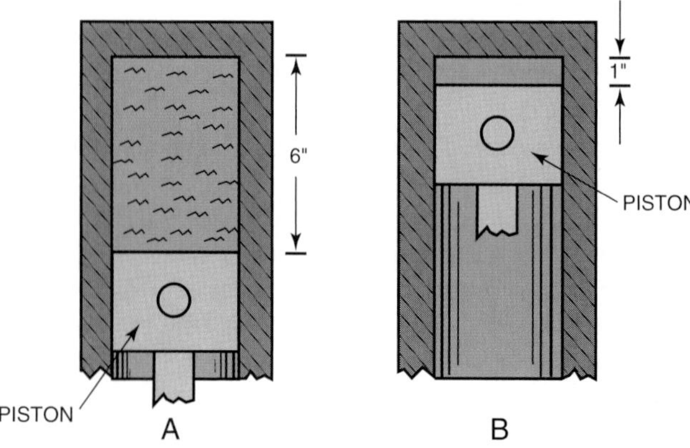

Figure 1-25. The compression ratio of an engine is the volume of a cylinder at the bottom of its stroke compared to its volume at the top of its stroke.

Stroke No. 3—The Power Stroke

As the piston reaches the top of the compression stroke, the mixture is broken into tiny particles and heated up. When ignited, it will explode with great force.

This is the right time to explode the mixture. A spark plug provides a spark inside the combustion chamber. The spark produced at the plug is formed by the ignition system. This will be discussed in Chapter 8.

Just imagine that a hot spark has been provided in the fuel mixture. The mixture will explode and, in turn, force the piston down into the cylinder. This gives the crankshaft a quick and forceful push. This is the ***power stroke***. Both valves must be kept closed during the power stroke or the pressure of the burning fuel will squirt out through the valve ports. See **Figure 1-26.**

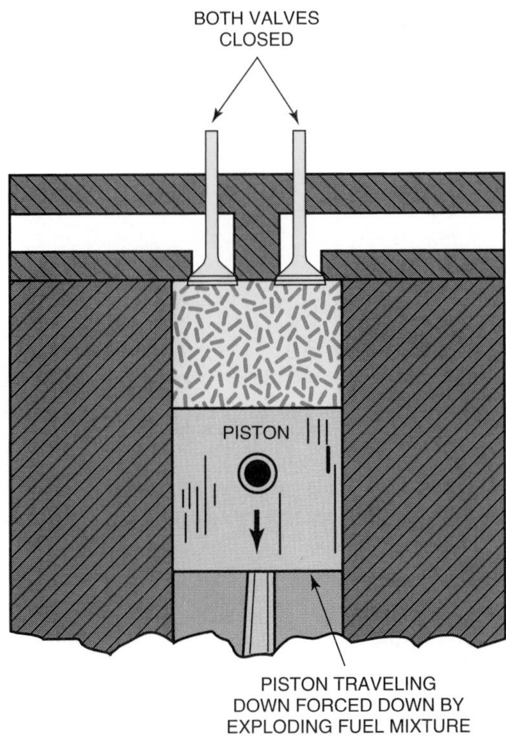

Figure 1-26. An engine's power stroke. The compressed air and fuel is ignited and the piston travels down.

Remember, the power stroke starts with the piston at the top of the cylinder (both valves closed) and stops with the piston at the bottom of the cylinder. This requires another half turn of the crankshaft.

Stroke No. 4—The Exhaust Stroke

When the piston reaches the bottom of the power stroke, the exhaust valve opens. The spinning crankshaft forces the piston up through the cylinder, pushing burned gases out, **Figure 1-27.** This is the ***exhaust stroke***.

Remember, the exhaust stroke starts with the piston at the bottom of the cylinder (exhaust valve open and intake valve closed). It stops with the piston at the top of the cylinder. This requires one more half turn of the crankshaft.

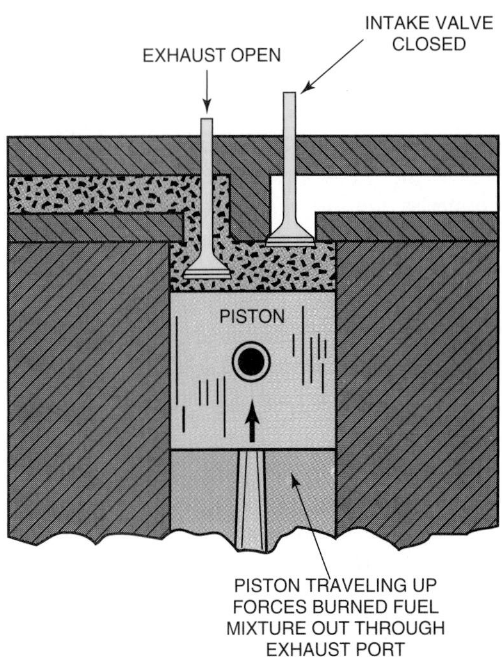

Figure 1-27. Engine cylinder exhaust stroke. As the piston travels up, it pushes the burned air and fuel exhaust gases out of the combustion chamber.

Completed Cycle

If you count the number of half turns in the intake, compression, power, and exhaust strokes, you will find you have a total of four. This gives you two complete turns, or ***revolutions***, of the crankshaft, **Figure 1-28.** While the crankshaft is turning around twice, it is receiving power only during one half turn, or one fourth of the time.

Cycle Repeated

As soon as the piston reaches the top of the exhaust stroke, it starts down on another intake, compression, power, and exhaust cycle. This cycle is repeated over and over. Each

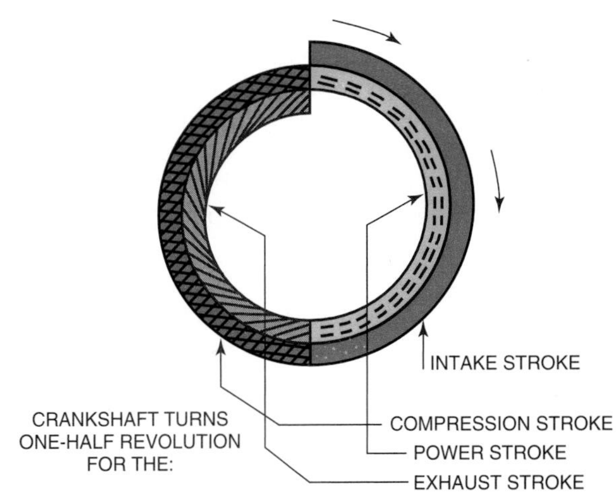

Figure 1-28. Four strokes are equal to two revolutions of the crankshaft.

complete cycle consists of four strokes of the piston, hence the name ***four-stroke cycle.*** There is also a two-stroke cycle, which will be covered later in this text.

What Opens and Closes the Valves?

You have seen that the intake valve must be opened for the intake stroke, both valves must remain closed during the compression and power strokes, and the exhaust valve opens during the exhaust stroke. You must now design a device to open and close the valves at the proper times.

Start by placing a round shaft above the valve stem. This shaft is supported in bearings placed in the front and the rear of the cylinder head. See **Figure 1-29.**

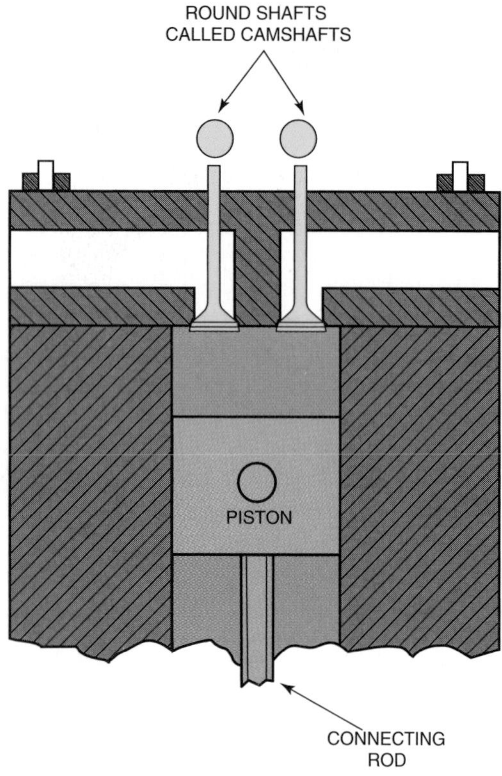

Figure 1-29. Engine section showing how the camshafts are positioned.

Camshaft

The shaft will have an egg-shaped bump called a ***cam lobe***. The cam lobe is machined as an integral part of the shaft. This shaft is called a ***camshaft,*** **Figure 1-30.**

The distance the valve will be raised, how long it will stay open, and how fast it opens and closes can all be controlled by the height and shape of the lobe. The lobe in this illustration is located directly over the valve stem, **Figure 1-31.** Other camshaft placements in the block and head will be discussed in later chapters.

As you will see later, it is impractical to have the cam lobe contact the end of the valve stem itself. You have placed the camshaft some distance above the end of the valve stem. When you turn the camshaft, the lobes will not even touch the valve stem.

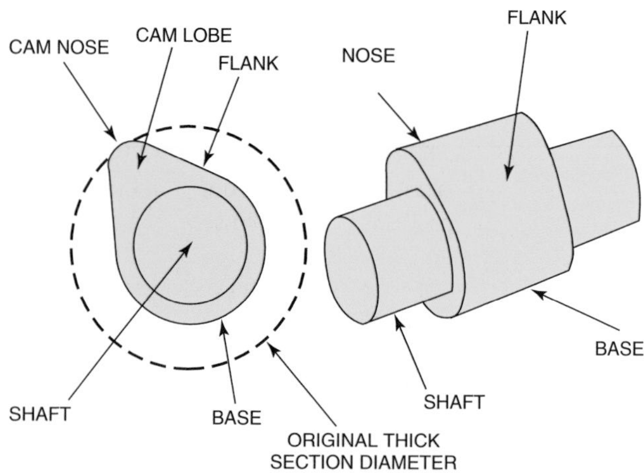

Figure 1-30. Section of a camshaft showing a cam lobe. Study the names of the various parts of the cam lobe.

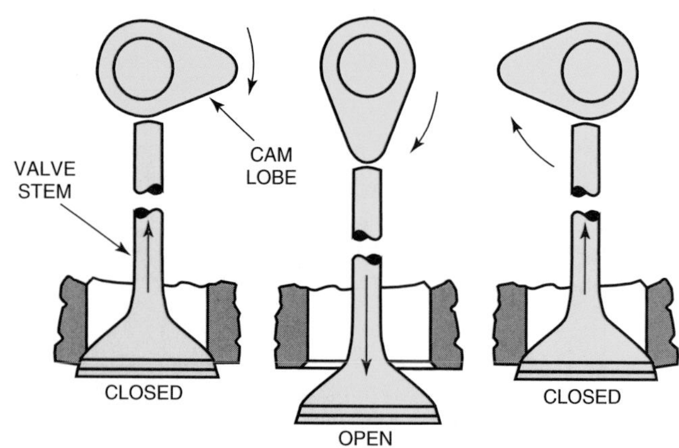

Figure 1-31. How the cam lobe opens a valve. The valve is closed by the valve spring assembly you studied earlier.

Valve Lifter

Your next step is to construct a cylindrical unit called a cam follower, or ***valve lifter***. The valve lifter in **Figure 1-32** is a solid lifter. Other lifters are hydraulic. Some lifters have an adjustment mechanism. These lifters are discussed in Chapter 2.

The lifter is installed between the cam lobe and the valve stem. The upper end rides on the lobe and the lower end almost touches the valve stem. The lifter slides up and down in a hole bored in the head metal that separates the valve stem from the camshaft. See **Figure 1-33.**

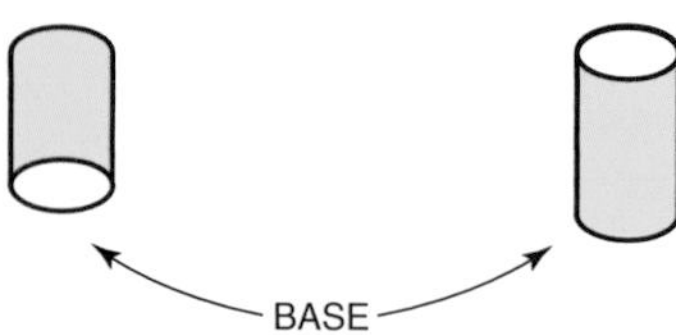

Figure 1-32. A simple valve lifter. Most valve lifters are more complex than this.

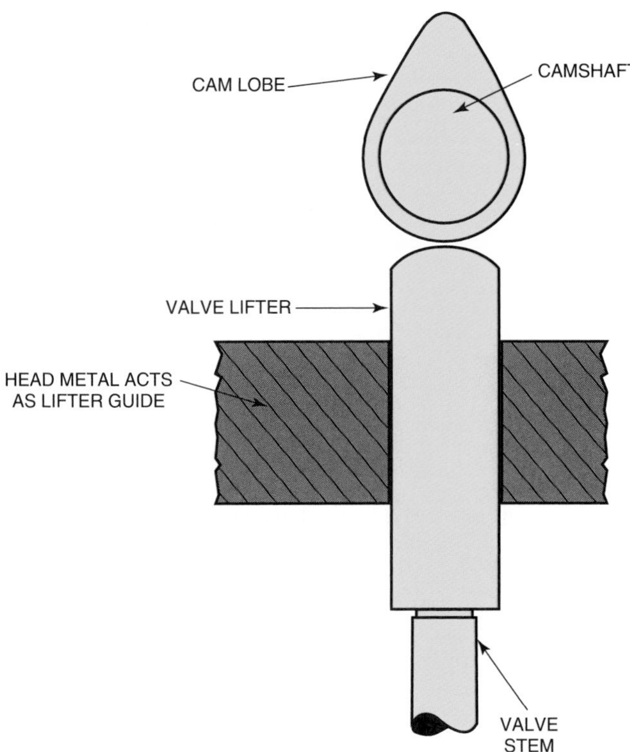

Figure 1-33. Exaggerated drawing showing valve lifter installation. The lifter is installed and moves inside the cylinder head metal. Note how the lifter would reduce wear and friction on the valve and camshaft lobe.

You have now built the essential parts of the valve system. The parts, in their proper positions, are referred to as the ***valve train***. See **Figure 1-34.**

Camshaft Speed

You have developed a method of opening and closing the valves. The next problem is how and at what speed to turn the camshafts. Stop a minute and think. Each valve must be open for one stroke. The intake valve is open during the intake stroke and remains closed during the compression, power, and exhaust strokes. This would indicate that the cam lobe must turn fast enough to raise the valve every fourth stroke. See **Figure 1-35.**

You can see from **Figure 1-35** that it takes one complete revolution of the cam lobe for every four strokes of the piston. Remember that four strokes of the piston requires two revolutions of the crankshaft. You can say that for every two revolutions of the crankshaft, the camshaft must turn once. If you are speaking of the speed of the camshaft, you can say that the camshaft must turn at one-half crankshaft speed.

Turning the Camshaft

If the crankshaft is turning and the camshaft must turn at one-half crankshaft speed, it seems logical to use the spinning crankshaft to turn the camshaft. One very simple way to drive the camshaft would be by means of gears and a belt. One gear is fastened on the end of the crankshaft, and the other is fastened on the end of the camshaft. The large camshaft gear drives the smaller crankshaft gear through the belt.

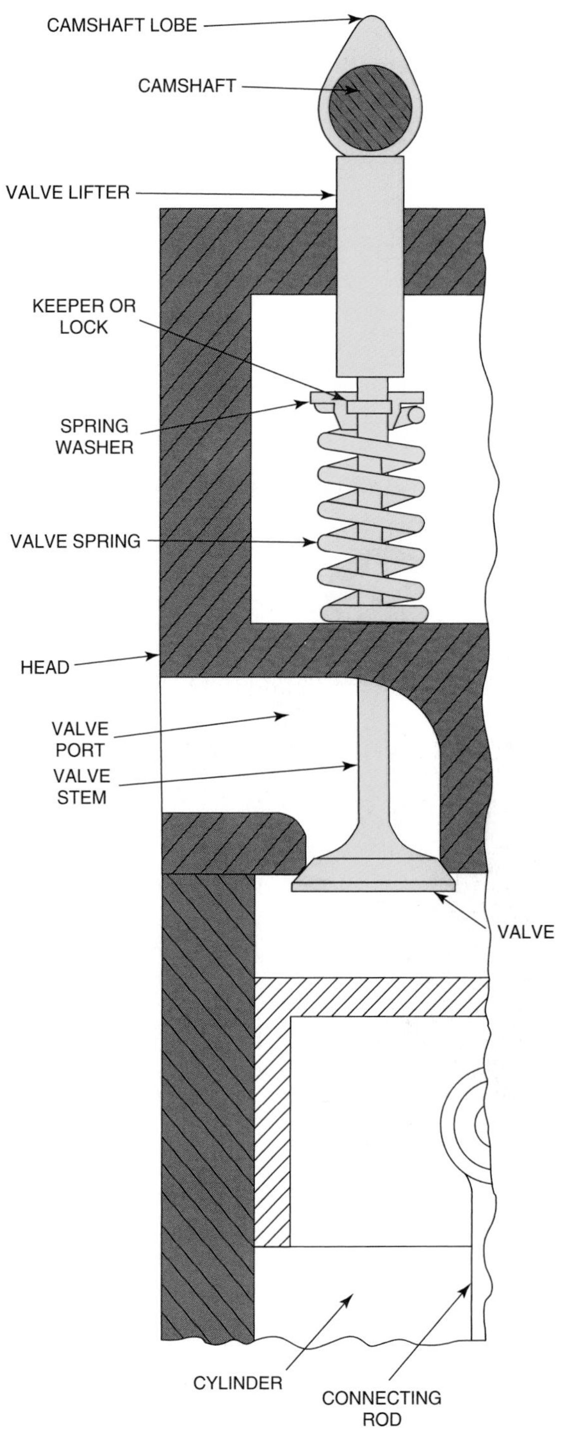

Figure 1-34. Complete valve train. Note how the cam lobe would force the valve open against valve spring pressure.

If, for instance, the small gear on the crankshaft has 10 teeth and the large gear on the camshaft has 20 teeth, the crankshaft will turn the camshaft at exactly one-half crankshaft speed. See **Figure 1-36. Figure 1-37** shows a front view of the ***crankshaft gear,*** both ***camshaft gears,*** and a ***timing belt*** as they would appear on the engine.

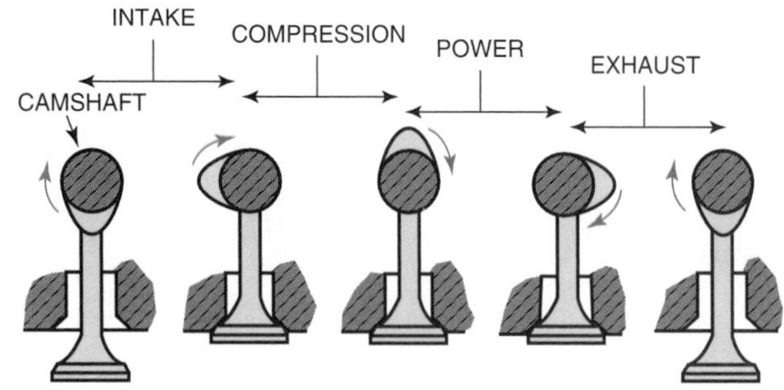

Figure 1-35. Each valve is opened once every fourth stroke. Each engine stroke turns the camshaft one-quarter turn.

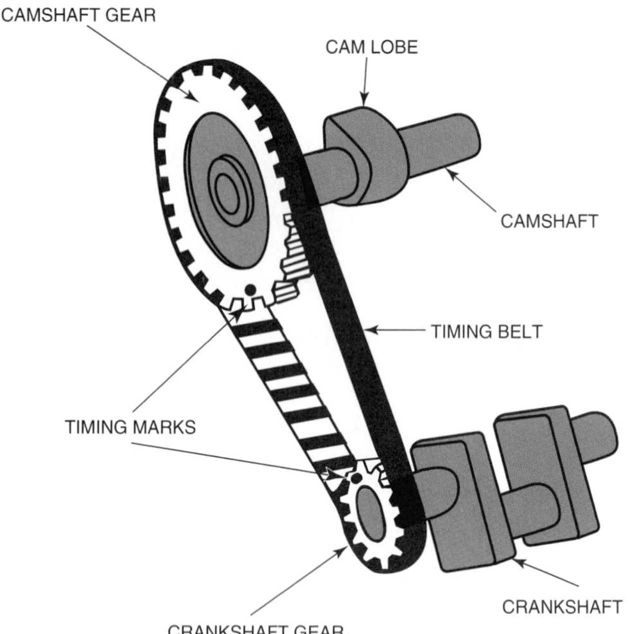

Figure 1-36. The crankshaft turns the camshaft at one-half crankshaft speed. The camshaft is driven through gears or sprockets and a chain or belt.

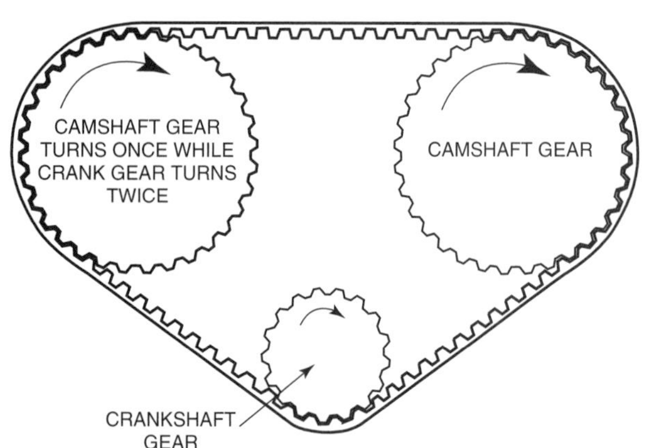

Figure 1-37. Front view. Crankshaft gear driving both camshaft gears through a belt.

Timing the Valves

You have now developed a method of turning the camshaft at the correct speed. The next problem that must be solved is how to get the valves to open at the proper time. This is called ***valve timing.***

Start with the intake stroke. As you have discovered, the intake valve must open as the piston starts down in the cylinder. Place your piston at top dead center (TDC topmost point of piston travel). See **Figure 1-38.**

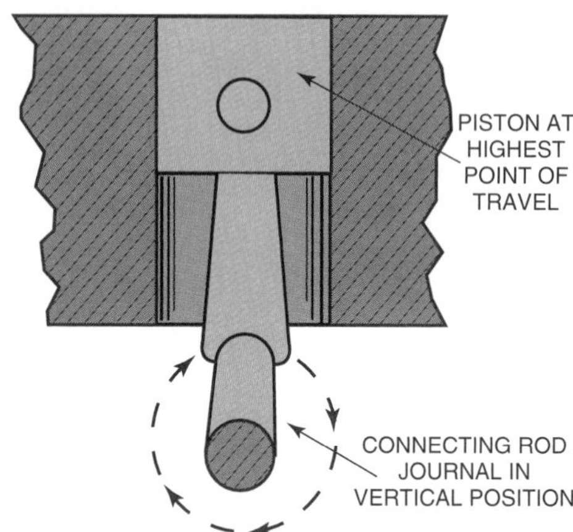

Figure 1-38. Engine piston at top dead center (TDC).

Insert the intake camshaft, then turn it in a counterclockwise direction until the flank of the cam lobe contacts the lifter. The timing gear should be connected to the crankshaft gear with a timing belt, **Figure 1-39.**

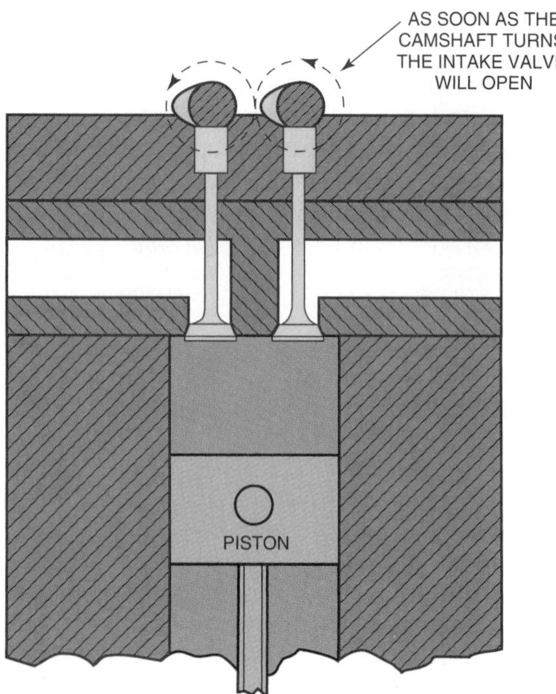

Figure 1-39. The position of the cam lobes at the start of the intake stroke. The camshafts are supported above the cylinder head surface. The lifters ride inside the cylinder head metal.

You should punch a mark on the crankshaft gear and a mark on the camshaft gear. These are called ***timing marks***. See **Figure 1-40.** If the camshaft is removed, it may be reinstalled by merely lining up the timing marks. Were you to crank the engine, the crankshaft would pull the piston down on the intake stroke. The camshaft would also rotate, causing the cam lobe to turn. The turning lobe will raise and lower the valve. During this stroke, the crankshaft would make a half turn and the camshaft would make a quarter turn. See **Figure 1-41.**

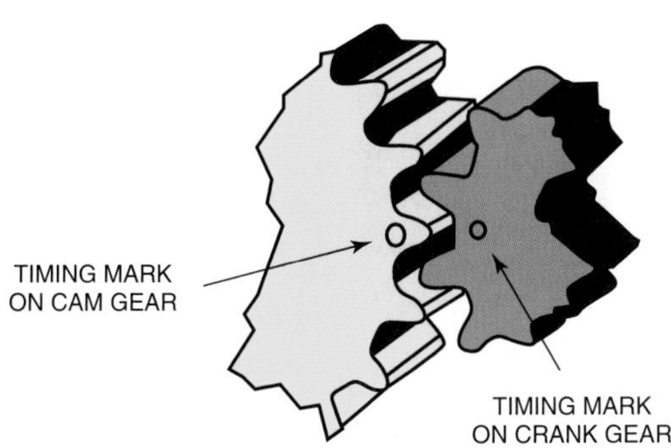

Figure 1-40. Timing gears meshed and marked. The gears shown here are meshed to illustrate proper timing mark alignment.

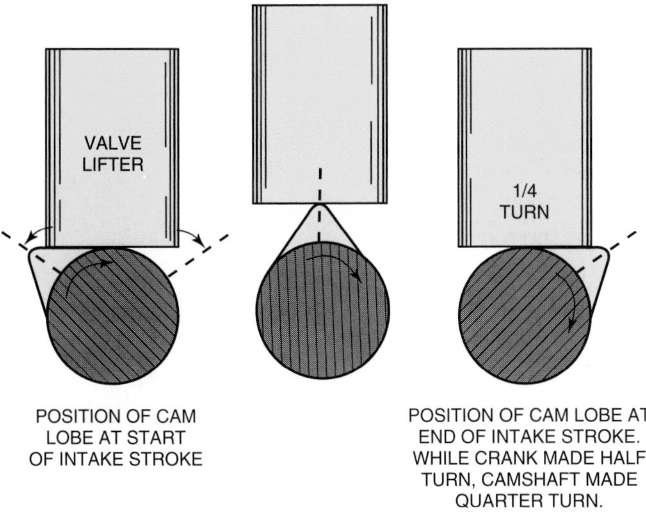

Figure 1-41. A camshaft makes one-quarter turn during each stroke.

Intake Remains Closed

As the crank continues to turn, it will push the piston up on compression, down on power, and up on exhaust. During these three strokes, the intake cam lobe continues to turn. When the piston reaches top dead center (TDC) on the exhaust stroke, the flank of the cam lobe will again be touching the intake lifter.

As your next stroke will be the intake, the piston is ready to start down and the cam lobe is ready to push on the valve lifter to provide proper timing. See **Figure 1-42.**

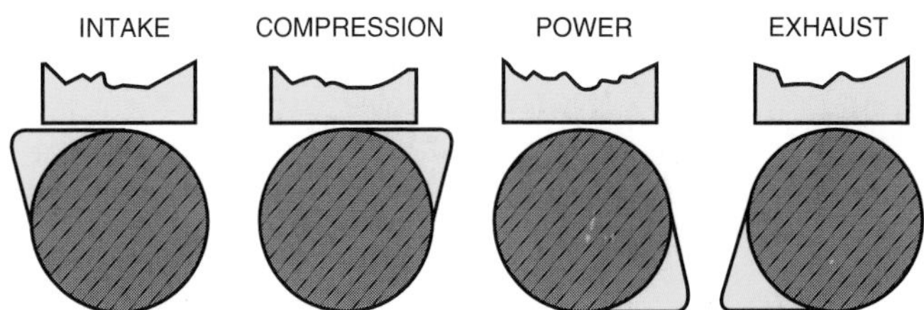

Figure 1-42. Position of an intake valve's camshaft lobe at the start of each stroke.

Exhaust Timed the Same Way

The only difference between the intake and exhaust timing is that you start by placing the piston on bottom dead center (BDC). Turn the exhaust camshaft until the lobe contacts the lifter, then mark the gears. Use two marks on the crankshaft gear and two on the camshaft gear for the exhaust camshaft. This prevents aligning with the wrong marks.

The Flywheel

When you crank an engine, it turns through all the necessary strokes. The only time the crankshaft receives power is during the power stroke. After the power stroke is completed, the crankshaft must continue to turn to exhaust the burned gases, take in fresh air and fuel, and then compress the mixture. One power stroke is not enough to keep the crankshaft turning during the four required strokes. See **Figure 1-43.**

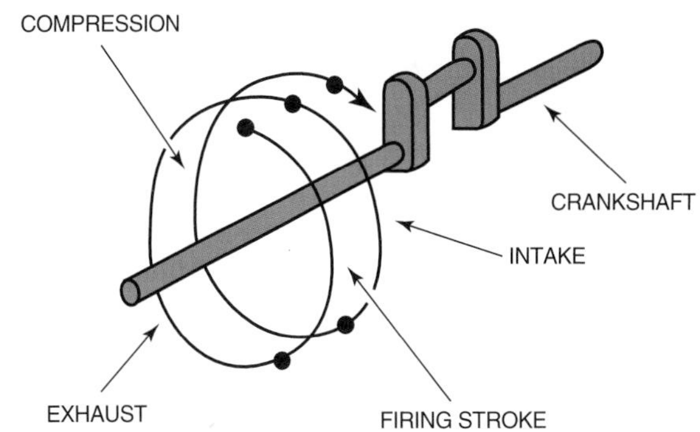

Figure 1-43. After the power stroke, the crankshaft must coast through the next three strokes.

You will recall that one end of the crankshaft has a timing gear attached. If you will fasten a fairly large, heavy wheel to the other end, the engine will run successfully. This wheel is called a flywheel.

The flywheel, which is caused to spin by the power stroke, will continue to spin because it is heavy. In other words, the inertia built up within the flywheel will cause it to keep turning. Because it is attached to the crankshaft, the flywheel will cause the shaft to continue to turn. The shaft will now spin long enough to reach the next power stroke. See **Figure 1-44.**

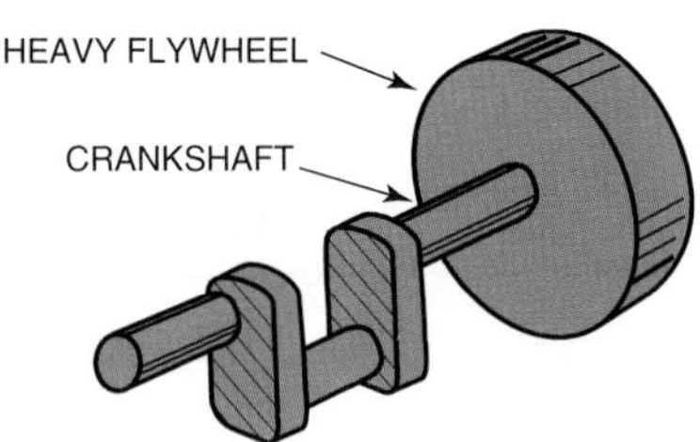

Figure 1-44. Flywheel inertia helps the engine to run.

Engine Will Run Smoothly

You can see that every time the engine fires, the crank receives a hard push. This will speed it up. As the power stroke is finished, the shaft coasts. This slows it down. The alternating fast and slow speeds would make for a very rough running engine.

With the heavy flywheel in place, the power stroke cannot increase the crankshaft speed as quickly since it must also speed up the flywheel. When the crankshaft is coasting, it cannot slow down as quickly because the flywheel velocity keeps it spinning. This will cause the crankshaft speed to become more constant and will, in turn, give a smooth-running engine.

Basic Engine Completed

You have now completed a basic engine. With the addition of fuel, spark, and oil, it will run. This is a one cylinder engine, and its uses would be confined to such small tasks as powering lawn mowers, scooters, portable generators, go-carts, and chainsaws. There are many modifications and improvements that can be made to your simple engine. These will be discussed in following chapters.

Regardless of the size, number of cylinders, and horsepower, all four-stroke cycle engines will contain the same basic parts as the small engine you have just built. The parts may be arranged in different ways and their shape may change, but they will do the same job in the same way. You will not find *different* basic parts, just more of them.

Summary

An engine operates by converting the heat given off by burning fuel into motion. The most convenient and widely used fuel is gasoline. Diesel and other fuels are also used. These fuels are able to burn efficiently in internal combustion engines. To deliver the maximum amount of power, the fuel is vaporized and mixed with air before it enters the engine.

As the piston is pulled down by the crankshaft, the camshaft opens the intake valve and the air-fuel mixture is drawn into the cylinder. When the piston reaches the bottom of its stroke, the intake valve closes and the crankshaft forces the piston up through the cylinder. This compresses the air-fuel mixture in the combustion chamber.

As the piston nears the top of the compression stroke, the air-fuel mixture is ignited by a spark from the spark plug. This explodes the mixture, and the pressure of the rapidly expanding gas drives the piston down through the cylinder. Both valves are closed during this power stroke.

After reaching the bottom of the power stroke, the exhaust valve opens, and the spinning crankshaft forces the piston up through the cylinder. This time, all the burned gases are driven (exhausted) from the cylinder and combustion chamber. When the piston reaches the top of the exhaust stroke, the exhaust valve closes and the intake opens. The piston is drawn down on another intake stroke.

Once the cylinder has fired, the heavy flywheel will keep the crankshaft spinning long enough to go through the exhaust, intake and compression strokes. It will then receive power from another power stroke. This cycle is repeated over and over, and the engine will run on its own power.

Know These Terms

Internal combustion
Engine
Fuel
Gasoline
Hydrocarbon
Octane rating
Unleaded gas
Tetraethyl lead
Energy
Heat
Reciprocating motion
Rotary motion
Block
Cylinder
Piston
Crankshaft
Connecting rod
Main bearings
Connecting rod bearing
Flywheel
Crankcase
Piston pin
Wrist pin
Combustion chamber
Cylinder head
Valve ports
Valve
Valve seat
Valve guide
Valve stem
Spring
Spring washer
Keeper
Vacuum
Intake stroke
Compression stroke
Compression ratio
Power stroke
Exhaust stroke
Revolutions
Four-stroke cycle
Cam lobe
Camshaft
Valve lifter
Valve train
Crankshaft gear
Camshaft gear
Timing belt
Valve timing
Timing marks

Review Questions—Chapter 1

Do not write in this book. Write your answers on a separate sheet of paper.

1. An engine converts heat into _____ .
   (A) exhaust gas
   (B) noise
   (C) useful work
   (D) All of the above.
2. Gasoline is obtained from __________ .
3. Gasoline is a mixture of hydrogen and carbon atoms and is therefore called a _____ fuel.
   (A) carbon
   (B) hydrogen

(C) hydrocarbon
(D) efficient

4. In order to burn, gasoline must be mixed with _____ .
   (A) oxygen
   (B) hydrogen
   (C) carbon
   (D) None of the above.
5. When the gasoline explodes in the cylinder, what movable part of the engine first receives the force?
   (A) Crankshaft.
   (B) Piston.
   (C) Flywheel.
   (D) Connecting rod.
6. The connecting rod must have a bearing at _____ .
   (A) the piston end
   (B) the crankshaft end
   (C) both ends
   (D) either end
7. On what stroke is a vacuum produced?
   (A) Intake.
   (B) Compression.
   (C) Power.
   (D) Exhaust.
8. What is the purpose of valve ports?
   (A) To let exhaust gases out of the cylinder.
   (B) To let air and gasoline into the cylinder.
   (C) To allow the valves to move up and down.
   (D) Both A & B.
9. The burning air-fuel mixture forces the piston down on what stroke?
   (A) Intake.
   (B) Compression.
   (C) Power.
   (D) Exhaust.
10. Describe what a valve does in a valve port.
11. The position of each valve during the various strokes is important. Give the position (open or closed) of each valve for each of the following strokes.

	Intake valve	Exhaust valve
(A) Intake Stroke	____________	____________
(B) Compression Stroke	____________	____________
(C) Power Stroke	____________	____________
(D) Exhaust Stroke	____________	____________

12. The part on a camshaft that moves the lifter is called a _____ .
13. The camshaft is turned by the _____ .
    (A) flywheel
    (B) crankshaft
    (C) lifter
    (D) connecting rod bearing
14. What is the speed of the camshaft in relation to that of the crankshaft?
    (A) One-half speed.
    (B) Same speed.
    (C) Twice as fast.
    (D) Depends on the type of engine.
15. In order to keep the engine running smoothly, it needs a large heavy _______ to keep the crankshaft turning.

NORTHEAST WISCONSIN TECHNICAL COLLEGE
LEARNING RESOURCE CENTER
GREEN BAY WI 54307

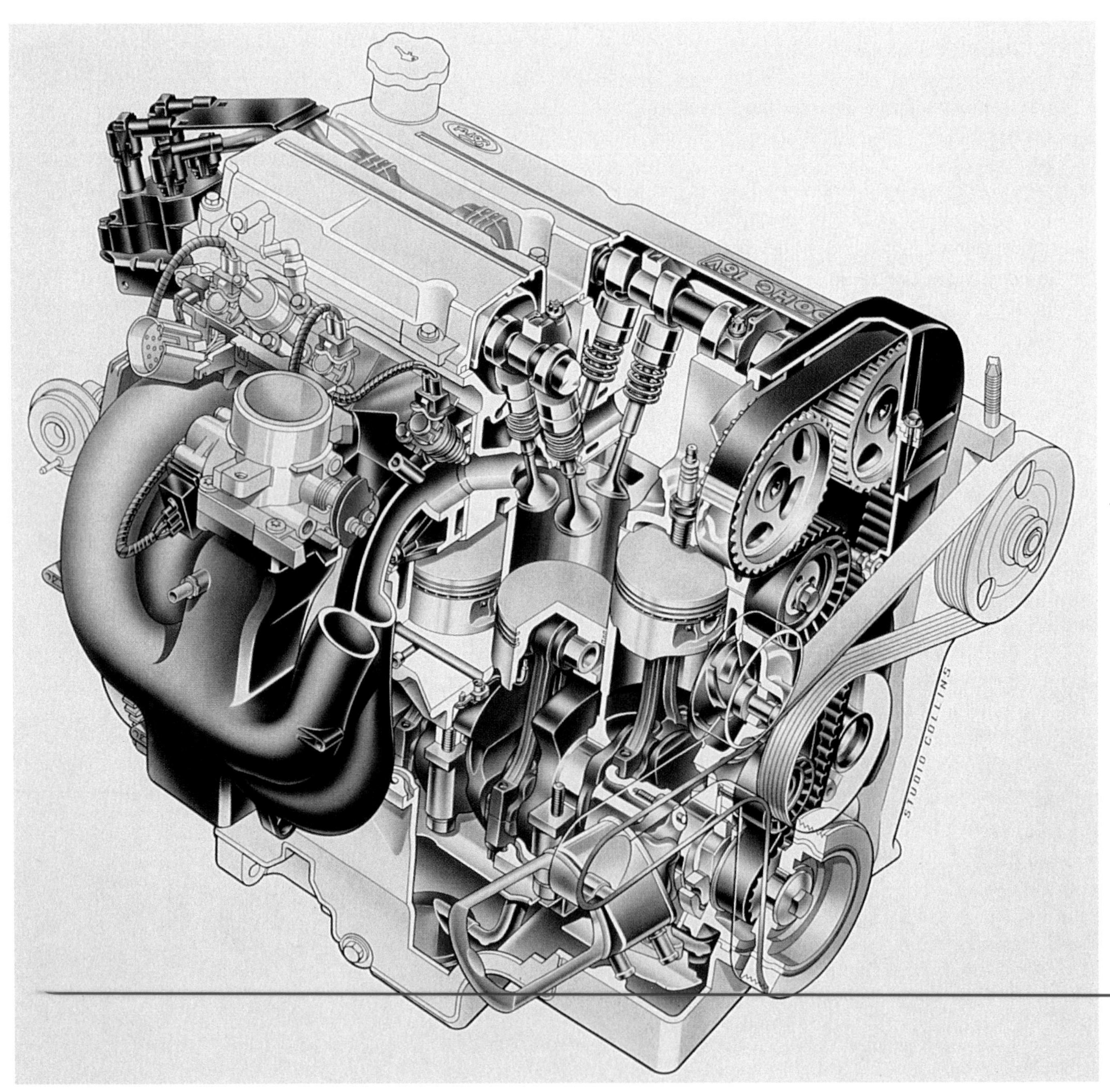

Cutaway of a 2.0 liter inline four-cylinder engine. Many of the parts in this engine are similar to those you just studied. (Ford Motor Co.)

2

Design, Construction, Application of Engine Components

After studying this chapter, you will be able to:

- Describe engine part design.
- Explain the construction of engine components.
- Define the purpose of each engine part.
- Identify engine part variations.

In Chapter 1, you designed and built an engine on paper. To give you a clear idea of how the engine works, it was necessary to move from step to step without studying the fine points of engine construction. Now that you are familiar with the general theory involved in engine design, you will be able to study the parts of the basic engine in greater detail. It is important for you to understand how these parts are built, of what materials they are constructed, why they are built the way they are, and the purpose each part serves.

The Engine Block

The ***engine block***, **Figure 2-1,** serves as a rigid metal foundation for all parts of an engine. It contains the cylinders and supports the crankshaft and camshaft. In older engines, the valve seats, ports, and guides are built into the block. Accessory units and the clutch housing are bolted to it. Note in **Figure 2-1** that the crankcase is formed with the block.

Blocks are made of either cast iron or aluminum. In some small one-cylinder engines, the material is die-cast metal. Die-cast metal is a relatively light, soft metal especially suited to the die casting process.

Blocks are commonly formed in two ways. One method is to pour molten iron or aluminum into a mold made of sand. A core is placed within the mold to form the cavities and passageways within the block. See **Figure 2-2.** After the casting has cooled, it is removed from the mold and the sand core is dissolved and washed out.

The second method forces molten aluminum or die-cast metal into a metal mold under pressure. The pressure-casting process has several advantages. It produces a block that is free of air bubbles (called voids), gives sharp corners, and is extremely accurate. This keeps machining operations to a minimum. The same mold can be used over and over. All parts of the aluminum or die-cast block that are subjected to wear will have iron or steel inserts either pressed in place or cast into the block.

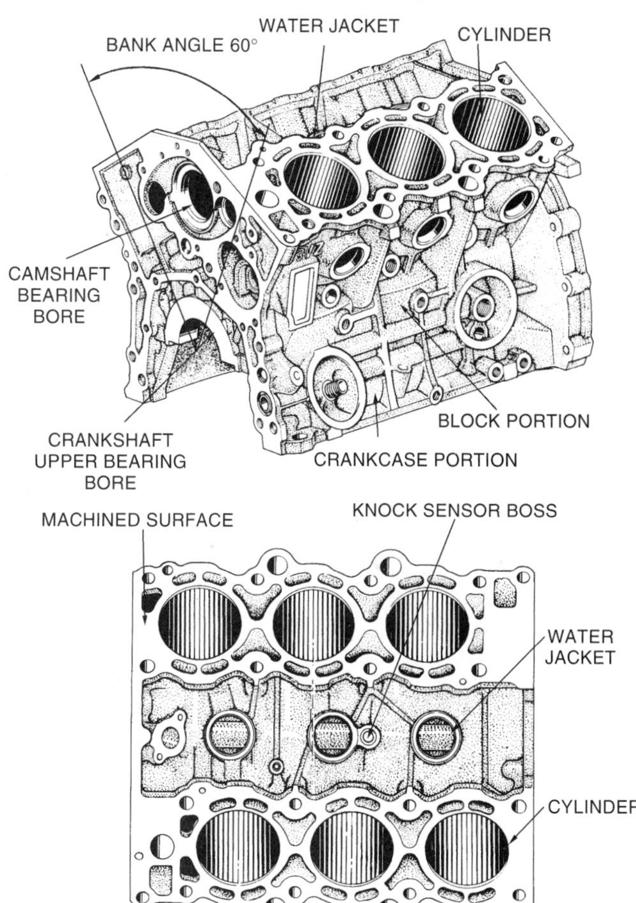

Figure 2-1. Typical block construction for a V-6, valve-in-head, gasoline engine. (Toyota)

The lighter the block (providing it has sufficient strength), the better. The modern thin-wall casting process controls core size and placement much more accurately than the older casting process. This permits casting the block walls much thinner, reducing the weight of the block. Since the block wall thickness is more uniform, block distortion during service is less severe.

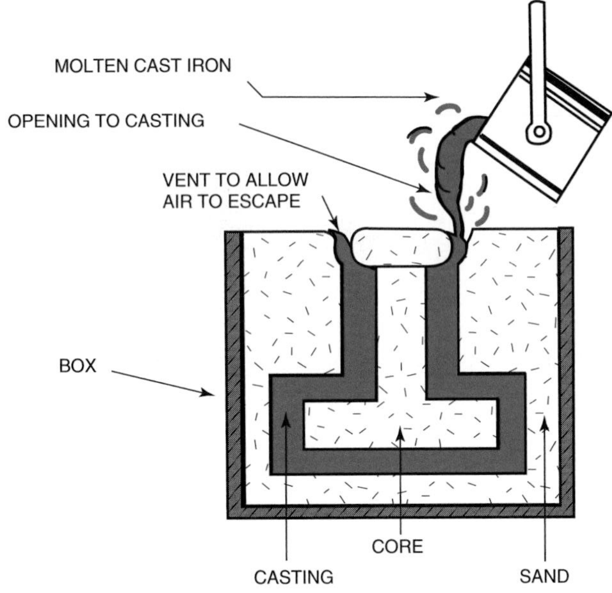

Figure 2-2. Simple mold used for block casting.

Cylinders

The cylinder is a round hole formed in the block. See **Figure 2-3.** It is first cast into the block and then bored and honed to a smooth finish. The cylinder dimensions must be extremely accurate. A good cylinder will not vary in diameter more than .0005″ (0.013 mm). The paper on which this text is printed is around .004″ (0.102 mm) thick.

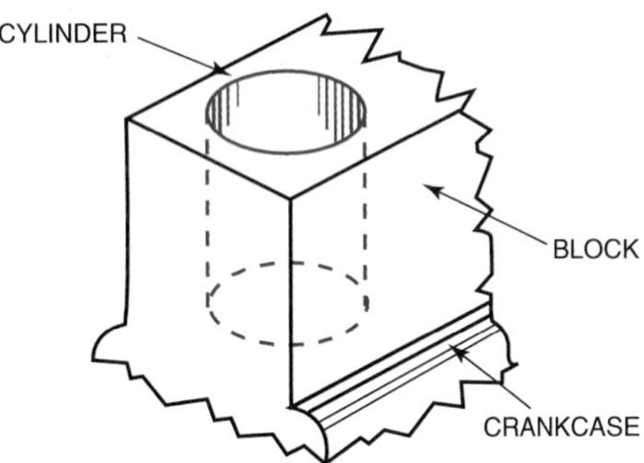

Figure 2-3. Section of an engine showing cylinder location.

The cylinder forms a guide for the piston and acts as a container for taking in, compressing, firing, and exhausting the air-fuel charge. Cylinders have been made of both steel and cast iron. Cast iron is by far the most popular.

Cylinder Sleeves

When steel cylinders are desired in an aluminum block, they are installed in the form of ***cylinder sleeves*** (round, pipe-like liners). These sleeves may be either cast or pressed into the block. Some engines use removable cylinder sleeves. When the cylinder becomes worn, the old sleeves can be pulled out and new sleeves can be pressed in. See **Figure 2-4.** The sleeves are pressed into oversize cylinder holes.

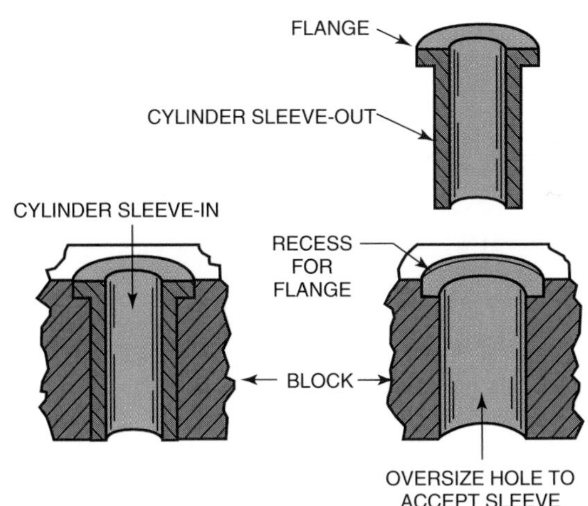

Figure 2-4. Section of a block showing typical cylinder sleeve installation.

Cylinder sleeves are widely used in heavy-duty truck and industrial engines. Sleeves can also be used to repair a worn or cracked cylinder in a cast iron block.

Cylinder sleeves are either the dry or wet type. The ***dry sleeve*** is pressed into a hole in the block and is surrounded over its full length by block metal, **Figure 2-5.** This type of sleeve can be quite thin because it utilizes the block metal to give it support over its full length. The ***wet sleeve*** is also pressed into a hole in the block. However, it is supported at the top and bottom only, **Figure 2-6.** The engine coolant is allowed to directly contact the sleeve. The wet sleeve must be heavier than the dry sleeve because it receives no central support from the block.

Securing the Sleeve

Sleeves can be secured in the block in several ways. Where a cast iron or steel sleeve is placed in an aluminum block, it can be cast into place. See **Figure 2-7.** The removable sleeve may be held in place by friction alone, **Figure 2-8.** However, this requires a very tight fit and is not dependable.

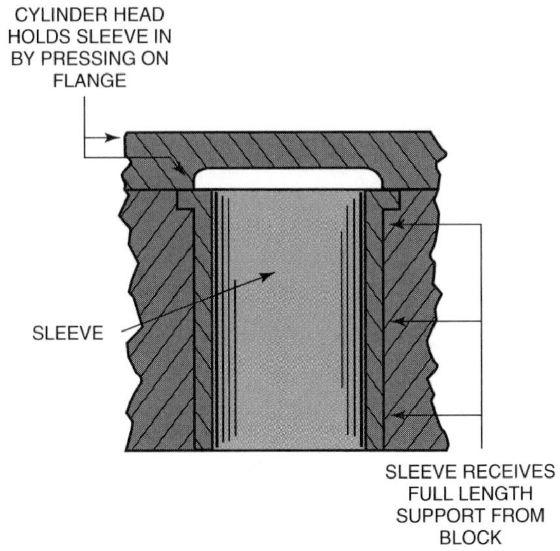

Figure 2-5. A dry sleeve in place in cylinder hole.

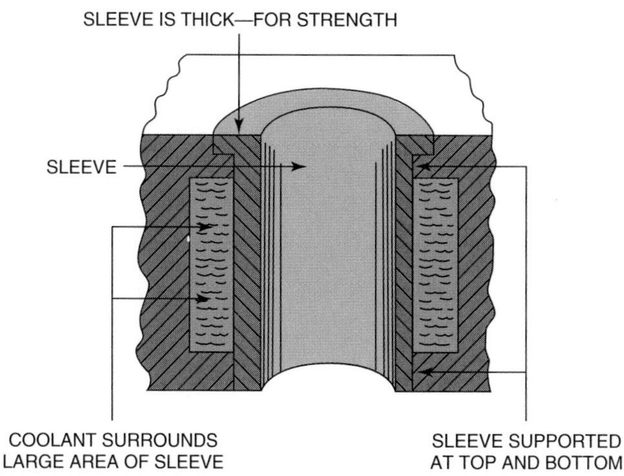

Figure 2-6. A wet sleeve in place in cylinder hole. Note that the sleeve is surrounded by coolant.

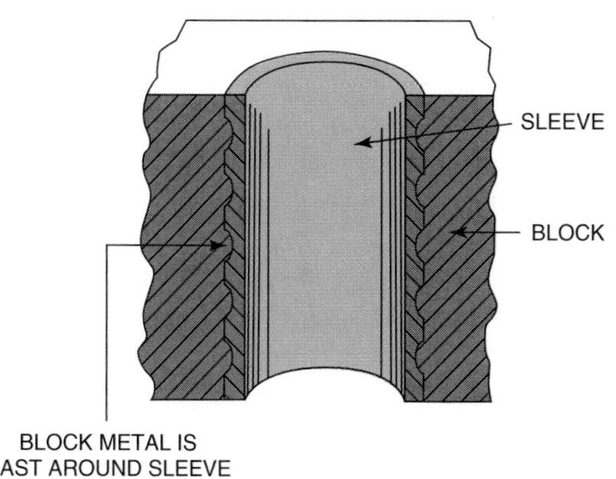

Figure 2-7. Some sleeves are held by the block casting. The grooves help prevent movement.

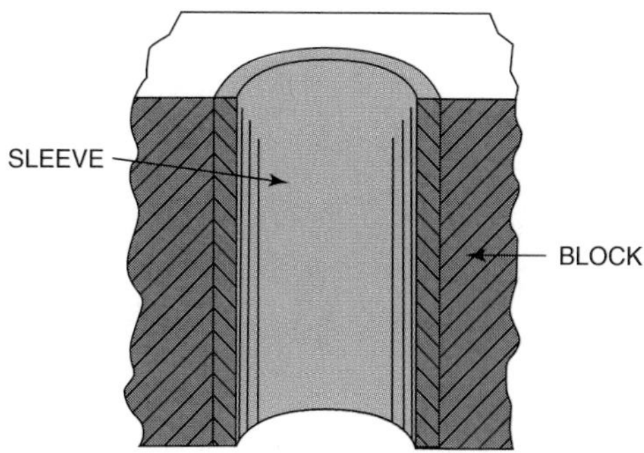

Figure 2-8. Sleeve held in place by friction between sleeve and block.

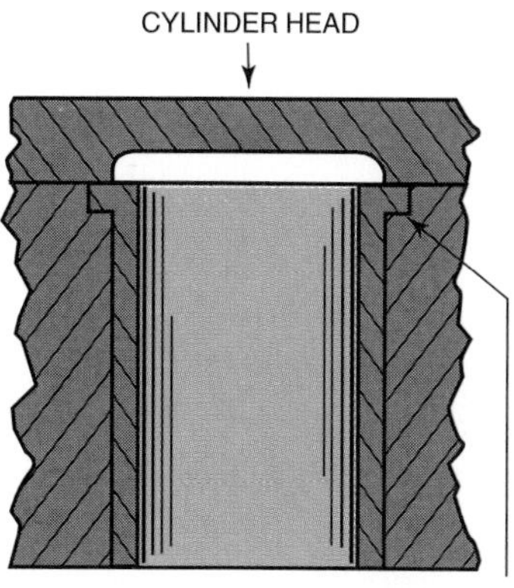

Figure 2-9. The flange on this sleeve fits a corresponding groove in the block. Sleeve is held in place by the cylinder head.

A better method is to have a flange on the top edge of the cylinder sleeve that drops into a corresponding groove in the block. See **Figure 2-9.** When the cylinder head is bolted on, it presses on the flange and holds the sleeve in place. This type of sleeve can be fitted with a greater degree of freedom. However, any cylinder sleeve must be a fairly snug fit so that heat is conducted away by the surrounding block material.

Pistons

The piston is literally a sliding plunger that rides up and down in the cylinder. It has several jobs to do in proper sequence.

The piston must move down through the cylinder to produce a vacuum to draw a fuel charge into the cylinder. It then travels up in the cylinder and compresses the mixture. When the mixture is fired, the pressure of the expanding gas is transmitted to the top of the piston. This drives the piston back down through the cylinder with great force, transmitting the energy of the expanding gas to the crankshaft. The piston then travels up through the cylinder and exhausts the burned fuel charge.

The overall job the piston performs is difficult. A piston is subjected to intense heat from the burning air-fuel mixture. It must change directions at high speed. It is subjected to friction against the cylinder walls. In addition to all this, the piston receives the tremendous thrust of power on the power stroke. Study the cross-sectional drawing of a piston in **Figure 2-10.** Learn the names of all the parts of the piston.

Piston Materials

Pistons are usually made of aluminum. Often, aluminum pistons are tin-plated to allow proper break-in when the engine is started. Aluminum pistons can be forged, but they are more commonly cast.

The aluminum piston is light. In most cases, this gives it an advantage over the cast iron type. A piston must change its direction of travel at the end of every stroke. Considering the fact that modern engines sometimes reach speeds in excess of 6000 revolutions per minute (RPM), it is obvious that lighter pistons help increase engine efficiency.

Cast iron is a good material for pistons used in a slow-speed engine. It has excellent wear characteristics and will provide good performance. Aluminum pistons that are designed to operate in aluminum cylinders are iron-plated.

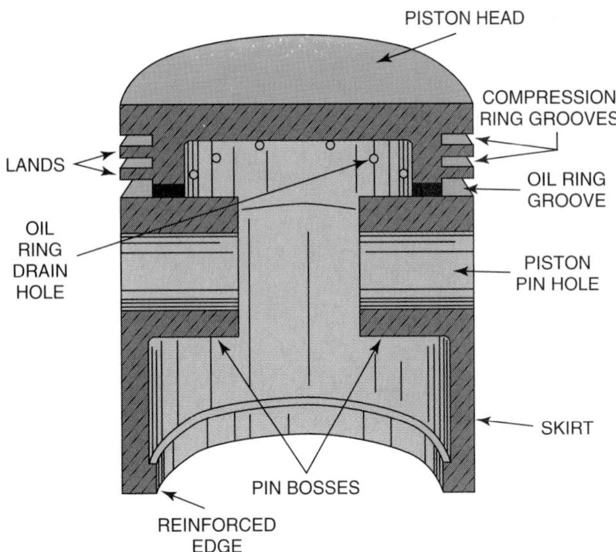

Figure 2-10. Sectional view of a piston. Note the names of each section of the piston.

Piston Expansion Problems

Pistons must be carefully fitted into engine cylinders to prevent them from tipping from side to side (slapping). They must hold the burning fuel charges above the piston heads and be tight enough against the cylinder walls to form a vacuum, compress the fuel charge, exhaust the burned gases, and prevent excess oil loss.

A piston will expand when it gets hot, so enough clearance must be left between the piston and cylinder wall to allow for this. Aluminum pistons expand more than cast iron pistons. See **Figure 2-11.** The problem of fitting an aluminum piston close enough to prevent tipping while leaving enough clearance for an oil film to separate the piston and the cylinder has been solved in several ways.

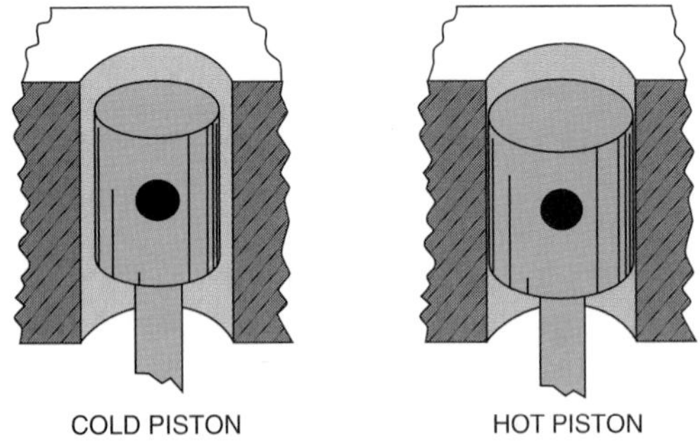

Figure 2-11. Exaggerated view shows the effect of heat on piston expansion.

Steel Strut Pistons

Steel struts and, in some cases, steel rings are cast into aluminum pistons. See **Figure 2-12.** Steel expands less than aluminum, and as a result, the steel struts tend to minimize piston expansion.

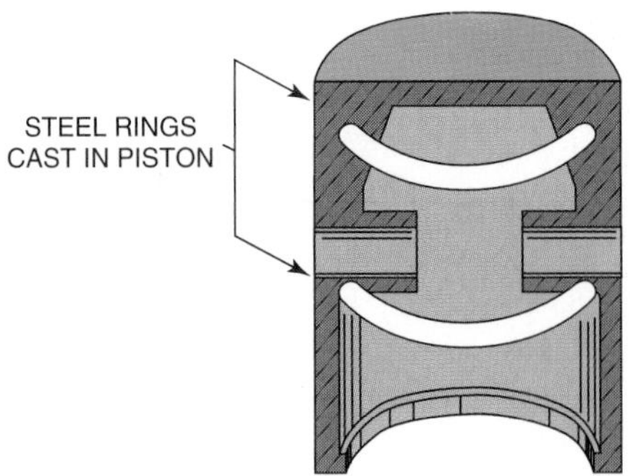

Figure 2-12. Steel rings control or minimize piston expansion.

Cam Ground Piston

The ***cam ground piston***, **Figure 2-13,** is a popular type. Instead of being made round, this piston is ground so that it is elliptical, or egg-shaped. Note in **Figure 2-13** that diameter A is larger than diameter B.

Diameter A is established so that the piston has a minimum amount of clearance in the cylinder. This clearance, around .001″ (0.025 mm), is necessary to allow oil to form a lubrication film between the piston and cylinder wall. The larger diameter at A in **Figure 2-13** is always at right angles to the block (in relation to the crankshaft centerline).

As the piston heats up, diameter A will not expand as much as diameter B. This will cause the piston to become round when fully heated. See **Figure 2-14.** Cam grinding will give a minimum clearance at the thrust surfaces. The thrust surfaces are the two sides of the piston that contact the cylinder walls at right angles to the crankshaft and the piston pin. See **Figure 2-15.** These surfaces support the piston and prevent tipping.

Partial Skirt Piston

To make ***partial skirt pistons***, (also called slipper skirt pistons), a large area of the ***piston skirt*** (the bottom section

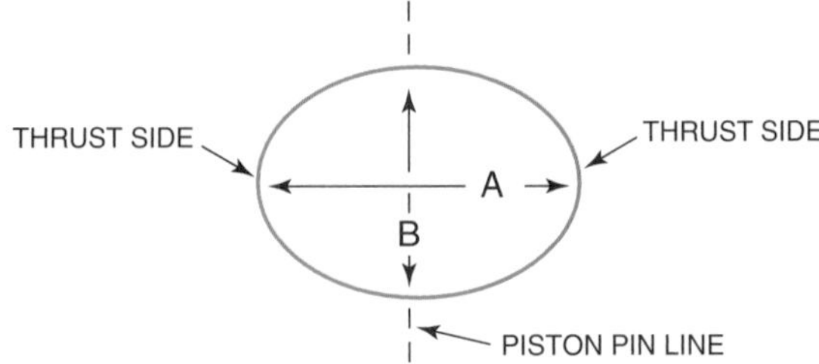

Figure 2-13. Exaggerated top view of a cam ground piston, which is designed to be wider across the thrust surfaces.

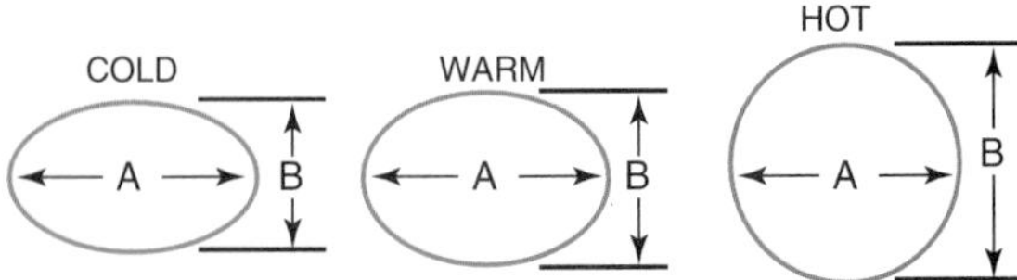

Figure 2-14. A cam ground piston becomes round when hot. Diameter A remains constant while diameter B expands.

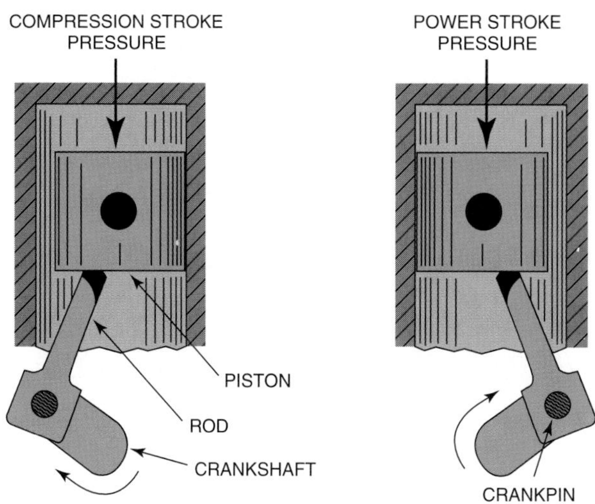

Figure 2-15. Exaggerated view shows how a loose piston is pushed from one side of the cylinder to the other. The sides that contact the cylinder wall are called thrust surfaces.

of the piston) is removed. This reduces piston weight and allows the piston to move closer to the crankshaft.

Since the non-thrust sides of the skirts are removed, the counterbalances on the crankshaft will not strike the piston, **Figure 2-16.** The non-thrust sides of a piston do not carry much of a load, and therefore, their removal is not detrimental. The partial skirt piston is also cam ground.

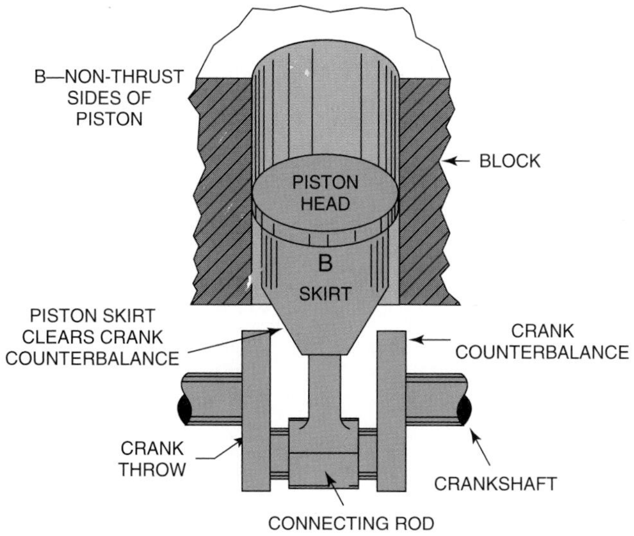

Figure 2-16. Non-thrust sides of the piston skirt are removed from partial skirt pistons. This allows the piston to approach the crankshaft closely without striking.

Split Skirt Piston

Some older engine designs make used of a ***split skirt piston***, **Figure 2-17,** the skirt is either partially or, in some cases, completely split. When the piston warms and begins to

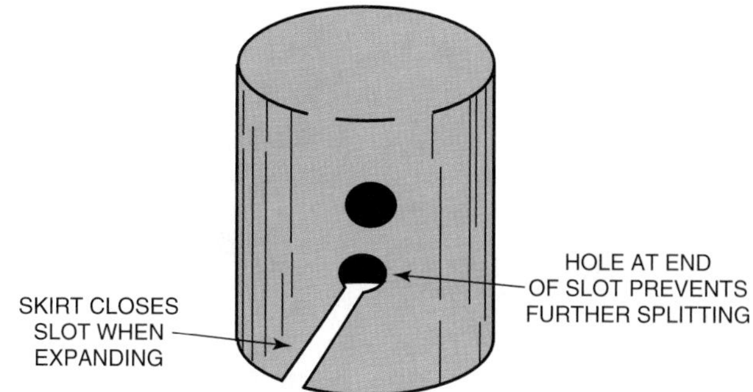

Figure 2-17. A split skirt piston is sometimes used to solve expansion problem.

expand, it cannot bind in the cylinder because the skirt merely closes the split.

T-Slot Piston

The ***t-slot piston***, **Figure 2-18,** is a variation of the split skirt. The top of the T tends to retard the transfer of heat from the piston head to the skirt. The vertical slot allows the skirt to close in when heated.

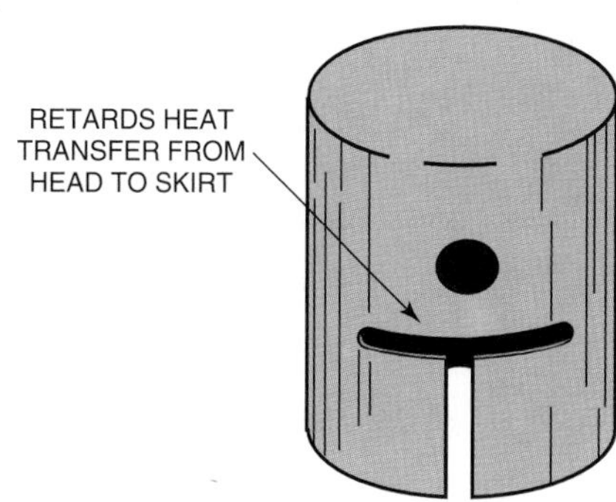

Figure 2-18. A T-slot piston is another type of piston used to control expansion.

Piston Temperature

The piston head is subjected to the direct heat of the exploding fuel. This heat can raise the temperature of the piston crown (very top) to over 600°F (316°C). The temperature will be lower as you go down the piston. The bottom of the skirt will be about 300°F (149°C), **Figure 2-19.**

The temperatures will vary according to engine design and work application. Because the bottom of the skirt is the coolest, the skirt on some pistons is slightly larger in diameter at the bottom. The top area of the skirt would be smaller in diameter.

Piston Construction

It is obvious that the piston head is by far the hottest part of the piston. As a result, it expands more. In order to avoid having the piston head grow tight in the cylinder, it is turned to a smaller diameter than the skirt (not cam ground). The head

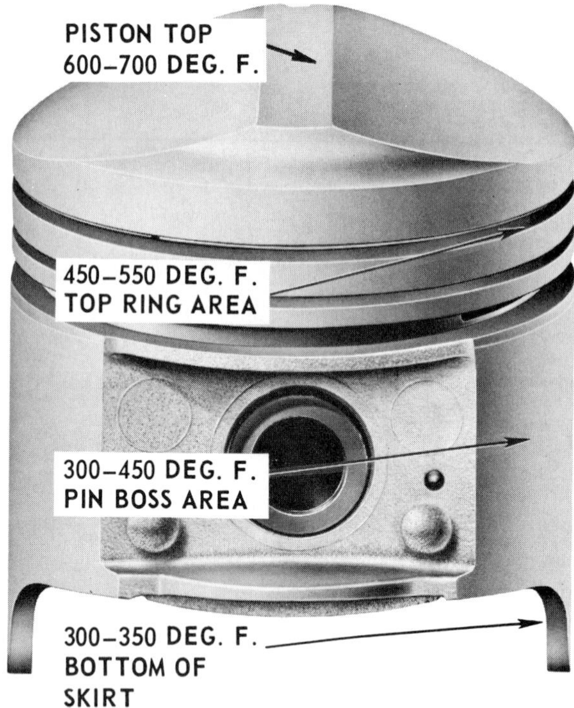

Figure 2-19. Approximate temperature range from piston head to skirt. Piston temperature can vary considerably based on engine type, design, and use.

will generally be .030″ to .040″. (0.76 to 1.02 mm) smaller than the skirt, **Figure 2-20.** The section of the piston that supports the pin must be thicker and stronger. This area is called the ***pin boss***. See **Figure 2-21.**

Piston Head Shape

Some pistons have flat-topped heads. Others are designed to assist in creating a rapid swirling and to help

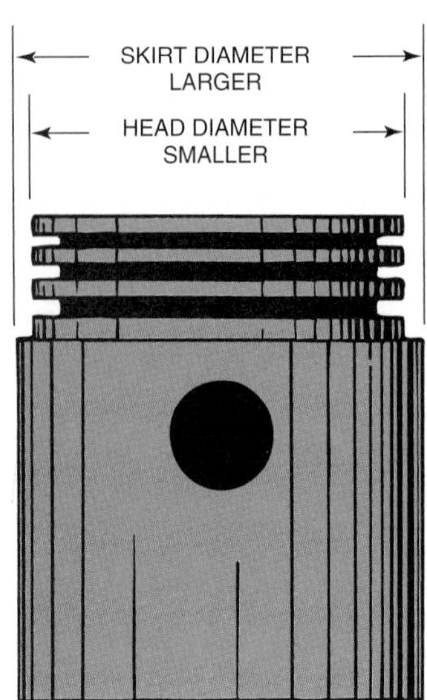

Figure 2-20. As the head is the hottest part of a piston, it must be ground approximately .030″-.040″ (0.76 mm to 1.02 mm) smaller than the skirt

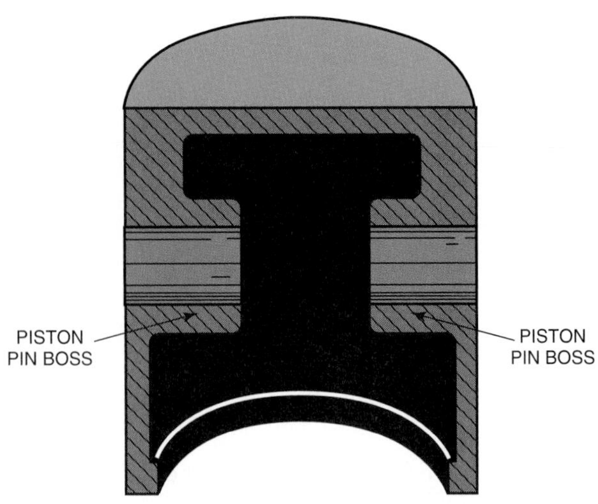

Figure 2-21. The pin boss must be strong to withstand directional changes during each stroke of the piston.

break up gasoline particles on the compression stroke. In one design, the shape of the combustion chamber is formed in the head of the piston, allowing the use of a flat-surfaced cylinder head. See **Figure 2-22.**

Piston Rings

The piston must have some clearance in the cylinder. If the skirt has .001″ to .002″ (0.025 to 0.05 mm) clearance and the head has .030″ to .040″ (0.76 to 1.02 mm) clearance, it is obvious that the piston cannot seal the cylinder effectively. See **Figure 2-23.**

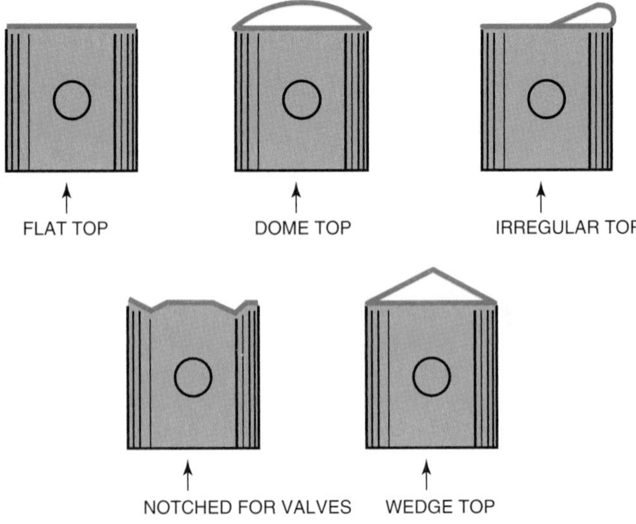

Figure 2-22. Several types of piston heads that have been used in engines. The flat top is the one most commonly used today.

To solve the leakage problem, ***piston rings*** are used. A properly constructed and fitted ring will rub against the cylinder wall with good contact all around the cylinder. The ring will ride in a groove that is cut into the piston head. There will be

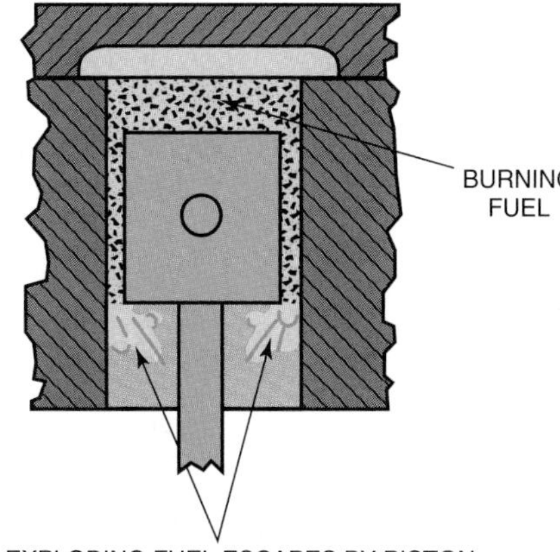

Figure 2-23. A piston cannot seal by itself because clearance with the cylinder wall must be maintained.

a slight clearance between the sides of the ring and the edges of the groove. This clearance, which is known as side clearance, is generally around .002″ (0.05 mm).

The rings do not contact the bottom of the ring grooves. Actually, the ring will rub the cylinder wall at all times but will not be solidly fastened to the piston at any one point. See **Figure 2-24.**

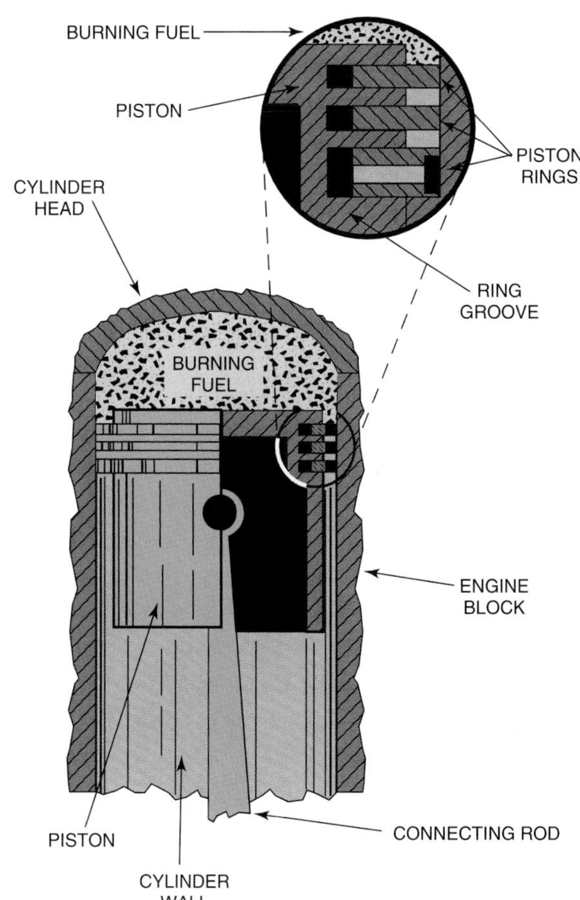

Figure 2-24. Piston rings are used to seal the gap between the piston and cylinder wall.

Ring Gap

The ring is built so it must be squeezed together to place it in the cylinder. This will cause the ring to exert an outward pressure, which keeps it tightly against the cylinder wall. See **Figure 2-25.** The ring is not solid all the way around, but is cut through in one spot. This cut spot forms what is called the ***ring gap***. See **Figure 2-26.**

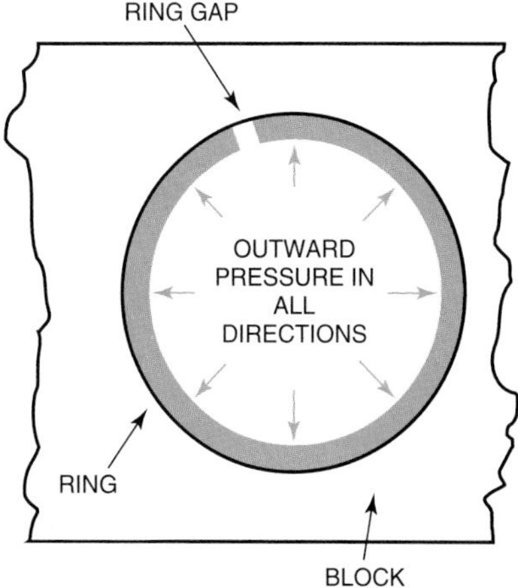

Figure 2-25. Top view of the cylinder shows the sealing action of a piston ring.

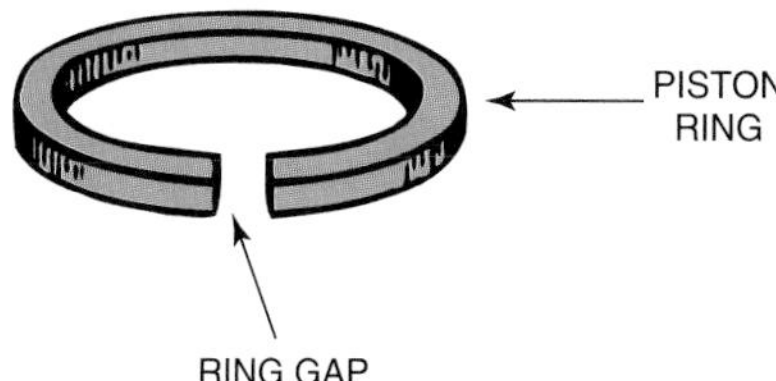

Figure 2-26. Simplified piston ring showing ring gap present on piston rings.

When the ring is in the cylinder, the cut ends must not touch. When the ring heats up, it will lengthen. Since it cannot expand outwardly, it will close the gap. See **Figure 2-27.** If there is not enough gap clearance, the ends will touch as the ring expands. As the ring continues to lengthen, it will break up into several pieces. This can quickly ruin an engine.

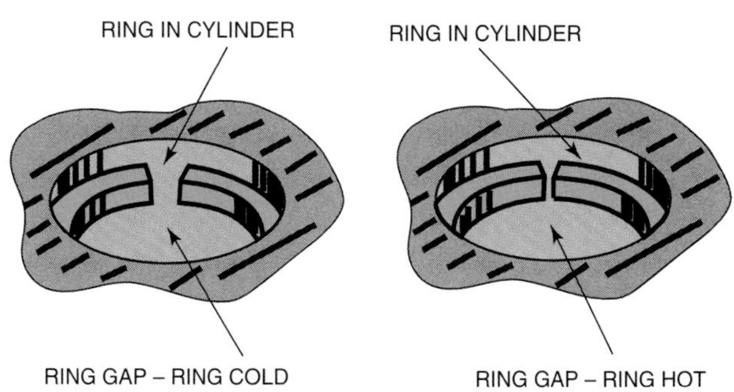

Figure 2-27. When the rings get hot, the gap closes.

A general rule for ring gap clearance is to allow .003″ to .004″ per inch (0.07 to 0.10 mm per 25.4 mm) of cylinder diameter. For example, a 4″ (101.6 mm) diameter cylinder would require from .012″ to .016″ (0.30 to 0.41 mm) clearance in the gap.

Many different types of joints have been used in an endeavor to stop leakage through the ring gap. This leakage is commonly referred to as ***blow-by***. It has been found that the common butt joint is effective and is simple to adjust. **Figure 2-28** illustrates a few of the joints that have been used.

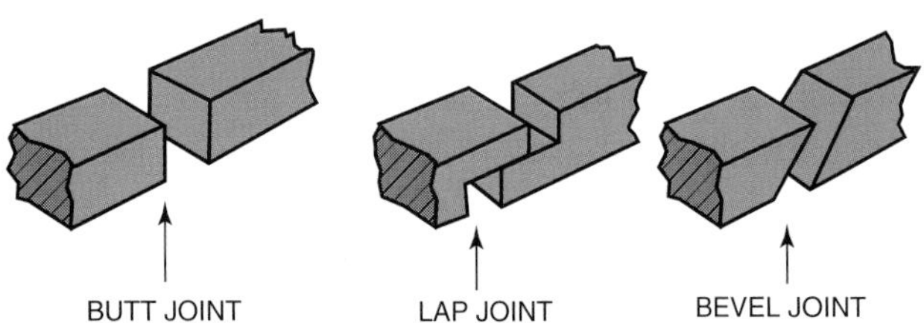

Figure 2-28. Types of piston ring gap joints. Most engines use three rings on each piston, two compression rings and one oil control ring. Others use two compression rings and two oil rings. Some diesel engines use five or more rings. All rings may be above the piston pin; or a second oil control ring may be set into a groove near the bottom of the skirt. The compression rings are always used in the top grooves and the oil control rings in the lower grooves.

The ring is placed in the groove by expanding it until it will slip over the piston head and slide down and into the ring groove. **Figure 2-29** illustrates how a ring fits the piston groove. Note that there is both side clearance and back clearance.

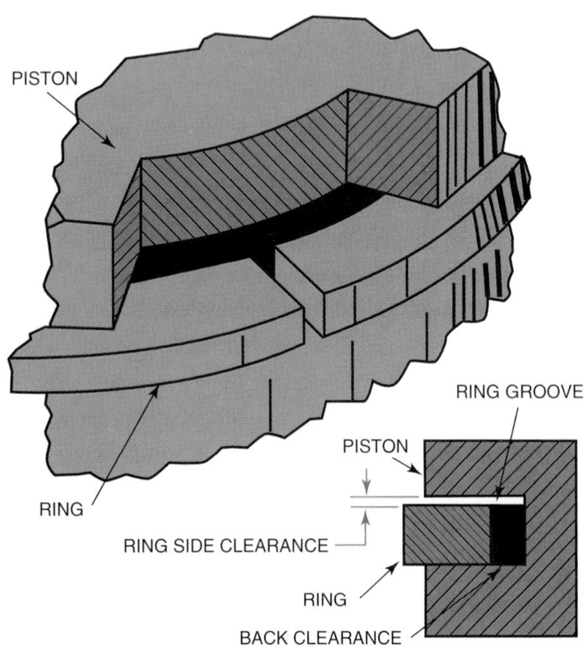

Figure 2-29. Compression ring-to-groove fit. Note that the ring has both side and back clearance. Side clearance is actually quite small, running around .0015″ to .002″ (0.04 to 0.05 mm).

Types of Rings

There are two distinct types of rings. One is called a ***compression ring,*** and the other is called an ***oil control ring***. Most engines have three rings: two compression rings at the top and one oil control ring at the bottom, **Figure 2-30.**

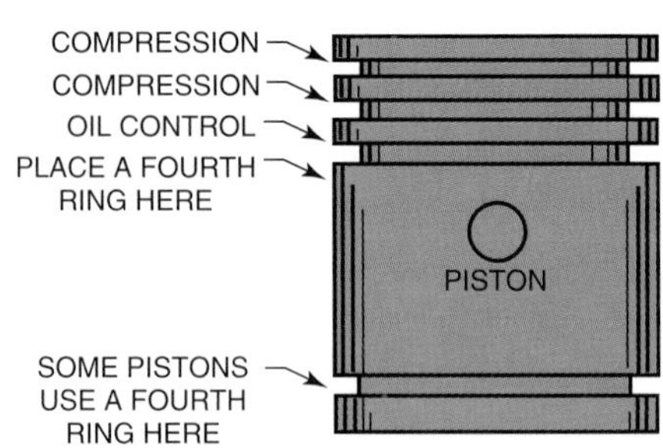

Figure 2-30. Location of compression and oil control rings. The locations and number of rings vary by application.

Compression Rings

Compression rings are designed to prevent leakage between the piston and the cylinder, **Figure 2-31.**

Various types of grooves, bevels, and chamfer shapes are used to achieve this goal. The idea is to create an internal stress within the ring. This stress will tend to cause the ring to twist in such a fashion that the lower edge of the ring is pressed against the cylinder wall on the intake stroke. This will cause the ring to act as a mild scraper. The scraping effect will assist in the removal of surplus oil that may have escaped the oil control rings, **Figure 2-32.**

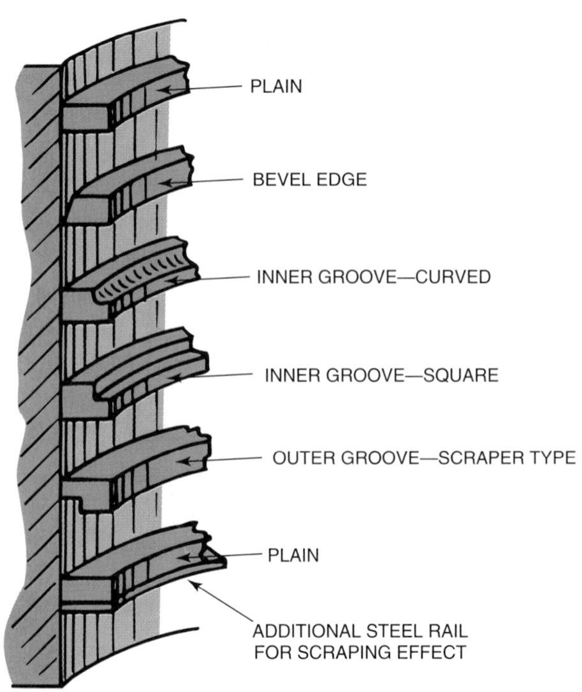

Figure 2-31. Typical compression ring shapes as they would look in a cylinder.

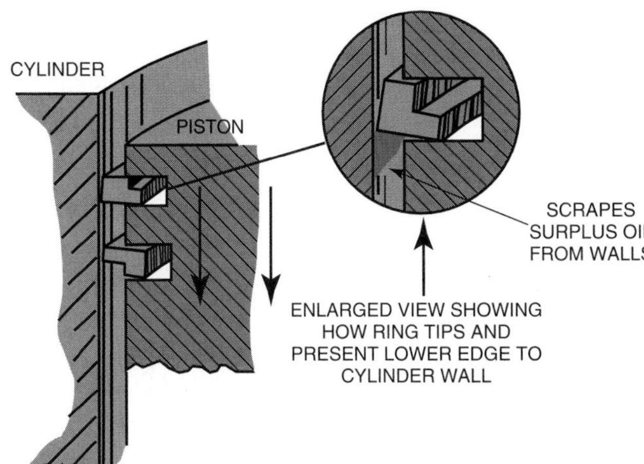

Figure 2-32. Internal stress causes the rings to tip and act as mild scrapers.

On compression and exhaust strokes, the rings will tend to slip lightly over the oil film. This will prolong the life of the ring, **Figure 2-33.** On the power stroke, pressure of the burning gases will force the top edge of the ring downward. This causes the ring to rub the wall with full face contact and provides a good seal for the enormous pressure generated by the power stroke. See **Figure 2-34.**

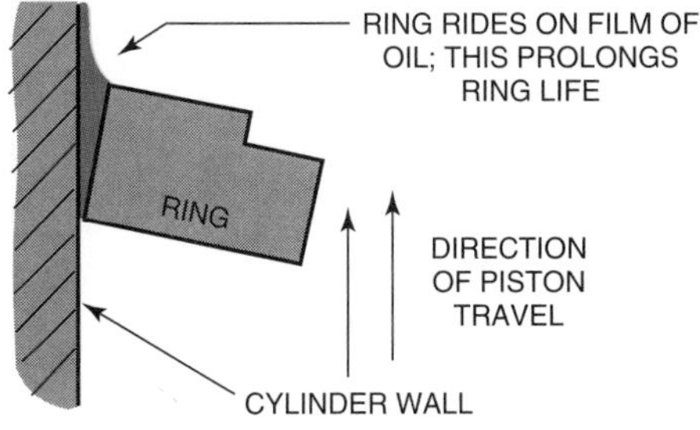

Figure 2-33. On compression and exhaust strokes, piston rings tip and slide easily on a film of oil.

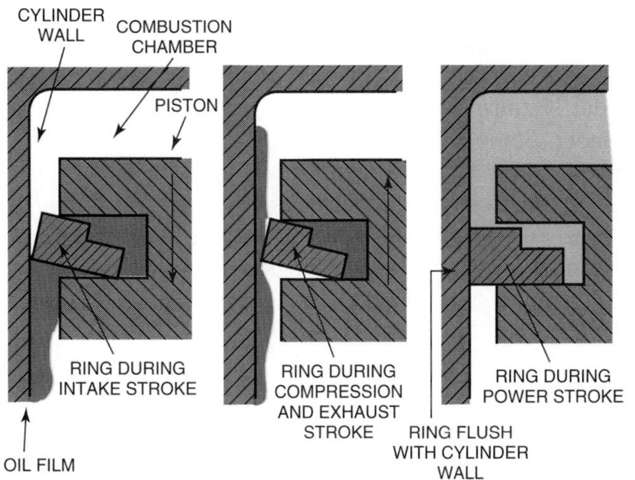

Figure 2-34. Pressure of burning gases during the power stroke force the ring down and out so that it is flush with the cylinder wall.

Heat Dam

The compression rings, especially the top ring, are subjected to intense heat. To minimize the transfer of heat from the head of the piston to the top ring, a ***heat dam*** is sometimes used. This is actually a thin groove cut into the head of the piston between the top ring groove and the top of the piston.

Instead of passing through the aluminum of the piston to the ring, the heat encounters the heat dam. The heat cannot move as readily through the empty space in the groove as it can through the piston metal. This helps minimize heat transfer, **Figure 2-35.**

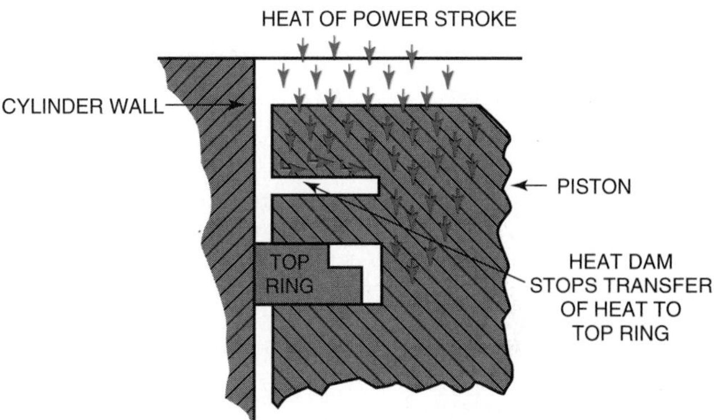

Figure 2-35. Heat dam are used to keep the top ring cool.

Top Ring Groove Insert

Some aluminum pistons have nickel-iron or comparable metal inserts cast into the piston heads. The top ring groove is cut in this metal. Since top ring grooves in aluminum pistons tend to be pounded out of shape during engine operation, this insert will prolong the useful life of the piston and ring, **Figure 2-36.**

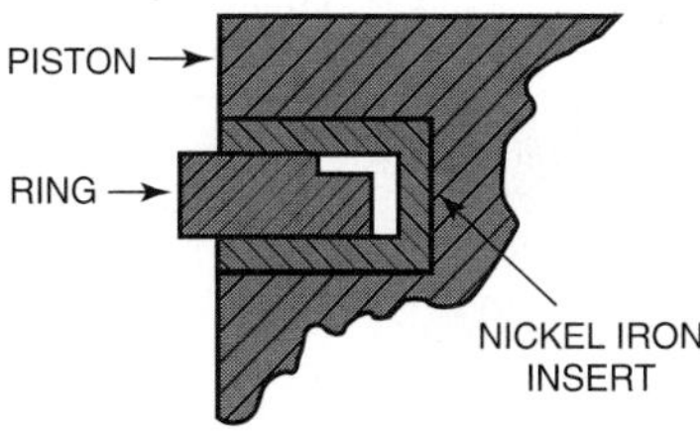

Figure 2-36. Top ring insert groove cast into the piston.

Oil Control Ring

The oil control ring is used to scrape the surplus oil from the cylinder walls. This prevents oil from escaping into the combustion chamber. As engine designs have changed, much research has been devoted to the design of oil rings. See **Figure 2-37.**

All oil rings are slotted and have edges that scrape the excess oil from the cylinder walls. The oil between the edges passes through slots on the ring. When two rings are used, the oil passes through slots in the bottom of the ring groove. From there, the oil drips down into the crankcase area, **Figure 2-38.**

Expander Devices

Some ring sets that are designed for worn cylinders utilize expanding springs between the bottom of the ring groove and the ring. This will force the ring outward against the cylinder wall. The top of a worn cylinder is usually wider than the bottom. When the head of the piston is on the bottom of its stroke, the rings will be squeezed in the smaller section of the cylinder. See **Figure 2-39.**

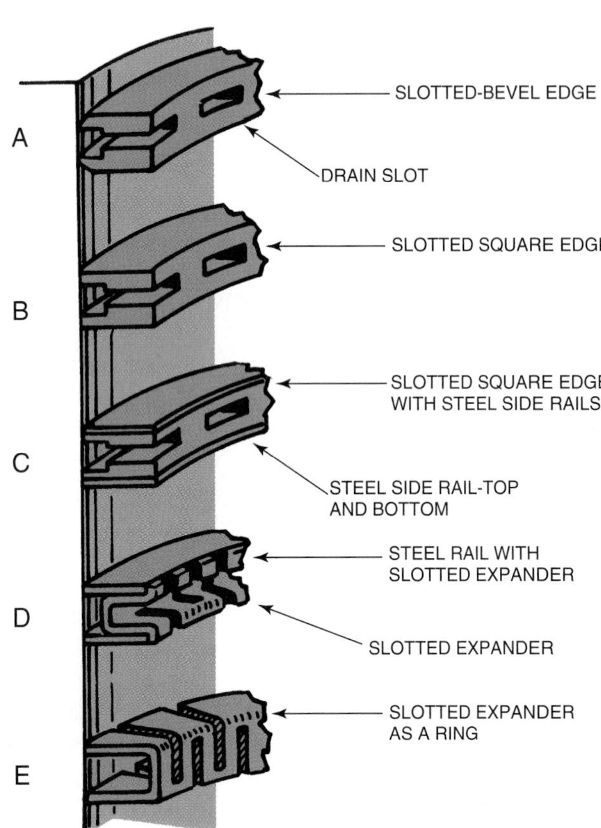

Figure 2-37. Several common types of oil control rings.

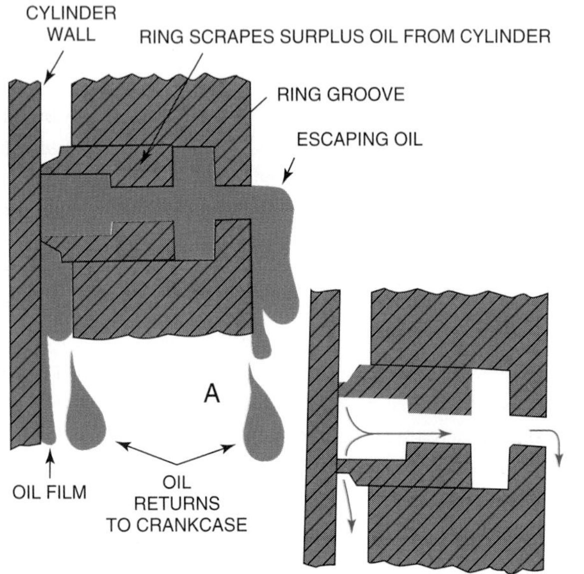

Figure 2-38. A—This shows the action of the oil ring as it travels down the cylinder wall. B—Note the path the oil takes through the ring.

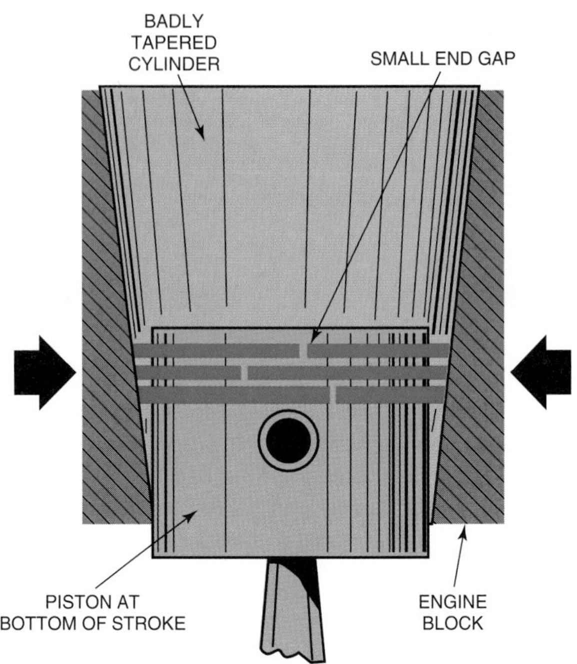

Figure 2-39. If a cylinder is badly tapered from wear, the rings are compressed in the bottom of the cylinder.

When the piston travels up the cylinder, the rings must expand outward to follow the ever-widening cylinder diameter, **Figure 2-40.** This makes it necessary for the rings to expand and contract during every stroke. If engine speed is high enough, the piston will travel up the cylinder and will be snapped back down before the rings have time to expand. This leaves the piston at the top of the stroke with the rings not touching the cylinder walls.

Expander devices used for rings designed for worn cylinders are stronger than those used for a new or rebored cylinder. **Figure 2-41** illustrates a common type of ring expander. Some expanders do not touch the bottom of the grooves. In this type, the ends butt together, and when the ring is compressed during piston installation, the expander is compressed within itself. Another type of expander is a round wire spring type. The ends butt together and even though the spring does not touch the bottom of the groove, it still pushes out on the ring, **Figure 2-42.**

Rings Must Fit the Cylinder

In addition to following the cylinder walls, rings must make perfect contact all the way around. **Figure 2-43** illustrates a ring that is poorly fitted and would do a poor job of sealing.

Even with accurately bored cylinders and new rings, it is impossible to secure an absolutely perfect fit when the rings

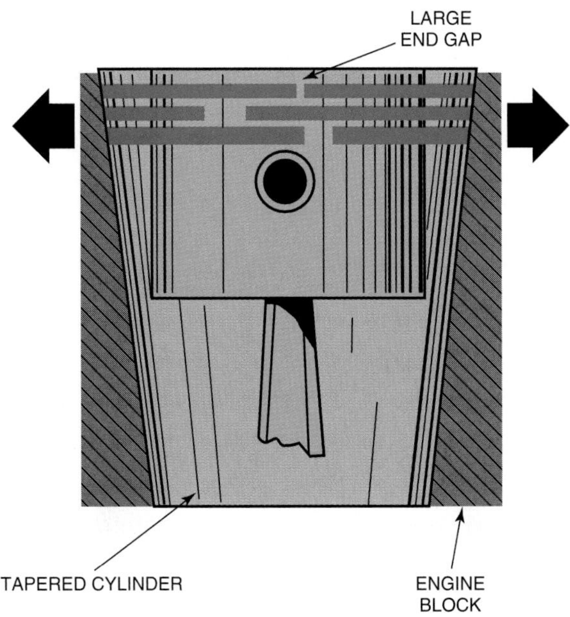

Figure 2-40. When the piston reaches the top of the tapered cylinder, the rings expand. Note the large end gaps.

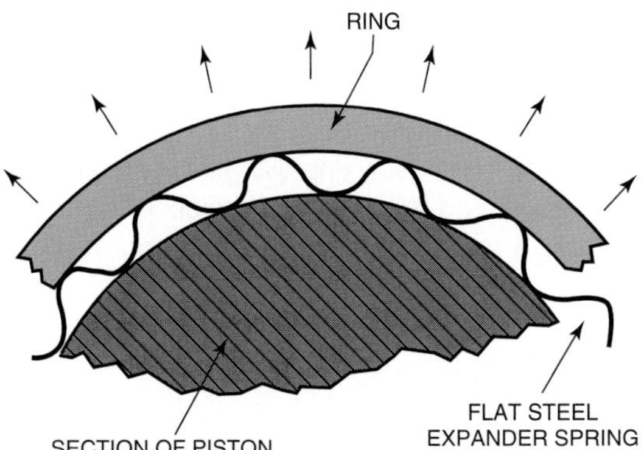

Figure 2-41. Spring expander rests on the bottom of the ring groove and forces the ring outward. This pressure will keep the ring in constant contact with the cylinder wall.

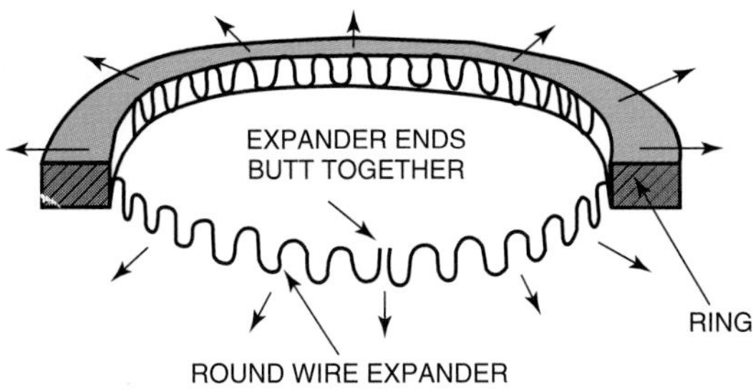

Figure 2-42. A round wire spring expander does not touch the bottom of the ring groove. When the ring is compressed to enter the cylinder, the spring is compressed within itself to keep constant tension on the ring.

are first installed. After the engine has operated for several hundred miles, the rings will wear into perfect contact with the cylinder walls.

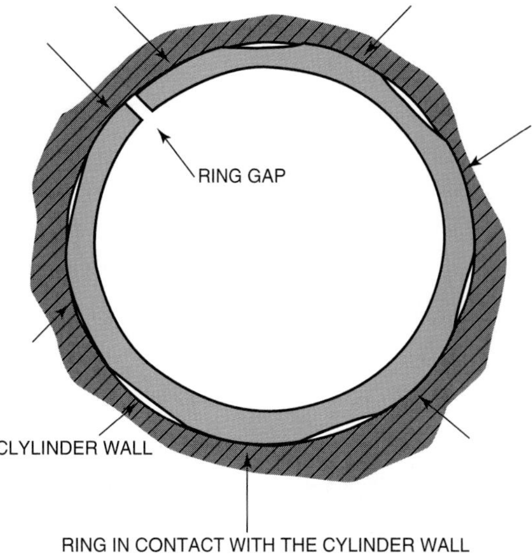

Figure 2-43. Poor ring contact with the cylinder wall. Arrows indicate high spots that touch, the area between the arrows do not touch.

Ring Break-In

To facilitate a fast job of ring break-in, the ring's outer face is left rough. Although this surface feels smooth to the touch, a close inspection will show that fine grooves are left in the ring surface. Similarly, the cylinders may look smooth, but final honing with fine stones leaves tiny surface scratches in the walls.

When the engine is started, the rings will be drawn up and down the cylinder. Since both the ring faces and the walls have fine scratches, some wear will occur. Any high spots on the ring faces will soon be worn off and the ring will fit properly, **Figure 2-44.**

The ring and wall surfaces must be designed with a degree of roughness that will cause the rings to wear in. By the time an acceptable fit is achieved, the initial roughness will be gone and excessive wear will cease.

Special Ring Coatings

To assist with a fast wearing-in period, ring faces are often coated with a soft, porous material. This material also absorbs some oil and allows a gentle wear-in. Materials such as graphite, phosphate, and molybdenum are used for this purpose. See **Figure 2-45.** Rings can also be tin coated.

All the expanding pressure of new rings is applied to the walls at the ring high spots. This can cause overheating and ***scuffing*** at these points. Scuffing is a roughening of the cylinder wall. It is caused when there is no oil film separating the moving parts and metal-to-metal contact is made. The porous coating wears quickly and since it holds oil, the danger of scuffing is lessened.

On some rings, the outer edge is chrome-plated. This chrome surface wears very well and stands up under severe operating conditions. Chrome-plated rings are generally finished smoother and to a higher degree of accuracy, giving fewer high points to retard wear-in. They may also have grooves to aid seating and break-in. See **Figure 2-46.**

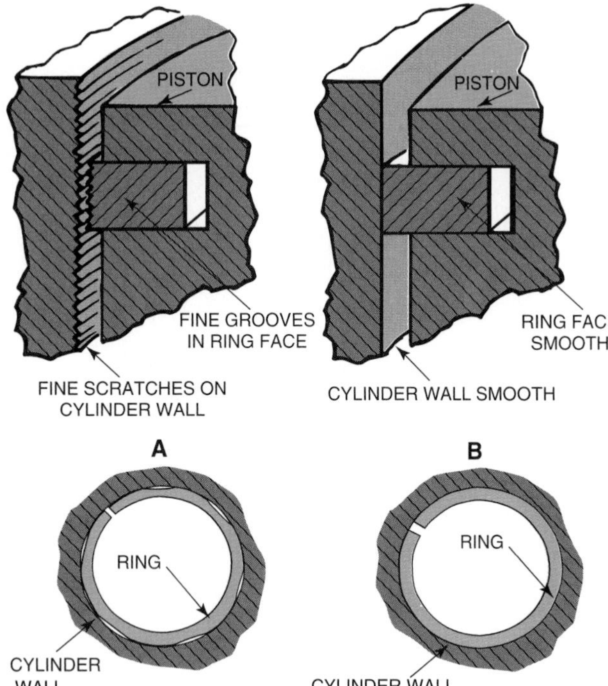

Figure 2-44. Fine scratches in the cylinder wall and grooves in the ring face rub against each other to assist in prompt and proper wear-in. A—Here, the ring touches cylinder only on high spots. B—This shows proper ring contact after wear-in.

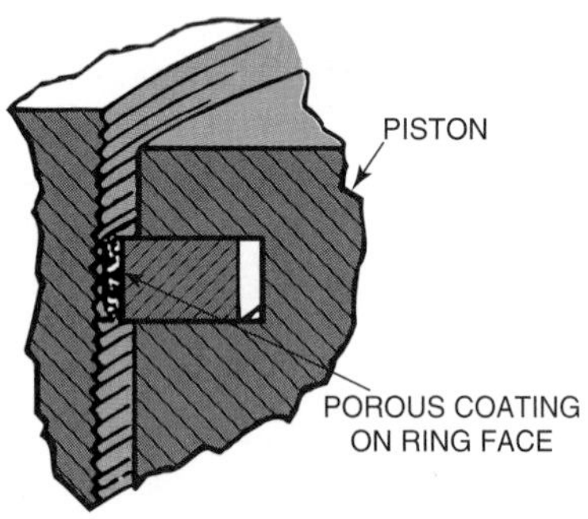

Figure 2-45. A porous coating on the ring face helps to ensure rapid and scuff-free break-in.

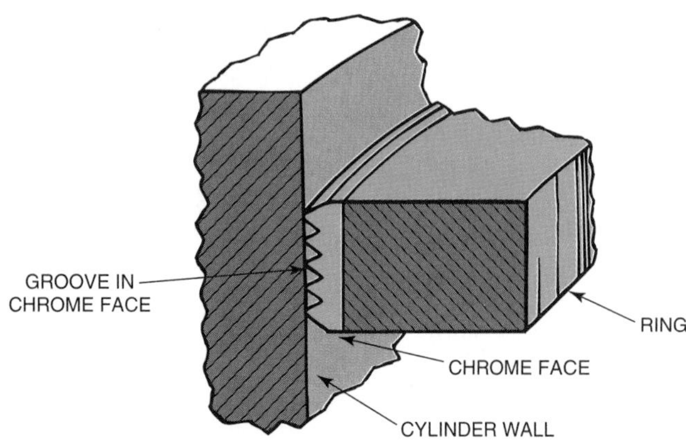

Figure 2-46. The chrome face of a chrome-plated ring is grooved to aid break-in.

Ring Material

Piston rings are made from high-quality cast iron. This material must have excellent wear properties and possess spring-like qualities that will hold it out against the walls. The installation and handling of cast iron must be done carefully to avoid breaking the brittle rings.

Thin oil ring rails are made of steel. Special expanding oil rings may also utilize steel in their construction, **Figure 2-47.** Some rings are made of stainless steel. Often, a molybdenum-filled cast iron ring is used in the top ring groove.

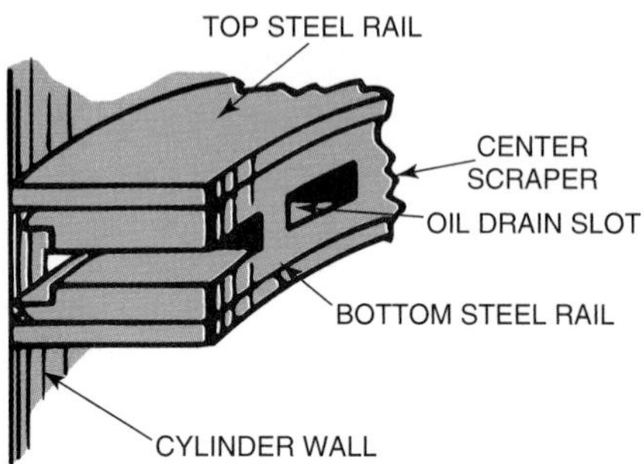

Figure 2-47. Side view of a three-piece oil ring. The top and bottom rails are made of steel.

Ring Types Are Numerous

Ring design is steadily improving. Many new designs are constantly appearing. All rings are designed to provide good sealing, long wear, quick break-in, excellent oil control, and freedom from breakage.

Piston Pins

Pistons are fastened to connecting rods with steel pins. These pins, called ***piston pins***, pass through one side of the piston, through the upper end of the connecting rod, and then through the other side of the piston. Look at **Figure 2-48.**

The piston pin is usually hollow to reduce weight. It is also case-hardened to provide a long-wearing surface. ***Case-hardening*** is a process that hardens the surface of steel, but leaves the inner part fairly soft and tough to prevent brittleness. This hardness penetrates from .004″ (0.10 mm) up to any depth desired. However, there would be little advantage in making the hard shell any deeper than a few thousandths of an inch, **Figure 2-49.** Piston pins are ground to a very accurate size and are highly polished.

Pin Installation

Piston pins are installed and secured to provide a bearing action. One method has the piston pin fastened to the rod and uses the piston bosses for bearings. At times, a separate bushing is pressed into the bosses to provide a bearing surface. Current practice favors a press fit (friction holds the two pieces together) between the piston pin and the connecting rod, with the pin oscillating in the aluminum pin bosses. See **Figure 2-50.**

Another method is to secure the pin at each end with a ***snap ring*** that rides in a shallow groove cut in each pin boss.

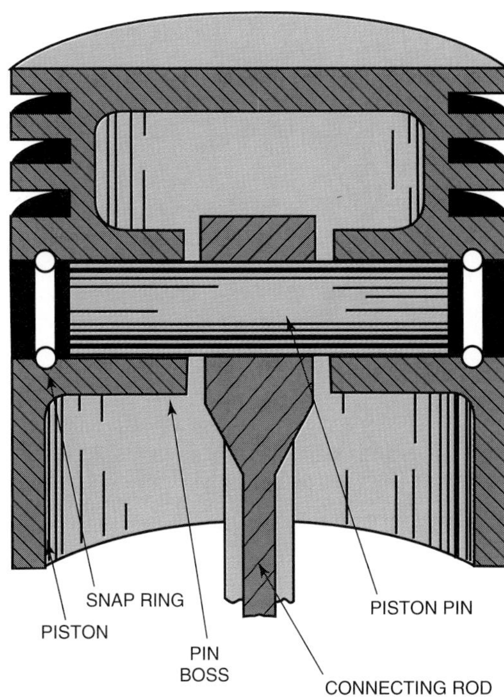

Figure 2-48. A piston is fastened to the connecting rod with a piston pin.

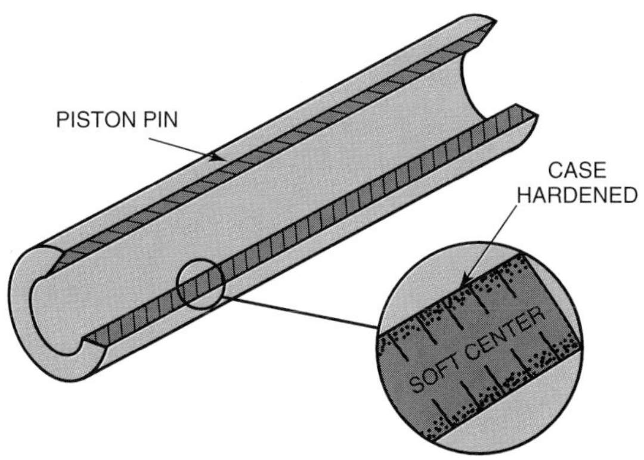

Figure 2-49. Sectioned piston pin shows the thin skin of case-hardening.

The pin is free to turn in the bosses and in the rod. This is called a free-floating pin, **Figure 2-51.**

Connecting Rods

As the name implies, connecting rods are used to connect pistons to the crankshaft **Figure 2-52.** The upper end of the rod oscillates (swings back and forth), while the lower, or big end, bearing rotates (turns).

Because there is very little bearing movement in the upper end, the bearing area can be reasonably small. The lower end rotates very fast, and the crankshaft bearing journal turns inside the connecting rod. This rotational speed tends to produce heat and wear. To make the rod wear well, a larger bearing area is required.

The upper end of the rod has a hole through it for the piston pin. The bottom of the large end of the connecting rod

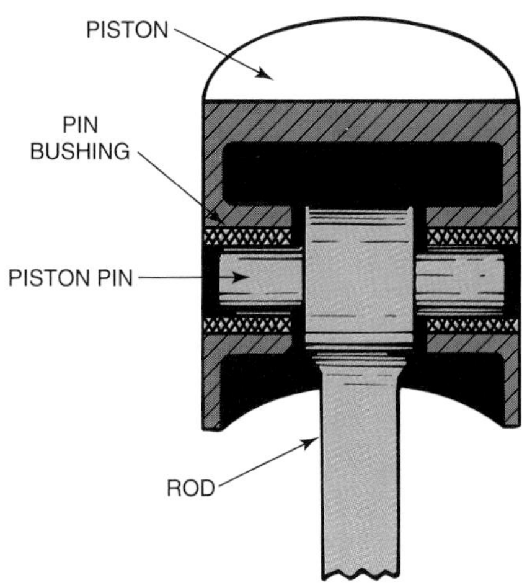

Figure 2-50. Piston pin locked to rod turns in a bushing. These are sometimes used on heavy duty diesel engines.

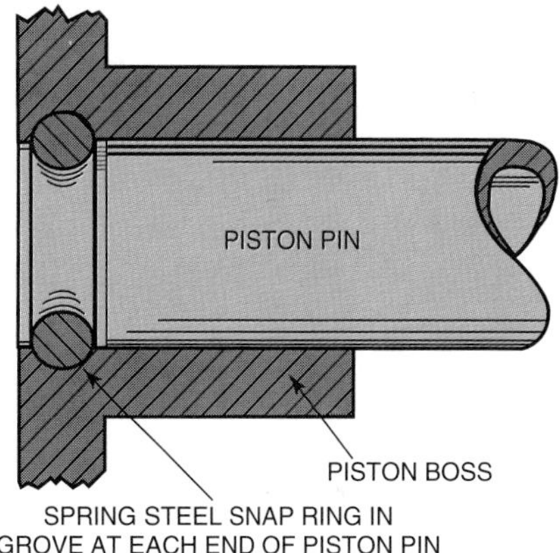

Figure 2-51. The piston pin is free to turn in the rod and in the pin bosses.

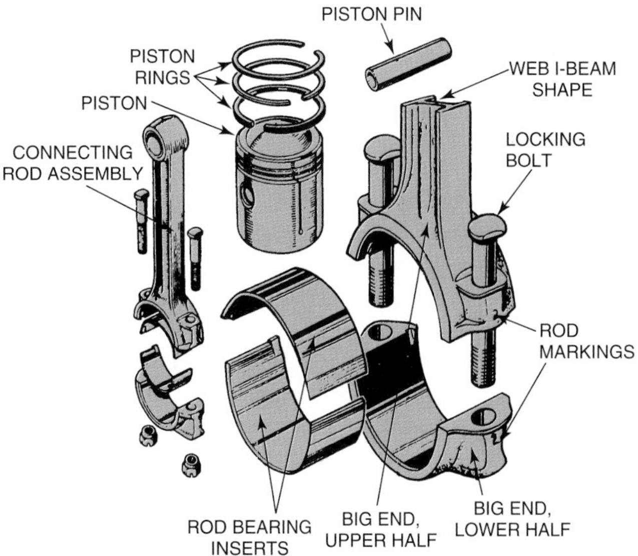

Figure 2-52. A typical connecting rod is made of forged steel rod. It is used in conjunction with an aluminum piston. (Jaguar)

must be removed so the rod can be installed on the crankshaft journal. The section that is removed is called the connecting rod cap.

The top of the rod and the connecting rod cap are bolted together. If removed, the rod and cap should be numbered. When installed, the numbers should be on the same side. This prevents turning the cap around when installing it on the rod.

Turning the connecting rod cap around would make the rod bearing hole out-of-round, **Figure 2-53.** When rods are manufactured, the upper and lower halves are bolted together and the holes are bored to an accurate size. The hole may be slightly off center. If the caps are crossed, the upper hole half may not line up with the lower hole.

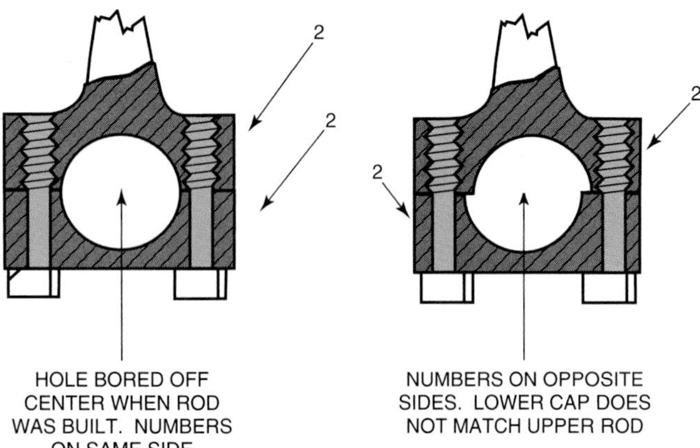

Figure 2-53. If a connecting rod is removed, they should be numbered and the rod and cap kept together.

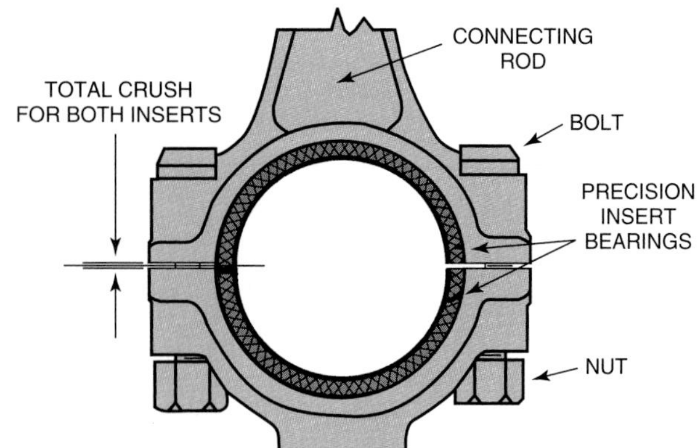

Figure 2-54. Connecting rod precision insert bearing is properly fitted when the bearing parting surfaces (ends) touch before the rod halves meet. This provides what is known as bearing crush. (Sunnen)

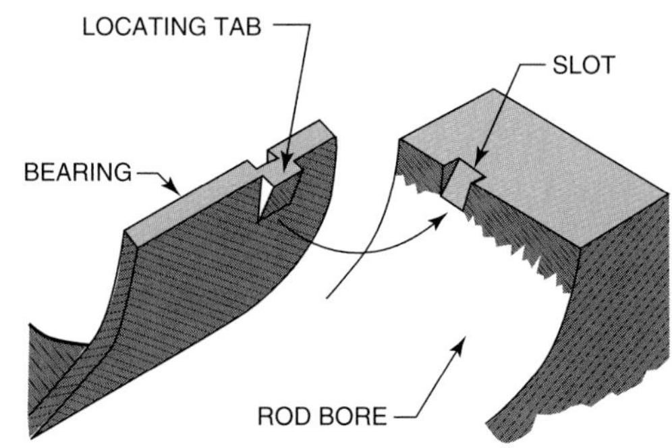

Figure 2-55. A connecting rod precision inserts are aligned and held in place by a bearing locating tab engaging a slot in the rod bore. (Clevite)

Connecting Rod Construction

Connecting rods are normally made of alloy steel. They are drop-forged to shape and then machined. The customary shape uses I-beam construction.

Some rods are made of aluminum. Generally, these are for small, light-duty engines. Small engines often utilize the rod material for both upper and lower bearing surfaces. Special aluminum rods for high-speed, high-performance engines are also available.

Connecting Rod Bearings

As mentioned, the upper end of the connecting rod may use a bushing for a bearing surface. If the rod is bolted or pressed to the piston pin, it will not use any special bearing. The bearings in this case would be in the piston boss holes. The lower end uses a precision insert. The rod bearing hole is bored larger than the crank journal and an ***insert bearing*** is placed between the rod and journal. See **Figure 2-54.**

A bearing insert does not turn in the rod. It is held in place by a locating tab (locking lip) on the insert that is placed in a corresponding notch in the rod. Look at **Figures 2-55** and **2-56.**

The insert must fit the rod snugly in order to transfer heat to the connecting rod. To ensure a proper fit, the insert will protrude a small amount above the rod bore parting surface.

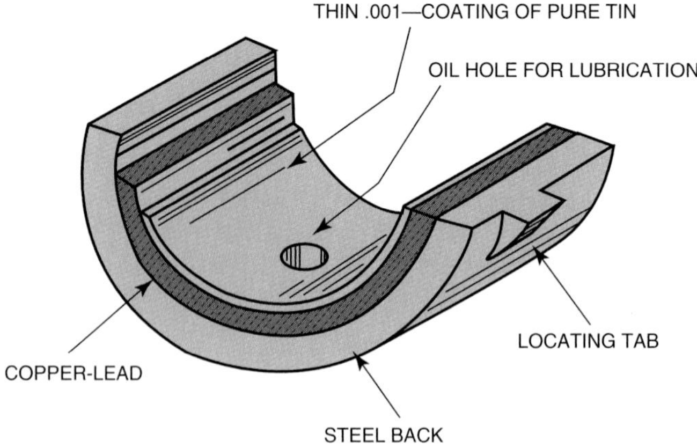

Figure 2-56. Cutaway showing construction of a precision insert bearing.

This distance (from less than .001″ to .002″ or less than 0.025 to 0.050 mm) is called ***crush height***. When the rod halves are drawn together, the inserts touch before the halves, forcing the insert tightly into place, **Figure 2-54.**

Insert Construction

An insert is started as a steel shell. This gives it shape and rigidity. The inner part of the shell that contacts the journal is then coated to form a lining. Some bearings have steel shells with a thin (.002″ to .005″ or 0.05 to 0.13 mm) babbitt lining. Others use steel shells coated with a copper-lead-tin matrix followed by a very thin (.001″ or 0.025 mm) coating of pure tin. Others use an aluminum coating. **Figure 2-56** illustrates a typical bearing insert.

Bearing Characteristics

The ideal bearing is not easy to construct. For every advantage, there seems to be a disadvantage that goes with it. A good bearing must have many characteristics. Refer to **Figure 2-57** as you read the next few paragraphs.

Load Strength

The bearing is subjected to tremendous stress from the power stroke. There is pounding as the crank first pushes up and then pulls down. The bearing must not fatigue, crack, or spread out from the pounding force.

Antiscuffing

The material in the bearing should be such that in the event the oil surface skin is destroyed, the shaft journal will not be damaged by scuffing and scratching.

Corrosion Resistance

The bearing material must resist any tendency to corrode when exposed to vapors and acids in the crankcase.

Conductivity

All bearings produce heat. It is essential that the bearing material be of a type that will conduct heat to the rod. This is one reason why insert bearings must be a near-perfect fit.

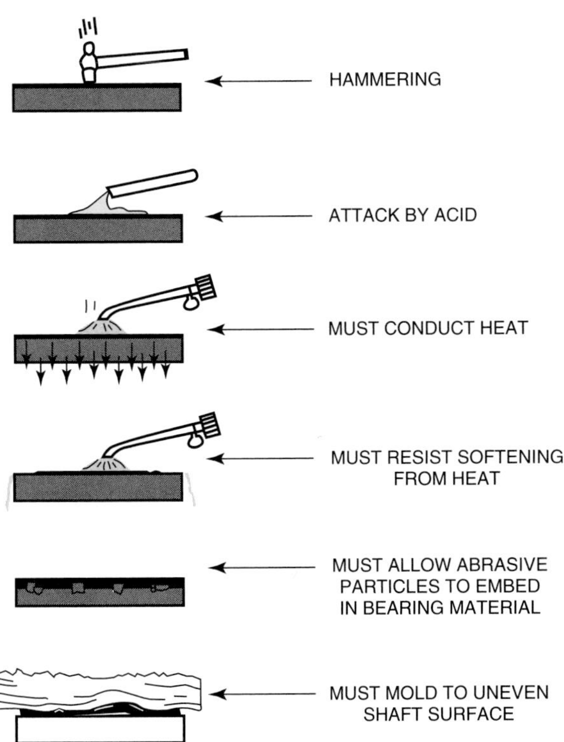

Figure 2-57. Bearing inserts must resist many destructive forces.

Temperature Change

The bearing strength must not be lessened when the bearing reaches its operating temperature. It must have reasonable strength when cold and when hot.

Each insert must have a small hole to allow oil to enter for lubrication. See **Figure 2-56.** Some inserts have shallow grooves in the surface to allow oil to spread out across the bearing surface.

Embedability

If a small abrasive particle enters the bearing area, the bearing should allow the particle to embed itself in the bearing material so it will not scratch the journal.

Conformation

No surface is perfectly true. Even with a highly accurate shaft, a new rod, and a new insert, the bearing will not fit the journal without some minute imperfection. A good bearing is soft and ductile enough to shape to the journal after it has been used for some time.

Crankshaft

The engine crankshaft provides a constant turning force to the wheels. It has throws to which connecting rods are attached, and its function is to change the reciprocating motion of the piston to a rotary motion to drive the wheels. Crankshafts are made of alloy steel or cast iron. **Figure 2-58** shows a typical crankshaft.

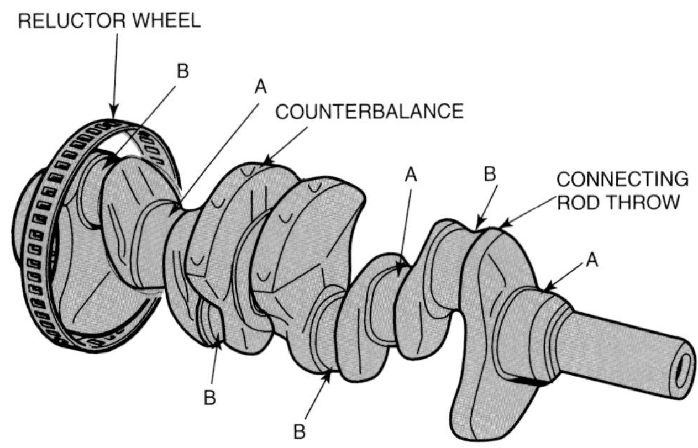

Figure 2-58. A typical four-cylinder engine crankshaft. A—Main bearing journals. B—Rod bearing journals. Note the notched reluctor wheel. A crankshaft position sensor reads the reluctor wheel (crankshaft) position and sends this data to the engine's computer or electronic control module.

Main Bearings

The crankshaft is held in position by a series of main bearings. The maximum number of main bearings for a crankshaft is one more than the number of cylinders. It may have fewer main bearings than cylinders. Most engines use precision insert bearings that are constructed like the connecting rod bearings, but are somewhat larger. See **Figure 2-59.**

In addition to supporting the crankshaft, one of the main bearings must control the forward and backward movement (end play) of the shaft. This bearing has flanges on the edges that rub against a ground surface on the edge of the crankshaft main journal. This bearing is referred to as a ***thrust bearing***. Look at the center bearing in **Figure 2-59.**

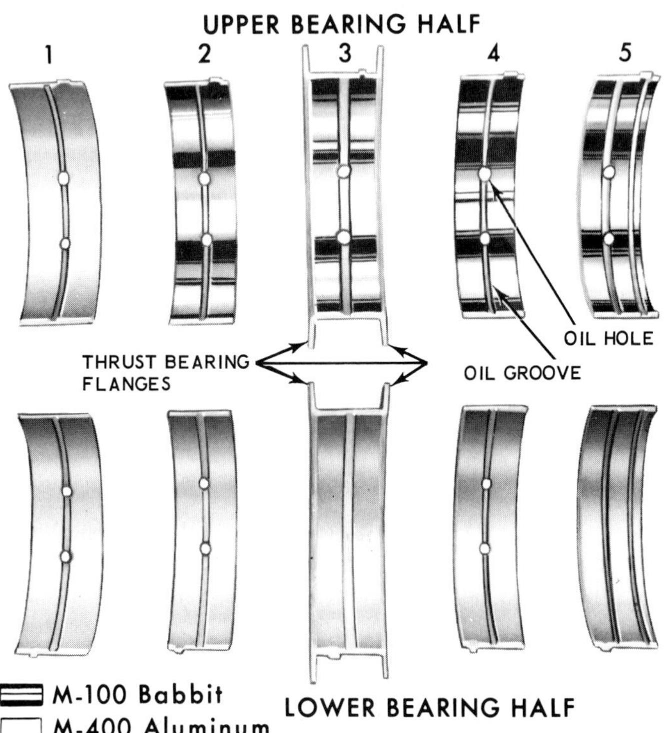

Figure 2-59. Set of main bearings. Some of these bearings use an aluminum coating, while others employ babbitt. Note the thrust flanges on thrust bearing. (Cadillac)

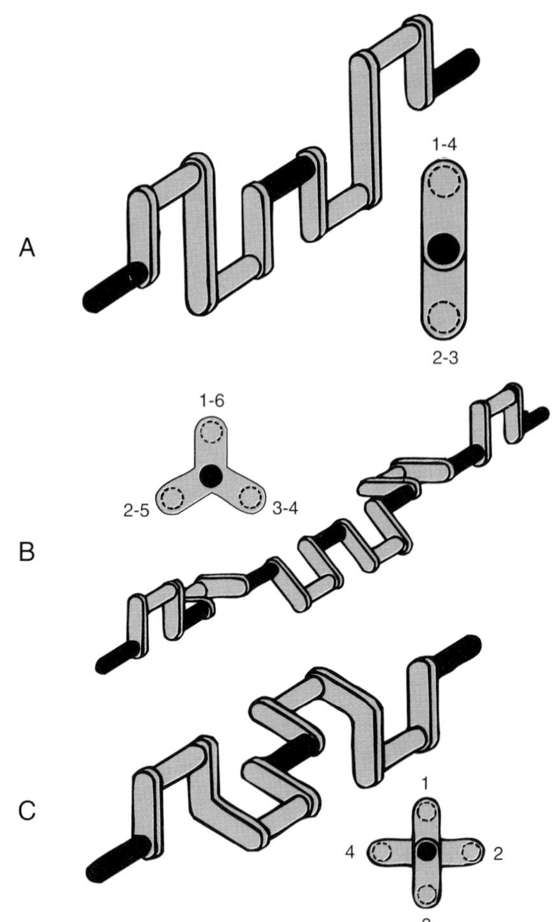

Figure 2-60. A—A four-cylinder crankshaft will normally have throws spaced 180° apart with cylinders 1 and 4 on the same side. B—A right hand crankshaft for a six-cylinder engine will have throws arranged like this. C—A V-8 engine may have the crankshaft throws arranged the same as a four-cylinder engine, or the method shown here which is widely used.

Crankshaft Throws

The ***crankshaft throws*** (part of the shaft the connecting rod fastens to) must be arranged in such a way as to bring the piston to top dead center at the proper time for the power stroke. As the pistons in a multicylinder (more than one) engine do not fire one after the other, but fire in a staggered sequence, the throw position is very important.

Figure 2-60 illustrates end views of four-, six-, and eight-cylinder crankshafts. **Figure 2-60A** and **Figure 2-60B** are for inline engines. An inline engine is one in which all the cylinders are arranged one after another in a straight row. An inline six-cylinder engine, **Figure 2-60B,** requires six connecting rod throws. A V-8 engine, **Figure 2-60C**, has four throws. Each throw services two connecting rods.

Crank Vibration

To offset the unbalanced condition caused by off-center throws, many crankshafts use counterbalances to stop vibration. They may be forged as part of the crankshaft, or they may be bolted to the crankshaft, **Figure 2-61.**

Vibration Damper

When the front cylinders fire, power is transmitted through the crankshaft. The pressure the piston applies can

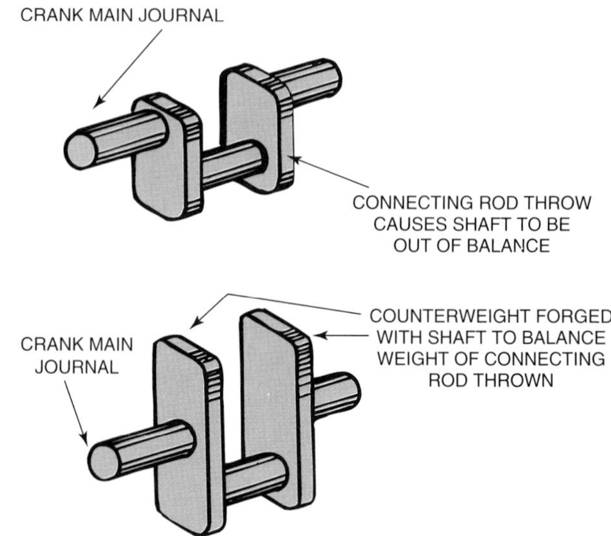

Figure 2-61. All crankshafts needs counterweights to bring them into proper balance.

exceed 3000 lbs. (1350 kg). The front of the crankshaft receiving this power tends to move before the rear, causing a twisting motion. When the torque is removed from the front, the partially twisted shaft will unwind and snap back in the other direction. This unwinding, although minute, causes what is known as torsional vibration.

To stop the vibration, a damper, sometimes called a ***harmonic balancer***, is attached to the front of the crankshaft. Basically the ***vibration damper*** is built in two pieces. These pieces may be connected by rubber plugs, spring-loaded friction discs, or a combination of the two, **Figure 2-62.**

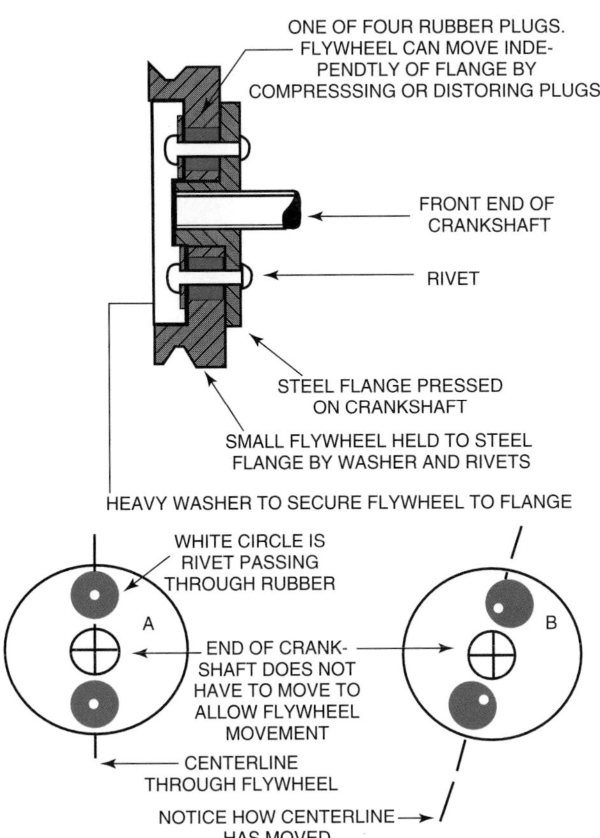

Figure 2-62. Vibration damper utilizes a small flywheel which moves a limited amount, bending the injected rubber plugs. See A and B.

When a front cylinder fires and the shaft tries to speed up, it tries to spin the heavy section of the damper. When this occurs, the rubber connecting the two parts of the damper is twisted. The shaft does not speed up as much with the damper attached. The force necessary to twist the rubber and to speed up the heavy damper wheel tends to smooth out the crankshaft operation. When the firing pressure is removed from the shaft, the shaft cannot spring back quickly because the twisted rubber attempts to keep the damper wheel turning. The unwinding force of the crankshaft cancels out the twist in the opposite direction.

Balance Shafts

Some engines use a ***balance shaft*** that is geared to the crankshaft or camshaft. This balance shaft helps to counterbalance vertical and torsional vibrations from the crank and camshafts. A balance shaft has offset bob weights that rotate in opposite directions, canceling out vibrations from the crankshaft and or camshaft. **Figure 2-63** shows a balance shaft used in a inline four-cylinder engine.

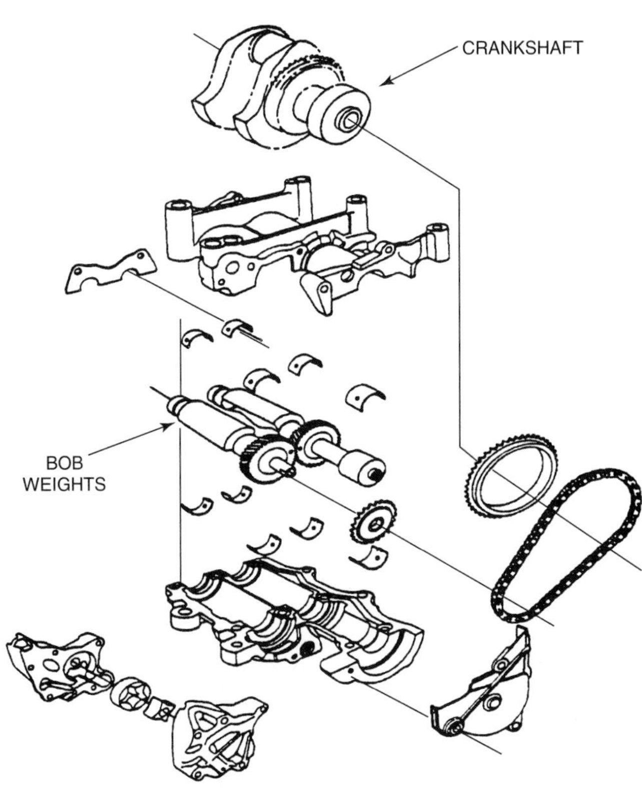

Figure 2-63. Some engines use balance shafts geared to the crankshaft or camshaft, depending on the design. Note that this particular setup uses a chain. The bob weights offset the vibrations created by the crankshaft. (Pontiac)

Drilled, Ground, and Polished

The crankshaft is drilled so oil being fed to the main bearings will pass through the shaft to the connecting rod throws, **Figure 2-64.** All bearing surfaces are precision ground and are highly polished.

Timing Gear or Sprocket

A ***timing gear*** or a ***chain sprocket*** is installed on the front end of the crankshaft. A chain sprocket is usually made of steel. In the case of the timing gear, the teeth are helical in shape. A timing gear is used to turn a camshaft timing gear. Where a sprocket is used, the sprocket will drive a ***timing chain*** or ***timing belt***, which turns a larger sprocket on the front of the camshaft, **Figure 2-65.** Some engines use a combination of chain and belt to drive the camshafts.

A timing gear or sprocket is secured with a metal key that rides in a slot cut in the crankshaft. The gear or sprocket also contains a slot or groove. See **Figure 2-66.**

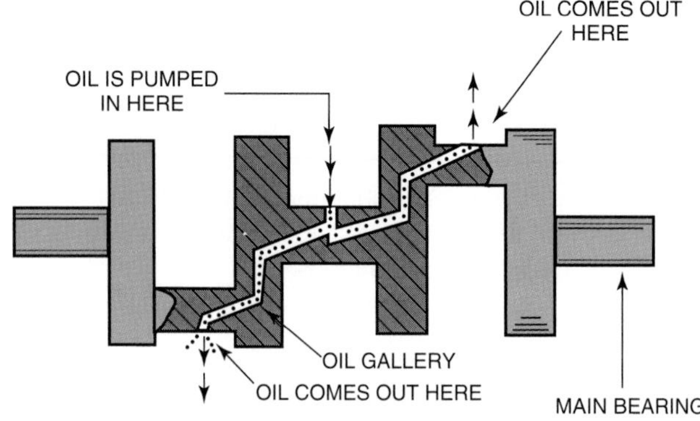

Figure 2-64. This crankshaft which is partially cut away shows how oil is pumped into the center main bearing. From there it enters an oil gallery drilled through the crankshaft to each rod journal.

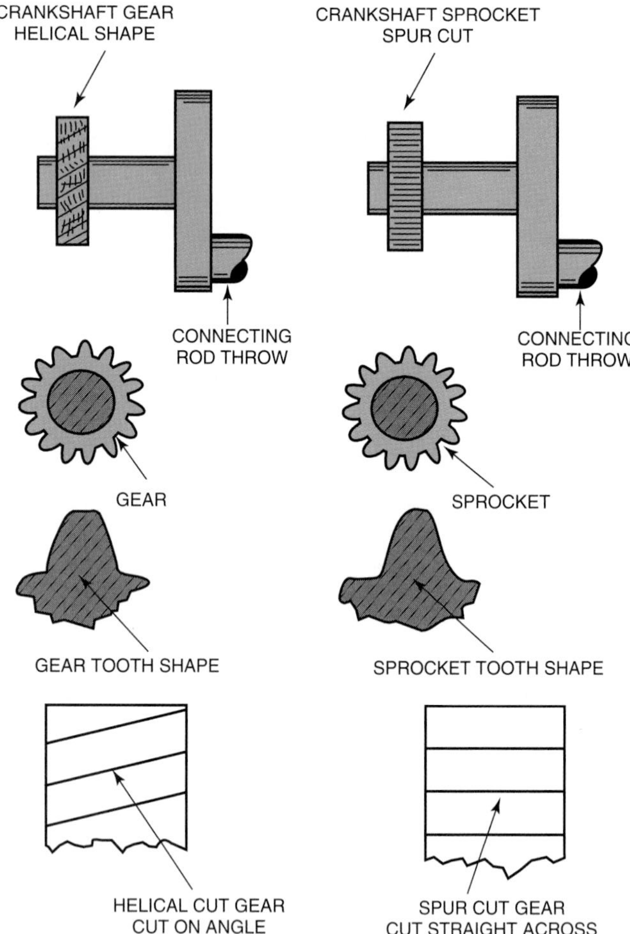

Figure 2-65. Timing gear drive on the crankshaft; also timing chain (or belt) drive sprocket on crankshaft.

Flywheel

A heavy flywheel is attached to the rear of the crankshaft with bolts. The function of the flywheel is to smooth out engine speed and keep the crankshaft spinning between power strokes. In some engines, the flywheel also serves as a mounting surface for the clutch. See **Figure 2-67.**

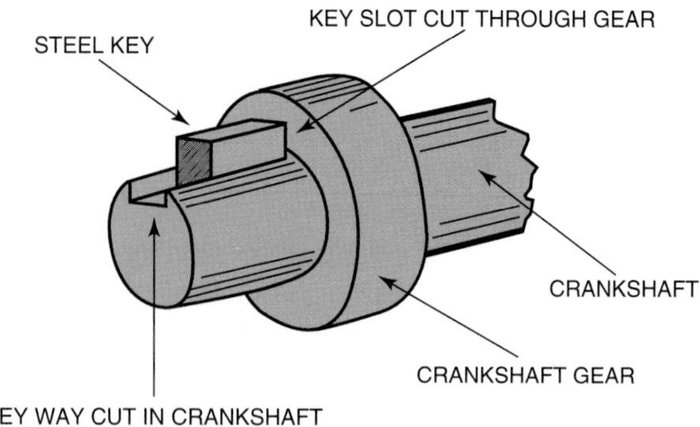

Figure 2-66. A steel key, which is installed in a slot in the crankshaft, passes through a similar slot in the gear to secure the gear to the shaft.

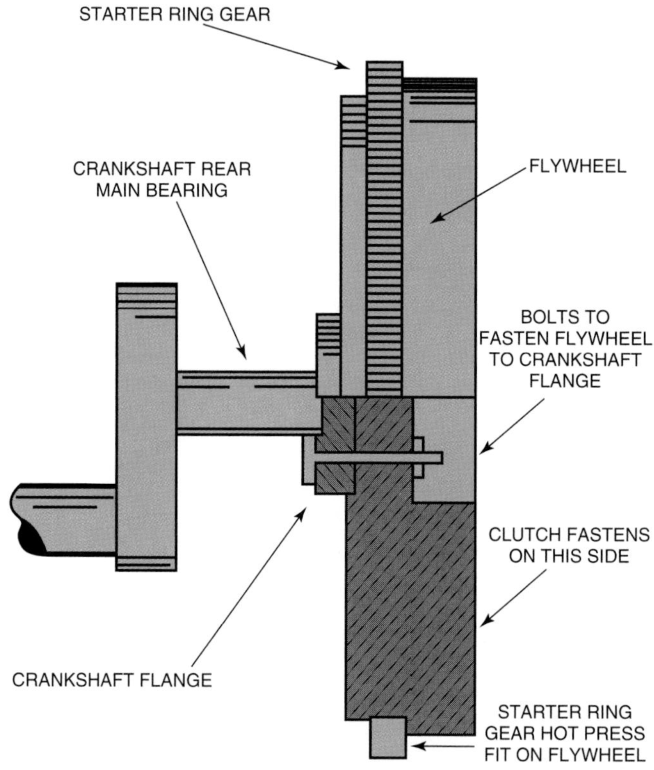

Figure 2-67. On all engines, a heavy flywheel is bolted to the crankshaft flange. The flywheel helps to keep the engine running at all speeds.

The outer rim of the flywheel has a large ring attached with gear teeth cut into it. The teeth of the starter motor engage these teeth and spin the flywheel to crank the engine. When an automatic transmission is used, the torque converter assembly works with the flywheel.

Camshaft

The camshaft is used to open and close the valves. There is one cam on the camshaft for each valve in the engine. Generally only one camshaft is used in most engine. Newer engines are increasingly equipped with two or more camshafts. (On the elementary engine you designed in Chapter 1, you used two camshafts and the valves were arranged to better illustrate the action in relation to an end view of the crankshaft.) A camshaft has a series of support bearings along its length. Camshafts turn at one-half crankshaft speed.

The cam lobes are usually not flat across the top as they might appear. The bottom of the valve lifter may be slightly crowned and the cam lobe tapered. This places the lobe-to-lifter contact to one side of center, **Figure 2-68.** A gear cut into the camshaft is used to drive the distributor and oil pump.

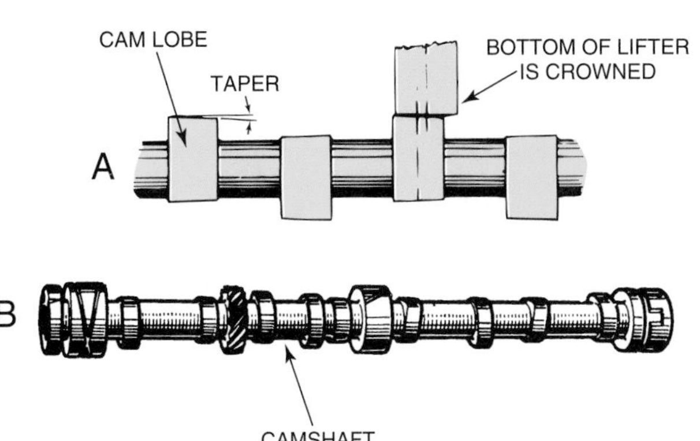

Figure 2-68. A—Section of a camshaft showing one method of tapering cam lobes. The lifter bottom is crowned. B—Overall view of typical camshaft design.

The camshaft is kept in place using a thrust washer behind the timing gear or by using a spring-loaded plunger to push on the end of the camshaft. An additional cam may be ground on the shaft to drive the fuel pump.

The typical camshaft is made of cast or forged steel. The cam surfaces are hardened for long life. **Figure 2-69** illustrates a typical camshaft with the gear and thrust washer in place. The camshaft may be turned by a gear, timing chain, or a toothed timing belt. See **Figure 2-70** through **Figure 2-72.** The camshaft gear can be made of cast steel, aluminum, or a special pressed fiber. Chain sprockets are made of cast steel.

Valves

Each engine cylinder ordinarily has two valves. However, modern engines often use four valves per cylinder (two intake and two exhaust). A few engines used in smaller vehicles have three valves per cylinder: two intake valves and one exhaust valve.

Because the head of an exhaust valve operates at temperatures up to 1300°F (704°C), valves are made of heat-resistant metal. It is obvious that the steel used in valve construction must be of high quality. In order to prevent burning, the valve must give off heat to the valve seat and to the valve guide. The valve must make good contact with the seat and must run with minimum clearance in the guide.

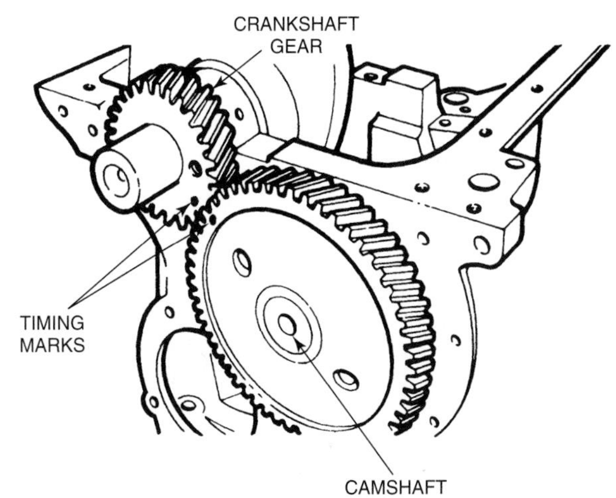

Figure 2-70. Timing gear drive for camshaft also requires alignment of timing marks. (Pontiac)

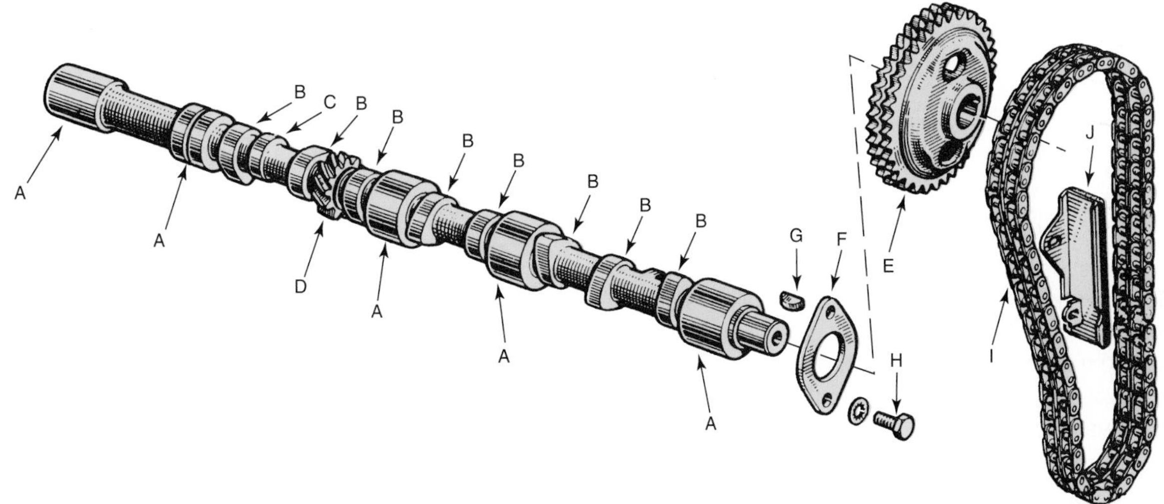

Figure 2-69. A camshaft and timing gear. A—Main camshaft bearing journals. B—Cam lobes. C—Eccentric to drive fuel pump. D—Gear to drive distributor. E—Camshaft sprocket. F—Thrust washer. G—Key. H—Retaining capscrew. I—Timing chain. J—Rubbing block.

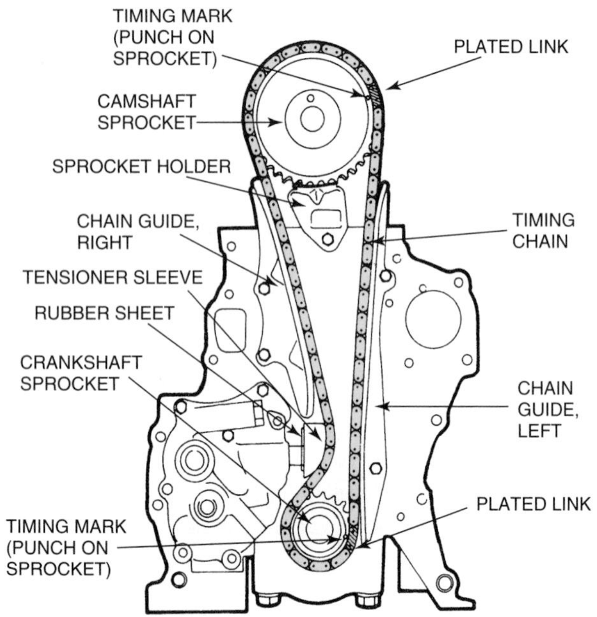

Figure 2-71. Front view of typical timing chain drive. Note the alignment of the gear timing marks. (Dodge)

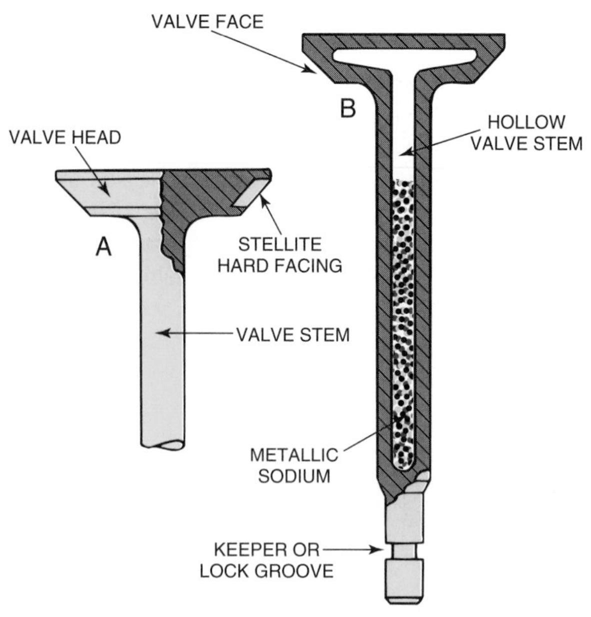

Figure 2-73. A—Valve with special hardfacing to lengthen useful life. B—Special heat resistant valve utilizing a hollow stem partially filled with metallic sodium. Never drill or cut into such a valve as the sodium inside the valve is dangerous. If worn out, dispose of carefully as recommended by the manufacturer.

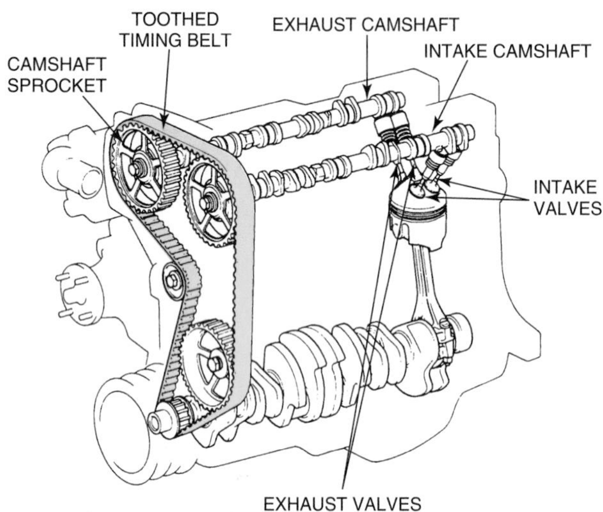

Figure 2-72. Dual overhead camshaft driven by a toothed drive belt. (Toyota)

Some valves have special hard coating on the face areas to increase their useful life. Others use hollow stems filled with metallic sodium, **Figure 2-73.** At operating temperatures, the sodium becomes a liquid and splashes up into the head of the valve. This draws the heat into the stem, where it transfers to the valve guide.

STOP Warning: Sodium is extremely toxic and reacts violently with any form of water, even the moisture in the air. Never cut, drill, melt, or burn a sodium-filled valve.

Valve Seats

Two angles commonly used for valve seats are 30° and 45°. Some manufacturers grind a 1° difference between the valve face and seat to permit fast seating, **Figure 2-74.** The seat may be ground to 45° and the valve to 44° or vice versa.

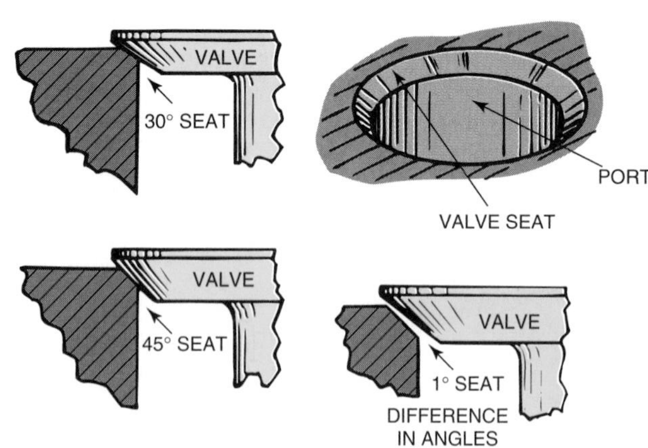

Figure 2-74. Common valve seat angles. 30° and 44° are the most common.

Valve seats can be cut in the cylinder head, or special, hard steel inserts can be pressed into the head, **Figure 2-75.** Seats can also be induction hardened, which is an operation that hardens the seat surface, increasing its useful service life. The valve face and seat must make perfect contact to ensure efficient operation.

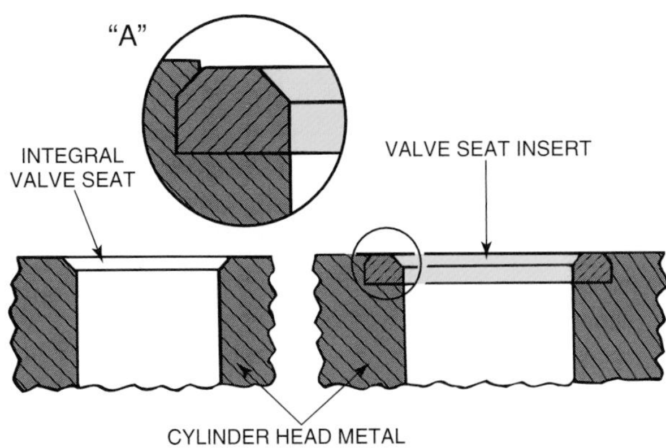

Figure 2-75. Most valve inserts are often frozen to shrink their diameter, then pressed into warm head to secure insert. The head metal, A, is sometimes peened over the edge of the insert to ensure a secure fit.

Valve Guides

The valve guide can be an integral part of the cylinder head, or it may be made as a separate unit that is pressed into a hole in the head or the block. The pressed-in type of valve guide is made of cast iron. The valve stem must fit this guide with about .002″ to .003″ (0.05 to 0.08 mm) clearance. See **Figure 2-76.**

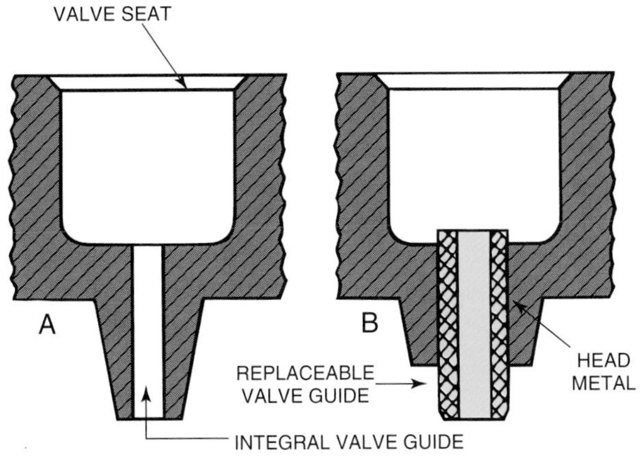

Figure 2-76. Examples of valve guides. A—Integral (not removable). B—Pressed-in (removable).

Valve Springs

Valve springs push the valves closed when the cams lower. Since the springs are compressed and expanded over 70,000 times per hour at 50 mph (80.5 kmh), they must be made of high-quality spring steel. You have already learned how the spring is fastened to the valve using a spring washer and split keepers.

Some springs have the coils closer together at one end than at the other. In an installation of this kind, the end with the more closely spaced coils must be placed against the head. Some engines use two springs per valve, one inside of the other.

Mechanical Valve Lifters

Mechanical valve lifters are usually made of cast iron. The bottom part that contacts the camshaft is hardened. Some lifters are hollow to reduce weight. Most valve trains that contain mechanical lifters have some provision for adjusting clearance. See **Figure 2-77.** Mechanical valve lifters were used in older engines.

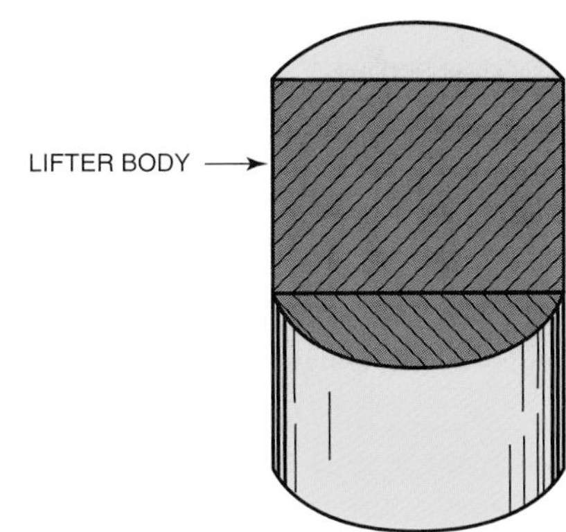

Figure 2-77. Mechanical valve lifters are usually solid. Some mechanical lifters are hollow, which reduces their weight.

Hydraulic Valve Lifters

Hydraulic valve lifters perform the same job as mechanical lifters. However, hydraulic lifters are self-adjusting, operate with no lifter-to-rocker arm clearance, and uses engine oil under pressure to operate. Hydraulic lifters are quiet in operation.

To operate, engine oil under pressure enters the hydraulic lifter body. The oil passes through a small opening in the bottom of an inner piston and into a cavity beneath the piston. The oil raises the piston up until it contacts the push rod (the oil pressure is not high enough to open the valve).

When the cam raises the lifter, pressure is applied to the inner piston. The piston tries to squirt the oil back through the small opening, but it cannot do so as a small check ball seals the opening.

As the cam raises, the lifter becomes solid and lifts the valve. When the cam lowers, the lifter will be pushed down by the push rod. The lifter will then automatically adjust to remove clearance. **Figure 2-78** shows one type of hydraulic valve lifter. See **Figure 2-79** for complete lifter action.

Turning Reduces Valve Wear

If the valve goes up and comes down in the same place time after time, carbon buildup between the valve face and seat may cause the valve to remain partially open. This will cause valve burning. If the valve turns even a few degrees on each opening, a wiping action between the face and seat will be developed. This keeps carbon from building up. Turning also helps to prevent localized hot spots since the valve will keep moving away from the hottest areas.

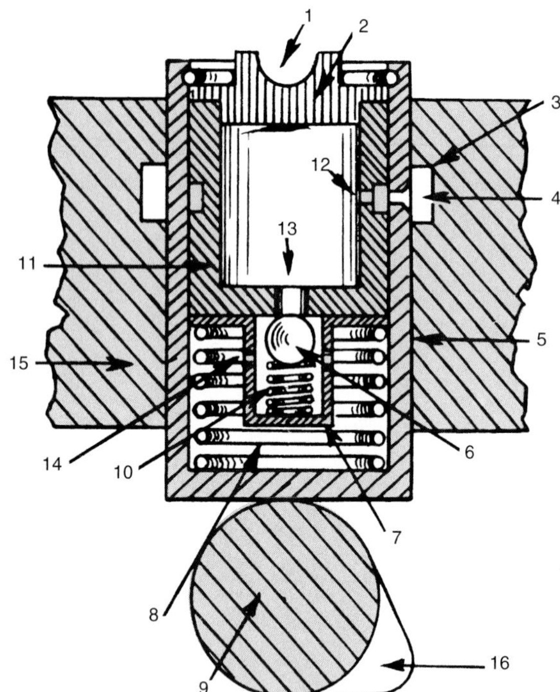

Figure 2-78. Typical hydraulic valve lifter. Oil enters channel (3) in the block, passes through hole (4) in the lifter body (5), then enters hole (12) in the inner piston. It fills the inner piston cavity and passes through hole (13). This pressure will push the check ball (6) off its seat. It now flows through hole (14) in the lifter cage (7). It fills the cavity under inner piston (11). The pressure raises the inner piston up until it contacts the valve push rod, or stem, whichever is used. Other parts include: 1—Push rod seat. 2—Inner piston cap. 15—Block. 8—Inner piston spring. 10—Ball check spring. 9—Camshaft. 16—Lobe.

Types of Rotators

There are several methods of causing valves to rotate as they open and close. One is through the use of a release-type mechanism. This removes the spring tension from the valve while it is open and induces rotation from engine vibration. See **Figure 2-80.** Another type of rotator is the positive-type, which uses a spring or rolling action to move the valve. This rolling action causes the retainer and the valve to turn a few degrees each time the valve opens.

Roller Lifters

Some late-model engines are equipped with hydraulic ***roller lifters***. The roller reduces wear on both the lifter and the camshaft, **Figure 2-81.** The roller lifter operates in the same

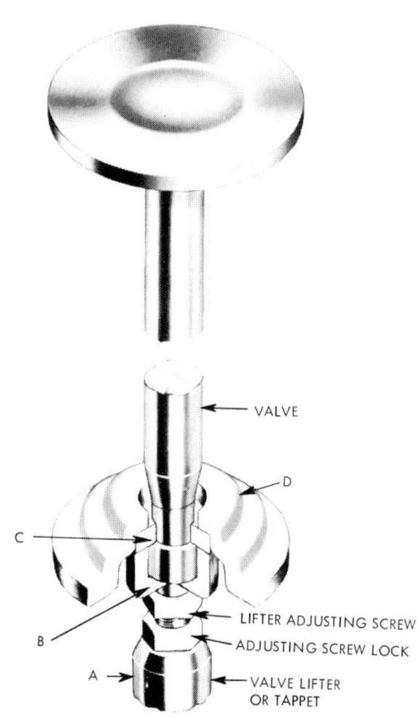

Figure 2-80. A release-type valve rotator. (Thompson Products)

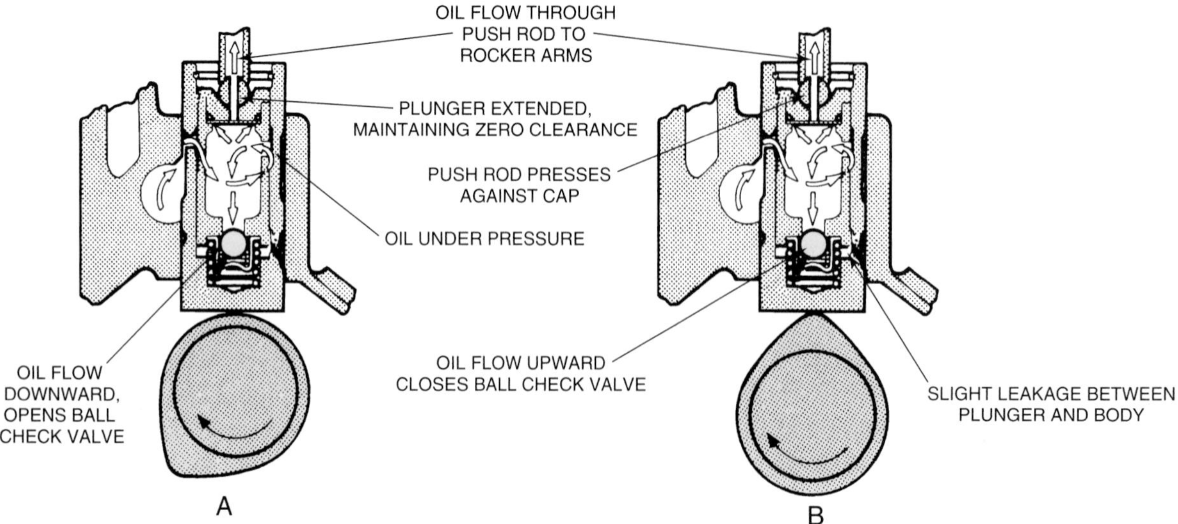

Figure 2-79. Hydraulic valve lifter operation during two stages. A—Valve closed. B—Valve open. (Cadillac)

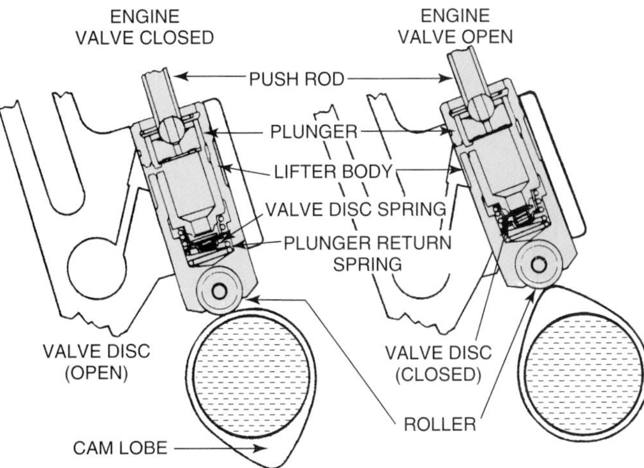

Figure 2-81. A roller lifter is used in some heavy-duty applications.

manner as a conventional hydraulic lifter with the exception of the roller, which engages the camshaft lobe. Instead of a sliding action, the lifter rolls on the cam surface.

Valve Timing

Both the intake and exhaust valves are open longer than it takes the piston to make a stroke. The exact number of degrees that a valve will open or close before top or bottom dead center varies, depending on engine design. The degrees shown in **Figure 2-82** are for one specific engine. Note that the intake valve opens about 20° before the piston starts down on the intake stroke. It closes about 67° after the piston reaches the bottom of its stroke.

The exhaust valve opens about 69° before the piston reaches bottom dead center (BDC) on the power stroke. It does not close until about 27° after top dead center (TDC) on the intake stroke. The early opening and late closing of both valves greatly improves the intake of fresh fuel mixture and the thorough exhausting of burned gases.

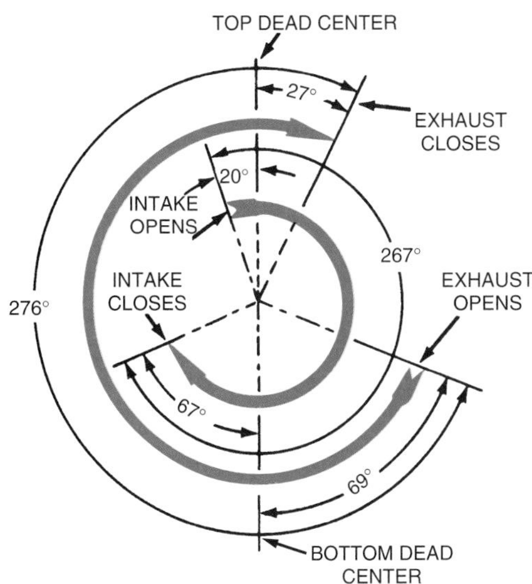

Figure 2-82. Valve timing diagram. The angles will vary widely depending upon engine design. The length of time, in degrees, that a valve is held open is referred to as valve duration.

The intake and exhaust valves in **Figure 2-82** are partially open at the same time. For example, the intake valve opens 20° before TDC and the exhaust valve closes 27° after TDC on the same intake stroke. This situation is termed ***valve overlap***. Overlap improves engine performance at higher speeds.

When a valve closes or opens, how fast it will rise, how long it will stay open, and how fast it will close depend on the shape of the cam lobe and the position of the camshaft in relation to the crankshaft.

The timing sprockets are generally marked to ensure correct valve timing. The camshaft is mounted to one side of the crankshaft on most inline engines (except engines using overhead camshafts), **Figure 2-83.** On V-type engines, the camshaft may be mounted above the crankshaft in the center region of the block. See **Figure 2-84.**

A few overhead cam engines have a hydraulic camshaft sprocket assembly that can vary the valve timing. These engines are said to have ***variable valve timing***. The variable valve timing sprocket assembly is operated by engine oil pressure. The assembly retards valve timing at low speeds for smooth operation and advances timing at high speeds for more power. Dual overhead cam engines may have a variable valve timing sprocket assembly at each camshaft.

Engine valves are not always operated directly from a camshaft. Some engines utilize additional linkage to operate the valves. This will be discussed in Chapter 3.

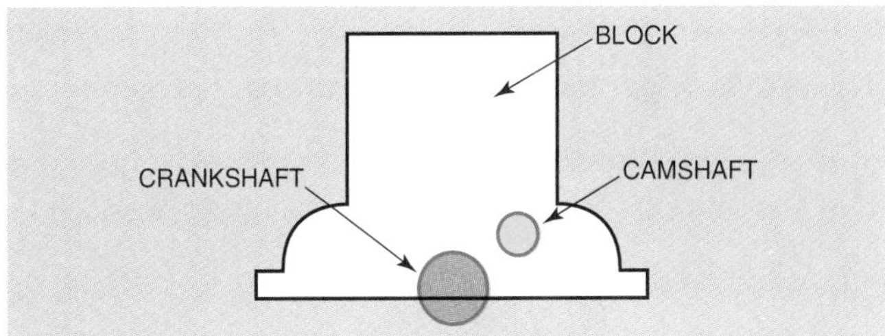

Figure 2-83. One location for the camshaft in an inline engine.

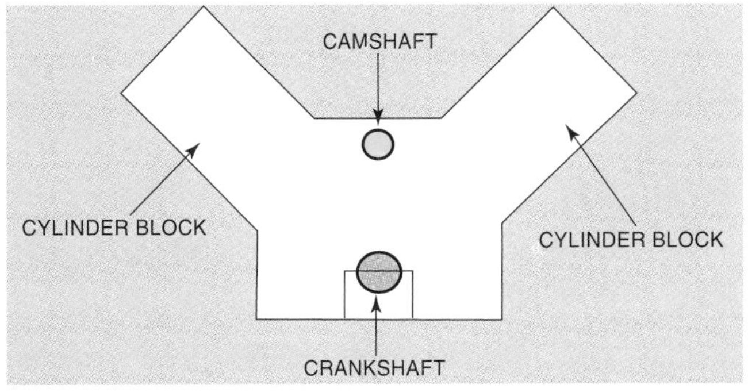

Figure 2-84. Camshaft location in a cam-in-block or valve-in-head (OHV) V-type engine.

Cylinder Head

The cylinder head serves as a cover for the cylinders and forms the top of the combustion chamber. The head also holds the spark plugs. All modern cylinder heads contain the valves. Engines with the valves located in the cylinder heads are called ***valve-in-head*** engines. Cylinder heads are usually made of cast iron or aluminum. They must be strong and rigid. They are bolted to the block with ***head bolts***. The surfaces of the head and block that make contact must be absolutely flat. See **Figure 2-85.**

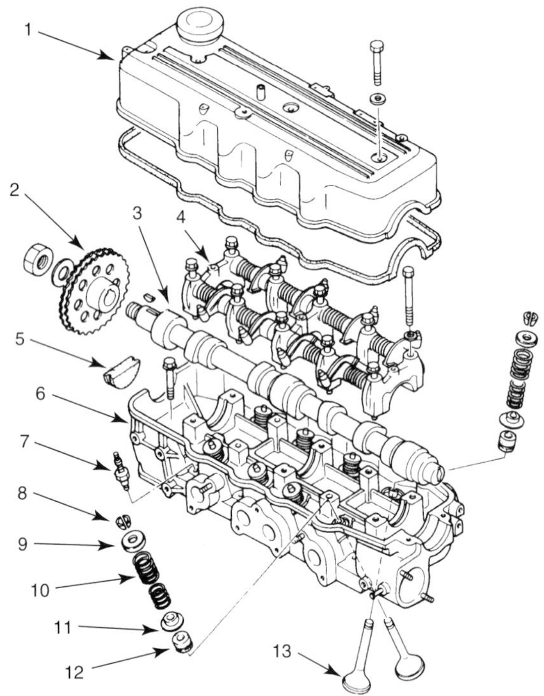

Figure 2-85. A typical overhead camshaft cylinder head arrangement. 1—Cylinder head cover. 2—Camshaft sprocket. 3—Camshaft. 4—Rocker arm assembly. 5—Semi-circular seal. 6—Cylinder head. 7—Spark plug. 8—Valve keeper. 9—Upper spring seat. 10—Spring. 11—Lower spring seat. 12—Valve seal 13—Valve. (Mazda)

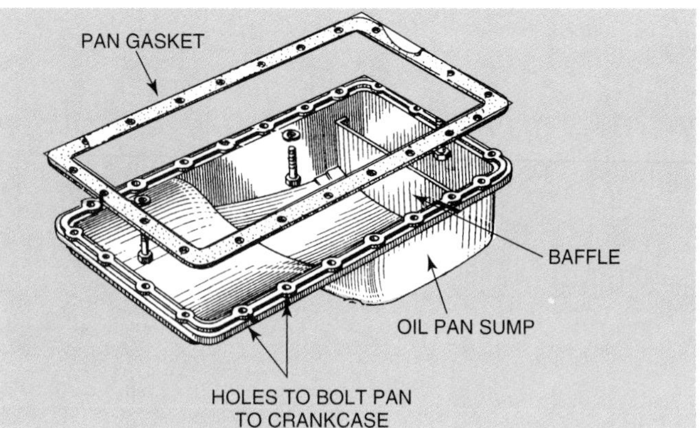

Figure 2-86. Engine oil pan size and shape will vary widely. (Nissan)

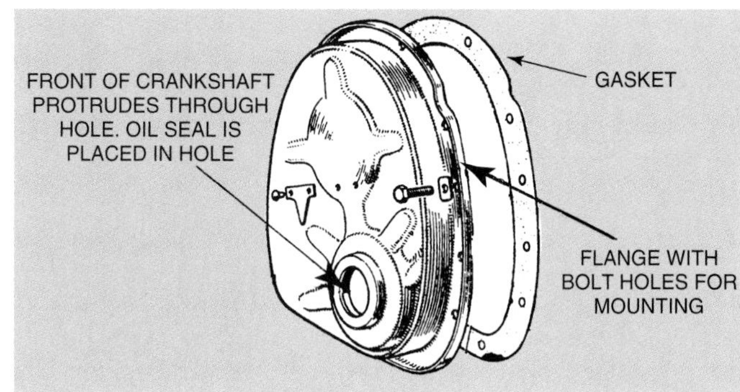

Figure 2-87. One type of stamped steel timing gear cover.

Oil Pan

The ***oil pan*** acts as a reservoir for oil and serves as a dust shield for the bottom of the engine. It is attached to the bottom of the block with cap screws. The pan is generally made of thin steel stamped to shape, **Figure 2-86.** Plastic and aluminum may also be used.

Timing Gear Cover

The timing gears, belts, and chains must be covered to prevent the entrance of dirt and to eliminate the loss of oil. In addition to this function, the cover often contains an oil seal that allows the crankshaft to protrude through the cover without leaking oil. Timing covers may be made of thin steel, plastic, aluminum, or cast iron. An example of a timing cover is shown in **Figure 2-87.**

Gaskets

In an engine where machined parts fit together, gaskets are used to make the joints tight and to prevent leakage of oil, water, and/or gasoline. The cylinder heads must seal in the water of the cooling system and must also contain the pressure of the exploding fuel. Thin steel, copper, and fiber ***gaskets*** are used between the head and engine block.

It is very difficult to machine metal parts to the degree of accuracy necessary for leakproof joints. As the engine expands and contracts during warm-up and cooling periods, there are minute shifts in the fastened parts. This, coupled with vibration, will loosen many parts to the point of leakage.

Gasket material is somewhat resilient (soft and springy) and will adapt itself to expansion and contraction. It will also conform to irregularities in the surfaces of the mating parts. Many materials are used in gasket construction, such as steel, aluminum, copper, cork, rubber (synthetic), paper, felt, and liquid silicone. **Figure 2-88** shows how the gasket material accomplishes this. **Figure 2-89** illustrates some of the common gaskets found in an engine.

Sealants

Some newer engines use chemical sealants in place of or along with a normal gasket. ***Sealant***, or gasket sealer, is a liquid or semiliquid material that is sprayed, brushed, or spread on the gasket surface. Various types, having different

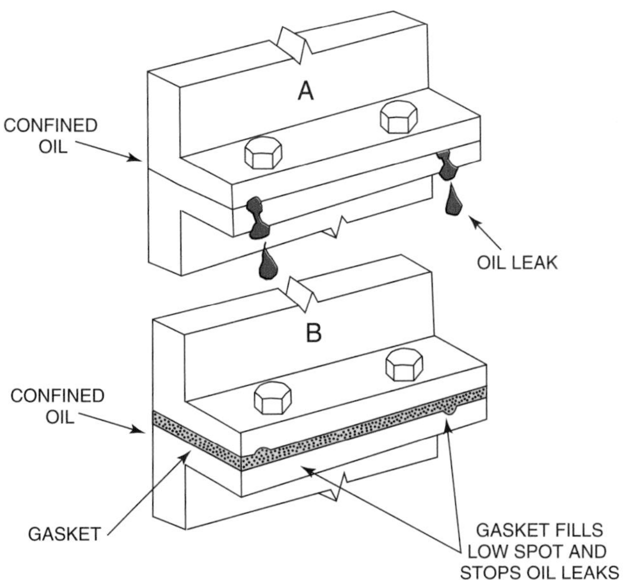

Figure 2-88. Gaskets are designed to compress and seal irregular surfaces. A—Note how oil is leaking out through the two areas. B—A gasket is used here to seal off leaks.

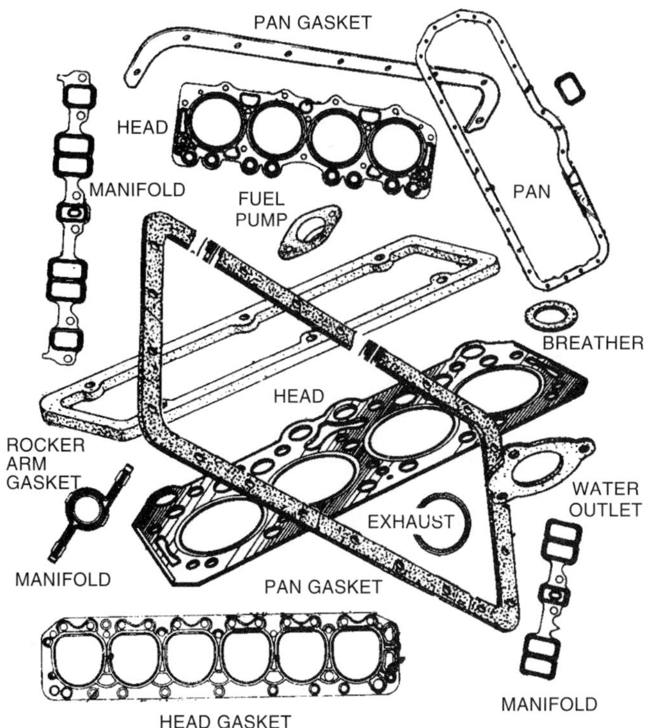

Figure 2-89. A few of the typical gaskets used in an engine.

properties, are available. Most sealants are highly resistant to oil, water, gas, diesel fuel, grease, antifreeze, mild acid, and salt solutions.

Two of the most popular are ***room temperature vulcanizing*** (RTV) and ***anaerobic*** sealants. RTV sealant hardens at room temperature and is used with gaskets or as a stand alone sealant. Anaerobic sealants are used as a stand alone replacement for a gasket. They have the capability of hardening in the absence of air.

Seals

A ***seal*** can be used to confine fluids, prevent the entry of foreign materials, and separate two different fluids. An oil seal is secured to one part, while the sealing lip allows the other part to rotate or reciprocate (move).

Sealing elements are usually made of synthetic rubber. Older engines used leather seals. Synthetic rubber seals are replacing leather in most applications. The rubber seal can be made to close tolerances. It can also be given special shapes and heat-resistant properties.

In the oil seal, a rubber sealing element is bonded to the case. The element rubs against the shaft. The case holds the element in place and in alignment. A garter spring forces the seal lip to conform to minor shaft runout (wobble) and maintains constant and controlled pressure on the lip.

Oil seals are used throughout the mechanical parts of the car. The engine, transmission, drive line, differential, wheels, steering, brakes, and accessories all use seals in their construction.

Other Parts

Other engine parts that are required by the various systems will be discussed in later chapters. **Figure 2-90** shows a cutaway of an engine. Note the relationship of each part to the entire engine.

Summary

The engine block forms the base on which all the other engine parts are mounted. Engine blocks are made of cast iron or aluminum. Cylinders are formed when the block is cast or cylinder sleeves are installed in the block after casting. Pistons ride up and down the length of the cylinders. The pistons must be carefully sized to prevent tipping or excessive friction. Pistons must also be constructed of rigid materials to withstand the wear associated with normal engine operation.

Cast iron rings are installed around the pistons to seal the combustion chamber from the crankcase. Piston rings are of two different types: compression and oil control. Expander devices are sometimes used with rings designed for worn cylinders. Rings are often coated with special materials to assist break-in and to give them long life.

The pistons are attached to the connecting rods through the use of piston pins. Most connecting rods use an I-beam construction. Connecting rod bearing inserts must be able to fit precisely while being rigid enough to withstand the tremendous forces in an engine. The crankshaft uses bearings similar to the connecting rod bearings. Most engines use dampers to offset vibration created by the crankshaft.

The camshaft is used to open the valves in precise time with the crankshaft. The camshaft is connected to the crankshaft through the use of timing gears or sprockets and a chain or belt. Mechanical and hydraulic lifters transfer the force from the camshaft to the valves. Valves are designed to turn in order to reduce heat and carbon build-up.

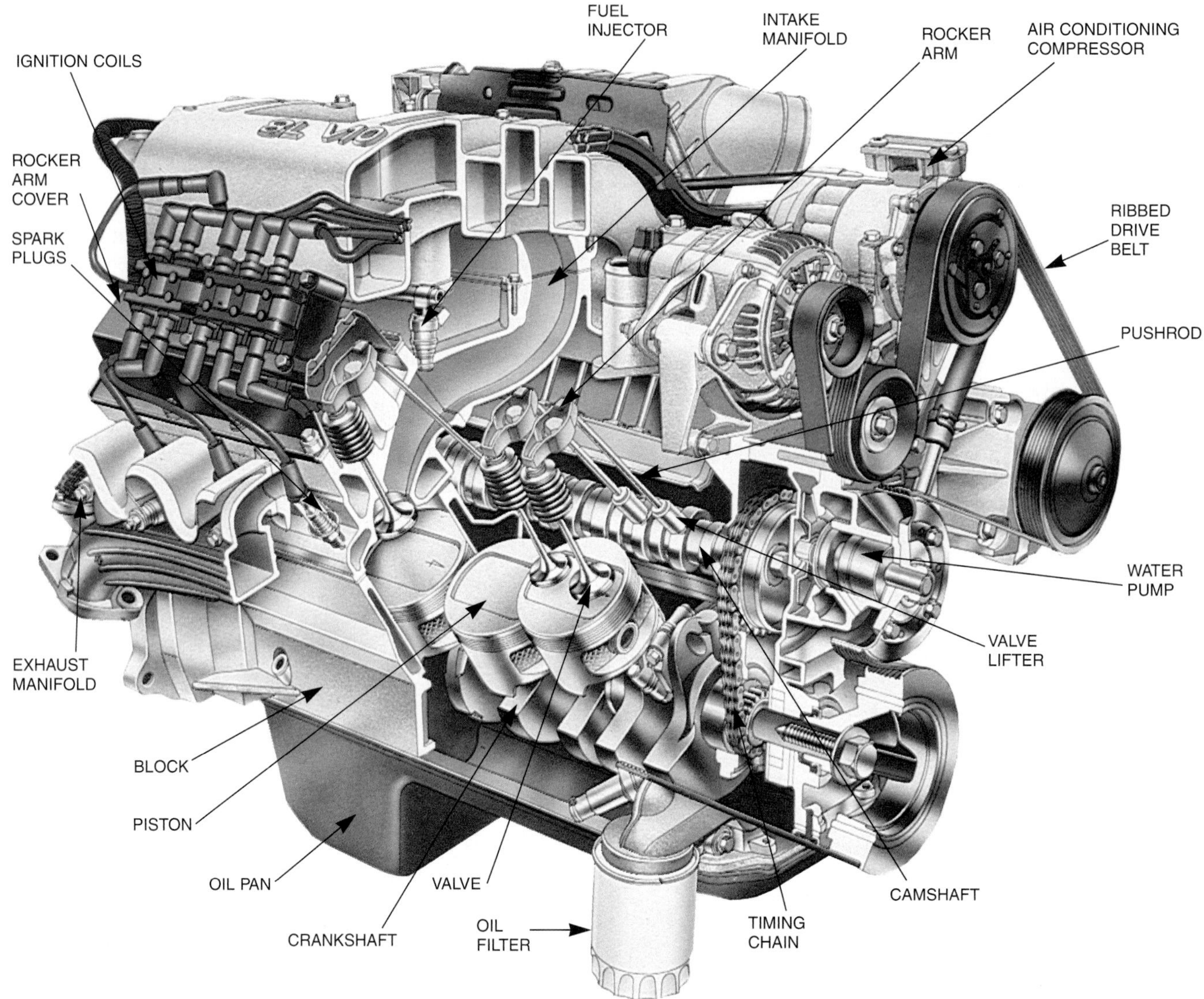

Figure 2-90. Cutaway of an 8.0 liter, V-10, fuel injected engine. Study all the parts.

The valves in most engines are housed in the cylinder head. The oil pan and timing cover house the crankcase and timing gears respectively. Gaskets, sealants, and seals are used to seal the engine and other vehicle components. They help to keep dirt and water out and vital fluids in.

Know These Terms

Engine block
Cylinder sleeves
Dry sleeve
Wet sleeve
Cam ground piston
Partial skirt piston
Piston skirt
Split skirt piston
T-slot piston
Pin boss
Piston rings
Ring gap
Blow-by
Compression ring
Oil control ring
Heat dam
Scuffing
Piston pins
Case-hardening
Snap ring
Insert bearing
Crush height
Thrust bearing
Crankshaft throws
Harmonic balancer
Vibration damper
Balance shaft
Timing gear
Chain sprocket
Timing chain
Timing belt
Mechanical valve lifters
Hydraulic valve lifters
Roller lifters
Valve overlap
Variable valve timing
Valve-in-head
Head bolts
Oil pan
Gaskets
Sealant
Room temperature vulcanizing
Anaerobic
Seal

Review Questions—Chapter 2

Do not write in this book. Write your answers on a separate sheet of paper.

1. All of the following statements about engine blocks are true, EXCEPT:
   (A) the block forms a basic foundation for the entire engine.
   (B) the heavier the block, the better it can control heat expansion.
   (C) blocks can be made of cast iron or aluminum.
   (D) blocks contain the engine cylinders.
2. Technician A says that a dry sleeve can be thin because the block metal supports it. Technician B says that coolant in the engine directly contacts the wet sleeve. Who is right?
   (A) A only.
   (B) B only.
   (C) Both A & B.
   (D) Neither A nor B.
3. The _____ the piston is, the more efficient it will be.
4. Technician A says that a cylinder that is round to within .100″ (2.54 mm) is accurate enough. Technician B says that a cylinder that is round to within .050″ (1.26 mm) is accurate enough. Who is right?
   (A) A only.
   (B) B only.
   (C) Both A & B.
   (D) Neither A nor B.
5. The main advantage of using aluminum instead of cast iron in engine construction is that it is _____ .
   (A) lighter
   (B) stronger
   (C) less likely to rust
   (D) cleaner
6. Name three common ways of controlling piston expansion.
7. Technician A says that the piston must fit tightly in the cylinder to prevent slapping. Technician B says that the piston must be loose enough in the cylinder to leave room for an oil film. Who is right?
   (A) A only.
   (B) B only.
   (C) Both A & B.
   (D) Neither A nor B.
8. A cam ground piston becomes round when it is _____ .
   (A) cold
   (B) partially heated
   (C) fully heated
   (D) None of the above.
9. The head of the piston will generally be .030″ to .040″ (0.76 to 1.02 mm) _____ than the skirt.
10. Piston heads are of different shapes. What are some of the reasons for these shapes?
11. What does a piston heat dam protect?
    (A) Piston.
    (B) Top ring.
    (C) Bottom ring.
    (D) Piston pin.
12. Name the two types of piston rings and state how many of each type are used on a modern piston.
13. Proper ring gap prevents ring _____ when the engine heats up.
14. All of the following statements about piston rings are true, EXCEPT:
    (A) chrome-plated rings are generally rough finished for quicker break in.
    (B) new ring and wall surfaces are designed with some roughness to cause the rings to wear in.
    (C) compression rings are made from cast iron.
    (D) oil ring rails are made of steel.
15. Ring expanders must be stronger when used on a piston installed with what type of cylinder?
    (A) new
    (B) used
    (C) rebored
    (D) None of the above.
16. During the break in period, new piston rings will wear at a _____ rate than after the break in.
17. The bottom rings are always _____ rings.
18. Piston pins fasten the _____ to the _____ .
19. Piston pins are hollow to reduce _____ .
    (A) wear
    (B) heat transfer
    (C) noise
    (D) weight
20. Case-hardening hardens what part of an engine component?
    (A) Outside.
    (B) Inside.
    (C) The sharp edges.
    (D) All of the component.
21. Connecting rods are used to connect the _____ to the _____ .
22. The smaller connecting rod upper end bearing is _____, but the big end bearing is _____ to facilitate installation.
23. Connecting rods use precision _____ for the big end bearings.
24. Name six characteristics of a good bearing material.
25. Crankshafts are usually made of _____ .
    (A) aluminum
    (B) stainless steel
    (C) wrought iron
    (D) None of the above.

26. Technician A says that crankshafts can vibrate from being off balance and also from twisting. Technician B says that the flow of power is smoother in a four-cylinder engine than in an eight-cylinder engine. Who is right?
    (A) A only.
    (B) B only.
    (C) Both A & B.
    (D) Neither A nor B.
27. Crankshafts have drilled passageways. What are they for?
28. Camshafts are driven in three ways. Name them.
29. Technician A says that all modern engines use one exhaust and one intake valve per cylinder. Technician B says that the exhaust valve sheds its heat into the exhaust stream leaving the cylinder.
    Who is right?
    (A) A only.
    (B) B only.
    (C) Both A & B.
    (D) Neither A nor B.
30. Two common valve seat angles are _____ degrees and _____ degrees.
31. The camshaft lobes are usually _____ .
32. Why is the valve rotated on some engines?
33. What are the advantages of a hydraulic lifter?
34. Which of the following parts is sometimes made of a hard metal and pressed into the head?
    (A) valve guide
    (B) valve seat.
    (C) valve spring
    (D) Both A & B.
35. Gaskets are used to seal _____ surfaces.

3

Engine Classification, Parts Identification

After studying this chapter, you will be able to:

- Compare four-stroke and two-stroke cycle engines.
- Explain the different engine classifications.
- Describe the operating principles of a rotary type engine.
- Identify the parts of various engine types.

In the first two chapters, you designed and constructed a basic engine and studied engine parts and their uses. This chapter studies the various methods in which basic engine theory and parts are utilized to produce multicylinder engines. Methods of engine classification as well as different kinds of engines will be discussed. Internal combustion engines can be classified in many ways. The most common methods used to classify engines are by cycle, valve location, cylinder arrangement, type of fuel used, cooling, and number of cylinders. This chapter will discuss each classification.

Cycle Classification

Engines are often classified according to cycle. Most internal combustion piston engines use a two- or four-stroke cycle. The basic engine which was discussed in the first two chapters is a ***four-stroke cycle*** engine. **Figure 3-1** illustrates that it takes two complete crankshaft revolutions to complete the cycle. As a result, each cylinder receives only one power stroke per cycle. All modern automobile engines use the four-stroke cycle engine.

Two-Stroke Cycle Engine

The ***two-stroke cycle*** engine performs the intake, compression, firing, and exhaust sequence in one revolution of the crankshaft. This is accomplished by eliminating the valves used in the four-stroke cycle engine. In place of the valves, two ports enter the cylinder wall. One is used for the intake of fuel and the other for exhaust.

A passage connects the intake port to the crankcase interior. The air-fuel mixture is drawn into the crankcase through a ***reed valve*** or a ***rotary valve***. Oil is usually mixed with the gasoline to provide lubrication. Some two-stroke cycle engines use an oil injection system. Upon entry into the crankcase, the oil provides ample lubrication for the bearings.

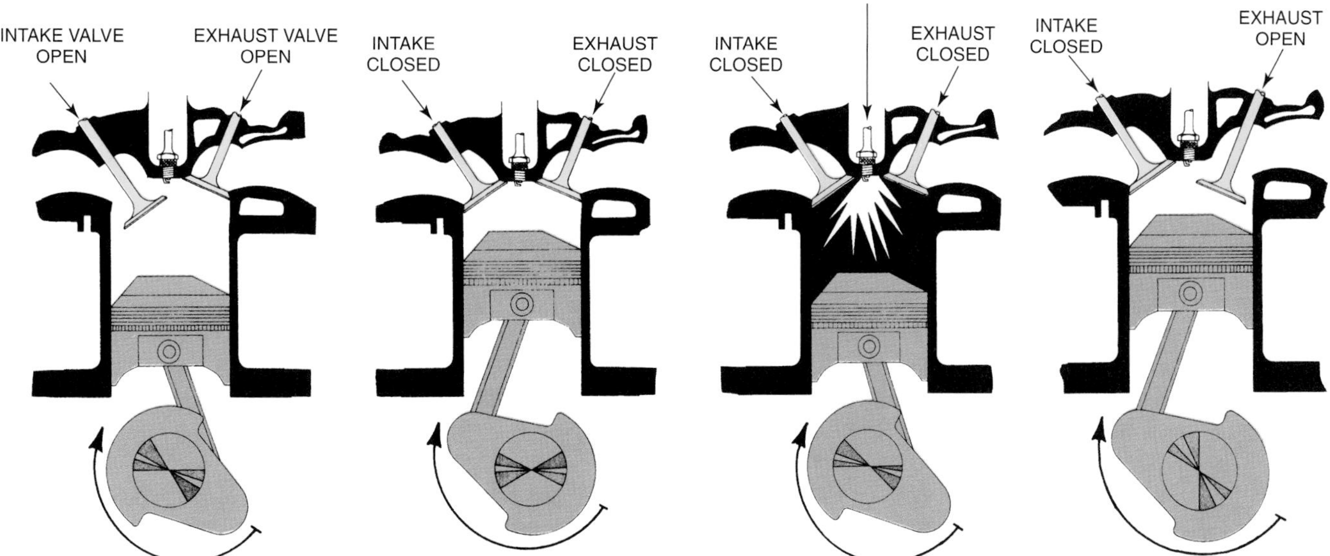

Figure 3-1. Four-stroke cycle principle. This series of events requires two complete revolutions (720° of rotation) of the crankshaft. Note that there is one power stroke per cycle. (TRW, Inc.)

Two-stroke cycle engines are used in chain saws, lawn mowers, marine outboard motors, go-carts, and other small engine applications. Due to emission control problems, gasoline powered two-stroke cycle engines are not used in automobiles.

Two-Stroke Cycle Operation

Imagine that a charge of air and fuel is in the cylinder and that the piston is at the bottom of its stroke. As the piston travels up, it closes off both intake and exhaust ports. As it continues upward, it compresses the air-fuel charge while forming a vacuum in the airtight crankcase. The vacuum pulls open a small flap-like reed valve and a mixture of air and fuel enters the crankcase. When the piston reaches the top of its stroke, it has compressed the charge in the cylinder and filled the crankcase with an air-fuel mixture. This has all happened in one-half revolution of the crankshaft.

The spark plug fires the cylinder charge and drives the piston down. Several things happen as the piston travels downward. The air-fuel mixture in the crankcase is compressed and the reed valve closes. The piston is also performing the power stroke at this time. When the piston has traveled down the cylinder far enough to expose the exhaust port, the exhaust will flow out of the port. As the piston continues to travel down, the intake port is uncovered. As the crankcase air-fuel mixture is compressed, it is forced to flow through the now uncovered intake port into the cylinder.

The piston top is usually shaped to deflect the incoming air-fuel charge upward. This fills the top of the cylinder with a fresh charge while simultaneously scavenging (cleaning out) the remaining exhaust gases. This has all taken place in one revolution of the crankshaft. Every time the piston reaches the top of its travel, it will be driven downward by a power stroke. Study **Figure 3-2**.

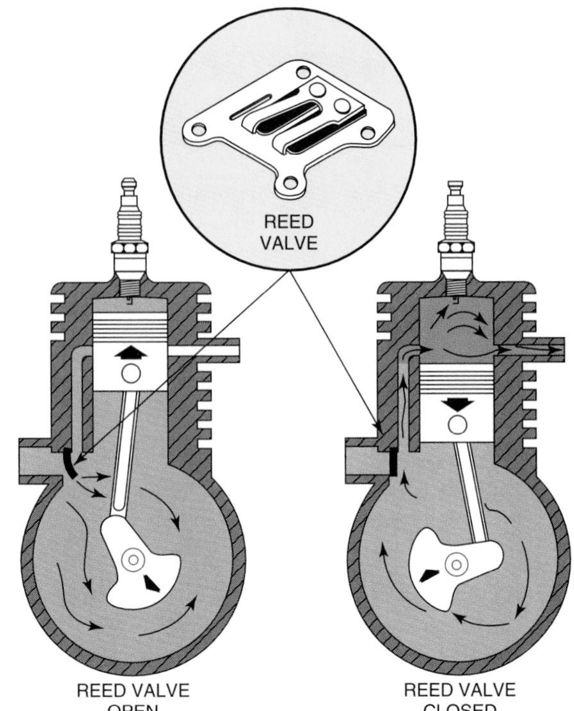

Figure 3-2. In the left-hand view, the piston is traveling upward and has almost reached the top of its stroke. It has compressed the air-fuel charge trapped above the piston. The spark plug has just fired and the mixture is starting to burn.

On its way upward, the piston has drawn a fresh fuel mixture into the crankcase through a reed valve. The fuel vapor contains suspended droplets of oil that lubricates the surfaces of moving parts.

In the right-hand view, the air-fuel mixture has fired and the piston is traveling down. It has uncovered the intake and exhaust ports, allowing burning gases to escape and fresh charge to enter. The air-fuel mixture flows through the intake port because the piston, while traveling down, compressed the mixture in the crankcase.

In two-strokes, the engine has performed all the necessary functions to enable it to receive a power stroke for every crankshaft revolution.

Advantages of the Two-Stroke Engine

The two-stroke cycle engine has several advantages that make it popular for use in small engine applications. By eliminating the valves, it also eliminates valve springs, cam followers, camshaft, and gears. Since there are no valves to "float," two-cycle engines can operate at very high speeds. This enables the engine to be quite compact and light. Machining costs are reduced.

Because the two-stroke engine fires on every stroke, more power per pound of engine weight is available. However, it does not produce the power of a four-stroke cycle engine of comparable size. A two-stroke engine has poor volumetric efficiency (ability to intake fuel) and will lose some power through compressing the crankcase mixture. In order to completely scavenge the cylinder, some of the intake air-fuel charge also leaves the cylinder, resulting in higher emissions and less fuel economy. A two-stroke cycle engine that utilizes an exhaust valve and no mixture in the crankcase will be shown in the section on diesel engines later in this chapter.

Valve Location Classifications

Engines are also classified according to the valve location. The various valve arrangements are known as the L-head, T-head, F-head, and I-head. The names are derived from the resemblance to letters of the alphabet. This will be demonstrated in forthcoming sketches. The I-head is almost universally used in cars and light trucks.

L-Head, T-Head, and F-Head Designs

These three head designs are no longer used in automotive applications, but it is possible to find them in some stationary power plants or small engine applications. The ***L-head*** engine has both valves in the block and on the same side of the cylinder. A line drawn through the cylinder and across to the valves will produce the letter *L* in an inverted (upside down) position. **Figure 3-3** illustrates the L-head design. This design may still be found on some small engines and forklifts.

In the ***T-head*** engine, both valves are in the block and on opposite sides of the cylinder, **Figure 3-4**. This design has been obsolete for many years and is no longer used. The ***F-head*** places one valve in the head and one in the block, **Figure 3-5**. Today, this design is only used on small air-cooled

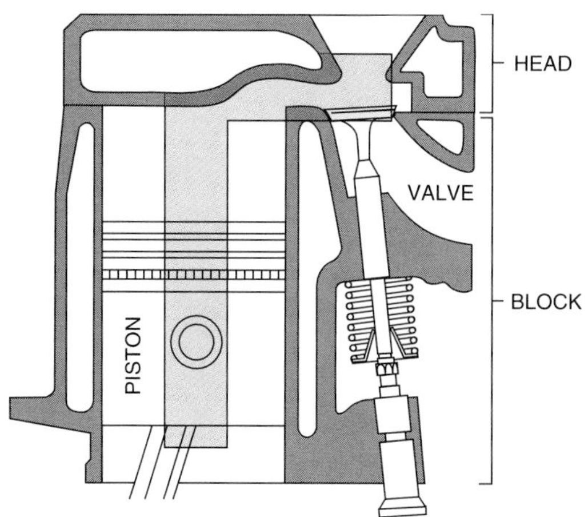

Figure 3-3. L-head engine. Valves are in the block, both on the same side. Note letter L. (TRW, Inc.)

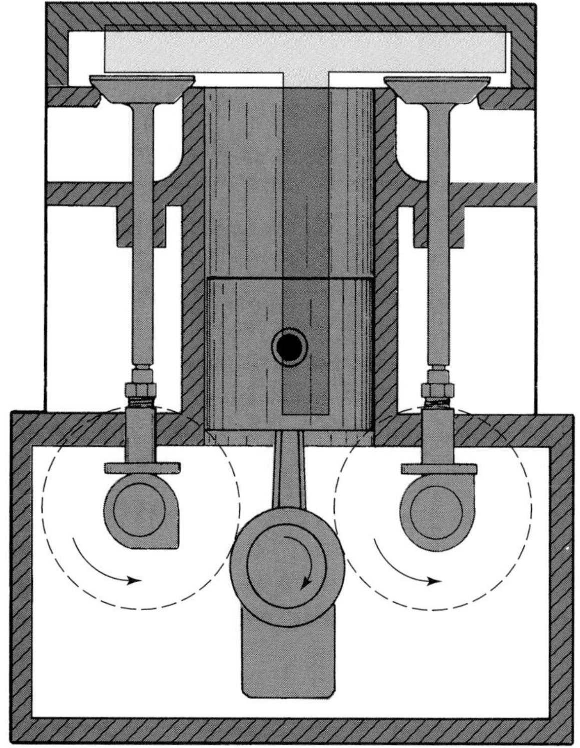

Figure 3-4. T-head configuration, like the L head engine, the valves are in the block. This configuration is not used anymore, but may be found on some older engines.

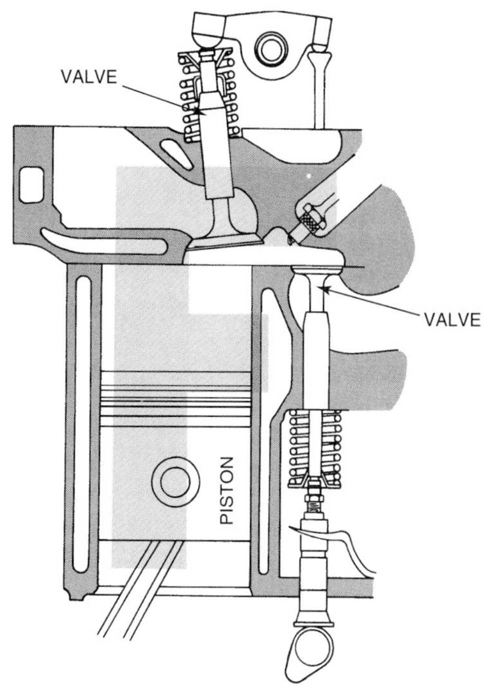

Figure 3-5. F-head engine. Note that there is one valve in the block and one in the head. (TRW, Inc.)

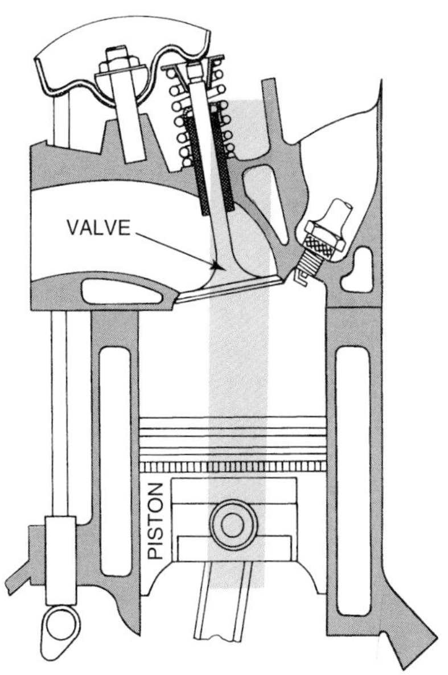

Figure 3-6. I-head engine. Both valves are located in the head. (TRW, Inc.)

engines, such as those found in most lawn mowers and garden tillers.

I-Head

In the ***I-head*** engine, both valves are located in the cylinder head. They may be arranged in a straight line, **Figure 3-6** or staggered, **Figure 3-7**. This is sometimes referred to as a valve-in-head engine.

Since the valves are placed in the I-head, a ***valve train*** must be used to operate them. **Figure 3-8** illustrates the use of a ***rocker arm*** and ***push rod*** setup to actuate the valves. The rocker arm is mounted on a ***rocker shaft***. Note how push rod force is transferred in the opposite direction by the rocker arm. **Figure 3-9** illustrates the use of a rocker arm-push rod valve train on an actual engine. **Figure 3-10** shows a different rocker arm design. Instead of oscillating on a rocker arm shaft, this type uses a ***ball pivot*** arrangement on which the rocker arm pivots.

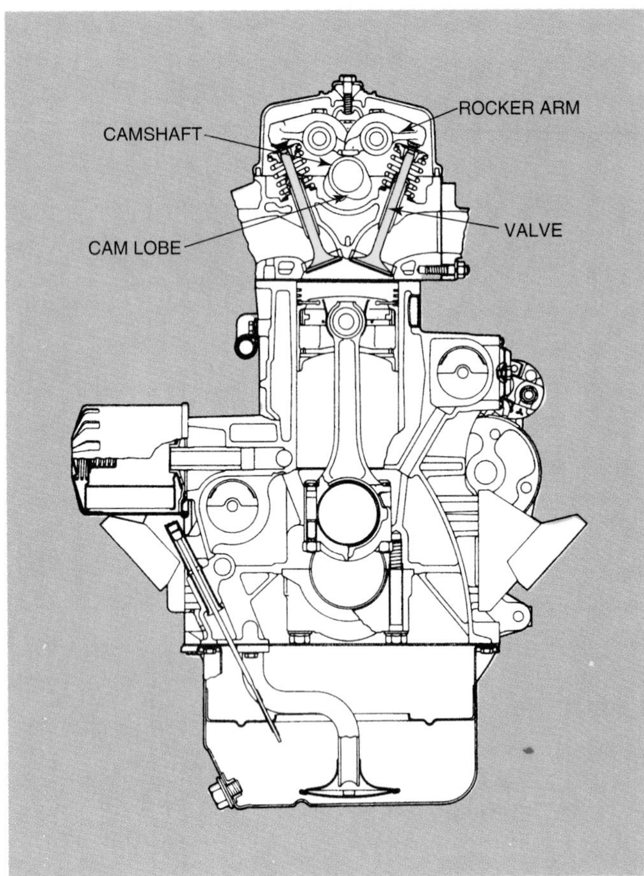

Figure 3-7. Transverse cross-section of an I-head (overhead valve) engine. Both intake and exhaust valves are located in the head. (Dodge)

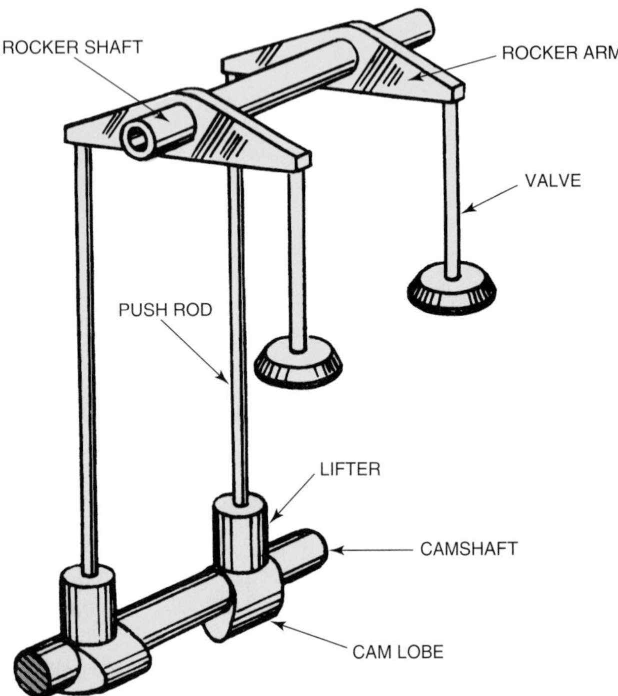

Figure 3-8. Schematic shows simplified valve train for one cylinder of an overhead valve engine.

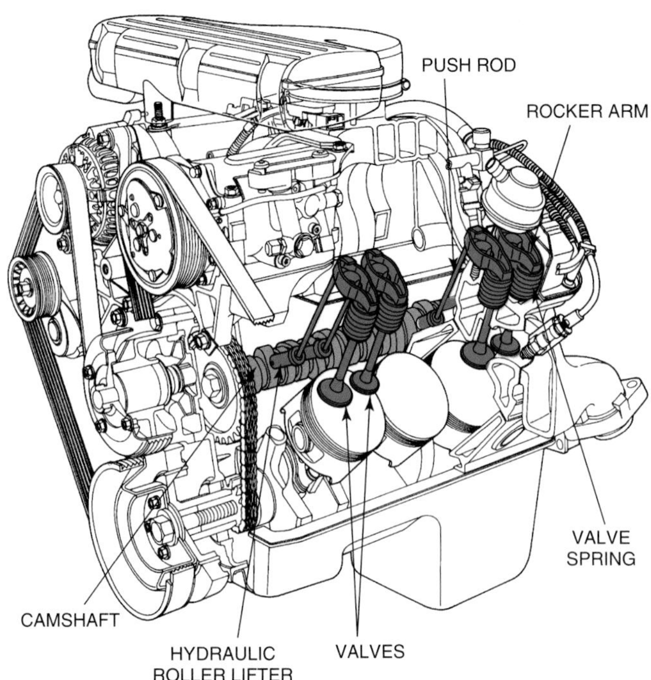

Figure 3-9. Cutaway view shows a typical valve train used to operate overhead valves on a V-6 engine. (Dodge.)

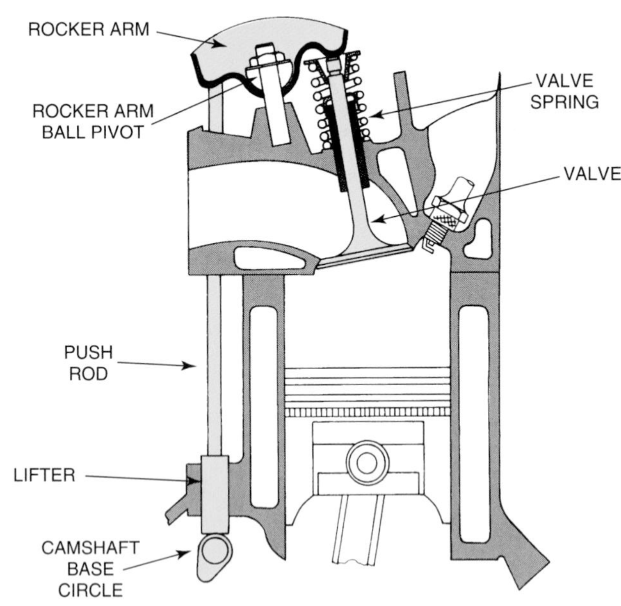

Figure 3-10. Ball type rocker arm. Note the shape of the ball pivot. (TRW, Inc.)

Overhead Camshaft I-Head

Some I-head engines mount the camshaft in the head. Either one or two shafts can be used. When an ***overhead camshaft*** is used, push rods are not required. The rocker arm setup shown in **Figure 3-11** is operated by the camshaft lobe rubbing directly on the rocker. Zero valve stem-to-rocker clearance is maintained by a hydraulic valve lash adjuster, **Figure 3-12**, which operates in the same manner as the hydraulic

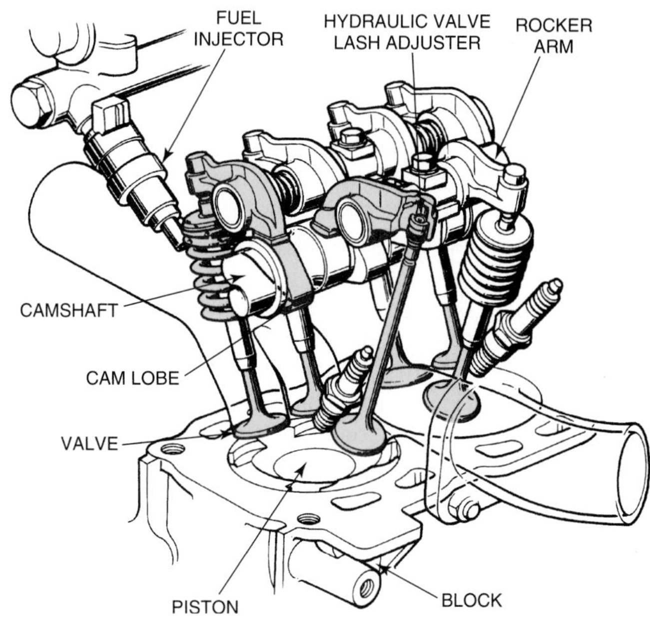

Figure 3-11. One type of overhead camshaft rocker arm arrangement. Note the hydraulic lash adjuster. (Ford)

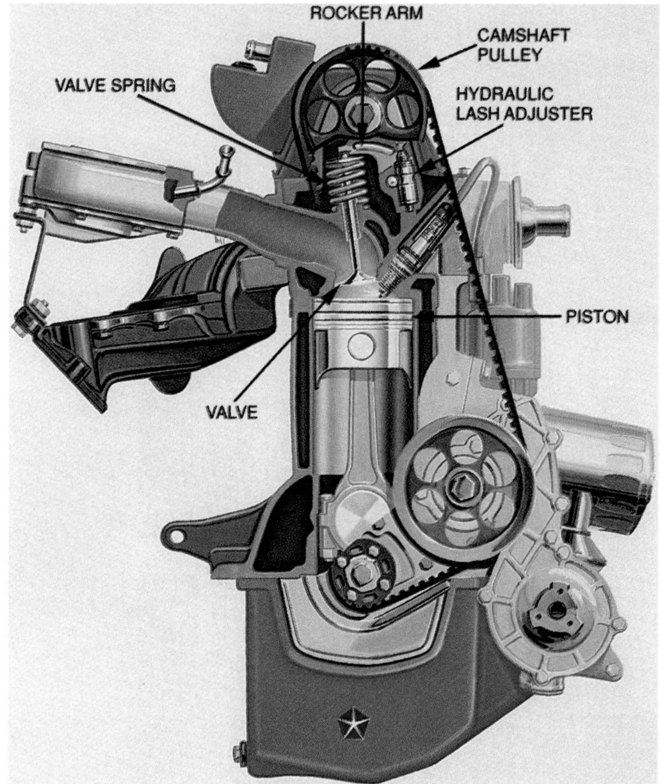

Figure 3-12. Cross-sectional view of an engine which incorporates a single overhead camshaft (SOHC) to operate the valves. Note the use of a hydraulic valve lash (clearance) adjuster. (Chrysler)

lifter discussed earlier in Chapter 2. The camshaft lobes can also be arranged to work directly on the valves as shown in **Figure 3-13**.

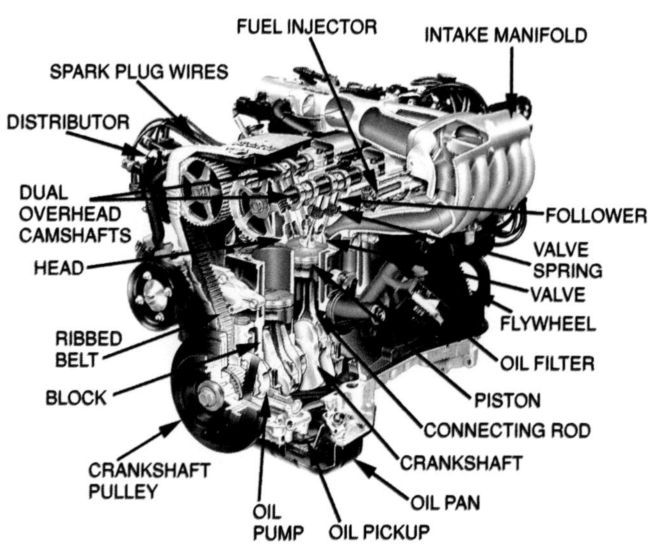

Figure 3-13. Cutaway view of a 3.0-liter, dual overhead camshaft (DOHC), 24-valve, inline six-cylinder, gasoline engine. Note the camshaft lobes act directly on the ends of the valves. No rocker arms are used. Study carefully. (Lexus)

Overhead Valve Adjustment

Some overhead valve installations provide an adjusting screw in the push rod end of the rocker arm. This screw is moved in or out until the recommended clearance is obtained between the valve stem and the rocker arm. See **Figure 3-14**. Some engines use hydraulic lifters, which eliminates the need to adjust clearance. The only adjustment necessary is the initial setting to place the lifter plunger at the halfway point in its travel. Some engines make the initial setting with adjustable

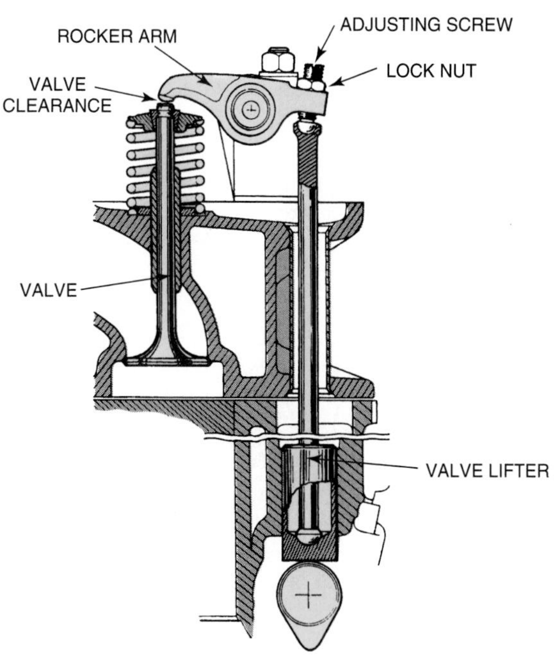

Figure 3-14. Rocker arm with adjustment screw used to set proper valve clearance or "lash." (British-Leyland)

rocker arms, while others use push rods that provide the proper setting when installed. **Figure 3-15** shows such a setup. Push rods of slightly different lengths are available to make minor changes.

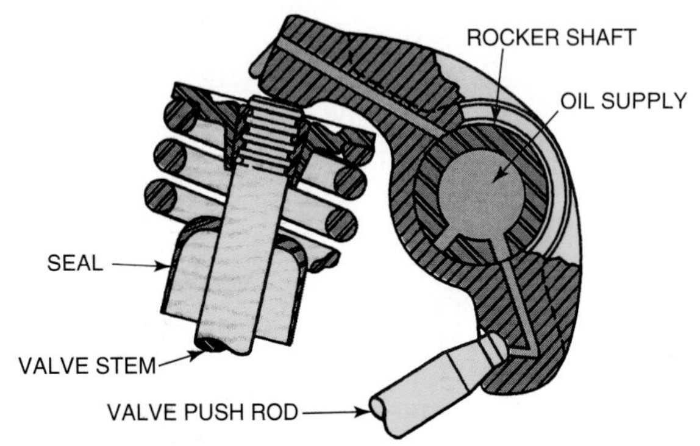

Figure 3-15. This non-adjustable rocker arm depends upon precise parts so that the hydraulic lifter plunger will be at midpoint of travel upon assembly. (Dodge)

Valve-to-Piston Clearance

Some engines utilize high compression, which squeezes the air-fuel charge into a smaller than normal area. In order to have this small area and still allow room for the valves to open, it is sometimes necessary for the pistons to have small indentations cut into their tops to allow the valves to clear the piston at top dead center (TDC), **Figure 3-16**. These indentations, or grooves, keep the piston head from striking the valves at the top of the exhaust stroke when both valves may be partially open. If the piston and valve(s) are allowed to make contact, extensive engine damage would result.

Detonation and Preignition

When a highly compressed air-fuel charge is fired, the flame travels from the plug in an outward direction. As the charge continues to burn, chamber pressure greatly increases. If the combustion chamber pressure increases faster in another part of the cylinder before the flame reaches it, this area will become hot enough to fire itself. This will produce two flame fronts, which increases the fuel charge burn rate. Instead of firing smoothly, there is a violent explosion that literally slams the piston down. This is called ***detonation***, which is very damaging to an engine.

Detonation can be caused by an excessively high compression ratio or by a low grade of gasoline. An overheated valve or piece of carbon can ignite the remaining air-fuel mixture and cause the formation of the second flame front, **Figure 3-17**. ***Preignition*** is caused when a glowing piece of carbon, an overheated spark plug, or a piece of metal extending into the combustion chamber ignites the fuel charge before the spark plug fires. When preignition occurs, the premature explosion attempts to drive the piston down against the direction of rotation. Detonation and preignition can destroy an engine if it is not corrected.

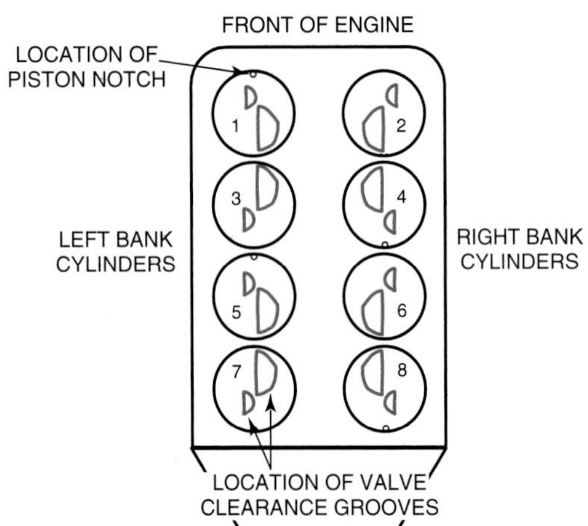

Figure 3-16. Piston indentations used for valve clearance. Each piston has two small indentations to allow the valves to clear. (Chevrolet)

Combustion Chamber Design

The size and shape of the combustion chamber is very important. A well designed chamber will allow high compression ratios without the occurrence of detonation or preignition.

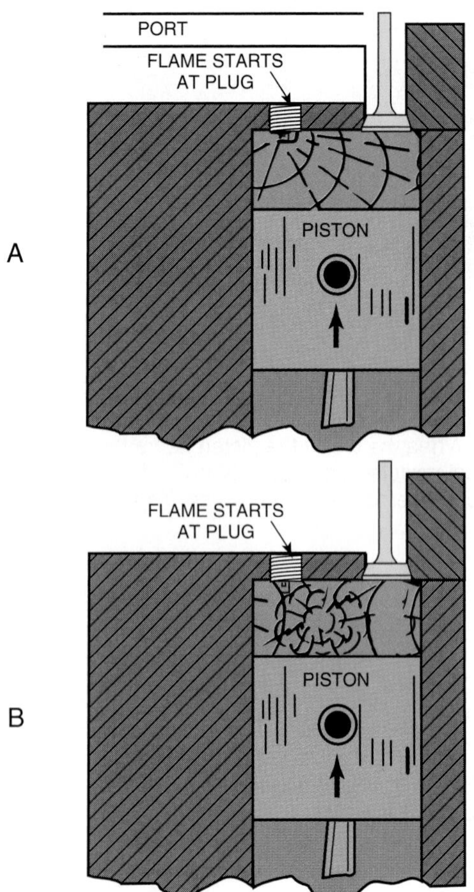

Figure 3-17. Detonation. A—Proper flame travel. B—Two flame fronts increase burning rate and cause detonation.

Compression ratios of 10.5 to 1 and higher were once widely used. Modern engines have compression ratios of around 8, 8.5, or 9 to 1, which provides both good performance and emission control.

Ricardo Combustion Chamber

Many years ago a man named ***Ricardo*** developed a combustion chamber for L-head engines that has been modified for use in many I-head engines. He found that by causing a violent turbulence in the compressed air-fuel charge and by locating the spark plug near the center of the turbulence, flame travel through the charge was smooth and compression could be raised considerably without detonation.

Modern designs locate the combustion chamber under the valves and allow the piston to come very close to the head. As the piston travels up on compression, it forces the fuel into the area below the valves. As the piston comes close to the head, the remaining mixture is rapidly squeezed, causing it to shoot into the combustion chamber, setting up a rapid turbulence. The area between the piston top and the head is referred to as the ***squish area***, **Figure 3-18**.

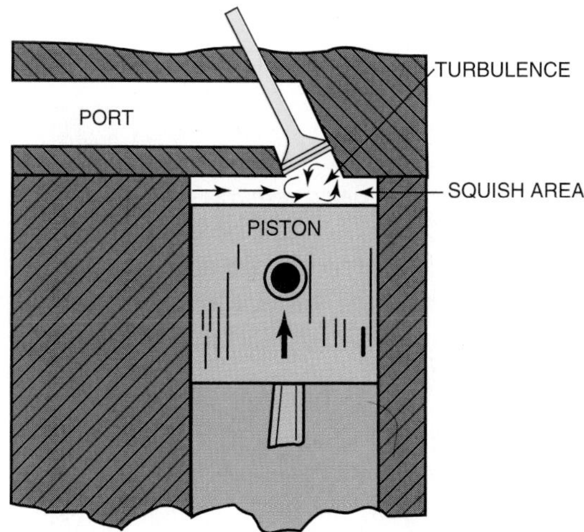

Figure 3-18. When the top of the piston nears the head, the trapped air-fuel mixture in the "squish" area shoots into the combustion chamber, causing turbulence. This is known as the Ricardo principle.

Hemispherical Combustion Chamber

The ***hemispherical combustion chamber*** is compact and allows high compression with very little detonation. By placing the valves in two planes, it is possible to use larger valves, thereby improving air-fuel intake and exhaust scavenging. The hemispherical combustion chamber was first used many years ago on high performance engines. Today, it is used on a number of modern engines. **Figure 3-19** illustrates the hemispherical combustion chamber and valve configuration.

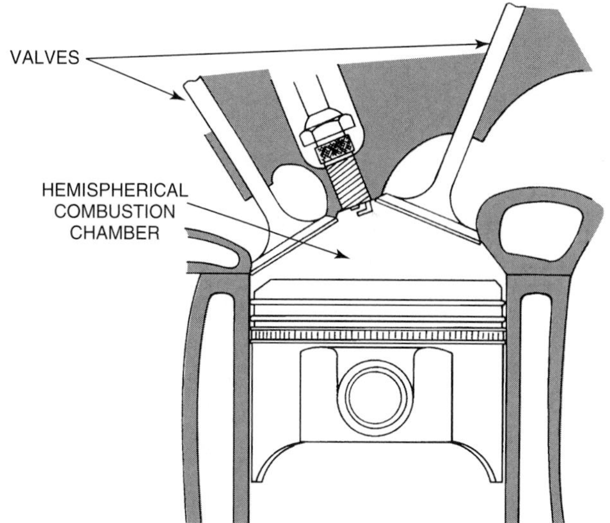

Figure 3-19. A cutaway view of a hemispherical combustion chamber. The valves are set in two places. (TRW)

Wedge Shape Combustion Chamber

Another combustion chamber shape is the ***wedge chamber***. When the wedge shape is incorporated in the head, a flat top piston will produce a desirable squish effect. When the top of the engine block is machined at an angle of around 10°, it is possible to build the cylinder head flat. The combustion chamber wedge is formed by the angled cylinder and a piston with an inverted V top. This setup will also produce a squish effect.

The spark plug is generally located at one corner of the wedge, **Figure 3-20**. This provides a smooth flame travel and loads (presses down) the piston gradually. When the charge is first ignited, it loads only a portion of the piston. This partial pressure starts the piston down smoothly. As the piston

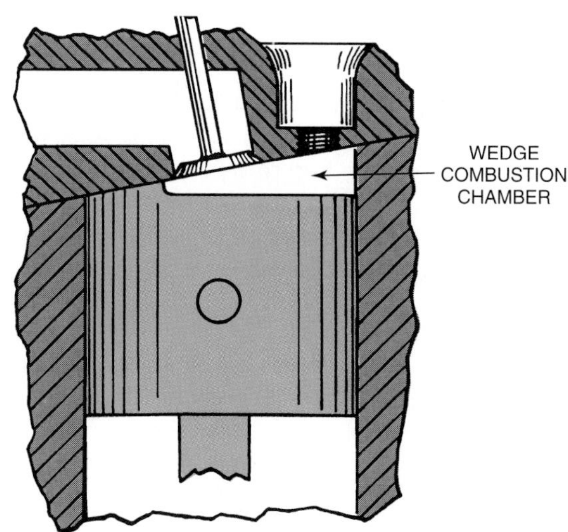

Figure 3-20. Notice that the head in this wedge combustion chamber is flat and cylinder block is machined at an angle — approximately 10°. The valves are set in a plane.

begins to travel down, the flame front spreads across the entire piston dome, providing smooth loading.

Cylinder Classification

Cylinder placement gives rise to one of the most common methods of engine classification. Many arrangements have been used, but only those in popular use today will be covered in this text. **Figure 3-21** shows one other type. This is a ***radial*** engine, which has been very popular for use in propeller-driven aircraft.

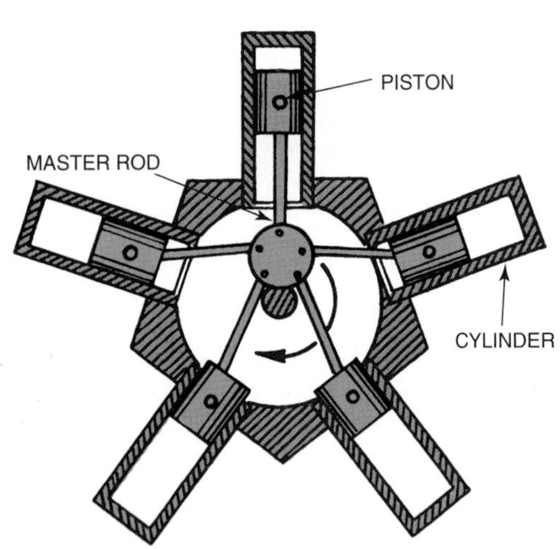

Figure 3-21. In this radial engine, all the connecting rods are fastened to the master rod. This design has been used for many years in aircraft.

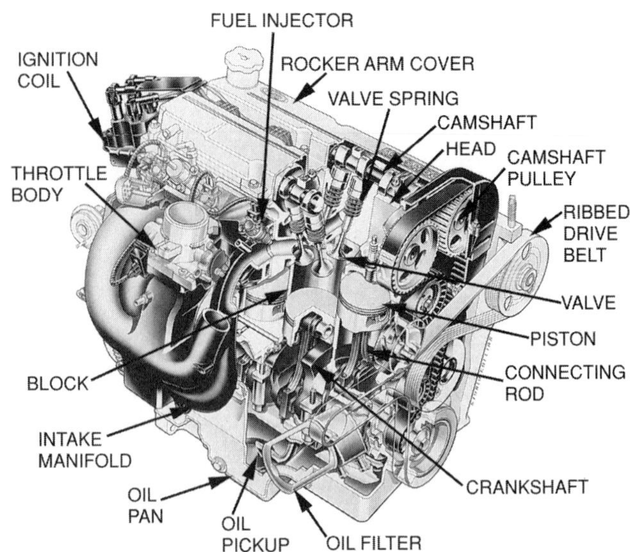

Figure 3-22. Cutaway view of an inline, four-cylinder, overhead valve, electronic fuel injected engine. Note the ribbed drive and timing belts. (Ford Motor Co.)

Inline Engine

The ***inline*** engine has its cylinders arranged one after the other in a straight line. They are in a vertical, or near vertical position, **Figure 3-22**. Most modern inline engines are four cylinders. Inline six-cylinder engines were used in older vehicles and can still be found in a few modern light trucks. Until the adoption of the V-8, eight cylinder inline engines were common. A few three- and five-cylinder inline engines are now produced. A variation of the inline engine is the ***slant*** or ***inclined engine***, **Figure 3-23**. An inclined engine is an inline engine with the cylinder block slanted to one side. This is done to fit the engine into a smaller engine compartment, allowing for a lower hood line.

V-Type Engine

A ***V-type*** engine places two banks or rows of cylinders at an angle to each other—generally at 60° or 90°. The V-type engine has several advantages: short length, extra block rigidity, a short, heavy crankshaft, and low profile that is conducive to low hood lines. The shorter block permits a reduction in vehicle length with no sacrifice in passenger room.

Higher compression ratios can be used in the V-type engine without block distortion under load. The short, heavy crankshaft is more resistant to torsional vibration. In the past, V-12, and V-16 engines were widely used. Current production V-type engines are the V-6 and V-8, along with a few V-10 and V-12 designs, **Figure 3-24**.

Horizontal-Opposed Engines

The ***horizontal opposed engine*** is like a V-type engine, except that both banks lie in a horizontal plane. The advantage here is an extremely low overall height, which makes this engine ideally suited to installations where space is limited. **Figure 3-25** shows a horizontal-opposed engine. It is an air-cooled, six-cylinder engine.

Firing Order

Firing order refers to the sequence in which the cylinders fire. Firing orders are given in order from the front of the engine to the rear. The order in which the engine must fire is determined by the number of cylinders and the arrangement of the crankshaft throws. Firing order can be found in the service manual for the vehicle being serviced. Firing orders vary between engines, and must be obtained from the service manual.

Cooling Classification

As you have learned, engines are either liquid-cooled or air-cooled. Most vehicles use liquid-cooled engines. Air-cooled engines are used in limited numbers on modern vehicles, but are widely used to operate chain saws, lawn mowers, scooters, and other equipment. Both types have advantages and disadvantages.

Air cooling is simple and has no coolant to leak or freeze in cold weather. However, air cooled vehicles are very uncomfortable in cold weather, since there is no efficient or safe way to transfer engine heat to the passenger compartment.

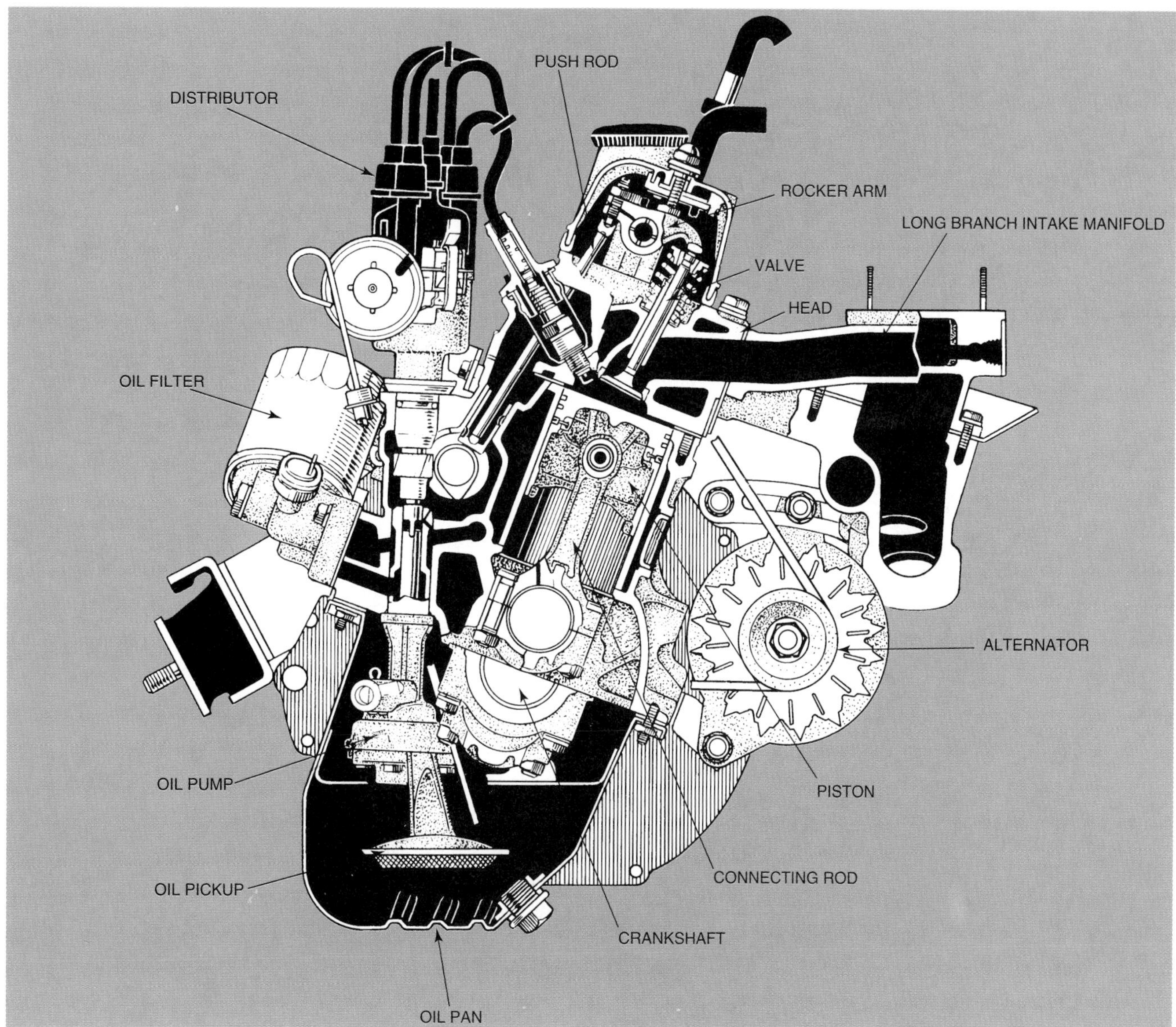

Figure 3-23. Cross-sectional view of an inclined four-cylinder engine. By setting the engine at an angle, overall height is reduced. This allows for a lower vehicle hood. (Toyota)

Fuel Classification

Automobile engines can use gasoline, diesel fuel, gasohol (mixture of gasoline and alcohol), alcohol, LNG (liquefied natural gas), CNG (compressed natural gas), or LPG (liquefied propane gas). Gasoline powers the majority of vehicles, but diesel fuel is used in some vehicles. Gasohol, LNG, CNG, and LPG are beginning to see wider use.

One of the principal differences in these engines is in method of fuel delivery and carburetion. Gasoline, LNG, CNG, and LPG utilize the same basic type of engine, but LNG, CNG, and LPG utilize a slightly different fuel delivery setup. Diesel engines do not use a carburetor or an ignition system. Due to the pressures and stresses involved, diesel engines are constructed very rigidly. The theory of diesel engines is covered in the diesel injection section in Chapter 9. Diesel engines are either two- or four-stroke cycle types.

Two-Stroke and Four-Stroke Cycle Diesel Engines

A two-stroke cycle diesel engine is somewhat different than the two-stroke gasoline engine discussed earlier in this chapter. The two-stroke cycle diesel engine utilizes an exhaust valve. No fuel mixture enters the crankcase. Due to the pressures involved, a supercharger is used to force air into the cylinder intake ports. **Figure 3-26** shows the two-stroke cycle operation in a diesel engine.

Notice that air rushing in through the cylinder liner ports helps scavenge the cylinder by flowing out the exhaust passage for a portion of the piston travel. Since only air enters the cylinder, there is no loss of fuel. **Figure 3-27** shows a typical four-stroke cycle, four-cylinder automotive diesel engine. Inline five-, six-, and V-8 diesels are also used, **Figure 3-28**. Diesel fuel injection system operation is discussed in Chapter 9.

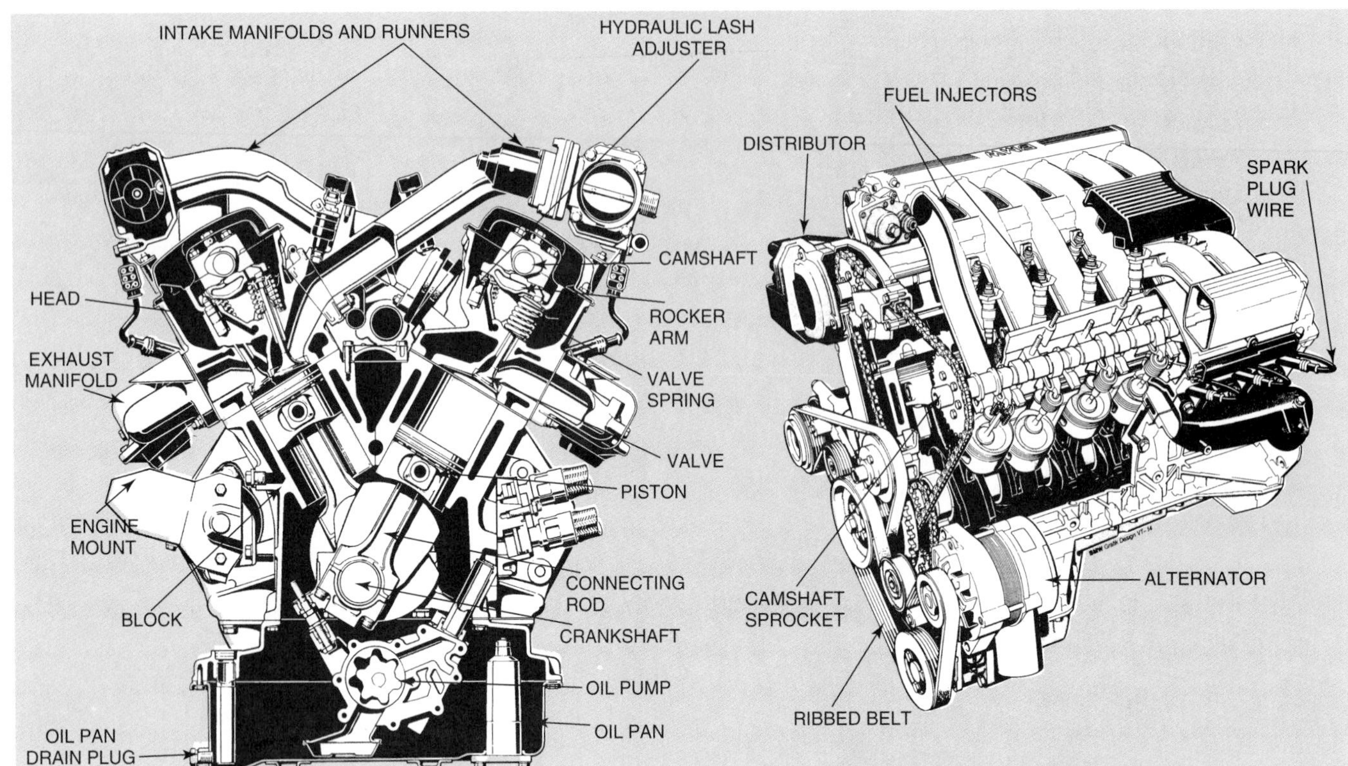

Figure 3-24. Two cutaway views of one particular V-12, electronic fuel injected, gasoline engine. Study the engine construction carefully in both views. (BMW)

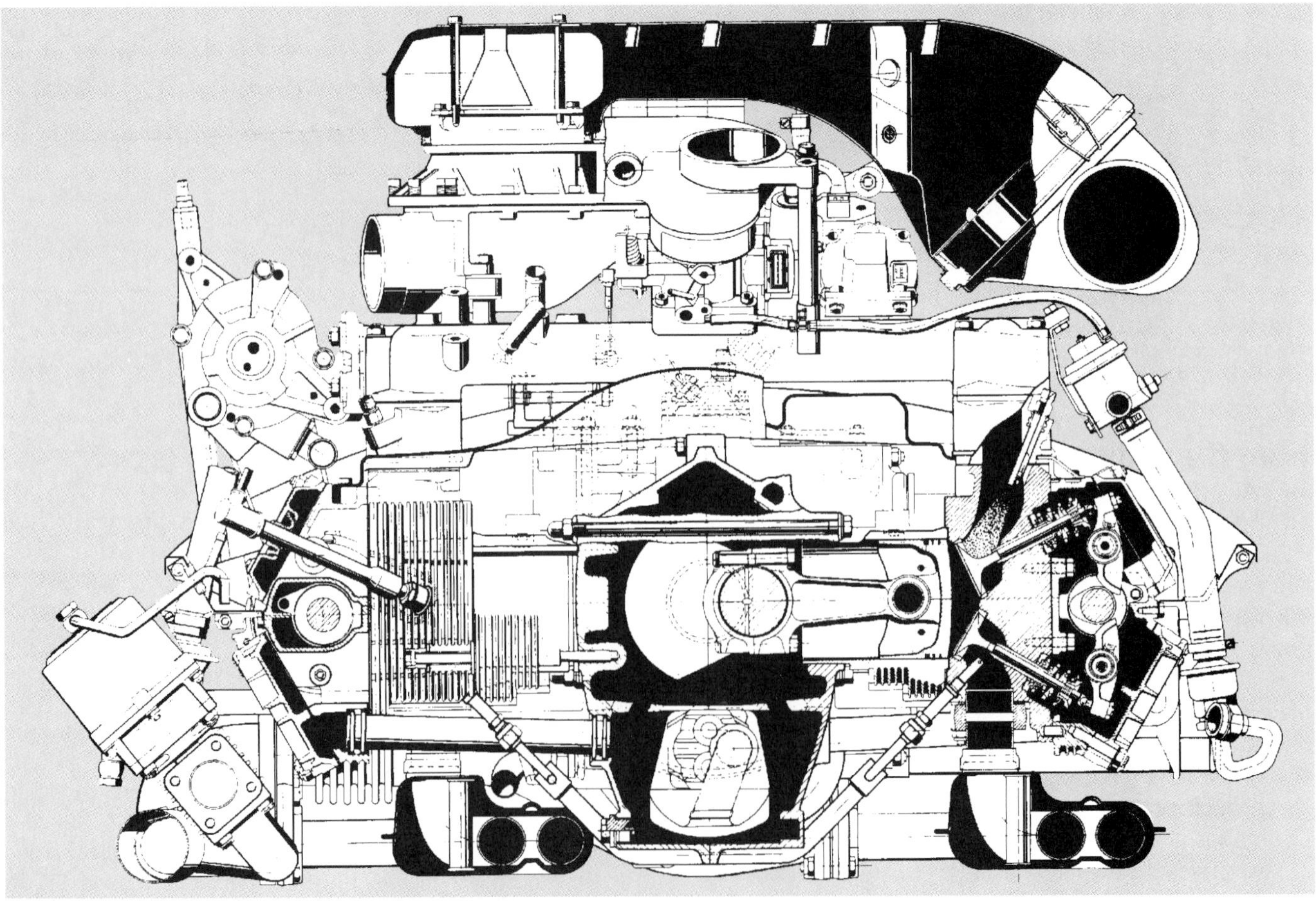

Figure 3-25. Cutaway view of a horizontally opposed 3.3 liter, six-cylinder, fuel injected engine. A low profile makes this engine suitable to limited spaces. (Porsche)

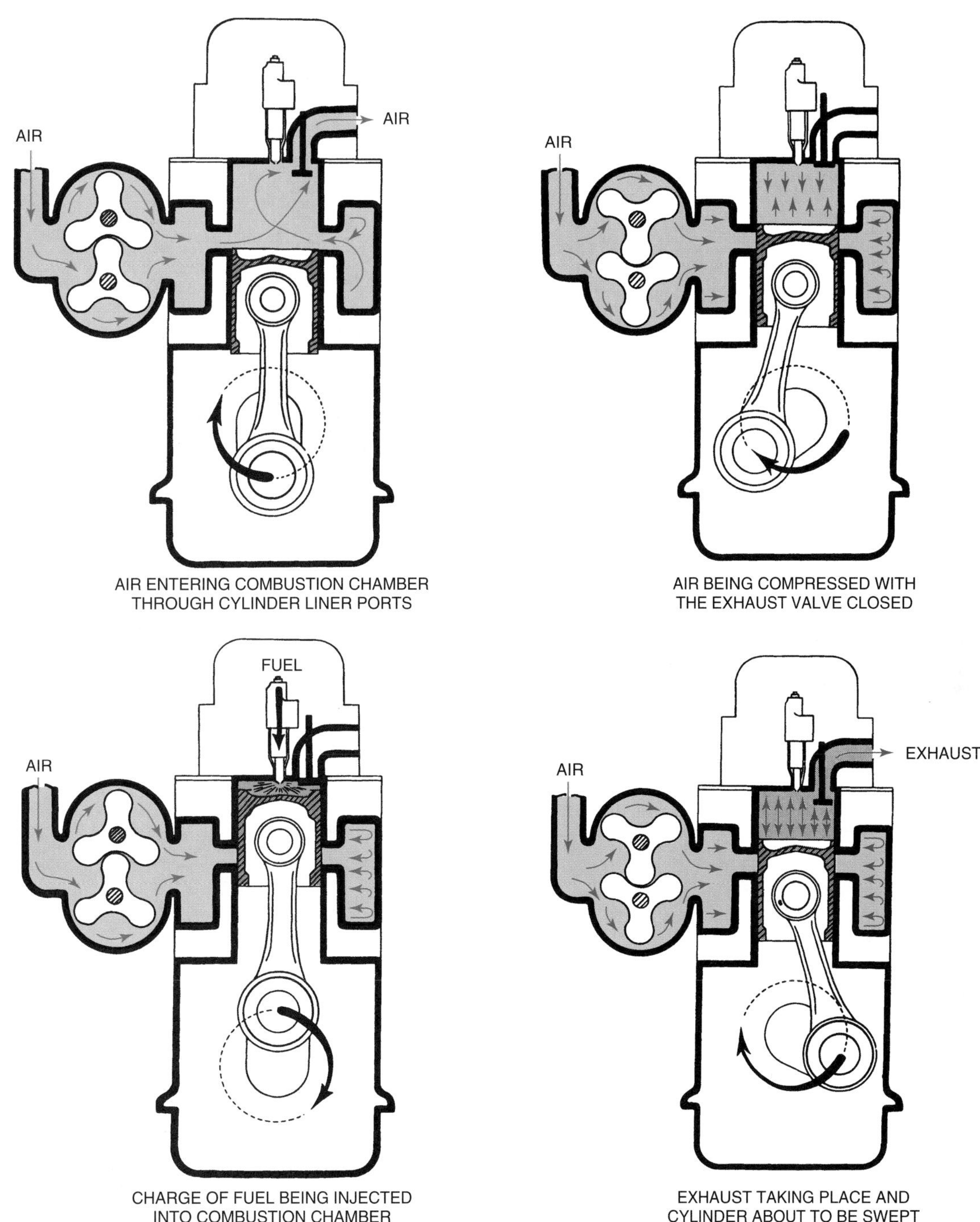
AIR
AIR
AIR ENTERING COMBUSTION CHAMBER THROUGH CYLINDER LINER PORTS
AIR
AIR BEING COMPRESSED WITH THE EXHAUST VALVE CLOSED
FUEL
AIR
CHARGE OF FUEL BEING INJECTED INTO COMBUSTION CHAMBER
AIR
EXHAUST
EXHAUST TAKING PLACE AND CYLINDER ABOUT TO BE SWEPT WITH CLEAN SCAVENGING AIR

Figure 3-26. Two-stroke cycle diesel engine operation. Note the use of a supercharger to force the air-fuel mixture into the combustion chamber. (Detroit Diesel Engine Div.—General Motors)

Rotary Engines

The ***rotary engine*** was used on a few vehicles for many years and was not produced in large numbers. The rotary engine is very powerful for its size. This engine is often referred to as a Wankel engine, **Figure 3-29**.

The rotary engine has fewer moving parts than a piston engine. Since it has no reciprocating motion, the rotary engine produces less vibration and offers more horsepower. Drawbacks of the rotary engine included difficulty meeting emission control standards, and poor fuel economy compared to a piston engine of similar power output.

Rotary Engine Construction

Rotary engines are classified according to the number of ***rotors*** used in the engine. The remainder of the engine consists of one or more ***rotor housings***, end housings, fixed gears, and a center housing all held rigidly together with a series of bolts.

There are just two or three moving parts in the rotary engine depending upon the number of rotors the engine has —the ***eccentric shaft*** (also called the mainshaft) and the cast iron rotor assemblies. The rotors have spring-loaded cast iron seals to close the very small running clearance between them

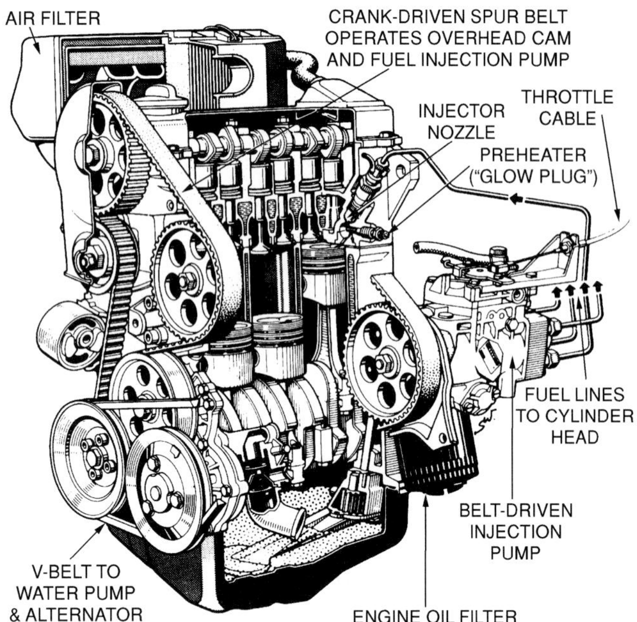

Figure 3-27. Four-cylinder diesel engine. Note the use of a timing belt instead of gears or chains. (VW)

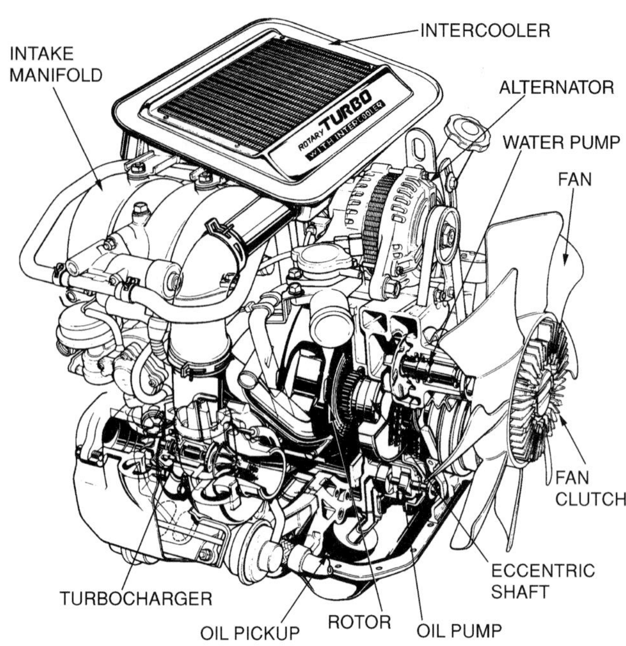

Figure 3-29. Cutaway view of a turbocharged rotary engine. This particular engine has a compression ratio of 8.5:1. Study the various parts. (Mazda)

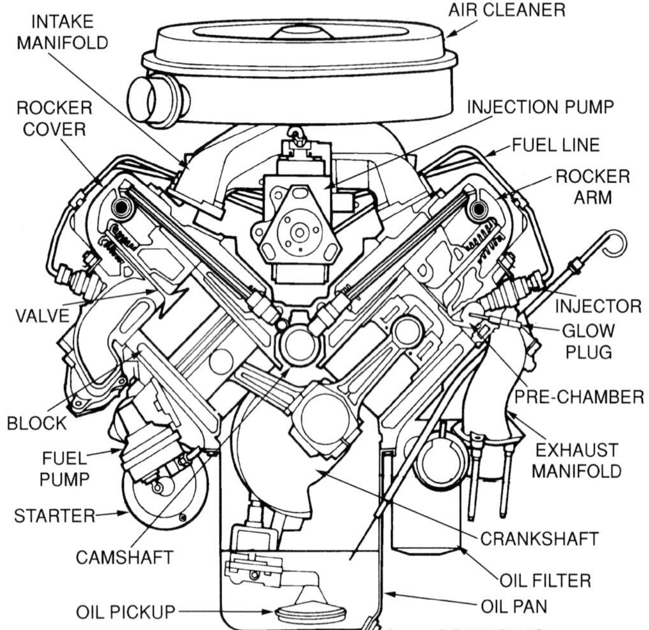

Figure 3-28. Cross-section of a 6.2 liter V-8 diesel engine. Note how similar this engine is to some gasoline engines. (General Motors)

and the housing surface while providing a barrier against compression loss and oil consumption.

Rotary Engine Operation

The rotary engine performs the four distinct strokes—intake, compression, power, and exhaust, **Figure 3-30**. Each face of the triangular rotor acts like a conventional piston. Instead of a reciprocating action, however, the rotor continually revolves in the same direction as the eccentric shaft.

By meshing the rotor internal gear with a fixed gear attached to the end housing, it forces the rotor to "walk" around the fixed gear when the eccentric shaft turns. This walking action causes the rotor to revolve around the eccentric shaft rotor journal in the same direction as the eccentric shaft. At the same time, the walking action keeps all three rotor tips constantly close to the wall of the rotor housing.

This planetary motion of the rotor constantly changes chamber volume between each rotor face and rotor housing. The change in volume is the principle used for engine operation. Study the rotor movement in **Figure 3-30A** through **Figure 3-30E**. We will follow one rotor face through the complete cycle. Note that this face lies between rotor tips A and B.

Rotary Engine Rotor Cycle

In **Figure 3-30A**, rotor face A and B has rotated past the intake port. As tip B passed the intake port, the chamber volume increased. This action forms a vacuum, causing the fuel mixture to flow into the enlarged chamber area. In **Figure 3-30B**, the rotor has turned further to the right and in so doing, chamber size has started to reduce, compressing the fuel. When rotor face A-B has turned to the position shown in C, maximum compression is applied.

When compression is at its maximum, the spark plug ignites the fuel charge, **Figure 3-30C**. When the charge ignites, the rapidly expanding gases apply pressure to the A-B face of the rotor, forcing it to continue walking around the fixed gear. As it rotates, it transmits the power of the burning gas to the eccentric shaft rotor journal, causing the eccentric shaft to spin, **Figure 3-30D**.

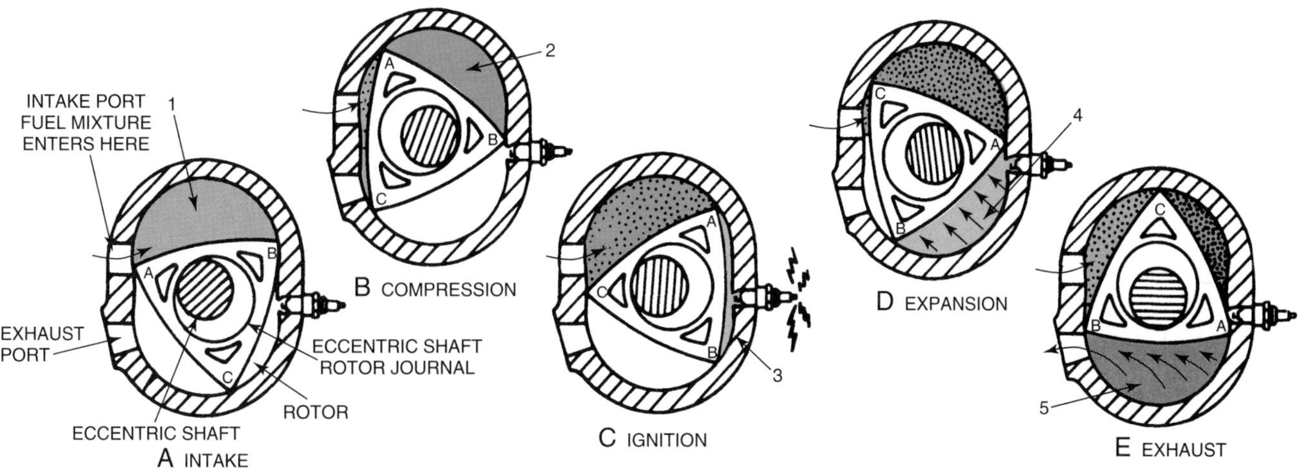

Figure 3-30. Rotor action within the rotor housing. The rotor internal gear and housing gear are not shown. (NSU)

The burning gas will continue to revolve the rotor until tip B uncovers the exhaust port, causing the burning gas to rush out the exhaust port as shown in **Figure 3-30E**. Follow rotor face A-B from **Figure 3-30A** to **Figure 3-30E** and note that as the ports are uncovered and the chamber volume changes, each phase of a four-stroke cycle is performed.

Miller-Cycle Engines

In a conventional engine, the intake valve is open during the piston's intake stroke, and closes almost as soon as the piston begins to move upward on the compression stroke. However, on a ***Miller-cycle*** engine, the intake valve is held open for a much longer time as the piston comes up on the compression stroke. To prevent the air-fuel mixture from exiting through the intake valve, a supercharger raises the intake manifold pressure. This forces more of the air-fuel mixture into the cylinder. The Miller-cycle design is said to produce more power and economy over a wider RPM range. **Figure 3-31** shows a cutaway of a Miller-cycle engine.

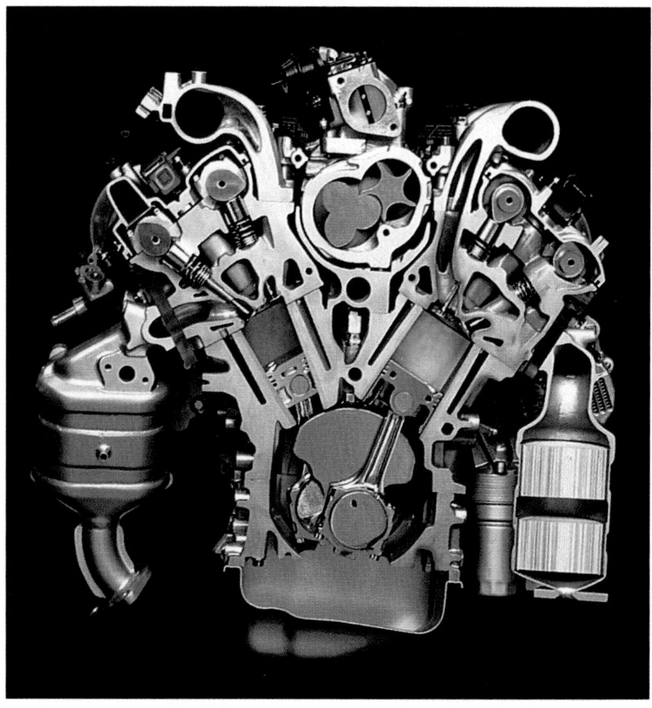

Figure 3-31. This Miller-cycle engine design allows the intake valve to open sooner. A supercharger is used to force the air-fuel mixture into the cylinders. (Mazda)

Natural Gas Engines

Natural gas is piped to millions of homes, where it is burned in water heaters, furnaces, and stoves. In addition, some vehicles operate on natural gas. The vehicle holds the gas in a high pressure storage tank, since it must be under high pressure in order to stay liquid. The gas must be maintained in a liquid state, as the vehicle could not carry enough to go very far if the gas were in its vapor state. The high pressure tank gives the natural gas vehicle a cruising range similar to that of a gasoline powered engine.

A natural gas engine is an internal combustion engine similar to a gasoline engine. In fact, most natural gas engines are production gasoline engines with a modified fuel system. The fuel system is modified to allow the natural gas to vaporize and mix with the incoming air. A simple mixing valve arrangement replaces the carburetor or fuel injectors. The natural gas system can even be adapted to work with the existing computer control system.

The advantage of natural gas is that it produces much less pollution and is readily available in large quantities. The disadvantage of natural gas is that the high pressure tank adds weight and price to the vehicle. Also, since most vehicles are gasoline powered, refueling stations are hard to find. Refueling time is slightly longer than for gasoline. If natural gas powered vehicles become more common, it may be possible in the future for homeowners to refuel their vehicles from their home gas system.

Experimental Engines

The search for better engines is never ending. Many ideas have been tried and adopted, others have failed. Currently, manufacturers are experimenting with engines that can operate on more than one type of fuel. Manufacturers are also developing vehicles driven by electricity. Whether or not these types of engines will find successful application as automobile power plants, only time will tell.

Engine Mounting

It is common practice to mount the engine on fluid-filled or rubber pads, usually called ***engine mounts***. This helps isolate engine vibrations from the vehicle body. It also eases torque (twisting strain) loading of the drive line. Mounts are used to support the engine and to minimize torque deflections. Small shock absorbers are sometimes employed, especially on diesel engines, to further control vibration.

The three point suspension, used mostly on rear-wheel drive vehicles, is one of the most popular. This method uses one mount on either side of the engine near the front and one at the back, usually under the transmission. However, other setups use two, three, four, or five engine mount arrangements. This varies from each model and manufacturer. **Figure 3-32** shows a five point mounting pattern.

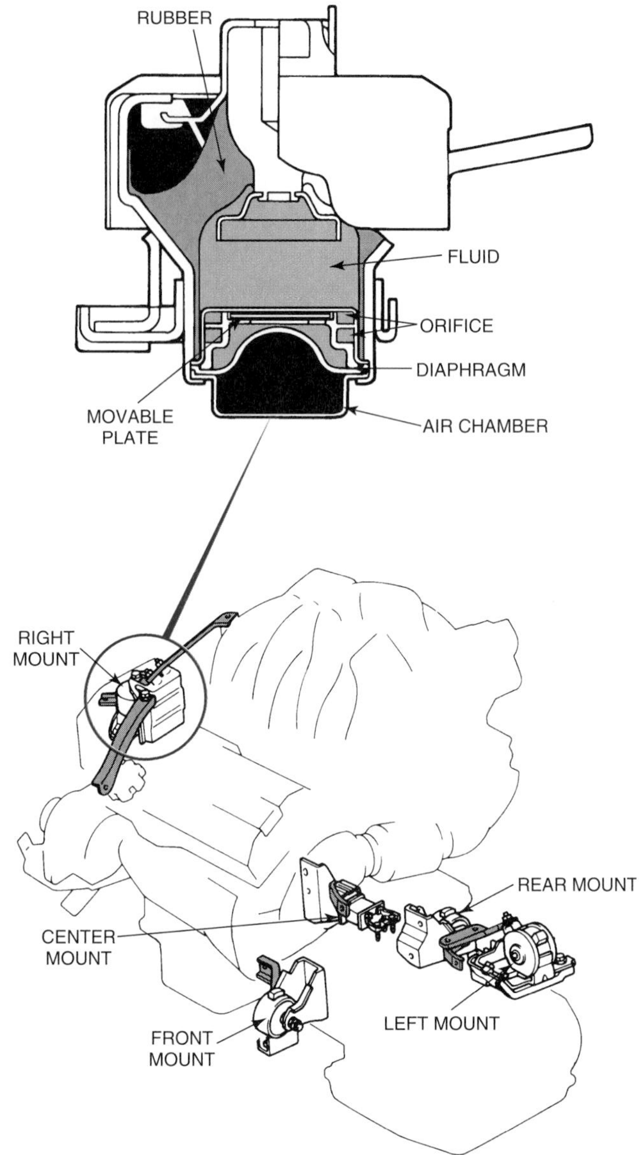

Figure 3-32. One type of front-wheel drive engine mount. The fluid sealed in the mount absorbs low speed vibrations. (Toyota)

Identification and Classification

The remainder of this chapter is devoted to engine classification and parts identification. Study all the engine figures in this chapter very carefully, especially **Figure 3-33** through **Figure 3-38**. Make certain you know the names and uses of all parts. It is most important that you build a complete and accurate mental image of the various types of engines, and the relationship of one part to the other.

As you study each figure, ask these questions:

- Is it a four-stroke cycle or a two-stroke cycle engine?
- Is this engine air- or liquid-cooled?
- Is this a V-type, inline, or opposed engine?
- How many cylinders does it have?
- Does it burn gasoline, diesel, or other fuel?
- If gasoline, is it carbureted or fuel injected?
- What type of head is on it?
- What type of combustion chamber design is used?
- Do I know the names and uses of all the parts?
- Do I know the working relationship of one part to another?

In addition to knowing the names and uses of all parts, be certain to test your knowledge of the various systems by identifying the cooling, lubrication, ignition, and fuel systems.

Summary

Engines can be classified in many ways. The most commonly used classifications are by cycle, valve location, cylinder arrangement, fuel used, cooling and number of cylinders. The vast majority of auto engines are of the four-stroke cycle type. The two-stroke cycle is generally used in small engines. Both have advantages and disadvantages. The I-head (overhead valve or valve-in-head) engine is in almost universal use today.

The most commonly used combustion chamber shapes are the wedge and the hemispherical. Detonation during the compression stroke can be stopped by proper fuel and combustion chamber design. Ricardo, in early experiments, set forth some basic principles of combustion chamber design that are still incorporated in modern engines. He found that by providing turbulence in the compressed charge, compression could be raised without detonation.

The V-8, V-6, and inline four-cylinder engines dominate the field of engine design and use. Some inline six-cylinder, V-10 and V-12, and four- or six-cylinder horizontal-opposed engines, are occasionally found. The diesel two-stroke cycle engine differs from the small gasoline engine of the same type. The diesel engine incorporates an exhaust valve, a supercharger and a conventional crankcase. The Wankel rotary piston engine uses a triangular shaped rotor that revolves inside a housing. It operates on the four-cycle principle. It has no reciprocating action and as a result, is very

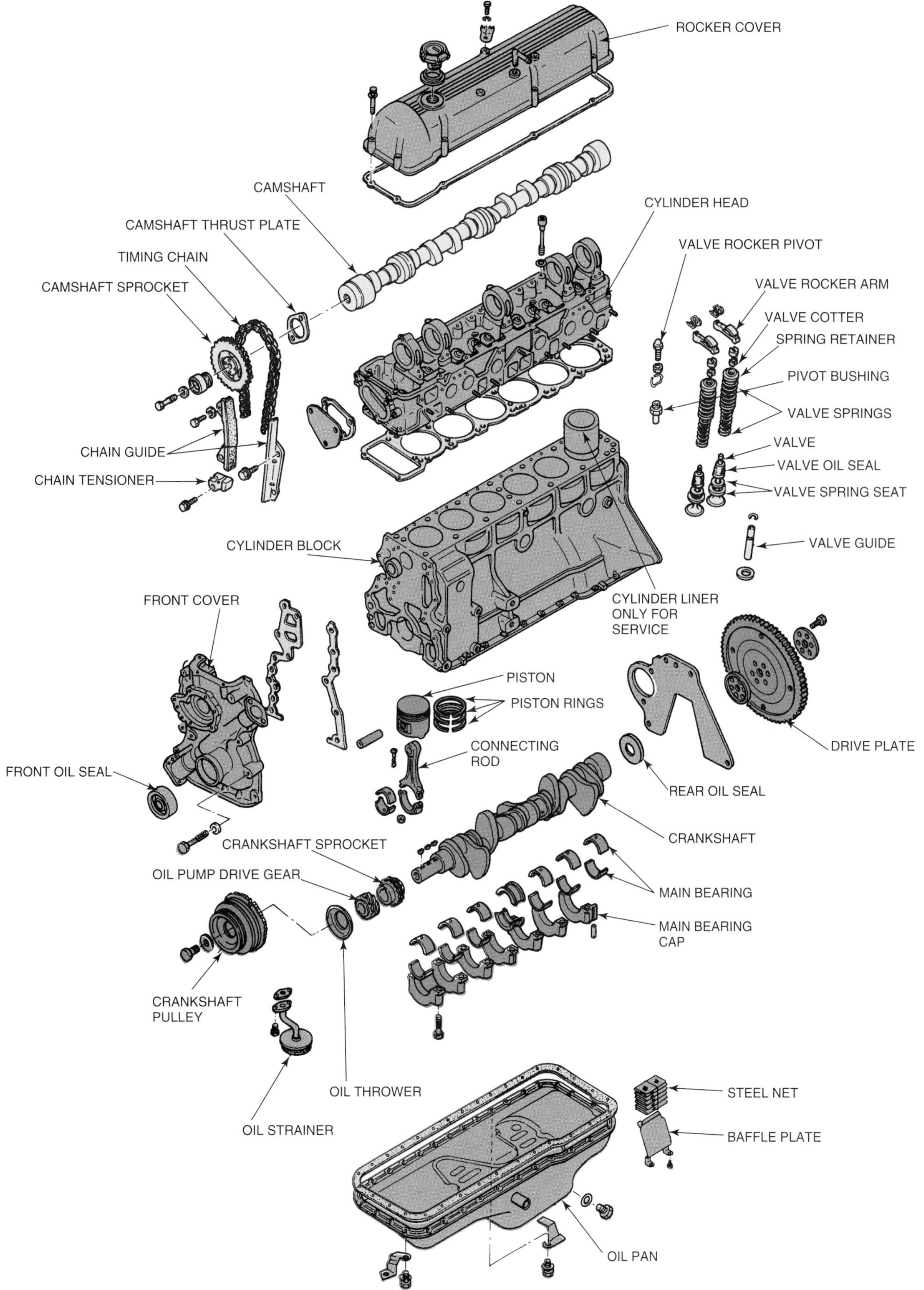

Figure 3-33. An exploded view of an inline six-cylinder engine. Study the construction. (Ford Motor Company)

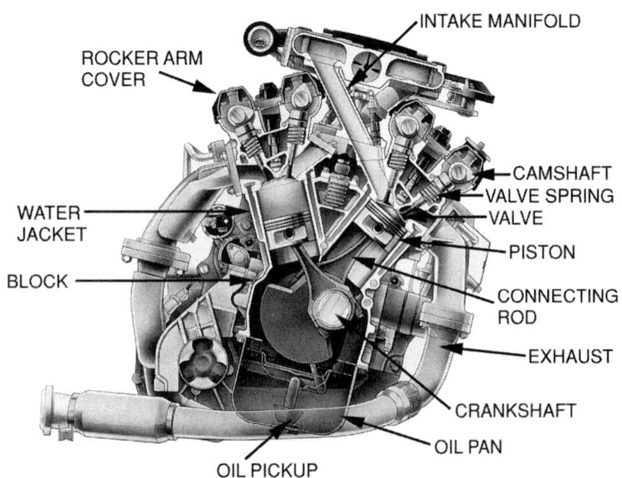

Figure 3-34. Cross-sectional view of a slant V-6 engine. By slanting the engine, overall height is reduced. This allows for a lower vehicle hood. (Saab)

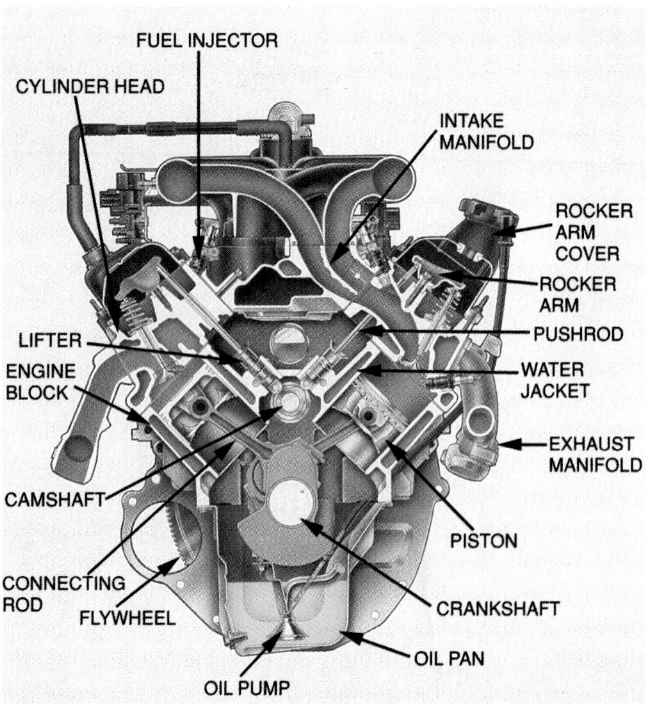

Figure 3-36. A cross-sectional view of a 3.8 liter V-6, split port induction (SPI) gasoline engine. Study the construction. (Ford Motor Company)

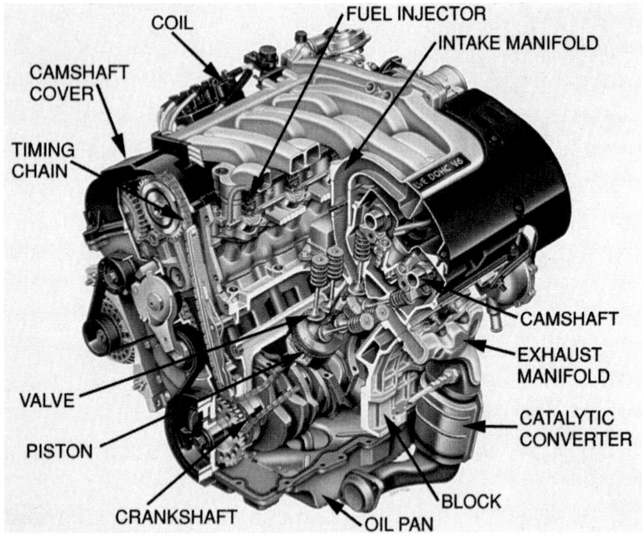

Figure 3-35. A cutaway view of an all-aluminum, low maintenance 2.5 liter dual overhead camshaft (DOHC) V-6 electronic fuel injected gasoline engine. This engine will not need a tune-up for 100,000 miles (160 000 km) under normal conditions. It produces 170 horsepower. (Ford Motor Company)

Figure 3-37. Overall view of a fuel injected, 6.0 liter V-12 gasoline engine. (Mercedes-Benz)

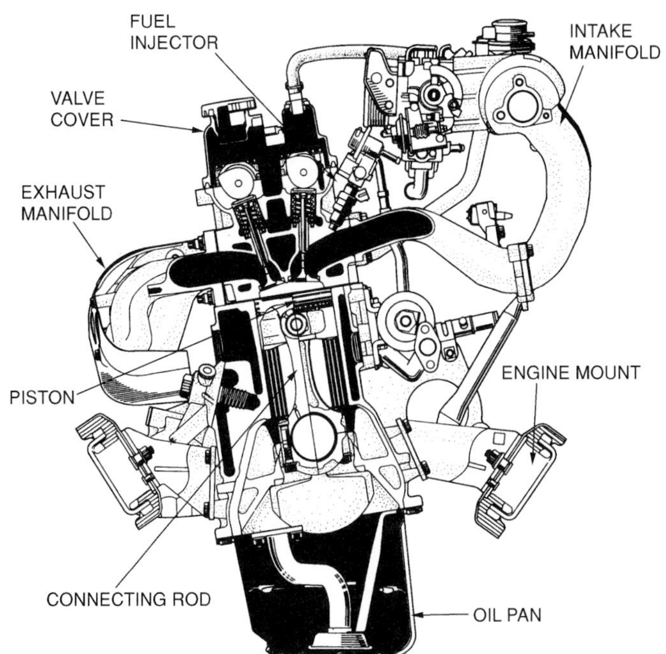

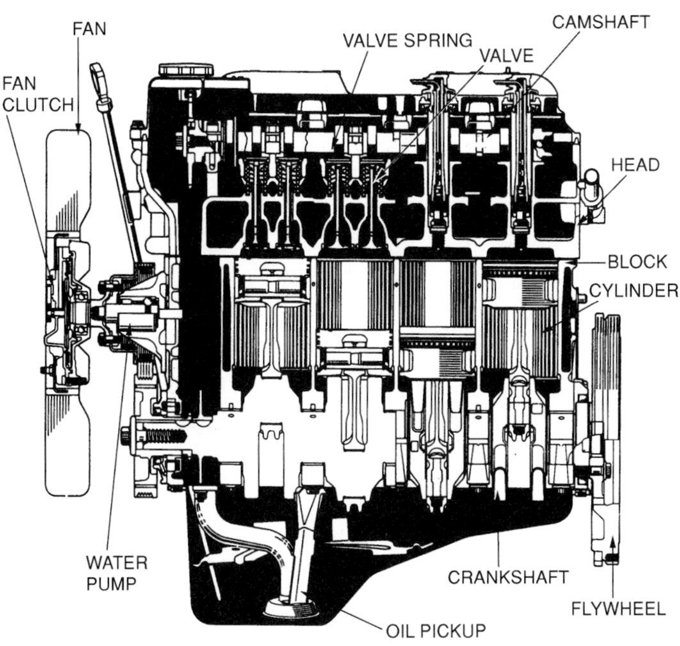

Figure 3-38. A cross-sectional view of an inline, four-cylinder, 2.7 liter (164.4 cu. in.), 16-valve, dual overhead camshaft, electronic fuel injected gasoline engine. The engine has a compression ratio of 9.5:1, and produces 150 horsepower at 4800 RPM and 177 lbs. ft of torque at 4000 RPM. (Toyota Motor Corp.)

smooth running. The rotary is a simple engine with very few moving parts (rotor and eccentric shaft) in the basic engine.

Experiments with new and different types of engines are always in progress. Further development of existing engines is also taking place. Considerable research is being done on natural gas and electric drive autos. Engines are mounted on rubber, generally using the three point suspension system, but some vehicles can utilize up to five mounts.

Know These Terms

Four-stroke cycle
Two-stroke cycle
Reed valve
Rotary valve
L-head
T-head
F-head
I-head
Valve train
Rocker arm
Push rod
Rocker shaft
Ball pivot
Overhead camshaft
Detonation
Preignition
Ricardo
Squish area
Hemispherical combustion chamber
Wedge chamber
Radial
Inline
Slant engine
Inclined engine
V-type
Horizontal-opposed engine
Firing order
Rotary engine
Rotor
Rotor housing
Eccentric shaft
Miller-cycle
Natural gas
Engine mounts

Review Questions—Chapter 3

Do not write in this book. Write your answers on a separate sheet of paper.

1. List the strokes in a four-cycle engine in their proper order.
2. The strokes of a four-cycle engine take place during how many revolutions of the crankshaft?
3. The two-strokes of a two-cycle engine take place during how many revolutions of the crankshaft?
4. What type of valves does the two-cycle gasoline engine have?
5. In a two-cycle engine, what opens and closes the reed valve(s)?
   (A) Air pressure and vacuum.
   (B) A camshaft lobe.
   (C) A crankshaft lobe.
   (D) All of the above, depending on the manufacturer.
6. Explain where the valves are located in the following engines.
   I-head
   F-head
   L-head
   T-head
7. Using cylinder arrangement, list the three most popular engines in use today.
8. A horizontally opposed engine has ______ banks of cylinders.
9. Where are the valves located in an overhead camshaft I-head engine?
10. Valve-to-piston clearance is most important on ________ compression engines.

11. Additional compression can be obtained without causing detonation if the air-fuel __________ is increased inside of the combustion chamber.
12. What is detonation and how is it caused?
13. All of the following could cause preignition, EXCEPT:
    (A) glowing carbon.
    (B) overheated spark plug.
    (C) burned valve.
    (D) piece of metal extending into the combustion chamber.
14. What determines an engine's firing order?
15. Which valve does a supercharged two-cycle diesel engine use?
    (A) Intake.
    (B) Exhaust.
    (C) Reed.
    (D) None.
16. Technician A says that engine mounts are used to reduce noise in the engine compartment. Technician B says that engine mounts can be made of solid rubber or fluid filled. Who is right?
    (A) A only.
    (B) B only.
    (C) Both A & B.
    (D) Neither A nor B.
17. Name two advantages of natural gas as a vehicle fuel.
18. In fuel classification, what are the three principal fuels used today?
19. In the rotary engine, the rotor turns _____ to every _____ turns of the eccentric shaft.
20. Technician A says that rotary engines have fewer moving parts than piston engines. Technician B says that piston engines operate more smoothly than rotary engines. Who is right?
    (A) A only.
    (B) B only.
    (C) A and B.
    (D) Neither A nor B.

4

Safety, Lab Procedures

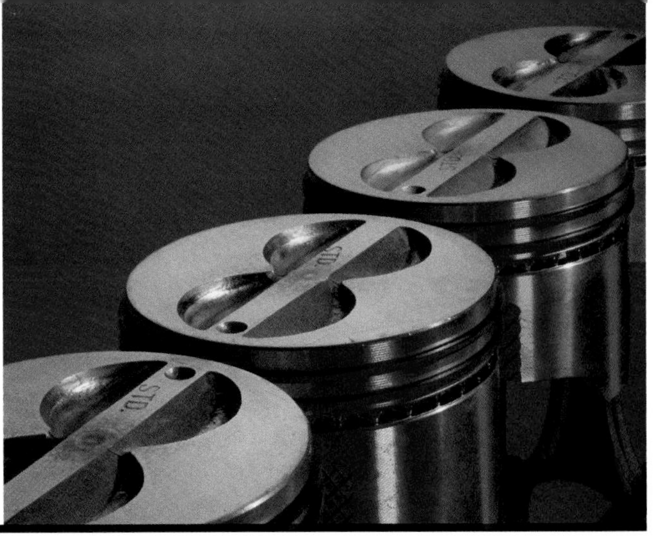

Upon completion of this chapter you will be able to:

- Identify the major types of accidents and their causes.
- Identify the consequences of accidents.
- Explain how to dress safely in the shop.
- Identify personal protective equipment.
- Explain how to identify and correct unsafe conditions in the shop.
- Identify unsafe work procedures.

This chapter covers the importance of safety. You will learn how to dress safely, work safely, and maintain a safe shop. The safety precautions given here are intended to help you avoid accidents as you perform laboratory jobs under your instructor's supervision. Learning good safety practices now will give you the safety habit at all jobs in the future. This chapter examines various types of unsafe shop conditions and work practices, and discusses ways to correct or avoid them.

Preventing Accidents

Safe working conditions and safety procedures protect the automotive technician, the vehicle, and the shop. ***Accidents*** are unplanned events that can occur without anyone intending for them to happen. Also unintended are the consequences of accidents, such as property damage or injury. The following sections show how a little foresight and common sense can eliminate the causes of most accidents.

Kinds of Accidents

There are many kinds of accidents. The most common accidents in the automotive shop include:

- ***Falls*** from slipping on a wet or oily floor.
- ***Fires*** and ***explosions*** caused by flammable substances contacting a heat source.
- ***Asphyxiation*** caused by exhaust fumes or refrigerants building up in a closed space.
- ***Injury to body parts***, such as hands, arms, face, head, or feet from moving parts, or from being caught under heavy parts.
- ***Strains and sprains*** from lifting heavy parts improperly.
- ***Burns*** from contacting hot coolant, oil, or metal.
- ***Chemical burns*** from battery acid or parts cleaner.
- ***Electrical shocks*** from ungrounded electrical tools or outlets.

In addition to the above accidents, the careless technician can be exposed to substances which cause long-term health problems including:

- ***Emphysema*** and ***lung cancer*** from inhaling brake dust over a period of years.
- ***Skin cancer*** from prolonged exposure to used motor oil and some cleaners.

As you can see, there are many ways to be injured or killed when working in an automotive shop. Therefore, it is good practice to find out how accidents happen and how to avoid them.

How Accidents Happen

Some of the accidents described earlier are caused when technicians try to take shortcuts instead of following proper repair procedures. Other accidents are caused when shop personnel fail to correct dangerous conditions in the work area. Accidents are often caused by a series of ***unsafe acts*** or ***unsafe conditions***, which lead to other actions. An example of an unsafe act is leaving spilled gasoline on the shop floor.

In some shops, this spill would remain unattended indefinitely because everybody is too busy to clean it up. Another unsafe act is using a standard light bulb in a drop light, which many technicians do because it is too much trouble to obtain a rough service bulb. Neither of these unsafe acts becomes an accident until someone drops the light to the shop floor. It breaks, and the spark ignites the gasoline. This causes a large fire, which could result in severe damage to the shop and any vehicles in the shop. It may also severely burn, or even kill someone in the shop. In this example, no one deliberately set out to cause an accident. Instead, it was just a series of things that were too much trouble to do correctly. The end result of this chain or series of unsafe acts was a fire, **Figure 4-1**.

Results of Accidents

An accident may result in injuries that keep you from working or enjoying your time off. Some accidents can kill. Even slight injuries are painful and annoying, and may impair

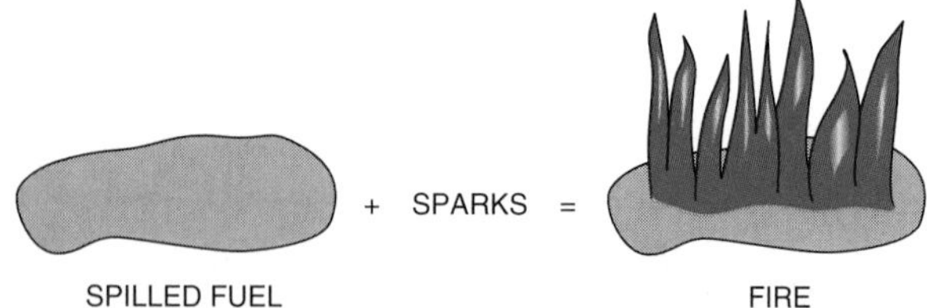

Figure 4-1. Spilled fuel and a stray spark or flame can cause a fire. Never leave spilled fuel or any other substance for someone else to clean. Clean it up immediately.

your ability to work and play. Even if an accident causes no personal injury, it can result in property damage. Damage to vehicles or shop equipment can be expensive and time consuming to fix, and could even cost you your job.

No mature and competent automotive technician wants to be injured or cause property damage. Even if immediate problems do not occur, remember that carelessness in the shop can also lead to long-term bodily harm, such as emphysema and cancer from prolonged contact with harmful liquids, vapors, and dust.

Keep Safety in Mind

The most experienced technician can become rushed and careless when they are under pressure. Pressure to complete a job on time eventually occurs in every shop. Falls, fires, injuries to hands and feet, explosions, electric shocks, and asphyxiation occur in auto repair shops every day. For these reasons, the technician must keep safety in mind at all times. This is especially true when pressure to complete a job tends to make safety the last thing on his or her mind.

Reducing the Possibility of Accidents

While it is true that some accidents are unavoidable, it makes sense to reduce the possibility of their occurrence as much as possible. Most accidents occur because of the following reasons:

- Wearing improper clothing.
- Not using protective equipment.
- Failure to keep the shop neat.
- Improper service procedures.
- A combination of all of these.

Prevent accidents by maintaining a neat work place, using safe methods and common sense when making repairs, and wearing protective equipment when needed. The following sections give some suggestions for reducing the possibility of accidents.

Dressing Safely for Work

The technician who keeps himself neat and clean is usually a safe worker. However, the most basic way to dress safely is to dress appropriately and avoid dangerous clothing. Do not wear open jackets or sweaters, scarves, or shirts with long, loose sleeves. They can get caught in the moving parts of an engine or machine and pull you in. Ties, belts, and suspenders, if they must be worn, should be kept away from moving parts. If you wear your hair long, keep it away from moving parts by tying it up or securing it under a hat.

Remove any rings, watches, or other jewelry. Jewelry can get caught in moving parts, resulting in a severe injury to your hands or fingers. If the jewelry should contact a short circuit, or is caught between a positive terminal and ground, severe burns will result.

Safety shoes, preferably with steel toe inserts, should be worn at all times. Most good quality safety shoes are constructed using materials which are oil and chemical resistant, giving the shoe a longer life expectancy in a harsh shop environment. Safety shoes have soles that are not only slip resistant, but also provide support and comfort. This is an important feature, since most technicians must stand on a concrete floor for an entire work day. In any shop, there is a constant danger of falling parts or tools. Since such an incident can occur at any time, it is a good idea to wear safety shoes whenever you are in the shop.

Personal Protective Equipment

Some ***personal protective equipment*** must be used in certain situations. This equipment protects you from personal injury, and if used regularly, also protects you from long-term illnesses, such as cancer. Long-term exposure to some common shop byproducts such as brake dust or oil fumes can cause diseases such as skin and lung cancer, and emphysema.

Eye protection is vital when working in any situation that could result in dirt, metal, or liquids being thrown into your face. This includes working around running engines; when using drills, saws, welders, or grinders; and when working around batteries and hot cooling system parts, **Figure 4-2**.

Figure 4-2. Always wear eye protection when working with grinders, air tools, hammers, chisels, etc. Remember, you can replace safety glasses, but not your eyesight. (Deere & Co.)

Respiratory protection, such as face masks, should be worn whenever you are working on brake systems or clutches. The dust from the friction lining materials used in these devices contain ***asbestos***, which can cause lung damage or cancer, **Figure 4-3**. Respiratory protection is also a good idea when working around any equipment that gives off fumes, such as a hot tank or steam cleaner. **Figure 4-4** shows a special brake washer tool, which should be used to clean brakes before service work begins.

a
WARNING
CONTAINS
ASBESTOS
Breathing asbestos dust is dangerous to health
Follow safety instructions

Figure 4-3. A typical asbestos warning label. Follow all safety rules when working with parts containing asbestos. Do not inhale the dust. (Sterling & Author)

Rubber gloves are necessary whenever you are working with chemicals such as parts cleaning solvents or carburetor cleaner. If you spill oil, gasoline, cleaning solvents, or any other substance on your skin, clean it off immediately. Prolonged exposure to even mild solvents or petroleum products can cause severe skin rashes or chemical burns. Material Safety Data Sheets, which are discussed later in this chapter, should be located and read before handling any chemical that is unknown to you.

Keep a Neat Shop

Neatness is the number one accident preventer. Whenever you are assigned projects in the school shop or lab, clean and return all tools and equipment to their proper places when the job is completed. Do not allow dirt, oil, old parts, boxes, or other refuse to pile up.

Keep all workbenches clean. This reduces the chance of tools or parts falling from the bench, where they could be lost, damaged, or possibly land on your foot. A clean workbench also reduces the possibility that critical parts will be lost in the clutter. It also reduces the chance of a fire from oily debris.

Return all tools and equipment to their proper storage places. This saves time in the long run, as well as reducing the chance of accidents and theft. Do not leave any pieces of equipment out where others could trip on them. If tools are dirty or oily, clean them before you put them away. This will not only extend the life of the tool, but is also a courtesy to others who use them.

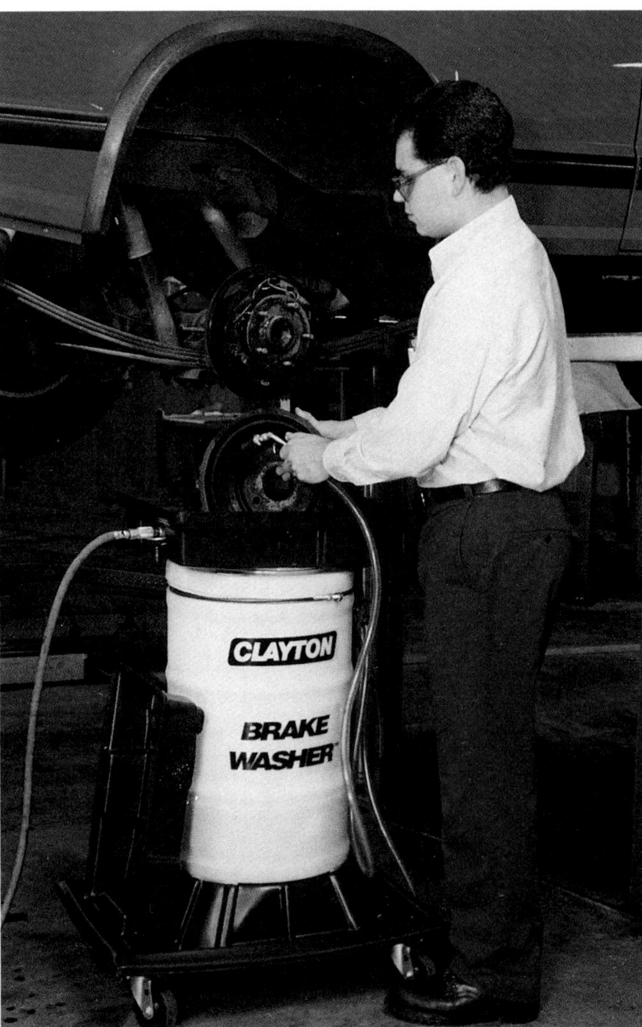

Figure 4-4. Use an approved brake washing tool in a well ventilated area of the shop. Note the safety glasses and protective shoes worn by the technician. (Clayton Mfg. Co.)

Clean up spills immediately, before they get tracked all over the shop. Many people are injured when they slip on floors coated with oil, antifreeze, or water. Gasoline spills can be extremely dangerous, since a flame or the smallest spark can ignite the vapors, causing a major explosion and fire. Do not leave open containers of any chemical in the shop or outside. Antifreeze, for example, contains ethylene glycol. While ethylene glycol is not flammable, it is harmful to the environment and will poison any animal (or person) who accidentally drinks it.

Fire Extinguishers

Although a shop fire is one of the most frightening things that you may encounter, it is important to keep your head and not panic. If you are familiar with the layout of your shop, you should be able to quickly locate and use the nearest **fire extinguisher**. The correct fire extinguisher will quickly smother a fire by cutting off its supply of oxygen. Fire extinguishers are identified by the type of fire that they will put out. See **Figure 4-5.**

Fire Extinguishers and Fire Classifications

Fires
Class A Fires Ordinary Combustibles (Materials such as wood, paper, textiles.) *Requires... cooling-quenching* A
Class B Fires Flammable Liquids (Liquids such as grease, gasoline, oils, and paints.) *Requires...blanketing or smothering* B
Class C Fires Electrical Equipment (Motors, switches, and so forth.) *Requires... a nonconducting agent.* C
Class D Fires Combustible Metals (Flammable metals such as magnesium and lithium.) *Requires...blanketing or smothering.* D

Type	Use	Operation
Soda-acid Bicarbonate of soda solution and sulfuric acid	Okay for use on: A Not for use on: B C D	Direct stream at base of flame.
Pressurized Water Water under pressure	Okay for use on: A Not for use on: B C D	Direct stream at base of flame.
Carbon Dioxide (CO_2) Carbon dioxide (CO_2) gas under pressure	Okay for use on: B C Not for use on: A D	Direct discharge as close to fire as possible, first at edge of flames and gradually forward and upward.
Foam Solution of aluminum sulfate and bicarbonate of soda	Okay for use on: A B Not for use on: C D	Direct stream into the burning material or liquid. Allow foam to fall lightly on tire.
Dry Chemical	Multi-purpose type: Okay for A B C; Not okay for D Ordinary BC type: Okay for B C; Not okay for A D	Direct stream at base of flames. Use rapid left-to-right motion toward flames.
Dry Chemical *Granular type material*	Okay for use on: D Not for use on: A B C	Smother flames by scooping granular material from bucket onto burning metal.

Figure 4-5. This chart illustrates the various fire classifications and fire extinguisher types. In the automotive repair shop, extinguishers designed for chemical and electrical fires (multi-purpose or BC types) should be used.

The three basic classes of fire extinguisher are:

- Class A: Will put out fires from ordinary combustibles, such as wood and paper.
- Class B: Will extinguish fires caused by flammable liquids, such as gasoline and oil.
- Class C: Designed to put out electrical fires.

The uses of these various types of fire extinguishers are clearly marked on the tank. Today, most shops have multi-purpose (A, B, and C class) dry chemical fire extinguishers, which will put out all of the fires described earlier. A fourth class of fire extinguisher is used on flammable metals, but this type is not encountered in the average automotive shop.

Never use water to put out a fire. If gasoline is involved, it will float on the water and spread the fire. Never depend on a fire extinguisher to any more than slow down the spread of the fire. Call the fire department first and then attempt to extinguish the fire. Stay low to avoid inhaling the smoke. If the fire begins to spread or if there is too much smoke or heat, *get out.* Do not open doors or windows as this will give the fire more oxygen. Never go back into a burning building for any reason.

Identifying Unsafe Conditions

In addition to keeping the shop neat, the student should identify and take steps to correct other unsafe conditions which may develop in the shop. Many of the statements below will seem to be just a matter of common sense, but they are often disregarded.

If you detect a problem in any shop equipment or tools, notify your instructor immediately. Assist your instructor in tool repair and maintenance. This may involve such things as replacing damaged leads on test equipment, checking and adding oil to hydraulic jacks, and regrinding the tips on chisels.

When servicing any shop equipment, be sure that the equipment is turned off and unplugged. Read the equipmentís service literature before beginning any repairs. Ensure that all shop equipment, such as grinders and drill presses, are equipped with safety guards. All of these devices are installed with guards by the manufacturer. These guards should never be removed, except for service operations such as changing the grinding wheels.

Know what type of chemicals are stored on the shop premises. Chemicals include carburetor cleaners, hot tank solutions, parts cleaner, and even motor oil and antifreeze. Chemical manufacturers provide ***Material Safety Data Sheets***, often called ***MSDS***, for every chemical that they produce. These sheets list all of the known dangers of the chemical, as well as first aid procedures for skin or respiratory system contact. There should be an MSDS for every type of chemical present in the shop. Locate the MSDS and read it before working with any unfamiliar chemical.

Ensure that the shop is well ventilated. A sealed shop, or a room in the shop, can permit the buildup of exhaust fumes or chemical vapors which can be fatal. Even a relatively harmless gas like air conditioner refrigerant can kill if it displaces all of the oxygen in a closed room.

Make sure that the shop is well lighted. Poor lighting makes it hard to see what you are doing. Not only does it make the job more time-consuming, it can lead to accidental contact with moving parts or hot surfaces. Overhead lights should be bright and centrally located. Portable trouble lights or "drop" lights, which are usually small and easy to use, should be used in close operating conditions.

Always use a rough service bulb in drop lights. These bulbs are more rugged than normal light bulbs and will not shatter if they are dropped. Do not use a high wattage bulb in a drop light. Light bulbs get very hot and can melt the light socket or burn anyone who touches the light safety cover.

Do not overload electrical outlets or extension cords by operating several electrical devices from one outlet. Do not pair up high current electrical devices, or operate them through extension cords. Examples of high current drawing electrical devices are drills, grinders, and electric heaters. Do not, under any circumstances, cut or break off the ground prong of a three-prong plug.

Waste Chemical Disposal

Probably the most common way in which auto shops cause environmental damage is the methods by which they dispose of used oil and other chemicals. These used chemicals are classified as hazardous waste. Pouring motor oil, transmission or brake fluid, antifreeze, and used cleaning solutions on the ground immediately contaminates the soil. These liquids sink further into the ground with every rainstorm. Eventually, it could contaminate your local source for drinking water. Another common way in which liquids can contaminate the ground and water is through leaking storage tanks. Although this problem is usually confined to underground gasoline tanks, any kind of tank can begin leaking. The shop is responsible for their safe disposal.

While the shop is using these chemicals, they are only subject to safety regulations concerning employee exposure to toxic chemicals. One the shop is done with them, however, they must be disposed of in a manner approved by federal, state, and local guidelines. Liquid waste should never be disposed of by pouring them into the local drainage system. Municipal waste treatment plants cannot handle the petroleum products, or their many additives, nor the heavy metals absorbed from the vehicle during their use. The Environmental Protection Agency has established strict guidelines for disposing of toxic waste. In some cases, you may be liable for cleaning up a contaminated area years after the actual violation occurred.

In many areas, local companies will recycle toxic wastes. These companies will accept used oil and antifreeze for recycling. The used oil and antifreeze is then refined and reused. Some shops use heaters that burn used oil as a fuel, eliminating the oil. Some used oil is burned by power plants to produce electricity. Additional information on waste disposal and vehicle emissions can be obtained from the Environmental Protection Agency. For more information you can write to:

Automotive/Emissions Division
United States Environmental Protection Agency
401 M Street SW
Washington, DC 29460

Follow Proper Work Procedures

Work procedures are the actual diagnosis and repair operations. You may be called on to do all sorts of procedures in your shop classes or labs. It is vital that you do them safely and correctly. Improper procedures include using the wrong tools or methods to perform repairs, using defective or otherwise inappropriate tools, not wearing protective equipment when necessary, and not paying close attention while performing the job. Whenever any work is performed, it should be performed in a safe manner, before any other considerations. The following safe work procedures may seem to be simple. However, they are often disregarded, with tragic results.

- Wear proper clothing and use protective equipment when called for.
- Study work procedures before beginning any job that is unfamiliar to you. Do not assume that the procedure you have used in the past will work with a different type of vehicle.
- Always work carefully. Speed is not nearly as important as doing the job right and avoiding injury. Avoid people who will not work carefully.
- Use the right tool for the job. For example, using a screwdriver as a chisel or a wrench as a hammer, is asking for an accident, or at least a broken tool.
- Inspect electrical cords and compressed air lines to ensure that they are in good condition. Do not run over, or close vehicle doors on, electric cords or air lines. Do not run electrical cords through water puddles or use them outside when it is raining.
- Learn how to use new equipment properly before using it. This is especially true of air-operated tools, such as impact wrenches or chisels, and large electrical devices such as drill presses, boring bars, and flywheel surfacers. These tools are very powerful and can hurt you if they are used improperly. A good way to start learning about new equipment is by reading the manufacturer's instructions.
- When working on electrical systems, avoid creating a short circuit with a jumper wire or metal tool. Not only will this damage the vehicle components or wiring, it will develop enough heat to cause a severe burn or start a fire.
- Lift safely. Make sure that you are strong enough to lift the object to be moved. Always lift with your legs, not your back. If an object is too heavy to lift by yourself, get someone to help you or use a jack.
- Do not smoke in the shop. You may accidentally ignite an unnoticed gasoline leak. A burning cigarette may also ignite oily rags or paper cartons.
- Do not attempt to raise a vehicle with an unsafe or undercapacity jack. Always support a raised vehicle with good quality jackstands. Never use boards or cement blocks to support a vehicle.
- Do not run any type of engine, even for a short time, in a closed area without good ventilation. ***Carbon monoxide***, which cannot be seen or smelled, can build up very quickly. This is true even of lawn mowers and other small engines.
- When working on or near a running engine, keep away from all moving parts. Never reach between moving engine parts for any reason. A seemingly harmless part such as the drive belts and fan can seriously injure you.
- Do not leave a running vehicle unattended. The vehicle may slip into gear, or overheat while you are away. Whenever you must work on a running vehicle, shift the transmission into park or neutral and set the parking brake.
- When road testing a vehicle, be alert and obey all traffic laws. Do not become so absorbed in diagnosing a problem that you forget to watch the road. Be alert for the actions of other drivers.

Do the Job Right

It is important to correctly perform repairs to the vehicle and to make sure that your repairs are not the cause of an accident. For example, if you follow every shop safety procedure when doing a brake job, but allow the vehicle to leave the shop with a leaking brake hose, an accident is sure to occur when all the brake fluid leaks out. The driver, along with any passengers, could be injured or killed. Making sure that the vehicle leaves the shop in a safe condition is just as important as working safely in the shop.

After all other factors have been figured in, it is still up to you to notice and correct safety hazards, work safely, and prevent accidents. Always use common sense when working on vehicles. Avoid people who do not follow good work and safety practices.

Summary

Each unsafe act may not cause an accident, but a combination of two or more unsafe acts may result in a severe accident. Not taking the time to do things correctly almost always results in an accident. Many accidents are caused when technicians try to take shortcuts instead of following proper repair procedures. Another common cause of accidents is failing to correct dangerous conditions in the work area, such as defects in equipment or oil spills. An accident may result in damage to equipment or property, personal injuries, long-term bodily harm, or death. No mature and competent automotive technician wants to be injured or cause property damage.

Always dress for the shop with an eye toward safety. Avoid loose clothing, dangling hair, and jewelry. Various types of protective equipment are needed to protect the eyes, feet, lungs, and skin. Protective equipment should protect against immediate injury and the long-term effects of exposure to toxic substances.

The best way to prevent accidents is to maintain a neat work place, use proper repair methods, and use protective equipment when needed. It is up to the technician to study the job beforehand, to work safely, and to prevent accidents. Do not allow a vehicle to leave the shop in unsafe condition. Always use common sense when working on vehicles, and avoid people who do not.

Know These Terms

Accident
Falls
Fires
Explosions
Asphyxiation
Injury to body parts
Strains and sprains
Burns
Chemical burns
Electrical shocks
Emphysema
Lung cancer
Skin cancer
Unsafe act
Unsafe condition
Personal protective equipment
Asbestos
Fire extinguisher
Material Safety Data Sheets (MSDS)
Carbon monoxide

Review Questions—Chapter 4

Do not write in this book. Write your answers on a separate sheet of paper.

1. An accident is an _____ event.
2. What two kinds of accidents could occur when a flammable substance contacts a heat source?
3. What kind of injury could be caused by battery acid or parts cleaner? _____ .
4. Accidents are often caused by a series of unsafe _____ which could be avoided.
   (A) conditions
   (B) acts
   (C) vehicles
   (D) Both A & B.
5. Name the five major causes of accidents.
6. Technician A says that, since air conditioner refrigerant is not poisonous, it cannot cause asphyxiation, no matter how much of it is present. Technician B says that it is OK to run an engine in a closed shop, as long as it is not run for more than 10 minutes.
   Who is right?
   (A) A only.
   (B) B only.
   (C) Both A & B.
   (D) Neither A nor B.
7. You should always wear respiratory protection when working on what two vehicle components?
8. Each shop electrical outlet should be used to operate how many high current draw tools?
   (A) 1
   (B) 2
   (C) 3
   (D) 4 or more
9. Material Safety Data Sheets are provided for all dangerous _____ .
   (A) procedures
   (B) tools
   (C) chemicals
   (D) working conditions
10. If the safety guards have been removed from a grinder, what should you do?
   (A) Be very careful when using the grinder.
   (B) Let someone else do the grinding.
   (C) Do not use the grinder until the guards are replaced.
   (D) Wear eye protection.
11. Always use a "rough service" bulb in _____ .
   (A) fluorescent drop lights
   (B) incandescent drop lights
   (C) every light in the shop
   (D) the ceiling lights
12. A high wattage bulb in a drop light can cause what type of personal injury?
   (A) Blindness.
   (B) Burns.
   (C) Electric shock.
   (D) All of the above.
13. Spilled ethylene glycol antifreeze can cause all of the following, EXCEPT:
   (A) slipping.
   (B) poisoning.
   (C) fires.
   (D) environmental damage.
14. The technician should wear eye protection when _____.
   (A) working around running engines
   (B) using drills, saws, or grinders
   (C) working around batteries
   (D) all of the above
15. Improperly fixing a vehicle could result in _____ .
   (A) repeat business for the technician
   (B) an accident
   (C) injury to the driver
   (D) Both B & C.

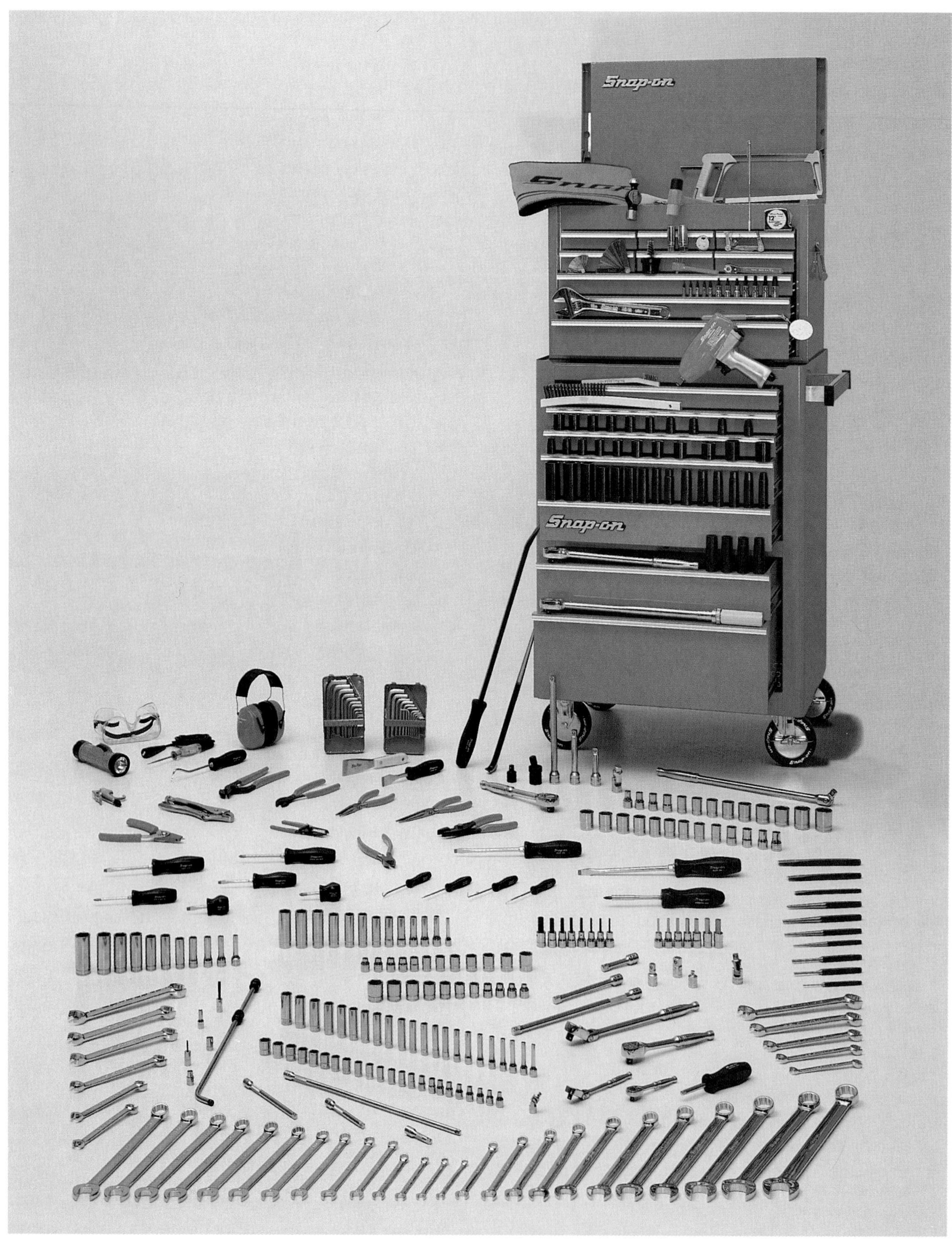

All technicians must own a good basic set of tools and a tool chest to keep them in. Many of the tools shown here are discussed in this chapter. (Snap-on Tools)

5

Tool Identification and Use

After studying this chapter, you will be able to:

- Identify the most common automotive tools.
- Describe commonly used measuring tools.
- Describe commonly used test equipment.
- Select the correct tool for a given job.

Today's technician must be familiar with a large number of tools. Proper tool selection and use will improve both the quality and speed of any repair operation. Many repair jobs are extremely difficult to perform without the right tools. Good technicians own a wide selection of quality tools. They constantly strive to add to their collection and take great pride in keeping their tools clean and in order.

What to Look for in a Tool

Some technicians prefer one certain brand of tools, while others prefer more than one brand. You will find, however, that technicians agree on several important features that are found in quality tools.

Tool Material

Quality tools are made of high-strength alloy steel. Consequently, they can be made without a great deal of bulk. These tools are light and easy to use in tight quarters. Heavy, "fat" tools are useless on many jobs. The alloy steel gives the tool great strength. Its working areas will withstand abuse, greatly extending its useful life.

Tool Construction

Quality tools receive superior heat treating. Their working surfaces are held to tight tolerances and sharp edges are removed. Additionally, quality tools are carefully polished. Polishing produces a finish that is easy to clean and comfortable to handle.

Tools worthy of consideration will be slim, strong, easy to clean, and a pleasure to use. Quality material and construction makes it possible for the manufacturer to offer a good (often a lifetime) guarantee on the tool. Efficient service should also be offered. In the case of some tools, replacement parts should be readily available. This can be an important feature. Usable tools often have to be discarded due to the loss or failure of some minor part that is no longer available.

Remember: When buying tools, you get what you pay for. Fine quality tools are a good investment.

Tool Handling and Storage

Use tools designed for the job at hand. Keep your tools orderly and clean. Store them in a good, roll-type cabinet tool chest and tote tray (a small tray that can be carried to the job to keep a few selected tools close at hand). Proper storage will help keep tools in good shape and readily available. All tools should have your name, initials, or a number that is unique to you engraved on the tool body. This will allow you to identify that a specific tool is your property.

Any tool subject to rust should be cleaned and lightly oiled. Separate cutting tools, such as files and chisels, to preserve their cutting edges. Place delicate measuring tools and electronic tools in protective containers. Store heavy tools by themselves and arrange tools so that those most often used are handy.

Remember to keep your tools clean. You cannot hope to successfully assemble a fine piece of machinery with dirty tools. The slightest bit of dirt or abrasive material that finds its way into moving parts can cause damage when the tool is used. Tools are a good indication of a technician's worth. If a technician uses dirty, worn, or disorganized tools, the quality of this technician's work is likely to be as poor as the tools.

Types of Tools

This chapter does not attempt to cover all the tools used in the automotive trade. Large shops and technicians that specialize in certain areas of repair will often utilize hundreds of special tools designed for specific jobs, models, and units. Therefore, the basic tools common to automotive shops and owned by every technician will be discussed in this chapter.

Hammers

Technicians frequently use *ball peen, brass-tipped,* and *plastic-tipped* **hammers.** The ball peen hammer is used for general striking work and is available in weights ranging from a couple of ounces to several pounds. Brass- and plastic-tipped hammers are used when there is danger of the steel ball peen marring the work surface. Some plastic-tipped hammers have their heads filled with lead shot. This prevents the hammer from rebounding when striking a hard object. See **Figure 5-1.** The handles should be kept clean and dry.

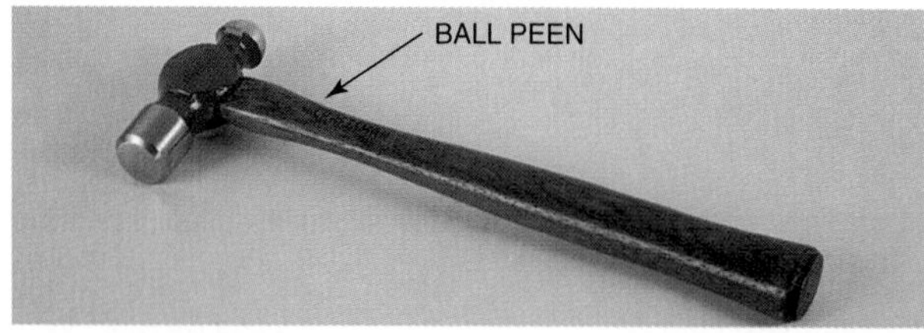

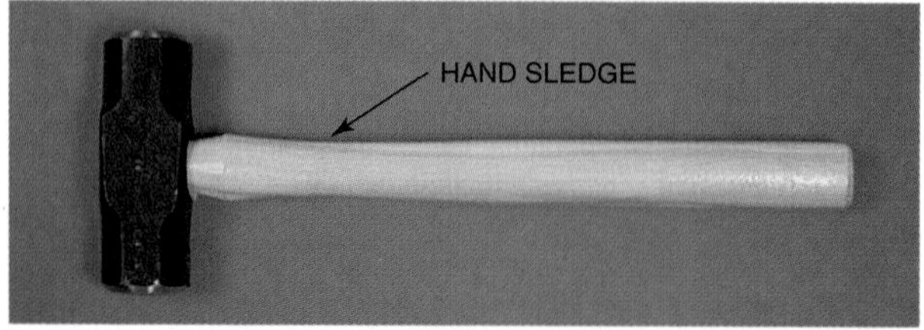

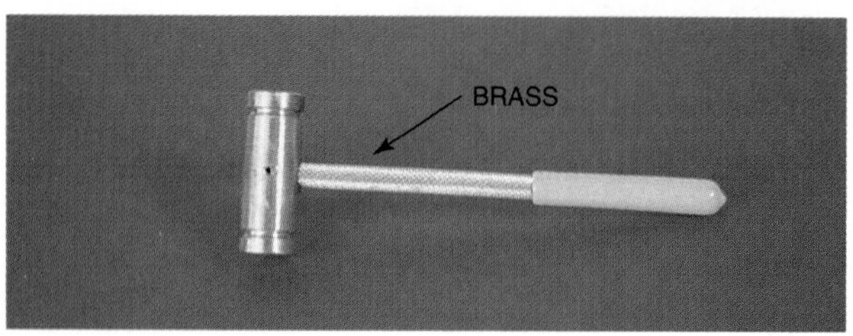

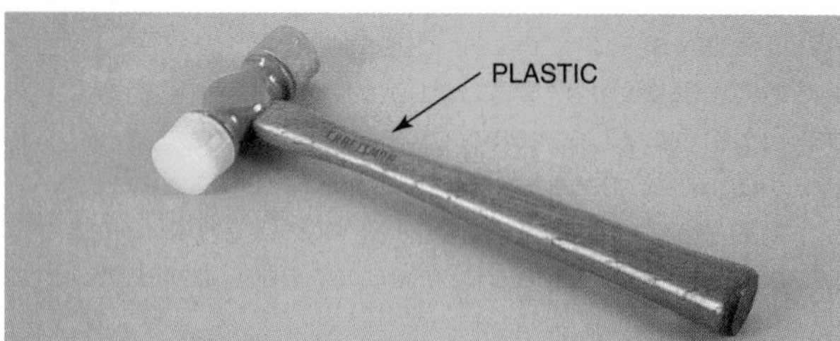

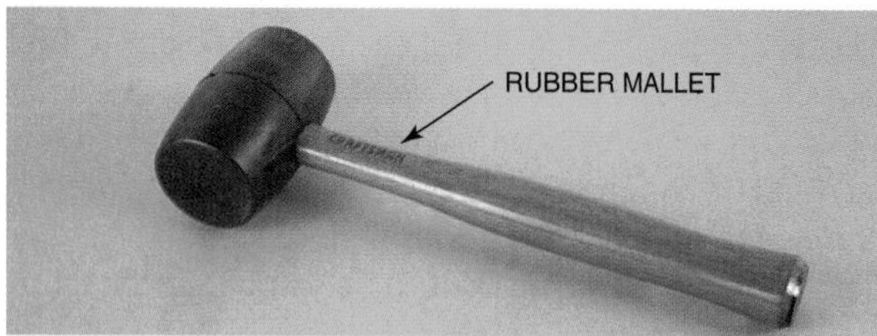

Figure 5-1. Several types of hammers. Each hammer is designed to perform a specific task.

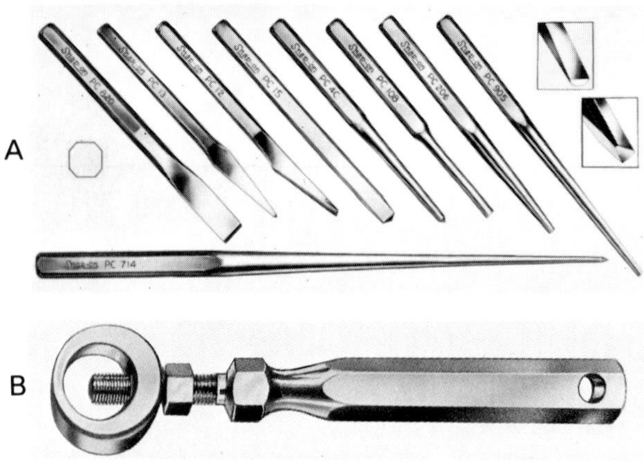

Figure 5-2. A—A chisel and punch assortment. B—A chisel holder. By holding the punch or chisel with this holder, it protects the hand when striking with a hammer. (Snap-on Tools)

heads have been cut off. Once a pin has been started, the starting punch can no longer be used (because of its taper). A *drift punch,* which has the same diameter for most of its length, is used to complete the job. See **Figure 5-3.** A *pin punch* is similar to a drift punch, but it is smaller in diameter.

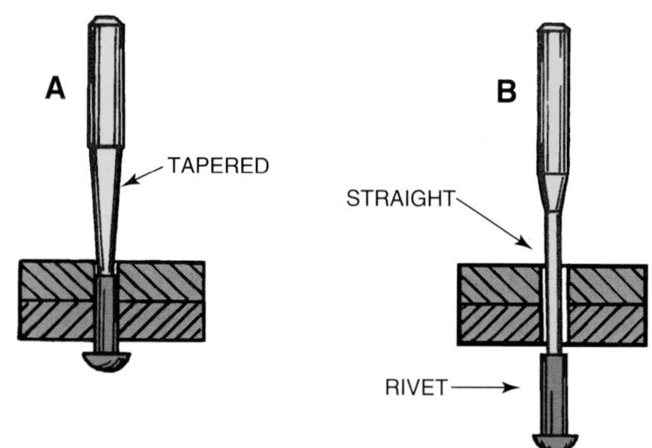

Figure 5-3. Starting and drift punches. A—The starting punch is used to start the rivet from the hole. B—The drift punch is used to drive the rivet from the hole.

Chisels

Several sizes and types of ***chisels*** are essential for cutting bolts or rivets, **Figure 5-2A.** Chisels can be held in your hand or secured in a holder. The holder will allow the use of large hammers and greater striking force without the chance of injuring your fingers or hand. One type of chisel holder is shown in **Figure 5-2B.** The *flat cold chisel* is used for general cutting. Special chisels, such as *cape, half-round,* and *diamond-point chisels,* are used when their shapes fit a definite need.

Punches

A variety of ***punches*** are used when servicing automobiles. A starting punch tapers to a flat tip. It is used when starting to punch out pins and when driving out rivets after the

An *aligning punch,* which has a long, gradual taper, is used to shift parts and bring corresponding holes into alignment, **Figure 5-4.** A *center punch* is used to mark materials before drilling. It leaves a small, V-shaped impression that assists in aligning the drill bit. A center punch may also be used to mark parts so they can be assembled in the correct positions. See **Figure 5-5.**

Files

Commonly used ***files*** include the *flat mill, half-round, round, square,* and *triangular files.* All come in different sizes and with cutting edges that range from fine to coarse. Some files have one or more cutting edges. Some files have one or more safe sides (surface with no cutting edges).

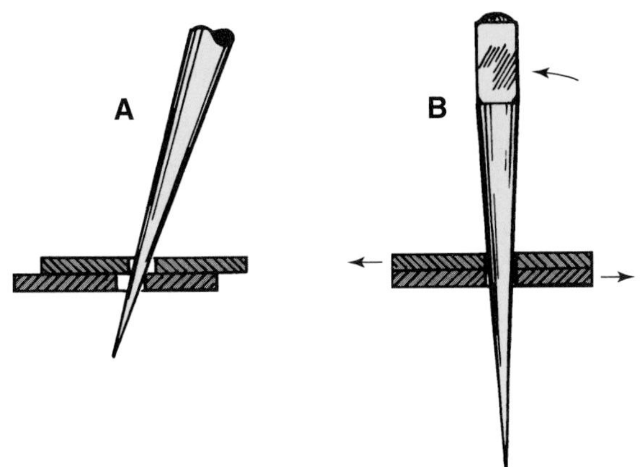

Figure 5-4. Aligning punch. A—This punch is pushed through the holes and then is pulled upright and pushed deeper into the holes. B—This lines up the two holes.

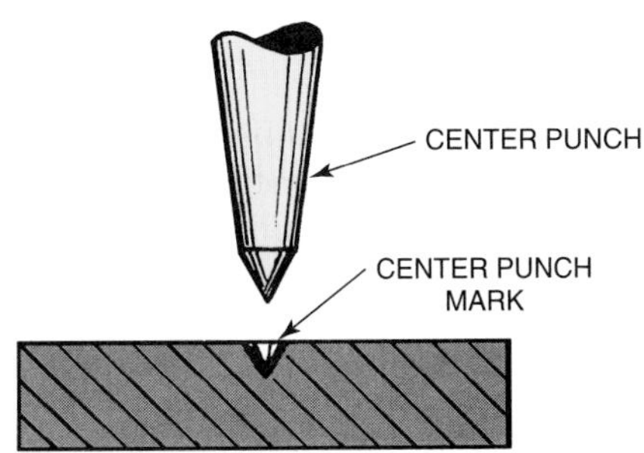

Figure 5-5. A center punch is used for marking parts and starting drill bits.

A file should have a handle firmly affixed to the tang. The handle gives a firm grip and eliminates the danger of the sharp tang piercing the hand. A typical mill file and nomenclature is shown in **Figure 5-6.**

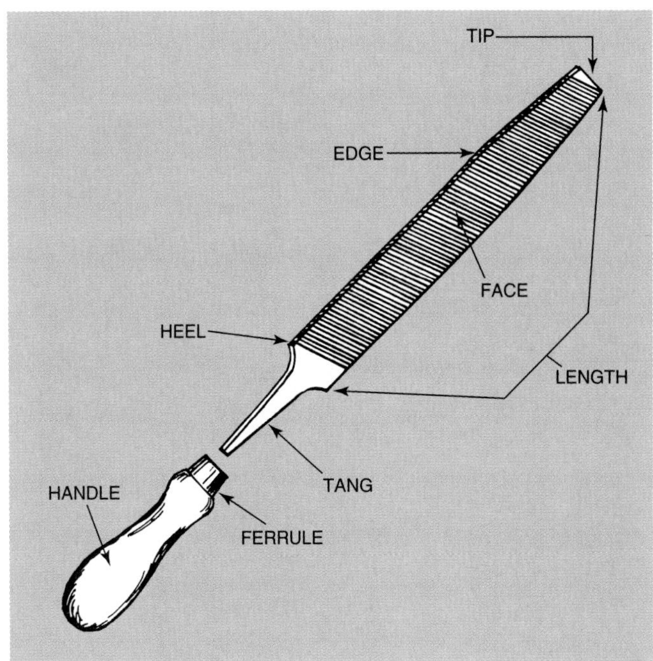

Figure 5-6. A typical single-cut mill file.

File Cut and Shape

When a file has a single series of cutting edges that are parallel to each other, the file is referred to as a *single-cut file.* A file with two sets of cutting edges that cross at an angle is called a *double-cut file.* See **Figure 5-7.**

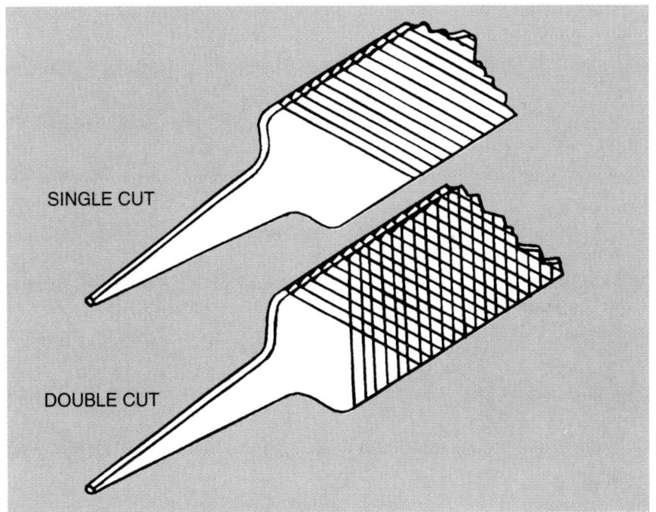

Figure 5-7. Single- and double-cut files.

File cut—from rough to smooth—depends on the number and size of the cutting edges. **Figure 5-8** illustrates several classifications of cut for both single- and double-cut files. Common file shapes are illustrated in **Figure 5-9.**

Rotary Files

A selection of *rotary files* is often found in the technician's toolbox. These files are designed to be chucked in an hand drill or hand-held air grinder. They are useful in working in "blind" holes, where a normal file is useless. See **Figure 5-10.**

Drills

Most drilling today is done with power operated hand ***drills***. Power drills are operated by compressed air or electric motors. Either type of power drill has a chuck, which is tightened to securely hold the ***drill bit.*** See **Figure 5-11.**

The most common hand drill sizes are 3/8″ and 1/2″. The drill size refers to the largest drill bit that the chuck can handle. 3/8″ hand drills are the most common and are used for drilling light metals, light honing, and operating wire brushes. 1/2″ drills are used for drilling through thick or hard metals and for turning engine cylinder hones. Both electric and air drills are common. The compressed air drill is used, however, in situations where a long drilling session would overheat an electric motor.

Cordless Drills

Cordless drills are beginning to see greater use in the modern shop. These compact and powerful tools can fit into

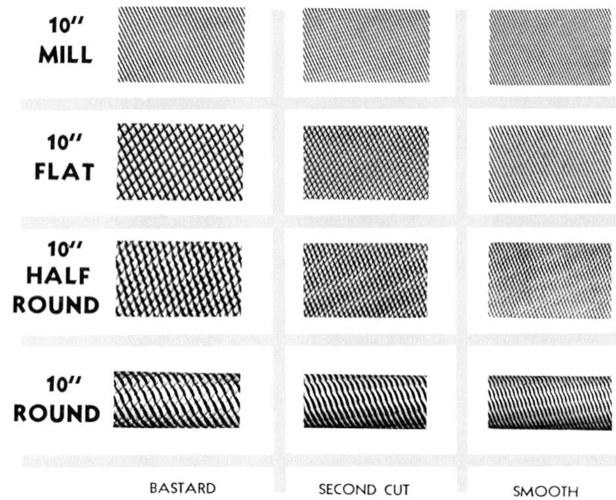

Figure 5-8. Three different file cuts: bastard, second-cut and smooth. (Simonds File Co.)

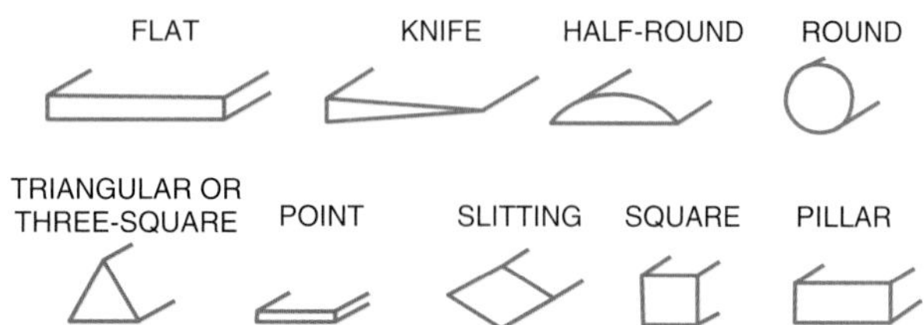

Figure 5-9. Nine common file shapes that are handy for the technician.

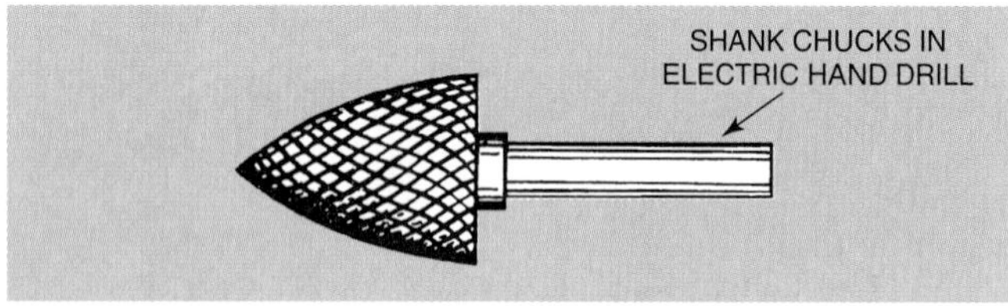

Figure 5-10. A rotary file. These come in a variety of sizes and shapes.

tight spots where the power cord or air hose to a conventional drill cannot go. By adding adapters, they can be used as power screw and nut drivers. This setup can make short work of removing a component that has many screws or bolts retaining it. Otherwise, this would normally be a very time consuming task.

Drill Press

Another type of drill is the drill press. The drill motor and chuck are mounted on a stand and are able to move up and down on a central shaft. The work to be drilled is mounted on a table directly under the drill motor. The drill bit, chuck, and motor are lowered by turning a lever to contact the work to be drilled.

The drill press eliminates the need to hold the drill and ensures that the drill bit will produce a properly aligned and

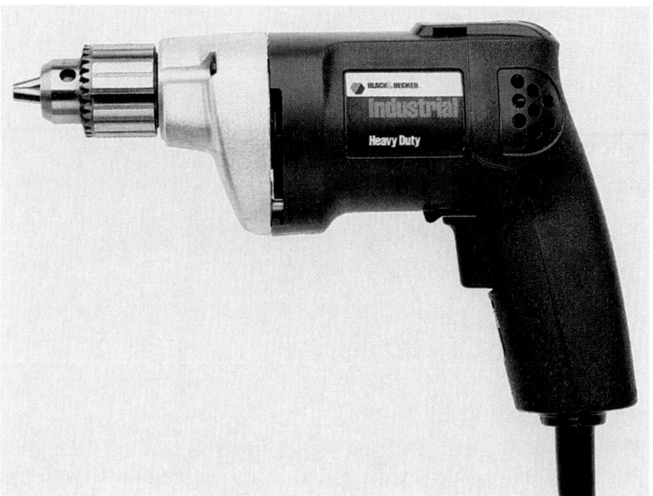

Figure 5-11. A heavy-duty electric drill. (Black & Decker)

sized hole. Drill presses can only be used on parts that will fit on the mounting table and cannot be taken to the vehicle for drilling any part that cannot be removed.

Drill Bits

A drill bit is chucked in a hand or power drill and does the actual cutting when drilling holes. Drill bits are available in fractional inch and metric sizes. For general auto shop use, a set of fractional size bits (from 1/16″ to 1/2″ or 1.5 to 13.0 mm) is sufficient to start. Other size drill bits may be purchased as needed.

Drill bits with a straight shank are used in hand drills and portable power drills. For heavy duty power drills, the taper shank bit is sometimes used. See **Figure 5-12.** Bits are commonly furnished in carbon steel and high-speed steel. Carbon bits will require more frequent sharpening, and they will not last nearly as long as high-speed steel drills. Cobalt drill bits, which are designed for heavy duty drilling, are also available.

Reamers

Reamers are used to enlarge or shape holes. They produce holes that are more accurate in size and smoother than those produced by drill bits. The reamer should not be used to make deep cuts in metal. A cut of a few thousandths of an inch at a time is all that should be attempted.

Some reamers are adjustable in size, while others are of a fixed size. Tapered reamers are handy for removing burrs and assisting in tap starting. Several different types of reamers are shown in **Figures 5-13** and **5-14.**

Impact Wrench

A power ***impact wrench*** is a must for fast work. Most impact wrenches are air operated, though electric impact tools are available. Parts can be removed and replaced in a fraction of the time required with hand wrenches. Although the impact wrench in no way replaces hand wrenches, it can be used often and should be included in the tool collection.

Heavy-duty impact sockets should always be used with an impact wrench. These sockets are much stronger than

Figure 5-12. Technicians use a variety of drill bit sizes and types.

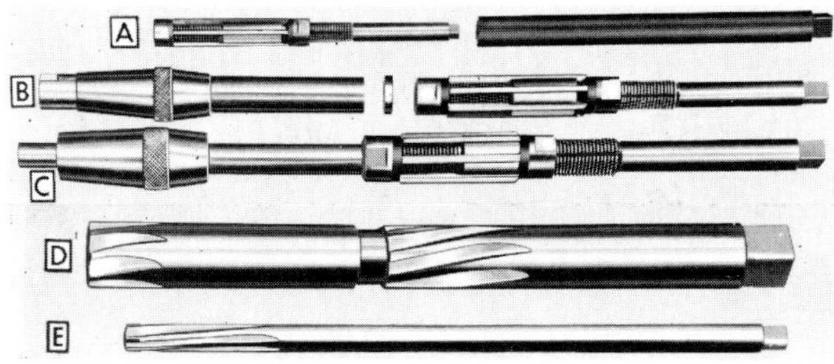

Figure 5-13. Reamers. A—Adjustable. B and C—Adjustable with a pilot to aid in reamer alignment. D—Kingpin reamer. E—Valve guide reamer. (Snap-on Tools)

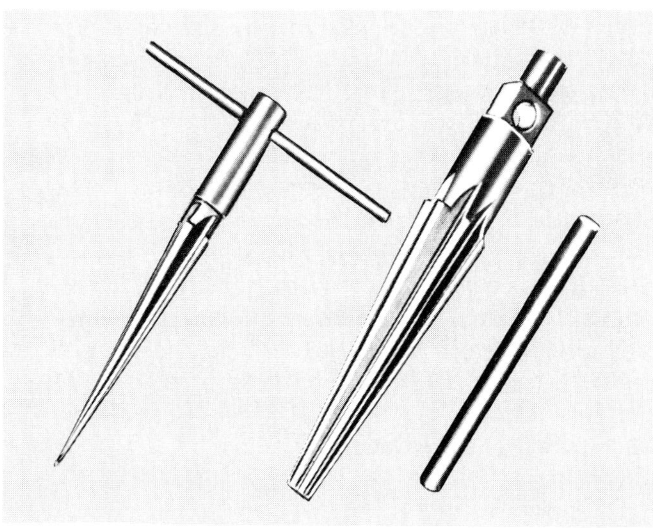

Figure 5-14. Tapered reamers. (Snap-on Tools)

standard sockets. When used with an impact wrench, standard sockets may break. **Figure 5-15** shows a pneumatic (air-operated) impact wrench. Sizes run from 1/4″ to 1″. Power ratchets, drills, chisels, hand-held grinders, sanders, and shears in a variety of designs are also available.

Figure 5-15. An impact (power) wrench is a must for every technician. They are available in a variety of sizes and shapes. (Snap-on Tools)

Hacksaws

The hand ***hacksaw*** is a much-used tool. It is excellent for cutting bolts, tubing, or light metal. Hacksaw blades are generally furnished with 14, 18, 24, or 32 teeth per inch. The 14-tooth blade is handy for cutting fairly thick articles. The 18-tooth blade handles work of medium thickness. The 24-tooth blade is best used on brass, copper, heavy sheet metal, and medium tubing. For thin sheet metal, use a 32-tooth blade. **Figure 5-16** illustrates typical work for each type of blade.

Hacksaw blades are made of various materials. Some blades are hardened across the full blade width; others are hardened only along the edge of the teeth. When cutting alloy steels, select one of the better quality blades, such as a high-speed (tungsten) steel blade. A conventional hacksaw is shown in **Figure 5-17.**

Special Hacksaws

The automotive technician will sometimes find use for a special hacksaw that allows cutting in restricted quarters. A hole saw, which is operated by a drill, is also handy for drilling large holes in thin sheet metal, **Figure 5-18.**

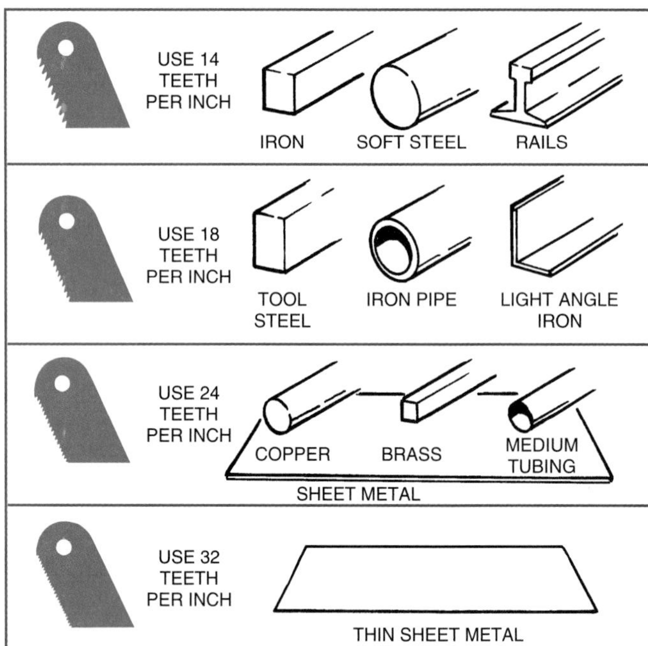

Figure 5-16. The type of work to be cut determines the blade selection. These are typical. (Crescent Tools)

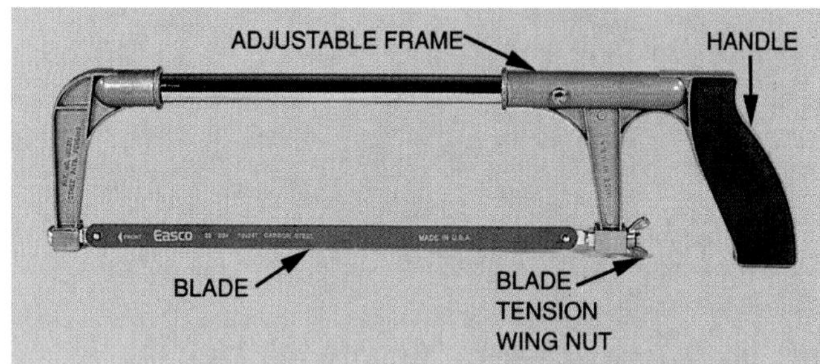

Figure 5-17. One type of hacksaw with an adjustable frame. This allows the saw to handle blades of different lengths. (Easco Tools)

Figure 5-18. Hole saws are handy for cutting openings in thin metal and for making large diameter holes.

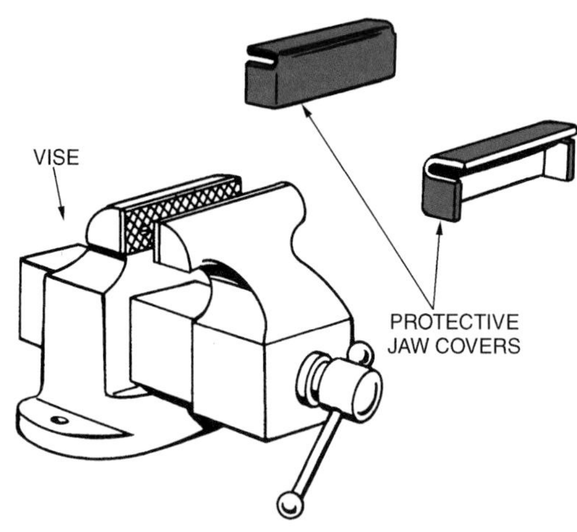

Figure 5-19. A typical bench vise.

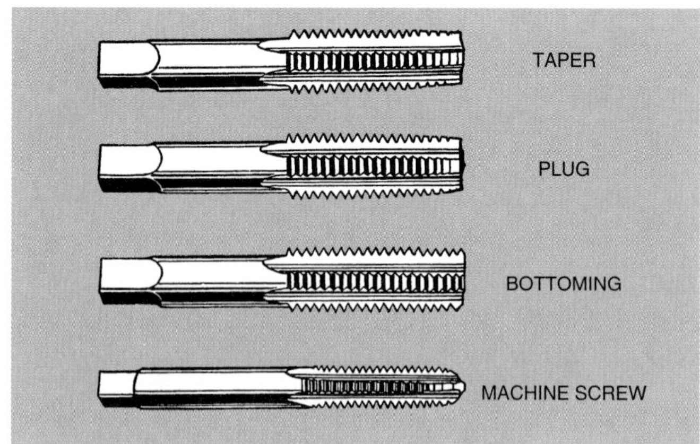

Figure 5-20. Four different types of taps.

Vise

A ***vise*** is a holding device. A typical bench vise is shown in **Figure 5-19.** If the work is easily marred, place special copper or nylon covers over the serrated steel jaws. Keep the vise clean. Oil all working parts. Do not use a vise for an anvil and do not hammer the handle to tighten the jaws on the workpiece.

Taps

Taps are used for cutting internal threads, **Figure 5-20.** For general use, a set of *taper, plug, bottom,* and *machine screw taps* should be available. Each type of tap will have to be provided in both National Fine and National Coarse threads, as well as common metric sizes.

The taper tap is used to tap completely through the hole. Notice that this tap has a long, gradual taper that allows it to start easily. The bottoming tap is used to cut threads all the way to the bottom of a blind hole. The plug tap should be used before using the bottoming tap because the bottoming tap will not start well. The plug tap is the most widely used tap. It will work satisfactorily in most cases. However, it cannot be used to run threads completely to the bottom of a blind hole. The machine screw tap is used for small diameter holes with fine threads. **Figure 5-21** shows tap nomenclature.

Dies

Dies are used to cut external threads. A die of the correct size is placed in a diestock (handle) and turned. In some

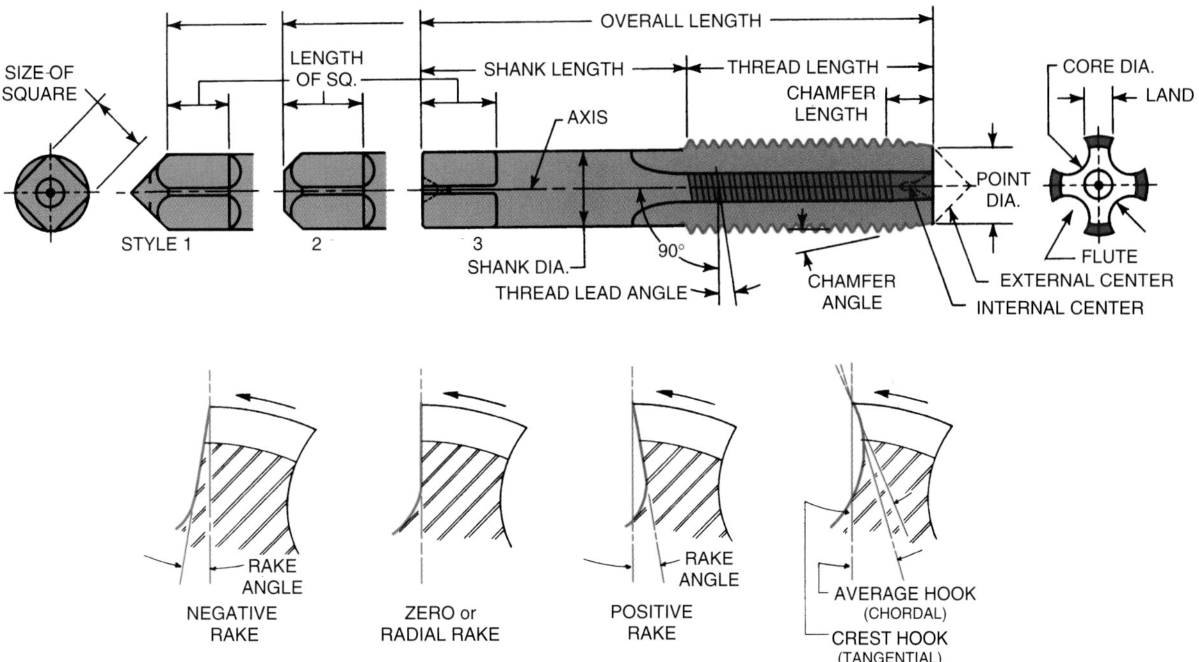

Figure 5-21. Tap nomenclature. (Winter Brothers)

cases, dies are adjustable in size, so you can enlarge or reduce (slightly) the outside diameter of a threaded area. Both taps and dies should be cleaned, lightly oiled, and placed in a protective box after each use.

Cleaning Tools

Various cleaning tools are illustrated in **Figure 5-22.** These tools provide valuable assistance in the thorough cleaning of parts. A wire wheel that can be used on a grinder mandrel is shown in **Figure 5-23.** Carbon cleaning brushes that can be operated in a 3/8″ drill are also available.

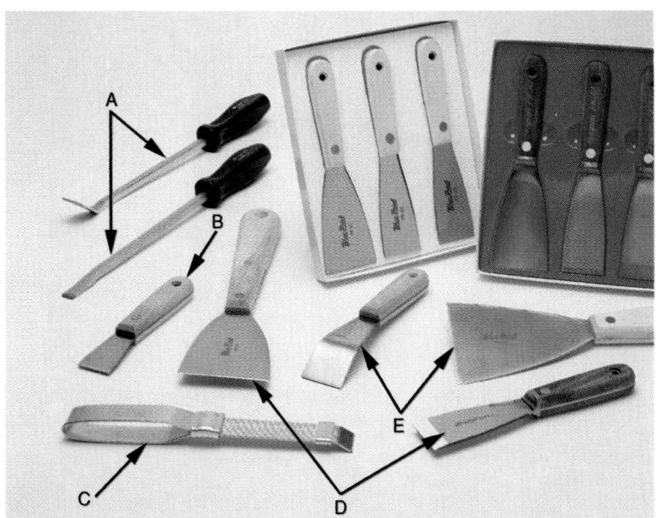

Figure 5-22. Cleaning tools. A—Rigid scrapers. B—Gasket scraper. C—Flexible carbon scraper. D—General purpose scrapers. E—Putty knives. (Snap-on tools)

Pliers

A technician needs a variety of ***pliers***. Study the pliers shown in **Figures 5-24** to **5-26.** Never use pliers in place of a wrench. Use pliers designed for the job at hand. Avoid cutting hardened parts with lineman, diagonal, or other cutting pliers. Needle nose pliers are delicate; use them carefully.

Figure 5-23. A power driven wire wheel. Wear eye protection.

Screwdrivers

Screwdrivers of various lengths and types are required. *Standard, Phillips, Reed and Prince, Clutch,* and *Torx®* screwdrivers will cover the various types of jobs ordinarily encountered by the automotive technician. It is wise to have several large screwdrivers that will stand some prying and hammering. The various types of screw head shapes used in most vehicles are shown in **Figure 5-27**.

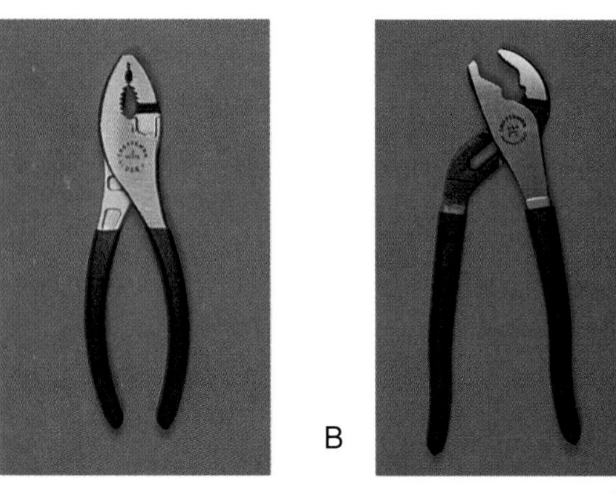

A B

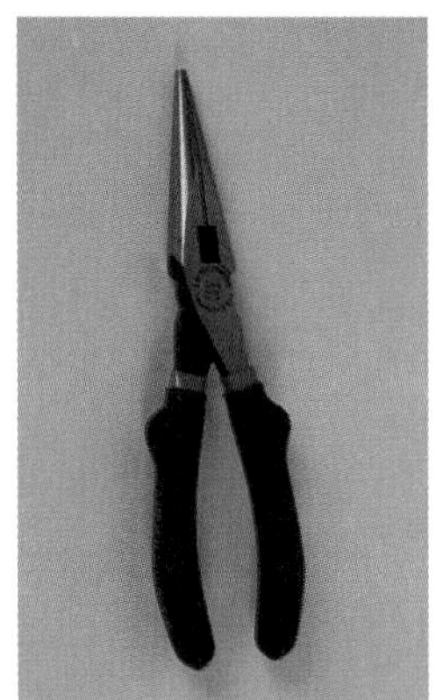

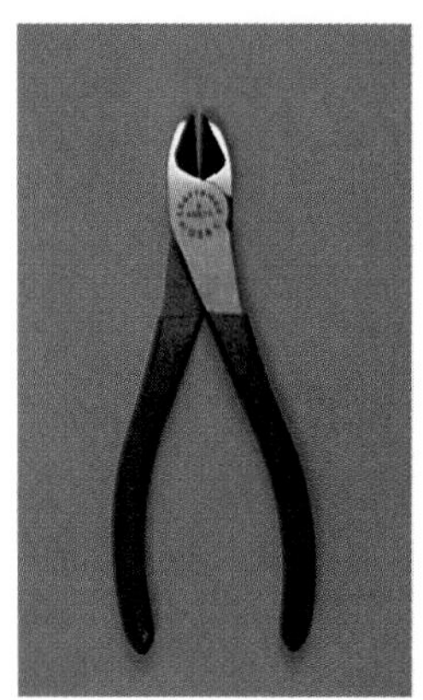

C D

Figure 5-24. Different types of pliers. A—Slip joint. B—Rib joint. C—Needle nose. D—Diagonal cutting.

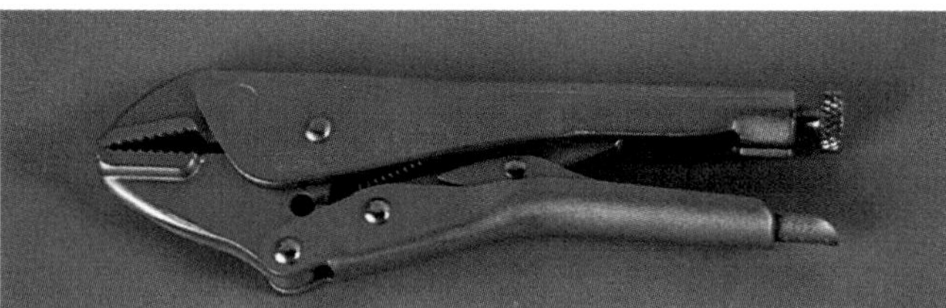

Figure 5-25. Locking pliers, or Vise grips

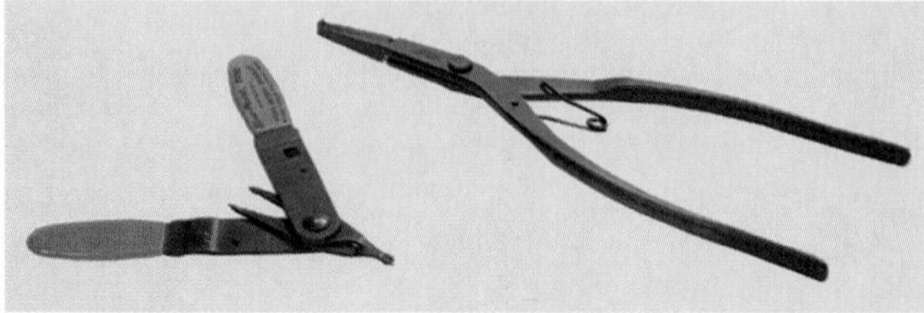

Figure 5-26. Two pairs of snap ring pliers. (Snap-on Tools)

The standard, or slotted, screwdriver can have a round or square shank. A set of slotted and Phillips screwdrivers is shown in **Figure 5-28.** Reed and Prince tip screwdrivers are similar to Phillips screwdrivers. A Pozidrive® screwdriver and

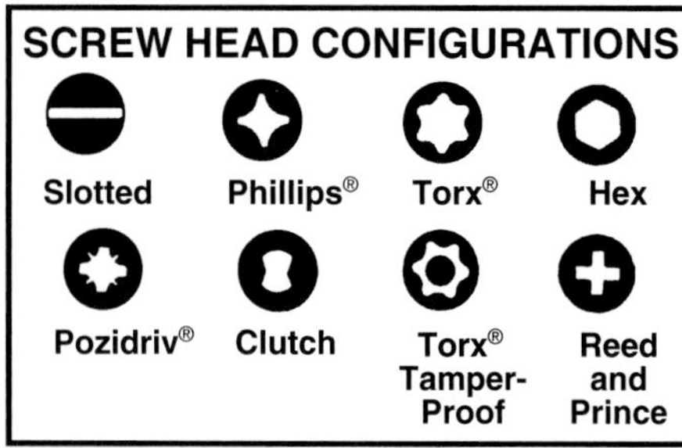

Figure 5-27. Configurations for screw heads used in the typical automobile.

screw head are shown in **Figure 5-29A.** A Torx® screw head and driver are illustrated in **Figure 5-29B.** Some Torx® screw heads have a stud protruding from their center and are tamper resistant. An assortment of offset screwdrivers makes it easy to remove and drive screws in difficult areas. Another essential screwdriver is the magnetic type designed to hold screws while they are being installed.

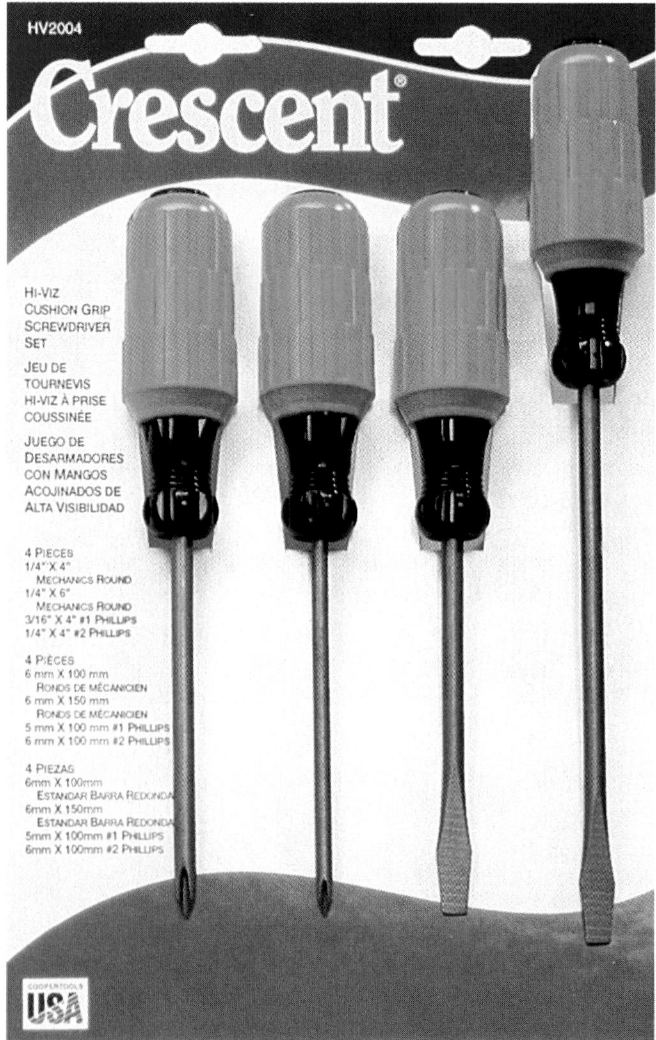

Figure 5-28. Screwdrivers are often best purchased in sets. (Cooper Tools)

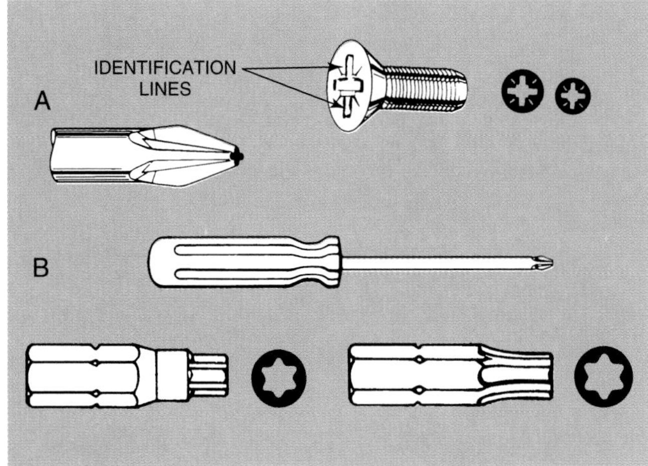

Figure 5-29. A—Similar to the Phillips design, the engagement tip of a pozidrive screwdriver is narrower and longer. A pozidrive fastener is identified by lines between the openings. B—Torx bits and screw head. (Vaco Tools)

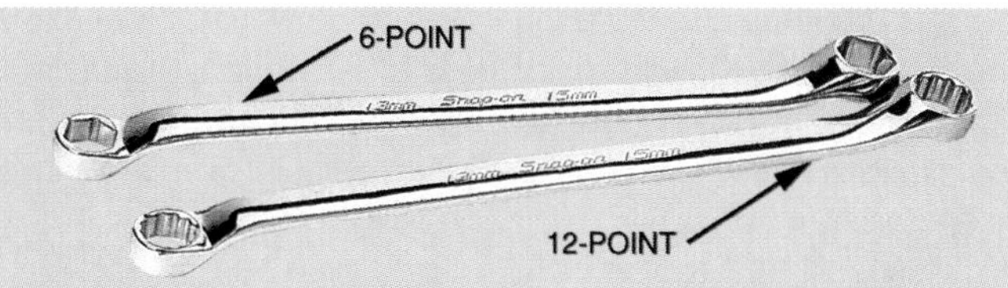

Figure 5-30. 10° offset box-end wrench in six- and twelve-point openings (Snap-on Tools)

Wrenches

All technicians must have a wide variety of both standard and metric ***wrenches*** at their disposal. A technician should have wrench sizes from 1/4″ to 1 1/8″ for standard and 5 mm to 24 mm for metric. A few of the wide variety of wrenches are discussed here.

Box-End Wrenches

The *box-end wrench* is an excellent tool because it grips the nut or bolt head on all sides. This reduces the chances of slipping, with resultant damage to the fastener and possibly the hand.

A box-end wrench is designed with either a 6- or 12-point opening. For general use, the 12-point opening works well. It allows the wrench to be removed and replaced without moving the handle over a long arc. When removing stubborn or damaged nuts, or when there is a danger of collapsing the nut, the 6-point will do a better job. It grips the nut across each flat, minimizing slippage and fastener damage.

The box-end wrench is available in either a double offset or a 15° offset. Sets in standard size openings, as well as standard lengths, will handle most jobs. Sets are available in midget lengths for work in cramped areas. Each end on the box-end wrench is a different size.

Warning: When using a wrench, pull on the handle whenever possible. If you must push on the wrench, push with the palm of the hand with the fingers outspread to avoid smashing them if the wrench slips.

Figure 5-30 shows two box-end wrenches. Both wrenches have an extra deep offset for additional clearance. Can you see why the wrench with the six-point opening will grasp the nut or cap screw more securely? These are standard length box-ends. Short and longer length wrenches are of the same design.

Open-End Wrenches

Open-end wrenches are handy, but they are not as dependable as box-end wrenches. They grasp the nut or bolt head on only two of its flats and are subject to slipping under a heavy pull. There are tight places, however, where they must be used. One example would be a threaded adjusting nut on a shaft or closed-end cable, such as a brake cable. See **Figure 5-31.**

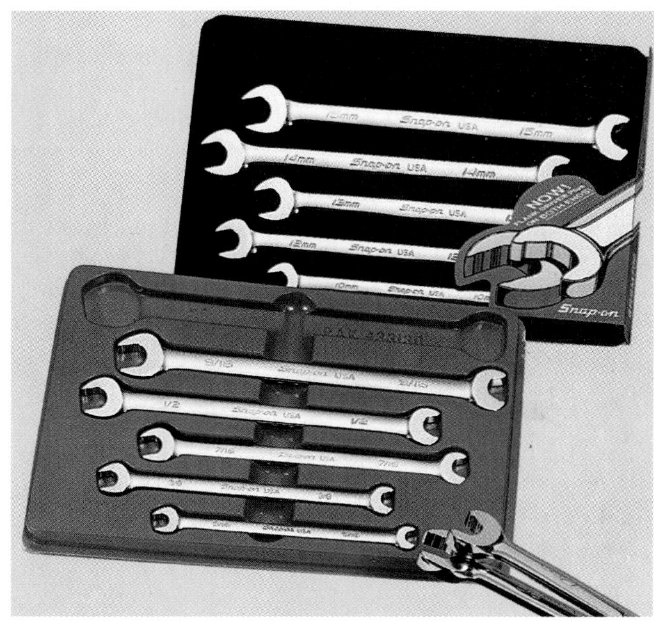

Figure 5-31. This set of open-end wrenches contains both standard and metric sizes. (Snap-on Tools)

The head of an open-end wrench is generally offset at an angle of 10-15°. When the wrench has traveled as far as allowable, it can be flipped over and placed on the nut in an arc of 30°. Open-end wrenches are also available with the heads offset more than 15°. Each end of an open-end wrench has a different size opening.

Combination Wrench

The *combination wrench* has a box-end head on one end and an open-end on the other. Generally, the open-end is offset 15° from the wrench centerline. Both ends are the same size.

The combination wrench is a very convenient tool since the box-end can be used for both breaking loose and final tightening. The open-end is for faster fastener removal or installation. See **Figure 5-32.** Some combination wrenches have flare-nut heads.

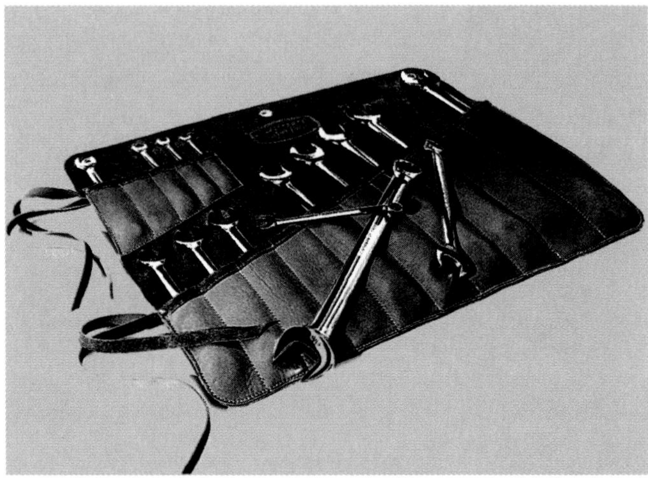

Figure 5-32. A combination wrench set in a roll-up tool pouch. (Proto Tool Division of Ingersoll-Rand Co.)

Flare-nut Wrench

All tube fittings should be removed and replaced with special *flare-nut wrenches*. See **Figure 5-33.** These wrenches are designed to provide a maximum amount of contact. They can also be used to remove soft fasteners without damage.

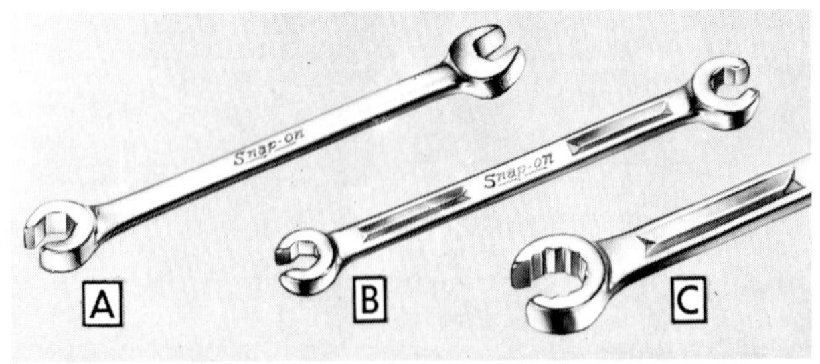

Figure 5-33. Flare-nut wrenches. A and B—Six-point openings C—Twelve point opening.

Adjustable Wrenches

The *adjustable wrench* is handy in that it can be adjusted for size. However, it tends to slip, so it is a poor wrench to use when other tools are available for a given job. When an adjustable wrench is used, adjust the jaws firmly. Also, make certain the wrench is placed so the pull on the handle is toward the bottom side. This relieves heavy pressure on the adjustable jaw. Some adjustable wrenches are spring-loaded and adjust themselves to the work.

Other Useful Wrenches

The flexible head socket wrench, **Figure 5-34A,** and the ratcheting box-end wrench, **Figure 5-34B,** are useful additions to the technician's toolbox. Allen and fluted wrenches are often used in tightening or removing setscrews and cap screws. They are strong and provide a good grip. See **Figure 5-35.**

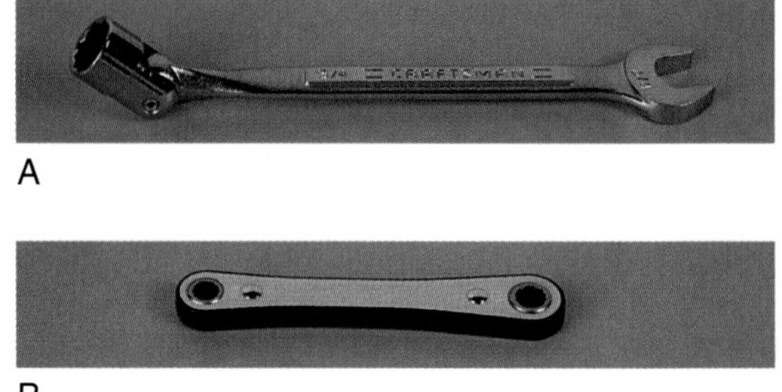

Figure 5-34. Flex head socket and ratchet box-end wrenches. A—Flex head socket. B—Ratchet box.

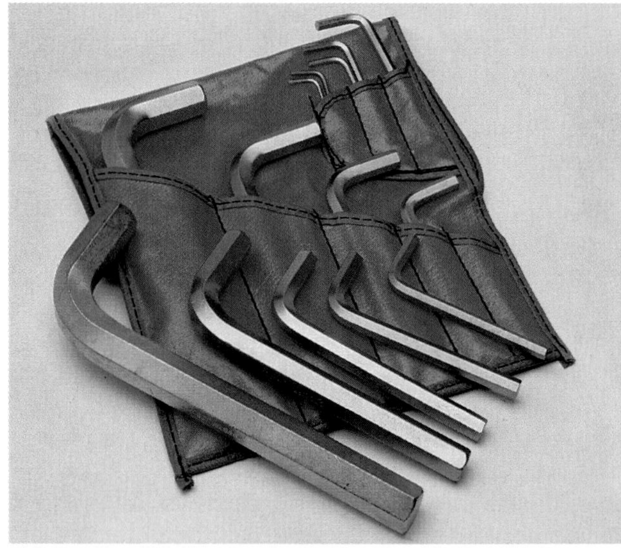

Figure 5-35. Allen, or hex, wrenches.

Torque Wrenches

The ***torque wrench*** allows technicians to tighten fasteners to exact torque (tightness) specifications supplied by the manufacturer. It is of utmost importance that bolts be tightened to specifications. Improper and varying torque on units or assemblies may cause distortion.

A torque wrench is shown in **Figure 5-36.** Torque wrenches are available in both foot-pound and inch-pound types. The inch-pound wrench is used for delicate, low-torque applications. Some torque wrenches are available with dual scales on the same wrench. One scale is for foot-pounds and the other is for inch-pounds. Still others with dual scales can have one scale for foot-pounds and one scale for metric values (Newton-meters). Torque wrenches are available in special shapes and sizes.

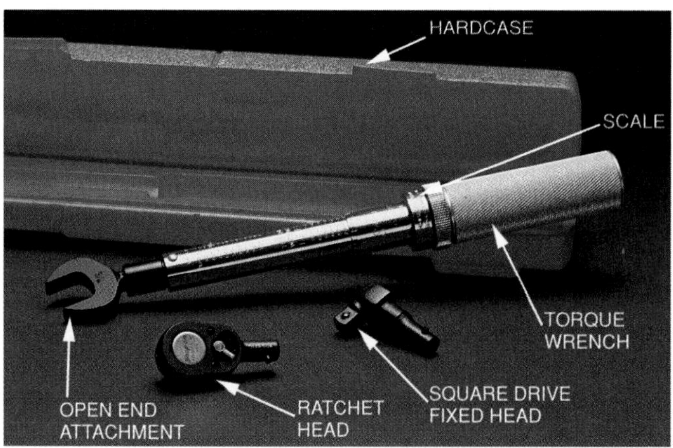

Figure 5-36. A click type interchangeable torque wrench. The assortment of heads permits ratcheting and open-end use. (Snap-on Tools)

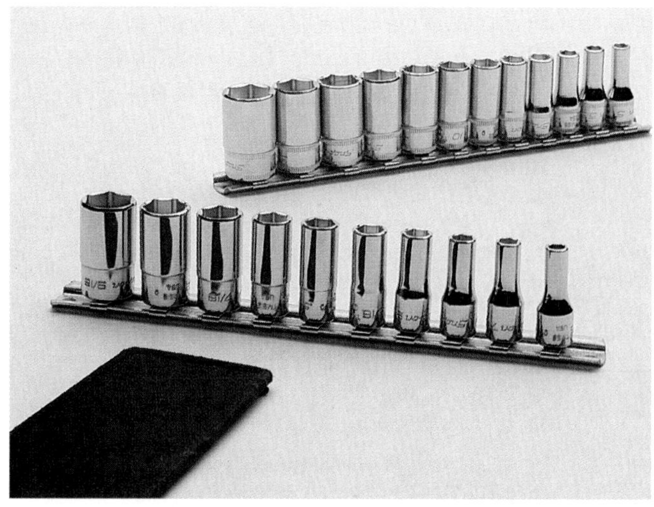

Figure 5-38. A set of deep six-point sockets in standard and metric sizes. Sockets should be stored on rails to prevent loss. (Snap-on Tools)

Sockets and Ratchets

Sockets and ratchets are very convenient and, in most instances, faster than normal wrenches. Both sockets and ratchets are available in a variety of sizes and types.

Sockets

Most ***sockets*** are available with 6-point, 12-point, and a variety of other openings and driver shapes. The 1/4″, 3/8″, 1/2″, 3/4″, and 1″ drive sizes will cover a wide range of jobs. The drive size refers to the size of the square hole into which the socket handle fits. The larger the drive, the heavier and bulkier the socket.

The 1/4″ drive is for small work in difficult areas. The 3/8″ drive will handle most general work where the torque (tightness) requirements are not too high. The 1/2″ drive is for all-around service. The 3/4″ and 1″ drives are for heavy duty work with high torque settings.

Standard-length sockets are available for general work, **Figure 5-37.** Deep sockets, **Figure 5-38,** are available for work requiring a longer than ordinary reach, such as spark plug removal and installation. Swivel sockets are also available for angle work.

Socket Handles and Ratchets

Various handles are available for sockets. See **Figure 5-39.** The speed handle is used for fast operation because it can be turned rapidly. Flex handles in varying lengths allow the socket to be turned with great force and at odd angles. ***Extension bars*** of different lengths allow the technician to lengthen the socket setup to reach difficult areas. The sliding T-handle varies the handle length.

The ***ratchet*** is versatile. It allows the user to either tighten or remove a nut or bolt without removing the socket from the fastener. On the backstroke, the handle ratchets. See **Figure 5-40.**

Figure 5-37. Six-point sockets. These sockets are slightly longer than the normal standard socket, which allows them to access recessed bolts or bolts with protruding studs. (Snap-on Tools)

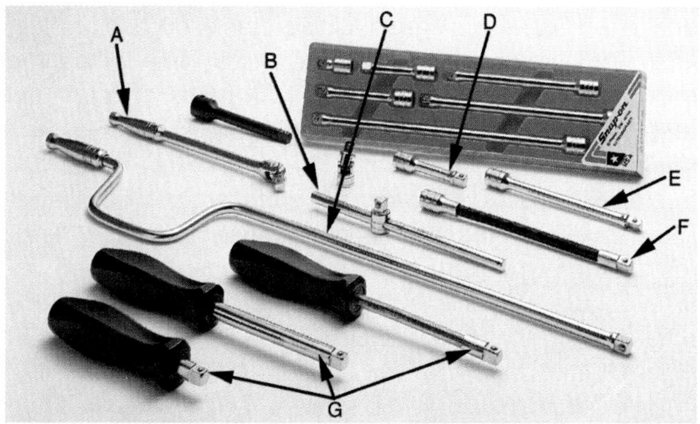

Figure 5-39. Accessory handles and extensions. A—Breaker bar. B—Sliding T-handle. C—Speed handle. D—Short extension. E—Long extension. F—Flexible extension. G—Socket drivers. (Snap-on Tools)

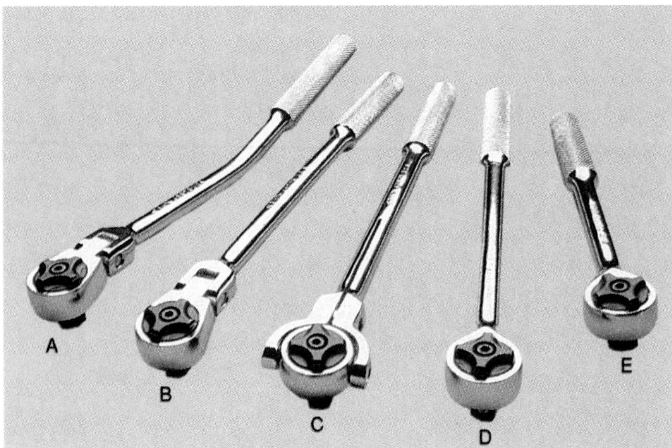

Figure 5-40. A selection of socket ratchet handles. A—Bent handle ratchet. B—Straight handle with flex head. C—Straight handle ratchet with 360° swivel head. D—Straight handle with fixed head. E—Short handle straight fixed head ratchet. (S-K Tools)

Socket Attachments

Many socket attachments are available. Such items as screwdriver heads, drag link sockets, pan screw sockets, crowfoot attachments, and Allen wrench heads combine to make a socket set a fine tool.

Pullers

Automotive technicians will find use for a wide variety of ***pullers.*** These tools are used for pulling gears, bearings, hubs, and many other parts. An assortment of pullers is shown in **Figure 5-41.** Note the various sizes.

Another excellent puller is the slide hammer type. The puller jaws grasp the work, and the weighted slide is shoved forcefully against the stop. This type of puller is fast and efficient on many types of jobs, **Figure 5-42.**

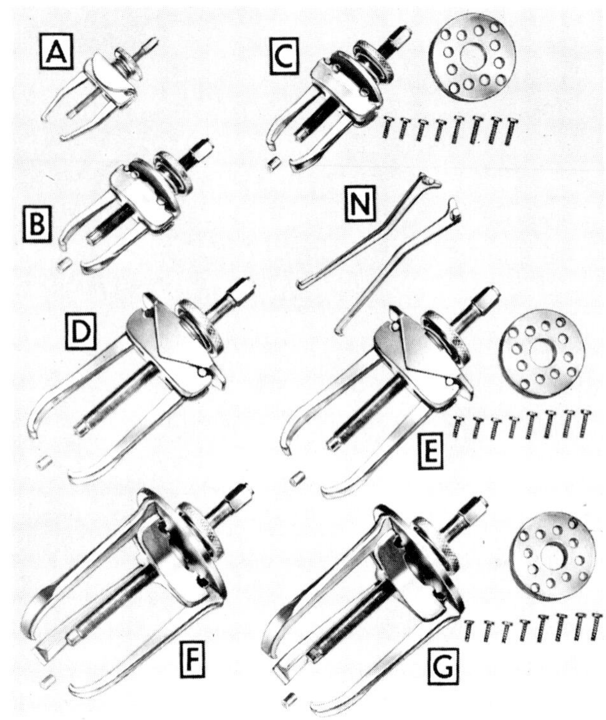

Figure 5-41. Pullers are available in various styles and sizes.

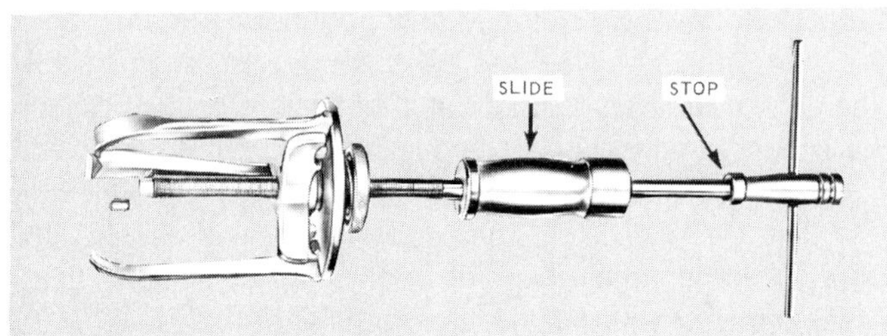

Figure 5-42. A slide hammer puller. The slide is "hammered" against the stop, developing a powerful pulling force.

Soldering Equipment

Electric ***soldering irons*** are used for most automotive soldering jobs, such as soldering wires. Modern 35 to 100 watt irons will handle most wire soldering jobs. The tip of an electric soldering iron will quickly reach soldering heat after it is plugged in. Soldering irons should be large enough for the job. An iron that is too small will require excess time to heat the work and may never heat it properly. A proper size iron will bring the metal up to the correct soldering heat quickly and result in a good soldering job.

Multitemperature ***soldering guns*** are also available. These tools allow the user to hold and use them easily. Their tips heat the work very quickly, ensuring a good solder job. Small lights are sometimes built into the body to provide additional light on the work. A typical soldering gun is shown in **Figure 5-43.**

For large soldering jobs such as radiator tanks, a large soldering iron, sometimes called a *copper* is used. These irons are heavy and must be placed in a gas flame or in an electric furnace for heating. They are too cumbersome and heavy for wire soldering.

Solder

For soldering all automotive wiring, use rosin-core solder only. Acid-core solder leaves a residue that will cause corrosion in electrical units and should never be used for wiring. Acid-core solder is satisfactory for radiator work.

Measuring Tools

Technicians are often called on to make precision measurements. To do this, they must be able to use an outside micrometer, an inside micrometer, dial gauges, calipers, depth gauges, and a combination square. In addition, technicians must often measure electrical properties, various pressures, and vacuum.

Micrometers

Outside and inside ***micrometers*** are used to make linear measurements that are accurate to a fraction of a thousandth of an inch. See **Figures 5-44** and **5-45.** Outside

Figure 5-43. A pistol grip type soldering gun. These develop heat quickly, easily heating the work for good soldering. (Cooper Tools)

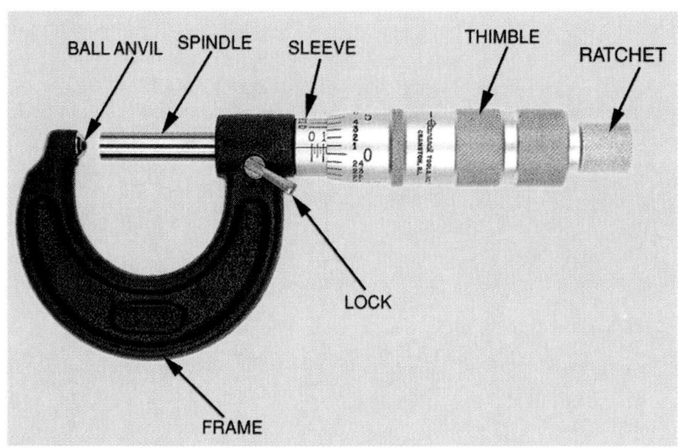

Figure 5-44. Typical outside ball anvil micrometer. This tool is handy for measuring curved surfaces. Handle with care. (Central Tools)

micrometers are used to measure outside diameters and thicknesses. Inside micrometers are used to measure internal dimensions, such as the inside diameter of a cylinder. Micrometers are available in a variety of sizes. Both standard and metric micrometers are available. The metric micrometer is graduated in millimeters instead of thousandths of an inch.

Inside and Outside Calipers

Calipers are useful tools for rough measurements. **Figure 5-46** illustrates a pair of outside calipers being used to measure the diameter of a shaft. **Figure 5-47** shows how the inside calipers can be used to measure the size of a hole. To determine the reading, hold the calipers on an accurate steel rule.

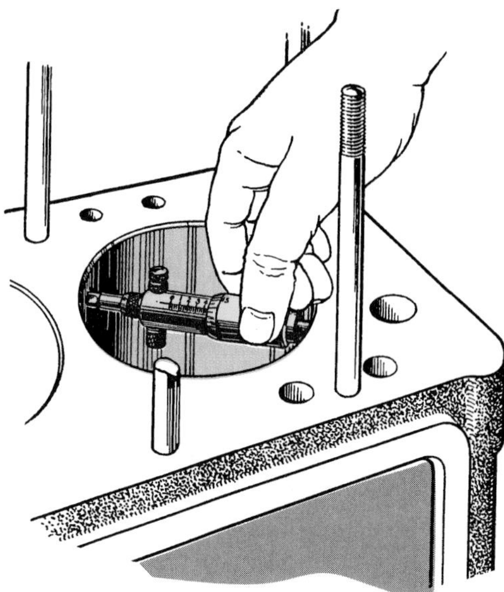

Figure 5-45. Using an inside micrometer to check cylinder bore size.

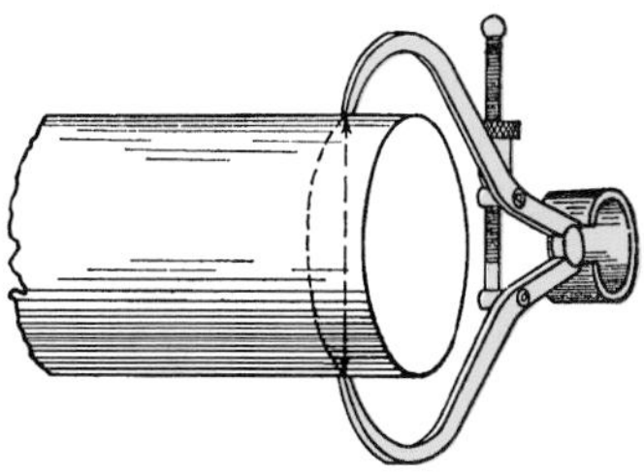

Figure 5-46. Using an outside caliper to measure the diameter of a shaft. (South Bend Lathe)

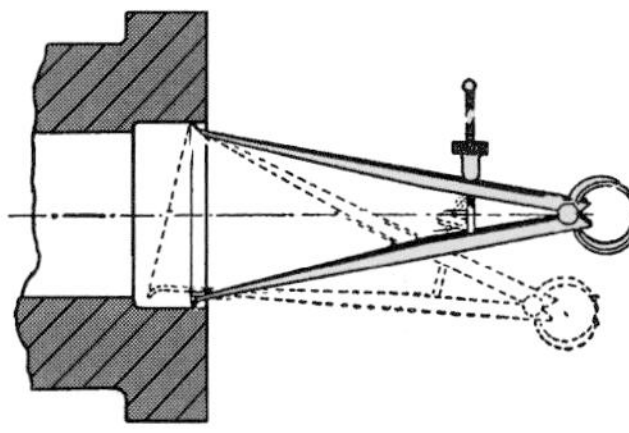

Figure 5-47. Using an inside caliper to measure the diameter of a counterbore.

Dividers

Dividers are made somewhat like calipers, but have straight shanks and pointed ends. They are handy for marking circles and making surface measurements. **Figure 5-48** shows dividers being used to find the center of a steel shaft.

Dial Indicator

A ***dial indicator*** is used to read movement in thousandths of an inch. Common uses are checking end play in shafts, backlash in gear teeth, valve lift, shaft runout, wheel runout, and taper in cylinders. **Figure 5-49** shows how a dial

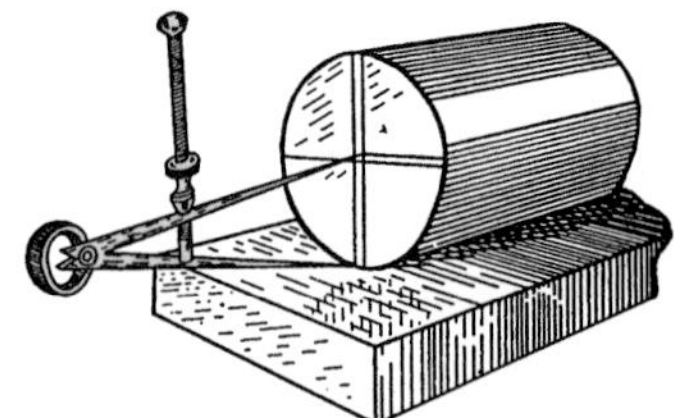

Figure 5-48. Dividers being used to find the center of a steel shaft. (South Bend Lathe)

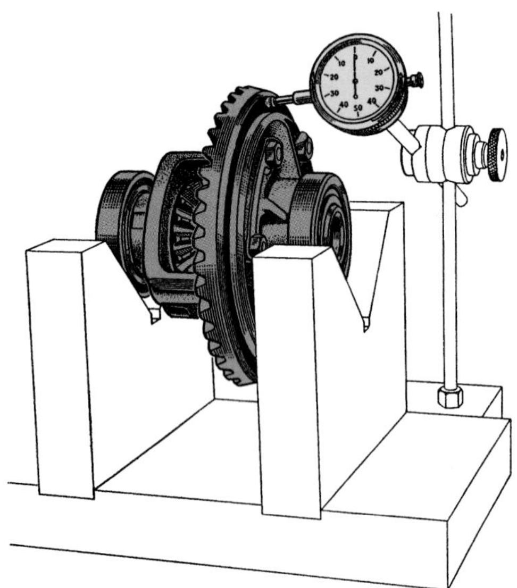

Figure 5-49. Checking ring gear runout with a dial indicator.

indicator is used to check a ring gear for runout. As the ring gear revolves in the V-block stand, any runout (wobble) of the gear will cause the dial needle to move. This movement is read in thousandths of an inch.

Dial Calipers

A ***dial caliper*** is a very handy and useful measuring tool. It is capable of taking inside, outside, and depth measurements. These instruments are highly accurate and should be handled with care.

The dial caliper shown in **Figure 5-50A** is capable of measuring objects up to 6″ (152.4 mm). Each graduation on the bar scale (body) is equal to .100″ (2.54 mm). The dial is calibrated in .001″ (0.025 mm). One full revolution of the dial needle equals .100″ on the bar scale. A thumb-operated roll knob is used to make fine adjustments. Any given adjustment can be secured by turning the lockscrew. This prevents the caliper from opening or closing. An electronic dial caliper is shown in **Figure 5-50B.**

Feeler Gauges

Feeler gauges are thin strips of specially hardened and ground steel. The thickness of each strip is marked in thousandths of an inch. Feeler gauges are used to check clearances between two parts or surfaces. See **Figure 5-51.**

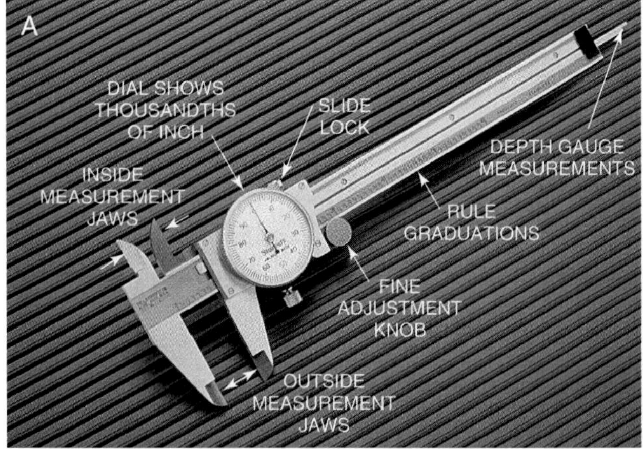

Figure 5-50. Two types of dial calipers. A—Standard dial indicator. B—Electronic digital caliper. Handle with care. (Starrett)

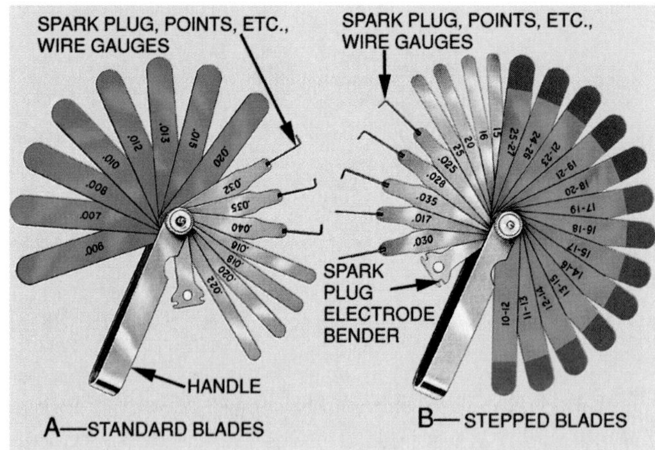

Figure 5-51. Feeler gauges. A—Wire gauges and standard flat blades. B—Wire gauges and stepped blades. (L.S. Starrett)

Electrical Testers

To troubleshoot modern electrical systems, the technician must be familiar with many pieces of electrical test equipment. The test light, voltmeter, ammeter, ohmmeter, and tachometer are all very important to proper diagnosis, service, and repair.

Some electrical testers have a needle or pointer to indicate the electrical reading. These are called analog gauges. Many modern electrical testers are digital meters, which display a number to indicate the reading.

Test Light

A ***test light*** is a simple electrical tester composed of a 12-volt light bulb, two terminals, and connecting wiring, as shown in **Figure 5-52.** Using a test light is a quick way to check an electrical device or circuit. However, test lights cannot be used to measure exact electrical units—they only tell if electricity is or is not present. Test lights should not be used to check computers or other solid-state devices, since these units can be ruined by the current flowing through the light.

There are two types of test lights: powered test lights and non-powered test lights. The powered test light contains a battery and is used to check for a complete circuit when no current is flowing in the circuit. The non-powered test light is connected between a powered circuit and ground. It will light up if power is present in the circuit.

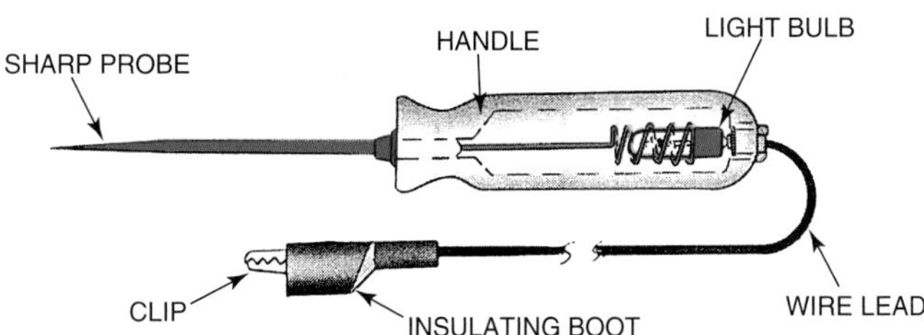

Figure 5-52. Phantom view of a 12-volt test lamp with an internal light bulb. The sharp probe will fit into tiny circuit connectors. Never use it to pierce wire insulation. (General Motors)

Ohmmeter

Ohmmeters are used to measure electrical resistance. Resistance, which exists in any electrical circuit or device, is the opposition to current flow. This resistance is measured in ohms. Resistance is discussed in more detail in Chapter 7. Ohmmeters can only be used to check an electrical circuit when no current is flowing in the circuit.

An ohmmeter has two leads, which are connected to each side of the unit or circuit to be tested. Polarity (direction of current flow) is not important when checking resistance, except when checking computer circuits and certain electronic components. Most ohmmeters have selector knobs for checking various ranges of resistance values. Analog ohmmeters have a special knob to adjust the needle to zero before checking resistance.

Voltmeter

Voltmeters are used to check voltage between two points in an energized circuit. On some voltmeters, different scales can be selected, depending on the voltage level to be measured. Always observe proper polarity when attaching voltmeter leads. When a voltmeter is used on a modern vehicle with a negative ground, the negative lead (usually black) should always be connected to the frame or other ground and the positive lead (usually red) should be attached to the positive part of the circuit.

Ammeter

An ***ammeter*** is used to check the amperage (current) flow in a circuit. The ammeter is used to check the amperage draw of starters or other motors and when checking battery condition. Ammeters can also be used to check the amperage draw of ignition coils, solenoids, and other electrical devices.

Multimeter

A ***multimeter*** is essentially a voltmeter, an ammeter, an ohmmeter, and other testers housed in one unit. Multimeters usually have at least one positive (red) lead and one negative (black) lead. Some multimeters are equipped with other leads for special functions. Polarity should be carefully noted when making certain tests with multimeters. A typical multimeter is shown in **Figure 5-53.**

Tachometer

Tachometers are used to measure engine speed. Modern tachometers are available in either analog or digital versions. Tachometers have at least two leads. One lead is connected to the distributor side of the ignition coil and the other lead is connected to ground. Some tachometers have an inductive clamp-on pickup that is placed around a plug wire to obtain the speed reading. Others use an electric "eye" setup.

The tachometer may have a low range to check idle speeds and a high speed range for making various tests at

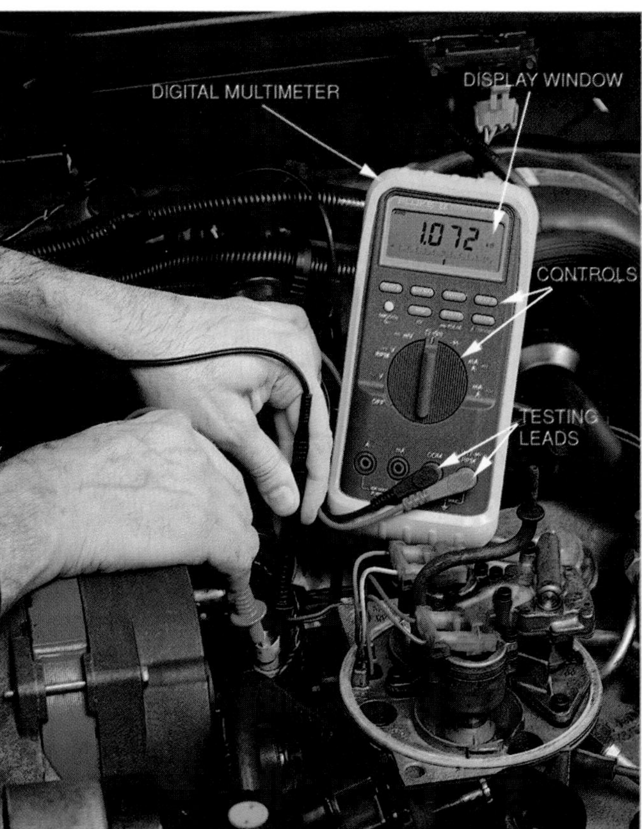

Figure 5-53. This technician is using a digital multimeter to test an electrical connection. Multimeters are precision instruments. They must be handled with care. Note the red and black leads. (Fluke)

high engine speeds. The ranges may be selected with a control knob, or they may automatically switch as engine speed varies. Some tachometers have a provision for checking distributor dwell (the amount of time that primary current is flowing in the ignition coil). This is useful when working on older vehicles with point-type ignition systems, and it has other applications as well. Some multimeters have a provision for a tachometer.

Pressure Gauges

The automotive technician must often measure pressures when diagnosing various problems. The technician needs several pressure gauges. These gauges measure air pressure or the pressure of several automotive liquids, including engine oil, transmission fluid, power steering fluid, and fuel. The most common pressure gauges are discussed below.

Compression Gauge

The ***compression gauge,*** **Figure 5-54,** measures the pressure developed in the engine cylinder when the piston comes up on the compression stroke. The compression gauge measures this compression in pounds per square inch (PSI) or kilopascals (kPa). By using a compression gauge to check the condition of each cylinder, the technician can identify many engine problems. The gauge is connected to the cylinder through a high-pressure hose threaded into the spark plug hole of a gasoline engine or the fuel injector hole of a diesel engine.

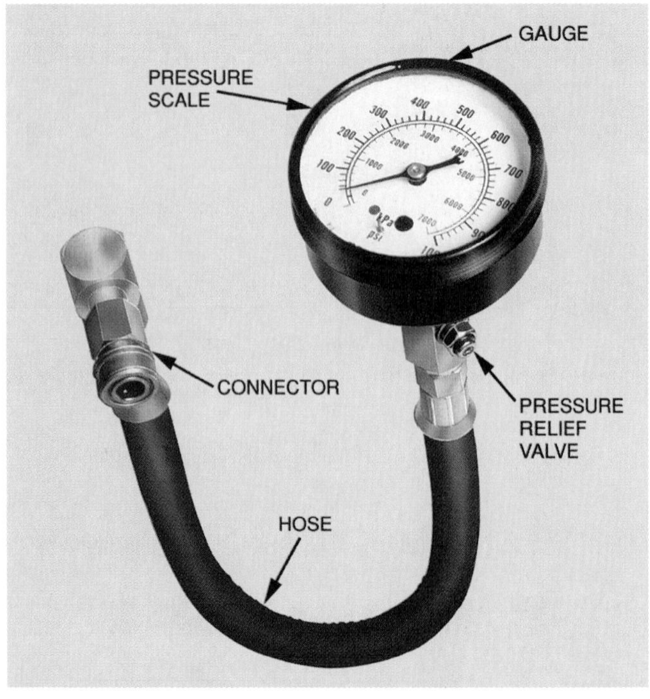

Figure 5-54. One type of engine compression gauge and connecting hose. Note the pressure relief valve. This valve allows relief of compression pressure so that a second reading can be obtained without disconnecting the gauge. It also aids in gauge removal from the engine. (Lisle Tools)

Vacuum Gauge

Vacuum gauges, **Figure 5-55,** measure negative air pressure, or vacuum. In an engine, vacuum is the difference in air pressure between the engine intake manifold and the outside atmosphere. Vacuum is a reliable indicator of engine load and condition. The vacuum gauge is connected by a length of hose to a port on the intake manifold. Variations in vacuum readings indicate various engine problems. Vacuum gauges can also be used to test for proper vacuum to and from vacuum-operated components, such as heater doors, brake boosters, etc.

The vacuum gauge may also be used with a hand vacuum pump to test various vacuum-operated devices. The pump creates a vacuum on the device to be tested while the gauge measures the vacuum and indicates whether or not the unit can hold the vacuum. The technician may also be able to observe the action of the vacuum device as the vacuum is applied.

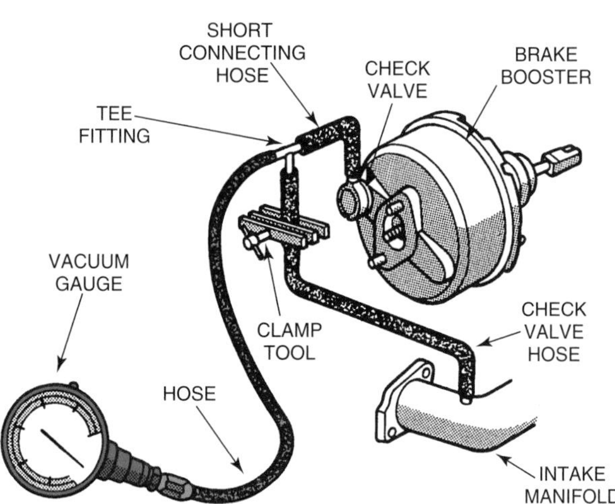

Figure 5-55. A standard vacuum gauge with one particular hookup to test the vacuum for a brake booster. This instrument will provide vacuum or pressure readings. (Dodge)

Oil Pressure Gauge

An ***oil pressure gauge*** measures the pressure developed by the engine oil pump, automatic transmission pump, brake fluid, or power steering pump. The oil pressure test gauge is used to determine whether the pump is providing the specified pressure. The pressure gauge is attached to the unit to be tested through a high-pressure hose that threads into a pressure port, **Figure 5-56.**

The pressure gauge scale is calibrated according to the output of the system being tested. Engine oil pressure seldom exceeds 80 PSI (552 kPa), while automatic transmission pressures are in the hundreds of PSI (or kPa), and brake and power steering pressures may reach over 2000 PSI (13 789 kPa).

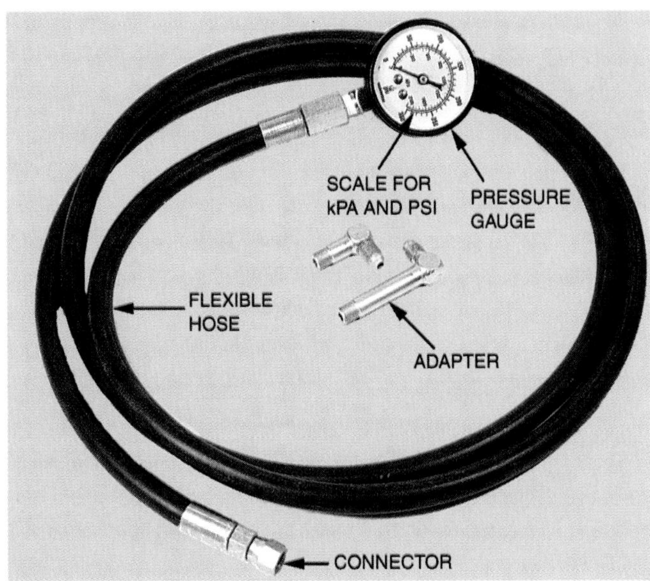

Figure 5-56. One type of engine oil pressure gauge with a flexible connection hose and adapters. (Kent-Moore Tools)

Fuel Pressure Tester

The ***fuel pressure gauge*** measures the pressure developed by the fuel pump. It is used to determine whether the pump is providing enough pressure and flow to keep the engine supplied with fuel. It can also be used to detect clogged fuel filters or other fuel system problems. Pressure testers can be used on both carbureted and fuel-injected engines. A typical fuel pressure tester for a fuel-injected engine is shown in **Figure 5-57.**

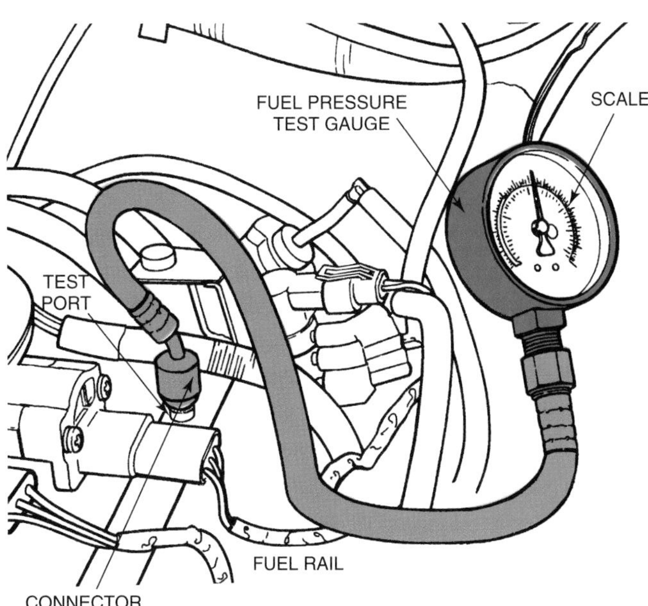

Figure 5-57. A fuel pressure gauge and connection to one type of electronic fuel injection fuel rail. Use caution when working with fuel. (Dodge)

Other Diagnostic Equipment

There are several other pieces of diagnostic equipment commonly used in the automotive repair shop.

A ***scan tool*** is a hand-held electronic device that can communicate with a vehicle's computer, **Figure 5-58.** The scan tool is connected to the computer through a special electrical connector known as the ***data link connector.***

Early scan tools could only be used to retrieve diagnostic trouble codes from the computer. Modern scan tools can also be used to check input sensor signals and verify the operation of the output devices. Some scan tools can be used to reprogram the computer on late-model vehicles.

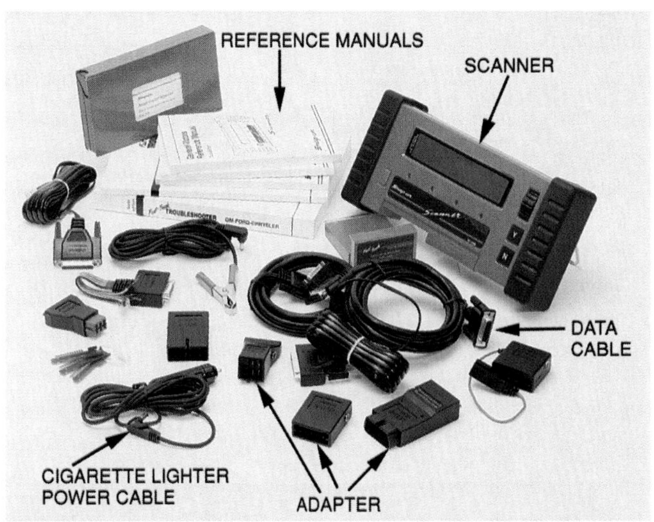

Figure 5-58. A hand-held diagnostic scan tool with related adapters and equipment. Such a tool is useful for diagnosing computer-controlled vehicles. (Snap-on Tools)

Although they are expensive, scan tools are widely used because they save diagnostic time. They also reduce the chance of misdiagnosis. On most vehicles made after 1996, the proper scan tool *must* be used to retrieve trouble codes.

The ***exhaust gas analyzer*** is used to measure the amount of various chemicals in the vehicle's exhaust. To use an exhaust gas analyzer, a hose is inserted into the vehicle's tailpipe. A small blower motor draws an exhaust gas sample from the tailpipe as the vehicle operates.

Early exhaust gas analyzers were able to measure carbon monoxide (CO) and unburned hydrocarbons (HC) only. In addition to CO and HC, the newest exhaust gas analyzers can measure oxides of nitrogen (NO_x), free oxygen (O_2), and carbon dioxide (CO_2).

Another commonly used tester is the ***oscilloscope.*** The most obvious component of the oscilloscope is the display screen, which resembles a television screen. The oscilloscope uses this screen to display a picture of the electrical activity in the vehicle's ignition system and other electrical systems. The picture is usually called a ***scope pattern.***

A ***vehicle analyzer*** combines the functions of many of the smaller testers previously discussed, including a multimeter, an oscilloscope, an exhaust gas analyzer, and a scan tool. In many cases, these testers are mounted in a large, roll-around cabinet. Using an analyzer eliminates the need for separate pieces of test equipment. See **Figure 5-59.**

Figure 5-59. This engine analyzer incorporates a digital oscilloscope and can act as a database for technical information. This analyzer can interface with hand-held scan tools to download information gathered during a test drive. (Ford Motor Co.)

Summary

Good quality tools are necessary for working on today's vehicles. Quality tools are also the sign of technicians who takes pride in their work. You get what you pay for, so only purchase quality tools. Take good care of your tools by keeping them clean and dry and by storing them in a tool cabinet.

A good set of quality tools not only makes the job easier, but it will also make the job go faster. Hammers, screwdrivers, chisels, punches, wrenches, pliers, and a socket set are a few of the basic tools common to every technician's tool collection. Other tools can be added as your experience and the need for them arises.

Measuring tools are important in making precision measurements and diagnosing problems. calipers, dial indicators, and micrometers are used by technicians to measure engine and other parts to diagnose problems or to ensure proper clearances during assembly. Scan tools, multimeters, test lights, and pressure gauges help the technician to properly diagnose the problem the first time.

Know These Terms

Hammer
Chisel
Punch
File
Drill
Drill bit
Reamer
Impact wrench
Hacksaw
Vise
Tap
Die
Pliers
Screwdriver
Wrenches
Torque wrench
Socket
Extension bar
Ratchet
Puller
Soldering iron
Soldering gun
Micrometer
Calipers
Dividers
Dial indicator
Dial caliper
Feeler gauge
Scan tool
Test light
Ohmmeter
Voltmeter
Ammeter
Multimeter
Tachometer
Compression gauge
Vacuum gauge
Oil pressure gauge
Fuel pressure gauge
Scan tool
Data link connector
Exhaust gas analyzer
Oscilloscope
Vehicle analyzer

Review Questions—Chapter 5

Do not write in this book. Write your answers on a separate sheet of paper.

1. All of the following statements about good quality tools are true, EXCEPT:
   (A) they are heavy and thick.
   (B) they are easy to use in tight quarters.
   (C) they are made of high-strength alloy steel.
   (D) they are guaranteed.
2. Any tool subjected to rust should be cleaned and _____.
3. Keep _____ tools by themselves.
4. You should arrange your tools in a tool box so that the tools that are the most often used are _____ .
   (A) in precise order according to size
   (B) at the top
   (C) easy to get to
   (D) securely locked up
5. How are the following punches shaped?
   starting punch _____
   drift punch _____
   aligning punch _____
6. A file with two sets of cutting edges at angles to each other is referred to as a _____ file.

7. A hammer handles should be _____ and _____ .
8. Technician A says that carbon steel drill bits will last longer than high-speed steel drill bits. Technician B says that carbon steel drill bits will require more frequent sharpening. Who is right?
   (A) A only.
   (B) B only.
   (C) Both A & B.
   (D) Neither A nor B.
9. A 24-tooth hacksaw blade is best for cutting all of the following, EXCEPT:
   (A) heavy sheet metal.
   (B) brass.
   (C) thinwall tubing.
   (D) copper.
10. Technician A says that reamers are used for drilling large holes. Technician B says that a husky vise is useful as an anvil. Who is right?
   (A) A only.
   (B) B only.
   (C) Both A & B.
   (D) Neither A nor B.
11. If you wish to cut internal threads in a hole, you would use a _____ .
12. If you wish to cut threads on a bolt, you would use a _____ .
13. Pliers can be used for all of the following, EXCEPT:
   (A) tightening tubing fittings.
   (B) bending cotter pins.
   (C) cutting wire.
   (D) crimping connections.
14. Briefly describe the following items:
   Box-end wrench.
   Open-end wrench.
   Adjustable wrench.
15. What advantage does a 6-point opening have over a 12-point opening?
16. What advantage does a 12-point opening have over a 6-point opening?
17. Technician A says that using an impact wrench will reduce the amount of time that a job takes. Technician B says that impact sockets should always be used with an impact wrench. Who is right?
   (A) A only.
   (B) B only.
   (C) Both A & B.
   (D) Neither A nor B.
18. Technician A says that a box-end wrench is preferred to an open-end wrench in most cases. Technician B says that combination box-end and open-end wrenches have a different size opening on each end. Who is right?
   (A) A only.
   (B) B only.
   (C) Both A & B.
   (D) Neither A nor B.
19. To remove a spark plug, the technician would need a _____ socket.
20. Standard size screwdrivers can be used for which of the following purposes?
   (A) Removing and installing screws.
   (B) Prying of oil pans.
   (C) Hammering on tight parts.
   (D) All of the above.
21. Technician A says that an adjustable wrench can be used to turn a crankshaft journal. Technician B says that a torque wrench should be used to tighten head bolts. Who is right?
   (A) A only.
   (B) B only.
   (C) Both A & B.
   (D) Neither A nor B.
22. Gears are usually removed by using a _____ .
23. Technician A says that an ohmmeter measures resistance to the flow of electricity. Technician B says that a multimeter can measure many kinds of electrical properties. Who is right?
   (A) A only.
   (B) B only.
   (C) Both A & B.
   (D) Neither A nor B.
24. Technician A says that a vacuum gauge measures negative air pressure. Technician B says that engine compression is measured with a vacuum gauge. Who is right?
   (A) A only.
   (B) B only.
   (C) Both A & B.
   (D) Neither A nor B.
25. Name three vehicle systems where an oil pressure tester could be used
26. Technician A says a fuel pessure gauge can be used to check for clogged fuel filters. Technician B says a fuel pressure gauge should not be used when testing carburated fuel systems. Who is right?
   (A) A only.
   (B) B only.
   (C) Both A & B.
   (D) Neither A nor B.

Environmental test chambers are used to test the entire vehicle as well as engines for proper operation in all kinds of weather conditions. (Chrysler)

27. A scan tool is connected to a vehicle's computer system through the _____ _____ connector.
28. A scan tool can be used for each of the following, EXCEPT:
    (A) retrieving trouble codes.
    (B) checking input sensor signals.
    (C) reprogramming the computer.
    (D) measuring exhaust emissions.
29. Technician A says exhaust gas analyzers can measure the amount of carbon monoxide in a vehicle's exhaust gases. Technician B says some exhaust gas analyzers can measure the amount of carbon dioxide in a vehicle's exhaust gases. Who is right?
    (A) A only.
    (B) B only.
    (C) Both A & B.
    (D) Neither A nor B.
30. A(n) _____ can be used to display a picture of the electrical activity in a vehicle's ignition system.

6

Engine Tests and Measurements

After studying this chapter, you will be able to:

- Recall engine measurement formulas.
- Solve basic engine measurement problems.
- Develop an understanding of engine ratings.

Everyone has heard the word horsepower, but very few people, including many technicians, understand exactly what it means and how it is computed. The science of engine testing and measurement is very complicated and, in its true sense, lies in the realm of engineering. Nevertheless, it is important that the student of automotive technology be familiar with basic formulas, terms, and how they are computed.

What Is Horsepower?

Many years ago, people recognized the need for a yardstick with which to measure the work-producing ability of an engine. Since horses had been used to do the work for so long, it was natural to compare the power developed by the early engine to that produced by a horse.

The work ability of an average draft horse measured in foot-pounds had long been the standard unit of work. ***Work*** can be defined as a force applied to a body that causes the body to move. The amount of work done can be computed by multiplying the distance the body is moved by the weight of the body. This answer would be in ***foot-pounds***.

It was found, for example, that the average draft horse could lift 100 pounds 330 feet in one minute. If one pound is lifted one foot in one minute, one foot-pound of work would be done. The horse lifted 100 pounds (lb.) 330 feet (ft.) in one minute. The amount of work performed can be found by using the formula:

Rate of work = Distance moved x Force weight

=33,000ft.-lb.=330ft.x100lb.

We find the horse performed work at the rate of 33,000 foot-pounds per minute. This became the standard measure for one ***horsepower***, **Figure 6-1.** The ***rate*** at which work is performed is called ***power***. Horsepower is sometimes called brake horsepower. This term stems from an early method of measuring horsepower.

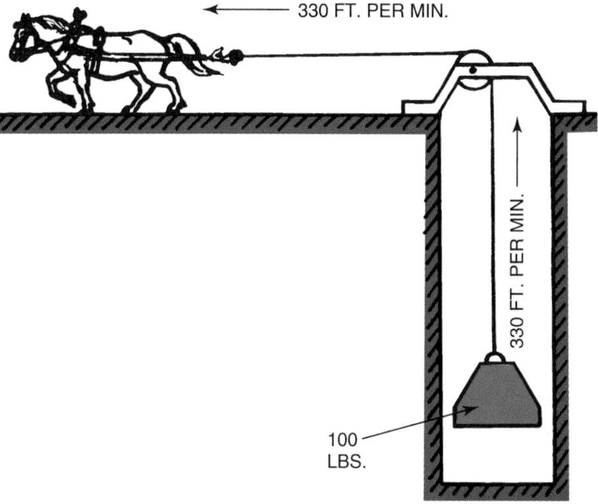

Figure 6-1. Horsepower is a measure of the rate at which work is performed. The ability to perform work at a rate of 33,000 ft.-lb. per min. is equal to 1 horsepower.

Horsepower Formula

To find engine horsepower, the total rate of work accomplished (in foot-pounds) is divided by 33,000. If a machine lifts 100 lb. 660 ft. per minute, its total rate of work would be 66,000 ft.-lb. Dividing this by 33,000 ft.-lb. (1 horsepower), you will find that the machine is rated as 2 horsepower (hp). Your formula would be:

$$\frac{\text{Rate of work in ft.-lbs.}}{233{,}000(1\text{hp})} = \frac{\text{Distance moved x force}}{33{,}000} = \frac{\text{D x W}}{33{,}000} = \text{hp}$$

This formula may also be used to determine the horsepower required to perform a specific task. Let us assume that you want to lift buckets of earth from an excavation site. The loaded bucket weight is 5000 lb., and you want to raise it 15 ft. per minute. Using the formula:

$$\text{hp} = \frac{\text{D x W}}{33{,}000} = \frac{15 \text{ x } 5000}{33{,}000} = \frac{75{,}000}{33{,}000} = 2.27 \text{ hp}$$

The minimum amount of horsepower (disregarding friction) required would be 2.27.

Potential Horsepower

If a certain engine burns a specific fuel at a certain rate, and if all the heat of the burning fuel could be converted into useful power, you would have the ideal engine. Unfortunately, engines have not reached this level of efficiency. In fact, even the best engines are relatively inefficient in that the usable power falls far short of the potential.

Engine efficiency can be broken down into two groups: ***thermal efficiency*** and ***mechanical efficiency***. Any comparison between different engines on an efficiency basis must be conducted under identical situations. Weather, altitude, temperature, engine speed, and humidity all affect the test results.

Thermal Efficiency

Engine thermal efficiency (heat efficiency) is based on how much of the energy (ability to do work) of the burning fuel is converted into useful horsepower.

The heat generated by the burning fuel drives the piston down during the power stroke. Much of this heat is lost to the cooling system, some to the lubrication system, and a great deal to the exhaust system. The thermal efficiency of an average engine is around 25%. This means that the engine is losing approximately 75% of the fuel's heat energy, **Figure 6-2.**

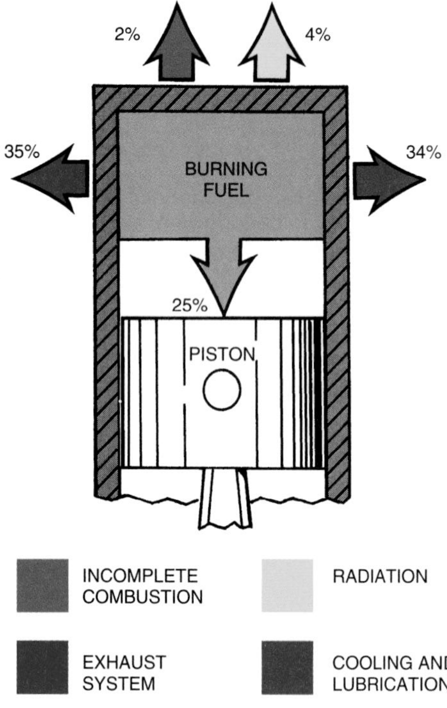

Figure 6-2. Average engine thermal efficiency is around 25%. Note the losses due to incomplete combustion, exhaust, radiation, cooling, and lubrication.

The generally accepted formula for computing brake thermal efficiency is:

$$\text{Brake Thermal Efficiency} = \frac{\text{Brake Horsepower (bhp)} \times 33{,}000}{778 \times \text{Fuel Heat Value} \times \text{Weight of Fuel Burned per Minute}}$$

The fuel heat value is rated in Btus (British thermal units) per pound (the number 778 is a constant).

Mechanical Efficiency

Mechanical efficiency is based on the relationship between the power developed within the engine and actual brake horsepower delivered at the crankshaft. Fortunately, mechanical efficiency is better than thermal efficiency—generally aroung 90%. This means that 10% is lost to friction within the engine. Engine friction increases with speed.

The formula for mechanical efficiency is:

$$\text{Mechanical Efficiency} = \frac{\text{Brake Horsepower}}{\text{Indicated Horsepower}} = \frac{\text{bhp}}{\text{ihp}}$$

Practical Efficiency

When dealing with vehicle engines, you are primarily interested in horsepower delivered at the drive wheels.

Every pound of gasoline that enters the engine has the ability to do a certain amount of work. The capacity of any material to do work is called its energy. However, there are many factors that rob energy before it gets to the drive wheels. As little as 15% of the potential energy in the fuel is actually delivered to the drive wheels, **Figure 6-3** illustrates some typical energy losses.

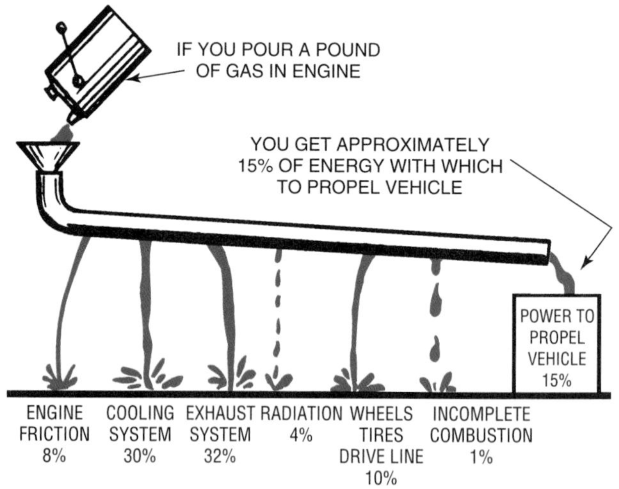

Figure 6-3. Practical engine efficiency. Energy is lost between the engine and the drive wheels. The percentages shown are approximate and depend on engine design and the conditions under which engine is operated.

Indicated Horsepower

Indicated horsepower (ihp) is a measure of the power developed by the burning fuel in the cylinders. In order to compute ihp, it is necessary to find the pressures within the cylinder during the intake, compression, power, and exhaust strokes. A measuring device gives a constant reading of the pressures. These pressures are placed on an indicator graph.

The next step is to establish a mean effective pressure (mep). This is done by taking the average pressure during the power stroke and subtracting the average pressure during the other three strokes. Once the mep is determined, the following formula is used to determine the indicated horsepower:

$$\text{Indicated Horsepower} = \frac{PLANK}{33{,}000}$$

P - mep in lb. per sq. in.

L - length of the stroke in feet

A - cylinder area in sq. in.

N - power strokes per min. $\frac{\text{rpm}}{2}$

K - number of cylinders

Indicated horsepower ratings do not reflect losses from friction.

Brake Horsepower

Brake horsepower (bhp) is a measure of actual usable horsepower delivered at the engine crankshaft. Brake horsepower is not constant: it increases with speed. An early method of measuring crankshaft horsepower made use of a device called a Prony brake, **Figure 6-4.**

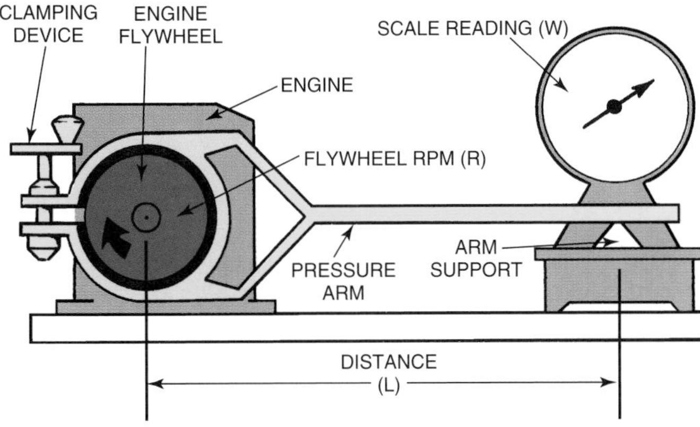

Figure 6-4. The Prony brake is used to measure brake horsepower (bhp). A clamping device grasps the spinning flywheel. The clamp is tightened until a specific rpm is reached with the throttle wide open. At this point, a weight reading on the scale is taken. By knowing the scale reading (W), the distance from the center of the crank to the arm support (L), and the rpm of the flywheel (R), it is possible to compute bhp.

The formula for determining brake horsepower on a Prony Brake is:

$$\text{bph} = \frac{2\pi \times R \times L \times W}{33{,}000}$$

$$\text{bph} = \frac{R \times L \times W}{5252}$$

R = Engine rpm.

L = Length from the center of the drive shaft to the point where the beam presses on the scale.

W = Weight as registered on the scale.

The modern method of testing and determining brake horsepower is by using an ***engine dynamometer*** or a ***chassis dynamometer***.

In one type of engine dynamometer setup, the engine drives a large generator (dynamo). By placing an electrical load in the circuit, bhp at various rpm can be determined. See **Figure 6-5.** An engine dynamometer that attaches directly to the engine is shown in **Figure 6-6.**

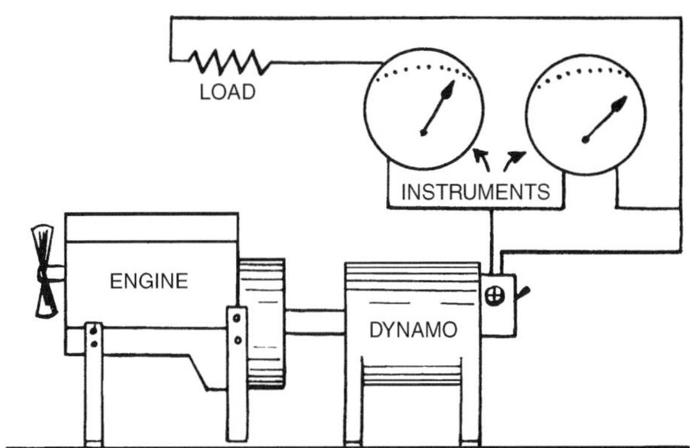

Figure 6-5. This schematic demonstrates the general idea of an engine dynamometer. The generator dynamo can be loaded until the engine rpm is at a certain point with the throttle wide open. By using instrument readings, horsepower can be calculated.

Figure 6-6. This engine dynamometer is located inside an altitude simulation chamber. This particular setup allows engineers to measure engine horsepower at different altitudes. The engine is operated at various atmospheric conditions and throttle positions, including wide open throttle. (Chrysler)

The chassis dynamometer is used to measure bhp at the drive wheels. The vehicle's drive wheels are placed on rollers, and the engine is engaged. As the wheels turn the rollers, the rollers are loaded (made increasingly difficult to turn). By computing the amount of loading the wheels can handle, it is possible to determine bhp at the drive wheels. This is practical efficiency that is meaningful to the average motorist. Chassis dynamometers are installed in some larger shops for use in engine tuning and testing, **Figure 6-7.**

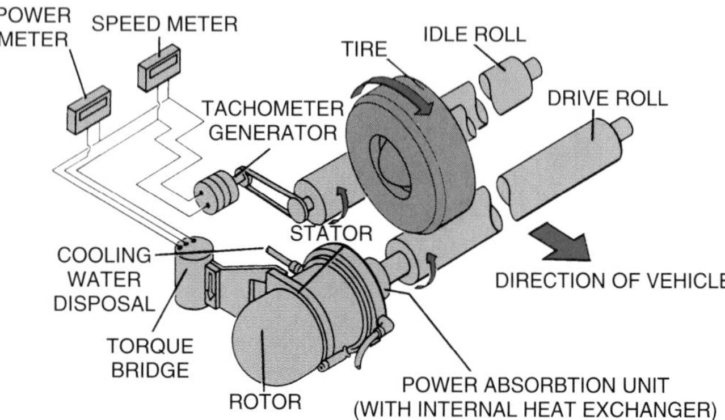

Figure 6-7. Schematic showing the operation of a typical chassis dynamometer. This unit measures the horsepower delivered to the drive wheels. (Clayton)

Gross and Net Horsepower Ratings

Under ***SAE*** (Society of Automotive Engineers) specifications, a ***gross horsepower*** rating is the maximum horsepower developed by an engine equipped with only the basic accessories needed for its operation. This would include the built-in emission controls, oil pump, coolant pump, and fuel pump. The fan, alternator, air cleaner, catalytic converter, and muffler are removed. Other engine-driven accessories, such as the power steering pump and air conditioner compressor, are also left off.

The ***net horsepower*** rating is the maximum horsepower developed by the engine when equipped with all the accessories it would have when installed in the vehicle, such as the fan, alternator, catalytic converter, muffler, air pump, air cleaner, fuel pump, and coolant pump. The net horsepower rating is a more informative figure for the consumer because it shows the horsepower that will actually be produced in normal usage.

The specified test conditions (29.38″ Hg. or 99.01 kPa barometric pressure, 85°F (29.46°C) ambient air temperature, 0.38″ Hg. or 1.28 kPa water vapor pressure or humidity) are the same for both gross and net horsepower ratings.

Some other nations use a ***din*** horsepower rating. This is a net rating measured under slightly different conditions and specifications. Therefore, it will vary from ratings used in the United States.

Frictional Horsepower

Frictional horsepower (fhp) is a measurement of the amount of horsepower lost to engine friction. It is computed by this formula:

$$fhp = ihp – bhp$$

As mentioned, frictional horsepower increases with engine speed. See **Figure 6-8.**

Torque

Engine torque (twisting action) is the ability of the crankshaft to impart a twisting or turning force. It is measured in pound-feet (Newton-meters). If a pipe wrench is placed on a pipe and a pressure of one pound is exerted one foot from the center of the pipe, the torque (torque = force x distance) applied to the pipe would be one pound-foot. See **Figure 6-9.** Actually, when measuring torque, the reading is given in lb.-ft. When measuring *work,* the reading is given in ft.-lb.

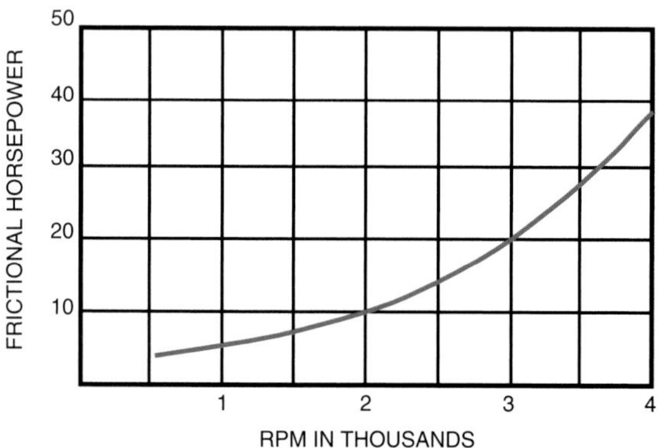

Figure 6-8. Graph showing frictional horsepower. Notice how an increase in engine rpm causes more friction.

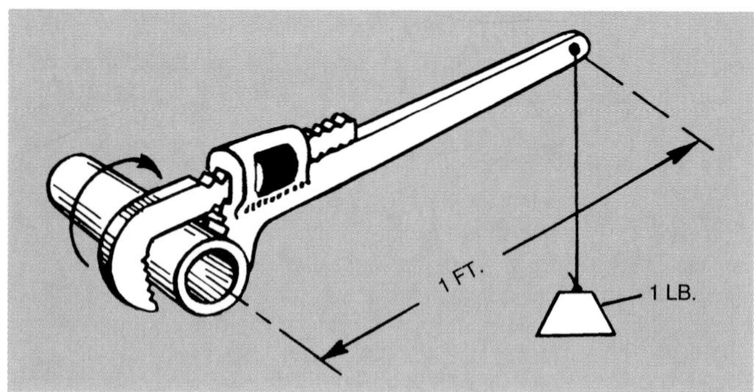

Figure 6-9. The torque, or twist, applied to this pipe would be one pound-foot. This is torque in excess of wrench weight.

Torque and Brake Horsepower Are Not the Same

Engine brake horsepower increases with engine speed. Engine torque increases with engine speed up to the point where the engine is drawing in the maximum amount of fuel mixture, after all factors are considered. Torque is greatest at this point, and any additional increase in rpm will cause torque to diminish. **Figure 6-10** compares torque and brake horsepower.

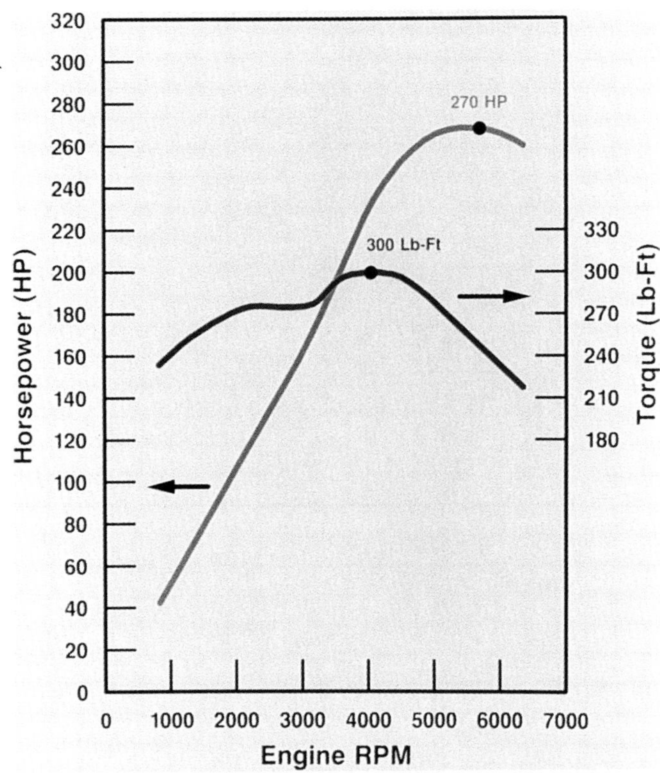

Figure 6-10. This graph shows brake horsepower and torque for one specific engine. Notice how horsepower increases almost to maximum speed, while torque drops off at a much lower rpm. The relationship between torque and bhp will vary depending on engine design. (Cadillac)

Volumetric Efficiency

Volumetric efficiency (sometimes referred to as breathing ability) is the measure of an engine's ability to draw the fuel mixture into the cylinders. It is determined by the ratio between what is actually drawn in and what could be drawn in if all cylinders were completely filled.

As engine speed increases beyond a certain point, the piston speed becomes so fast that the intake stroke is of such short duration that less and less fuel mixture is drawn in. As torque is greatest when the cylinders receive the largest amount of fuel mixture, torque will drop off as volumetric efficiency decreases.

Many factors influence volumetric efficiency, including engine speed, temperature, throttle position, intake system design, valve size and position, exhaust system configuration, and atmospheric pressure.

Volumetric efficiency can be improved by adding a supercharger; a straighter, smoother, and ram-length intake manifold (sometimes called a tuned intake manifold); larger intake valves; a more efficient exhaust system, and other modifications to the induction and exhaust systems.

A formula used to determine volumetric efficiency is:

$$\text{Volumetric Efficiency} = \frac{\text{Total Volume of the Charge}}{\text{Total Cylinder Volume (displacement)}}$$

All tests between engines must be under identical conditions. See **Figure 6-11** and **Figure 6-12**.

Engine Size (Displacement)

Engine size (not physical dimensions) is related to piston displacement. Piston displacement refers to the total number of cubic inches of space (liters) in a cylinder when the piston moves from the top of its stroke (TDC) to the bottom (BDC), **Figure 6-13.**

All related factors (such as compression, head design, and fuel system) being equal, the greater the engine displacement, the more power produced. To calculate displacement, find the area of the cylinder (0.7854 x D^2). Multiply the area by the total piston travel (stroke) from TDC to BDC. This answer is then multiplied by the number of cylinders to calculate the total engine displacement in cubic inches. The formula for total displacement is:

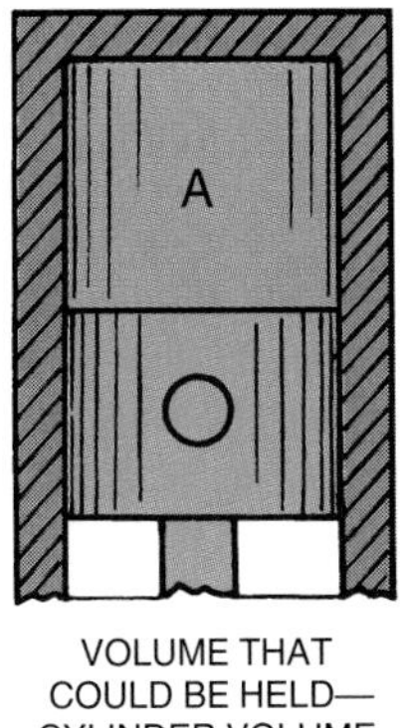

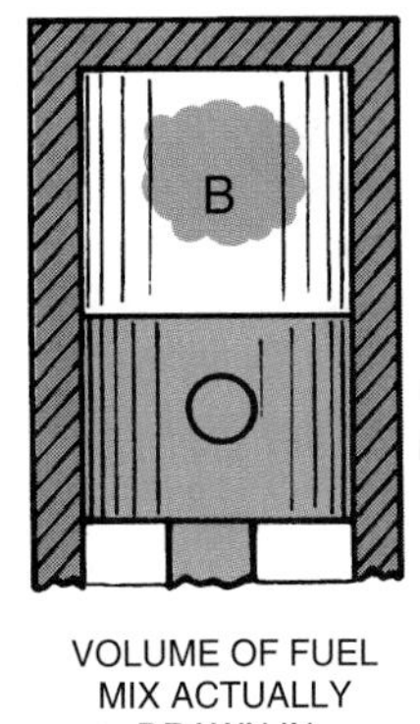

Figure 6-11. Comparison of volumetric efficiency. A—This shows the volume that the cylinder can actually hold if it were completely filled. B—This shows the volume of air-fuel mixture actually drawn in by vacuum.

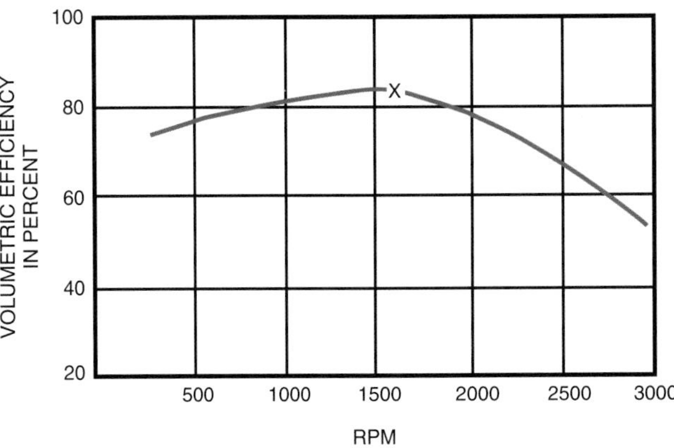

Figure 6-12. This graph shows volumetric efficiency at various speeds. The graph is only approximate as volumetric efficiency will vary from engine to engine. The X marks the point of highest efficiency and highest torque.

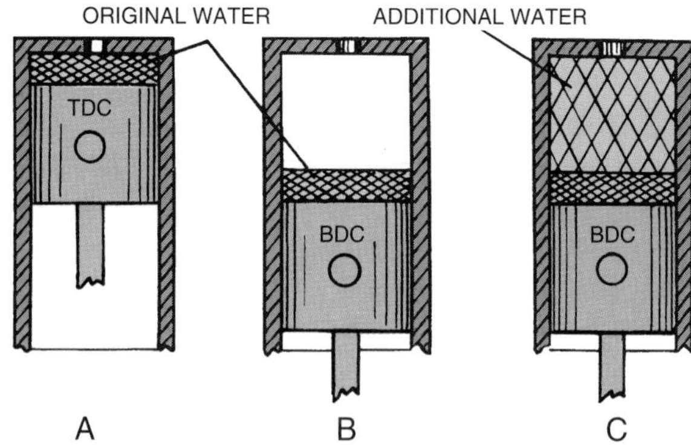

Figure 6-13. Measurement of piston displacement. If you were to fill the cylinder with water when the piston is at TDC (A), move the piston to BDC (B), and measure the additional water required to fill the cylinder (C), you would be able to determine the cylinder displacement in cubic inches. When the piston is at TDC, the space from the top of the piston to the cylinder head is called clearance volume. This space is occupied by the water in A.

Total Engine Displacement $= 0.7854 \times D^2 \times \text{Travel} \times \text{Number of Cylinders}$

To find the displacement of a V-8 engine having a cylinder diameter (bore) of 4.000 inches and a piston travel (stroke) of 3.000 inches, use the formula as follows.

$0.7854 \times D^2$ (4^2=16 in^2) x Travel (3″) x Number of Cylinders (8) = 301.5936 cu. in,

This is close enough to 302 cu. in. to tell you that you have a 302, or 5 liter, engine.

This formula can also be used to find out how many cubic inches an engine gained when it was bored out. If a 173 cubic inch (2.8 liter) V-6 was bored .030 over, add .030 to the original bore (3.503 inches) to get a new bore of 3.533. The travel is 2.992, so you can use the formula as described earlier:

$0.7854 \times D^2$ (3.503^2 = 12.48 square inches) x Travel (2.992 inches) x Number of Cylinders (6) = 175.99 cu. in,

This makes the bored 173 cubic inch (2.8 liter) engine a 176 cubic inch (2.9 liter) engine.

Cylinder Bore and Piston Stroke

As you noticed in the previous section, displacement hinges on the cylinder ***bore*** (diameter) and the piston ***stroke***. The bore is the diameter of the cylinder. The stroke is the distance the piston moves from TDC to BDC, **Figure 6-14**.

When bore and stroke are equal, the engine is referred to as being ***square***. Modern engines use a stroke that is shorter than bore diameter. This is referred to as an ***over square*** engine. By reducing the stroke, piston speed is decreased. This prolongs the life of the cylinders, pistons, and rings.

For example, say that you want to compute the distance a piston ring slides up and down during one mile of driving.

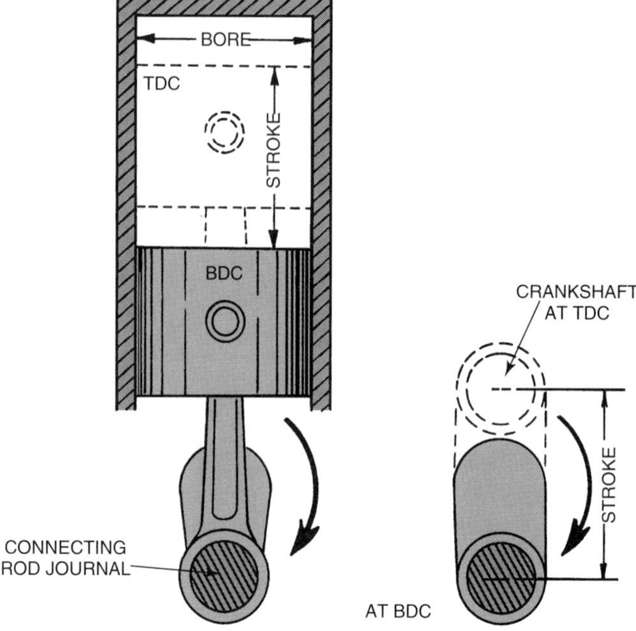

Figure 6-14. Engine bore (cylinder diameter) and stroke (distance the piston moves in traveling from top dead center to bottom dead center). The connecting rod journal offset, or throw, determines the length of the stroke.

Your hypothetical (make believe) car will have a tire diameter of 30′, the engine drives 1 to 1 through the transmission (standard transmission with no overdrive), and the differential (rear end) reduces driveline speed by a ratio of 4 to 1 (driveline turns four times for each revolution of the wheels).

The circumference of the tire (6.283 x radius) would be 94.245″. Converting this circumference to feet, you would have 7.85′. If the tire rolls 7.85′ per revolution, it will make 672.61 revolutions per mile.

The engine rpm would be four times the wheel rpm. This would give 2690.44 engine rpm per mile. The rings slide 3″ during each piston stroke, or 6″ per engine revolution. With 2690.44 rpm per mile, the rings would slide 1345.22′ for every mile traveled.

If the engine had a 6″ stroke, the rings would slide twice as far. This shows that the short stroke decreases wear. Not only is the sliding distance decreased, but the speed at which the rings slide is greatly reduced.

Compression Ratio

The ***compression ratio*** is the relationship between the cylinder volume (clearance volume) when the piston is on TDC and the cylinder volume when the piston is on BDC, **Figure 6-15**.

Compression ratios range from 5:1 to slightly over 11:1 for gasoline engines (8.5:1 is about average). Diesel engine compression ratios range from 17.5:1 to 22.5:1. To a certain degree, the higher the compression ratio, the more power the engine will develop. Such being the case, you may wonder why the compression ratio is not increased a great deal on all engines.

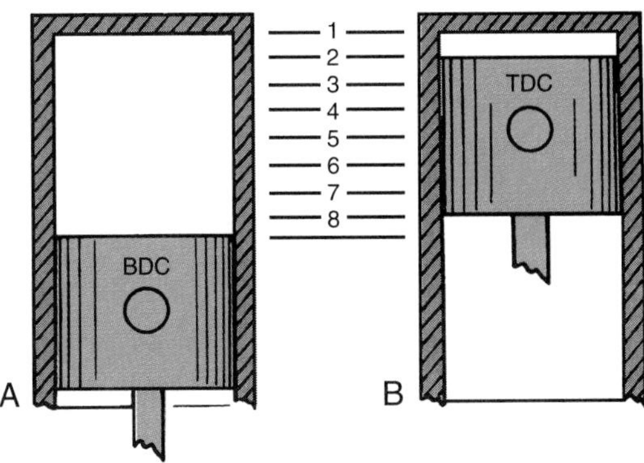

Figure 6-15. To determine engine compression ratio, the volume of the cylinder at both BDC and TDC is measured. A—When the piston is at BDC, the cylinder volume is 8. B—When the piston is at TDC, the cylinder volume is 1. therefore, the compression ration is 8:1.

One reason is that modern fuels will not allow increased compression beyond a certain point. When a fuel mixture is compressed, heat energy is developed. The temperature at which the air-fuel mixture will ignite itself limits the amount it can be satisfactorily compressed. Another problem is that when the compression ratio of an engine is too high, detonation, or knocking, will occur in the engine. Additionally, effective emission control has required that compression ratios be lowered from 10 or 11:1 down to about 8:1.

Summary

Horsepower is a measure of the rate at which work is performed. One horsepower is the ability to perform work at the rate of 33,000 foot-pounds per minute.

Engines are relatively inefficient. Efficiency is based on thermal and mechanical efficiency. Most engines are about 25% efficient on a thermal basis and about 90% efficient on a mechanical basis. On a practical efficiency basis, only about 15% of the total potential energy of the fuel is used to drive the wheels. Energy is lost to the cooling, lubrication, and exhaust systems, as well as to engine, driveline, and wheel friction.

Indicated horsepower is a measure of the power developed by the burning fuel within the cylinders. Ihp is not the power delivered by the crankshaft. Crankshaft horsepower is called brake horsepower and is measured with a dynamometer. Frictional horsepower is a measure of the horsepower lost to engine friction. Net brake horsepower ratings are taken with all normal accessories on the engine. Gross brake horsepower figures are calculated with accessories such as the fan, air cleaner, and alternator removed.

Engine brake horsepower and torque are not the same. Brake horsepower increases with engine speed until maximum rpm is reached. Torque increases to the point where volumetric efficiency is greatest, and any additional speed will cause it to diminish. Torque is a measure of the engine's turning force at the crankshaft.

Volumetric efficiency refers to how well the engine is able to draw in the air-fuel mixture. It is influenced by engine design, engine speed, and a number of other factors.

Engine size is usually given in cubic inches or liters. The number of cubic inches or liters of cylinder space is the total displacement of the engine cylinders. Cylinder bore and stroke are determining factors in cubic inch or liter computations. Compression ratios are important. High compression ratios improve power and economy. However, current ratios have been lowered somewhat to reduce emissions.

Know These Terms

Work	Gross horsepower
Foot-pounds	Net horsepower
Horsepower	Din
Rate	Frictional horsepower
Power	Engine torque
Thermal efficiency	Volumetric efficiency
Mechanical efficiency	Bore
Indicated horsepower	Stroke
Brake horsepower	Square
Engine dynamometer	Over square
Chassis dynamometer	Compression ratio
SAE	

Review Questions—Chapter 6

Do not write in this book. Write your answers on a separate sheet of paper.

1. What is horsepower?
2. Give the formula for computing horsepower.
3. Engine efficiency can be classified as _____ .
   (A) thermal
   (B) mechanical
   (C) volumetric
   (D) Both A & B.
4. What is indicated horsepower?
5. How thermally efficient is the average internal combustion engine?
   (A) 10%
   (B) 25%
   (C) 90%
   (D) 100%
6. How much mechanical efficiency does the average internal combustion engine lose to internal friction?
   (A) 10%
   (B) 25%
   (C) 90%
   (D) 100%
7. The __________ is now used to determine brake horsepower.
   (A) Prony brake
   (B) engine dynamometer
   (C) chassis dynamometer
   (D) Both A & B.

8. If the potential energy in a pound of fuel is 100%, how much of this energy lost before reaching the drive wheels?
   (A) 10%
   (B) 35%
   (C) 85%
   (D) 95%
9. Technician A says that horsepower increases with speed. Technician B says that torque decreases beyond a certain speed. Who is right?
   (A) A only.
   (B) B only.
   (C) Both A & B.
   (D) Neither A nor B.
10. The din horsepower rating is used in other countries. It is a _____ .
   (A) net rating with metric outputs
   (B) net rating determined under different operating conditions
   (C) gross rating with metric outputs
   (D) gross rating determined different operating conditions
11. The difference between indicated horsepower and brake horsepower is classified as _____ horsepower.
12. Torque is a _____ motion.
13. Technician A says that horsepower and torque are the same. Technician B says that volumetric efficiency and cubic inch displacement are the same. Who is right?
   (A) A only.
   (B) B only.
   (C) Both A & B.
   (D) Neither A nor B.
14. Torque will drop off with a decrease in _____ .
15. Piston area can be found by squaring the _____ and multiplying this value by _____ .
16. If a cylinder is bored out, the amount of oversize is added to the original _____ to determine the new displacement.
   (A) bore
   (B) stroke
   (C) compression ratio
   (D) number of cylinders
17. Compression ratio is figured by comparing the volume in the cylinder at _____ with the volume at _____ .
18. What is meant by the cubic inch or cubic liter rating of an engine?
19. What is energy?
20. Explain the difference between *gross* brake horsepower and *net* brake horsepower.

7

Electrical System Fundamentals

After studying this chapter, you will be able to:

- Explain the principles of electricity and electronics.
- Name the values used to measure electricity.
- List the basic types of electrical circuits.
- Explain the effects of magnetism.
- Explain how diodes work.
- Explain how transistors work.
- Explain the construction of an integrated circuit.

The field of electricity and electronics covers a wide and complicated range. We know much about electricity, however, much more remains to be discovered. A thorough coverage of electricity and electronics is not within the scope of this text. However, a basic understanding of electrical and electronic fundamentals and their application to the vehicle is a must for all students of automotive technology.

Basic Electricity

We have learned to control electricity, predict its actions, and harness it in many ways. No person has literally seen electricity flow through a wire or a device. For many years no one was even sure of the direction it moved or what, if anything, actually did move. There are two theories of how current flows through a circuit.

The Electron Theory

The accepted theory regarding electricity is termed the ***electron theory***. This theory has explained many of the strange behavior patterns as well as opening the way to many new discoveries, not only in electricity, but in other fields as well.

The electron theory is based on the assumption that the operation of all electrical equipment is produced by a flow of electrons from one area to another. It also states that the flow of electrons is caused by having an excess number in one area, and a shortage in the other. The area with the surplus electrons is referred to as the ***negative*** and the area in short supply, the ***positive***.

Conventional Theory

The conventional theory says that electrons flow from the positive to the negative. Electrons leave the positive terminal and flow to the negative. It is the opposite of the electron theory. Some believe that this is how electricity actually works, however, the electron theory described earlier is more widely accepted.

Matter

Anything that takes up space and has mass is considered ***matter***. Everything on Earth is composed of matter. Matter can be a solid, a liquid, or a gas. Matter is made up of different combinations of atoms which are called ***molecules***. Some substances, such as water can exist as a solid (ice), liquid, or gas (steam). Other substances will remain in the same state under the normal range of Earth temperatures.

Atoms

There are over one hundred different types of known ***atoms***. Various combinations of these atoms will produce different kinds of matter. Some atoms exist in their pure state (uncombined with other kinds of atoms), and are called ***elements***. A molecule of water is made up of one atom of oxygen and two atoms of hydrogen. Both oxygen and hydrogen are gases. All matter is made up of different atoms, or the same type of atoms. **Figure 7-1** illustrates a progressive breakdown of matter.

Structure of the Atom

The atom itself is made up of small subatomic particles called ***electrons***, ***protons***, and ***neutrons***. The electron is a negatively (-) charged particle. The proton is a positively (+) charged particle. The neutron is a neutral particle with no electrical charge. Protons and neutrons form the center, or nucleus, of the atom. The electrons rotate rapidly around this nucleus. As unlike charges attract each other, the electrons are held in their orbital path. They will not draw in any closer, because the centrifugal force built up by their orbital flight counterbalances the mutual attraction force, **Figure 7-2**.

There can be many different atoms, each with a different number of protons, neutrons, and electrons. Every atom has the same number of electrons as it has protons. Protons and neutrons are alike in any type of atom. The number of neutrons, protons, and electrons, and their arrangement produce different types of matter. A simple atom is that of hydrogen. The hydrogen atom consists of a single neutron and a single proton. A more complex atom is that of copper. This atom consists of 34 neutrons, 29 protons, and 29 electrons, **Figure 7-3**.

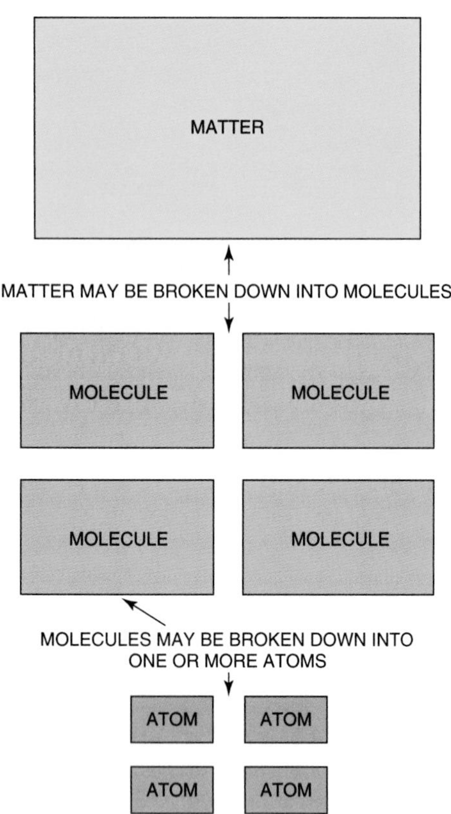

Figure 7-1. Breakdown of matter. This can be applied to any substance, solid, liquid, or gas.

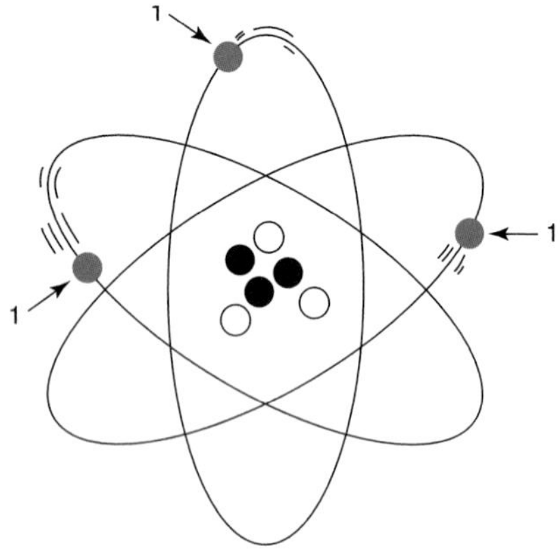

Figure 7-2. Structure of an atom. Electrons, 1, are whirling about the nucleus. Solid dots in the nucleus represent neutrons; white dots are photons.

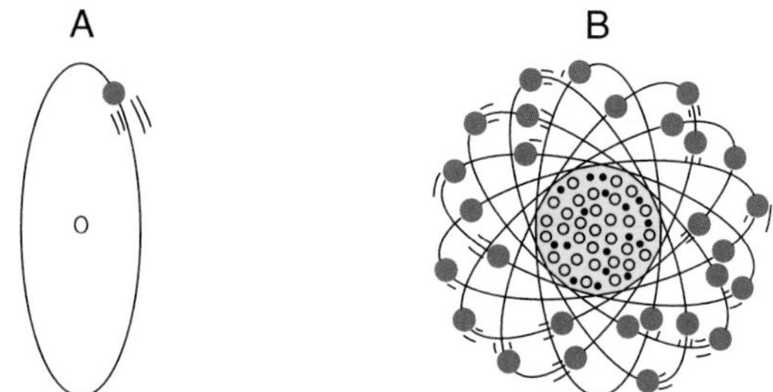

Figure 7-3. Two different atoms. A—A simple hydrogen atom consisting of one proton and one electron. B—The more complex copper atom has 34 neutrons, 29 protons, and 29 electrons.

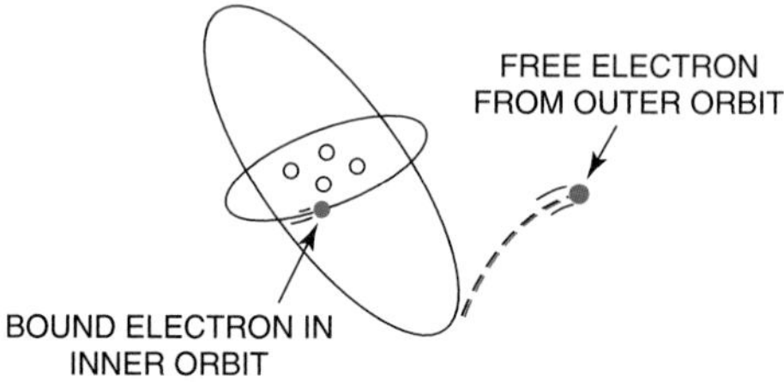

Figure 7-4. Free and bound electrons. A free electron has been knocked from its outer orbit.

Positive and Negative Charge Atoms

You will notice in **Figure 7-4** that the electrons move in different size orbits. The inner orbits contain what is referred to as ***bound electrons***. The outer orbits contain the ***free electrons***. The bound electrons are so named because it is difficult to force them from their orbits. The outer free electrons can be more readily moved from their orbits.

If some of the free electrons are moved out of their orbits, the atom will have more protons than electrons. This will give the atom a positive charge. If the electrons are returned until they equal the number of protons, the atom will become electrically neutral. If a surplus of electrons are present, the atom is negatively charged. A negatively charged body will repel electrons, while the positively charged body will attract electrons, **Figure 7-5**.

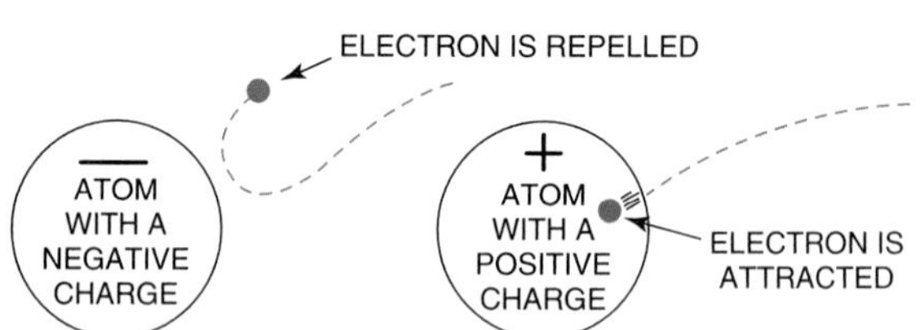

Figure 7-5. Negative and positive charge atoms. Atoms with a negative charge repel electrons. Positive charged atoms attract electrons.

Electrical Current

An electrical ***current*** is a movement of electrons through a ***conductor***, such as a wire. Electrical current may be likened to water flowing through a pipe. The amount of electrons passing any given point in the circuit in one second determines the amount of current flowing. The amount of current flow is measured in ***amperes***. Before current will flow, there must be a surplus of electrons at one end of the conductor and a scarcity at the other end.

Current Flow

Figure 7-6 illustrates a schematic of a simple circuit utilizing a battery as the source. Note that the electron flow is from negative to positive. The battery, through chemical action (which will be covered in Chapter 16), has built up a surplus of electrons at the negative post and a lack of electrons at the positive post. As soon as the copper wire conductor is connected to both posts, current will flow from the negative to the positive post.

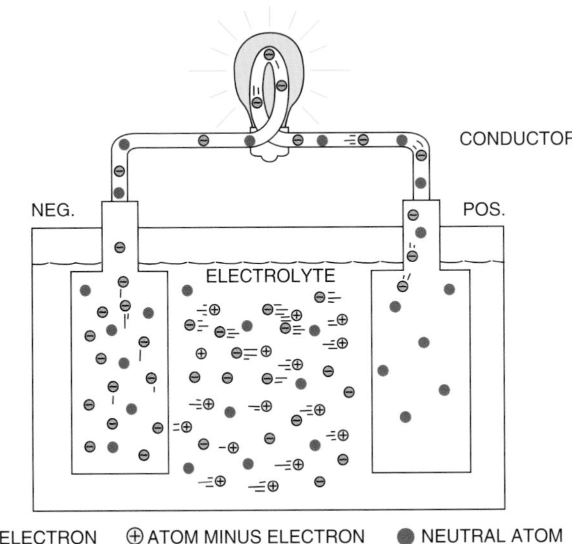

Figure 7-6. Current flow in a simple circuit. Electrons move from the negative post through a conductor and on to positive post.

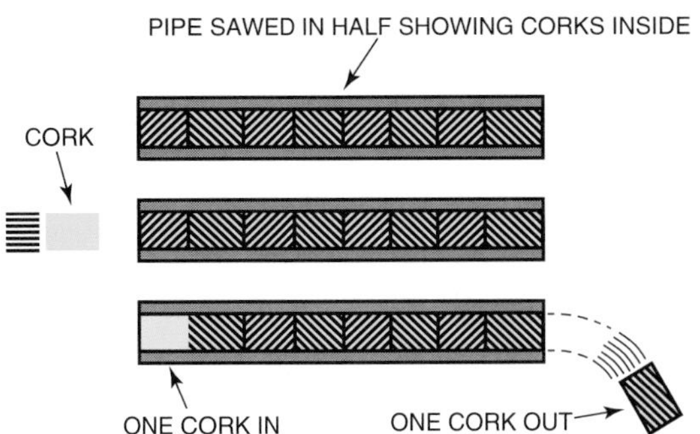

Figure 7-7. Cork (electron) in a pipe (conductor). Even though the cork has just entered the pipe, it causes a chain of pressure that results in a different cork being forced out of the pipe's opposite end.

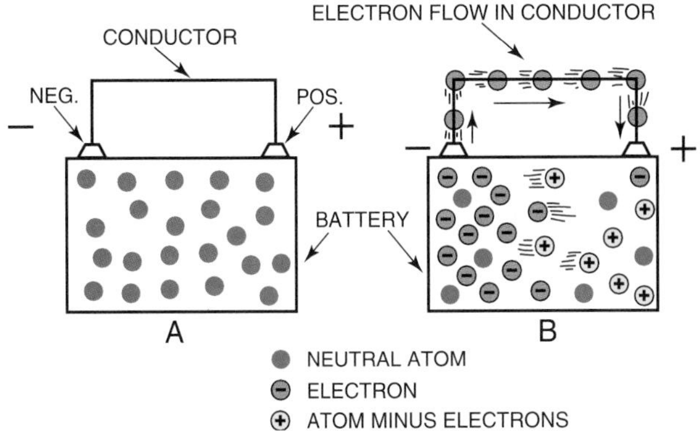

Figure 7-8. A—This battery is completely discharged. There is no voltage or current flow. B—Once the battery is charged, normal voltage returns and electrons are massed on the negative side. Current (electrons) flow through the conductor.

On their way from the negative post, the electrons will pass through the lightbulb. As the filament in the bulb creates a resistance to the flow of electrons, it will heat up until it is white hot and glows. After passing through the filament, the electrons continue on to the positive post. When an electron leaves the negative source, this single electron does not speed through the conductor. It strikes an adjacent electron and knocks it from its orbit. This electron strikes its neighbor and knocks another electron from orbit.

To help illustrate this action, imagine placing enough corks in a pipe to fill it from end to end. If you push one cork (electron) into one end of the pipe (conductor), one cork (electron) will pop out the other. As soon as one cork presses on the column of corks, its movement is transmitted through the column. Even though it enters only a small distance, its influence is felt at the other end almost at once. See **Figure 7-7**. In the circuit in **Figure 7-8**, as one electron knocks another out of its orbit, it replaces the one it knocked out.

All automotive electricity is known as ***direct current***. Direct current always flows in the same direction. Some automotive systems use ***alternating current***. This type of current is normally used in homes and has the ability to change direction.

Voltage

As mentioned earlier, current flow is produced by having a difference in electrical potential, or pressure, at each end of a conductor (one end with a surplus of electrons and other lacking electrons). This pressure differential causes the current to flow. To keep the current flowing, it is necessary to maintain the electrical pressure or, as it is commonly called, ***voltage***. This pressure is sometimes called ***electromotive force***, or ***EMF***.

In the automotive electrical system, initial voltage is produced by a battery, which uses electrochemical action. Operating voltage is produced by an alternator, which uses electromagnetic action. The amount of pressure or voltage is measured in ***volts***. The higher the voltage, the more current flow it can create. Modern vehicles use a 12-volt system. Some large trucks have 24-volt systems. Very old cars have 6-volt systems, while some farm and construction equipment use 8-volt systems.

The battery in **Figure 7-8A** is discharged (dead). Both the negative and positive sides have a balanced electron condition. There is no voltage or pressure difference. In **Figure 7-8B**, the battery is charged. The negative side has a great number of surplus electrons, while the positive side has a deficiency. When a wire is connected across the terminals, the voltage will cause current to flow from negative to positive.

Electrical Resistance

When current flows through a conductor, it meets ***resistance***. A relatively poor conductor offers greater resistance than a good conductor. As the electrons attempt to flow, they bump into other atoms in the conductor. See **Figure 7-9**. This causes a heating effect. As the conductor heats up, the movement of the atoms becomes more agitated. This further increases the number of collisions. If current flow is increased, the conductor will become hotter and hotter until it will literally burn up.

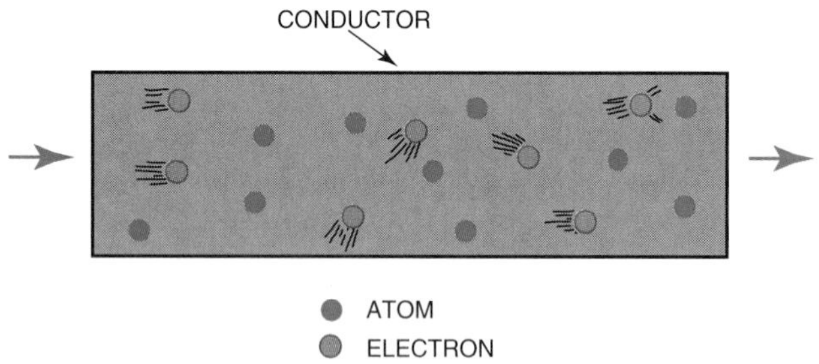

Figure 7-9. As electrons attempt to move through a conductor, they strike atoms. This deflects them and in effect resists their efforts to pass through the conductor.

This heating action is one of the great uses for electricity. Such things as lights, electric burners, and heaters make use of the resistance found in conductors. In order for the wire to reach a specific desired heat, the size and type of conductor and amount of voltage and current flow must be carefully balanced. A large diameter conductor of a given material will offer less resistance than a smaller one. The longer the conductor, the greater its resistance. The amount of resistance in a conductor is measured in ***ohms***.

Test Meters

There are several tools which can be used to measure the flow of electricity. You studied the test light in Chapter 5, which can be used to check for the presence of electricity. To measure the rate of current flow in amperes, a gauge called an ***ammeter*** is used. The ammeter is connected in series with the circuit, **Figure 7-10**. Voltage is determined by using a ***voltmeter***. The voltmeter is connected in parallel across the circuit. See **Figure 7-11**. Resistance in ohms is calculated by using an ***ohmmeter***. The ohmmeter test leads are connected across the resistance being measured, **Figure 7-12**. These meters are often combined into one tool called a ***multimeter***.

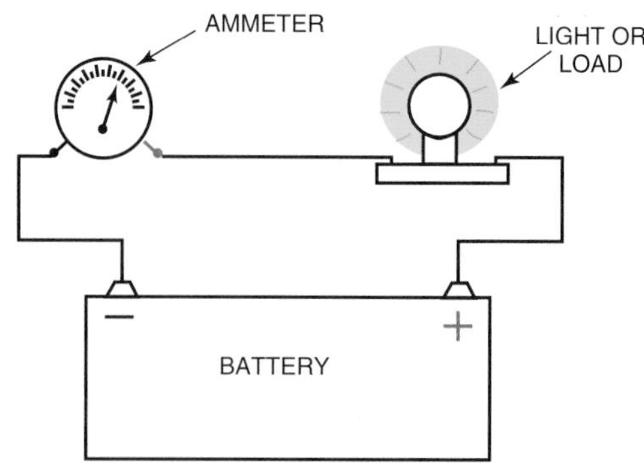

Figure 7-10. Using an ammeter to check current flow. The ammeter must be connected in series with a circuit.

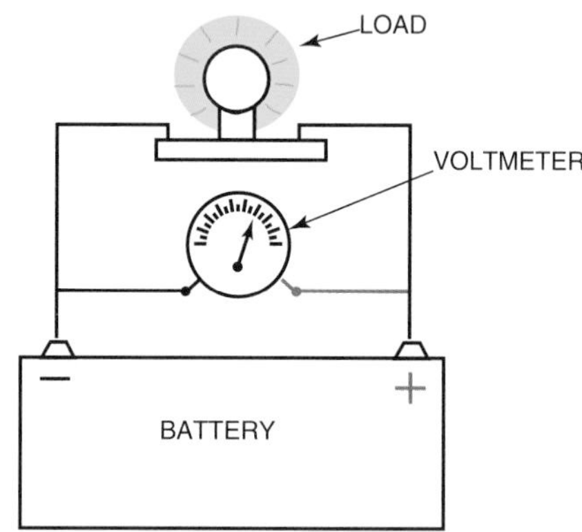

Figure 7-11. Using a voltmeter to check circuit voltage. Voltmeter is connected in parallel with a circuit.

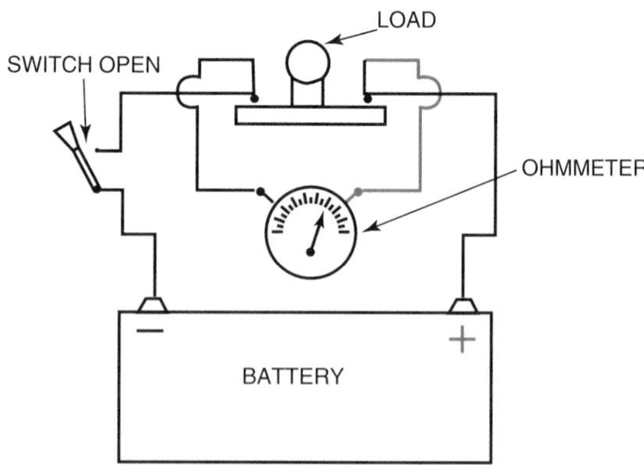

Figure 7-12. Using an ohmmeter to check load resistance. Note that the circuit is open to prevent damage to the ohmmeter. The ohmmeter has its own source of current and is hooked directly to the load.

Materials Used for Conductors

In order for electrons to flow, they must have a path or conductor. If more than one path is available, current will flow in the path with the least resistance. Because it has a great number of free electrons and is relatively inexpensive, copper is widely used as a conductor. Aluminum has also been used as a conductor. The air itself will act as a conductor when voltage is high enough. There are other materials, such as gold and platinum, that are better conductors than copper. Since they are expensive, their use in vehicles is limited. To make certain current flows where desired, the conductors in the vehicle electrical system must be insulated.

Insulators

Any material in which the atoms have more than four electrons in their outer orbit will make a good ***insulator***. Since the electrons in the insulator's atoms tend to be bound and will resist being knocked from their orbits, current cannot flow. Glass, ceramics, plastics, rubber, fiber, and paper are excellent insulators. All are used in automotive electrical systems. The wiring itself uses a rubber or plastic coating to prevent the escape of electrons to adjacent metal parts or to other wires. Spark plugs use ceramic insulators. Switches and relays use fiber, rubber, and glass for specific points where insulation is required.

Electrical Circuits

An ***electrical circuit*** consists of a power source, a unit to be operated, and the necessary wiring to provide a path for electron flow. The circuit must be complete (connected to create a path for current) in order to function. In a ***complete circuit***, all of the components and wiring are electrically connected.

A very basic circuit, **Figure 7-13**, consists of a power source (battery), a single unit to be operated (bulb) and connecting wires. The current flows from the negative terminal through the conductor, through the unit, and then on through the second conductor to the positive terminal of the battery. It then passes from the positive terminal to the negative terminal through the inside of the battery.

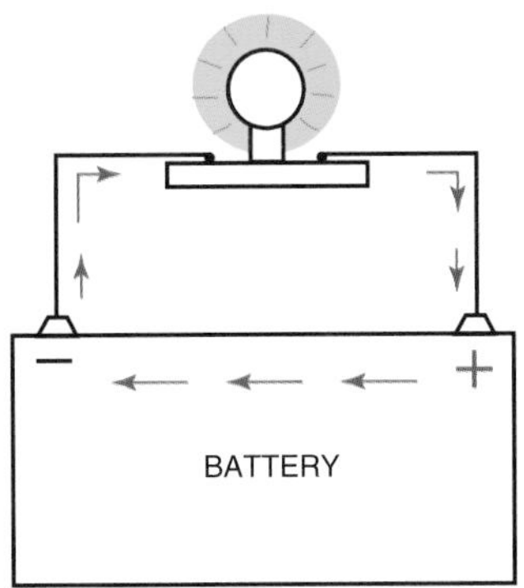

Figure 7-13. A simple circuit. Current flows from negative to positive through the lamp.

Series Circuit

A ***series circuit*** consists of one or more electrical devices (sometimes called resistance units or loads) that are wired so that current has but one path to follow. To complete the circuit, the current must pass through each device. As each electrical device impedes the rate of current flow, the total amount of resistance offered by all devices in the circuit will determine the amperage flow. The more units added to a specific circuit, the greater the resistance, with a corresponding reduction in current flow. If amperage flow is checked at several points in a series circuit, all readings will be the same.

Voltage Drop in a Series Circuit

As the available voltage is lowered by each resistance unit, the voltage will drop as the units occur. The voltage may be up to a full rated 12 volts at the battery, but will become progressively lower farther out on the circuit. Study **Figure 7-14**. Notice that the current must pass through one unit to get to another. Also note that the voltage is dropping while the current flow is the same throughout, and that if any one resistance unit is broken or burned out (wire severed), the current cannot flow.

Any type of circuit will have a certain amount of normal ***voltage drop***. Even the conductor itself can somewhat impede electron flow. When voltage drop becomes excessive, poor unit operation will result. Voltage drop is controlled by using the proper size wires, ample source voltage, proper insulation, and clean connections.

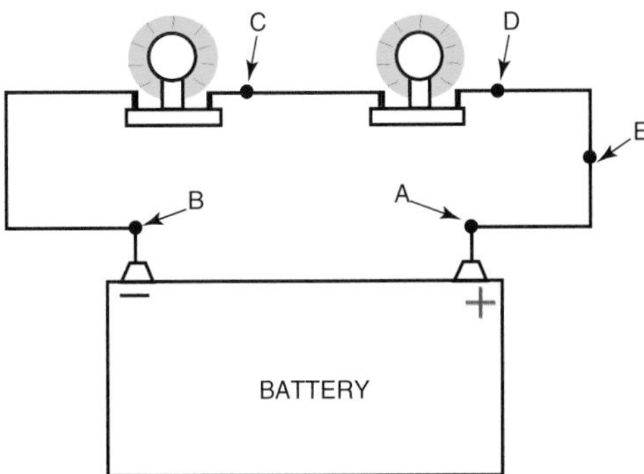

Figure 7-14. A series circuit. If an ammeter were inserted in the circuit at points B, C, D, and E, all readings would be the same. If a voltmeter were installed across or between points B and A, it would show a voltage drop (less voltage). If connected between points B and D, it would show still more voltage drop.

Parallel Circuit

The ***parallel circuit*** consists of two or more electrical devices connected in such a way as to allow current flow to pass through all of the devices simultaneously. This eliminates the necessity of the current going through one before reaching the other. Also, if one device burns out, the others will still function.

If you measure the total resistance in a parallel circuit, you would find it to be less than any one device's resistance. You will recall that in the series circuit, the current has to pass through all of the electrical devices, one after the other. This gave a resistance to current movement that was equal to the sum of all the devices. In a parallel circuit, the current has several devices that it can pass through, actually reducing the total resistance in relation to any one device's resistance value.

Equal voltage is applied to all electrical devices in a parallel circuit. The amount of current flow through each device will vary according to its individual resistance, but the total current flowing will equal the sum of the current flowing through all the devices. See **Figure 7-15**.

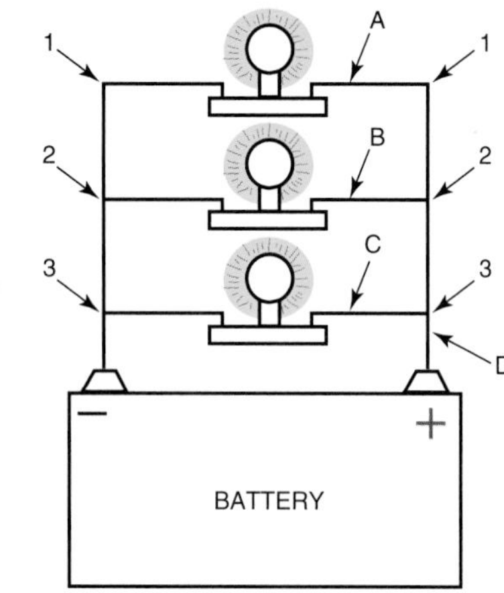

Figure 7-15. A parallel circuit. Voltage across 1-1, 2-2, and 3-3 would be equal. Current flow in amperes at A, B, and C would depend upon the resistance unit. Current flow at D would equal the sum of current flowing through A, B, and C. Total resistance of A, B, and C would be less than any single one.

Series-Parallel Circuit

In many automotive systems, series and parallel circuits are combined to form a ***series-parallel circuit***. It requires the connecting of three or more electrical devices in a combination series and parallel hookup. To determine the resistance of this circuit, the total resistance of the parallel circuit is added to that of the series portion. Current flow will be determined by the total resistance. Voltage drop will be the drop across the parallel circuit, plus that of any series resistance unit. See **Figure 7-16**.

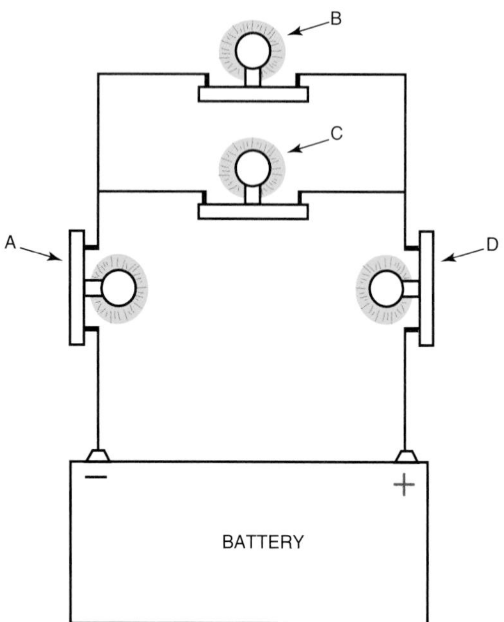

Figure 7-16. A series-parallel circuit. Units A and D are in series. Units B and C are in parallel.

Vehicle Chassis Can Act as a Ground Wire

A complete circuit requires one wire from the source to the unit and one wire from the unit back to the source. In a vehicle where the body, engine, and/or drive train are made of iron and steel, it is not necessary to have a two wire circuit. The metal structure of the vehicle can act as one of the wires. The battery's negative post is connected to the steel framework of the vehicle. The positive terminal is connected to the various electrical units with an insulated wire. **Figure 7-17** illustrates the frame being used in place of a wire.

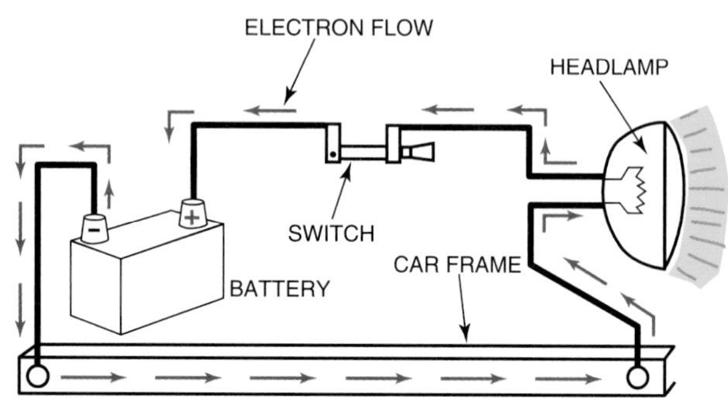

Figure 7-17. Using a vehicle frame as one wire in a circuit. The battery and electrical units are grounded to the frame. Electron flow is from the negative (–) battery terminal to the frame, to the headlamp, and back to the battery through the closed switch.

Some older vehicles used positive ground. However, most vehicleís today are negatively grounded. When looking at wiring diagrams, you will see a ***ground*** symbol where the circuit is connected to the vehicle's chassis. This symbol usually indicates that the circuit is grounded to the chassis, and is not physically connected by a separate wire to the battery. See **Figure 7-18**.

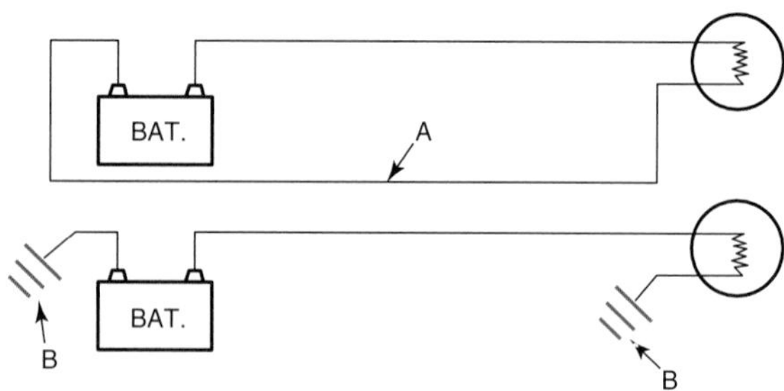

Figure 7-18. Schematic symbol for ground. In the top circuit, wire A, (indicating steel structure of car) has been drawn in. In the lower circuit, ground symbols B have eliminated drawing wire A portion of circuit.

Ohm's Law

All electrical theory is based on ***Ohm's law***. Ohm's Law can be expressed in three different ways:

Amperes equals volts divided by ohms ($I = E \div R$).
Volts equals amperes multiplied by ohms ($E = I \times R$).
Ohms equals volts divided by amperes ($R = E \div I$).

Ohm's law can be used to find any one unknown factor when the other two are known. A handy form in which to remember Ohm's law in its three applications is illustrated in **Figure 7-19**. To use the table in **Figure 7-19**, cover the unknown value and do the appropriate math using the other two values. This will give you the unknown value.

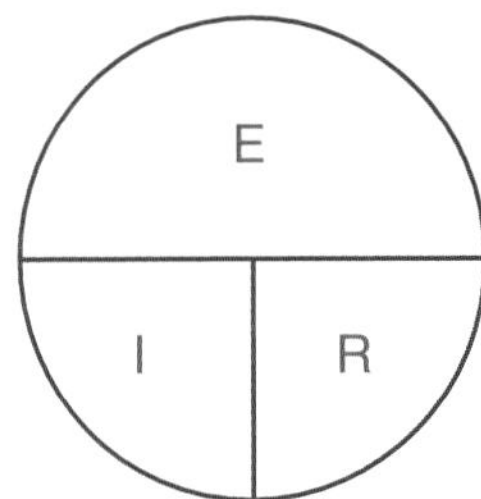

Figure 7-19. Handy form of Ohm's law. Try three combinations using this form. Simply cover the unknown value and compute using the known values.

Magnetism

A great deal remains to be learned about the strange force termed ***magnetism***. It is probably best understood by observing some of its effects and powers. One theory regarding magnetism states that electrons rotating about the nucleus contain the essentials of magnetism. These electrons not only rotate about the nucleus, but also spin on their own axes. Even though each electron creates a very weak magnetic field, the electrons are orbiting in so many directions that the fields tend to cancel each other out.

Magnetic Domains

In an unmagnetized material, there are many minute sections in which the atoms line up to produce a magnetic field. As these magnetic domains are scattered through the material in all directions, one domain tends to cancel out the force of the next one.

To magnetize materials such as steel, it is necessary to orient or line up these domains so that most of them point in the same direction. This is done by placing the steel in a strong magnetic field produced by a current passing through a coil of wire (covered later). The domains will grow in size as they move into alignment. This growth involves lining up neighboring atoms so the magnetic effect is increased. Look at **Figure 7-20**.

Lines of Force

The first thing to remember about the term ***magnetic lines of force*** is that it is used to describe the invisible force involved in magnetism. If a simple bar magnet is placed under a sheet of glass, and fine iron filings are sprinkled on the surface of the glass, the filings will arrange themselves in accordance with the lines of force, **Figure 7-21**.

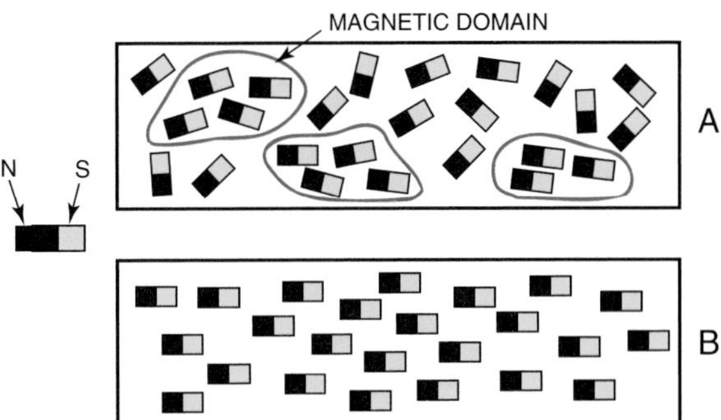

Figure 7-20. A—Magnetic domains are small and are surrounded by atoms that are not aligned. B—When metal is magnetized, it enlarges the domains and causes them to line up. The combined force creates a strong magnetic field. Note that N and S poles are aligned.

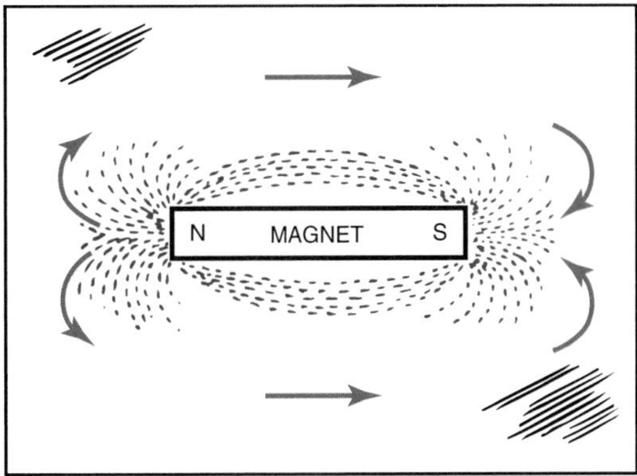

Figure 7-21. Lines of force around a bar magnet. Lines of force leave the N pole and travel along the magnet, then enter the S pole.

The lines of force leave the north pole of the magnet and enter the south pole. They then pass from the south to the north pole through the body of the magnet. All magnetized objects will have north and south ***magnetic poles***. Notice the effect produced on the iron filings when a horseshoe magnet is placed under the glass. The glass should be tapped to allow the filings to line up, **Figure 7-22**. You will note that these fine lines of force do not cross. The lines tend to be drawn in as nearly a straight line as possible, but push each other apart at the same time.

Principles of Magnetism

If two bar magnets are shoved together with unlike poles on the approaching ends, they will snap together. If the approaching ends have similar poles, they will repel each other. If forced together, they will snap apart when released. Therefore, we can say that unlike poles attract and like poles repel.

By studying **Figure 7-23**, you will see that the magnetic force lines are leaving the north pole. Remember that these lines tend to draw into as straight a line as possible, yet they will not cross as the lines repel each other. As the two magnets near each other, the force lines leaving the north pole of

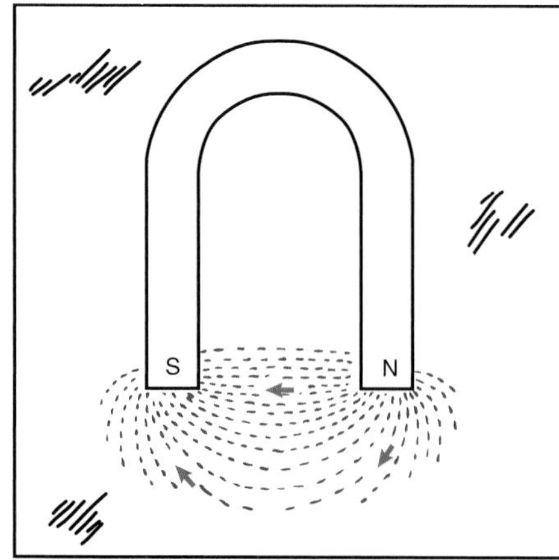

Figure 7-22. Lines of force around a horseshoe magnet.

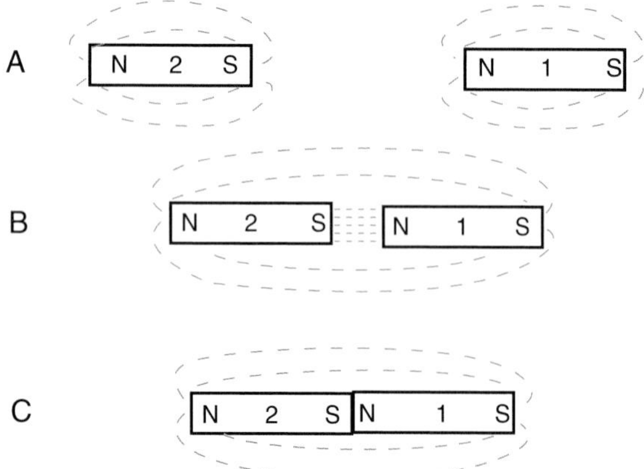

Figure 7-23. Remember that unlike magnetic poles attract. A—Both magnets are apart and the lines of force assume their normal state. B—The magnets have approached each other, causing the lines of force from N pole of magnet No. 1 to enter S pole of magnet No. 2. This draws them together, as in C, and the two magnets act as one long bar magnet.

magnet number one, instead of swinging up and back to its own south pole, leap across the gap to the other closer south pole of magnet number two. The lines leaving the north pole of magnet number two will now swing along both magnets to reach the south pole of magnet number one. As the lines stretch as tightly as possible, the magnets cling tightly to each other. They have, in effect, become one long bar magnet.

When like poles near each other, the magnetic lines entering each pole will repel each other and refuse to cross. This causes each magnet to push away from the other, until the magnetic fields are far enough apart to become too weak to cause further movement. Magnetic lines of force will follow metal more readily than air. If a metal bar is placed across the ends of a horseshoe magnet, the force lines follow the bar from the north pole to the south pole.

Electromagnetism

In the previous sections, you have studied both electricity and magnetism. This section will explain how electricity and magnetism interact. The relationship of electricity to magnetism must be explained to better understand almost any type of automotive electrical equipment.

Magnetism can be used to create electricity. If a bar magnet is moved in the vicinity of a wire, the magnetic field causes movement of the electrons in the wire. An electric current is created, or ***induced***, in the wire. If the wire is wound into a coil and the bar magnet is then moved repeatedly through it, a great amount of electricity can be produced. This is the basic principle of the vehicle alternator.

On the other hand, electricity can be used to create magnetism. If an electric current is moving in a wire, the electrons flowing in the wire create a magnetic field by energizing the electrons around the wire. If the wire is wound into a coil around a metal bar, a much stronger field is created. This device is known as an ***electromagnet***. When current is passing through the electromagnet, it is said to be energized. The energized electromagnet will attract any iron or steel objects placed near it. Automotive applications of electromagnets include solenoids, relays, and ignition coils.

Like a bar magnet, the electromagnet has both north and south poles. Unlike the bar magnet, however, the polarity of the magnet can be changed by reversing the direction of current flow. Polarity is important when installing ignition coils, as a coil with reversed polarity will produce a weaker spark.

Basic Automotive Electronics

The following sections provide an explanation of electronic devices as they are applied to automobiles and light trucks. All modern automotive electronics depend on the use of ***semiconductors***. Semiconductors are substances that are sometimes conductors and sometimes insulators. Semiconductors make this change at the atomic level; in other words, the atoms themselves change between conducting and insulating. Semiconductor materials are used to make diodes and transistors, which are explained below. Semiconductor devices are small and durable, making them ideal for use in vehicles.

Diodes

Diodes are made of a semiconductor material specially designed to allow current to flow in one direction only. When the current flow reverses and attempts to flow through the diode in the opposite direction, it is stopped. A diode is essentially a one-way check valve for electricity, turning off or on as the semiconductor material changes between being a conductor and an insulator, depending on which way the current tries to flow.

Diode Construction and Operation

The diode is made up of two layers of semiconductor material. One layer is positive, or P material, while the other layer is negative, or N material. See **Figure 7-24**. When current tries to flow in one direction, the boundary layer between the P and N materials changes to act like an insulator and no current can flow. When current tries to flow in the other direction, the boundary layer changes into a conductor and current flows.

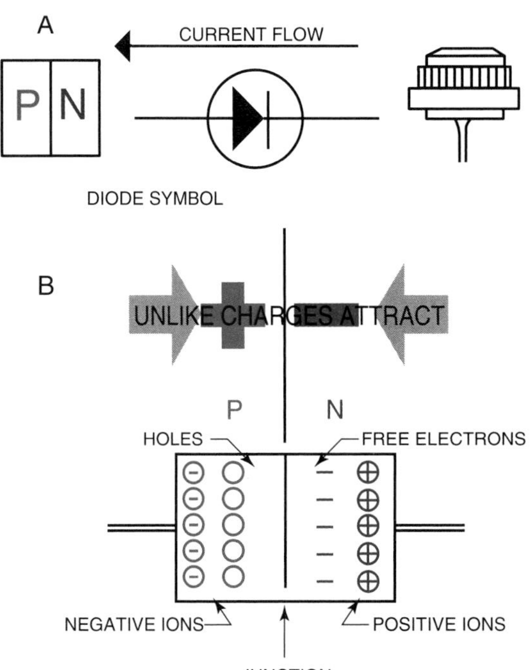

Figure 7-24. A diode is an electrical unit which will allow electric current to travel through itself in one direction only. A—Diode symbol. B—Diode construction.

Diodes are used where alternating current (ac) must be rectified (changed) to direct current (dc). They are used in alternators, which charges the battery and operates other vehicle systems. Alternator diodes are arranged to permit the current to leave the alternator in one direction only. Modern alternators contain six diodes to allow all of the alternator output to be processed into direct current. Diodes are also used in air conditioning compressor and other vehicle circuits.

Another type of diode is the ***Zener diode***, which will not allow current to flow until a certain voltage is reached. When the triggering voltage is reached, the diode will become a conductor, allowing current to pass. Zener diodes are often used in electronic voltage regulators.

Transistors

A ***transistor*** is a solid state electronic switching device. A transistor is operated by very low current flows, but can carry heavy current. Transistors are used in almost every part of the modern vehicle. Common parts that have been replaced by the transistor include mechanical relays and switches. Unlike mechanical devices, transistors have no moving parts to wear out, go out of adjustment, or contact points to pit or corrode.

Transistor Construction and Operation

Many modern transistors are so small that they cannot be seen. The thickness of the parts shown is exaggerated for clarity. The transistor is made up of two outer layers of P material with a center section of N material. Other transistors have a center of P material, between two pieces of N material. The three materials are fused together. An easy way to remember is to think of the transistor as two diodes sandwiched around a common base material. The method of combining the two materials determines the polarity of the transistor. The one shown in **Figure 7-25A** would be a PNP type. **Figure 7-25B** illustrates how the transistor connections are made and **Figure 7-25C** shows the conventional wiring symbol.

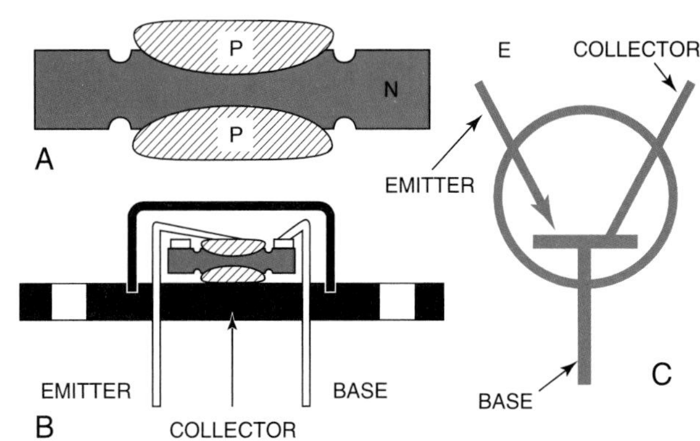

Figure 7-25. A—PNP transistor construction. B—Typical transistor connections. C—Wiring schematic symbol for a transistor. (Delco-Remy)

By connecting wires to the transistor, as shown in **Figure 7-26** and **Figure 7-27**, we find that when a battery is placed in the circuit, current will flow through the transistor in two ways. A small ***base*** current (about 2% of the total) will pass through the base circuit. When the base current is flowing, the transistor will allow a much heavier ***collector*** current (98% of the total) to pass through the collector circuit. The two circuits combine, and 100% of the current leaves through the ***emitter*** circuit. When the base circuit is broken, collector current instantly stops. Most of the transistors in use today require extremely small amounts of current to operate.

By making and breaking the primary circuit with a transistor instead of distributor breaker points, it is possible to increase primary current flow and thus produce higher available voltage at the coil. In addition, there are no physical parts, such as contact points, that can wear out or go out of adjustment. This makes the transistor an excellent switch for operating the ignition system primary circuit, the voltage regulator, and performing other tasks that were formerly done by mechanical relays. Transistors which are designed to carry heavy current loads are called ***power transistors***. Some provision is usually made for cooling a power transistor.

Integrated Circuits

Many small transistors can be combined onto a large complex electronic circuit by etching the transistor circuitry on a single piece of semiconductor material. A circuit made in this manner is called an ***integrated circuit***, or ***IC***. The IC contains resistors, diodes, Zener diodes, and other electronic devices all in a device the size of a dime. A single IC can operate an entire system, such as controlling alternator voltage, making

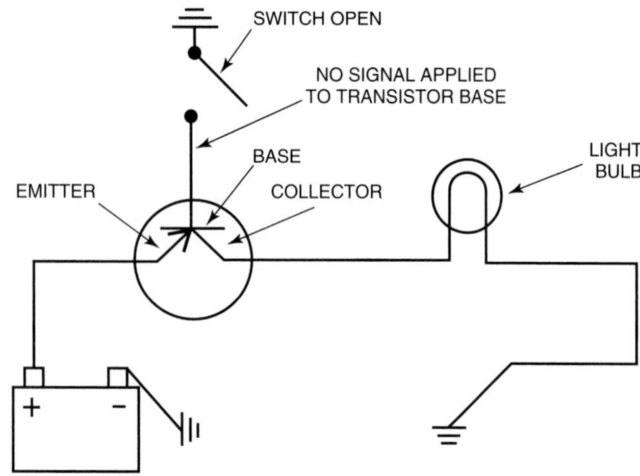

Figure 7-26. Since no voltage is applied to the transistor's base, current does not flow and the light bulb remains off.

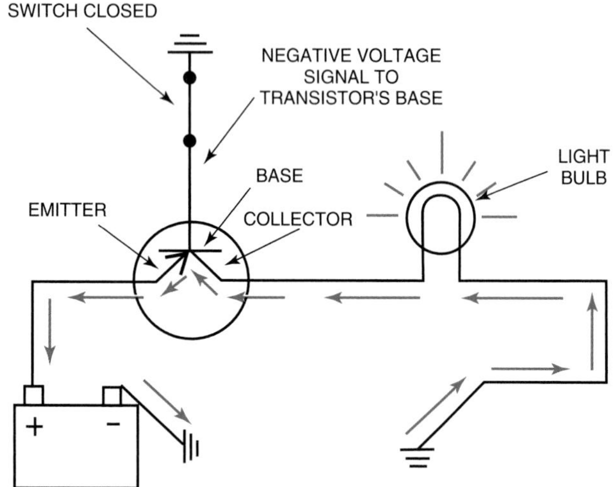

Figure 7-27. Colored arrows indicate current flow through a PNP transistor and circuit when negative voltage is applied to its base. When the transistor is turned on, the circuit is complete, which causes the lightbulb to turn on.

and breaking the ignition primary circuit, or adjusting fuel injector opening. A series of ICs can control the operation of the entire engine and vehicle electrical system.

All modern vehicles have integrated circuits as part of the engine control module. IC's are also used in the vehicle radio and tape/CD player, air bag systems, anti-lock brake system, traction control systems, automatic ride controllers, and any other place where electronic circuitry is used to monitor and control vehicle operation. An IC is often called a ***chip***. Chips are used in many electronic devices. The use of integrated circuits is discussed in more detail in Chapter 15.

Summary

The electron theory states that electricity is caused by a flow of electrons through a conductor. There is a shortage of electrons at one end of the conductor and a surplus at the other. Whenever this condition is present, current will flow.

An atom is made of neutrons, protons, and electrons. The electron is a negative charge of electricity, while the proton is positively charged. The neutron is electrically neutral. The free electrons in an atom are those that are relatively easy to take out of orbit. The free electrons are those that cause the current flow.

The pressure that causes current to flow is voltage. The amount of current flowing is measured in amperes. Every conductor will have some resistance to the movement of electrons. The amount of resistance is measured in ohms. Insulators are made of materials containing atoms with very few free electrons. The bound electrons resist current flow.

In order for electricity to flow, it must have a circuit or path that is complete. Three circuits are series, parallel, and series-parallel. The lining up or orienting of the magnetic domains in a material is what causes the material to become magnetized. A magnet is surrounded by lines of force. They leave the magnet at the north pole, and enter the south pole. Like poles repel while unlike poles attract.

Semiconductors are substances that can be conductors or insulators, depending on current flow through them. Semiconductors make this change at the atomic level. A diode is made of two layers of semiconductor material, one positive, one negative. Diodes are one-way check valves for electricity. A transistor is an electronic switching device composed of two layers of P material with one piece of N material in the center. Other transistors have a center of P material, between two pieces of N material. The transistor is operated by changes in current flow through its layers.

An integrated circuit (IC) contains many small transistors, resistors, diodes, Zener diodes, and other electronic devices in a single piece of semiconductor material. An IC can operate a system in a vehicle and a series of ICs can operate the entire vehicle.

Know These Terms

Electron theory
Negative
Positive
Matter
Molecule
Atoms
Elements
Electron
Proton
Neutron
Bound electrons
Free electrons
Current
Conductor
Ampere
Direct current
Alternating current
Voltage
Electromotive force
EMF
Volts
Resistance
Ohms
Ammeter
Voltmeter
Ohmmeter
Multimeter
Insulator
Electrical circuit
Complete circuit
Series circuit
Voltage drop
Parallel circuit
Series-parallel circuit
Ground
Ohm's law
Magnetism
Magnetic lines of force

Magnetic poles	Base
Induced	Collector
Electromagnet	Emitter
Semiconductor	Power transistor
Diode	Integrated circuit
Zener diode	IC
Transistor	Chip

Review Questions—Chapter 7

Do not write in this book. Write your answers on a separate sheet of paper.

1. According to the electron theory, what causes the operation of electrical devices?
2. What causes the action described in the answer to question 1?
3. Atoms can combine to form _____ .
4. What two subatomic particles are at the center of the atom?
5. Technician A says that the electron revolves around the center of the atom. Technician B says that bound electrons are loosely held to the atom.
   Who is right?
   (A) A only.
   (B) B only.
   (C) Both A & B.
   (D) Neither A nor B.
6. If a surplus of electrons are present, the atom becomes _____ charged.
7. Electrical current is a mass movement of electrons through a _____ .
8. Current flow is measured in _____ .
   (A) volts
   (B) amperes
   (C) ohms
   (D) watts
9. Technician A says that current flows from negative to positive. Technician B says that an excess of electrons on one side of a complete circuit will cause current to flow.
   Who is right?
   (A) A only.
   (B) B only.
   (C) Both A & B.
   (D) Neither A nor B.
10. List the voltages of the following systems.
    Modern vehicles:
    Large trucks:
    Very old cars:
    Farm equipment:
11. What is another name for voltage?
12. Resistance to the flow of electricity is measured in _____.
13. Name the instrument used to check each of the following:
    (A) Voltage.
    (B) Amperage.
    (C) Resistance.
14. What material is widely used as a conductor in vehicle electrical systems?
    (A) Copper.
    (B) Lead.
    (C) Glass.
    (D) Gold.
15. In a series circuit, the current must pass through _____ .
    (A) a portion of the electrical devices
    (B) the high resistance electrical devices
    (C) every electrical device in the circuit
    (D) the light units only
16. The total resistance in a parallel circuit is _____ than any one of the electrical devices in the circuit.
17. A series parallel circuit requires at least _____ electrical devices in a combination series and parallel hookup.
18. If a 12-volt circuit is drawing 6 amperes, what is the resistance in the circuit?
19. If a circuit has 3 ohms of resistance, and the system is operated on 12 volts, what is the amperage draw?
20. If a circuit with 12 ohms of resistance is drawing .5 amperes, what is the voltage of the circuit?
21. Unlike magnetic poles _____, like poles _____ .
22. All of the following statements about semiconductors are true, EXCEPT:
    (A) semiconductors are sometimes conductors, and sometimes insulators.
    (B) semiconductors change at the atomic level.
    (C) semiconductors are being replaced with mechanical relays.
    (D) semiconductors are used in diodes and transistors.
23. A Zener diode will permit current to flow when _____ .
    (A) amperage reaches a certain level
    (B) voltage reaches a certain level
    (C) resistance heat reaches a certain level
    (D) Either A or B.
24. All of the transistor current flows through the _____ .
    (A) base
    (B) collector
    (C) emitter
    (D) Both A & C.
25. An integrated circuit contains many small _____ .
    (A) transistors
    (B) diodes
    (C) resistors
    (D) All of the above.

Three computer-controlled ignition coils are used by this engine. Each coil fires the spark plugs in two opposing cylinders. This engine is part of the same family as the one shown at the top of the next page. (General Motors)

8

Ignition Systems

After studying this chapter, you will be able to:

- Explain why battery voltage must be increased.
- Describe the primary circuits of an ignition system.
- Describe the secondary circuits of an ignition system.
- Describe the construction of ignition system components.
- Summarize the basic function and operating principles of ignition system parts.
- List the differences between a contact point and an electronic ignition system.
- Explain the action of spark advance mechanisms.
- Explain how a distributorless ignition system operates.

Chapters 1, 2, and 3 made frequent reference to the use of a spark to ignite the air-fuel mixture. This chapter will provide you with a clear and concise description of the various units in the ignition system. The theory, design, and construction of the parts will be discussed. You will also learn how they are combined to produce, control, and distribute the spark. Before beginning this chapter, be sure that you are familiar with the material in Chapter 7. This will provide you with a basic introduction to electricity and electronics. Such knowledge is essential to understanding ignition systems.

High Voltage Is Necessary

Jumping a spark plug gap requires thousands of volts. However, the vehicle's battery can produce only 12 volts. Since this is not enough voltage to jump across the electrodes of a spark plug in the combustion chamber, a way to raise the voltage must be found. In modern vehicles, the voltage needed at the spark plugs can exceed 60,000 volts. This means that the battery's 12 volts is increased many times. However, if you touch a plug and are shocked, there is little real danger involved. Current, which is the part of electricity that does the actual damage, is very low in an automotive ignition system.

STOP **Warning: Do not confuse vehicle electrical system voltage with the voltage in your home electrical system. Household current at 120 or 240 volts can, and often does, electrocute people. Always treat electricity with caution and respect.**

The ignition system is divided into two separate circuits, which are called the primary and secondary circuits. As you read this chapter, refer to this and other illustrations to learn the names of the various units and where they are used in the primary and secondary circuits, which are shown in **Figure 8-1**. The primary circuit will be covered first.

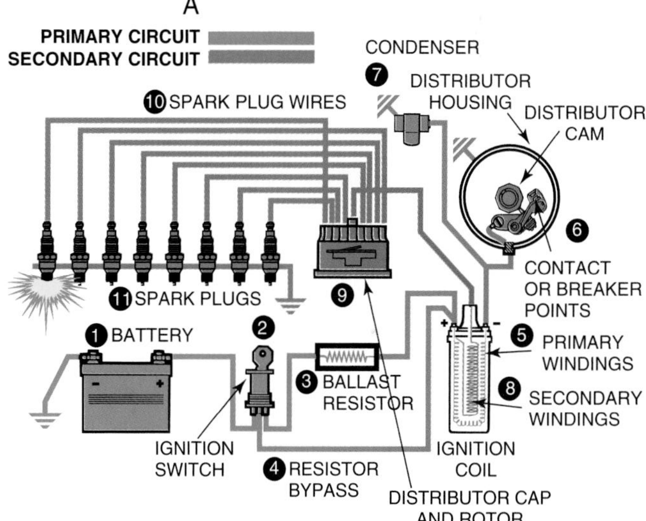

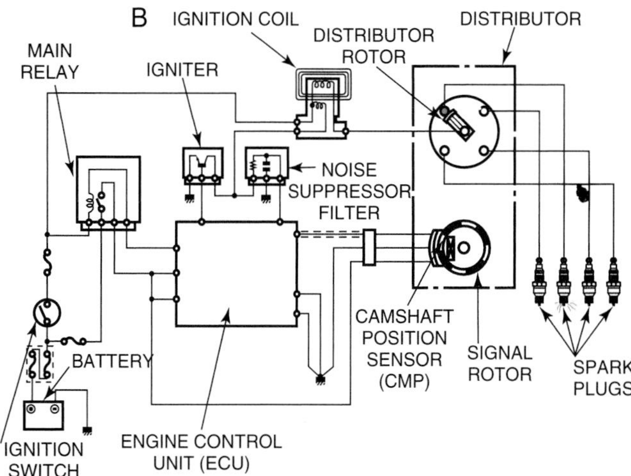

Figure 8-1. Two different types of ignition systems. A—A conventional contact or breaker point type. B—An electronic ignition system which incorporates a breakerless style distributor. (Sun, Geo)

Primary Circuit

The ***primary circuit*** consists of the battery, ignition switch, resistor (where used), ignition module or contact points, and coil primary wiring. They are covered in the order that electricity flows through them. The primary circuit voltage is low, operating on the battery's 12 volts. The wiring in this circuit is covered with a thin layer of insulation to prevent short circuits.

Battery

To better understand the operation of the ignition system's primary circuits, we will start at the battery and trace the flow of electricity through the system. The ***battery***, **Figure 8-2**, is the source of electrical energy needed to operate the ignition system. The battery stores and produces electricity through chemical action. When it is being charged, it converts electricity into chemical energy. When it is discharging (producing current), the battery converts chemical energy into electricity. See Chapter 16 for more information about batteries. To function properly, the battery must be in condition or ***charged***, to produce the highest output of electricity.

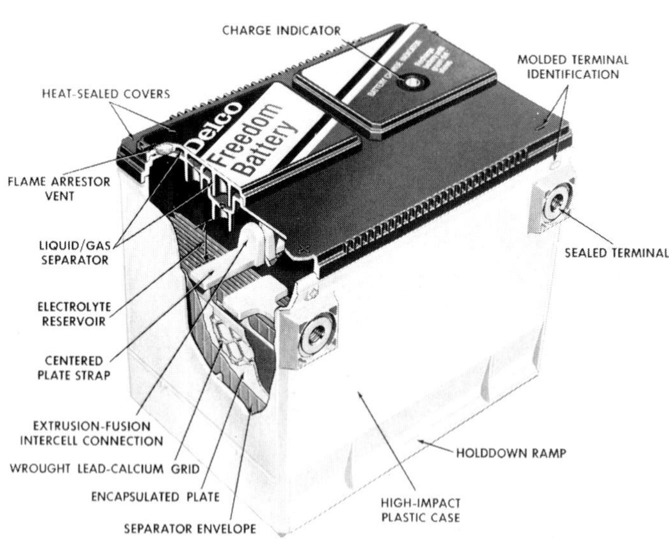

Figure 8-2. Typical maintenance free battery construction. Note the use of a charge indicator. (Delco Remy)

Ignition Switch

The primary circuit starts at the battery and flows to the ***ignition switch***. The ignition switch is operated by the ignition key. It controls the flow of electricity across the terminals. The ignition switch may have additional terminals that supply electricity to other vehicle systems when the key is turned on. Most ignition switches are installed on the steering column, **Figure 8-3**.

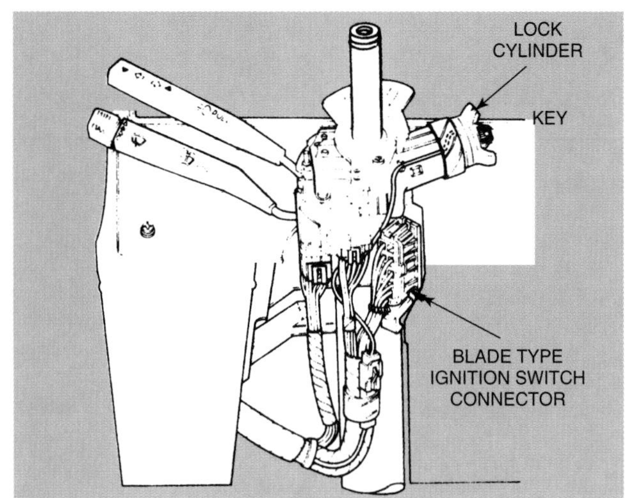

Figure 8-3. One type of ignition switch mounted on the steering column. Note the various wire connections. (Ford)

Resistors

Some ignition systems include a ***resistor*** in their primary circuits. Electricity flows from the ignition switch to the resistor. A simple resistor is shown in **Figure 8-4**. The resistor controls the amount of current reaching the coil. The resistor may be either the calibrated resistance wire or the ballast type.

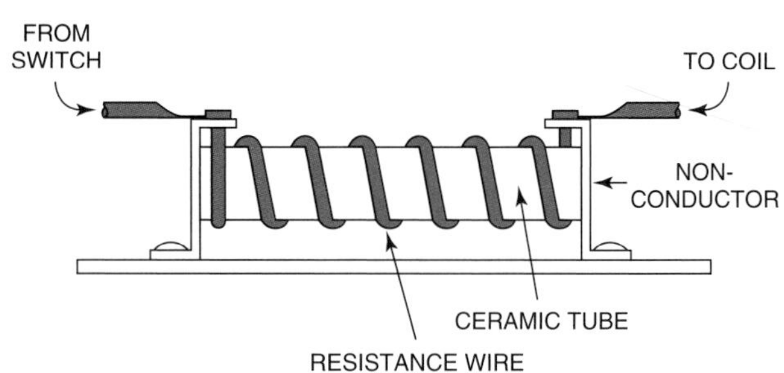

Figure 8-4. Simple resistor used in the primary circuit of an ignition system.

Most resistors simply consist of a calibrated resistance wire built into the wiring harness between the ignition switch and coil. The resistance wire lowers battery voltage to around 9.5 volts during normal engine operation. However, when the engine is started, the coil receives full battery voltage from a bypass wire. The bypass wire supplies the coil with full battery voltage from the ignition switch or starter solenoid while the engine is cranking. When the key is released, the circuit receives its power through the resistance wire.

The ballast resistor, which is used on some vehicles, is a temperature sensitive, variable resistance unit. A ballast resistor is designed to heat up at low engine speeds as more current attempts to flow through the coil. As it heats up, its resistance value increases, causing lower voltage to pass into the coil. As engine speed increases, the duration of current flow lessens. This causes a lowering of temperature. As the temperature drops, the resistor allows the voltage to the coil to increase.

At high speeds, when a hotter spark is needed, the coil receives full battery voltage. The ballast resistor is a coil of nickel-chrome or nichrome wire. The nichrome wire's properties tend to increase or decrease the voltage in direct proportion to the heat of the wire, **Figure 8-5**. Some early transistor ignition systems use two ballast resistors to control coil voltage. From the resistor, the current travels to the coil. Most modern vehicles with electronic ignition do not use a resistor in the ignition circuit. The majority of modern electronic ignition systems use full battery voltage at all times.

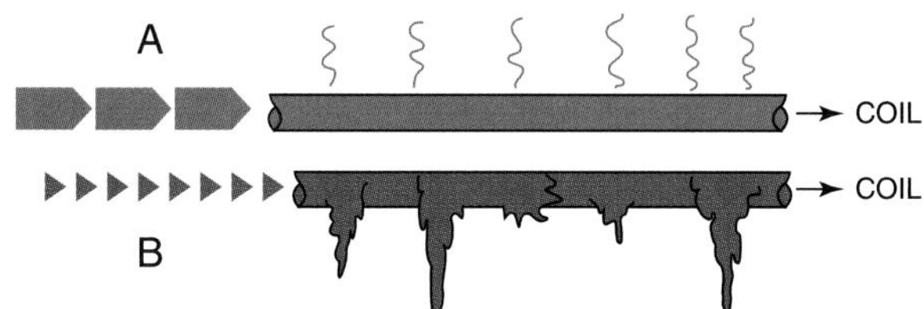

Figure 8-5. Ballast resistor principle. A—This Illustrates long pulsations of current passing through a special ballast resistor wire at slow engine speeds. The current heats the special wire and lowers amount of current reaching coil. B—This illustration shows short pulsations at high speeds. This allows wire to cool, and a heavier current to flow to the coil.

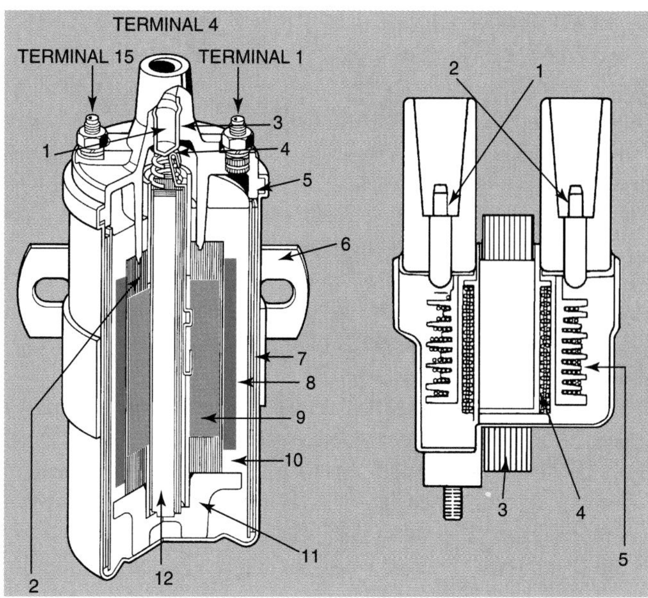

Figure 8-6. Cutaways of two ignition coils. A—1-High voltage terminal. 2-Winding layers with insulating paper. 3-Coil cap. 4-High voltage connection via spring. 5-Coil case. 6-Mounting bracket. 7-Metal plate jacketing (magnetic). 8-Primary winding, (heavy wire). 9-Secondary winding, (fine wire). 10-Sealing compound. 11-Insulation. 12-Laminated iron core. B—1-Secondary output (+). 2-Secondary output (–). 3-Iron core. 4-Primary winding. 5-Secondary winding. (Mercedes-Benz, Deere & Co.)

Ignition Coil

The primary circuit leads from the ignition switch or resistor to the ***ignition coil***. An ignition coil is actually a transformer that is capable of increasing battery voltage to as much as 100,000 volts, although most coils produce about 50,000-60,000 volts. Coils vary in size and shape to meet the demands of different vehicles.

Coil Construction

The coil is constructed with a special laminated iron ***core***. Around this central core, many thousands of turns of very fine copper wire are wound. This fine wire is covered by a thin coating of high temperature insulating varnish. One end of the fine wire is connected to the high tension terminal and the other is connected to the primary circuit wire within the coil. All these turns of fine wire form what is called the ***secondary winding***.

Several hundred turns of heavier copper wire are wrapped around the secondary coil windings. Each end is connected to a primary circuit terminal on the coil. These windings are also insulated. The turns of heavier wire form the ***primary winding***. Study the winding connections in the cutaways shown in **Figure 8-6**.

The core, with both the secondary and primary windings attached, is placed inside a laminated iron shell. The job of the shell is to help concentrate the magnetic lines of force that will be developed by the windings. This entire unit is then placed inside a steel, aluminum, or Bakelite case. In some coil designs, the case is filled with oil or a paraffin-like material. In other designs, the coil windings are encased in heavy plastic.

The coil is sealed to prevent the entrance of dirt or moisture. The primary and secondary coil terminals are carefully sealed to withstand vibration, heat, moisture, and the stresses of high induced voltages. **Figure 8-7** shows some types of coils found on modern vehicles. Although their shapes are different, they all work in the same way.

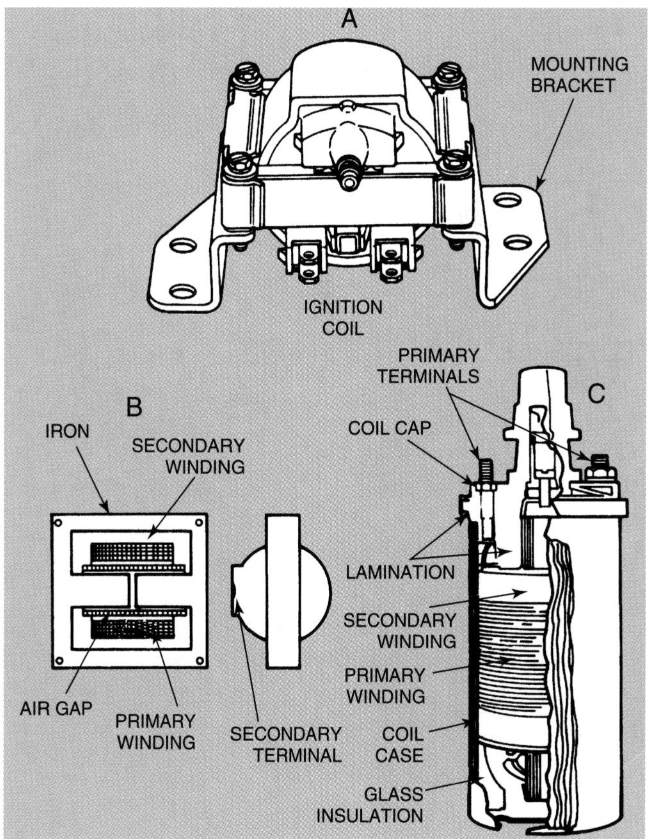

Figure 8-7. Several different ignition coils and their construction, A—Remote mounted high energy ignition (HEI) coil. B—Cutaway view of HEI coil construction. C—Cutaway view of a conventional style coil showing its construction. (AC-Delco)

Coil Operation

When the ignition switch is turned on, the current flows through the primary windings of the coil to ground. See **Figure 8-8**. As you learned from Chapter 7, when a current flows through a wire, a ***magnetic field*** is built up around the conductor. Since there are several hundred turns of wire in the primary windings, a strong magnetic field is produced. This magnetic field surrounds the secondary as well as the primary windings. See **Figure 8-8B**. If there is a quick and clean interruption of current flow on its way to ground after passing through the coil, the magnetic field will ***collapse*** into the laminated iron core, **Figure 8-8B**.

As the field collapses through the primary winding, the voltage in the primary windings will be increased. This is called ***self-induction***, since the primary windings produce its own voltage increase. The voltage induced in the primary windings is about 200 volts, since it consists of only several hundred turns of wire. Self-induction does not affect secondary winding operation, but can cause point arcing on contact point systems.

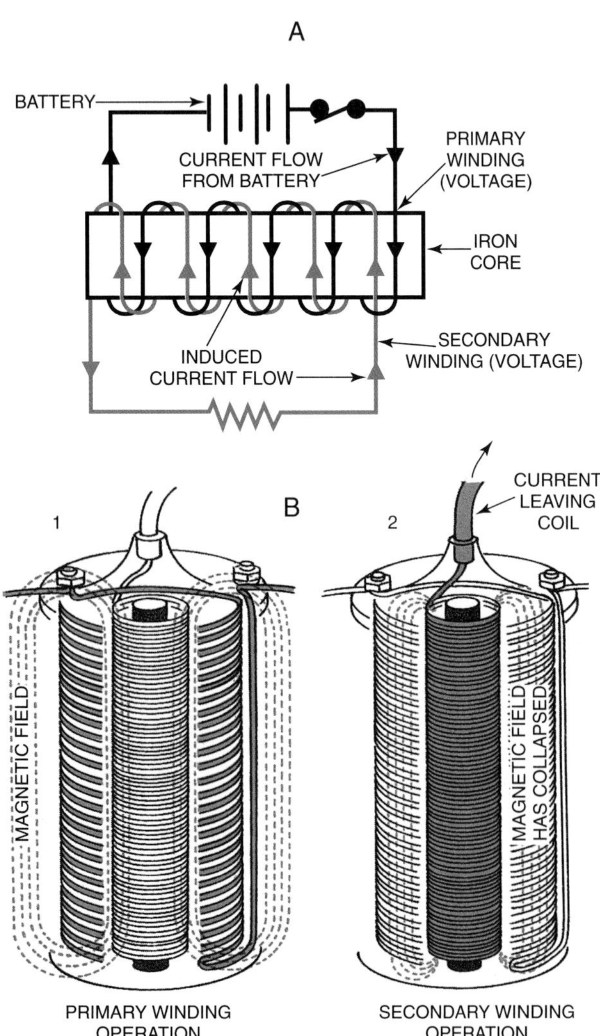

Figure 8-8. A—Wiring schematic illustrating how the coil induces current flow in a secondary coil. B—Ignition coil operaiton. 1-Primary winding. 2-Secondary winding. Current is now leaving the coil on its way to the spark plugs via the distributor. (Deere and Sun Electric)

As the magnetic field collapses, it passes through the secondary winding, producing a tiny current in each turn. The secondary windings possess thousands of turns of wire. Since they are in series, the voltage of each turn of wire is multiplied by the number of turns. This can produce a voltage exceeding 100,000 volts. This is known as ***induction***. The high voltage produced by the secondary windings exits the high tension coil terminal and is directed to the spark plugs.

Most coils have primary terminals marked with a (+) and (–). The plus sign indicates positive and the minus indicates negative. The coil must be installed in the primary circuit according to the way the battery is grounded. This alignment of the positive and negative terminals is called ***polarity***. If the battery's negative terminal is grounded, the negative terminal of the coil must be connected through the ignition module or distributor to ground as applicable. This is done to ensure the correct polarity at the spark plug.

Actual Coil Output

Even though the voltage output of some coils can exceed 100,000 volts, the coil will only build up enough voltage to produce a spark. This may be as low as 2000 volts at idle on an older vehicle without emission controls, or as high as 60,000 volts on a new vehicle with the leanest possible mixture and under a load. See **Figure 8-9**.

To control the coil's output, most engines have a ***distributor***. The job of the distributor it to trigger the coil and to distribute the high voltage current to the right spark plug at the right time.

Methods of Current Interruption

To cause the coil's magnetic field to collapse, the current flow through the primary windings must be interrupted instantly and cleanly with no ***flashover*** (current jumps or arcs across space) at the point of disconnection. For about 75

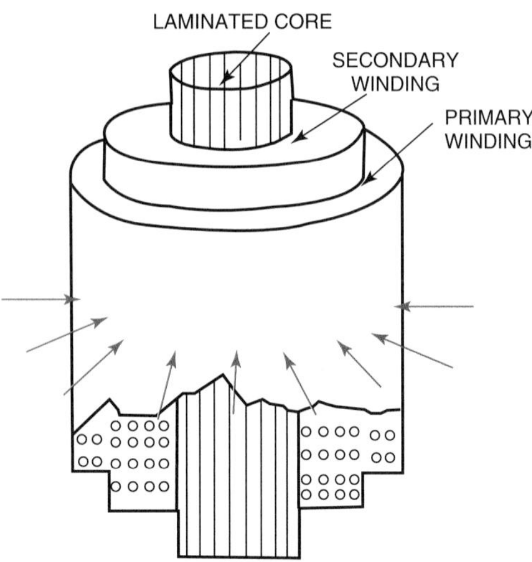

Figure 8-9. Collapse of primary field. When the primary circuit is broken, the magnetic field will collapse through secondary winding to the core.

years, the primary current flow was controlled by using a set of contact points to break current flow and collapse the coil primary field. Over the last 20 years, contact point systems have been replaced by electronic ignition systems, which uses transistors to operate the primary circuit. Transistor construction and operation were covered in Chapter 7.

Electronic ignitions can produce the high voltage spark needed to fire the leaner mixtures used on modern vehicles. While the old contact point system could produce no more than 20,000 or 30,000 volts, electronic ignition systems allow as much as 100,000 volts to be used. All modern vehicles use ignition systems with electronic primary circuit control. The basic difference between contact point ignition systems and electronic ignition systems is the method employed to interrupt the coil primary circuit.

Contact Points

The ***contact points*** used on older vehicles were a simple mechanical way of making and breaking the coil primary circuit. Note the contact point set in **Figure 8-10**. The stationary piece is grounded through the distributor contact point mounting plate. This section does not move other than for an initial point adjustment.

The second piece is the movable contact point. It is pivoted on a steel post. A fiber bushing is used as a bearing on the pivot post. A thin steel spring presses the movable contact arm against the stationary unit, causing the two contact points to touch each other. The movable arm is pushed outward by the ***distributor cam lobes***, which are turned by the distributor shaft. The cam lobe on the distributor shaft opens and closes the points as it revolves. The number of lobes corresponds to the number of cylinders in the engine.

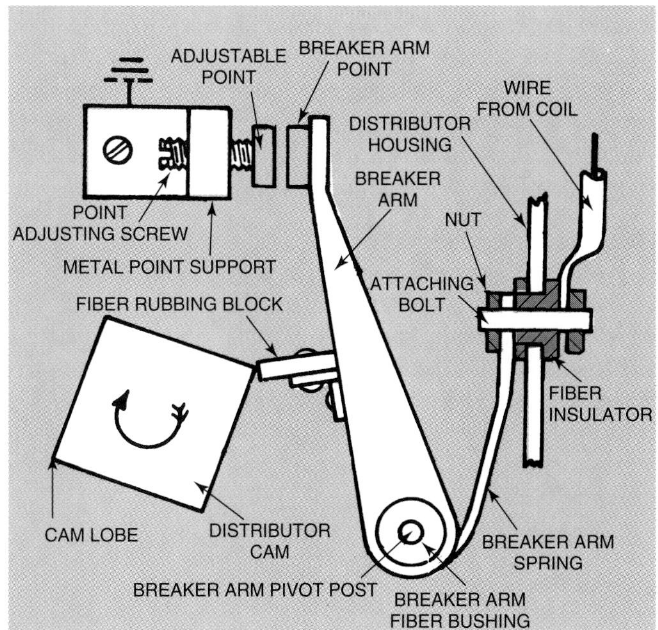

Figure 8-10. Typical contact point construction. Most incorporate the adjustable point into an adjustable support base. Average point gap specifications run around .018″ to .022″ (0.457 mm to 0.559 mm).

The cam moves the contact arm through a fiber rubbing block. This block is fastened to the contact arm and rubs against the cam. High temperature lubricant is used on the block to reduce wear. The movable contact arm is insulated so that when the primary lead from the coil is attached to it, the primary circuit will not be grounded unless the contact points are touching, **Figure 8-10**.

Contact Point Dwell

The number of degrees the distributor cam rotates from the time the points close until they open again is called ***dwell***, and is sometimes referred to as cam angle. See **Figure 8-11**. Dwell is important as it affects the magnetic buildup of the primary windings. The longer the points are closed, the greater the magnetic buildup. However, too much dwell can result in point arcing and burning. If the dwell is too small, the points will open and collapse the field before it has built up enough voltage to produce a satisfactory spark.

When setting contact point gaps, as the gap is reduced, dwell is increased. When the gap is enlarged, dwell is decreased. The dwell cannot be adjusted on electronic ignition systems, but can be measured as an aid to diagnosis. Always check the manufacturer's specifications for dwell when setting points.

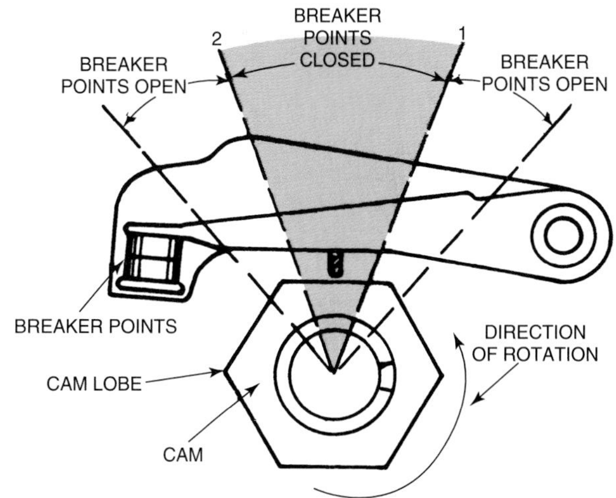

Figure 8-11. These ignition points close at 1 and remain closed as cam rotates to 2. The number of degrees formed by this angle determines dwell.

Condenser

The ***condenser***, sometimes called a capacitor, absorbs excess primary current when the contact points are opened. The condenser prevents point arcing and resulting overheating, pitting, and excessive wear. In addition to increasing contact point service life, the condenser allows the coil's magnetic field to collapse quickly, producing a strong, instant spark.

Most condensers are constructed of two sheets of very thin foil separated by two or three layers of insulation. The foil and insulation are wound together into a cylindrical shape.

The cylinder is then placed in a small metal case and sealed to prevent the entrance of moisture. **Figure 8-12** illustrates the construction of the typical condenser. The close placement of the foil strips creates ***capacitance***, or the ability to attract electrons.

When the points are closed, the condenser is inactive as the coil's magnetic field begins to build in strength. When the points open, the magnetic field begins to collapse and voltage in the primary windings increase due to self induction. If a condenser was not used, the voltage in the primary circuit would arc across the points, consuming the coil's energy before the magnetic field passes through the secondary windings.

However, the condenser attracts the excess primary voltage, preventing an arc across the points. By the time the condenser has become fully charged, the points have opened too far for the current to arc. The magnetic field collapses through the secondary windings, producing a quick, strong spark.

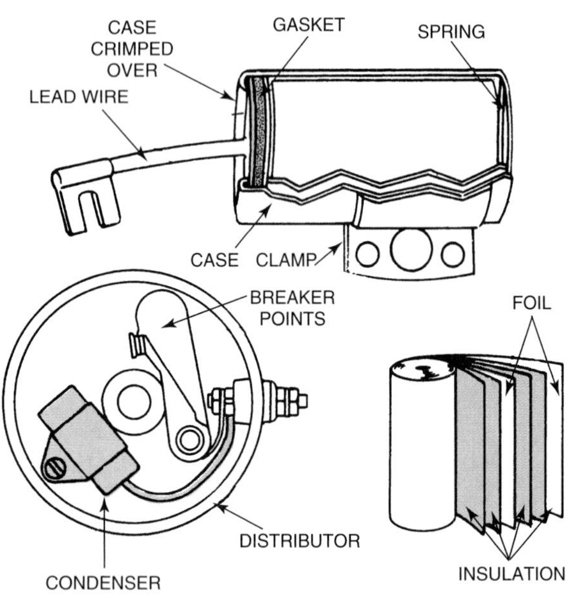

Figure 8-12. This condenser unit is hermetically sealed in a metal case. Note how the condenser is attached in the distributor. (AC Delco, Deere & Co.)

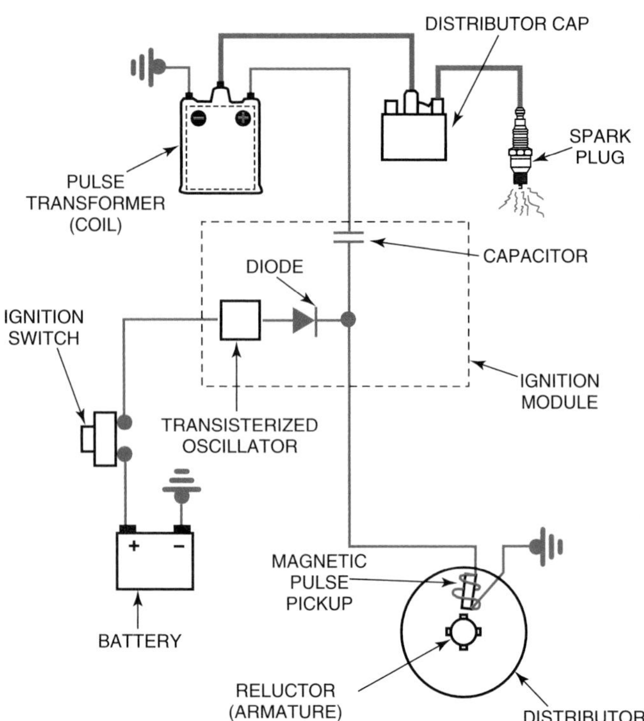

Figure 8-13. Schematic showing power flow through one type of electronic ignition circuit. (Sun Electric Corporation)

Electronic Ignition

The schematic in **Figure 8-13** is a simple electronic ignition circuit. Note that there are no mechanical devices to make and break the circuit. The entire process is done electronically. Current flows from the ignition switch, through the ***ignition module***, to the coil. The ignition module contains the electronic components which cause the coil to produce a high voltage spark. Ignition modules process the inputs from other ignition components.

Ignition modules are sometimes installed on the engine firewall or inner fender to protect them from excessive engine heat. Other modules are located in the distributor, installed outside on the distributor body, or as part of the coil assembly. Some typical electronic ignition modules are shown in **Figure 8-14**. Current from the ignition switch enters the module and passes through a power transistor before reaching the coil, as in **Figure 8-14C**. The power transistor acts like a conductor, allowing full current to flow in the circuit. This begins the build up of the magnetic field in the coil.

When the power transistor is signaled by the triggering device (explained below) and other module circuitry, it becomes an insulator. Since current cannot flow through an insulator, this stops current flow through the coil primary circuit. When current flow stops, the magnetic field collapses, creating the high voltage current in the secondary windings. After the coil collapse is complete, the process is repeated as current flow through the power transistor begins again.

Electronic Triggering Devices

Electronic ***triggering devices*** send a signal current to the ignition module, which then breaks the primary circuit. The parts of the triggering device do not wear, which gives them a much longer life expectancy than contact points. Since the triggering device does not wear, engine timing does not change. This improves engine performance, emissions output, and reliability. There are three types of triggering devices currently in use:

- Magnetic
- Hall effect
- Optical

Most of the triggering devices are operated by rotation of the distributor shaft. Some triggering devices are installed into or on the engine block and are operated by the rotation of the crankshaft and/or camshaft.

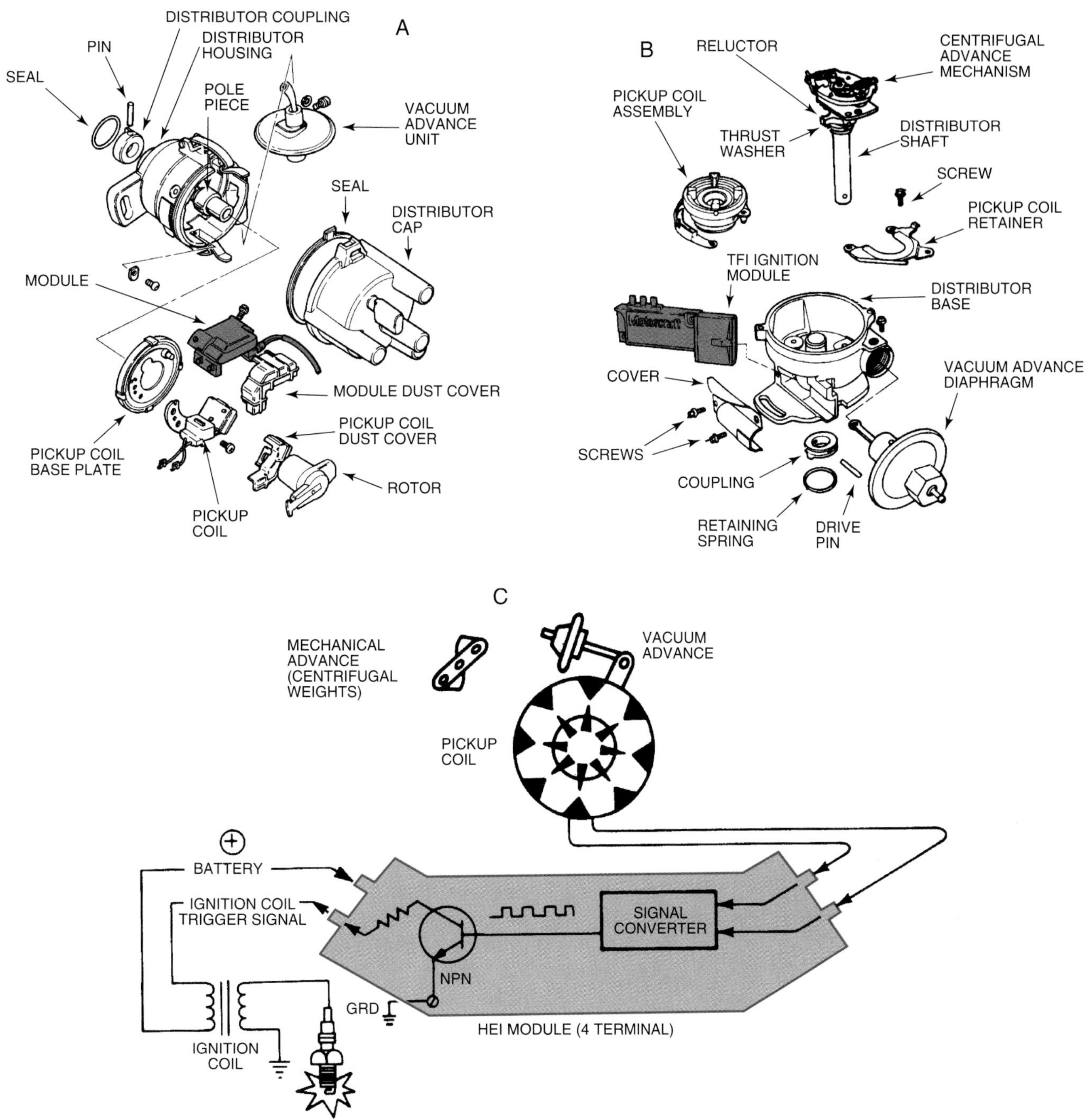

Figure 8-14. A & B—Exploded views of distributor assemblies which house electronic ignition modules. C—Ignition system schematic showing an electronic ignition module. (General Motors & Ford)

Magnetic Pickup

The ***magnetic pickup*** shown in **Figure 8-15A** is mounted in the distributor and reacts to distributor speed, which is one half of crankshaft speed. This sensor generates alternating current, **Figure 8-15B**. The current produced is small, (about 250 millivolts) but can be read easily by the ignition module. The rotating tooth assembly is called a ***reluctor***, or ***trigger wheel***. The stationary assembly is called the pickup coil or stator.

An air gap between the rotating and stationary teeth prevents physical contact and eliminates wear. When a tooth of the reluctor aligns with the tooth of the pick-up coil, a voltage signal is sent to the ignition module, which turns the power transistor off and interrupts primary current to the ignition coil, causing it to fire a spark plug.

Some sensors are mounted near the crankshaft, as shown in **Figure 8-15C**. A reluctor wheel is part of the crankshaft and is placed at its midpoint. An air gap also exists

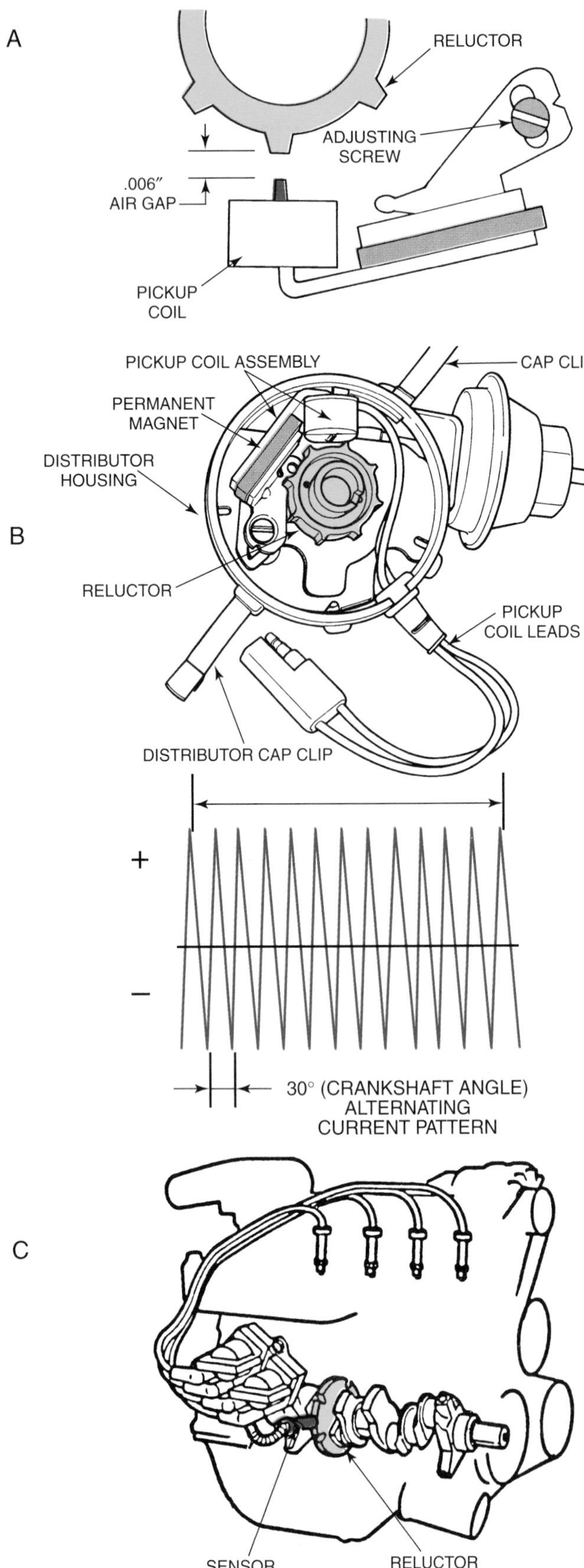

Figure 8-15. Several different magnetic crankshaft position sensors. A—An air gap exists between the reluctor and pickup coil mounted on the distributor. B—An alternating current pattern is produced by this sensor. C—A position sensor and reluctor located at the crankshaft. (Chrysler, Toyota, and Oldsmobile)

between this sensor and the reluctor. When the sensor is in the middle of each slot, the transistor is turned off and interrupts current flow to the ignition coil, causing a spark plug to fire. The air gap is critical on all magnetic sensors and must be set to specifications.

Hall-Effect Switch

The ***Hall-effect switch*** can be mounted in the distributor or at the crankshaft, **Figure 8-16**. The Hall-effect sensor is a thin wafer of semiconductor material with voltage applied to it constantly. A magnet is located opposite the sensor. There is an air gap between the sensor and magnet.

The magnetic field acts on the sensor until a metal tab, usually called a ***shutter***, is placed between the sensor and magnet. This metal tab does not touch the magnet or sensor. When contact between the magnetic field and sensor is interrupted, it causes its output voltage to be reduced. This signals the ignition module to turn the power transistor off. This interrupts primary current to the ignition coil, causing it to fire.

Optical Sensor

The ***optical sensor*** is usually located in the distributor, as shown in **Figure 8-17A**. The rotor plate, **Figure 8-17B**, has many slits in it through which light passes from the ***light emitting diode*** (LED) to the ***photo sensitive diode*** (light receiving). As the rotor plate turns, it interrupts the light beam from the LED to the photo diode. When the photo diode does not detect light, it sends a voltage signal to the ignition module, causing it to fire the coil.

Distributorless Ignition System

The ***distributorless ignition system*** has no distributor, as seen in the schematic in **Figure 8-18**. It uses a ***crankshaft position sensor***, which is a magnetic pickup or Hall-effect switch. The crankshaft sensor is mounted on or in the engine block. Some distributorless systems have a second sensor on the camshaft. The sensor performs the same job that the pickup coil or Hall-effect switch does in the distributor, matching the firing of the spark plug to the piston compression stroke. The advantage of this system is the elimination of the distributor assembly, rotor, and distributor cap.

An electrical signal is generated whenever the crankshaft is rotating and the signal is sent to the ignition module and/or on-board computer. This signal allows the computer to determine the position of each piston in the engine. On systems with crankshaft and camshaft sensors, both sensor readings are used to determine piston position. The sensor input may also be used by the computer to determine engine RPM and the amount of ignition timing advance.

The distributorless ignition creates a high voltage spark using multiple ignition coils. There is one ignition coil for every two cylinders. A four-cylinder version has two coils, a six-cylinder has three coils, and a V-8 uses four coils. Multiple coils must be used since there is no distributor cap and rotor to distribute the spark.

All distributorless ignition coils have two discharge terminals. These terminals are connected to two of the engine spark plugs through conventional resistor plug wires. When

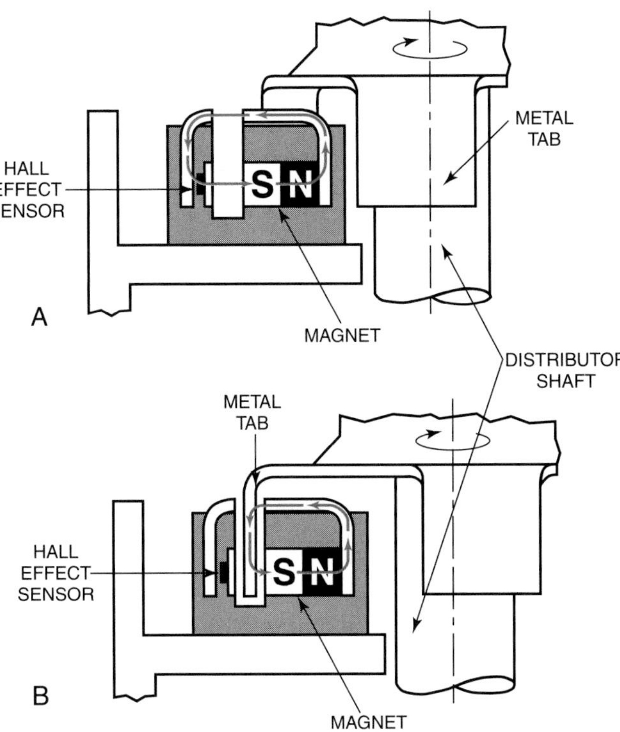

Figure 8-16. A—A magnetic field can act on a Hall-effect sensor. B—When the metal tab, attached to the distributor shaft rotates between the magnet and Hall-effect sensor, the magnetic field is interrupted. The ignition coil sends high voltage to the distributor any time the magnetic field is interrupted. (Robert Bosch)

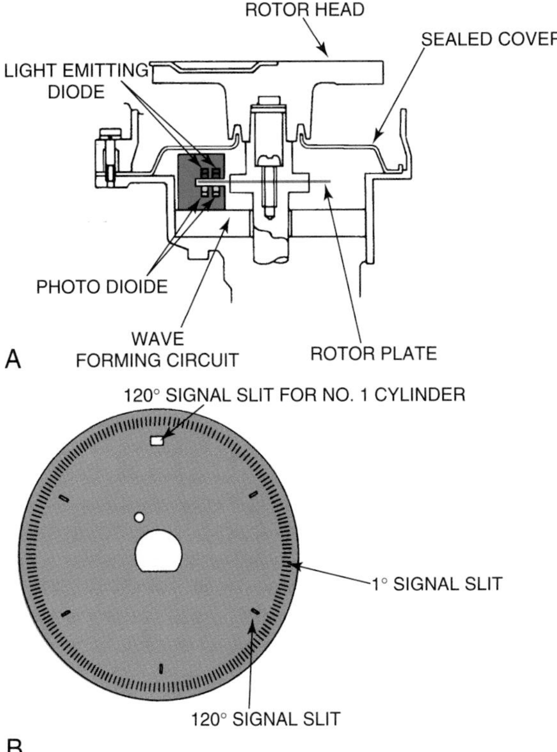

Figure 8-17. A—An optical crankshaft position sensor uses an LED to send a beam of light to a photo diode through slits in a rotor plate. B—A rotor plate used with the optical sensor. Notice the spacing of the slits. (Nissan)

the coil fires, the spark exits one terminal, travels through the plug wire to fire the plug, and returns to the other coil terminal through the engine block, the other spark plug, and the other plug wire. In effect, the coil fires both plugs at the same time.

The coil wires are arranged so that the coil fires one plug on the top of the compression stroke and the other plug on the top of the exhaust stroke. The plug firing on the top of the exhaust stroke has no effect on the operation of the engine and is often called a ***waste spark***. Since it takes very little voltage to jump the spark plug gap on the exhaust stroke, the coil is powerful enough to fire both plugs.

An ***integrated direct ignition system*** is a variation of the distributorless ignition system. This system uses conductor strips instead of spark plug wires to transfer electricity from the coils to the spark plugs. As in all distributorless systems, each coil serves two spark plugs. See **Figure 8-19**.

Direct Ignition System

A ***direct ignition system*** is similar to a distributorless ignition system. However, there is one coil for each spark plug in a direct ignition system. Spark plug wires or other conductors are not used between the coils and the plugs. Instead, the coil towers are connected directly to the spark plugs. See **Figure 8-20**.

Primary Circuit Complete

You have now traced the flow of current through the primary system. After going through the control module or contact points, it returns to the battery through the metal parts of the vehicle to which it is grounded. Be sure you understand how each unit works and its relationship to the other parts.

Secondary Circuit

You have seen how the coil produces high voltage current. The job of the ***secondary circuit*** is to send this current to the proper spark plug at the proper time. The current jumps from the center electrode of the plug to the side, or ground electrode. When it jumps across, it produces a hot spark that ignites the air-fuel mixture. This section explains how this is accomplished.

Spark Plug Construction

All internal combustion gasoline engines have one thing in common: they use spark plugs. The ***spark plug*** is made up of three major parts: the electrodes, insulator, and shell. See **Figure 8-21**.

Spark Plug Electrodes

The typical modern spark plug may be expected to last from 15,000-30,000 miles (24 000-48 000 km) or more. Some spark plugs can last as long as 100,000 miles (160 000 km). To accomplish this, the plug ***electrodes*** must be constructed of material that will be resistant to heat, oxidization, and burning. Typical materials used to make spark plug center electrodes include copper and nickel alloy. Platinum, although expensive, is sometimes used. The center electrode is insulated from the rest of the plug by a ceramic insulator. The top terminates in a push-on terminal to which the plug wire may be attached, **Figure 8-21**.

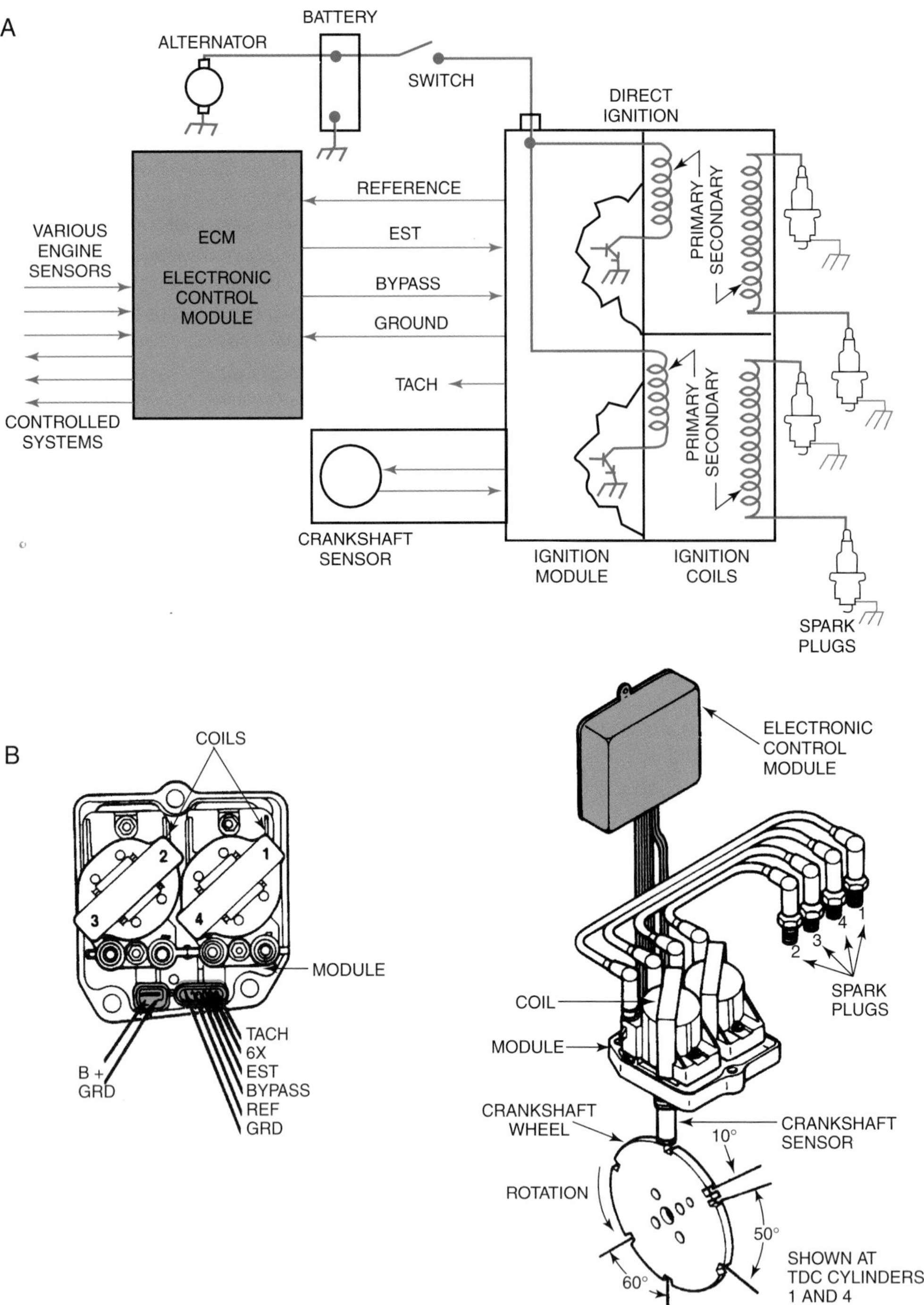

Figure 8-18. A—An electronic distributorless ignition system schematic. B—One possible arrangement of components for a distributorless ignition system. (AC-Delco)

In a modern spark plug there are two electrodes, the center electrode and the side electrode. The space between the two is called the plug ***gap***. On older vehicles, the plug gap may be as small as .025″ (0.63 mm). The spark plugs in newer vehicles may require gaps of as much as .095″ (2.41 mm). The majority of plug gaps are between .045″ to .065″ (1.1 to 1.65 mm). Always check the service manual for the exact plug gap.

Electrons will flow more easily from a hot surface to a cooler one. As the center electrode of a spark plug is always hotter than its side electrode, the plug's center electrode has negative polarity. This ensures a better flow of current from the center electrode. Air is more easily ionized (broken down from a nonconductor to a conductor) near the hotter center electrode. As a result, less voltage is needed to create a spark, which lightens the load on the coil.

In distributorless ignition systems, the waste spark travels from the side electrode to the center electrode. Since the cylinder is not under compression, an extra load is not placed on the coil. However, reverse firing can cause electrode

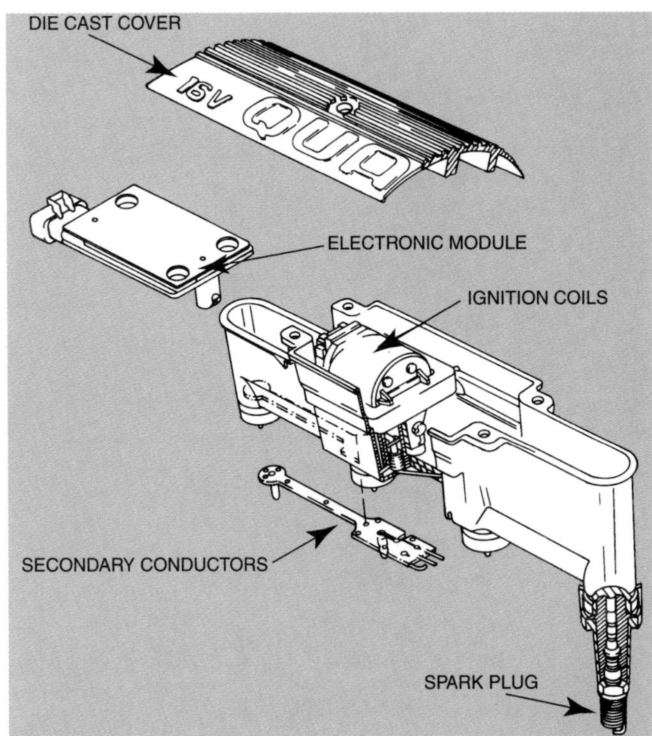

Figure 8-19. Exploded view of a direct ignition system. This two-coil setup is for use with a four-cylinder engine. (Oldsmobile)

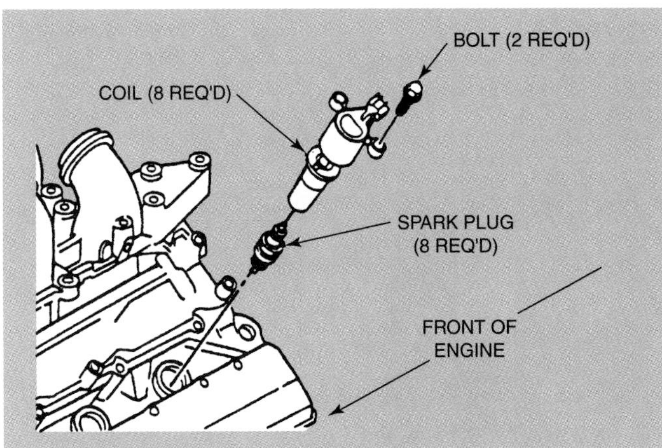

Figure 8-20. Exploded view showing the coil and spark plug arrangement for one cylinder of a V-8 engine with direct ignition. Each plug in this engine has its own coil. (Ford)

damage to certain types of spark plugs. The technician must always make sure that the replacement plugs are designed for use in a distributorless system.

Spark Plug Insulators

Spark plug ***insulators*** must have special properties. They must resist heat, cold, chemical corrosion, and sudden voltage changes. Insulators must also be resistant to vibration and physical shock. A common material used for making spark plug insulators is aluminum oxide. The aluminum oxide is fired at high temperature to produce a glassy smooth, dense, and very hard insulator. The length, diameter, and location of the insulator has a direct bearing on the heat range of the particular plug, which is covered later in this chapter. The top end of the insulator is often ribbed, or grooved, to prevent flashover, **Figure 8-21**.

Spark Plug Shells

The center electrode, surrounded by the insulator, is placed in a steel ***shell.*** The steel shell top is generally crimped over to bear against a seal. The crimping process grips the insulator tightly and also forms a pressure seal at both the top and bottom of the insulator. This prevents combustion leaks and arcing.

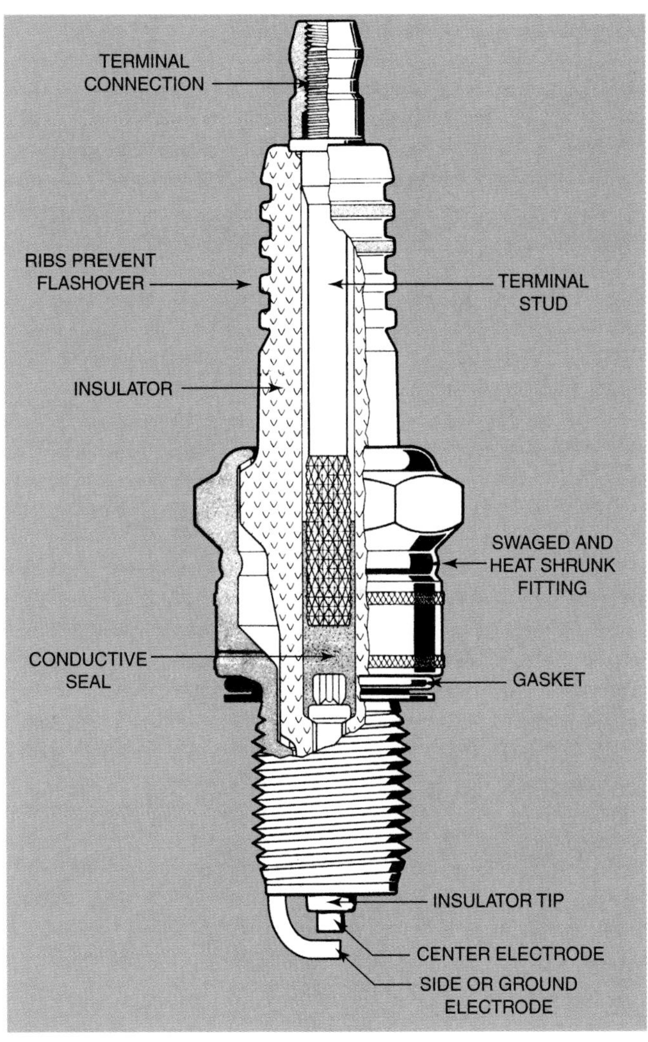

Figure 8-21. Cutaway of a spark plug. Study and learn the names of all the parts. (Bosch)

The side electrode is welded to the steel shell. The shell is threaded so it will screw into a threaded hole in the cylinder head. The shell's threaded area will vary in length and thread pattern to conform to the cylinder head in which it is installed. The shell forms a seal with the head by means of a copper or aluminum gasket, or by a beveled edge that wedges against a similar bevel in the cylinder head, **Figure 8-22**. The thread seal is important since this is an area through which a great deal of the plug's heat is transferred to the head metal. The insulator is subjected to tremendous temperatures. In order to prevent burning, it must get rid of surplus heat.

Spark Plug Heat Range

Spark plugs are designed to operate at a certain temperature, neither too hot nor too cold. If a plug runs too cold, it will collect deposits and become ***fouled***. If it runs too hot (much above 1,700°F or 967°C), it will cause ***preignition*** from hot carbon deposits igniting the fuel charge before the plug fires. It is important that the plug extend into the combustion chamber just the right amount. The ***heat range*** is determined by the diameter and length of the insulator as measured from the sealing ring down to the tip, **Figure 8-23**.

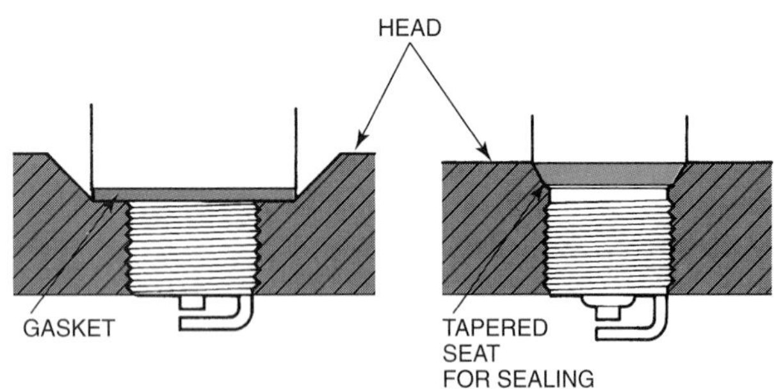

Figure 8-22. Spark plug thread seals. Not all spark plugs use a gasket for sealing purposes.

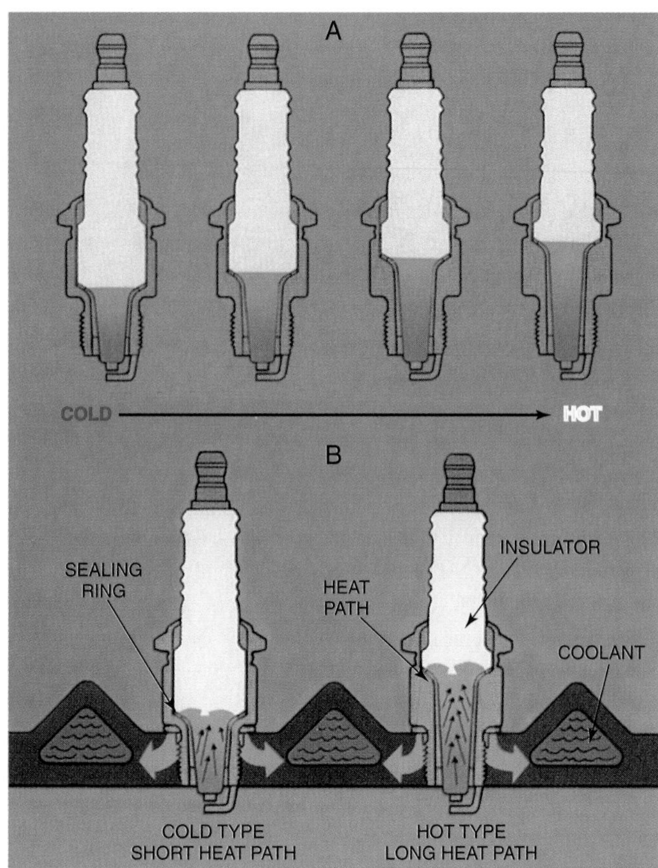

Figure 8-23. A—Cold-to-hot spark plug heat range. B—A cold type, short heat path and a hot type, longer heat path. Study the heat flow through both types. (Champion Spark Plug Company)

Notice in **Figure 8-23** that the heat generated in the insulator must travel up until it can escape through the seal to the shell and then to the head. The longer and thinner the insulator tip, the less efficiently it can transfer heat. As a result, it will run hotter (hot plug). The short, heavy insulator carries heat well and will operate much cooler (cold plug). All engine manufacturers specify heat ranges for the plugs to be used in their engines.

Projected Nose Plugs

One spark plug design extends the electrode tip and insulator further down into the combustion chamber. This ***projected nose*** places the insulator more directly in the path of the incoming fuel charge, which cools the tip. See **Figure 8-24**. The projected nose plug, due to its long tip, tends to run hot at low engine speeds because the incoming fuel charge is moving slowly and provides little cooling effect. This prevents low speed fouling and assists in producing a spark without excessive voltage from the coil.

At high engine speeds, the insulator temperature is kept from exceeding acceptable limits by the rapid flow of the incoming fuel mixture. The washing effect of the hot escaping exhaust tends to keep the plug clean. The projected nose plug will perform satisfactorily within acceptable temperature limits over a wider vehicle speed and load range. See **Figure 8-25**.

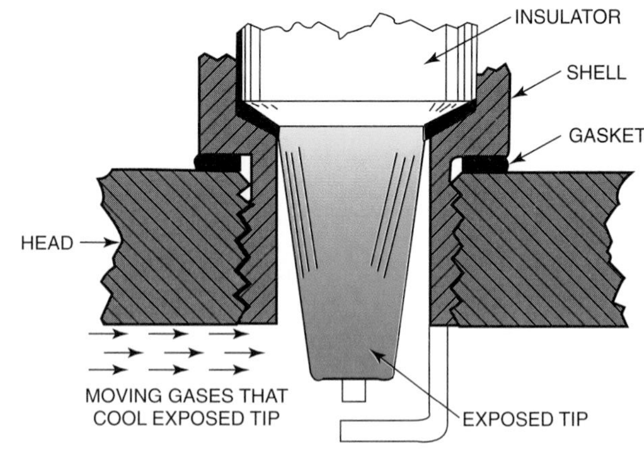

Figure 8-24. Projected core nose plug. Not all spark plugs use this type of nose plug.

Resistor Spark Plugs

The spark at the plug electrodes is delivered in two stages. The voltage at the plug's center electrode will rise rapidly until the voltage is sufficient to ionize the gap and cause the plug to fire. This is the first stage. The second stage is longer and follows the first. It is produced by the remaining residual voltage in the coil.

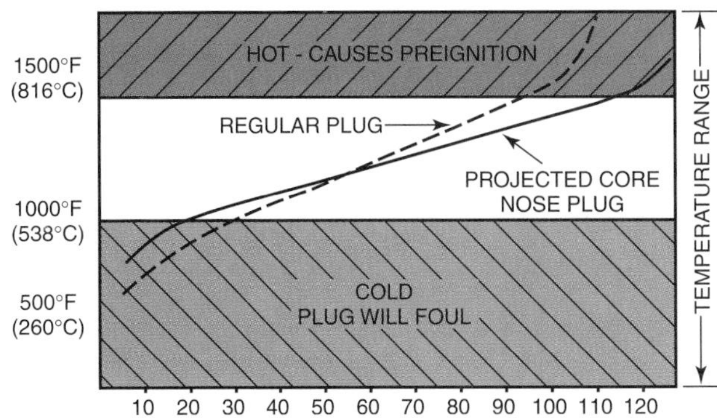

Figure 8-25. Speed/temperature range performance graph.

The combustion process takes place during the first stage. The second stage causes radio interference and can cause the plug electrodes to wear. To shorten the second stage, a ***resistor*** of around 10,000 ohms is placed in the ceramic insulator. The resistor shortens both the stages. It does not require any higher voltage and will lengthen electrode life as well as suppress radio interference. A resistor plug is shown In **Figure 8-26**. **Figure 8-27** illustrates the difference between the spark produced by a standard plug and a resistor type plug. No additional voltage is required for the resistor plug. You have seen how the necessary voltage is produced. How this voltage is delivered to each plug at the proper time will now be covered.

Distributor Cap

High voltage from the coil is carried by an insulated wire to the center terminal of a ***distributor cap***. On some vehicles, the coil is installed in the distributor cap. Additional terminals, one per cylinder, will be arranged in a circle around the center terminal. Each one of these will have a heavily insulated wire connecting it with a spark plug. The distributor cap is made of a plastic material, sometimes mica-filled to reduce flashover tendency. The materials used must provide excellent insulation. **Figure 8-28** shows the relation of the distributor cap to the other distributor components.

On the inside of the cup-like cap, the plug terminals have metal (aluminum, brass, etc.) lugs extended down past the cap material. Note the fixed carbon rod used in **Figure 8-29**. The distributor cap shown in **Figure 8-30** contains the coil terminal and a small metal connector or brush which forms the connection between the coil and rotor.

Rotor

To carry the secondary voltage from the center cap terminal to any one side terminal requires what is known as a ***rotor***. In cases where the coil is installed in the distributor cap, voltage is discharged directly to the rotor through a metal connector in the cap. The rotor body is made of an insulating material, usually plastic. It is attached to the top of the distributor shaft. Its outer edge has a brass terminal that passes, as it turns, very close to the side terminals. This outer edge terminal is connected to a spring that rubs against the center coil terminal carbon post.

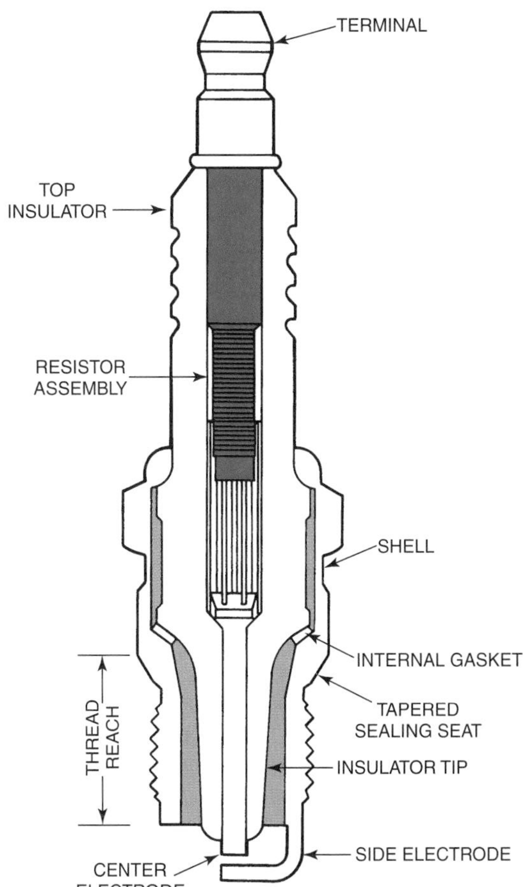

Figure 8-26. A cutaway view of one type of resistor spark plug. (AC-Delco)

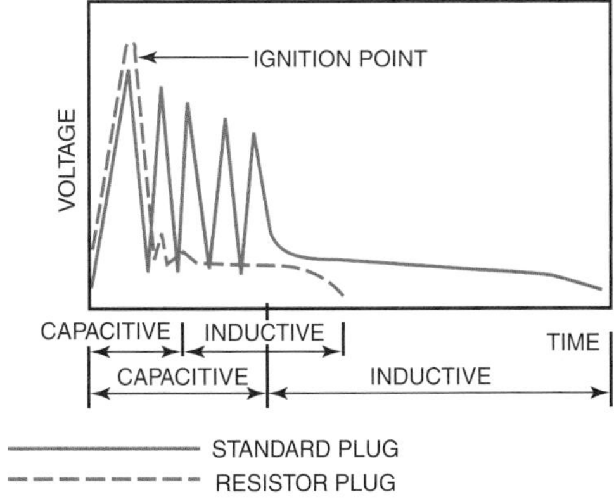

Figure 8-27. Spark plug discharge graph. Notice how the resistor plug reduces undesirable inductive portion of spark.

When current from the coil arrives at the center terminal, it travels down the carbon rod, through the spring, and out to the rotor's outer edge terminal. From this point, it jumps the small gap between the rotor and side terminal and goes on to the plug. A typical rotor is shown in **Figure 8-30**. Jumping the gap between the rotor and distributor cap terminal requires about 3000 extra volts, well within the coil's capacity.

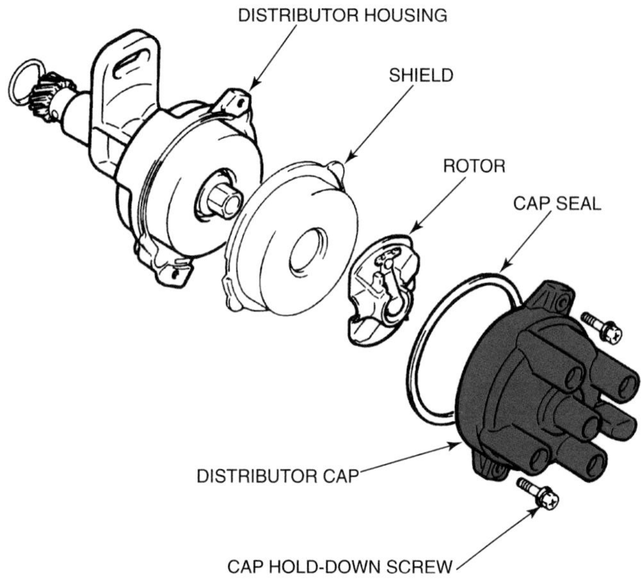

Figure 8-28. A distributor cap for a four-cylinder engine. (Geo)

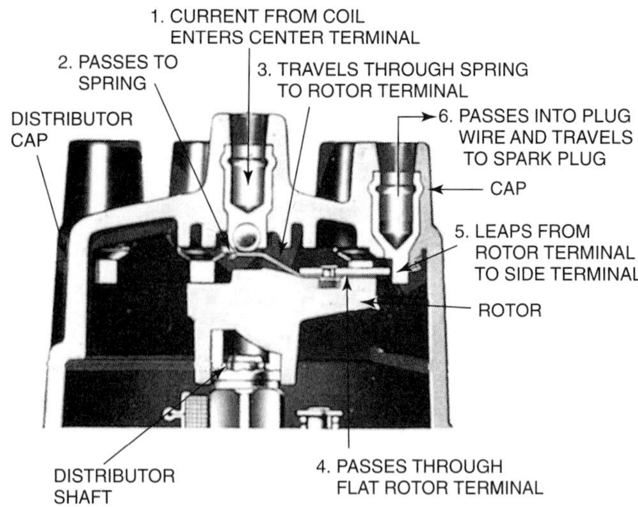

Figure 8-30. Cutaway view of a distributor cap and rotor. The rotor transfers current from the center terminal to the outer terminal. (Oldsmobile)

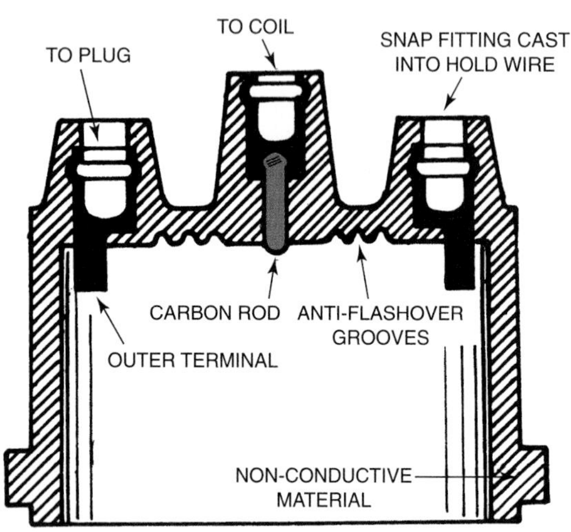

Figure 8-29. Cross-sectional view of a distributor cap. Note the carbon rod.

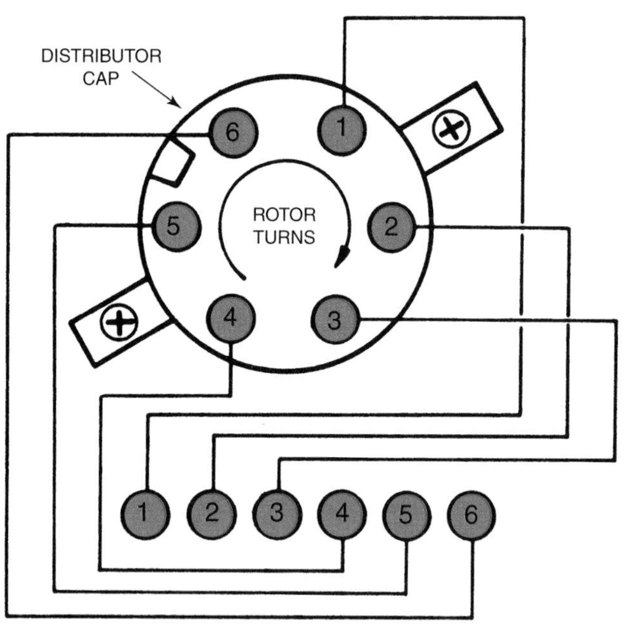

Figure 8-31. A simplified firing order for a six-cylinder engine: 1,2,3,4,5,6. Notice how the plug wires are arranged in the distributor cap to produce this firing order. (Chevrolet)

Spark Plug Wires

The ***spark plug wires*** carry the high voltage current from the distributor cap to the spark plugs. Spark plug wires are covered with a thick layer of insulation to protect the wire and to reduce the possibility of arcing. Original equipment plug wires contain a built-in resistance to reduce radio interference. Most replacement wires are also resistor types. Resistor wires should always be used in late model vehicles to prevent radio and computer system interference. The plug wires are arranged in distributor cap in that order. Direction of rotor rotation dictates whether a clockwise or counterclockwise order is required, **Figure 8-31**.

The cap, rotor, and wires distribute the spark to each plug at the proper time. It is essential that the plug wires be arranged in the proper sequence. The firing sequence in **Figure 8-31** is never used. An example of a typical firing order (order in which pistons reach TDC on the compression stroke) in a six-cylinder engine is 1, 6, 5, 2, 4, 3.

Ignition Timing and Advance

For correct ***ignition timing***, each cylinder should receive a spark at the plug electrodes as the piston nears the top of its compression stroke (a few degrees before TDC). This is made possible by driving the distributor shaft so that it turns at one-half crankshaft speed. The distributor shaft may be turned by one-to-one gearing with the camshaft, which is already turning at one-half of the engine's rotational speed. On some engines using a timing belt, the distributor is driven by the belt.

The distributor shaft gear is timed so that the spark is produced when the cylinder is ready to fire. The rotor will then point toward the cylinder's cap plug terminal. A plug wire is attached to this terminal. The wires are attached to the cap starting at the number one cylinder and following the firing order in the direction of distributor shaft rotation.

As the engine turns, the distributor shaft revolves. Each time the distributor shaft has turned enough to cause the rotor to point to a plug terminal, the ignition system produces a spark. This cycle is repeated over and over. The engine manufacturer specifies timing in regard to the number of degrees before top dead center (TDC) that the number one cylinder should fire. All other cylinders will fire at the same number of degrees before TDC. If the plug fires later than the specified setting, the timing is said to be ***retarded***. If the plug fires earlier than specified, the timing is referred to as ***advanced***.

Setting Basic Timing

Most older engines, and many newer ones, have timing marks in the form of a line marked on the rim of the vibration damper. Some engines in front-wheel drive vehicles have timing marks on the flywheel. A pointer is attached to the timing cover. When the mark is exactly under the pointer, as in **Figure 8-32**, the engine is ready to fire number one cylinder. The spark will occur with the rotor pointing to the number one cap terminal. The timing is generally set by using a strobe lamp, which is a light that is operated by high voltage surges from the spark plug wire. See **Figure 8-33**. The strobe lamp is usually referred to simply as a ***timing light***.

To time the engine, the timing light pickup is clamped over the number one (or other cylinder as may be specified) plug wire. On most engines, special steps must be taken before setting the initial timing, such as disconnecting the vacuum line to the distributor, or grounding an electrical connector to the computer. On older engines with contact points, the point gap should be set before timing the engine. The engine is then started and operated at idle. Many modern vehicles do not have a provision for setting timing. Check the emissions sticker in the engine compartment before looking for the timing marks.

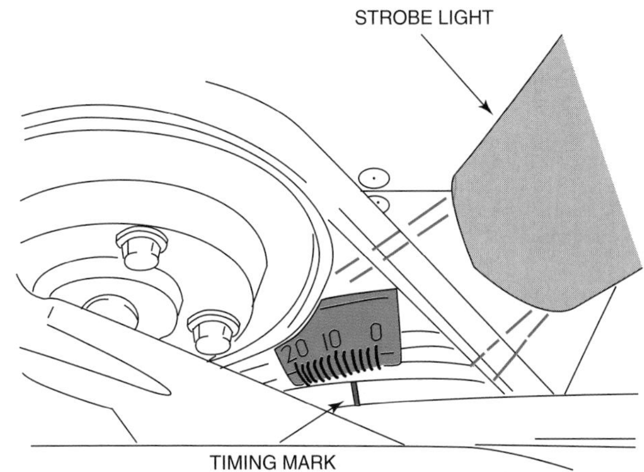

Figure 8-33. Using a strobe light to time the ignition. Every time the No. 1 plug fires, the strobe light will illuminate timing marks. (General Motors)

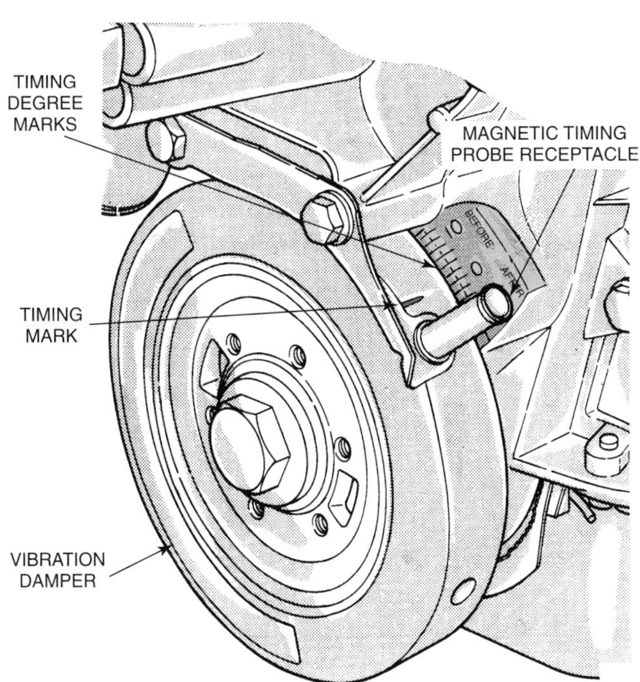

Figure 8-32. Typical ignition timing marks have degrees marked before and after top dead center. This setup also incorporates a magnetic timing probe receptacle. (Chrysler)

Using a Timing Light

The timing light will illuminate the pointer over the vibration damper. The timing is checked by pointing the light at the timing marks. Every time the number one plug fires, the strobe lamp lights. Each time it fires with the damper in the same position in relation to the pointer, the damper timing mark looks as though it were standing still.

To adjust the timing, the distributor clamp is loosened and the distributor is turned by hand. As it is turned, the timing mark will move. When turned in the proper direction, the mark will line up with the pointer. When the two are aligned, the engine is properly timed and the distributor clamp can be tightened. Remember to reconnect any vacuum lines or electrical connectors as applicable.

Magnetic Timing Meter

Many late model engines can be timed with a ***magnetic timing meter***. This meter has a timing probe which is installed in a magnetic timing receptacle near the conventional timing marks, such as the one shown in **Figure 8-32**. The timing meter also uses an inductive pickup which clamps over the number one spark plug. Once all connections are made, the engine is started, and timing can be read directly from the meter dial.

Timing Advance Mechanisms

As engine speed increases, it is necessary to fire the mixture sooner. If this is not done, the piston would reach TDC and start down before the air-fuel mixture can be properly ignited. To properly fire the air-fuel charge, a device is needed to advance the engine timing (firing more degrees before TDC) as the engine speed increases. It is also necessary to retard the timing to control exhaust emissions and prevent spark knock.

When the engine is at idle, very little advance is necessary. At higher engine speeds, it is necessary to fire the mixture somewhat sooner. To see this concept, look at **Figure 8-34**. In this example, the pressure of a burning fuel charge will end when the piston reaches 23° after TDC. Notice in **Figure 8-34A**, that the combustion cycle must start at 18° before TDC in order to be complete by 23° after TDC. In **Figure 8-34B**, engine speed has tripled. It is now necessary to ignite the charge at 40° before TDC in order to complete combustion by 23° after TDC. The three common methods of advancing the ignition timing are centrifugal advance, vacuum advance, and electronic advance.

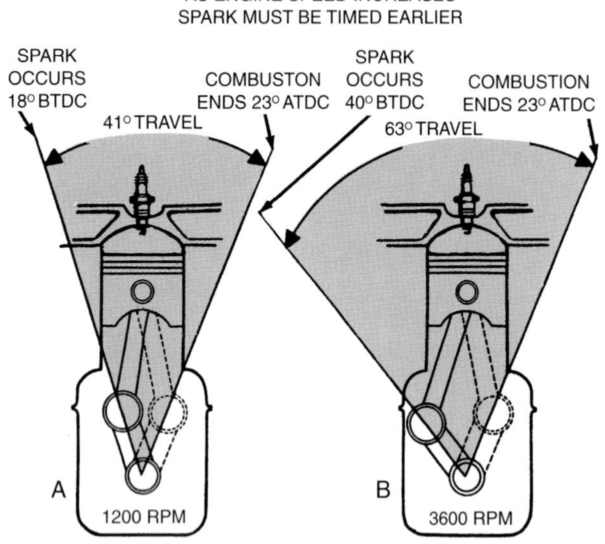

Figure 8-34. As engine speed increases, the spark must be timed sooner. Notice in A that only 41(of crankshaft travel is required while in B, at 3,600 RPM, 63° (is necessary. (Ford)

Centrifugal Advance

One method of advancing the timing is through the use of a ***centrifugal advance mechanism***, which is assembled in the distributor shaft. In effect, the distributor shaft is divided into upper and lower parts, with the upper part able to advance in relation to the lower part, **Figures 8-35** and **8-36**. When the distributor shaft turns, it turns the centrifugal unit which turns either the cam (contact point ignition) or the reluctor or shutter (electronic ignition). Centrifugal advance will advance engine timing in relation to engine speed.

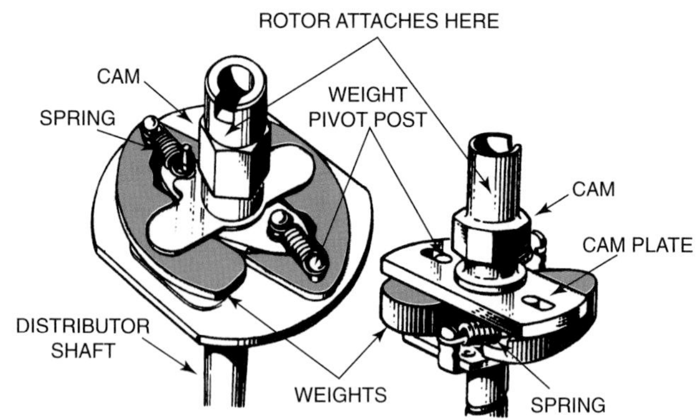

Figure 8-35. Two different types of distributor centrifugal advance mechanisms. (Automotive Electric Association)

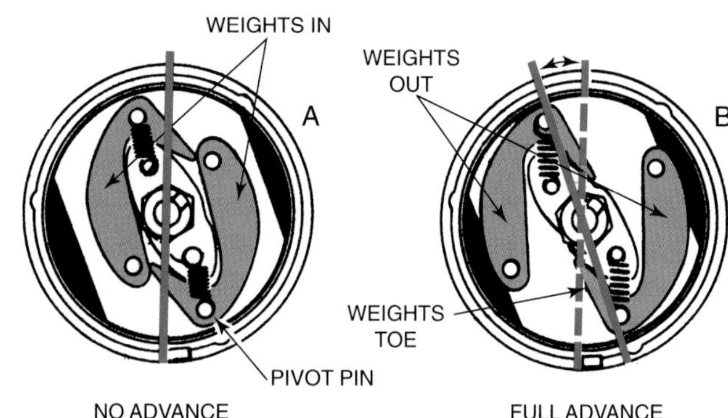

Figure 8-36. Distributor centrifugal advance unit in action. A—The engine is idling and the springs have drawn the weights in. Timing has no advance. B—The engine is running at high speed. Centrifugal force has drawn the weights outward. As they pivot, the weight toe ends force the cam plate to turn, advancing the timing.

When the engine is idling, spring pressure keeps the two weights drawn together and the shaft remains at the position for low speed timing. As the engine speeds up, the weights are drawn out by centrifugal force. As the weights move apart, they force the upper part of the shaft to move in an advance direction in relation to the lower part of the shaft. As the upper part of the shaft is advanced, it causes the triggering device to fire the coil sooner, causing the plugs to fire more degrees before TDC. The faster the engine turns, the farther apart the weights move, until they finally reach the limit of their travel.

As engine speed is decreased, the centrifugal pull on the weights is lessened and the springs pull the weights together, retarding the timing. By calculating the pull of the springs and the size of the weights, it is possible to properly advance the timing over a long RPM range. **Figure 8-36** illustrates how the weights control advance. It is possible to change the amount of centrifugal advance by changing the weights and springs. This should be done very carefully to reduce the chance of engine damage.

Vacuum Advance

It has been found that at the partial throttle open position, additional advance over and above that provided by a centrifugal mechanism is desirable. This is due to the fact that there is high vacuum in the intake manifold when the throttle valve is partially open. This high vacuum draws in less air and fuel. The smaller air-fuel mixture will be compressed less and will burn slower.

To maximize economy from this part of the fuel charge, it is necessary to advance the timing beyond that provided by centrifugal weights. This is provided by the ***vacuum advance mechanism***, **Figure 8-37**. Vacuum advance is used to advance the timing in relation to engine load. Any benefit from additional advance applies only to the partial open throttle position. During hard acceleration or wide open throttle operation, there is no manifold vacuum to operate the vacuum advance mechanism.

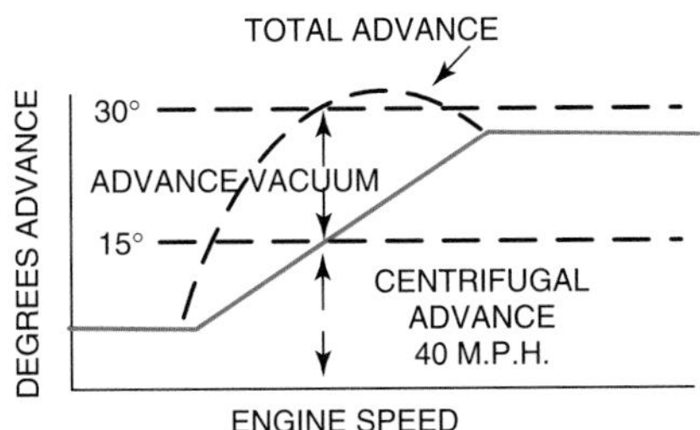

Figure 8-37. A vacuum advance chart. Notice that part throttle vacuum advance is in addition to regular centrifugal advance.

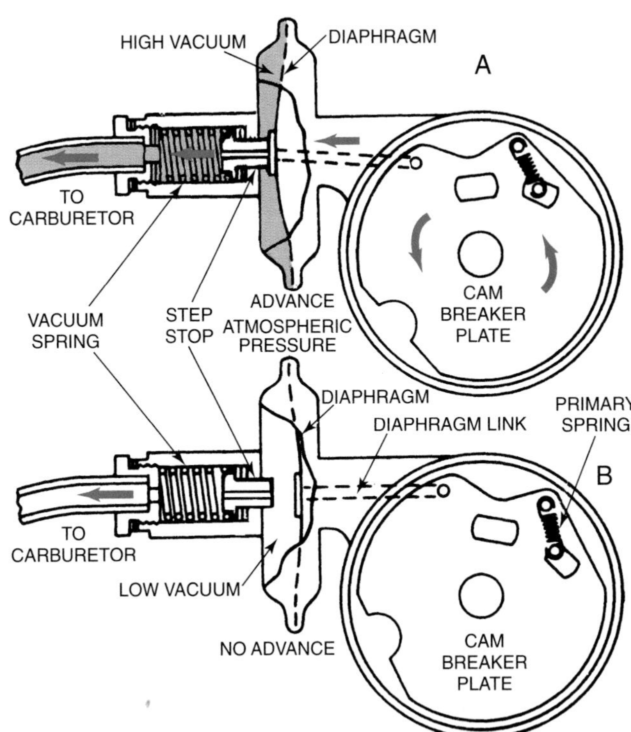

Figure 8-38. Vacuum advance mechanism operation. A—The carburetor throttle valve is in the part throttle position, drawing a heavy vacuum. With a vacuum on the left side, atmospheric pressure forces the diaphragm to the left. The diaphragm link will pull the cam contact plate around and advance the timing. B—When the throttle is opened and vacuum is lowered, the contact plate primary spring will pull the contact plate back and retard the timing. The vacuum spring also controls the limits of advance. (Ford)

Vacuum Advance Operation

The electronic pickup or contact points are mounted on a movable plate. This plate can revolve either on a center bushing, or on ball bearings on its outer edge. With either type of plate, an advance in timing may be obtained by turning the plate against distributor shaft rotation. See **Figure 8-38**.

The plate is rotated by means of a vacuum advance diaphragm. This is a stamped steel container with a neoprene-coated cloth diaphragm stretched across the center. One end is airtight and is connected to the carburetor, below or slightly above the closed position of the throttle valve. The other end is open to the atmosphere. The diaphragm has a lever rod attached to its center. The lever rod is connected to the moveable plate. See **Figure 8-38**.

When the throttle valve is partially open, as shown in **Figure 8-38A**, there is high vacuum in the intake manifold. The vacuum pulls the diaphragm back toward the vacuum side. This in turn pulls the plate around and advances the timing. When the throttle is opened, the vacuum drops and the spring draws the diaphragm back toward the distributor. This rotates the moveable plate in the retard direction, **Figure 8-38B**. When the engine is idling, the throttle valve closes below the vacuum advance opening. This removes vacuum pull and the spark will be retarded for idling. The vacuum advance mechanism is constantly moving as vacuum varies with the movement of the throttle valve.

The vacuum advance is adjustable on some vehicles. If the vacuum advance is not operating, it can affect engine performance and fuel mileage. On some older vehicles, the vacuum may be ported, or not activated until the throttle is partially opened. See **Figure 8-39**. This is covered in more detail in Chapter 14.

Computer-Controlled Ignition

The triggering device for electronic ignition may be inside the distributor, or it may be a camshaft or crankshaft position sensor. Since the entire spark production process is performed electronically, it follows that the spark timing can also be modified electronically. While many older electronic ignitions use a centrifugal and vacuum advance, most modern systems use the on-board engine control computer to create ***electronic advance***.

The latest on-board computers monitor all engine and external variables, such as engine RPM and temperature, engine accessory operation, manifold vacuum, barometric pressure, airflow rate, air temperature, throttle opening, exhaust gas oxygen, transmission gear, vehicle speed, system voltage, and whether the engine is knocking. The computer advances or retards the timing to exactly match the needs of the engine and vehicle. There are no vacuum or

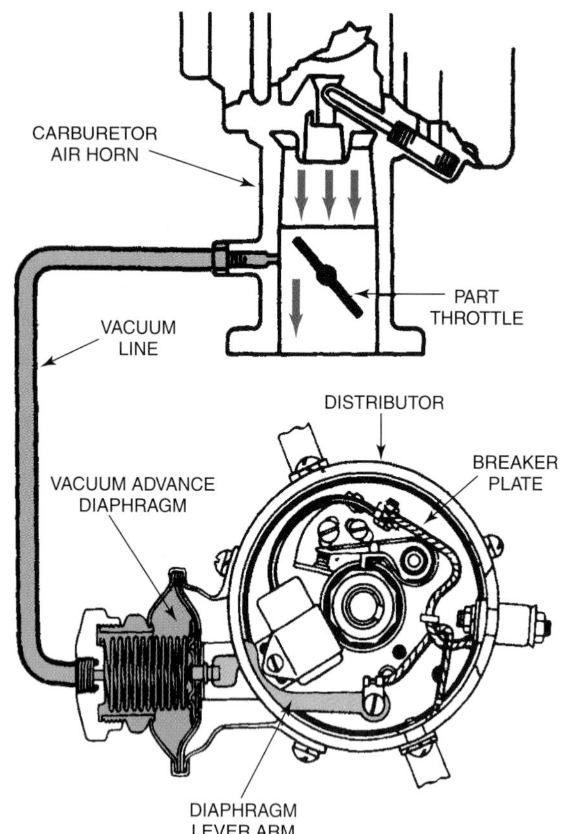

Figure 8-39. This distributor vacuum advance unit utilizes engine vacuum to move the distributor contact plate. (Buick)

centrifugal advance mechanisms on computerized ignition systems. This will be covered in more detail in Chapter 15.

On some systems, the computer contains the ignition module, and controls the coil directly. On other systems the ignition control module is separate, and interacts with the engine control computer. In either case, the amount of advance is set by the computer, and cannot be adjusted.

Summary

Current from the battery flows to the ignition switch. When the switch is turned on, current flows through a resistor to the primary windings in the coil. From the primary coil windings, the current travels to the ignition module or contact points and on to ground where it returns to the battery via the metal parts of the vehicle.

Current flow through the primary windings builds up a magnetic field that surrounds the secondary windings. When the primary ignition control system breaks the primary circuit, the magnetic field around the secondary windings collapses and induces a high voltage in the secondary circuit. The primary circuit can be broken by an ignition module or by contact points. The electronic ignition module is signaled by one of three types of triggering devices.

Direct and integrated direct ignition systems do not use a distributor. Sensors are located on the crankshaft and/or camshaft. These systems use multiple coils. The secondary system consists of the coil secondary winding, distributor cap, rotor, plug wires, and spark plugs. The plugs must last at least 15,000-30,000 miles and sometimes up to 100,000 miles. Most spark plugs contain a resistor for radio static suppression. The cap and rotor distribute the spark to the right plug. Plug wires are usually resistor types for radio interference suppression.

Initial timing is set by the use of a timing light or timing advance meter. The timing is advanced by vacuum, centrifugal, or electronic means. Computerized ignition has the computer advance the spark after measuring all possible engine and vehicle variables.

Know These Terms

Primary circuit
Battery
Charged
Ignition switch
Resistor
Ignition coil
Core
Secondary winding
Primary winding
Magnetic field
Field Collapse
Self-induction
Induction
Polarity
Distributor
Flashover
Contact points
Distributor cam lobes
Dwell
Condenser
Capacitance
Ignition module
Triggering device
Magnetic pickup
Reluctor
Trigger wheel
Hall-effect switch
Shutter
Optical sensor
Light emitting diode
Photo sensitive diode
Distributorless ignition system
Crankshaft position sensor
Waste spark
Integrated direct ignition system
Direct ignition system
Secondary circuit
Spark plug
Electrodes
Gap
Insulator
Shell
Fouled
Preignition
Heat range
Projected nose
Resistor
Distributor cap
Rotor
Spark plug wires
Ignition timing
Retarded
Advanced
Timing light
Magnetic timing meter
Centrifugal advance mechanism
Vacuum advance mechanism
Electronic advance

Review Questions—Chapter 8

Do not write in this book. Write your answers on a separate sheet of paper.

1. What is the maximum voltage that some modern electronic ignition systems can produce?
   (A) 20,000
   (B) 50,000
   (C) 60,000
   (D) 100,000

2. Normal voltage requirements for successful plug firing can range from _____ volts to _____ volts.
3. What is the major difference between electronic and contact point ignition systems?
4. What is the primary job of the ignition coil?
   (A) Convert low voltage to high voltage.
   (B) Convert high voltage to low voltage.
   (C) Reduce primary voltage.
   (D) Reduce secondary voltage.
5. Optical sensors rely on the use of _____ .
6. What causes the electronic ignition module to stop the flow of primary current?
   (A) A signal from the contact points.
   (B) A signal from the triggering device.
   (C) Coil self-induction.
   (D) Coil over-voltage.
7. Dwell is controlled by the _____ .
   (A) vacuum advance
   (B) centrifugal advance
   (C) distributor cap
   (D) None of the above.
8. The magnetic crankshaft sensor generates a(n) _____ that is sent to the ignition module or computer.
9. The ignition control module contains _____ so small that they cannot be seen without a microscope.
10. Where is the power transistor located?
    (A) Pickup coil assembly.
    (B) Ignition module.
    (C) Coil.
    (D) Ignition switch.
11. Triggering devices are operated by rotation of the _____.
    (A) distributor shaft
    (B) crankshaft
    (C) camshaft
    (D) All of the above, depending on the design.
12. What opens and closes the contact points?
13. When the contact point gap is increased, the dwell is _____ .
14. Contact point arcing is prevented by using a _____ .
15. What are the three major parts of the spark plug?
16. If the spark plug heat range is too cold, what could happen?
    (A) Fouling.
    (B) Preignition.
    (C) Detonation.
    (D) Weak spark.
17. A spark plug gap of .025″ (.635 mm) would be found on _____ .
    (A) the newest vehicles
    (B) vehicles with electronic ignition
    (C) older vehicles
    (D) Both A & B.
18. The most common spark plug gaps used today are about _____ .
    (A) .025″ (.635 mm) to .045″ (1.1 mm)
    (B) .045″ (1.1 mm) to .065″ (1.65 mm)
    (C) .065″ (1.65 mm) to .095″ (2.413 mm)
    (D) .095″ (2.413 mm) to .120″ (3.04 mm)
19. Using a resistor spark plug reduces what undesirable condition?
    (A) Static.
    (B) Radio interference.
    (C) Electrode overheating.
    (D) Both A & B.
20. What two components does the center tower of the distributor cap connect?
21. Name the two common methods of advancing the timing in a non-computer controlled ignition system while the engine is operating.
22. All transistor ignition systems use mechanical vacuum and centrifugal advance units. True or False?
23. A six cylinder engine with direct ignition has how many ignition coils?
    (A) 1
    (B) 2
    (C) 3
    (D) 6
24. Take a blank sheet of paper and sketch an electronic ignition system with a magnetic pickup.
25. Take a blank piece of paper and sketch an entire battery ignition (conventional contact point) system on it.

This fuel injected engine uses individual intake ports to channel air to each cylinder. (Ford Motor Co.)

9

Fuel Injection Systems

After studying this chapter, you will be able to:

- Describe the major parts of a gasoline fuel injection system.
- Describe the major parts of a diesel injection system.
- Identify the parts of gasoline and diesel injection systems.
- Summarize the operating principles of gasoline and diesel injection systems.
- Compare the different types of gasoline and diesel injection systems.
- Identify the parts of superchargers and turbochargers.
- Compare supercharger and turbocharger operation.

This chapter covers the principles of gasoline and diesel engine fuel injection. Operating principles and major components of the various types of fuel injection are discussed. Differences and similarities between the major types of fuel injection systems are covered. Studying this chapter will give you the basic information that you need to recognize the different types of fuel injection systems and begin to troubleshoot them.

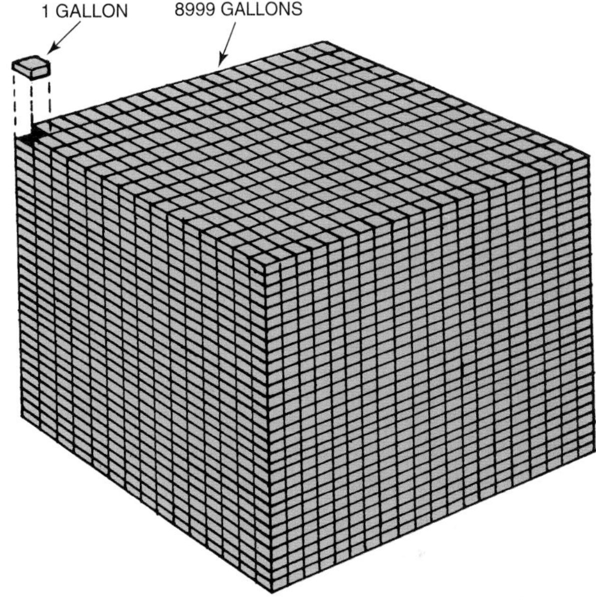

Figure 9-1. The ratio of gasoline to air, by volume. If this were 9000 one gallon cans, 8999 would be filled with air and 1 with gasoline. This is an average fuel mixture ratio.

Providing the Correct Air-Fuel Mixture

As you have learned in the preceding chapters, an engine uses a mixture of air and gasoline or other fuel. This mixture is drawn into the engine by the vacuum formed during the intake stroke. It is then compressed, fired, and exhausted. This cycle is repeated over and over with each crankshaft revolution.

The proportion of air and fuel necessary for proper combustion will vary according to engine speed, load, temperature, and design of the engine. An average mixture for a moderate cruising speed could be 14 to 16 parts of air to 1 part of fuel by weight. For a mixture that produces more power, 12 to 13 parts of air to 1 of fuel might be needed.

Gasoline is 600 times heavier than air. Therefore, if we consider the air-fuel ratio in terms of volume instead of weight, one gallon of gasoline would require around 8999 gallons (34 061 L) of air at an air-fuel ratio of 15 to 1. The relative proportions of gasoline and air by volume are shown in **Figure 9-1**. It is the job of the fuel system to constantly deliver the proper air-fuel mixture to the engine. The mixture and amount must be variable to meet the everchanging needs of the engine, as well as government requirements for mileage and emissions.

Fuel Injection

Fuel injection is the process of spraying fuel directly into the engine. Fuel is sprayed in a ***cone shaped-pattern*** for maximum distribution and atomization, **Figure 9-2**. The fuel is always pressurized. The fuel injection system controls the air-fuel mixture by modifying either the system pressure or the opening of the injectors. The spraying action of the injectors ***atomizes*** the fuel, allowing it to better mix with the air. Although the fuel injection system does not use a carburetor, the system does have a throttle body to control airflow. A throttle body or plate is used on all fuel injection systems.

Direct and Indirect Injection

The pressurized fuel is sprayed through the injector nozzles. Placement of the nozzles is what determines the type of injection. The fuel may be sprayed directly into the cylinder.

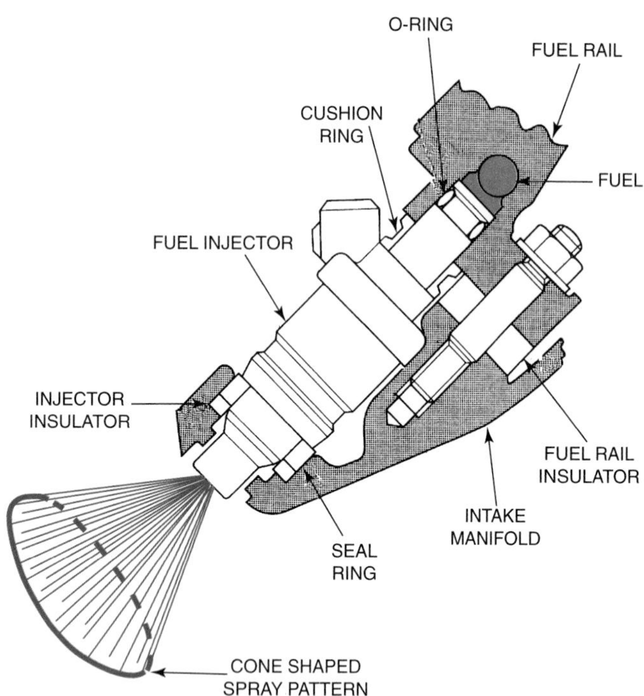

Figure 9-2. An electronic fuel injector produces a "cone" shaped fuel spray pattern in the intake manifold. (Honda Motor Co.)

This is called ***direct fuel injection***, and is used in diesel engines and a few older gasoline engines.

The fuel may also be sprayed into the intake manifold just ahead of the intake valve, or into the throttle body at the entrance to the intake manifold. This is called ***indirect fuel injection***. Modern gasoline injection systems use the indirect injection method. **Figure 9-3** shows one indirect setup. Note how fuel enters the intake manifold just ahead of the intake valve.

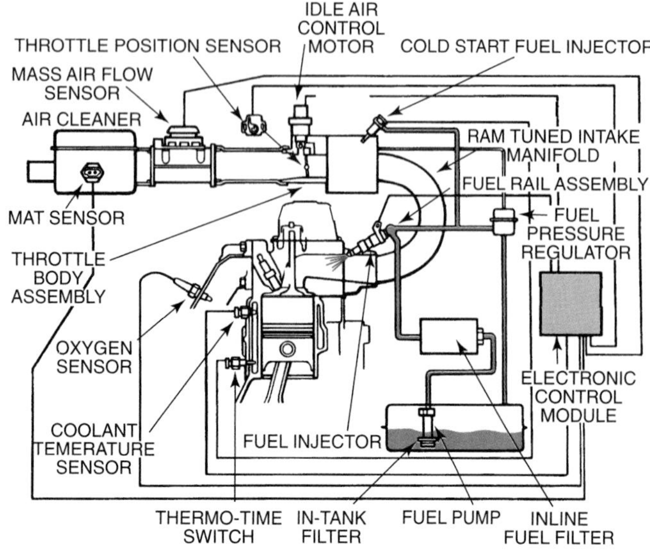

Figure 9-3. Schematic of a computer-controlled electronic fuel injection system. Study the various components and their relationship to the overall system. (AC-Delco)

Mechanical and Electronic Injection

Mechanical fuel injection systems were the first, and, for many years, the only type of fuel injection. These systems used an engine-driven injector pump that distributed pressurized fuel to the injection nozzles. The fuel was then sprayed into the cylinder or into the intake manifold. The pump was precisely timed to the engine intake strokes, in a manner similar to timing an ignition distributor.

The amount of fuel delivered was controlled by the position of the accelerator linkage that in turn actuated a valve or other mechanism within the pump body used to control the amount of fuel injected. A similar system using a fuel rail is discussed in the diesel fuel injection section.

The ***electronic fuel injection*** system was made possible by the introduction of compact, durable, and inexpensive electronic components and computers for automotive use. Like the mechanical type, the electronic fuel injection system delivers fuel, under pressure, to the injectors. Unlike the mechanical type, however, the electronic system does not use an injector pump. An electric fuel pump provides sufficient pressure to produce the proper injector spray pattern. This pressure can vary with the type of system, from about 7 psi (48.27 kPa) up to around 65 psi (448.1 kPa) for others. The computer controls when the fuel is sprayed into the engine.

Computer-Controlled Electronic Fuel Injection System

The operation of most modern fuel injection systems is sensitive to the oxygen content of the exhaust. The system determines the proper air-fuel ratio and then constantly monitors the engine exhaust to verify the accuracy of the mixture setting. Whenever the exhaust sensor determines the oxygen content is off, the system corrects itself to bring the oxygen content back to proper levels.

The system tries to maintain as nearly a perfect air-fuel mixture as possible. A ***stoichiometric*** air-fuel mixture will burn very cleanly with a minimum of pollutants. The ideal mixture is 14.7 parts of air to 1 part of fuel by weight. The computer receives signals from sensors that monitor such things as:

- Throttle position.
- Engine RPM.
- Intake air temperature.
- Coolant temperature.
- Intake airflow.
- Intake manifold vacuum.
- Barometric pressure.
- Exhaust gas oxygen.
- Vehicle speed.

Based on the sensor inputs, the computer determines exactly how much fuel is needed. Changes in sensor signals allow the computer to instantly change injector operation to correspond with engine fuel needs. **Figure 9-3** is a schematic of a computer-controlled fuel injection system. Operation of the computer is covered in detail in Chapter 15.

Pulse and Continuous Electronic Fuel Injection

In the electronic injection system, the air-fuel mixture is controlled in one of two ways. The injector may be opened and closed by electrical signals. This is a ***pulse fuel injection*** system. In the other type of system, the injector is forced open by fuel pressure. Fuel pressure is controlled by an electronic control device, or by an airflow metering valve, depending on the system. This is the ***continuous fuel injection*** system.

Pulse Fuel Injection

In pulse injection systems, the rate of fuel flow through the injectors remains constant. The total amount of fuel delivered is determined by the length of time the injectors are held open. Systems are classified by the type of timing sequence used and the location of the injectors.

Timed and Non-Timed Pulse Fuel Injection

Some pulse fuel injection systems inject fuel ahead of each valve just before and/or during the intake stroke. Each injector is open for only several milliseconds and is carefully timed to each cylinder intake stroke. This is the ***timed fuel injection*** system.

The injector is opened by means of an electric solenoid (magnetic winding). When an electrical current is passed through the solenoid winding, it forms a magnetic field that draws the sealing needle off its seat. This admits a stream of fuel for as long as the injector remains energized. The injector must emit an even, smooth stream. Remember, the injector is only open while energized.

Many pulse fuel injection systems do not attempt to precisely time each injector to a specific cylinder intake stroke. Some systems energize half of the injectors all at once, then the other half on an alternating basis. Other systems use two injectors at the entrance to the intake manifold and operate them either at the same time or on an alternating basis. These are called ***non-timed fuel injection*** systems.

These non-timed injection systems spray fuel into the manifold where it will remain until the intake stroke draws it into the cylinder. The injector sprays fuel at regular intervals. Although such systems might appear poorly designed, in actual use, they work very well.

Injector Pulse Width

Injector ***pulse width*** refers to the amount of time during which the injector electronic solenoid is energized, causing the injector to open. Pulse width is a measurement of how long the injector is kept open; the wider the pulse width, the longer the injector stays open. The amount of fuel delivered by a given injector depends upon the pulse width. Pulse width is controlled by the computer.

Types of Pulse Fuel Injection Systems

The next section will discuss common fuel injection systems which are typical of the pulse systems now in use. The pulse injection system is usually electronically controlled. Although there are a number of design differences between various pulse fuel injection systems, the operating principles are very similar.

Throttle Body Injection System

The ***throttle body injection*** system uses an assembly containing one or two fuel injectors. The assembly is mounted at the entrance to the intake manifold and injects fuel ahead of the throttle valve. These systems are also called single-point injection or central fuel injection. A typical throttle body injection system is shown in **Figure 9-4**.

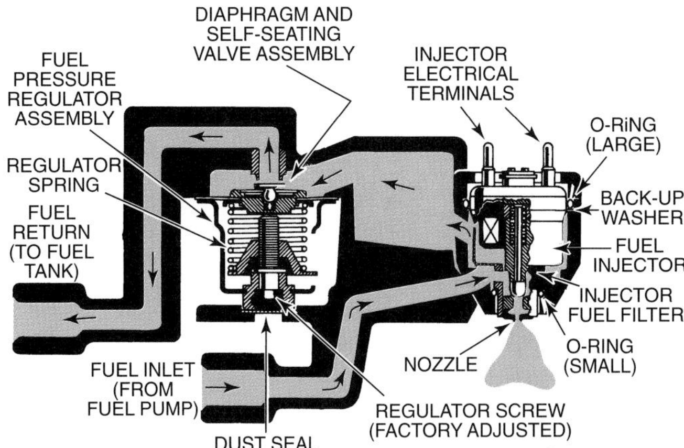

Figure 9-4. This unit will regulate fuel pressure to the injector. Excess pressure opens the regulator valve and extra fuel flows back to the tank. (Cadillac)

The throttle body injection system shown in **Figure 9-5** employs two injectors. Both injectors are mounted on the throttle body directly above the throttle valves. When the engine is running, the injectors are energized alternately. The throttle body mounts on top of the intake manifold much like a carburetor and contains both injectors, throttle valves, idle air control valve, throttle position sensor, and pressure regulator.

In this system, the pump supplies fuel flow. Fuel pressure to the injectors in this system is approximately 10 psi (68.9 kPa). Fuel pressure is controlled by a vacuum-operated regulator. Any excess fuel is returned to the tank. The throttle valves, actuated by the accelerator, determines airflow. The computer, acting on signals from the sensors, determines and constantly adjusts injector pulse width and energizes the injector solenoids accordingly.

Multiport Fuel Injection

When the system uses one injector per cylinder, the individual injector nozzles are located inside the intake manifold or in the cylinder head, close to the intake valve. These systems are called ***multiport fuel injection***. Some other names for these systems include multipoint fuel injection, sequential fuel injection (timed), and port fuel injection. Typical multiport injectors are shown in **Figure 9-6**.

A version of the multiport injector system uses one central fuel injector with lines connecting it to outlet nozzles at each intake passage. This system, shown in **Figure 9-7**, combines the simplicity of the single injector with the more precise fuel control of the multiport system.

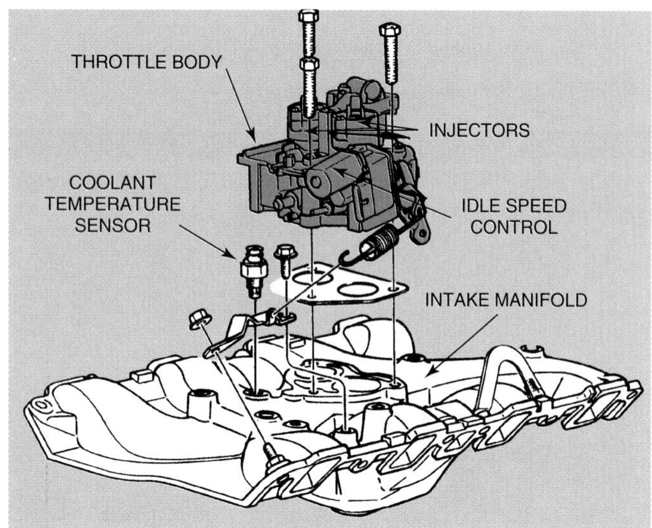

Figure 9-5. Throttle body with injectors. The injectors are energized at the same time during cranking and alternately during normal engine operation. (Cadillac)

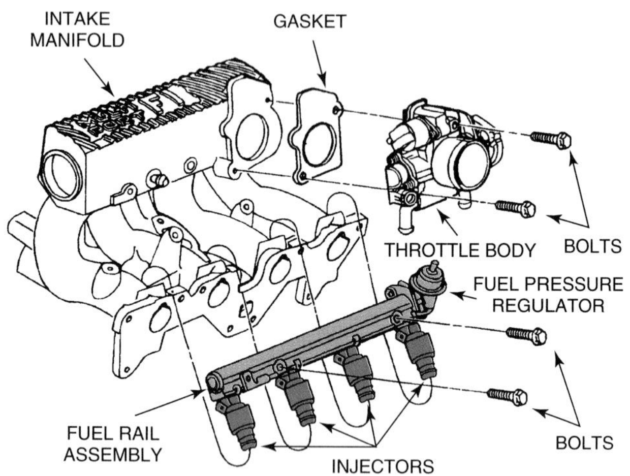

Figure 9-6. One particular multiport fuel injector system arrangement for a four-cylinder engine. Each individual injector is connected to the fuel rail. (Chevrolet)

Control of the multiport system resembles the throttle body fuel injection system. The electric fuel pump supplies fuel and creates fuel system pressure. The fuel pressure regulator controls the system pressure. Movement of the throttle valve determines airflow. The computer receives signals from the sensors and determines proper injector pulse width. The computer constantly controls the injector pulse width to maintain the proper air-fuel ratio.

Pulse Fuel Injection System Components

The typical pulse system can be broken into four basic parts:

- Air induction system.
- Fuel delivery system.
- Engine control computer.
- Electronic sensors.

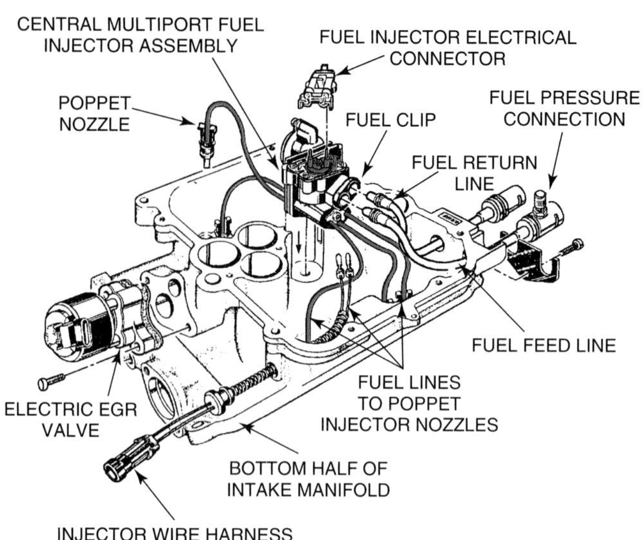

Figure 9-7. A central, multiport fuel injector layout with individual fuel lines to the poppet nozzles. (Chevrolet)

Air Induction System

The air induction system consists of an ***air cleaner assembly***, ***throttle body***, and ***intake manifold***. Some systems may have a thermostatic air cleaner, which is covered in Chapter 14. The throttle body section contains the ***throttle valve***, which is opened and closed by the driver to control the amount of air entering the intake manifold.

On throttle body fuel injection systems, the throttle body also contains the fuel injection components as discussed earlier. The intake manifold forms a closed passageway between the throttle body and the cylinder head. One type of intake manifold, air cleaner, and throttle body is shown in **Figure 9-8**.

Fuel Delivery System

The fuel delivery system provides the fuel which mixes with the air. Pressure is provided by an electric fuel pump. Some systems use two pumps: a low-pressure fuel pump that delivers fuel to another pump, which develops the pressure

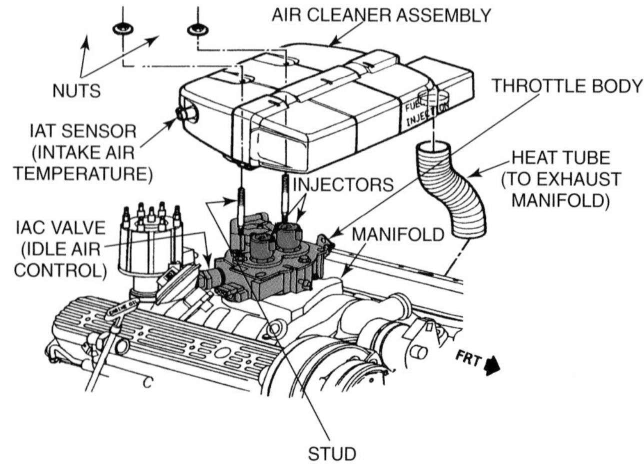

Figure 9-8. A dual-injector, throttle body assembly mounted on an intake manifold for a V-8 engine. (General Motors)

needed at the injectors. However, most systems use only one pump.

The fuel delivery system also contains at least one ***filter*** and in the case of diesel systems, a ***water trap*** which removes any water from the fuel. A fuel delivery system using a single in-tank pump is shown in **Figure 9-9**. The in-tank electric pump forces fuel from the tank, through a filter, and into the fuel rail.

A

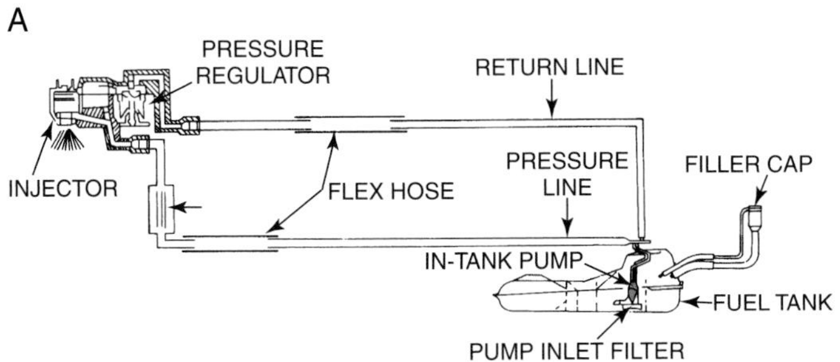

B

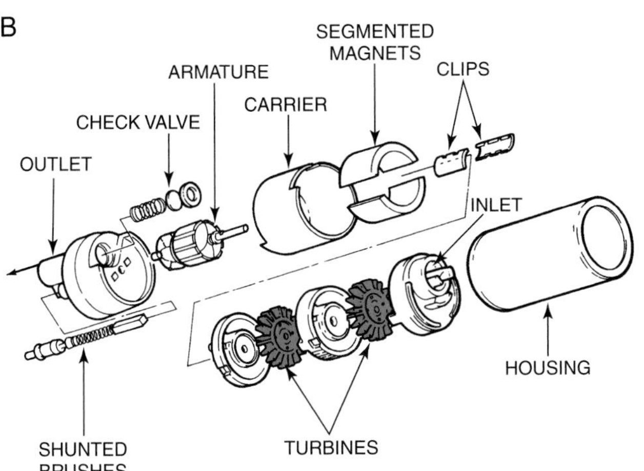

Figure 9-9. A—An electronic fuel injection system which uses a fuel pump located in the fuel tank. B—Exploded view of the in-tank, electric fuel pump from illustration A. This pump is a twin turbine, low pressure type. (AC-Delco)

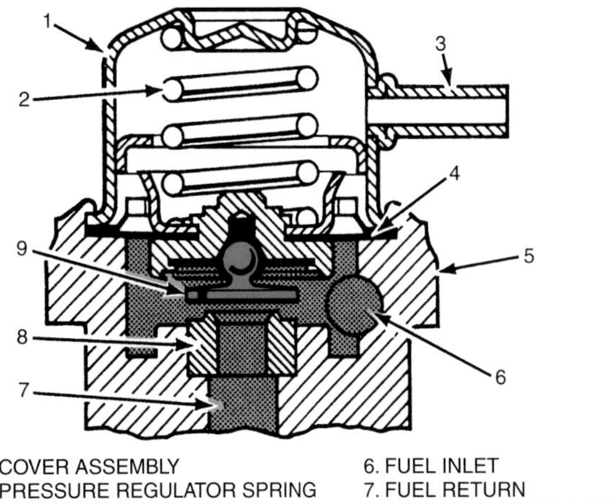

Figure 9-10. A small pressure regulator used with a fuel rail setup. (Chevrolet)

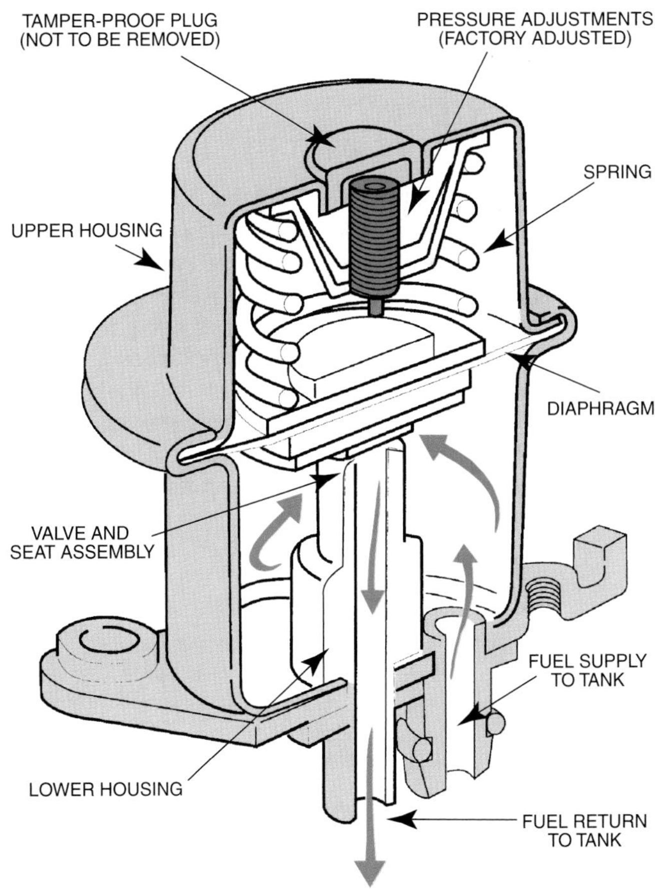

Figure 9-11. Fuel injector pressure regulator action with a non-vacuum assist regulator. Note that this unit is factory adjusted and is sealed with a tamper-proof plug. This regulator, like the vacuum assist type in Figure 9-10, will maintain a constant pressure drop across the injectors, regardless of changes in engine speed, load, etc. (OTC Tools)

Fuel Injection Pressure Regulators

Pressure in some throttle body fuel injection systems is as low as 7 psi, while some multiport systems can reach 60 psi (55 kPa to 380 kPa) or more. Injection system pressure is controlled by a ***pressure regulator***. The regulator controls pressure by bleeding excess fuel back into the fuel inlet line or the fuel tank.

Pressure regulators on multiport fuel injection systems are usually connected by a hose to the intake manifold. A typical pressure regulator is shown in **Figure 9-10**. Note that regulator pressure is controlled by a spring.

Changes in engine load affect intake manifold vacuum, which is then used to modify the fuel pressure setting. High vacuum caused by low engine load allows fuel pressure to unseat the spring easily, keeping fuel pressures low. Low vacuum caused by high engine loads allows the spring to apply full pressure to the fuel return valve, and keep fuel pressures high.

The pressure regulators used on most throttle body fuel injection systems are operated by spring pressure, **Figure 9-11**. The purpose of all fuel pressure regulators is to maintain a constant fuel pressure at the injectors at all times by controlling fuel flow.

Returnless Fuel Injection Systems

Returnless fuel injection systems do not use a fuel return line between the engine and the tank. Instead, the vehicle's computer adjusts pressure at the injectors by increasing or reducing the output of the electric fuel pump. The computer receives inputs from a fuel pressure sensor located at fuel rail. Based on this input, as well as input from other sensors, the computer applies more or less current to the pump motor. Some returnless systems use a conventional electric fuel pump with a pressure sensor mounted on or near the pump assembly. In these systems, a constant voltage is applied to the fuel pump and the regulator bypasses fuel back to the tank when pressure exceeds specifications. Again, a separate return line is not needed.

Fuel Injectors

The fuel injectors receive fuel from the pump or pumps and pressure regulator and spray it into the intake manifold. Injectors can be part of a throttle body, **Figure 9-12**, or installed in the intake manifold and connected to the fuel system through a ***fuel rail***, as in **Figure 9-13**. The fuel rail is a rigid piece of steel tubing that feeds fuel to the injectors. Some systems use flexible hoses to connect the injectors to the fuel rail.

One type of electronic fuel injector is illustrated in **Figure 9-14**. Fuel passes from the fuel rail to the injector. Filter (A) is built into the injector. The fuel moves down through the injector until stopped at the ***nozzle*** (injector outlet) by the ***sealing pintle*** (needle) (B). When the injector is energized by the computer, the ***solenoid winding*** (C) forms a strong magnetic field that attracts the sealing needle armature (D) and draws it upward against spring pressure (E). In this injector, the sealing pintle (B) is lifted about .006″ (0.15 mm.), allowing fuel to spray out. When de-energized, ***spring pressure*** forces the needle closed. The length of time that the injector remains open is very short, usually measured in thousandths of a second.

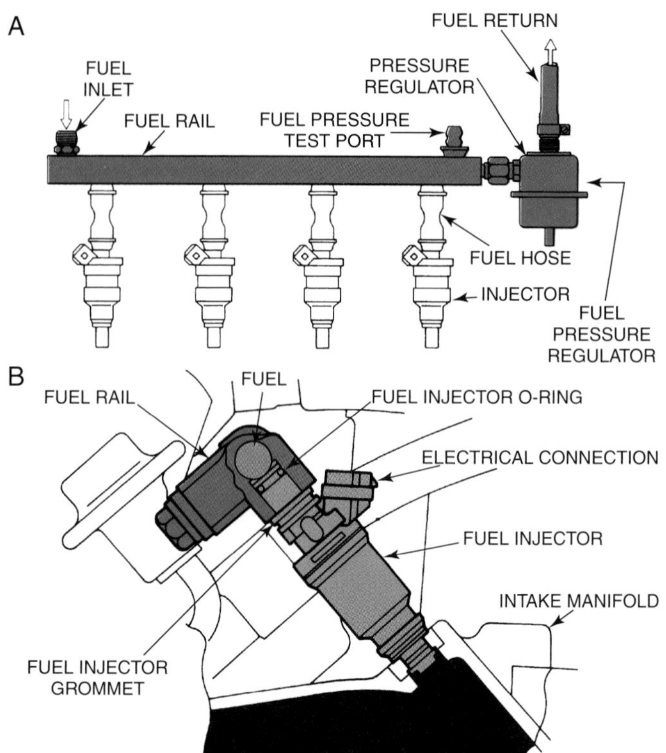

Figure 9-13. A—Manifold mounted fuel injectors that are connected to the fuel rail with high pressure fuel hose and clamps. B—Manifold mounted fuel injector that is attached directly to the fuel rail. Note the O-ring seal. (Geo & Buick)

Cold Start Aids

In addition to the sensors described, the system can incorporate other units for special functions, such as cold starts. Some systems use a ***cold start valve*** to add extra fuel to the intake manifold when starting a cold engine. The cold start valve is an extra injector which sprays extra fuel into the intake manifold for a temperature controlled period of time.

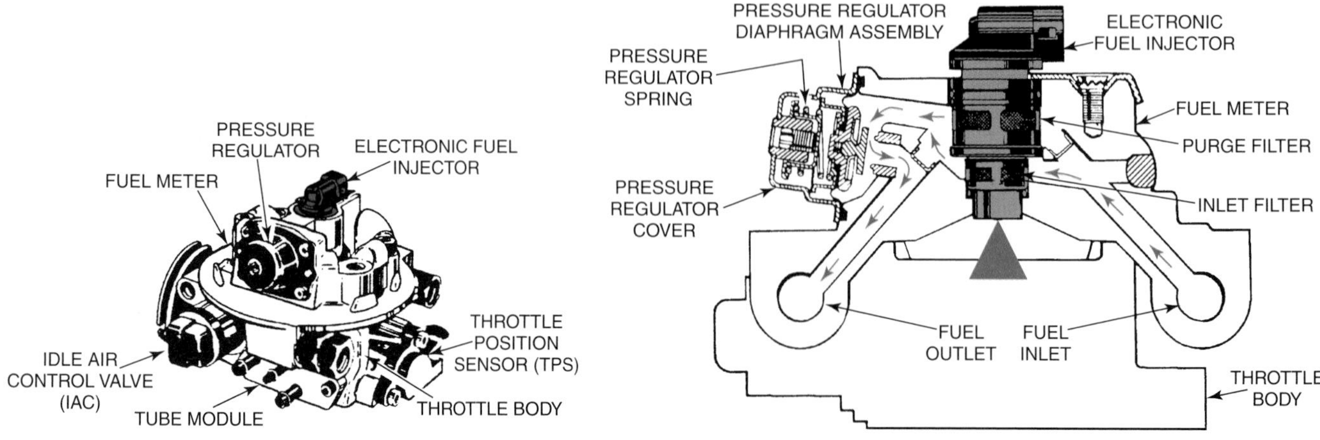

Figure 9-12. Central fuel system injector setup. Study the layout very carefully. (General Motors)

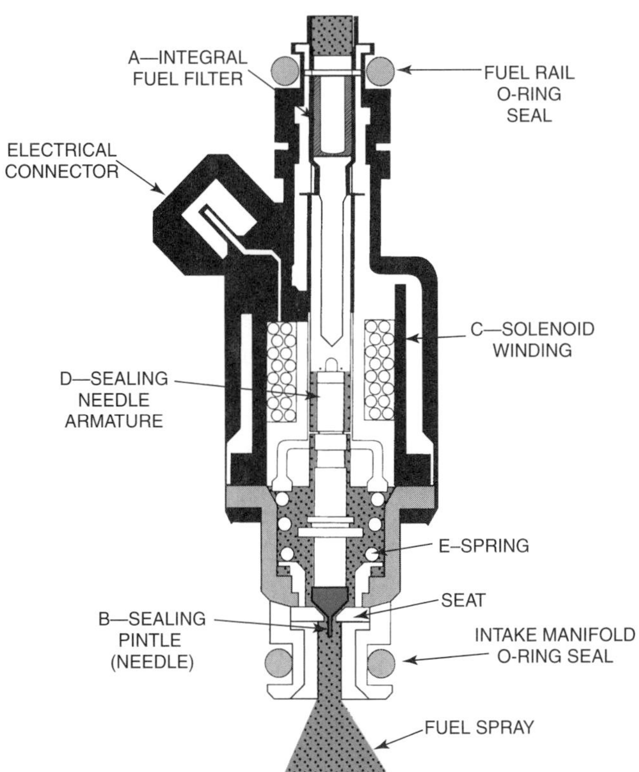

Figure 9-14. Cutaway view of an electronic fuel injector with an internal integral fuel filter. Study the various parts. Note the intake manifold O-ring seal.

The amount of fuel injected is controlled by a thermal switch which regulates the amount of time that the valve is energized (opened), depending on the engine temperature. The valve will be energized for the maximum amount of time when the engine temperature is below -5°F (-21°C). As ambient temperatures increase, the amount of time that the valve is energized will decrease until an engine temperature of 95°F (35°C) is reached. Above this point, the cold start valve will not be energized.

During cold startup on some systems, an ***auxiliary air regulator*** admits additional air into the intake manifold to increase the idle speed. The auxiliary air regulator is controlled by a thermostatic switch located in the engine water jacket. At a coolant temperature of -13°F (-25°C), the thermostatic switch allows maximum airflow through the air regulator. As the coolant warms up, the air regulator flow is gradually decreased. When the temperature reaches 140°F (60°C), airflow is completely shut off.

Engine Control Computer

Engine control computers, such as the one shown in **Figure 9-15**, are constructed using many different electronic circuits and components. Size and complexity varies depending upon the system. Modern computers control many other engine systems in addition to the fuel injectors, as shown in **Figure 9-16**. The computer is usually located in a protected area away from engine vibration and heat, and is connected to the rest of the injection system by means of a sealed wiring harness plug.

The computer receives signals from a number of sensors whenever the engine is running. From this input, the computer evaluates engine fuel needs and adjusts injector pulse width accordingly. Some computers also energize the fuel pump or pumps for 2-5 seconds to pressurize the fuel system before a cold start. Many fuel injection systems do this in place of a cold start valve.

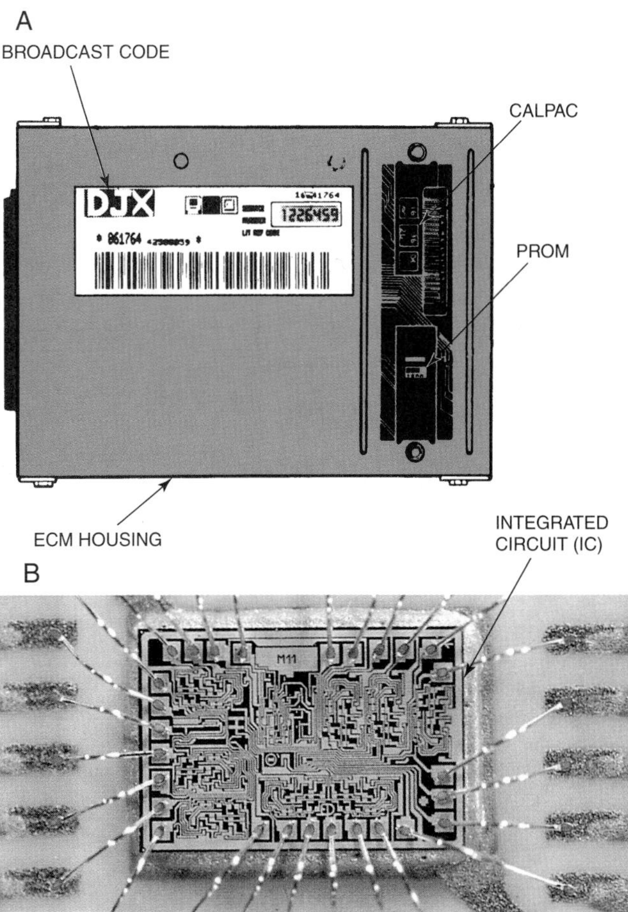

Figure 9-15. A—An ECM (electronic control module) with its PROM chip access cover removed. B—A 30 power magnification of an ECM integrated circuit (IC). (Bosch & AC-Delco)

Electronic Sensors

The electronic ***sensors*** monitor (check) various engine functions and feed this information to the computer. The number and types of sensors vary with the system. **Figure 9-16** illustrates some of the sensors used by one system. Note that information from the sensors enters the computer where it is processed into commands to the fuel injectors and other engine devices as well as the transmission. Operation of these sensors is discussed in Chapter 15.

Oxygen Sensor

The ***oxygen sensor*** monitors the amount of oxygen in the engine's exhaust gases. As the oxygen content in the exhaust gases changes, the voltage signal produced by the sensor also changes. The computer uses signals from the oxygen sensor to control the air-fuel mixture. The oxygen sensor is generally mounted in the exhaust manifold. Most electronic fuel injection systems rely on this sensor for much of the engine's operating information, **Figure 9-17**.

Most computer-controlled engines made after 1994 have a second O_2 sensor installed behind the catalytic converter. The purpose of this sensor is to monitor the operation of the catalytic converter. Input from this O_2 sensor, along with the input from other sensors, provides the computer with information about vehicle emissions and the condition of the converter. Computer controls are discussed in more detail in later chapters.

Engine Speed Sensor

The ***engine speed sensor*** monitors engine RPM. Many speed sensors are mounted in the distributor, **Figure 9-18**, where they obtain a signal from the rotating distributor shaft. In some cases, the ignition coil or Hall-effect switch provides

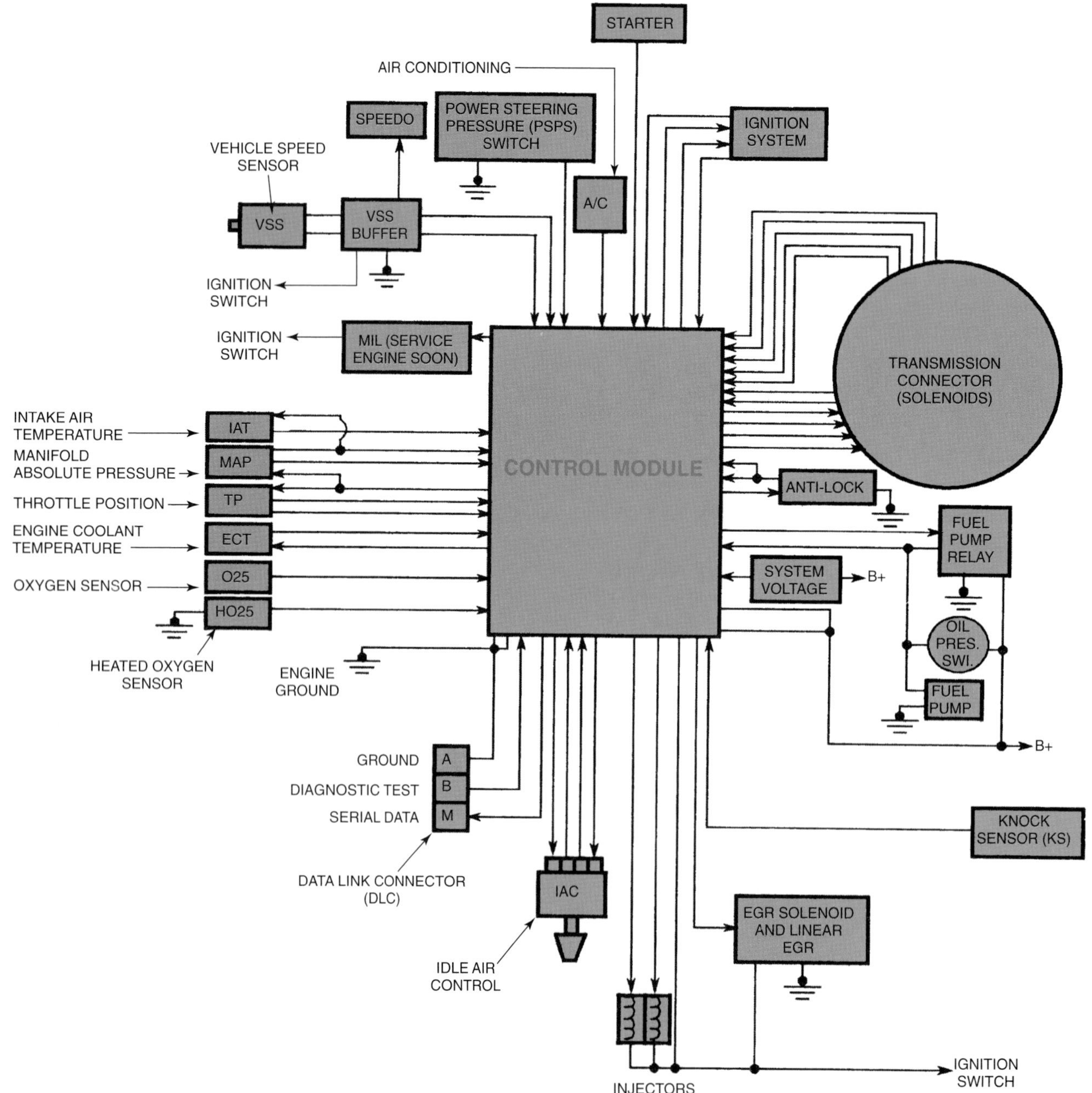

Figure 9-16. An electronic control module (ECM) schematic with various sensors, switches, solenoids, etc., for one particular vehicle. (Chevrolet)

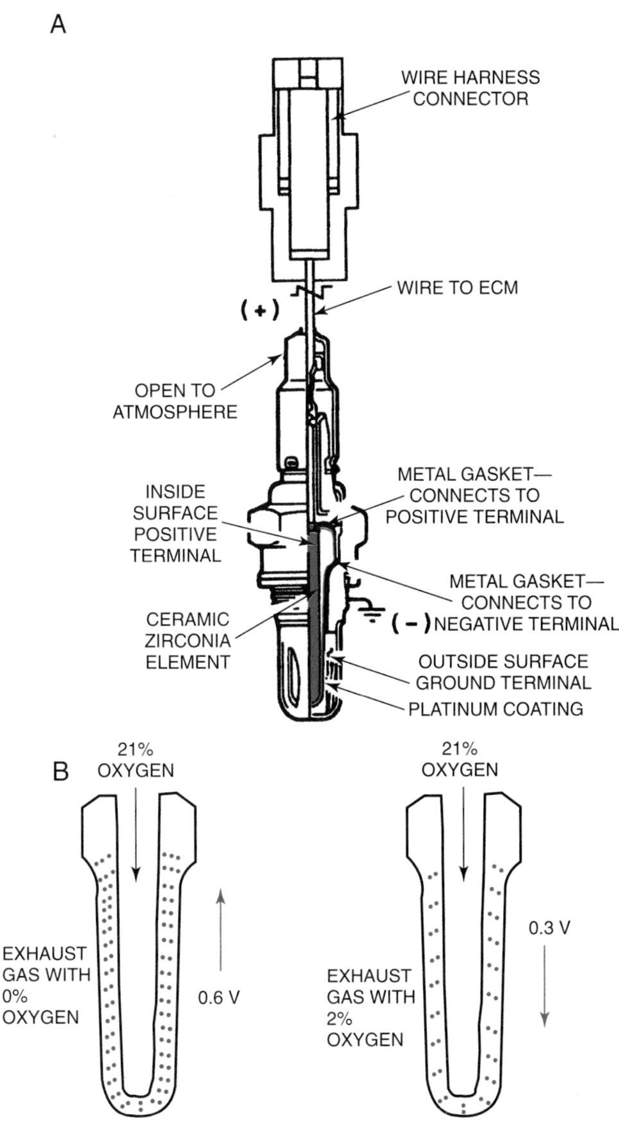

Figure 9-17. A—A cutaway view of a non-heated oxygen sensor assembly. The oxygen sensor produces a low voltage signal which sends information on the oxygen content in the exhaust gas to the electronic control module (ECM). B—This shows the oxygen sensor element. The element begins to produce a signal voltage at an operating temperature of 600°F (315°C). The amount of voltage produced depends on the oxygen content in the exhaust gas. (Chevrolet)

the signal to the computer. In other systems, the speed sensor is mounted so that it can monitor crankshaft or camshaft rotation.

These speed sensors also indicate the crankshaft and camshaft position so that the computer can open the injector before the intake valve opens. This information is utilized by the computer (along with other sensor inputs) to help determine injector pulse timing and pulse width.

Throttle Position Sensor

Throttle position is relayed to the computer by the ***throttle position sensor***. Throttle position sensors are installed on

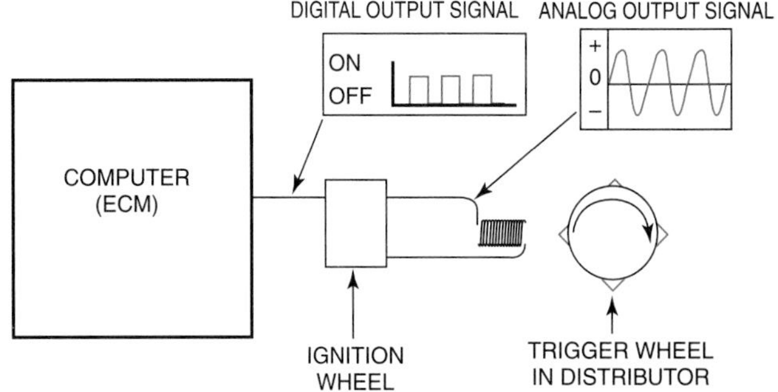

Figure 9-18. A schematic showing an alternating current (AC) signal generator (triggering device) as used with a distributor. Types and construction vary. (OTC Tools)

the throttle body, and are used to monitor the movement of the throttle valve. They can be resistance types, **Figure 9-19**, or transducers, **Figure 9-20**. Resistance sensors contain variable resistance units which send a varying signal to the computer, depending on throttle position. Transducers vary the input signal by creating a magnetic field which is then modified by movement of a metal rod attached to the throttle linkage.

Manifold Vacuum Sensor

Engine load is transmitted to the computer by means of an intake ***manifold vacuum sensor.*** The sensor shown in **Figure 9-21** converts manifold vacuum into a small electrical signal. This input allows the computer to increase fuel supply when the engine is under load and needs a richer mixture, and decrease fuel supply when engine load is light.

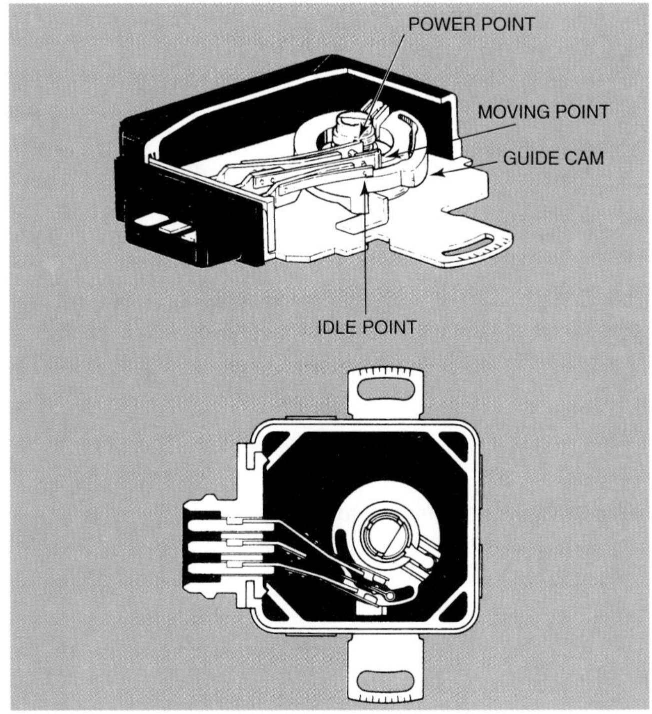

Figure 9-19. Throttle position sensor. The throttle shaft is attached to and operates the switch. (Toyota)

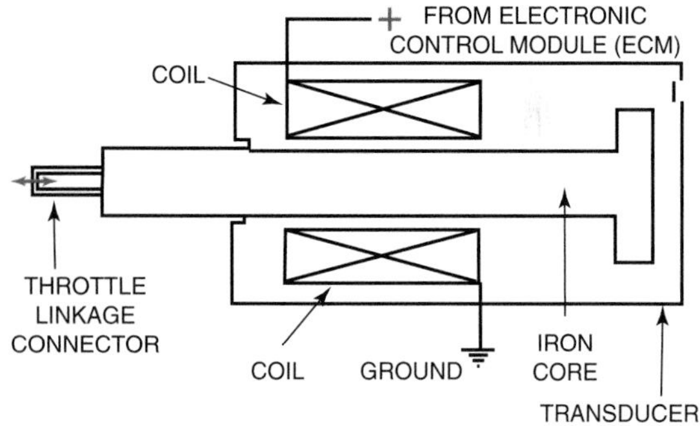

Figure 9-20. A cross-sectional view of a transducer type, throttle position sensor (TPS).

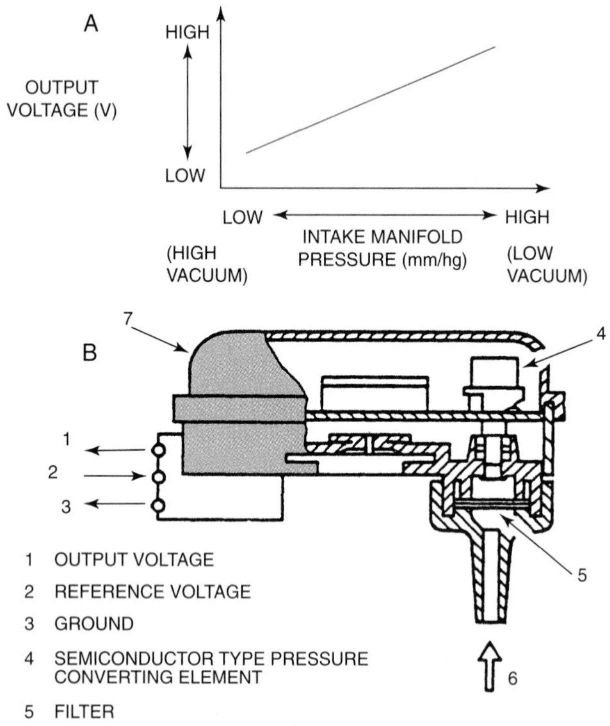

Figure 9-21. A—This graph illustrates intake manifold pressure (low to high vacuum) in relation to the sensor output voltage. B—A cutaway of a manifold vacuum sensor assembly. (Geo)

Barometric Pressure Sensor

Many injection systems have a sensor to measure the pressure of the outside air, usually called atmospheric or ***barometric pressure.*** Barometric pressure is compared with manifold vacuum by the computer to more closely monitor engine load. The input of this sensor is important when the vehicle is driven to higher and lower altitudes. This sensor is sometimes combined with the manifold vacuum sensor into a single unit, as in **Figure 9-22**.

Temperature Sensor

Temperature has a great effect on the operation of the fuel injection system. To overcome the tendency of cold fuel to condense into an noncombustible liquid, cold engines must have a richer mixture if they are to run properly. Every injection system has a temperature sensor to measure engine coolant temperature. Many injection systems have an additional sensor to measure the temperature of the incoming air. A typical temperature sensor is shown in **Figure 9-23**.

Airflow Sensor

Some late model fuel injection systems monitor the amount of air entering the engine by using an ***airflow sensor***. The computer takes this input and compares it with engine RPM and manifold vacuum to determine the amount of fuel to inject. There are three types of airflow sensors, the heated wire, **Figure 9-24A**, the air valve, **Figure 9-24B**, and the Karmann vortex.

The heated wire sensor uses a resistance wire which extends into the incoming airstream. As the wire heats up, its resistance increases, and the amount of current which flows through it decreases. More airflow carries away more heat, reducing the wire's resistance. Less airflow causes the wire to heat up, increasing resistance and reducing current flow. The computer can read this change in current as changes in airflow. A heated film is sometimes used in place of the wire.

The Karmann vortex system uses a restrictor placed in the air stream. The turbulence caused by this restriction can be translated into a signal to the computer.

The air valve consists of a flap which extends into the incoming airstream. Changes in air movement cause the flap to move, moving the contacts in an electrical resistance unit. Changes in current flow through the resistance unit are read by the computer as changes in airflow.

Speed Density

Most throttle body and some multiport injection systems do not use an airflow sensor. Instead, these systems use a combination of the other sensors described earlier to determine air flow. This method is referred to as ***speed density***. The computer receives information from these sensors and controls the injector pulse width, timing and other outputs based on a calculated airflow.

Overall System

The complete electronic fuel injection system is shown in **Figure 9-25**. This is a schematic showing the overall system including sensors, controls, and other components similar to those just discussed. Study **Figure 9-25** and become familiar with the units and the part they play in the overall operation.

Continuous Fuel Injection System

On the continuous injection system, the injectors are always open and fuel pressure is varied to match the fuel flow with the airflow. These systems contain a mechanically operated pressure control valve. This valve increases pressure based on the amount of air flowing into the engine. On later versions of these systems, the fuel pressure is electronically controlled and bases its pressure decisions on inputs from various sensors, **Figure 9-26**.

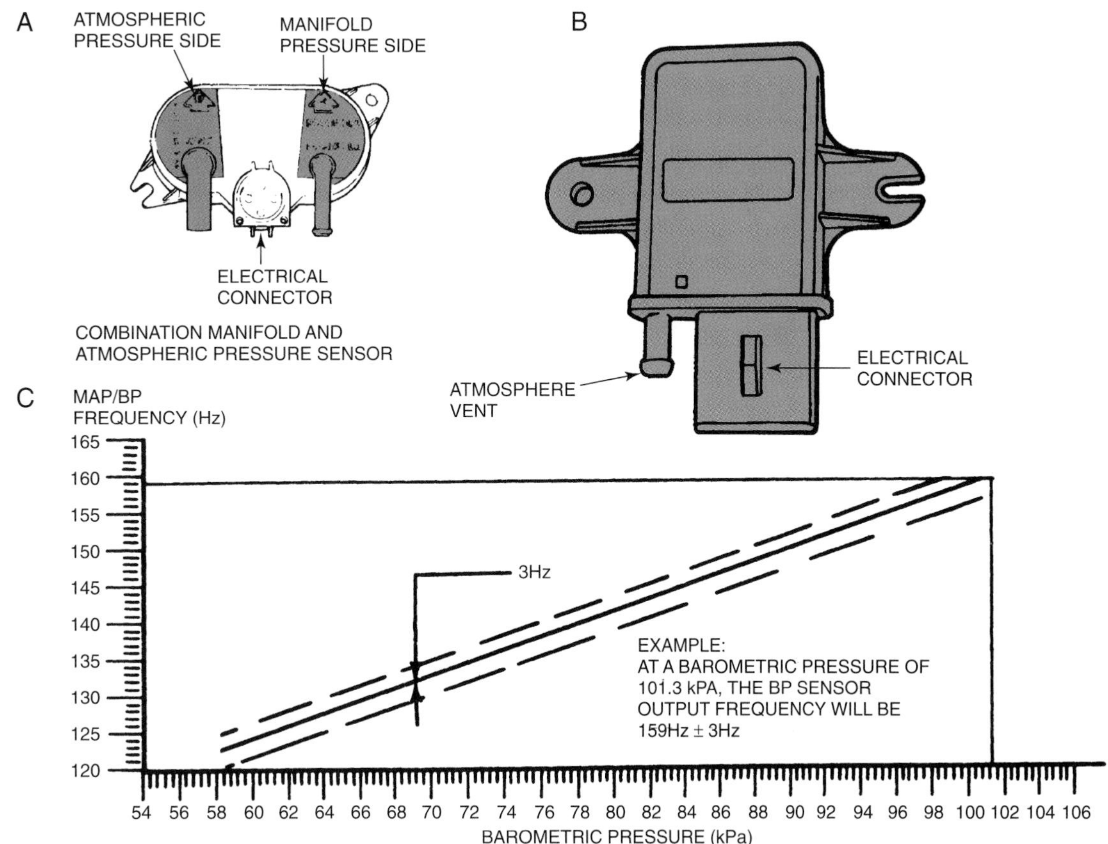

Figure 9-22. A—A combination atmospheric pressure and manifold absolute pressure sensor. B—An atmospheric pressure sensor. This unit converts atmospheric pressure (air density) into electric signals which are sent to and used by the electronic control module. The ECM processes the signals and adjusts the engine fuel mixture according to altitude. C—This table illustrates how barometric pressure changes will alter the frequency range. (Sun Electric Co. & Ford Motor Co.)

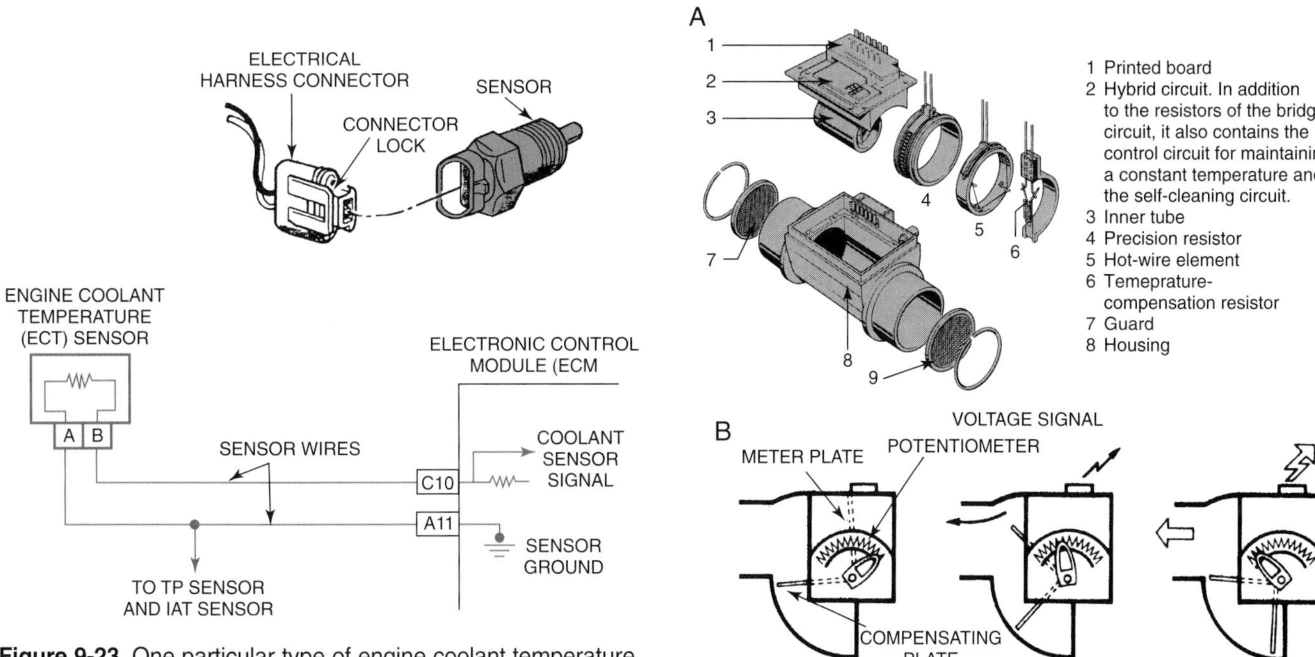

Figure 9-23. One particular type of engine coolant temperature sensor and its electrical schematic. The sensor is a thermistor (a resistor which changes its electrical resistance values in response to heat). Cold engine coolant creates a high resistance, while hot coolant produces a low resistance. (General Motors)

Figure 9-24. A—An exploded view of a heated (hot) wire airflow meter. B—Potentiometer action as an air meter hinged plate is forced open by airflow. Study the various parts (Robert Bosch)

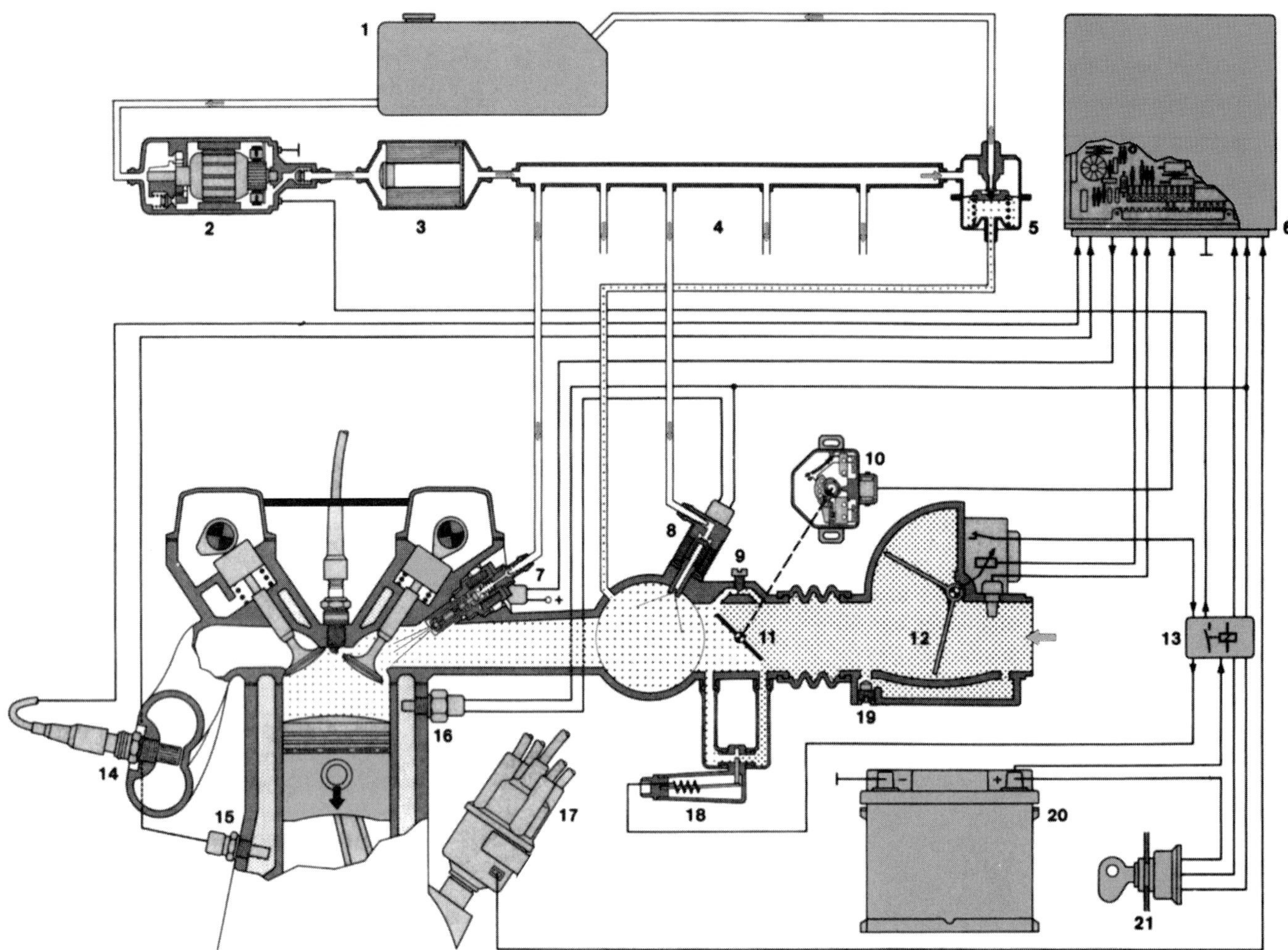

Figure 9-25. Overall cutaway view of one particular complete electronic fuel (gasoline) injection system schematic. Study it very carefully. 1–Fuel tank. 2–Electric fuel pump. 3–Fuel filter. 4–Fuel rail (distributor pipe). 5–Pressure regulator. 6–Electronic control unit (ECU). 7–Fuel injector. 8–Cold start injector. 9–Idle speed adjusting screw. 10–Throttle position sensor. 11–Throttle plate. 12–Airflow sensor. 13–Relay. 14–Oxygen sensor. 15–Coolant temperature sensor. 16–Thermal time switch. 17–Distributor. 18–Auxiliary air valve. 19–Idle mixture adjustment screw. 20–Battery. 21–Ignition switch. (Bosch)

During engine operation, the continuous fuel injection system feeds atomized fuel from the injectors at all times. The amount will vary to suit engine needs. In most systems, the injectors have a valve or check ball, which is closed by a spring when there is no pressure in the injection system. This keeps fuel from dripping out of the nozzles and causing hard starting conditions when the engine is hot. The continuous system injectors do not use a solenoid to open or close them. Fuel pump pressure forces the spring-loaded injector valve open, allowing fuel to spray into the intake manifold.

The injector is opened at minimum fuel pressure. The injectors in **Figure 9-26** open at 47 psi (324 kPa). As long as the engine is operating, pressure will exceed this amount and fuel will flow continuously from the injector nozzles. The amount of fuel will vary, depending upon system pressure. The injector is designed to atomize fuel properly at any pressure or flow rate. A continuous injector is shown in **Figure 9-27**.

Fuel Flow Control

Air passing into the intake manifold flows through a mechanical airflow sensor, **Figure 9-28**. The sensor utilizes a hinged lever that pivots on a pivot rod. An airflow sensor plate attached to one end of the lever rides up and down in the center of the air venturi, **Figure 9-28A**. A balance weight is used to balance the lever and plate assembly, which allows the sensor plate to "float" in the venturi.

During part load operation, **Figure 9-28B**, airflow has forced the sensor plate to rise a small amount. Note that the fuel distributor control valve installed in the distributor block has been raised somewhat by the lever. The airflow sensor controls the fuel distributor valve. As the engine is accelerated, more and more air flows through the venturi, raising the sensor plate higher and higher. At full load, the sensor plate is raised to its highest position, **Figure 9-28C**. The lever has moved the fuel distributor control valve to the wide open position, permitting maximum fuel flow to the injectors.

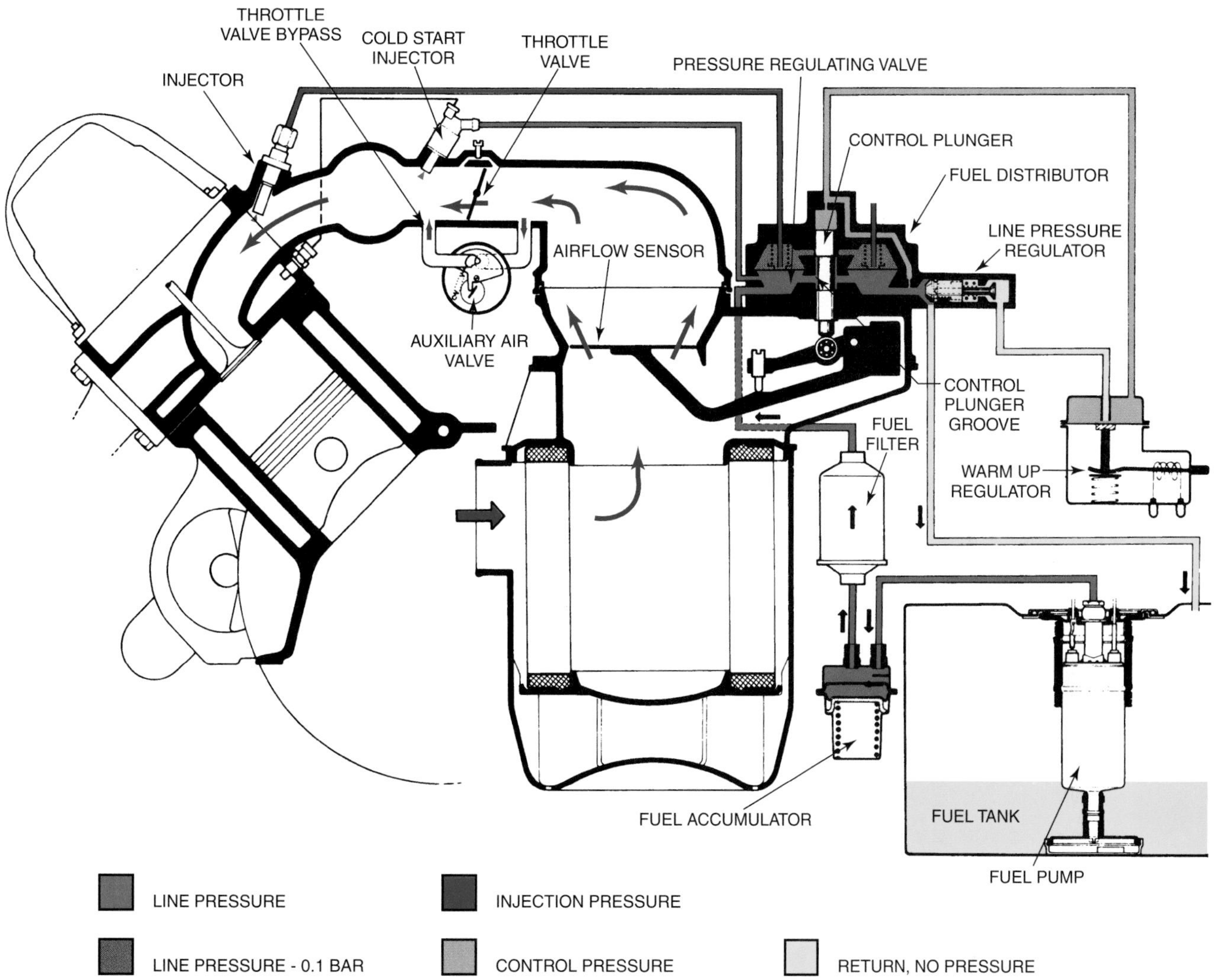

Figure 9-26. Schematic of a typical mechanically controlled continuous fuel injection system. Main controls are the airflow sensor and fuel distributor. (Saab)

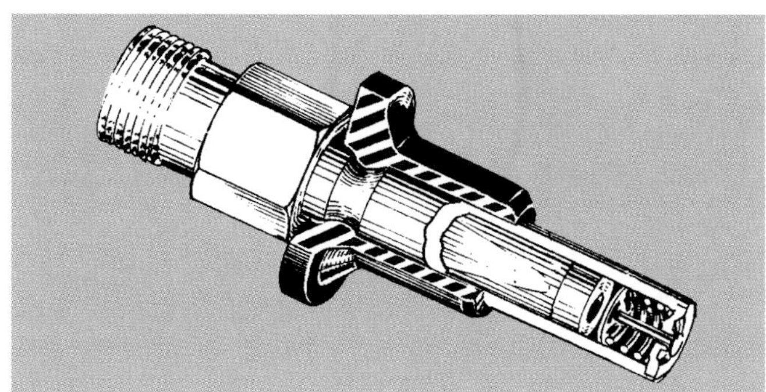

Figure 9-27. Fuel injector used on a continuous injection system. (Volvo)

In order to provide a constant pressure drop through the fuel distributor metering slots, a pressure regulating valve is used for each injector outlet. The valves are built into the ***fuel distributor*** housing. The constant pressure drop is needed to keep the amount of fuel injected directly proportional to the size of the metering slot opening. See **Figure 9-29**.

Complete Air-Fuel Control Unit

Figure 9-30 shows the operation of the fuel control plunger. Note that there is pressure in the control chamber above the control plunger. This controlled pressure (around 52.2 psi or 360 kPa) is needed to dampen the sensor plate lever movement so that it will not be overly sensitive to sudden acceleration.

Control Pressure Temperature Regulator

The control pressure temperature regulator maintains a steady pressure in the chamber above the control plunger when the engine is at normal operating temperature. If the engine is started cold, a heating coil causes a bimetallic spring to alter the pressure on the diaphragm valve spring. This changes spring pressure on the diaphragm valve, causing a lowering of control pressure. This allows the sensor plate lever to shove the control plunger farther up, allowing extra fuel to

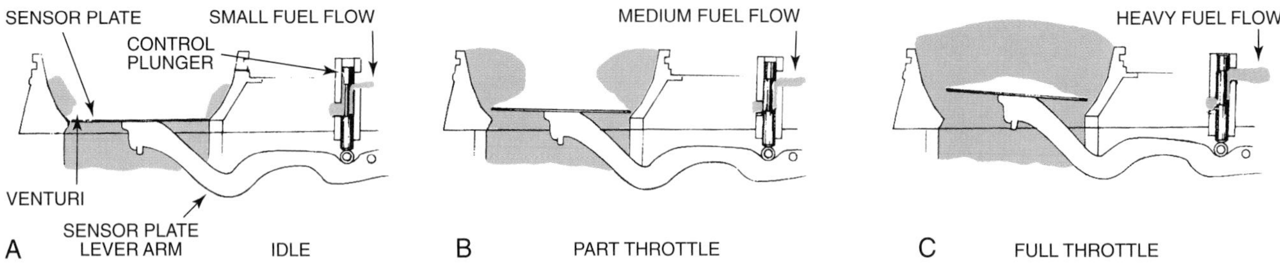

Figure 9-28. Airflow sensor plate action. As airflow increases through the venturi, the plate is forced upward. The lever arm will raise the control plunger, increasing fuel flow. (Volvo)

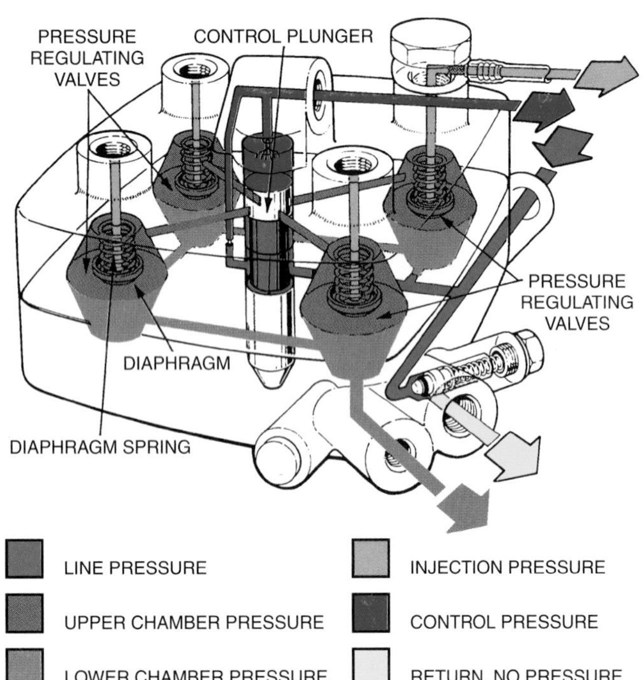

Figure 9-29. Fuel distributor for a mechanically controlled continuous injection system. There is a pressure regulating valve for each injector. (Saab)

flow through the metering slots. As the engine warms, normal fuel pressure is resumed.

Diesel Fuel Injection

Diesel engines are used on some cars and light trucks. The diesel engine uses the heat of compression, instead of a spark plug, to ignite the fuel. ***Diesel fuel injection*** is similar to gasoline injection, but requires higher injection pressures. Air is drawn into the cylinder and highly compressed. The compression is so high that the compressed air will reach temperatures of 1000°F (538.2°C). At the precise time the piston has completed the compression stroke, diesel fuel is sprayed into the combustion chamber, or an offset ***prechamber***, **Figure 9-31**. The intense heat of the compressed air ignites the fuel, and the power stroke follows.

Diesel Fuel

The two common types of diesel fuel are grades 1-D and 2-D. 2-D fuel is specified for most engines operating under normal conditions and within normal temperature ranges. 1-D fuel is used only for extreme cold weather (below 9°F or -7°C) or in very special cases of frequent stop and start driving. Automotive diesel fuel is the same as diesel used in boat engines and construction equipment. However, diesel fuel made for non-highway use is dyed red to prevent its use in automotive diesel engines.

Diesel fuel has a ***cetane number***. The cetane number is determined by comparing the burning characteristics of a sample of diesel fuel against a test fuel called cetane. It should not be confused with the ***octane*** (resistance to knocking) rating of gasoline. It is very important that only clean, fresh, water free diesel fuel be used.

Overall Diesel Fuel System

A schematic of a diesel fuel system layout is shown in **Figure 9-32**. Typical diesel system components include the tank, fuel pump, filter, injection pump, injectors, and the necessary fuel lines. The system in **Figure 9-32** is typical of newer systems that utilize a computer-controlled electronic fuel injection pump.

Water In Fuel Sensor

Some vehicles are equipped with a ***water in fuel sensor*** that will alert the driver to the fact that water is present in the fuel, **Figure 9-33**. The water may then be drawn off through a valve in the bottom of a water separator. Water is extremely damaging to the injection system.

Fuel Filter

Diesel fuel must be kept clean to avoid injector system trouble. Very small particles are enough to clog injector nozzles or jam the injector. Diesel fuel systems typically use one or more filters to trap dirt and other deposits. One type of diesel fuel filter is shown in **Figure 9-34**.

Diesel Injection Pump

Diesel fuel is injected directly into the cylinder at the top of the compression stroke. Due to the high compression (around 22:1) used in diesel engines, the ***diesel injection pump*** must be capable of producing high pressures. Injector pump pressures vary greatly depending on the pump and

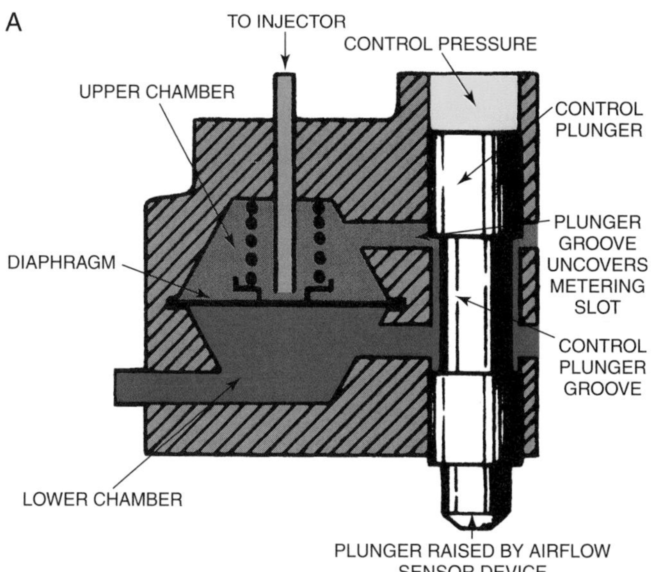

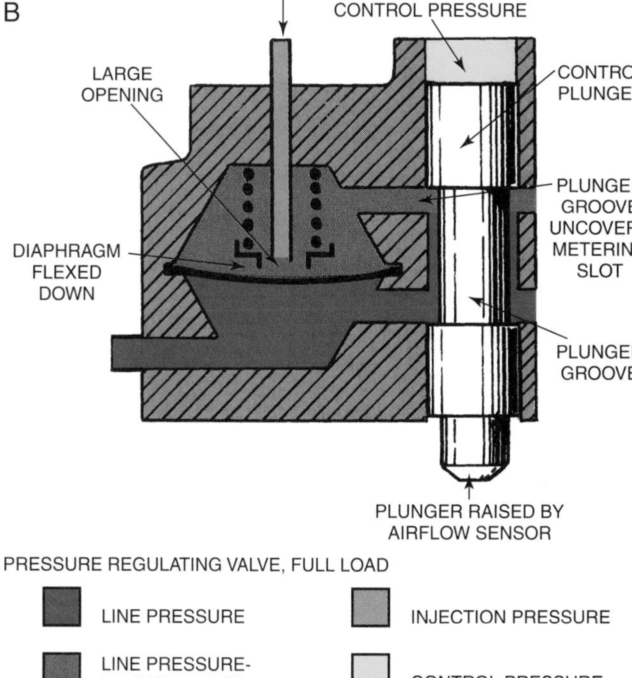

Figure 9-30. A—Control plunger action-partial engine load condition. The diaphragm has flexed downward increasing fuel flow to the injector. B—Control plunger action-heavy engine load. Diaphragms forced down, increasing fuel flow to injector. (Saab)

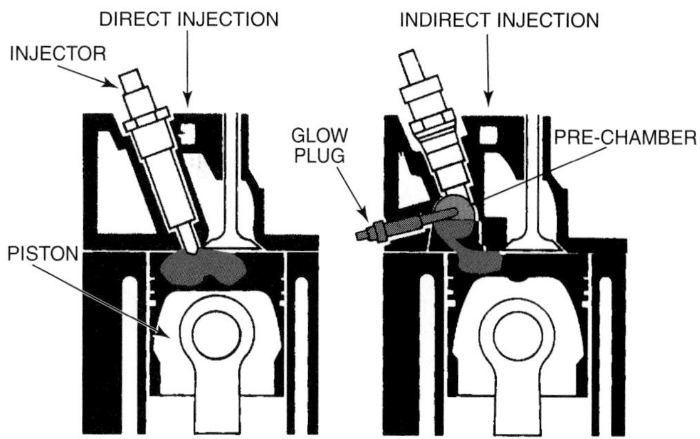

Figure 9-31. This Illustration compares diesel fuel direct injection to indirect injection. Note the pre-chamber on the indirect injection engine. (Cummins Engine Co.)

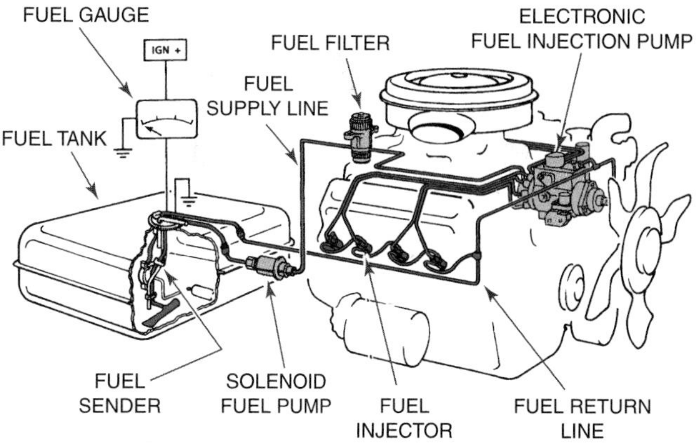

Figure 9-32. An overall view of one particular diesel fuel injection system as used with a V-8 engine. This system incorporates an electronic fuel injection pump equipped with various sensors and solenoids which are controlled by the powertrain control module (PCM). (General Motors)

engine. Pressures can range from 3000-15,000 psi (20 685-103 425 kPa).

How the Injector Pump Produces Pressure

By using a plunger operating in a cylinder, the pump traps fuel ahead of the plunger. When the plunger is forced up, it builds pressure on the fuel. Pressure will be increased to the point at which the injector will open and spray fuel into the engine. **Figure 9-35** shows the essential parts. Note that the shape of the cam influences the timing, quantity of fuel, and the rate of injection—all very important to proper operation. Pumps can use a single plunger for each injector or a single pump (often consisting of two or more opposing plungers) to serve all injectors.

Injector Pump Types

There are a number of injector pump designs. The most commonly used automobile and light truck diesel injector pumps are classified as inline or distributor. Both are mechanical pumps that pressurize the fuel and deliver it to the cylinders at the proper time for each.

Inline Injector Pump

The ***inline injector pump*** uses a separate cylinder plunger-cam for each cylinder. They are contained in a

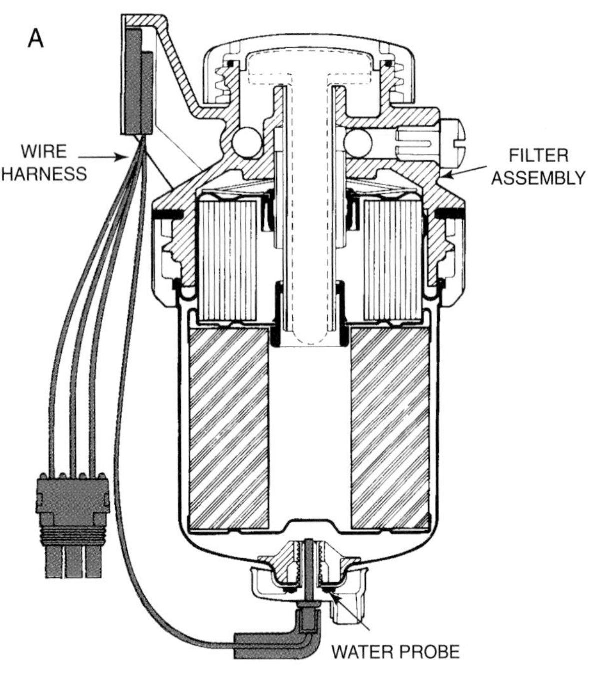

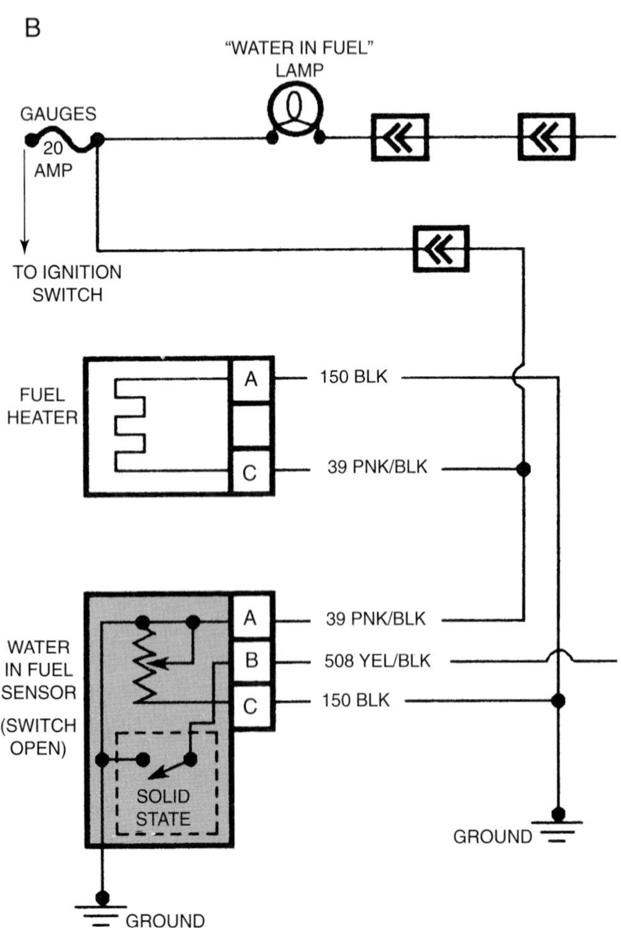

Figure 9-33. A—An electronic water in fuel sensor assembly. B—Water in fuel electrical circuit schematic. As the water reaches the sensor probe it closes a switch. This completes an electrical circuit path, which turns on the "water in fuel" lamp on the vehicle instrument panel. (Stanadyne & General Motors)

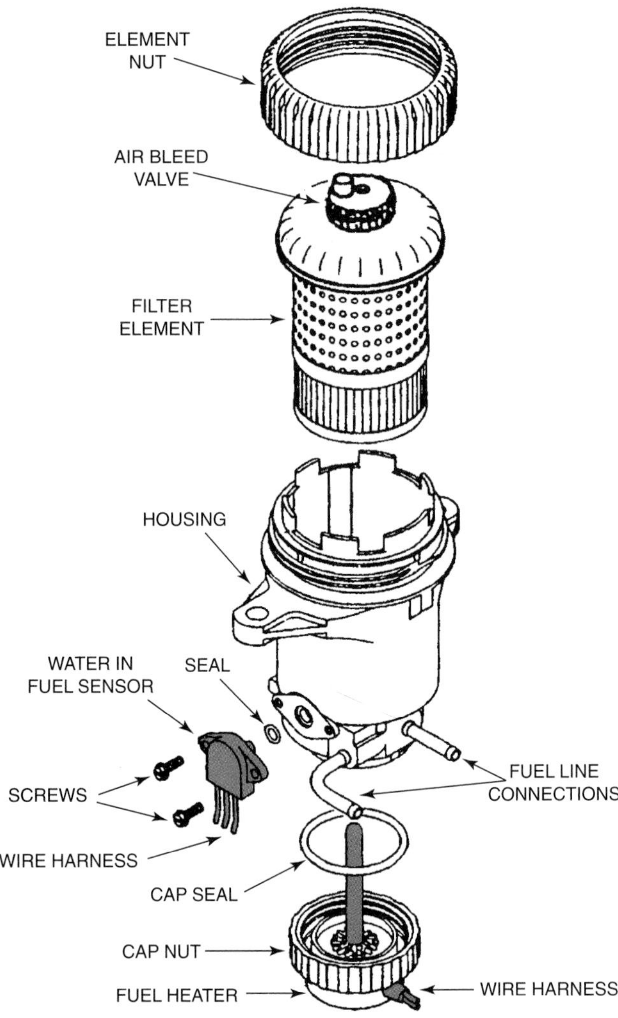

Figure 9-34. An exploded view of one type of diesel fuel filter assembly. Note the water in fuel sensor and the fuel heater. (General Motors)

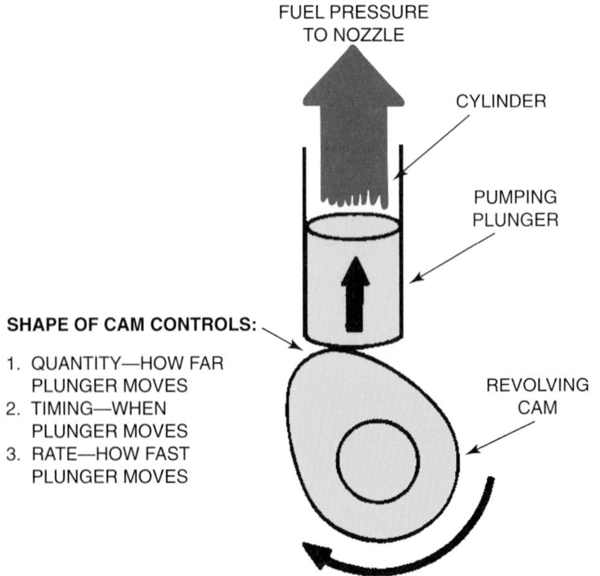

Figure 9-35. The cam operated plunger working in a cylinder, is used to produce pressure in a diesel injector pump. (Deere & Co.)

straight row in a common housing. The cams are part of a common camshaft driven by the engine at one-half engine speed. Drive may be by chain, belt, or gear. Fuel from the fuel pump is fed to all of the plunger assemblies. As the camshaft rotates, each of the plungers are moved up and down in their cylinders.

A single control rack or rail with gear teeth along one side is meshed with each plunger assembly. By moving the control rack, the effective pump action (amount of fuel pumped) of all plungers can be adjusted. A typical inline pump is illustrated in **Figure 9-36**. Study it carefully and note the location of all parts.

Inline Injector Pump Metering Control

Each plunger operates in a separate cylinder or barrel. The top section of the plunger has a wide band or groove cutaway. One edge of the cut forms a helix (spiral shape). The plunger also has a vertical slot reaching from the top of the plunger down into the cutaway area.

The bottom of the plunger is connected, by means of a flange, to the toothed control sleeve that meshes with the ***control rack*** teeth. When the control rack is moved back and forth, the plunger will rotate in the cylinder. Study the arrangement in **Figure 9-37**. Plunger action during the entire pumping cycle is shown in **Figure 9-38**. Note how fuel is trapped above the plunger in **Figure 9-38B**, forced out of cylinder in **Figure 9-38C**, and how the helix uncovers the port in **Figure 9-38D**, stopping pump action.

Injector Pump Delivery Valve

Fuel flowing from the pump cylinder is forced through a delivery valve on its way to the injector. The fuel flow forces the delivery valve from its seat in **Figure 9-39**. When the helix uncovers the cylinder ports, fuel pressure in the pump cylinder drops instantly. This causes the delivery valve plunger to quickly return to its seat. In so doing, the volume in the delivery line (injector housing, line to injector and delivery valve spring chamber) is suddenly increased by the amount of the delivery valve plunger travel. This increase in volume causes a sudden reduction in delivery line pressure that allows the injector nozzle valve to snap shut quickly and cleanly without dripping. A delivery valve is used on both inline and distributor type pumps.

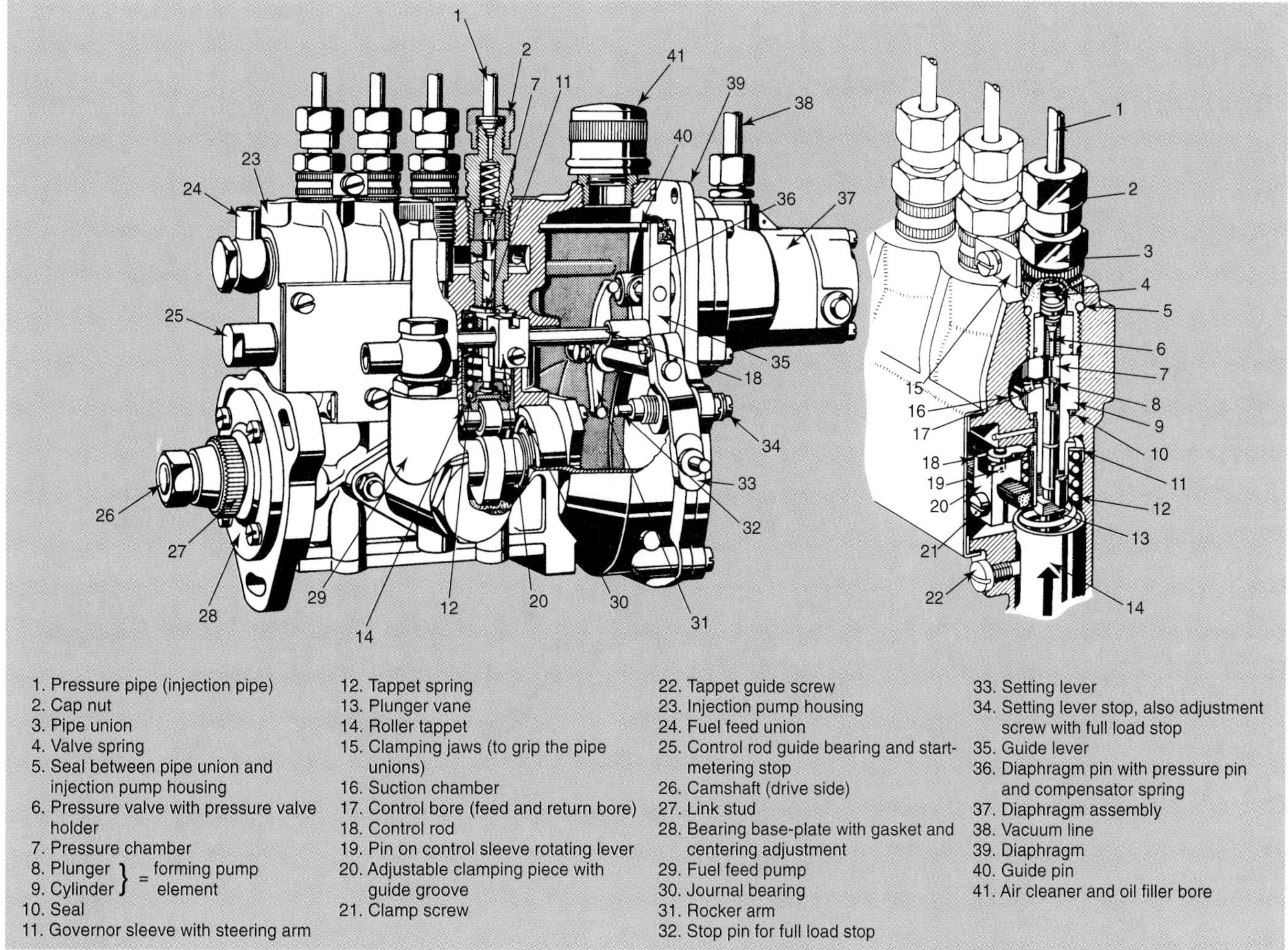

Figure 9-36. Typical diesel engine inline injection pump. Study the cutaway carefully and learn part names. (Mercedes-Benz)

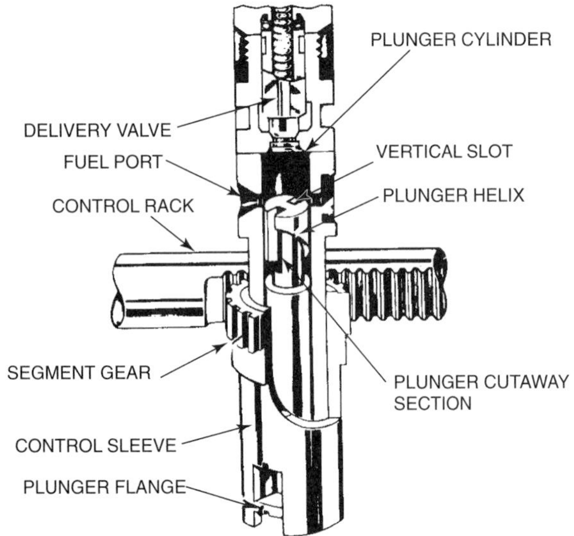

Figure 9-37. Pump plunger construction. When the control rack is moved, the plunger will be forced to rotate by the segmented gear sleeve acting on the plunger flange. (Bosch)

Distributor Type Injection Pump

The ***distributor injection pump*** is shown in **Figure 9-40**. You will note that a single, double-plunger type pump serves all injector outlets. A single delivery valve is located in the rotor. Study the pump thoroughly and learn all part names.

Pump Charging Cycle

Fuel enters the transfer pump from the inline fuel pump. From the transfer pump, fuel moves through the rotor passages into the pumping chamber and fuel metering valve area. Excess fuel flows through the transfer pump pressure regulator back to the inlet side.

For fuel to flow through the rotor passages, the rotor must turn far enough to align the rotor passages with the fuel

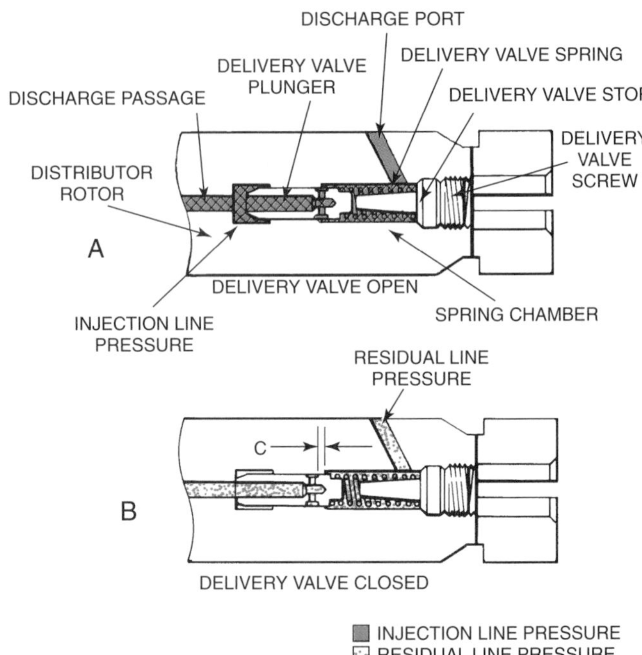

Figure 9-39. Injector pump fuel delivery valve. This valve produces a sudden drop in injector line pressure as soon as fuel delivery ceases. This helps the injector nozzle valve to close without dripping. (Oldsmobile)

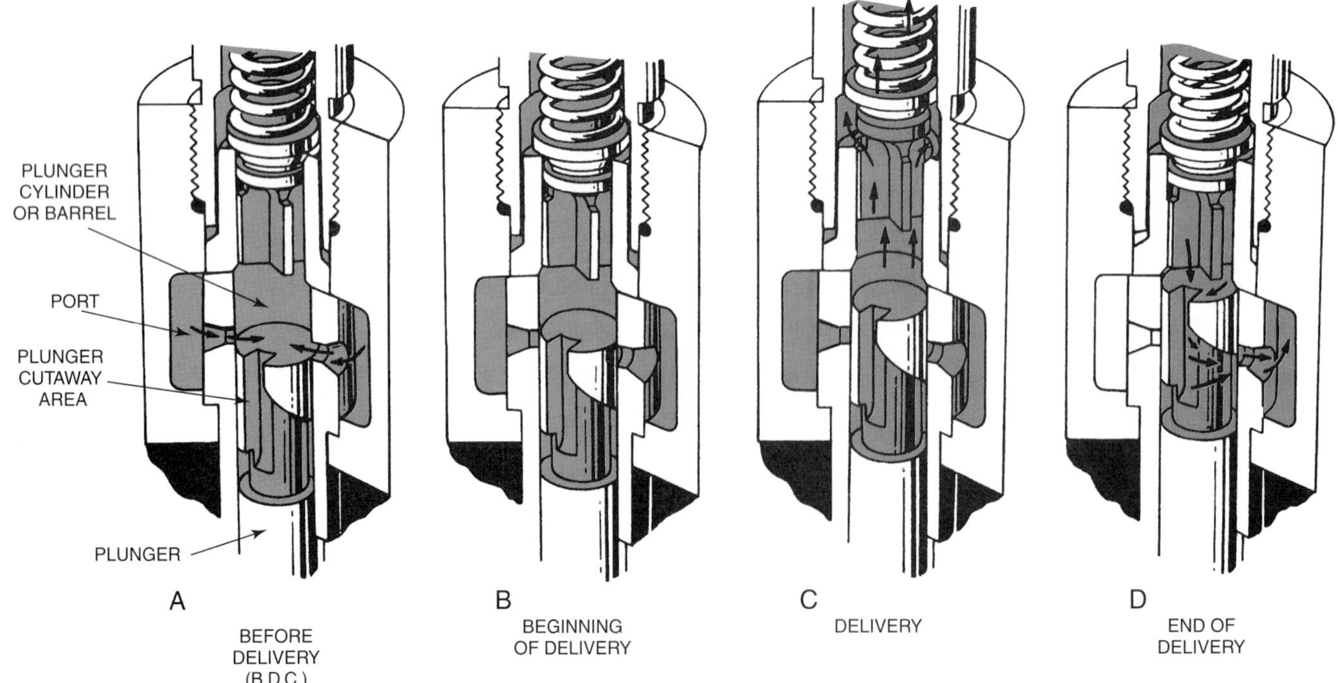

Figure 9-38. Pump plunger action during its entire pumping cycle. Note how the helix has uncovered port in D, stopping fuel delivery even though full plunger lift travel may raise it still higher. (Deere & Co.)

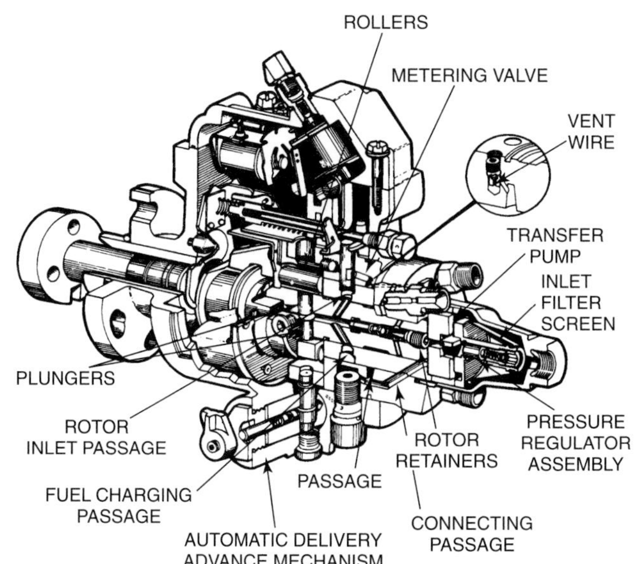

Figure 9-40. Cutaway view of a distributor type injection pump. This pump creates pressure and distributes fuel from a single rotary pump to all injectors. Learn the part names and their relationship to overall pump operation. (Ford)

charging ports. The cam ring is positioned to permit the pump plungers to be forced apart by the fuel entering the chamber. The distance the pump plungers move apart determines how much fuel is delivered. The distance the plungers move apart is controlled by the governor operated metering valve.

Pump Discharge Cycle

When the angled inlet passages no longer line up with the charging ports, fuel is trapped in the pumping chamber. The cam ring rollers have been squeezed inward, forcing the two plungers toward each other. As the plungers are forced together, the trapped fuel is forced past the delivery valve, into the injector line which opens the injector nozzle, spraying fuel into the cylinder.

Pump Fuel Return Circuit

Transfer pump pressure forces fuel through a vent passage in the hydraulic head. Flow through the passage is partially blocked by a wire to prevent excessive flow and pressure loss. Any air from the transfer pump flows, with some fuel, through the vent into the governor linkage compartment and then back to the fuel tank.

Electrical Fuel Shutoff

An electrical ***fuel shutoff***, **Figure 9-41**, is used on diesel engines to stop the engine. It rotates the metering valve to the closed position when the shutoff solenoid is de-energized. This stops fuel flow.

Other Pump Features

The pump can also offer a ***viscosity compensating valve*** that tends to keep the injector pump functioning the same with either light or heavy (or cold) fuels. Fuel delivery timing can be automatically advanced or retarded to compensate for engine speed and/or load changes. A fuel automatic ***delivery advance mechanism*** is shown in **Figure 9-40**.

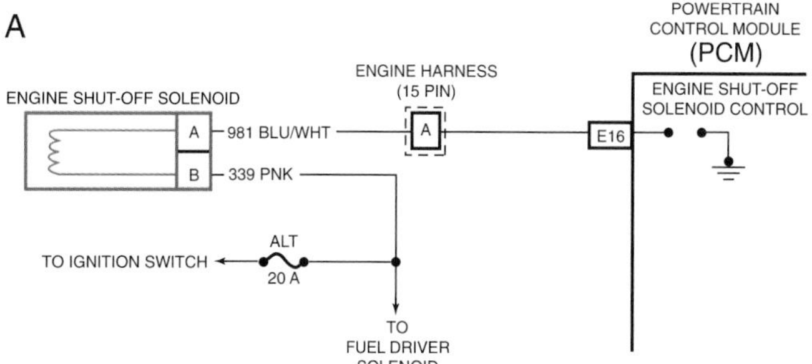

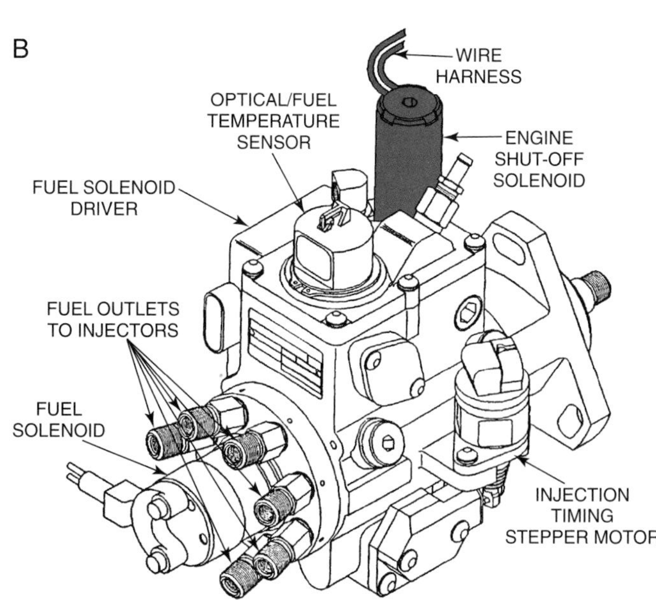

Figure 9-41. A—An electrical schematic for a fuel shutoff circuit. B—A distributor type fuel injection pump showing the location of one particular fuel shutoff solenoid. This solenoid is controlled by the powertrain control module (PCM). When energized, it is designed to completely stop the fuel flow into the injection pump. (Chevrolet)

A schematic of the entire fuel system is shown in **Figure 9-42**. Start at the fuel tank and study the various components and their relationship to the overall system.

Diesel Injector Types

Fuel pressure, fuel viscosity, nozzle design, and nozzle condition must all be correct so that a proper spray pattern is formed. There are many different injector designs, but all are a variation of two basic types—***inward opening*** and ***outward opening***. Study the action of the two nozzle designs in **Figure 9-43**. Diesel spray patterns will vary depending upon engine design.

One type of inward opening injector nozzle is illustrated in **Figure 9-44**. Note the heavy spring (10), keeping the jet needle (12), against its seat. Injection pressure will force fuel down canal (1), to pressure passage (4) where it will act against the shoulder of the jet needle. When fuel pressure reaches a point high enough to overcome spring pressure, the needle will lift off its seat and fuel will be sprayed into the cylinder.

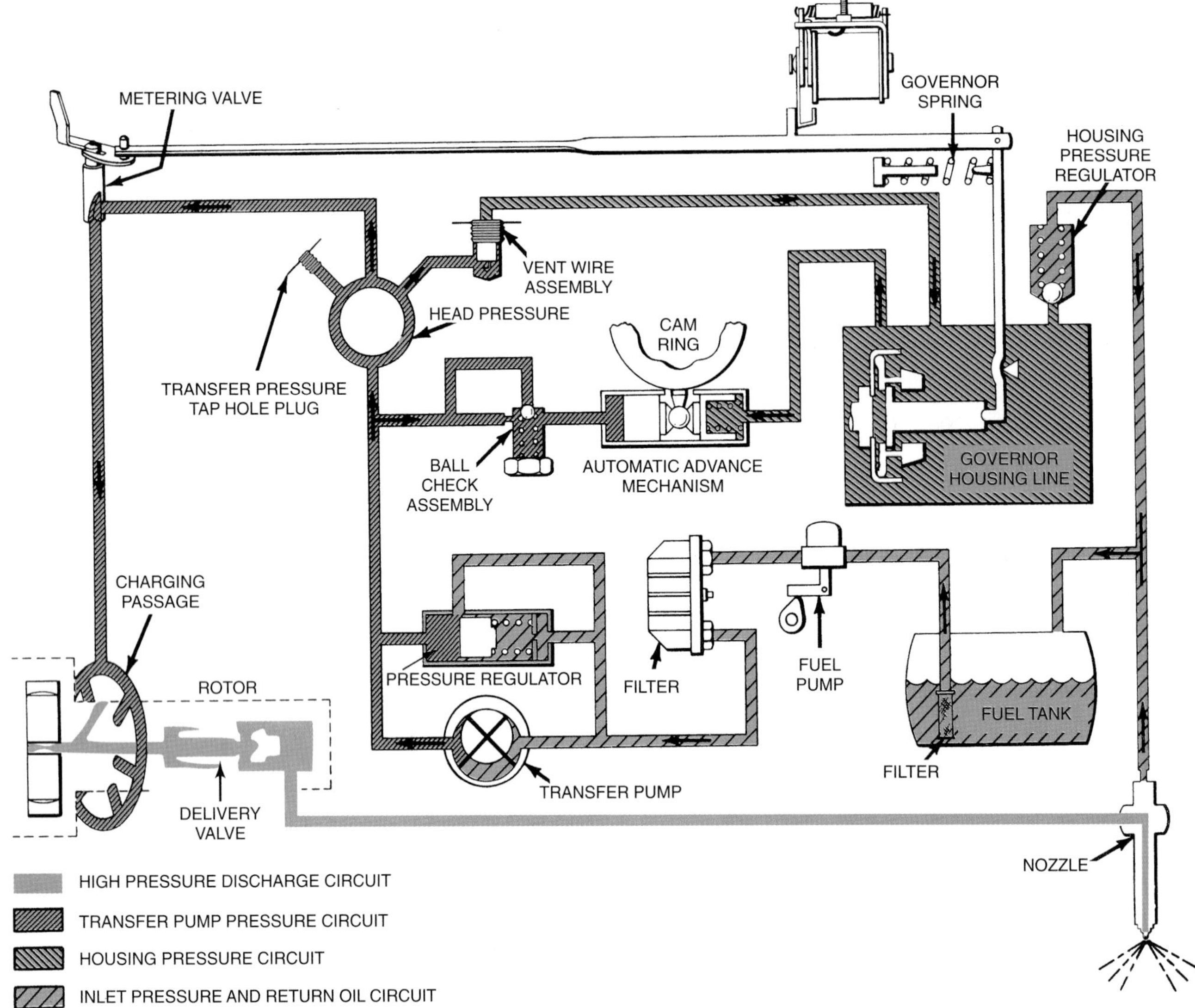

Figure 9-42. A schematic of a diesel fuel system employing a rotary type injector pump. (General Motors)

Glow Plugs and Intake Air Heaters

To aid in igniting the compressed air-fuel mix during cold starting, diesel engines use ***glow plugs*** and ***intake air heaters***. The glow plug is an electrically operated device that protrudes into the precombustion chamber. When energized, the glow plug tip heats to a dull red. The swirling, highly compressed air-fuel mixture is heated by the glow plug tip. This additional heat helps the engine to start by causing the mixture to ignite more readily.

On older vehicles, the glow plugs must be turned on by the vehicle operator and allowed to heat the combustion chamber for several seconds before the engine can be started. On later engines, the glow plugs are turned on and off automatically. Under most starting conditions, the plugs are energized for a very short period. The engine can be started in seconds after the key is turned on, **Figure 9-45**. The glow plugs may still be energized for a very short while following engine startup.

Intake air heaters warm the air as it passes through on its way to the intake manifold. The heater element is a wire mesh that is electrically heated. The computer energizes the heater element before cranking, **Figure 9-46**.

Superchargers and Turbochargers

The purpose of superchargers and turbochargers is to force more of the air-fuel mixture into the engine cylinders. This extra air and fuel results in more power output. Superchargers are driven by a belt or chain from the engine crankshaft. Turbochargers are a type of supercharger that is driven by the engine exhaust.

Superchargers

A ***supercharger*** assembly is shown in **Figure 9-47**. It consists of a drive belt, drive and driven gears, and a centrifugal ***compressor wheel***. The belt drives the gear and compressor wheel. The compressor wheel pumps air into the engine. Superchargers are installed on top of the intake manifold on V-type engines, and on the side of inline engines.

Supercharger speed is dependent on engine speed and is most efficient at higher engine speeds. Another type of

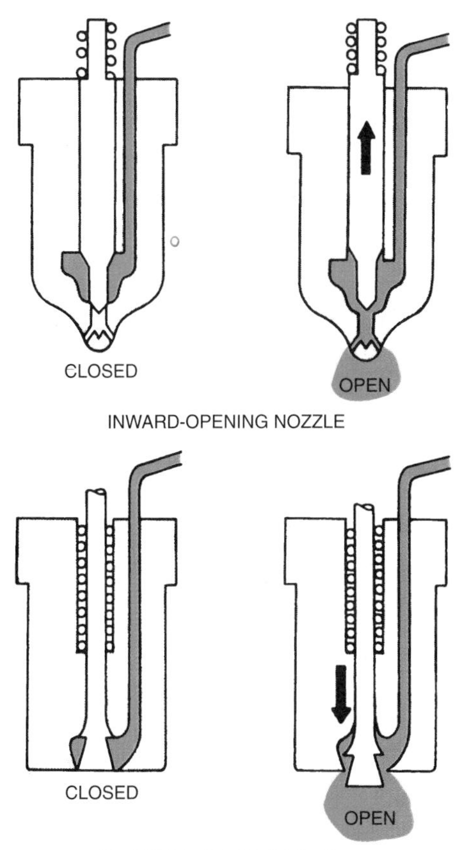

Figure 9-43 Diesel fuel injector nozzles are of either inward opening or outward opening design. (Deere & Co.)

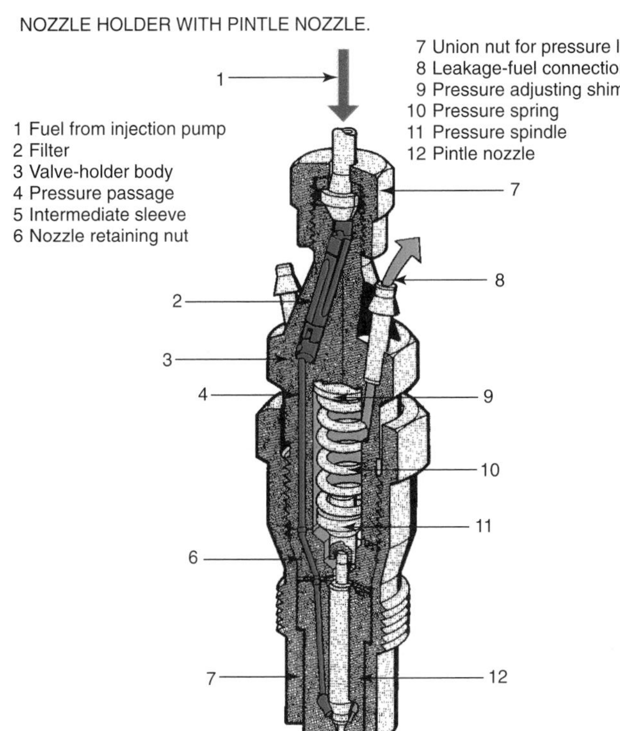

Figure 9-44. Cutaway view of an inward opening, diesel fuel injector. (Robert Bosch)

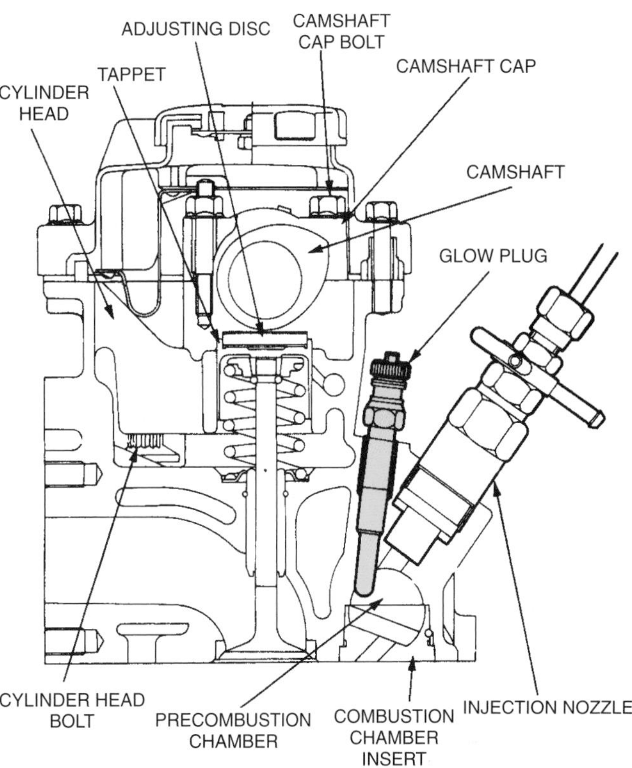

Figure 9-45. Diesel engine glow plug. The glow plug heats the air-fuel mixture in a precombustion chamber to assist in cold engine startup and initial driveaway. Note how the glow plug tip protrudes into the precombustion chamber. Injector spray enters this area. (Mazda)

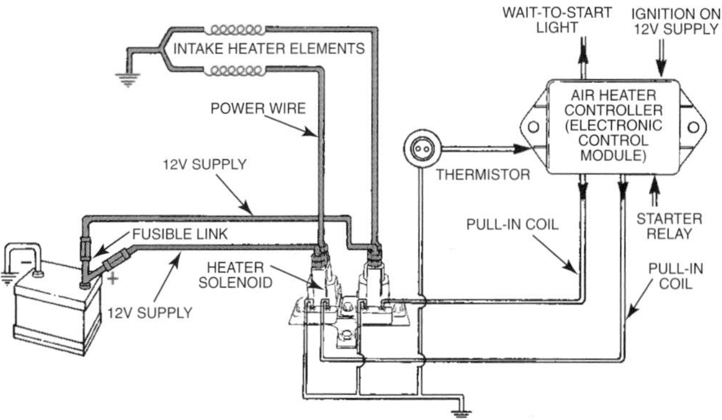

Figure 9-46. An electrical schematic for an intake air heater circuit. This unit takes the place of glow plugs to aid in cold starting. (Dodge)

supercharger is a positive displacement vane type and works like a vane oil pump. Its operational speed is also dependent on engine speed.

Turbochargers

Turbochargers use the exhaust gases to drive a ***turbine wheel***. It is connected to the compressor wheel by a shaft, **Figure 9-48**. Shaft speeds may approach 150,000 RPM. The shaft is supported by ball or tapered roller bearings that are pressure lubricated and cooled by the engine oiling system.

The turbine wheel is installed after the exhaust manifold. Exhaust gases leave the engine at high speed, and spin the turbine wheel. The turbine wheel spins the compressor wheel, which is connected to the intake manifold passages after the air cleaner and before the intake valve. The compressor wheel forces more of the air-fuel mixture into the engine.

Turbocharger Boost Control

If turbocharger ***boost pressure*** (pressure delivered to the intake manifold) becomes too high, it can cause detonation and other engine damage. Boost pressure is controlled by a ***waste gate***, **Figure 9-49**. The waste gate controls the engine exhaust gas flow.

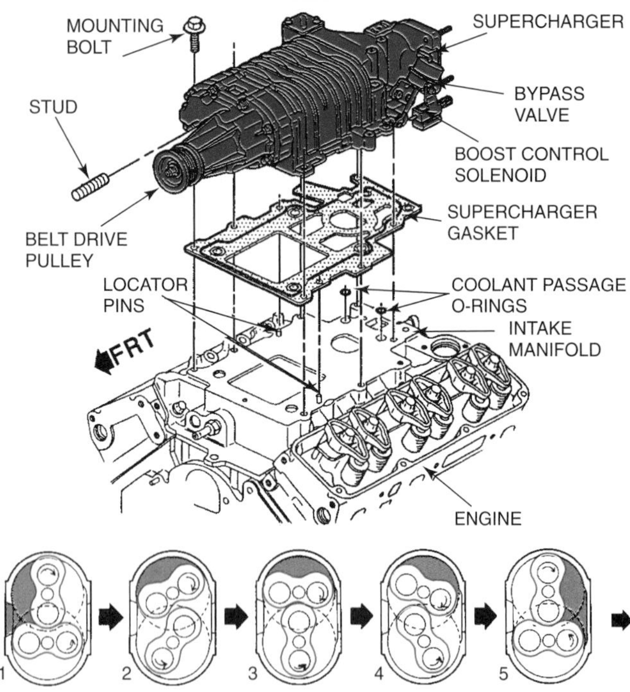

Figure 9-47. A—A V-6 engine that uses a belt-driven, positive displacement supercharger. The supercharger contains two counter-rotating rotors that are connected by spur gears. They produce a boost pressure of 7 psi -11 psi (48.3 kPa - 75.8 kPa). The unit is lubricated with its own supply of synthetic oil. It is sealed for life, and does not require service. B—Supercharger rotor positions. 1–Taking air in. 2, 3 and 4–Compressing the air. 5–Forcing the air out. The air is pumped out four times per rotor revolution. (Pontiac and Toyota)

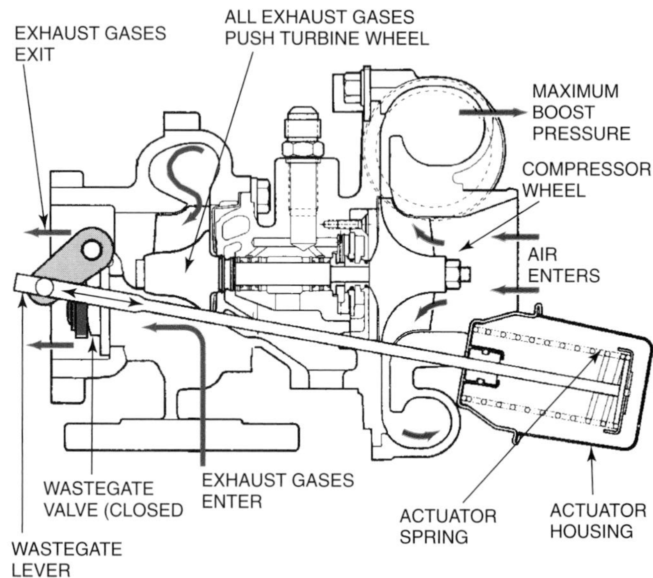

Figure 9-49. A cross-sectional view of a turbocharger which incorporates a mechanically operated waste gate. When a preset boost pressure is reached and exceeded by the turbocharger, the waste gate valve will overcome spring pressure in the actuator, allowing it to open and vent off excessive exhaust gas pressure. Study the air and exhaust flow paths.

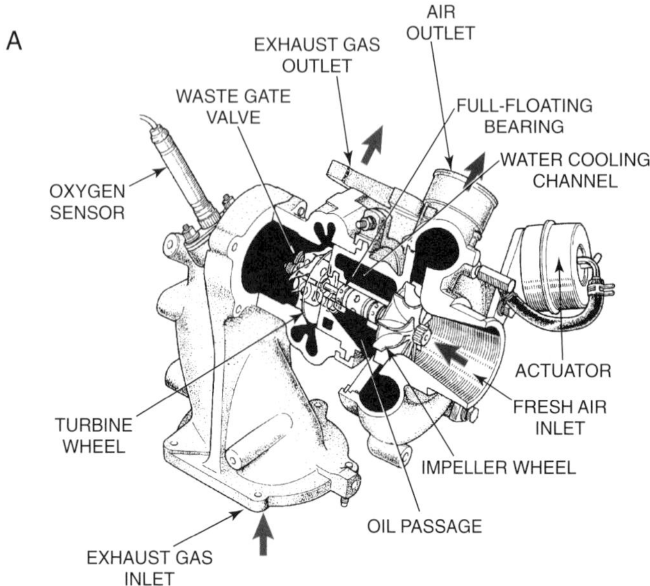

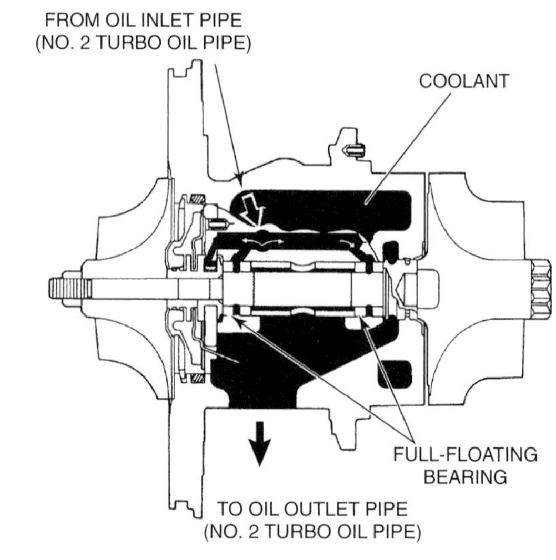

Figure 9-48. An exhaust gas driven turbocharger. A—Cutaway view of an exhaust gas driven turbocharger unit. B—A cutaway view of the turbocharger bearings, oil, and coolant passages. An intercooler is used to cool the incoming air, helping to create a denser air/fuel mixture. (Toyota)

The waste gate is operated mechanically or by a vacuum diaphragm, which will move when the intake manifold pressure rises above a set level. When the pressure exceeds this level, the diaphragm opens the waste gate. When the waste gate is open, exhaust gas is permitted to bypass the turbine wheel. This causes the wheel to slow down, reducing the turbocharger boost pressure to a more suitable level.

Due to the increased air-fuel mixture delivered to the engine (with a corresponding increase in pressure), it is essential that the engine compression be lower than that of a non-supercharged engine. Modern supercharged and turbocharged engines have electronic controls which can adjust the ignition timing to prevent detonation.

Summary

Gasoline fuel injection systems, both electronic and mechanically controlled systems, have completely replaced the carburetor on the newest vehicles. Some systems use pulse injection while others employ continuous injection. The major types of pulsed injection systems are the throttle body injection system, which takes the place formerly occupied by the carburetor; and the multiport injection system which uses one injector per cylinder. Fuel injection systems provide extremely accurate control of the air-fuel mixture thus increasing engine efficiency while at the same time reducing emission levels.

Older fuel injection systems were controlled by mechanical linkage. The new injection systems are controlled by the engine control computer based on inputs from sensors.

Diesel engines are highly efficient and are available in trucks and some cars. They use high pressure injector pumps and inject fuel directly into the precombustion chamber.

Superchargers and turbochargers are simply air pumps which force more of the air-fuel mixture into the engine cylinders to create extra power. Superchargers are driven by a belt or chain from the engine crankshaft and are always turning in relation to engine speed.

Turbochargers are a type of supercharger that is driven by the engine exhaust. The turbocharger has no effect until engine speed provides enough exhaust flow to spin the turbine. A waste gate prevents excessive boost pressure in the intake manifold.

Know These Terms

Fuel injection
Cone shaped pattern
Atomizes
Direct fuel injection
Indirect fuel injection
Mechanical fuel injection
Electronic fuel injection
Stoichiometric
Pulse fuel injection
Continuous fuel injection
Timed fuel injection
Non-timed fuel injection
Pulse width
Throttle body fuel injection
Multiport fuel injection
Air cleaner assembly
Throttle body
Intake manifold
Throttle valve
Filter
Water trap
Pressure regulator
Returnless fuel injection systems
Fuel rail
Nozzle
Sealing pintle
Solenoid winding
Spring pressure
Cold start valve
Auxiliary air regulator
Engine control computer
Sensors
Oxygen sensor
Engine speed sensor
Throttle position sensor
Manifold vacuum sensor
Barometric pressure
Airflow sensor
Speed density
Fuel distributor
Diesel fuel injection
Prechamber
Cetane number
Octane
Water in fuel sensor
Diesel injection pump
Inline injector pump
Control rack
Distributor injection pump
Fuel shutoff
Viscosity compensator valve
Delivery advance mechanism
Inward opening injectors
Outward opening injectors
Glow plugs
Intake air heaters
Supercharger
Compressor wheel
Turbocharger
Turbine wheel
Boost pressure
Waste gate

Review Questions—Chapter 9

Do not write in this book. Write your answers on a separate sheet of paper.

1. Technician A says that most gasoline fuel injection systems inject the fuel directly into the cylinders. Technician B says that most gasoline injection systems are continuous flow types.
   Who is right?
   (A) A only.
   (B) B only.
   (C) Both A & B.
   (D) Neither A nor B.
2. Gasoline fuel injection systems employing a mechanical pump operate on the _____ flow principle.
   (A) continuous
   (B) pulse
   (C) rotating
   (D) timed
3. The fuel injector solenoid is energized by the _____ .
   (A) ignition switch
   (B) computer
   (C) airflow sensor
   (D) ignition module
4. The two kinds of electronic gasoline injection systems are _____ and _____ .
5. Technician A says that central fuel injection systems have one injector for each cylinder. Technician B says that electronic injection systems use an electric fuel pump to produce the needed injector pressure. Who is right?

(A) A only.
(B) B only.
(C) Both A & B.
(D) Neither A nor B.

6. Mechanical injection system injectors are opened by _____ .
   (A) electric solenoids
   (B) airflow
   (C) fuel pressure
   (D) compression pressure
7. In the continuous flow fuel injection system, basic air-fuel ratio control is provided by a(n) _____ sensor.
   (A) pressure
   (B) engine speed
   (C) manifold vacuum
   (D) airflow
8. List the four basic parts of the pulse fuel injection system.
9. Electronic fuel injection systems can use either:
   (A) one injector per cylinder.
   (B) two injectors for the entire engine.
   (C) one injector for the entire engine.
   (D) All of the above.
10. On some fuel pressure regulators, the spring loaded valve is assisted by _________.
11. The cold start injector supplies _________ for cold starts.
12. All of the following statements about gasoline fuel injectors are true, EXCEPT:
    (A) injectors spray gasoline into the intake manifold.
    (B) injectors can be part of a central fuel system assembly.
    (C) injectors are operated by manifold vacuum.
    (D) injectors can be installed in the intake manifold.
13. Multiport injectors are connected to the fuel pump by a __________ .
14. Technician A says that all gasoline injection systems use higher pressures than diesel injection systems. Technician B says that some gasoline injection systems use higher pressures than diesel injection systems. Who is right?
    (A) A only.
    (B) B only.
    (C) Both A & B.
    (D) Neither A nor B.
15. All of the following statements about diesel injectors are true, EXCEPT:
    (A) diesel injectors are operated by fuel pressure.
    (B) diesel injectors inject fuel directly into the cylinder.
    (C) diesel injectors are closed electrically.
    (D) diesel injectors are closed by spring pressure.
16. Glow plugs are used to _____ .
    (A) open the injectors
    (B) heat the injectors
    (C) assist in cold starting
    (D) assist in hot starting
17. Diesel injection pumps are of two general types. Name them.
18. Technician A says that diesel injection systems feed some fuel from the injector nozzle at all times. Technician B says that some diesel injection pumps inject fuel at any time.
    Who is right?
    (A) A only.
    (B) B only.
    (C) Both A & B.
    (D) Neither A nor B.
19. Some diesel injection pumps use a control rack to adjust fuel flow. The rack is used to rotate the __________ .
20. Rewrite the following sentence, leaving out the wrong words. "Diesel engines use a glow plug that protrudes into the cylinder, precombustion chamber, crankcase."
21. The function of the delivery valve is to _____ .
    (A) prevent system leaks
    (B) regulate system pressure
    (C) inject fuel into the hollow control rack
    (D) None of the above.
22. Diesel injector nozzle designs are of two general types. Name them.
23. Superchargers are driven by the crankshaft through _____ .
    (A) a chain
    (B) a belt
    (C) gears
    (D) Both A & B.
24. Maximum turbocharger speed is around _____ RPM.
    (A) 5000
    (B) 150,000
    (C) 20,000
    (D) 2000
25. The turbocharger waste gate is operated by _____ .
    (A) a diaphragm
    (B) manifold pressure
    (C) a detonation sensor
    (D) Both A & B.

10

Fuel Supply and Carburetors

After studying this chapter, you will be able to:

- Describe the components of the fuel supply system.
- Explain the basic function of the fuel pump.
- Describe the major parts of the fuel pump.
- Identify the basic parts of a carburetor.
- Summarize carburetor operating principles.
- List the seven basic carburetor circuits.

This chapter will cover the methods by which fuel is transferred from the fuel tank to the engine, and the principles of carburetion. Modern engines must be provided with an air-fuel mixture in the right proportions for proper engine operation, and to meet mileage and emission requirements. In Chapter 9 we discussed the need to atomize and mix gasoline for proper combustion and how this was accomplished by the fuel injection system. This chapter will describe how gasoline is atomized and mixed with air by the carburetor. The fuel supply system will also be discussed in this chapter.

Fuel Supply System

If fuel does not reach the carburetor and fuel injection system, the vehicle will not run. Therefore, the fuel supply system is vitally important to the operation of the vehicle.

Fuel Tank

The ***fuel tank*** has been located in just about every conceivable spot on a vehicle—in the rear, behind the seats, even in the front. An ideal location is a spot where the tank is protected from flying stones and debris, where it will not bottom on the road, and where it will not easily rupture in the event of an accident. The overall design features of a car often dictate the location of the tank.

The tank is usually constructed of thin sheet steel and coated with a lead-tin alloy to protect against rusting. Tanks can also be made of aluminum or plastic such as polyethylene. Capacity can be anywhere from 9 to 26 gallons (34 to 98 L). Internal baffles are arranged to prevent sloshing, **Figure 10-1**. A filler neck, reaching to a convenient spot, is attached either directly or by a neoprene hose. The filler cap normally seals the tank, but does have a vacuum and pressure relief valve to allow air to either enter or leave the tank, in the event the emission control vent lines were to become inoperative.

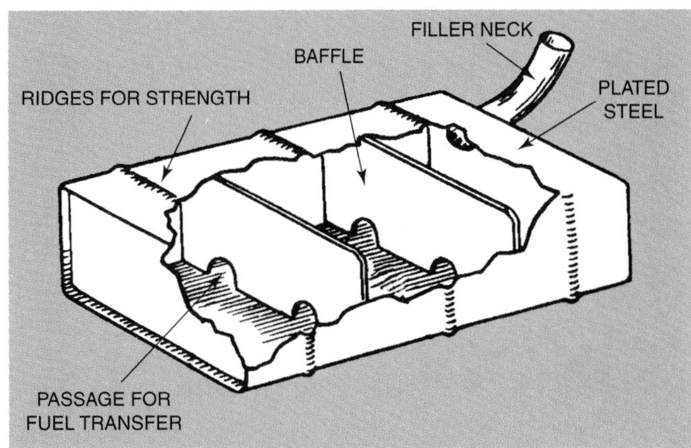

Figure 10-1. Construction of a typical fuel tank.

The tank is vented to a charcoal canister (covered in Chapter 14) which traps gasoline vapors before they enter the atmosphere. The tank must be vented to allow fuel to be drawn out. The tank is securely strapped into place. It is often protected on the outside by an application of undercoating material.

Warning: A fuel tank, even when apparently empty, is as dangerous as a bomb! Keep all sparks, heat, and open flames away from the tank.

Gasoline Pickup Tube

A ***pickup tube*** enters the tank to draw off fuel for the engine. This tube can enter from the top, side, or bottom. The tube end is generally located about one-half inch from the tank bottom. This allows considerable water and sediment to form before being drawn into the pickup tube.

Several types of filters are used on the ends of pickup tubes, including plastic, metal mesh, and fiberglass. An effort is made to design screens that will not clog while filtering out sediment and most water, **Figure 10-2**.

Water Forms in Tank

When a fuel tank is only partially full, water tends to form within the tank. Any moisture within the tank condenses on the

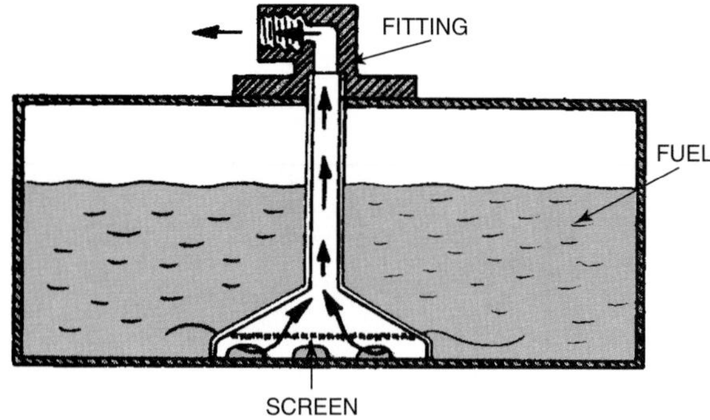

Figure 10-2. Fuel pickup pipe. Notice that the pickup screen is above the bottom of the tank.

tank wall, then runs under the fuel. Since water is heavier than gasoline or diesel fuel, it will remain at the bottom where it will fail to evaporate as it is not exposed to the air. Water buildup is a slow process, but can eventually be drawn into other parts of the fuel system or can cause the tank to rust out, **Figure 10-3**. In cold weather, water in the fuel system can cause the fuel lines to freeze.

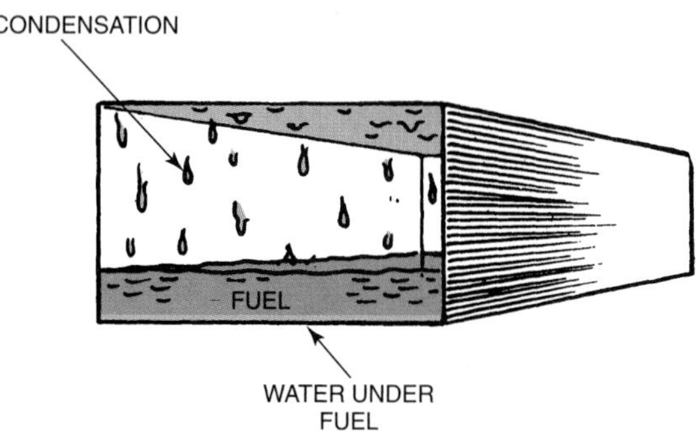

Figure 10-3. Fuel tank showing condensation when the fuel level is low.

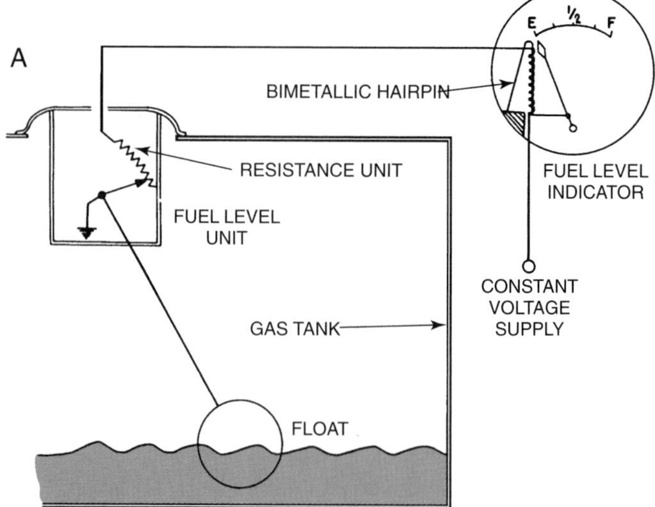

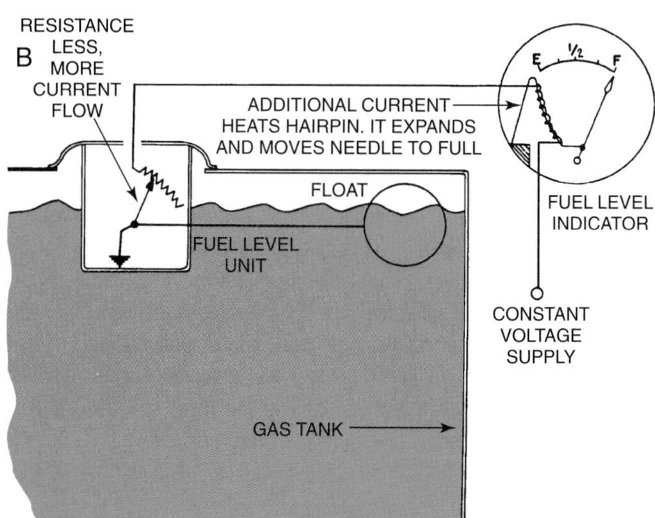

Figure 10-4 A—Fuel gauge operation with tank empty. When the tank is empty, the hairpin expands, causing the needle to move toward the full mark. B—Fuel gauge operation with the tank full. (Plymouth)

Tank Gauge Unit

The fuel tank contains a float that is attached to a resistance device similar to that used in an electric oil pressure sender. In this case, the up-and-down travel of the float causes the resistance to vary, depending on the depth to which the tank is filled. This is the ***tank gauge unit***.

Most gauges are designed to read empty when one or two gallons of fuel are left in the tank. This is a concession to human nature as some people are inclined to drive until the gauge does read empty. **Figure 10-4** illustrates a typical tank sender and bimetallic receiver. When the float drops, current flow to ground through the bimetal wire coil will be less, because it must travel through more resistance wire. This cools the bimetallic hairpin and pulls it together. When the tank is filled, the contact slides up, cutting out resistance and current flow increases. This heats the hairpin and causes it to separate.

How Gasoline Leaves the Tank

Fuel tanks usually are lower than the rest of the fuel system, so gravity must be overcome before the fuel will reach the engine. The fuel must be drawn from the tank and delivered to the rest of the fuel system. This is accomplished by the vehicle fuel pump. Fuel pumps can be operated mechanically or electrically.

Mechanical Fuel Pump

Mechanical fuel pumps are mounted on the engine and operated by the camshaft. The pump consists of an air chamber divided in the center by a flexible ***diaphragm*** made of gasoline-resistant material. The diaphragm is pinched between the halves, either by a series of screws that hold the pump halves together, **Figure 10-5**, or by crimping the halves together (on modern pumps).

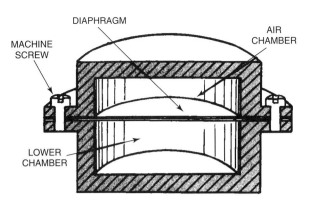

Figure 10-5. An air chamber and diaphragm typical of a basic fuel pump.

Check Valves

Two ***check valves***, one inlet and one outlet, are added to the top chamber. The inlet is connected to the fuel tank and the outlet to the carburetor via copper, steel, or plastic tubing. These check valves use flat washer valves held closed by spring pressure, **Figure 10-6**. The purpose of the check valves is to ensure that the fuel flows in one direction only.

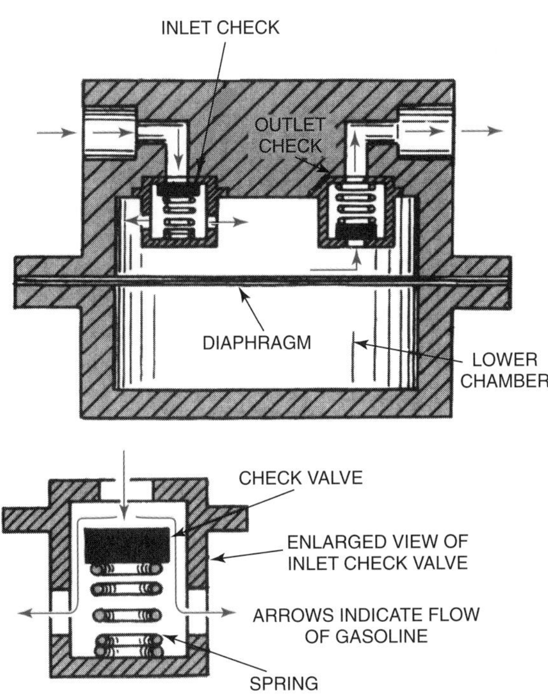

Figure 10-6. Check valves added to the top chamber. Small springs hold the valves shut until vacuum or pressure overcomes spring tension.

Pull Rod and Return Spring

A ***pull rod*** is now fastened to the center of the diaphragm and a strong ***return spring*** pushes the diaphragm in an upward direction. If you were to connect the inlet of this fuel pump to a container of gasoline, then run the outlet to another container, the pump could be worked by pulling down on the pull rod by hand. Fuel pump operation is shown in **Figure 10-7**.

Rocker Arm

In order to pull the diaphragm down, it is necessary to install a ***rocker arm***. The rocker arm hooks in the slotted end of the pull rod and extends out where it rides on a cam on or is attached to the front of the camshaft. Note that the cam lobe used for this is different from the cam lobes used to operate the valves. When the cam lobe pushes up on the rocker arm, it pulls the pull rod down. When the cam turns, the rocker follows the lobe down.

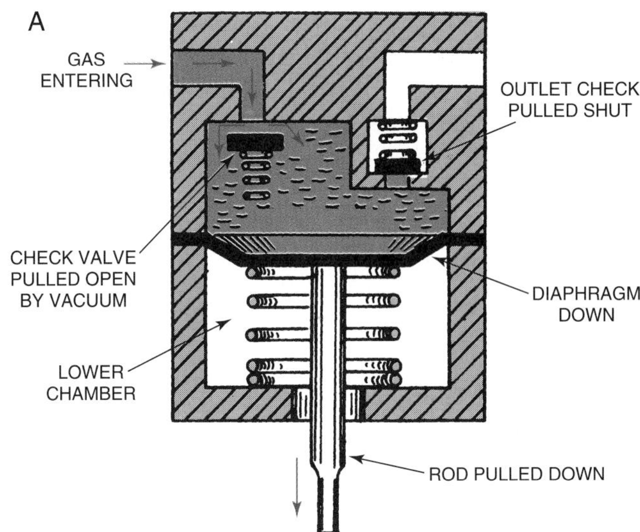

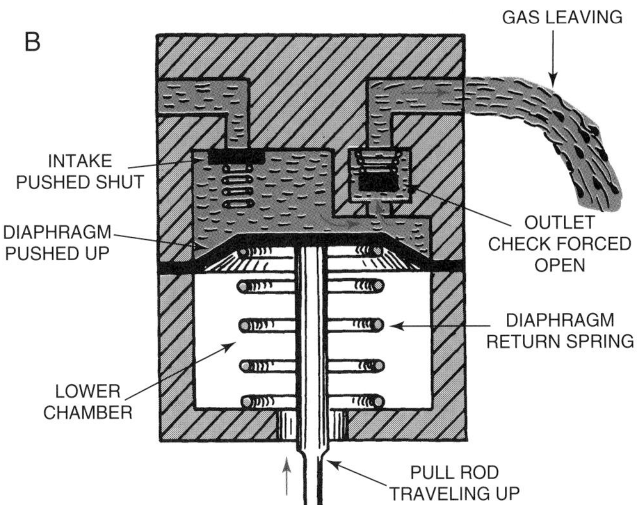

Figure 10-7 A—Fuel pump intake stroke. When the pull rod is drawn down, diaphragm is flexed downward. This forms a vacuum in the air chamber. The intake check valve will open and gasoline will be drawn into the pump. B—Fuel pump outlet stroke. When the pull rod is released, the diaphragm is pushed up by the return spring. This creates pressure, forcing the inlet check valve shut and opening the outlet check valve. Gasoline will be forced from the outlet.

The pull rod is raised by the spring, not the rocker arm. A small spring keeps the rocker firmly against the cam lobe. The rocker arm is contained in a housing which also forms the mounting flange. This flange, along with a gasket and/or sealer prevents engine oil from leaking out where the rocker arm enters the engine. A complete pump assembly is shown in **Figure 10-8**.

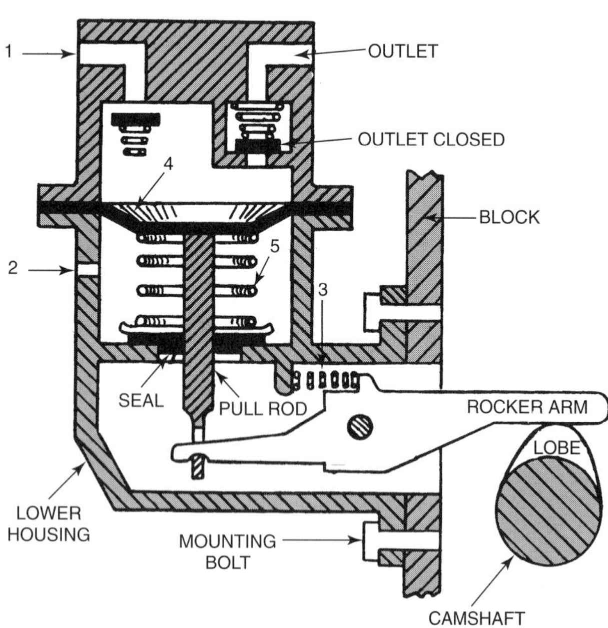

Figure 10-8. Lower housing, rocker and link added. Notice that the cam lobe has raised the rocker arm, pulling diaphragm (4) down. This opens the inlet check valve, pulling in fuel. When the cam lobe turns, spring (3) will force the rocker arm to follow. The diaphragm spring will push the diaphragm up, close the inlet, and expel gasoline through the outlet. Vent (2) is provided so the diaphragm can flex without compressing air in the lower chamber. A seal keeps engine oil from entering the lower chamber. All link and rocker rubbing surfaces are hardened to retard wear.

The seal around the pull rod keeps engine oil and fumes from entering the lower chamber and prevents any gasoline, which leaks from the diaphragm, from entering the engine. A vent is provided in the lower chamber so the diaphragm can be pulled down without compressing the air in the chamber. Note that when the diaphragm is held down by fuel pressure in the carburetor, the rocker arm can slide up and down on the diaphragm pull rod. Study the parts and learn their names and functions. As illustrated, this fuel pump will move gasoline efficiently, but will not filter nor dampen the pulsating movement of gasoline leaving the outlet.

Fuel Pulsation Damper

Fuel pressure rises and falls, or pulsates, as the fuel pump operates. This pulsation can create a slamming effect against the carburetor float needle valve, causing it to come off its seat and flood the carburetor. **Figure 10-9** illustrates this effect.

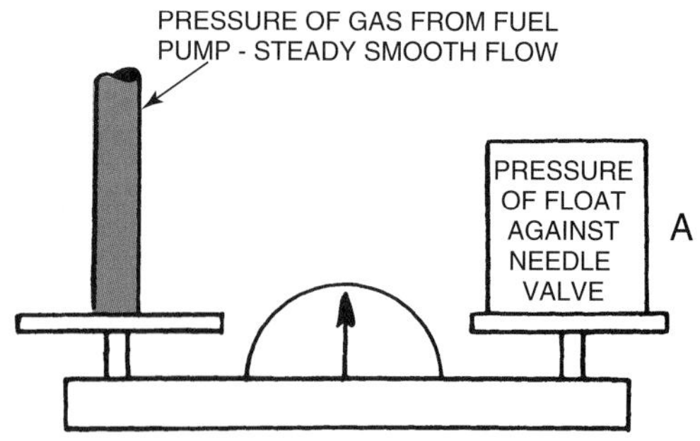

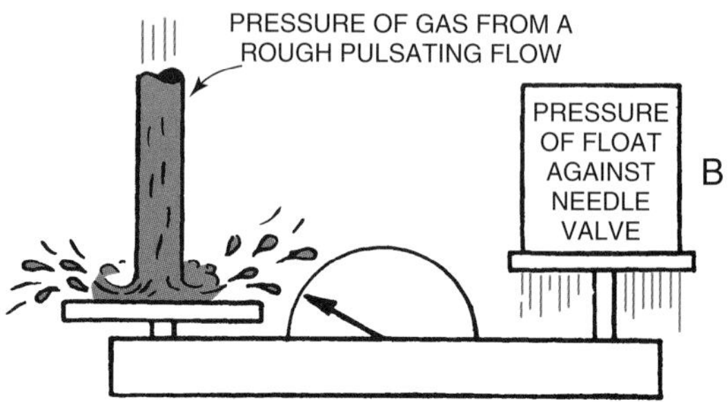

Figure 10-9. Pulsation effect. A—Notice how float pressure controls the flow of gasoline. B—Here, fuel is being delivered in a jerky manner, called pulsation. See how it slams against the scale and upsets needle pressure.

To smooth out the flow of fuel, many systems are equipped with a ***pulsation damper***. A pulsation damper is a flexible diaphragm stretched across an air compartment. The trapped air acts as a spring against which the diaphragm can flex. The damper can be built into the pump, mounted on the outlet of the pump, or line as a separate unit.

When the pump injects fuel into the line, the pulsation bulges the diaphragm outward, absorbing the shock. While the pump is in its intake stroke, the diaphragm will return to its normal position, **Figure 10-10**.

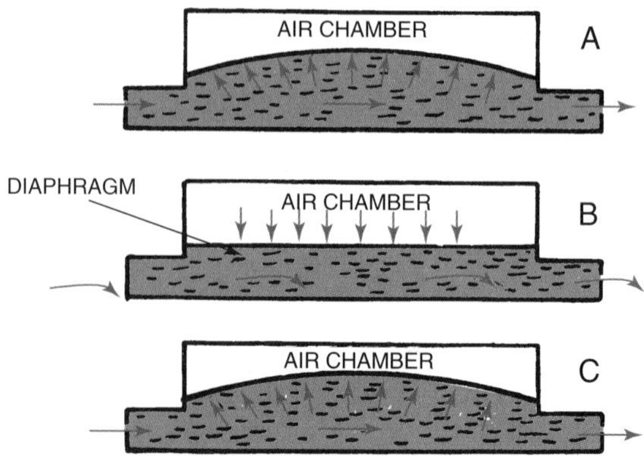

Figure 10-10. Pulsation damper principle. A—When the pump injects fuel, the diaphragm pushes up against the trapped air inside the air chamber. This absorbs the slamming effect. B—No fuel from the pump. Trapped air pushes the diaphragm down, continuing some fuel pressure during the lull. C—When the pump ejects more fuel, it is absorbed again by the diaphragm.

Mechanical Fuel Pump Efficiency Measurements

The amount of pressure the mechanical fuel pump exerts on the fuel line to the carburetor depends on the strength of the diaphragm spring. Pump pressures can be designed into the pump by the manufacturer according to need. An average would be around 4 to 10 psi (28 to 41 kPa). The normal pump can discharge over a quart per minute.

Serviceable and Non-Serviceable Fuel Pumps

The majority of modern mechanical fuel pumps are non-repairable, and must be replaced as a unit. If the pump is defective, it is discarded. A repairable fuel pump is shown in **Figure 10-11**. Note the various parts making up the pump.

Electric Fuel Pump

An ***electric fuel pump*** is always used with fuel injection and on some carbureted engines. The electric pump will fill the fuel lines merely by turning on the key. Another feature is the electric pump's adaptability to almost any location. This allows the pump to be mounted away from the heat of the engine to reduce the chance of vapor lock. Some electric pumps are submerged inside the fuel tank. A few fuel injected vehicles have two fuel pumps. A low pressure transfer pump feeds the high pressure pump for the fuel injectors.

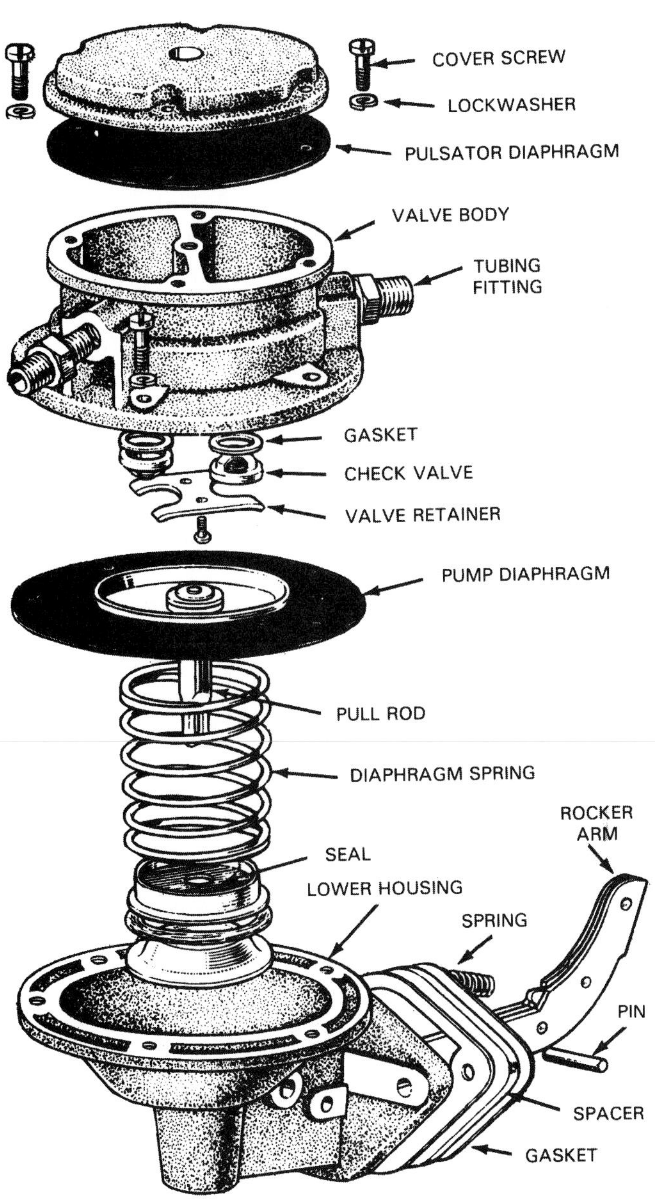

Figure 10-11. Typical serviceable fuel pump construction. Learn the part names. (Nissan)

Older electric fuel pumps used an electromagnet (a magnet produced by electricity flowing through a coil) to operate a metal bellows that alternately formed a vacuum and then pressure. It resembled a mechanical fuel pump, complete with one-way check valves. When the key was turned on, the electromagnet was energized and attracted a metal armature that was attached to the bellows. This stretched the bellows and formed a vacuum, which drew in gasoline through the inlet check valve.

As the armature continued downward, it would strike a pair of contact points, opening them. This would break the circuit to the electromagnet and it would release its pull on the bellows. A return spring shoved the armature and bellows up. As the bellows closed, gasoline was forced out the outlet check valve. When the armature traveled all the way up, the points closed and it would be drawn down again. When the carburetor's float bowl was full, the return spring could not collapse the bellows and the pump would be inoperative until some of the gasoline was drawn from the bowl.

Other electric fuel pumps use a small electric motor to drive a rotating pump. Modern rotating pumps are one-piece units. The motor is usually surrounded by the incoming gasoline for cooling. These fuel pumps can be the ***positive displacement*** type, which delivers a certain amount of fuel every time the pump makes one revolution. Or they may be a ***non-positive displacement*** type, which can vary the output on each revolution. Electric fuel pumps can use positive displacement vanes or rotors to pump the fuel, or a non-positive impeller. **Figure 10-12** shows a cutaway of one type of electric rotating pump.

Fuel Filter

The tank pickup screen will not filter out small particles. It is necessary to have an additional ***fuel filter*** either between the tank and the pump or between the pump and the carburetor (or fuel rail). Some pumps have a filter built into the pump assembly.

Types of Fuel Filters

Filters can be sealed units which are replaced as a unit, or may have a replaceable element inside of a reusable housing. The filter element may be a fine screen, or be made of ceramic, treated paper, or sintered bronze. All filters need periodic replacement. **Figure 10-13** illustrates an inline (placed in the fuel line) filter. All fuel must pass through the cleaning element. A special pleated paper is widely used for cleaning elements.

Fuel Lines

Gasoline is carried from the tank to the carburetor in plated steel, neoprene, or plastic ***fuel lines***. When plastic is

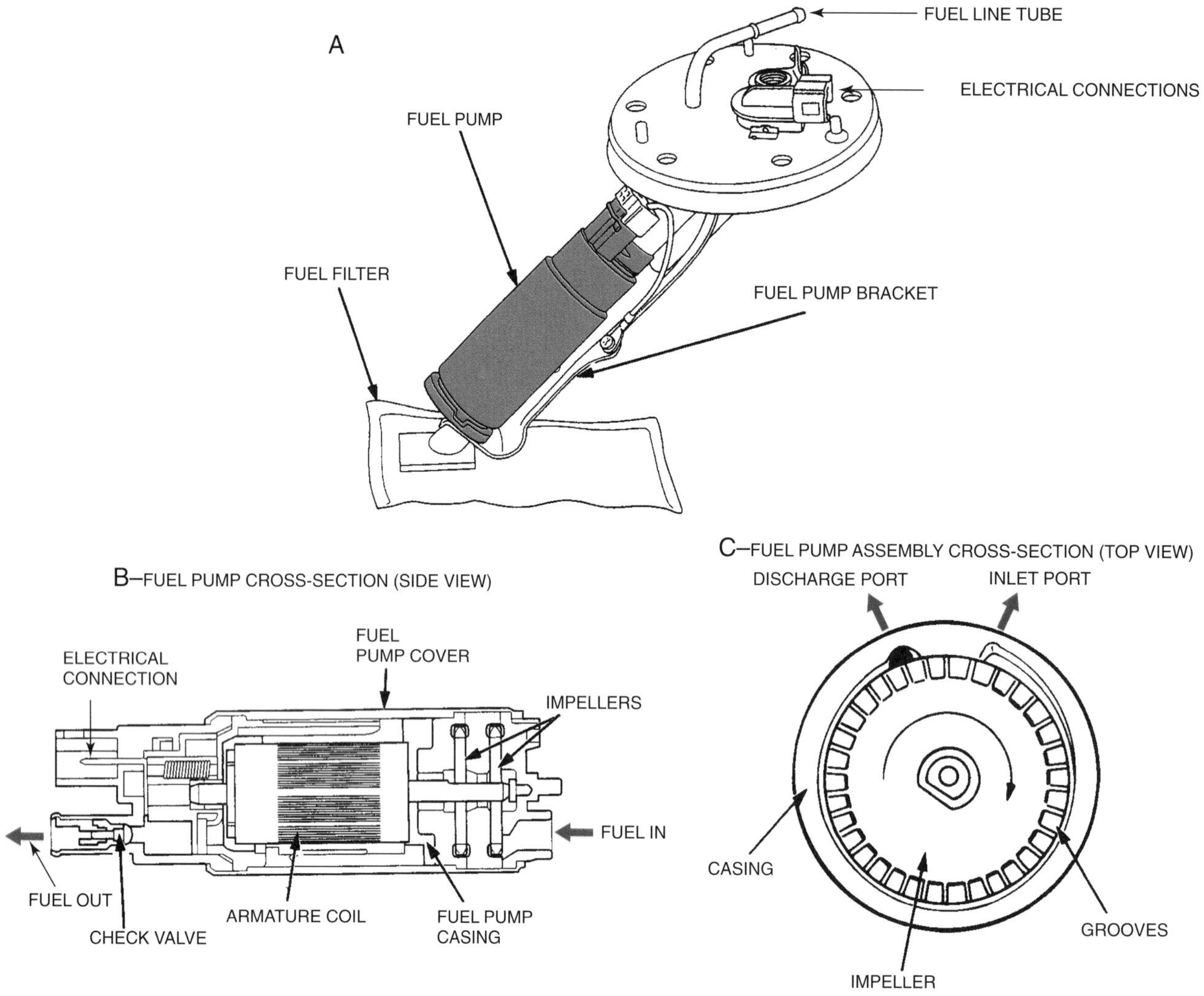

Figure 10-12. A—An in-tank electric impeller type fuel pump assembly. B—Cross-sectional view of a dc (direct current) fuel pump. C—Cross-section top view of an in-tank pump. When the engine is started, the computer main relay turns on the pump—turning the impellers. Fuel pressure is produced by the many grooves around the impellers. Fuel entering the inlet port travels inside the motor from the pumping chamber and is forced out the discharge port check valve. If the fuel flow is obstructed at the discharge side, the relief valve will open and bypass fuel to the inlet port to prevent excessive fuel pressure. When the engine is turned off, the fuel pump will stop automatically. A check valve closes by gravity to help retain residual line pressure. This aids in restarting the engine.

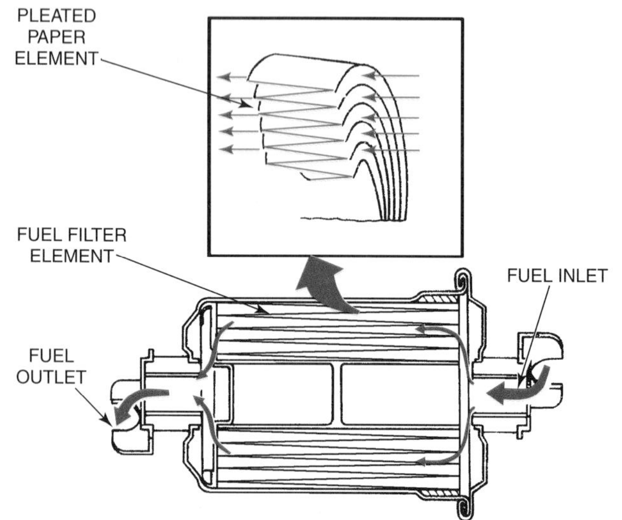

Figure 10-13. Cutaway view of a paper or throwaway fuel filter. (Chevrolet)

used, it often runs from the pump to the carburetor. All lines and joints must be tight and secure to prevent destructive vibration and rubbing.

Warning: Fire is a constant hazard around any fuel system. A loose connection or unsecured fuel line can break or leak. If this happens, fuel can be sprayed over the engine, which will cause a dangerous fire if ignited.

Vapor Lock Eliminators

It is important to shield fuel lines, pump, and carburetor from excessive heat. When gasoline reaches a certain temperature, it will turn to a vapor. Since the fuel pump is not designed to pump vapor, the fuel flow can either be reduced or completely stopped. This condition is called ***vapor lock***. Upon cooling, fuel flow will resume at the normal rate.

Metal plates, heat shields, and heat resistant line covers are used to prevent vapor lock. Chemical additives are blended into gasoline to raise its boiling point to eliminate this tendency as much as possible. A fuel pump vapor discharge (return) valve can be incorporated to prevent vapor lock when underhood temperatures are high. The vapors return, via a separate line, to the fuel tank where they condense and rejoin the liquid fuel. Cars equipped with air conditioning often utilize vapor return devices to offset the danger of vapor lock due to increased underhood temperatures.

Carburetors

The job of the fuel system is to constantly deliver an ***air-fuel ratio*** of the proper proportions to the engine. The proportion must be variable to meet the ever changing needs of the engine. For many years this was effectively accomplished by the ***carburetor***.

While the carburetor has been replaced by fuel injection on late model vehicles, there are still millions of vehicles with carburetors on the road. The latest carburetors are highly developed, complex systems. There are many types and sizes, however, they all share the same basic concepts.

Build a Carburetor

As you have done with an engine, you will now build a basic carburetor on paper. In Chapter 1, you learned that gasoline must be mixed with air to burn. The proportions of gasoline and air can vary, depending on engine temperature, speed, load, and design. A rapidly moving stream of air enters each cylinder on the intake stroke. You will use this airstream to make your carburetor work. Each intake port feeds into the intake manifold. The intake manifold runs from the ports to the carburetor.

Air Horn

Start with a plain round metal tube. This tube is called the ***air horn***. The air enters the engine through the air horn. The tube has a flange that bolts to the intake manifold, **Figure 10-14**.

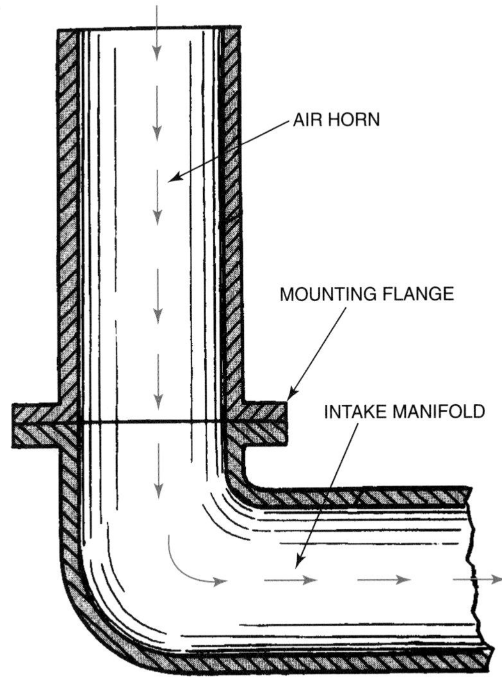

Figure 10-14. Air horn. Arrows indicate flow of air through the horn into the intake manifold and on into the cylinders.

Fuel Bowl

A ***fuel bowl*** is attached to the air horn to hold a supply of gasoline. Inside of the bowl, a ***float*** is hinged. The float controls a tapered pin called a ***needle valve*** that will open and close in relation to its ***needle seat*** to admit gasoline. When the bowl fills, the float will rise and shove the needle valve against its seat to stop the flow of gasoline. When the fuel level drops, the float releases its pressure, the needle valve lifts from the seat and more gasoline enters. The float and needle valve will keep the fuel level constant.

The fuel level in the bowl is vital for maintaining the proper air-fuel ratio. A high float level in the carburetor will make the engine run ***rich*** (too much gasoline reaching the engine). A low float level in the carburetor will make the engine run ***lean*** (too little gasoline reaching the engine). The float level can be adjusted to maintain the proper level. The bowl must have a hole (vent) to allow air to enter or leave as the fuel level changes. The bowl may be vented to the air horn as in **Figure 10-15**, to the atmosphere, or to both.

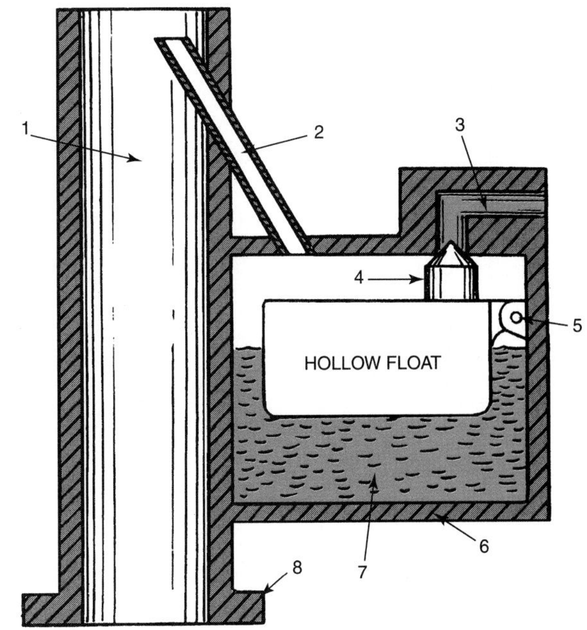

Figure 10-15. Float bowl, float, needle valve and bowl vent added to the carburetor air horn. Notice how the float has pushed the needle valve into its seat to shut off the flow of gasoline. 1—Air horn. 2—Bowl vent or balance tube. 3—Fuel inlet. 4—Needle valve. 5—Hinge. 6—Float bowl. 7—Gasoline. 8—Flange.

Fuel Bowl Vent

If the air cleaner becomes clogged, a stronger vacuum will result in the air horn. This tends to draw in more fuel. To prevent an overly rich mixture, a ***bowl vent*** is made between the bowl and air horn, **Figure 10-15**. If it is vented, the pressure on the fuel in the bowl will decrease in the same proportion and fuel intake will tend to remain constant.

A common setup employs the internal bowl vent plus an external vent that opens during periods of engine idle. This type of external vent helps prevent hot (engine overheated) starting problems by venting fuel vapors that would otherwise be forced into the carburetor air horn. The external vents are channeled to the emission control charcoal canister, which absorbs the fuel vapor. The operation of the canister will be covered in Chapter 14.

Main Discharge Tube

In order to allow the gasoline to be drawn into the air horn, it is necessary to connect the fuel bowl and horn with a small tube. This tube is called the ***main discharge tube***, **Figure 10-16**.

Notice in **Figure 10-16** that the gasoline level in the discharge tube is slightly below the nozzle. In order to get the gasoline to flow out the nozzle and mix with the airstream, a vacuum is necessary. Some vacuum is created by the moving air, but it is not strong enough.

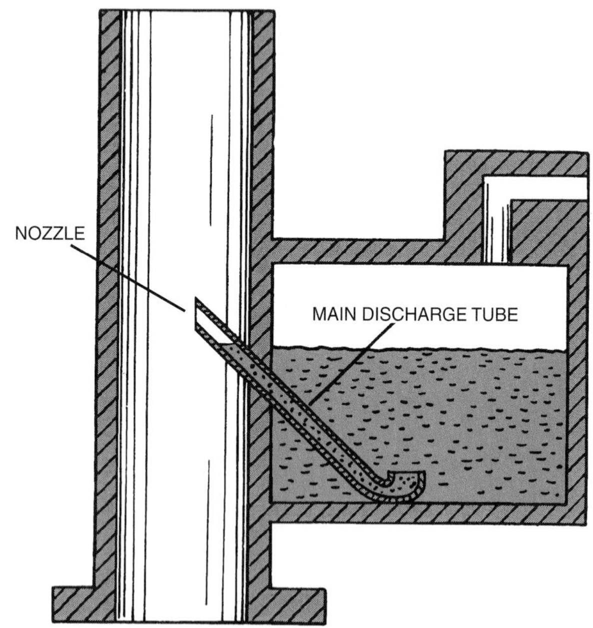

Figure 10-16. Main discharge tube. Note the fuel level in discharge tube.

Creating Venturi Vacuum

By building a restriction, or ***venturi***, in the air horn, the air will be forced to speed up where it strikes the venturi. As the air suddenly speeds up, it tends to be "stretched." In other words, a vacuum is created. The more the air speeds up, the stronger the vacuum. This ***venturi vacuum*** will be strongest in the throat of the venturi, but will still exist for a short distance below it.

Figure 10-17 illustrates an air horn with a venturi. Imagine three tubes containing mercury enter through the horn at different points. Notice the variations in vacuum at different spots in the horn. The higher the air velocity, the more vacuum formed. This is evident by examining the distance the atmospheric pressure has pushed the mercury up the tube.

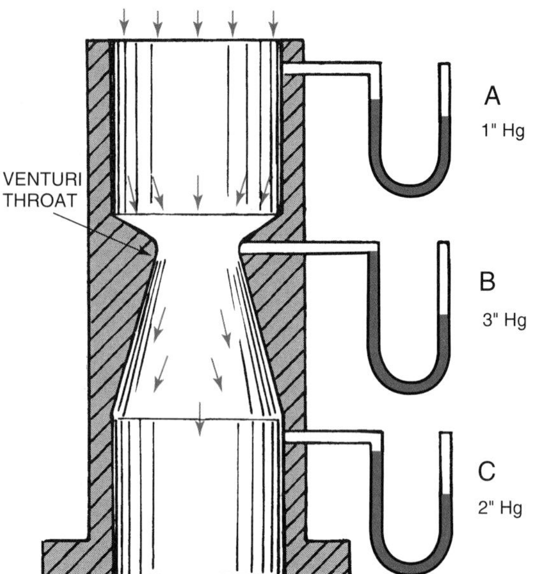

Figure 10-17. A venturi increases the speed of vacuum. A—Air rushing into the air horn near the top produces some vacuum. B—As it travels down into the venturi throat, it speeds up. This causes an increase in vacuum. C—After leaving the venturi throat, the air expands and slows down somewhat. This reduces vacuum.

Secondary Venturi

The vacuum can still be increased even more by constructing a smaller venturi in the center of the air horn and positioning it so that its outlet end is even with the throat of the main, or primary venturi. The primary venturi forms a vacuum at the outlet of the secondary venturi, causing the air to flow quickly through the secondary venturi. This produces a sizable increase in vacuum.

The main discharge tube nozzle enters the secondary venturi. When the engine starts, rushing air draws off drops of gasoline that join the passing air. See **Figure 10-18**. One carburetor uses a venturi valve that can change size in relation to engine vacuum and throttle position.

Air Bleed

The setup so far would add gasoline to the passing air, but the drops would be rather large and would tend to stick to the nozzle. Adding a tiny stream of air to the fuel as it travels through the nozzle causes the drops to combine with the air and break up. The smaller particles will move quickly and vaporize easily. The tube through which the air travels is called an ***air bleed***. See **Figure 10-19**.

Jets

The carburetor must be able to control the amount of gasoline that can travel through a passageway. The carburetor uses ***jets*** to accomplish this task. The jets form a restriction to certain passageways. Jet sizes are very carefully

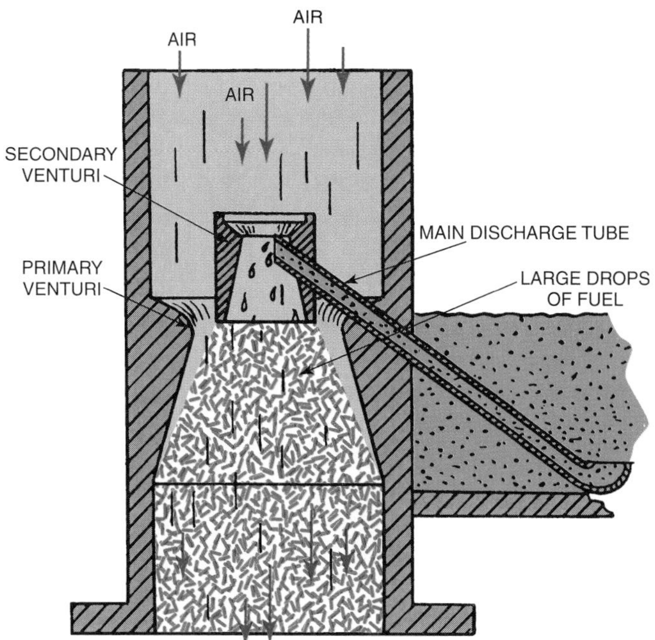

Figure 10-18. A secondary venturi can increase the vacuum speed. Note large drops of fuel being drawn from the nozzle.

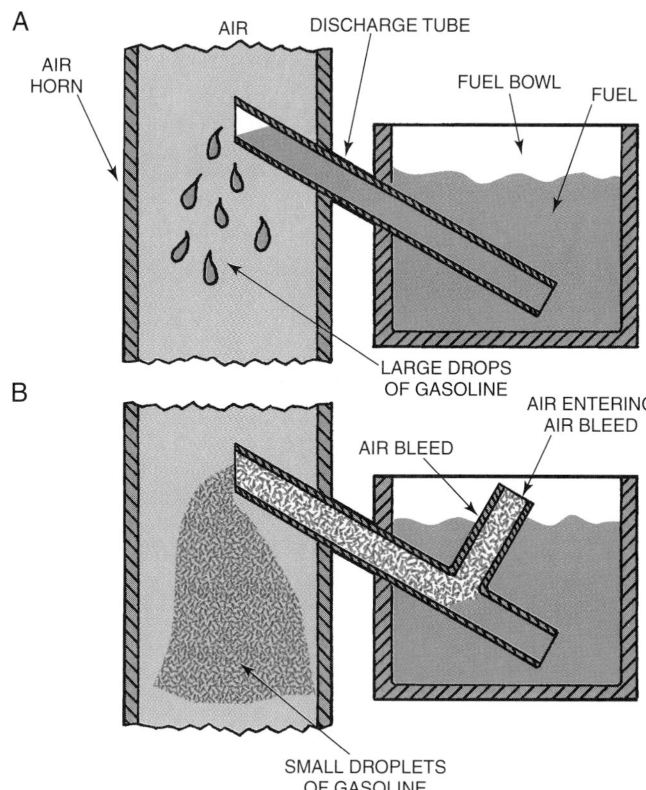

Figure 10-19. An air bleed is used to break up the large drops. A—If no air bleed is used, large fuel drops result. B—An air bleed mixes air into fuel by breaking the gasoline into small droplets.

calculated to control the amount of gasoline that can travel to these passageways.

Jets are constructed of brass and screw into the ends of the various passageways in the carburetor. Some jets are pressed into place and are not removed for routine maintenance. A jet is placed at the lower end of the main discharge tube to control the amount of gasoline that can pass through the tube. See **Figure 10-20**.

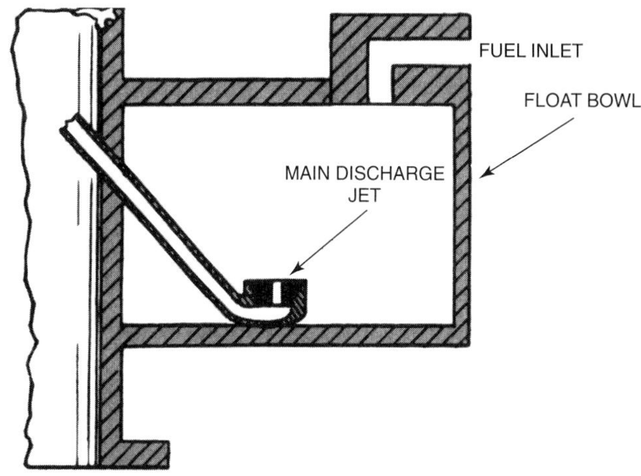

Figure 10-20. Main discharge jet. This jet controls the amount of fuel entering the discharge tube.

Throttle Valve

An engine could start and run with the carburetor described above, but would run wild or wide open. We have not provided a way to control the speed. A valve of some sort is needed to control the idle speed. The ***throttle valve*** is a circular metal plate placed between the bottom of the primary venturi and thc mounting flange. In the closed position, it will almost completely seal off the air horn.

Moving the throttle valve will give us control of the amount of air-fuel mixture reaching the cylinders. It is controlled by linkage connected to the accelerator pedal. **Figure 10-21** illustrates a throttle valve, and **Figure 10-22** shows the necessary linkage.

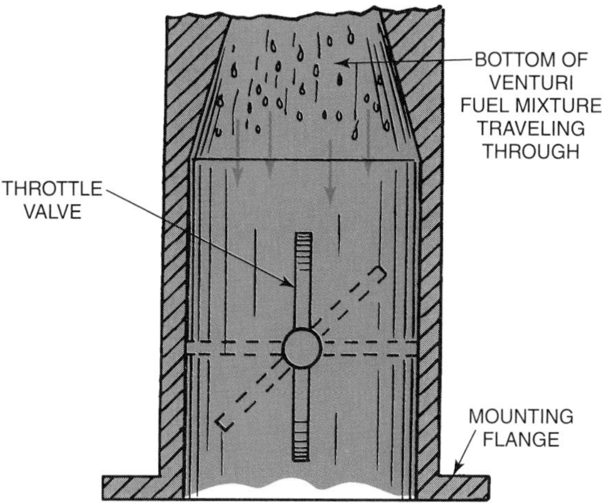

Figure 10-21. Throttle valve. Throttle valve controls the flow of air-fuel mixture. Shown in wide open, half open, and closed positions.

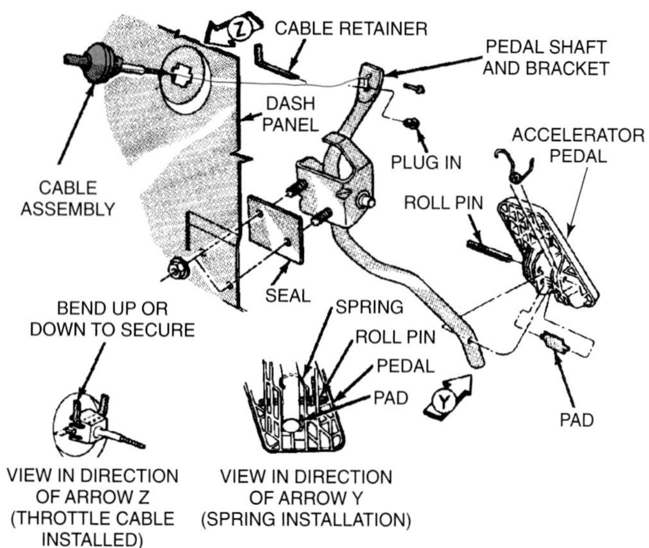

Figure 10-22. One type of carburetor throttle control, cable linkage. Linkage types and arrangements vary widely. (Chrysler)

Basic Carburetor Is Complete

Study the basic carburetor in **Figure 10-23**. Notice the position of the air horn, float bowl, float, needle and seat, fuel passages, and throttle valve. Study the venturi design and notice the relationship of the secondary to the primary venturi.

A basic carburetor as shown could be used to operate an engine. The throttle is opened to about one-half speed and the main discharge nozzle meters fuel into the airstream. If the main discharge jet is the correct size, the engine will run efficiently as long as this certain engine speed was maintained. The system designed so far is called the ***main metering system*** and is used in all carburetors for cruising at highway speeds. The ***float system*** is also in place for maintaining the proper level in the fuel bowl.

Correcting the Remaining Faults

This basic carburetor, however, has many faults. The engine will not idle without stalling and will have little power at low speed. Whenever the throttle valve is quickly opened to full throttle, the engine will have a flat spot (engine will seem to die, power fades, and for a brief second there will be no response) until the main metering system begins working.

The engine will not get enough gasoline to operate under a heavy load. There must be some provision for adding extra fuel when the engine is cold. The carburetor would be subject to vapor lock. This basic carburetor seems rather hopeless, but it is the foundation upon which all carburetors are built. By adding a few more components, we can make it operate very well.

Carburetor Circuits

The job of the various ***carburetor circuits*** is to produce the proper air-fuel ratio to meet all possible engine needs. The

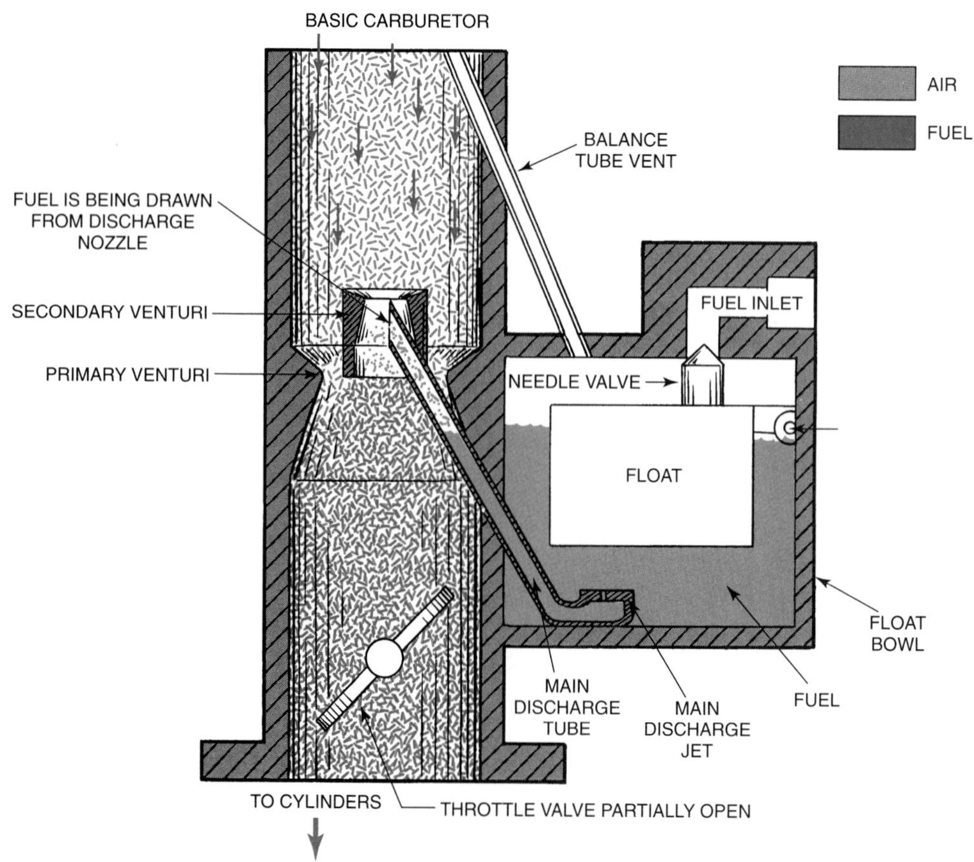

Figure 10-23. A cutaway view of a basic single-barrel carburetor.

ideal ratio for fuel economy and emission control is 14.7:1. However, ratios must vary for other conditions. When starting a cold engine, the ratio can be a low as 8:1 for a brief period. When accelerating from a stop, and when the engine is under a heavy load, the mixture might be around 12:1. When cruising at highway speed, the mixture might be as high as 15:1. These ratios are averages and will vary from engine to engine.

We have already covered the float circuit and high speed, or main metering circuit as we built our basic carburetor. The other basic circuits (often called systems) which most carburetors have can be listed as:

- Idle circuit.
- Low speed circuit.
- High speed full power circuit.
- Acceleration circuit.
- Choke circuit.
- A secondary circuit (on some carburetors).

There are other devices that are used in the construction of carburetors, but they are not classified as circuits. We will now see how each circuit overcomes a specific problem or fault. The problem will be given first, followed by the manner in which a particular circuit functions.

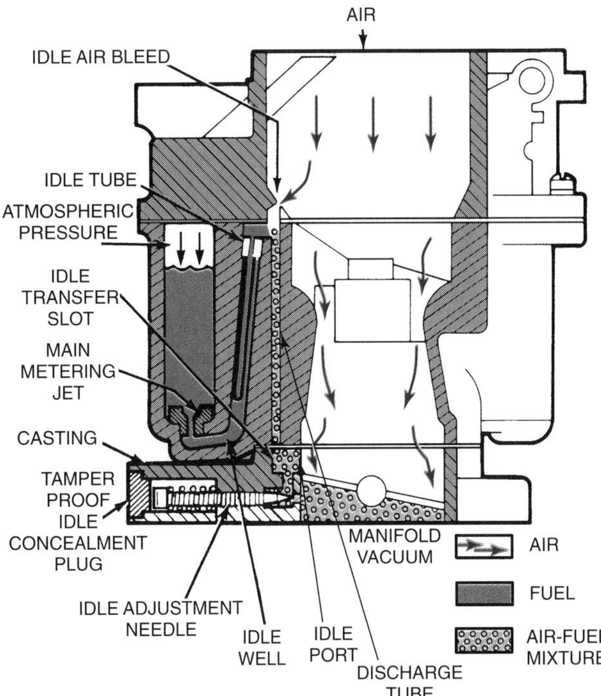

Figure 10-24. Carburetor idle circuit. Follow the flow of fuel from the metering jet to the idle screw port. This idle circuit uses a factory set idle adjustment needle. The access hole to the needle is plugged. (Ford)

Problem—Engine Will Not Idle Properly

When the throttle valve is closed to slow the engine to idle speed, the main discharge nozzle stops supplying fuel, and the engine will stall. The nozzle fails to feed gasoline because the rush of air through the air horn is cut off when the throttle is closed.

Solution—Idle Circuit

Notice in **Figure 10-24** that a passage branches off of the main discharge tube, and leads to the idle mixture screw. This is the ***idle circuit***. Since there is no vacuum above the throttle but a strong vacuum below, the vacuum will draw fuel from the bowl in through the passageway to the area underneath the throttle plate. The vacuum is also drawing air in at the idle air bleed openings. The fuel is thoroughly mixed with this air and travels to the mixture screw.

The amount of the mixture that can be drawn into the engine is determined by the position of the ***idle mixture adjustment screw***. Turning it in reduces the amount and turning it out increases the amount. In the past the screw could be adjusted to produce the smoothest idle. Today, mixture adjustment procedures must be followed exactly to produce the lowest possible exhaust emissions. Note that the idle adjustment screw illustrated in **Figure 10-24** is set at the factory and is sealed to prevent anyone from changing the setting.

To maintain the proper idle speed, an idle speed screw is installed at the base of the carburetor, and touches a piece of linkage attached to the throttle shaft. Turning this screw adjusts the idle speed.

Problem—Engine Will Not Operate at Part Throttle

As soon as the throttle valve is opened past the idle position, more fuel mixture is necessary. The idle screw port will not furnish enough fuel mixture and airflow through the main metering system is not strong enough to cause the fuel to be drawn from the main discharge nozzle. The engine will not operate properly in this range.

Solution—Low Speed Circuit

Going back to **Figure 10-24**, you will see a slot-shaped opening just above the throttle. This is called the idle port. It opens into the idle fuel passage. When the throttle valve starts to open, this port is uncovered and it will begin to supply fuel mixture to the engine. This is in addition to what is already supplied through the idle screw port.

This additional mixture will furnish enough extra fuel to allow the engine speed to increase to the point where the main discharge nozzle will function. Instead of the slot-shaped opening, some carburetors use a series of small round holes. As the throttle opening grows wider, it keeps uncovering extra holes. This type is illustrated in **Figure 10-25**. The individual holes perform the same function as the slot-shaped idle port. This is the ***low speed circuit***, sometimes called the off-idle or idle transfer circuit. **Figure 10-25** illustrates the low speed circuit in action.

Problem—Not Enough Fuel for High Speed, Full Power

At normal cruising speeds, the main metering system can provide enough fuel. When the throttle reaches the wide open position, venturi vacuum peaks off (reaches its highest point). Since not enough fuel can pass the metering rod jet under this condition, engine performance will be weak.

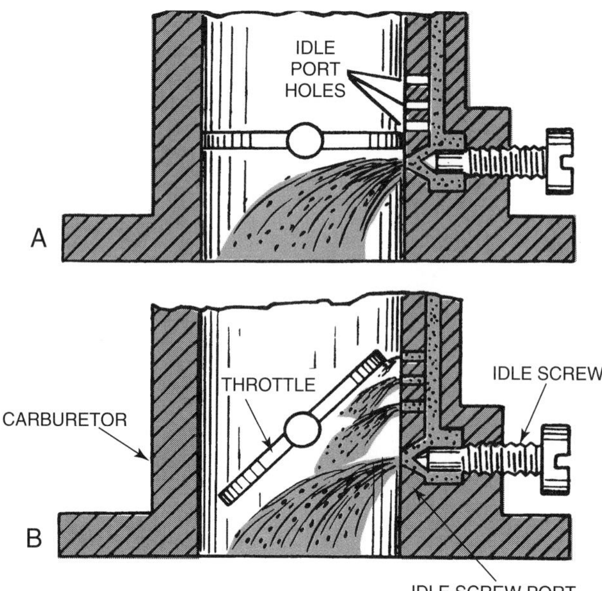

Figure 10-25. Low speed circuit—part throttle. A—The throttle is closed. The idle screw port is feeding fuel and the idle port holes are inoperative. B—When the throttle is part way open, the idle port holes are uncovered and begin to feed additional fuel to the mixture. The idle screw port is also feeding the mixture.

Solution—Full Power Circuit

The solution is to allow extra fuel to bypass the main metering system. The power enrichment system is operated by engine manifold vacuum and spring tension. It enriches the fuel mixture by either bypassing, or increasing, the size of the main jet. Either system allows more gasoline to enter the main metering system when the engine is under load. The two main types of enrichment systems are the ***power valve*** and the ***step-up rod***.

The power valve, **Figure 10-26**, is controlled by the manifold vacuum opposing the valve's spring pressure. High manifold vacuum holds the valve closed against spring pressure. When the engine is under a heavy load, the manifold vacuum drops. The lowered manifold vacuum cannot hold the valve closed against spring tension, and the spring pulls the valve open. When the valve opens, extra gasoline enters the engine through a passageway separate from the main jets.

The step-up rod, **Figure 10-27**, is a stepped, or tapered needle placed in the main jet. Manifold vacuum opposes spring pressure which tries to pull the needle out of the jet. Under light loads, manifold vacuum is high enough to push the needle far into the main jet, restricting fuel flow. When the engine is under a heavy load, vacuum drops off and spring tension can pull the needle out of the jet. The needle is tapered to match fuel flow to increases in air flow; the further it is pulled from the jet, the more fuel can flow through the jet. On some older carburetors, the needle is operated by linkage from the throttle plates.

Problem—Flat Spot on Acceleration

The carburetor, as discussed so far, would still allow a "flat spot" when sudden acceleration is attempted. This is especially true at speeds below 30 mph (48 kmh). When the throttle valve is thrown open, the engine will hesitate, seem to die, and may even backfire through the intake manifold, then catch and accelerate. The momentary flat spot is the problem.

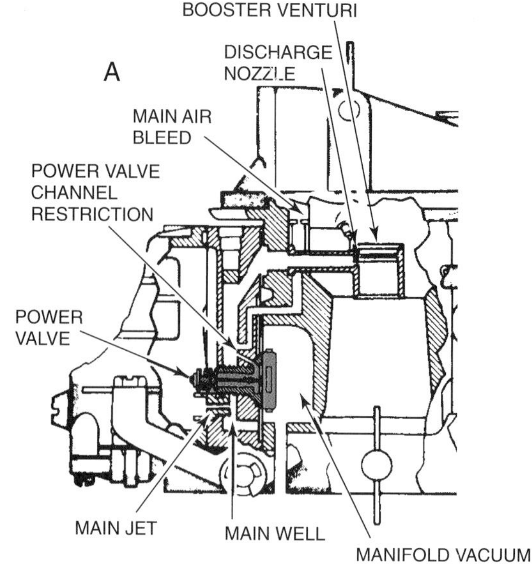

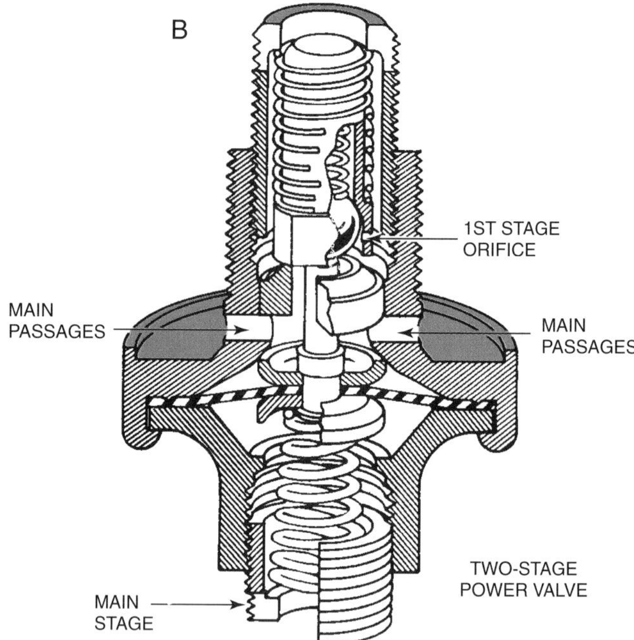

Figure 10-26. A—A cutaway view of a diaphragm power valve with routing passages for vacuum and fuel. Using the power valve on the secondary side allows a smaller main jet to improve fuel management during deceleration and braking. B—A cutaway view of a two-stage power valve. (Holley)

Solution—Accelerator Pump Circuit

An ***accelerator pump***, using either a piston or a diaphragm is built into the carburetor. The accelerator pump is a small fuel pump which injects gasoline into the carburetor throat. In the diaphragm type, **Figure 10-28**, the diaphragm is operated by linkage connected to the throttle valve.

When the throttle is opened, the linkage causes the diaphragm to push fuel into the carburetor air horn. The pump

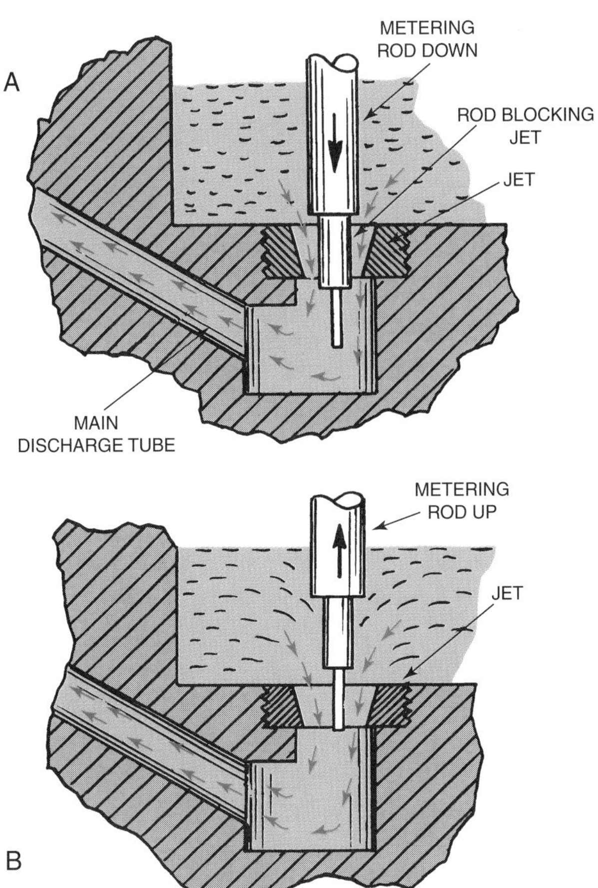

Figure 10-27. A—A metering rod is used to control the size of the jet opening. The large step of the metering rod is now in the jet. When the rod is raised, the small step will block the jet opening. B—When the metering rod is in the raised position, the small step will allow more fuel to pass.

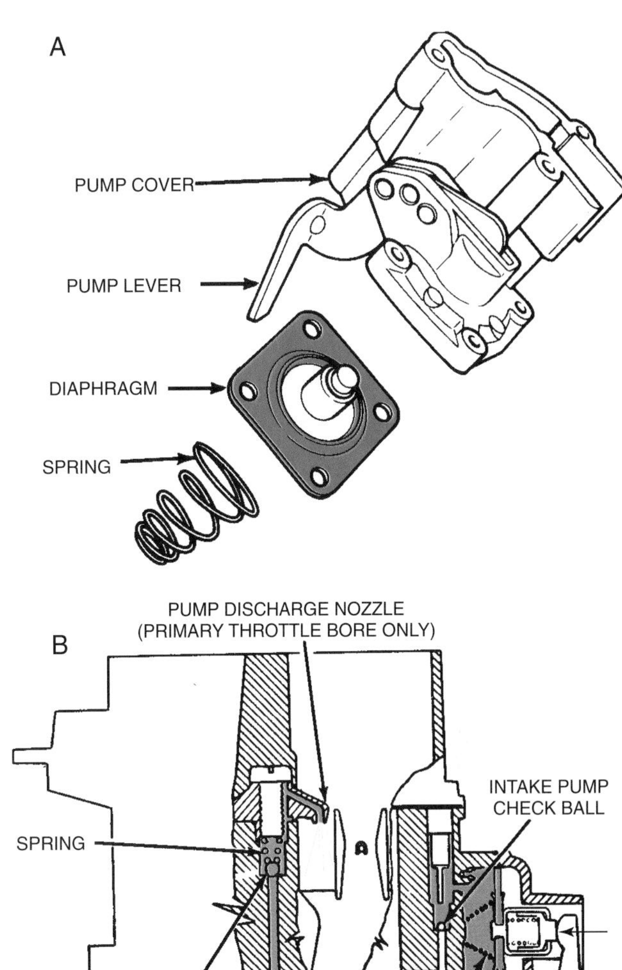

Figure 10-28. A—An exploded view of one type of diaphragm accelerator pump. B—A cross-sectional view of the accelerator pump mounted on the carburetor. Note the pump discharge nozzle. (Pontiac)

can discharge fuel into the air horn through one or more ***pump discharge nozzles***. Check valves ensure that the fuel flows in the proper direction and that air is not drawn into the system when the throttle is released. The system may use pump discharge jets to control the amount of fuel into the air horn.

Figure 10-29 shows a piston type accelerator pump. It is held in the top position by linkage connected to the throttle valve. The reservoir below the piston is filled with fuel. When the throttle valve opens, the linkage pushes the piston downward. This closes the inlet check ball and forces gasoline past the discharge check valve, through the pump nozzle, and into the air horn.

When the throttle is closed, the linkage raises the piston and the reservoir refills through the inlet check valve. A spring placed above the piston compresses when the throttle is opened quickly, allowing the piston to move down steadily and produce a steady stream of gasoline.

Problem—Engine Needs a Rich Mixture for Cold Starting

When an engine is cold, some of the gasoline condenses in the intake manifold. Since liquid gasoline will not burn, the carburetor must provide a rich mixture to allow enough gasoline to stay in the vapor state. As it warms up, the mixture must lean out. When it is hot, the mixture must be even leaner.

Solution—Automatic Choke Circuit

To increase the amount of fuel entering the air horn, a ***choke plate*** is used. The choke plate is installed at the top of the air horn. When it is closed, vacuum builds up in the air horn. This draws extra fuel from the other carburetor systems, enriching the mixture.

On some older vehicles, a ***manual choke*** was used. This choke used a flexible control cable. One end of the cable was attached to the choke valve shaft lever and the other ended in a knob on the dash. Although some older cars used

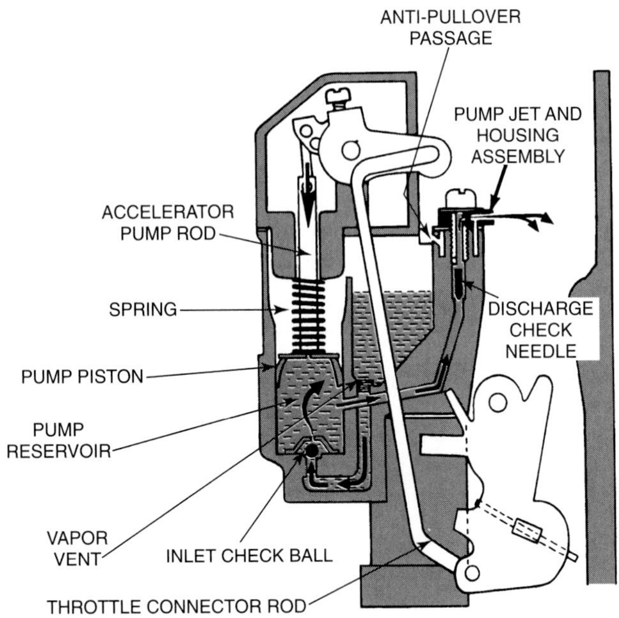

Figure 10-29. Mechanically operated accelerator pump piston. When the piston is forced down, it will compress gasoline in the pump reservoir, closing the inlet check valve and forcing gasoline by the discharge check needle. The pullover passage vents the discharge area to prevent vacuum from drawing gasoline from the accelerator discharge jet during normal operation. (Ford)

a manual choke, most modern carburetors are equipped with an ***automatic choke***.

Choke Thermostatic Coil

The choke valve shaft extends through the carburetor into a round housing. Inside the housing, there is a ***thermostatic coil spring*** (a bimetallic spring formed of two dissimilar metals that cause the spring to wind or unwind, depending on temperature change). One end of the spring is attached to the housing cover and the other end is attached to a lever on the choke shaft.

The thermostatic coil is installed in a well in either the exhaust manifold or in the exhaust crossover (V-type engine) section of the intake manifold. When it is cold, the coil will hold the choke valve in the closed position. When the coil is hot, it will tend to lose its tension and allow the choke valve to open, **Figure 10-30**. Another type of choke, called the ***divorced choke***, is installed on the manifold, and operates the choke plate through linkage.

Opening the Choke Plate

The choke plate must be fully closed to start the engine. However, as soon as the engine starts, it needs some air to continue running. Two methods are used to accomplish this.

Choke Vacuum Break

A small vacuum-operated piston or diaphragm is connected to the choke plate shaft by means of a link. As soon as the engine starts, the vacuum will pull on this piston or diaphragm and cause it to slightly open the choke plate

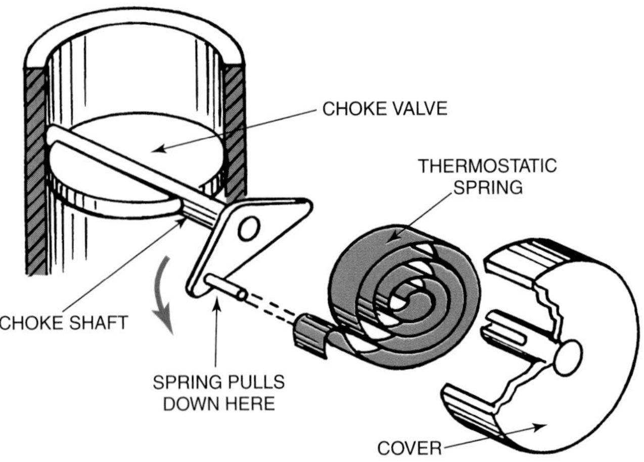

Figure 10-30. Thermostatic coil and cover. When cold, the spring pulls down in the direction of the arrow. The cover can be turned to increase or decrease spring tension.

against the pull of the thermostatic coil. This vacuum operated device is called a ***vacuum break***. **Figure 10-31** shows a vacuum break diaphragm that is used to open the choke on most modern carburetors.

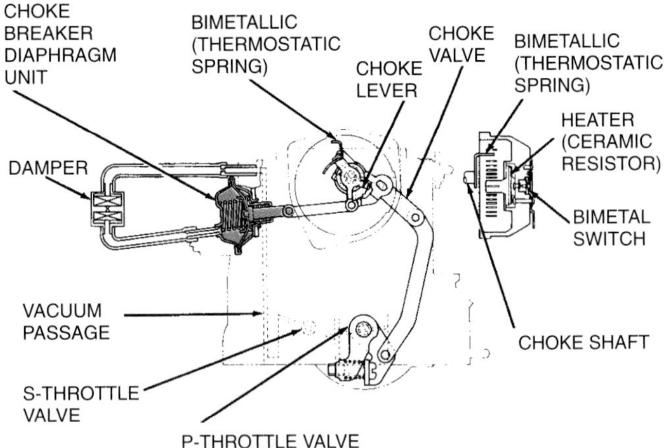

Figure 10-31. A cross-sectional view of a vacuum break diaphragm which is used to open the choke valve when needed. (Geo)

Offset Choke Valve

The choke shaft does not grasp the choke plate in the center, but is positioned so that it is slightly off center. When the engine starts, the velocity of air trying to enter the carburetor will push more on the long side and partially open the plate. See **Figure 10-32**. A small inlet valve is sometimes placed in the choke plate to admit extra air.

The cold thermostatic coil closes the offset choke plate. When the engine starts, air entering the carburetor will open the plate slightly. The vacuum piston or diaphragm will also pull to open the choke valve a little more. Between the two, they will offset the pull of the coil and open the valve enough

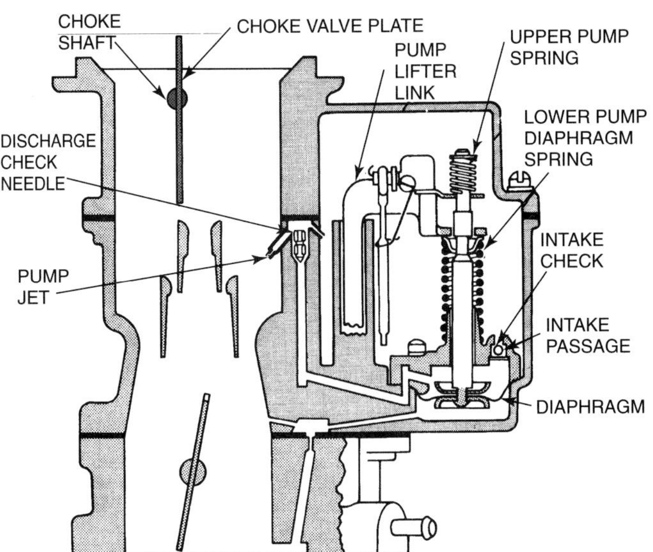

Figure 10-32. A cross-sectional view of a single barrel carburetor which places the choke valve plate in an "offset" position on the choke shaft. Also shown is a diaphragm accelerator pump circuit. (Carter)

to permit the engine to run. Note the choke action in **Figure 10-33**. Study the relationship of the parts. If the engine is accelerated, the vacuum pull on the piston will diminish and allow the choke valve to close more tightly and enrich the air-fuel mixture.

Choke Coil Heating

When the engine starts, the choke thermostatic coil must be heated so that it will gradually open the choke as the engine heats up. Older chokes used heated air drawn from the exhaust crossover passage by intake manifold vacuum. Passages and tubing allowed heated air to pass through the choke housing and heat the coil. This was called a ***choke stove***. Divorced choke coils are directly warmed by heat from the engine.

Note that the choke circuit shown in **Figure 10-34** utilizes a vacuum break diaphragm unit to open the choke slightly as soon as the engine starts. Newer vehicles use an electrically heated coil to speed up the opening of the choke thermostatic spring, **Figure 10-35.** This is usually referred to as an ***electric choke.*** An older design passes engine coolant around the choke housing for heating.

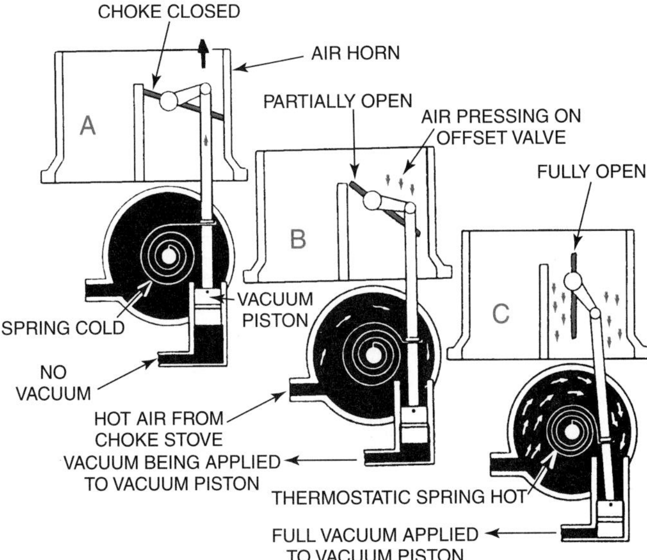

Figure 10-33. Thermostatically controlled choke action. A—when the engine is stopped, the choke is closed. B—As the engine is started, the vacuum piston and offset choke valve force the choke open a small amount. C—When the engine is at operating temperature, the thermostatic spring allows the choke plate to fully open. (Toyota)

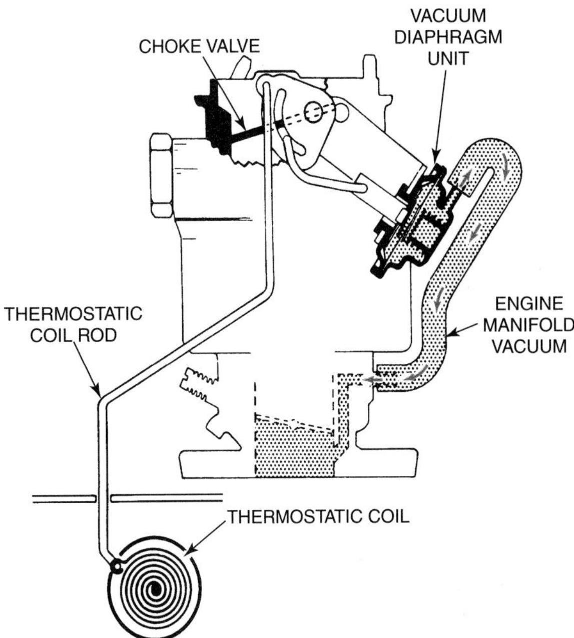

Figure 10-34. Automatic choke, well type. The thermostatic coil may be placed in the exhaust crossover (hot) area of the intake manifold. This setup uses a diaphragm unit instead of a piston to partially open the choke when the engine starts. (Chevrolet)

Preventing Flooding

In case the carburetor ***floods*** the engine (puts too rich a mixture in the cylinders), a choke ***unloader***, which is an additional lever attached to the choke shaft, is used to add more air to the mixture. In the event of flooding, the throttle valve is opened all the way, causing linkage from the throttle lever to strike the unloader trip lever and open the choke to allow air to enter the cylinders.

Fast Idle Cam

When the engine is warming up, it is necessary to increase the idle speed to prevent stalling. This is accomplished by connecting the choke shaft to a ***fast idle cam***. When the choke is on, the fast idle cam swings out in front of the idle speed adjustment screw. This holds the throttle valve partially open to speed up the engine. As the choke opens, the fast idle cam swings down and the throttle returns to its

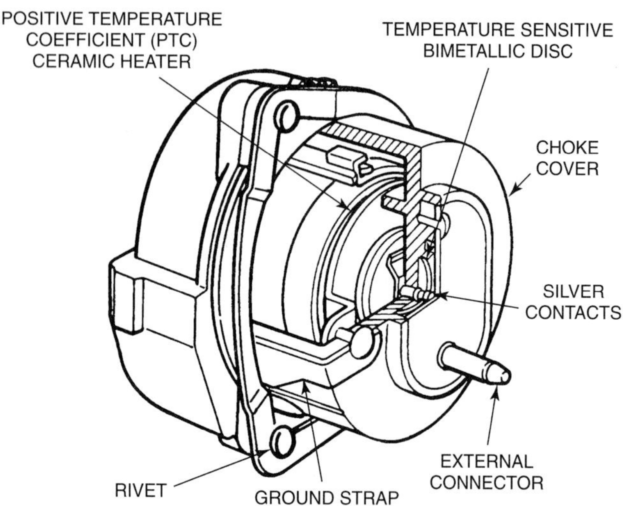

Figure 10-35. An electric assist choke. Note the rivets used to retain the choke cover. This prevents tampering with the choke setting. (Ford)

regular hot idle position. The fast idle cam is sometimes curved or notched, **Figure 10-36**.

Problem—Engine Needs More Power

Many engines are capable of producing more power than the carburetor will allow. To provide for smooth idle and low speed operation, the carburetor throats are relatively small. Small throats cause faster air speed, increasing the venturi vacuum and fuel mixing. However, these small throats will not let in enough fuel and air at high engine RPM. Making a large throat will make the engine faster at high speeds, but will increase roughness at low speeds.

Solution—Secondary Circuit

The ***secondary circuit*** consists of an additional carburetor throat or throats which allow extra air and gas into the engine when more power is needed. Not all carburetors have secondary systems. The secondary throats are equipped with throttle valves that are opened only when needed, as shown in **Figure 10-37.** The throat, or throats, used for normal driving are called the ***primary circuit.***

When the accelerator is depressed to completely open the primary throttle valve, linkage also opens the secondary throttle valve. When the secondary throttle valve opens, extra air enters the engine through the secondary throat. A venturi system in the secondary throat delivers extra gasoline. Some secondary throttle valves are vacuum-actuated. These respond to changes in engine load. Using a carburetor with a secondary system allows the engine to run smoothly and get good mileage at small throttle openings while still performing well when extra power is needed.

Notice that only the primary barrels have a choke valve. No choke valve is needed in the secondary barrels since they are closed when starting. Although secondary systems were once found only on four-barrel carburetors, many two-barrel carburetors, with one of the barrels functioning as a secondary system, are used on four- and six-cylinder engines.

Other Carburetor Features

Many other devices are installed on the carburetor. They are used to improve carburetor performance, operate other engine systems, or help meet emission regulations. Other devices, installed elsewhere on the engine, improve carburetor operation. Some of the most common are covered below.

Hot Idle Compensator

To prevent stalling when idling in very hot weather, some carburetors utilize a thermostatically-controlled air valve called a ***hot idle compensator***, **Figure 10-38**. When the carburetor becomes hot, the bimetallic strip will curl upward and open the hot idle air valve. This feeds extra air to the manifold below the throttle valve. The extra air compensates for an overly rich idle mixture due to heat-produced fuel vapor entering the carburetor. Hot idle compensators are used less on

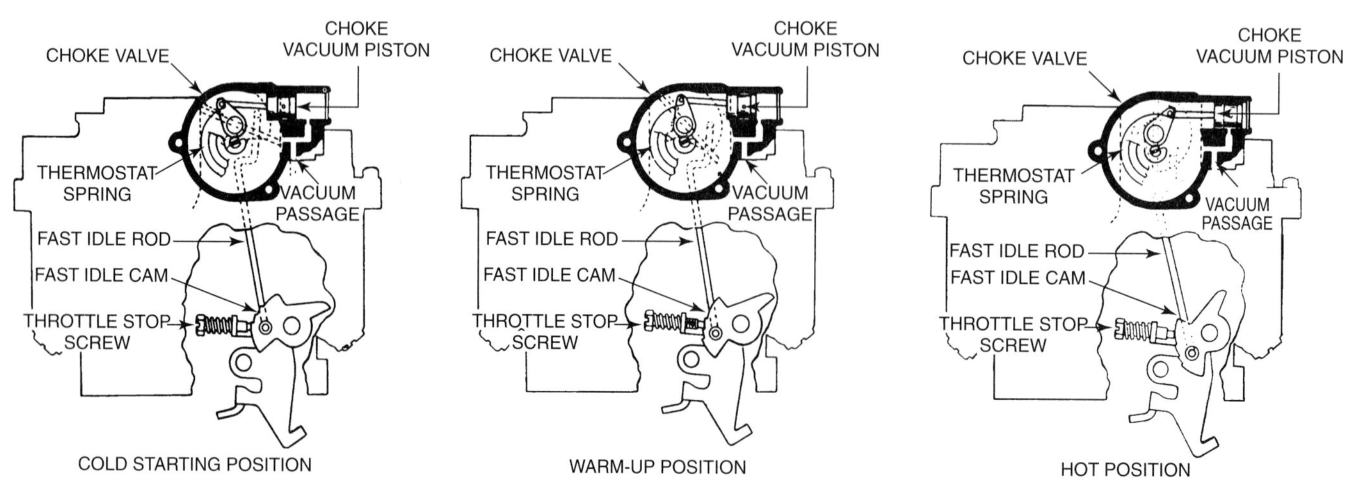

Figure 10-36. Action of fast idle cam during engine warm-up. As the choke opens, the fast idle rod will lower the fast idle cam. As the cam lowers, it will allow the idle throttle stop screw to close the throttle more and more. This will reduce engine speed. When the choke is fully open, the engine will idle at normal speed. (Bendix)

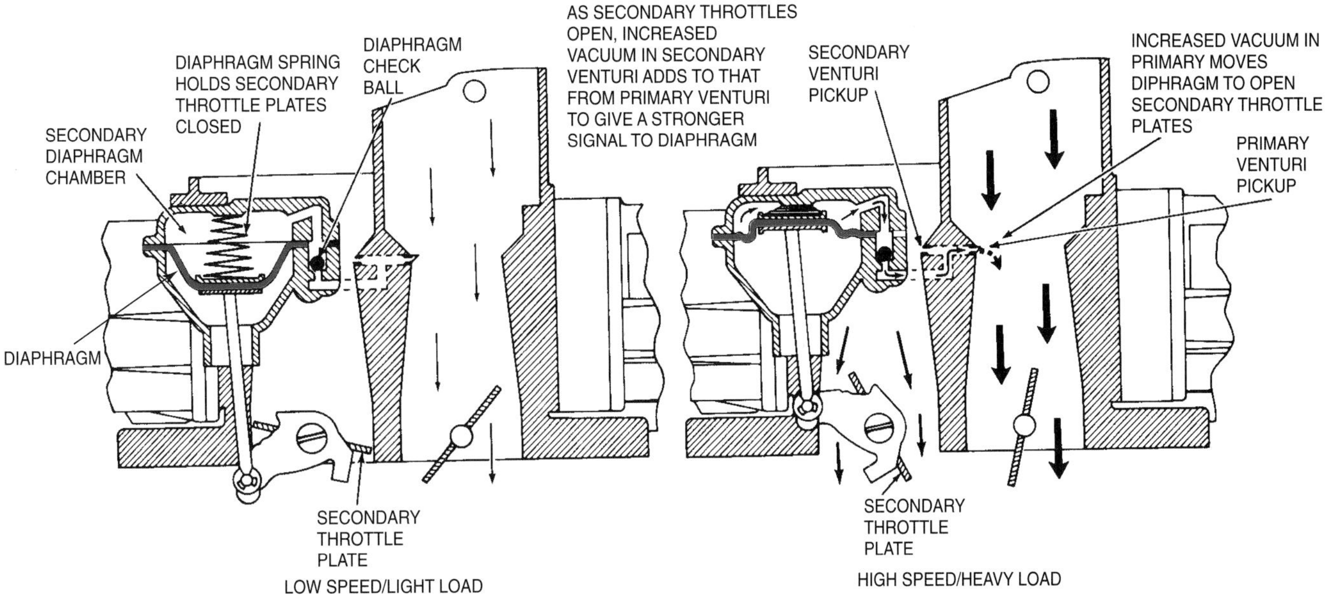

Figure 10-37. A schematic of a diaphragm-operated secondary throttle circuit. A vacuum signal from the primary venturi is aided by an increasing vacuum signal from the secondary venturi as the secondary throttle starts to open. Closing the primary throttle closes the secondary with mechanical override linkage.

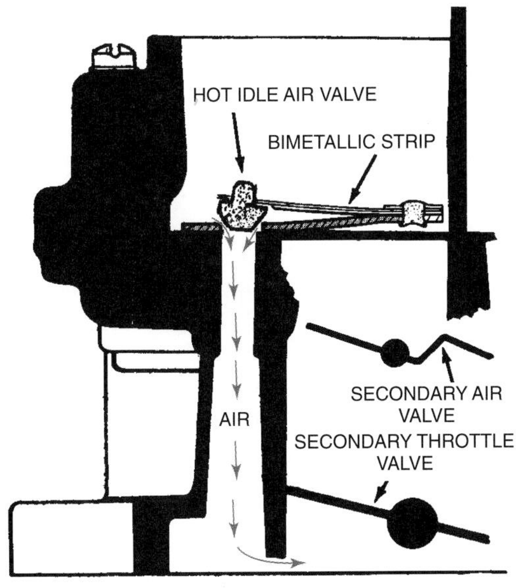

Figure 10-38. Hot idle compensator. Heat will cause the bimetallic strip to open the air valve. (Pontiac)

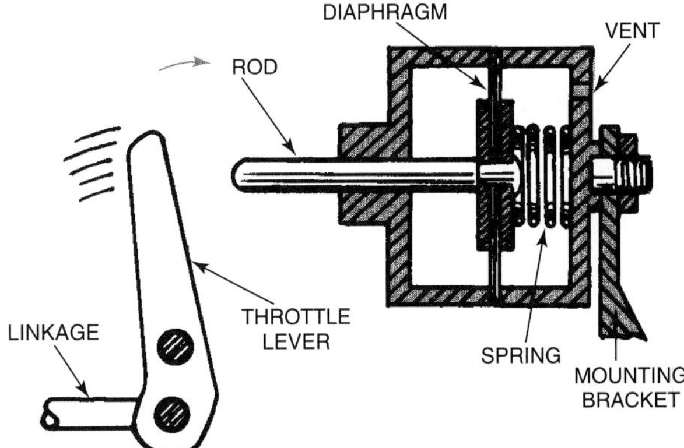

Figure 10-39. Throttle return dashpot. When the throttle strikes the dashpot rod, the rod pushes the diaphragm back. Before it can move back, air must be forced out of the vent. This causes the diaphragm to move slowly. When the throttle lever is moved away, a spring will force the diaphragm out. Air will be drawn in the vent and the dashpot is ready to work again.

modern engines, since their presence makes exact fuel mixture control more difficult.

Throttle Return Dashpot

On cars equipped with automatic transmissions, the drag placed on the engine when the throttle is opened and then suddenly returned to idle will result in engine stalling. The ***throttle return dashpot***, usually called the dashpot, is arranged so the throttle lever will strike the dashpot just before reaching idling range. The dashpot diaphragm rod will slowly collapse, allowing the throttle to slowly return to the idling position. This occurs when the car is stopped, **Figure 10-39.** Some dashpots are built as an integral part of the carburetor. Some dashpots are combined with other idle control devices.

Electrical Idle Speedup Control Devices

Some engines have electrical solenoid controlled devices to vary the idle for certain special operating conditions. Such units used to control idle speed for various reasons are often termed throttle positioners, **Figure 10-40.**

To compensate for the increased load on the engine when the air conditioning compressor clutch is engaged, an ***idle speed-up solenoid*** is often used. The increased engine speed during idle also provides increased airflow for better

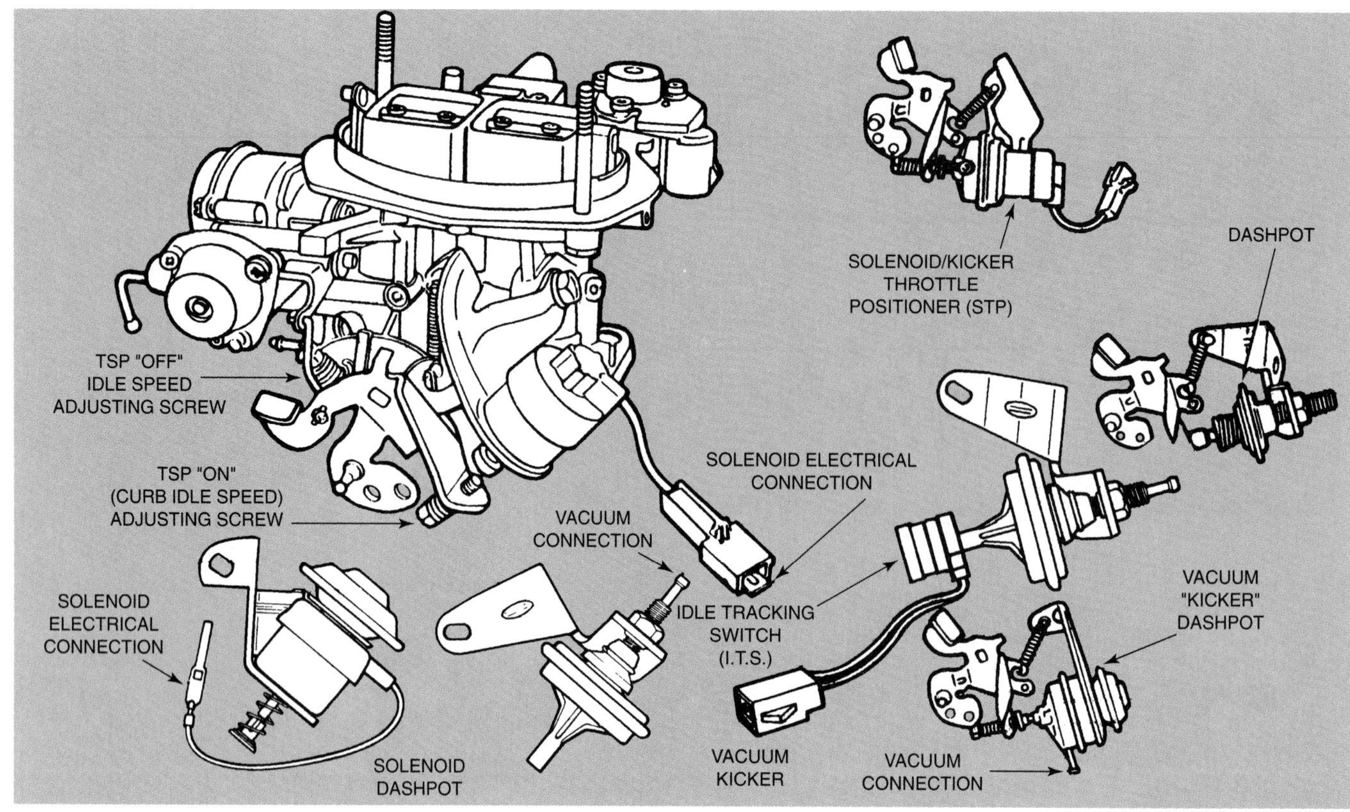

Figure 10-40. Various styles of vacuum and electric throttle positioners.

cooling. When the compressor clutch is energized, electricity is fed to a small solenoid on the side of the carburetor. The solenoid then opens an air valve, permitting increased airflow below the throttle plates.

On some engines, the idle speed-up solenoid is energized in third and fourth gears to hold the throttle plates slightly open when the engine is decelerating. This reduces engine emissions. **Figure 10-41** illustrates the idle speedup control.

Some engines are prone to ***dieseling*** (condition in which the engine continues to run when the ignition is turned off). Dieseling, sometimes called run on, is caused by the hotter combustion chamber temperatures in modern emission-controlled vehicles. To prevent dieseling, an ***anti-dieseling solenoid*** is used to set the hot idle. When the engine is turned off, the solenoid deenergizes, closing the throttle completely. The engine is deprived of air and cannot run.

Mixture Control Solenoid

A computer-controlled ***mixture control solenoid*** is used on most carburetors made recently. This solenoid is a type of electromagnet (coil of wire through which an electrical current is passed), used to move the rod. Electrical current to the solenoid is pulsed on and off to change rod position in response to engine fuel needs. The computer controls solenoid operation as needed. **Figure 10-42** shows one type of solenoid operated rod.

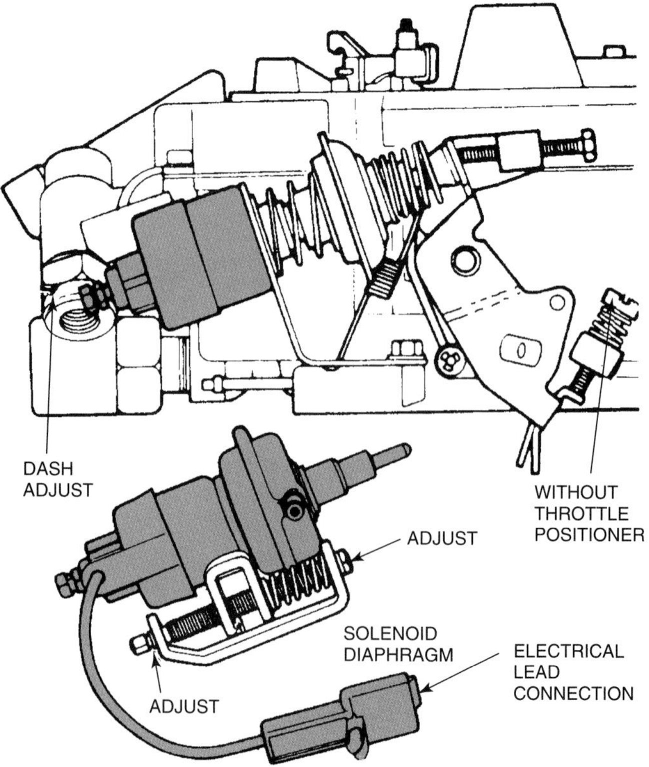

Figure 10-41. An idle speed-up solenoid assembly and its mounting location. (Ford)

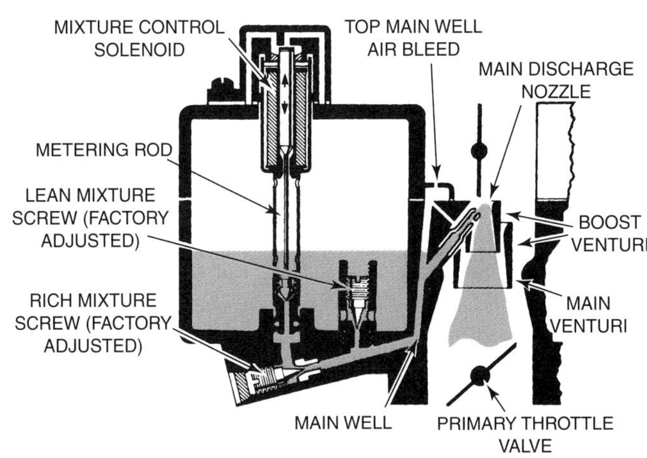

Figure 10-42. A solenoid operated metering rod provides the control needed by the "closed loop" computer system employing an exhaust sensor. (Chevrolet)

Throttle Position Sensor

Throttle position is relayed to the control module by the ***throttle position sensor***. As you learned in Chapter 9, throttle position sensors can be either resistance types or transducers. Throttle position sensors are usually installed on or in the carburetor bowl.

Devices for Preventing Carburetor Icing

It is possible for ice to form on the throttle valves. As the nozzles feed fuel into the air horn, it turns into a vapor. This vaporizing action chills the air-fuel mixture and the throttle bores in the throttle valve area. This condition, called ***carburetor icing***, happens most often when the outside air temperature is just above freezing and humidity is high.

When the moisture-laden mixture contacts the chilled throttle valve, it condenses, then freezes. This pileup of ice can reach the point where, when the accelerator is released, the engine will stall because no air can pass around the throttle valve. Icing occurs during the warm-up period, and stops after the engine is fully warmed up.

Manifold Heat Control Valve

When the air-fuel mixture enters the manifold during the warm-up period, gases from the exhaust manifold are directed to the base of the intake manifold where the carburetor is attached. This warms the incoming mixture. When the engine is hot, the gases are directed away from the intake manifold. This action is controlled by the ***manifold heat control valve***.

The manifold heat control valve can be operated by a thermostatic spring or by a vacuum diaphragm operated by a temperature switch. **Figure 10-43** shows the action of the heat control valve on an inline engine. On a V-type engine, the manifold heat control valve is located in the exhaust manifold on one bank. When cold, the heat control closes and directs some of the exhaust gases up through the intake manifold. After passing through, the gases flow out the exhaust manifold on the other side. When hot, the manifold heat control valve opens and both exhaust manifolds expel gas directly into the exhaust system.

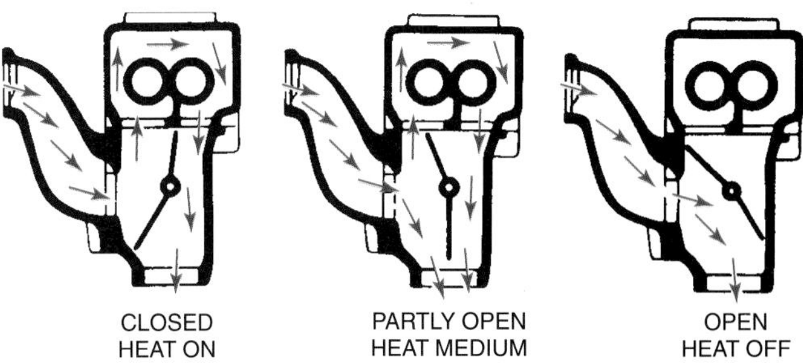

Figure 10-43. Operation of a manifold heat control valve. Notice how hot gases are directed up and around the intake manifold passages.

On most modern carbureted engines, the intake air is heated by means of a thermostatically-controlled air cleaner, shown in **Figure 10-44**. See Chapter 14 for additional information about manifold heat control valves and thermostatic air cleaners.

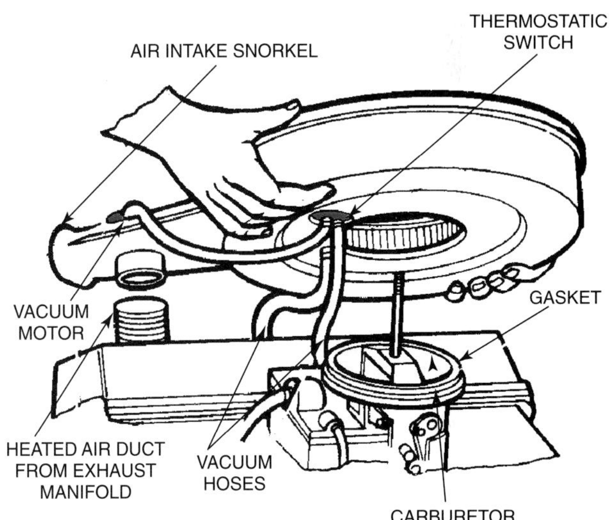

Figure 10-44. A thermostatically controlled, heated air cleaner assembly being removed from a carburetor. There are many different styles and control systems.

Coolant Heated Manifold

Some engines are designed so that coolant is pumped through passages in the intake manifold. This ***coolant heated manifold*** provides a controlled temperature for the fuel mixture. **Figure 10-45** illustrates a coolant-heated intake manifold for a V-8 engine. The top view shows the complete manifold for a four-barrel carburetor. The shaded portion in the second sketch shows the intake passages. The bottom sketch shows the flow of coolant through the manifold.

Electrical Heating of Fuel Mixture

Some older engines utilize a thin ***ceramic heater*** built as part of the carburetor base-to-intake manifold spacer. The heating grids are located under the primary bores and provide rapid heating of the fuel mixture during cold startup and while

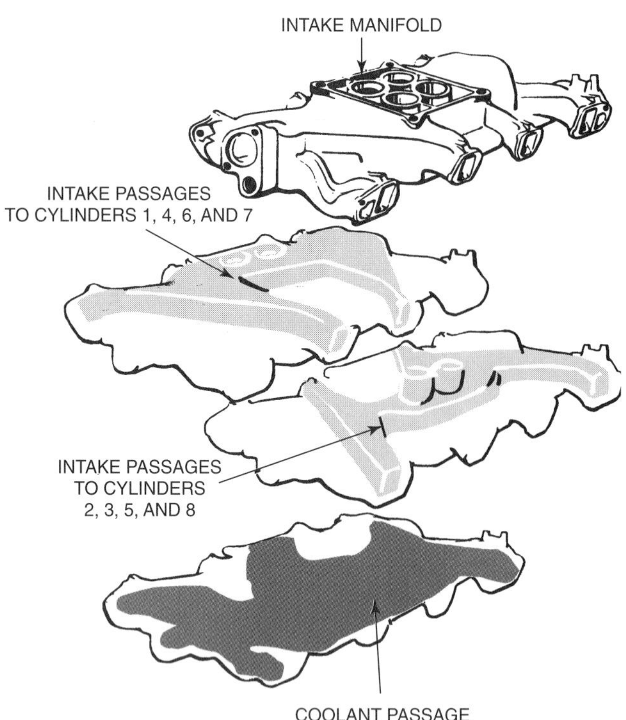

Figure 10-45. A coolant heated intake manifold. Coolant replaces hot gases in this installation.

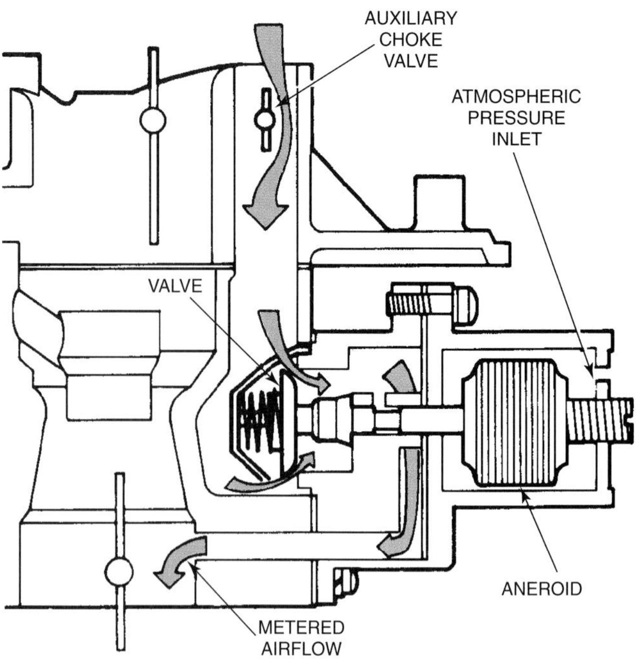

Figure 10-46. An aneroid controlled, spring loaded valve compensates for changes in altitude. As altitude increases, the aneroid expands, opening the valve. When the engine is cold and the regular choke valve is closed, the auxiliary choke will also close to prevent excess air delivery. (Chrysler)

the engine is warming up. This improves fuel vaporization and reduces the length of time the choke must be closed. The heater is shut off by a thermostatic switch when the engine reaches a specified temperature.

Altitude Compensators

When operating at higher elevations, the fuel mixture tends to become overly rich as the air becomes thinner and contains less oxygen to sustain burning. On older carburetors, the mixture was simply adjusted or jets were changed to compensate for the thin air. Today, this problem is overcome by either adjusting the amount of fuel or the amount of air in the mixture using an ***altitude compensator.***

On older engines, this adjustment was made by using an ***aneroid***, which is a pressure sensitive sealed bellows. The aneroid can control either a fuel or airflow valve so as to maintain the proper air-fuel ratio at various altitudes. **Figure 10-46** illustrates the use of an aneroid to admit extra air as altitude increases. At lower elevations, atmospheric pressure collapses the aneroid which in turn holds the valve closed. As elevation increases, atmospheric pressure drops and the aneroid expands, pushing the valve open. This admits extra air into the stream of fuel, which leans the air-fuel ratio.

On computer-controlled vehicles, the oxygen sensor reads the decrease in oxygen content caused by higher altitude operation. Some computer control systems with carburetors are equipped with a barometric pressure sensor. The computer reads this and other inputs and makes adjustments accordingly. The operation of the computer control system is covered in detail in Chapter 15.

Other Carburetor Functions

In addition to its job of mixing air and gasoline in the proper proportions, the carburetor performs other functions. The carburetor may furnish an opening for the distributor vacuum advance line, or for various vacuum operated emission controls such as the EGR (exhaust gas recirculation) valve. The throttle linkage may be attached to kickdown switches for automatic transmissions, or air conditioner full throttle cutout switches. The carburetor may be connected to a throttle valve (TV) linkage or cable to operate the automatic transmission throttle valve.

Intake Manifold Design

The ***intake manifold*** is designed to carry fuel from the carburetor to the valve ports. They can be made of cast iron or aluminum. The V-8 engine mounts the manifold between the cylinder heads. Inline engines have the manifold bolted to the cylinder head. Intake manifolds usually have fittings for attaching vacuum lines and may have a mounting pad for the EGR valve or other emission controls.

It is not enough just to carry fuel from the carburetor to the cylinders. Each cylinder should get the same type and amount of fuel mixture. To design an intake manifold to deliver equal amounts of fuel to all cylinders is not easy. It is essential that the passages in the manifold be as nearly equal in length as possible. In addition, they should be smooth with gentle curves. **Figure 10-47** shows one specific V-8 intake manifold design for a dual carburetor.

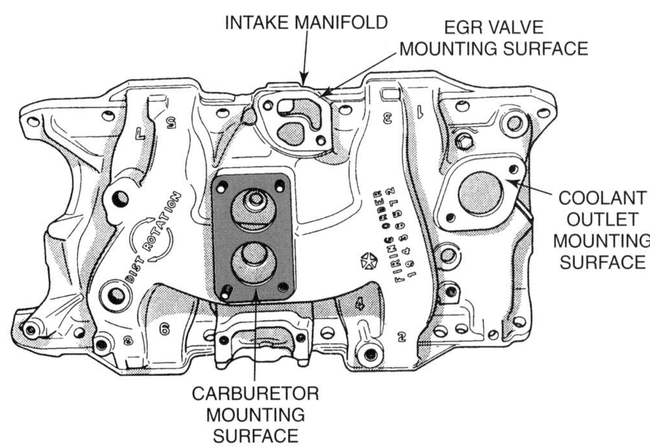

Figure 10-47. An intake manifold for a V-8 engine which uses a two-barrel carburetor. (Chrysler)

Air Cleaners

It is essential that air entering the carburetor or throttle body is clean. A good ***air cleaner,*** properly cleaned or replaced at the proper intervals, will add greatly to the useful life of the engine.

Air Cleaner Element

An air cleaner draws air through a treated paper ***air cleaner element*** which will allow air to flow, but traps dirt particles, **Figure 10-48**. The paper is pleated accordion style for added filtering area. Air cleaner elements can be round (radial airflow) or flat (linear airflow), **Figure 10-49**.

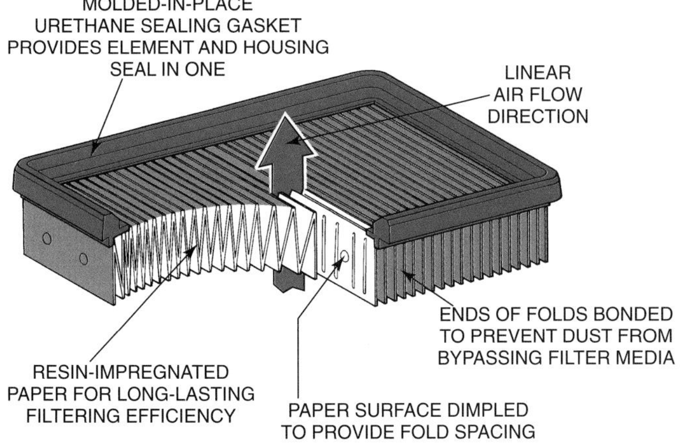

Figure 10-48. A cutaway view of the linear flow air filter. These elements are generally used in remote housing locations. (Wix Corp.)

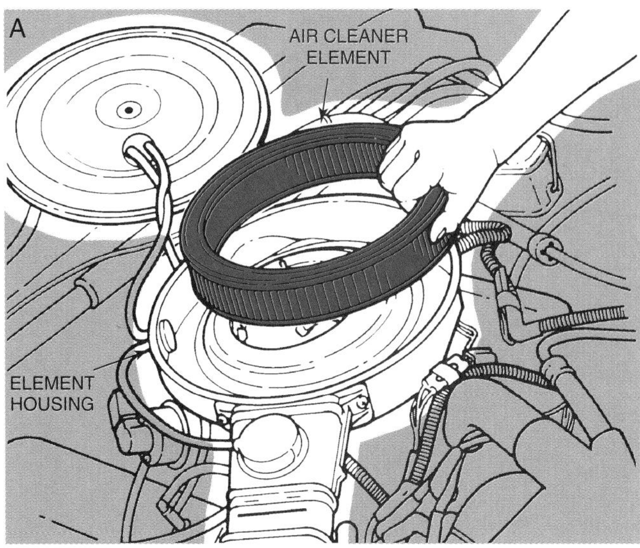

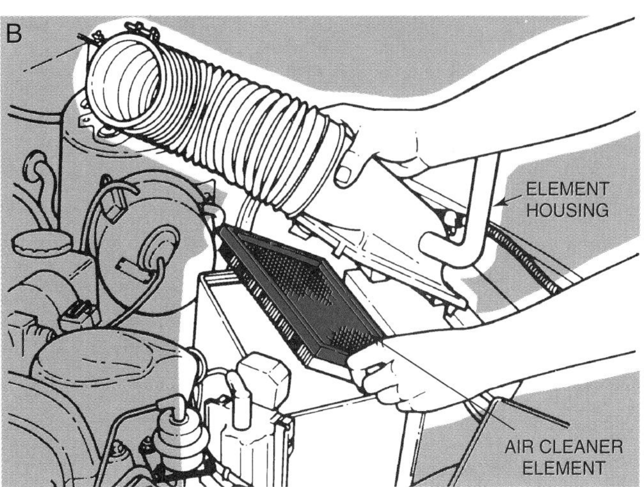

Figure 10-49. A—A round, radial air flow, dry type, air cleaner element being removed from its housing. B—A flat, linear air flow, air cleaner element being removed from its housing. (AC-Delco)

Carburetor Classifications

Since carburetors have been used for over 100 years, many different types have been made. However, all carburetors perform the same function and every carburetor can be placed into a specific classification. Carburetors are classified according to several criteria. While discussing carburetor classifications, it will be helpful to note the three sections of a carburetor that almost all carburetors have:

Air horn, the upper section of the carburetor containing the air inlet and air cleaner mounting bracket, and sometimes the fuel inlet, float assembly, and accelerator pump linkage and plunger.

Main body, the center section of the carburetor containing the float bowl (chamber) and some of the internal passages. The main body usually contains the venturi cluster and accelerator pump outlet nozzle.

Throttle plate or base, the lower section containing the throttle valve and associated linkage, and usually the idle mixture screw(s).

Some carburetors incorporate the main body and throttle body into one unit, but these three major sections are distinct on all carburetors.

Airflow Direction

Incoming air can enter the carburetors in three directions. The three carburetor designs based on these flows are ***downdraft***, ***updraft***, and ***sidedraft***. Your basic carburetor was a downdraft type, **Figure 10-50.** While sidedraft and updraft carburetors are sometimes found on small engines and off-road equipment, they have not been used on cars and light trucks for many years.

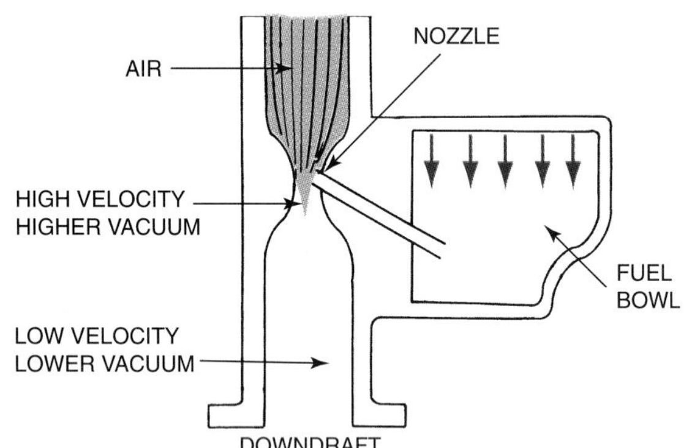

Figure 10-50. A cutaway of a downdraft carburetor. (Deere)

Number of Throats

The number of throats, often called ***barrels,*** is a common way of identifying carburetors. Carburetors can be single barrel, two-barrel, or four-barrel designs. In the past, multiple carburetors (three two-barrel carburetors or two four-barrel carburetors) have been placed on engines.

Single-Barrel Carburetor

The basic carburetor illustrated earlier was a ***single-barrel*** model. A typical downdraft, single-barrel carburetor is shown in **Figure 10-51.** The single-barrel carburetor is used on three- and four-cylinder engines. Study the various parts.

Two-Barrel Carburetor

Many V-type engines and some inline engines use the ***two-barrel*** carburetor. This carburetor is basically two single barrel carburetors mounted side by side. Each barrel has its own discharge nozzle, idle port, accelerator discharge jet, and throttle valve. Some two-barrel carburetors make use of a secondary system, with one barrel as the primary and the other the secondary, **Figure 10-52**. On some V-type engines, each barrel feeds into a separate intake passage.

Four-Barrel Carburetor

A ***four-barrel*** carburetor can be thought of as a cluster of four, single-barrel carburetors. One group of two barrels is the primary side and the other two are the secondary side. Four-barrel carburetors are primarily used on V-8 engines, although some have been used on other engines with fewer cylinders. There is wide variation in four-barrel carburetor design. The carburetor illustrated in **Figure 10-53** is typical.

Some four-barrel carburetors have separate primary and secondary, or separate right and left float circuits, but most modern four-barrel carburetors have only one float. Most four-barrel carburetors have an airflow-operated auxiliary valve. The auxiliary valves are installed on top of the secondary throttle plates. They are mounted off-center and are pulled open by manifold vacuum only when engine load is sufficient to use the extra air and fuel. Therefore, even when the secondary throttle valves are opened at low speeds, the secondary system does not operate until the engine can use it.

Summary

The fuel tank holds the gasoline or diesel fuel until it is needed. A mechanical or electric fuel pump draws, fuel from the tank. The fuel is drawn out through a pickup tube with a screen on the end. The fuel flows through lines, then passes through a filter on its way to the fuel delivery system.

For efficient operation an engine used in a passenger car requires an air-fuel ratio that is constantly changing. For many years, this varied mixture was produced and delivered by the carburetor.

The carburetor contains a number of circuits designed to produce a proper air-fuel mixture for all engine needs. When the engine first starts, the choke valve is closed. The vacuum is strong in the area beneath the choke valve, resulting in a very rich mixture by reducing the air supply and increasing the withdrawal of fuel from the bowl. As soon as the engine starts, the choke partially opens. As the engine is idled to warmup, fuel is fed into the passing airstream by the idle screw port. The choke continues to open. When the engine is hot, it will be fully opened.

When the accelerator opens the throttle valve, the idle screw port fuel delivery is assisted by additional fuel from the idle port or part throttle holes just above the idle screw port. As the throttle continues to open, the air speed through the air horn and venturi increases until fuel begins to feed from the main discharge nozzle. At very high speeds where the throttle is wide open, or when the engine is loaded, additional fuel is administered a power valve or by a metering rod.

During the cruising and full power range, no fuel is fed by the idle port or idle screw port. To assist in smooth acceleration, an accelerator pump is provided to feed additional gasoline during acceleration only. Engine speed is controlled by the position of the throttle valve in the carburetor.

The air and fuel mixture is delivered to the combustion chambers by means of an intake manifold. The temperature of this mixture is raised during engine warmup by heat crossover passages, heated air intake, and in some cases, by running the coolant through passages in the intake manifold. Even distribution of fuel to all cylinders is very important.

Air entering the carburetor must be cleaned. Failure to clean the air will result in rapid deterioration of the engine. Air is generally cleaned by a dry-type air cleaner that utilizes special paper as the filtering element. Cleaners must be serviced regularly to maintain efficiency.

Carburetors can be classified as downdraft, updraft, and sidedraft types. Some carburetors are single-barrel, some two-barrel, and others have four barrels. All carburetors—despite shape, size and type—attempt to provide as nearly perfect an air-fuel mixture as possible for all engine needs.

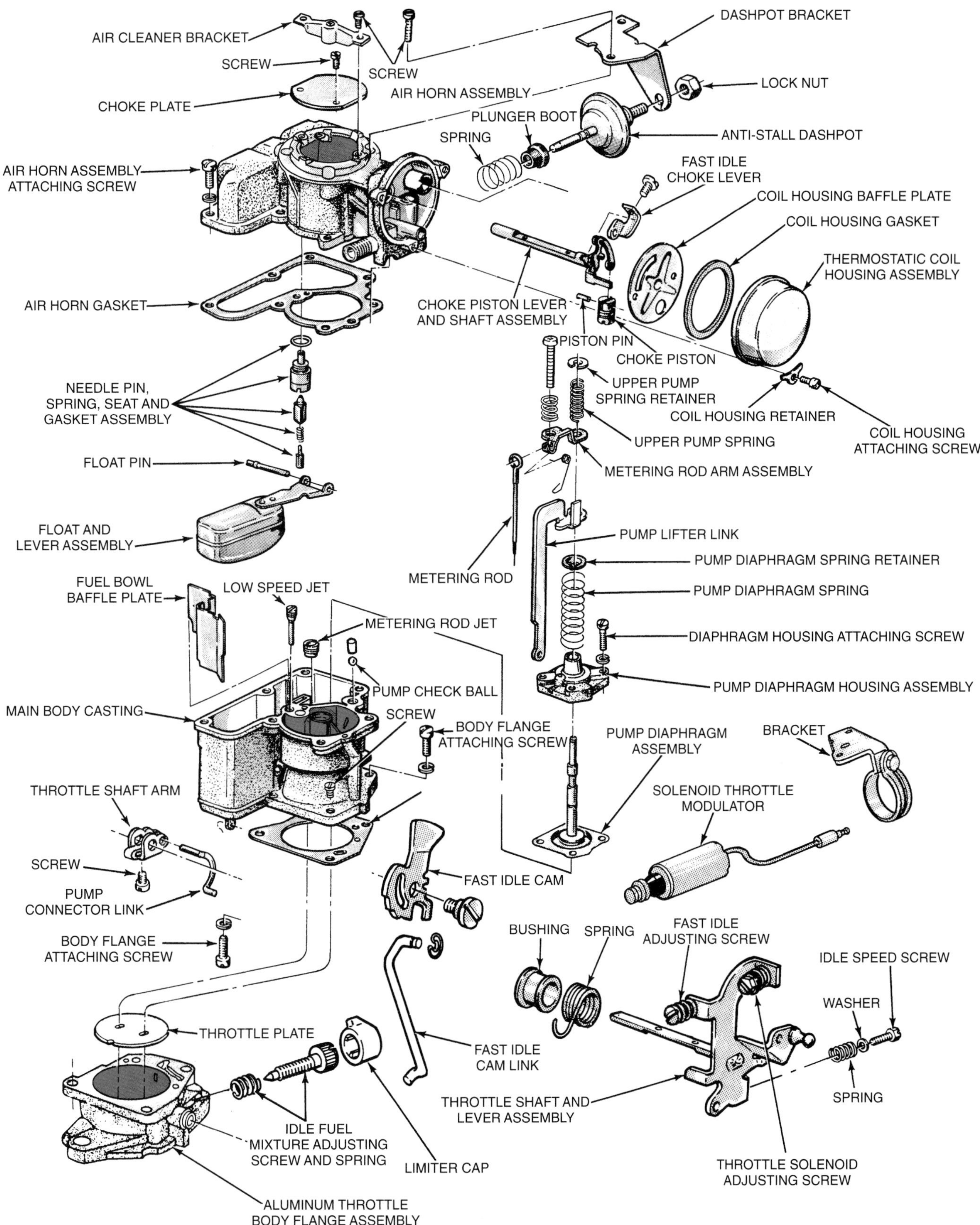

Figure 10-51. An exploded view of a single barrel, downdraft carburetor. Types and construction vary widely. Study all the individual parts carefully. (Ford)

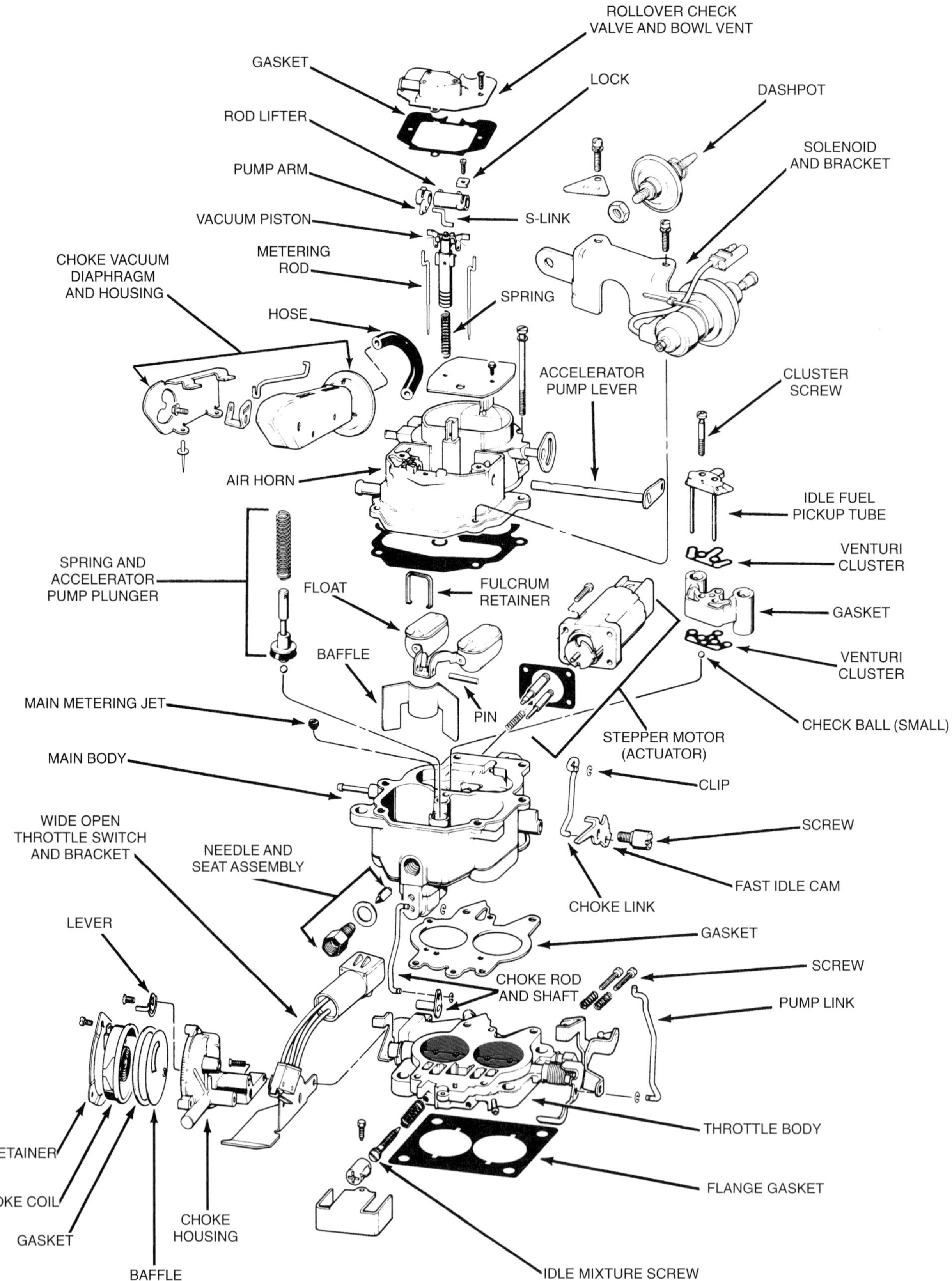

Figure 10-52. An exploded view of a two-barrel, downdraft carburetor. Note the main sections — air horn, main body, and the throttle. (Chrysler)

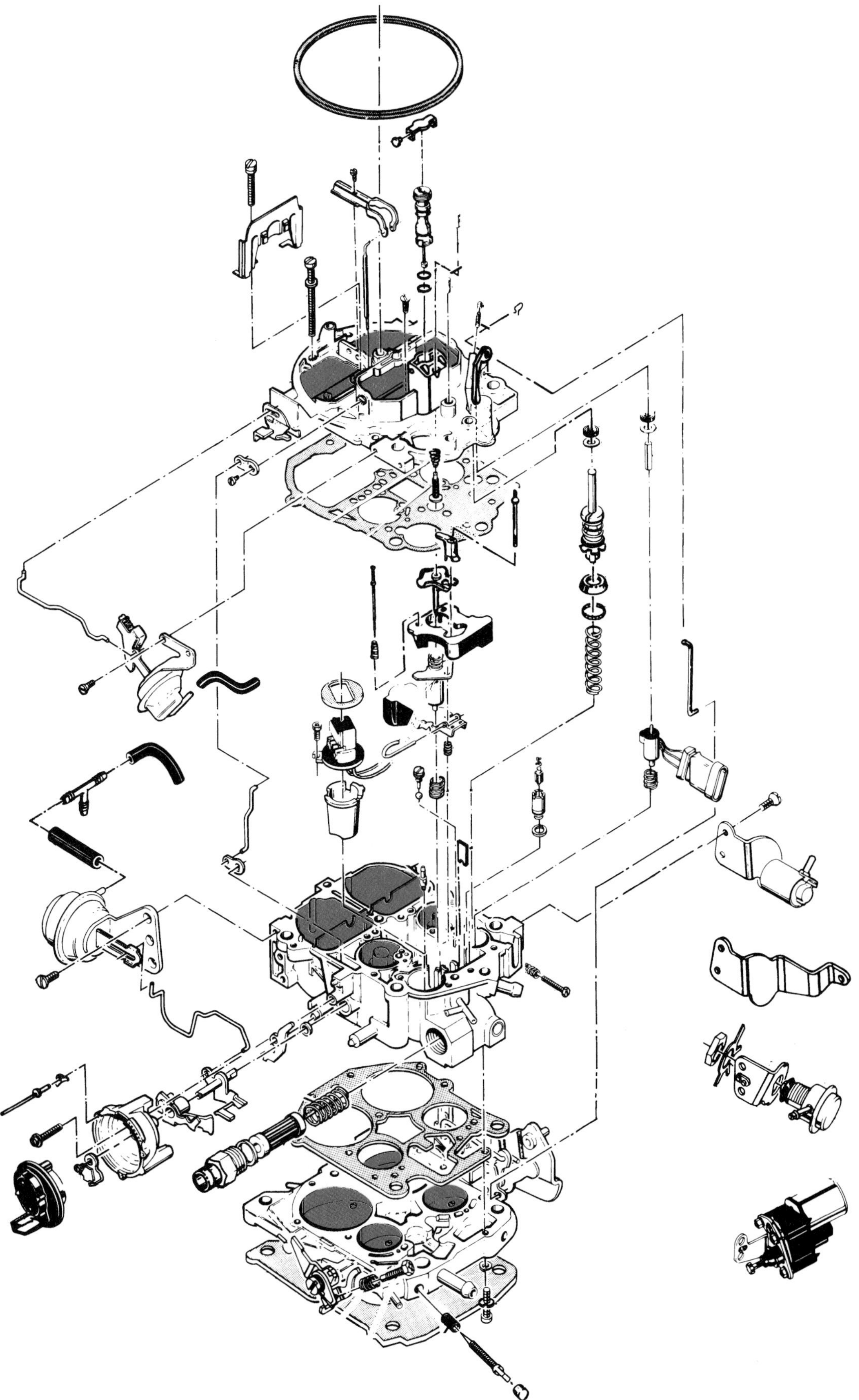

Figure 10-53. An exploded view of a four-barrel, downdraft carburetor. After studying carefully, how many parts are you able to correctly identify? (Pontiac)

Know These Terms

Fuel tank
Pickup tube
Tank gauge unit
Mechanical fuel pump
Diaphragm
Check valve
Pull rod
Return spring
Rocker arm
Pulsation damper
Electric fuel pump
Positive displacement
Non-positive displacement
Fuel filter
Fuel lines
Vapor lock
Air-fuel ratio
Carburetor
Air horn
Fuel bowl
Float
Needle valve
Needle seat
Rich
Lean
Bowl vent
Main discharge tube
Venturi
Venturi vacuum
Air bleed
Jets
Throttle valve
Main metering system
Float system
Carburetor circuits
Idle circuit
Idle mixture adjustment screw
Low speed circuit
Power valve
Step-up rod
Accelerator pump
Pump discharge nozzles
Choke plate
Manual choke
Automatic choke
Thermostatic coil spring
Divorced choke
Vacuum break
Choke stove
Electric choke
Flooding
Unloader
Fast idle cam
Secondary circuit
Primary circuit
Hot idle compensator
Throttle return dashpot
Idle speed-up solenoid
Dieseling
Anti-dieseling solenoid
Mixture control solenoid
Throttle position sensor
Carburetor icing
Manifold heat control valve
Coolant heated manifold
Ceramic heater
Altitude compensator
Aneroid
Intake manifold
Air cleaner
Air cleaner element
Main body
Throttle plate
Downdraft
Updraft
Sidedraft
Barrels
Single-barrel
Two-barrel
Four-barrel

Review Question—Chapter 10

Do not write in this text. Write your answers on a separate sheet of paper.

1. Fuel tanks are made from _____ .
   (A) thin sheet steel
   (B) aluminum
   (C) plastic
   (D) All of the above.
2. Internal baffles in the fuel tank prevent fuel _____ .
3. All of the following statements about mechanical fuel pumps are true, EXCEPT:
   (A) check valves are used to ensure that fuel flows through the pump in one direction.
   (B) the pump is operated by the camshaft.
   (C) a pull rod is fastened to the center of the diaphragm.
   (D) the return spring assists the rocker arm in moving the diaphragm.
4. An electric fuel pump that delivers a certain amount of fuel every time the pump makes one revolution is a _____ type.
5. How many check valves does an mechanical fuel pump use?
   (A) 2
   (B) 3
   (C) 4
   (D) One for every cylinder.
6. Name three materials commonly used to make fuel filters.
7. What is vapor lock?
8. The air bleed makes the gasoline particles _____ .
9. The fuel level in the carburetor is controlled by the _____ system.
10. A high fuel level will make the carburetor run _____ .
11. What does the use of carburetor jets prevent?
    (A) Too little gas going through a passageway.
    (B) Too much gas going through a passageway.
    (C) Lean carburetor operation.
    (D) Both A & B.
12. Engine speed is controlled by the _____ .
13. How is the idle mixture adjusted on late model engines?
    (A) Turning the mixture screw at the throttle body.
    (B) Turning the speed screw at the throttle body.
    (C) Adjusting the float level.
    (D) The mixture on late model engines cannot be adjusted.
14. What circuit is provided to assist the idle circuit when the throttle is opened just beyond an idle?
15. What circuit provides fuel for normal cruising speeds?
16. What operates the power valve or step up rod when full power is needed?
17. What circuit is installed in the carburetor to cure a "flat spot?"
18. What closes the automatic choke when the engine is cold?
19. During the warmup period, engines must idle fairly fast. What is the name of the device provided for this purpose?
20. When does the choke pull operate?

(A) Before the engine is started.
(B) As soon as the engine starts.
(C) As the engine warms up.
(D) When the engine is fully warmed up.

21. If a cold engine becomes "flooded," what mechanism will force the choke valve open?
22. What causes ice to form in a carburetor?
    (A) Gasoline vaporization.
    (B) Low temperatures.
    (C) High humidity.
    (D) All of the above.
23. The manifold heat control valve directs _________ under the carburetor to raise the temperature and reduce icing.
24. Which of the following types of carburetor could have a secondary circuit?
    (A) Single-barrel.
    (B) Two-barrel.
    (C) Four-barrel.
    (D) Both B & C.
25. List two safety precautions to remember when dealing with the fuel system.

No matter what fuel or delivery system a vehicle uses, it must be in good operating condition for optimum engine performance. (Ford Motor Co.)

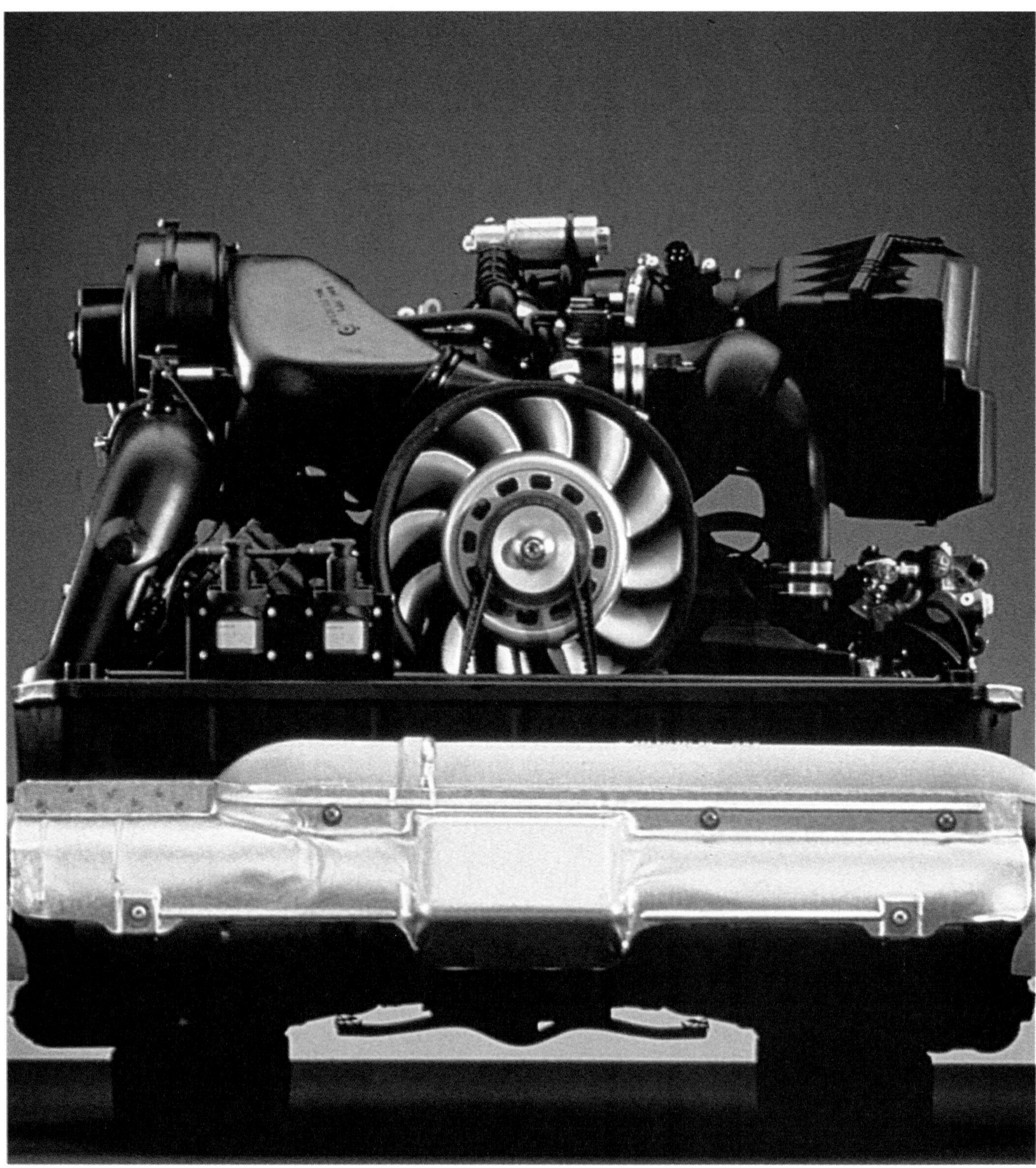

This opposed six-cylinder engine uses air for cooling. The fan is driven by the engine and air is channeled by the shroud surrounding the fan. (Porsche)

11

Cooling Systems

After studying this chapter, you will be able to:

- List the functions of an engine cooling system.
- Identify the two major kinds of cooling systems.
- Identify the major components of a liquid cooling system.
- Trace the flow of coolant through the liquid cooling system.
- Identify the basic parts of an air cooling system.

This chapter will cover the principles of cooling systems, why a cooling system is needed, and why the engine must operate within a certain temperature range. The different types of cooling systems and cooling system components will also be discussed.

The Need for a Cooling System

When fuel is burned in an engine, heat is produced. Modern engines can have combustion chamber temperatures in excess of 4000°F (2204°C). Of this heat, only about one-third is used to power the engine. The rest is unwanted heat. This unwanted heat must be removed. About one-third of the total heat leaves with the exhaust gases. This leaves one-third to be removed by the cooling system. The engine cooling system actually has three jobs:

- It must remove surplus or unwanted heat.
- It must maintain an efficient temperature under all operating conditions.
- It must bring a cold engine up to operating temperature as soon as possible.

As the temperature of the engine parts increases, a point may be reached where the oil will break down if the remaining heat is not dissipated. When this happens, lubrication is no longer possible and the engine will be ruined. On the other hand, a cold engine is very inefficient. A cold engine will run poorly, which can result in oil contamination, formation of engine deposits, increased wear, decreased horsepower, higher exhaust emissions, and poor fuel mileage. This is shown in **Figure 11-1**.

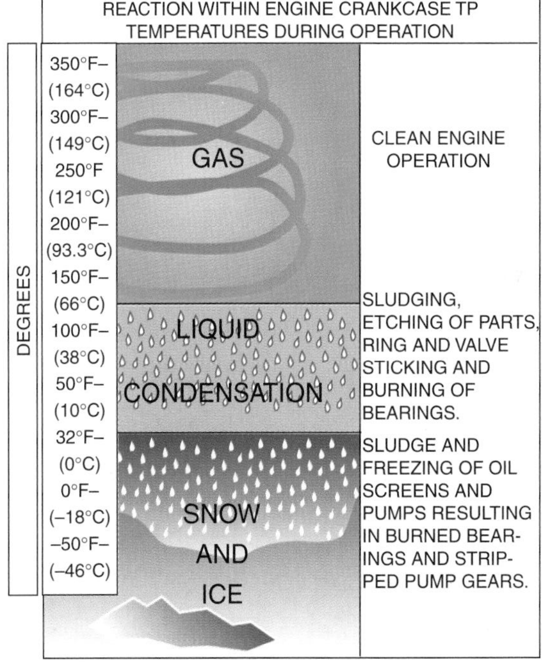

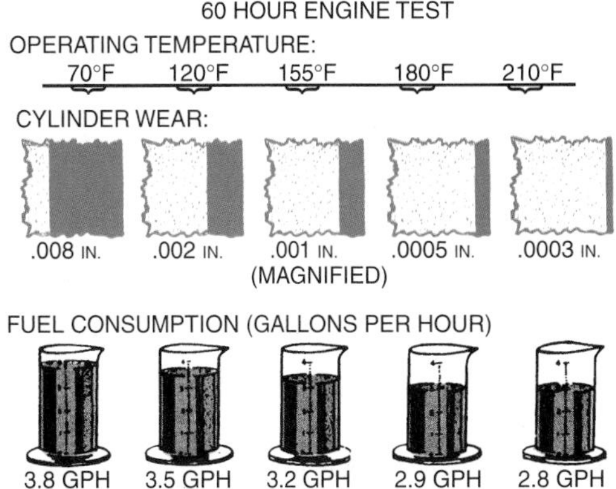

Figure 11-1. The relationship between engine temperature and performance, engine wear, life, and fuel economy. Note that engine performance increases and wear decreases as the engine becomes hotter.

Two Methods of Cooling

When something is too hot, we either blow on it or pour water or some other liquid on it. These same two methods, highly refined, are utilized in cooling an internal combustion engine. Most car and light truck engines use a liquid cooling system. Some vehicle engines and many small one- and two-cylinder engines use air cooling.

Liquid-Cooled Engines

The ***liquid-cooled engine*** must have passages for the flow of liquid through the cylinder block and head. Indirectly, the liquid must contact the cylinder walls, combustion chamber, valve seats, valve guides, and all other engine parts. This is accomplished by taking the basic engine and building a container around it.

In actual practice, liquid circulation passages, called ***cooling passages***, are cast into the engine block and cylinder head. When these passages are filled with a liquid, the engine is surrounded by a jacket of liquid. These cooling passages are often referred to as ***water jackets***. See **Figure 11-2**. Although the liquid-cooled engine is sometimes referred to as a water-cooled engine, all modern engines must use a mixture of 50% water and 50% ethylene glycol antifreeze, usually called ***coolant***. Antifreeze will be discussed in more detail later in this chapter.

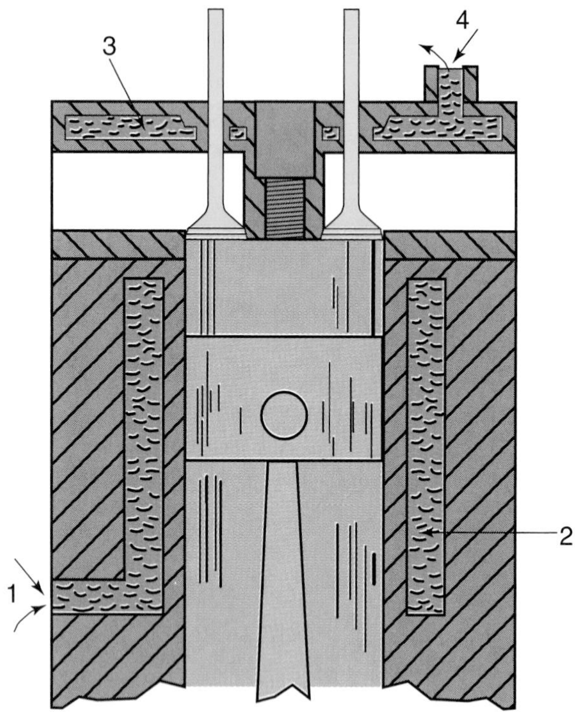

Figure 11-2. Coolant passages. 1–Coolant enters the block here, circulates around the cylinder to 2, and then flows up into the head 3, where it leaves at 4. Notice how coolant surrounds the cylinder, valve guides, head, spark plug seat, and top of the combustion chamber.

Holes are left in the block during the casting process. These holes are sealed by using stamped metal or brass plugs called ***freeze plugs***. They are designed to allow for expansion should the coolant freeze. However, in spite of their name, block damage usually occurs if the coolant freezes.

Heat Transfer

Engine parts, such as the block and cylinder head, absorb the heat of the burning fuel. This heat is conducted through the part and into the coolant contained in the cooling passages. The coolant absorbs some of the heat from the part and transfers it to the outside of the engine, **Figure 11-3**.

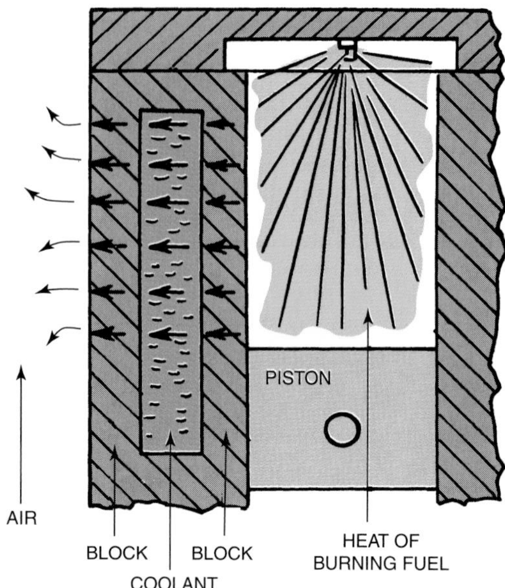

Figure 11-3. If coolant does not circulate, there is no transfer of heat. If this coolant is allowed to sit without circulating, the engine will literally burn up.

The engine coolant is the medium of heat transfer. It is able to absorb and release heat, and will not damage any other cooling system parts. However, the coolant will absorb just so much heat, then it will boil away. If the coolant stands still, it is of no value for cooling purposes. In order to function as a cooling agent, the coolant must move, or circulate through the engine. As the coolant is heated, it must be moved out of the engine and replaced with coolant that is not as hot. There must be no dead spots (areas in which there is little or no circulation), **Figure 11-4**.

A Pump Is Needed

To circulate coolant through the engine, a ***coolant pump*** is needed. Most pumps are ***centrifugal pumps*** utilizing an ***impeller***, **Figure 11-5**. The impeller consists of a round plate fastened to a rotating shaft to which a series of flat or curved vanes are attached. When the impeller turns, the coolant is thrown outward by centrifugal force. By placing the impeller in a closed housing with an inlet to allow coolant to contact the front face of the vanes and an outlet to carry away coolant thrown outward, such a pump will move coolant efficiently, **Figure 11-6**.

The pump housing is made of cast iron or aluminum. The impeller shaft is supported on a double row of sealed ball bearings. A spring-loaded seal is used to prevent coolant leakage. For the pump to operate at engine speed, a way must be found to turn the impeller. To accomplish this task, a hub is cast on the front of the impeller shaft. A pulley is then bolted to this hub. Sometimes a fan is also bolted or attached to the

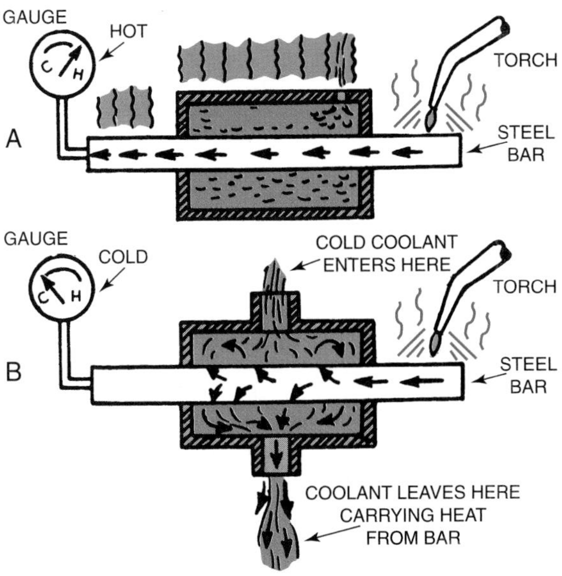

Figure 11-4. Circulation is vital. A—Notice that a torch has heated the coolant and the entire steel bar. The gauge reads hot. B—Coolant is circulating through the passages, carrying heat away from the bar. The gauge now reads cool.

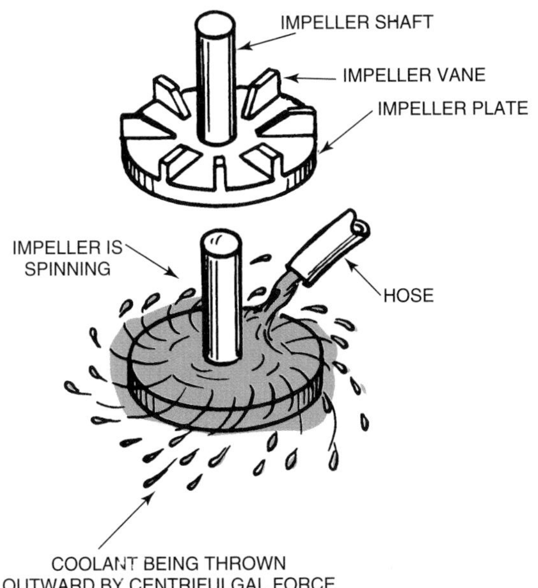

Figure 11-5. Normal action of a coolant pump impeller. Note how coolant is thrown off in all directions. This impeller is inefficient in this state.

hub or pulley. One particular cooling pump is shown in **Figure 11-7**.

On some pumps, the impeller extends into the coolant passages of the engine, as in **Figure 11-7**. Some pumps are driven directly by the engine, **Figure 11-8**. Other pumps have the impeller enclosed in the pump housing, as in **Figure 11-9**. The coolant pump is usually secured to the front of the engine by bolts. Coolant pumps are usually referred to as ***water pumps***.

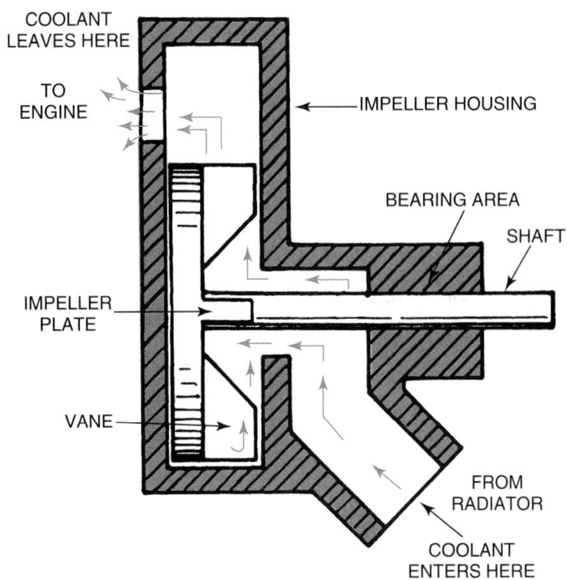

Figure 11-6. If the impeller is placed inside of a housing, coolant directed to the center of the impeller is thrown outward into the housing, and an outlet directs the flow out of the housing.

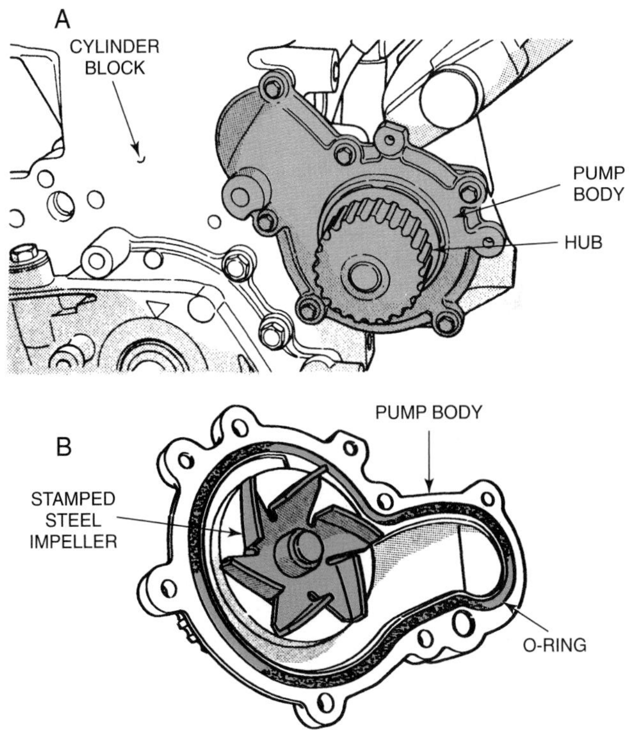

Figure 11-7. A—Water (coolant) pump mounted on the engine with the impeller inside the water jacket. B—Water pump removed. Note the six-blade, stamped steel impeller, and the O-ring pump body-to-block seal (Chrysler Corp.)

Belt Drive

Figure 11-9 shows a ***V-belt*** being used to drive a coolant pump. The belt is driven by a pulley on the front of the crankshaft. This same belt may be used to drive other engine

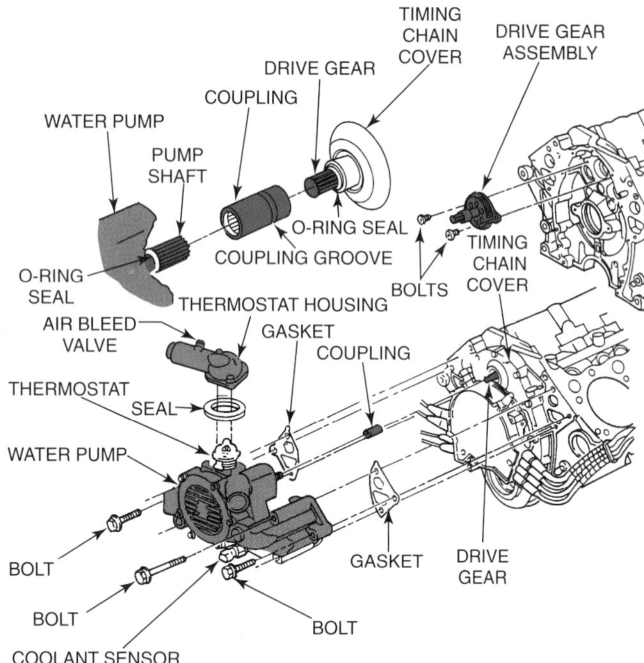

Figure 11-8. Water pump which is gear-driven by the engine. Note that the thermostat is housed in the water pump body and that an air bleed is used to aid in removing all air from the system. Electric cooling fans are used with this arrangement. (Chevrolet)

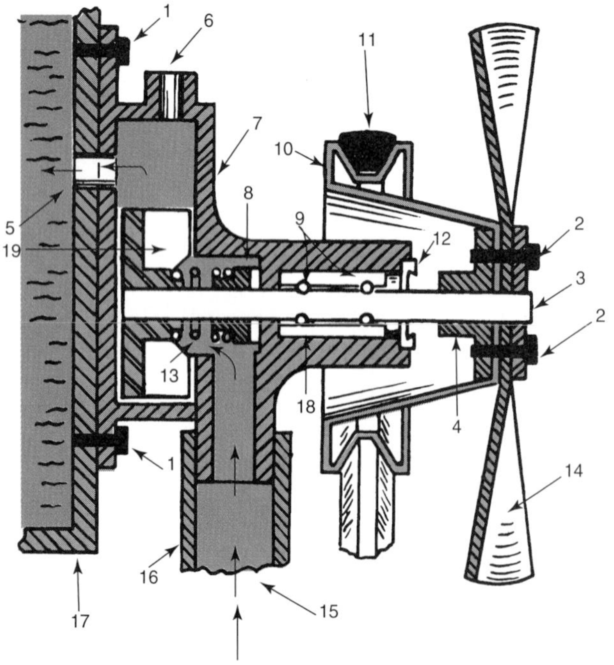

accessories. Typical V-belt construction is shown in **Figure 11-10**. A modification of the V-belt is the ***ribbed belt***, which is sometimes referred to as a ***serpentine belt***, **Figure 11-11**. It is a wide belt with several ribs that act as several small V-belts.

All belts must be in good condition and properly adjusted. Excessive tightness will place a heavy load on the coolant pump's bearings, causing premature wear. Belt looseness will permit slippage, which will reduce pump and fan speed, possibly causing overheating. A loose belt will have a

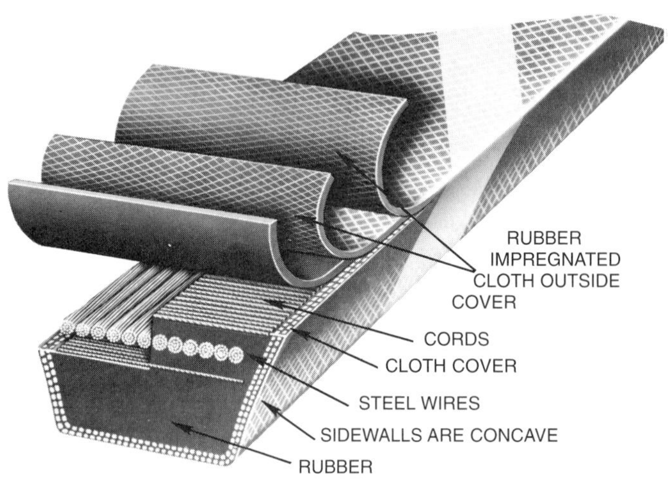

Figure 11-10. This particular V-belt construction uses steel wires and is designed to drive the air conditioning compressor. (Gates)

Figure 11-9. Cutaway of a typical coolant pump. 1–Bolts-pump to block. 2–Bolts-fan and pulley. 3–Pump shaft. 4–Hub. 5–Coolant outlet. 6–Bypass. 7–Pump housing. 8–Seal. 9–Ball bearings. 10–Pulley. 11–V-belt. 12–Dust seal. 13–Seal spring. 14–Fan. 15–Coolant inlet. 16–Hose. 17–Block. 18–Bearing race. 19–Impeller.

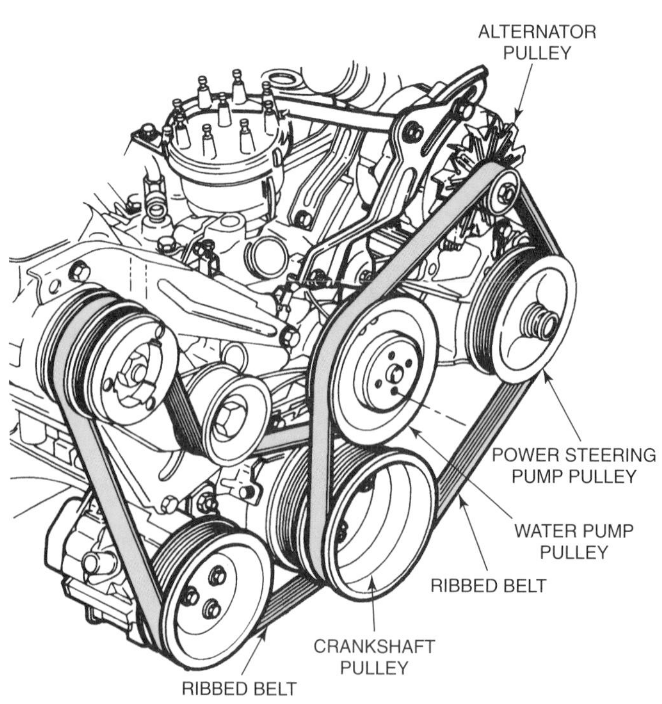

Figure 11-11. This engine uses a ribbed belt to drive the coolant pump. Note the difference between this belt and the regular V-belt in Figure 11-10. (Ford)

tendency to whip or flap, which can cause intermittent heavy loading of the bearings and possible pump failure.

Controlled Circulation

The coolant must flow around the cylinders and pass up through the cylinder head on its way out. It must reach every part of the system. It is not enough to pump coolant into the bottom of an engine and let it flow out the top. If this were done, the coolant would rapidly circulate through some areas and would move slowly, or not at all through others, causing localized overheating.

The coolant may be channeled into parts of the engine by means of ***distribution tubes***. **Figure 11-12** illustrates a typical distribution tube. The tube is placed in the coolant passages. The outlet holes are lined up to circulate coolant into the hard-to-cool areas in the engine.

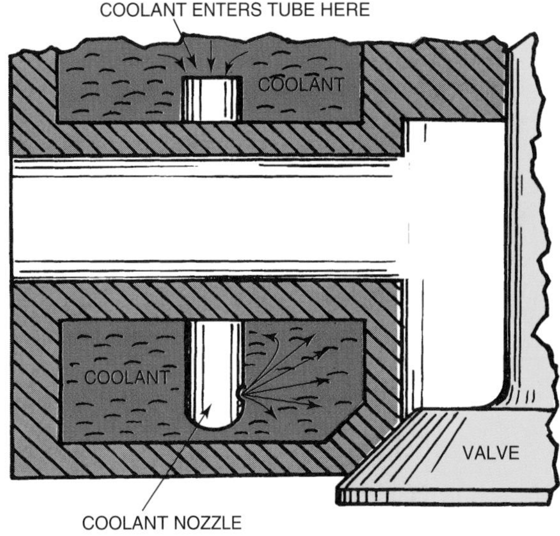

Figure 11-13. Notice how the coolant nozzle pumps coolant. As the coolant leaves the small hole in top, it causes a strong circulation around the valve seat area.

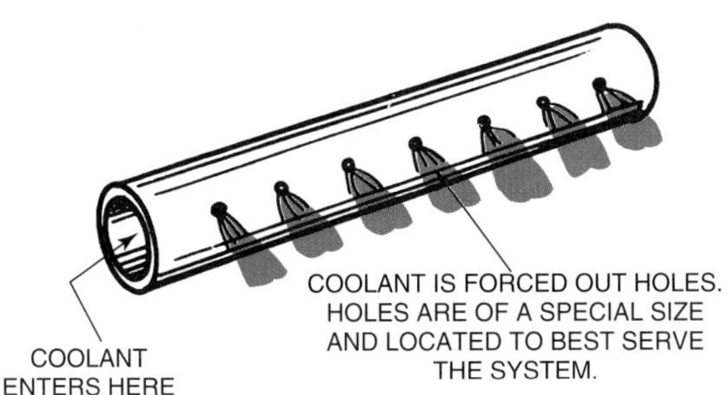

Figure 11-12. This distribution tube may be located in the block, head, or both. It allows coolant to reach areas of the block which would not be readily accessible by the water jackets.

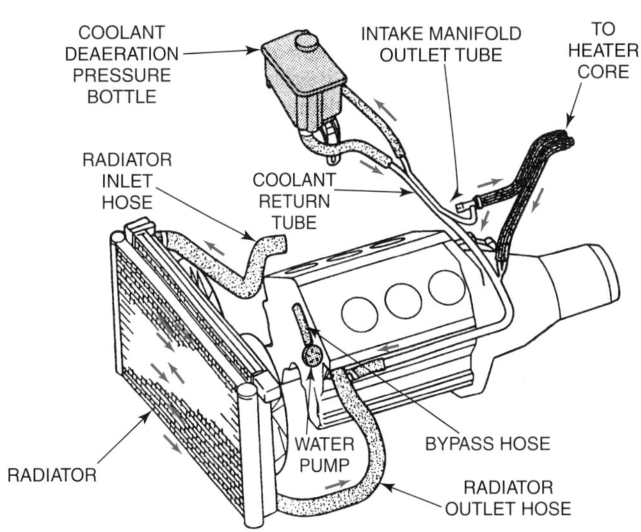

Figure 11-14. Cooling system operation with one particular V-6 engine. Follow the coolant flow path arrows. (Chrysler Corp.)

Some engines employ ***coolant nozzles*** to speed up circulation around the valve seats. A nozzle is shown in action in **Figure 11-13**. On most modern engines, the distribution tubes and nozzles are precision cast into the cylinder head, eliminating the need for separate tubes and nozzles. **Figure 11-14** illustrates the path the coolant takes on its way through the engine.

The Radiator

When the coolant leaves the engine, it is quite hot—often well over 200°F (93.3°C). If it were immediately pumped back into the engine, it would be too hot to absorb additional heat. Before the coolant can be reused, the heat must be removed. Cooling is accomplished by passing the coolant, as it leaves the engine, through a ***heat exchanger***. This heat exchanger is usually referred to as a ***radiator***.

Some radiators are made of brass and copper to facilitate cooling and to protect it from rust and corrosion. All joints are soldered. However, most modern radiators use aluminum and plastic in their construction to reduce the weight of the unit. The radiator is placed so that air can pass through it when the vehicle is in motion.

From the top- or side-mounted entry tank, coolant flows down or across through small copper or aluminum tubes. The tubes are flattened to provide the least resistance to airflow and have thin metal fins soldered over their entire length. The tube and fin assembly is known as the ***radiator core***. As the coolant makes its way through the core, it gives off heat to the tubes and fins. Since both copper and aluminum are excellent conductors, the tubes give off their heat, via the fins, to the air passing around them.

By the time the coolant reaches the outlet tank, it is cool enough to reuse. **Figure 11-15** shows a cross-section of a simple radiator with top and bottom tanks. Modern vehicles use cross-flow radiators, such as the one shown in **Figure 11-14**. The coolant enters one side tank and then passes through the horizontal tubes to the opposite tank, sometimes called the return tank.

The radiator usually contains a filler neck for adding coolant when necessary. Some filler necks are mounted on

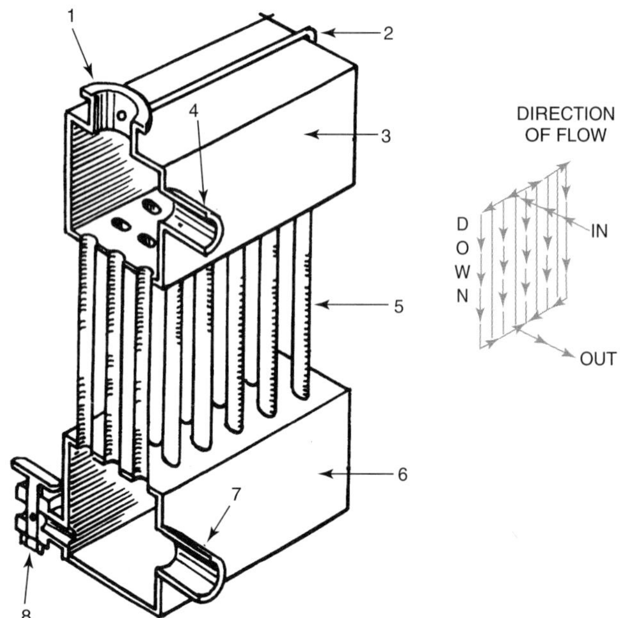

Figure 11-15. In this downflow radiator, coolant enters the top hose connection 4, then passes into the top tank 3. From there it flows down through the core tubes 5. When it reaches the bottom tank 6, it has been cooled. 1–Filler neck. 2–Overflow tube. 7–Lower hose connections. 8–Drain petcock. Not shown are the thin fins normally soldered to the core tubes to help dissipate heat.

the vehicle's engine. The neck is sealed by a pressure cap, which is discussed later. The radiator may also contain another heat exchanger to provide cooling for the automatic transmission fluid. The ***transmission oil cooler*** can be a separate unit or placed in the return tank to give its heat up to the coolant after the coolant temperature has been reduced in the radiator core. Transmission oil coolers are shown in **Figure 11-16**.

Forced Air

To speed up the cooling action of a radiator, one or more power-operated ***fans*** are used to draw air through the radiator core. The fan can be driven by the engine, or by an electric motor, **Figure 11-17**. Engine-driven fans are an integral part of the water pump and are driven by the water pump's drive belt. Electric fans are mounted as independent units.

If an engine-driven fan is set too far back from the radiator core, it will recirculate the same air, **Figure 11-18A**. To avoid the loss of fan efficiency, engine-driven fans require a radiator shroud. This prevents recirculation of air, **Figure 11-18B**. Electric fans can be placed close to the radiator and do not require a shroud.

Fan Blades

Fans may have as few as two blades and as many as ten, depending on specific requirements. Blades are usually rounded and arranged in an uneven manner in an attempt to lessen noise. One type of fan employs flexible blades that reduce the pitch (angle at which the blade is mounted) at higher speeds, permitting the fan to turn more easily.

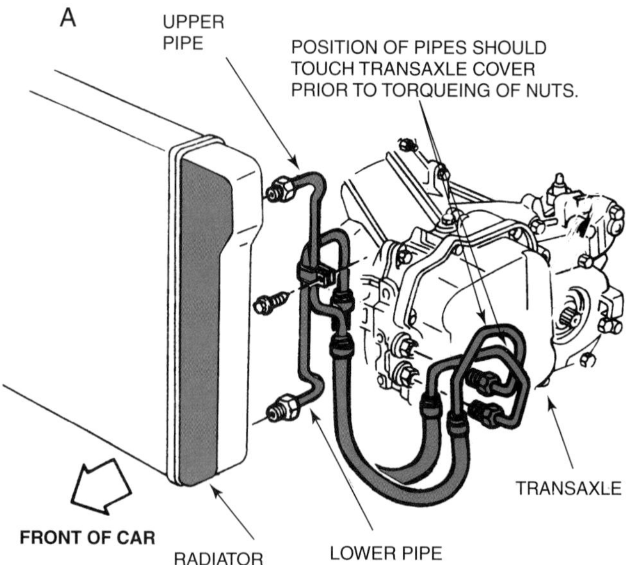

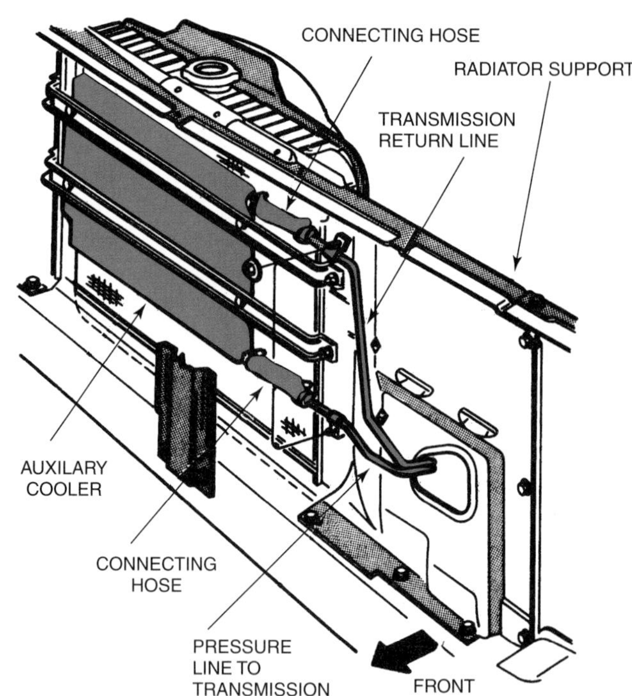

Figure 11-16. Two types of transmission fluid coolers. A—A cooler installed inside the radiator side tank. B—An auxiliary cooler mounted in front of the radiator assembly, with lines running to the transmission.

STOP **Warning: A fan can inflict serious injury. Keep hands and other objects away from the fan. Electric fans can come on automatically, even when the vehicle is shut off. Do not stand in line with a revolving fan—a blade can fly out with lethal force.**

Fan Clutches

As the vehicle gains speed on the road, air is forced through the radiator core. As a result, the fan becomes less and less useful as road speed increases. Since they require

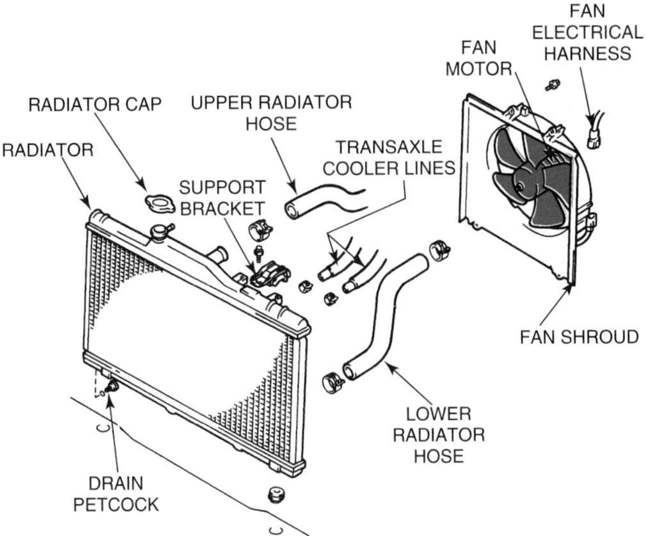

Figure 11-17. This electric cooling fan assembly is used instead of the engine-driven fan. Many vehicles have more than one electric cooling fan. (Geo)

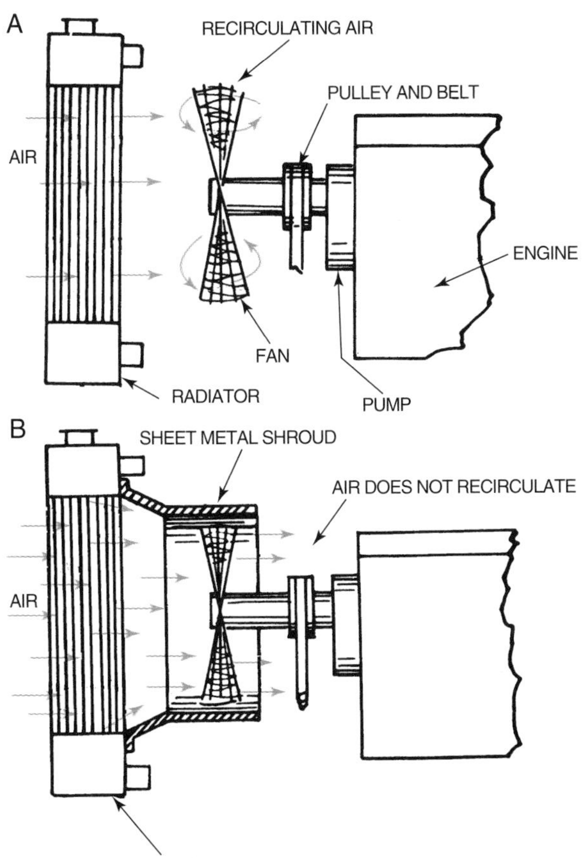

Figure 11-18 A—This fan is too far from the radiator which allows air recirculation. Some air is being pulled through the radiator, but a great deal is being recirculated through the fan. B—This fan shroud stops recirculation which improves airflow. Most vehicles have a thin sheet metal or plastic shroud attached to the radiator.

engine power to spin, some engine-driven fans have a device that causes the fan to slip, or freewheel above a certain engine RPM. See **Figure 11-19**. This device is called a ***fan clutch***. Fan clutches are designed to save horsepower and to cut down on objectionable fan noise. Some fan clutches have a thermostatic control that will allow the fan to slip until the engine warms up and the fan is needed.

Electric Cooling Fans

Electric fans are driven by small electric motors. They are commonly used on front-wheel drive vehicles. The fan motor is usually controlled by a thermostatic switch installed in a coolant passage or in the radiator tank. Some vehicles have more than one cooling fan.

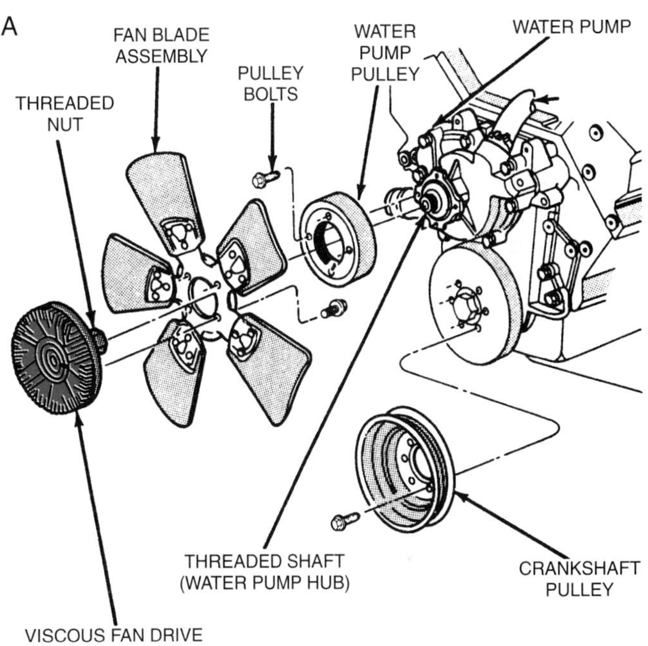

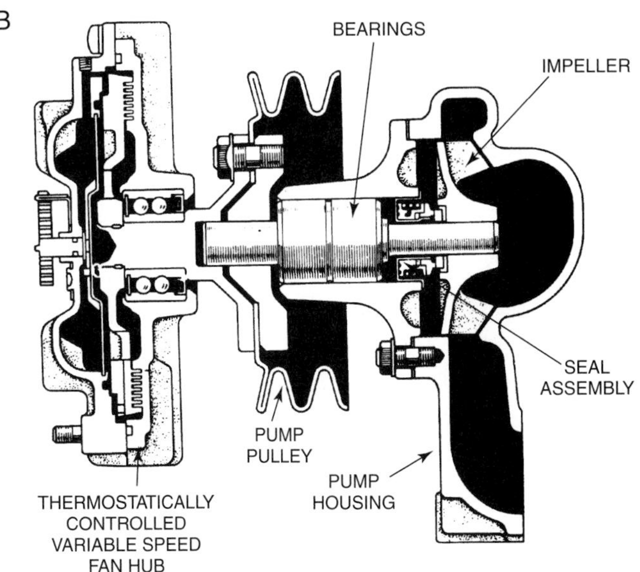

Figure 11-19. A thermostatically controlled, variable speed viscous fan drive unit. A—An exploded view of one particular setup. B—A cutaway of a viscous fan drive unit. (Dodge & Toyota)

When the engine is first stated, the electric fan does not run. As the coolant heats up, the fan comes on as needed to keep the engine from overheating. Some electric fans are operated by the engine control computer, which can also vary the fan's speed in some cases. The cooling fan will also stay on at all times while the vehicle's air conditioning system is operating.

Hoses

The radiator is connected to the engine and pump by means of rubber ***hoses***. Hoses must be strong enough to withstand cooling system heat and pressure, while remaining flexible to withstand engine vibration. These hoses are of different diameters, depending upon the particular engine. There are three common types of hoses, as shown in **Figure 11-20**.

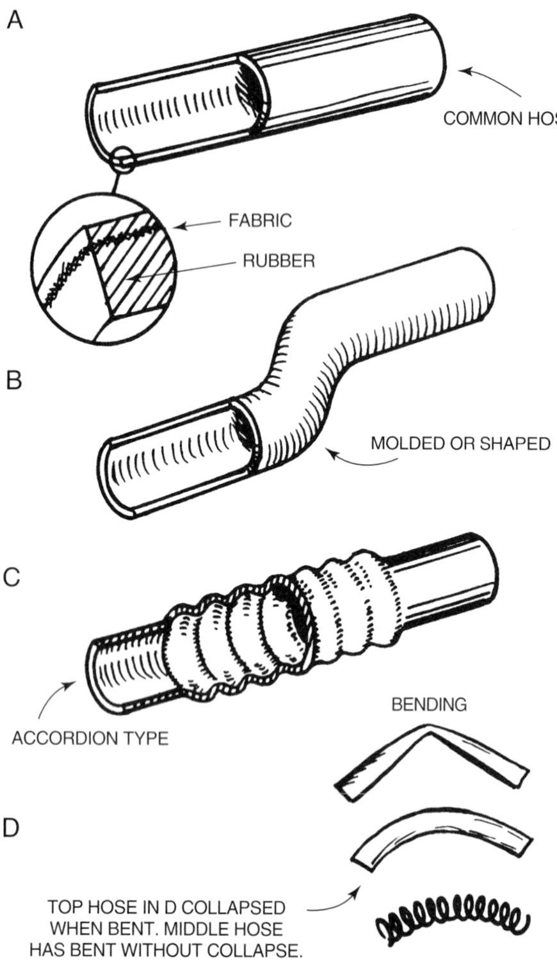

Figure 11-20. Radiator hose types. A—This common straight hose is made of rubber with one or two layers of fabric for strength. This hose will not stand much bending without collapsing. B—The molded or shaped hose shares the same construction as A. Since all the necessary bends are molded in, this hose will not collapse. C—The accordion hose will stand severe bending without collapse. It will also reduce the transfer of vibration from the engine to the radiator. D—The top hose has collapsed with bent. Middle hose has bent with no sign of collapse as it has a spiral spring-like wire inserted into it before bending. The wire is shown beneath the middle hose. All three types of hose may use the spiral wire in their construction. It can be molded in or placed in.

To bring hot coolant to the heater core, ***heater hoses*** are used. These hoses are smaller than the radiator hoses, and carry a smaller amount of coolant. A ***heater shut-off valve*** may be installed in the heater hoses. Heater construction and operation will be covered in more detail in Chapter 28. Some cooling systems make use of metal tubing as part of the coolant transfer system.

Hose Clamps

Hoses are secured to their fittings by ***hose clamps***. There are two popular types. One is the adjustable, screw tightened type, and the other is the snap, or ring type. Both the screw and snap types have many adaptations. Four different styles of hose clamps are shown in **Figure 11-21**.

First Job Complete

As mentioned earlier, the cooling system has three jobs. The first job is to remove surplus heat from the engine. The second job is to maintain an even, efficient heat level. The third job is to bring a cold engine up to proper operating temperature as soon as possible after starting. The first job is completed by the coolant passages, pump, radiator, hoses, and fan, **Figure 11-22**. We will now discuss how the cooling system performs the last two functions.

The Thermostat

The system, as studied so far, would cool at a constant rate. If the capacity is sufficient to cool the engine in hot weather under heavy loads, it would obviously overcool in cold weather. At some engine speeds, the coolant may pass through the radiator too quickly to properly remove the excess heat, resulting in overheating. Some type of temperature control is necessary for good cooling system operation in hot and cold weather. To provide constant temperature control, a ***thermostat*** is installed in the cooling system. It is often placed directly beneath the top coolant outlet. The coolant outlet may have a bowl-shaped bottom to house the thermostat.

The Purpose of the Thermostat

The thermostat is designed to open at a predetermined temperature. When the engine is cold, the thermostat shuts off the flow of coolant from the engine to the radiator. As the coolant is confined to the engine, it heats quickly. When it reaches the thermostat's predetermined rated temperature, it opens and allows coolant to circulate. The thermostat will remain open until the temperature falls below its heat setting. The thermostat, therefore, has no effect on the maximum temperature of the engine.

If it is extremely cold, the thermostat may not open all the way. Coolant can circulate, but slowly, keeping the engine temperature at the proper level. The thermostat constantly changes the size of its opening in relation to coolant temperature. This action will bring a cold engine up to proper heat in a minimum amount of time. It will also maintain a minimum, efficient temperature level under all cold weather driving

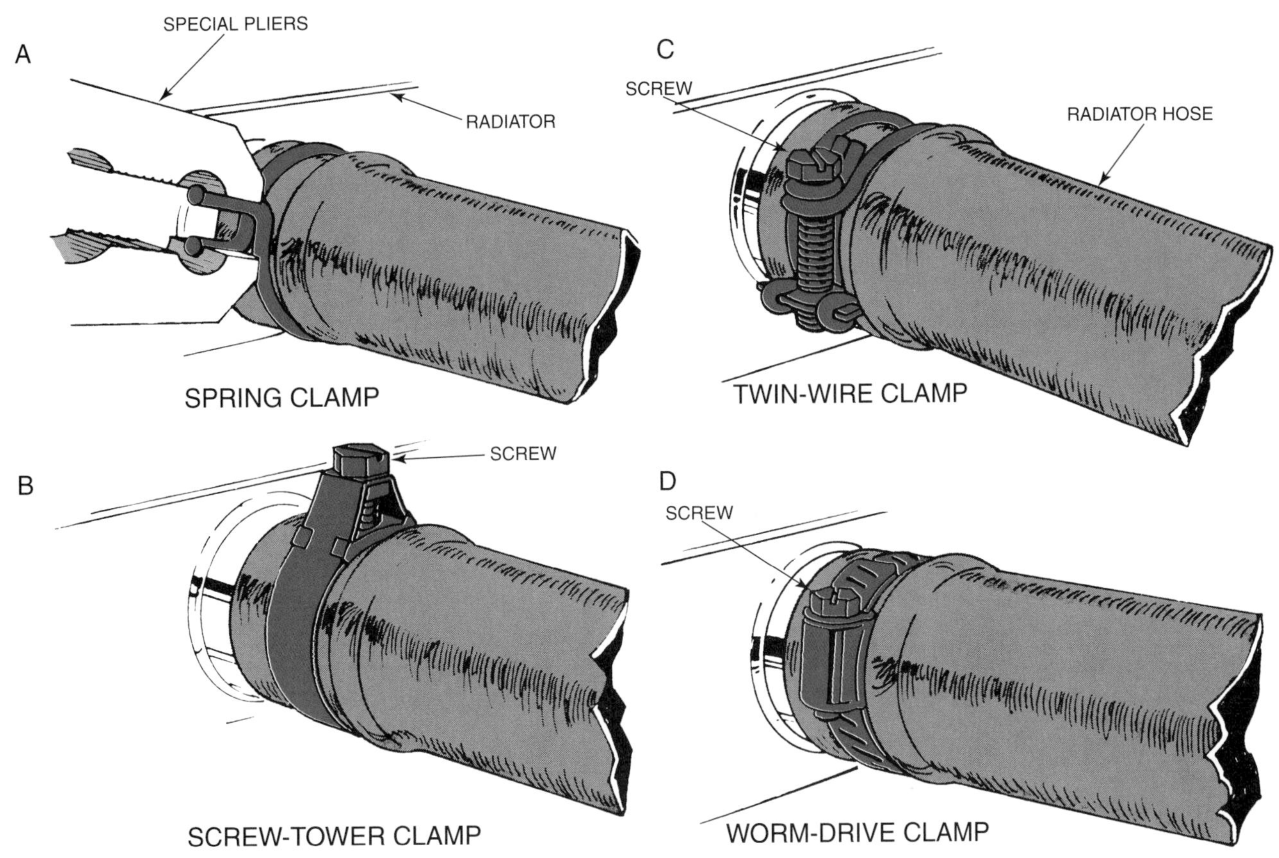

Figure 11-21. A—A constant tension, spring-type hose clamp. B, C, and D—Three different screw-type hose clamps. D is the most commonly used. (Gates Rubber Co.)

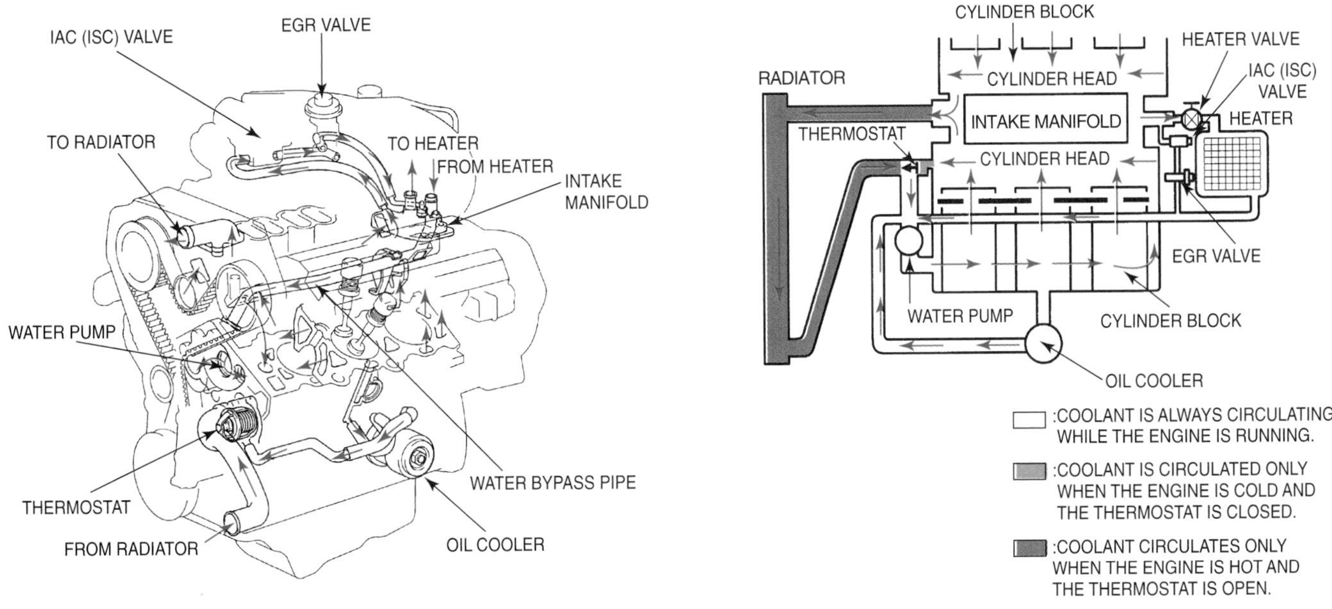

Figure 11-22. Schematics illustrating the basic coolant flow through the various parts of a V-6 engine. Study both illustrations carefully. (Toyota)

conditions. See **Figures 11-23A** and **11-23B**. The thermostat will also allow the vehicle's heater to produce heat more quickly in cold weather.

Note in **Figure 11-24** how the coolant is heated in three stages. **Figure 11-24A** shows the thermostat closed, no flow to the radiator, and with the top engine area (cylinder heads) heating. **Figure 11-24B** illustrates stage 2, with the entire engine heating up. **Figure 11-24C** shows the cooling system with the thermostat open and the coolant circulating throughout the system.

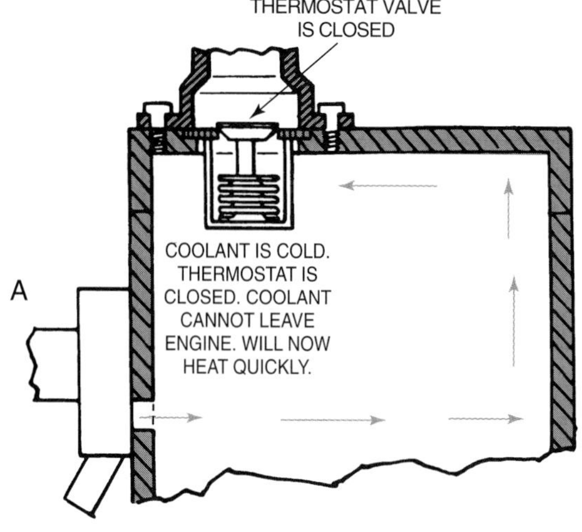

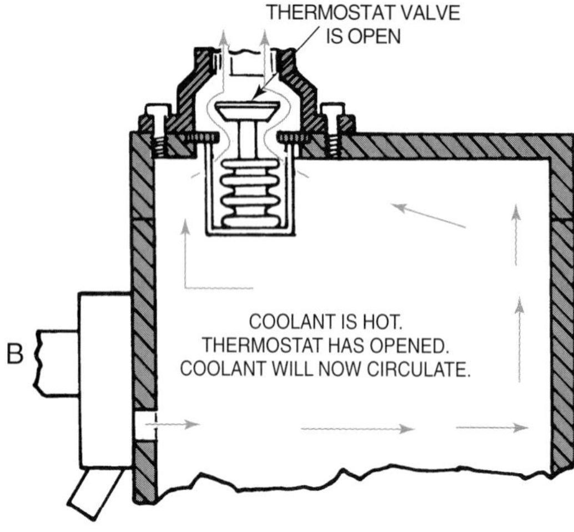

Figure 11-23. A—In a cold engine, a thermostat will remain closed to allow the coolant to heat quickly. B—As the engine begins to approach operating temperature, the thermostat will open and allow hot coolant to circulate.

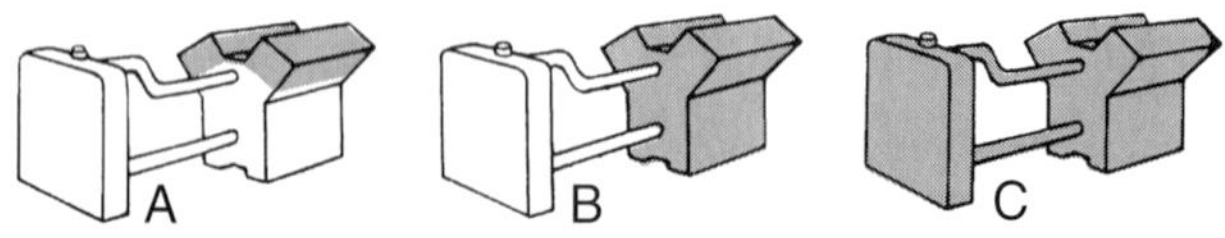

Figure 11-24. The cooling system comes up to operating temperature in three stages. A—The coolant around the top of the engine is heated first. B—The coolant in the entire block has begun to heat and the thermostat will begin to open. C—This illustration shows that the coolant has begun to circulate through the radiator.

Thermostat Operation

The modern thermostat is operated by a wax pellet contained in a housing. Also in the housing is a valve that is operated by the wax pellet, **Figure 11-25A**. The thermostat is placed in the coolant outlet, so that the pellet case rests in the engine block coolant. When the engine is cold, there is no pellet action and the spring holds the thermostat valve closed.

As the engine coolant warms, the wax pellet is heated. This causes the wax to expand, forcing the rubber tightly against the steel piston or pin, **Figure 11-25B**. When the coolant temperature reaches a predetermined level, the wax will have expanded to the point that the pressure on the rubber is so great that it will force the case downward against spring pressure. This pressure will pull the valve open and allow coolant to flow through the thermostat on into the radiator.

There are numerous pellet thermostat designs. Many have a pull-push valve similar to that shown while others operate a flap or a butterfly valve. The pellet thermostat is not pressure sensitive and works well in a fully pressurized cooling system. A small hole may be drilled in the thermostat valve. This acts as an aid when refilling the system as it permits air trapped in the block to escape as the coolant flows upward. In some vehicles where the radiator is lower than the engine, a ***manual bleeder valve*** is installed on top of the thermostat housing to help remove trapped air from the block. See **Figure 11-26**.

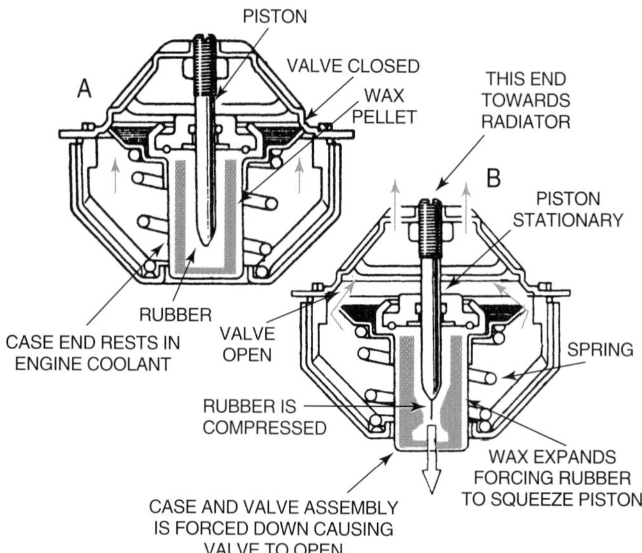

Figure 11-25. In a pellet type thermostat, the pellet end rests in block coolant. A—At this point, the engine coolant is cold and cannot pass by the closed thermostat valve. B—Coolant has now reached operating temperature. The wax pellet expands, squeezing the rubber against the piston hard enough to force the pellet case downward to open the valve. (Toyota)

Thermostat Heat Ranges

Thermostats are available in different heat ranges. Common heat ranges are the 180° and 195°F (82.2°-90.6°C) thermostats. The 180°F thermostat opens near this temperature and is fully opened around 200°F (93.3°C). The 195°F type starts opening near 195°F and is fully open around 218°F (103.3°C). Some thermostats range above 200°F.

These higher range thermostats raise the engine operating temperature and, in so doing, aids in emission (exhaust pollution) control. The exact thermostat temperature range used will depend on the type of engine, vehicle load, and the weather where the vehicle is operated.

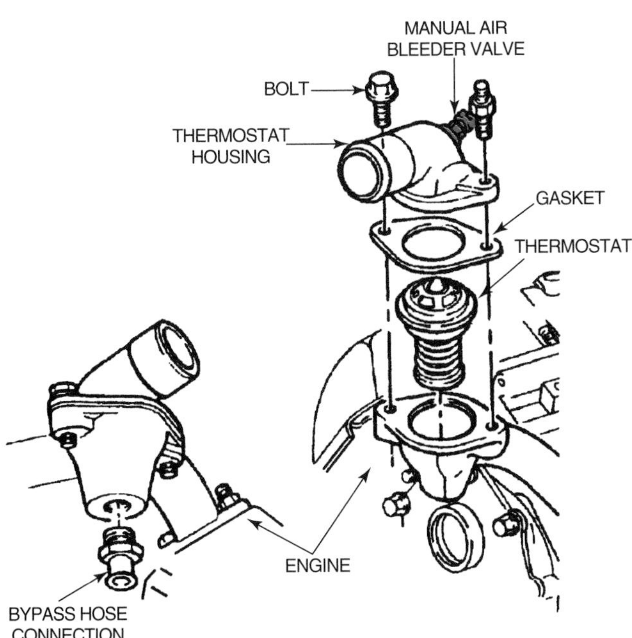

Figure 11-26. A manual bleeder valve assembly which is located in the upper half of the thermostat housing. When the valve is opened, it will allow trapped air in the cooling system to escape. (General Motors)

Internal Coolant Circulation

A thermostat will remain closed until the coolant in the engine reaches a specific temperature. If the coolant did not circulate until this temperature was reached, localized overheating (at the exhaust valves for instance) could occur. Therefore, the coolant in the engine must circulate within the engine cooling passages.

To allow the pump to circulate coolant through the engine when the thermostat is closed, a ***bypass*** is used. The bypass may be a fixed opening that allows coolant to recirculate within the block and heads during warmup. Coolant circulates from the cylinder head, back through the pump, and through the block and heads again. Most vehicles use a fixed bypass that allows continuous recirculation of a small portion of the coolant. When the thermostat opens, the pump can draw coolant from the radiator, and draws only a small amount of coolant through the bypass.

On some systems, the bypass is set up as shown in **Figures 11-27A** and **11-27B**. In **Figure 11-27A**, the thermostat is closed, the bypass cover finger is raised and coolant moves from the pump through the distribution tube to the open bypass pipe and into the pump. Coolant continues to recirculate until the thermostat is fully open. In **Figure 11-27B**, the pellet thermostat case has moved down, pulling the valve open and at the same time, forcing the finger to close the bypass pipe. Coolant now moves through the thermostat to the radiator.

Pressurized Cooling System

The modern engine produces tremendous heat. This heat could cause the coolant to exceed its boiling point and boil away. If the cooling system is pressurized (placed under pressure), the boiling point of the coolant is raised. A ***pressurized cooling system*** allows the coolant to circulate at a temperature that is closer to the engine's optimum operating temperature. Higher coolant temperatures allow the engine to run hotter which results in less exhaust pollutants.

Every one PSI (6.895 kPa) increase in cooling system pressure raises the boiling point by 3°F (1.66°C). By pressurizing the system to 15 PSI (103 kPa), the boiling point of the coolant will increase by 45°F (25°C). Since the boiling point of water is 212°F (100°C), pressurization makes it possible to operate an engine at temperatures exceeding 250°F (121°C) without boiling. Pressurization also increases coolant pump efficiency by reducing cavitation (coolant in the low-pressure area near the pump impeller blades turning to steam, allowing the pump impeller blades to spin in the air bubbles).

How System is Pressurized

Pressurizing a cooling system is done by placing a ***pressure cap*** on the radiator filler neck. The pressure cap presses firmly against a seat in the filler neck. As the coolant in the system heats, it also expands. As it continues heating and expanding, it squeezes the air on the top of the coolant.

Since the pressure cap has sealed the opening, the air cannot escape. When the pressure overcomes the spring holding the pressure cap seal valve closed, the seal valve will rise. As soon as the excess pressure (over what the cap is designed for) escapes, the spring closes the seal and keeps the predetermined system pressure constant. The escaping air passes out through the radiator overflow pipe.

When the engine is shut off and cools down, another small valve in the cap opens. As the coolant cools, it contracts and forms a vacuum in the radiator. This draws the vacuum valve open and allows air from the overflow pipe to enter the radiator. Without a vacuum valve, the radiator hoses could collapse.

Pressure Cap Operation

The radiator filler neck has a flange near the bottom. When the pressure cap is installed, it has a spring-loaded valve that seats against the flange. The spring-loaded valve has a valve in the center that opens when a vacuum is formed in the tank. This vacuum valve can also be spring-loaded. The large pressure release control spring is calibrated to open at a specific pressure (15-16 PSI or 103-110 kPa). **Figures 11-28A** and **11-28B** shows the operation of a pressure cap.

Coolant Recovery System

On older pressurized cooling systems, the cooling system could never be completely filled, since coolant expelled through the pressure cap was replaced by air when the engine cooled off. This meant that the cooling system could never be operated at full capacity. The ***closed cooling system***, which had the addition of a ***coolant recovery system***, provided vehicles with a cooling system that could be operated at full capacity. The major features are a ***coolant reservoir*** and a different pressure cap design. Some systems use the coolant reservoir as part of the actual cooling system. The pressure caps in these systems are mounted on the reservoir itself.

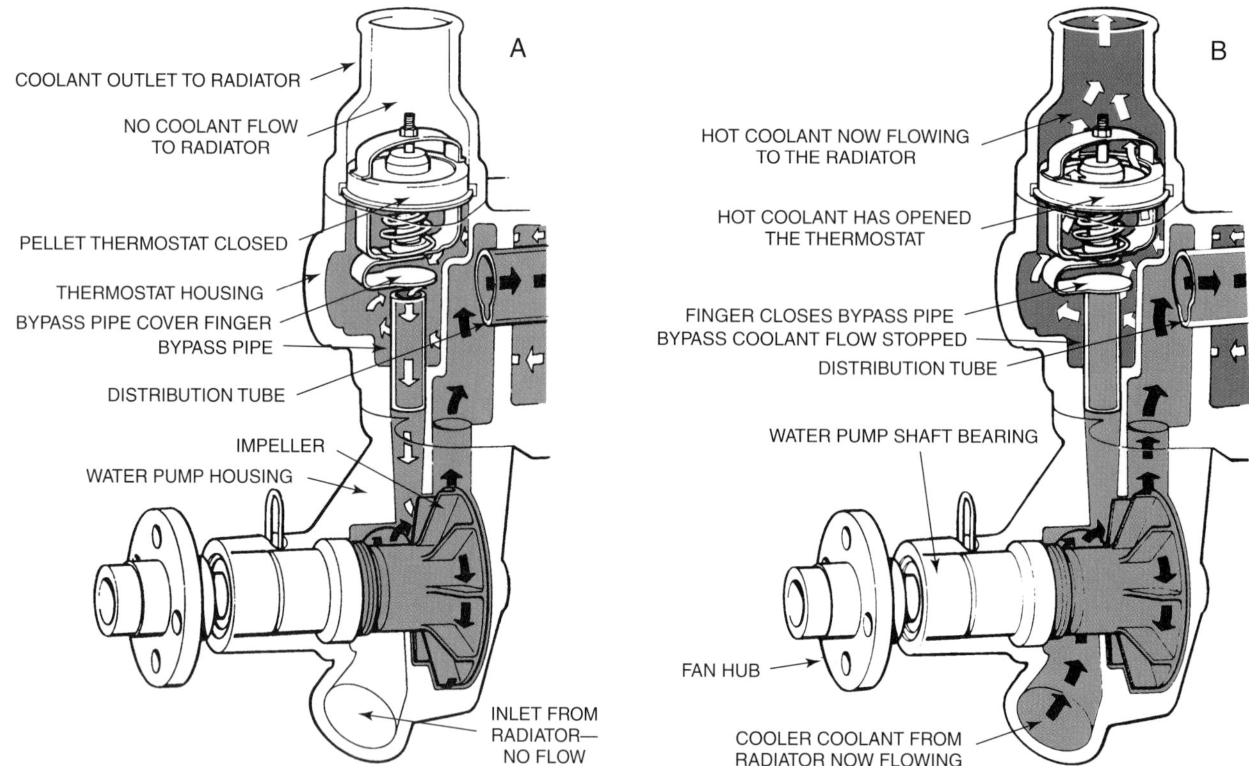

Figure 11-27. Thermostat and bypass housing action. A—The coolant is still cold and the thermostat is closed. The bypass pipe cover finger is up, exposing the bypass pipe opening. Coolant circulates from the pump, through the distribution tube, down through the bypass pipe to the pump. B—The coolant has heated, causing the pellet thermostat to open. This closes the bypass pipe and coolant flows through the thermostat valve to the radiator. (Volvo)

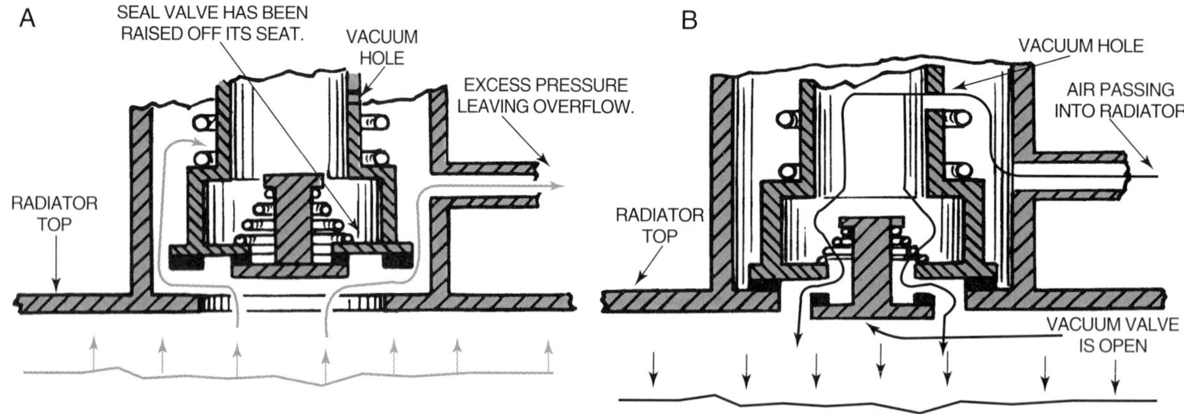

Figure 11-28. A—Coolant will expand until the pressure cap limit is reached. At that point, the cap will open and excess pressure will be released out of the overflow. B—When the coolant has cooled and contracted, this will form a vacuum in the radiator. The vacuum valve is drawn open and air passes into the tank through the overflow pipe. In a closed cooling system, coolant from the reservoir would flow back into the tank.

Coolant reservoirs are used to provide a storage place for coolant expelled from the radiator as the engine heats up. Coolant reservoirs are sometimes called ***expansion tanks***, or ***overflow tanks***. The pressure cap used with the coolant recovery system seals the radiator filler neck lower flange as well as the top flange of the neck, as illustrated in **Figure 11-28**. **Figure 11-29** shows the typical non-closed system cap. Note that the upper flange is sealed only by a thin metal seal. This will prevent leakage if the system overheats, but does not form a positive airtight seal.

The coolant recovery system cap illustrated in **Figure 11-30** provides an additional seal ring for the filler neck upper flange. This cap provides a seal at both the upper and lower neck flanges. Note the nonloaded (no spring pressure) center vent valve. The overflow pipe is located between the two sealed areas and is connected to the coolant reservoir via a

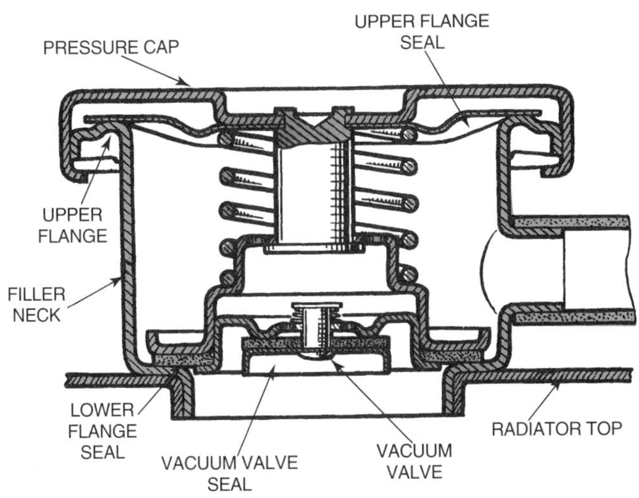

Figure 11-29. This is a pressure cap used on many open cooling systems. Upper flange seal does not use a soft seal insert common on the lower flange seal and vacuum valve seal. (Chrysler)

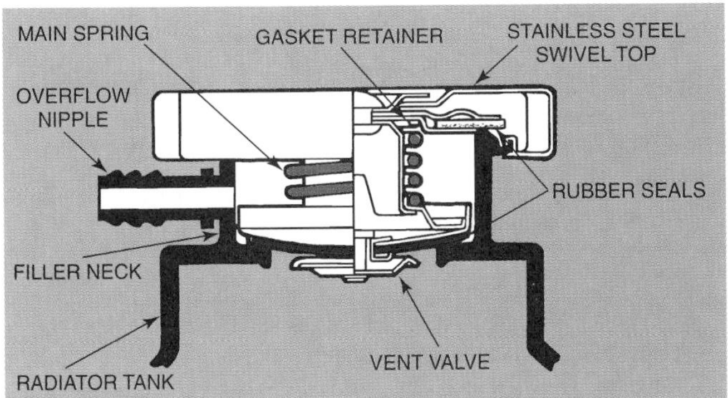

Figure 11-30. Closed cooling system pressure cap. Note that this cap uses an additional soft seal to provide an airtight seal at the top flange. (Chrysler)

rubber hose. The hose (or a rigid pipe) passes through the cover of the coolant reservoir and extends to the bottom of the reservoir. On some vehicles, the hose is connected externally to the base of the reservoir. The reservoir usually has a cold and hot full mark, **Figure 11-31**. The radiator, unlike the older open system, is completely filled with coolant.

Coolant Recovery System Operation

When the coolant heats, it expands and flows up through the nonloaded center vent valve. This displaced coolant flows through the overflow pipe and on into the coolant reservoir. Any air that is in the system rises to the top of the coolant in the coolant reservoir and escapes out the coolant reservoir vent hose. At this point, the entire cooling system is completely filled with coolant and all air has been displaced.

As the coolant temperature nears the boiling point, the vent will be forced closed to create pressure on the system. At extremely high temperatures, the system may exceed the maximum cap pressure. When this happens, system pressure forces the bottom cap seal valve to raise off its seat and

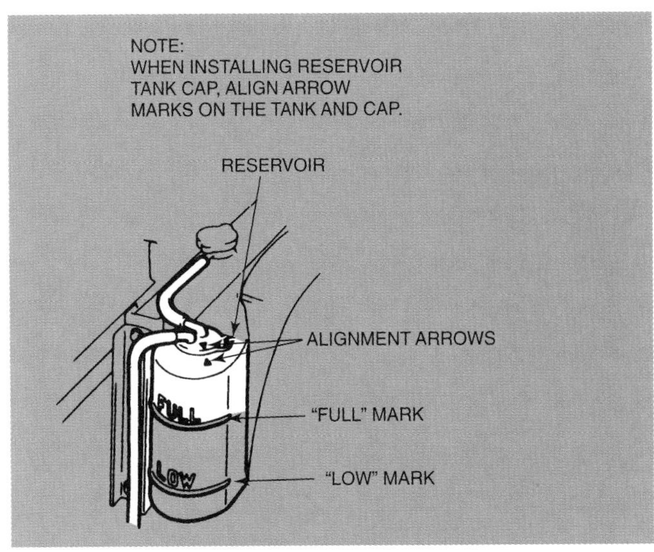

Figure 11-31. A closed cooling system reservoir tank. Excess coolant moves to the reservoir through a rubber hose. When system cools, coolant is drawn back into radiator from the reservoir.

dissipate the pressure in the form of coolant flow into the reservoir.

When the engine cools, the coolant contracts, forming a mild vacuum. This draws the cap center vent valve open and coolant, instead of air, is drawn into the system from the coolant reservoir. The recovery system allows the cooling system to operate more efficiently by eliminating air, which also reduces corrosion and overheating. The coolant level can be visually inspected without removing the pressure cap. To replace slight coolant losses, do not remove the pressure cap, merely bring the coolant in the reservoir up to the indicated level.

Temperature Indicators

The temperature of the coolant can be monitored by observing the ***temperature gauge*** or ***warning light*** on the dashboard. Temperature gauges use an electric sender and a gauge, as in **Figure 11-32**. Warning lights are operated by an on-off type of temperature switch, **Figure 11-33**. Many modern temperature gauge senders make use of electronic circuitry and also send an engine temperature signal to the on-board computer.

Using Antifreeze

To protect against engine and radiator freeze up in the winter and overheating in the summer, as well as to protect the system from rust and corrosion, manufacturers install ***antifreeze***. Antifreeze is a chemical which contains ***ethylene glycol***, with some ***corrosion inhibitors*** to reduce rust, and a small amount of ***water soluble oils*** to lubricate seals and moving parts.

Pure ethylene glycol freezes at about 9°F (-12°C), and water freezes at 32°F (0°C). However, when ethylene glycol and water are mixed, the freezing point of the mixture is lower than either liquid by itself. A two-third ethylene glycol and

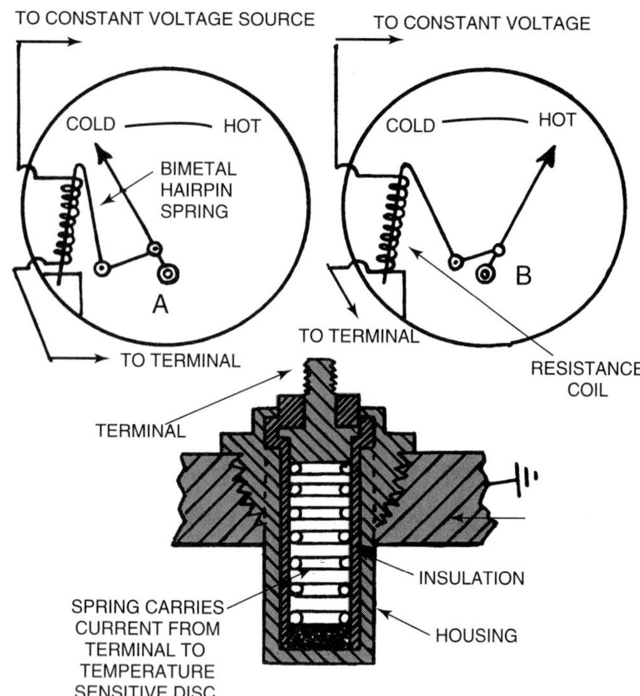

Figure 11-32. An electric temperature gauge. Current from the battery flows through a wire coil that surrounds a bimetal (two different metals fastened together) hairpin. It then passes on to the terminal of the sender unit, traveling from the terminal through the spring to the temperature sensitive disc.

As coolant heats the disc, it will carry more electricity to ground. The resistance coil on the hairpin will start to heat up. This will transfer heat to the hairpin.

The expansion of two dissimilar metals in the hairpin causes it to bend as it heats. This bending actuates the gauge needle and causes it to move to hot.

As the coolant cools, the disc will slow down flow of electricity and the hairpin will cool, bending back to its original shape.

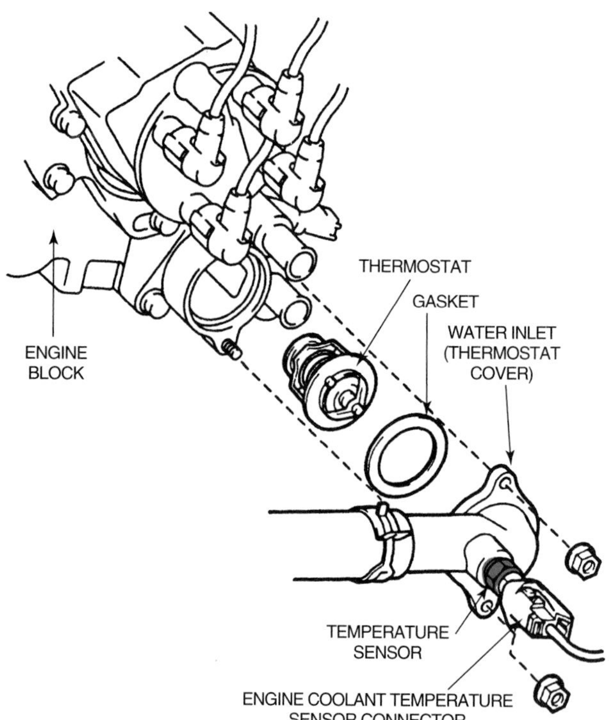

Figure 11-33. An engine coolant temperature sensor which is mounted in the thermostat cover. The sensor tip extends through the cover where it is submerged in the coolant stream. (Toyota)

one-third water mixture will not freeze until the temperature reaches approximately -67°F (-55°C). A 50-50 mixture will freeze at about -35°F (-37°C).

A mixture of antifreeze and water has a higher boiling point than plain water. Water boils at 212°F (100°C). A 50-50 mixture will not boil until its temperature is about 11°F (6°C) higher than the boiling point of water. Therefore an antifreeze/water solution will not boil until the temperature reaches about 223°F (106°C). If a pressure cap is used, the boiling point of the coolant mixture can be raised to almost 275°F (135°C).

Most manufacturers recommend a 50-50 mix of ethylene glycol antifreeze and water. A mixture of up to 70% ethylene glycol antifreeze and 30% water can be used in extreme hot or cold climates. Most manufacturers recommend a one or two year drain, flush, and refill cycle. Some manufacturers produce coolant that can last up to five years (100,000 miles [160,000 km]). This long-life coolant is orange or red for ease of identification. It is important to note that mixing long-life coolant with conventional antifreeze will reduce its effective service life.

Air-Cooled Engines

Many small engines are ***air-cooled***. Air cooling was used on many automotive engines in the past. Vehicles with air-cooled engines were manufactured in great numbers well into the 1970s. A few modern vehicles continue to use air-cooled engines.

Air cooling is efficient and dependable. Generally an air-cooled engine warms up more quickly and operates at a somewhat higher temperature than the liquid-cooled engine. The disadvantage of air cooling is that there is no effective way to heat the passenger compartment during cold weather.

Basically, an air cooling system consists of ***cooling fins*** on the engine head and block, a fan to force air, a shroud to direct the air over the fins, and in some cases a thermostatically adjusted air control. Air cooling is most efficient when applied to engines that are constructed of aluminum. Aluminum is a better conductor of heat than cast iron. In fact, aluminum will conduct heat about three times better than cast iron. However, air cooling has been successfully used on cast iron engines.

It is essential that the engine block and head have a number of rather deep cooling fins. The fins are cast into the engine parts. The theory behind the use of fins is simple. Heat generated inside a metal container must pass through the walls of the container and be carried away by the airflow.

Assume that you have a square cylinder block of aluminum 6″ (15.24 cm) wide and 6″ (15.24 cm) high. Inside the block is your engine combustion chamber. As heat is generated by the burning fuel mixture, it will be conducted through the walls of the block and carried away by the air, **Figure 11-34**. If the engine produces enough heat in the combustion chamber, it is obvious that the block will not be able to dissipate the heat into the air fast enough and will eventually melt.

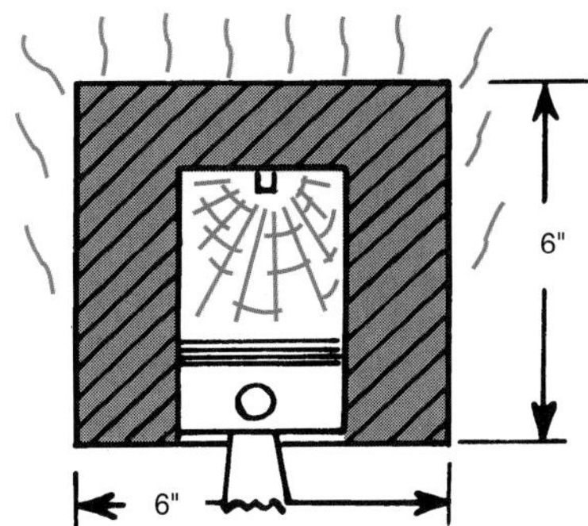

Figure 11-34. Heat is conducted through the walls of the block and is carried away by surrounding air. All engines will give off some of their heat in this manner.

The rate at which the aluminum block will be able to cool itself will, to a considerable degree, be determined by the amount of block surface that is exposed to the air. Measuring and adding the lengths of the top and two sides of the block in **Figure 11-35**, you will find that three sides at 6″ (15.24 cm) per side equals 18 lineal inches. In other words, the air moving around the block can draw heat only from these 18 lineal inches of exposed surface. By adding fins to the sides and top, **Figure 11-35**, you can see how much more surface the air can strike. By adding 2″ (5.08 cm) fins to the top and sides, the air can now circulate over 162 lineal inches of block surface. The block now has nine times the cooling surface.

So far we have considered lineal (straight line) inches as viewed on three sides. If we consider our cylinder block to be square, this would give us five sides upon which to calculate the number of square inches of contact area. You will ignore the bottom side, as this would be fastened to the crankcase and would not be finned, **Figure 11-36**. By adding fins as

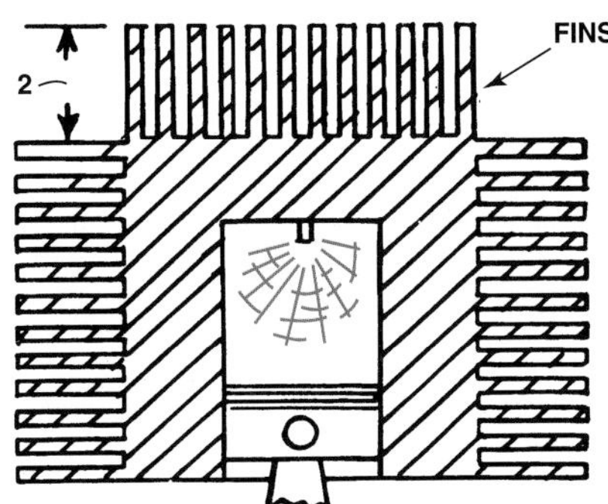

Figure 11-35. If you add fins to the block, it will increase the amount of surface area exposed to the air.

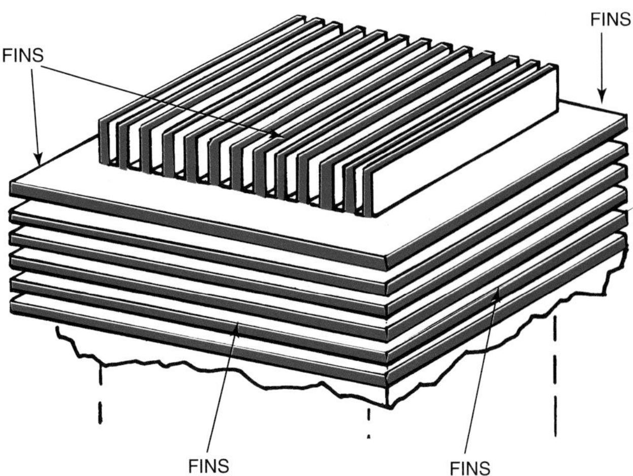

Figure 11-36. A block with finned surfaces on five sides would be able to give off a great deal of heat. This is the principle for the air-cooled engine.

shown in **Figure 11-36**, you would now have a total of about 1800 in^2. (11 700 cm^2) of contact area. This would be compared to five sides of the unfinned block at 180 in^2. (1170 cm^2). This represents a sizable increase in cooling efficiency.

Moving The Air

In addition to adding fins, it is essential that you provide a fan or blower to produce a moving stream of air. The air, if not moved, will soon heat up and be unsuitable for further cooling of the finned surfaces. Air movement may be accomplished in several ways. One way is to have a separate fan or blower. Another method is to use the flywheel as a fan by casting fins on it. Engines designed to move rapidly through the air, such as motorcycles and aircraft, often use no fan, depending entirely upon forward momentum to provide a cooling stream of air.

Any vehicle with an air-cooled engine should not be operated for prolonged periods while standing still. Automotive engines use a separate fan, while most small engines use the flywheel fan, **Figures 11-37** and **11-38**. The separate fan in **Figure 11-37** does a better job, but it is not always possible to line up the fan with the fins. **Figure 11-38** demonstrates that even though the fan is throwing air, it is not directed onto the fins. From **Figure 11-38**, you can see that the flywheel fan throws air in a fashion similar to the way a coolant pump impeller throws coolant.

Both fans can lose efficiency by recirculating the air near the outer edges of the blades or by losing some of it. These problems are overcome by the use of an air shroud. The shroud prevents recirculation and also guides the air into the fins. **Figures 11-39** and **11-40** illustrate the use of shrouds. Some shrouds are designed to form a plenum chamber (an enclosed area in which the air pressure is higher than the surrounding atmospheric pressure).

A Thermostat is Sometimes Used

An air-cooled engine can run too cool. To prevent this, a thermostatically controlled shutter, or valve, is sometimes

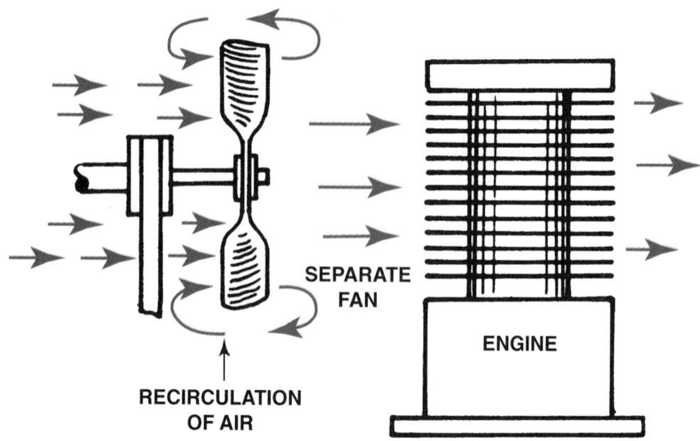

Figure 11-37. An air-cooled engine uses a fan much like liquid-cooled engines to circulate air.

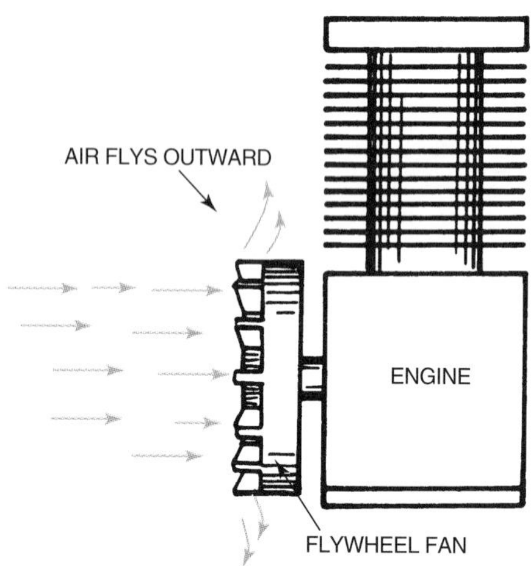

Figure 11-38. Like fans used on liquid-cooled engines, this flywheel fan is very inefficient without a shroud.

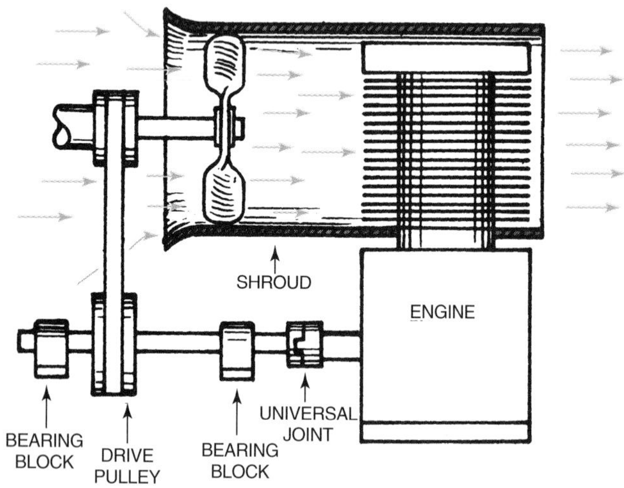

Figure 11-39. A separate fan with a sheet metal shroud. Notice that the shroud prevents recirculation and also guides air to the fins.

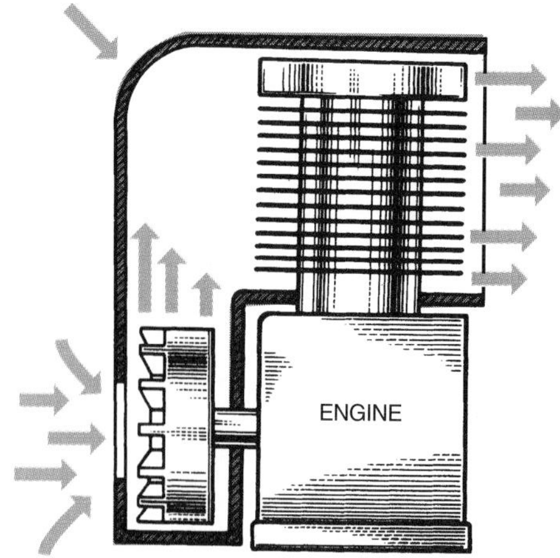

Figure 11-40. Flywheel fan with a sheet metal shroud. The shroud collects air thrown off by fan and guides it to fins. This fan acts as a centrifugal blower.

used to limit airflow. The air-cooled engine uses a bellows thermostat located in a spot that will allow it to accurately measure engine temperature.

The bellows is filled with liquid under a vacuum. If the bellows leaks, or in any other way becomes inoperative, the shutter will then return to the full open position. This eliminates the possibility of the shutter remaining in the closed position which could result in engine failure. **Figure 11-41** demonstrates the use of such a bellows.

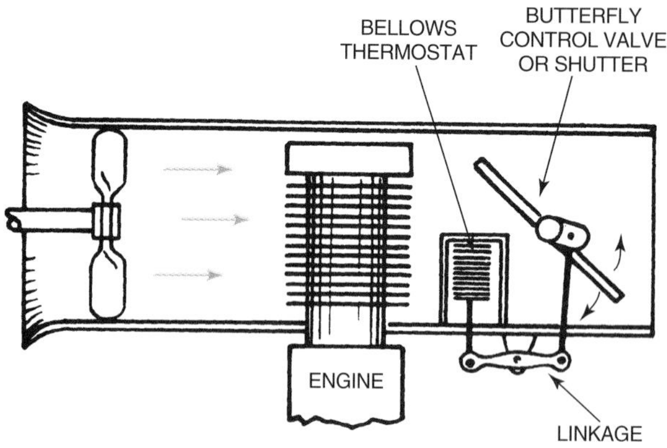

Figure 11-41. A temperature control system for an air-cooled engine. When the engine is cold, the bellows thermostat contracts, closing the shutter. As the engine heats up, air passing over the engine will warm the thermostat, causing it to expand and open the shutter.

Summary

All internal combustion gasoline engines must be cooled, either with a liquid or air. The liquid-cooled engine commonly uses a 50-50 mix of water and ethylene glycol. In the liquid-cooled engine, the coolant must surround all the areas that are subjected to heat from the burning fuel charge.

The coolant is contained in an area referred to as the coolant passages or water jackets. When the engine is first started, the thermostat is closed and the coolant in the jacket cannot circulate. Pressure from the coolant pump then forces coolant through the bypass. This allows coolant to circulate from the pump into the block, then back into the pump.

As soon as the coolant in the engine reaches the predetermined operating temperature, the thermostat will open and allow the coolant to be pumped from the engine into the top or side radiator tank, depending on the type of radiator used. From this tank, the coolant flows through the radiator core to the other radiator tank. The coolant is then pumped back into the engine. The coolant repeats this circulation pattern over and over.

When the engine is started, the block heats quickly. The heat is transferred by conduction to the fins. Air, moved by a fan, passes through a shroud surrounding the fins. Moving air removes the heat from the cooling fins. The amount of air can be controlled by a thermostatic shutter.

Know These Terms

Liquid cooled engines
Cooling passages
Water jackets
Coolant
Freeze plugs
Coolant pump
Centrifugal pump
Impeller
Water pumps
V-belt
Ribbed belt
Serpentine belt
Distribution tubes
Coolant nozzles
Heat exchanger
Radiator
Radiator core
Transmission oil cooler
Fans
Fan clutch
Hoses
Heater hoses
Heater shut-off valve
Hose clamps
Thermostat
Manual bleeder valve
Bypass
Pressurized cooling system
Pressure cap
Closed cooling system
Coolant recovery system
Coolant reservoir
Expansion tank
Overflow tank
Temperature gauge
Warning light
Antifreeze
Ethylene glycol
Corrosion inhibitors
Water soluble oils
Air-cooled
Cooling fins

Review Questions—Chapter 11

Do not write in this book. Write your answers on a separate sheet of paper.

1. If an engine runs too cool, all of the following will happen, EXCEPT:
   (A) oil contamination.
   (B) deposit formation.
   (C) increased wear.
   (D) good fuel mileage.
2. What are the three jobs that the cooling system must perform?
3. There are two popular methods of cooling the engine. Name them.
4. To prevent dead spots and to produce a better than normal circulation in some vital areas, cooling systems use _____ .
   (A) distribution tubes
   (B) coolant nozzles
   (C) thermostats
   (D) Both A & B.
5. Modern radiators are constructed from _____ .
   (A) copper
   (B) aluminum
   (C) plastic
   (D) All of the above.
6. Fan clutches are designed to reduce _____ .
   (A) horsepower
   (B) noise
   (C) torque
   (D) All of the above.
7. The radiator heat exchanger removes heat from the _____ .
8. The thermostat should be completely closed when the engine is _____ .
   (A) cold
   (B) partially warm
   (C) fully warmed up
   (D) overheating
9. A 180° thermostat would just begin to open at _____.
   (A) 160°
   (B) 180°
   (C) 195°
   (D) 200°
10. The bypass valve prevents what problem?
    (A) Overcooling.
    (B) Overpressure.
    (C) Localized overheating.
    (D) Rust and corrosion.
11. Pressurizing the cooling system raises _____ .
    (A) engine load
    (B) coolant boiling point
    (C) thermostat opening temperature
    (D) pressure cap sealing
12. All of the following statements about coolant recovery systems are true, EXCEPT:
    (A) it allows the cooling system to be operated at full capacity.
    (B) the center vent valve of the pressure cap is non-loaded.
    (C) coolant is expelled into the coolant reservoir when the engine cools off.
    (D) the pressure cap has an extra seal at the top of the filler neck.

13. The freezing point a 50-50 mixture of water and ethylene glycol antifreeze is _____.
    (A) 9°F (-12°C)
    (B) 32°F (0°C)
    (C) –67°F (-55°C)
    (D) –35°F (-37°C)
14. A cooling system with a 50-50 mixture of antifreeze and water and a 12 pound pressure cap will begin to boil at _____ .
15. What is the major disadvantage of air-cooled engines for automobiles?
    (A) They warm up too quickly.
    (B) They are lighter.
    (C) There is no effective way to heat the passenger compartment.
    (D) They operate at higher temperatures than the liquid-cooled engine.

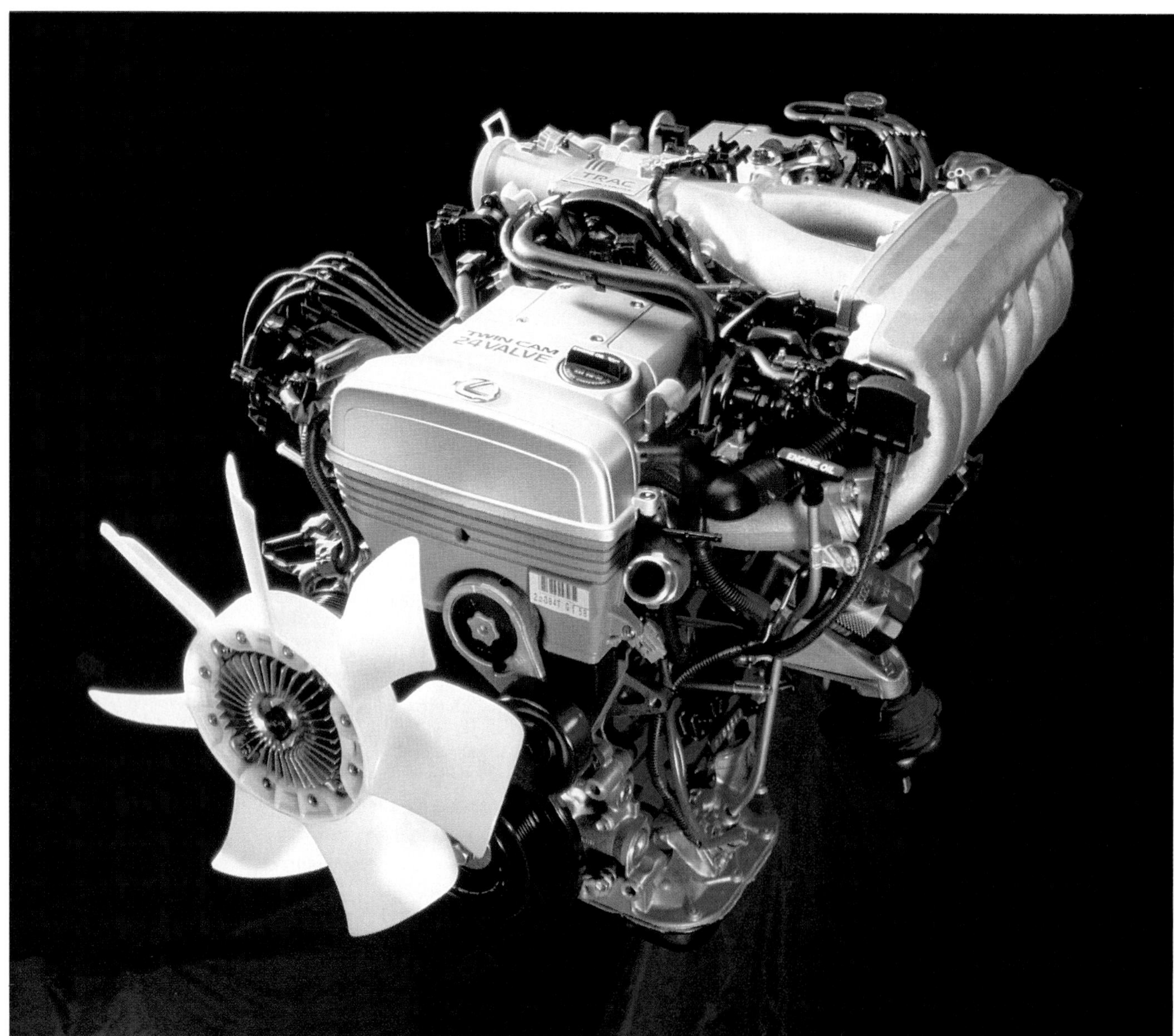

In order to operate at the high temperatures needed for good economy, modern engines need cooling and lubrication systems that are in good condition. Note the use of plastic and aluminum parts to further reduce heat. (Lexus)

12

Lubrication Systems

After studying this chapter, you will be able to:

- Explain the need for a lubrication system.
- Identify the major parts of a lubrication system.
- Explain engine oil classifications.
- Summarize the operation of an engine lubrication system.
- Identify the components of an engine lubrication system.
- List the different types of oil pumps.
- List the different types of oil filters.

This chapter will cover the causes of engine friction and reasons for a lubrication system. Also discussed will be the design and construction of the modern lubrication system and its major components. This chapter will also identify the jobs of engine oil, oil classifications, and how oil is cleaned and cooled.

The Need for a Lubrication System

If you were to start and run an engine without lubrication, it would last only a short time. The bearing surfaces in the engine would heat up and scuff. Eventually the engine would seize, or lock up. This occurs when lack of lubrication causes the moving parts in an engine to expand and overheat, tearing bits of metal from each other in the process, **Figure 12-1**. Eventually the parts will seize and refuse to move. The force that causes wear and heat and robs the engine of a portion of its potential power is called ***friction***.

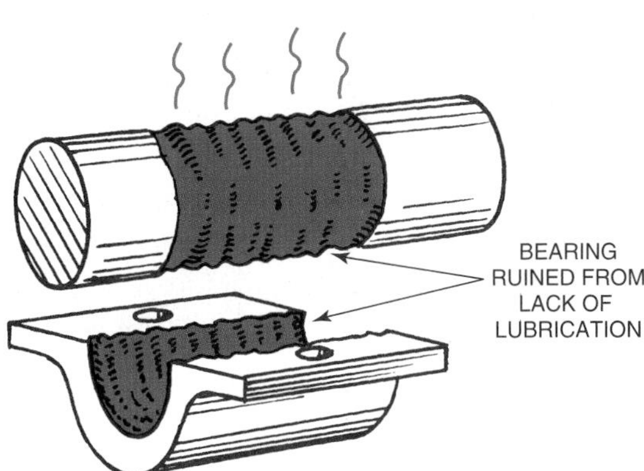

Figure 12-1. The size of this illustration is exaggerated to show the ruined journal and bearing. Oil would have prevented this.

Friction

Friction is defined as the force that opposes movement between any two objects in contact with each other. The amount of friction depends on the type of material, surface finish, amount of pressure holding the two objects together, and relative amount of movement between the objects. You can see in **Figure 12-2**, that it takes a force to pull block A over block B.

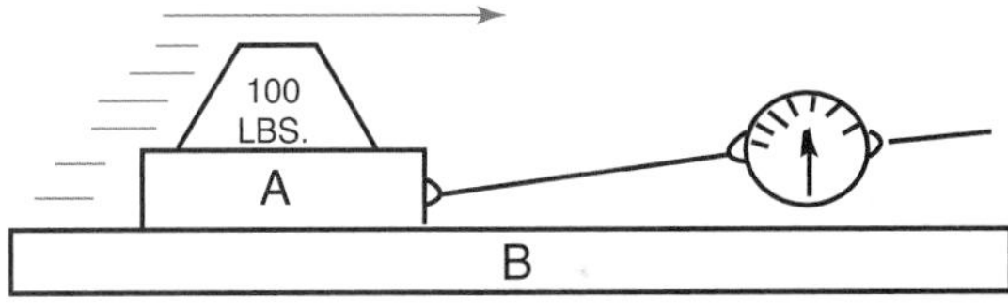

Figure 12-2. It takes force to pull block A along the surface of block B. The force which counters the pulling force is called friction.

If you were to examine the contact surfaces of the two blocks under a microscope, you would find a series of sharp points, grooves, and other imperfections, regardless of how smooth they were polished. When the two objects are placed together, these jagged surfaces touch each other, **Figure 12-3**.

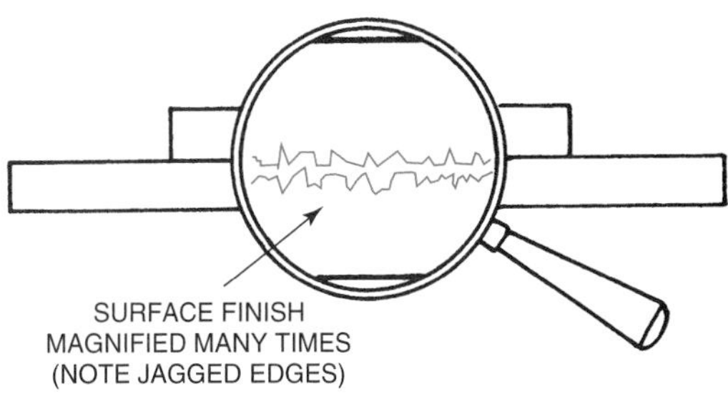

Figure 12-3. Microscopic view (exaggerated) of block contact surfaces. All surfaces look like this, no matter how smooth they might feel or appear.

Friction Generates Wear and Heat

As soon as block A starts to move, the edges collide and impede the movement. If there is considerable pressure holding the two blocks together (bearing pressures in a normal engine can reach 1000 lbs. or 450 kg) and the movement is continued, two things happen. Each block will literally tear pieces from both itself and the other block. These tiny particles roll along under block A until they work out the back edge. Removal of these pieces is continuous and will result in wear.

Another byproduct of friction is heat. Put your hands together and rub them briskly back and forth. A fire can be started by rubbing two sticks together. These are examples of heat produced by friction. In an engine, the rubbing speeds and pressures are tremendous. Without lubrication, friction will build up enough heat to melt the bearings in a short time. It is impossible to completely eliminate friction between moving parts. If this were possible, perpetual motion machines could be easily constructed. The most that can be done is to reduce friction to an absolute minimum.

How Friction is Reduced

There are two kinds of friction, ***dry friction*** and ***wet friction***. So far we have dealt with dry friction. Dry friction is when both rubbing parts are clean and free of other materials. Wet friction is when rubbing parts have other material or materials placed between them. Have you ever tried playing a fast game of basketball on a well-waxed gym floor? This is a good example of dry friction being changed into wet friction by placing something between your shoes and the floor—in this case, wax. It is obvious that the engine must operate under conditions of wet friction.

Bearing Construction and Friction

Bearing construction also plays a part in determining the amount of friction involved. There are two principal types of bearings used in engines. One is called a ***friction bearing***. The friction bearing is literally two smooth surfaces sliding against each other. The only motion involved is a sliding effect. The babbitt insert and the crankshaft journal are examples of this type. The friction bearing is inexpensive and works well enough to be used in all automotive engines.

The other type is called an ***antifriction bearing***. This bearing utilizes balls or rollers between the moving parts. One or more antifriction bearings is used in almost every part of a vehicle. Both types of bearings must be used wet. Ball and roller bearings are used in many small, high speed engines in motorcycles and outboard motors. **Figure 12-4** illustrates both bearing types.

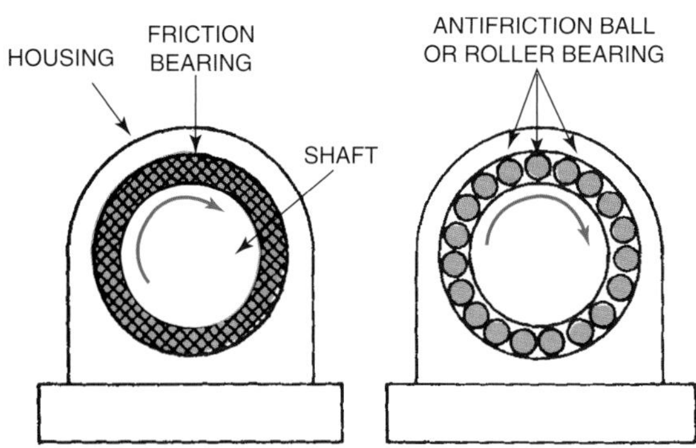

Figure 12-4. Both friction and antifriction bearings need lubrication for long, trouble-free service.

Engine Oil

It has been found that high quality lubricating oil successfully reduces friction to an acceptable level. The oils used to lubricate engines are called ***engine oil***, or sometimes ***motor oil***. Since vehicle engines are not really motors, the term motor oil may seem incorrect. The term comes from the early days of the automobile, when the entire vehicle was called a motor.

Engine Oil Has Several Jobs

The most obvious job of oil is ***lubricating*** the engine's internal moving parts. But in addition to lubrication, the oil has other jobs in the engine. Engine oil must also assist in ***cooling***. The steady flow of oil through the bearings carries away a great deal of the heat that is generated. Oil strikes the piston head and carries heat with it. As the oil is returned to the sump before being recirculated, it is much cooler by the time it reaches the bearings again.

Oil must also ***seal***. Although the piston rings are closely fitted to the cylinder, they would not form much of a seal without oil as a final sealing agent. Engine oil also assists in sealing around valve stems. A good oil will also ***clean***. The detergents in the oil assist in removing dirt and other impurities. They hold these impurities in suspension until they can be filtered out.

Refining Engine Oil

Engine oils are ***distillated*** from crude oil or petroleum that is removed from the earth through wells. The crude oil is heated and is then pumped into a tall ***fractionating tower***, which contains trays at different heights. As each component of crude oil has a different ***relative volatility***, they will evaporate at different temperatures. As the crude oil is heated, the lighter or more volatile (easily evaporated) components turn to a vapor or "flash" and begin to travel to the top of the tower.

As the crude oil vapor travels up the tower, the heavier or less volatile parts will condense and settle onto the trays. The lighter fractions are able to travel to the very top of the tower. A series of pipes are connected to the fractionating tower at the different tray heights. If the tower is operating properly and operating conditions are correct, a different fractionated liquid or gas will travel out of each pipe. The fractionated liquid or gas will enter a cooling tank and then will run down into storage tanks.

After the lubricating oil leaves the distillation tower, it still requires further refining and processing. Impurities are removed in a process that takes several steps. Various additives may also be added to produce oil for specific requirements. **Figure 12-5** shows the basic steps in engine oil refining.

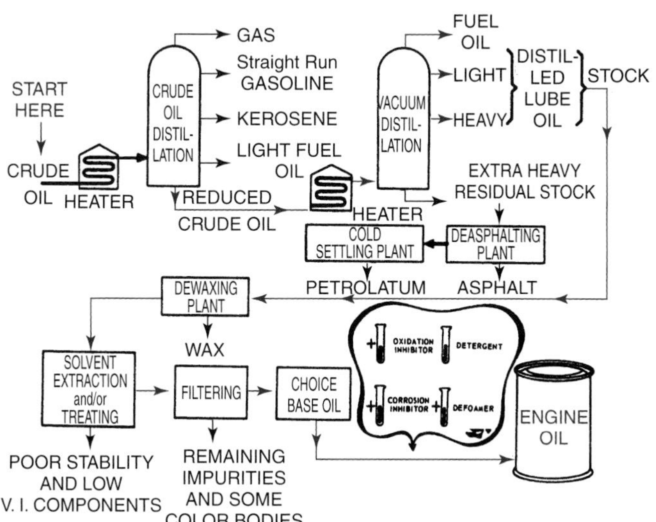

Figure 12-5. The refining of lubricating oil is a somewhat complicated task. Study the various steps involved. (John Deere)

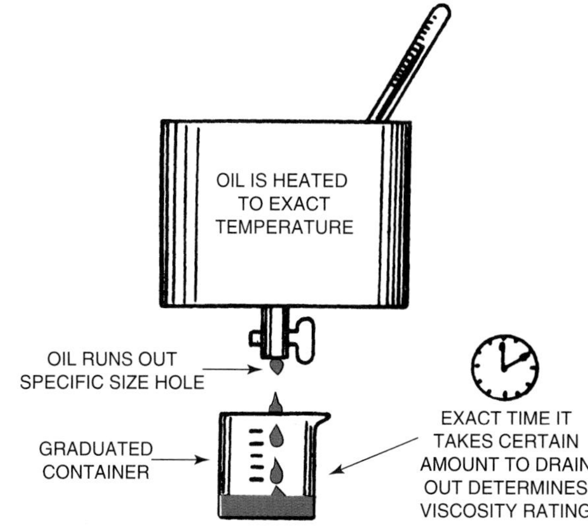

Figure 12-6. The viscometer is used to measure the viscosity of engine oil. While appearing to be quite simple, it is a very important indicator of an oil's quality.

Oil Additives

Engine oils have many extra compounds called ***additives*** added to them during the refining process. Various additives such as oxidation inhibitors, detergents, defoamers, corrosion inhibitors, pour point depressants, and viscosity improvers (VI), may be added to produce oil for specific requirements. These additives are designed to improve the oil by keeping small foreign particles in suspension, fighting corrosion, reducing oxidization, minimizing carbon, water, lacquer, and gum formation, and reducing engine wear.

Modern oils have a high detergent quality that helps to keep the engine clean by preventing the formation of acids, carbon, lacquer, and gum deposits. These are all caused by the fact that some unburned gases and raw fuel can get past the piston rings and into the crankcase. Therefore, it is vital that the detergent additives in the oil hold these impurities in suspension until they can be trapped by the oil filter.

Oil Viscosity

Oil ***viscosity*** refers to the thickness or fluidity of the oil. In other words, it is a measure of an oil's ability to resist flowing. The Society of Automotive Engineers (SAE) has set low (0°F or -18°C) and high (210°F or 99°C) temperature requirements for oil. Oils that meet special SAE low temperature requirements have the letter W following the viscosity rating, such as 10W. Oils that meet the high temperature requirements have no letter, but are designated by a number such as 30.

Viscosity is determined by taking a sample of oil and heating it to a certain point. The heated oil is then allowed to flow out of a very precise hole. The length of time it takes to flow determines its viscosity rating. The faster it flows, the lower the rating. This testing device is referred to as a viscosimeter, **Figure 12-6**.

Viscosity Index

Oils thin out when heated and thicken when cooled. As a good engine oil must be reasonably thin for cold starts and thick at high temperatures, the ***viscosity index*** (VI) is important. The VI is a measure of an oil's ability to resist changes in viscosity with changing temperatures. When oils are heated, they tend to lose their body (resistance to oil film failure). If a given oil is thin at cold temperatures and does not thin out much more when heated, it is said to have a high VI. If the heating produces a great change in viscosity, the oil has a low VI.

Multi-Viscosity Oils

A ***multi-viscosity oil*** contains special hydrocarbon molecules commonly known as ***polymers***. Polymers make the oil thinner when cold and thicker when hot. Polymers allow a multi-viscosity oil to meet the low temperature requirements of a light oil and the high temperature requirements of a heavy oil. Since they meet the viscosity and performance requirements of two or more SAE grades, they are marketed in such weights as 5W-30, 10W-30, 10W-40, and 20W-50. They are often referred to as multi-grade, multi-weight, all-season, and all-weather oil.

Choosing Oil Viscosity

Always follow the manufacturer's recommendation for viscosity rating. The most commonly used oils are 30, 10W-30, 10W-40, and 20W-50. 5W, 10W, or 5W-30 are recommended in areas where the temperature is frequently lower than 0°F (-18°C). These recommendations are based on engine design and the lowest anticipated temperatures.

Engine Oil Service Classification

Presently, there are nine oil ***service grades*** for automotive gasoline engines. The American Petroleum Institute (API) has classified these oils as SA, SB, SC, SD, SE, SF, SG, SH, and SJ for use in gasoline engines. The letter S indicates the oil is for use in ***spark ignition*** (gasoline) engines. **Figure 12-7** illustrates the typical API marking found on most oils intended for gasoline engines.

Figure 12-7. This symbol or "logo" is used on containers of engine oil to assist the buyer in selecting the proper oil. The oil's quality and service recommendation is given by API (American Petroleum Institute) numbers. In this example, Service SJ (oil for gasoline engines), the viscosity is indicated by the SAE (Society of Automotive Engineers) number 5W-30. The term "Energy Conserving II" indicates that this oil has additional friction reducers for improved fuel economy.

The API has classified oils for use in diesel engines as CA, CB, CC, CD, and CE. The C indicates that the oil is for use in ***compression ignition*** (diesel) engines. The second letter indicates the amount of anti-wear additives, oxidation stabilizers, and detergents. The eight current gasoline engine oil grades are discussed below.

SA and SB oils were the first engine oils to be classified by the API. They are recommended for light loads, moderate speeds, and clean conditions. They contain no additives and are known as ***non-detergent oils***. These oils are still available for use in very old vehicle and stationary engines and are sometimes used in gearboxes and in two-stroke cycle engines. They should never be used in late model vehicle engines.

The first ***detergent oil*** was grade SC, followed over the years by SD, SE, SF, SG, SH, and SJ. The latest oil designation is SJ. This oil will also replace all former oil classifications. Most manufacturers do not bother to put the earlier grade designations on their oil, as they have been superseded by grade SJ. CF, CG, and CH are the latest oil grades for diesel engines and supersede CA, CB, CC, CD, and CE oils. Many oils have more than one classification.

Detergent oil is used to prevent the buildup of dirt, soot, and sludge in the engine. They hold impurities in suspension in the oil until they can be trapped by the oil filter. However, detergent oils may not be suitable replacements for non-detergent SA or SB, since the detergent qualities may loosen internal engine deposits, clogging the filter or engine oil passages.

Refer to the chart in **Figure 12-8**. Note the relationship between the previously used API classifications and the new ones in use today. Engine oils may also be classified according to manufacturer's specifications or to military specifications (standards developed by the US military for oils to be used by them).

	NEW API ENGINE SERVICE CLASSIFICATIONS	PREVIOUS API ENGINE SERVICE CLASSIFICATIONS	RELATED DESIGNATIONS MILITARY AND INDUSTRY
GASOLINE ENGINES	SA	ML	Straight Mineral Oil
	SB	MM	Inhibited Oil
	SC	MS (1964)	1964 MS Warranty Approved. M2C101-A
	SD	MS (1968)	1968 MS Warranty Approved, MC2C101-B, 6041 (Prior to July, 1970)
	SE	None	1972 Warranty Approved, M2C101-C, 6041-M (July 1970)
	SF	None	1980 Warranty Approved
	SG	None	1987 Warranty Approved
	SH	None	1994 Warranty Approved
	SJ	None	1998 Warranty Approved
DIESEL ENGINES	CA	DG	MIL-L-2104A
	CB	DM	Supp. 1
	CC	DM	MIL-L-2104B
	CD	DS	MIL-L-45199B, Series 3
	CE		
	CF		
	CG		
	CH		

Figure 12-8. Chart shows the relationship between the old and the new API Service classifications for gasoline and diesel engines.

Oil Change Intervals

The answer to this frequently asked question depends on so many variables that a general recommendation is very difficult to make. Many manufacturers state that oil does not have to be changed before 7500 miles (12 000 km) or every 6 months under normal conditions. However, the definition of normal conditions varies greatly. Any grade of engine oil will retain proper lubricating properties longer if:

- The engine is operated for reasonably long periods each time it is started.
- The engine is in good mechanical condition.
- The engine runs the majority of the time at its normal operating temperature (over 180°F (93.3°C) or higher).
- The engine has a good oil filter.
- The positive crankcase ventilation (PCV) system is functioning properly.
- The engine is operated at moderate to fairly high speeds.
- The engine is provided with an efficient air cleaner.
- The fuel system is properly adjusted.
- The engine is operated under clean conditions.

The conditions given are ideal. If an engine operates under them, the oil can be used for the maximum recommended time. Engines that are operated in dusty or dirty conditions, that are in poor mechanical condition, or that do not have a filtration system, should have the oil changed at least every 3000 miles (1800 km) or 3 months. Turbocharged and supercharged engines should have their oil changed every 3000 miles (1800 km) under all conditions.

Engine Lubrication Systems

Even the best oil is useless unless it can reach the engine's moving parts. The job of the lubrication system is to get the oil to the moving parts of the engine. How it does this is covered below.

How Oil Works

When oil is injected into a bearing, **Figure 12-9**, it is distributed between the bearing surface and the shaft, creating a film that separates the parts. The parts then revolve without actually touching each other. When the engine is shut off and oil is no longer supplied, the shaft will settle down on the bearing, squeezing the oil film apart. Some residual oil (oil that clings to the parts) will be retained.

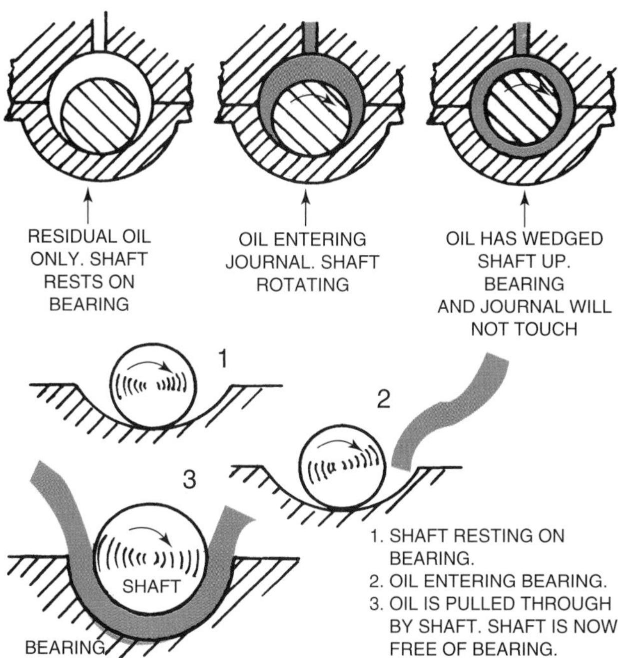

RESIDUAL OIL ONLY. SHAFT RESTS ON BEARING
OIL ENTERING JOURNAL. SHAFT ROTATING
OIL HAS WEDGED SHAFT UP. BEARING AND JOURNAL WILL NOT TOUCH
1
2
3
SHAFT
BEARING
1. SHAFT RESTING ON BEARING.
2. OIL ENTERING BEARING.
3. OIL IS PULLED THROUGH BY SHAFT. SHAFT IS NOW FREE OF BEARING.

Figure 12-9. Oil enters a bearing by pressure and rotation. The pressurized oil enters the bearing and journal area as the shaft is rotating. The rotation of the shaft pulls the oil around. As more oil is pulled around, the shaft begins to ride on a thin film of oil, away from the bearing and journal.

The top of the film that contacts the shaft journal and the bottom of the film that contacts the bearing, resist movement to some extent. The sliding effect of the oil is achieved near the center of the film. See **Figure 12-10**. This slipping action allows the journal to spin easily and still not contact the bearing surface.

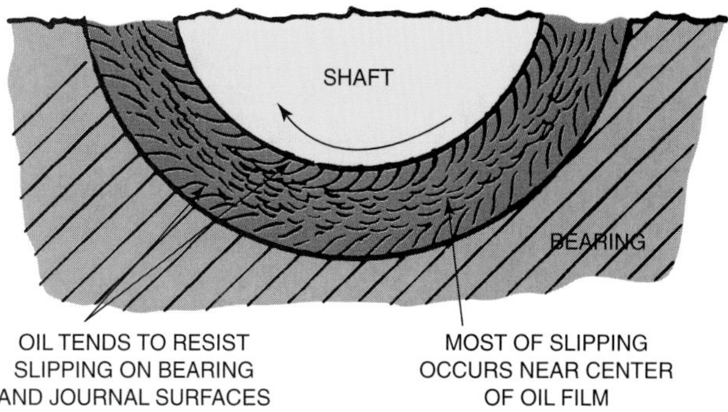

SHAFT
BEARING
OIL TENDS TO RESIST SLIPPING ON BEARING AND JOURNAL SURFACES
MOST OF SLIPPING OCCURS NEAR CENTER OF OIL FILM

Figure 12-10. The oil provides a thin film on which the shaft rides. This keeps the shaft separated from the bearing.

When the engine is started, the shaft will have only residual oil for the first few seconds of operation. This is why engines should be allowed to idle, in park or neutral, until the oil light goes out or the oil pressure gauge needle begins to rise. In cold conditions, additional time should be given to allow the oil to circulate. Other engine parts, such as the valve train and cylinders, are under the same load at all times and will not be unduly damaged if the engine is driven before they begin to receive oil.

Bearing Clearance

It is important that all bearings have the proper clearances. The ideal bearing clearance is loose enough for oil to enter, but tight enough to resist pounding. Oil clearances are very small, around .002″ (0.05 mm).

Three Types of Lubrication Systems

The engine bearings and all other moving parts are lubricated by one of three systems:

- Full pressure system.
- Splash system.
- Combination pressure-splash system.

The full pressure system will be discussed first.

Full Pressure System

The ***full pressure system*** draws oil from the ***oil pan*** by means of an ***oil pump***. The pump then forces oil through drilled and cast passages, called ***oil galleries***, to the crankshaft and camshaft journals. The crankshaft is drilled to permit oil to flow to the connecting rod journals. In some engines, the rods are drilled their full length to allow oil to pass up to the wrist pin bushings.

Bearing throw-off may be helped by spurt holes in the rod to lubricate the cylinder walls and camshaft lobes. Timing gears and chains, lifters, pushrods, and rocker arms are also oiled. In the pressure system, all bearings are oiled by either pumping oil into the bearings, or by squirting or dripping it on, **Figure 12-11**. The lubrication system also provides the pressure to fill the hydraulic lifters.

Splash System

The most basic ***splash system*** supplies oil to moving parts by attaching ***dippers*** to the bottom of the connecting rods. These dippers can either dip into shallow trays or into the sump itself. The spinning dippers splash oil over the inside of the engine. The dipper is usually drilled so that oil is forced into the connecting rod bearings. An oil pump may be used to keep the oil trays full at all times. The basic splash system is used primarily on small one-cylinder engines, **Figure 12-12**.

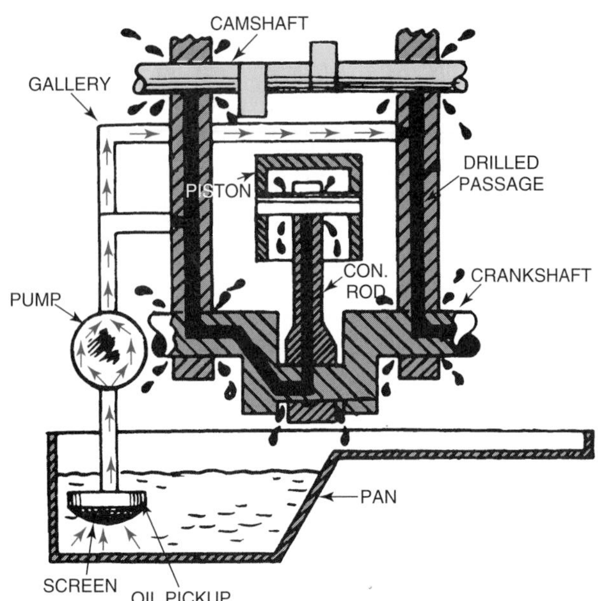

Figure 12-11. Notice how oil in the pressure system travels to all major bearings. When it passes through the bearings, it is thrown around, creating an oil spray and mist. The oil eventually returns to the oil pan sump.

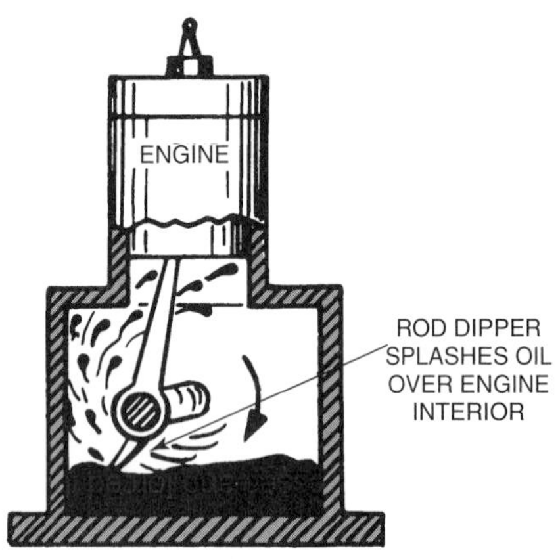

Figure 12-12. This simple splash oiling system, while used mostly on small engines, is combined with the pressure system on some engines to provide additional oiling.

Combination Pressure-Splash System

This system uses an oil pump to supply oil to the camshaft and crankshaft bearings. The movement of the crankshaft splashes oil onto the cylinder walls and other nearby parts. Oil spurt holes in the connecting rods are used to spray oil onto the cylinder walls and piston pins.

Oil is also pumped to the cylinder head to oil the camshaft, rocker arms, and valves as applicable. Pressurized oil leaving the pushrods sprays oil around the valve cover, lubricating the rocker arms and valve stems. A nozzle is sometimes used to squirt oil onto the timing gears and chain. Some combination systems make use of a dipper to splash the oil, as in **Figure 12-13**.

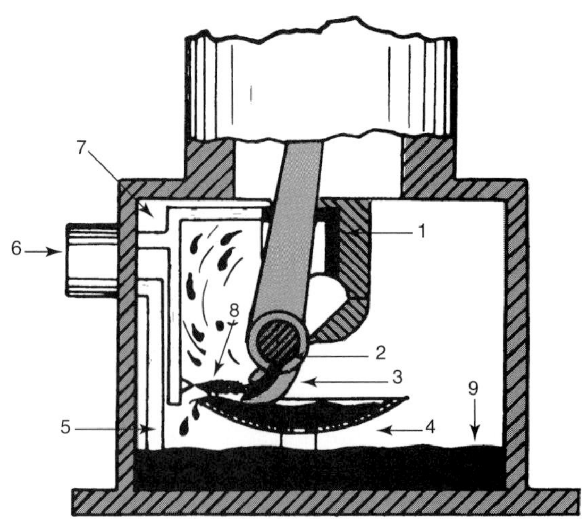

Figure 12-13. Combination splash and pressure oiling system. Oil pump 6 draws oil up pipe 5 from the oil sump 9. It then pumps the oil to main bearings 1 through the discharge pipe 7. Oil is also pumped down discharge nozzle 8 where it sprays out and strikes dipper 3. This feeds oil into rod bearing 2. Tray 4 also holds oil for the dipper to strike.

Components of the Pressure-Splash System

The combination ***pressure-splash system*** best meets the needs of the modern engine and is therefore in almost universal use. Oil is constantly drawn from the bottom of the engine, pressurized, and distributed to the engine parts. The oil then drips to the bottom of the oil pan sump and begins the circuit again. The rest of this chapter will deal primarily with the components which develop pressure.

Oil Pan Sump

Oil is poured into the engine and flows down into the oil pan. One end of the pan is lowered to form a reservoir called a ***sump***. Baffle plates and screens are sometimes used to prevent the oil from sloshing around and striking the crankshaft. The drain plug is placed in the side or bottom of the sump. See **Figure 12-14**

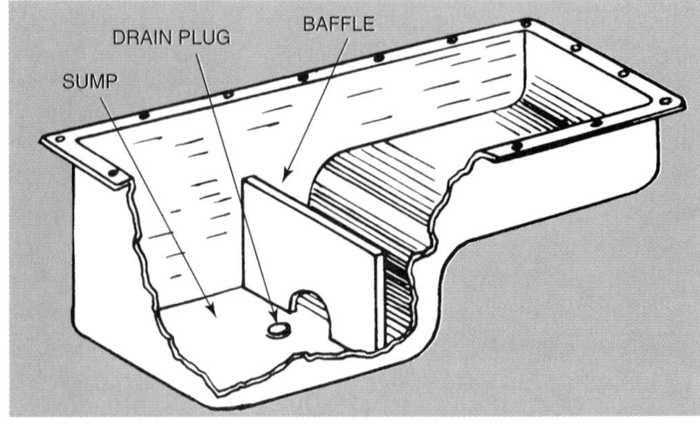

Figure 12-14. Oil pan with sump and baffle. There are many different designs for oil pans, sumps, and baffles.

Oil Pickup

Most engines have a rigid pipe leading down into the oil in the sump. The ***oil pickup*** does not touch the bottom to avoid picking up sludge. Some engines use a floating pickup. This pickup actually floats on top of the oil, drawing only from the top of the sump. As the oil level drops, the pickup will also drop to maintain a constant supply of oil.

All oil pickups use screens to exclude the larger particles that may be drawn into the system. In the event the screen becomes clogged, there may be a small valve that will be drawn open, allowing the pump to bypass the screen and obtain oil, **Figure 12-15**.

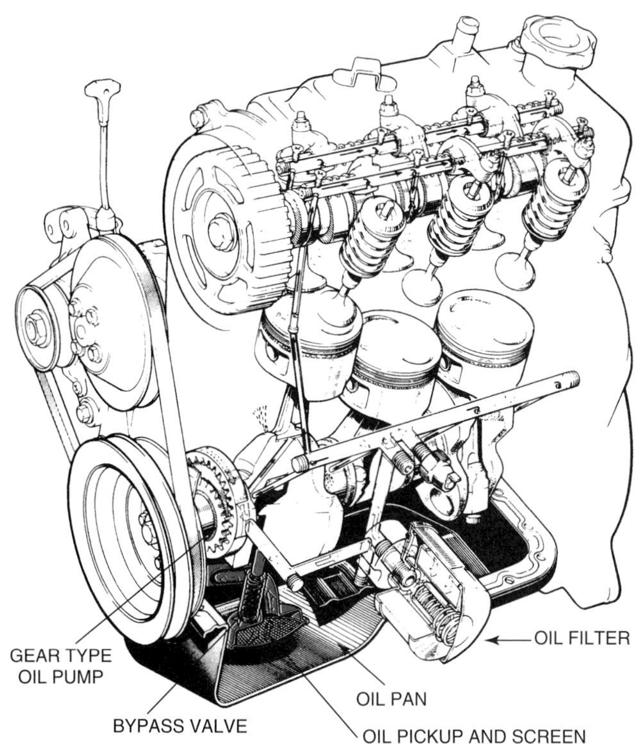

Figure 12-15. Oil pickup with filtering screen and a bypass valve. Note the gear type oil pump. (Chevrolet)

Oil Pumps

There are three principal types of oil pumps that have been used on engines. They are the gear, rotary, and vane pumps. The two most widely used are the gear and the rotary types.

Gear Pump

The ***gear pump*** uses driving and driven gears as the moving force to pump oil. These gears are placed in a compact housing. The driven gear revolves on a short shaft, while the driving gear is fastened to a long shaft. This shaft is turned by a spiral gear meshed with a similar gear on the camshaft. The teeth of the two gears must mesh with a minimum amount of clearance. The housing just clears the top, bottom, and sides of both gears.

An inlet pipe allows oil to be drawn in by the turning gear teeth. As the teeth move apart in the inlet side, they create a vacuum which draws in oil. The oil is carried around to the other side. As the teeth move together, they pressurize the oil, which is then discharged from the pump outlet. The pressurized oil then enters the rest of the lubrication system. The gear pump is efficient and produces high pressure and flow when needed. Gear pump operation is shown in **Figure 12-16**. An exploded view of a gear pump is shown in **Figure 12-17**.

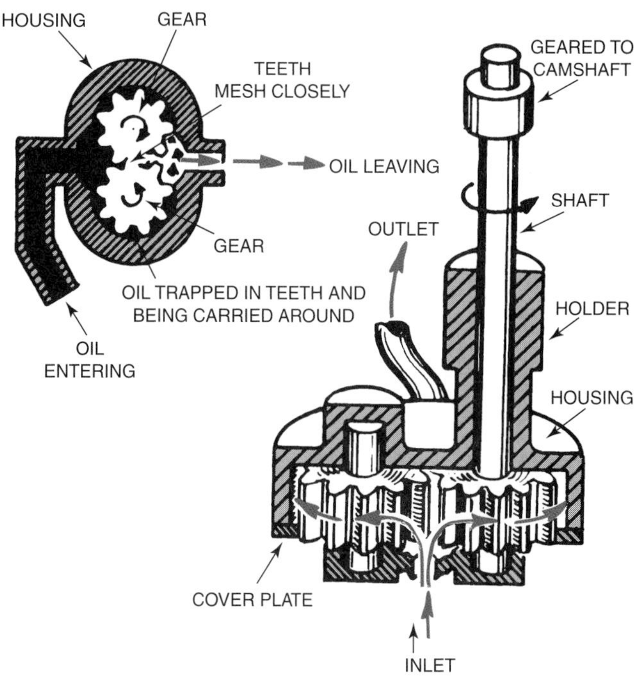

Figure 12-16. Gear oil pump. Notice how oil is caught in the gear teeth and carried around. As the pump spins rapidly, it will produce a steady stream of oil.

A different adaptation of the gear pump is shown in **Figure 12-18**. The driving gear (center) and driven gear (outer) teeth carry oil around to the outlet area. As the teeth mesh, the oil is compressed and forced through the outlet. An exploded view of a similar pump is shown in **Figure 12-19**.

Rotary Pump

A ***rotary pump*** uses a housing, an inner rotor, and an outer rotor. The outer rotor is cut away in the form of a star with rounded points. The inner rotor is shaped in the form of a cross with rounded points that fit in the outer rotor star. The inner rotor is much smaller. It is driven by a shaft turned by the camshaft.

The inner rotor is mounted off center so that when it turns, the rounded points walk around in the star-shaped outer rotor. This causes the outer rotor to revolve also. As the inner rotor walks the outer rotor around, the outer rotor openings pick up oil at the inlet pipe and pull it around until it lines up with the outlet. It is then forced out as the inner and outer rotor points close together. The oil cannot make a circuit around, as the inner rotor fits snugly against the outer rotor at

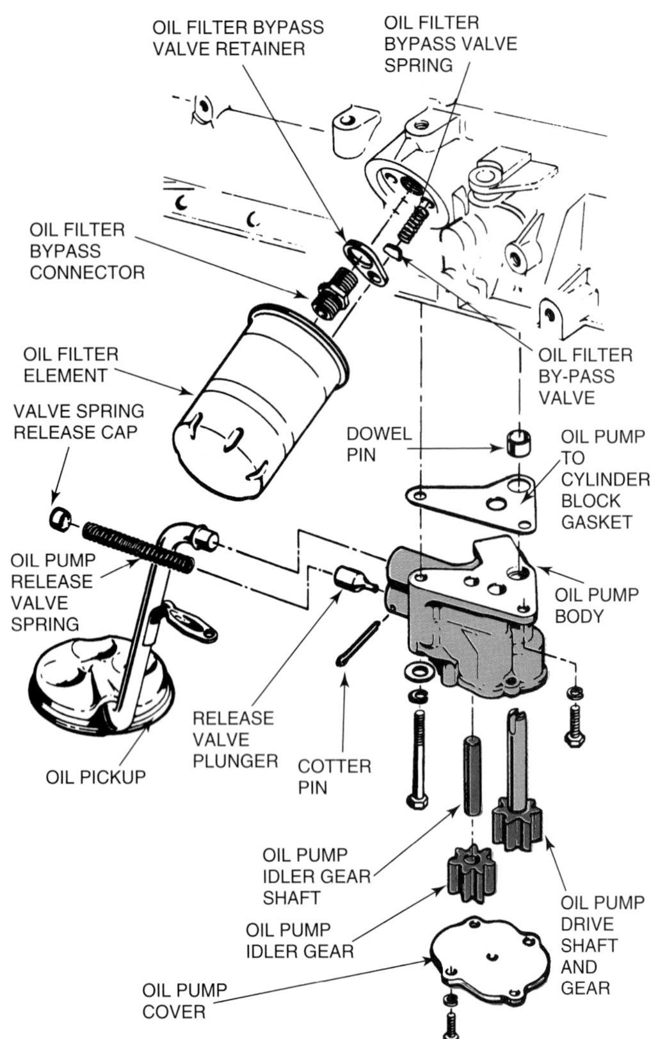

Figure 12-17. An exploded view of a gear type oil pump. Note the slot in the pump drive shaft which connects to a shaft from the distributor. (Jeep)

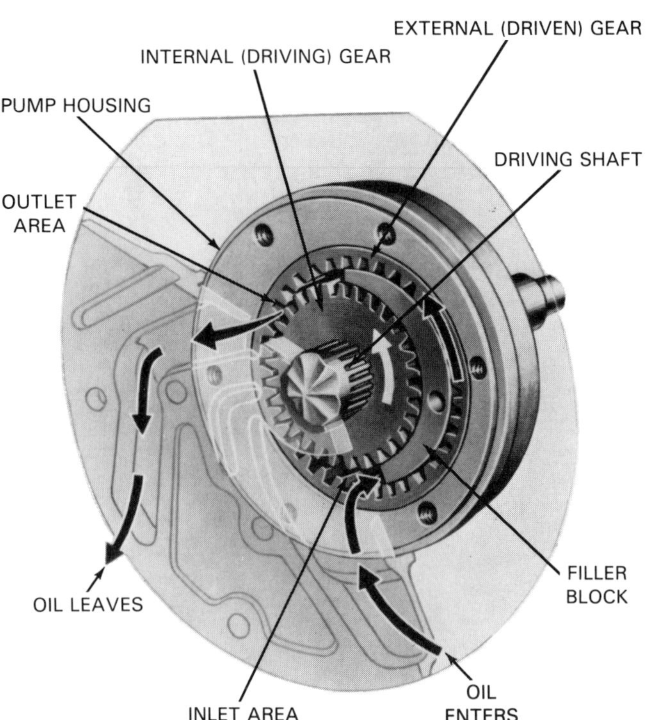

Figure 12-18. Gear oil pump using one gear with internal teeth and another with external teeth. As gears rotate, oil is carried past the filler block in the teeth of each gear. This oil is deposited in the outlet area. Since the gears mesh, the oil has no place to escape and pressure is built up, forcing oil through the outlet. (Deere & Co.)

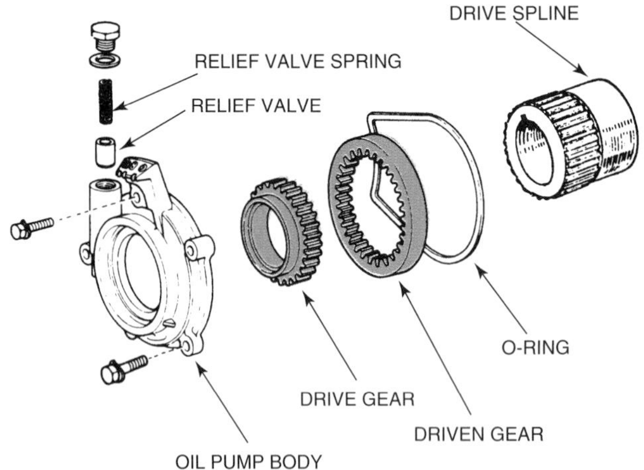

Figure 12-19. One form of gear oil pump. The principle of operation is same as that in Figure 12-18. (Toyota)

one spot. **Figure 12-20** shows rotor pump operation. **Figure 12-21** is an exploded view of a rotor pump.

Vane Pump

The ***vane pump*** uses a round housing with a rotor placed off center. This rotor contains two or more vanes, or rotor blades, that fit into slots in the rotor. The vanes are held out against the round housing by springs between the vanes. Some designs feed oil under pressure to the inner vane ends. This pressure, along with centrifugal force, helps hold the vanes in firm contact with the housing wall.

As the shaft turns, the vanes are forced in and out to keep contact with the housing. As they move outward, they create a vacuum which draws oil in the inlet port. As the vane is forced in on the other side, it pressurizes the oil and pushes it out of the pump. As a steady charge is picked up, carried around, and squeezed out, the pump produces a good supply of oil. **Figure 12-22** shows the operation and an exploded view of a vane pump.

Pressure Relief Valve

An oil pump in good condition will produce pressures far beyond those necessary for lubrication. The pumps are designed to carry a large flow of oil, but when the usable pressure limits are reached, a ***pressure relief valve*** opens and allows oil to return to the sump. See **Figure 12-23**.

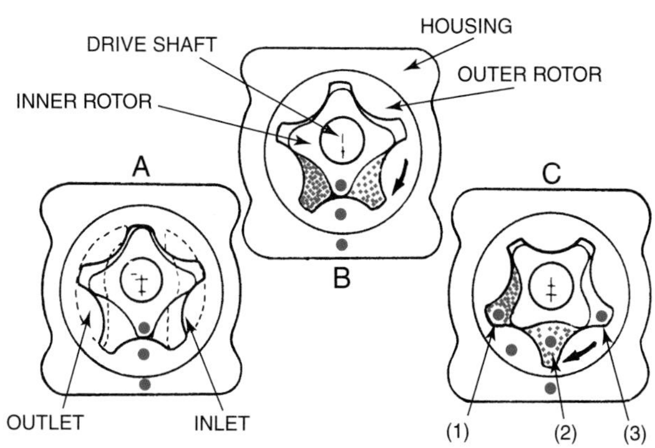

Figure 12-20. Rotor oil pump action. A—Pump inlet and outlet areas. B—Drive shaft turns the inner rotor, causing it to walk around in the outer rotor. The outer rotor will also turn in response. C—The rotor pulls around a charge of oil and forces it through the outlet. (1). While this is happening, another charge is moving around (2) and another is being picked up at the inlet (3). (Jaguar)

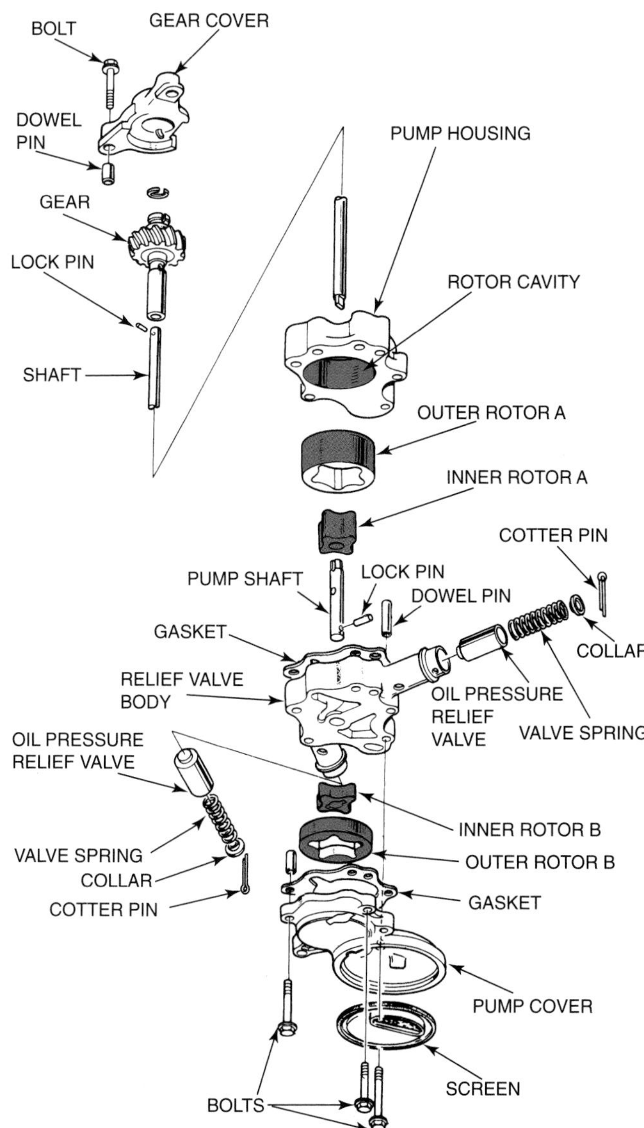

Figure 12-21. An exploded view of a dual rotor type engine oil pump, and related parts. (Honda Motor Co.)

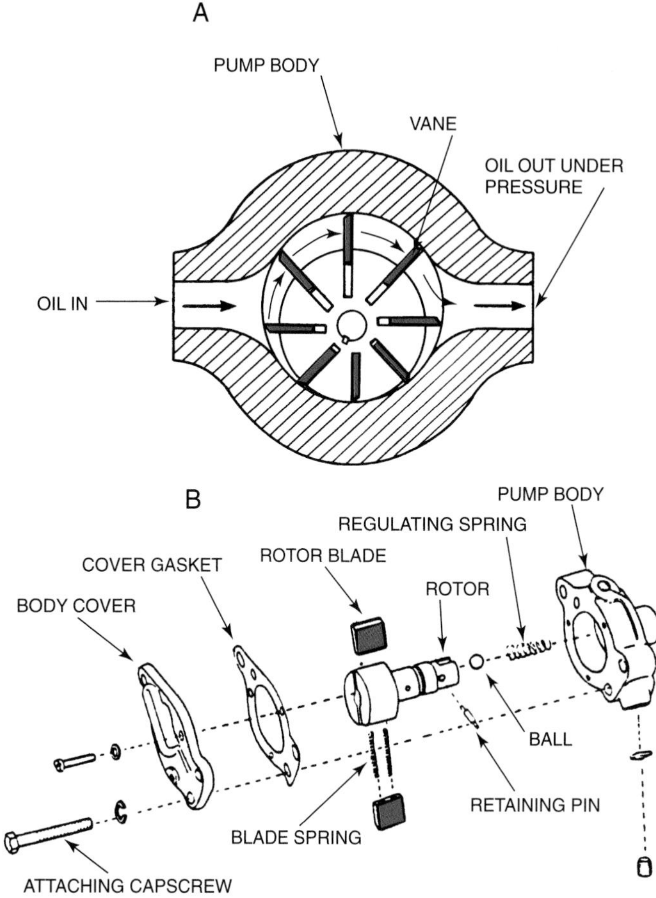

Figure 12-22. A vane type oil pump. A—Vane pump cross-sectional view. Follow the oil flow. B—Exploded view of a vane style oil pump. Note the blade (vane) springs which will keep the vane in contact with the pump chamber wall as they spin. (Deere & Co. and Allis Chalmers)

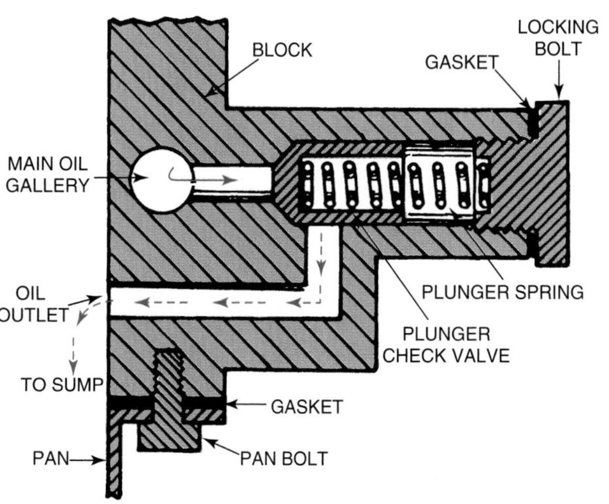

Figure 12-23. Oil in the main oil gallery is under pressure from the oil pump. When pressure exceeds the setting for the oil pressure relief valve, it will push the plunger check valve back against the plunger spring pressure. When the valve is forced back far enough to expose the oil outlet channel, oil will pass from the main oil gallery, through the oil outlet, and back into the pan sump.

The relief valve will allow the pressure reaching the bearings to remain at a predetermined level. By varying spring tension, the pressure can be raised or lowered. The relief valve can be located right at the pump, **Figure 12-24**, or at any spot between the pump and the bearings.

Oil Galleries and Lines

Oil that is allowed to pass the pressure relief valve is carried to the bearing through the oil galleries. The main oil gallery runs lengthwise in the block and is connected to most of the other galleries. Most engines have two or three plugs in the back of the block at the end of the main gallery. During an engine overhaul, the plugs should be removed and the gallery cleaned.

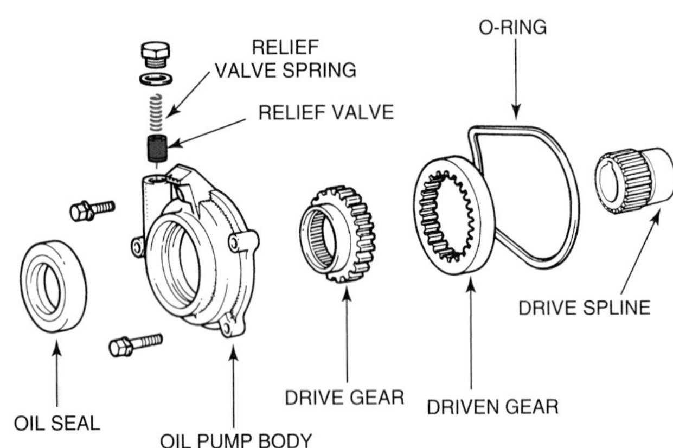

Figure 12-24. An exploded view of an oil pressure relief valve which is housed in the oil pump body. The relief valves can be located in many different areas, depending upon lubrication system design and construction. (Toyota)

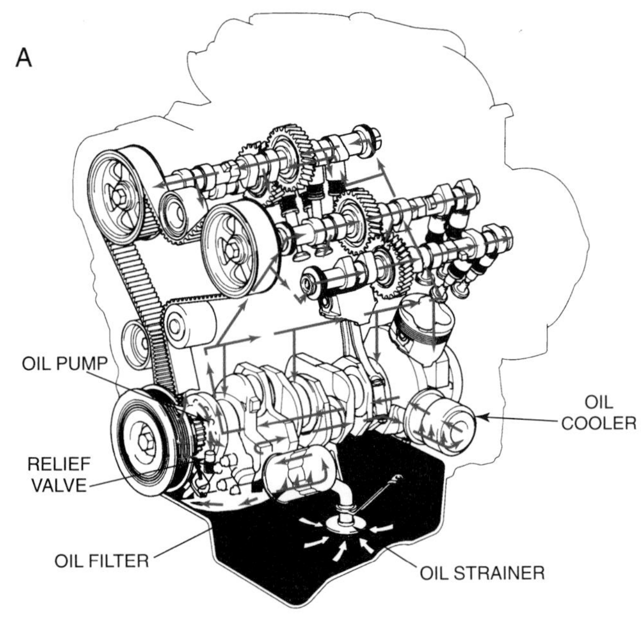

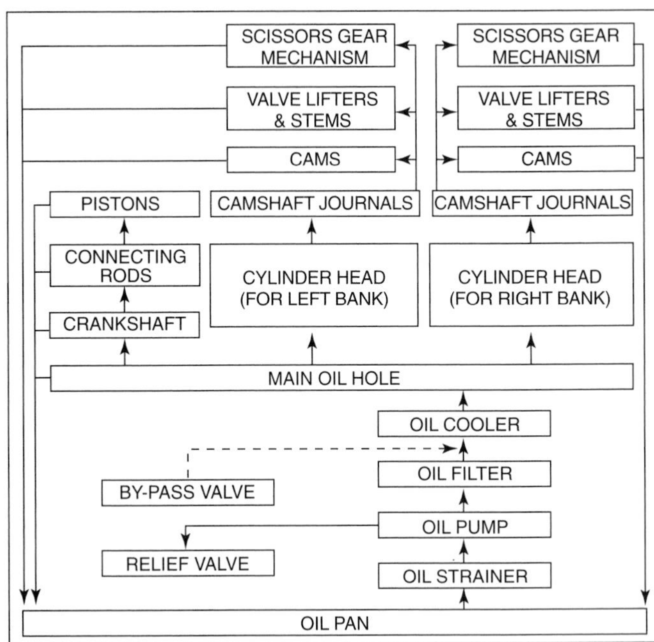

Figure 12-25. Oil flow through a V-6 engine. A—Cutaway showing the oil flow through the engine itself. B—An oil flow system schematic. Note that this engine uses an oil cooler. (Toyota)

Passages from the main gallery are drilled to the main bearings. The crankshaft is drilled to carry oil from the main journals to the rod journals. Some engines use separate lines to carry part of the oil flow. See **Figures 12-25** and **12-26**. Some rods are drilled through their full length to carry oil to the wrist pins. Others have tiny holes that line up with the crank journal once every revolution. When the two holes align, oil is squirted through the spit hole and onto the cylinder wall. Spit holes are also used to lubricate the camshaft lobes in some engines.

The timing chain is oiled either by spray, by oil discharged from the No. 1 camshaft bearing, or by oil discharged from an oil nozzle. The valve train is lubricated by oil spray and in the case of overhead valve engines, by oil pumped into the rocker arm shaft, through the rockers to the valves. Oil can also reach the rocker arms via hollow push rods. Some engines do not employ rocker arms, but operate the valves directly from the camshaft lobes, **Figure 12-26**. Camshafts receive oil under pressure to their journals. The lobes are lubricated by oil spray.

Oil Enters Bearing

The bearing receives oil from a hole drilled through the bearing support. The insert has a hole that aligns with the one in the bearing support. The oil passes through and engages the turning journal and is pulled around between the insert and journal. Some inserts have shallow grooves cut into them to assist in spreading the oil.

Oil is constantly forced through the bearings. Oil leaving a bearing is thrown violently outward. This helps to produce the fine oil mist inside the engine that is useful in lubricating hard-to-reach areas. All oil eventually flows back into the sump.

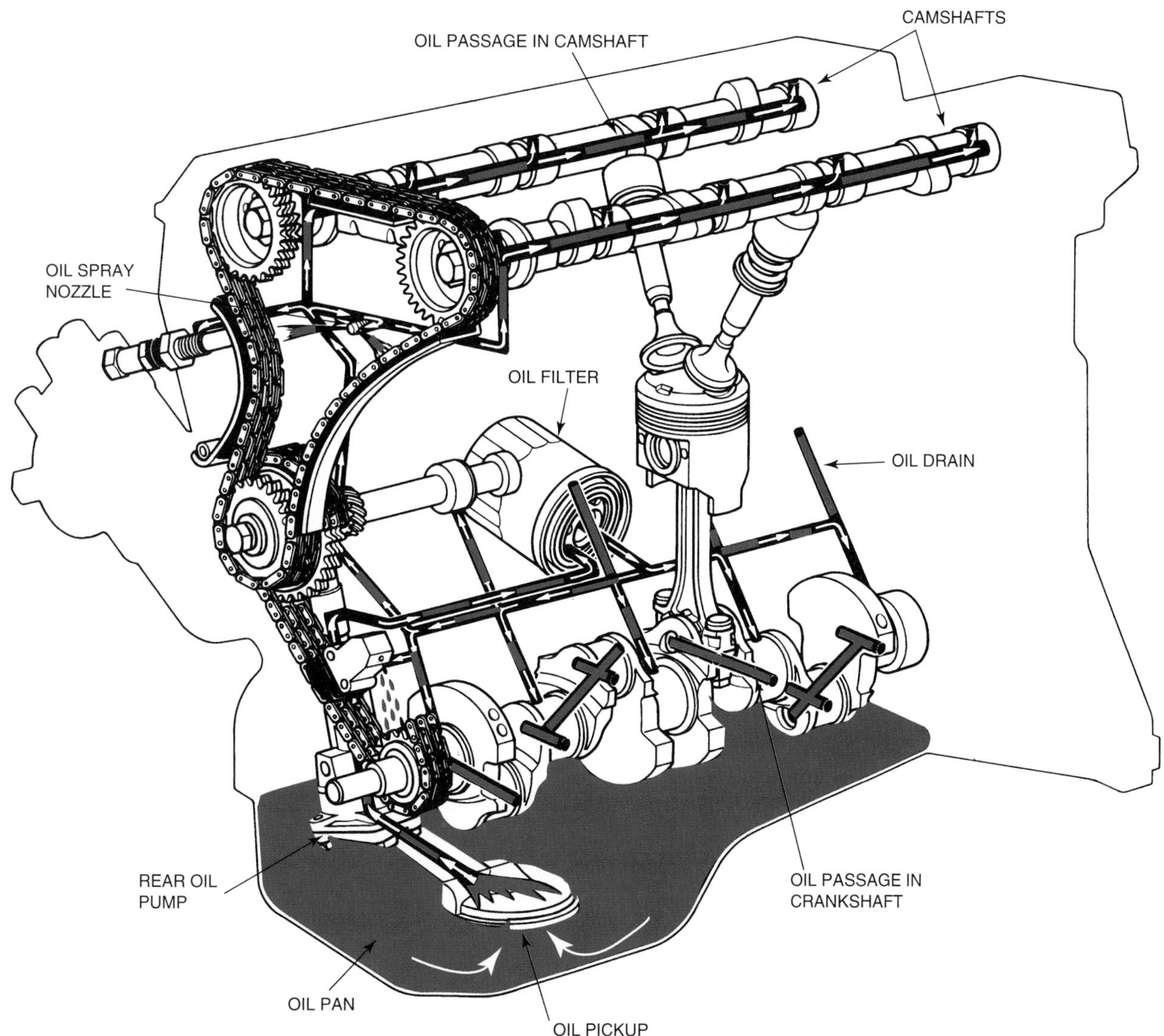

Figure 12-26. A full-pressure engine oiling system that uses a rotor oil pump. (Toyota)

Hydraulic Valve Lifters

Hydraulic valve lifters are used on many engines. One type of hydraulic lifter is shown in **Figure 12-27**. In this lifter, oil under pressure is forced into the tappet when the valve is closed. This extends the plunger in the tappet, so valve clearance is eliminated.

When the camshaft lobe moves around and starts to raise the tappet, oil is forced upward in the lower chamber of the tappet. This closes the ball check valve and oil cannot escape. Since oil does not compress, the valve is actually lifted on the column of oil and is supported by the oil while the valve is open.

When the cam lobe moves from under the tappet and the spring causes the valve to close, pressure in the lower chamber of the tappet is relieved. Oil loss from the lower chamber is replaced by oil pressure from the lubricating system.

Oil Filtration

After it has been in the engine a while, the oil becomes contaminated with dust, bits of carbon, metal dust, and sludge. Modern detergent oils keep these contaminants in suspension instead of allowing them to build up on the inside of the engine. If these impurities are allowed to remain in the oil, it will hasten engine wear. Engines are equipped with an oil filtering system to remove these harmful impurities.

Oil Filters

Many materials have been tried in an effort to find a satisfactory filtering medium. Cotton waste, plant fibers, metal, clay, and paper have all been used. Today, filters are grouped into two main classifications, ***depth filters*** and ***surface filters***. Depth filters are made of various fibers, while surface filters are made of treated paper. A few filters used in heavy trucks and construction equipment are made of closely spaced metal plates. **Figure 12-28** illustrates the filtering principle used in cotton, paper, and metal filters.

Some oil filters use a permanent metal container that holds a replaceable filtering element, or cartridge. Most filters are one-piece disposable units in which the entire filter

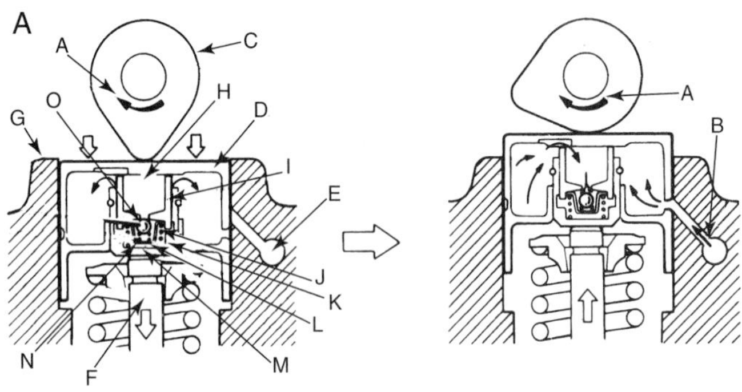

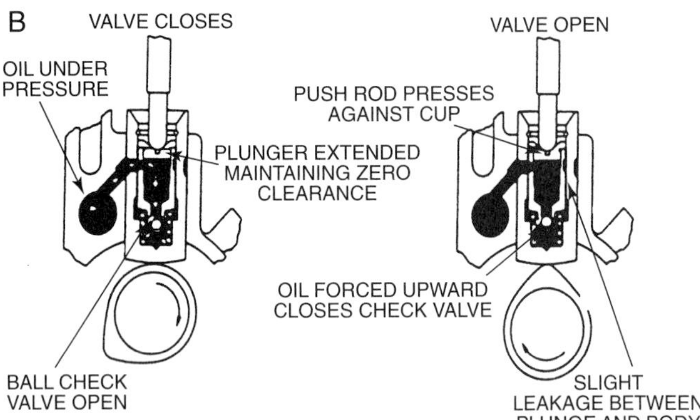

Figure 12-27. A—The action of this hydraulic valve lash adjuster allows it to receive oil during normal camshaft rotation. B—Oil flow. C—Camshaft. D—Hydraulic valve adjuster body. E—Oil gallery. F—Valve stem. G—Cylinder head. H—"A" chamber. I—Plunger. J—Plunger spring. K—Body. L—Check ball cage. M—"B" chamber. N—Check ball spring. O—Check ball. B—Hydraulic valve lifter (non-roller type), showing the tappet action. Note that the hydraulic valve lash adjuster (1), is in direct contact with the overhead camshaft. (General Motors)

assembly is discarded. The filter assembly is usually installed on the bottom or side of the engine. One filter design is installed inside the oil pan.

Filter Operation

Oil is pumped into the container and surrounds the filter. The oil passes through the filtering element to the center outlet pipe. It is then returned to the sump or to the bearings, depending on the type of system. **Figures 12-29** and **12-30** illustrate full-flow and depth filters.

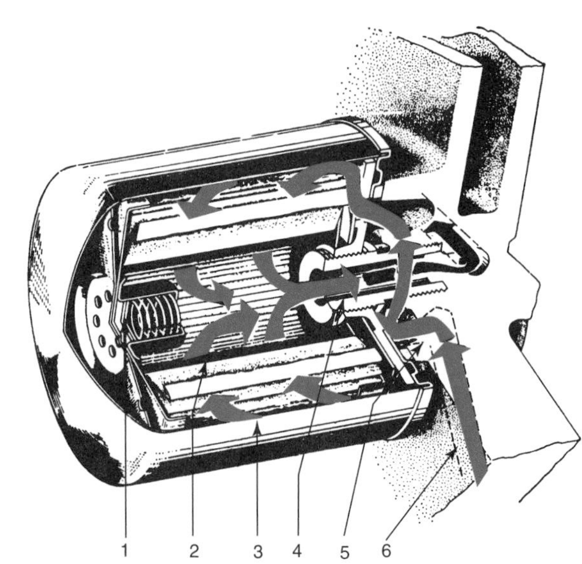

Figure 12-29. When a full-flow oil filter must be changed, the entire filter is replaced. 1–Relief valve. 2–Filter element (folded paper). 3–Body. 4–Connecting pipe. 5–Gasket. 6–Engine block. This pleated paper filter provides surface filtration in that all particles are trapped on the outer surface. (Volvo)

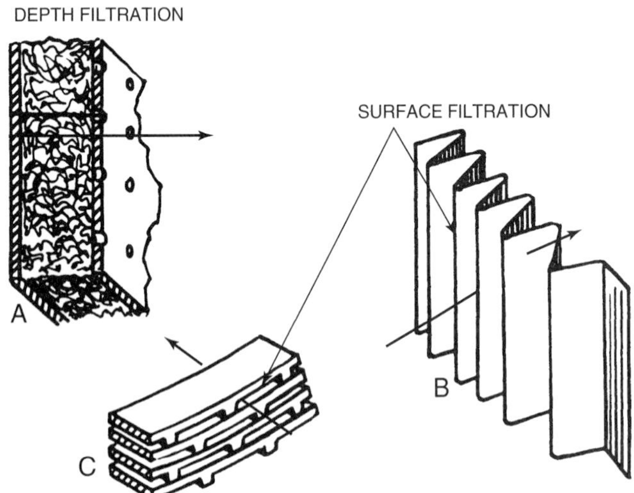

Figure 12-28. These materials are used to construct different types of oil filters. A—Cotton waste between thin perforated metal sheets. B—Paper treated with resins. C—Uno type uses thin layers of metal sheet held apart by bumps or bars.

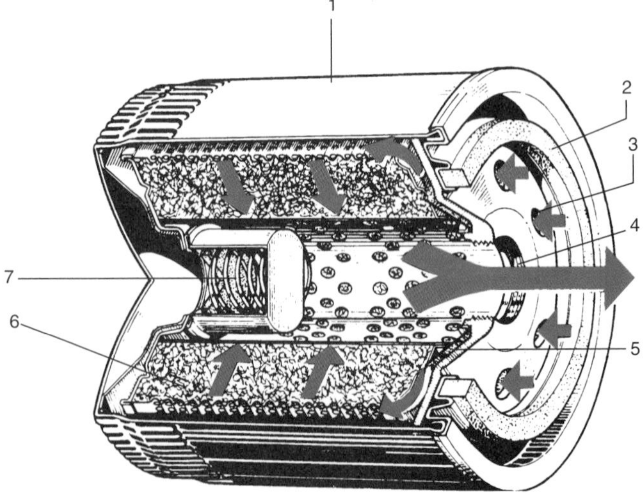

Figure 12-30. Depth type filtration oil filter. 1–Housing. 2–Seal. 3–Oil inlet. 4–Oil outlet. 5–Check valve to prevent oil drain back when filter is mounted in either a horizontal or inverted (upside down) position. 6–Filter material. 7–Relief valve opens if filter clogs. (Volvo)

Full-Flow Filter System

On modern engines, the entire output of the pump goes through the filter. This is called the ***full-flow filter system***. After passing through the filter, the oil goes to the oil galleries. In this way, all oil is filtered before reaching the bearings. In the event the filter becomes clogged, a small relief valve opens and allows the oil to flow directly to the bearings. This protects the bearings from damage due to lack of oil. See **Figure 12-31**.

Bypass Filter System

The ***bypass*** system is used on older vehicles. It differs from the full-flow system in that it does not filter the oil before it reaches the bearings. However, it is constantly filtering some of the oil in the system and then returning it to the sump. See **Figure 12-32**.

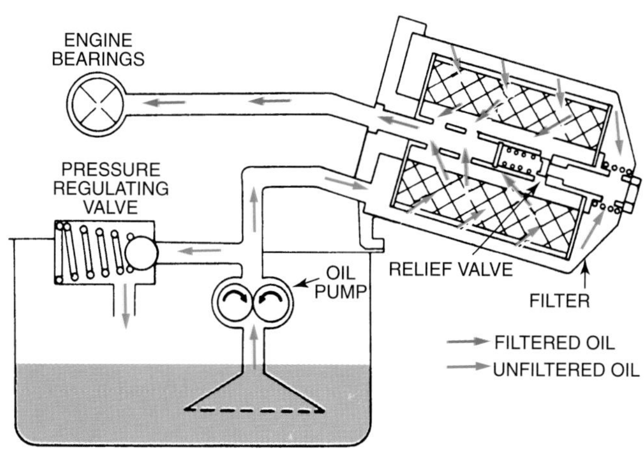

Figure 12-31. Note built-in relief valve in this full flow oil filter system. This valve allows oil to flow even if the filter becomes clogged. This system filters all oil before it reaches the bearings. (John Deere)

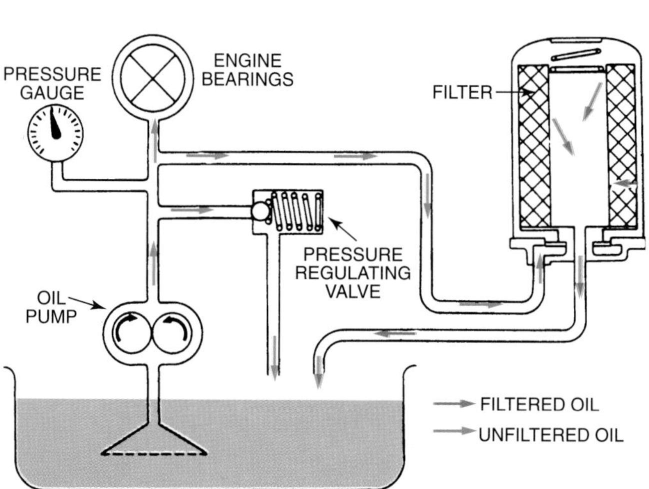

Figure 12-32. This bypass filter system filters some of the oil all of the time, but does not filter oil before it reaches the bearings. (John Deere)

Filter Replacement Intervals

All oil filters, if not replaced, will eventually become clogged. The better the filter can remove contaminants from the oil, the sooner it can become clogged. Filter change intervals are subject to the same variables as were discussed in the section on engine oil changing. Replacement varies, but will average between 1000 and 6000 miles (1600 to 9600 km) per filter. Most manufacturers recommend changing the filter at each oil change or at every other oil change.

Crankcase Ventilation

Even in new engines, a certain amount of exhaust gas and unburned fuel gets by the piston rings and into the crankcase. This is called ***blowby***, which causes pressure to build up in the crankcase. Unless it is vented, blowby will force oil to escape by the oil seals.

It is also important to remove the gases and raw fuel before they form moisture and acids in the oil. When the engine heats up, the unwanted blowby gases will evaporate and the ventilation system will remove them. If they are not removed, the oil will be quickly contaminated and the engine will be damaged.

On older engines, a vent pipe called a road draft tube was used to vent the crankcase. The road draft tube ran down below the engine. When the vehicle was in motion, a partial vacuum was formed at the end of the pipe, which drew air through a breather cap.

The cap was filled with an oil-soaked metal wool to prevent the entrance of dust and dirt. The road draft tube contained a baffle, which prevented oil vapor from being drawn through, **Figure 12-33**. Unfortunately, the road draft ventilation system allowed air pollutants to enter the atmosphere and did not do a thorough job of removing blowby. All modern engines use a ***positive crankcase ventilation***, or ***PCV*** system.

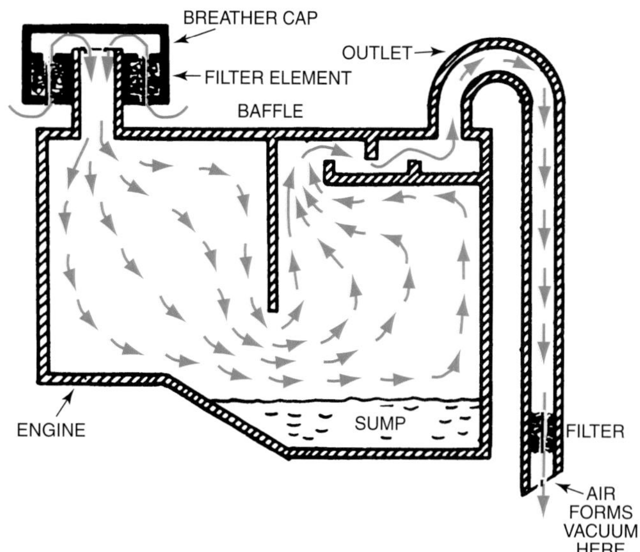

Figure 12-33. Airflow in a road draft crankcase ventilation system. Notice the filter material in both inlet and outlet.

Positive Crankcase Ventilation (PCV)

The PCV system uses intake manifold vacuum to draw fresh air through the engine. Clean air is taken from either the breather cap (open system) or the air cleaner (closed system). All engines now use the closed system to prevent fumes from leaving the engine during periods of high crankcase pressure or when the engine is stopped.

The air is pulled through the engine where it mixes with gases in the crankcase. The mixture of air and gases is then drawn into the intake manifold, where it is burned in the cylinders. A PCV valve controls air flow through the crankcase. **Figure 12-34** shows a cross-sectional view of a closed positive crankcase ventilation system. **Figure 12-35** illustrates the difference between the open and closed ventilation systems. The complete operation of the positive crankcase ventilation system will be covered in Chapter 14.

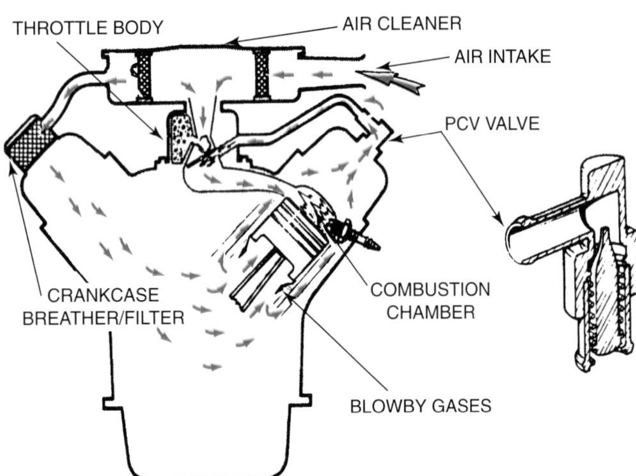

Figure 12-34. Vacuum created in the intake manifold draws air from the air cleaner and passes it through the engine. (Dodge)

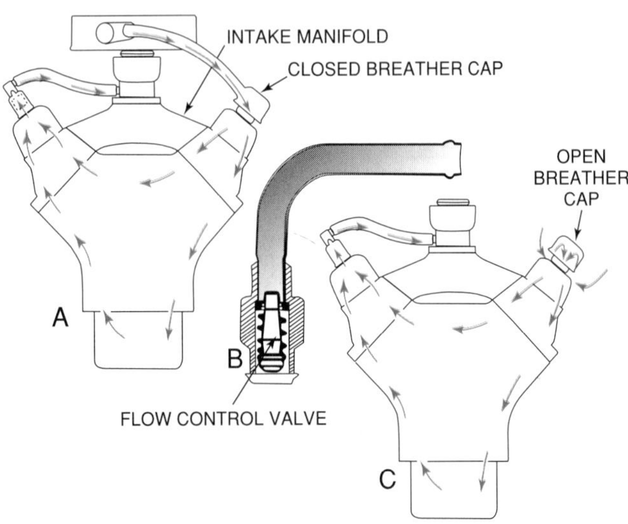

Figure 12-35. Two types of positive crankcase ventilation systems. A—Closed type that draws air in through the air cleaner. B—Cutaway of the air flow control valve. C—Open type that draws air in through the breather cap. (Plymouth)

Oil Seals

Oil is retained in the engine by ***gaskets***, ***sealants***, and ***oil seals***. Gaskets, such as the one in **Figure 12-36**, are used throughout the engine. Room temperature vulcanizing (RTV) and anaerobic sealants are used with or in place of some gaskets on modern engines. Seals are placed at the front and the rear of the crankshaft, **Figure 12-37**. Some crankshafts have an ***oil slinger*** cast into them, **Figure 12-38**. The slinger catches what oil may escape past the bearing and throws it outward into a catch trough that returns the oil to the sump.

Oil Pressure Indicators

Oil pressure can be determined by a dashboard mounted ***oil pressure gauge*** or ***warning light***. An electric oil pressure gauge uses a different sender unit than that used by an electric water temperature gauge. **Figure 12-39** shows a

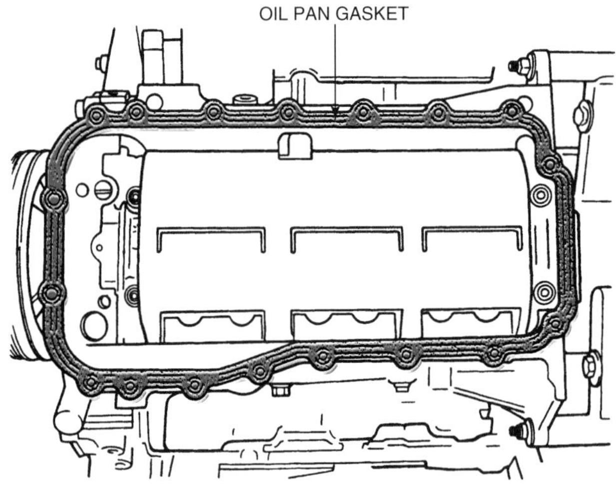

Figure 12-36. A silicone rubber oil pan gasket is used to prevent oil leakage between the block and oil pan. (Chrysler)

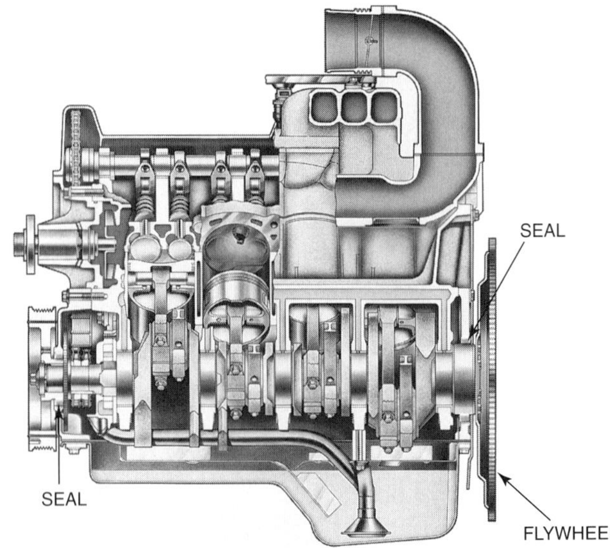

Figure 12-37. A cross-sectional view of a V-8 engine showing the front and rear crankshaft oil seals. (Ford)

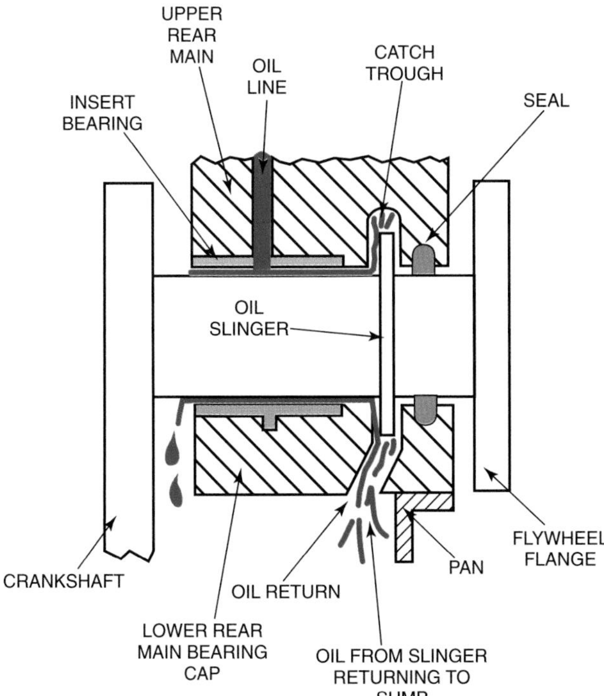

Figure 12-38. Oil that escapes past the bearing is stopped by the slinger and seal. Oil on the slinger is thrown out into the catch trough and returned to the sump. Some engines use a seal only. One type of seal is made of rope, oil impregnated and covered with graphite. Other crankshaft seals are made of synthetic rubber.

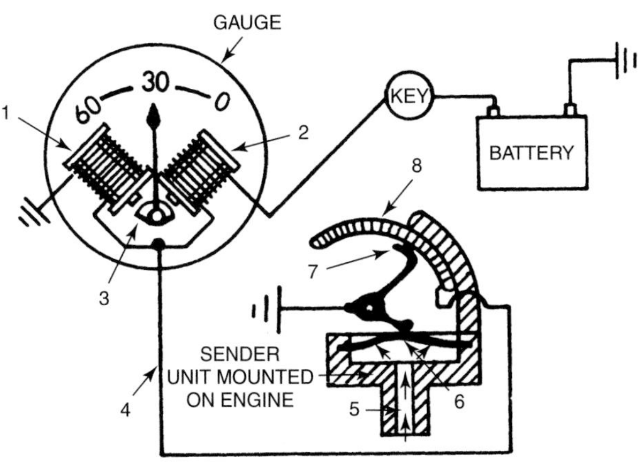

Figure 12-39. Study the schematic diagram of this electric oil gauge carefully. Electricity travels from the battery through the key switch to coil 2; then across wire 4 to coil 1 and on to ground. Both coils are connected by wire 4 to resistance unit 8, mounted on the engine.

Oil entering at 5 will raise diaphragm 6 causing sliding contact 7 to slide further out on the resistance unit, which will increase resistance.

When resistance is increased, coil 1 will strengthen its magnetic pull on armature 3. This will pull the armature and indicating needle over to read more pressure.

When oil pressure drops, the sliding contact will come back and decrease resistance. This strengthens coil 2 and weakens coil 1. Coil 2 will now pull the armature and needle down to read less pressure.

typical electric oil pressure gauge setup using the balancing magnetic coil principle. **Figure 12-40** shows a heating coil electric pressure gauge setup. Note the difference between this unit and the one in **Figure 12-39**. Some vehicles use a warning light to indicate when the pressure drops below a certain point. **Figure 12-41** illustrates a warning light system that warns of low crankcase oil level. The light warns of system failure, but does not indicate actual oil pressure.

Dipstick

A long metal rod, commonly referred to as a ***dipstick***, is used to check the amount of oil in the crankcase sump. The

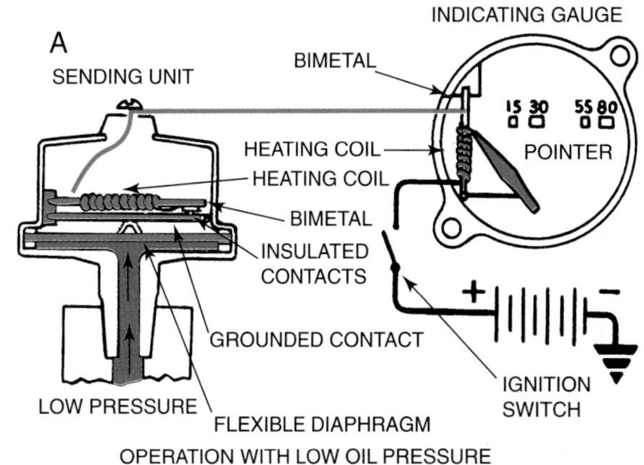

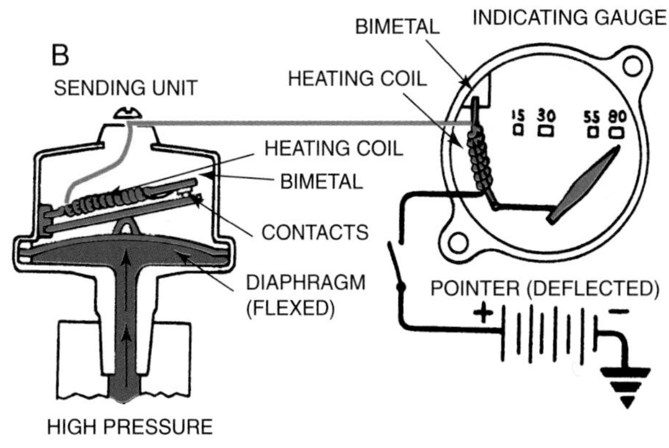

Figure 12-40. In a heated coil oil pressure gauge, current flows through both the gauge coil and the sender heating coil. Both coils heat up and cause the bimetal strips to bend. The gauge bimetal strips will move the indicator needle when it bends. How far the needle moves depends on how hot the bimetal strips get. The gauge bimetal temperature is controlled by the sender's bimetal strips. As the sender heats up, the bimetal strips bend away from ground, breaking the circuit. When the circuit is broken, the sender coil cools and the bimetal strips make contact again, reestablishing current flow. This process occurs over and over.

A—The sender diaphragm is not pushing the insulated contact up. The bimetal breaks current easily, so the coils heat for a short period. B—Oil pressure is high, which forces the diaphragm up, bending the grounded contact up. The sender coil will now have to stay on much longer to heat the bimetal enough to break contact. This keeps the gauge coil heated longer, imparting a greater bend to its bimetal strip. The indicator needle now moves further across the gauge indicating more oil pressure. (Deere & Co.)

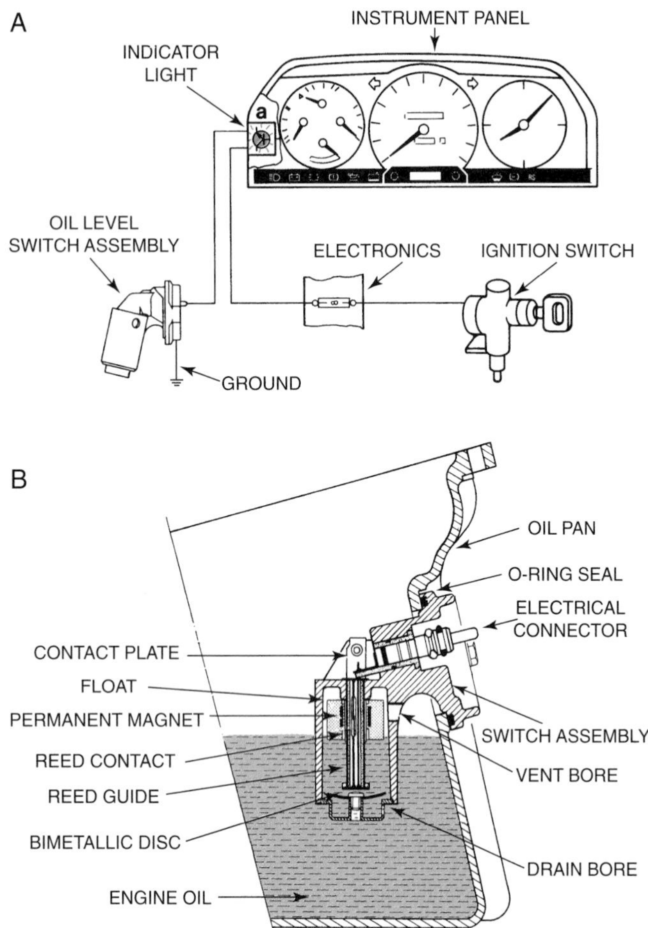

Figure 12-41. A—Low oil level indicator light electrical circuit schematic. B—Cross-sectional view of the low oil level switch and its mounting location in the oil pan. As the floats travel down (from oil loss) on the reed guide, it will open a contact switch as it reaches a predetermined minimum oil level in the pan. This will turn on the indicator light on the instrument panel. (Mercedes-Benz)

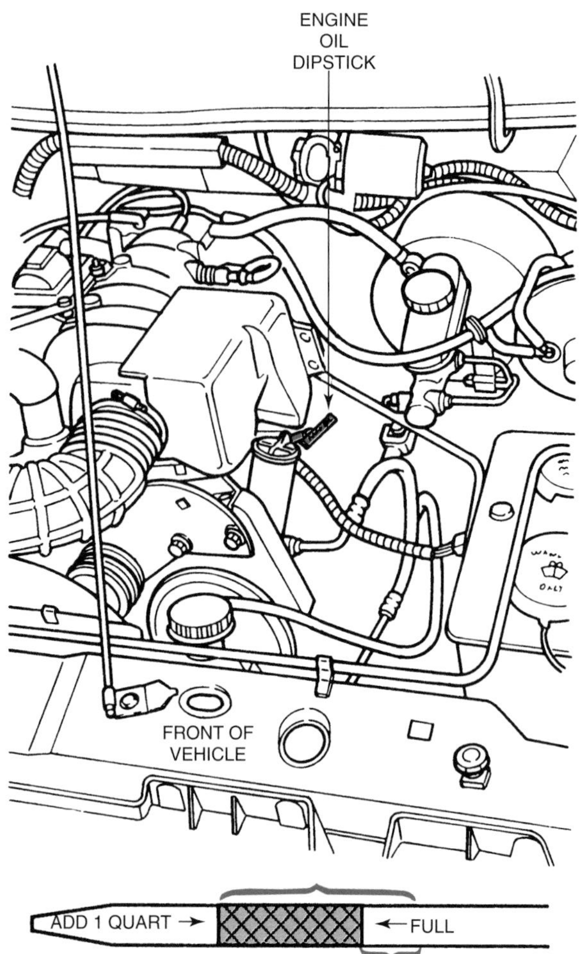

Figure 12-42. Engine oil dipstick location as used by one vehicle manufacturer. Note the dipstick markings. (Ford)

dipstick is held in a tube, just long enough for the end of the dipstick to protrude into the sump. Some dipsticks are connected to the oil fill cap. It may be withdrawn and the oil level in the sump checked by examining the dipstick to see where the oil reaches. **Figure 12-42** shows a dipstick and its markings.

Cooling the Oil

Oil removes much heat from the pistons and cylinder walls and becomes quite hot. It is important that the oil have some way to rid itself of excess heat. In the average engine, this is accomplished by the airstream passing over the thin oil pan, thus drawing much heat from the oil. Some oil pans have fins cast into them to aid in the cooling. The cooling system keeps the cylinder block and head area within an acceptable temperature range thus limiting oil heat exposure. Oil moving through the filter gives off some heat to the airstream and the entire engine surface gives off heat. All these sources of heat removal will normally prevent excess heating of the oil.

Many high performance and heavy duty engines as well as some normal duty engines that may occasionally be exposed to heavy operating loads (trailer towing, steep roads, or extremely hot weather), are equipped with an ***auxiliary oil cooler***. Numerous designs have been used. The oil can be cooled by air or coolant depending on the design.

One type of engine oil cooler is shown in **Figure 12-43**. This particular cooler fits between the oil filter and the engine block. Engine coolant circulates around the thin discs making up the oil compartment. As the oil passes through the cooler on its way to the filter, it is forced through the thin disc compartments where it gives off heat to the passing coolant.

Summary

Friction, the enemy of all engines, is controlled through the use of lubricating oil. Lubricating oil must cool, clean, seal, and lubricate. The viscosity, or weight, and the service rating, such as SH or CE, are important factors in choosing the proper oil for a specific engine.

Engine oil is drawn from the oil pan sump through a strainer by the oil pump. It is then pumped to all the major bearings through a network of oil galleries and tubes. Maximum oil pressure is controlled by a pressure relief valve.

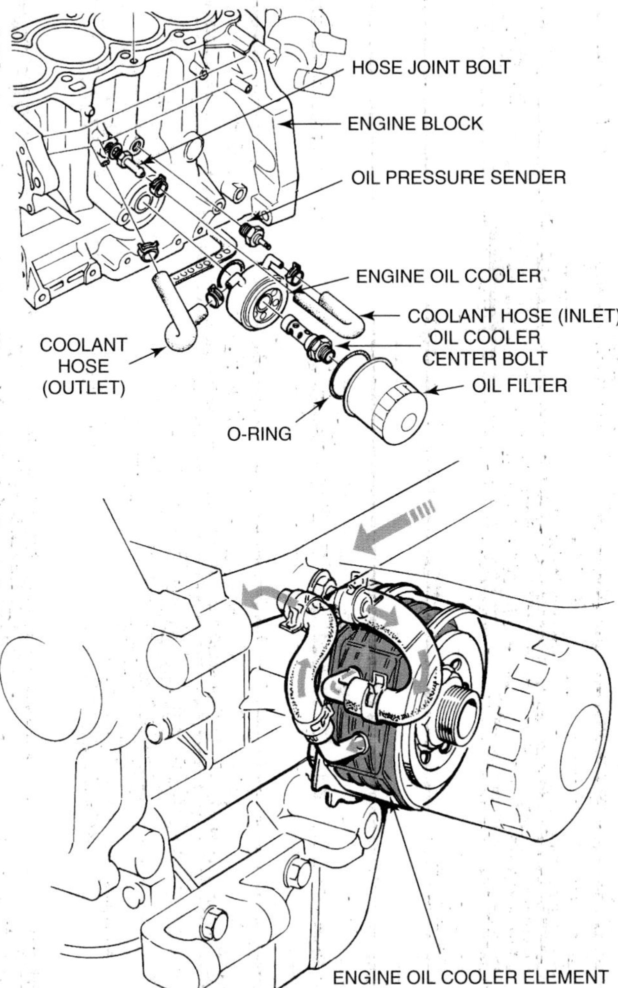

Figure 12-43. One type of engine oil cooler that utilizes circulating engine coolant to help reduce the oil operating temperature. In this unit, the oil travels through the cooler on its way to the oil filter. (Honda)

The most commonly used oil pumps are the gear and the rotor types that are either turned or actuated by the camshaft.

Upon entering the bearings, the engine oil separates the moving parts with a fine film. As it passes through the bearings, it is thrown outward in every direction. Oil drops down on some parts and is sprayed against others. The spinning parts create an oily mist that lubricates the other areas. It is then returned to the sump for recirculation. The system is ventilated to remove moisture, hot gases, and excessive pressure.

Modern engines filter the oil through the use a full-flow system. Even with filtering, oil must be changed periodically. The length of time between changes depends on driving conditions and the number of miles driven. Oil pressure is measured by means of a gauge or a warning light. The amount of oil in the sump can be checked by using the dipstick. Some engines are equipped with oil coolers.

Know These Terms

Friction
Dry friction
Wet friction
Friction bearing
Antifriction bearing
Engine oil
Motor oil
Lubricating
Cooling
Seal
Clean
Distillated
Fractionating tower
Relative volatility
Additives
Viscosity
Viscosity index
Multi-viscosity oil
Polymers
Service grades
Spark ignition
Compression ignition
Non-detergent oil
Detergent oil
Full pressure system
Oil pan
Oil pump
Oil galleries
Splash system
Dippers
Pressure-splash system
Sump
Oil pickup
Gear pump
Rotary pump
Vane pump
Pressure relief valve
Hydraulic valve lifters
Depth filter
Surface filter
Full-flow filter system
Bypass
Blowby
Positive crankcase ventilation (PCV)
Gaskets
Sealants
Oil seals
Oil slinger
Oil pressure gauge
Warning light
Dipstick
Auxiliary oil cooler

Review Questions—Chapter 12

Do not write in this book. Write your answers on a separate sheet of paper.

1. The amount of friction developed between moving parts depends on _____ .
   (A) the type of material
   (B) surface finish
   (C) pressure between moving parts
   (D) All of the above.
2. For purposes of illustration, friction was defined or divided into two kinds. Name them.
3. What is a frozen bearing?
   (A) A very cold bearing.
   (B) A very oily bearing.
   (C) A stuck bearing.
   (D) A loose bearing.
4. Name two kinds of bearings that can be used in an engine.
5. Name the two organizations which classify engine oil according to viscosity and grade.
6. A detergent oil prevents or reduces the formation of _____ .

7. A good engine oil has four tasks. What are they?
8. Polymers are used to make what kind of oil?
9. Technician A says that SAE 30 oil should be used for cold weather operation. Technician B says that SAE 5W-30 oil should be used for cold weather operation. Who is right?
   (A) A only.
   (B) B only.
   (C) Both A & B.
   (D) Neither A nor B.
10. The most common method of lubricating an automotive engine is _____ .
   (A) pressure
   (B) splash
   (C) combination pressure-splash
   (D) adding oil to the fuel
11. Name three types of oil pumps.
12. Which two of the above oil pumps are the most widely used?
13. The oil pressure in a system cannot exceed a predetermined pressure. How is this pressure controlled?
14. What is an oil gallery?
15. If only part of the oil leaving the oil pump is filtered, what kind of filter system is it?
16. Depth and surface are the two most popular types of _____ .
   (A) oil pumps
   (B) oil filters
   (C) bearings
   (D) hydraulic lifters
17. In the full-flow filtering system, what happens when the filter is clogged?
   (A) The engine is ruined.
   (B) The oil must be changed.
   (C) The bypass valve opens.
   (D) Both A & B.
18. What does the positive crankcase ventilation system (PCV) prevent?
   (A) Oil contamination.
   (B) Sludge buildup.
   (C) Crankcase pressure buildup.
   (D) All of the above.
19. In addition to gaskets, oil _____ are placed on the front and rear of the crankshaft.
20. What do you use to tell how much oil is in the sump?
   (A) An oil pressure gauge.
   (B) A warning light.
   (C) A dipstick.
   (D) The amount of time since last oil change.

13

Exhaust Systems

After studying this chapter, you will be able to:

- Explain the purpose of the exhaust system.
- Describe the operation of the exhaust system.
- Identify exhaust system components.

This chapter outlines the purpose of the exhaust system, describes various exhaust system designs, and identifies the major exhaust system components. While the exhaust system appears to be relatively simple, it is vitally important to the efficient operation of a vehicle's engine. Studying this chapter will prepare you for the material in Chapter 14, which discusses emission control systems.

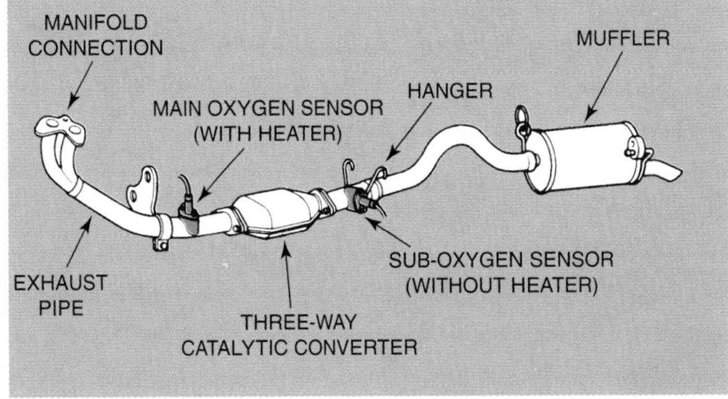

Figure 13-1. A single exhaust system. Note the main oxygen sensor and the sub-oxygen sensor. The sub-oxygen sensor rechecks the exhaust content and sends a signal to the electronic control unit to make any needed corrections to the air-fuel mixture. It also acts as a back up should the main oxygen sensor fail.(Toyota)

Purpose of the Exhaust System

The purpose of the exhaust system is to route the exhaust gases from the engine, reduce exhaust noise, and keep the gases from entering the passenger compartment. The exhaust system must be free from leaks and vibration. It must be designed to prevent excessive ***back pressure***. Back pressure occurs when the flow of exhaust gases is restricted, making it harder for cylinders to clear themselves during the exhaust stroke. Excessive back pressure can seriously reduce engine power and mileage. In extreme cases, back pressure can cause the exhaust valves to burn.

Exhaust System Designs

Every exhaust system functions in the same manner. Burned gases leave the engine through the exhaust ports and enter the exhaust manifold. From the manifold, gases travel through the exhaust pipe, catalytic converter, and muffler. The gases leave the vehicle via the resonator, if used, and the tailpipe.

There are two basic types of exhaust systems: single exhaust systems and dual exhaust systems. The following sections describe each type of exhaust system in detail.

Single Exhaust System

In the ***single exhaust system,*** only one exhaust pipe is used. See **Figure 13-1.** Note that this system has two oxygen sensors installed in the pipes. The purpose of the oxygen sensor is covered in Chapter 15. The pipe, converter, and muffler are sized in proportion to the engine displacement. Single exhaust systems are the most common and are used on all types of engines.

Dual Exhaust System

The ***dual exhaust system*** is used primarily on V-type engines. It consists of two separate exhaust systems, each equipped with a catalytic converter and muffler. **Figure 13-2** shows a typical dual exhaust system. Because dual exhaust systems offer twice as much exhaust system capacity, they reduce back pressure. However, the components in a dual exhaust system do not get as hot as those in a single exhaust system. Therefore, dual exhaust components tend to corrode more easily.

A dual exhaust system also has a higher noise level since the noise reduction potential of self canceling action is lost. ***Self canceling*** occurs when the noise and pressure pulses of the engine cylinders cancel each other out when they are channeled through a single exhaust system.

There are many types of dual exhaust setups. Some systems have dual exhaust pipes from the engine to the catalytic converters and then join the pipes so that they use a single muffler and tailpipe. Many dual exhaust systems have an ***equalizer pipe*** that connects each exhaust pipe ahead of the mufflers or converters. See **Figure 13-3.**

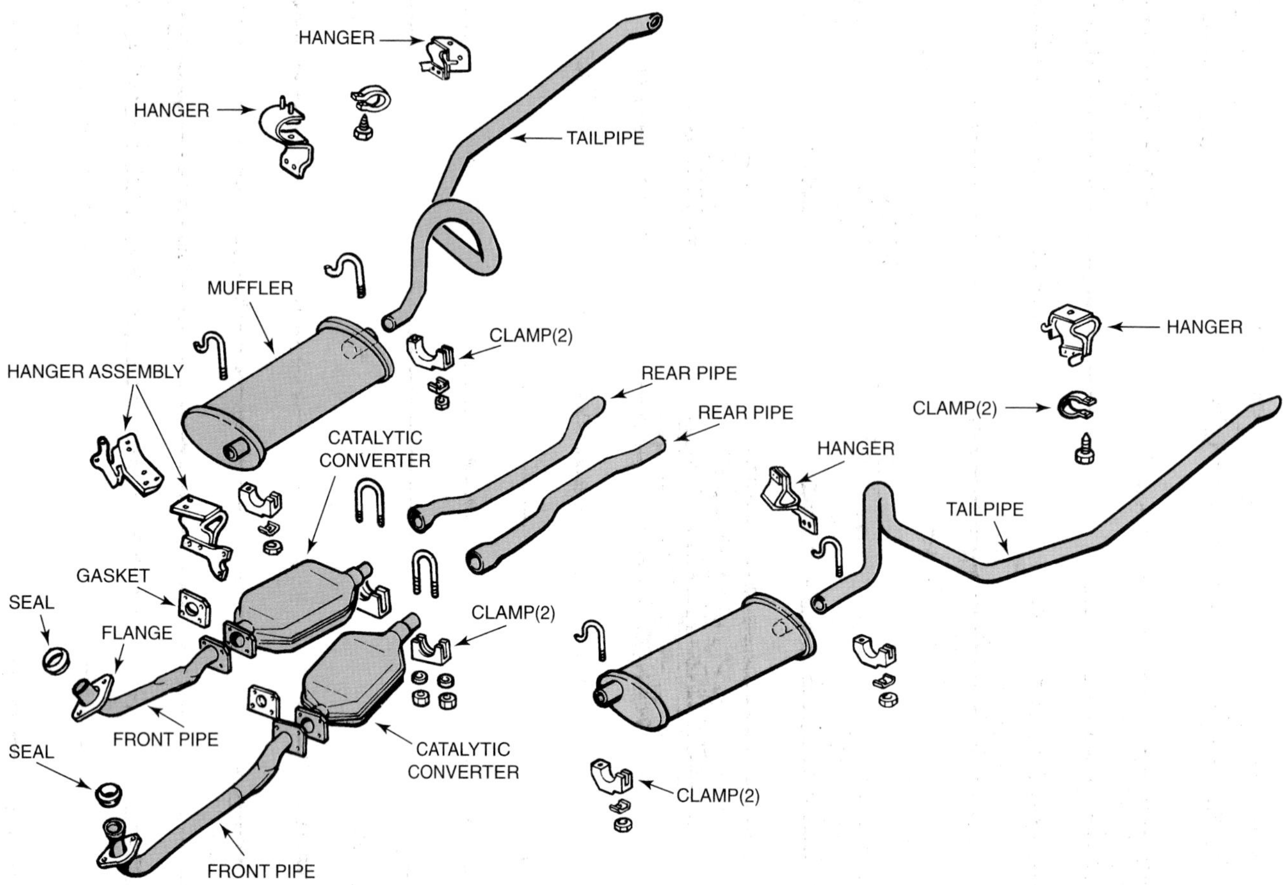

Figure 13-2. One particular exhaust system with hangers, clamps, gaskets, etc., for a V-8 engine. Note the dual catalytic converters. (Chrysler)

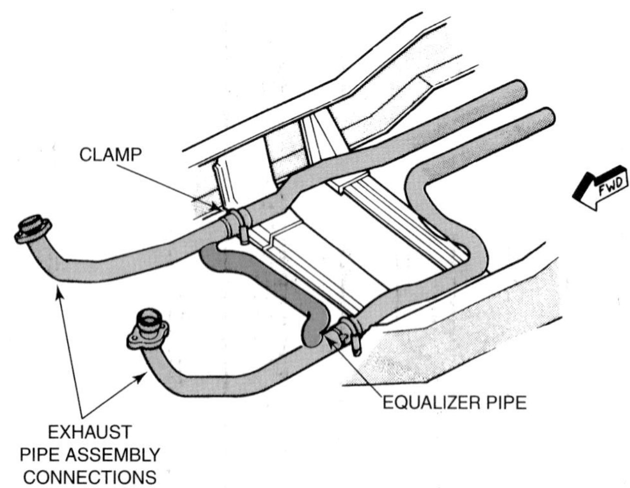

Figure 13-3. Dual exhaust pipe assemblies with an equalizer pipe connecting the two pipes before the mufflers. The equalizer pipe helps to even out the pressure between the two sides. (Chrysler)

Exhaust System Components

The same basic components are found in all exhaust systems. The sections below explain the construction and operation of these components.

Exhaust Manifolds

Exhaust manifolds channel exhaust gases from the exhaust ports to the rest of the exhaust system. Exhaust manifolds are constructed of cast iron or stainless steel. Manifolds are manufactured with smooth, gentle bends to improve exhaust flow. **Figure 13-4** illustrates exhaust manifolds for a V-6 engine.

Some engines use steel tube exhaust manifolds, often called ***headers***. They are carefully engineered to reduce restrictions. Each header tube is the same length between the exhaust port in the cylinder head and the collector box where the tubes connect to a single large exhaust tube, **Figure 13-5.** This allows the exhaust pulses to interact with each other to reduce back pressure. This interaction is called exhaust ***scavenging***. Modern manifolds are accurately designed to take advantage of the exhaust scavenging effect. These manifolds are sometimes called ***cast iron headers***.

The exhaust manifold is bolted to the cylinder head. Most manifolds use a gasket at the head attaching point, while some older manifolds rely on a metal-to-metal contact for sealing. The manifold may contain a heat control valve and passageways for air injection nozzles or an oxygen sensor. These are covered in detail in Chapter 14.

Exhaust Pipes

The ***exhaust pipes*** connect the exhaust manifold, muffler, converter, and resonator. Any pipe other than the tailpipe can be called an exhaust pipe. They are produced using coated and stainless steel. Exhaust pipes are often ***double-walled*** (made of two pipes, one pressed inside of the other) for added sound deadening. See **Figure 13-6.**

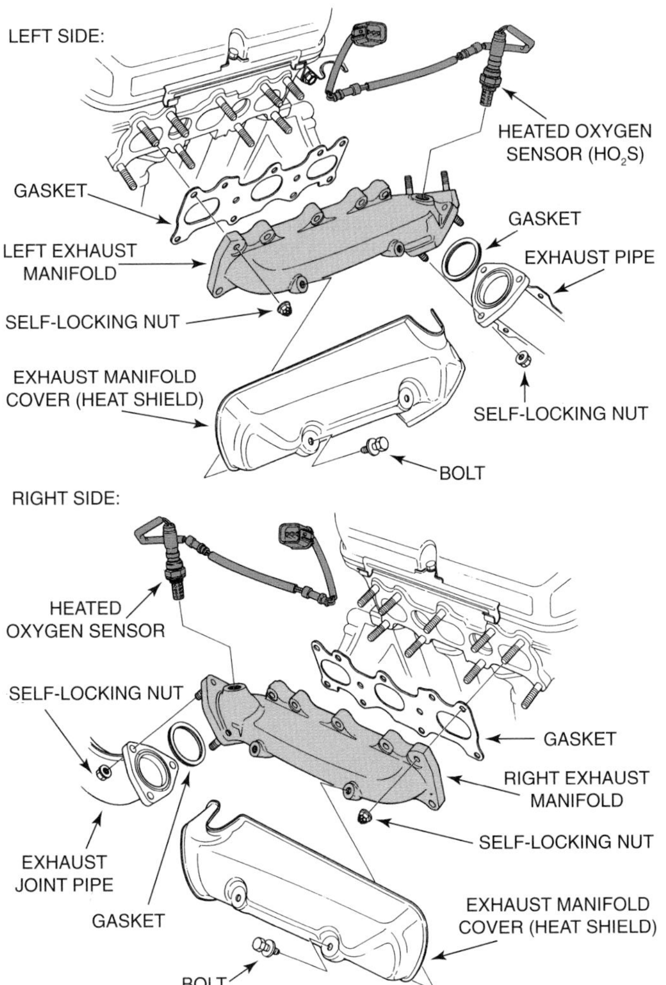

Figure 13-4. Right and left-hand side exhaust manifolds as used on a V-6 engine. Note that each manifold has its own heated oxygen sensor (HO_2S). (Honda Motor Company)

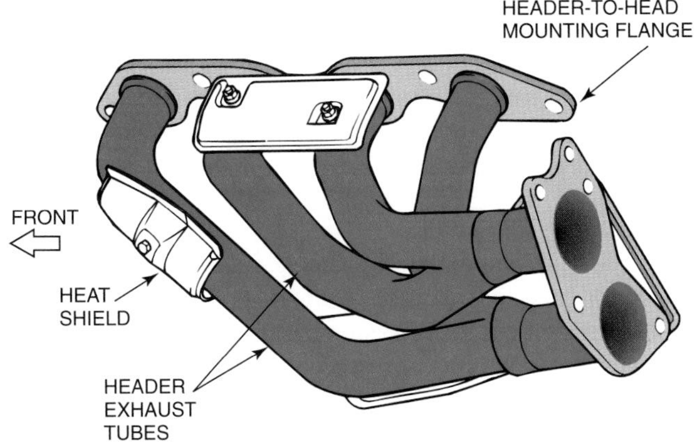

Figure 13-5. A stainless steel tube exhaust header. This header will increase the exhaust gas flow efficiency, reduce weight, and aid in catalytic converter warm-up. (Toyota)

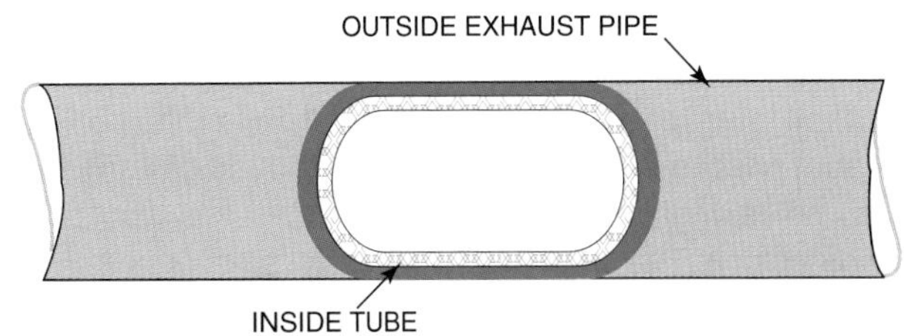

Figure 13-6. A cutaway of a double-walled exhaust pipe.

Some exhaust systems make use of a ***flexible tube*** to allow for expansion and contraction as the system temperature changes. The tube is usually an accordion-shaped pipe covered with braided wire.

Catalytic Converters

All modern vehicles contain a catalytic converter. It is placed in the exhaust system before the muffler to clean exhaust gases. Most converters are clamped or bolted to the exhaust system. Catalytic converters are discussed in more detail in Chapter 14.

Mufflers

The muffler has the job of reducing engine noise without restricting exhaust flow enough to cause excessive back pressure. There are two basic muffler designs: the ***reverse-flow muffler*** and the ***straight-through muffler***.

Reverse-Flow Muffler

The reverse-flow muffler uses a series of ***muffler chambers*** to cause the exhaust gases to reverse direction. This change in direction reduces the noise level, **Figure 13-7.** Small openings and slits between the various chambers allow the sound waves in each chamber to cancel each other. Internal ***baffles*** absorb and dampen more of the exhaust sounds. Many reverse-flow mufflers have a small hole at the bottom of the case to allow moisture and acids to drip out.

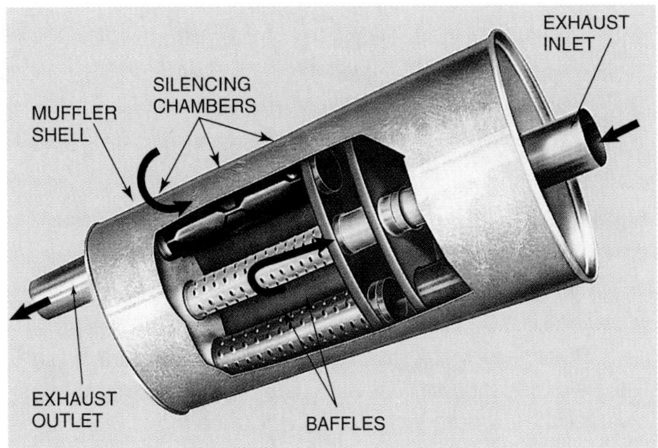

Figure 13-7. Cutaway of one type of reverse-flow muffler. (AP Parts)

Straight-Through Muffler

The straight-through muffler is sometimes called a "*glass pack*" or "*steel pack*" muffler, **Figure 13-8**. In this muffler, the exhaust passes straight through a perforated pipe (pipe with many holes or splits) that is surrounded by steel wool or fiberglass packing. The steel wool or fiberglass absorbs the noise.

The straight-through muffler is designed to minimize back pressure. Straight-through mufflers also require less space under the vehicle. A disadvantage of the straight-through muffler is the tendency of the packing to settle and break up, requiring early muffler replacement.

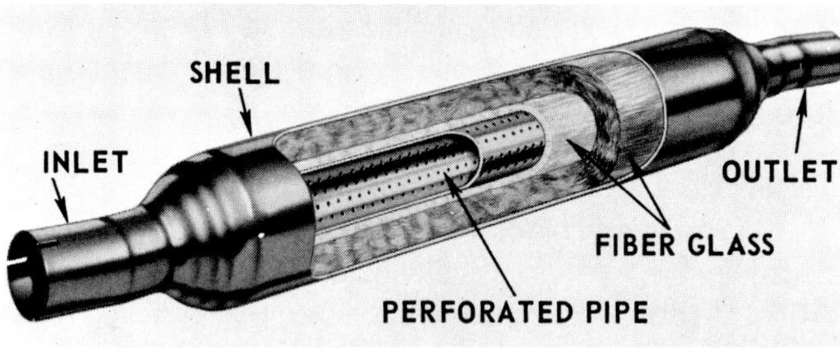

Figure 13-8. A straight-through muffler. Heavy fiberglass packing is used around the perforated pipe.

Another kind of straight-through muffler is shown in **Figure 13-9.** Instead of using fiberglass, the perforated sections of the pipe are surrounded with sound-deadening chambers. This design, termed a "*chambered pipe*," requires a minimum of space beneath the vehicle.

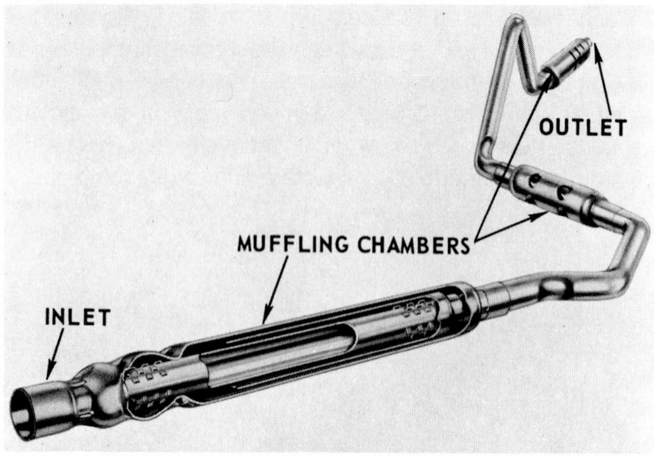

Figure 13-9. A chambered pipe exhaust system. A series of muffling chambers are incorporated into this system.

Resonators

Resonators are small extra mufflers designed to further reduce exhaust noise. They are usually found on more expensive vehicles, where the goal is to produce the quietest possible exhaust. Resonators may be reverse-flow or straight-through designs. They are generally installed near the end of the exhaust system to silence any resonance. However, some newer vehicles have the resonator before the mufflers, **Figure 13-10.** ***Resonance*** is the vibration sound created or sustained within the exhaust system after the quieting action of the regular exhaust system.

Tailpipes

The ***tailpipe*** is the last pipe in the exhaust system. Most tailpipes are of single-wall construction. Many late-model tailpipes are welded to the muffler and may contain the resonator, if used.

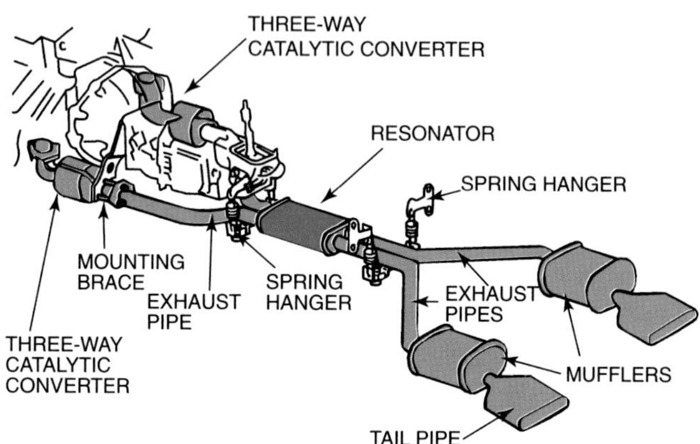

Figure 13-10. This dual exhaust system contains a resonator. Note the spring hangers, which allow additional system flexibility. (General Motors)

Hangers

The exhaust system is suspended on ***hangers*** designed to provide support while allowing the system to flex. The hangers must not allow exhaust system noise and vibration to be transmitted to the vehicle body. They must provide shock protection and sound deadening and stand up to exhaust system heat and road debris. See **Figure 13-11.** On many modern vehicles, rubber retaining rings are used to attach the exhaust system to the hanger assemblies. They allow a greater degree of flexibility than a conventional hanger. **Figure 13-12** shows one type of retaining ring arrangement.

Some vehicle exhaust systems use rods that are welded onto the pipes. The ends of the rods fit into bushings or retaining rings, **Figure 13-13.** The use of these rods allows the exhaust system components to expand and contract as their temperature changes.

Heat Shields

The modern exhaust system, especially the catalytic converter, produces a great deal of heat. To provide sufficient ground clearance, these systems must be mounted close to the vehicle's floor pan. To prevent heat damage to various components under the vehicle and to prevent the transfer of heat to the passenger compartment, ***heat shielding*** is used on modern vehicles.

Heat shields generally consist of sheet metal housings placed around the hottest parts of the system, metal shields

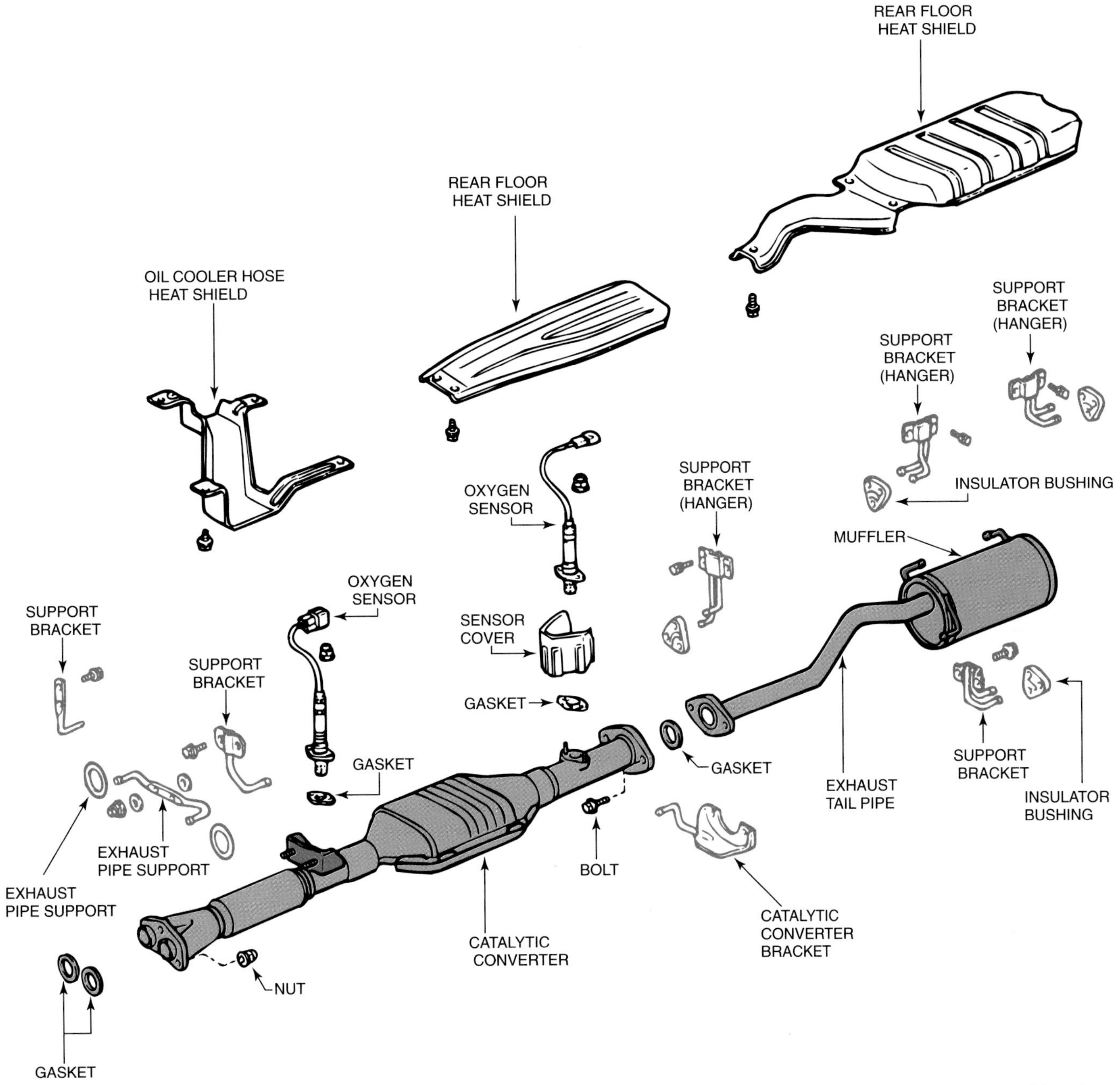

Figure 13-11. Exploded view of an exhaust system's various support hangers and rubber insulator bushings. The style and location of the hangers vary by vehicle and manufacturer. (Toyota)

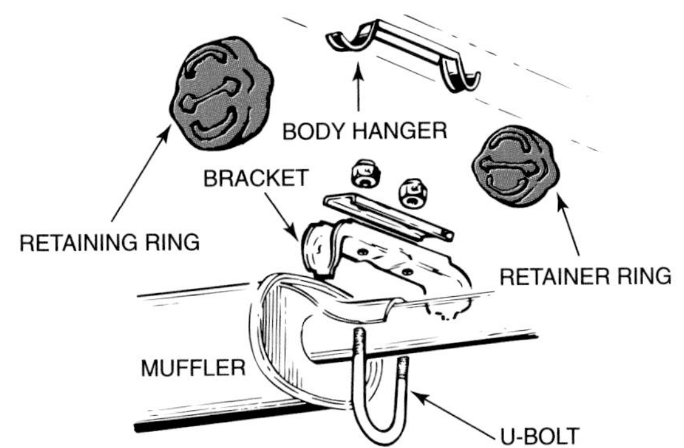

Figure 13-12. This exhaust system hanger allows flexing and reduces noise. Note the rubber retaining rings, or biscuits. (Volkswagen)

located between the floor pan and exhaust components, or heavy insulation placed under the carpet. Heat shields are also used under the catalytic converter to protect items under the vehicle from the heat produced by the converter. If this were not done, it would be possible for the converter to ignite dry grass or other flammable materials. **Figure 13-14** shows the heat shields used on one particular single exhaust system. Also note the flexible tube described earlier in this chapter.

Exhaust System Materials

Exhaust components are subjected to intense heat (as much as 1600°F or 871°C), corrosive gases on their inner surfaces, and water, dirt, salt, and debris on their outer surfaces. To provide proper service life, stainless steel is used for the

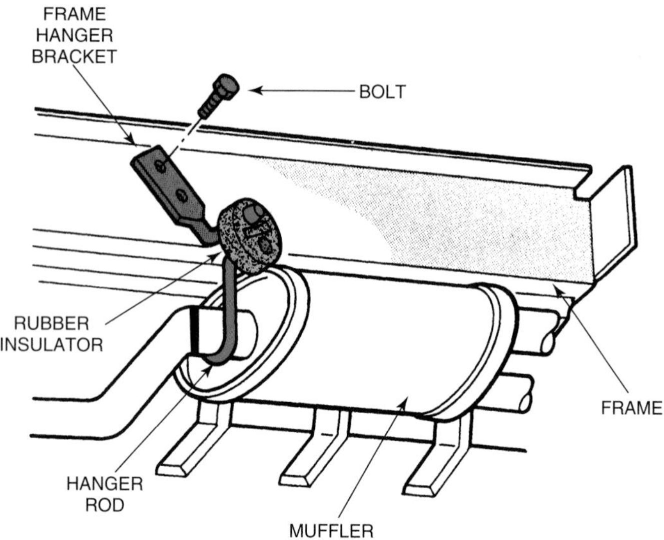

Figure 13-13. A metal hanger rod is welded to this muffler. It attaches to a frame hanger bracket with a rubber insulator. (Dodge)

converters and occasionally for some of the piping. Aluminized steel (coated with a layer of aluminum, inside and out) is widely used for exhaust pipes, tailpipes, and mufflers. These metals are highly resistant to corrosion. Because the catalytic converter produces tremendous heat, which also helps to drive out moisture, exhaust systems on late-model vehicles tend to last longer than systems on older vehicles

Sealing the Exhaust System

As mentioned earlier, it is vitally important that the exhaust system be sealed, both to reduce exhaust noise and to prevent exhaust gas entry into the passenger compartment. Important sealing principles are covered in the following sections.

Exhaust Pipe-to-Manifold Connections

An exhaust pipe-to-manifold connection may be one of several types. A flat-surface connection which utilizes a gasket is shown in **Figure 13-15.** The ***semi-ball connector,*** **Figure 13-16,** uses the ball seal on one side only and allows

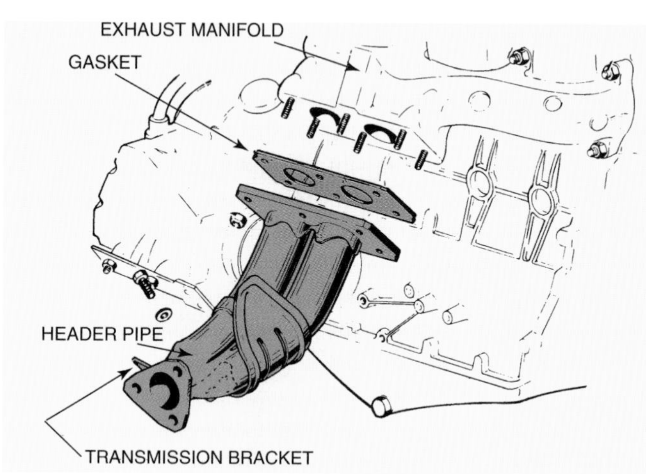

Figure 13-15. Flat, gasketed exhaust manifold-to-pipe connection. (Volkswagen)

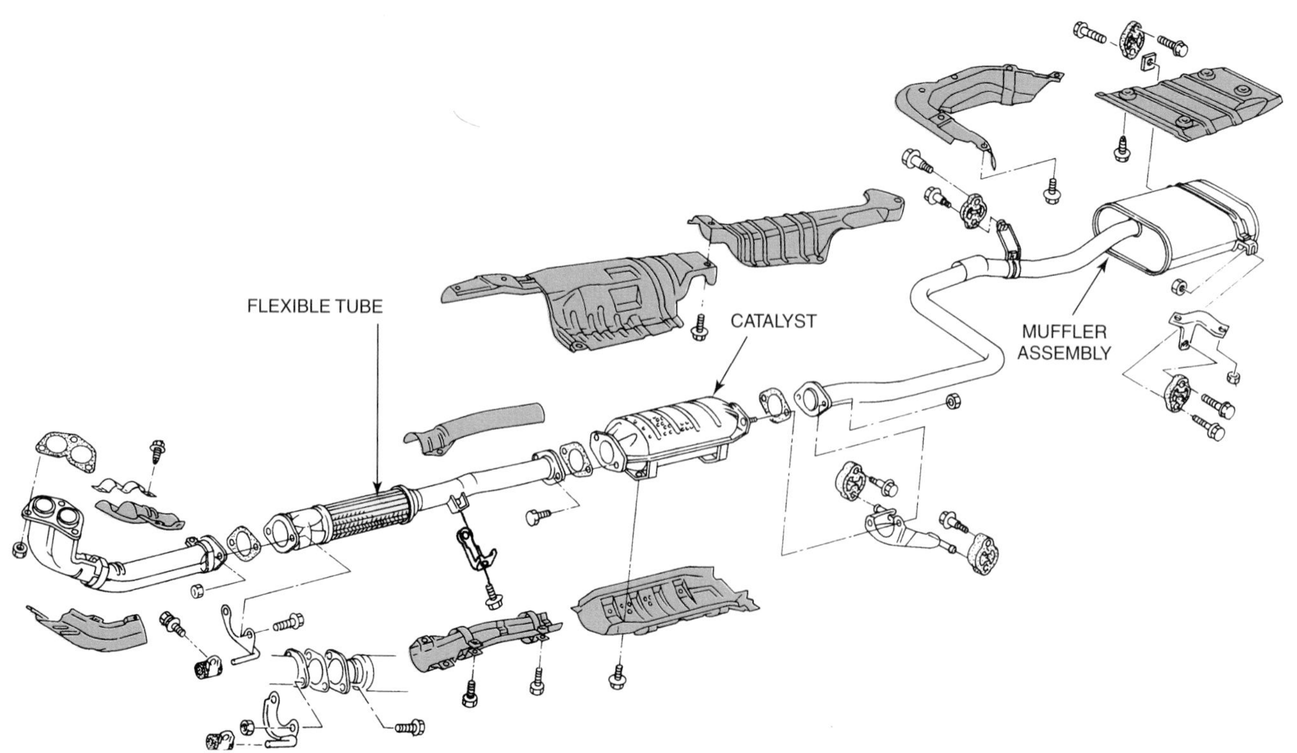

Figure 13-14. A complete exhaust system with heat shielding (in color) and a flexible tube. (Nissan)

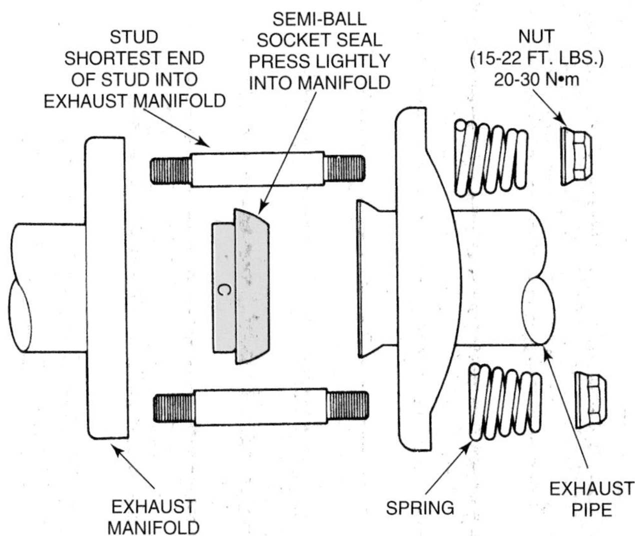

Figure 13-16. Semi-ball manifold-to-pipe connection. This connection allows a degree of flexibility in response to engine and vehicle movement. (Cadillac)

some flexing without breaking either the seal or the manifold-pipe assembly. A gasket is not needed. Note the use of springs to maintain pressure under vibration and temperature changes.

A similar fitting makes use of a semi-ball gasket made of high-temperature fiber. The advantage of this type is that the gasket can compress to fit uneven surfaces. A ***full-ball connector*** having ball seals on both sides is shown in **Figure 13-17.** This design also provides some flexing without leaking.

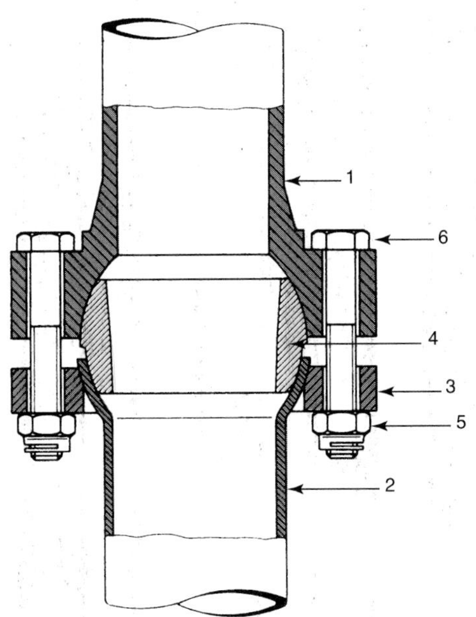

Figure 13-17. Full-ball manifold-to-pipe connection. 1–Exhaust manifold. 2–Spherically flared exhaust pipe. 3–Flange. 4–Spherical conneciton adapter. 5–Self-locking hex head nut. 6–Hex head bolt. (Mercedes-Benz)

Pipe-to-Pipe Connections

Most connections between the exhaust pipes, tailpipes, mufflers, and resonators are pipe-to-pipe connections. Pipe-to-pipe connections are made by sliding one pipe into another and securing the connection with an ***exhaust system clamp***, or ***muffler clamp***, **Figure 13-18.**

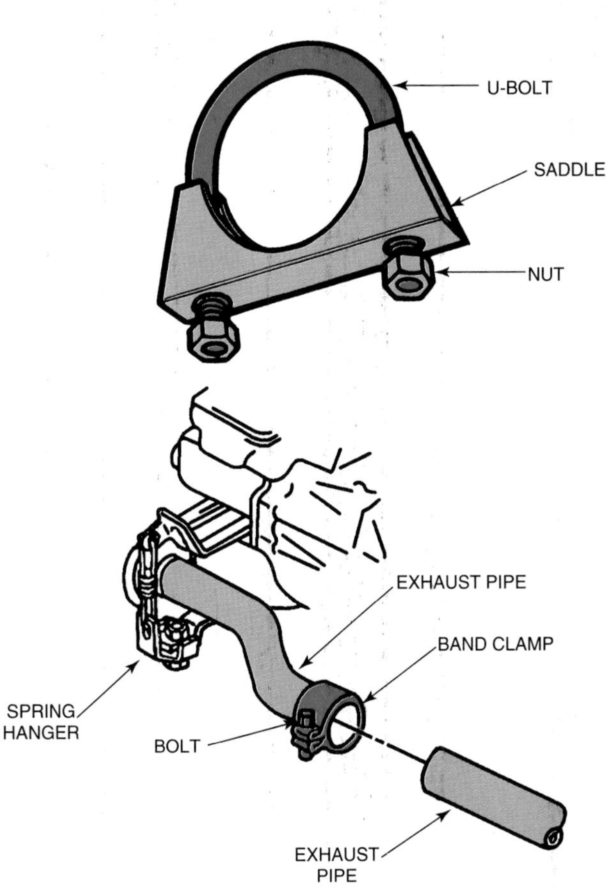

Figure 13-18. A—U-bolt exhaust clamp. B—Band clamp. All clamps will provide a 360° leakproof seal when correctly installed. (Chevrolet, Goerlich Co.)

The pipes are carefully manufactured so that they can be assembled easily and will seal well when clamped. Typical clamp connections are shown in **Figure 13-19.** In many cases, an air tube from the air injection pump is attached to the catalytic converter using a small clamp, **Figure 13-20.**

Many catalytic converters are connected to the rest of the exhaust system with the semi-ball or full-ball connections discussed earlier. In many cases, pipe-to-pipe connections are welded together. When the connection is welded, the pipes must be cut off to replace the component. All connections must be tight and absolutely leak free.

Other Connections

Many other devices are attached to the exhaust system on modern vehicles. These include one or more oxygen sensors, air injector nozzles, and heat control devices. Oxygen sensors and air injector nozzles are usually threaded into the

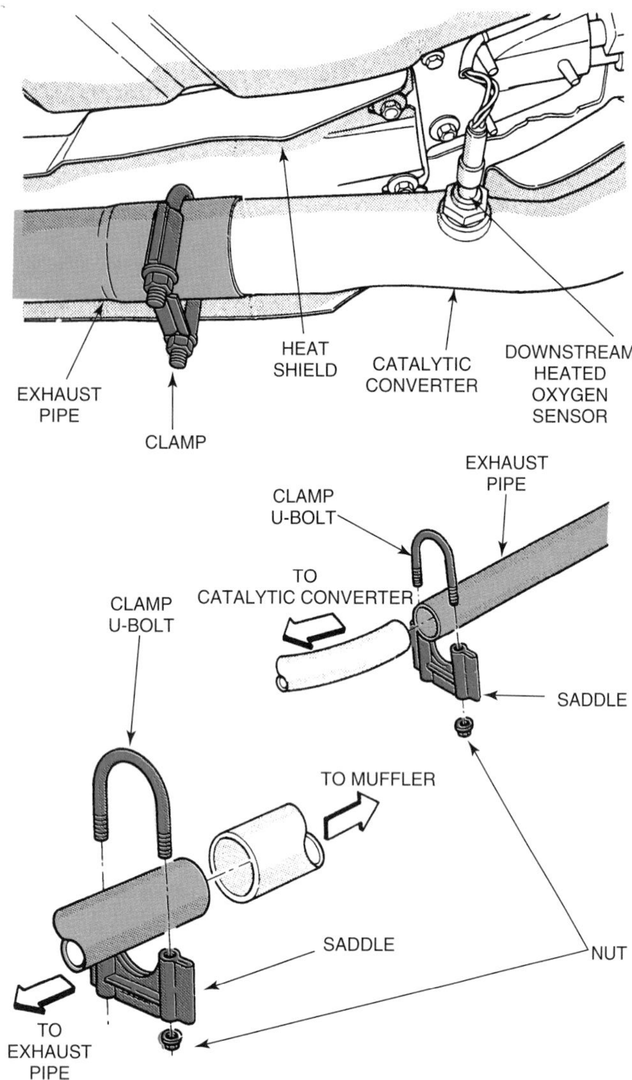

Figure 13-19. Several typical exhaust system clamps and their positions in relationship to the pipes, muffler, etc. (Chrysler)

appropriate exhaust system component and require anti-seize compound on their threads. The manifold heat control device must move in its shaft. If the shaft opening begins to leak, the entire component should be replaced. Sometimes weights are clamped to the pipes to further reduce exhaust resonance.

Summary

Exhaust systems usually consist of an exhaust manifold, catalytic converter, muffler, resonator, exhaust pipes, tailpipe, and all the necessary connections, fittings, heat shields, and hangers.

Stainless steel is used for catalytic converters and some exhaust pipes. Aluminized steel is widely used for the remainder of the system. Exhaust systems on late-model vehicles last longer than systems on older vehicles because the catalytic converter produces more heat, which drives out additional moisture

Exhaust systems must not leak. They should quiet the exhaust and discharge it properly to the passing air stream. Exhaust pipe-to-manifold connections can be of many types. Gasketed flat-surface, semi-ball, and full-ball connections are often used. Pipe-to-pipe connections often slip together and are secured with clamps.

Know These Terms

Back pressure
Single exhaust system
Dual exhaust system
Self canceling
Equalizer pipe
Exhaust manifold
Headers
Scavenging
Cast iron headers
Exhaust pipes
Double-walled
Flexible tube
Reverse-flow muffler
Straight-through muffler
Muffler chambers
Baffles
Resonators
Resonance
Tailpipe
Hangers

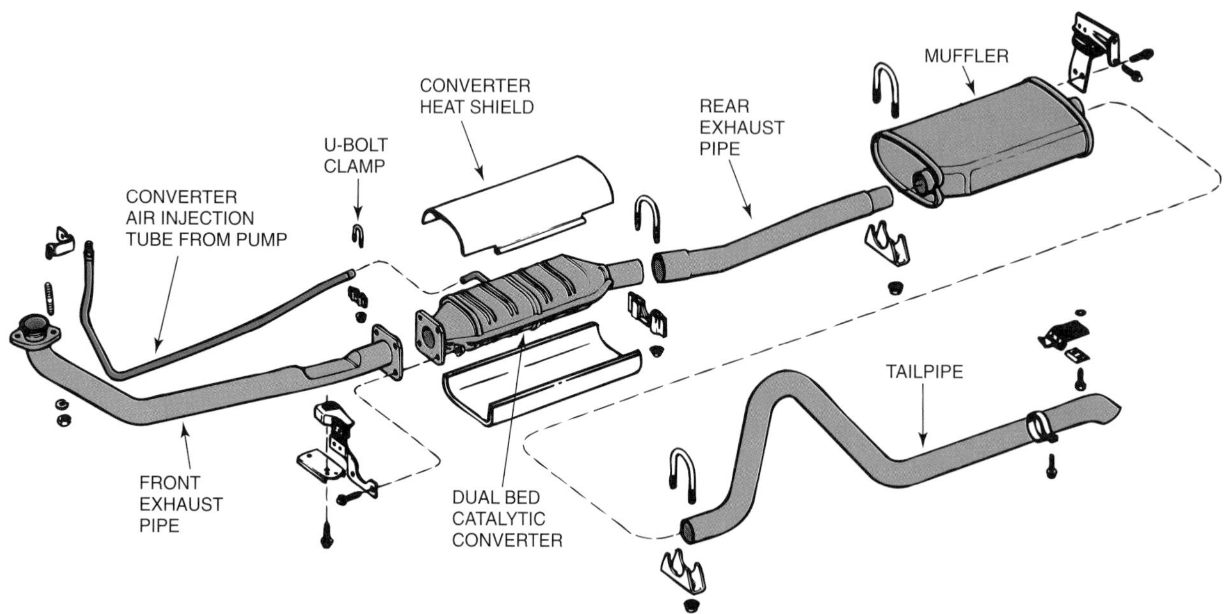

Figure 13-20. This exhaust system has an air injection pump tube that is secured to the catalytic converter with a U-bolt clamp. (Dodge)

Heat shielding
Semi-ball connector
Full-ball connector
Exhaust system clamp
Muffler clamp

Review Questions—Chapter 13

Do not write in this book. Write your answers on a separate sheet of paper.

1. All of the following statements about exhaust system back pressure are true, EXCEPT:
   (A) excessive back pressure can reduce engine power and mileage.
   (B) back pressure occurs when the exhaust gas flow is restricted.
   (C) complete lack of back pressure can cause exhaust valves to burn.
   (D) excessive back pressure makes it harder for cylinders to clear themselves during the exhaust stroke.
2. List two types of muffler construction.
3. Where is double-wall construction used?
   (A) Exhaust pipes.
   (B) Mufflers.
   (C) Resonators.
   (D) Tailpipes.
4. A resonator is used to _____ .
   (A) reduce exhaust sounds
   (B) force air into the converter
   (C) prevent exhaust back pressure
   (D) cool the exhaust gases
5. Explain why exhaust systems on late-model vehicles last longer than those on older vehicles.
6. Technician A says that exhaust system hangers hold the exhaust system rigidly to provide support. Technician B says that the hangers cannot allow exhaust system noise and vibration to be transmitted to the vehicle body. Who is right?
   (A) A only.
   (B) B only.
   (C) Both A & B.
   (D) Neither A nor B.
7. The modern exhaust system produces a great deal of heat. When installing catalytic converters and exhaust pipes, all _________ must be in place to prevent damage to adjacent parts.
8. Modern manifolds are designed with tubes of equal length to make use of the exhaust _____ effect.
9. Many pipe-to-pipe connections are held together with _____ .
   (A) clamps
   (B) welds
   (C) gaskets
   (D) Both A & B.
10. Welded exhaust system components must be _________ if they need to be replaced.

Manufacturers are experimenting with zero emissions vehicles, such as this electric car. A three-phase alternating current motor delivers the driving force to the wheels. (General Motors)

14

Emission Controls

After studying this chapter, you will be able to:

- Define automotive emissions.
- List the three most harmful automotive emissions.
- List the three major classes of emission controls.
- Explain the operating principles of common emission control systems.
- Identify the major parts of common emission control systems.

This chapter covers the various types of emission controls used on modern vehicles. Many emission control devices have been developed over the last 30 years. Many early devices were developed as the need arose. Most modern emission controls are descendants of these devices. These modern systems are fully integrated into the overall engine system and are usually monitored and controlled by the engine control computer. After studying this chapter, you will understand how these systems work and how they are coordinated to keep emissions as low as possible. This chapter will also prepare you for the computer control information in Chapter 15.

The Need for Emission Controls

Many exhaust emissions, such as water, oxygen, nitrogen, and carbon dioxide, are of little concern. Others, such as ***hydrocarbons*** (HC), ***oxides of nitrogen*** (NO_x), and ***carbon monoxide*** (CO), can pose health and environmental problems if they are not controlled. For example, hydrocarbon emission into the atmosphere tends to produce photochemical (sunlight causing chemical reactions) smog.

Emission Control Systems

Progress has been made in the development of emission control devices. When properly maintained, these units do an excellent job of reducing harmful emissions. Emission controls may be divided into three general types:

- Engine modifications and controls.
- External cleaning systems.
- Fuel vapor controls.

Engine modifications and controls include the basic design of internal engine parts, such as combustion chamber shape, camshaft lift and duration, intake manifold design, and the controls used to alter ignition timing, fuel mixture, and combustion chamber temperature.

External cleaning systems are used to help the exhaust gases continue the burning process as they move through the exhaust system. These controls include the air injection system and the catalytic converter.

The ***fuel vapor controls*** are designed to prevent the escape of gasoline vapor, a common source of unburned hydrocarbons, from the tank filler cap, tank, carburetor, and other vehicle components. Also falling under vapor control is the control of engine crankcase fumes caused by unburned fuel leaking past the piston rings.

Engine Modifications and Controls

Engine modifications and controls are designed to more completely burn the fuel charge within the combustion chamber and to ensure that all gases leaving the chamber are as environmentally safe as possible.

Engine Modifications

In modern vehicles, internal engine parts are designed to balance the needs of performance, emissions, and driveability. The most common of these internal modifications are explained in the following sections.

Combustion Chamber Shape

The combustion chamber in modern cylinder heads is designed to reduce the amount of hydrocarbons in the exhaust. Older head designs had complex combustion chamber shapes, with areas that almost touched the piston head when the piston was at the top of the cylinder. This mixed the air and fuel more thoroughly, producing more power with less detonation. However, it caused unburned fuel to condense and cling to the sides of the combustion chamber without burning.

Combustion chambers in newer heads have simple shapes and small surface areas to reduce fuel condensation. They also have a larger volume, which lowers compression ratios. The combustion chamber area is reduced, minimizing the chance for fuel to become trapped.

Lower Compression Ratios

Modern engines have lower compression ratios than older engines. This allows modern engines to operate on low-octane unleaded gasoline without detonating or dieseling. Since lower compression also lowers the temperature of the combustion process, the oxygen and nitrogen are less likely to combine into NO_x. In addition, the lower compression allows the engine to operate on unleaded gas, which must be used if the vehicle has a catalytic converter.

Camshaft Designs

Modern camshafts have less ***overlap*** (the time when both the intake and exhaust valves are open) than older camshaft designs. Camshafts with large amounts of overlap give the engine good high speed performance. However, this overlap causes the incoming air-fuel mixture to be diluted by exhaust gases at idle and low speeds. On older engines, this dilution was overcome by rich idle mixtures. Since it is no longer desirable to have rich idle mixtures, overlap has been reduced to prevent exhaust dilution and roughness at low speeds. These "milder" camshafts reduce engine power to increase smoothness.

Hotter Thermostats

Cooling system thermostats used on modern engines do not begin to open until the coolant temperature has reached at least 190°F (88°C). Thermostats used on older vehicles opened between 160° and 180°F (71° and 82°C), depending on the vehicle.

The modern engine cooling system warms up quickly and operates at higher temperatures than older systems. This keeps the combustion chamber at higher temperatures for most of the time that the engine is operating. Higher temperatures reduce gasoline condensation on the combustion chamber surfaces and cylinder walls, reducing the levels of HC.

Fuel System Controls

Most fuel system controls involve producing a leaner air-fuel ratio—one in which there is a higher percentage of air.

Lean Carburetors

Carburetors and fuel injection systems on newer vehicles operate at an air-fuel ratio as close to 14.7:1 as possible. This is the ideal ratio for mileage and emissions, but not necessarily for power and driveability. Newer carburetors have smaller main jets, reduced accelerator pump output, and power valve springs that do not allow the power valve to open until the engine is under very heavy loads. The idle mixture screws are sealed to prevent readjustment. See **Figure 14-1.**

Modern carburetors are controlled by the engine computer through a mixture control solenoid, **Figure 14-2.** When the engine is warming up, or when wide open throttle operation is necessary, the air-fuel ratio may be enriched by the choke, accelerator pump, or power valve. Rich ratios are used as little as possible.

Other ways in which the carburetor has been modified to lower emissions are faster-opening chokes, more precise idle speed control, and the use of throttle positioners. These devices were covered in detail in Chapter 10.

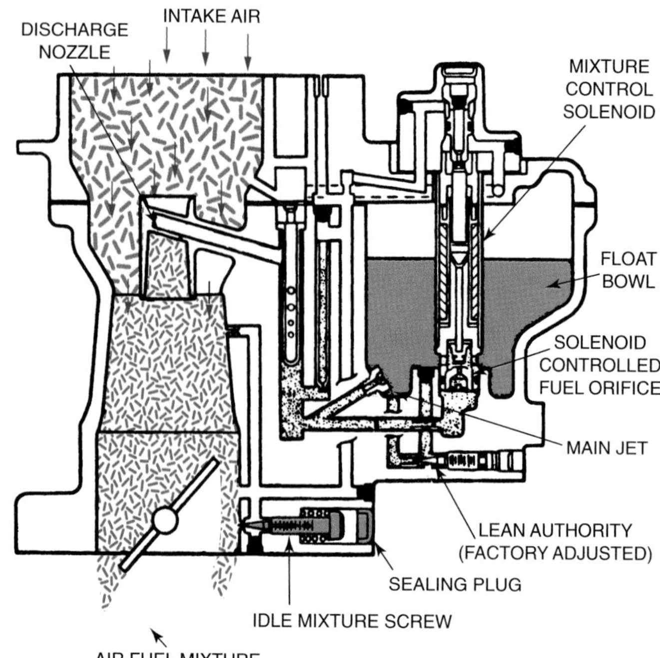

Figure 14-1. Cross-sectional view of a carburetor with a sealed idle mixture adjustment screw and sealing plug. (Pontiac)

Lean Carburetor Dieseling

Due to lean carburetor mixtures, high operating temperatures, and fast idle speeds, some modern engines develop hot spots in the combustion chambers that ignite the fuel mixture and tend to keep the engine turning after the ignition key is turned off. This is commonly known as ***dieseling*** or ***run-on***. Besides being extremely annoying, dieseling can produce a considerable amount of pollution.

To prevent dieseling, some engines employ an electric anti-dieseling solenoid. When the key is on, the solenoid keeps the throttle valve open far enough for normal idle. When the key is turned off, the solenoid allows the throttle to close completely. This shuts off all air, and the engine cannot diesel. Another setup automatically engages the air conditioning compressor drive clutch for several seconds whenever the key is turned off. This places the engine under a small load, causing it to slow down and stop without dieseling.

The ultimate solution to air-fuel ratio control is computer-controlled electronic fuel injection. Fuel injection, discussed in Chapter 9, has replaced the carburetor on all late model engines.

Quick Air-Fuel Mixture Heatup

To allow a cold engine to run properly with lean mixtures, the incoming air must be heated to prevent fuel condensation in the intake manifold. Heating the incoming air allows the choke to be opened sooner, reducing the amount of HC. It also helps prevent ***carburetor icing***. Carburetor icing occurs when the lower air pressure in the carburetor venturi causes water vapor to freeze in the carburetor throat. This ice formation can cause the engine to run roughly or stall in cold, damp weather.

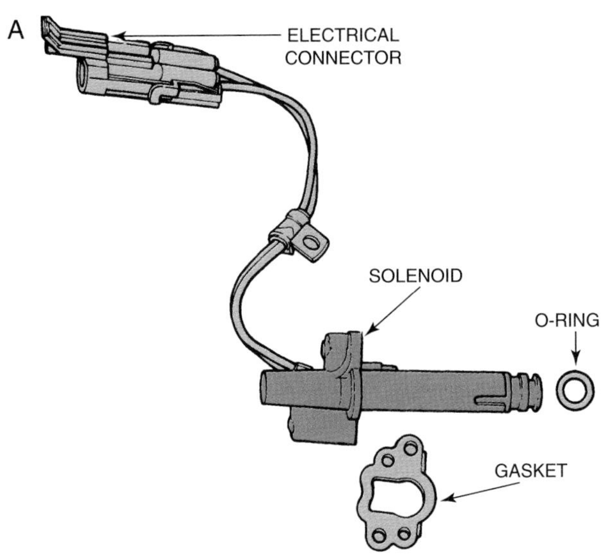

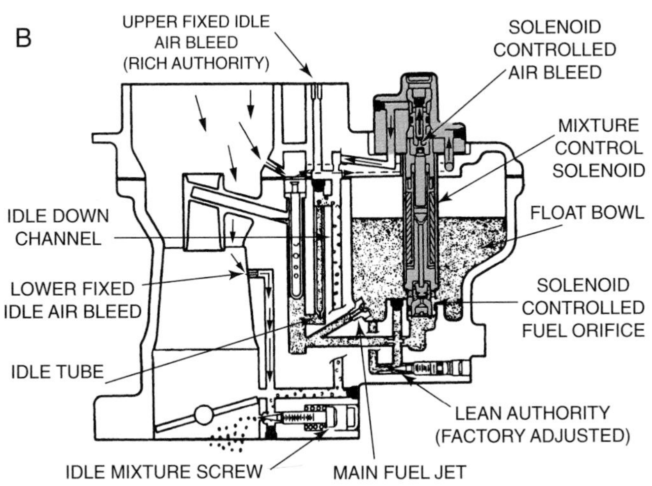

Figure 14-2. A—Electrically operated mixture control solenoid with O-ring seal, gasket, and electrical conneciton harness. B—Cross-section of a carburetor and mixture control solenoid mounted in place. This mixture solenoid is cycled up and down ten times a second to carefully control the air-fuel mixture. The solenoid is preset and sealed at the factory. (Pontiac)

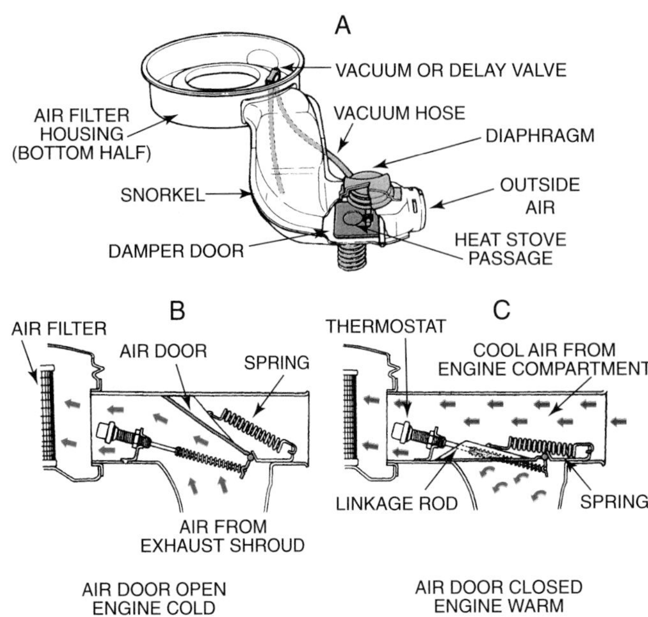

Figure 14-3. Two different types of thermostatic air cleaners (Thermac). A—This air cleaner uses a vacuum-operated diaphragm to control the heat stove. B & C—This illustrates the air door position on an air cleaner which uses a thermostat linkage assembly to open the air door. Study the air flow arrows. (AC-Delco & Sun Electric Corp.)

Thermostatic Air Cleaner

The ***thermostatic air cleaner*** is one of the devices used to heat the incoming air. It employs a special valve in the air inlet section of the air cleaner housing. See **Figure 14-3.** When the valve is closed, ambient temperature air is drawn in. When the valve is fully opened, heated air (drawn from a shroud around the exhaust manifold) is admitted and the cooler air is excluded. At positions between fully open and fully closed, a blend of both heated air and ambient air is admitted. This permits smooth engine operation on relatively lean fuel mixtures.

The modern air control door is actuated by a vacuum valve or motor, **Figure 14-4.** When the engine is started cold, a vacuum motor is employed. The cold engine start position is shown in **Figure 14-4B.** The thermal sensor bleed valve is closed, and full vacuum is applied to the vacuum motor. This pulls the diaphragm up against spring pressure and fully opens the air valve to allow the entrance of heated air.

When the thermal sensor warms, it opens the air bleed. This stops vacuum to the vacuum motor, allowing the spring to force the diaphragm down and close the valve, **Figure 14-4C.** A cutaway view of the complete thermostatically controlled air cleaner assembly is illustrated in **Figure 14-5.**

Early Fuel Evaporation System

The ***early fuel evaporation system*** (EFE) warms the incoming air by transmitting exhaust heat to the exhaust crossover under the intake manifold. A vacuum-operated valve provides this enriched flow of hot gases during cold engine operation. This improves engine performance and reduces emissions. On older vehicles, a thermal vacuum switch controls the EFE valve. Later EFE valves are controlled by a solenoid that is operated by the engine control computer.

When coolant temperature is below a certain point, the vacuum switch or solenoid applies vacuum to the EFE valve, which, in turn, diverts the gases up under the intake manifold before they exit through the exhaust system. See **Figure 14-6.** EFE systems were discussed in detail in Chapter 10.

Spark Timing Controls

A number of systems have been developed to alter ignition spark advance to meet most engine operating conditions. Most of these systems retard the operation of the vacuum advance unit, when used, or use the engine computer to modify the timing.

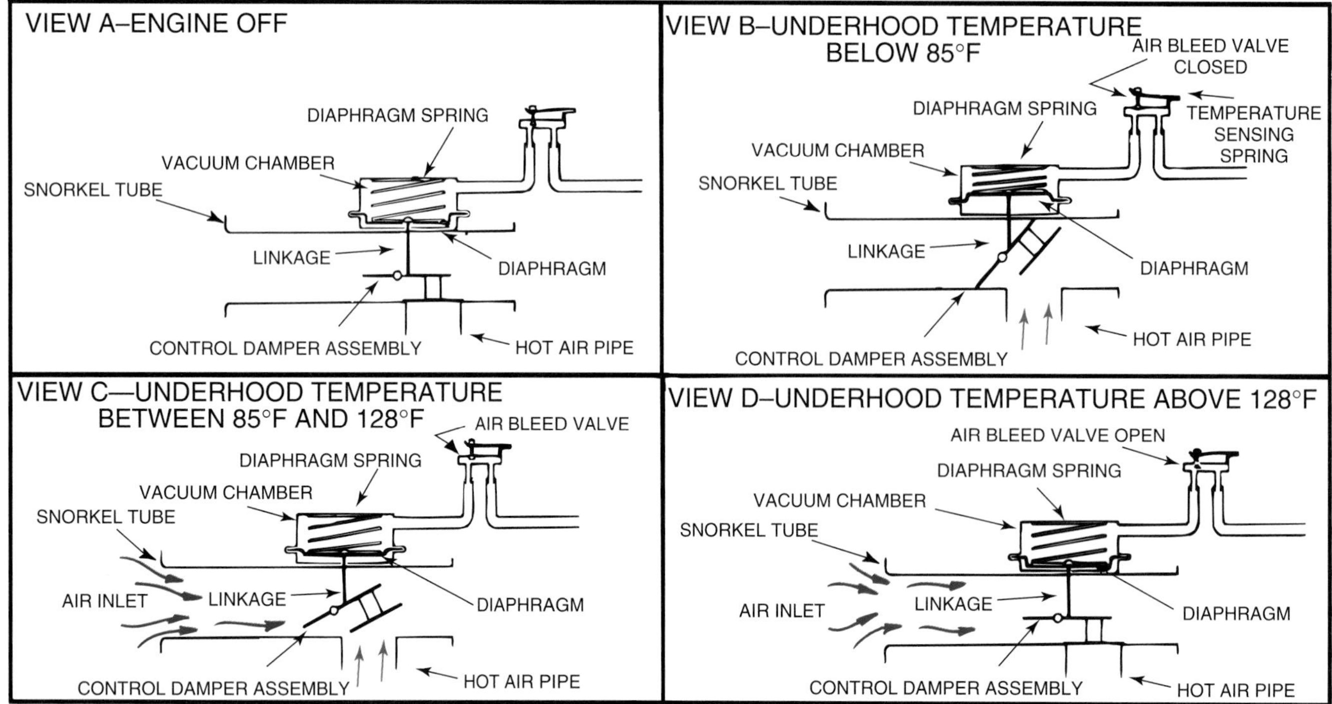

Figure 14-4. Operation of the thermostatically controlled air cleaner. Study each view and note the action of various controls in determining the position of the control damper.

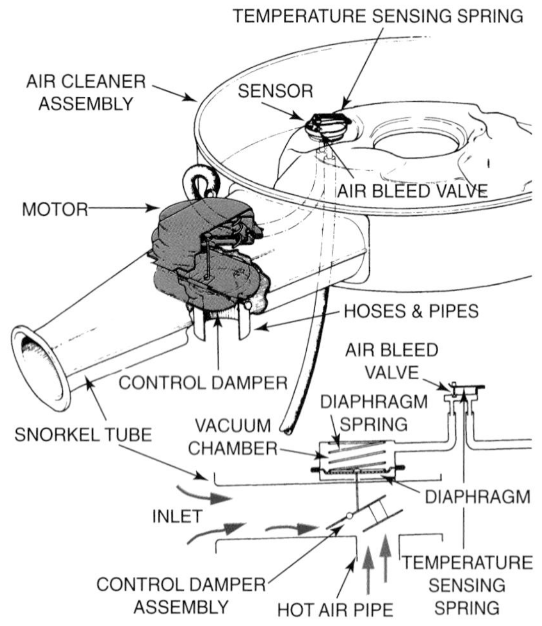

Figure 14-5. Construction details of a thermostatically controlled air cleaner. The damper is blending both hot and cold air. (Buick)

Older vehicles may have a ***vacuum advance restrictor,*** which is attached between the vacuum advance unit and the intake manifold. On these systems, the advance unit is connected to manifold vacuum through a ported vacuum opening on the carburetor. The ported vacuum connection ensures that no vacuum reaches the advance unit at idle speeds.

A vacuum restrictor is inserted into the vacuum line. When the throttle is opened, ported vacuum is forced to pass through the restrictor. The restrictor prevents full vacuum from reaching the advance unit for up to 30 seconds after the throttle is opened. This lowers HC emissions while providing good highway mileage.

On other older vehicles, solenoid-operated vacuum valves are installed in the vacuum line from the manifold to the advance unit. These valves are energized by electrical switches on the transmission or speedometer. The vacuum advance cannot receive full vacuum until the vehicle is in high gear or has reached a certain speed. Some systems allow for overriding of the control system when engine or outside air temperature is below a certain point. When a temperature override switch opens, the solenoid is de-energized, providing full advance until the engine warms up. A vacuum advance control device is shown in **Figure 14-7.**

Since the introduction of computer engine controls, ignition timing has been precisely controlled by the engine computer. Modern systems use a spark control computer and a number of engine sensors to provide instantaneous timing control. This permits smooth engine operation on lean fuel mixtures.

Exhaust Gas Recirculation (EGR)

When peak combustion temperature exceeds 2500°F (1372°C), the nitrogen in the air mixes with the oxygen to produce oxides of nitrogen (NO_x). By introducing a metered portion of the hot exhaust gases into the intake manifold, the ***exhaust gas recirculation system,*** or EGR system, dilutes

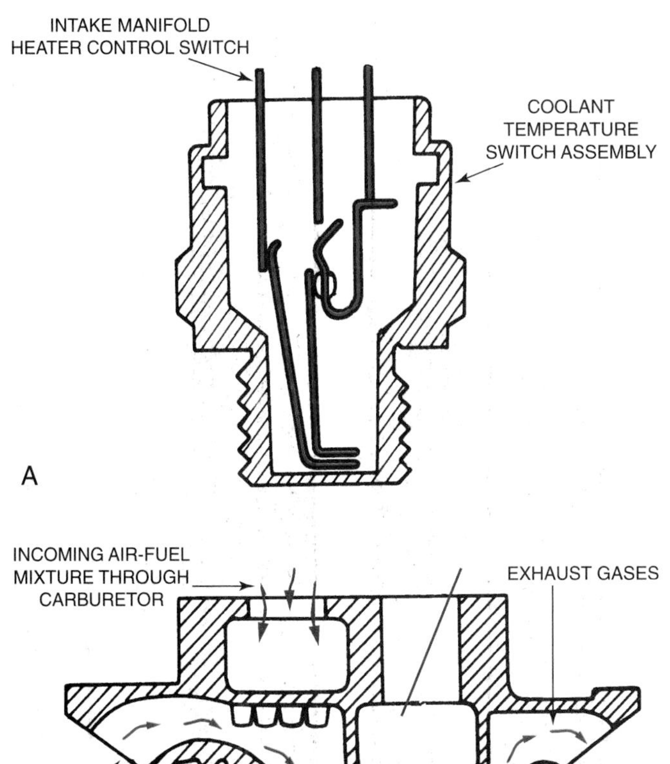

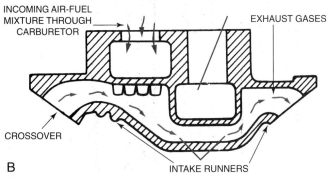

Figure 14-6. A—Early fuel evaporation (EFE) coolant temperature operating switch schematic. B—Exhaust gases crossing under the intake manifold to warm the incoming air-fuel mixture. (OTC, Sun)

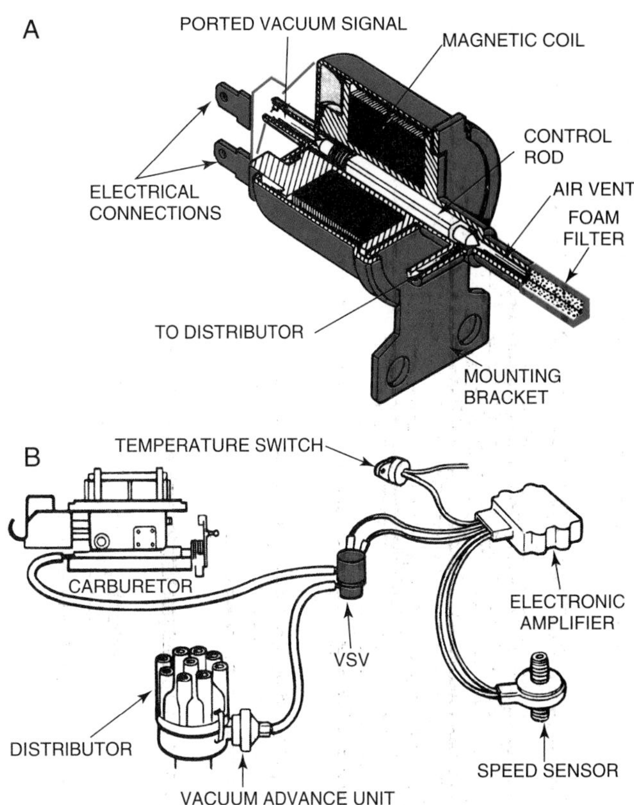

Figure 14-7. Various electrical solenoid-operated vacuum valves. A—Cutaway of an electrically operated vacuum valve. B—Overall system using a vacuum solenoid valve (VSV). This system uses the speedometer cable sensor to indicate vehicle speed. The amplifier boosts the weak speed sensor electrical signal so that it is strong enough to operate the vacuum solenoid valve. (Sun)

the incoming fuel charge and lowers combustion temperatures. This reduces the level of NO_x.

The amount of exhaust gas that is recirculated is controlled by an EGR valve. Such valves are widely used, as they can be designed to admit carefully controlled amounts of exhaust gases into the intake air stream. The exact amount required will vary according to engine speed and loading.

There are three basic types of EGR valves:

- Vacuum modulated (ported) valves.
- Back pressure modulated valves.
- Electronic-vacuum modulated valves.

Vacuum Modulated (Ported) EGR Valve

The ***vacuum modulated EGR valve*** uses a vacuum diaphragm to operate the gas flow control valve, or pintle. The vacuum tube is connected to a timed port in the carburetor or throttle body. The port is located just above the throttle valve in the idle position. When in the idle position, the throttle valve covers the port and no vacuum is applied to the EGR valve, **Figure 14-8A.** Note the diaphragm spring forces the valve closed and exhaust gas does not flow.

When the throttle valve is opened beyond the idle position, the port is uncovered and engine vacuum is applied to the EGR valve. The vacuum causes the diaphragm to raise upward against spring tension. This draws the valve (pintle) upward, allowing exhaust gas to flow through the valve into the intake manifold. See **Figure 14-8B.**

The EGR valve remains closed during periods of idle or heavy deceleration (accelerator completely released). Flow is reduced or stopped during heavy acceleration because low engine vacuum is not strong enough to operate the diaphragm.

The EGR valve should remain closed when the engine is cold because a cold engine will run rough if the EGR valve is opened. On some engines, a temperature-sensitive control is placed between the EGR valve and the carburetor or throttle body. Below certain temperatures, the EGR vacuum line is closed, preventing the EGR valve from operating. A schematic showing how the timed port on an engine with a carburetor admits vacuum to the EGR valve diaphragm is shown in **Figure 14-9.**

Back Pressure Modulated EGR Valve

Some ***back pressure modulated EGR valves*** use two separate diaphragms, while others, such as the valve illustrated in **Figure 14-10,** utilize a single diaphragm that serves two separate functions.

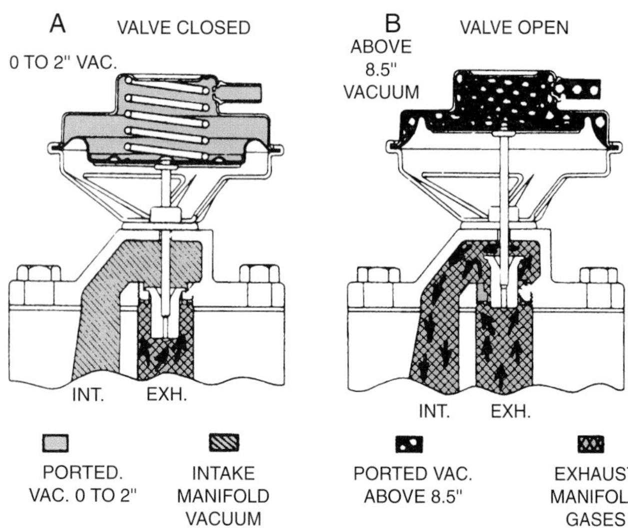

Figure 14-8. Action of the vacuum diaphragm operated EGR valve. Note how the engine vacuum is used to open the valve in B.

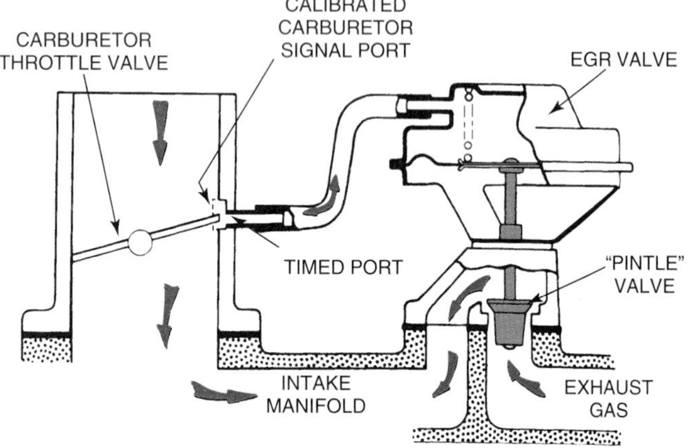

Figure 14-9. When the throttle valve starts to open, the timed port is uncovered, admitting engine vacuum to the EGR vacuum diaphragm. This opens the pintle valve. (General Motors)

The power diaphragm portion, shown in **Figure 14-10A,** is secured between the outer edges of the diaphragm plate and the valve stem support housing. The center, or control diaphragm portion divides the chamber formed between the diaphragm plate and the valve stem support housing. The power diaphragm plate is spring-loaded. The spring closes the pintle valve when the engine is stopped or when intake vacuum cannot be applied to the power diaphragm.

In **Figure 14-10A,** the engine is operating. Timed vacuum is applied to the power diaphragm. The control valve is open as the control diaphragm is held down by the valve spring. As long as it remains open, air is allowed to enter and a vacuum cannot be formed above the power diaphragm. Until a sufficient vacuum is formed, the power diaphragm will remain down, keeping the pintle valve closed. Although exhaust gases cannot pass the pintle valve, they move up through the hollow valve stem and enter the chamber beneath the control diaphragm. Pressure builds up in the chamber and attempts to force the control diaphragm upward.

In **Figure 14-10B,** the exhaust gas back pressure has risen to the point that it has forced the control diaphragm upward, causing it to seal off the control valve opening. Because air can no longer pass, vacuum is formed above the power diaphragm. Atmospheric pressure acting on the bottom of the power diaphragm housing assembly moved the entire unit up, opening the pintle valve. Exhaust gas now passes into the intake manifold.

Electronic EGR Valve

The operation of the ***electronic EGR valve*** is similar to that of a ported EGR with a vacuum regulator. The electronic EGR valve has a sealed cover. A solenoid valve opens and closes the vacuum signal, which in turn controls the amount of vacuum that is vented into the atmosphere. The solenoid is controlled by the engine control computer, which is discussed in more detail in Chapter 15. The EGR valve is cycled on and off many times per second. See **Figure 14-11.**

External Cleaning Systems

Unlike the systems discussed above, that reduce the amount of harmful gases produced by the engine, external cleaning systems clean up the exhaust gases after they leave the engine. The most common external cleaning systems are the air injection system and catalytic converter.

Air Injection System

Hydrocarbon and carbon monoxide emission through the exhaust system can be reduced by thoroughly burning the fuel remaining in the exhaust gases. This is performed by the ***air injection system***. The air injection system is sometimes called an Air Injection Reactor, Thermactor, Air Guard, or simply, a smog pump.

The exhaust gases leaving the engine on the exhaust stroke contain unburned hydrocarbons (HC). The exhaust stream is still hot enough to continue burning these hydrocarbons, but there is not enough oxygen to do so. Injecting outside air into the burning gases as they pass the exhaust valve adds more oxygen, causing the burning process to continue. This consumes a great deal of the normally unburned hydrocarbons and changes a significant portion of the carbon monoxide gas into harmless carbon dioxide. **Figure 14-12** illustrates one setup for injecting air into the area near the exhaust valve.

Air Injection System Action

A typical air injection system is illustrated in **Figure 14-13.** Note that the air pump draws fresh air in through the air cleaner. Some other setups draw air through a separate filter. The air is compressed to a slightly higher value than that of the exhaust gases in the manifold. The compressed air is then directed, via a hose, to a check valve and into a distribution manifold. An air bypass valve, or diverter valve, and a switching valve may also be used.

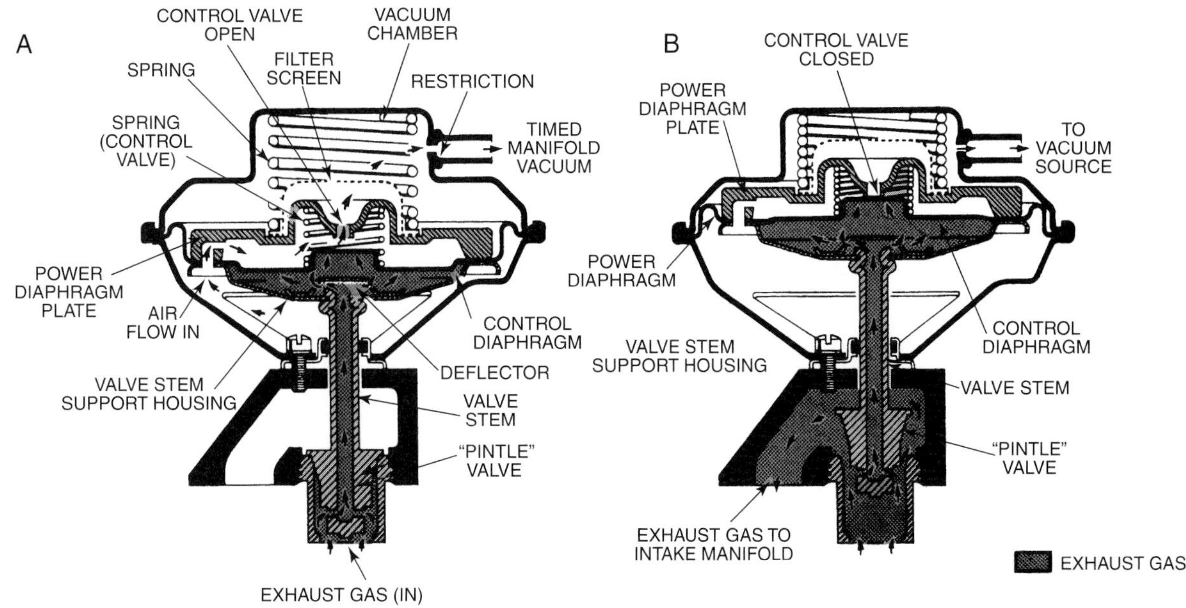

Figure 14-10. An exhaust back pressure modulated EGR valve. A—When there is low back pressure, the valve remains closed. B—The exhaust back pressure closes the control valve. A vacuum forms above the power diaphragm plate, drawing the pintle valve up to the open position. (Buick)

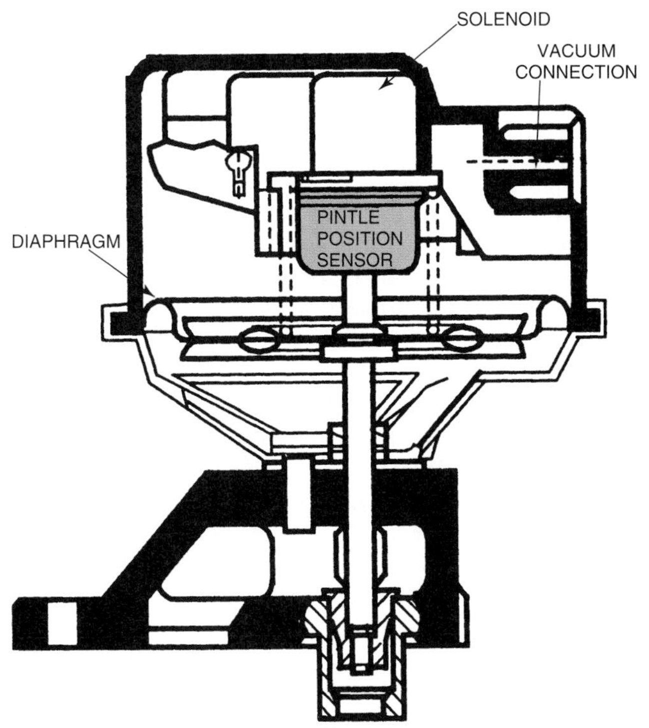

Figure 14-11. An electronic EGR valve. An electrical signal is sent to the ECM by the pintle position sensor. The ECM then knows how much exhaust is flowing. This valve also uses vacuum to modulate EGR operation. (Buick)

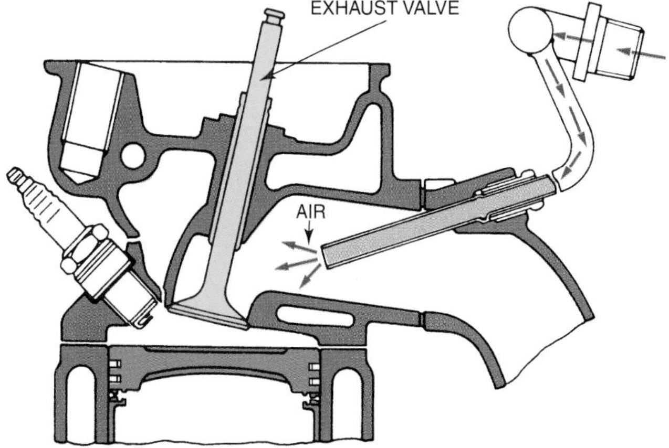

Figure 14-12. When air is injected into the exhaust stream, it provides the oxygen needed to keep exhaust gases burning. This helps reduce emissions of HC and CO. (Nissan)

Air Pump

The air pump, or compressor, is usually of the positive displacement vane type. A relief valve limits the amount of pressure the pump can develop. Spring-loaded carbon shoes permit sliding contact between the vanes and the rotor. The pump is belt driven from the engine. **Figure 14-14** shows a cutaway of a typical air pump.

As each rotating pump vane passes the intake chamber, **Figure 14-15,** air is drawn in. The next vane pushes the air around into the compression chamber. The compression chamber reduces the volume of the air, pressurizing it. Pressurized air is then discharged through the exhaust chamber. Note that the intake and exhaust chambers are separated by the section of the housing called the stripper.

Check Valve

The one-way check valve, **Figure 14-16,** is forced open by the air passing through to the distribution manifold. In the

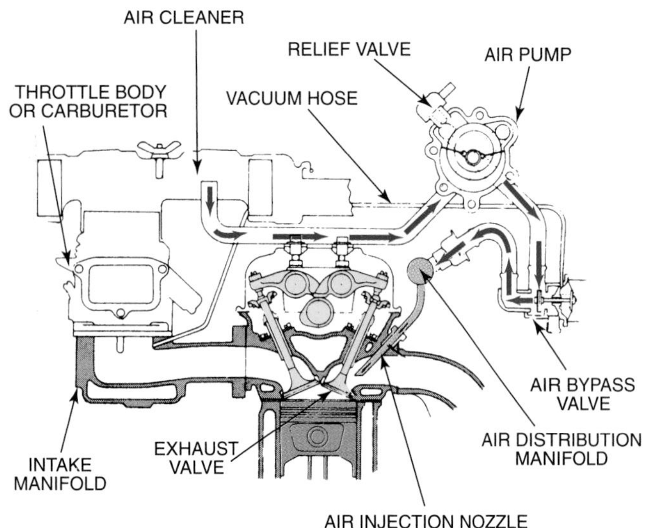

Figure 14-13. Typical air injection exhaust emission control system. (Chevrolet)

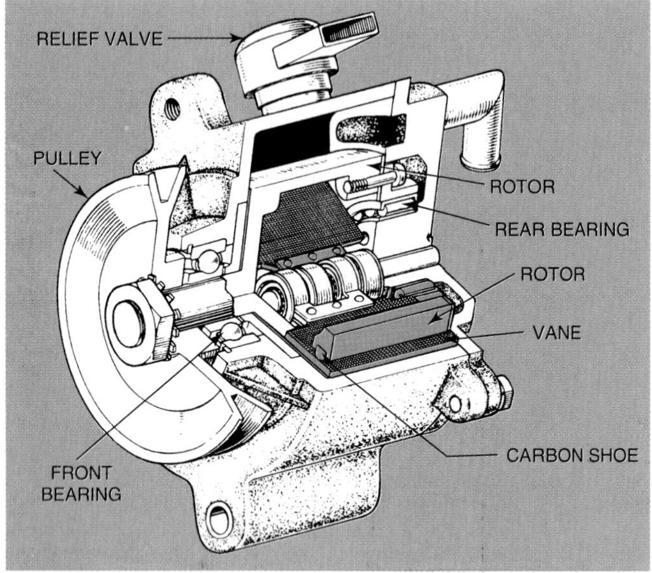

Figure 14-14. Cutaway of an air injection pump. (Toyota)

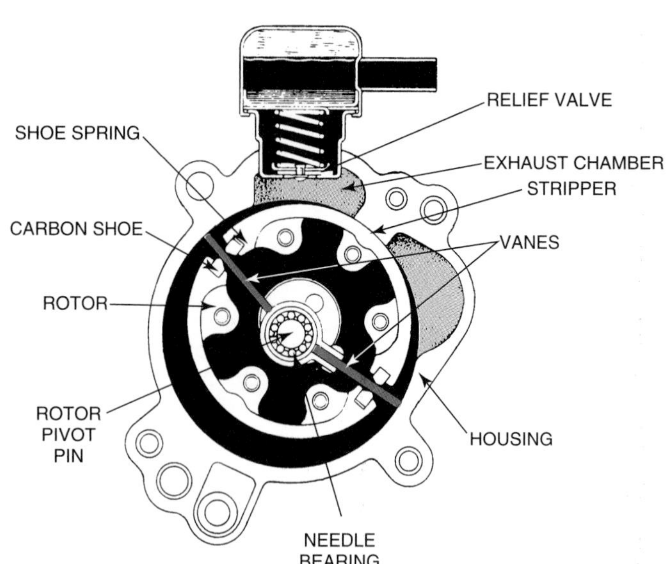

Figure 14-15. End view of an air injection pump. Note how the outer edges of the vanes maintain constant contact with the housing wall.

event of pump failure, or any time that exhaust manifold pressure equals or exceeds pump pressure, the check valve will seat and prevent back feeding of hot exhaust into the hoses or pump.

Distribution Manifold and Air Injection Tubes

An air distribution manifold and air injection tube setup is shown in **Figure 14-16.** Some engines incorporate the distribution manifold and injection tubes as an integral part of the cylinder head.

Diverter Valve

When the throttle valve is closed quickly, gasoline continues to flow. This is most common with carbureted engines, where the manifold vacuum can draw fuel through the idle system, but it can occur on some fuel injected engines. Since little or no air is entering the engine during this period, the flow of gasoline produces a rich mixture. The rich mixture will leave a considerable amount of unburned gas following the power stroke. When this unburned gas enters the exhaust manifold and meets the injected stream of oxygen, it will begin to burn again with explosive force, causing backfiring.

The ***diverter valve*** is designed to momentarily divert the air stream away from the injection nozzles. It does this by means of a vacuum diaphragm or an electrically controlled metering valve. See **Figures 14-16** and **14-17.**

Diverter valve action is shown in **Figures 14-18** and **14-19.** Airflow is passing through the diverter valve and on to the air injectors. When the throttle is suddenly released, heavy intake manifold vacuum is applied to the metering valve control diaphragm. The diaphragm is drawn downward, forcing the metering valve to block off the passage to the air injection manifold. This downward movement opens up the diverter passage and pump air is momentarily discharged into the atmosphere. See **Figure 14-19.**

Air Switching Valve

Some air injection emission control setups use ***air switching valves***. In some systems, these valves can be made as an integral, combination unit. In other systems, separate valves are used to perform specific functions.

The switching valve can be used to perform the normal diverter valve functions. It can also be used to channel pump air to the intake manifold, exhaust manifold, or catalytic converter, as determined by system needs. **Figure 14-20** shows one switching valve that, during engine warm-up, channels pump air to the exhaust port area. When a specific engine temperature is reached, the valve diverts a large percentage of the air to the catalytic converter. Some air continues to be

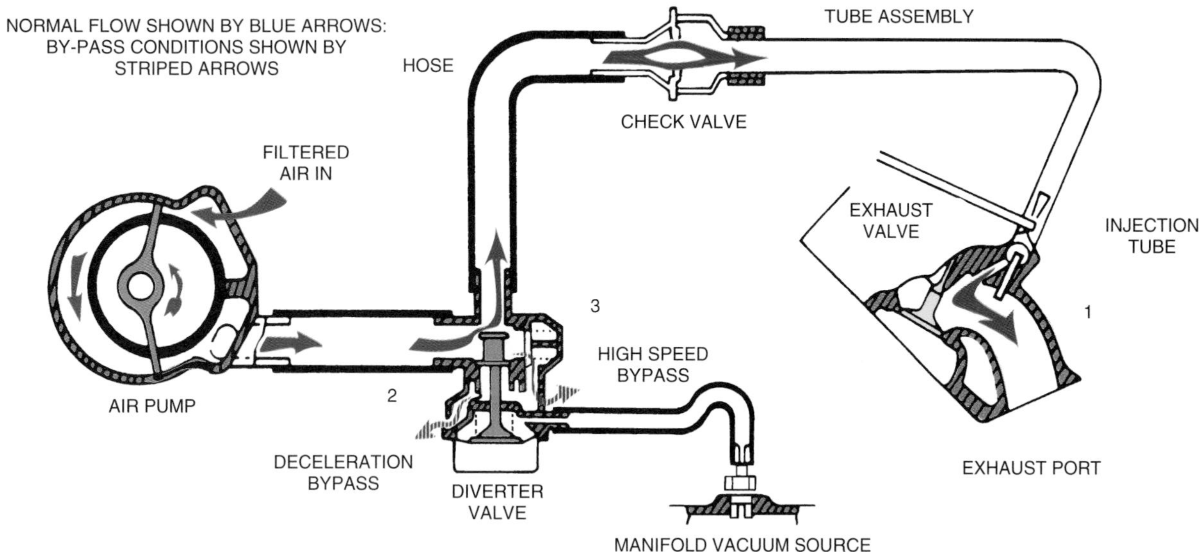

Figure 14-16. Air injection system diverter valve functions. 1–Admitting air to the exhaust port. 2–Diverting air during deceleration. 3–Relieving excessive air pressure during high-speed operation. (General Motors)

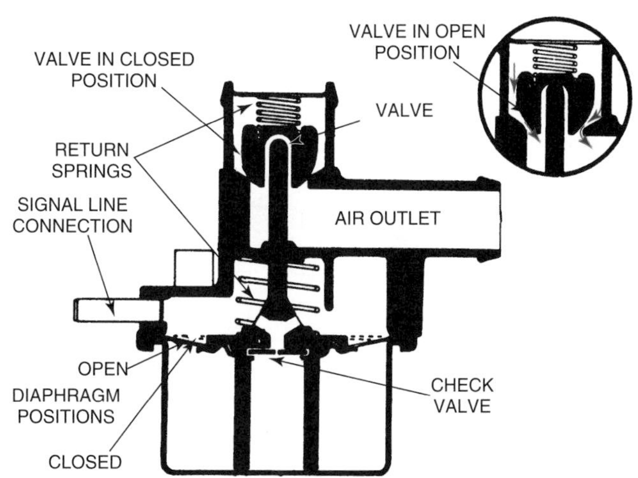

Figure 14-17. The anti-backfire (diverter) valve admits air to intake manifold upon deceleration. The vacuum diaphragm forces the diverter valve open.

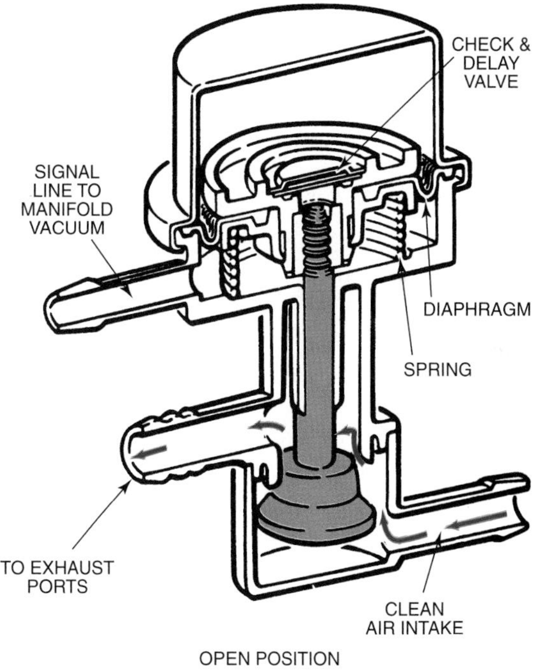

Figure 14-18. Deceleration diverter valve. This valve is shown in the open position. (General Motors)

sent to the exhaust manifold. The switching valve improves the operation of the air injector pump, converter, and EGR system to reduce emissions.

Pulse Air Injection

The ***pulse air injection system*** does not use an air pump. Instead, it employs the normal exhaust pressure pulsations to draw air into the exhaust manifold. This is similar to the exhaust scavenging effect discussed in Chapter 13. The momentary vacuum pulse draws fresh air in through an air injection tube. Continual alternating vacuum and pressure pulses produce a sufficient flow of fresh air to significantly increase exhaust burning (oxidization), thus reducing CO and HC emissions.

One setup employing the pulse air principle is shown in **Figure 14-21.** Note that a separate air pipe is used for each cylinder on this specific engine. Each pipe has a separate air check valve.

Pulse Air Injection Operation

The air injection tubes are connected to distribution (manifold) pipes that lead to an air filter, **Figure 14-22**. Some

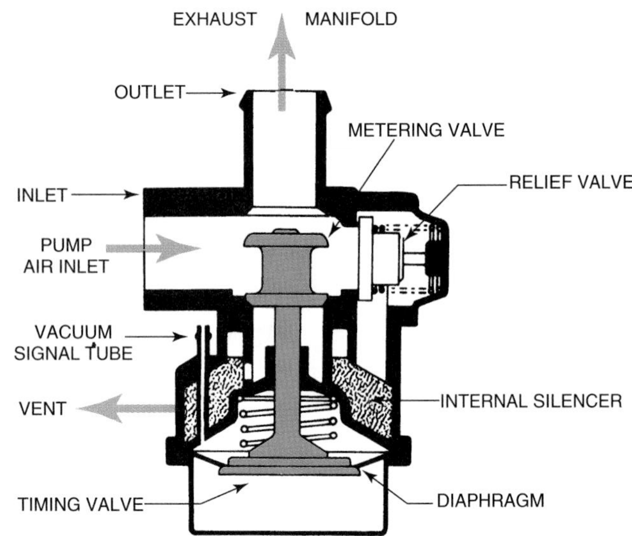

Figure 14-19. Cutaway of a diverter valve. Note the construction. (Oldsmobile)

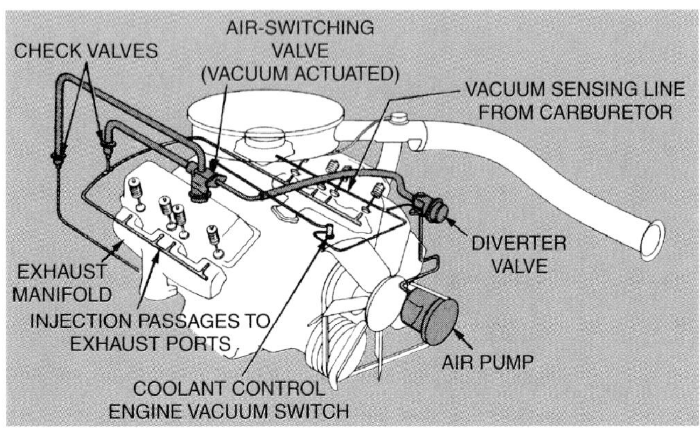

Figure 14-20. An air injection system using an air-switching valve for dual-flow injection control. (Chrysler)

installations use the air cleaner, while others use a separate filter. A one-way air check valve is used in the lines. Note that this arrangement also uses four injection tubes.

A different design pulse air system is shown in **Figure 14-23.** Instead of admitting fresh air to all the exhaust port areas, air is drawn into ports 3, 4, and 5 only. Air is fed continually (once per vacuum pulse) through pipe C. Air pipe D will allow air to be drawn into exhaust ports 4 and 5 only when the intake manifold vacuum is high enough to operate the diaphragm on check valve B. The system is shown with all pipes feeding air.

Pulse Air Check Valves Are Important

You will note that the various pulse air systems shown in **Figures 14-21** to **14-23** all use air check valves. The check vaives open during the vacuum pulse phase but snap shut during the pressure phase. This prevents back-feeding of exhaust gases and ensures that fresh air is drawn in.

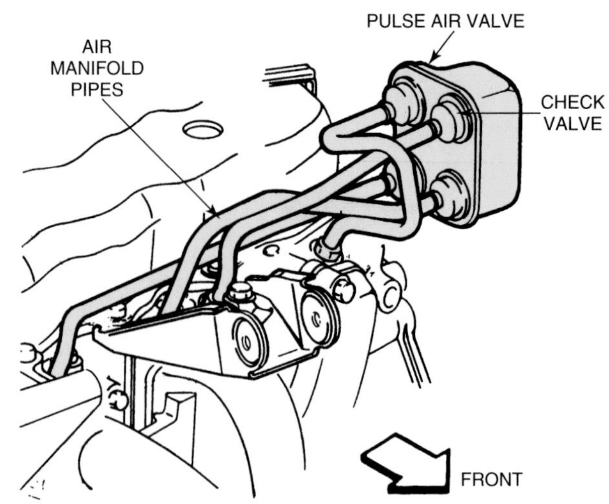

Figure 14-21. A pulse air injection system does not use an air pump. Exhaust pressure pulsations cause fresh air to be drawn in during the low pressure (vacuum) pulsations. (Buick)

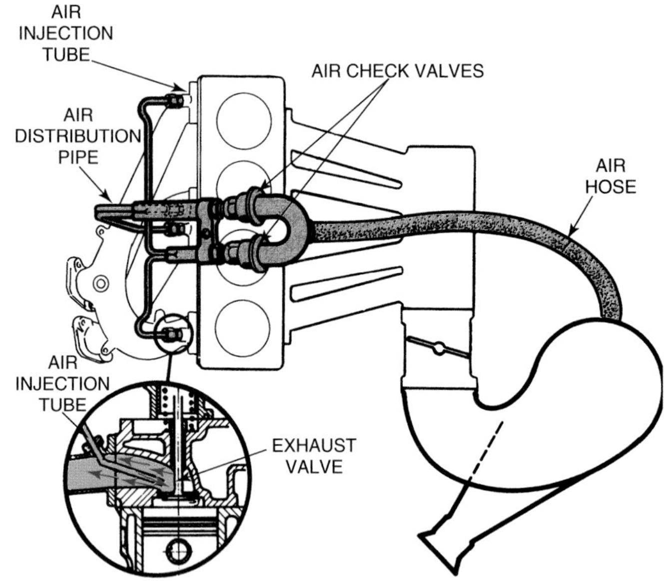

Figure 14-22. Pulse air injection system. Note how vacuum pulsation draws fresh air in through the injection tube. Airflow is from the air cleaner. (Saab)

Catalytic Converter

A ***catalytic converter*** (reactor) is a device installed in the exhaust line to significantly lower levels of hydrocarbons (HC), carbon monoxide (CO), and in the case of some converters, oxides of nitrogen (NO_x).

Converters are of two basic catalyst types: the ***two-way converter*** (oxidation) and the ***three-way converter*** (oxidation and reduction). Both types employ either a pellet or monolith design.

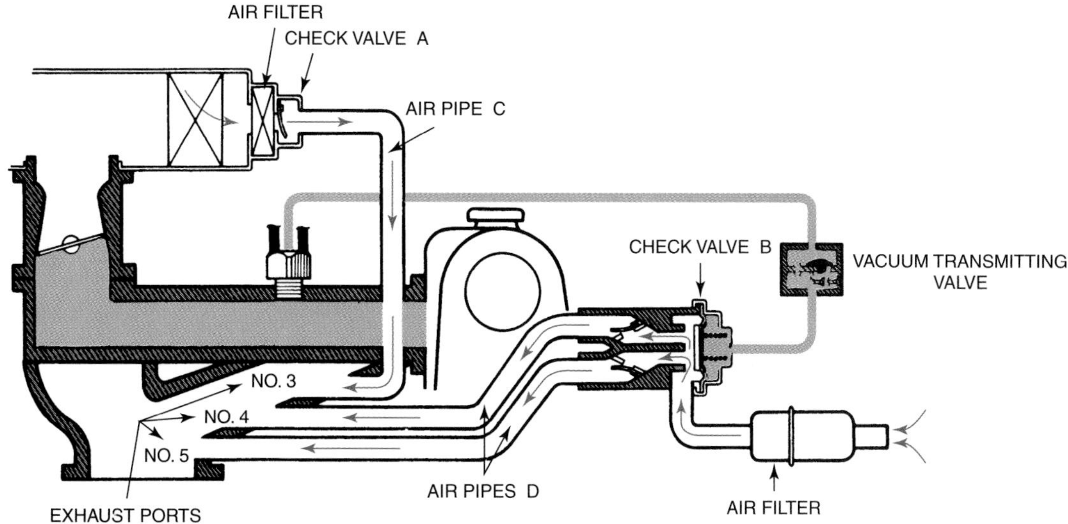

Figure 14-23. Another pulse air injection setup. Air is fed into number 3 port constantly. Additional air is fed into number 4 port and into number 5 port during periods of high vacuum. (Toyota)

Two-Way Converter

The two-way converter is used to reduce CO and HC. All early converters and some later converters are two-way designs. They can be subdivided into the ***pellet*** and the ***monolith***, or ***honeycomb,*** types. The pellet type is being replaced with the monolith type, which produces less back pressure in the exhaust system.

Pellet-Type Converter

In a pellet-type converter, thousands of tiny beads, or pellets, of porous aluminum oxide covered with a thin coating of platinum (Pt) and palladium (Pd) are placed inside a perforated container. The container, or "core," is then suspended inside the converter shell. Baffles are arranged in such a way that the exhaust gases enter from one side of the perforated core, flow through the pellet bed, and exit through the other end of the core.

Converters are made of stainless steel because of the very high operating temperatures, which can exceed 1600°F (872°C). The outer shell is lined with a layer of insulation to prevent heat transfer to other areas and to maintain high catalyst temperatures. A two-way, pellet-type converter is illustrated in **Figure 14-24.** Study the flow of exhaust through the converter.

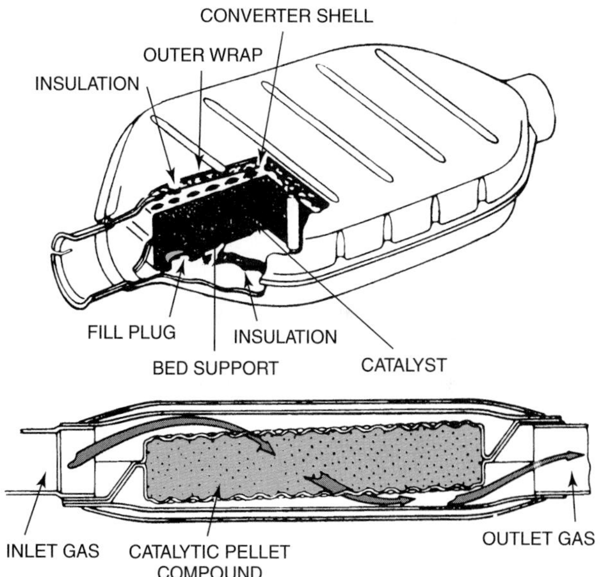

Figure 14-24. Two-way, pellet-type catalytic converter. Note how the exhaust gases are forced to travel through the catalyst bed (pellets). (Pontiac)

Monolith Converter

In place of porous beads, the monolith, or honeycomb, converter uses a ceramic honeycomb-like block that is coated with alumina and treated with a very thin coating of platinum and palladium. All exhaust gases flow through this treated honeycomb mass. A monolith converter is shown in **Figure 14-25.**

Two-Way Catalytic Converter Operation

The two-way catalytic converter (in some cases two are employed) is placed in the exhaust system between the exhaust manifold and muffler. When the hot gases are forced through the converter, they contact the catalyst-coated pellets or honeycomb, depending on the type. The catalyst causes a rapid increase in the exhaust temperature. This, in turn, causes the carbon monoxide and hydrocarbons to change (by means of an oxidizing process) into water (H_20) vapor and carbon dioxide (CO_2). The two-way oxidizing converter does not reduce oxides of nitrogen (NO_x). The catalyst itself will not be altered or consumed by this process.

Three-Way Converter

The three-way converter, sometimes called the dual bed converter, uses an additional catalyst bed coated with

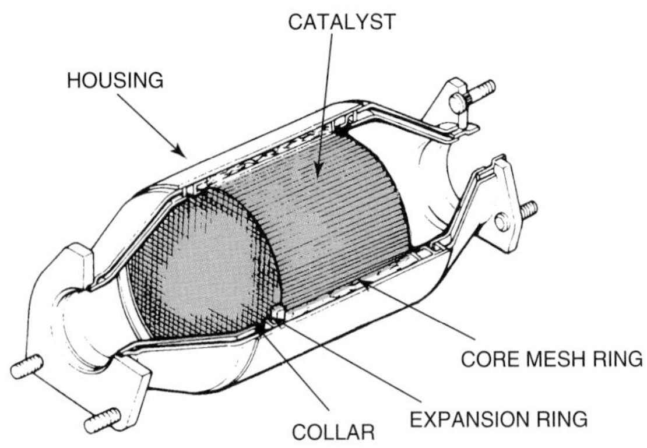

Figure 14-25. Two-way, monolith-type catalytic converter. The catalyst bed is of honeycomb construction to allow exhaust to flow through. (Honda)

platinum and rhodium (Rh). This bed not only helps reduce hydrocarbons and carbon monoxide, but it also lowers the levels of oxides of nitrogen.

One type of three-way, dual-bed monolith converter is shown in **Figure 14-26.** Note the use of two separate catalyst beds. They have a space between them into which air from the air injection pump may be forced.

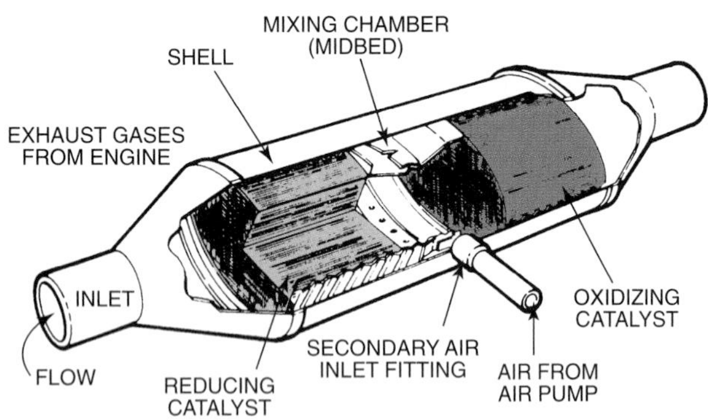

Figure 14-26. Three-way, dual-bed monolith catalyst converter. Note the secondary air pipe (air from air injection pump). This converter lowers the levels of hydrocarbons, oxides of nitrogen, and carbon monoxide. (Ford)

The front bed (inlet end) is treated with platinum and rhodium and is termed a ***reducing catalyst*** (catalyzes reduction of NO_x). The rear bed is coated with palladium and platinum and is referred to as an ***oxidizing catalyst***. Study the construction of the three-way converter in **Figure 14-26.** Note the air inlet from the air injection pump and the exhaust flow pattern.

Operation of a Three-Way Converter

Exhaust gases pass first into the reducing catalyst. Levels of oxides of nitrogen are reduced. There is also some oxidizing of hydrocarbons and carbon monoxide. The air injection system forces air into the passage between the catalyst beds. This adds oxygen to the exhaust gases, allowing a high oxidization heat level.

As the exhaust gases encounter the second oxidizing catalyst, tremendous heat is created and the resultant oxidizing process changes the carbon monoxide and hydrocarbons into water vapor and carbon dioxide. **Figure 14-27** illustrates the operation of the dual-bed, three-way converter system. Note how the air pump forces air into the center section of the converter.

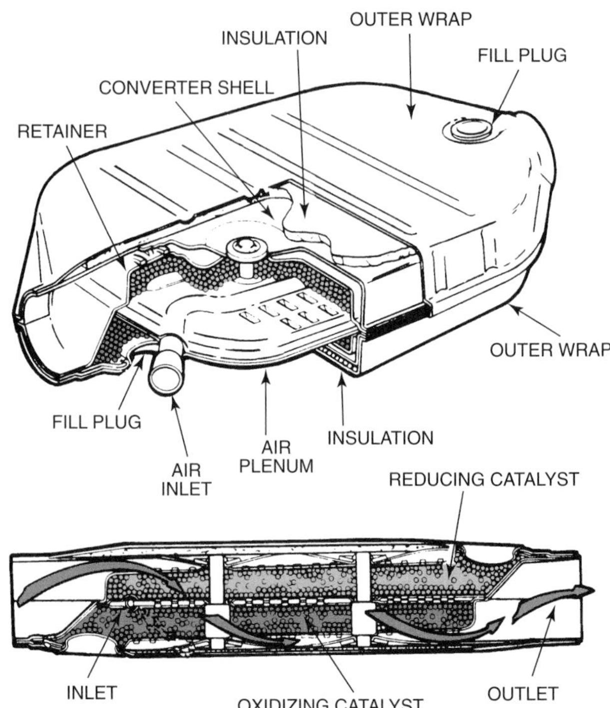

Figure 14-27. Three-way, dual-bed, pellet-type catalytic converter. The air inlet pipe admits airflow from the air injection pump. Note the exhaust flow pattern. (Buick)

In order to prevent excessive heating of the three-way converter, the air injection system air switching valve will divert air from the converter under deceleration conditions, during wide open throttle, when engine temperature becomes excessive during a prolonged idle, and whenever the engine is cold and the choke or cold enrichment system causes a rich mixture. **Figure 14-28** shows the action of the air switching valve when engine temperature is relatively cool. Note how the air is diverted to the exhaust manifold to help in the reduction of carbon monoxide and hydrocarbons, which are produced in large amounts during warm-up.

Multiple Converters

Some vehicles have two catalytic converters. In these systems, one converter, which is often quite small, is located very close to the exhaust manifold. The other converter is larger and is located downstream in the exhaust system between the manifold and muffler. Vehicles with V-type engines may have two identical converters (one near each exhaust manifold) instead of a single converter located after the pipes join.

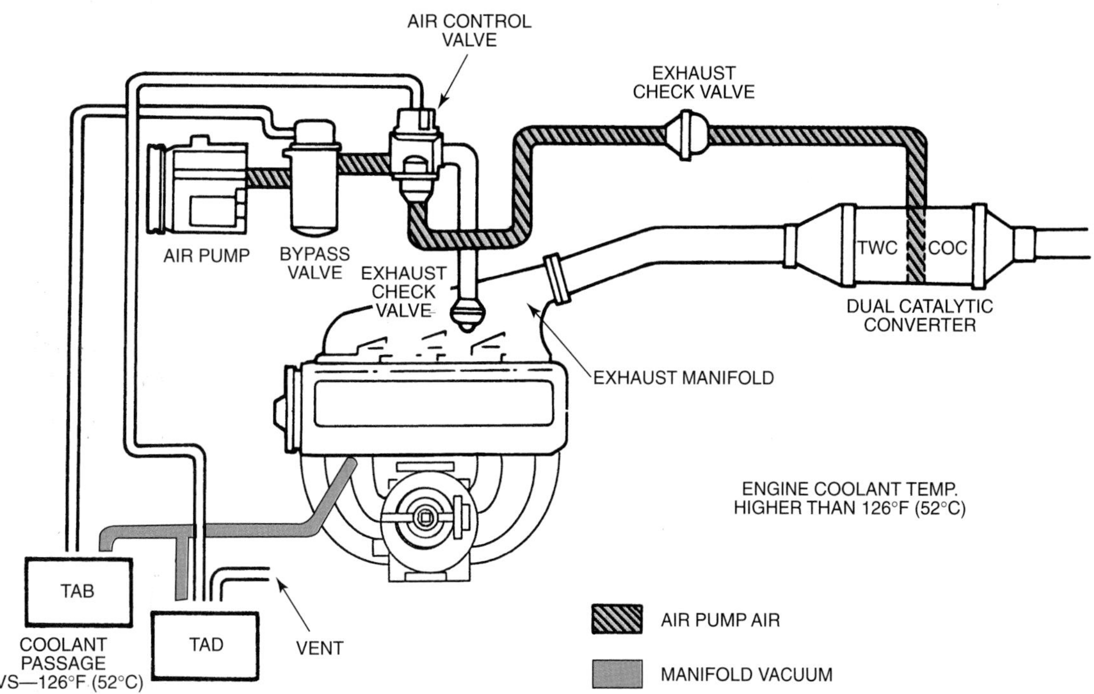

Figure 14-28. Dual-bed, air-injected, three-way catalytic converter system. When engine temperature exceeds 126°F (52°C), manifold vacuum is not applied to the air control valve, and the valve is passing pump air into the center section of the converter. Note the exhaust check valve, which prevents exhaust gases from feeding back into the air supply system. (Ford)

Diesel Catalytic Converters

This oxidizing catalytic converter is used to reduce exhaust smoke particulates. The oxidizing converter operates at normal exhaust system temperatures. The converter is not serviceable and sizes and shapes vary.

Fuel Vapor Controls

One of the major sources of automotive pollution in the past was the release of unburned hydrocarbons in the form of raw gas into the atmosphere. Major sources of these unburned gases were the gas tank vent and the crankcase breather tube. The following systems prevent the release of unburned hydrocarbons into the atmosphere from either of these sources.

Positive Crankcase Ventilation

The combustion process produces many gases. Some of these gases are very corrosive. A great deal of water (almost a gallon of water for each gallon of gasoline) is also formed. The high pressure in the cylinder during the firing stroke forces some of these gases and water vapor by the piston rings and into the crankcase area. To prevent acid and sludge formation in the crankcase and the dilution of the engine oil, these blowby gases must be removed.

For many years a simple road draft tube pulled the gases out of the crankcase and into the atmosphere. This created significant amounts of air pollutants. The ***positive crankcase ventilation system***, or PCV system, removes the gases and allows them to be burned in the engine instead of being discharged into the atmosphere.

By connecting a hose between the engine interior and the intake manifold, engine vacuum will draw the crankcase fumes out of the crankcase and into the cylinders, where the gases will be burned along with the regular fuel charge. A schematic illustrating a typical PCV system is shown in **Figure 14-29.**

PCV Control Valve Is Needed

If allowed to flow into the intake manifold at all times and in any amount, the crankcase gases would upset the operation of the fuel system. To prevent this, it is essential that a PCV flow control valve, usually called a ***PCV valve***, be incorporated into the system. The flow pattern for blowby gases through a PCV valve is shown in **Figure 14-30.**

PCV Control Valve Action

When the engine is stopped, no intake manifold vacuum exists and the PCV valve is held closed by spring action, **Figure 14-31A.**

During periods of idle or deceleration, vacuum is high. The tapered valve plunger is drawn into the metering opening. This reduces the size of the passageway, allowing only a small flow of crankcase gases to the intake manifold. See **Figure 14-31B.**

When the vehicle is operating at a normal load, the vacuum drops off somewhat, allowing the spring to push the plunger part way out of the opening. This increases the opening, and gas flow is reasonably heavy, **Figure 14-31C.**

During acceleration or heavy loading, manifold vacuum drops greatly. The spring instantly pushes the plunger further

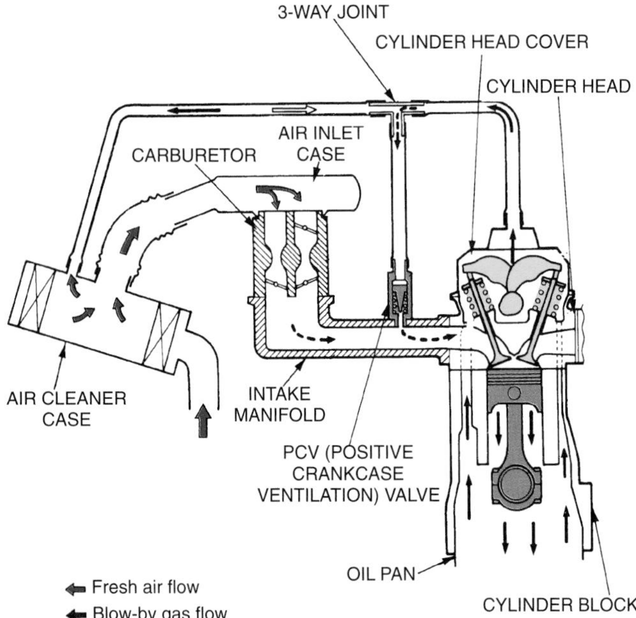

Figure 14-29. Schematic illustrating positive crankcase ventilation system. Note how crankcase blowby gases are drawn into the intake manifold, where they mix with the fuel charge for burning. (Suzuki)

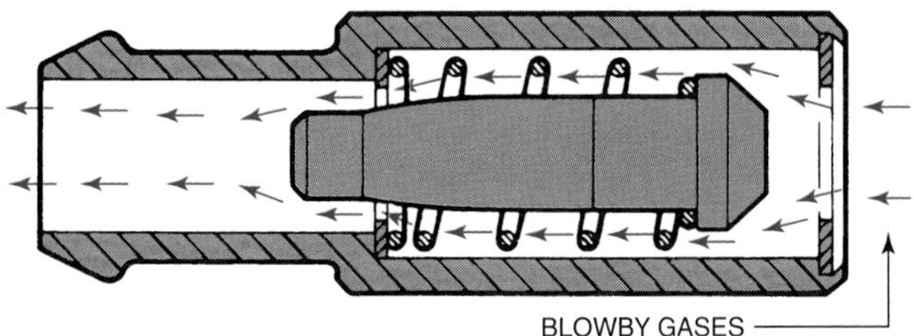

Figure 14-30. Typical PCV valve. Note the flow of blowby gases through the valve. (Jeep)

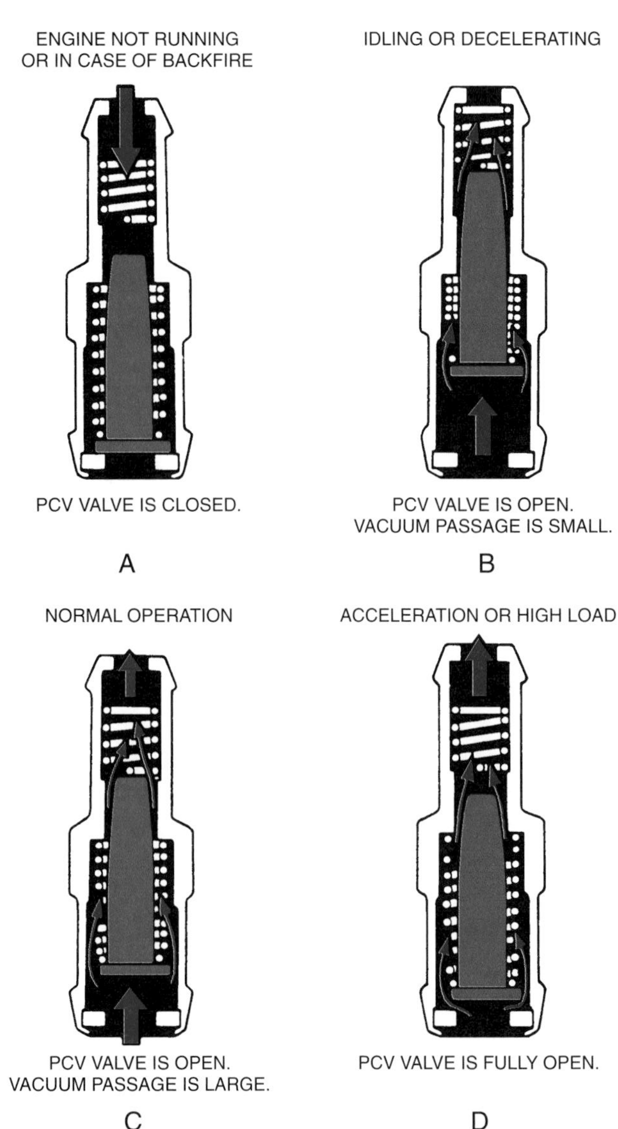

Figure 14-31. PCV control valve action. As the tapered plunger moves out of the metering opening, the gas flow area is increased. (Toyota)

out of the opening. Crankcase gas flow is now at its heaviest. Look at **Figure 14-31D.**

In the event of a backfire in the intake manifold, the reverse pressure allows the spring to fully close the valve, **Figure 14-31A,** preventing the flame from traveling into the crankcase. This prevents a possible explosion in the crankcase area.

Modern PCV systems are ***closed systems.*** In the closed system, fresh air enters the crankcase through the air cleaner. This keeps any vapors from escaping from the crankcase under heavy loads or when the engine is off. Older open systems drew air through a separate breather cap, which was usually mounted on the valve cover.

Evaporation Control System (ECS)

The ***evaporation control system*** (also called evaporative emission control system) is designed to prevent the release of either liquid gasoline or gasoline vapor into the atmosphere.

A typical evaporation control system is comprised of a nonvented (closed to the atmosphere) fuel tank with a sealed filler cap, a tank fill limiter, and a vapor separator. A vent line connects the tank to a charcoal canister. Other lines extend to the carburetor (when used) and intake manifold. The system may have a purge valve and an excess fuel and vapor return line to the tank. A rollover valve may also be used to prevent fuel leakage from the tank in the event the vehicle rolls over. A schematic of one specific evaporative control setup is illustrated in **Figure 14-32.** Study the various parts and their relationships to the overall system.

Evaporation Control System Operation

The fuel tank is sealed with a ***pressure-vacuum cap.*** This cap uses both a pressure relief valve and a vacuum relief valve to prevent excessive pressure or vacuum in the tank in the event of system failure. The valves remain closed under

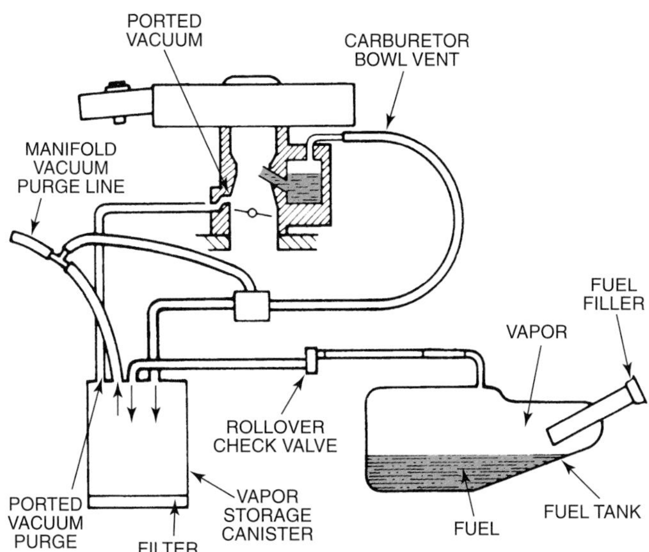

Figure 14-32. An evaporation control system. Study and learn the names of all the parts. (Chrysler)

normal operating conditions. When pressure exceeds around 0.8 psi (5.52 kPa), or when a vacuum forms in excess of 0.1 Hg (0.69 kPa), the valves are activated to vent the tank. See **Figure 14-33.** The fuel tank breathes through the vent tube as fuel is consumed or when it expands or contracts from heating and cooling.

The tank shown in **Figure 14-34** is designed so that when it is filled to capacity, there is a sufficient air dome above the fuel to permit thermal expansion without forcing liquid gasoline into the vent lines. When filling the tank, the small orifice (hole) in the vent restrictor slows down the venting of the tank so that an air dome is present, thus limiting fill capacity.

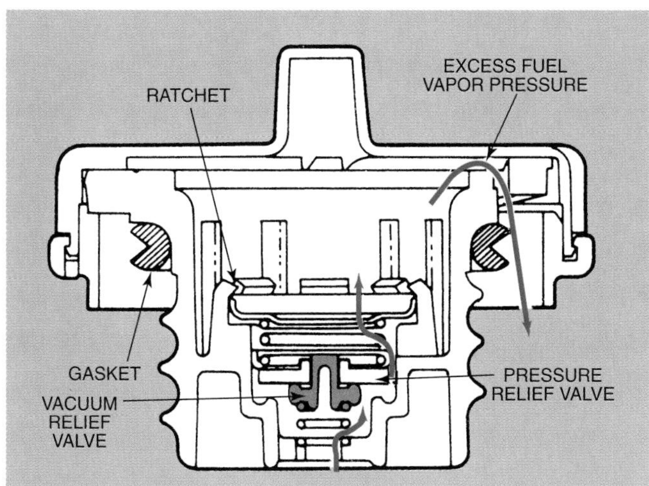

Figure 14-33. Pressure-vacuum fuel tank cap. This cap will relieve either excessive pressure or vacuum, preventing tank or other system damage. (Chrysler)

Any vapors in the tank travel through the fuel tank vent tube to the ***charcoal canister***. On carbureted engines, the float bowl is also vented to the canister through a bowl vent valve. Vapors generated by "engine off heat soak" (heat from engine transferred to the carburetor after the engine is stopped) will be forced into the canister.

The canister is filled with activated charcoal granules (small particles). When the vapors touch the charcoal, the granules absorb and store the vapors. As long as the engine remains off, the vapors will be retained in the canister. When the engine is started, vacuum under the throttle valve will draw fresh air through the charcoal bed. As the air passes over the charcoal, it will pick up the stored vapors and draw them into the intake manifold, where they will be mixed with the fuel charge for burning. As the engine continues to run, the canister will be purged (cleaned) of vapors.

When the engine is stopped, tank and carburetor bowl (when used) vapor will again enter the canister. Study the system shown in **Figure 14-34.**

Canister Construction

All charcoal canisters contain a bed of activated charcoal (carbon), **Figure 14-35A.** Early designs have a replaceable fiberglass filter at the fresh air intake of the canister, **Figure 14-35B**. Canister vents vary with the type of system. Some canisters used with carbureted engines vent the canister directly to the carburetor, as in **Figure 14-34.** Some vents are contained in the canister and open when manifold vacuum builds up as the engine begins to run.

Another canister design has a canister purge valve, **Figure 14-36.** The purge airflow rate is controlled by the size of the orifice in the purge control valve. A thermal (heat controlled) vacuum switch controls vacuum to the purge valve. When engine operating temperature reaches a specified level, the thermal vacuum valve will apply vacuum to the purge valve and purge airflow will start. Note that this too is a closed system because clean purge air is drawn in from the air cleaner.

Liquid Fuel Control

In the closed fuel system, liquid fuel must never enter the canister. Evaporative control systems may use an air dome in the tank, a vapor separator, or a liquid check valve to prevent liquid fuel from leaving the tank.

Liquid-Vapor Separator

The liquid-vapor separator unit collects any liquid fuel and drains it back to the tank, preventing it from entering the charcoal canister. The separator is connected to as many as three vent lines. The rear vent standpipe in the separator has a hole near the bottom that permits any liquid fuel that may have been forced into the separator to drain back into the tank.

The liquid-vapor separator is mounted above the gas tank so that normal vehicle angles will always leave one vent above the fuel level and free to pass vapor to the canister. **Figure 14-37** illustrates an evaporation control system using a liquid-vapor separator.

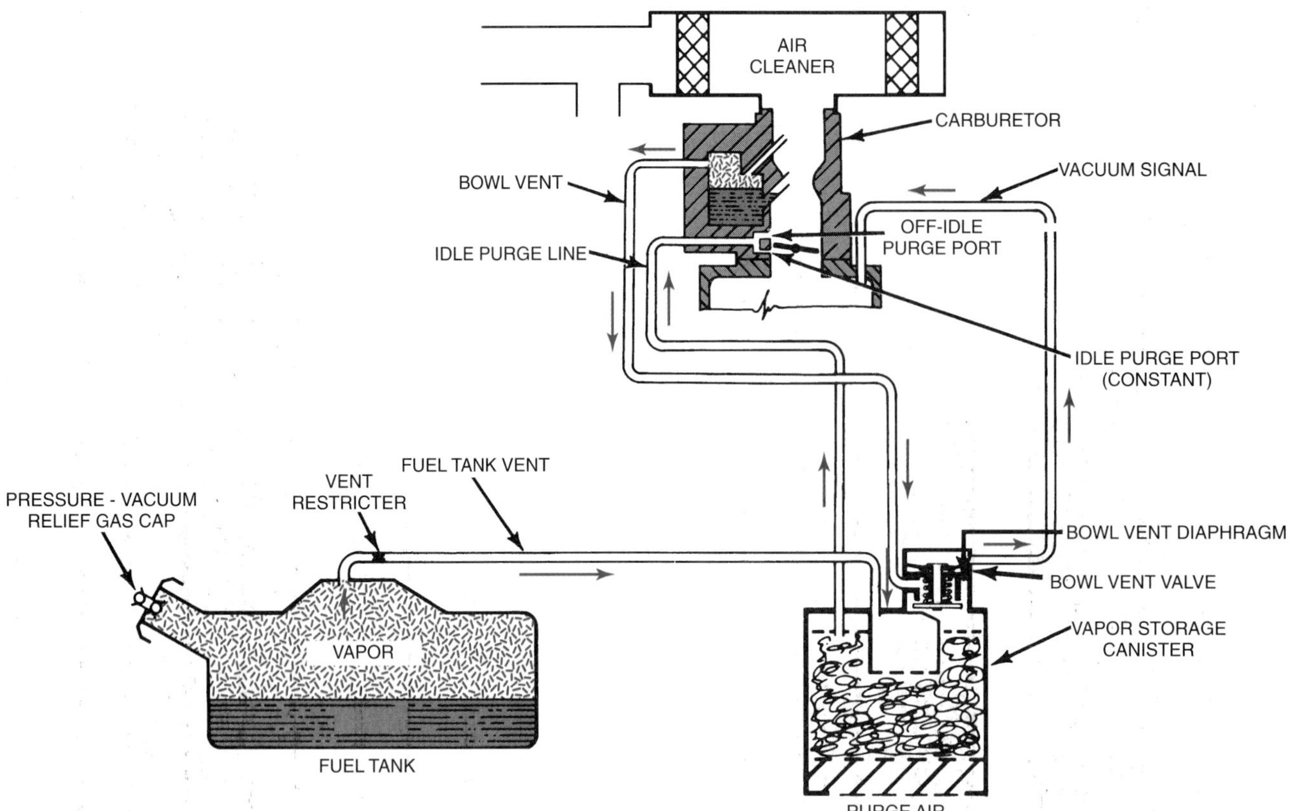

Figure 14-34. Schematic of an evaporation control system. This setup uses a bowl vent valve. (Buick)

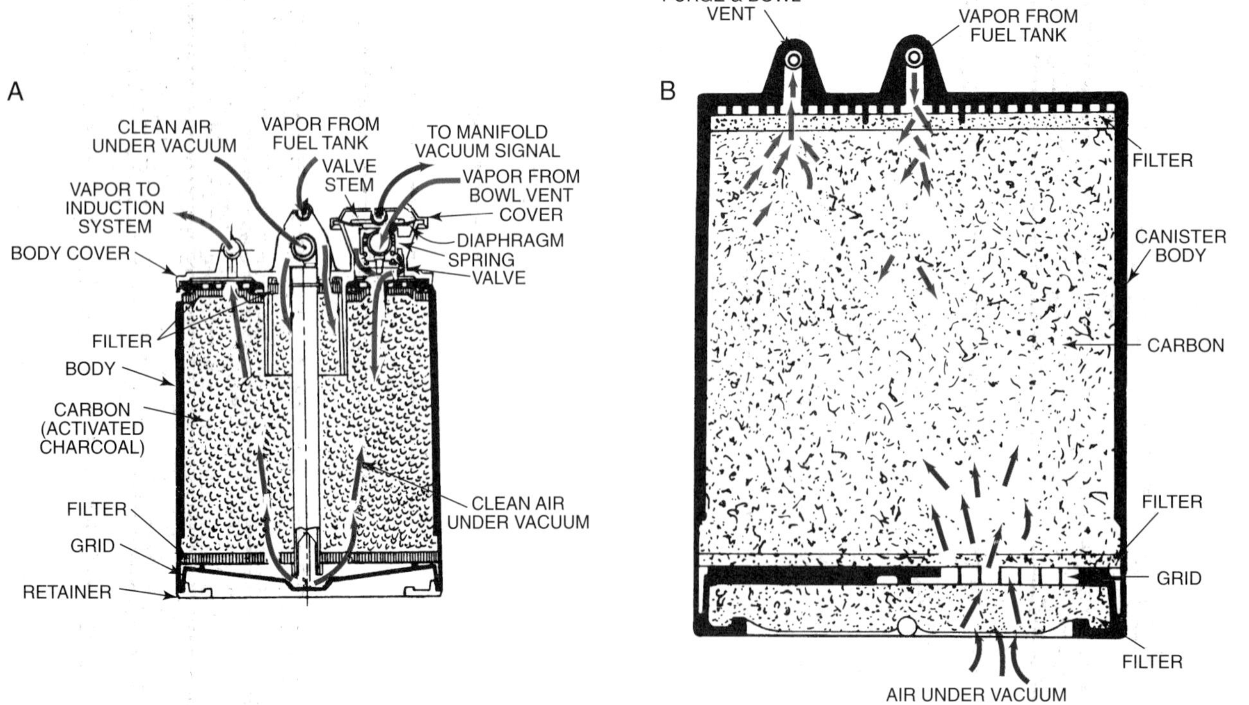

Figure 14-35. A—Cutaway of a vapor storage canister using only a bowl vent valve. Canister purge air is vented directly to the intake (induction) system. B—This vapor canister draws purge air up through an external filter at the bottom. The control valve is remotely located. This is an open-type canister. (Pontiac)

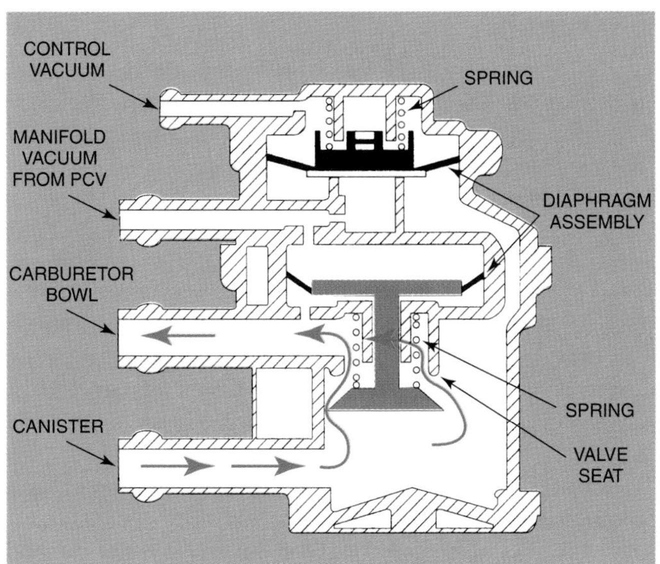

Figure 14-36. Cutaway of one type of charcoal canister purge valve. Study the construction. (OTC Tools)

Liquid Check Valve

The liquid check valve allows tank vapors to pass into the canister under normal conditions. If, however, liquid fuel starts filling up the check valve, the float will rise. This forces the needle into its seat, closing off the vent line until the liquid fuel drains back into the tank, **Figure 14-38.**

Rollover Check Valve

A rollover check valve can be used to seal off the tank vent line in the event a vehicle is turned upside down in an accident. Some rollover valves also incorporate a liquid separator valve as part of their construction. One form of rollover valve is show in **Figure 14-39.** When the valve is inverted, the steel ball forces the plunger to close the vent line.

Computer Monitoring of the Evaporation Control System

On most vehicles built after 1996, the computer operates the evaporation control system. The purge valve is operated by a computer-controlled solenoid. Other solenoids may be used to control the flow of vapor through the system.

The computer also monitors the evaporation control system. On many vehicles, the computer controls a small air pump. When energized by the computer, the pump pressurizes the fuel tank and related system components. After the desired pressure is achieved, the pump stops and the computer holds the purge valve closed while monitoring tank

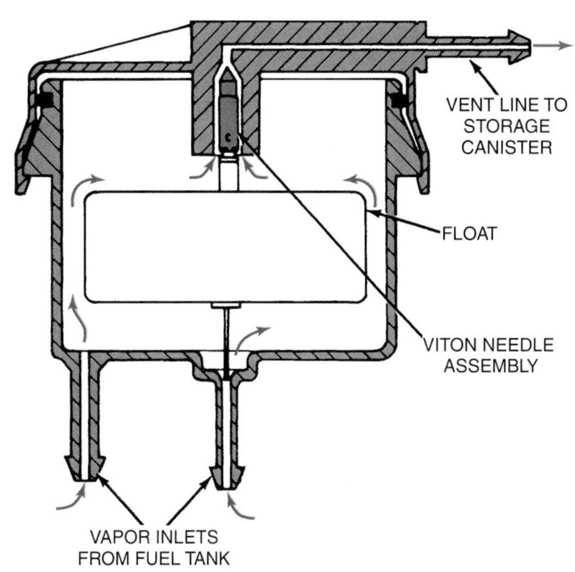

Figure 14-38. Typical liquid check valve. Fuel entering the valve raises the float, forcing the needle to seal off the vent line. (Chrysler)

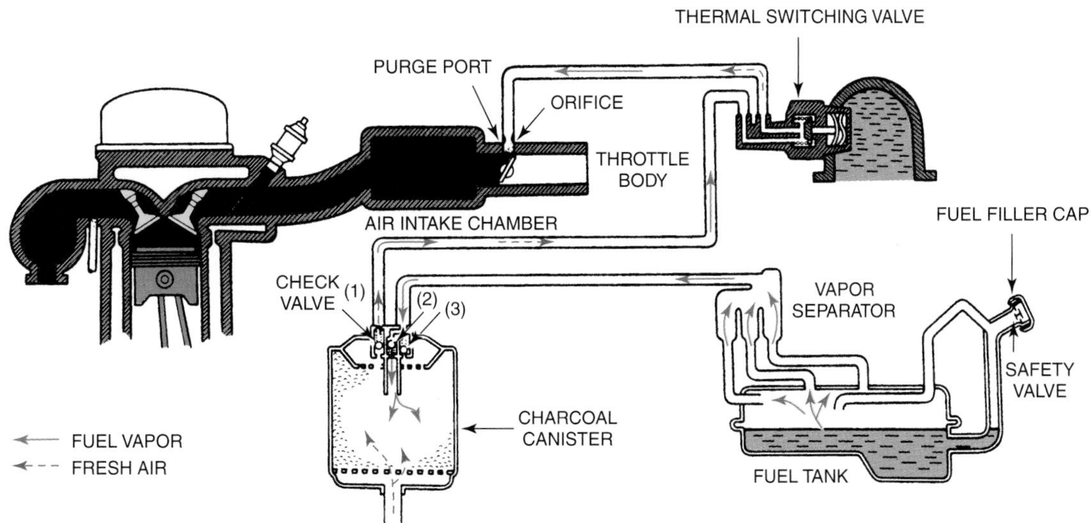

Figure 14-37. This evaporative control system employs a liquid-vapor separator to prevent liquid fuel from being forced into the charcoal (carbon) canister. (Toyota)

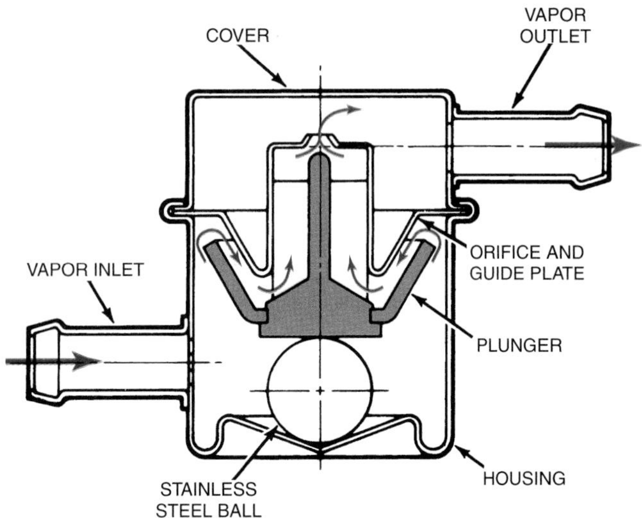

Figure 14-39. One type of rollover check valve. In the event of a rollover, the ball will depress the plunger, sealing the vapor line and preventing fuel leakage. (Chrysler)

pressure with a pressure sensor. If the pressure drops too quickly, the sensor reading alerts the computer to the fact that the fuel system is leaking to the atmosphere. The computer then sets a trouble code.

Some evaporation control systems do not pressurize the tank to check for leaks. Instead, the computer will cycle the purge valve under certain conditions. The computer then monitors the engine airflow and air-fuel ratio to determine whether the system is operating properly. If a problem is detected, the computer will set a diagnostic trouble code.

Computer-Controlled Emissions Systems

The computer-controlled emission system is covered extensively in Chapter 15. Using a computer provides optimum air-fuel ratios for a variety of operating conditions. In many cases, the computer controls the air-fuel ratio so precisely that some emission controls can be eliminated. Other emission controls are operated directly by the computer. Modern vehicles have low emissions with reasonable performance and good economy.

Emission Systems Caution

Proper emission reduction depends on a carefully balanced and integrated number of emission control devices and systems. When performing any emission-related adjustment or repair, keep in mind that an adjustment or alteration in one system will very often affect other systems.

Also, remember that it is against federal and state law to remove or disable any emission control device. Always follow the manufacturer's recommendations and specifications. Do not alter, disconnect, or override emission controls for any reason.

Summary

Automotive exhaust emissions of major concern include hydrocarbons, oxides of nitrogen, and carbon monoxide. Emission controls are of three basic types: engine modifications and controls, external cleaning systems, and fuel system vapor control.

Many engine modifications are used to reduce emissions, including reduced camshaft overlap, lowered compression, hotter thermostats, and modified combustion chamber shapes. Fuel systems are designed to produce an air-fuel ratio as close to 14.7:1 as possible. Proper choke control during engine warmup and driveaway is very important. Electric chokes may be used to speed up opening under certain conditions.

Heated air is drawn into the air cleaner during engine warmup. The thermostatically controlled air cleaner can admit all hot or all cold air, or it can blend the two together, depending on engine temperature.

Ignition timing must be kept within specified limits to reduce emissions. Many early devices used a restrictor or other controls to reduce the vacuum to the vacuum advance unit. Most modern vehicles have electronic advance mechanisms, which change timing based on inputs such as engine temperature, vacuum, and RPM.

To lower NO_x, a portion of the exhaust gas is allowed to enter the intake manifold. This lowers combustion temperatures. The exhaust gas recirculation (EGR) valve controls the flow of exhaust to the manifold. EGR valves can be the back pressure modulated, vacuum modulated, or electronic-vacuum modulated design.

Air injection systems force a stream of fresh air into the burning exhaust as it leaves the cylinder. This prolongs burning to reduce hydrocarbon and carbon monoxide emission. An engine-driven vane-type pump provides the air. Some setups use pulse air instead of a pump. A check valve prevents exhaust from entering the air system in the event of excessive exhaust pressure. A diverter valve is used to prevent engine backfiring during deceleration.

Catalytic converters reduce exhaust emissions. Converters can use either a pellet bed or a monolith catalyst. The oxidizing catalyst is treated with platinum and palladium and will change carbon monoxide and hydrocarbons into water and carbon dioxide. The reducing catalyst is treated with platinum and rhodium. It will reduce the amount of nitrates of oxygen. It will also oxidize hydrocarbons and carbon monoxide to some degree. Two-way converters use an oxidizing catalyst. Three-way converters contain both reducing and oxidizing catalysts.

Some converters utilize a controlled stream of injected air to improve their efficiency. The air injection system can be used as the air source. Systems can use more than one converter. The positive crankcase ventilation system removes fumes from the crankcase and burns them in the cylinders. The PCV valve controls the flow of gases.

The evaporative control system prevents the escape of liquid gasoline or gasoline vapor into the atmosphere. Vapors are stored in a charcoal canister until the engine is started.

They are then drawn off and burned in the engine. Canisters can be of open or closed design.

Computer-controlled emission systems offer the ultimate in emission control. The electronic system instantly makes adjustments needed to maintain the best air-fuel ratio. Care must be used when making adjustments on any engine system, as adjustments on one system may affect other systems. Follow manufacturer's specifications. Do not disconnect or alter emission controls.

Know These Terms

Hydrocarbons (HC)
Oxides of nitrogen (NO_x)
Carbon monoxide (CO)
Engine modifications and controls
External cleaning systems
Fuel vapor controls
Overlap
Dieseling
Run-on
Carburetor icing
Thermostatic air cleaner
Early fuel evaporation system (EFE)
Vacuum advance restrictor
Exhaust gas recirculation (EGR) system
Vacuum modulated EGR valve
Back pressure modulated EGR valve
Electronic EGR valve
Air injection system
Diverter valve
Air switching valve
Pulse air injection system
Catalytic converter
Two-way converter
Three-way converter
Pellet converter
Monolith converter
Honeycomb converter
Reducing catalyst
Oxidizing catalyst
Positive crankcase ventilation system (PCV)
PCV valve
Closed system
Evaporation control system
Pressure-vacuum cap
Charcoal canister

Review Questions—Chapter 14

Do not write in this book. Write your answers on a separate sheet of paper.

1. The three major exhaust pollutants are _____, _____, and _____ .
2. List the three general types of emission controls.
3. The combustion chamber in modern cylinder heads is designed to reduce the amount of _____ in the exhaust.
   (A) CO
   (B) NO_x
   (C) HC
   (D) O_2
4. Camshaft overlap is _____ to improve idling with lean mixtures.
5. All of the following statements about lower engine compression ratios are true, EXCEPT:
   (A) lower compression ratios allow the use of low-octane gasoline.
   (B) NO_x formation is greater with low compression ratios.
   (C) lower compression allows the engine to operate on unleaded gas.
   (D) detonation is less likely with lower compression ratios.
6. Dieseling is caused by hot spots in the _____ .
   (A) block
   (B) intake manifold
   (C) combustion chamber
   (D) exhaust manifold
7. Technician A says that carburetor icing is reduced by the use of a thermostatic air cleaner. Technician B says that carburetor icing is reduced by the use of early fuel evaporation systems. Who is right?
   (A) A only.
   (B) B only.
   (C) Both A & B.
   (D) Neither A nor B.
8. Lowering the combustion flame temperature reduces the formation of _____ .
   (A) HC
   (B) NO_x
   (C) CO
   (D) O_2
9. Technician A says that the EGR valve admits exhaust gas to the intake manifold at all times. Technician B says that the PCV valve admits crankcase gases into the intake manifold at all times. Who is right?
   (A) A only.
   (B) B only.
   (C) Both A & B.
   (D) Neither A nor B.
10. The air injection system forces air into the _____ .
    (A) intake manifold near the intake valve
    (B) exhaust manifold near the exhaust valve
    (C) catalytic converter
    (D) All of the above, depending on the system.
11. The diverter valve is used on the air injection system to prevent _____ .
12. What unit provides air for the air injection system?
    (A) Fan.
    (B) Air pump.
    (C) Air cleaner snorkel ram pressure.
    (D) Intake system back pressure.

13. All of the following statements about air switching valves are true, EXCEPT:
    (A) the switching valve can channel pump air to the intake manifold, exhaust manifold, or the catalytic converter.
    (B) the switching valve can be used to perform the normal diverter valve functions.
    (C) some switching valves channel pump air between the exhaust port area and the catalytic converter.
    (D) the switching valve reduces the efficiency of the catalytic converter.
14. One air injection system uses the pressure pulsations of the exhaust system to draw air into the exhaust system. What is this system called?
15. Technician A says that the thermostatically controlled air cleaner admits cool air to the engine during engine warmup. Technician B says that the thermostatically controlled air cleaner allows a leaner air-fuel mixture during the warmup period. Who is right?
    (A) A only.
    (B) B only.
    (C) Both A & B.
    (D) Neither A nor B.
16. The pellet-type converter catalyst bed is being replaced because it creates more _____ than the monolith type.
    (A) heat
    (B) back pressure
    (C) noise
    (D) vibration
17. A reducing catalyst can lower _____ emissions.
    (A) HC
    (B) NO_x
    (C) CO
    (D) All of the above.
18. The operation of a three-way, air-injected catalytic converter requires all of the below, EXCEPT:
    (A) a catalyst bed containing rhodium.
    (B) an air injection pump.
    (C) an oxidizing catalyst bed.
    (D) a vacuum-operated EGR valve.
19. To speed up choke release during engine warmup, some carburetors are equipped with a(n) _____ heated choke.
20. Many vehicles with emission controls reduce ignition timing advance by placing a restrictor on the _____ advance mechanism.
    (A) vacuum
    (B) centrifugal
    (C) electronic
    (D) None of the above.
21. To prevent crankcase fumes from escaping to the atmosphere, all modern engines are equipped with a(n) _____ system.
    (A) EGR
    (B) PCV
    (C) diverter valve
    (D) thermostatic air cleaner
22. Technician A says that crankcase gases are allowed to flow heavily into the intake manifold during engine idle. Technician B says that the vapor control canister is filled with fiberglass. Who is right?
    (A) A only.
    (B) B only.
    (C) Both A & B.
    (D) Neither A nor B.
23. Evaporative control systems may use one of three systems to prevent liquid fuel from leaving the tank. Name these systems.
24. In the closed (fuel vapor control) fuel system, liquid fuel must not enter the _____ .
    (A) fuel lines
    (B) fuel return lines
    (C) carbon canister
    (D) All of the above.
25. The pressure-vacuum cap used on a sealed fuel tank has a _____ valve.
    (A) pressure relief
    (B) vacuum relief
    (C) rollover
    (D) Both A & B.

15

Computer Systems

Upon completion of this chapter, you will be able to:

- Explain why computers are used in modern vehicles.
- Describe computer control system operation.
- Describe the major components of an on-board computer.
- Identify and describe computer input sensors.
- Identify and describe computer output devices.

In previous chapters, we have discussed computer-controlled systems on modern vehicles. In this chapter we will see how the computer operates to control the fuel and ignition systems, as well as other vehicle systems. This chapter will concentrate on the engine and powertrain control computer. Other computer systems will be discussed in the chapters dealing with the system that they control.

Computers on Modern Vehicles

Almost every vehicle built in the last 10-15 years has at least one on-board ***computer***. There may be up to six different computers on some vehicles. On-board computers control the ignition and fuel systems, emission controls, cooling fans, air conditioner compressor, torque converter clutch, and transmission shift points.

Other vehicle computers control the air bags, anti-lock brake and traction control systems, interior air temperature and distribution, suspension and steering systems, and the anti-theft system. Many modern vehicles combine the engine control computer and other vehicle computers into one large microprocessor, which monitors and controls most vehicle functions.

The primary reason for using on-board computers in modern vehicles is to meet fuel economy and emissions regulations. The use of computers allows the ignition, fuel, and emission systems operations to be controlled precisely. The end result is a stoichiometric air-fuel ratio that is as close to 14.7:1 as possible, **Figure 15-1**.

This air-fuel ratio is necessary for optimum catalytic converter operation as well as good fuel economy and emissions performance. Newer on-board computers also control the torque converter clutch, transmission shift points, and air conditioner operation to increase fuel mileage as well as lower emissions. The computer control system can be divided into three major subsystems:

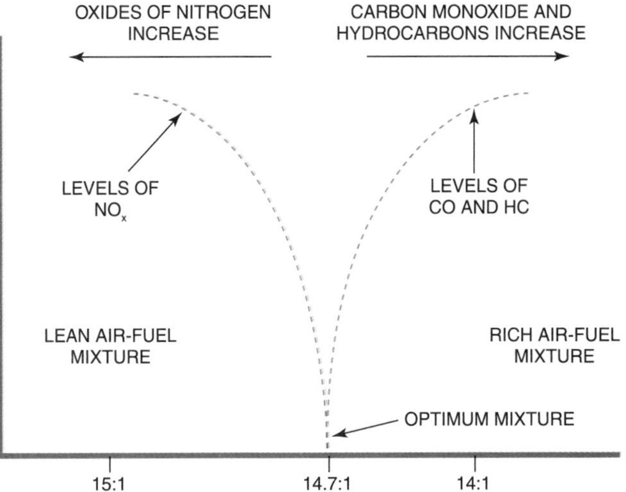

Figure 15-1. A stoichiometric air-fuel ratio allows for optimum performance. Note how the levels of carbon monoxide (CO) and hydrocarbons (HC) increase as the mixture becomes richer. However, if the mixture is allowed to become too lean, oxides of nitrogen (NO_x), increase.

- Computer.
- Input sensors.
- Output devices.

Figure 15-2 shows the interrelation of these three subsystems used to operate and monitor the engine and drive train. The following sections discuss each of these areas in detail.

Electronic Control Unit (ECU)

A vehicle's on-board computer is normally referred to as an ***electronic control unit*** or ***ECU***. In the first fourteen chapters of this text, we used the term computer or on-board computer for the ECU. The ECU has many different names. A vehicle's ECU may be referred to by one of the following terms:

- controller
- microprocessor
- on-board computer
- engine control computer (ECC)
- engine or electronic control module (ECM)
- powertrain control module (PCM)

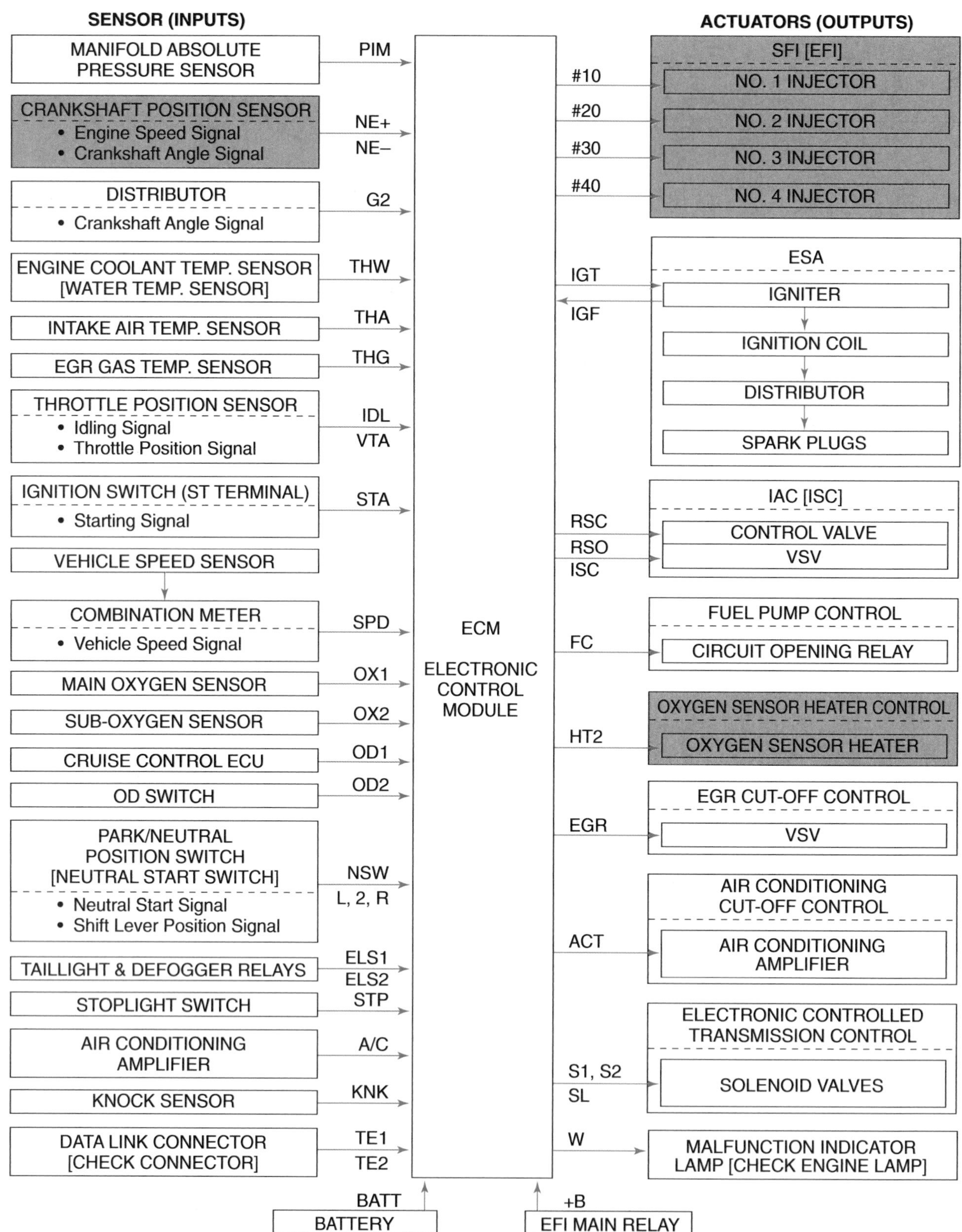

Figure 15-2. The computer (electronic control module) monitors inputs from the many sensors before adjusting the outputs (actuators). Study carefully. (Toyota)

These names are the most common and depending on the manufacturer and vehicle system, other specific names may be used. ECU's are usually contained in one casing. Most ECU's are located inside the vehicle's interior away from the heat and vibration of the engine compartment, but can be located almost anywhere depending on the vehicle, system, and manufacturer.

A few ECU's are made in two parts, with the processor located in the passenger compartment and the output part under the hood. Whatever their physical location, all ECU's contain two main sections: the central processing unit and the memory section. The memory section has two separate sections of memory—permanent and temporary.

Central Processing Unit

The ***central processing unit (CPU)*** is the section of the electronic control unit that performs calculations and makes decisions. The CPU is sometimes called the controller. The CPU is constructed using one or more integrated circuits, or ICs, which were covered in Chapter 7. Incoming data from the input sensors must go through the CPU, where the data is processed and adjustments made to the output sensors. The ECU must be replaced if the CPU is defective, as it is affixed to the ECU's circuit board.

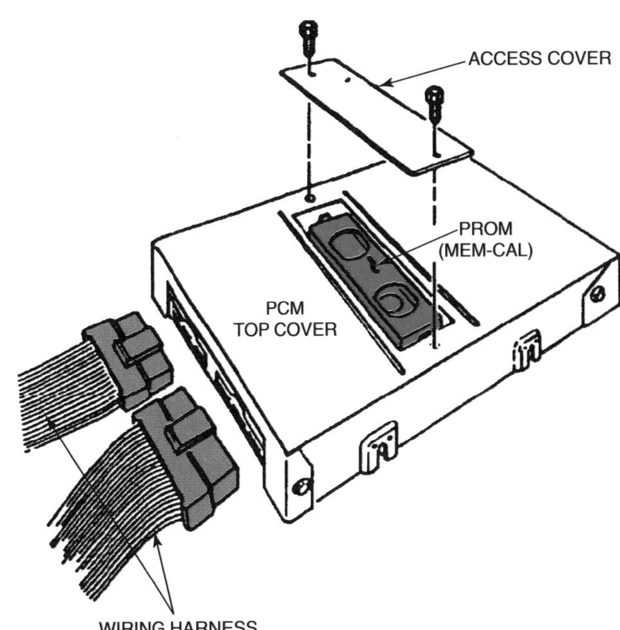

Figure 15-3. The PROM (Mem-Cal) is plugged into the ECU. To gain access to the PROM, the metal cover must first be removed. (Chevrolet)

Permanent Computer Memory

There are four basic types of permanent computer memory, ***read only memory (ROM), programmable read only memory (PROM), erasable programmable read only memory (EPROM),*** and ***electronically erasable programmable read only memory (EEPROM).*** A program that tells the ECU what to do and when to do it is stored in ROM. The ROM also contains general information that is used as a standard of operation under various conditions. Once this information is placed in memory, it cannot be modified. The ROM is an integral part of the ECU. If the ROM is defective, the ECU must be replaced.

Programmable read only memory (PROM) is a variation of ROM. On some vehicles, the PROM is plugged into the ECU and can be replaced if it is defective or if a revised PROM is needed, **Figure 15-3**. This eliminates the expense of replacing the entire ECU. On most newer vehicles, the PROM is permanently affixed to the ECU. The ROM and PROM are nonvolatile memories. This means that information stored in these memories will remain even if the battery is disconnected.

The EPROM can be reprogrammed by exposing it to ultraviolet light and then reprogramming it. This can only be done by the manufacturer. The EEPROM can be erased electronically and reprogrammed or "burned in", using either electronic scan tools or computerized diagnostic equipment.

Temporary Computer Memory

The ECU uses ***random access memory (RAM)*** as a temporary storage place for data from the sensors. As new information is received from the sensors, it is written over the old information inside the RAM. The ECU constantly receives new signals from the sensors while the engine is running. If no signal or the wrong signal is received by the ECU, a diagnostic trouble code is stored in the RAM.

The RAM is contained inside the ECU. If the RAM is defective, the ECU must be replaced. The RAM has a volatile memory. This means that if the battery, fuse, or the power source at the ECU is disconnected, all data in the RAM will be erased, including any stored trouble codes. Some ECUs use another type of memory called ***keep alive memory (KAM).*** KAM is a memory chip that allows the ECU to maintain normal vehicle performance by compensating for sensor and parts wear. This is sometimes called an adaptive strategy. Any data in KAM is also lost if power to the ECU is disconnected.

Computer Control Operation

Although it is far simpler, the computer control system operates in much the same way as the human nervous system. If you touch a hot engine part, the nerves in your hand send a signal to your brain. Based on the signal, your brain decides that your hand may get burned and tells your arm muscles to pull your hand away from the part.

An ECU receives input data in the form of electrical signals from the sensors (nerves). The ECU (brain) compares the input data to the standards in the ROM. Based on its calculations, the ECU sends commands to the output devices (arm muscles), **Figure 15-4**. This process is performed over and over as engine, environmental, and vehicle conditions change.

Open and Closed Loop Operation

When the engine is cold, it cannot provide optimum fuel economy and emissions control. Also, several of the input

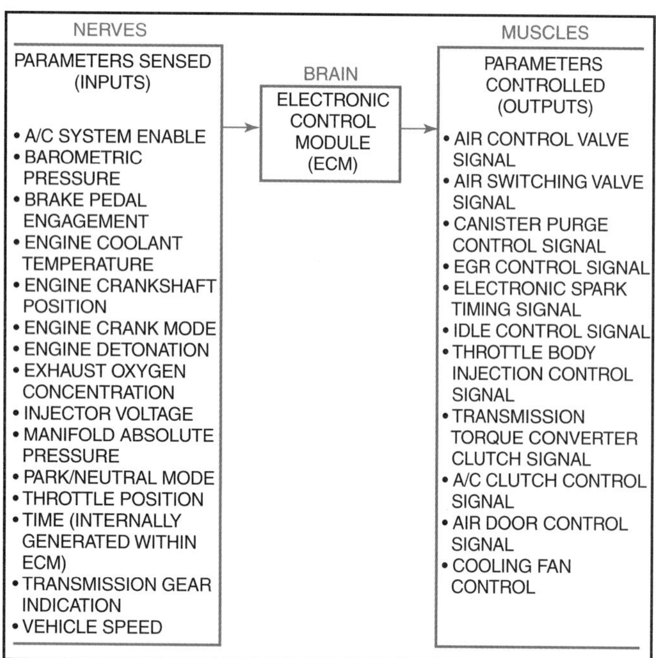

Figure 15-4. The computer operating input parameters are the nerves. The electronic control module is the brain, and the output parameters are the muscles. (AC-Delco)

sensors must warm up to operating temperature before they can provide reliable readings. Therefore, while one or more of the input sensors tells the ECU that the engine is below a certain temperature, the ECU operates in ***open loop*** mode.

In open loop mode, the ECU controls the air-fuel ratio based on preset information while monitoring the input sensor readings. Some computer control systems also go into open loop mode when the throttle is opened completely, often called wide open throttle mode.

When the input sensors tell the ECU that the engine has warmed up sufficiently, the ECU goes into ***closed loop*** mode. In the closed loop mode, the ECU begins processing the information from the input sensors and bases its commands to the output devices on the sensor information. The result is optimum fuel economy and engine performance with minimal emissions output.

Diagnostic Trouble Codes

When the computer control system develops a problem, the ECU illuminates a warning light on the dashbord. This light is sometimes called the ***check engine light*** or the ***service engine soon light***. In late-model vehicles, the warning light is usually called the ***malfunction indicator light,*** or ***MIL***. Some MILs can flash ***diagnostic trouble codes*** when the technician puts the computer into the self-test mode. Trouble codes are defined in the service manual. Trouble codes indicate which component and circuit the technician uses as a starting point for diagnosis.

The ECU's diagnostic system may have an internal chip which will allow the vehicle to operate even if the ECU fails. This chip will operate the vehicle on preset information in what is usually referred to as the limp-home mode. This allows the vehicle to be driven until the computer control system can be serviced.

Input Sensors

The electronic control unit receives hundreds of input signals every second. These signals are created by ***input sensors***. Each sensor measures one operating condition of the engine or drive train. The sensor then sends an electrical signal that represents the measurement to the ECU. Some sensors produce their own voltage signal and other sensors adjust the voltage sent to it by the ECU. The voltage sent to these sensors by the ECU is called the ***reference voltage***.

The amount of resistance at the sensor determines how much output voltage is sent to the ECU. The ECU then compares its reference voltage to the sensor output voltage. This determines the sensor measurement. Some inputs to the ECU are electrical in nature and do not require a separate sensor to operate them. Examples are the charging system voltage, and voltage to apply the air conditioner clutch.

On-Off Switches

On-off switches are the simplest type of sensor. They are turned on and off by the movement of mechanical linkage, just as you flip a switch to turn on a light, **Figure 15-5**. The ECU receives their input as either full voltage (on) or no voltage (off). An example of a common on-off switch used in a vehicle is the brake light switch. On-off switches are also operated by transmission or transfer case linkage.

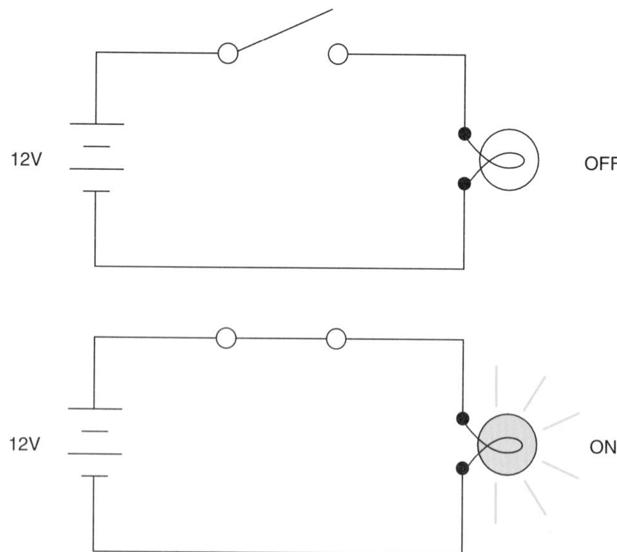

Figure 15-5. This simple light circuit is an example of the operation of an on-off switch. On-off switches are the simplest types of sensors.

Pressure Sensors

Pressure sensors are on-off switches which send a full voltage (on) or no voltage (off) signal to the ECU. **Figure 15-6** shows a simple power steering pressure sensor. Most pressure sensors in a vehicle are operated by hydraulic (liquid) pressure of some kind. Pressure sensors are often used

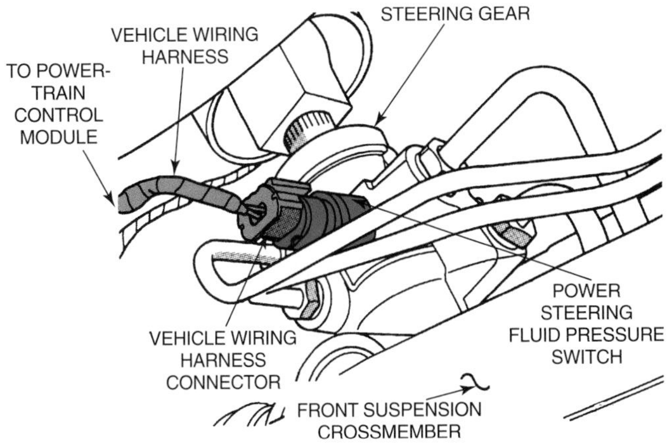

Figure 15-6. One type of power steering electrical fluid pressure sensor (switch). This switch helps with vehicle engine idle speed through the powertrain control module (PCM). It senses high pressure fluid conditions, such as turning the wheels while parking. An electrical signal from the switch travels to the PCM, which in turn will increase the engine idle speed and make other adjustments to prevent the engine from stalling. (Chrysler)

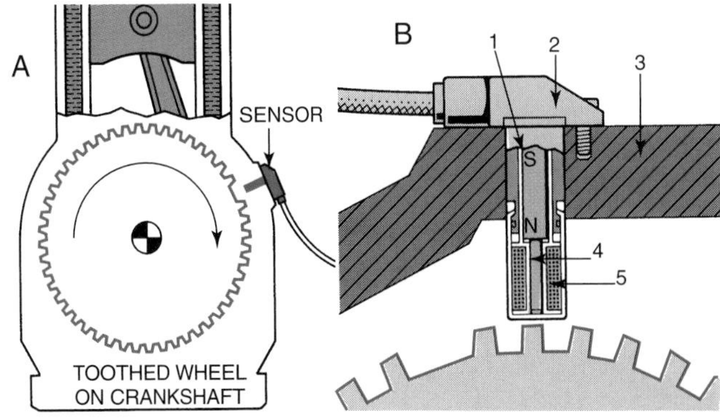

Figure 15-7. A crankshaft position sensor assembly. A—Sensor mounted in the engine block. B—Cutaway view of the sensor. 1—Permanent magnet. 2—Sensor housing. 3—Engine block. 4—Soft iron core. 5—Winding. 6—Toothed wheel with reference mark (wider gap spacing than the other teeth). Study closely. (Bosch)

in automatic transmissions. They are installed in hydraulic pressure passages and tell the ECU which transmission gear has been selected, either by the driver, or by the hydraulic control system. Pressure switches are also used in power steering systems.

Crankshaft Speed and Position Sensors

The ***crankshaft speed and position sensor*** monitors engine speed (RPM) and position of the piston. This sensor may be part of the distributor or it may be mounted on the engine block to monitor the position of the rotating crankshaft. Some systems use a ***camshaft position sensor*** to provide an additional reference signal to the ECU. Some computer control systems use the triggering device in the distributor as a speed and position sensor.

The ECU uses the crankshaft sensor input, along with the camshaft sensor and other sensor inputs, to determine engine timing. These sensor inputs are used by the ECU to tell the ignition module when to fire the ignition coil and to energize the fuel injectors. Many of the crankshaft sensors operate on the same principles as the sensors used in electronic ignition systems. There are three types of crankshaft position sensors. They are the ***magnetic pickup coil***, the ***Hall-effect switch***, and the ***optical sensor***. Refer to Chapter 8 for more information about these sensors. **Figure 15-7** shows a crankshaft position sensor.

The signal from the ignition system triggering device may pass through the ignition module before going to the ECU. In this sense, the ignition module itself can be an input sensor. Ignition modules were discussed in Chapter 8.

Other Speed Sensors

Another type of speed sensor is the ***reed switch***, **Figure 15-8**. Reed switches are used to measure speed by measuring rotation. The reed switch consists of two thin metal blades, or reeds, which contact each other inside a closed chamber. A small amount of current flowing in the reeds creates a magnetic field which attracts the reeds to each other. Any outside magnetic field, however, will cause the reeds to repel each other and move apart.

The reed switch is placed near a magnet attached to the moving part to be monitored. Movement of the magnet causes the two reeds to move apart, and then spring back together, starting and stopping the current flow through the switch. The ECU can read this current variation as a speed signal.

Some vehicles have speed sensors mounted behind the speedometer, **Figure 15-8**, on the rear of the transmission, at the axle housing, or at the wheels. These sensors allow the ECU to determine vehicle speed and sometimes the relative speeds of each wheel. The ECU uses these inputs to control engine settings. The individual wheel speed sensors signals are used as inputs to the anti-lock brake and traction control system ECU.

Knock Sensor

The ***knock sensor***, **Figure 15-9**, is installed in the intake manifold or engine block. The purpose of this sensor is to cause the ECU to retard the timing when the engine begins to knock. Engine knock is usually associated with detonation or preignition, which can cause damage to the piston and valves.

The sensor uses a piezoelectric (pressure-produced electricity) ceramic element that produces a voltage proportionate to the strength of the knock. This means a mild knock will send a low voltage signal to the ECU, while a heavy knock will send a high voltage signal to the ECU. The ECU prevents the engine damaging knock from occurring by adjusting the ignition timing.

Temperature Sensors

Temperature sensors are used to notify the ECU when the engine is warm enough to begin normal computer control operations. Other temperature sensors monitor outside air temperature. There are two major types of engine temperature sensors—the coolant temperature sensor and the

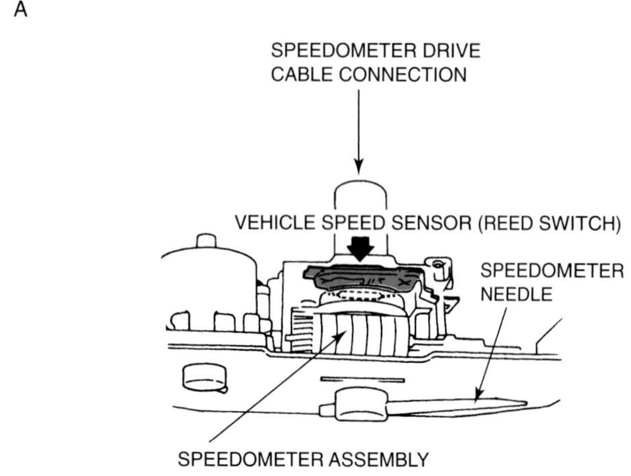

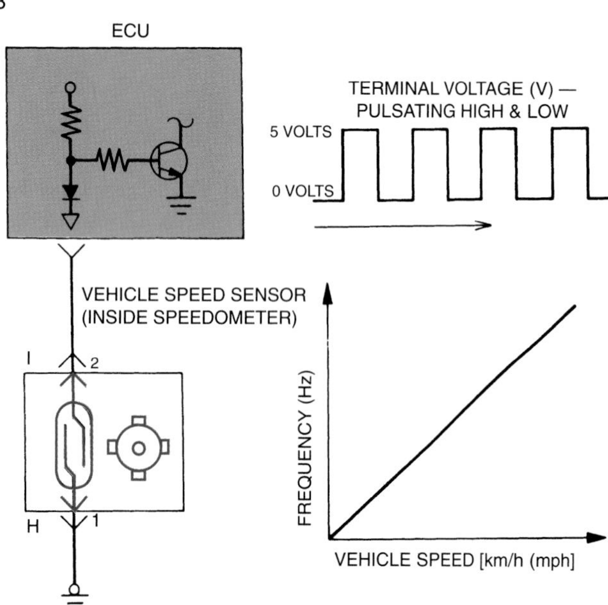

Figure 15-8. One type of vehicle speed sensor (VSS). A—A reed switch vehicle speed sensor built into a speedometer assembly. This sensor is driven by the speedometer cable from the transmission/transaxle. B—The reed switch opens and closes four times for every one revolution of the speedometer cable. The electronic control unit (ECU) furnishes the speed sensor with a five volt reference signal. This voltage is cycled from low to high by the operation of the reed switch. The ECU can read this returning pulsating voltage as vehicle speed. (Hyundai)

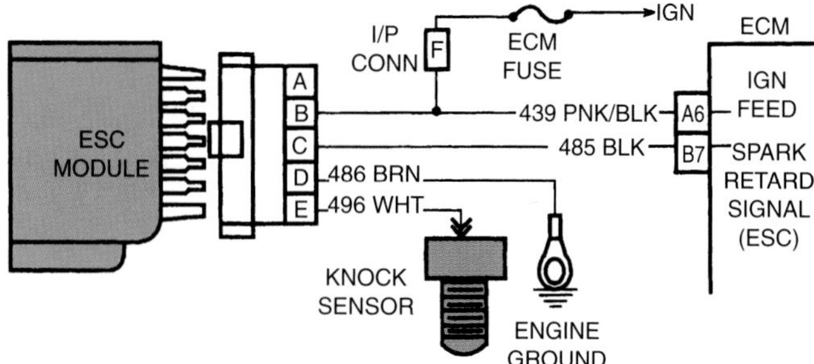

Figure 15-9. The engine control module (ECM) monitors the engine for preignition or detonation (knock) by using a spark control module and one or more knock sensors. When a knock sensor detects engine knock, it sends an electrical signal to the spark control module. The engine timing can then be retarded to eliminate the engine knock. (General Motors)

incoming air temperature sensor. Temperature sensors are usually ***thermistors***. A thermistor is made of electrically conductive material which changes its resistance in relation to temperature. The two types are discussed below.

Coolant Temperature Sensor

The engine ***coolant temperature sensor*** is mounted where its tip can contact the engine coolant, **Figure 15-10**. The purpose of this sensor is to monitor engine temperature. In most computer control systems, the coolant sensor input is used to determine when the ECU should operate in open or closed loop mode. If equipped, the ECU can turn the electric engine fan on when the coolant reaches a specified temperature.

Intake Air Temperature Sensor

The ***intake air temperature sensor*** or IAT sensor, **Figure 15-11**, like the coolant temperature sensor, is a thermistor. This sensor is placed in the air stream of the incoming air, on the air cleaner housing, intake air ductwork, or intake manifold. Since cold air is denser than hot air, a richer fuel mixture is required to provide the correct air-fuel ratio. This sensor is also called a ***manifold air temperature***, or MAT sensor.

Manifold Vacuum Sensor

The ***manifold vacuum sensor*** measures the negative pressure, or vacuum, in the intake manifold when the engine is running. Since vacuum is directly related to engine load, the ECU can use this signal to determine engine load, **Figure 15-12A**. At low engine loads, vacuum is high. The voltage signal sent to the ECU at idle is relatively low, **Figure 15-12B**. However, when the engine load is high, manifold vacuum is low and the voltage signal sent to the ECU is high.

Sometimes, vacuum is referred to as ***manifold absolute pressure***, or MAP. Manifold absolute pressure is the pressure developed inside of the manifold unrelated to outside air pressure. Even though the intake manifold is under a vacuum when the engine is running, it still has pressure in relation to an absolute vacuum. The changes in manifold absolute pressure correspond exactly to changes in vacuum. Therefore, a ***MAP sensor*** is simply a vacuum sensor. Some systems use an airflow meter in combination with a manifold vacuum or MAP sensor.

Barometric Pressure Sensor

The ***barometric pressure sensor*** contains sensors for both the manifold vacuum and outside air pressure in one housing. The barometric pressure sensor operates in the same manner as the vacuum sensor. The barometric pressure sensor allows the ECU to make changes in engine operating settings to compensate for changing weather conditions and changes in altitude. Barometric pressure sensors are usually incorporated into MAP or manifold vacuum sensors.

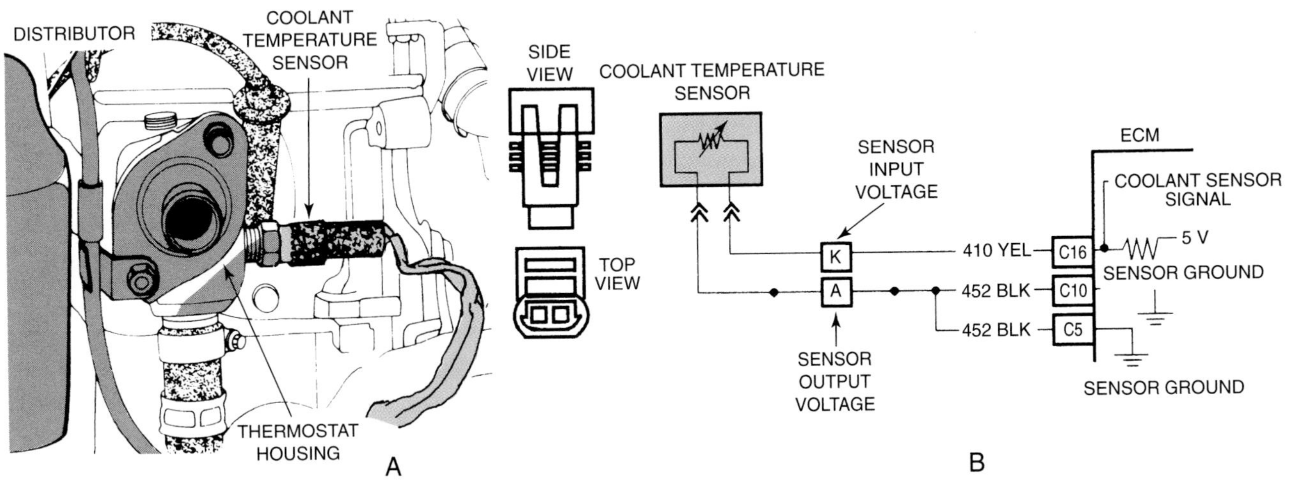

Figure 15-10. A coolant temperature sensor. A—Mounting location. B—Reference or sensor input voltage is adjusted by the amount of resistance at the sensor. Temperature of the coolant determines the amount of resistance. Sensor output voltage (signal) informs the computer of the engine's operating temperature. (Oldsmobile & Chrysler)

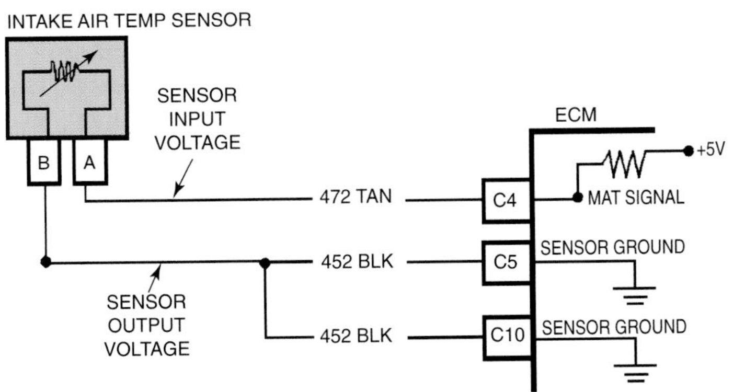

Figure 15-11. A sensor input voltage reference signal is sent to the intake air temperature sensor. The air temperature determines the amount of resistance and an output voltage signal is sent to the computer. (Oldsmobile)

Airflow Sensors

Airflow sensors are used with some computer control systems to measure the amount of air entering the engine. They can be divided into two groups, the flap-type airflow meter and the mass airflow sensor. The purpose of this sensor is to cause the ECU to advance the timing and richen the fuel mixture in relation to an increase in airflow. As you learned in Chapter 9, vehicles without airflow sensors use speed density, which uses the other sensor inputs to calculate airflow rate.

Airflow Meter

The ***airflow meter***, **Figure 15-13**, uses a spring-loaded flap to measure the rate of air entering the engine. The spring-loaded sensor flap is pushed to an increasingly smaller angle due to the increased airflow. Mechanical linkage

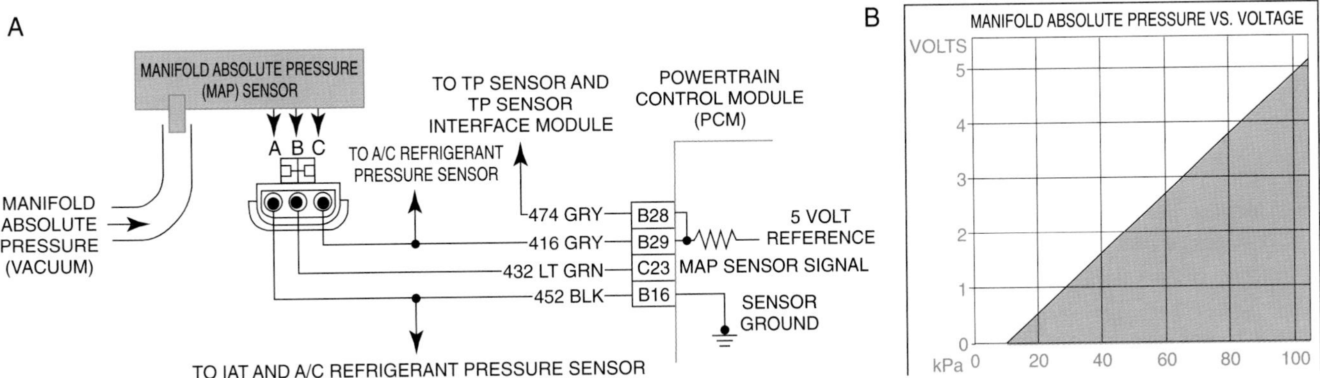

Figure 15-12. A—A manifold absolute pressure (MAP) sensor system schematic. The MAP sensor reacts to manifold pressure (vacuum) changes and sends a return electrical signal back to the PCM (powertrain control module). This particular systems signal voltage varies from 1V–1.5 V at idle, to 4.0 V–4.8 V at wide open throttle (WOT). (Cadillac) B—A manifold absolute pressure (MAP) sensor versus output voltage graph. The powertrain control module (PCM) furnishes a 5 V reference signal (electrical current) to the MAP sensor, which is then returned to the PCM by the sensor. The return voltage varies depending upon manifold pressure (vacuum). When manifold pressure increases, the voltage signal increases, and vice-versa. (Cadillac)

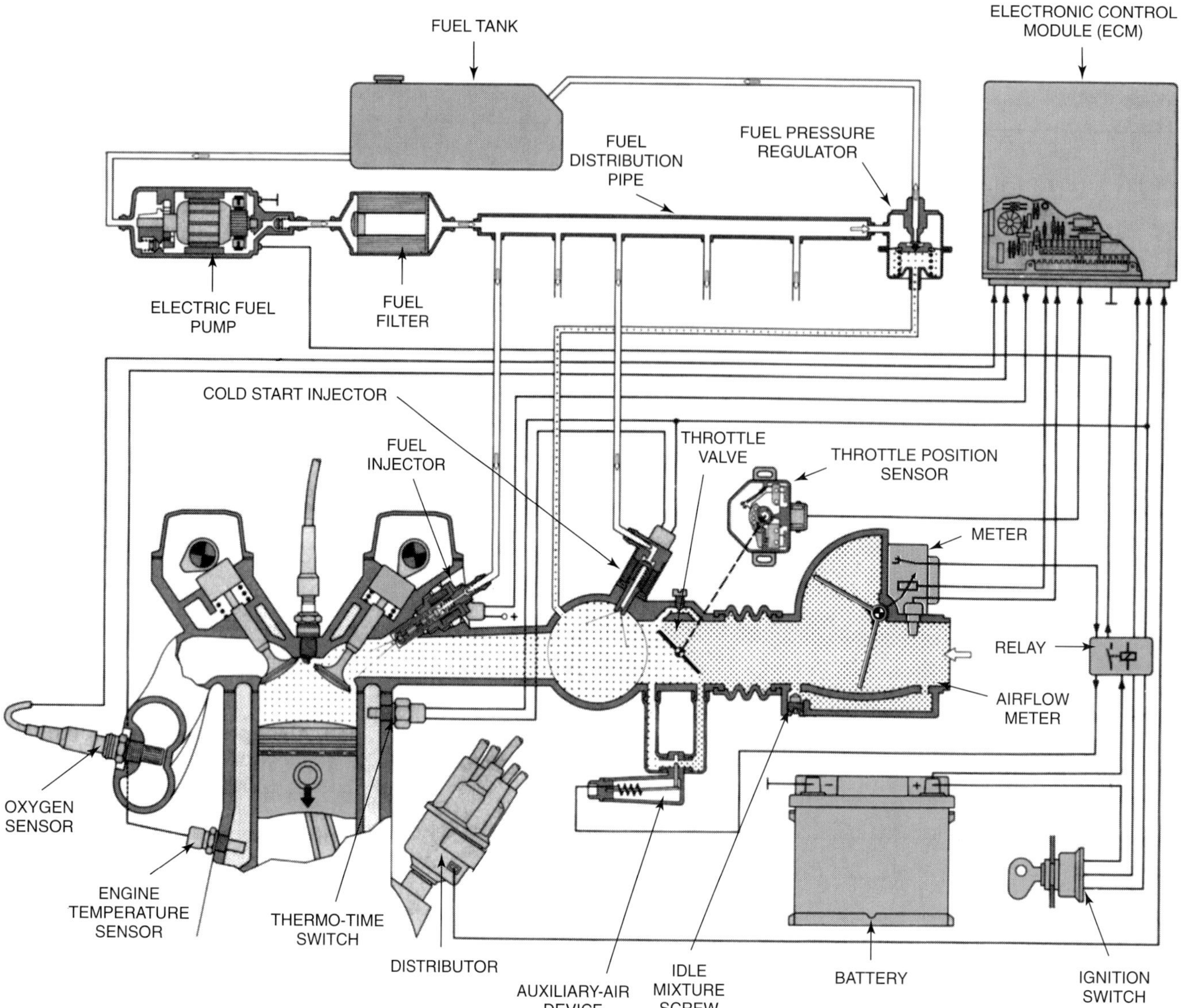

Figure 15-13. Parts of the airflow meter and computer control system. The wiper arm is connected to the sensor flap at one end and rides against a variable resistor at the other end. As the sensor flap is deflected by an increased airflow, the voltage signal sent to the electronic control unit is adjusted. (Bosch)

connects the sensor flap to the wiper arm which rides on a variable resistor. As the sensor flap moves, so does the wiper arm against the resistor. This adjusts the sensor output voltage which represents the amount of airflow. This output voltage allows the ECU to monitor sensor flap angle and therefore, air intake by the engine.

Mass Airflow Sensor

The ***mass airflow sensor***, or ***MAF sensor,*** combines the function of the manifold air temperature sensor and airflow meter, **Figure 15-14**. However, the MAF sensor measures the airflow electronically instead of using mechanical linkage. One advantage of this type of airflow meter is that it can compensate somewhat for the amount of humidity, or amount of moisture in the air.

A wire is placed across the path of the incoming air. The wire is heated to a specified temperature, while the incoming air cools the wire. The control circuit monitors this and increases the voltage to maintain the specified temperature of the wire. The ECU monitors the amount of voltage the control circuit sends to keep the hot wire at the specified temperature. The voltage signal that the ECU monitors at the control circuit represents the amount of incoming air. Some mass airflow sensors use a heated film which works on the same principle as the wire sensor.

Oxygen Sensor

The ***oxygen sensor*** is located in the exhaust manifold and is usually referred to as the O_2 sensor, **Figure 15-15**. It senses the amount of oxygen in the exhaust gas. A lean air-fuel mixture will contain excess oxygen after combustion. A rich air-fuel mixture will contain very little oxygen after combustion. The difference between the oxygen in the exhaust and the outside air causes a chemical reaction which produces a very small amount of electricity. This results in a voltage signal to the ECU.

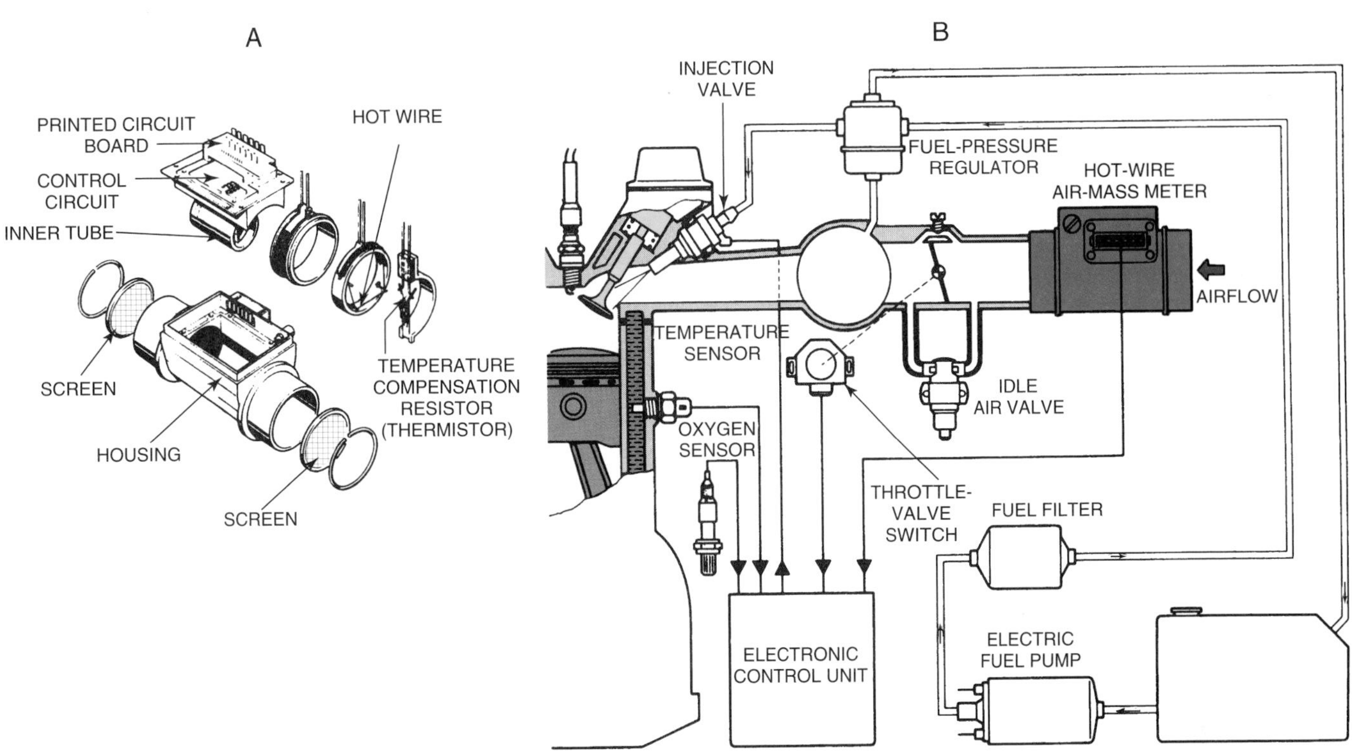

Figure 15-14. Parts of one particular mass airflow sensor. It is located between the throttle body and air filter. A—Exploded view of the sensor assembly. B—Airflow sensor and its location in one particular system. (Bosch)

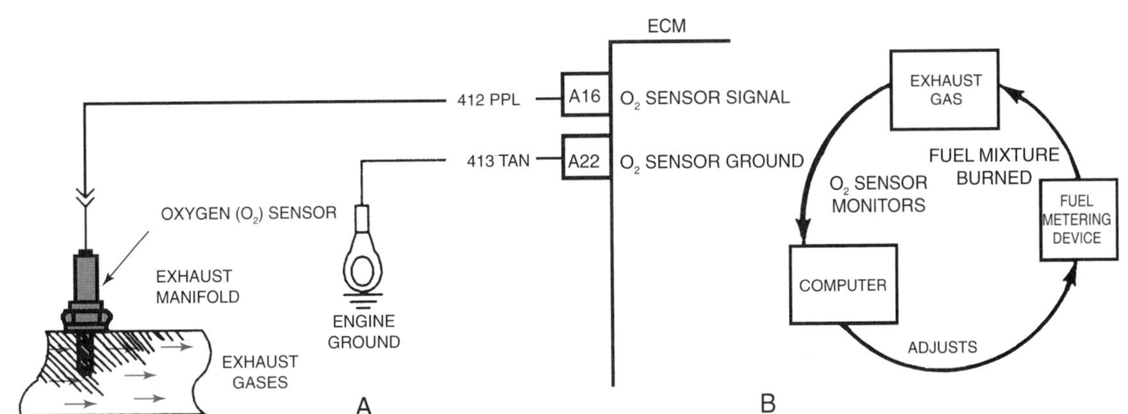

Figure 15-15. A—An oxygen sensor is mounted in the exhaust manifold (on this system), to monitor the oxygen content of the exhaust gas. B—When the exhaust temperatures reach 600°F (315.5°C), the sensor produces voltage inversely related to the oxygen content in the exhaust gas. When the electronic unit uses this voltage signal to adjust the fuel mixture, it is termed closed loop. (Oldsmobile)

The oxygen sensor is unique in that it can produce its own voltage signal. Some computer control systems send a reference voltage from the ECU to the O_2 sensor. The ECU compares this reference voltage to the voltage sent by the oxygen sensor to determine a lean or rich fuel mixture. In many computer control systems, the oxygen sensor is used almost exclusively by the ECU to determine the air-fuel mixture during closed loop operation.

When the oxygen sensor detects an absence of oxygen in the exhaust gas, it generates a relatively high voltage signal to the ECU. The ECU interprets this as an excessively rich fuel mixture and makes adjustments to produce a leaner mixture. However, when there is excess oxygen in the exhaust gas, the oxygen sensor generates a lower voltage signal to the ECU. The ECU interprets this as an excessively lean fuel mixture and makes adjustments to produce a richer mixture. When the sensor produces a voltage signal between the two extremes, the ECU interprets the fuel mixture as correct and no action is taken. This happens rapidly in the normal operation of an engine.

Some computer control systems used with V-type engines have an oxygen sensor in each exhaust manifold.

Other systems have a second sensor further along in the exhaust system as a backup to the first sensor. Some vehicles have oxygen sensors that contain a heated element. This element allows the O_2 sensor to quickly reach and maintain its normal operating temperature, shortening the time necessary for the vehicle to enter closed loop mode. This sensor is referred to as a ***heated oxygen sensor*** or HO_2S.

Throttle Position Sensor

The ***throttle position sensor*** or TPS, monitors the angle change of the throttle valve, **Figure 15-16**. As the angle of the throttle valve increases, the timing must advance proportionately. Also, the air-fuel mixture must be richened. The position sensor also senses the rate of change to determine how quickly the vehicle is being accelerated.

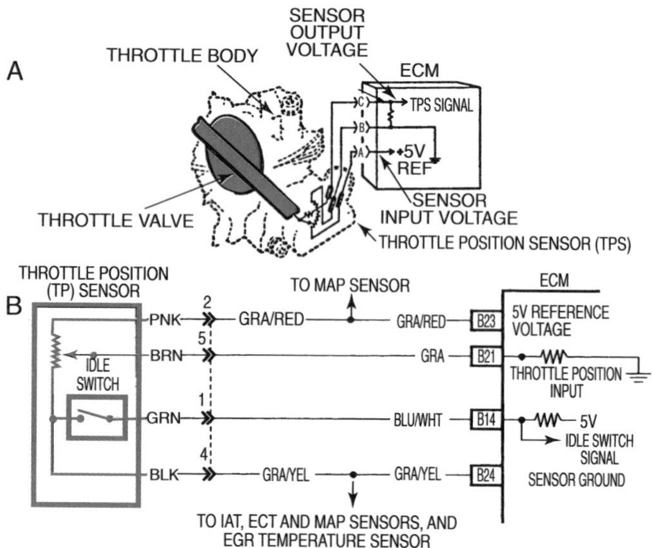

Figure 15-16. The shaft on this particular throttle valve moves the throttle position sensor. The sensor input voltage is adjusted by the variable resistor before being sent to the ECM. The ECM will use this sensor input to correctly determine the angle (position) of the throttle valve. A—Throttle valve and sensor assembly. B—Wiring schematic. (General Motors)

On one type, the throttle shaft connects directly to an arm that rides against a ***variable resistor***. As the arm varies the resistance, the reference voltage is adjusted and sent to the ECU. Another type is called a ***transducer***, which consists of a wire coil surrounding a metal rod. A small current flows through the coil and into the ECU whenever the engine is running. The rod is connected to the throttle shaft. Movement of the throttle shaft causes the rod inside the coil to move. This affects the magnetic field, thereby modifying the electrical signal to the ECU.

EGR Position Sensor

The ***EGR position sensor*** monitors the opening of the EGR (exhaust gas recirculation) valve. The ECU uses this input to determine whether the EGR control solenoid should be activated to give the engine more or less exhaust gas flow.

Other ECU Inputs

Some inputs to the ECU are made using on-off switches, pressure switches, or electrical inputs. For example, when the vehicle air conditioner is turned on, a wire from the air conditioner relay delivers a voltage signal to the ECU. The ECU then can adjust fuel, ignition timing, and idle speed to compensate for the extra load on the engine. Some vehicle mechanical switches, such as the brake light switch, send an on-off voltage signal to the ECU. Pressure switches in computer-controlled automatic transmissions send an on-off voltage signal to the ECU when gear changes are made.

Some inputs are not from any type of sensor. The ECU monitors vehicle system voltage and can compensate for low voltage or high demands on the vehicle's charging system. If the ECU reads low voltage, it can compensate by raising the idle speed to increase alternator output. A few ECU's contain the charging system voltage regulator and can control the alternator's field current in response to output voltage levels.

Output Devices

An electrical unit controlled by the ECU is called an ***output device***. Output devices control the fuel, ignition, emission control, and other parts of the vehicle. Output devices can be solenoids, relays, electric motors, or other electronic devices. Examples of output devices are the idle speed motor or solenoid, mixture control solenoid, fuel injector, ignition coil, ignition module, EGR system, air injection system, fuel pump relay, and electric fan motor relay.

Idle Speed Controls

Idle speed is carefully controlled by the ECU to maintain proper control of emissions at idle. Idle speed is no longer set by adjusting a screw. There are two major kinds of idle speed controls, the ***idle speed solenoid***, and the ***idle speed motor***. The two types are discussed below.

Idle Speed Solenoids

Idle speed solenoids are used on many carbureted vehicles. A solenoid is an iron or steel rod operated by a magnetic coil, **Figure 15-17A**. When the coil is energized, a magnetic field is created and moves the rod. When the coil is de-energized, an internal spring returns the plunger to its original position. The operation of the solenoid is controlled by the ECU.

An example of the use of an idle speed solenoid is increasing engine idle speed when the air conditioner is in operation. The rod is positioned to contact a part of the throttle linkage when the coil is energized. The rod pushes on the throttle linkage, opening the throttle to increase idle. The coil can be energized directly from the ECU, or by a relay operated by the engine or air conditioner control ECU.

Idle Speed Motors

The idle speed motor, **Figure 15-17B**, automatically increases idle speed to prevent engine stalling. The idle speed is increased by the motor extending its plunger. Idle speed motors of this type are no longer used. If the load on the engine is increased (air conditioner turned on) or the engine is cold on start up, the idle speed increases. Once the

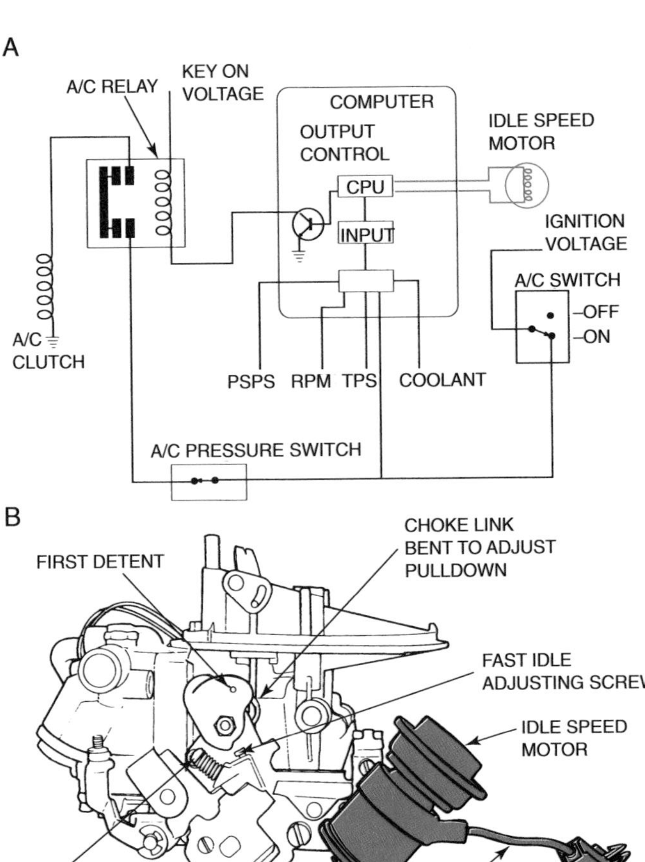

Figure 15-17. A—Idle speed motor operational wiring schematic. Just before the air conditioning clutch is engaged, the computer will send an electrical signal to the idle speed motor, increasing engine rpm. B—Idle speed motor and linkage attached to a two-barrel carburetor. (Ford & OTC Tools)

load is removed or the engine reaches operating temperature, the plunger is retracted and idle returns to normal.

Idle Air Control Valves

A variation of the idle speed motor is the ***idle air control valve***, or IAC valve, **Figure 15-18**. The idle air control valve, unlike the other idle control devices, does not move the throttle blade to change idle speed. Instead, the IAC valve is installed in a passage which allows incoming air to bypass the throttle blade. It regulates the amount of air entering the engine without affecting throttle blade position.

If the load on the engine increases, the valve is retracted and more air flows into the engine, increasing the idle speed. However, when the load on the engine is reduced, the valve is extended, which reduces the airflow past the throttle blades when the increased idle speed is no longer needed. IAC valves are used with fuel injection systems.

Fuel Injector

The ***fuel injector***, **Figure 15-19**, is a solenoid through which fuel is metered. It discharges pressurized fuel to the combustion chamber. The ECU pulses the injectors on and off. When the fuel injector is energized, an electromagnet causes the plunger to move upward. The needle valve, attached to the plunger, is also pulled up which causes the fuel to spray in a fine mist. Fuel injector design and operation were covered in Chapter 9.

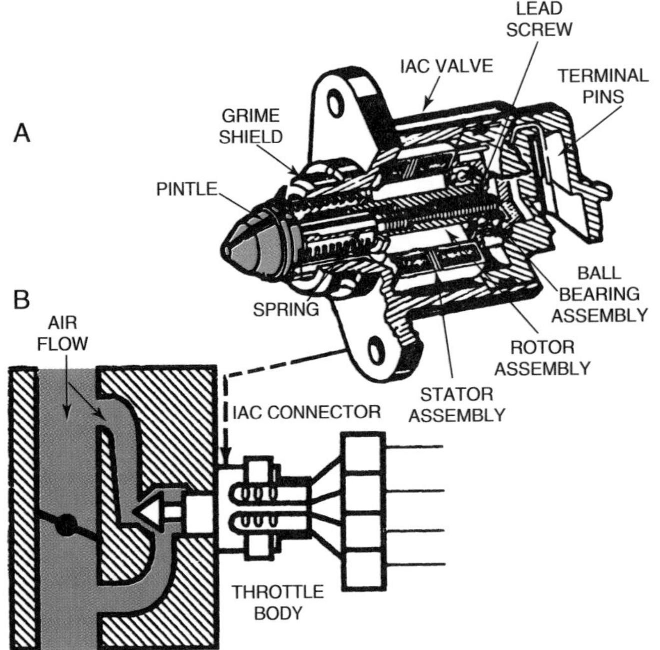

Figure 15-18. A—Cutaway of an idle air control (IAC) valve B—Schematic of the idle air control valve and its location in the throttle body. This valve regulates airflow past the throttle blades to control idle speed. (Oldsmobile)

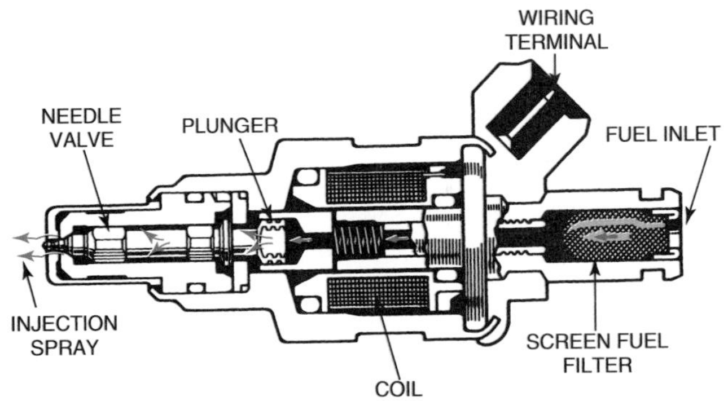

Figure 15-19. Cutaway view of an electronic fuel injector. Note the fuel inlet screen filter. Follow the fuel flow through the injector. (Toyota Motor Corp.)

Mixture Control Solenoid

The ***mixture control solenoid*** is used only on carbureted engines, **Figure 15-20**. When the solenoid is energized by the ECU, the metering rod is forced into the metering jet. This reduces the flow of fuel (lean mixture) to the engine. When the solenoid is de-energized, the metering rod is forced upward. This allows more fuel to flow (rich mixture) to the

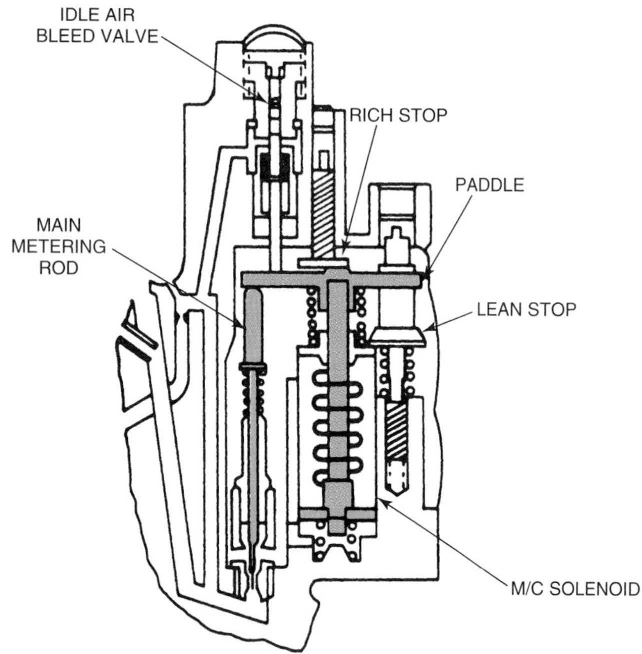

Figure 15-20. Cutaway view of a mixture control solenoid assembly which adjusts the fuel mixture on carbureted engines. The mixture control solenoid responds to electrical signals sent to it by the ECU.

engine. Some mixture control solenoids adjust air-fuel ratio by modifying the amount of air flowing through the main metering system. Mixture control solenoids were discussed in Chapter 10.

Ignition Devices

The ECU can control the operation of the ignition system by sending output commands to one of several ignition devices. The ECU can control the coil directly, or through a separate ignition module. The design of the computer control system dictates which device will be controlled.

Ignition Module

Many computer controlled ignitions have a separate ***ignition module***, **Figure 15-21**. The module receives the triggering signal from the ECU based on inputs from the sensors. The module uses this triggering signal to create the spark. The module may also function as a sensor, passing the triggering signal from the distributor pickup or crankshaft position sensor on to the ECU.

Some ignition modules have a fail-safe provision as part of the ECU's limp-home mode. If the ECU fails, the ignition system will continue to operate with fixed timing settings until the vehicle can reach a service facility. Ignition modules were discussed in Chapter 8.

Ignition Coil

On some computer control systems, the ***ignition coil*** is directly operated by the ECU, **Figure 15-22**. The ECU interrupts the primary current flow to the ignition coil based on crankshaft sensor input. When the primary current to the coil

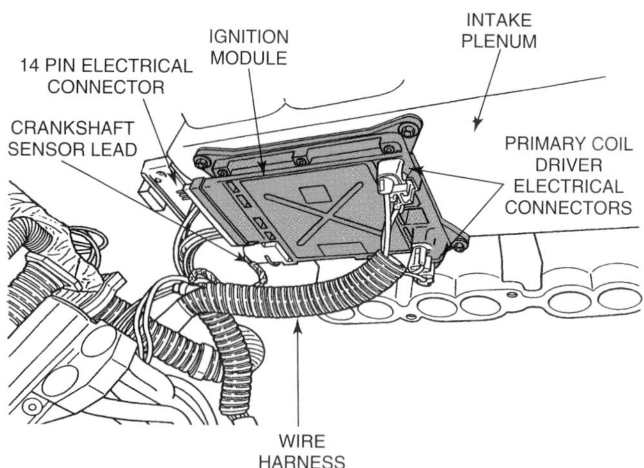

Figure 15-21. An ignition module and its mounting location as used by one vehicle manufacturer. (General Motors)

is interrupted, the ignition coil fires. This voltage is sent to the distributor and rotor for routing to the correct spark plug. The ECU can also control spark timing for maximum engine performance and efficiency.

Distributorless Ignition System

Distributorless ignition systems used on many modern vehicles do not use a conventional distributor, rotor, or coil. Distributorless ignition is comprised of one ignition coil for every two cylinders, an ignition module and a crankshaft and/or camshaft sensor. Distributorless ignition was discussed in Chapter 8. See **Figure 15-23**.

EGR System Solenoid

The ***EGR system solenoid*** controls vacuum between the intake manifold and EGR valve diaphragm, **Figure 15-24**. Another type has the solenoid inside and on top of the EGR valve, **Figure 15-25**. Also, note the valve position sensor, which is a variable resistor. This provides feedback on the position of the valve which tells the ECU the exact amount of exhaust gas flow recirculating.

If the exhaust flow is too much for operating conditions, the ECU then opens the solenoid valve. This vents atmospheric pressure to the EGR valve's diaphragm and the valve returns to its seat. When the engine is cold, the ECU determines that no vacuum is needed at the EGR valve diaphragm and vents the vacuum to the atmosphere.

Air Injection System

The ***air injection system*** may have a diverter valve and switching valve controlled by vacuum diaphragms, usually operated by the ECU through solenoid valves. These valves were discussed in Chapter 14. The solenoid valves are energized by the ECU. The ECU controls the diverter and switching valves to allow the incoming oxygen to combine most efficiently with the unburned gasoline.

Electric Fan Relay

Relays are used when an electrical component uses more current than can be handled by the ECU's internal

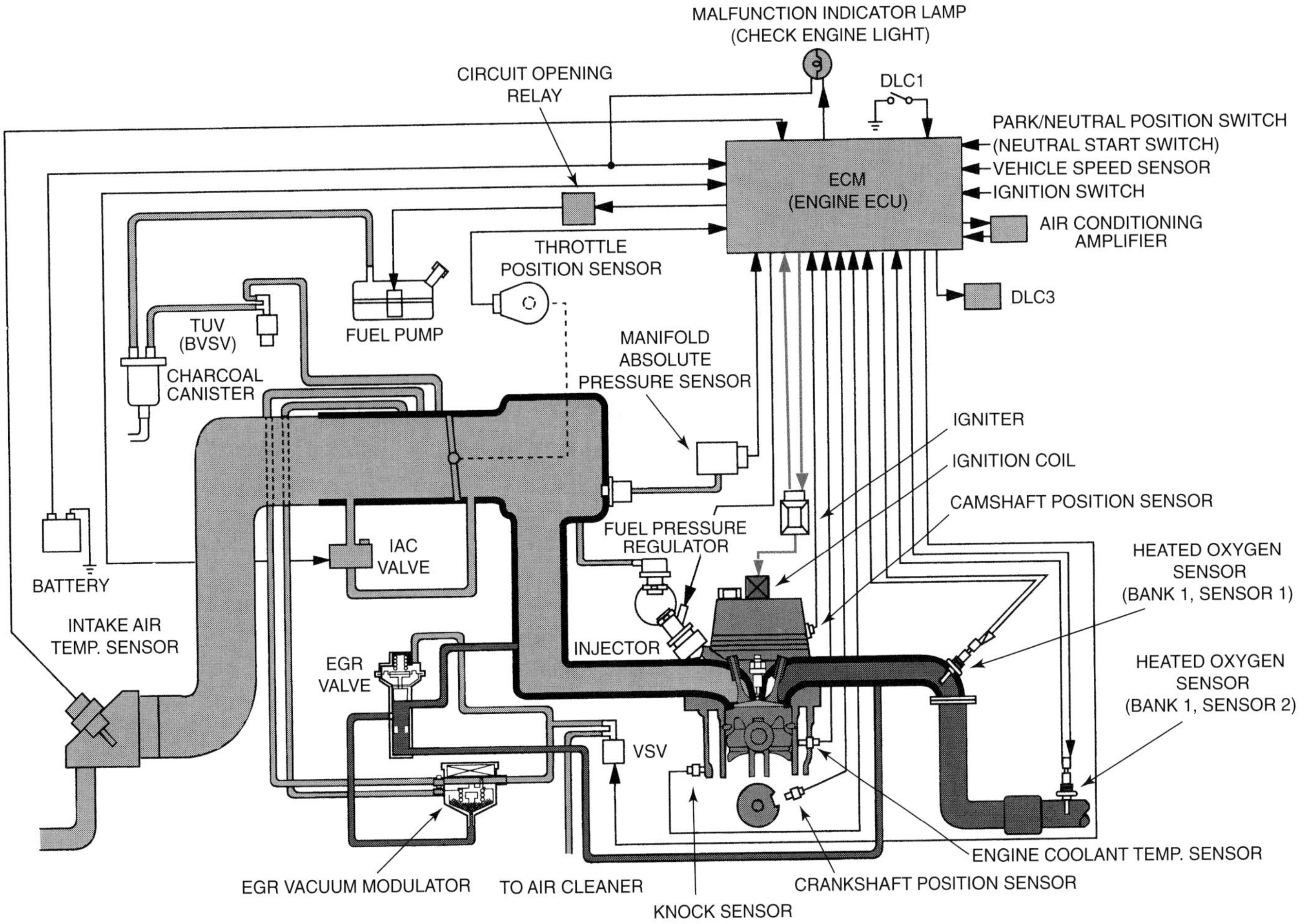

Figure 15-22. One particular electronic engine control system schematic. This system uses a computer-controlled ignition coil. Study the system carefully. (Toyota)

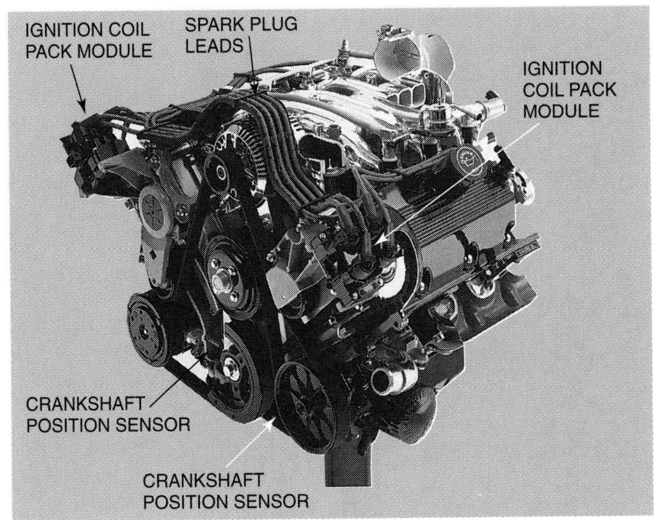

Figure 15-23. A 4.6-liter, electronic fuel injected (EFI) V-8 engine which uses a direct ignition system. There are two ignition coil pack modules, one on each side of the engine. Each module pack contains four ignition coils—one per spark plug. Note the position of the crankshaft and camshaft position sensors. (Ford Motor Company)

circuits. If the relay is the mechanical type, a small amount of current from the ECU energizes the relay solenoid, closing a set of contact points. In an electronic relay, the ECU current turns on a power transistor. A large amount of current can flow through the relay to the electrical component. Relays are discussed in detail in Chapter 17.

When an electric cooling fan is used, it remains off until the coolant temperature sensor signals the ECU that the engine has reached a predetermined temperature. Once the coolant temperature reaches around 227°F (108°C), the ECU energizes the ***fan relay***, **Figure 15-26**, which turns the engine cooling fan on. Also, the fan relay is energized any time the air conditioner is turned on regardless of engine temperature.

Electric Fuel Pump Relay

On many vehicles, the fuel pump is controlled by the ECU through a relay. The ECU energizes the ***fuel pump relay*** when the key is turned to the on position, **Figure 15-27**. This causes the electric fuel pump to deliver fuel through the system. When the ECU detects that the engine is not running (loss of RPM signal pulses) after three seconds, the ECU

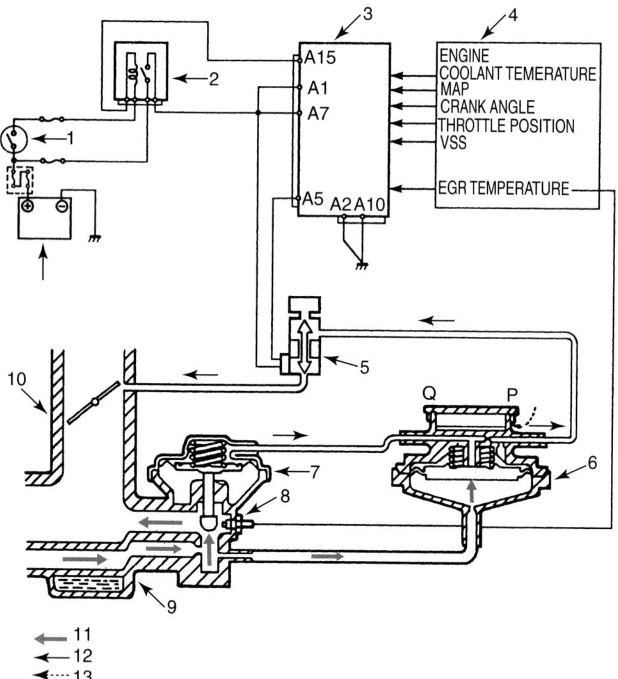

Figure 15-24. An exhaust gas recirculation (EGR) valve which is controlled through a solenoid vacuum valve by the electronic control module. Operation of the EGR valve is constantly monitored by the electronic control module which receives an electrical signal from the exhaust gas recirculation temperature sensor. If an abnormal temperature condition is detected, the ECM will turn the malfunction indicator lamp (MIL) on to alert the driver. 1—Ignition switch. 2—Main relay. 3—Electronic control module. 4—Sensed information. 5—EGR solenoid vacuum valve. 6—EGR modulator. 7—EGR valve. 8—Exhaust gas recirculation temperature sensor. 9—Intake manifold. 10—throttle body. 11—Exhaust gas. 12—Vacuum. 13—Air. 14—Battery (Geo)

de-energizes the relay. This reduces the chance of fire in the event of an accident.

Note that the oil pressure switch is wired parallel from relay to electric fuel pump. The oil pressure switch acts as a backup in the event the relay fails allowing the vehicle to start. However, it will take longer to start the engine as the oil must pressurize enough to close the contacts in the switch. Electrical current then passes through the switch and energizes the electric fuel pump.

Transmission Control Solenoids

On many automatic transmissions and transaxles, solenoids control some of the operations of the hydraulic system. The ***transmission control solenoid*** is attached to flow control valves in the hydraulic system. The solenoid moves the valve, which then modifies the flow of transmission fluid. Solenoids have been used to operate the torque converter lockup clutch on many older transmissions. In the transmissions and transaxles installed in newer vehicles, the solenoids operate the shift valves, selecting transmission shift points according to instructions from the ECU.

On-Board Diagnostic Systems

Although computer systems precisely control various operating conditions in modern vehicles, they are often very difficult to troubleshoot. To simplify the diagnosis and service of these systems, manufacturers have designed on-board diagnostic systems into their products.

Early on-board diagnostic systems were designed to monitor computer system parts, such as sensors and actuators. If a part failed, the ECU received an out-of-range electrical reading. The ECU would then set one or more trouble codes and turn on the MIL. These early systems are now called on-board diagnostics generation I, or ***OBD I***, systems.

The latest on-board diagnostic systems not only detect failed parts, but they also monitor engine operating conditions that indicate the production of excessive emissions. These new systems are called on-board diagnostics generation II, or ***OBD II***, systems. A typical OBD II system will monitor air-fuel ratio, engine misfires, engine temperature, catalyst efficiency, and other operating conditions. If the OBD II system determines that the vehicle is producing, or is about to produce, excess emissions, it will set a diagnostic trouble code and turn on the MIL.

Another important difference between OBD I and OBD II systems is the number of possible diagnostic trouble codes each system can produce. OBD I systems use a two-number trouble code format. Therefore, an OBD I system can have only 100 possible trouble codes. OBD II systems use a five-character alphanumeric code. The combination of letters and numbers allows for 8000 possible trouble codes.

Static Electricity

A technician can become charged with ***static electricity*** by friction or induction. Sliding across a seat, a technician can become charged by friction. A technician can become charged by induction by touching a ground while standing near a highly charged object. Many electronic components, including the ECU, can be destroyed by static electricity, **Figure 15-28**.

The static electricity charge while sliding across a seat can be as high as 25,000 V. The change remains static (at rest) in the body until the technician touches something made of metal or other conductive material. A high voltage spark jumps from the body to the object. If that object is an ECU, the transistors and IC's inside it will be destroyed. It only takes about 15 V, in some cases, to destroy the inner workings of an ECU and the static electrical discharge is far greater. Always ground yourself by touching the vehicleís body or a ground connection before handling the ECU.

Electromagnetic Interference

With the increasing amount of electronics and wiring in the modern automobile, it is difficult to keep signals from each circuit from crossing. For example, if sensor wires are routed too close to the ignition or other wires or are allowed to touch in some cases, ***electromagnetic interference*** or EMI, can result.

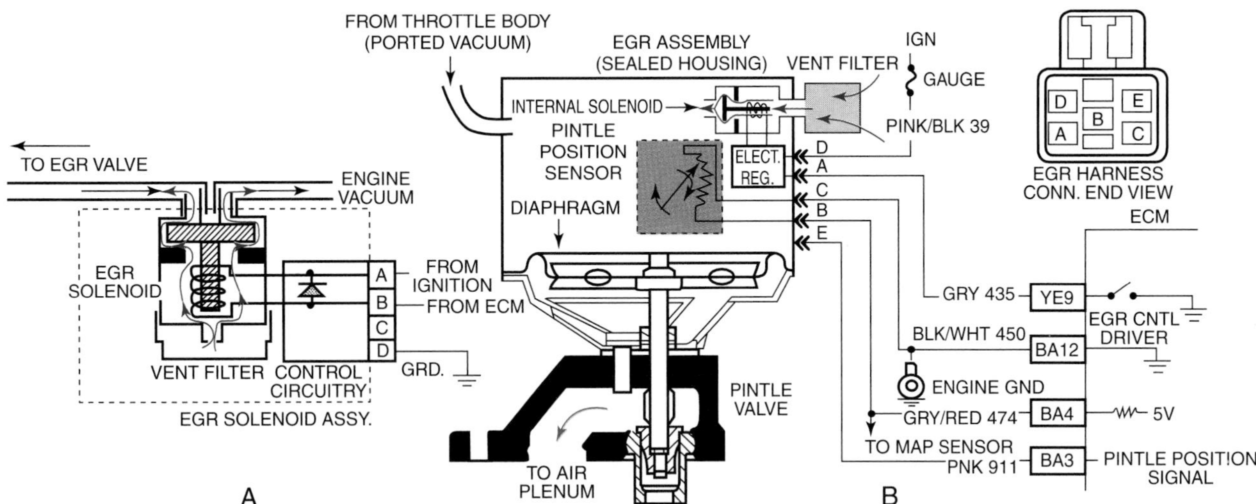

Figure 15-25. A—This computer-controlled EGR solenoid allows either vacuum to act on the valve's diaphragm or vents it to the atmosphere. Colored arrows indicate atmospheric pressure allowed to enter the lines when the solenoid valve is open, which prevents operation of the EGR valve. B—Sometimes solenoids are located on top of the EGR valve inside a sealed housing. Colored arrows indicate atmospheric pressure vented to the housing when the solenoid valve is open, preventing EGR valve operation. (Chevrolet & Oldsmobile)

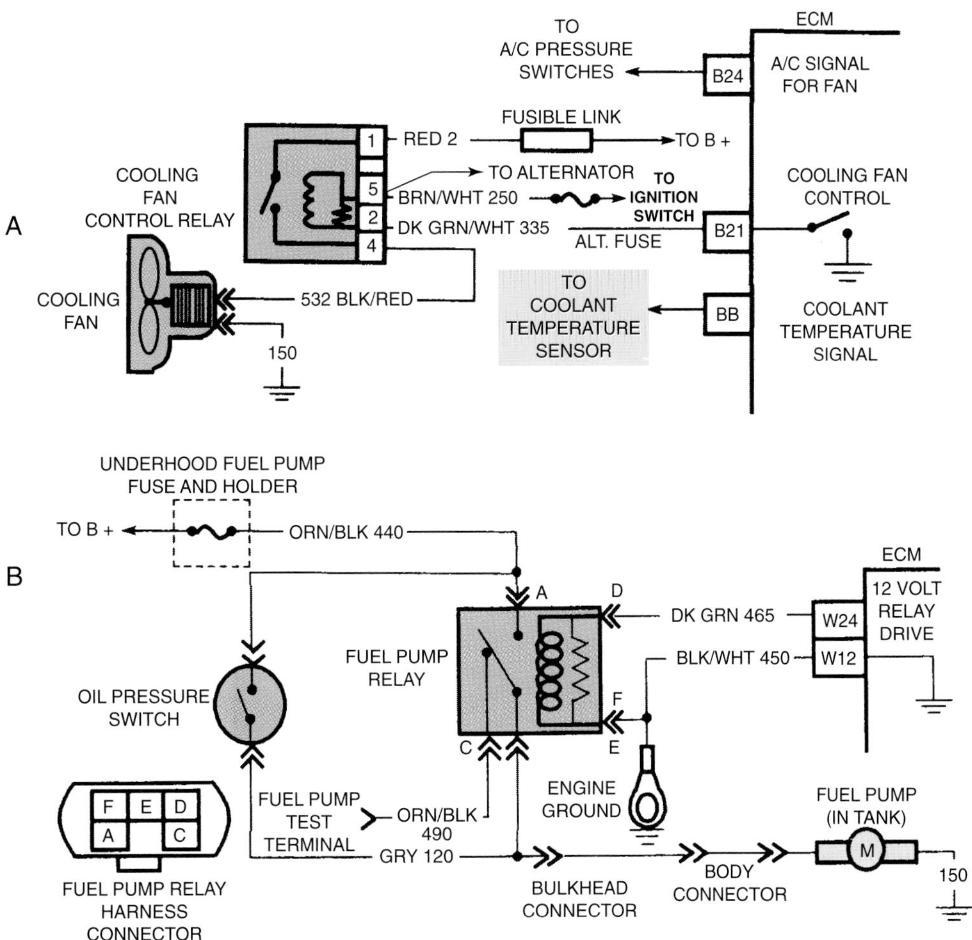

Figure 15-26. A—Computer-controlled relays are used to energize the cooling fan, based on data from the coolant temperature sensor. B—Computer can energize the electric fuel pump relay when the ignition is turned to the run position, which sends current to the fuel pump motor. The relay will remain energized if the computer detects that the engine is running (engine rpm input). Note the oil pressure switch which is wired parallel to the relay. This acts as a bypass if the relay should fail, allowing the engine to be started. (Oldsmobile)

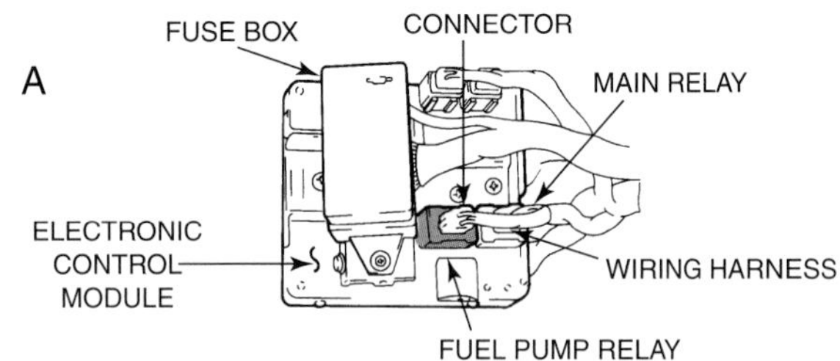

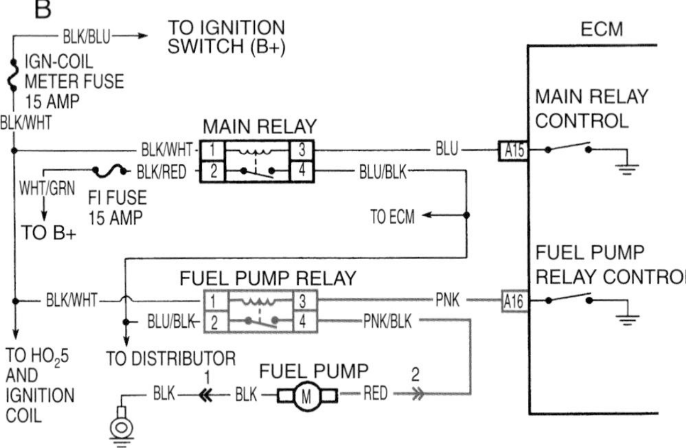

Figure 15-27. An electric fuel pump relay which is controlled by the electronic control module (ECM). A—Fuel pump mounting location as used by one manufacturer. B—Fuel pump relay and related wiring schematic. (Geo)

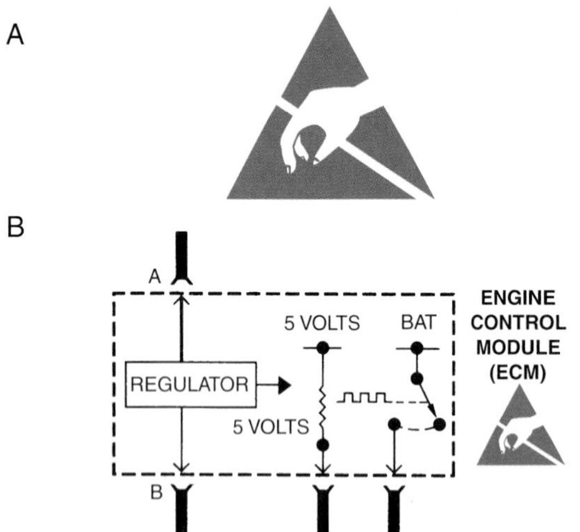

Figure 15-28. All electrostatic discharge (ESD) parts contain solid state circuits which are sensitive to static electricity. A—An electrostatic discharge symbol. This symbol is placed on the outside packing of any part that can be damaged by static electricity. B—Electrical part on an electrical schematic that is shown as electrostatic discharge sensitive. (General Motors)

In the past, electromagnetic interference was limited to ignition feedback that could be heard through the radio speakers. Today, electromagnetic interference from cellular telephones, police and CB radios, and other electronic accessories have been the cause of many problems in automotive systems. Wire harnesses on modern vehicles are carefully routed (and shielded in some cases) to reduce and prevent EMI.

Summary

An electronic control unit is used on modern vehicles to meet stringent fuel economy and emission control regulations. The ECU precisely controls the operation of the fuel, ignition, and other engine vehicle components to produce an air-fuel ratio as close to 14.7 to 1 as possible. Many modern vehicles combine the engine control unit and other system controllers into one large microprocessor which monitors and control most vehicle functions. The automotive computer control system is divided into three major areas, the ECU, the input sensors, and the output devices.

An ECU contains a central processing unit and four types of memory. The central processing unit controls the operation of the ECU. The memory is a set of instructions or programs that tells the CPU what to do. The program is stored in the read only memory (ROM), programmable read only memory (PROM), electronically erasable programmable read only memory (EEPROM), or erasable programmable read only memory (EPROM). The ROM or PROM have nonvolatile memories which means it will not lose any data when the battery is disconnected.

The random access memory (RAM) is a temporary storage place for data from the sensors including trouble codes. The RAM has a volatile memory which means its data is erased when the battery is disconnected. Keep alive memory (KAM) allows the ECU to compensate for engine component and sensor wear.

The input sensors measure the operating conditions of the engine or drive train, and convert them to an electrical signal that is then sent to the ECU. Some sensors produce their own voltage signal, while other sensors adjust the reference voltage sent to them. An output device is an electrical unit that is controlled by the ECU. The output device in turn controls some facet of engine or drive train operation. Output devices can be solenoids, relays, electric motors, or other electronic devices such as ignition modules.

Know These Terms

Computer
Electronic control unit (ECU)
Central processing unit (CPU)
Read only memory (ROM)
Programmable read only memory (PROM)
Erasable programmable read only memory (EPROM)
Electronically erasable programmable read only memory (EEPROM)
Random access memory (RAM)
Keep alive memory (KAM)
Open loop
Closed loop
Check engine light
Service engine soon light
Malfunction indicator light
Diagnostic trouble codes
Input sensors
Reference voltage
On-off switches
Pressure sensors
Crankshaft speed and position sensor
Camshaft position sensor
Magnetic pickup coil
Hall-effect switch
Optical sensor
Reed switch

Knock sensor
Thermistor
Coolant temperature sensor
Intake air temperature sensor
Manifold air temperature
Manifold vacuum sensor
Manifold absolute pressure
MAP sensor
Barometric pressure sensor
Airflow meter
Mass airflow sensor (MAF)
Oxygen sensor
Heated oxygen sensor
Throttle position sensor
Variable resistor
Transducer
EGR position sensor
Output device
Idle speed solenoid
Idle speed motor
Idle air control valve (IAC)
Fuel injector
Mixture control solenoid
Ignition module
Ignition coil
Distributorless ignition system
EGR system solenoid
Air injection system
Relay
Fan relay
Fuel pump relay
Transmission control solenoid
OBD I
OBD II
Static electricity
Electromagnetic interference (EMI)

Review Questions—Chapter 15

Do not write in this book. Write your answers on a separate sheet of paper.

1. Technician A says that the primary use of engine control computers on modern vehicles is to control exhaust emissions.Technician B says that the primary use of engine control computers on modern vehicles is to improve performance. Who is right?
   (A) A only.
   (B) B only.
   (C) Both A & B.
   (D) Neither A nor B.
2. The computer control system can be divided into three major subsystems. What are they?
3. What could be considered the fourth major subsystem?
4. Which of the following types of computer memory stores data temporarily?
   (A) Read only memory (ROM).
   (B) Programmable read only memory (PROM).
   (C) Random access memory (RAM).
   (D) All of the above.
5. When does the ECU go into open loop mode?
   (A) When the engine is cold.
   (B) When the engine is hot.
   (C) When the engine is at wide open throttle.
   (D) Both A & C.
6. Of read only memory (ROM), and programmable read only memory (PROM), which is contained on a chip? Which one can be removed from the main computer?
7. Computer system trouble codes are stored in _____ .
   (A) ROM
   (B) PROM
   (C) RAM
   (D) All of the above.
8. The simplest type of input sensor is the _____ .
   (A) on-off switch
   (B) pressure switch
   (C) magnetic pickup coil
   (D) reed switch
9. The ignition module can be an _____ .
   (A) output device
   (B) input sensor
   (C) knock sensor
   (D) Both A & B.
10. Reed switches are used to measure speed by measuring _____ .
11. In a piezoelectric element, electricity is produced by _____ .
    (A) heat
    (B) electricity
    (C) rotation
    (D) pressure
12. Temperature sensors measure the temperature of _____ and _____ .
13. Technician A says that manifold vacuum is the negative pressure in the intake manifold when the engine is running. Technician B says that manifold absolute pressure changes correspond to changes in manifold vacuum. Who is right?
    (A) A only.
    (B) B only.
    (C) Both A & B.
    (D) Neither A nor B.
14. The two types of throttle position sensors are _____ and _____ .
15. In a mass airflow sensor, the wire is heated by _____ .
    (A) the incoming air
    (B) electricity
    (C) engine heat
    (D) exhaust gases
16. The difference between the oxygen in the exhaust and the outside air causes _____ in the oxygen sensor.
    (A) heat
    (B) pressure
    (C) a chemical reaction
    (D) a short circuit

17. The mixture control solenoid is used only on engines with _____ .
    (A) electronic ignition
    (B) carburetors
    (C) fuel injection
    (D) air cooling
18. A relay is used when an electrical component uses more current than the _____ can handle.
    (A) battery
    (B) sensors
    (C) ignition system
    (D) internal computer circuits
19. The ECU energizes the electric fuel pump relay when _____ .
    (A) the engine starts
    (B) more fuel is needed for acceleration
    (C) the key is turned to the on position
    (D) the mechanical fuel pump stops operating
20. On a late model vehicle, the transmission control solenoids are used to control the automatic transmission _____ .
    (A) fluid temperature
    (B) shift points
    (C) noise
    (D) pump output

16

Charging and Starting Systems

After studying this chapter, you will be able to:

- Describe the construction and operation of an automotive battery.
- Explain the operating principles of a charging system.
- Identify the major parts of a charging system.
- Explain the operating principles of a starting system.
- Identify the major parts of a starting system.

This chapter will cover the fundamentals of vehicle charging and starting systems. The charging and starting systems are vital to the operation of almost every system on the vehicle. Without the electrical components discussed here, the ignition system, fuel injection, emission controls, electronic control unit, air conditioner, lights, and radio would not be functional. Studying this chapter will give you an understanding of the battery and charging and starting systems. These systems are the foundation for all other electrical equipment in the vehicle.

Battery

An electric current can be produced by means of a ***lead-acid battery***. The battery stores this energy in chemical form until it is needed. The main purpose of the battery is to supply current for the ignition system and starter motor until the vehicle is running. The battery also acts as a source of extra power for the vehicle's electrical system. The various components of the battery are discussed below.

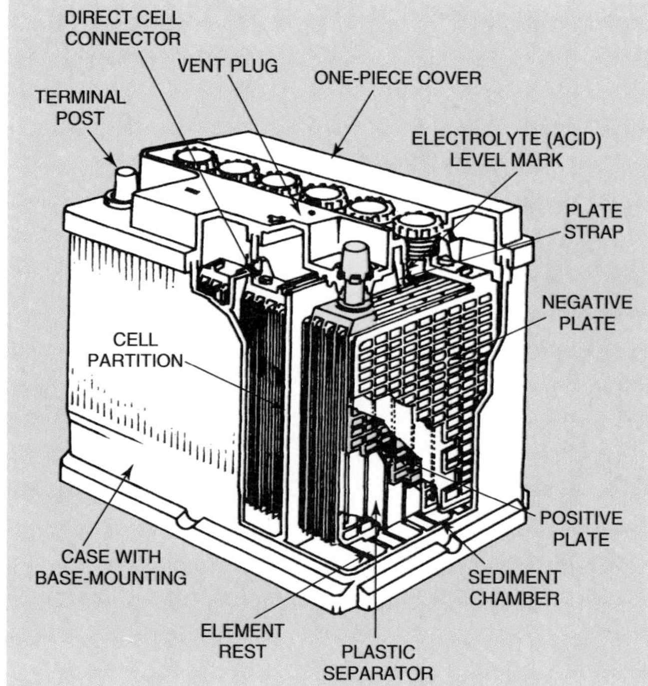

Figure 16-1. Cutaway of a 12 volt battery. Note the arrangement of cells (each one supplies approximately 2 volts), cell connectors, and sediment chamber.

Battery Cells

A battery is constructed of separate elements, or ***cells***, **Figure 16-1**. Each cell is made up of two groups of plates. Each battery cell has an open circuit voltage of 2 volts. Total battery voltage is determined by the number of cells. One group of plates forms the positive group, the other forms the negative group. The number of plates does not affect open circuit voltage.

Battery Cell Plates

The ***positive plate*** is a grid (wire-like, skeletal framework, made of lead and antimony or of lead and calcium), filled with lead peroxide (PbO_2) in sponge form. The ***negative plate*** is filled with porous lead (Pb) and expanders to prevent the lead from returning to an inactive solid state. Both are sandwiched together so that a negative group is next to a positive group. The plates are arranged negative, positive, negative, positive, and so on.

Insulating ***separators*** of nonconducting plastic, rubber, glass, cellulose fiber, or other types of material are placed between each set of plates. The separators keep the plates from touching each other and short circuiting. **Figure 16-2** illustrates a battery cell's positive and negative plates with separators. Note how the elements are kept away from the bottom of the container. This allows room for shedding plate material to deposit in the form of sediment. If this sediment touches the plates, it will cause a short circuit.

Electrolyte

The cell components are assembled in a hard rubber or plastic (such as polypropylene) battery box or case. The case has partitions to divide it into six compartments or cells. The

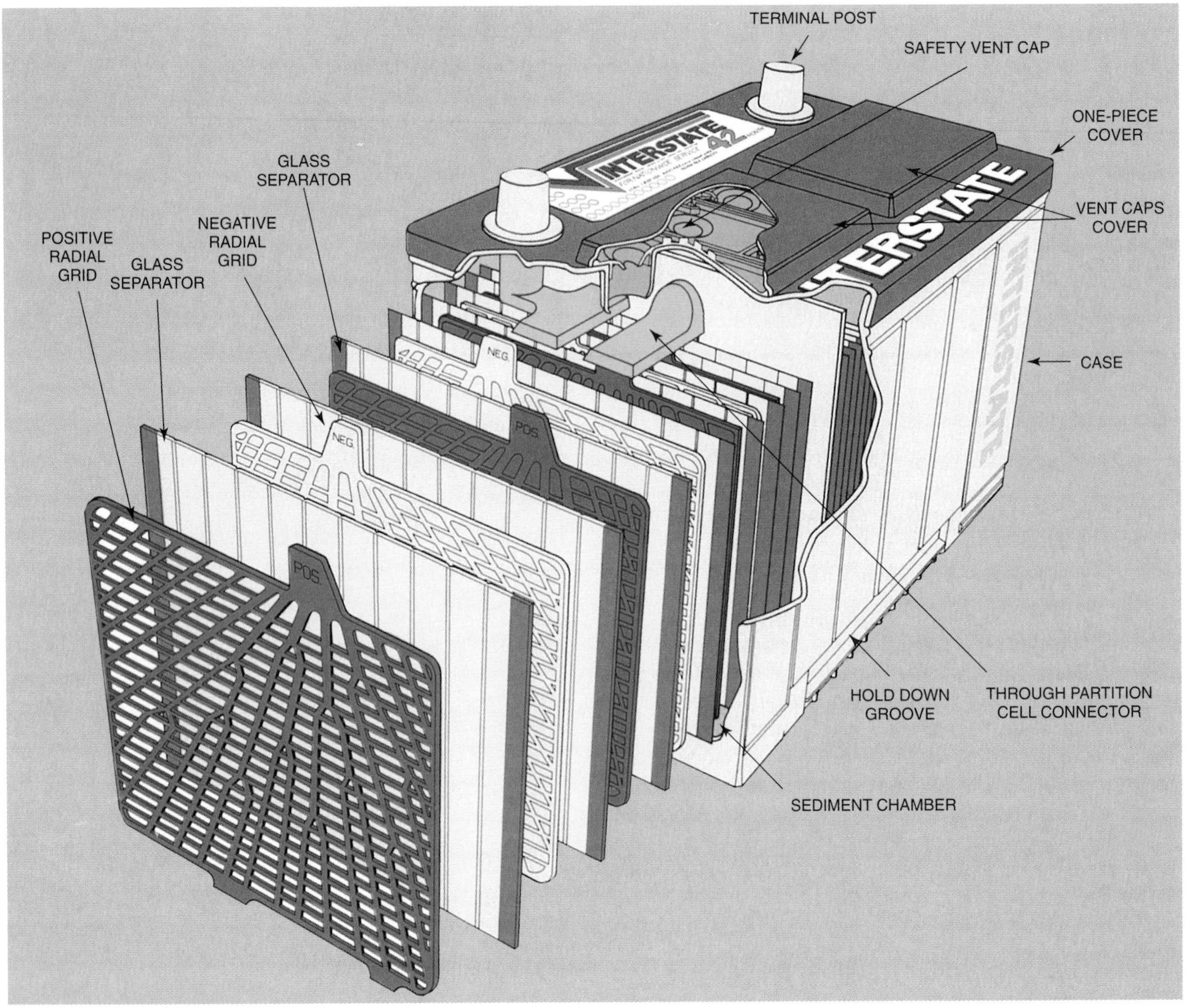

Figure 16-2. Typical 12-volt, maintenance free battery construction. Study the various parts and how they relate and connect to each other. (Interstate Battery)

cells are filled with an ***electrolyte*** solution made up of about 64% distilled water (H_2O) and 36% sulfuric acid (H_2SO_4), **Figure 16-3**. The electrolyte will slightly cover the top of the plates. The sponge form lead peroxide and porous lead allows the electrolyte to penetrate the positive and negative plates.

Battery Construction

Figure 16-4 shows how each cell is fitted into the individual compartments. As each element produces around 2 volts, 12 volts will be produced when they are connected in

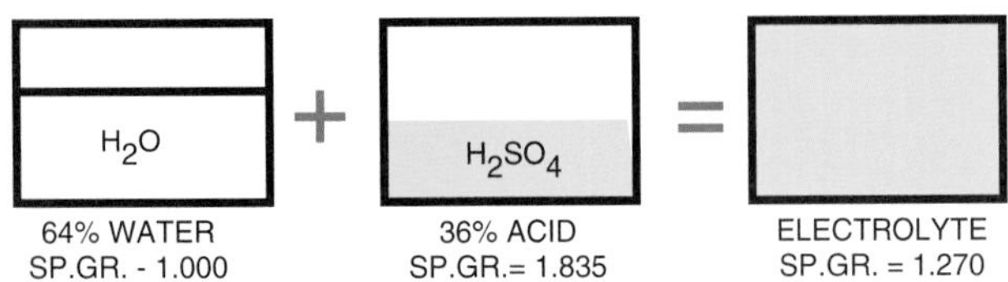

Figure 16-3. Chemical composition of battery electrolyte. Note the specific gravity of water compared to the sulfuric acid and the final electrolyte solution. (Deere & Co.)

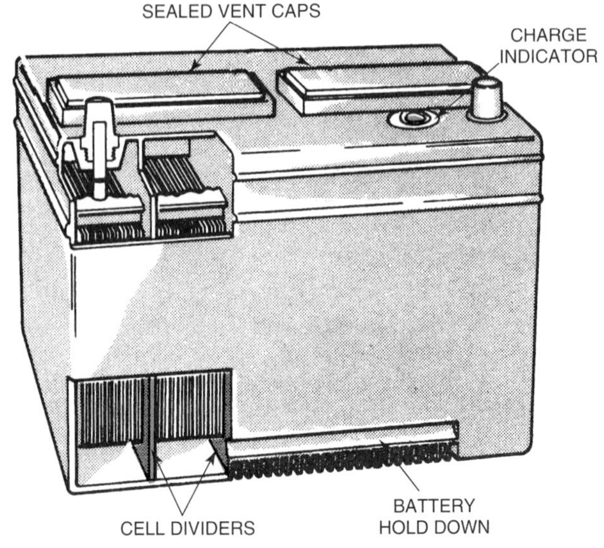

Figure 16-4. Battery container showing placement of cell dividers. Note the use of a built-in charge indicator. (Chrysler)

series. **Figure 16-5** shows the elements in place in the battery box. After installation, each cell is connected to its neighbor by a lead cell connector or strap. The end cells are connected to the battery ***terminals***.

Figure 16-6 illustrates all the various battery components and how they are joined together to produce the battery. Modern cars and light trucks have 12 volt batteries. Some older cars have 6 volt systems, while some agricultural vehicles have 8 volt systems. A few large trucks have 24 volt systems.

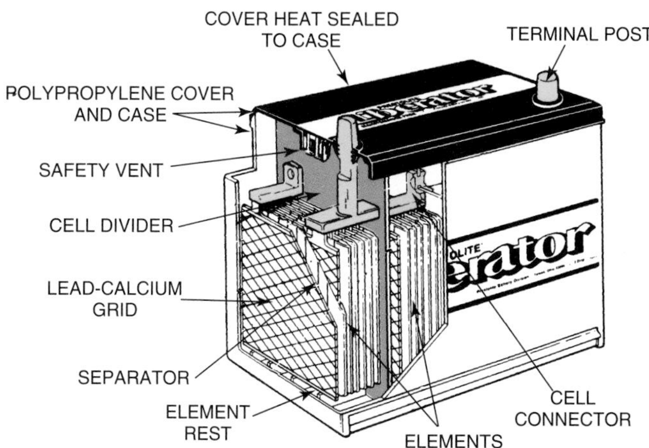

Figure 16-5. Sectioned view of a complete 12 volt battery. Learn the names of all parts. (Prestolite)

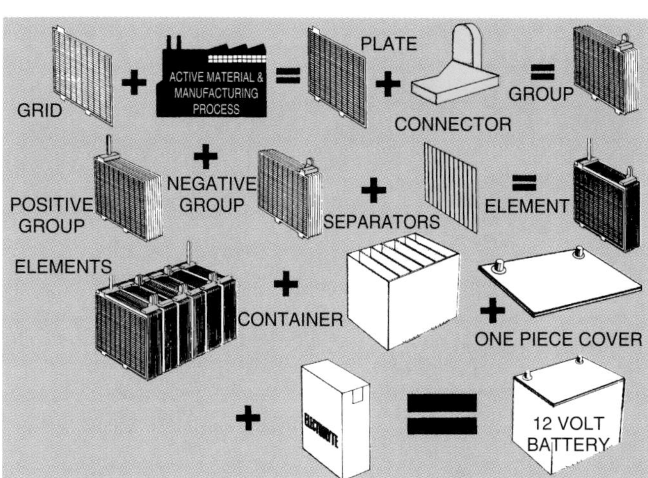

Figure 16-6. Study the manufacturing sequence that occurs in a battery's construction. Note in the final battery that the plates are joined by cell connector straps. (Firestone)

Some batteries may occasionally need water added to the electrolyte. Distilled water should be used whenever possible. Tap water can be used, however, in some areas tap water can contain chemicals that will reduce the life and efficiency of a battery. In the past, rainwater was collected and used to refill batteries. Freezer and refrigerator frost, when melted, is a good source of distilled water.

The electrolyte evaporated from older batteries, which required water to be added every few weeks. Modern batteries are sealed to reduce the amount of evaporation. Water cannot escape and condenses back into the electrolyte. These batteries are known as maintenance free batteries.

Warning: The sulfuric acid in electrolyte can cause severe burns. Avoid contact and wear protective clothing. A battery will sometimes release explosive hydrogen gas as part of its chemical action. Do not smoke or create any sparks or flame near a battery, Figure 16-7.

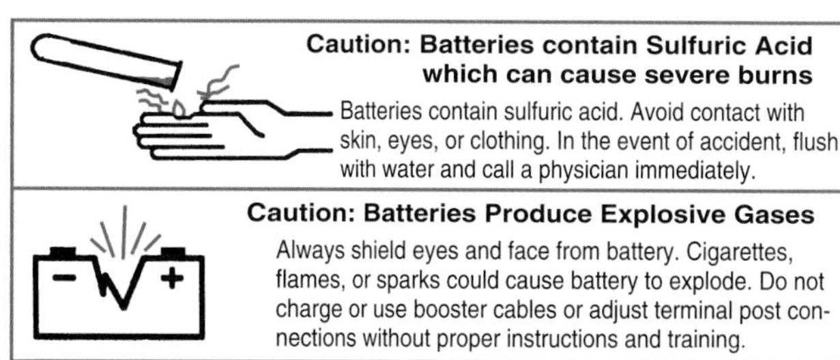

Figure 16-7. Typical warning label placed on batteries. Use extreme care whenever you are handling or near batteries. Severe burns to skin and eyes, as well as battery explosions are possible.

Battery Chemical Action

The battery does not actually store electricity. When charging, it converts electricity into chemical energy. Chemical action of electrolyte, working on active material in the plates, causes a transfer of electrons from the positive plates to the negative plates.

The amount of chemical energy in a battery is not inexhaustible. If the battery is not recharged, it will eventually be unable to operate. **Figure 16-8** shows the typical current load for modern vehicles. The charging system must produce sufficient current output to cover all possible electrical needs.

Battery Discharging

When the battery is placed in a closed circuit, such as when the starter is being operated, a surplus of electrons at the negative post will flow to the positive post. An electric current is produced by converting chemical energy into electricity, **Figure 16-9**.

As the current flows, the battery starts discharging. The sulfate ion (SO_4) in the sulfuric acid (H_2SO_4) combines with the lead (Pb) in the plate materials to form lead sulfate ($PbSO_4$). The hydrogen ion (H_2) combines with the oxygen in the lead peroxide to form water (H_2O). As the current continues flowing, the rate of current flow and the electrolyte become weaker and weaker. Finally the battery is completely discharged.

Battery Charging

The battery can be recharged by passing an electric current back into the battery (with a battery charger or the

TYPICAL CURRENT LOAD OF MODERN CARS
(IN AMPERES)

SWITCH	ACCESSORY SWITCH LOAD	MAX. VEHICLE OPERATING LOAD
Parking	4 to 8	–
Low-Beam Headlamp	8 to 14	–
High-Beam (4) Headlamp	12 to 18	18
Heater	6 to 7	–
Windshield Wiper	2 to 3	3
Air Conditioner	10 to 15	15
Radio	0.4 to 1.8	1.8
Electronic Ignition	8 to 12	12
Alternator Field	3 to 5	5
Total		54.8
Summer Starting	100 - 400 Amperes*	
Winter Starting	225 - 500 Amperes*	

*Values vary with engine size, engine temperature, and oil viscosity.

Figure 16-8. Typical current load for various electrical units. The load will vary somewhat depending upon the type of system and vehicle. (Prestolite)

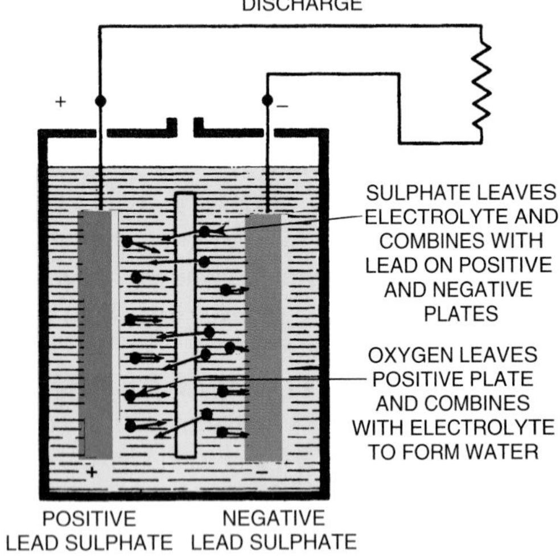

Figure 16-9. Normal chemical reaction during battery discharge. Note that the sulfate combines with the lead on the cell plates. This is why a battery is sometimes said to be sulfated. (Ford)

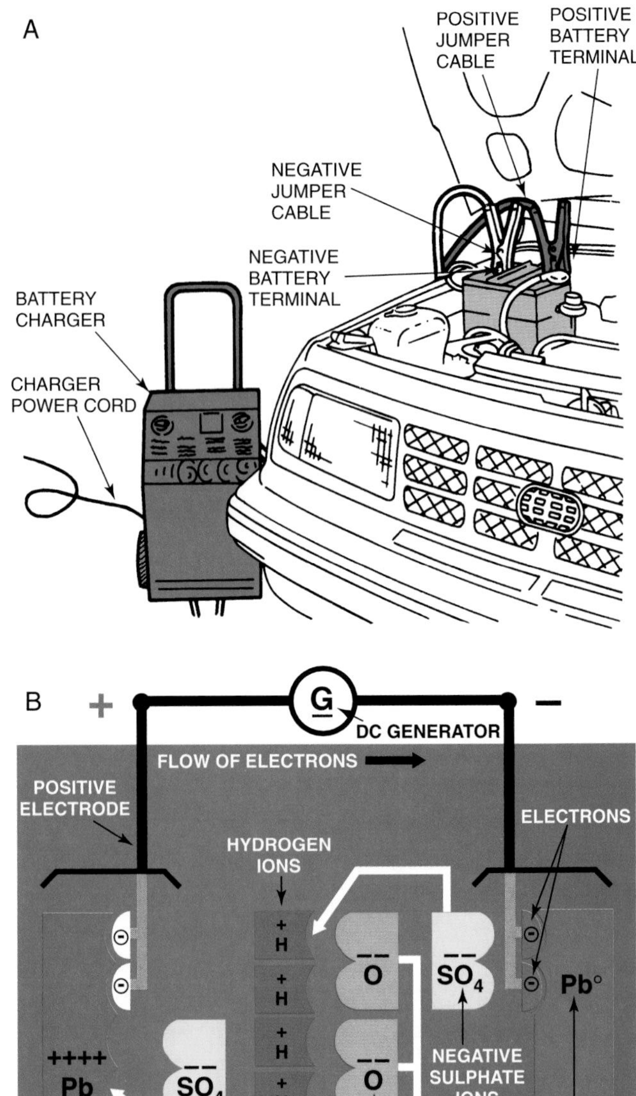

Figure 16-10. A—A battery charger can be used to charge a vehicle battery or to provide extra current for starting. B—Schematic of a battery cell and its chemical action during charging. Note that during charging the flow of electrons is from the positive electrode to the negative electrode. This is in reverse to battery current drain (discharge). (Bosch & Geo)

vehicle alternator) in a reverse direction. The sulfate (SO_4) ions leave the plates and combine with the hydrogen (H_2) from the water to form sulfuric acid (H_2SO_4). The lead (Pb) in the positive plates combine with the free oxygen (O_2) ions to form lead peroxide (PbO_2). If the battery plates and electrolyte are in good condition, the battery will eventually become fully charged, **Figure 16-10**.

Specific Gravity of the Electrolyte

When the battery is fully charged, the ***specific gravity*** (relative weight of a given volume of a certain material as compared with an equal volume of water) of the battery is around 1.280. This is determined by drawing a sample of the electrolyte into a ***hydrometer***, **Figure 16-11**. The depth at which the marked float rests in the electrolyte determines its specific gravity. The mark at the top of the electrolyte is read; the higher the float, the higher the charge.

It is not possible to use a hydrometer on maintenance free batteries. Battery state of charge is often indicated by a temperature compensated hydrometer built into the battery. Such a unit is illustrated in **Figure 16-12**. It contains a small green or other color ball that floats when the electrolyte specific gravity is within a certain range. Below this range, the ball sinks.

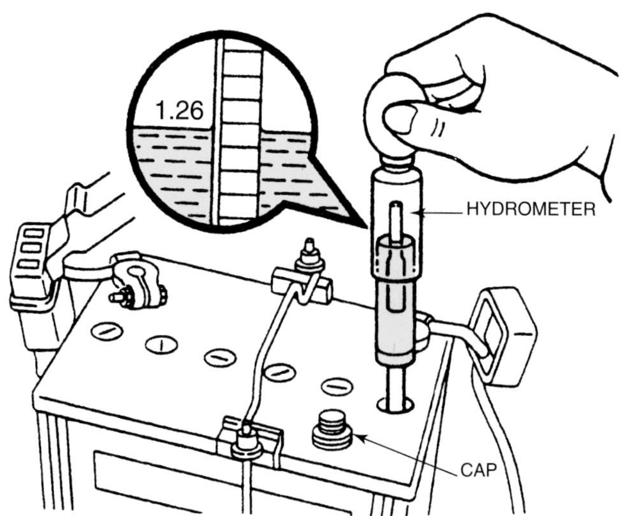

Figure 16-11. To check the specific gravity of electrolyte, it is drawn up into a hydrometer. The specific gravity is related to battery state of charge. As the battery charge falls off, the float will sink lower and lower into the electrolyte. (Daihatsu)

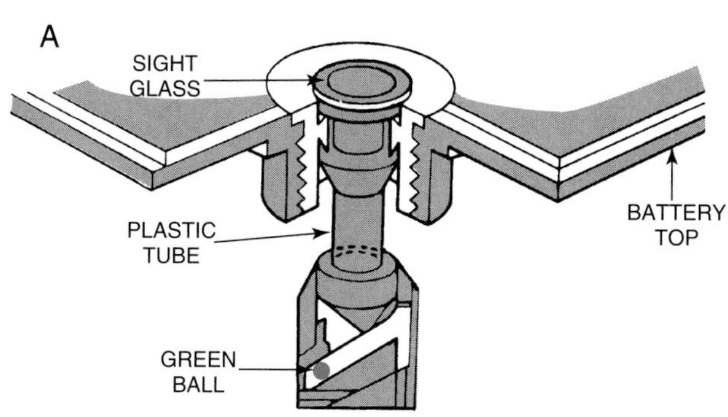

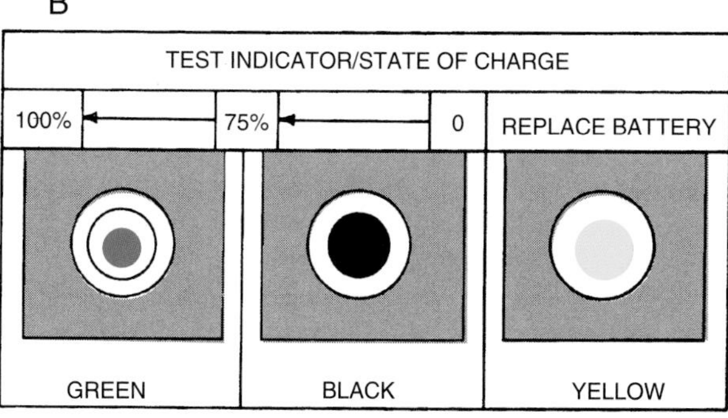

Figure 16-12. A—A cutaway of a built-in battery charge indicator. B—The sight glass or "eye" color indicates the battery's state of charge. (Dodge)

When the ball is floating, it shows as a bright green dot indicating the battery is 80% charged. When the ball sinks, the unit shows up dark, indicating the battery is discharged. If a light yellow color is visible, it indicates the electrolyte level is too low, **Figure 16-12**.

Temperature Correction

When reading a hydrometer, remember that the temperature of the electrolyte affects the specific gravity. Many hydrometers have built-in thermometers with temperature correction scales, **Figure 16-13**. The chart in **Figure 16-14** illustrates how temperature affects the efficiency of a battery. If a battery is discharged too much, the electrolyte can freeze and the battery will burst. The freezing point in relation to specific gravity is shown in **Figure 16-15**.

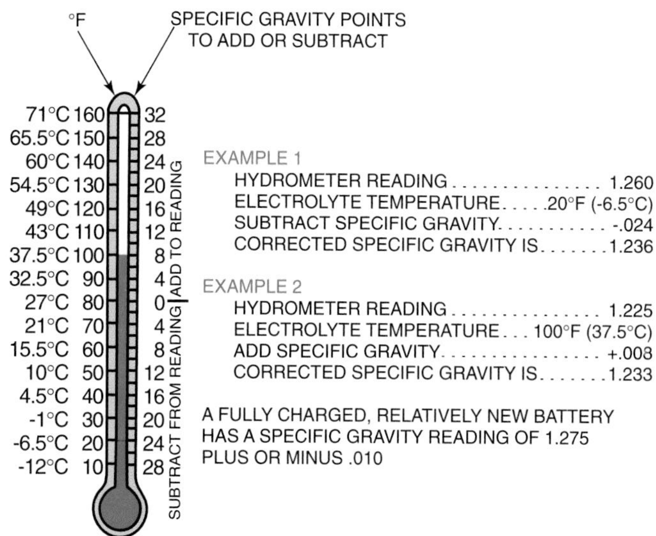

Figure 16-13. Temperature correction chart for hydrometer use. (Plymouth)

BATTERY EFFICIENCY AT VARIOUS TEMPERATURES	
TEMPERATURE	EFFICIENCY OF A FULLY CHARGED BATTERY
80°F – 26.7°C	100%
50°F – 10°C	82%
30°F – -1.1°C	64%
20°F – -6.7°C	58%
10°F – -12.2°C	50%
0°F – -17.8°C	40%
-10°F – -23.3°C	33%

Figure 16-14. The ambient air temperature can affect battery efficiency. It is obvious why a fully charged battery is a must for cold weather starting. (Ford)

SPECIFIC GRAVITY	FREEZING POINT DEGREES F	DEGREES C	SPECIFIC GRAVITY	FREEZING POINT DEGREES F	DEGREES C
1.100	+18	-8	1.220	-31	-35.0
1.120	+13	-11	1.240	-50	-46
1.140	+8	-13	1.260	-75	-59.4
1.160	+1	-17	1.280	-92	-69
1.180	-6	-21	1.300	-95	-71
1.200	-17	-27			

Figure 16-15. A battery's specific gravity can affect the electrolyte's freezing point. The electrolyte in a battery with a low charge can freeze and crack the case. (GMC)

Battery Ratings

Electrical size has almost no relationship to battery physical size. Often, a battery that is small in size can have a high cranking amp rating. This varies because of different battery materials and construction standards. Newer batteries are often smaller than the older batteries that they replace, but have higher electrical ratings. The list below shows the latest cold cranking and reserve capacity measurements:

- ***Cold cranking amps*** (CCA): The maximum amount of current that flows for 30 seconds at 7.2 volts, with the battery temperature at 0°F (-18°C). This measurement indicates how much current the battery can produce when cold, and is the standard measurement for modern batteries.
- ***Reserve capacity*** (RC): The number of minutes that the battery can produce 25 amps at 10.2 volts, with battery temperature at 80°F (26.7°C). Reserve capacity indicates how long the battery can operate the vehicle electrical system in the event of charging system failure.
- ***Cranking amps*** (CA): The maximum amount of current that flows for 30 seconds at 7.2 volts at a temperature of 32°F (0°C). This measurement is also called hot or marine cranking amps.

Parasitic Battery Loads

The electrical and electronic components found on modern vehicles can cause a small, continuous drain on the battery after the ignition is turned off. This is called ***parasitic battery load***. Parasitic battery loads do not cause the battery to go dead, due to the small current draw (measured in milliamps), and should be considered normal. The chart in **Figure 16-16** shows some typical parasitic loads. An ammeter must be used to determine whether current flow is excessive.

Identifying Battery Terminals

The positive post of the battery, which is usually marked *POS*, or (+), is somewhat larger than the negative post and its top is sometimes marked in red. The negative post is the smaller of the two, and is sometimes marked in green or black. It is usually marked with a *NEG*, or (–), **Figure 16-17**.

On side terminal batteries, the positive connector is surrounded by a red plastic collar. On some side terminal batteries the cables cannot be interchanged as the positive connector is a different size than the negative.

The battery is usually located as close to the engine starter as possible to minimize voltage drop. The close location also reduces the length and cost of the wire necessary to connect the battery to the starter. An underhood location also makes the battery accessible for maintenance and replacement.

	CURRENT DRAW IN MILLIAMPERES (mA)	
COMPONENT	TYPICAL PARASITIC DRAW	MAXIMUM PARASITIC DRAW
ABS-ECU	1.0	1.0
Alternator	1.5	1.5
Auto Door Locks	1.0	1.0
BCM	3.6	12.4
Chime	1.0	1.0
ECM	2.6	10.0
ELC	2.0	3.3
HVAC Power Module	1.0	1.0
Illuminated Entry	1.0	1.0
Keyless Entry	2.2	5.5
Radio	6.9	6.0
SRS	1.6	2.7
Theft Deterrent System	0.4	1.0

Figure 16-16. Parasitic loads (mA), of various electronic components. ABS-ECU-antilock brake system/electronic control unit. BCM-body computer module. ECM-electronic control module. ELC-electronic level control. SRS-supplemental restraint system.

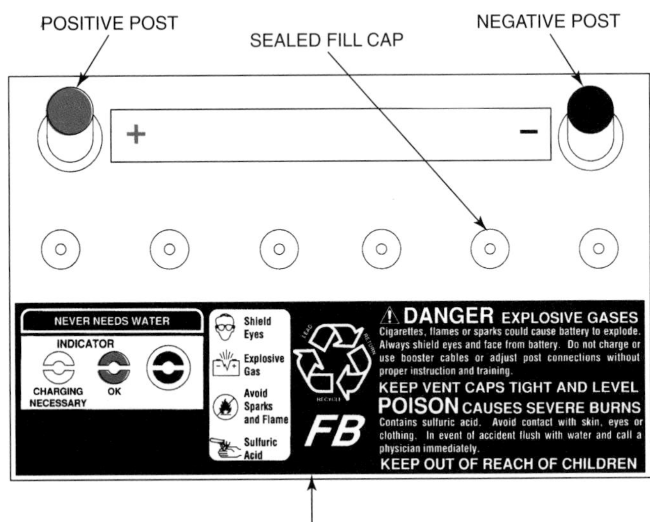

Figure 16-17. Typical features of a top post battery. Note the battery's specifications/warning label. (Geo)

The negative terminal is grounded on all vehicles, except for very old cars and trucks with 6 volt electrical systems and a few other older vehicles. The ground wire is fastened to the engine or to some other suitable metal location.

Dry-Charged Battery

Some batteries are shipped with the plates charged, but without electrolyte. This type of battery is considered to be ***dry-charged***. If this battery is kept in a cool, dry area, it will remain charged for a long time. Unlike the wet battery, dry-charged batteries do not utilize trickle charging (charging constantly at a very low rate). When the battery is delivered to the customer, the electrolyte is added. Dry-charged batteries are used in motorcycles, lawn equipment, and some cars.

Charging System

The charging system uses the rotation of the engine to create electricity. The electricity recharges the battery and provides power to operate the various vehicle electrical systems. The modern charging system consists of the alternator and regulator. On many vehicles, the regulator is built into the alternator. Alternator and regulator construction and operation are explained in this section.

Alternator

The ***alternator*** is a belt-driven, electromagnetic device. After the engine is started, the alternator produces electricity to meet the needs of the vehicle and to keep the battery charged. The modern vehicle contains many circuits that place a heavy load on the electrical system. Since many vehicles are used in stop-and-go city driving, this makes it difficult to maintain the battery in a fully charged state. Therefore, the alternator must be extremely efficient at all speeds. A typical alternator used in one charging system is shown in **Figure 16-18**.

Using Magnetism to Produce Electricity

The alternator uses mechanical energy from the engine to create electricity. It can do this because of a basic fact of electricity and magnetism. You may want to review the section on magnetic fields and magnetism in Chapter 7.

When a magnetic field moves in relation to a wire, voltage is generated in the wire. The wire can be moved through a stationary magnetic field, or the magnetic field can be moved through a stationary wire. If the wire is part of a closed circuit, current will flow, **Figure 16-19**. If the wire is passed through the magnetic field in the opposite direction, the current flow in the wire will be reversed. The amount of current produced in the wire will depend on two factors:

- The strength of the magnetic field.
- The difference in speed between the wire and field.

As the wire moves relative to the field, the lines of force are distorted or bent around the leading side. You will notice in **Figure 16-20** that the arrow around the wire indicates the direction in which the magnetic field moves around the wire. This excites the electrons in the wire, causing them to flow.

Inside an actual alternator, a field coil, or ***rotor***, with alternating (N) and (S) poles, spins inside a set of stationary ***stator*** windings. Current produced in the stator windings is delivered to the rest of the vehicle's electrical system. The alternator field coil is excited (current passed through it) by connecting it to a battery. The current draw is relatively light, around 2 amps.

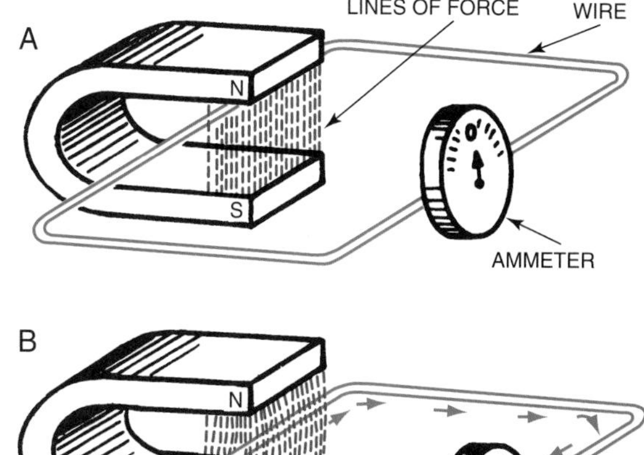

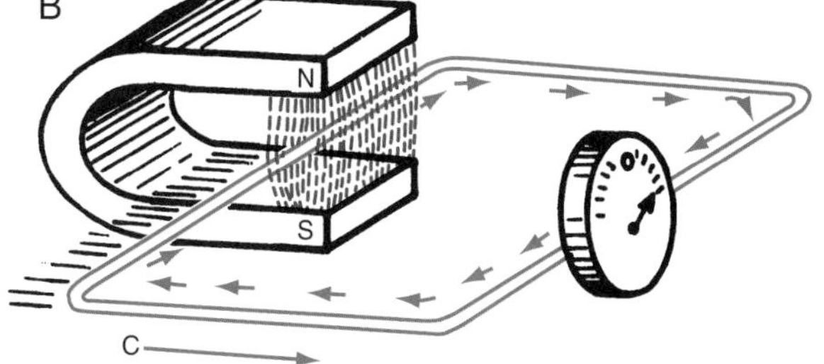

Figure 16-19. Inducing voltage in a wire. A—The wire is stationary and no current is flowing. B—The wire is being drawn swiftly through magnetic lines of force (in direction C). As the wire passes through the magnetic field, it will have voltage induced in it. As it makes a complete circuit, current flows (see needle on ammeter).

Alternator Construction

As mentioned earlier, the alternator spins a field coil with alternating (N) and (S) poles inside the stator windings. A round iron core is placed on a shaft, and the field winding is placed around the core. Two iron pole pieces are slid on the

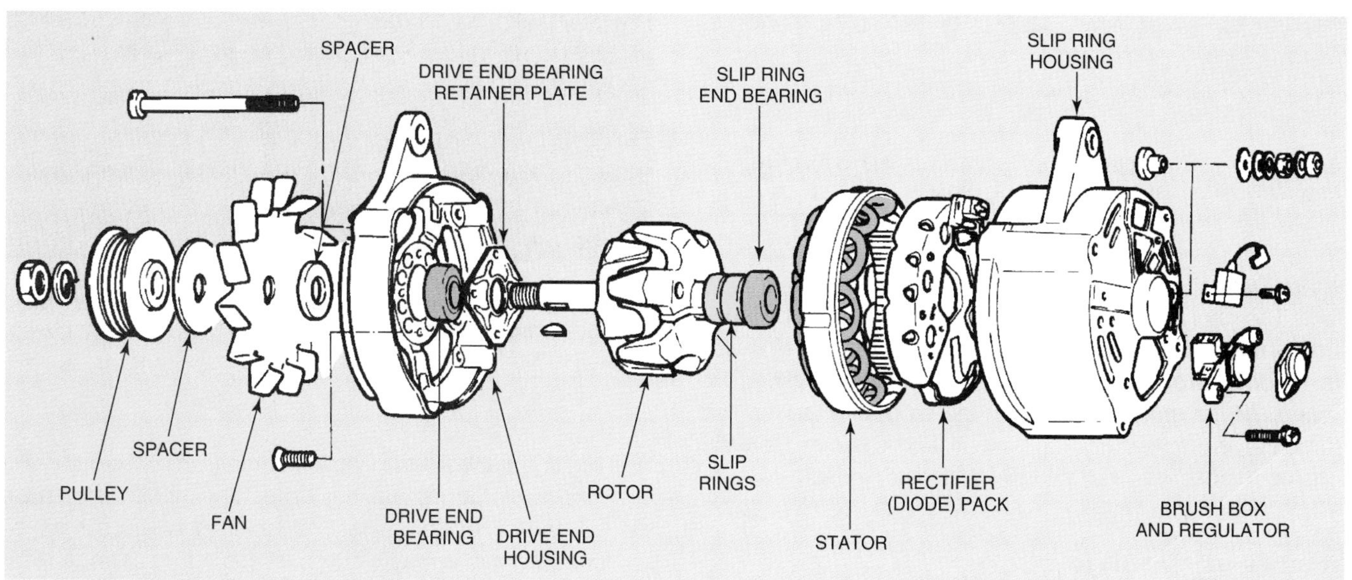

Figure 16-18. Typical parts of an alternator. Study the parts and learn their names and their relation to each other. (Hyundai)

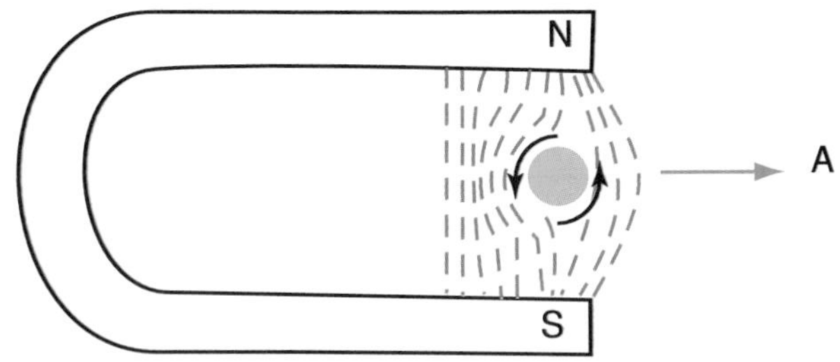

Figure 16-20. Wire passing through a magnetic field. The wire is passing through the magnetic field in direction A. Note how the lines of force are bent as the wire moves through the field. Two arrows around the wire indicate direction in which magnetic fields moves around wire.

shaft. They cover the field winding and are arranged so the fingers are interspersed. The fingers on one pole piece all form (N) poles. The fingers on the other form (S) poles. As the fingers are interspersed, they form alternate N-S-N-S- poles. **Figure 16-21** shows the complete rotor assembly in exploded form.

As the rotor magnetic fields cut through the conductor, voltage is generated and current flows in the complete circuit. As the alternating (N) and (S) pole fields pass around the conductor, the current will alternate, first one direction and then the other. See **Figure 16-22**.

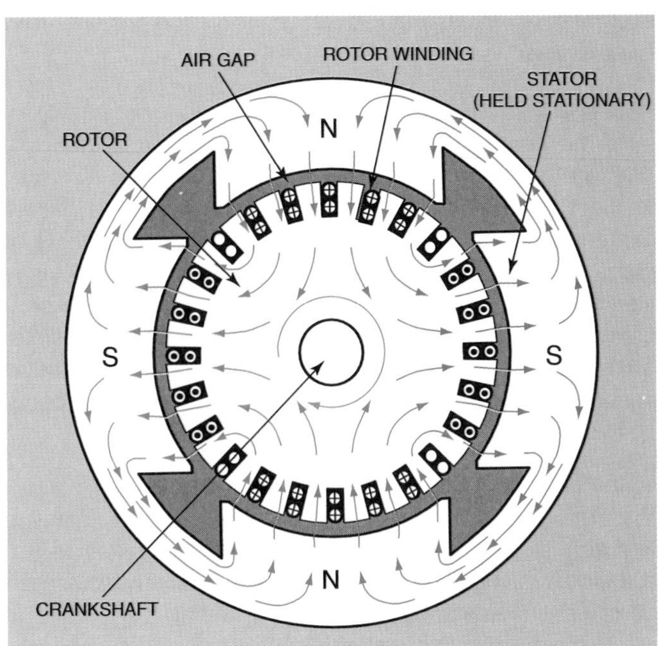

Figure 16-22. Rotor spinning inside the stationary stator assembly. Note the pattern of the magnetic lines of force. Note how they are similar to the magnetic fields you studied earlier. (Bosch)

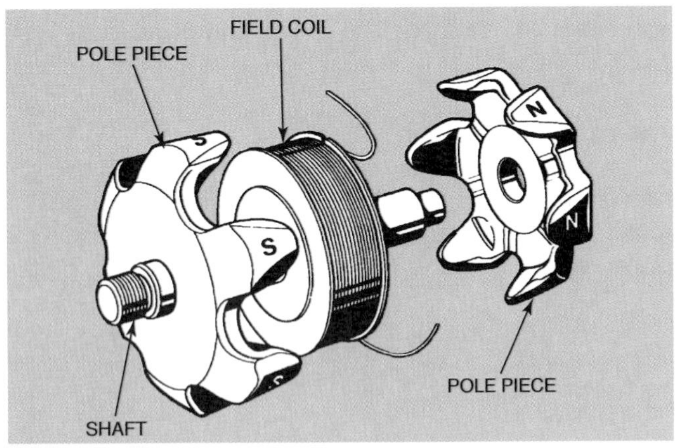

Figure 16-21. The pole pieces and field winding in this rotor assembly are pressed on the shaft. (Bosch)

Figure 16-23. A four-cylinder engine showing the diameter difference between the alternator pulley and the crankshaft pulley. (Toyota Motor Corp.)

Rotor Drive

The rotor is attached to a pulley which is driven by a belt from a pulley on the engine crankshaft. The crankshaft pulley is the ***drive pulley***, the rotor pulley is the ***driven pulley***. The rotor pulley is much smaller than the crankshaft pulley, and therefore the rotor turns at a much faster speed than the engine crankshaft, **Figure 16-23**. Rotor speeds of over 10,000 RPM are common.

Brushes and Slip Rings

Direct current from the battery is fed to the field coil by using ***brushes*** rubbing against ***slip rings***. One end of the field coil is fastened to the insulated brush, while the other end is attached to the grounded brush. As the pole fields pass through the conductor, voltage is imparted in the conductor, and current flows in one direction. When the rotor turns 180°, and the (N) pole and (S) pole are at opposite positions, this causes current to flow in the opposite direction.

A phantom view of a typical alternator is shown in **Figure 16-24**. Study the location of the various components. The use of diodes will be explained later in this section. The rear view of an assembled alternator is illustrated in **Figure 16-25A**. Note battery, field, and sensing terminals. A sectional view of the same alternator is shown in **Figure 16-25B**.

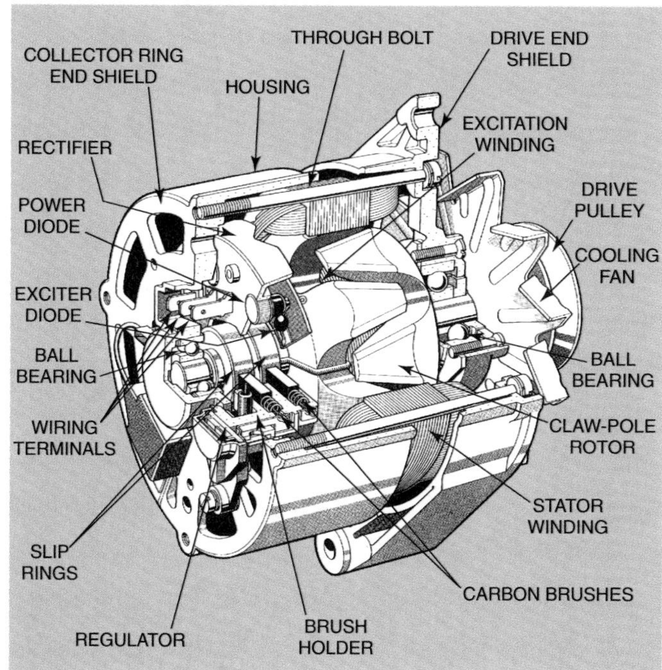

Figure 16-24. Cutaway view of an alternator showing the construction and layout of the various components. (Bosch)

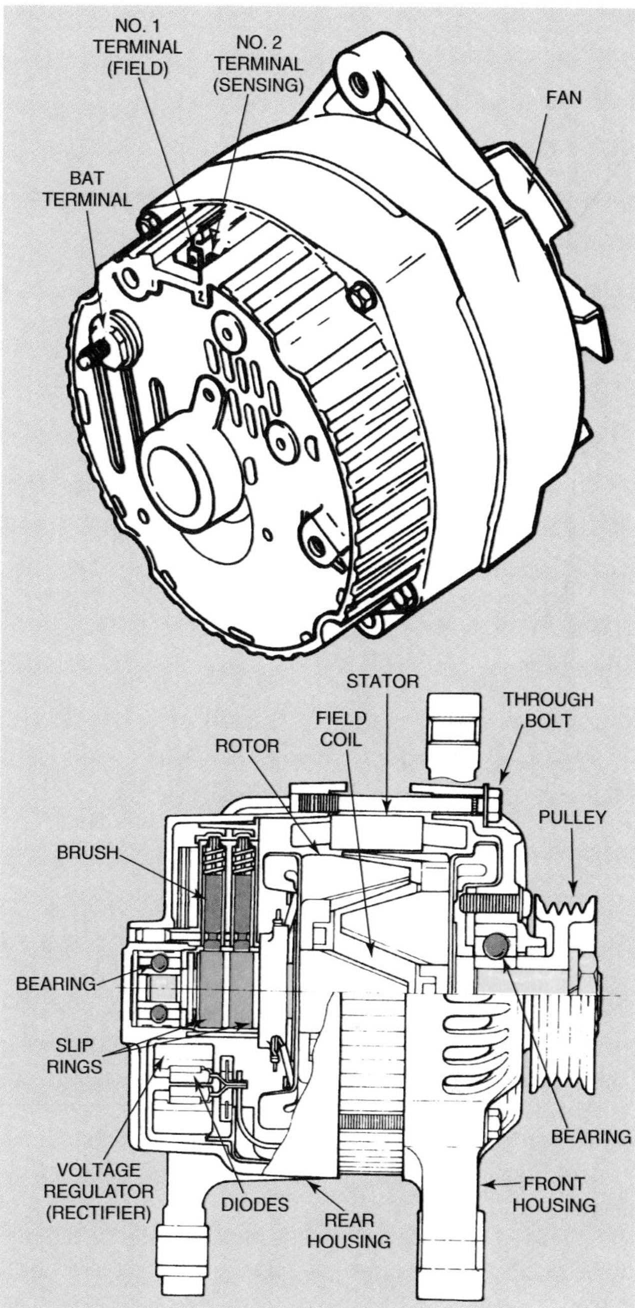

Figure 16-25. This alternator incorporates the voltage regulator into the alternator slip rings-end frame assembly. The regulator is a solid state electronic unit. (Chevrolet)

Alternator Output

The stator windings are made up of three separate windings. The windings produces what is known as three-phase alternating current. When only one winding is used, single-phase current results. A curve illustrating single-phase alternating current flow is shown in **Figure 16-26A**. **Figure 16-26B** illustrates three-phase alternating current.

Notice that the distance between high points in the output is shorter. The important thing to remember is that the stator produces alternating current. The vehicle battery requires direct current. Before it can be used, alternating current must be ***rectified*** (changed) to direct current.

Diodes

Alternating current is rectified by ***diodes***. Diodes have a peculiar ability to allow current to flow readily in one direction. When the current reverses and attempts to flow through in the other direction, current flow is stopped. Refer to Chapter 7 for more information about diodes.

The diodes are located in one end plate or shield. In the alternator illustrated in **Figure 16-27** four negative polarity diodes are pressed into the aluminum end plate. These are on the ground side. Four positive polarity diodes are pressed into a heat sink which is insulated from the end plate.

Remember that the stator has three windings. To rectify the current, each winding requires one negative and one positive diode. No matter which direction the current leaves any stator winding, the diodes are arranged to cause it to leave the alternator in the proper direction. The arrangement of the rectifiers will allow current to flow from the alternator to the battery, but will not allow current flow from the battery to the alternator.

The wiring arrangement of the stator and diodes is shown in **Figure 16-28**. A cross-sectional view of a diode is given in **Figure 16-29**. **Figure 16-30** shows a graphic representation of rectified current flow.

A

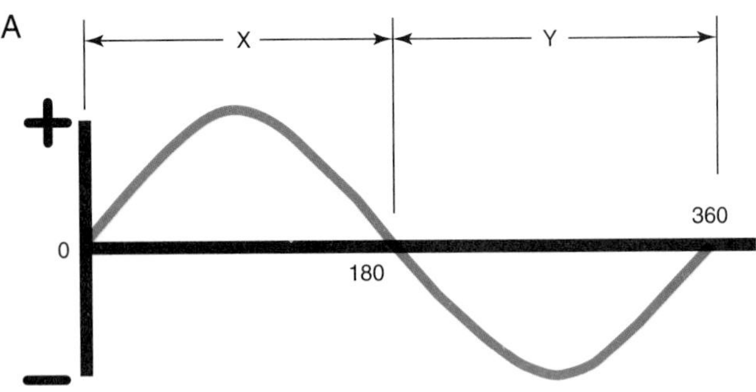

B

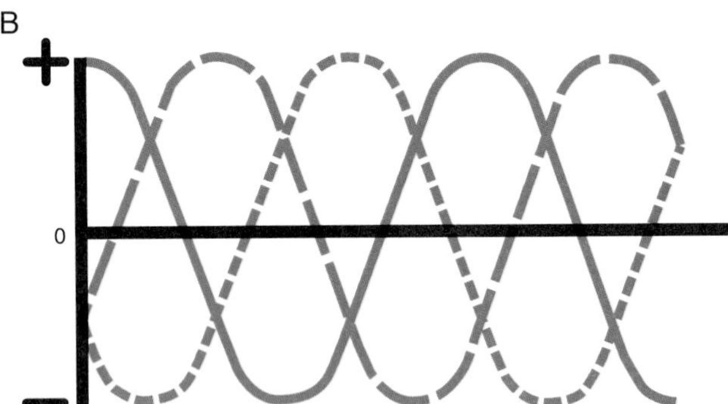

Figure 16-26. A—A single-phase alternating current wave. X represents positive half of wave and Y negative half. B—In three-phase alternating current, the phase outputs overlap, but do not coincide. (Delco-Remy)

A

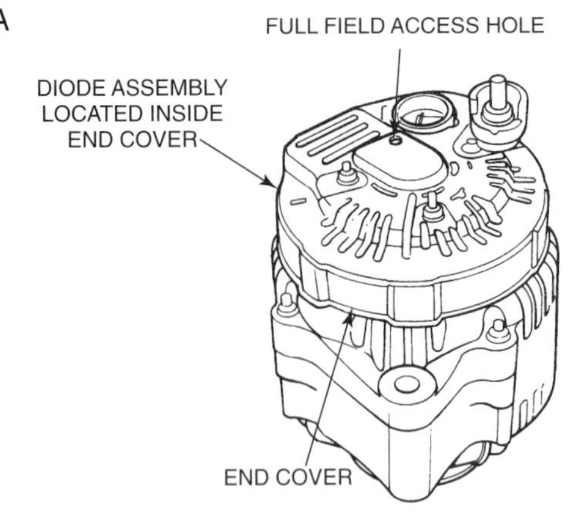

B

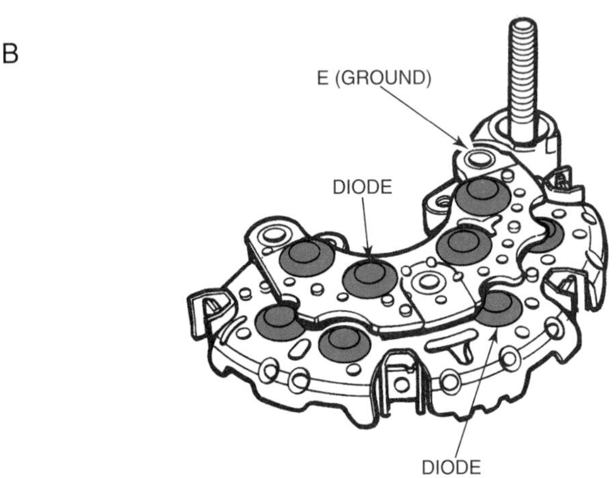

C

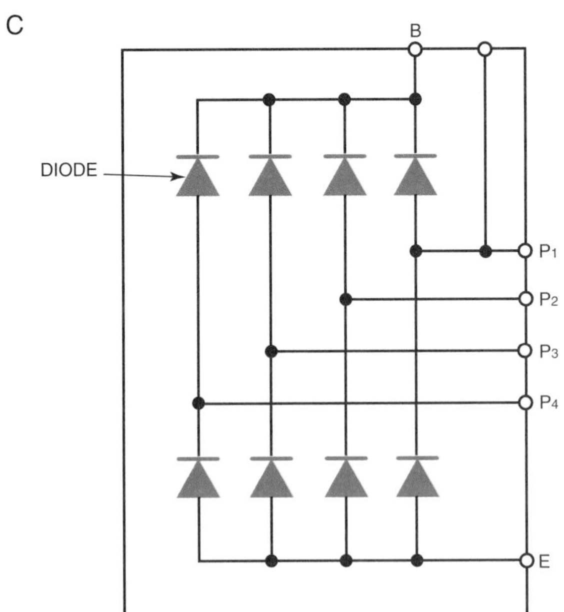

Figure 16-27. A—End view of an alternator. Note that the location of the diodes is next to an open vent in the alternator casing. B—This diode bridge contains eight diodes. Other diode bridges contain six. C—Electrical schematic for the diode bridge. (Acura)

Alternator Control

At idle, when the rotor is turning relatively slowly, the alternator can keep up with the demands of the vehicle electrical system. At normal vehicle speeds, the alternator can produce much more voltage and current than needed. To protect the alternator and the rest of the electrical system from damage, a means of reducing the output must be found.

As mentioned earlier in this chapter, there are two ways to control the amount of electricity produced in a wire: controlling the speed of rotor movement, or controlling the strength of the magnetic field. Since the rotor is driven directly by the engine crankshaft through belts and pulleys, controlling alternator speed would be difficult. The simplest way for controlling alternator output is to control the field strength.

To reduce alternator output, a ***voltage regulator*** is used. The voltage regulator controls the alternator output by adjusting the amount of current reaching the rotor. This controls the strength of the magnetic field passing through the stator windings. The regulator controls the field strength by reading voltage output from the alternator. If output voltage becomes too high, the regulator reduces field strength. If the output voltage drops, the regulator increases field strength.

Electromechanical Voltage Regulator

Older voltage regulators use contact points that are operated by electromagnetic coils. These were called ***electromechanical voltage regulators***, since they used a combination of electrical and mechanical parts. You learned earlier that electricity can be used to produce magnetism. If current is applied to a coil placed around a bar of steel, the bar will become magnetized, **Figure 16-31**. The more turns of wire and the stronger the current, the more powerful the magnet. This principle is used in electromechanical voltage regulators.

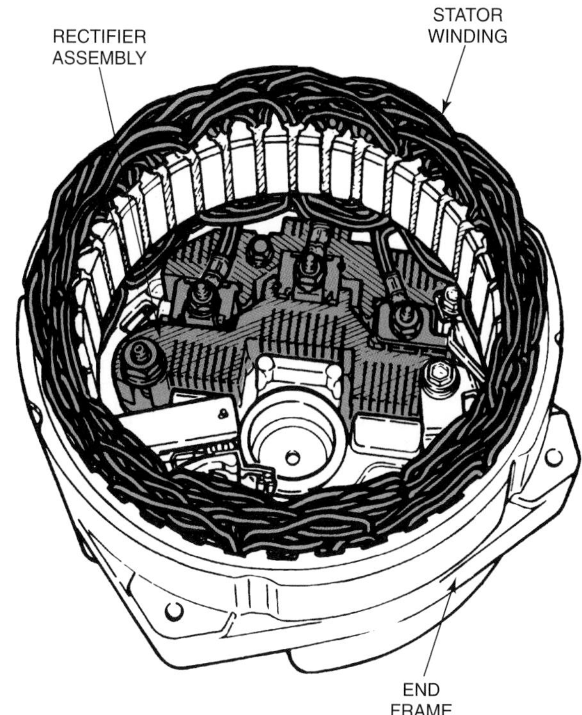

Figure 16-28. Typical rectifier wiring arrangement. Note that the stator windings are connected to the diode rectifier bridge. (General Motors)

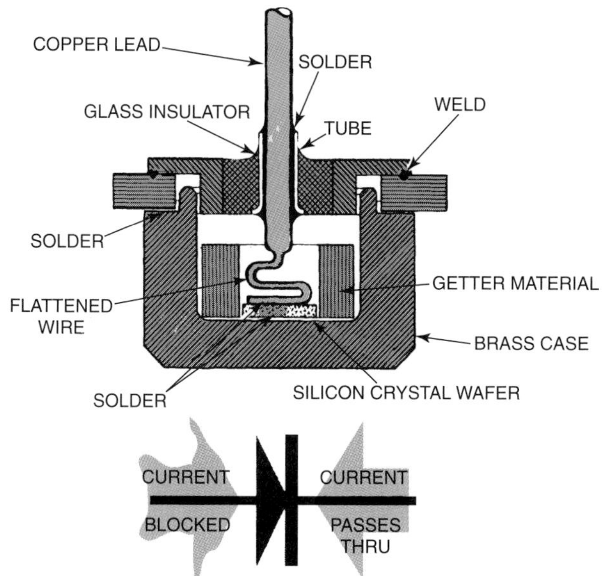

Figure 16-29. Cross-section of a diode (rectifier). Current is free to pass one way only. A diode, like a transistor, is a semiconductor (conducts electricity under certain conditions, acts as an insulator under others).

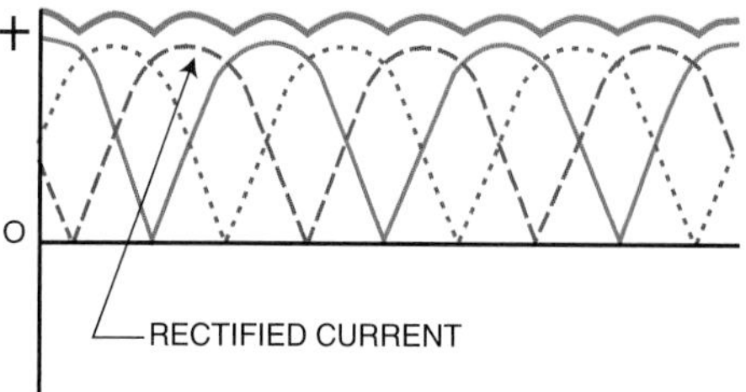

Figure 16-30. This rectified current flow wave pattern represents the three-phase alternating current depicted in Figure 16-26 after it has passed through the rectifier diodes. (Delco-Remy)

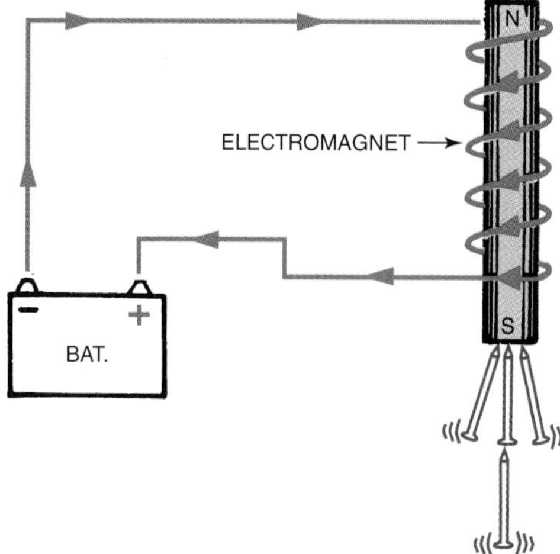

Figure 16-31. When current is passed through a coil surrounding an iron bar, the bar becomes an electromagnet. Notice how the bar attracts and holds nails while current is flowing.

The field current passes through a contact point set. Some voltage regulators cause the field current to go through a resistor which reduces, but does not stop current flow. The points are held closed by a spring. A small amount of output current from the alternator is delivered to the coil under the point set. If the voltage becomes too high, the coil's magnetic field become powerful enough to overcome spring tension, opening the points. This breaks the field circuit to the alternator. Alternator output stops, causing the electromagnetic coil to lose its field. This allows the spring to close the points, restoring current flow to the field.

This process occurs over and over many times per second. Alternator voltage output is determined by the tension of the point closing spring. On some older electromechanical regulators, the spring is adjustable. **Figure 16-32** shows a wiring diagram of a typical alternator charging circuit using a two unit regulator. When an ammeter is used instead of an indicator lamp, a single unit regulator, **Figure 16-33**, can be used. The single unit regulator contains a voltage regulator only.

Electronic Voltage Regulator

All vehicles made within the last 15 years use an ***electronic voltage regulator***. The electronic voltage regulator uses power transistors, integrated circuits, diodes, and other solid state parts to control alternator output. The electronic voltage regulator eliminates contact points and moving parts that can stick, oxidize, or wear. Some electronic regulators are remotely mounted. However, electronic voltage regulators are reliable, durable, and small enough to be incorporated into the alternator itself as shown in **Figure 16-34**.

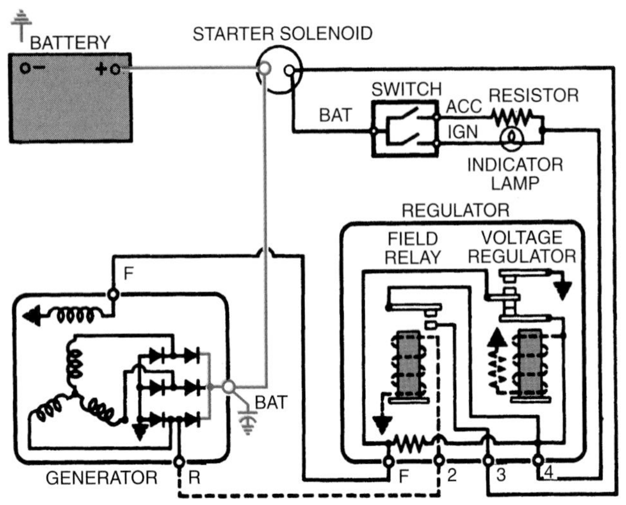

Figure 16-32. This alternator wiring diagram shows the use of a two unit regulator. Note how the field relay operates the "no charge" indicator lamp. (Pontiac)

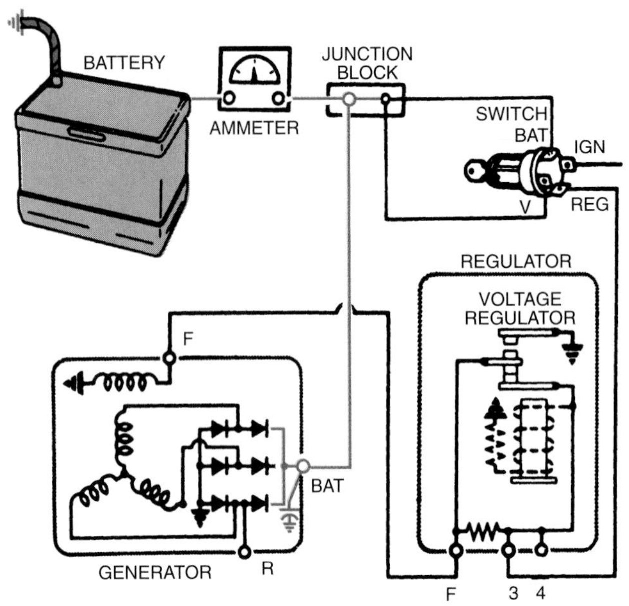

Figure 16-33. Schematic of a single unit voltage regulator. Note the use of an ammeter, eliminating the need for a field relay.

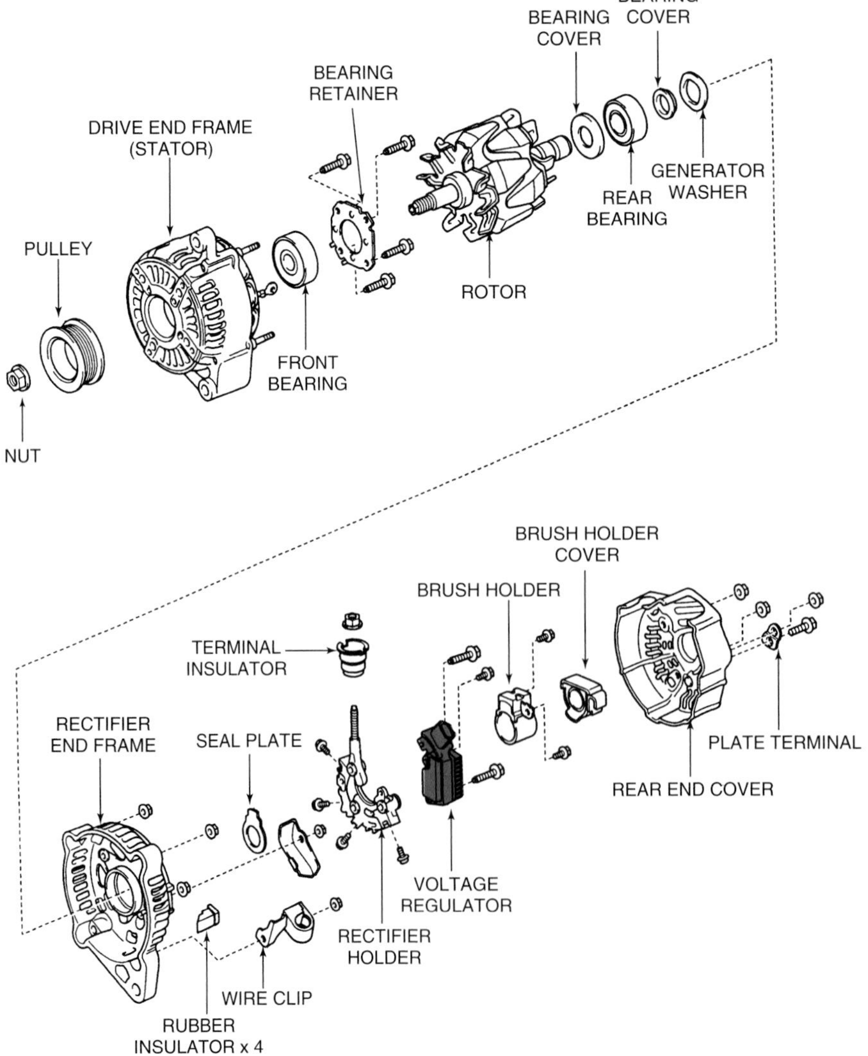

Figure 16-34. Exploded view of an alternator with a built-in-voltage regulator. (Toyota)

A charging circuit employing a transistor regulator is shown in **Figure 16-35**. The most important part in this regulator is the ***Zener diode***, discussed in Chapter 7. The breakdown point of the Zener diode is set at the maximum charging system voltage. The Zener diode is nonconducting until this voltage is reached. At this point, the diode breaks down and transmits electricity.

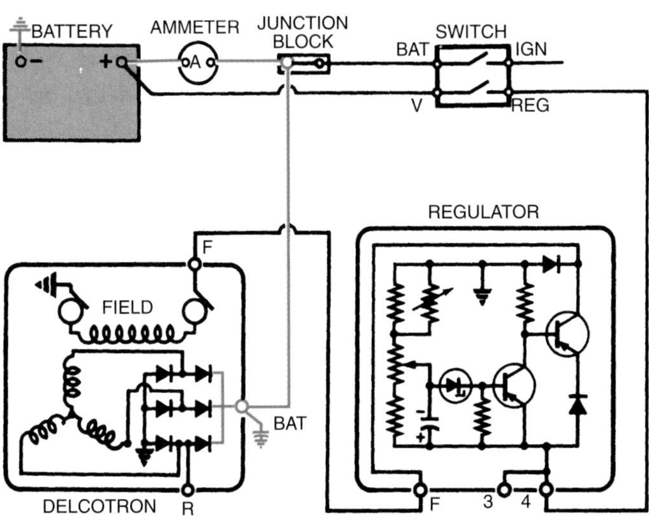

Figure 16-35. Schematic of a charging circuit employing a transistor regulator.

This current flow, acting through other transistors, turns the power transistor off and stops the flow of field current to the alternator. This causes the voltage output to fall below Zener diode break down voltage, at which time the diode once again becomes nonconductive. This stops the current flow to the transistors and allows resumption of field current flow. The cycle is repeated thousands of times per second. The external wiring circuit for one remotely mounted transistor regulator is in **Figure 16-36**.

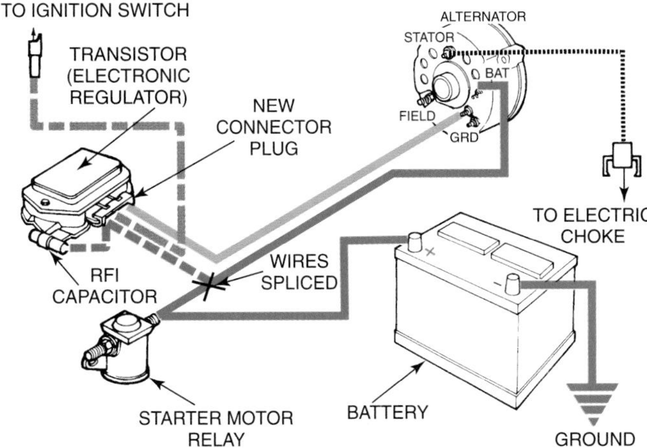

Figure 16-36. External wiring circuit as used on one charging system employing a remotely mounted transistor regulator. (Ford)

Built-In Electronic Regulator Operation

A common charging system circuit with a built-in regulator is illustrated in **Figure 16-37**. A small amount of current passes from the stator through the diode trio to the regulator. This equalizes the voltage on either side of the indicator light, putting it out, and tries to ground through resistor R3.

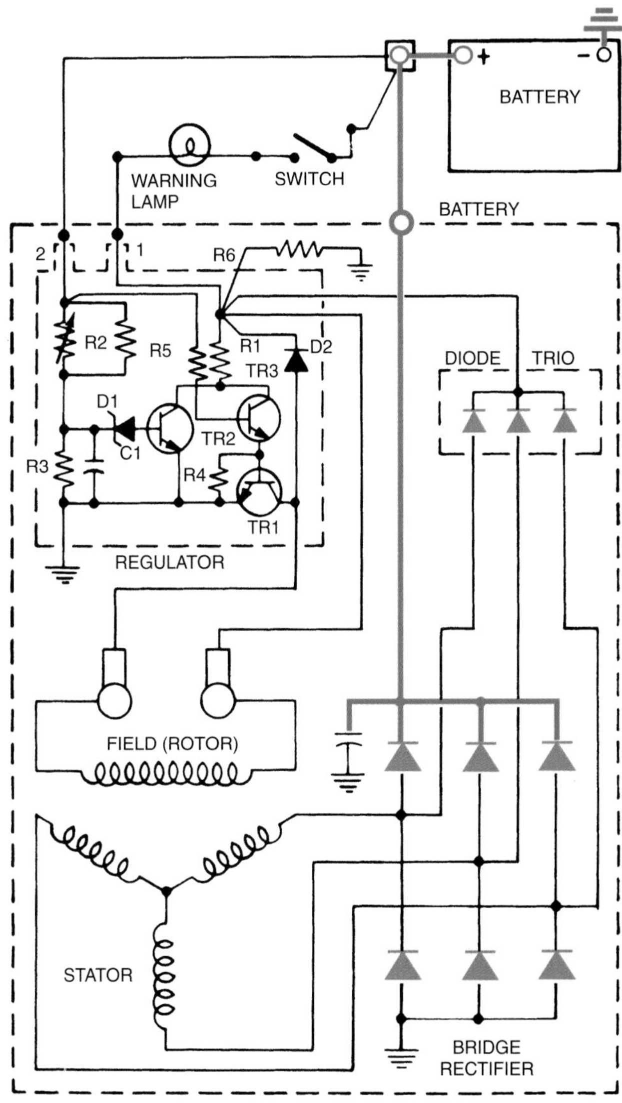

Figure 16-37. Charging system schematic showing a transistor voltage regulator. Note the diode trio. This unit rectifies current flow from ground through TR1 and the field windings to the stator windings. The rectifier diode bridge change stator ac voltage to dc output. (Chrysler)

When the stator voltage reaches a certain level, voltage going to ground through resistor R3 will increase to the point that it will cause Zener diode D1 to conduct. Forward-biased (voltage applied in direction causing current flow) transistor TR2 conducts and transistor TR3 is reverse-biased off. This turns transistor TR1 off. The field current and system voltage immediately decrease with TR1 off.

When system voltage decreases, voltage through R3 decreases and D1 stops conducting. This causes transistor TR2 to become reverse-biased off and transistors TR1 and

TR3 forward-biased on. Output voltage and field current will increase. This cycle is repeated thousands of times per second, holding alternator voltage output to a preset level.

Sudden voltage change across R3 is prevented by capacitor C1, and excessive back current through TR1 at high temperature is prevented by R4. TR1 is protected by D2 that prevents a high induced voltage in the field winding when TR1 is off. Voltage control temperature correction is provided by a thermistor (resistance decreases as temperature increases).

Computer-Controlled Voltage Regulation

Most vehicles today have several electronic control units (ECUs) that control the engine and other systems. ECU's can also be used to control voltage to the field winding, which eliminates the need for a separate regulator. The logic module tells the power module what to do in this charging system, **Figure 16-38**. The disadvantage of this system is if the portion of the ECU that controls voltage regulation becomes defective, the entire ECU must be replaced.

Alternator Operation

When the ignition key is turned on, the battery is connected to the field terminal through the voltage regulator. When the engine is started, it will drive the alternator rotor. As the turning rotor pole force fields cut through the stator windings, alternating current voltage is generated. The diodes rectify the alternating current into direct current. The alternator output leaves the positive terminal and travels to the battery. The return circuit is completed through the engine block and vehicle frame.

Brushless Alternator

Figure 16-39 illustrates a heavy duty, truck alternator with built-in voltage regulation. This particular unit, unlike light duty alternators, does not employ slip rings and brushes. All conductors carrying current are stationary (do not move). These are regulator circuit components, field winding, stator windings and rectifier diodes. Note the stationary (does not turn with rotor) field coil.

Starter System

To start an internal combustion engine, it must be cranked (rotated by an outside source). Small two-stroke engines are often cranked by pulling a rope which causes the internal engine parts to revolve. The first automobiles used a hand crank rod for starting. However, this method was dangerous and impractical. A method was soon developed to start the engine by using an electric starting motor, or ***starter***.

The starter is mounted on the transmission housing and operates a small gear that can be meshed with a large ring gear attached to the flywheel. To energize the starter, the driver turns the ignition switch to start, which completes an electrical circuit to the starter. When the starter motor armature begins to turn, the starter gear moves out and engages the ring gear, which spins the crankshaft. When the engine starts, the driver breaks the starter electrical circuit by releasing the ignition key switch. This causes the starter gear to move out of mesh with the ring gear.

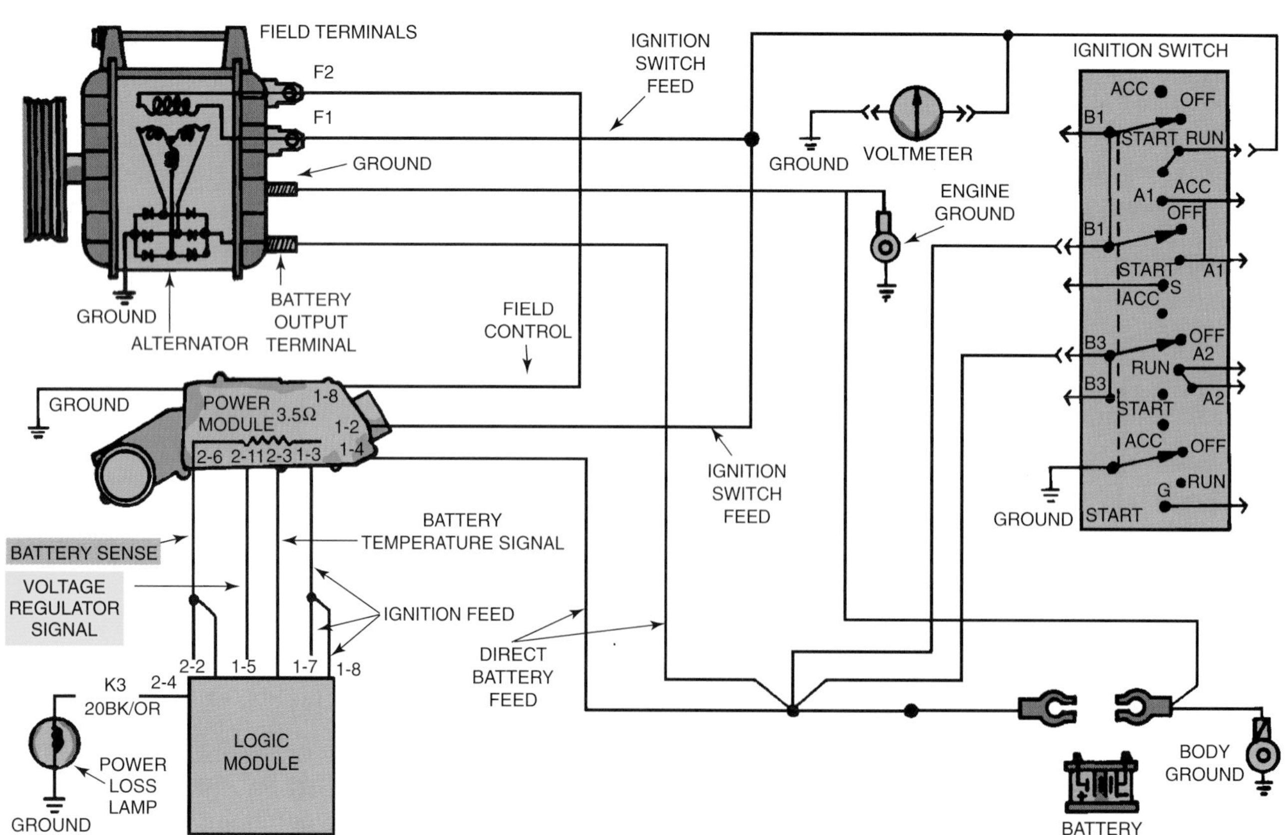

Figure 16-38. This power module controls voltage to the rotor windings, which is dependent on battery voltage. (Chrysler)

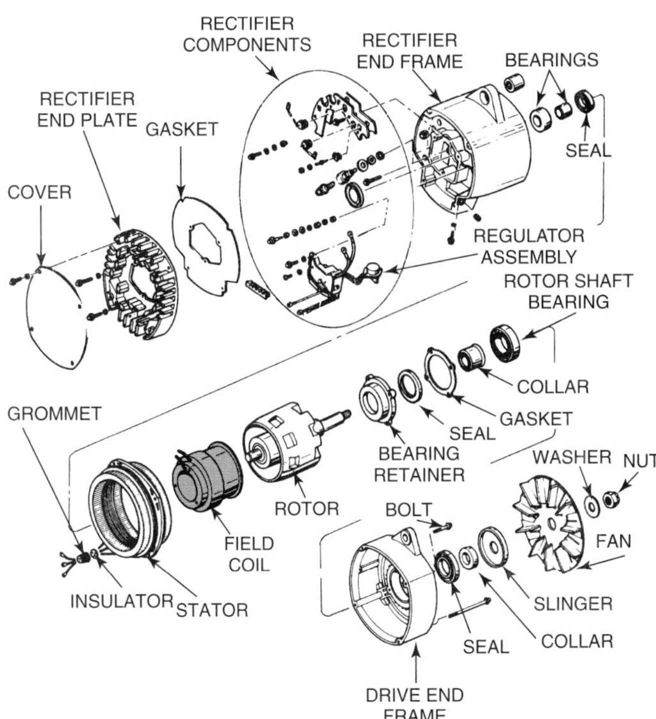

Figure 16-39. Heavy duty, truck type alternator with built-in electronic regulator. Note the stationary field coil and absence of slip rings and brushes. (Chevrolet)

Starter Motor Principles

All electric motors operate from the same fundamental principles. If a current carrying wire is placed in a magnetic field, the field set up around the wire by the moving current will oppose that of the magnetic field, **Figure 16-40**. If the wire is free to move, it will move out of the field.

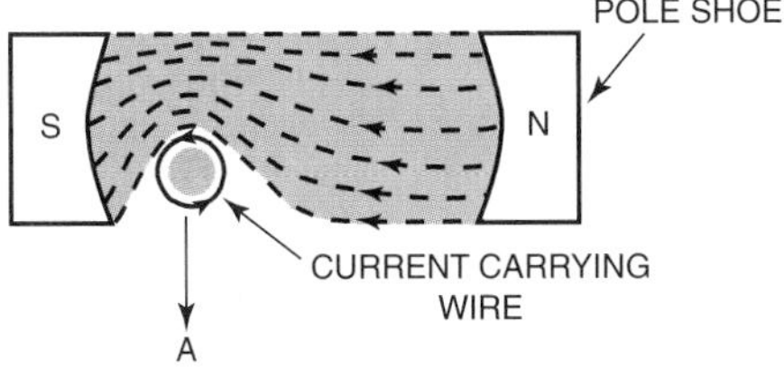

Figure 16-40. Current carrying conductor in a magnetic field. The magnetic field around the wire (see small arrows) will oppose the magnetic field from the poles. This will cause the wire to be forced out in direction of arrow A.

Armature Loop

If the conductor is bent in the form of a loop, each side of the loop will set up a magnetic field. Since the current is traveling in opposite directions, the two sides will have opposing fields, **Figure 16-41**. This is called an ***armature loop***. When the armature loop is placed in a magnetic field, both sides of the loop will oppose the magnetic field, but in opposite directions. This will cause the loop to rotate out of the field, **Figure 16-42**.

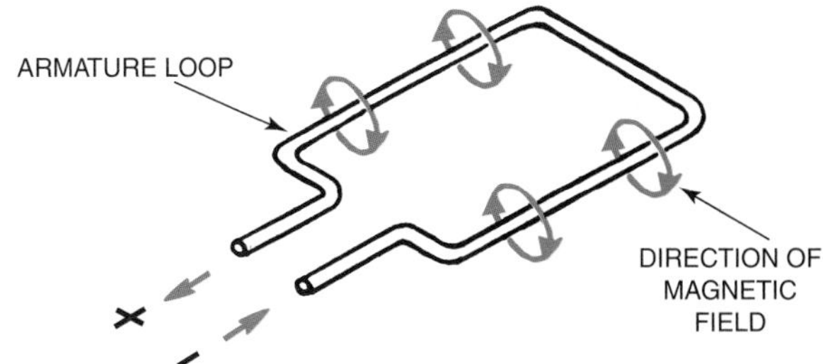

Figure 16-41. Note the direction of the magnetic field on each side of this armature loop.

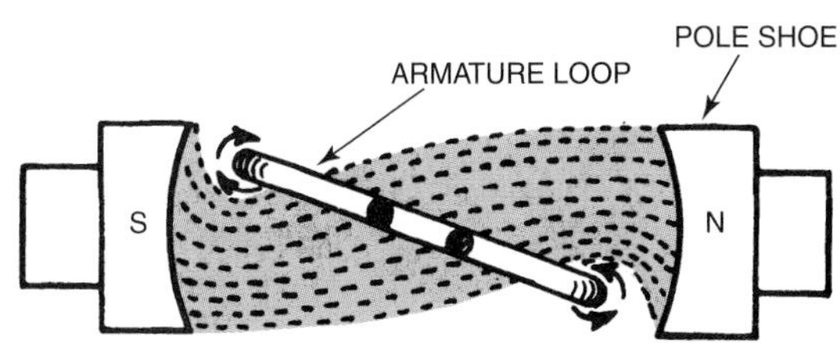

Figure 16-42. Each side of the armature loop has a field moving in opposite directions. These fields will oppose the pole shoe field, causing the loop to rotate.

Commutator Segments

In order to pass an electrical current into the revolving loop, the loop ends must be fastened to ***commutator segments***. Copper ***brushes*** rub against the turning segments, **Figure 16-43**. One loop would not give the starter motor sufficient power, so many loops are used. The ends of each loop are connected to a copper commutator segment, **Figure 16-44**.

The loops are insulated and formed over a laminated iron core. The ends of all loops are connected to the copper

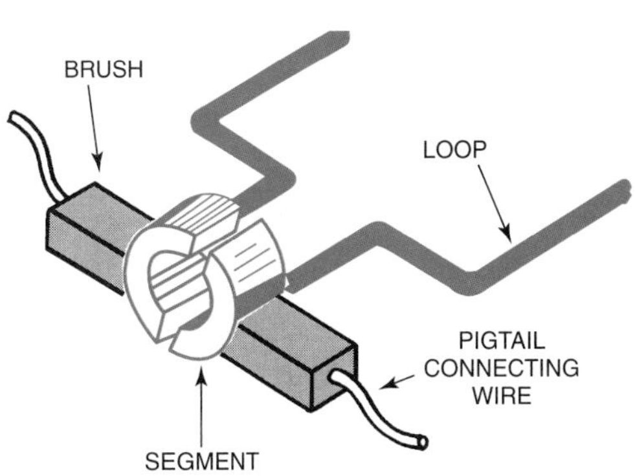

Figure 16-43. In an actual starter motor, many loops and commutator segments are used. Each commutator segment is insulated from the shaft and other segments.

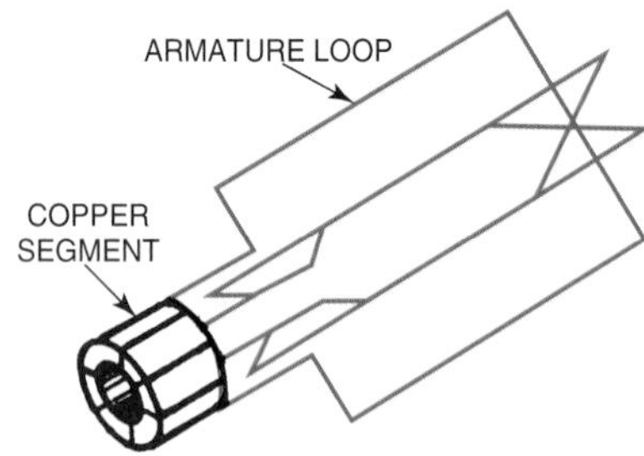

Figure 16-44. Many loops are required in a starter motor. Additional loops will increase the starter motor's torque.

commutator segments. The segments are insulated from the shaft and from each other. The core is supported on a steel shaft that, in turn, is supported in bronze bushings.

When current is fed into two of the four brushes, it flows through all the loops and out the other brushes. This creates a magnetic field around each loop. As the armature turns, the loop will move to a position where the current flow reverses. If this were not done, the magnetic field around the loops would push them out of one side of the field. As a result, they would be repelled when they entered the other side of the pole field. This constant reversal of current flow is accomplished by the commutator segments moving under the brushes. Two brushes are insulated and two are grounded.

Field Pole Shoes

Field pole shoes, usually called poles, are placed inside a steel cylinder (field frame). There may be four or more pole shoes. Two or more of the shoes are wound with heavy copper strips. Then, when current is passed through the coils, the pole shoes become powerful magnets. The field frame provides a low reluctance magnetic path, further strengthening the magnetic field.

Starter Electrical Circuits

The copper loops and field windings are heavy enough to carry a large amount of current with minimum resistance. Since they draw heavy amounts of current, they must not be operated on a continuous basis for longer than 15 seconds. After operating a starter for 15 seconds, wait a couple of minutes to let the starter motor dissipate some of its heat. A starter heats quickly, so prolonged use can cause serious damage.

Various wiring arrangements are used in starter construction. **Figure 16-45A** illustrates a four pole, two field winding. By wiring the field coil in parallel, resistance is lowered, and the field coils produce stronger magnetic fields. The other two poles have no windings, but serve to strengthen the field. **Figure 16-45B** shows a four pole, four winding armature wired series-parallel.

Figure 16-45C illustrates a four pole, three winding setup. Two of the windings are in series with themselves and the armature. One winding does not pass through the armature, but goes directly to ground. This ***shunt winding*** aids with additional starting torque. As starter speed increases, the shunt draws a heavy current and tends to keep starter speed within acceptable limits.

Figure 16-45D shows a four pole, four winding setup. Three windings are in series, one is a shunt winding. **Figure 16-45E** is a four pole, four winding setup with all windings in series. There are other setups, however, the circuits illustrated in **Figure 16-45** are the most commonly used.

Starter Motor Construction

Figure 16-46 shows an exploded view of the various parts of a typical starter motor. This particular starter has a four pole, three winding circuit. One of the windings is a shunt

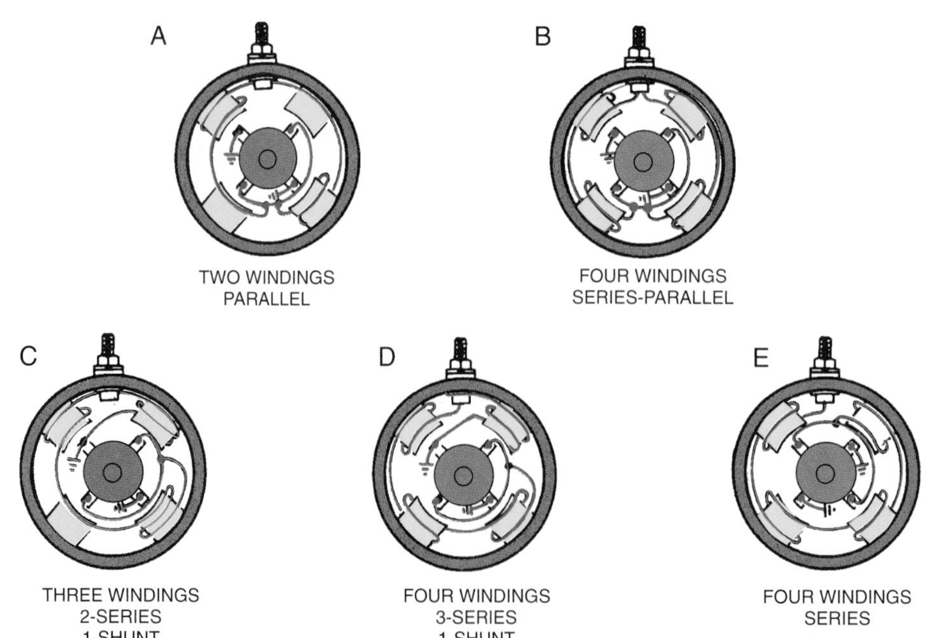

Figure 16-45. Study these various starter wiring circuits. Continue at it until you have learned the different types and their wiring patterns. (Motorcraft)

winding. **Figure 16-47** shows a cutaway view of a typical starter using an overrunning clutch starter drive (discussed below). Note how the armature shaft is supported at both ends by a bronze bushing.

Starter Drive System

To provide a means for the starting motor to turn the engine, a drive system is used. It consists of a ***ring gear*** and a ***starter pinion gear***. These are explained in the following sections.

Ring Gear

The ring gear is attached to the flywheel of vehicles with manual transmissions. The ring gear is either welded or shrink fit, **Figure 16-48**, to the flywheel. In shrink fitting, the ring gear is expanded by heat and fitted on the flywheel while hot. As it cools, the metal contracts and the ring grasps the flywheel securely. Vehicles with automatic transmissions and transaxles have a separate, lightweight ring gear assembly which bolts between the engine crankshaft and torque converter. The ring gear usually has 150 to 200 teeth.

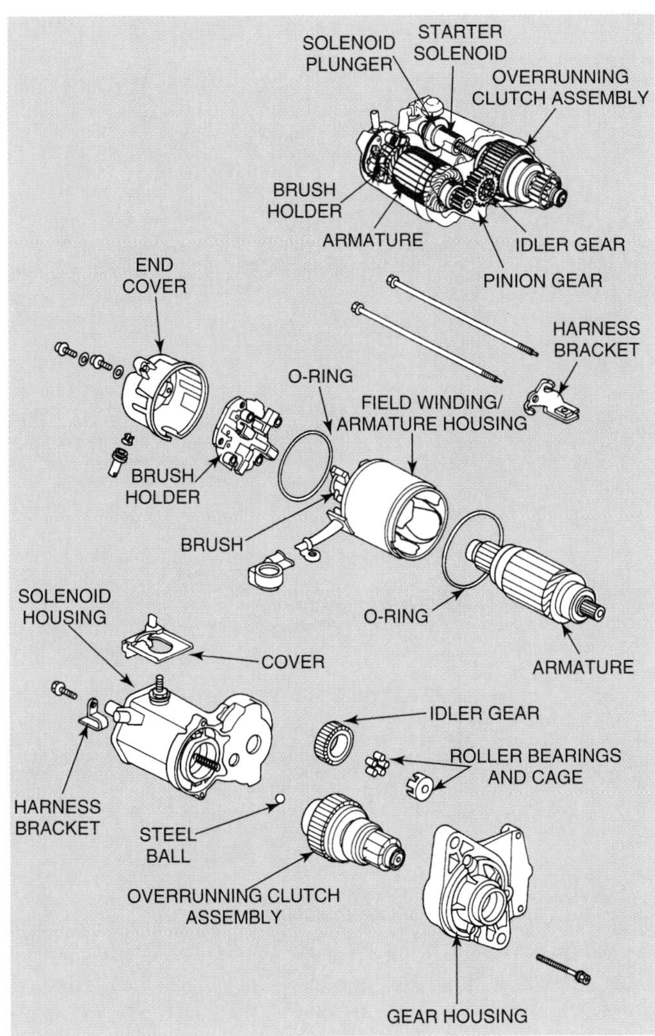

Figure 16-46. Exploded view of a typical starter motor. Learn the names and locations of all the parts. (Honda)

Starter Pinion Gear

A small gear, called a starter pinion gear, is attached to the starter armature shaft. The pinion gear is much smaller than the ring gear (about 10 to 15 teeth) and must turn 15 to 30 revolutions in order to crank the flywheel one revolution.

Pinion Engagement Devices

It is necessary to mesh the pinion gear for cranking and demesh (disengage) the pinion when the engine has started. The two basic types of pinion engagement devices are the inertia type, and the overrunning clutch type. There are several designs of each basic type.

Inertia Starter Drive

With the ***inertia drive***, sometimes called the ***Bendix drive*** or ***self-engaging drive***, the armature shaft spins and the drive pinion stands still while the threaded sleeve spins inside the pinion. As the sleeve spins, the pinion slides out and meshes with the ring gear. As soon as the pinion reaches its stop, the turning sleeve causes the pinion to turn with it to crank the engine.

The pinion sleeve fits loosely on the armature shaft and is connected to a large spring that is fastened to the armature shaft. The spring absorbs the sudden shock of pinion engagement. When the engine starts, the pinion will spin faster than the armature turns the pinion sleeve. This will reverse its travel on the threaded pinion, and it will spin backward, free of the ring gear. A small coil spring will prevent it from working back into mesh.

Figure 16-49 illustrates the action of the inertia type pinion engaging device. In **Figure 16-49A**, the armature shaft and pinion sleeve begin to spin. The stationary pinion is sliding to the left. In **Figure 16-49B**, the sliding action is stopped, and the pinion begins to turn and crank the engine.

When the engine starts, **Figure 16-49C**, the pinion is spinning faster than the threaded pinion sleeve. This spins the pinion back out of mesh with the ring gear. An inertia drive, in place on a starter motor, is shown in **Figure 16-50**. Notice the threaded pinion sleeve and pinion. The pinion is weighted on one side, so it will tend to resist being turned until it has slid into mesh with the ring gear.

Overrunning Clutch

The one-way or ***overrunning clutch drive*** is engaged with the ring gear by an electric solenoid that actuates the shift linkage. In operation, the pinion is turned by the overrunning clutch. When the pinion is slid into mesh with the ring, the shifting mechanism completes the starter circuit and the overrunning clutch drives the pinion.

When the engine starts, the ring will spin the pinion faster than the starter armature is turning. The pinion is free to run faster because the overrunning clutch engages the pinion only when the armature is driving it.

Figure 16-51 illustrates a typical overrunning clutch starter drive. Note the pinion, drive shell and sleeve assembly, coil spring, and shift collar. The coil spring comes into use when the pinion teeth mesh against the ring teeth. The actuating lever, or shift lever, will continue to move the shift collar, compressing the spring. When the lever has moved far

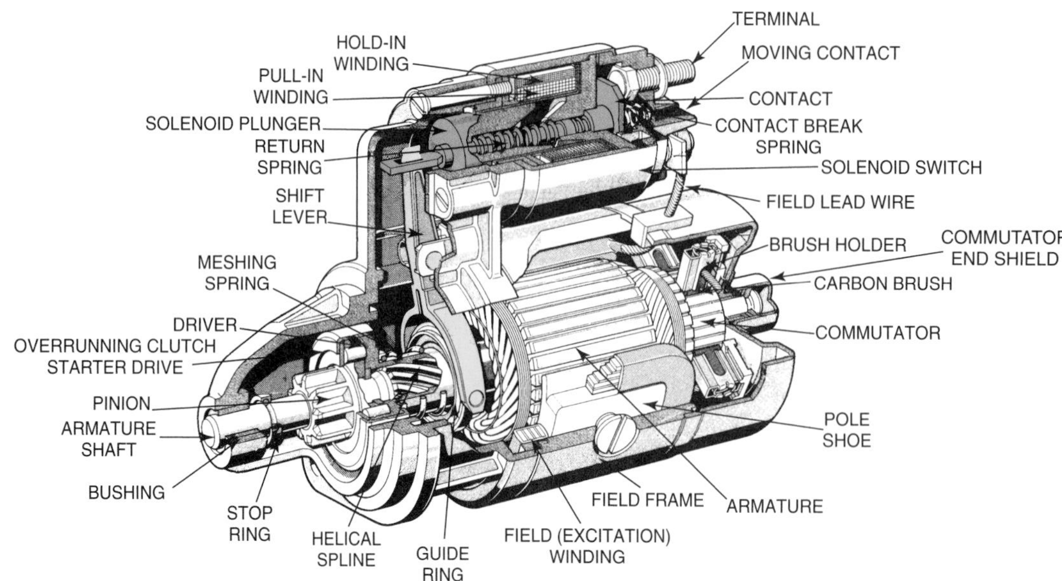

Figure 16-47. Cutaway of a starter motor using an overrunning clutch starter drive. Note the shift lever and solenoid. (Bosch)

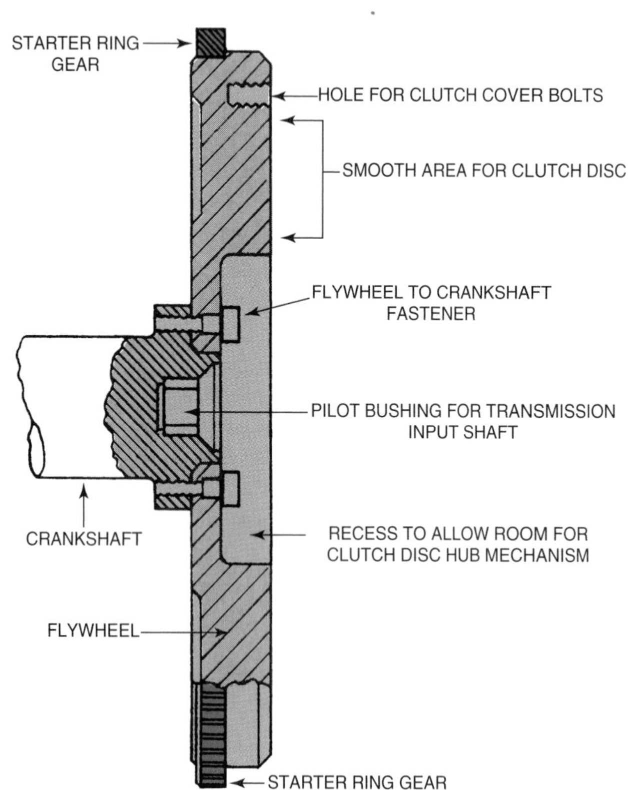

Figure 16-48. A cross-sectional view of a typical flywheel. Note how the ring gear is placed on the flywheel. This particular ring gear is pressed on.

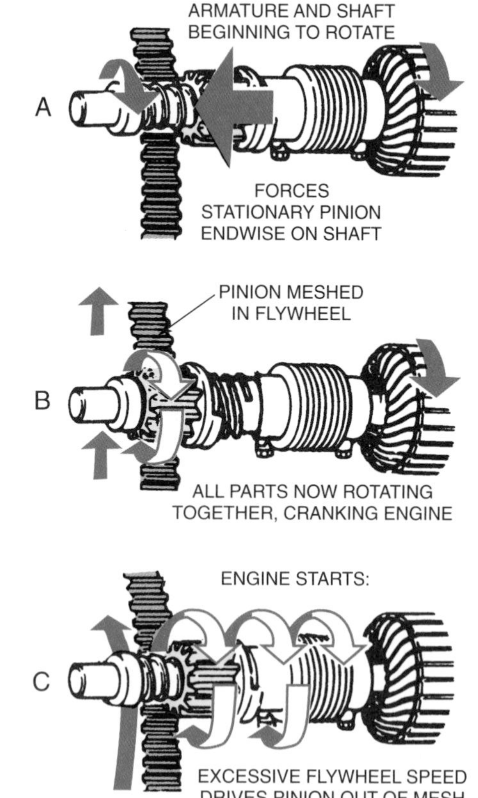

Figure 16-49. Bendix starter drive action. Note how the spring forces the starter gear back and forth. This allows the starter gear to mesh with the flywheel without gear clash. Another type of starter drive, called the overrunning clutch, are often mistakenly referred to as a Bendix drive. (Delco-Remy)

enough to close the starter switch, the pinion will turn and the compressed spring will force it into mesh with the ring gear.

Figure 16-51 illustrates an end view showing the spring-loaded clutch rollers. When the drive sleeve turns faster than the pinion and collar assembly, the rollers are wedged against the pinion, causing it to turn. When pinion speed exceeds that of the sleeve, the rollers retract and the pinion is free of the shell and sleeve assembly. **Figure 16-52** shows the overrunning clutch drive in place on a starter.

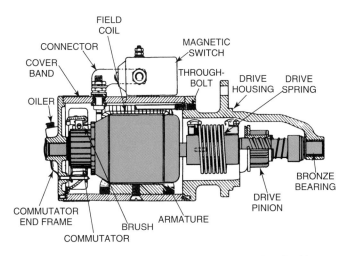

Figure 16-50. Cutaway of a starter motor with a Bendix drive.

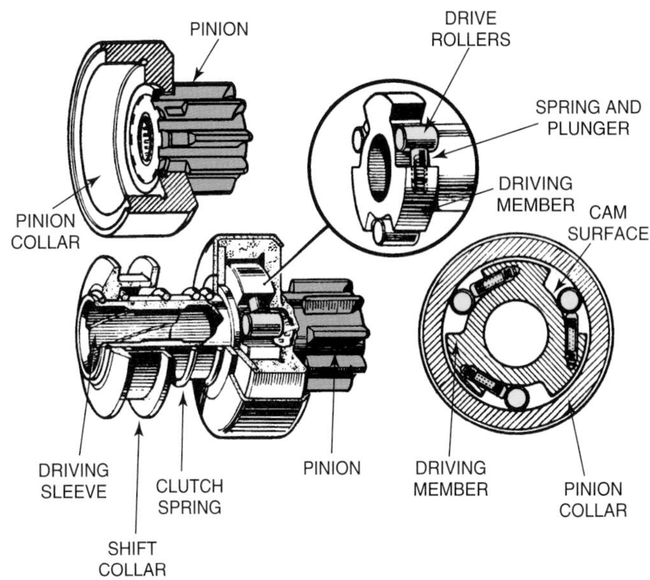

Figure 16-51. In this overrunning clutch starter drive, the rollers lock the pinion and collar assembly to the drive member when the armature shaft is spinning the drive member faster than the pinion. (British-Leyland)

Starter Solenoid

If an iron core is placed inside a coil of wire, an electric current is passed through the coil, the iron core will be drawn into the coil. Such a unit is called a ***solenoid***. A solenoid has many applications and can be used where a push or pull motion is desired. The primary job of the solenoid is to make the electrical contact between the battery and starter motor. A typical solenoid is shown in **Figure 16-53**. The solenoid is energized from the start position on the ignition switch.

Figure 16-54 shows how a solenoid can be used to pull on a starter shift lever. You will note that the solenoid in **Figure 16-55** not only pulled the starter shift lever until the pinion meshed, but it also closed the starter switch after the meshing was almost complete.

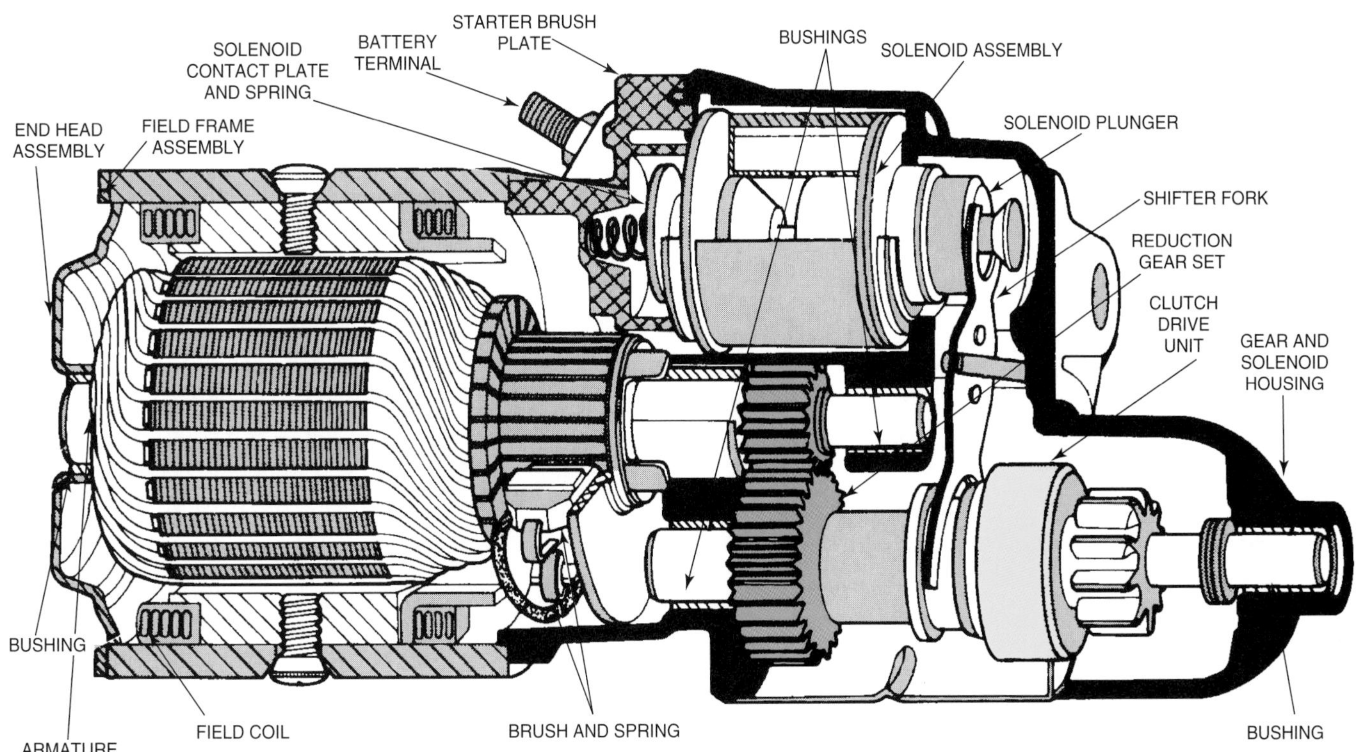

Figure 16-52. Cutaway of a starter motor using an overrunning clutch starter drive. This starter is engaged by an electric solenoid. It increases cranking torque by using reduction gears. (Chrysler)

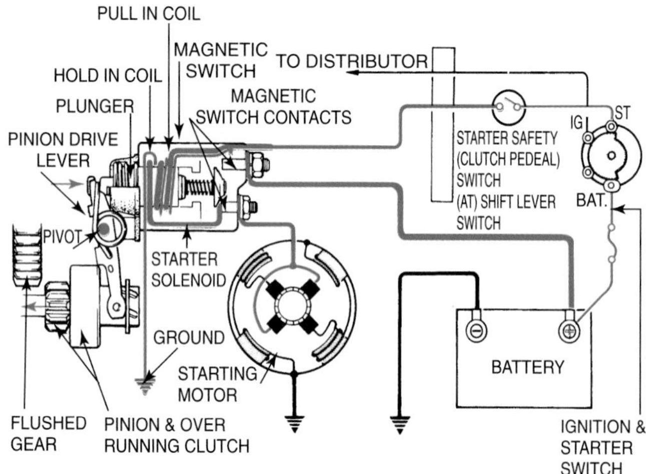

Figure 16-53. A starting circuit solenoid schematic illustrating parts and related electrical wiring. When the starter solenoid is energized, the plunger is drawn (pulled) in, bringing the top of the pinion drive lever in. This forces the bottom to move out, moving the pinion gear into engagement with the flywheel gear and engine cranking begins. Note the starter safety switch. (Chevrolet)

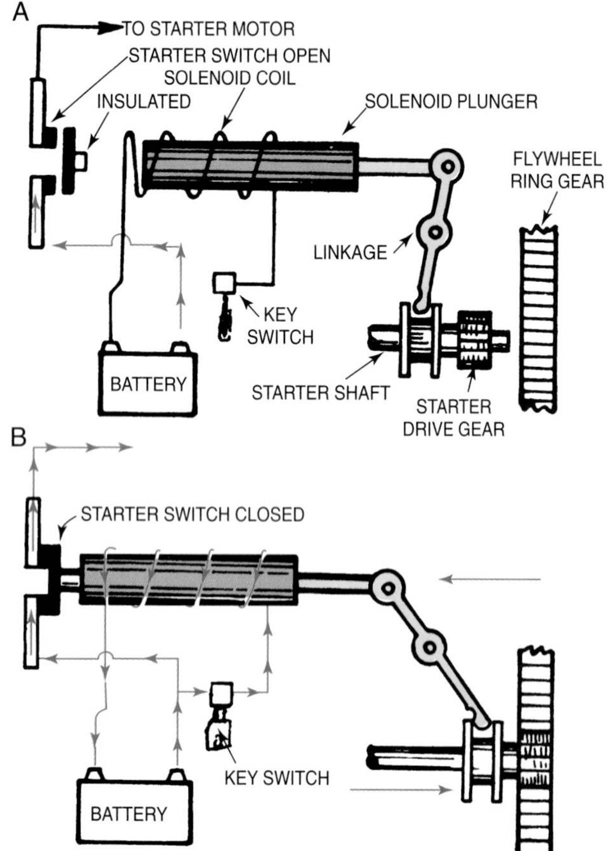

Figure 16-54. Starter drive solenoid operation. A—The solenoid plunger has moved to the right, opening the starter switch and retracting the starter drive gear. The ignition switch is off. B—The ignition switch is on, energizing the solenoid, which draws the plunger into the coil. The plunger has engaged the drive gear and closed the starter switch. The starter will now crank the engine.

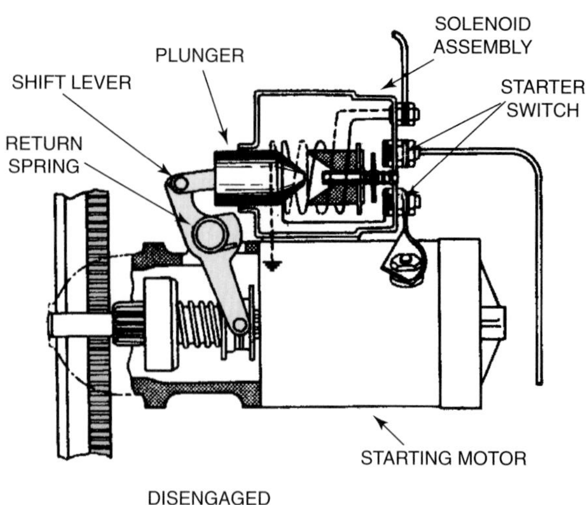

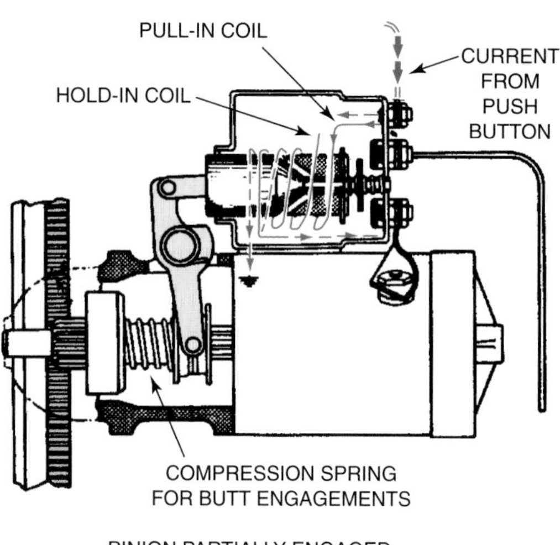

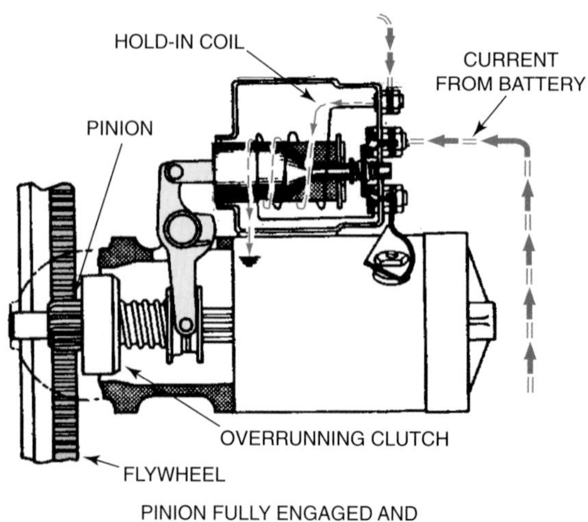

Figure 16-55. Starter solenoid operation stages. Note the electric solenoid operation in each stage. (Chevrolet)

Starter Solenoid Operation

Figure 16-55 shows the starter solenoid in three stages of operation. In the disengaged position, the pinion is demeshed. The solenoid plunger is released and no current is flowing. The partially engaged stage shows current from the starter switch flowing through both the pull-in and the hold-in coils. Both coils are drawing the plunger inward, causing it to pull the shift lever. The pinion is just starting to engage.

The cranking stage shows the plunger pulled into the coil. The pinion is fully engaged. The hold-in coil is operating, but the pull-in winding has cut out. The plunger has operated the starting switch and current is flowing from the battery through the switch and to the starter. When the engine has started, the hold-in coil will cut out and the plunger will move out, retracting the pinion and opening the starter switch, **Figure 16-55**. Starters with self-engaging inertia drives generally use a solenoid to open and close the starter switch.

A starter motor employing a solenoid and overrunning clutch with permanent magnets is shown in **Figure 16-56**. A somewhat different method of actuating the starter drive is illustrated in **Figure 16-57**. In this setup, a movable pole shoe is connected to the starter drive shift lever. When the starter is in the off position, one of the field coils is grounded through a set of contacts.

When the starter is first actuated, a very heavy flow of current through this field coil pulls on the movable pole shoe. The shoe, in turn, engages the starter drive. When fully engaged, the field coil contacts open and normal starter operation takes place. A holding coil keeps the sliding pole shoe in the engaged position. When current to the starter is interrupted, the pole shoe moves back and disengages the starter drive.

Reduction Gear Starters

To increase starter torque, some starter drives are driven through a ***reduction gear***. The reduction gear is used in high speed starter motors. The reduction gear reduces speed to increase starter torque. A common reduction gear starter is shown in **Figure 16-58**.

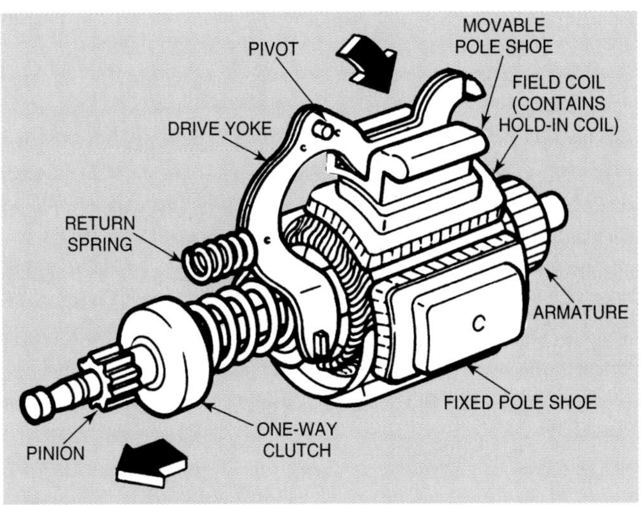

Figure 16-57. Starter design using one movable pole shoe as a means of actuating the starter drive. (Jeep)

Planetary Gear Reduction Starter

Planetary gears have been used in automatic transmissions to obtain desired gear reduction for many years. Planetary gearsets are also used for reduction in some starter motors, **Figure 16-59**. This design decreases the overall size and weight of the starter by approximately 35% while maintaining the needed starter torque.

The planetary gears, located on the shaft, mesh with the teeth of the internal or ring gear. Most reduction gear starters use permanent magnets which eliminates the field coil windings. There are six permanent magnets mounted inside the armature housing which reduce current draw while starting,

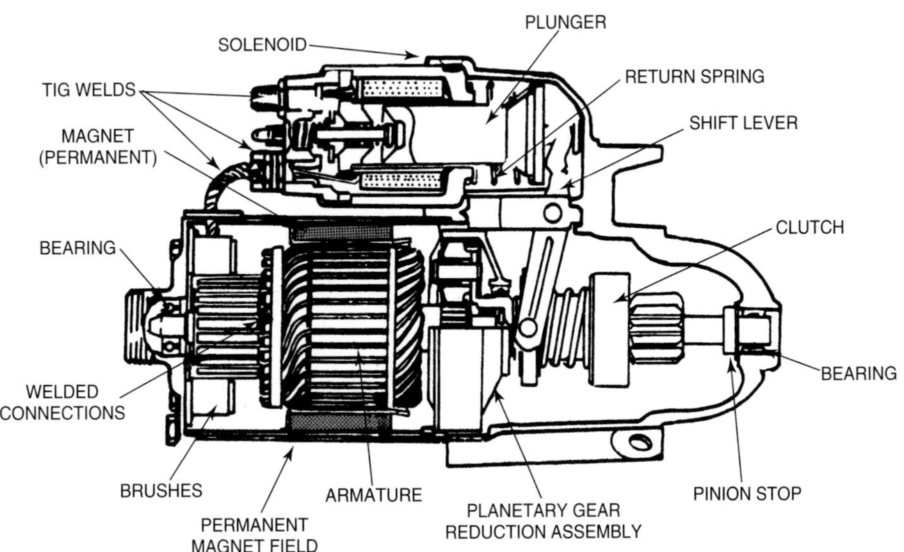

Figure 16-56. A cutaway view of a permanent magnet starter motor. Six small magnets are located inside the field frame. They take the place of the current carrying field coils. This type of starter is lighter than those using field coils. (Chevrolet)

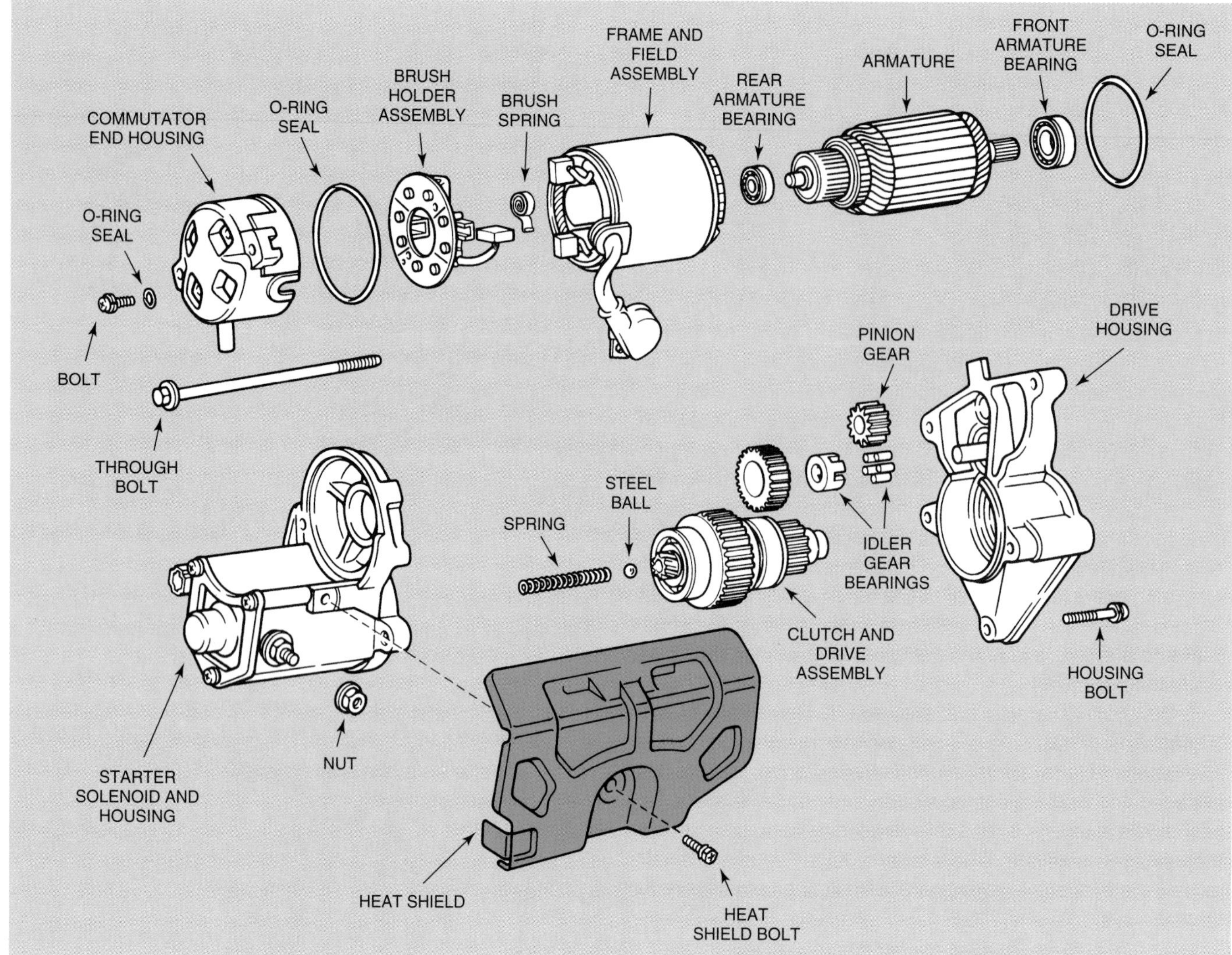

Figure 16-58. Exploded view of a typical reduction gear starter motor assembly. Note the heat shield used to help protect the starter from excessive engine and/or exhaust heat. (Geo)

Figure 16-59. The magnets are aligned according to polarity and cannot be removed from the housing. The magnets are permanently affixed to the housing.

Starter Actuating Switches

In addition to the solenoid-operated switch that directs battery current to the starter, it is necessary to furnish current to the solenoid. On most vehicles, turning the key to the fullest travel closes a switch to the solenoid. However, current must first pass through a ***neutral safety switch***, **Figure 16-60**, before energizing the solenoid.

On some vehicles equipped with manual transmissions, the clutch must be depressed to close the neutral safety switch. On vehicles equipped with automatic transmissions, the gear selector must be placed in park or neutral for current to flow through the neutral safety switch to the solenoid. This prevents the vehicle from being started when the transmission is in gear.

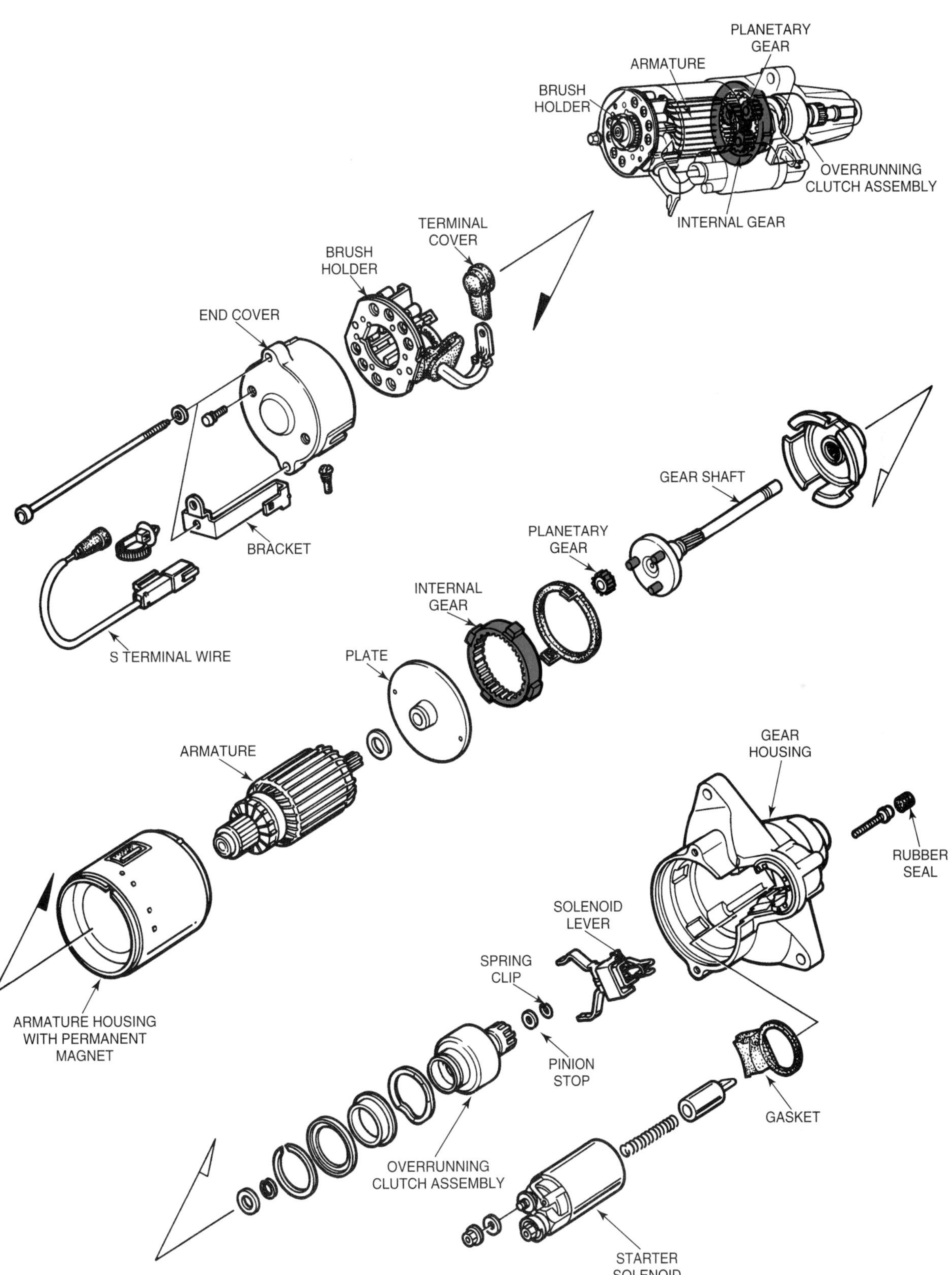

Figure 16-59. Exploded view of a planetary gear reduction starter motor assembly. Note the use of an overrunning clutch. (Honda)

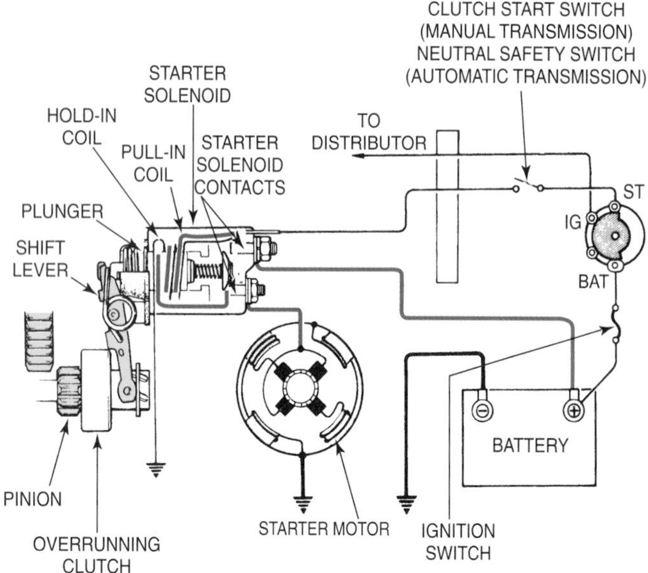

Figure 16-60. Starting circuit including neutral safety or clutch start switch. If the switch remains open, the vehicle will not start. (Geo/General Motors)

Summary

A battery is an electrochemical device used to supply current for starting the vehicle. It contains positive and negative plates that are connected in such a way as to produce six groups of cells. The plates are covered with electrolyte. A charged battery will gather a surplus of electrons at the negative post. When the battery is placed in a completed circuit, the electrons will flow from the negative post, through the circuit, and on to the positive post.

The battery is charged by passing electricity through it in the reverse direction of battery current flow. The battery's state of charge is checked by using a hydrometer to determine the specific gravity of the electrolyte. Upon starting, the alternator will recharge the battery. Vehicle batteries are usually 12 volt, though other battery sizes are used by trucks and commercial vehicles. Be careful of battery electrolyte—it is dangerous. Batteries can explode and should never be exposed to sparks or open flame.

The primary job of the alternator is to supply the necessary current for the electrical needs of the vehicle and to keep the battery charged. The alternator operates by using magnetism to change mechanical energy (rotation) into electricity. It is necessary to control the maximum voltage and current output developed by the alternator. This is done by using the regulator.

On older regulators, an electromagnet breaks a set of contact points and interrupts current flow to the field circuit when system voltage exceeds the alternator's designed output. As the points vibrate, they produce a smooth, controlled alternator output. The battery, alternator, and regulator work together and the function of any one unit is dependent upon the action of the other two.

Battery condition and electrical load determine the current and voltage requirements. The alternator produces the required voltage and current and the regulator controls the alternator output on a level consistent with system needs.

The starter is used to crank the engine for starting purposes. The starter motor utilizes a current carrying series of armature loops placed in a strong magnetic field produced by field coils. As the armature loops are repelled by the magnetic field, the armature is forced to spin. By arranging a sufficient number of loops and connecting them to commutator bars or segments, current in the loops keeps reversing so the repelling force will remain constant. This allows the armature to continue spinning.

The starter uses large copper conductors in both the field and armature circuits. The field and armature are connected in series via four brushes. Since the field strength is very strong, the starter motor will produce enough torque to crank the engine. Starters draw heavy current loads and therefore heat quite rapidly. They must not be operated continuously for longer than 15 seconds without a brief rest in which to dissipate the heat.

The starter pinion that engages the flywheel ring gear uses one of two different designs. One uses the Bendix drive. This type spins a pinion sleeve that causes the pinion to slide out on the sleeve until it engages the ring. The overrunning clutch type moves the starter pinion into mesh by means of a shift lever actuated by a solenoid. When this starter drive is used, the overrunning clutch is used to allow the drive pinion to freewheel when the engine starts, until it is drawn out of mesh. The starter is energized by the solenoid. The solenoid is energized by operating the ignition switch, through a neutral safety switch.

Know These Terms

Lead-acid battery
Cells
Positive plate
Negative plate
Separator
Electrolyte
Terminals
Specific gravity
Hydrometer
Cold cranking amps
Cranking amps
Reserve capacity
Parasitic battery load
Dry-charge
Alternator
Rotor
Stator
Drive pulley
Driven pulley
Brushes
Slip rings
Rectified
Diode
Voltage regulator
Electromechanical voltage regulator
Electronic voltage regulator
Zener diode
Starter
Armature loop
Commutator segments
Brushes
Field pole shoes
Shunt winding
Ring gear
Starter pinion gear
Inertia starter drive
Bendix drive
Self-engaging drive
Overrunning clutch drive
Solenoid
Reduction gear
Planetary gears
Neutral safety switch

Review Questions—Chapter 16

Do not write in this book. Write your answers on a separate sheet of paper.

1. Technician A says that each battery cell of a 12 volt battery produces 12 volts. Technician B says that most of the battery electrolyte is sulfuric acid. Who is right?
   (A) A only.
   (B) B only.
   (C) Both A & B.
   (D) Neither A nor B.
2. All of the following statements about batteries are true, EXCEPT:
   (A) an automotive battery contains positive and negative plates.
   (B) a 12 volt battery will contain three cells.
   (C) the battery electrolyte contains water.
   (D) a battery stores chemical energy.
3. What two devices are able to charge a battery?
4. How do you check the specific gravity of a battery?
5. How are the battery posts identified?
6. A battery that is low on electrolyte should be filled with _____ .
   (A) rainwater
   (B) distilled water
   (C) sulfuric acid
   (D) hydrochloric acid
7. A dry-charged battery is shipped without _____ .
   (A) plates
   (B) terminals
   (C) individual cells
   (D) electrolyte
8. Which of the following alternator parts rotates?
   (A) Rotor.
   (B) Stator.
   (C) Diodes.
   (D) Brushes.
9. Which of the following alternator parts change alternating current into direct current?
   (A) Rotor.
   (B) Stator.
   (C) Diodes.
   (D) Pulley.
10. The alternator slip rings are used with _____ to transfer current.
11. What are the two methods of controlling field strength in the alternator rotor?
12. Of the answers to question 11, which is controlled to control the output of the alternator?
13. Modern voltage regulators use _____ to control the alternator.
    (A) contact points
    (B) electromagnetic coils
    (C) brushes
    (D) electronic components
14. A Zener diode allows electricity to flow when the _____ passes a certain point.
    (A) temperature
    (B) amperage
    (C) voltage
    (D) phase shift
15. When the voltage on either side of the indicator light is equal, the light _____ .
    (A) goes out
    (B) comes on
    (C) flashes
    (D) flickers
16. An internal combustion engine must be _____ by an external source before it will start.
17. If a magnetic field surrounds a wire, in what direction will the wire try to move?
    (A) Away from the field.
    (B) Toward the field.
    (C) At right angles to the field.
    (D) It does not move.
18. What are the two purposes of the shunt circuit in the starter motor?
19. Which of the following starter parts do not rotate during starter operation?
    (A) Armature windings.
    (B) Commutator bars.
    (C) Drive pinion.
    (D) Pole shoes.
20. Technician A says that starter motors can be used continuously for up to ten minutes before stopping. Technician B says that the starter motors used with reduction gears are capable of higher speeds than conventional starter motors. Who is right?
    (A) A only.
    (B) B only.
    (C) Both A & B.
    (D) Neither A nor B.
21. Name the two general classes of starter drive mechanisms.
22. Some starters have a movable pole shoe to _____ .
    (A) increase starter torque
    (B) reduce starter speed
    (C) engage the starter drive
    (D) prevent starter overheating

23. The overrunning clutch engages the pinion only when the _____ is driving it.
    (A) starter motor armature
    (B) engine ring gear
    (C) starter solenoid
    (D) Both A & B.

24. The primary job of the starter solenoid is to provide an electrical connection between the battery and the _____.
    (A) ignition switch
    (B) starter motor
    (C) alternator
    (D) drive mechanism

25. Many starter solenoids are attached to linkage which engages the _____ with the flywheel during cranking.

Modern vehicles are thoroughly tested to ensure that their electronic systems will not be affected by electromagnetic interference. The antenna array is capable of generating radio waves across a wide spectrum. The blue material on the chamber walls is for electromagnetic absorption. (Chrysler)

17

Chassis Electrical

Upon completion of this chapter, you will be able to:

- Identify and define chassis wiring and related electronic components.
- Identify and define circuit protection devices.
- Identify and define common vehicle chassis lights.
- Identify common chassis electrical equipment.

This chapter will cover the basic principles of chassis electrical systems, such as chassis wiring, vehicle lights, safety devices such as horns and windshield wipers, and comfort items such as power windows and door locks. Devices such as radios and CD players are not within the scope of this text. Other chassis electrical systems, such as blower motors and anti-lock brakes, are covered in the chapters dealing with these systems. Studying this chapter will allow you to become familiar with the electrical components and associated wiring that are used on the vehicle body and chassis.

Chassis Wiring

Chassis wiring consists of all of the vehicle wiring that is not directly connected with the engine and drive train. This wiring extends throughout the vehicle body. Wiring to devices that will not be operated except when the vehicle is being driven, such as the windshield wipers and radio, is routed through the ignition switch. Wiring to vehicle systems which can be operated at any time, such as lights and horns, is powered directly from the vehicle battery, **Figure 17-1**.

Wire Sizes

The size of a particular vehicle wire will vary with the amount of current that is expected to flow through it. Wire size is measured by its gage. The higher the gage number, the smaller the wire diameter. Wire to a single bulb might have a gage of 16 or 18 while battery cable is 4 or 6. The gage standard used to measure most automotive wire is called the ***American Wire Gage*** or AWG.

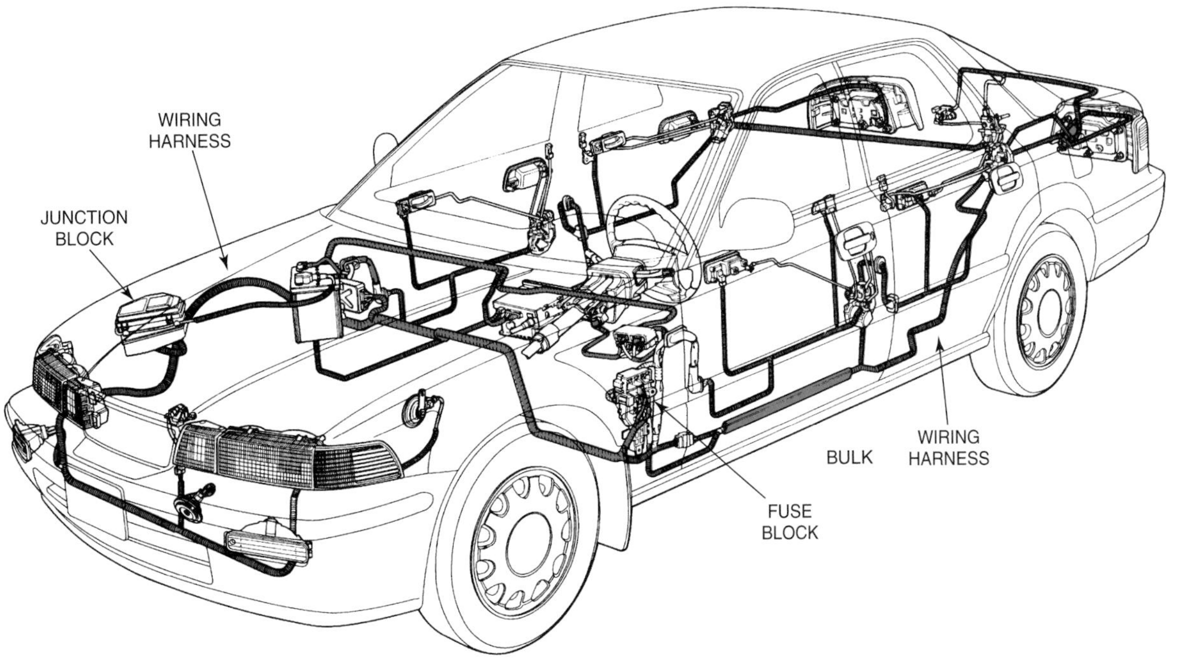

Figure 17-1. Chassis wiring is located throughout the vehicle. Each assembly contains many wires that connect to a variety of electrical components. This illustration shows a wiring assembly for a theft prevention system only. (Honda)

Wiring Harness

Almost all vehicle wiring is installed as part of a ***wiring harness***. Wiring harnesses are groups of vehicle wires wrapped together for ease of installation. The actual wires are encased in ***looms***. The looms help reduce chafing and other damage and keep the wire in a neat package. Wiring harnesses have molded electrical connectors that attach to electrical components or other harnesses.

Some vehicles use a ***printed circuit*** for at least part of their wiring needs. The printed circuit shown in **Figure 17-2** is a vinyl plastic sheet with the circuits etched or printed on it. Since the printed material is electrically conductive, power flows through the paths of the printed circuit in the same way that it flows through standard wiring.

Many electrical circuits operate on very low voltages. These low-voltage circuits must be protected from stray electrical impulses and magnetic fields. To accomplish this, many wires and electrical devices are ***shielded*** (surrounded with grounded metal covers). In some cases, the wire or device that produces a magnetic field is shielded. More commonly, the low-voltage wiring that could be affected is shielded.

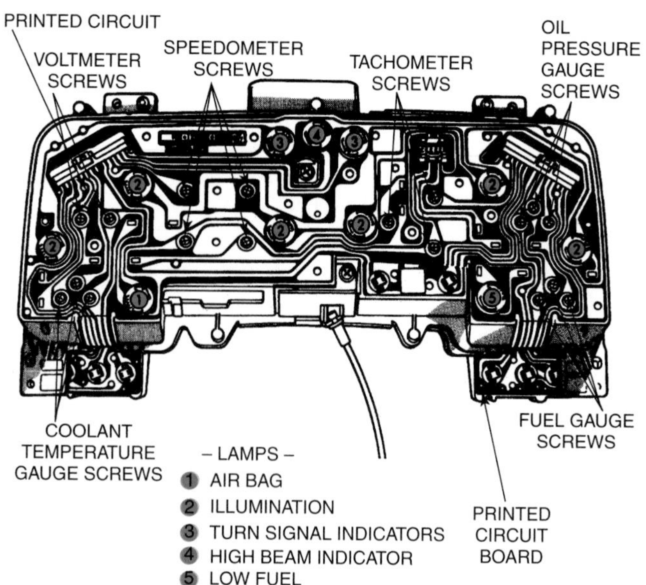

Figure 17-2. An instrument cluster printed circuit board. (Dodge)

Many Circuits

The modern automobile has a large number of electrical circuits. A typical wiring diagram, **Figure 17-3**, shows a cruise control circuit. The electrical circuits, with the addition of many accessory items, are becoming so complex that wiring diagrams are often broken down by sections. Some common wiring circuit symbols are shown in **Figure 17-4**.

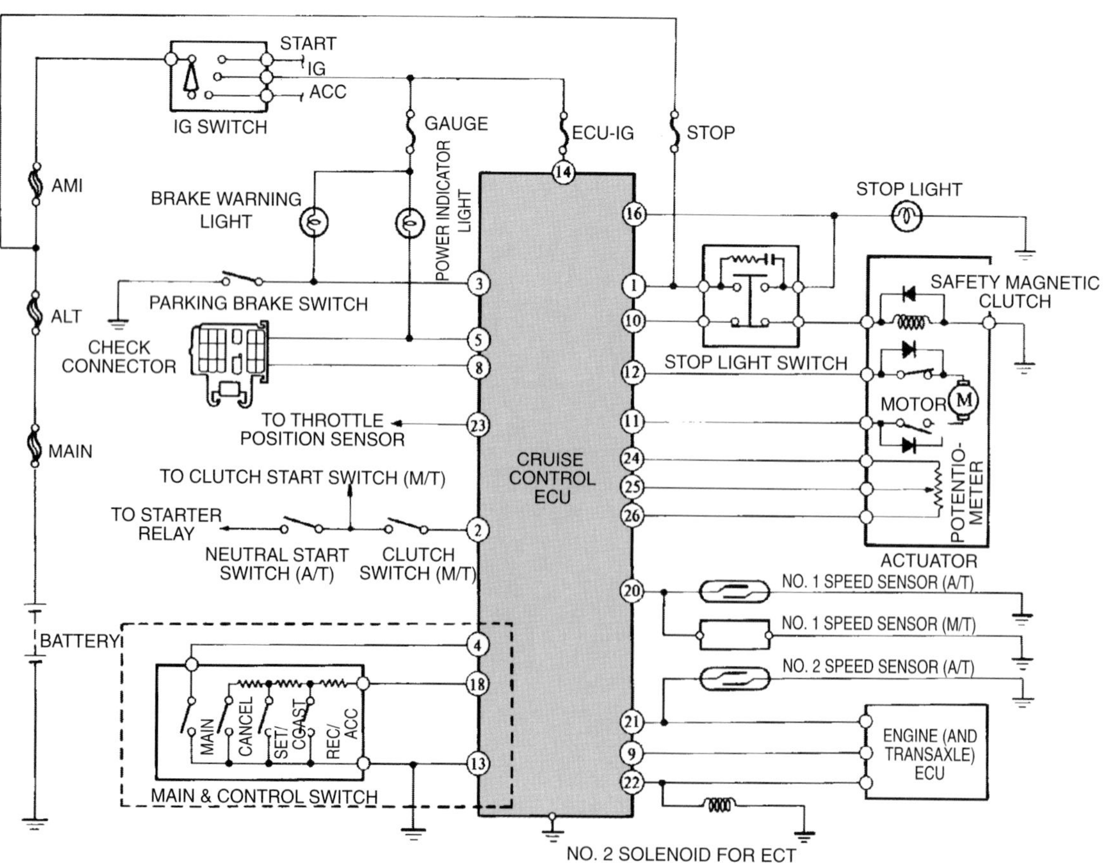

Figure 17-3. A typical electrical system diagram showing a cruise control circuit as used by one manufacturer. (Toyota)

Legends of Symbols Used on Wiring Diagrams			
+	Positive		Connector
−	Negative		Male Connector
	Ground		Female Connector
	Fuse		Denotes Wire Continues Elsewhere
	Gang Fuses with Bus Bar		Denotes Wires Goes to One of Two Circuits
	Circuit Breaker		Splice
	Capacitor	J2 2	Splice Identification
	Ohms		Thermal Element
	Resistor	TIMER	Timer
	Variable Resistor		Multiple Connector
	Series Resistor		Optional — Wiring with / Wiring without
	Coil		"Y" Windings
	Step Up Coil	88:88	Digital Readout
	Open Contact		Single Filament Lamp
	Closed Contact		Dual Filament Lamp
	Closed Switch		L.E.D.—Light Emitting Diode
	Open Switch		Thermistor
	Closed Ganged Switch		Gauge
	Open Ganged Switch		Sensor
	Two Pole Single Throw Switch		Fuel Injector
	Pressure Switch	#36	Denotes Wire Goes Through Bulkhead Disconnect
	Solenoid Switch	#19 STRG COLUMN	Denotes Wire Goes Through Steering Column Connector
	Mercury Switch	INST PANEL #14	Denotes Wire Goes Through Instrument Panel Connector
	Diode or Rectifier	ENG #7	Denotes Wire Goes Through Grommet to Engine Compartment
	Bi-Directional Zener Diode		Denotes Wire Goes Through Grommet
	Motor		Heated Grid Elements
	Armature and Brushes		

Figure 17-4. An assortment of electrical symbols for one vehicle manufacturer's wiring diagrams. Learning these symbols will greatly improve your ability to use and understand electrical diagrams. (Chrysler)

Color Coding

All chassis wires are ***color coded*** to make identification of individual wires easy. This is true of wires installed in wiring harnesses. Color codes can be single wire colors, or colors with contrasting stripes or bands. The use of stripes and bands makes hundreds of color combinations possible, **Figure 17-5**.

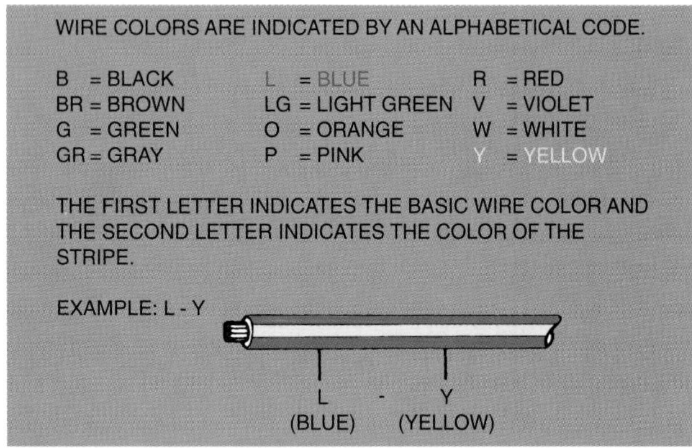

Figure 17-5. One example of a wiring color code as used in a wiring diagram. These will vary between manufacturers. (Toyota)

Terminal Blocks and Junction Blocks

To connect separate wires and harnesses, terminal blocks and junction blocks are used. They consist of plastic blocks with metal screw terminals to attach each wire. The wire ends have eye terminals to connect to the terminal screws.

The ***terminal block*** is used to supply current to several circuits from one feeder source. The hot wire (wire connected to source of electricity) is attached to one terminal. This terminal is connected to all others by a bus bar (metal plate).

The ***junction block*** serves as a common connection point for a number of wires. It may be of the terminal screw or the plug-in type. Unlike the terminal block, the junction block connects one wire to a corresponding wire on the other side with no common bus bar. Junction blocks are also known as power distribution centers.

Circuit Protection Devices

It is necessary to protect electrical circuits against ***shorting*** (an exposed part of the circuit accidentally touching ground) that could burn up the wiring. The electrical units must be protected against overloads also. ***Circuit protection devices*** are installed in series with the electrical device or wiring to be protected. They are used to break the circuit to prevent damage to the other equipment in the circuit. The most common types of circuit protection devices are the fuse, fusible link, and the circuit breaker, discussed below.

Fuses

When the current in a circuit becomes excessive (exceeds the amount of current that it can safely handle), a ***fuse*** will blow and break the circuit. To restore circuit operation, the fuse must be replaced. Older fuses consists of a glass tube with metal end caps. A built-in conductor carries the current from one cap to the other. This conductor is designed to carry a specific maximum load.

Most fuses in current use look like the fuses shown in **Figure 17-6**. Both use a soft conductor that melts when fuse capacity is exceeded, **Figure 17-7**. Most are color coded to identify amperage. All of the vehicle fuses are usually contained in a single holder known as a ***fuse block***. A fuse block is shown in **Figure 17-8**. Some add-on electrical devices may have a fuse in the wiring to it. This is called an ***inline fuse***.

Circuit Breaker

A ***circuit breaker*** feeds current through a bimetallic strip and a set of points to the remainder of the circuit. When circuit amperage exceeds the breaker's rating (the amount of current it is designed to handle), the bimetallic strip heats and bends to separate the points. Once it cools, it will close and reestablish the circuit. If the overload condition still exists, it will heat and reopen.

The circuit breaker can open and close quite rapidly. This gives it an advantage over a fuse. An overload will burn out a fuse and the circuit will not function until the fuse is replaced. This could cause an accident if a fuse were used in a headlight circuit. However, a circuit breaker will cause the lights to flicker on and off rapidly, but will still produce enough light to allow the driver to pull safely off the road, **Figure 17-9**.

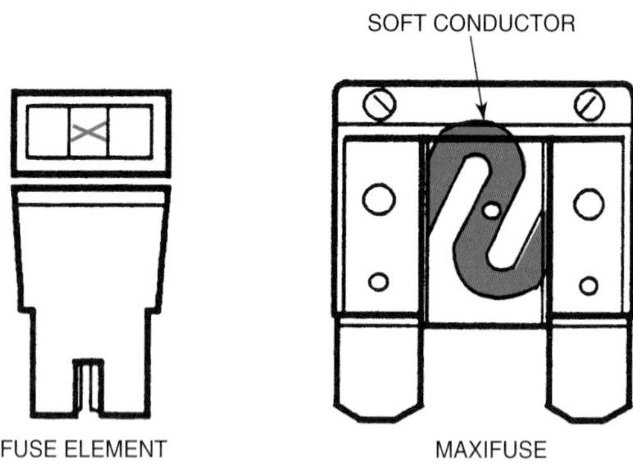

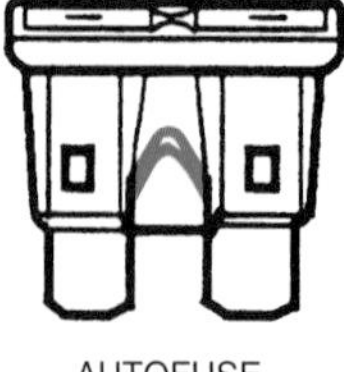

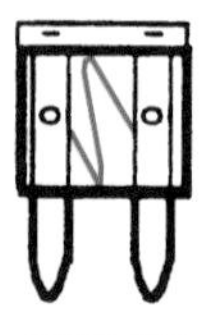

Figure 17-6. An assortment of fuses. All of these fuses give specific amperage rating and use. They are all color coded to indicate their current (amperage) rating. The color tan is used for a 5 amp rating, blue for 15 amps, green for 30 amps, etc. (Chevrolet)

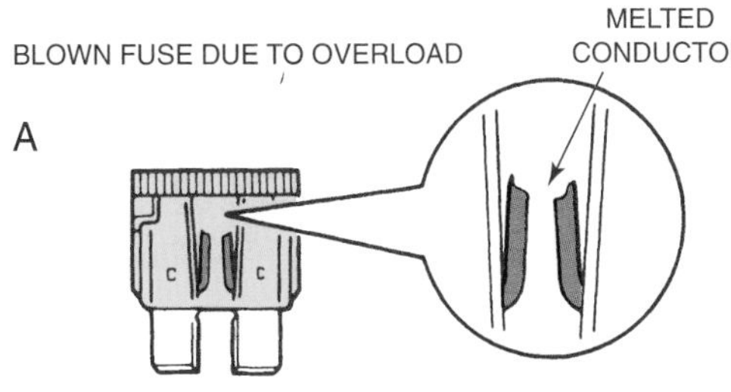

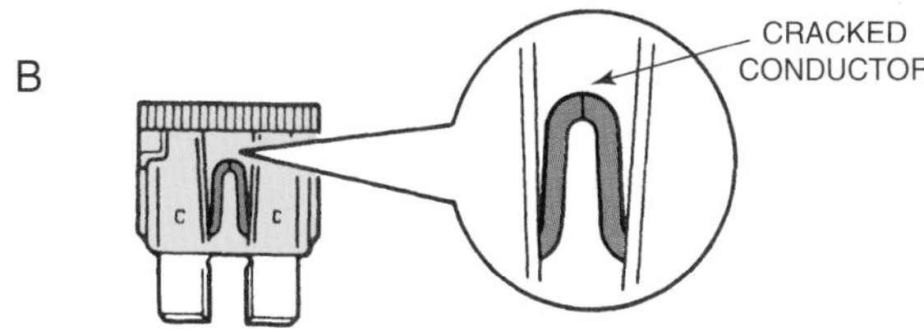

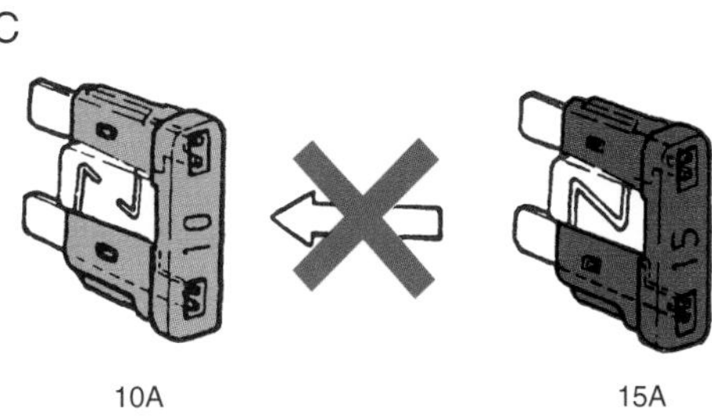

Figure 17-7. A and B—These illustrations show two common failures common in fuses. C—*Never* replace a blown fuse with a fuse of a higher amperage rating. Extensive electrical system and/or part damage can result. (Hyundai)

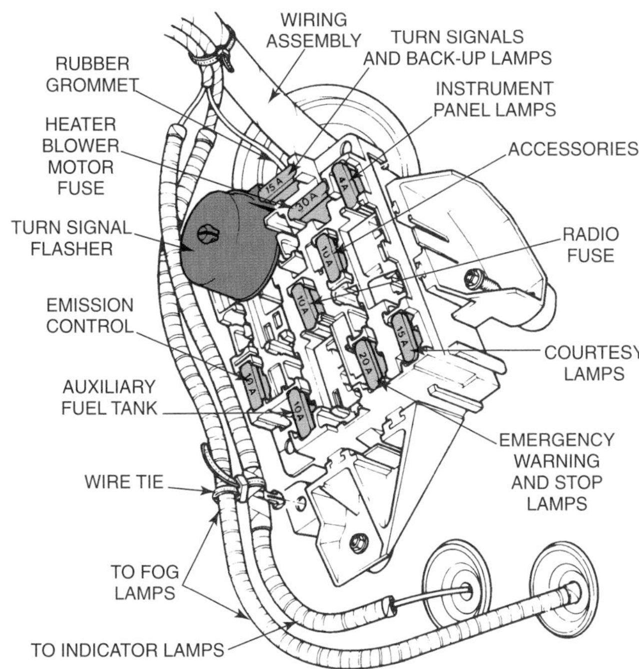

Figure 17-8. A fuse block that utilizes miniature fuses. Some vehicles have more than one fuse block. (Ford)

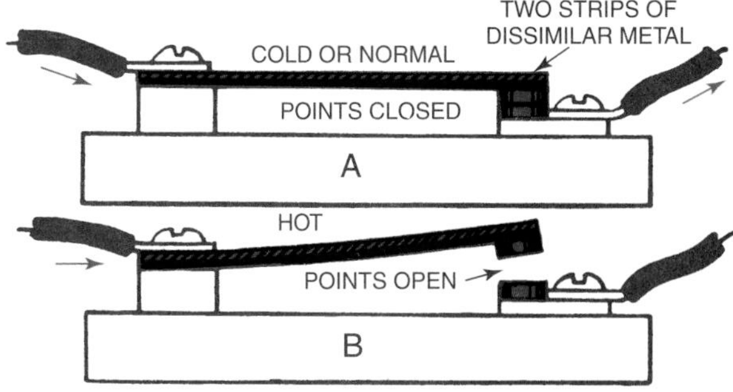

Figure 17-9. Normal operation of a circuit breaker. A—When the breaker is carrying a normal load, the contact points remain closed. B—If the breaker is overloaded, the bimetallic strip heats up. This causes the strip to curve upward, separating the contact points and breaking the circuit.

Fusible Link

A circuit may be protected by the use of a special wire that acts as a fuse. This wire is called a ***fusible link***. A fusible link uses a soft conductor and is one or two sizes smaller than the rest of the circuit wiring. When the fusible link is subjected to an overload condition, it begins to heat, causing the insulation to blister and smoke. If the overload is continued, the insulation ruptures and the fusible link eventually burns out, breaking the circuit, **Figure 17-10**. The fusible link must be replaced once the cause of the overload condition is repaired.

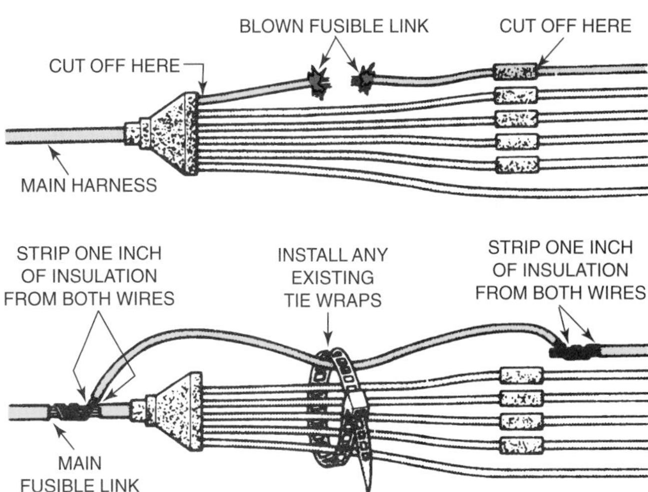

Figure 17-10. The fusible link prevents system damage due to shorts and overloads. Note, at top, how the outer insulation has melted off the link. Fusible link repair requires cutting off the burnt link at the solder joint and installing new fusible link. (Chrysler)

Vehicle Lights and Light Switches

The modern vehicle contains many lights in addition to the normal headlights, brake lights, turn signal lights, and taillights. Lights installed in a modern vehicle include several different types of interior lights, trunk light, glove compartment light, underhood light, and dashboard indicator lights. All of these lights will be discussed in the following sections.

A wide variety of bulb shapes are used on modern cars and trucks. All lamps contain one or more ***filaments***. The filament is a resistance element that glows when current passes through it. Lamps that have two filaments can be used for two separate lighting jobs. The filament is contained inside a vacuum-filled bulb to increase its service life. Lamps must be of the proper candlepower or lumen rating (measurement of light intensity) and voltage. Lamps are also rated by wattage and amperage.

Headlights

The headlights are one of the most important safety features on any vehicle. Obviously, the headlight must be constructed to provide a high intensity light beam, much brighter than that provided by normal service bulbs. The modern headlight can be either the sealed beam or the halogen light. Each type is discussed in the following sections.

Some late-model vehicles have ***daytime running lamps.*** These lamps, which are generally part of the headlight assembly, make the vehicle more visible during daylight hours. Daytime running lamps are connected to the ignition switch and illuminate whenever the ignition is on.

Sealed Beam Headlight

The ***sealed beam*** headlight is an airtight unit containing the lens and mirrored interior surfaces to reflect the light produced by the filament. See **Figure 17-11.** The sealed beams used on vehicles with two headlights and the outer lights on four headlight systems contain two filaments, one each for low and high beams. The inner lights on four headlight systems contain only one high beam filament.

Halogen Headlight

The ***halogen*** headlight is common on modern vehicles. The halogen gas allows the filament to be much brighter than older sealed beam designs, with about the same service life. Some halogen headlights are manufactured as direct replacements for the old sealed beam headlights, **Figure 17-12**.

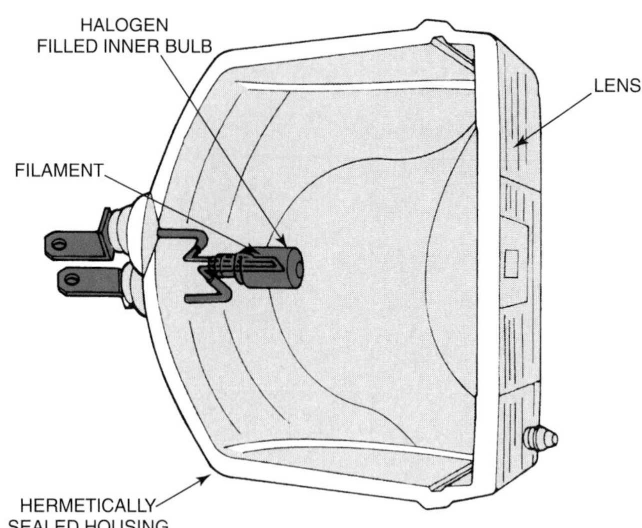

Figure 17-12. One type of halogen headlamp. Some designs use removable bulbs. (Chrysler)

On some newer vehicles, instead of replacing a complete sealed beam headlight, only the lamp is replaced. The filament is contained in a halogen gas-filled bulb that installs in the rear of a ***composite headlight assembly***. An exploded view of the halogen bulb and composite assembly is shown in **Figure 17-13**.

Headlight Switch

Headlights are operated by the driver through the ***headlight switch***. **Figure 17-14** shows a knob style headlight switch, while **Figure 17-15** shows a rocker style headlight switch. On some newer vehicles, the headlight switch is included as part of the turn signal lever, **Figure 17-16**.

The headlight switch also operates the taillights and marker lights. The headlight switch also contains a ***rheostat***,

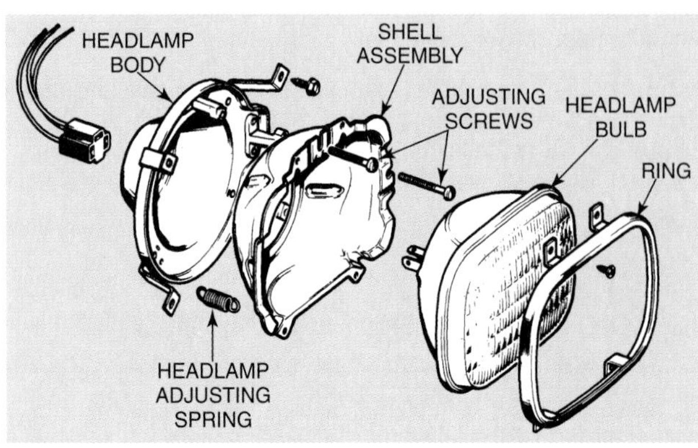

Figure 17-11. Exploded view of a sealed beam headlamp assembly. Note the adjusting screws for headlamp aiming. (Jeep)

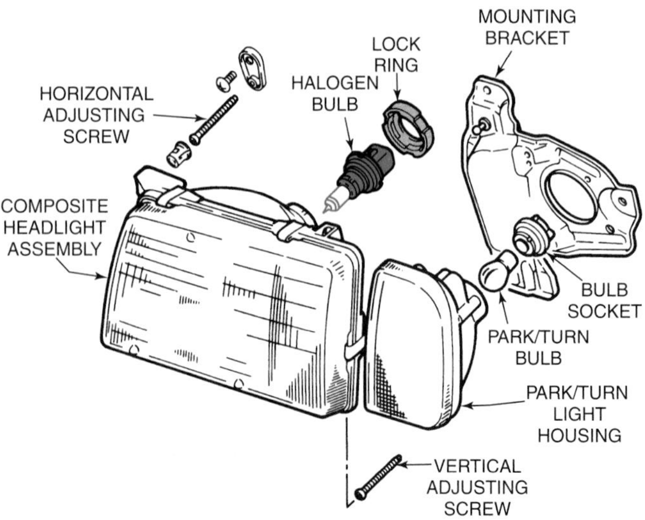

Figure 17-13. An exploded view of one type of halogen bulb headlight assembly. This bulb may be replaced by unscrewing the lock ring. The lock ring and bulb may then be removed from the composite assembly. (Geo)

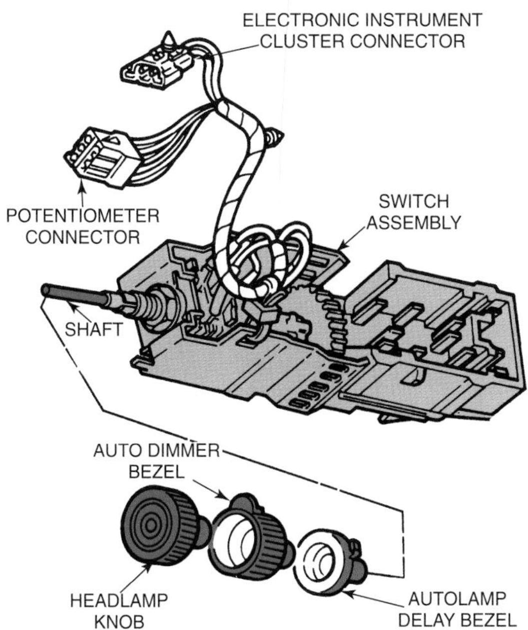

Figure 17-14. One type of headlamp switch and mounting arrangement. (Ford)

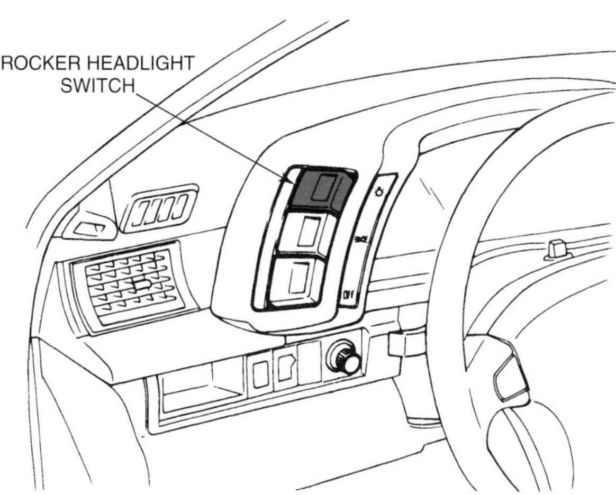

Figure 17-15. A "rocker" style headlight switch is mounted on the instrument cluster. (Geo)

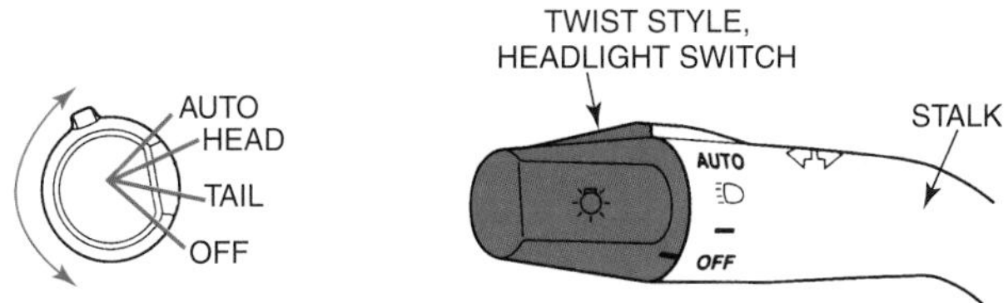

Figure 17-16. A "twist" style headlight switch. This unit is mounted on a stalk (usually the turn signal lever) attached to the steering column. Note the auto setting. In the auto position, an automatic light control sensor will determine the lighting conditions of the vehicle's surroundings and will turn on the headlights and taillights automatically if the sensor reads a low light condition. (Toyota)

which is used to adjust the brightness of the dashboard lights. The headlight switch may also be used to operate the interior lights when the doors are closed. Most headlight switches contain an overload breaker, instead of a fuse.

Dimmer Switch

A ***dimmer switch*** is located in series with the headlight circuit. It is a simple switch that directs current to one of the two terminals, but never to both. One terminal leads to the high beam filament, the other to the low beam. The high beam terminal also feeds current to the high beam indicator lamp on the dash.

The dimmer switch on older vehicles was located on the floorboard close to where the driver's left foot would rest, but on late-model vehicles, the dimmer switch is installed on the turn signal lever. On some vehicles, the dimmer switch wiring passes through the headlight switch. This additional safety feature allows the driver to flash the headlights at other vehicles before passing, **Figure 17-17**.

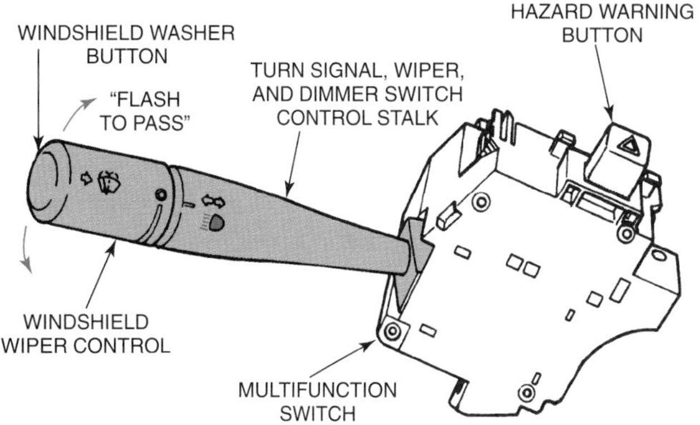

Figure 17-17. A steering column mounted multifunction switch with a built-in turn signal, headlight dimmer, and "flash to pass" function. (Chrysler)

Stoplights and Switches

A brake pedal operated ***stoplight switch*** is provided to operate the brake lights, that are either single or dual filament lamps. The switch is a simple on-off type. When the brake pedal is depressed, the lights come on. When the pedal is released, the lights go off.

Older stoplight switches were operated by hydraulic pressure in the brake system. Newer stoplight switches are mechanical and are energized when the brake pedal is pressed. On some older vehicles, the brake light wiring passes through the turn signal assembly so that the brake lights are bypassed when the turn signals are being used. Modern vehicles usually have separate lamps for the brake and turn signals, **Figure 17-18**.

Turn Signals and Switches

The ***turn signal selector*** lever mounted on the steering column contains contacts which send current to the ***turn signal lights*** on the left or right side of the vehicle, as well as

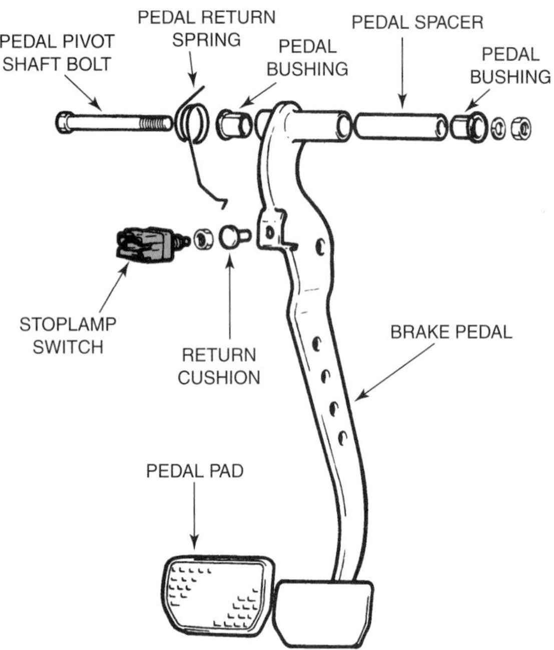

Figure 17-18. One particular type of brake pedal stoplamp switch. Some stoplamp switches are part of the cruise control and anti-lock brake systems. (Geo)

left-right dashboard indicators. The lever is operated by the vehicle driver. A ***canceling cam*** in the steering column returns the lever to the centered position when the steering wheel is returned to the straight-ahead position, **Figure 17-19**.

The turn signal wiring contains a ***flasher unit***, **Figure 17-20**, which causes the turn signal lights to flash on and off rapidly. The turn signal flasher unit is usually installed on the fuse block or in a special holder under the vehicle dashboard. At one time, the turn signals and brake lights used the same filament, usually in the same bulb as the taillight filament. Modern practice, however, is to use separate filaments in separate bulbs for each function.

The ***hazard warning flasher*** switch will send current to the turn signal lights on both sides of the vehicle when turned on. The flasher switch is usually installed on the steering column and is part of the turn signal switch assembly. The

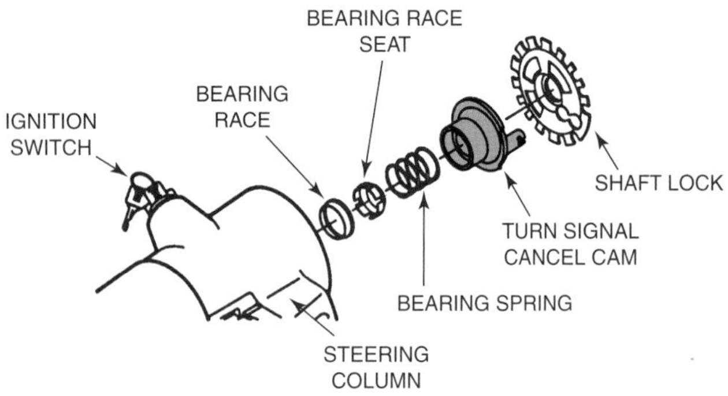

Figure 17-19. An exploded view of the top of a steering column showing the turn signal cancel cam. (General Motors)

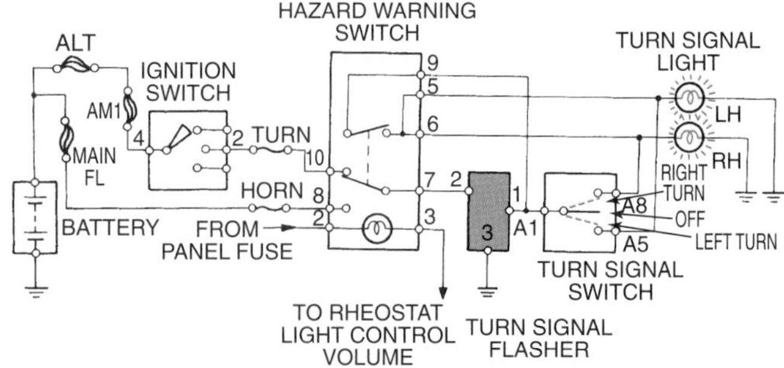

Figure 17-20. A turn signal flasher and wiring schematic. Study carefully. (Lexus)

hazard flasher unit is separate from the turn signal flasher, and is usually installed on the fuse block, or under the vehicle dashboard. Some flashers are called combo-flashers and contain the turn signal and hazard warning flasher in one unit. See **Figure 17-21**.

Backup Lights and Switches

Backup lights are single filament lamps with no connection to the other vehicle lights. ***Backup light switches*** are located on the transmission, in the neutral safety switch assembly (covered later), or the transmission linkage. The backup lights operate only when the vehicle is placed in reverse gear.

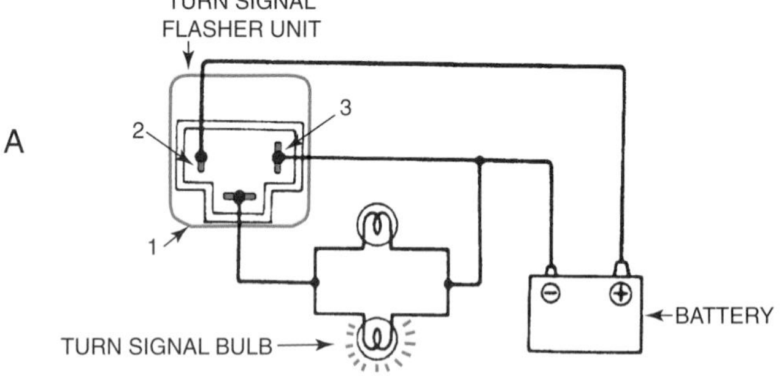

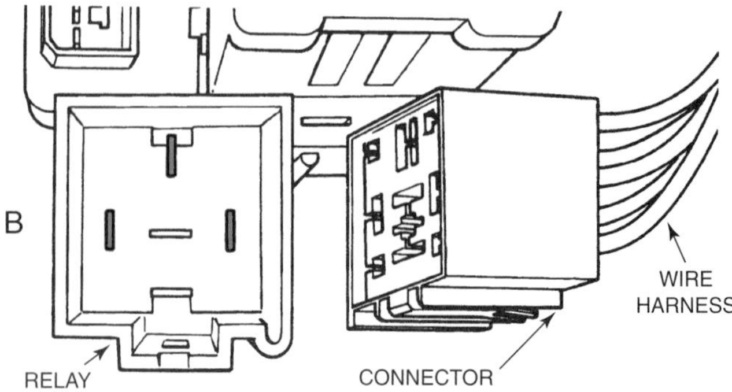

Figure 17-21. A—Turn signal flasher unit and a simple wiring circuit. B—Turn signal/hazard warning flasher assembly with wiring connector and harness. (Toyota & Chrysler)

Dashboard Lights

There are two major groups of vehicle dashboard lights. The first, ***illuminating lights***, light up the speedometer and other gauges if used, and the wiper, headlight, and heater-air conditioner controls. The second group, ***indicator lights***, warn of engine problems, such as charging system problems, overheating, low oil pressure, and computer system problems, **Figure 17-22**.

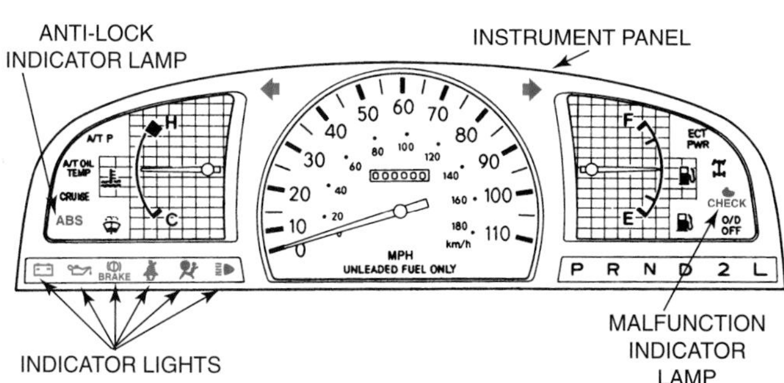

Figure 17-22. This instrument panel shows a variety of warning type indicator lights. The anti-lock brake (ABS) and malfunction indicator lamp (MIL) are computer-controlled. (Toyota)

The illuminating lights are controlled through the headlight switch. A rheostat is used to vary their brightness. There are usually several lights, although a few modern vehicles use only a few bulbs and distribute their light through a ***fiber optic*** (light carrying) harness.

Indicator lights are installed on the dashboard and are either computer-controlled or operated by ***senders***. Typical indicator lights monitor engine temperature, oil pressure, and charging system condition. These lights are either red- or amber-colored. Indicator light senders are simple on-off switches, operated by temperature changes (engine temperature lights), or pressure changes (oil pressure lights). The charging system light is usually operated by the alternator voltage regulator. Typical computer-controlled indicator lights include the ECU check engine and the anti-lock brake malfunction indicator.

Convenience Lights

This group of lights includes all of the lights that are not required for vehicle safety, but are installed to illuminate the passenger sections of the vehicle. The vast majority of these lights are single filament lights operated by on-off switches installed in the chassis, doors, glove compartment, or trunk.

Some courtesy lights are operated by a mercury switch, **Figure 17-23**. Examples are the trunk and underhood lights. The mercury in the switch will not make contact when the

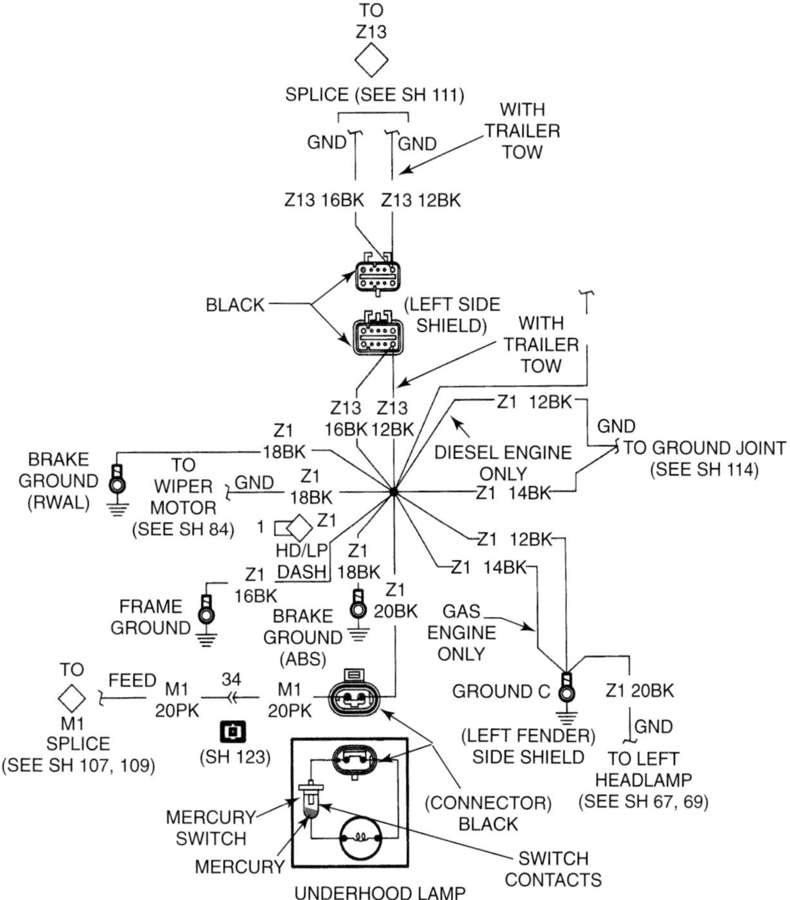

Figure 17-23. A wiring schematic showing an underhood lamp that is turned on and off with a mercury switch. Here, the switch is shown in the off position. When you open the hood, the mercury will flow over the switch contacts, completing the circuit and the light will come on. Closing the hood causes the mercury to flow back to the off position. (Chrysler)

trunk or hood is down. However, when the hood or trunk lid is raised, the mercury flows into contact with the electrical connectors, closing the circuit and allowing the light to come on.

Chassis Mounted Solenoids and Relays

The ***solenoids*** used in the vehicle chassis are similar to those used to energize the vehicle starter. They are used to operate door locks and trunk releases. Relays are a type of solenoid that closes electrical contacts instead of operating a mechanical device. They are used to direct large current flows that might overheat and damage a switch. Switches used to control most solenoids are spring-loaded to return to the off position, preventing accidental operation.

Motor Control Relays

Control relays are similar to starter solenoids. A small current flow energizes a magnetic winding, closing a set of contact points to control high current to another unit, such as a motor or heating coil. Use of relays eliminates having to pass heavy current through the dashboard controls or the ECU, **Figure 17-24**.

Many relays used on late-model vehicles are power transistors with no moving parts. Relays prevent circuit overheating them and voltage drop from the resistance of extra wiring. This reduces electricity consumption and increases reliability. Relays on the modern vehicle are used to operate many high current electrical devices.

Trunk Release Solenoids

Trunk release solenoids are single position solenoids operated by a switch installed in the vehicle glove compartment or instrument cluster, **Figure 17-25**. Power to the trunk release switch passes through the ignition switch and keyless remote module, if equipped. This keeps unauthorized persons from opening the trunk.

Power Door Lock Solenoids and Switches

Power door lock solenoids are two position solenoids, **Figure 17-26**. The solenoids have two windings, allowing them to move a control rod in two directions. When the control switch is moved in one direction, the solenoid moves the door lock to the unlocked position. When the switch is moved in the other direction, the solenoid moves the door lock to the

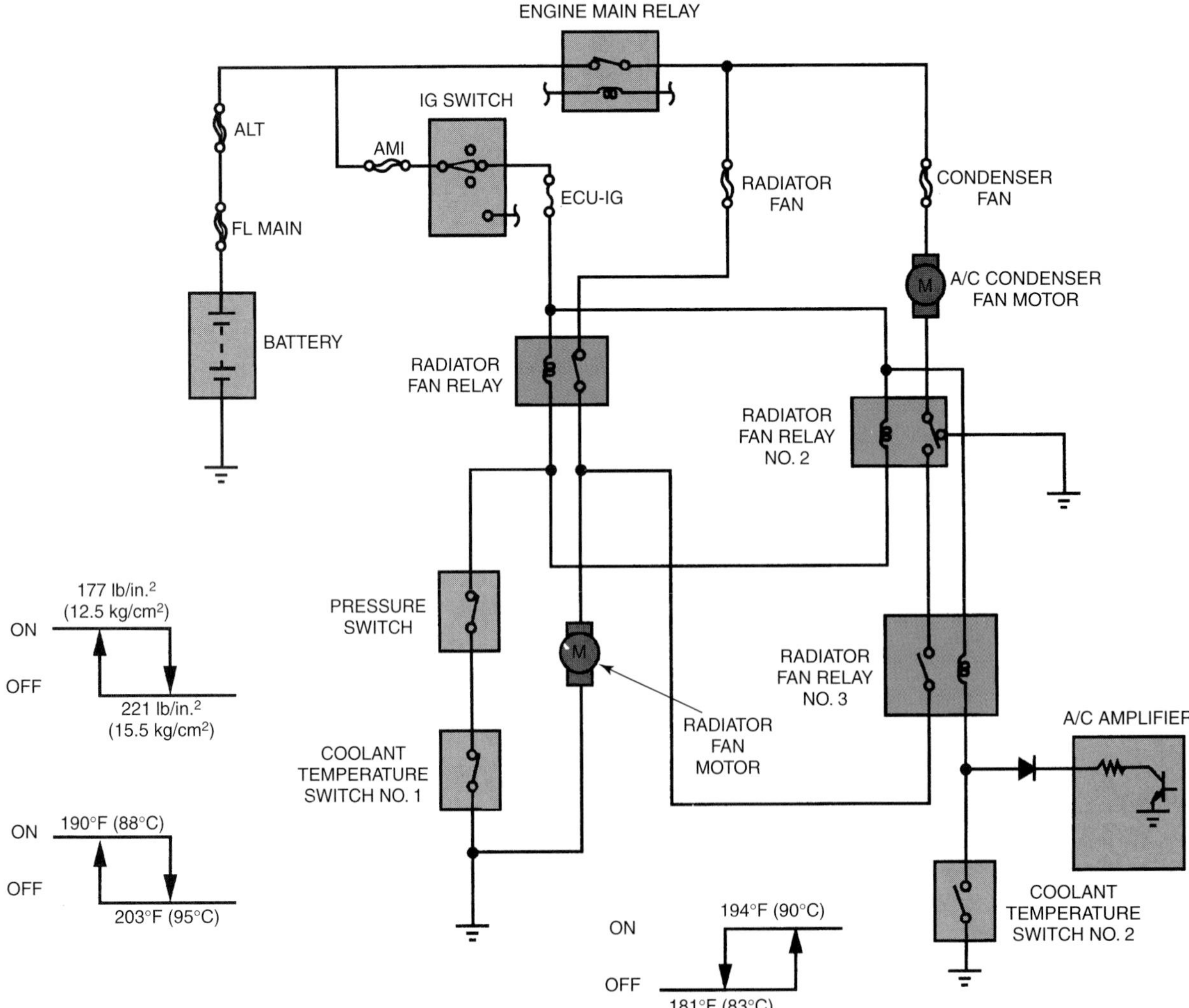

Figure 17-24. An electric cooling fan schematic circuit showing several motor control fan relays. Note how the coolant temperature switches turn the fan on and off at specific coolant temperatures. (Toyota)

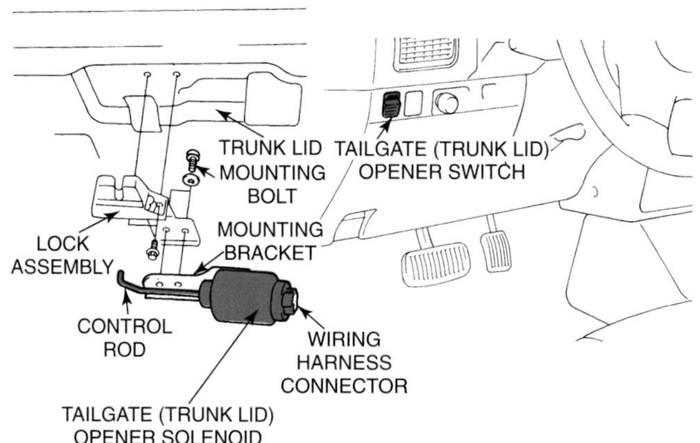

Figure 17-25. An electric trunk release solenoid and its dash-mounted operating switch. Some trunk release switches are located in the glove box or in other locations in the passenger compartment. (Hyundai)

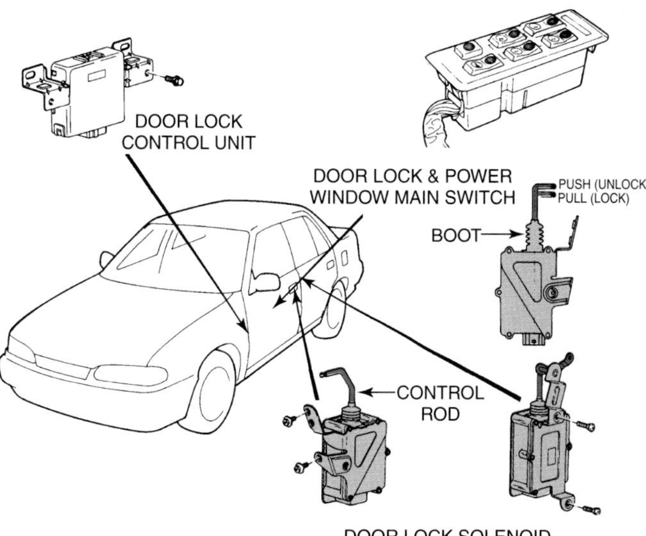

Figure 17-26. Several door lock solenoids and their control rods. The solenoids can be controlled by the master switch, which is usually mounted in the driver's door panel, or by separate switches at each door. (Hyundai)

locked position. Power door locks use simple wiring with separate circuits for the lock and unlock positions. See **Figure 17-27**.

Some power door lock systems have an electronic module added to the circuit as part of a keyless entry system. This system can be used to lock or unlock the vehicle or open the trunk by the driver using a key fob, **Figure 17-28**.

Horns and Horn Relays

The ***horn*** makes use of an electromagnet, vibrator points, and a diaphragm. Referring to **Figure 17-29**, you will see that the electromagnet field coil will attract the metal armature whenever the contact points are closed. Current flows through the horn relay contact points, energizing the electromagnet. The electromagnet draws the armature up, the diaphragm (thin metal disc) will be bent upward.

When the armature moves upward, it causes the points to open and break the circuit. The electromagnet releases the armature and it moves downward, flexing the diaphragm in the other direction. As it moves down, the points close and the armature is again drawn upward. This cycle is repeated very fast. These small up-and-down pulsations of the diaphragm cause the sound. Sound is magnified by passing it through a long or a spiral trumpet, **Figure 17-29**.

Horn Relay

By using a relay between the battery and the horn, it is possible to run a wire from the battery to the relay and from the relay to the horn. When the driver actuates the relay with the horn button, it will complete the circuit and the horn will blow. As this is the shortest possible route, resistance and voltage drop will be low. **Figure 17-30** shows a typical horn circuit using a relay. This particular circuit is operational at all times.

When the driver depresses the horn button, the circuit is completed and current will flow. Current will flow through a heavy wire to the relay, across the closed points, on to the horn and to ground. As soon as the horn button is released, the armature pops up, breaking the points, and the horn stops blowing.

By using the relay, it is possible to use a short run of heavy wire to avoid voltage drop. It is also possible to actuate the relay from a distance. Headlight relays are constructed in a similar fashion.

Rear Window and Mirror Defrosters

The heater ***grid*** consists of a series of fine wires applied to the surface of a rear window or mirror, **Figure 17-31**. The grid material has a calibrated resistance that opposes current flow, creating heat. Current flows through the grid, producing heat to defrost the window or to melt ice or snow. The temperature flow through the grid is kept constant by a load sensing device. Most glass defrosters are controlled by a dashboard switch working through a relay.

Chassis Mounted Motors

Chassis mounted motors include the windshield wiper motor, power door and vent windows, tailgate windows, sliding door motors, and sunroof motors, as well as motors that operate convertible tops and power antennas. The modern vehicle may have as many as ten small dc motors to operate various systems.

Switches used to control these motors, except for the windshield wiper motor, are spring loaded so they return to the off position when the operator's hand is removed. This prevents accidental operation of the motor, with resulting motor overheating and damage.

Windshield Wiper Motors and Controls

Windshield wiper motors, **Figure 17-32**, are high torque motors capable of overcoming the binding effects of ice on the wiper blades and mechanisms. For this reason, wiper motors are very durable, requiring relatively little maintenance.

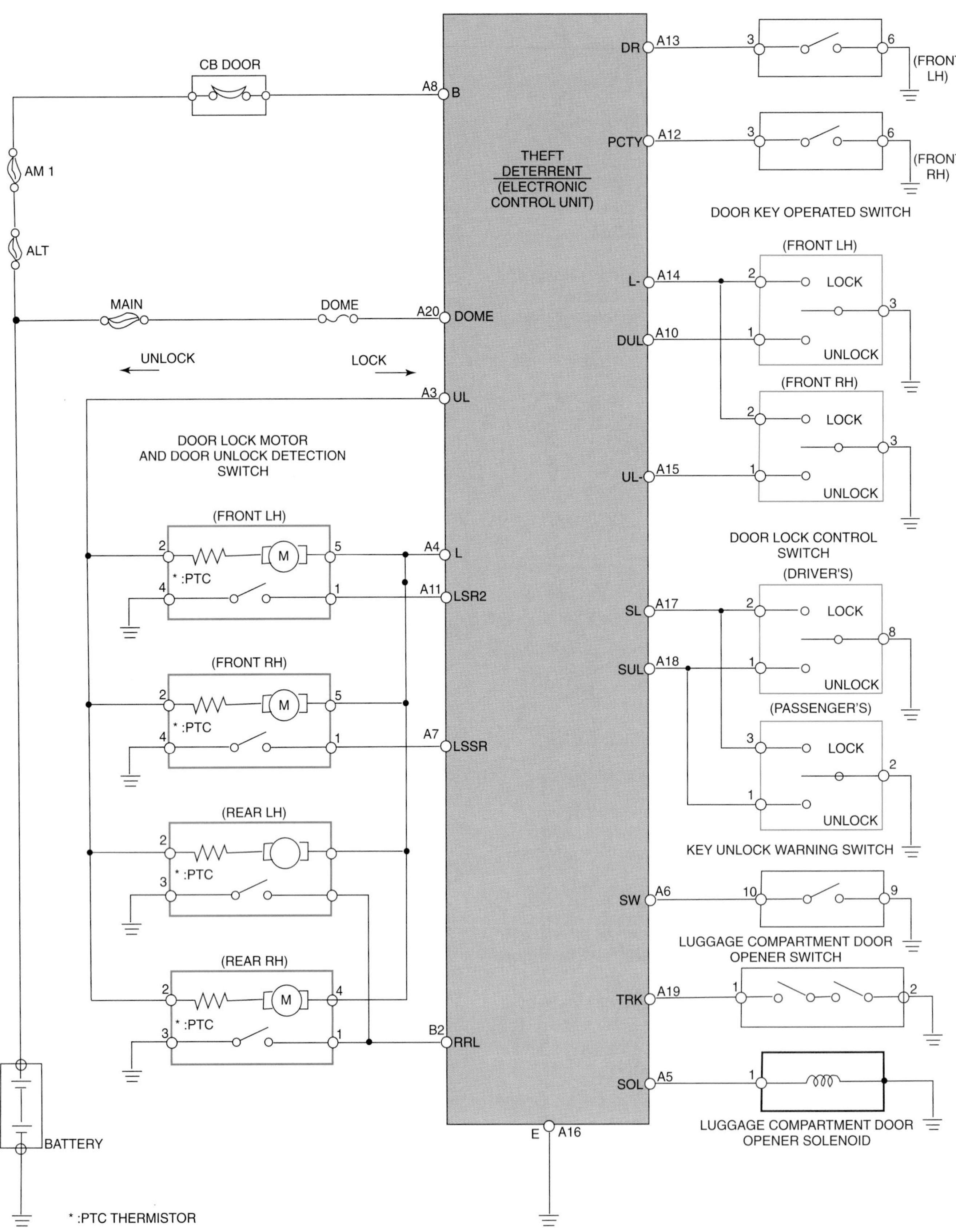

Figure 17-27. An electrical wiring schematic for a four-door vehicle equipped with power door and trunk locks. Study the schematic thoroughly. (Lexus)

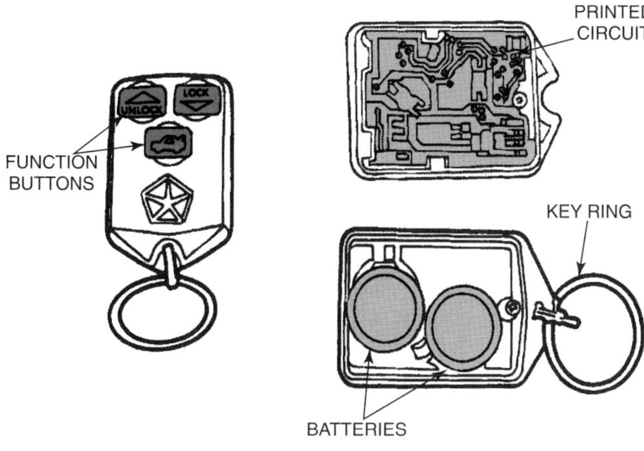

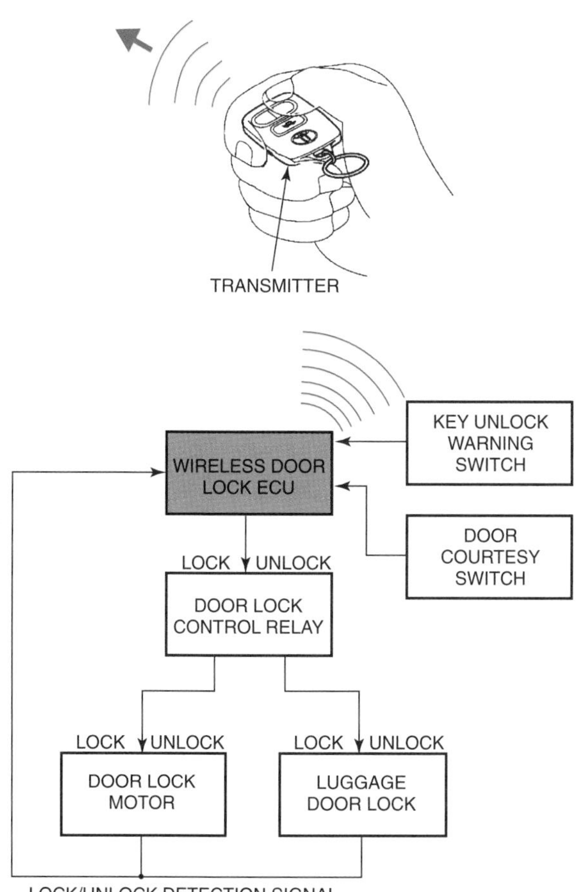

Figure 17-28. A—A wireless, battery operated, keyless remote transmitter. B—The transmitter sends a radio wave (signal) to the wireless door lock electronic control unit (ECU) receiver. The ECU will send the proper signal to the door lock control relay, which will trigger the door or trunk lock motors. (Toyota & Chrysler)

Windshield wiper motors have special park mechanisms which cause the wiper blades to return to the fully down position when they are turned off. Some older wiper motors were equipped to operate the windshield washer pump, but most modern designs use a separate pump located on or in the washer reservoir.

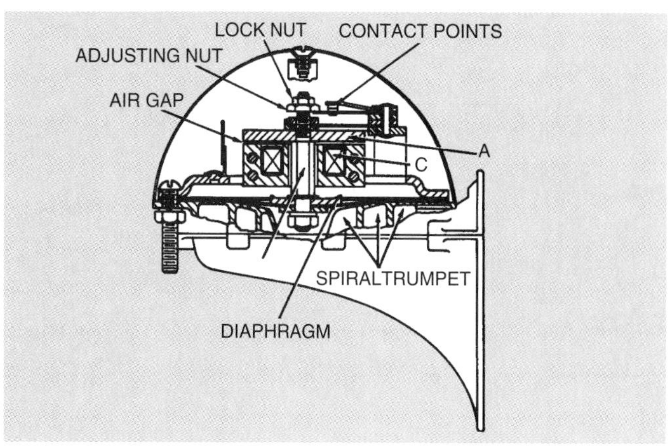

Figure 17-29. A cross-sectional view of an electric horn. The horn air chamber has a snail shell construction. This design will produce maximum sound volume from the rapidly vibrating diaphragm. (Deere & Co.)

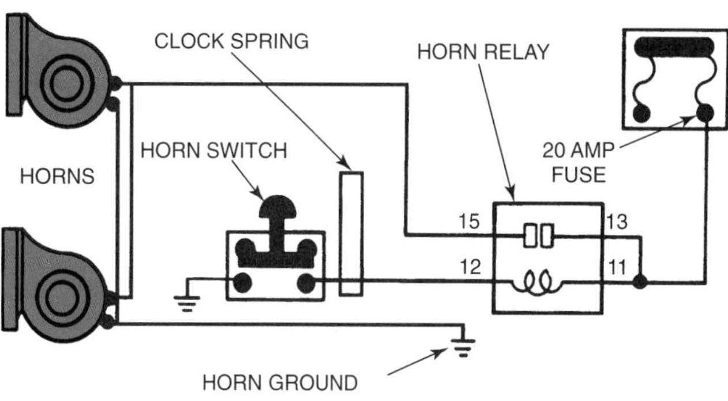

Figure 17-30. A dual horn electrical circuit schematic. The relay used here is similar to relays used in other systems. (Dodge)

Power Window, Tailgate, Sliding Door, and Sunroof Motors

These are generally high torque motors, able to overcome the drag of dry or tight operating mechanisms and glass channels. Most of these motors have built-in overload switches or circuit breakers to prevent overheating damage. The majority of these motors are operated by on-off switches located on the vehicle dashboard, door, or console. See **Figure 17-33**.

Convertible Top Motors and Switches

The convertible top motor on modern vehicles is used to drive a hydraulic pump to power hydraulic cylinders rather than working directly to raise and lower the top. These motors also have built-in overload protection and are operated by a spring-loaded on-off switch.

Power Antenna Motors and Switches

Power antenna motors are simple in construction and are usually operated through an automatic relay. Some power

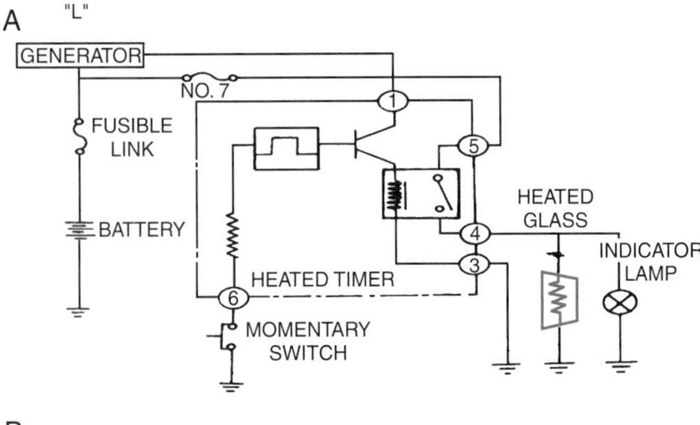

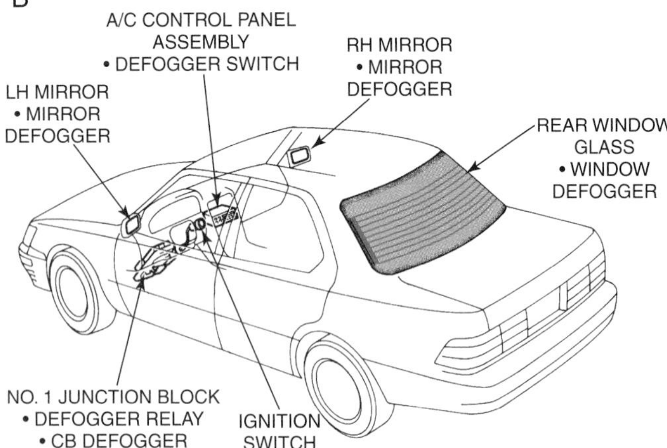

Figure 17-31. A—An electrical schematic for a rear window defogger. B—A vehicle with a rear window defogger. Note that this car also uses outside heated mirror defoggers, similar to the rear glass, even through no wires are exposed. Study the electrical circuit carefully (Hyundai & Lexus)

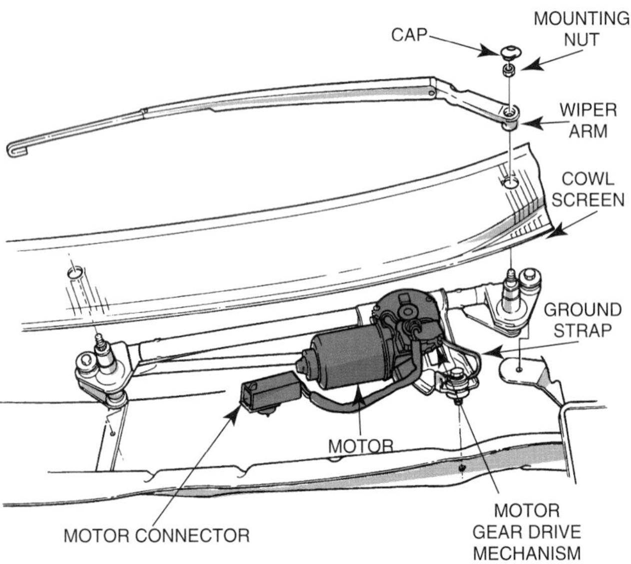

Figure 17-32. An electric windshield wiper motor, gear drive mechanism, and wiper linkage assembly. (Chrysler)

antennas are controlled by a manual 3-position switch. The relay is energized when the radio is turned on, or by the ignition switch when the radio is already on, **Figure 17-34**.

Electric Gauges

Electric gauges, including oil pressure, temperature, fuel, and ammeter, were discussed in the chapters covering systems using these gauges. Gauge illumination is provided by the vehicle dashboard lighting system.

Cruise Control

Most older ***cruise control*** systems were operated by various electromechanical devices. However, the latest cruise control systems are operated by ECU's. Cruise control systems consist of several common components which include:

- Vehicle speed sensor.
- Operator controls.
- Control module.
- Throttle actuator.

Vehicle Speed Sensor

The vehicle speed sensor uses a rotating magnet to generate a small electrical signal. It is usually mounted on the drive shaft or is driven by the transmission or transaxle governor assembly. It sends a speed signal to the control module. The speed sensor may also send a signal to the engine control computer.

Operator Controls and Control Module

The operator controls are used to set the desired speed. Control location varies between manufacturers, but is usually mounted on the turn signal lever. The operator can set the speed, and can also make minor adjustments to speed as necessary.

Another operator control is the brake pedal release switch, usually mounted on the brake pedal bracket. When the driver presses on the brake pedal, the switch sends a signal to the cruise control module to disengage the cruise control. The control module processes the inputs from the controls and speed sensor and produces an output signal to the throttle actuator.

Throttle Actuator

The throttle actuator opens and closes the vehicle throttle to maintain the operator set speed. Throttle actuators can be electric motors directly operated by the control module. In most cases, however, the throttle actuator is a vacuum diaphragm servo connected to a vacuum controller. The module operates the vacuum controller, which in turn operates the vacuum diaphragm.

Air Bag Systems

Air bag systems are becoming standard equipment on most new vehicles. These systems are designed to inflate or ***deploy*** when a vehicle is involved in a frontal collision of sufficient force. Most air bag systems consist of several common components, including a diagnostic control module, frontal impact sensors, coil assembly, and inflator module. These components work together to fully deploy the air bag within 50 milliseconds of impact. After deployment, the air bag will deflate in approximately 100 milliseconds.

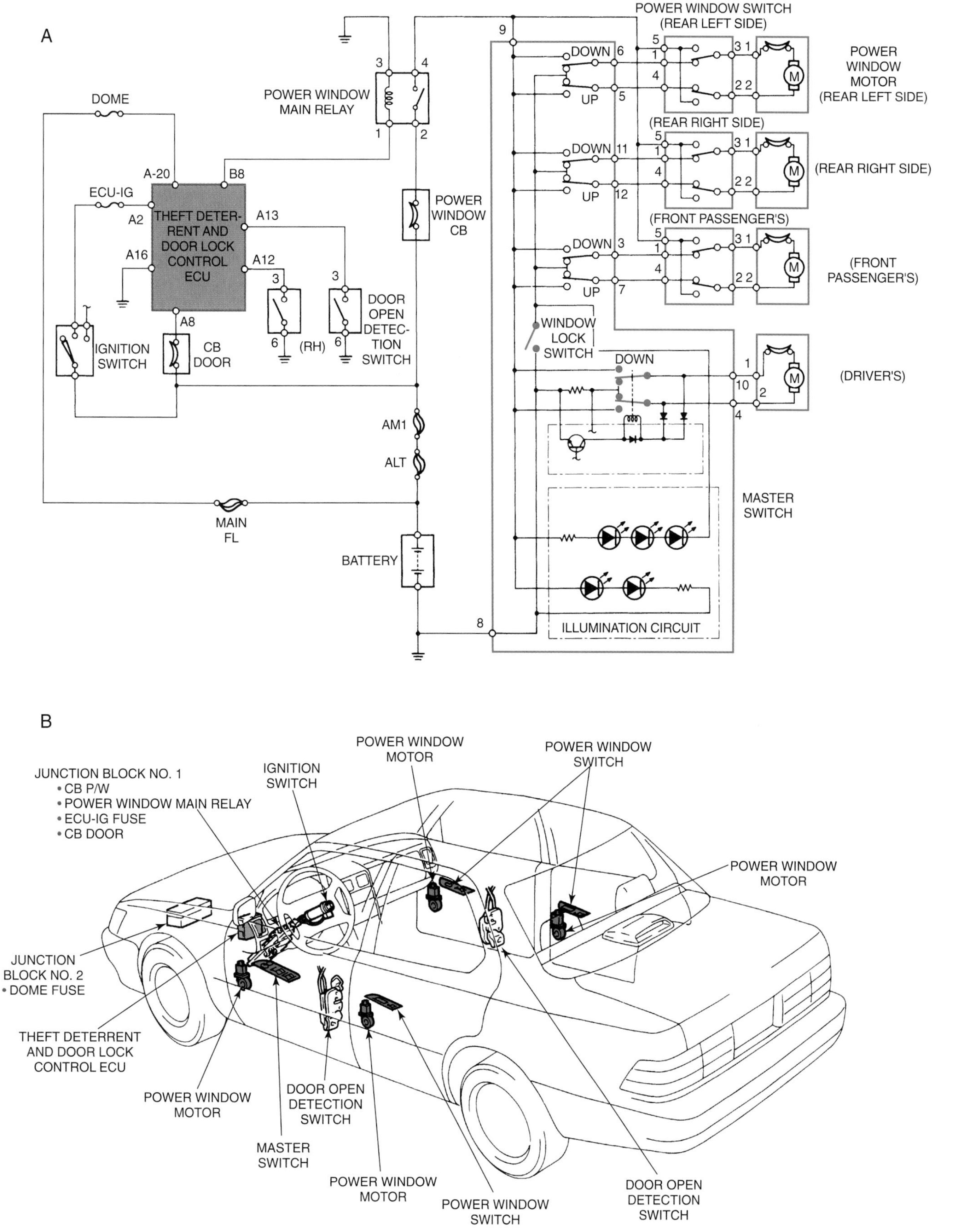

A

B

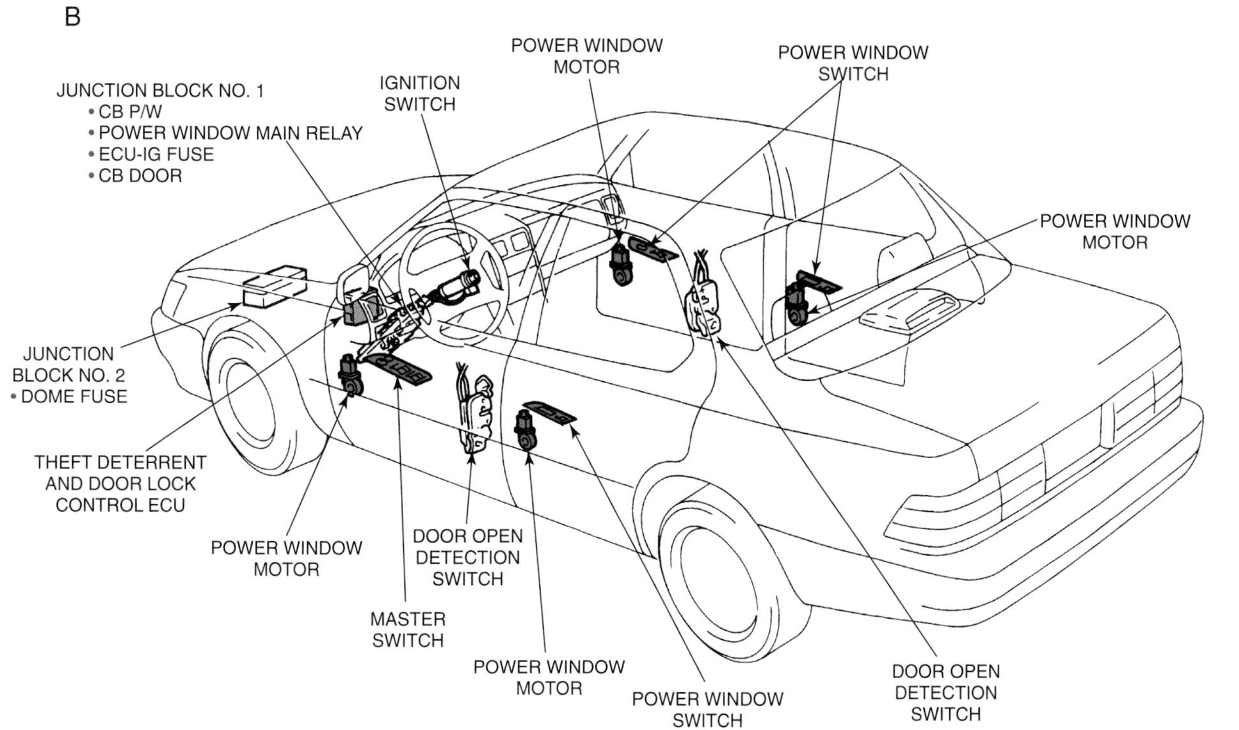

Figure 17-33. A—An electrical schematic for a power window motor and switch system. Note the theft deterrent and door lock ECU. B—A vehicle illustrating the various window motors, operating switches, and their locations. (Lexus)

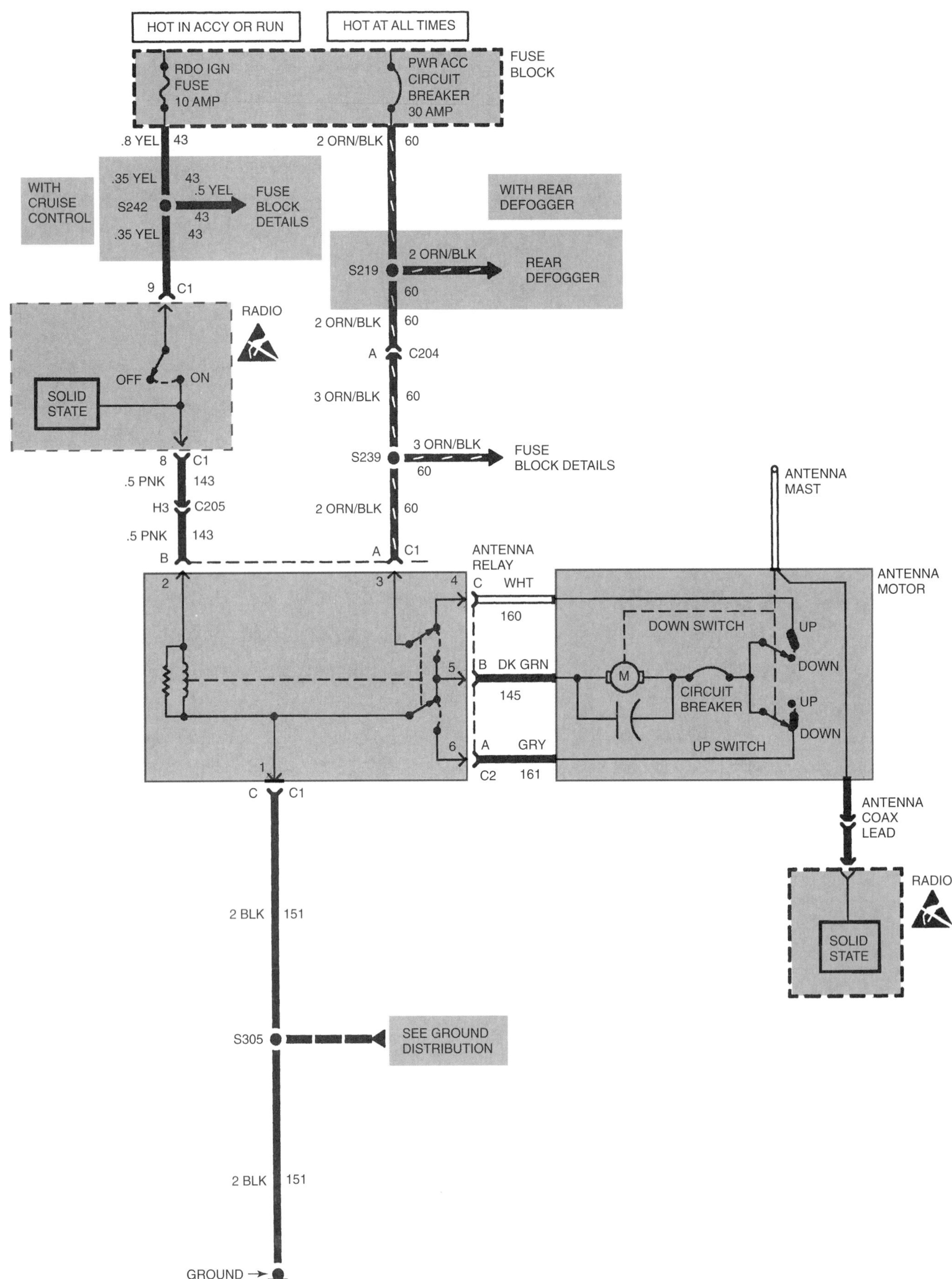

Figure 17-34. An electrical schematic illustrating one particular power radio antenna system. This antenna will go up and down by turning the radio on or off. (Honda)

Air bag systems are often called ***supplemental restraint systems***. It is important to remember that air bags are a supplement to seat belts, they *do not replace them*. Anyone driving or riding in any vehicle should always wear their seat belt.

Diagnostic Control Module

The ***diagnostic control module*** performs several functions. The module serves as a monitoring system for the air bag components. It can warn the driver of system malfunctions by activating the air bag indicator light, and stores trouble codes, used for system diagnosis.

Most diagnostic control modules provide the air bag system with an alternative source of power in case battery voltage is lost in an accident. This reserve power can last up to 30 minutes after the battery has been disconnected. The diagnostic control module also contains a ***safing sensor,*** designed to prevent accidental deployment. This sensor must close simultaneously with one of the impact sensors to trigger the inflator module. The safing sensor prevents a faulty impact sensor from activating the air bag system.

Impact Sensors

One or more ***impact sensors*** are located on the front sector of a vehicle's chassis or on each side of the radiator housing. An impact sensor is an open switch, designed to close on impact. It will close only when the impact is severe enough to warrant air bag deployment. A ball type impact sensor is shown in **Figure 17-35**.

Coil Assembly and Inflator Module

The ***coil assembly*** consist of two current-carrying coils attached to the steering column. As the steering wheel is rotated, the coils maintain a continuous electrical connection between the inflator module and diagnostic control module. The coil assembly is often referred to as a *clock spring*.

The ***inflator module*** is located in the center of the steering wheel and above the glove box on passenger side air bag systems. The module contains the inflatable fabric air bag, an initiator squib, and an inflator (gas generating material). The inflator module is covered with a trim piece. A cutaway of an inflator module is shown in **Figure 17-36**.

Air Bag Operation

When a vehicle is involved in a collision that is severe enough to warrant air bag deployment, the impact sensors close. If at least one impact sensor and the safing sensor in the diagnostic module close, the module sends a signal to the initiator squib. The squib then generates a thermal reaction which, in turn, ignites the gas generating material (sodium azide based propellant) in the inflator. The material releases a large amount of harmless gas, which quickly fills the air bag. The air bag deployment process is shown in **Figure 17-37**.

Side Air Bags

Some late-model vehicles have ***side air bags***, which are installed in the door panels or the seats, **Figure 17-38.** Side air bags operate in the same manner as conventional air bags. However, they are designed to deploy during a side impact.

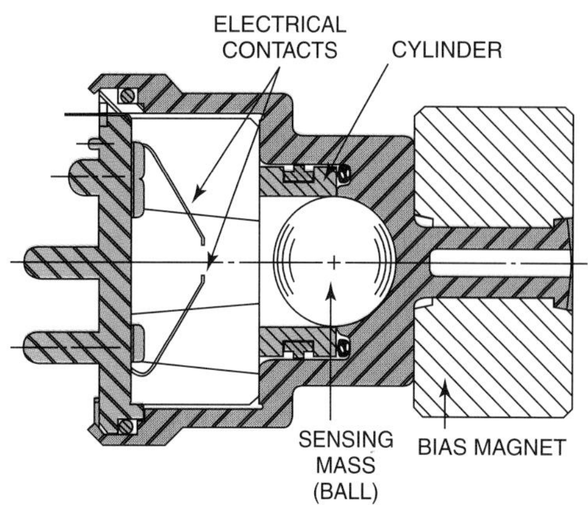

Figure 17-35. Cutaway view of an electromagnetic impact sensor. During vehicle impact, the gold plated sensing ball will move forward inside its stainless steel tube. The ball movement is controlled by a viscous air damping force and a bias magnet. The sensor can be produced to respond to various crash speeds by the ball and tube clearance, ball travel distance, and the bias magnet strength. (Breed Automotive Corporation)

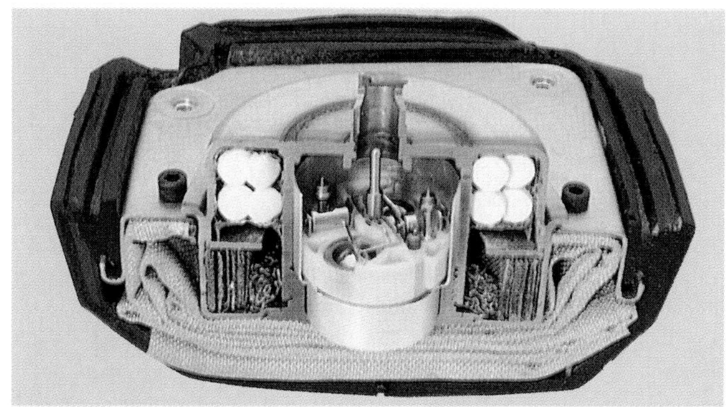

Figure 17-36. A cutaway view of an inflator module. Study all of the parts. (Breed Automotive Corporation)

Summary

Chassis wiring extends throughout the vehicle body. It consists of all vehicle wiring that is not directly connected with the engine and drive train. Wiring to electrical systems can pass through the ignition switch, or is powered directly from the vehicle battery.

Almost all vehicle wiring is part of wiring harnesses. All chassis wires are color coded. Circuit protection devices are used to break the circuit to prevent damage to the other equipment in the circuit. The most common types of circuit protection devices are the fuse, fusible link, and the circuit breaker.

The modern vehicle uses headlights, taillights, brake and turn signal lights, interior lights, trunk light, glove compartment light, underhood light, and dashboard indicator lights. A wide variety of lightbulb shapes are used on modern cars and trucks. Bulbs are composed of a filament inside of a glass bulb which glows when current passes through it. All of the air has been removed from the bulb to increase filament life. The modern headlight uses either sealed beams or halogen bulbs in a composite assembly.

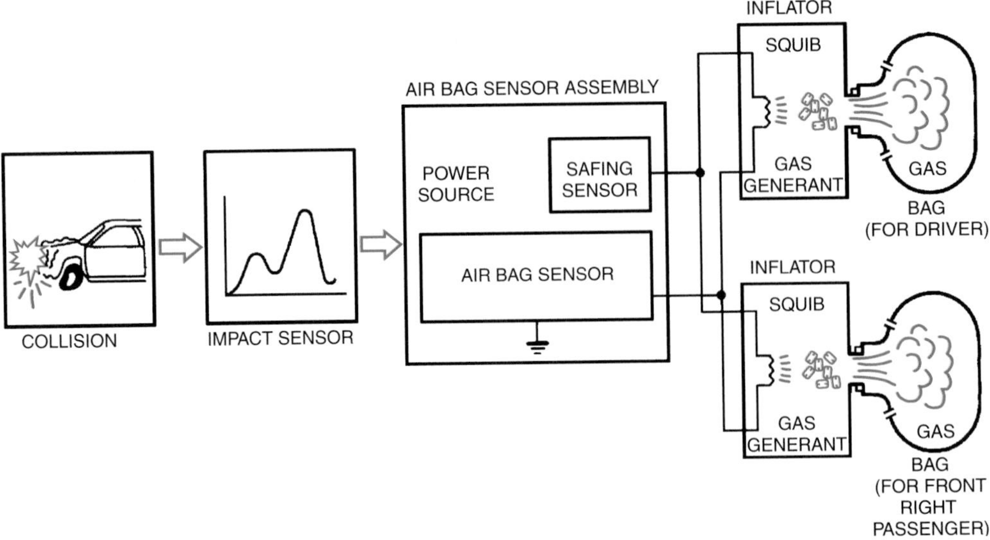

Figure 17-37. A schematic illustrating the air bag deployment process when a vehicle is involved in a frontal collision. Note the dual air bag arrangement. (Toyota Motor Corporation)

Figure 17-38. Side air bags help protect the vehicle's occupants in the event of a side impact. (Cadillac)

Solenoids are used to operate door locks and trunk releases. A relay is an electromagnetic unit designed to make and break a circuit. The relay can be a distance from the operator and will still function with only a ground wire and switch. Electric horns use an electromagnet, vibrator points, and a diaphragm to produce sounds. The horn is operated through a relay.

Rear window and mirror defrosters consist of a grid of fine resistance wires applied to the glass surface of the rear window or mirror. Current passing through the grid produces heat to melt any ice.

Chassis motors are used to operate the windshield wipers, power windows and vent windows, tailgate windows, sunroofs, convertible tops, and power antennas.

Disconnect the battery when working on the electrical system. Make sure that all electrical components are properly fused. Wires must be insulated, away from hot areas, and secured to prevent chafing. Wire terminals should be clean and tight.

Know These Terms

Chassis wiring
American Wire Gage (AWG)
Wiring harness
Looms
Printed circuit
Shielded
Color coded
Terminal block
Junction block
Shorting
Circuit protection device
Fuse
Fuse block
Inline fuse
Circuit breaker
Fusible link
Filament
Daytime running lamps
Sealed beam
Halogen
Composite headlight assembly
Headlight switch
Rheostat
Dimmer switch
Stoplight switch
Turn signal selector
Turn signal lights
Canceling cam
Flasher unit
Hazard warning flasher
Backup lights
Backup light switch
Illuminating lights
Indicator lights
Fiber optic
Senders
Solenoid
Horn
Grid
Cruise control
Air bag system
Deploy
Supplemental restraint system
Diagnostic control module
Safing sensor
Impact sensor
Coil assembly
Inflator module
Side air bags

Review Questions—Chapter 17

Do not write in this book. Write your answers on a separate sheet of paper.

1. Is a wire with a gage of 17 smaller or larger than a wire with a gage of 8?
2. What is the gage of a wire decided by?
   (A) Wire diameter.
   (B) Wire length.
   (C) The wire material.
   (D) Both A & B.
3. Technician A says that electrical devices that can not be operated except when the vehicle is being driven are wired through the ignition switch. Technician B says that units that can be operated at any time are powered directly from the vehicle battery. Who is right?
   (A) A only.
   (B) B only.
   (C) Both A & B.
   (D) Neither A nor B.
4. Chassis wiring consists of all vehicle wiring that is not directly connected with the vehicle _____ .
   (A) engine
   (B) lighting
   (C) drive train
   (D) Both A & C.
5. What are the three most common types of circuit protection devices?
6. Any new electrical component installed on a vehicle must have a _____ .
7. All of the following statements about chassis mounted solenoids and relays are true, EXCEPT:
   (A) chassis solenoids are similar to starter solenoids.
   (B) solenoids are used to operate convertible top hydraulic pumps.
   (C) relays are a type of solenoid that close electrical contacts.
   (D) using relays prevents damage to switches.
8. Technician A says that small up-and-down pulsations of an internal diaphragm produce the sound in an electric horn. Technician B says that the electric horn operates by flexing a strong spring. Who is right?
   (A) A only.
   (B) B only.
   (C) Both A & B.
   (D) Neither A nor B.
9. Which of the following motors are not controlled by a spring loaded switch?
   (A) Windshield wiper motor.
   (B) Power window motor.
   (C) Convertible top motor.
   (D) Power antenna motor.
10. What is a bulb filament?
11. Technician A says that the dimmer switch is connected in series with the headlights. Technician B says that the rheostat in the headlight switch is connected in series with the headlights. Who is right?
    (A) A only.
    (B) B only.
    (C) Both A & B.
    (D) Neither A nor B.
12. A blown fusible link is _____ .
    (A) repaired
    (B) replaced with a length of copper wire
    (C) replaced with the same gage of wire
    (D) Either A or C.
13. Technician A says that the convertible top motor on modern vehicles drives a hydraulic pump. Technician B says that power antenna motors are usually operated by a 3-position switch. Who is right?
    (A) A only.
    (B) B only.
    (C) Both A & B.
    (D) Neither A nor B.
14. What does the air bag system safing sensor prevent?
15. Briefly explain how the air bag is inflated once the diagnostic control module sends the signal.

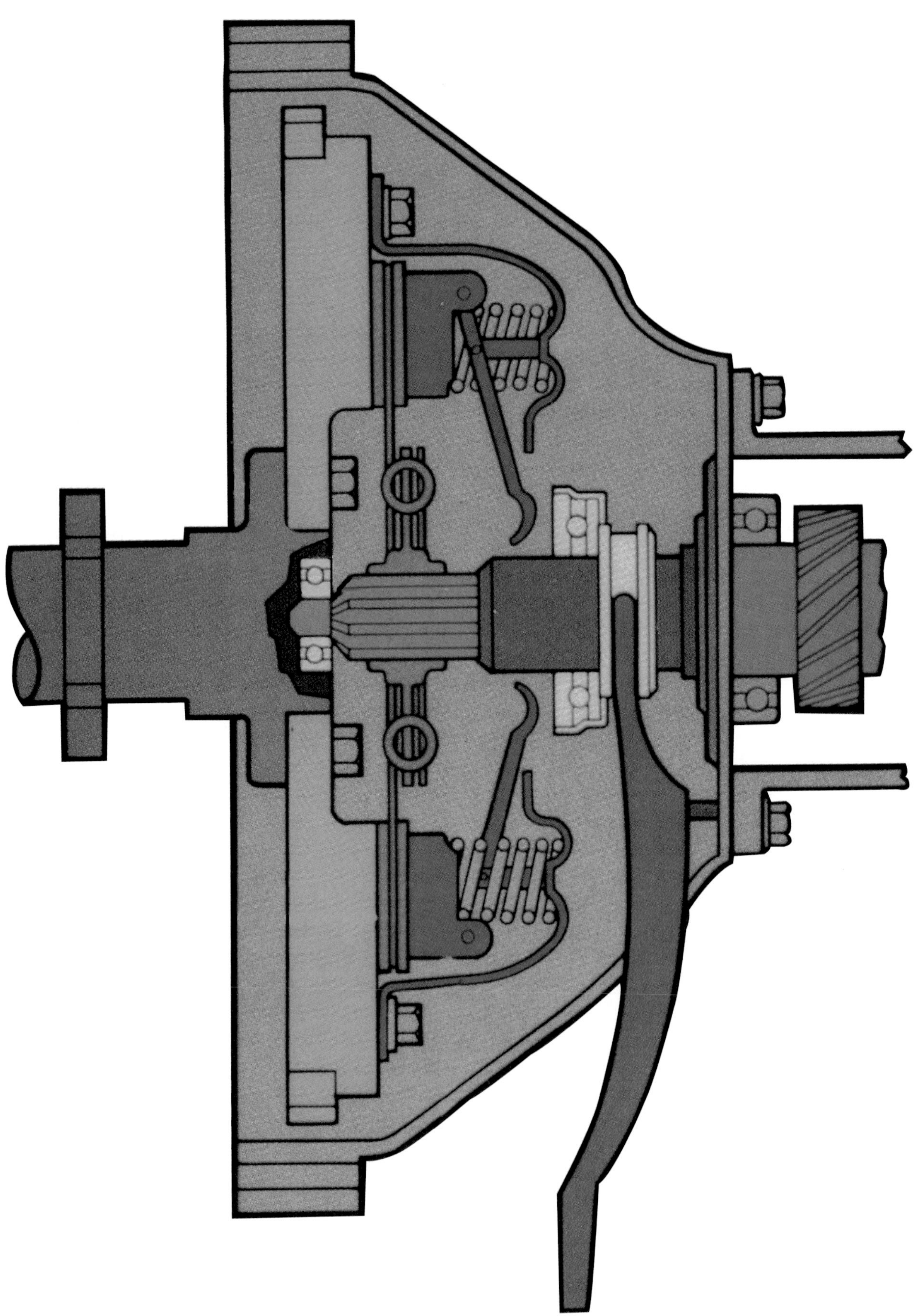

Cutaway of a manual transmission clutch assembly. Although it looks complicated, it is quite simple in design and operation. (Federal-Mogul)

18

Engine Clutches

After studying this chapter, you will be able to:

- Explain clutch operation.
- Name the different types of clutches.
- List the various clutch release mechanisms.
- Define the basic parts of a clutch.
- Identify the parts of a clutch.

The function of the clutch assembly in cars and trucks with manual transmissions or transaxles is to connect or disconnect the flow of power from the engine to the drive line. Although there are many variations in the design of clutches, all work on the same basic principle. This chapter concerns the clutch assemblies used on modern vehicles with manual transmissions and transaxles.

Clutch Purpose

A ***clutch*** is a mechanism designed to connect or disconnect power from one working part to another. In a vehicle, the clutch is used to transmit engine power and to disengage the engine and transmission when shifting gears. It also allows the engine to operate when the vehicle is stopped without placing the transmission in neutral (out of gear).

Clutch Construction

The modern clutch is a single plate, dry disc. It consists of five major parts: flywheel, clutch disc, pressure plate assembly, throw-out bearing, and clutch linkage. Other parts which make up the clutch assembly are the transmission input shaft and clutch housing. To best illustrate clutch construction and operation, each part will be discussed in order.

Flywheel

In addition to providing a base for the starter ring gear, the ***flywheel*** forms the foundation on which the other clutch parts are attached. The flywheel used with manual transmissions is thick to enable it to absorb a large amount of heat generated by clutch operation.

The clutch side is machined smooth to provide a friction surface. Holes are drilled into the flywheel to provide a means of mounting the clutch assembly. A hole is usually drilled directly into the rear of the crankshaft. This hole allows a bearing to be installed in the center of the flywheel.

The bearing in the center of the flywheel will act as a support for the outboard end of the transmission input shaft. It is referred to as a ***pilot bearing***. The pilot bearing may be either a ball bearing or a bronze bushing. Both must be provided with lubrication, **Figure 18-1**.

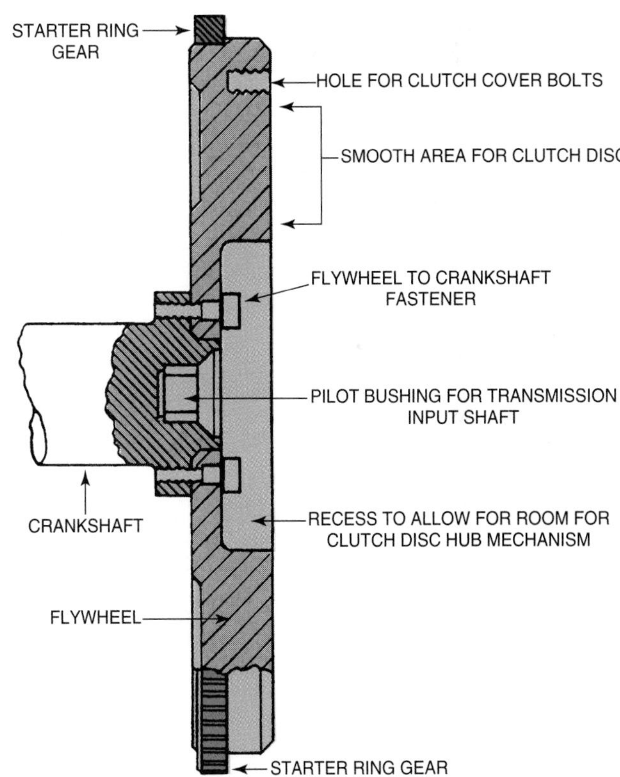

Figure 18-1. This cross-sectional view shows typical flywheel design. The flywheel provides the foundation on which the clutch is built.

Dual Mass Flywheel

A dual mass flywheel is sometimes used with diesel engines. To help absorb engine power stroke pulsations, springs mounted inside the flywheel act as a shock absorbing unit when the flywheel sections partially compress, smoothing out the power flow. They also help to reduce stress on the clutch and transmission parts, **Figure 18-2**.

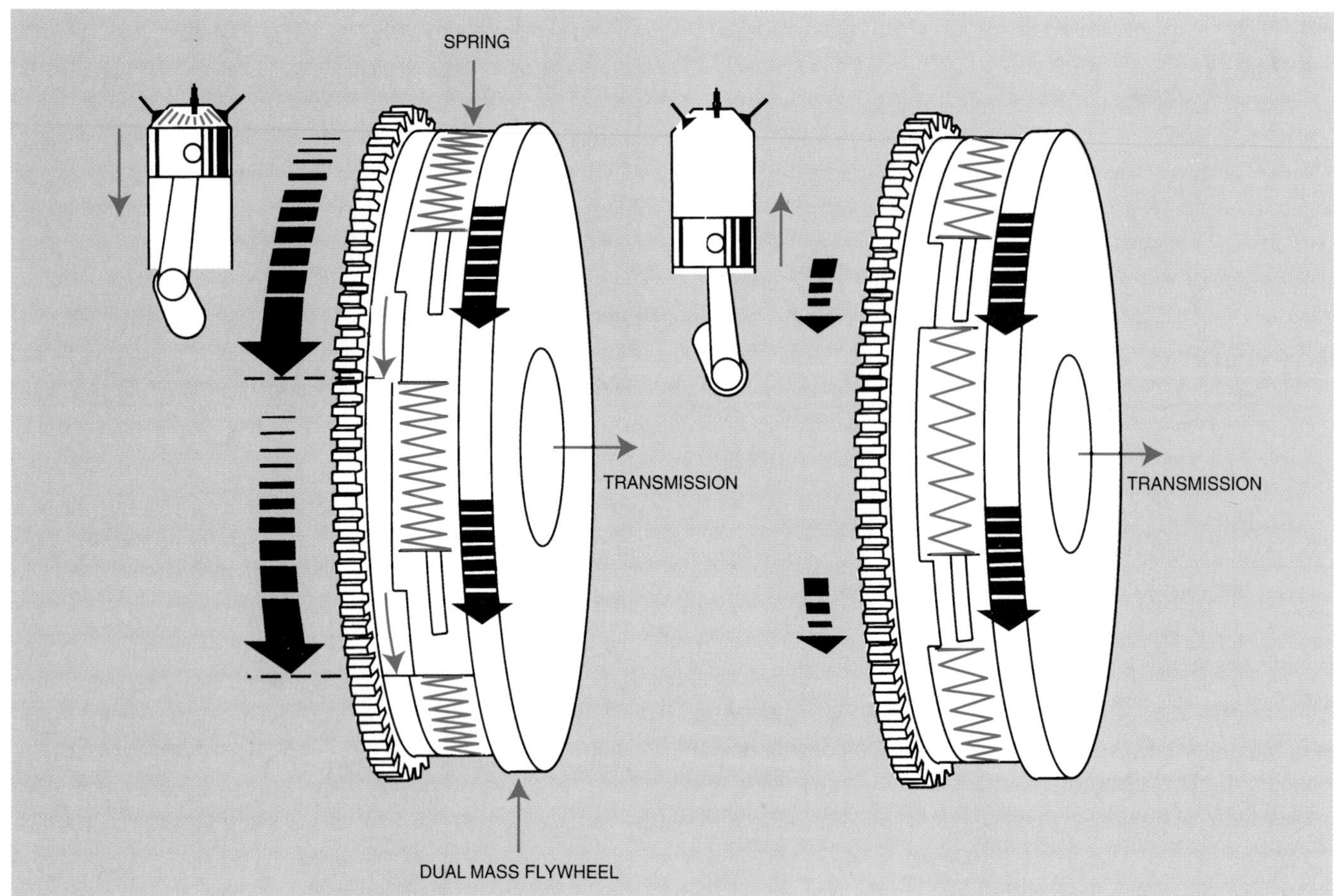

Figure 18-2. A dual mass flywheel assembly can absorb the power stroke pulsations of a diesel engine by using heavy coil springs mounted between the two-piece flywheel. (Luk Incorporated)

Clutch Disc

The ***clutch disc*** is round and constructed of thin, high quality steel with a splined (grooved) hub placed in the center. The hub ***splines*** engage splines on the transmission input shaft. The clutch disc can move back and forth on the shaft, but when the disc is turned, the shaft must turn also.

Both sides of its outer edge are covered with friction material. It is often made of asbestos or other high temperature material and copper wires either woven or molded together. It is riveted to the disc.

STOP **Warning: Clutch friction materials can contain asbestos—a known carcinogen (a substance that can cause cancer). Disassembling the clutch can produce small airborne particles of asbestos. Breathing these particles may cause cancer.**

To assist in smooth engagement, the disc outer edges are often split and each piece cupped. The friction material is riveted to these segments. When the disc is compressed, these cupped segments act as a spring-like cushion.

The inner hub and thin outer disc are fastened together in such a manner as to allow a certain amount of radial (circle around the center) movement between them. This movement is controlled by stop pins. Coil springs act as a drive unit between the hub flange and outer disc. These springs act as a dampening device, softening the torque thrust when the outer disc is pressed against the flywheel. It also helps to transmit this thrust to the hub. Vibration is controlled by a molded friction washer between the clutch hub flange and outer disc. **Figure 18-3** illustrates the edges of the cupped segments. Notice how the friction material is riveted to the segments. A typical clutch disc is shown in **Figure 18-4**.

Pressure Plate Assembly

There are two basic types of ***pressure plates***, the ***coil spring pressure plate*** and the ***diaphragm spring pressure plate***. The basic operating principles of each are the same. The difference lies in the type of apply spring used.

Pressure Plate Disc

The clutch pressure plate disc is the same in each type of pressure plate assembly. It is made of a thick piece of cast iron for maximum heat absorption. It is round and about the same diameter as the clutch disc. One side of the pressure plate is machined smooth. This side will press the clutch disc facing area against the flywheel. The outer side has various shapes to facilitate attachment of springs and release mechanisms, **Figure 18-5**.

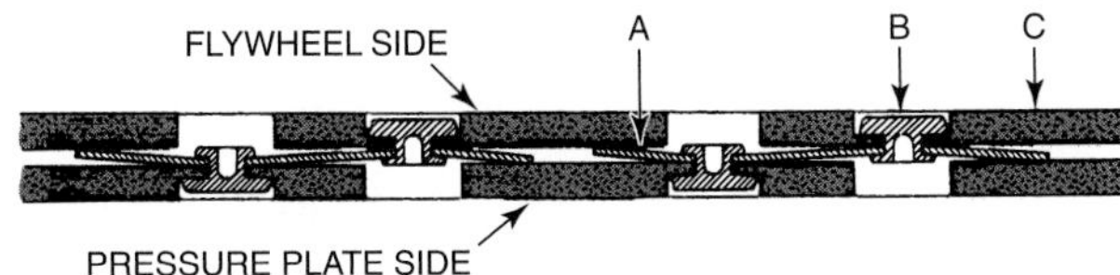

Figure 18-3. Side view of a clutch disc. Notice how the segments are cupped to produce a cushion effect that assists in smooth clutch engagement. A—Cupped segment. B—Rivet. C—Lining.

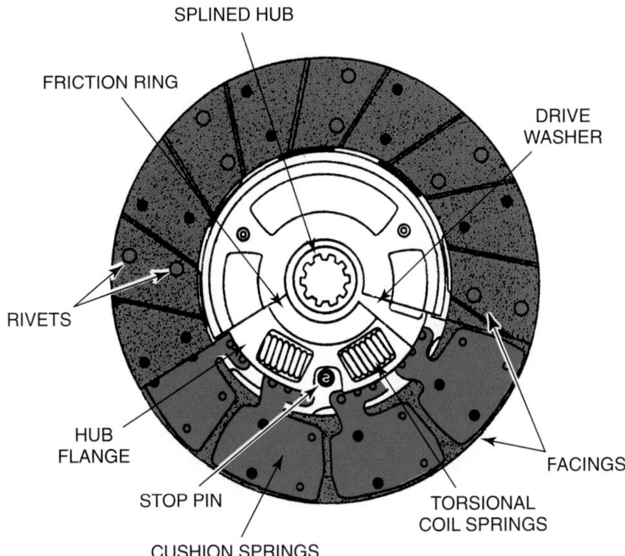

Figure 18-4. One type of clutch disc. Note the splined hub and arrangement of torsional coil springs that cushions the shock of engagement.

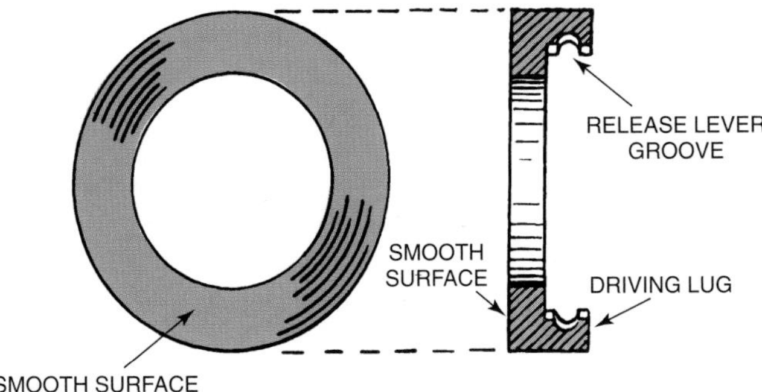

Figure 18-5. Pressure plate. The pressure plate is spring-loaded and presses the clutch disc against the flywheel.

Coil Spring Pressure Plate Assembly

In addition to the pressure plate itself, the coil spring pressure plate assembly contains a number of coil springs, a ***pressure plate cover***, and ***release levers***. The springs are used to provide pressure against the pressure plate. Various numbers of springs are used, depending on the type of service for which the pressure plate is designed. The springs push against the clutch cover and pressure plate, **Figure 18-6**.

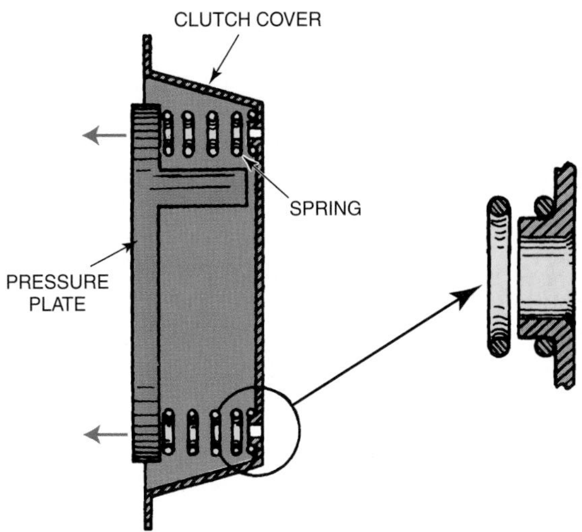

Figure 18-6. Clutch springs—coil type. Size and number of springs will vary with clutch design.

The release levers are designed to draw the pressure plate away from the clutch disc. One end of the release lever engages the pressure plate. The other end is free and is designed to be pressed inward. Between the two ends, the lever is hinged to the clutch cover by means of an eyebolt.

There are generally three release levers. Some levers are of the semi-centrifugal type. This type has a weight added to its outer end. As the clutch assembly spins, centrifugal force will act upon these weights and cause them to exert additional pressure against the pressure plate. This assists the springs. Release lever location and action is shown in **Figure 18-7**.

The clutch cover bolts to the flywheel and acts as a base for the springs. The pressure plate release lugs pass through slots in the cover, allowing the cover to drive the pressure plate.

Coil Spring Pressure Plate Operation

When the throw-out bearing (discussed later) pushes against the clutch release levers, and sufficient pressure is applied to the fork, the release levers will draw the pressure plate back, releasing the clutch disc. As the throw-out bearing contacts the whirling release fingers, the bearing will rotate. The sleeve is held by the clutch throw-out fork. The action of the entire throw-out assembly is shown in **Figure 18-8**.

Diaphragm Spring Pressure Plate Assembly

Instead of coil springs, the diaphragm spring pressure plate uses a single diaphragm spring. General clutch construction is the same. The difference is in the pressure plate spring. The diaphragm spring is round and thin. It is made of high quality, heat-treated steel and is constructed with a dished profile to produce the necessary spring effect. A number of fingers radiate from the edges where they terminate in an open center.

One application cuts six of the fingers somewhat shorter to assist in cooling. The remaining twelve are left full length, **Figure 18-9**. Another version bends the six fingers upward. Weights are attached to produce a centrifugal effect to assist spring pressure at high rpm.

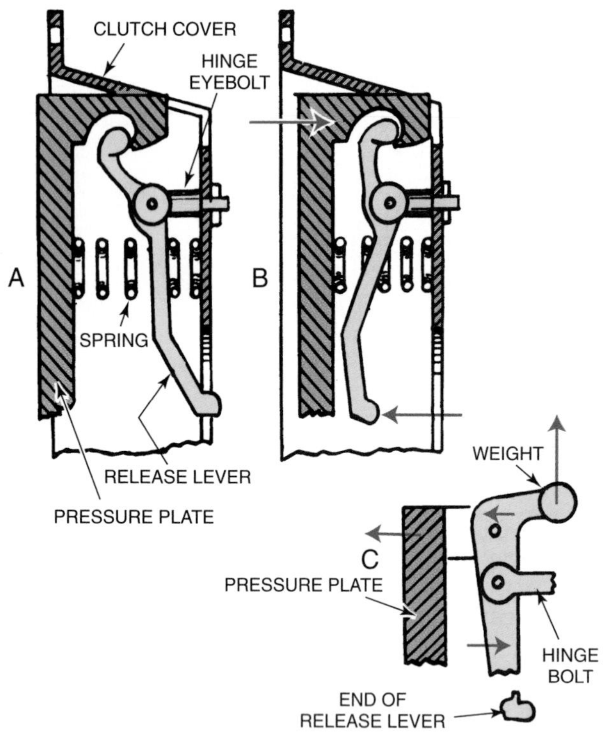

Figure 18-7. Release lever action. A—Here, the release lever is free and the pressure plate is engaged. B—When the release lever is pushed inward, it draws the pressure plate away from the flywheel. C—A weight is added to the centrifugal-type release lever. When the clutch revolves, the weight is thrown outward, adding to spring pressure.

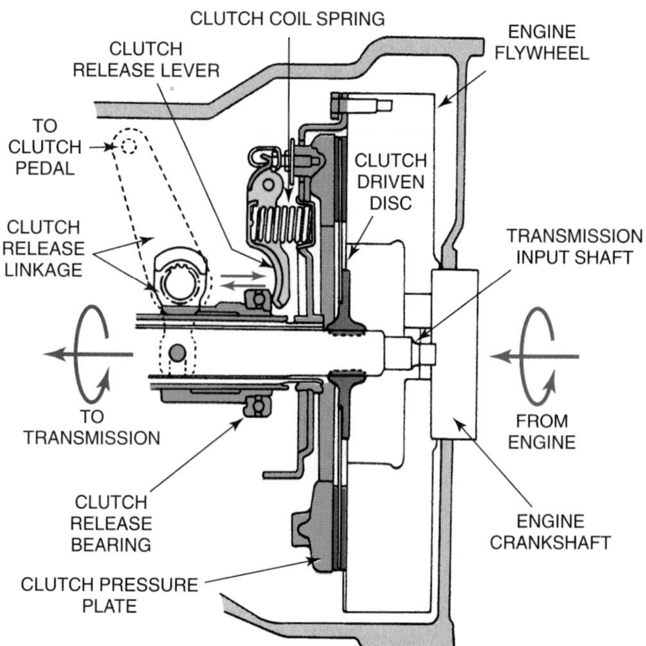

Figure 18-8. A cross-sectional view of a coil spring clutch assembly. As the clutch pedal is depressed, the throw-out bearing travels forward against the release levers. As the levers pivot, the pressure plate will pull backwards to collapse the clutch coil springs and release the pressure on the disc. The disc is now separated from the pressure plate and flywheel. This interrupts the power flow traveling from the engine to the transmission input shaft. As the clutch pedal is released, the collapsed coil springs force the pressure plate forward to apply pressure on the clutch disc, pinning it to the flywheel. Power flow to the transmission input shaft begins.

The outer edge of the diaphragm engages the pressure plate. The fingers point inward and dish out slightly. Two pivot rings are placed a short distance from the outer edge. The pivot rings are secured by means of a stud to the clutch cover. One pivot ring is on the outside of the diaphragm, the other is on the inside.

The pressure plate is driven by three double spring steel straps. The straps are riveted to the cover and bolted to the pressure plate. The throw-out bearing contacts the ends of the fingers. The fingers then act in place of release levers, **Figure 18-10**. An exploded view of a diaphragm clutch arrangement is illustrated in **Figure 18-11**.

Diaphragm Pressure Plate Operation

When the throw-out bearing applies pressure to the diaphragm fingers, the entire diaphragm is bent inward. The action can best be described as the same sort of effect as squeezing the bottom of an oil can. The inner pivot ring prevents the outer edge of the diaphragm from moving inward, so when the center of the diaphragm is pressed inward, the outer edges pivot outward. When the fingers are released, the diaphragm resumes its original position.

Figure 18-12 shows the position of the diaphragm when the clutch is engaged. Notice how the fingers point outward, causing the diaphragm to pivot on the outer pivot ring. This action presses the diaphragm outer edge tightly against the pressure plate. The clutch disc is held firmly.

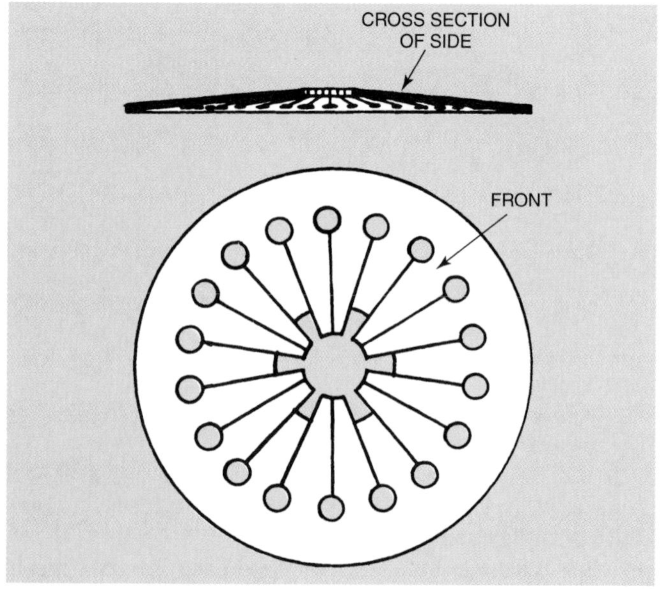

Figure 18-9. Diaphragm spring. Note the evenly spaced short fingers for cooling.

Now study **Figure 18-13**. The throw-out bearing has pressed the fingers in. The diaphragm now pivots on the inner pivot ring and the outer edges lift the pressure plate away from

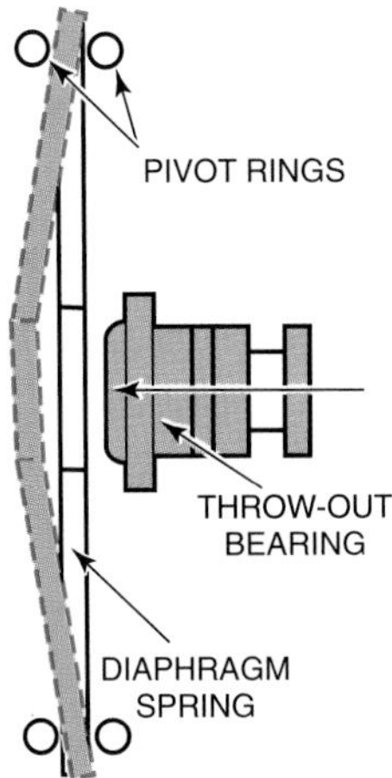

Figure 18-10. Diaphragm spring action. Dotted line shows the diaphragm in the disengaged position. Notice how the outer edge moves back and forth, depending on the position of the fingers. (Chevrolet)

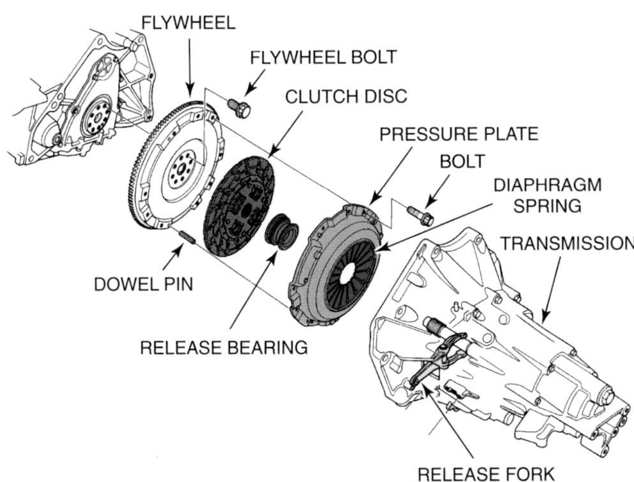

Figure 18-11. Exploded view of a diaphragm clutch assembly. Learn all the part names. (Honda)

the clutch disc. The disc is free and the clutch is disengaged. A complete diaphragm clutch assembly is shown in **Figure 18-14**.

Balance Marks

Most flywheel and pressure plate assemblies have ***balance marks***. At the factory, the flywheel and clutch pressure plate assembly are bolted together and balanced. After balancing, they are marked so that if the transmission or clutch assembly must be serviced, flywheel balance may be retained. The flywheel, clutch disc, pressure plate, release levers, springs, and cover are shown in their respective positions in **Figure 18-15**.

Clutch Release Lever Operating Mechanism

Another essential part of the clutch is the ***release lever operating mechanism***. This consists of a ball bearing normally referred to as a ***throw-out bearing***. This bearing is mounted on a sleeve or collar that slides back and forth on a hub that is an integral part of the transmission front bearing retainer.

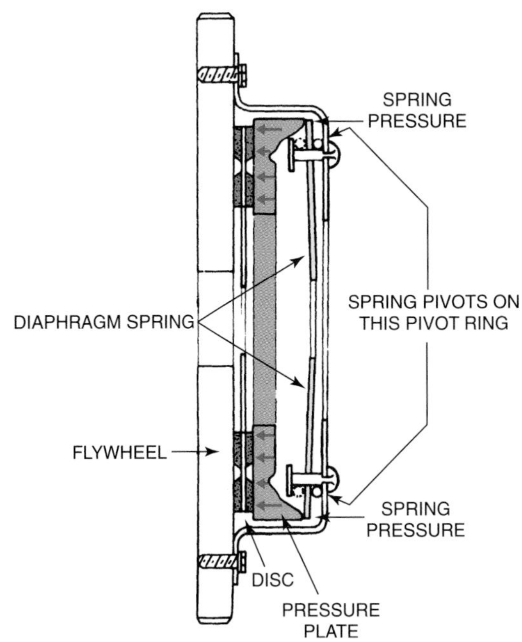

Figure 18-12. Diaphragm spring clutch in engaged position. The disc is pinched between the pressure plate and the flywheel. Notice the spring pressure produced by the diaphragm at the outer edges. (Chevrolet)

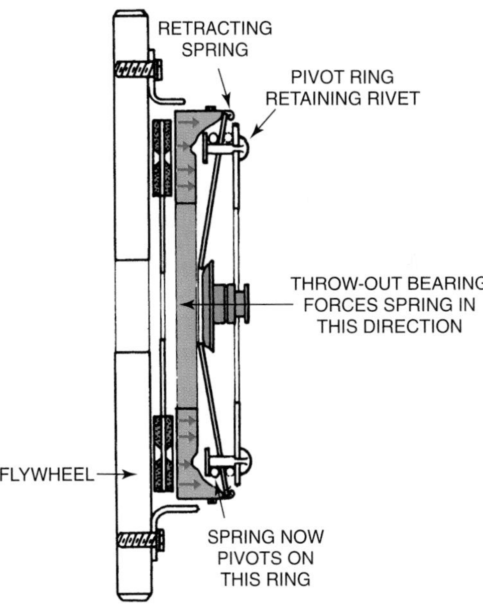

Figure 18-13. Diaphragm spring clutch in disengaged position. The throw-out bearing has moved the clutch diaphragm fingers inward. This pulls the outer edge back, retracting the pressure plate and leaving the clutch disc free. (Chevrolet)

The throw-out bearing is filled with grease at the factory and does not need service during its useful life. Another type of throw-out bearing that is used on many foreign vehicles is the graphite type. This throw-out bearing employs a ring of

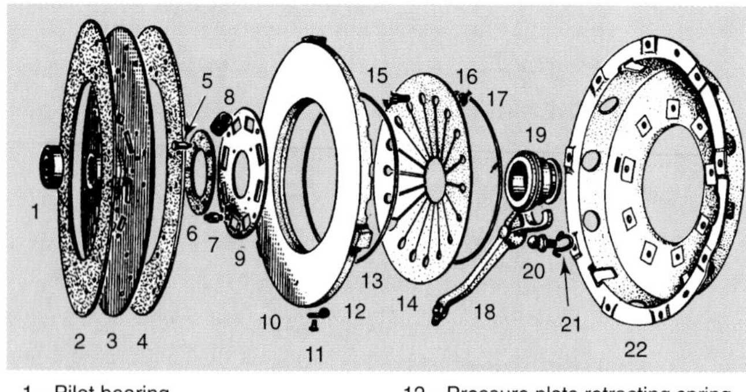

1—Pilot bearing.
2—Facing.
3—Disc.
4—Facing.
5—Dampener spring plate rivet.
6—Dampener spring outer plate.
7—Driven member stop pin.
8—Dampener spring.
9—Dampener spring plate.
10—Pressure plate.
11—Retracting spring bolt.
12—Pressure plate retracting spring.
13—Spring pivot ring.
14—Clutch spring.
15—Spring retainer bolt.
16—Spring retainer bolt nut.
17—Spring pivot ring.
18—Throw-out fork.
19—Release bearing and support assembly
20—Fork ball head stud.
21—Ball retainer
22—Cover

Figure 18-14. Exploded view—diaphragm spring clutch. This type of clutch uses the diaphragm fingers as release levers. (Luk, Inc.)

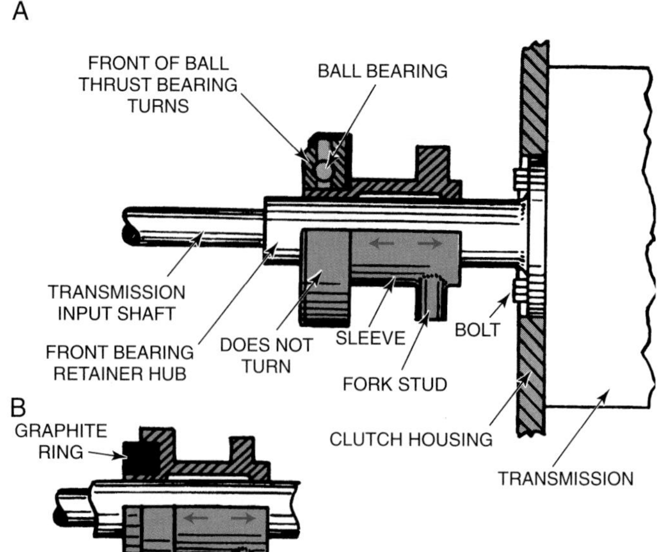

Figure 18-16. Clutch throw-out bearings. A—Ball bearing type. B—Graphite type.

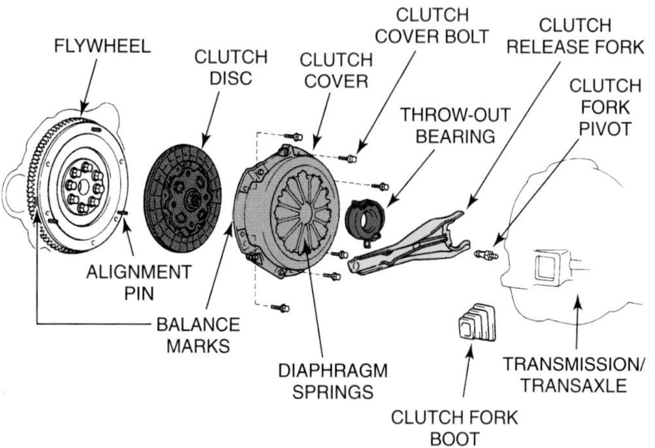

Figure 18-15. Various clutch parts in sequence. Note the balance marks and alignment pins which are used for reassembly. (Geo)

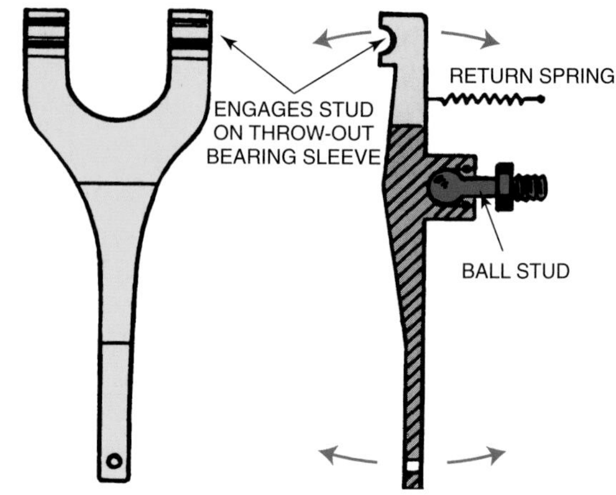

Figure 18-17. Clutch throw-out fork. The fork engages the throw-out bearing and pivots on a ball stud.

graphite to press against a smooth plate fastened to the clutch levers, **Figure 18-16**. The throw-out bearing sleeve is moved in and out by a ***throw-out fork*** or ***clutch fork***. This fork is usually pivoted on a ***ball head stud***. A ***return spring*** pulls the fork away from the pressure plate, **Figure 18-17**.

Clutch Housing and Input Shaft (Clutch Shaft)

The clutch housing bolts onto the engine and surrounds the flywheel and clutch mechanism. The housing is made of cast iron or aluminum. The bottom section of some housings are open, while other housings form a complete enclosure. Housings usually have openings to allow air circulation to cool the clutch and vent excess dust. The transmission can either bolt to the housing, or be an integral part of it.

The transmission input shaft passes through the clutch mechanism and its outboard (outer) end is supported by the clutch pilot bearing. The shaft is splined near the end, and the clutch disc rides on it at this point, **Figure 18-18**.

Complete Assembly

Figure 18-19 shows an exploded view of the entire clutch mechanism. Notice the use of the graphite throw-out bearing and the full enclosure clutch housing. The transmission input shaft is not shown. **Figure 18-20** shows a cross-section of a clutch fully assembled. The transmission input shaft is shown in place. Both pilot and throw-out bearings use an antifriction ball bearing.

Other Types of Clutches

There have been a great number of clutch types used over the years. The two shown in this chapter completely

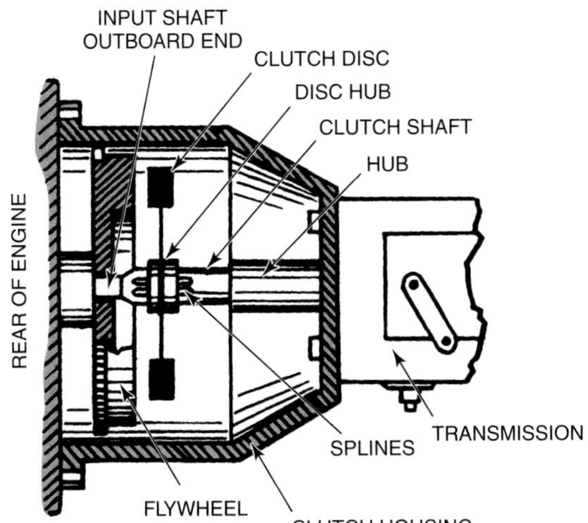

Figure 18-18. Clutch housing and input shaft, showing the flywheel, clutch disc, and hub. The remainder of the clutch mechanism is now shown. This housing forms the full enclosure. Notice how the outboard end of the input shaft is supported by the pilot bearing pressed into the flywheel.

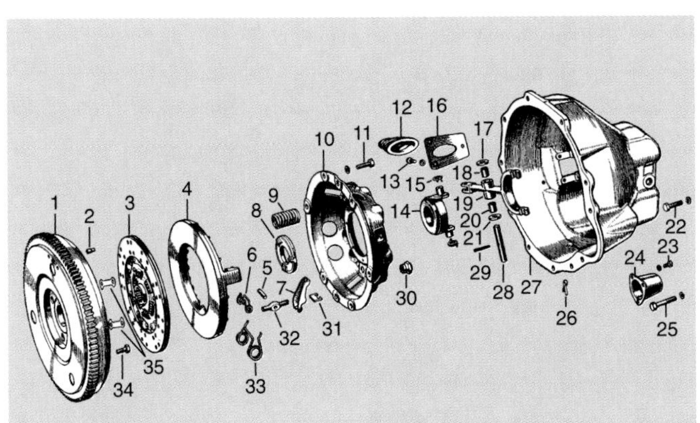

1—Flywheel.
2—Locating pin.
3—Clutch plate with lining.
4—Pressure plate.
5—Release lever pin.
6—Release lever retainer.
7—Release lever.
8—Release lever plate.
9—Pressure plate spring.
10—Clutch cover.
11—Cover bolt.
12—Fork and lever seal.
13—Retaining plate screw.
14—Release bearing.
15—Release bearing retainer spring.
16—Seal retaining plate.
17—Fork and lever thrust washer.
18—Fork and lever shaft bushing.
19—Clutch fork and lever.
20—Fork and lever shaft bushing.
21—Fork and lever thrust washer.
22—Clutch to gearbox bolt.
23—Starter cover screw.
24—Cover.
25—Clutch to gearbox bolt.
26—Split pin for drain hole.
27—Clutch housing.
28—Fork and lever shaft.
29—Taper pin.
30—Eyebolt nut.
31—Release lever strut.
32—Eyebolt.
33—Anti-rattle spring.
34—Flywheel to crankshaft bolt.
35—Lockwashers.

Figure 18-19. Exploded view—clutch assembly. Only one release lever is shown—there are actually three. The number of springs used will vary from five to seven in an average clutch.

dominate the automotive field today. Some heavy-duty clutches utilize extra springs and two clutch discs with a second pressure plate that is smooth on both sides sandwiched between them.

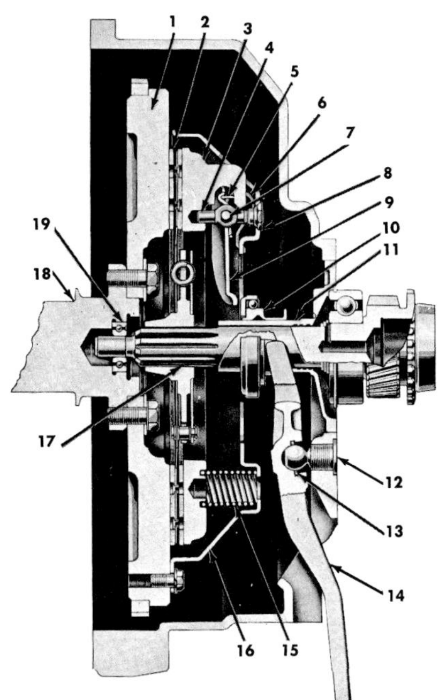

1—Engine flywheel.
2—Drive disc.
3—Pressure plate.
4—Eye bolt.
5—Strut.
6—Anti-rattle spring.
7—Release lever pin.
8—Adjusting nut.
9—Release lever.
10—Release bearing and support assembly.
11—Trans. drive gear bearing retainer.
12—Release fork ball stud.
13—Retainer.
14—Clutch release fork.
15—Pressure spring.
16—Cover.
17—Input shaft.
18—Engine crankshaft.
19—Pilot bearing.

Figure 18-20. Cross-section of an assembled clutch. Study and learn the names of the clutch parts. (GMC)

Clutch Operation

Remember that the flywheel, clutch cover, release levers, and pressure plate all revolve as a single unit. The transmission input shaft and the clutch disc are splined together, forming another unit.

The only time the clutch disc will turn (vehicle standing still) is when it is pinched between the flywheel and pressure plate. When the release levers draw the pressure plate away from the flywheel, the clutch disc will stand still while everything else continues to revolve. When the pressure plate travels back toward the flywheel, the disc is seized and forced to turn the transmission input shaft, **Figure 18-21**.

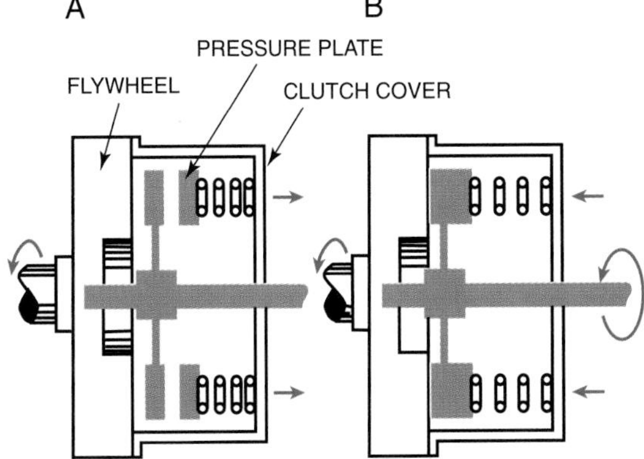

Figure 18-21. Clutch operation simplified. A—The flywheel, pressure plate, springs, and clutch cover are revolving. The clutch disc and clutch shaft are stopped as the pressure plate is disengaged. B—When the pressure plate is engaged, it seizes the clutch disc and the entire unit revolves.

Clutch Pedal Linkage

A way must be provided to connect the clutch fork to the clutch pedal. This is accomplished by the ***clutch linkage***. There are two principal kinds of clutch linkage used to actuate the clutch throw-out fork, the ***mechanical linkage*** and the ***hydraulic linkage***.

Mechanical Linkage

The mechanical linkage system incorporates the use of levers and rods or cables connected between the clutch pedal and throw-out fork. When the pedal is depressed, the force is transmitted to the fork by the mechanical linkage. **Figure 18-22** illustrates a simple rod and lever hookup. **Figure 18-23** shows a mechanical linkage using a cable.

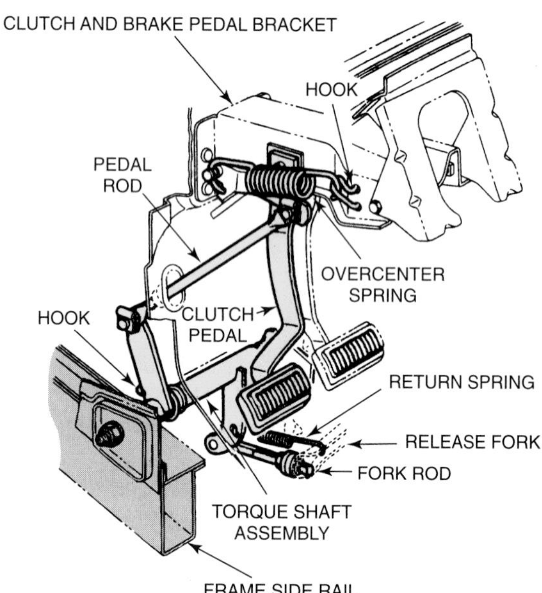

Figure 18-22. One mechanical clutch linkage arrangement. When the clutch pedal is depressed, the clutch throw-out fork is actuated. (Plymouth)

Hydraulic Linkage

The other system uses hydraulic power to control the fork. On many vehicles, the mechanical linkage required to operate the fork would be extremely complicated. The hydraulic control simplifies the job. It also reduces the amount of clutch pedal effort by the driver.

When the clutch is depressed, it actuates a small ***clutch master cylinder***, **Figure 18-24**. The pressure created in the master cylinder is transmitted to a ***slave cylinder*** bolted near the fork. The slave cylinder is connected to the fork with a short adjustable rod. When pressure is applied to the slave cylinder, it operates the fork. An exploded view of one type of hydraulically-operated clutch setup is shown in **Figure 18-25**. Both master and slave cylinders are of simple design. Regardless of position or obstructions, they are connected with hydraulic tubing. The principles of hydraulics will be discussed in more detail in Chapter 23.

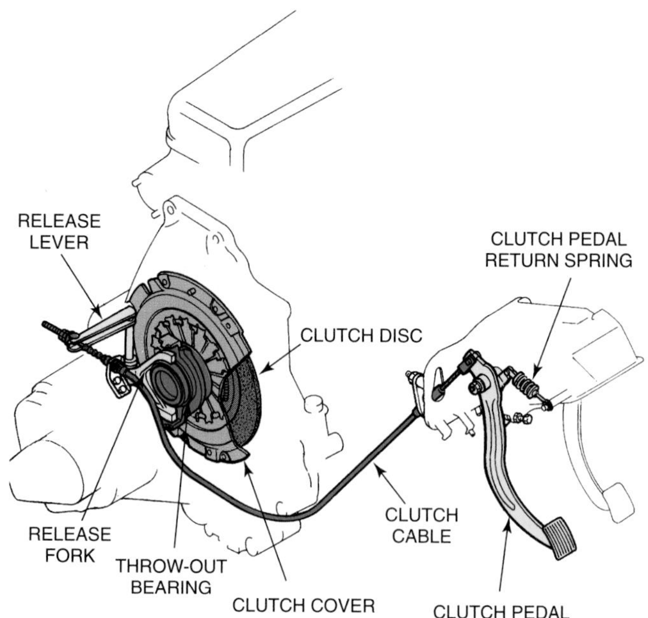

Figure 18-23. One type of mechanically operated clutch release system using a clutch cable to actuate the clutch release lever and fork. (Mazda)

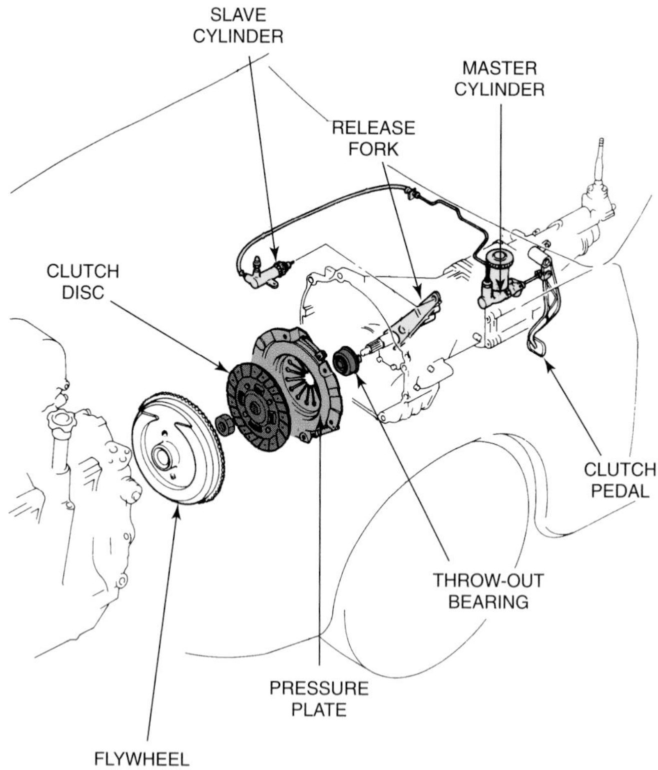

Figure 18-24. Hydraulic clutch control. Clutch pedal movement builds pressure in the master cylinder that in turn actuates the slave cylinder. (Mazda)

Another type of hydraulic clutch is shown in **Figure 18-26**. In this type, the throw-out bearing and slave cylinder are combined into a single unit. When the slave cylinder is actuated, it pushes the throw-out bearing into contact with the

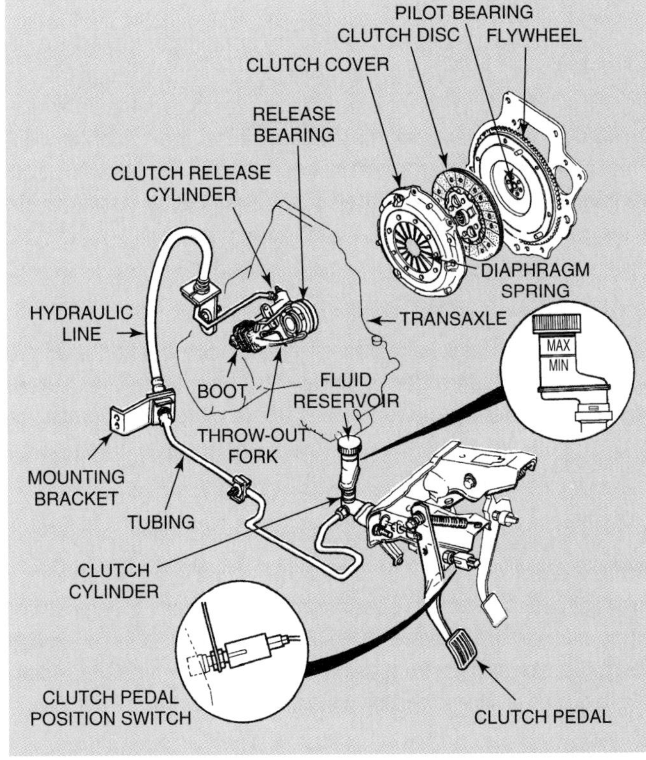

Figure 18-25. An exploded view of a clutch assembly which is used with a manual transaxle. Note the use of a hydraulic system to actuate the throw-out lever. (Kia Motors)

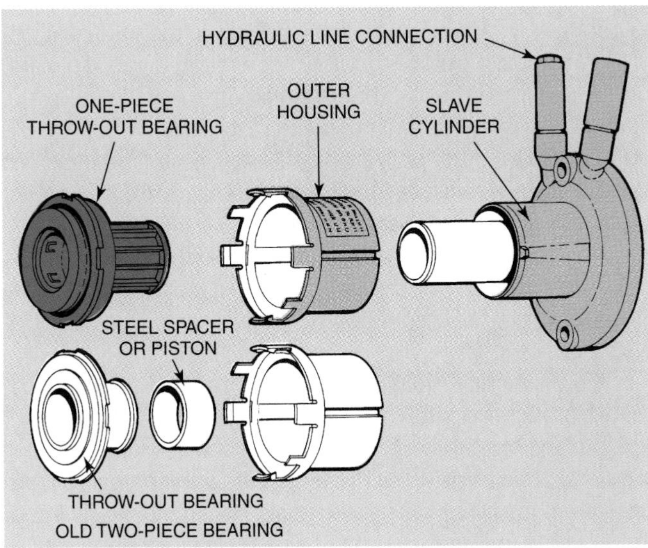

Figure 18-26. Exploded view of one style of hydraulic throw-out bearing and slave cylinder unit. Note the older style two-piece bearing assembly. (Luk Inc.)

release fingers of the pressure plate. This design eliminates the need for a clutch fork and related linkage.

Self-Adjusting Clutch

Most modern clutch assemblies have no adjustments and, with the exception of checking the hydraulic fluid level in the master cylinder, are maintenance free. Some older vehicles use a ***self-adjusting clutch linkage***. The self-adjusting mechanism is a mechanical arrangement that is attached to the clutch pedal and mounting bracket, **Figure 18-27**. The cable that is used with this clutch cannot be made shorter or longer by cutting.

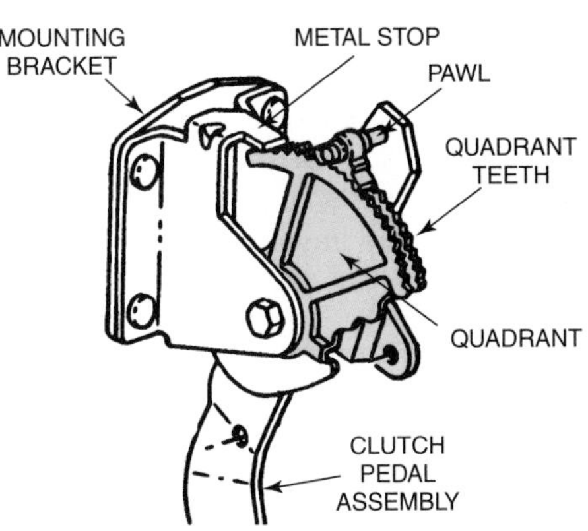

Figure 18-27. Self-adjusting clutch mechanism. Note the teeth on the quadrant and pawl. (Oldsmobile)

The position of the clutch cable may be altered by adjusting the position of the quadrant (toothed-type ratchet) in relation to the clutch pedal. This unit keeps proper cable tension and makes adjustments as the clutch lining wears down. The pawl is shown in engagement with the quadrant teeth in **Figure 18-28**.

When the clutch pedal is in the normal driving position (not depressed), the pawl is held away from the quadrant with a metal stop located on the bracket. In this position, the quadrant hub spring works against the spring force being applied

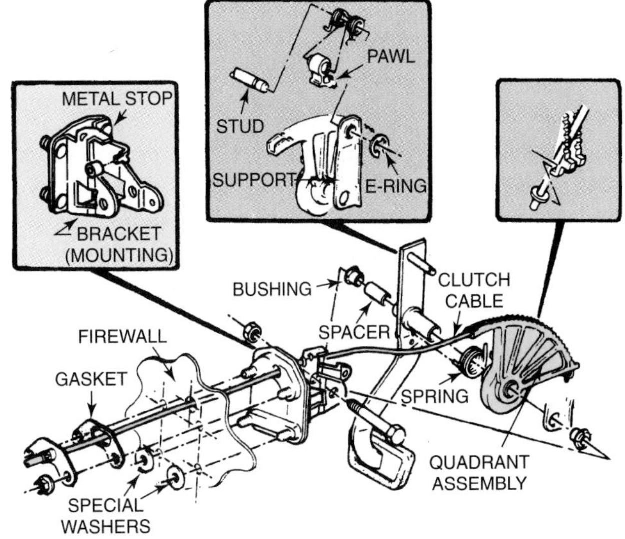

Figure 18-28. Exploded view of a self-adjusting clutch setup. Note the clutch cable and how it is attached to the quadrant. (Oldsmobile)

at the throw-out (release) bearing. The cable is now in a "balanced" state, with correct tension being applied.

As the clutch pedal is depressed, the pawl will travel away from the metal stop. At this point, the pawl spring will push the pawl down and into engagement with the quadrant teeth. Continued depression of the pedal pulls on the cable, which in turn, forces the throw-out bearing against the fingers or levers in the pressure plate, releasing the clutch.

As the pedal is released, the quadrant returns to its resting position and the pawl is disengaged from the quadrant by the metal stop. The cable is once again in a balanced state of even tension. As the lining wears off the clutch disc, the pawl will engage a different tooth on the quadrant, providing proper compensation for the extra travel.

Clutch Pedal Free Travel

The ***free travel***, or ***free play*** of the clutch pedal is most important. When the clutch pedal is released, the throw-out bearing must not touch the release levers. It must clear for two reasons. As long as it touches, it will continue to rotate. This will shorten its useful life. Also, if the throw-out bearing is not fully released, it may bear against the release levers hard enough to partially disengage the clutch. Any removal of pressure plate force will cause the clutch to slip. When slippage occurs, the clutch facing will overheat and burn up.

All manufacturers specify the amount of free travel that must be allowed. Free travel, or pedal travel, means how far the clutch pedal can be depressed before the throw-out bearing strikes the release levers. The distance the pedal moves from the full out position until it becomes hard to push is the free travel, **Figure 18-29**.

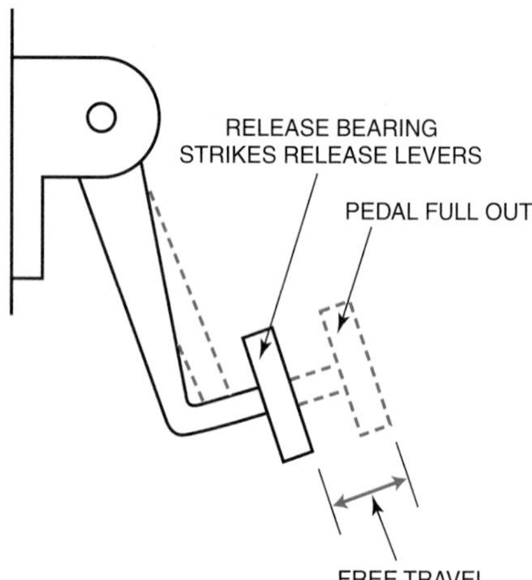

Figure 18-29. Clutch pedal free travel. Always refer to the manufacturer's recommendations for the proper pedal height.

Clutch Cooling

By their very nature, clutches generate considerable heat. You will notice that all clutch covers have openings and that the clutch housings are also provided with openings to facilitate air circulation.

Riding the Clutch

Some drivers develop the habit of riding the clutch (resting their foot on the clutch pedal). The natural weight of the leg will cause the throw-out bearing to contact the release levers. Any additional pressure may even cause the clutch to slip. This can cause premature clutch failure. The clutch pedal should only be touched when the transmission needs to be shifted or disengaged from the engine.

Will a Clutch Explode?

Clutches used in racing and other competition engines can, and sometimes do, explode. The high RPM and horsepower sets up tremendous forces. If the clutch fails, parts are thrown violently outward, which can shatter the clutch housing. The flying clutch parts and metal can endanger the driver and anyone standing nearby.

In any vehicle used for competition, it is imperative that a ***scatter shield*** be used to protect the driver and others in the event of clutch failure. A scatter shield is a heavy steel plate or fabric blanket that covers the clutch housing. It should extend down the sides to prevent parts from striking anyone standing nearby. It must be bolted or welded securely into place.

Summary

A clutch is necessary to assist in smooth starting, shifting gears, and allowing the vehicle to be stopped without putting the transmission in neutral. The clutch is designed to connect or disconnect the transmission of power from one working part to another—in this case, from the engine to the transmission.

Automotive clutches in use today are of the single plate, dry disc type. The flywheel, clutch disc, pressure plate, springs, clutch cover, and operating mechanism linkage make up the primary parts of the typical clutch. One type uses coil springs as a means of loading the pressure plate. Another type uses a diaphragm spring.

In the operating position (engaged), the clutch disc is held firmly between the flywheel and pressure plate. When the flywheel turns, the disc turns the transmission input shaft to which it is splined.

To disengage the clutch, the clutch pedal is depressed. Either through mechanical or hydraulic linkage, the throw-out fork pushes the throw-out bearing into contact with the pressure plate release levers. The levers draw the pressure plate away from the clutch disc to release it. When the disc is freed, the flywheel and pressure plate assembly continue to rotate, even though the disc and input shaft are not moving.

There are two principal types of throw-out bearings, the ball bearing type and the graphite type. Most vehicles use the ball bearing type. The clutch disc is designed to assist in smooth engagement by torsion springs in the hub assembly as well as cupped segments to which the friction facing is riveted.

Clutch free pedal should be adjusted according to the manufacturer's specifications. Some older clutches are self-adjusting. Others must be manually (by hand) adjusted as the lining wears down. Never ride the clutch.

Know These Terms

Clutch
Flywheel
Pilot bearing
Clutch disc
Splines
Pressure plate
Coil spring pressure plate
Diaphragm spring pressure plate
Pressure plate cover
Balance marks
Release lever
Release lever operating mechanism
Throw-out bearing
Throw-out fork
Clutch fork
Ball head stud
Return spring
Clutch linkage
Mechanical linkage
Hydraulic linkage
Clutch master cylinder
Slave cylinder
Self-adjusting clutch linkage
Free travel
Free play
Scatter shield

Review Questions—Chapter 18

Do not write in this book. Write your answers on a separate sheet of paper.

1. The clutch is used to disengage the engine and transmission when _____ .
   (A) the driver is shifting gears
   (B) the vehicle is stopped and the engine is running
   (C) the vehicle is being reversed
   (D) Both A & B.
2. Why is the flywheel thick?
3. Technician A says that the pilot bearing may be either a ball bearing or a bronze bushing. Technician B says that the pilot bearing does not require lubrication. Who is right?
   (A) A only.
   (B) B only.
   (C) Both A & B.
   (D) Neither A nor B.
4. The clutch disc hub splines engage splines on the _____.
   (A) pressure plate assembly
   (B) transmission input shaft
   (C) flywheel
   (D) rear of the crankshaft
5. What clutch part actually locks the clutch disc to the flywheel and pressure plate assembly?
   (A) Flywheel.
   (B) Clutch disc.
   (C) Pressure plate assembly.
   (D) Throw-out bearing.
6. What clutch part releases the clutch?
   (A) Flywheel.
   (B) Clutch disc.
   (C) Pressure plate assembly.
   (D) Throw-out bearing.
7. What are the two main classes of pressure plate springs?
8. The pressure plate cover is bolted to the _____ .
9. The _____ operates the throw-out bearing.
10. If the clutch pedal has no free travel, which of the following is least likely to occur?
    (A) The throw-out bearing will last longer.
    (B) The clutch will be partially disengaged.
    (C) The throw-out bearing will touch the release levers.
    (D) The clutch will slip and overheat.
11. Even though most competition engines use heavy-duty "beefed up" clutches, what additional unit must be installed to safeguard the driver and spectators?
12. Technician A says that clutches have aligning punch marks so that the pressure plate can be reinstalled in the same position as it was originally. Technician B says that aligning punch marks are used to preserve clutch assembly balance. Who is right?
    (A) A only.
    (B) B only.
    (C) Both A & B.
    (D) Neither A nor B.
13. The outboard end of the transmission input shaft is supported by the _____ .
    (A) clutch disc
    (B) clutch housing
    (C) pilot bearing
    (D) Both B & C.
14. Why does the clutch housing have openings?
    (A) To allow air circulation.
    (B) To allow oil to drip out.
    (C) To allow dust to fly out.
    (D) Both A & C.
15. Define the term "riding the clutch."

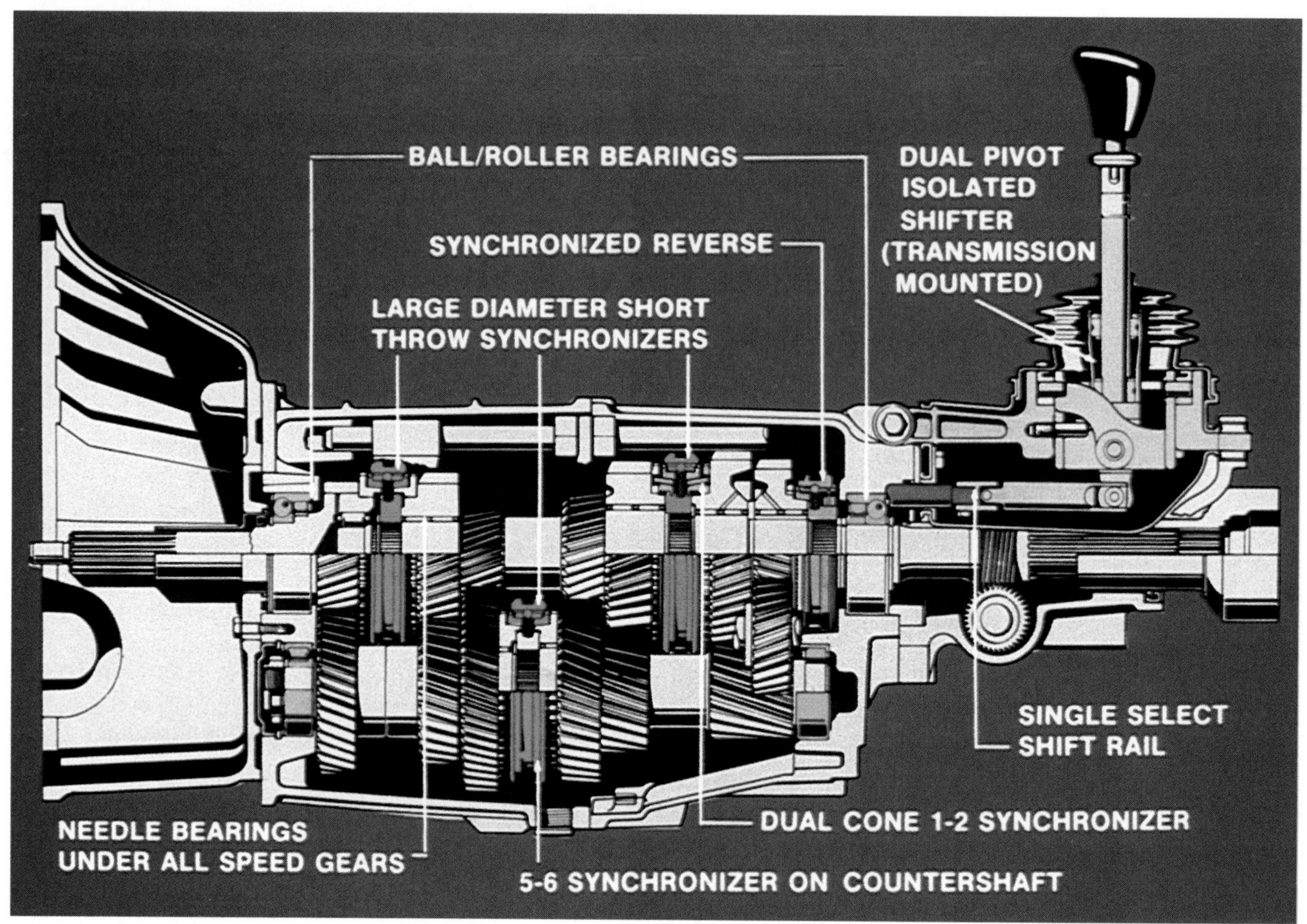

Cutaway view of a six-speed manual transmission. A unique feature of this transmission is that it has two overdrive gears. (General Motors)

19

Manual Transmissions and Transaxles

After studying this chapter, you will be able to:

- Explain how a manual transmission operates.
- Identify the basic parts of a manual transmission.
- Trace power flow through transmission gears.
- Identify the differences between a manual transmission and manual transaxle.
- Identify the similarities between a manual transmission and manual transaxle.
- Identify the basic parts of a manual transaxle.
- Trace power flow through transaxle gears.

This chapter covers the purpose, construction, and operating principles of manual transmissions and transaxles. Manual transmissions and transaxles are operated by the driver in conjunction with the manual clutch, discussed in Chapter 18. The basic operating principles and designs of all manual transmissions and transaxles have some similarities. Studying this chapter will familiarize you with the basic operating principles and component parts of manual transmissions and transaxles.

Manual Transmissions and Transaxles

Before beginning to discuss the principles of ***manual transmissions*** and ***manual transaxles***, it would be helpful to state the differences between the two. Transmissions are always used on vehicles with front engines and rear-wheel drive. Transaxles are used on vehicles with front engines and front-wheel drive. Some transaxles are used on vehicles with rear engines and rear-wheel drive. Although they look very different, the purposes and layout of all manual transmissions and transaxles is similar.

The Engine Needs Help

An internal combustion engine would have enough power to operate a vehicle without a transmission if the vehicle was operated on level roads and maintained sufficient speed. When the vehicle must be started from a standstill, or when attempting to negotiate steep grades, the engine would not provide sufficient power and the vehicle would stall.

Much less torque is required to drive a vehicle rolling on level ground than up a steep hill, **Figure 19-1**. To enable the engine to increase torque to the drive train, it is obvious that something must be used to multiply torque, **Figure 19-2**. The simplest torque multiplier is a manual transmission. By using a transmission as a torque multiplier, it is possible to adapt the available power of the engine to meet changing road and load conditions.

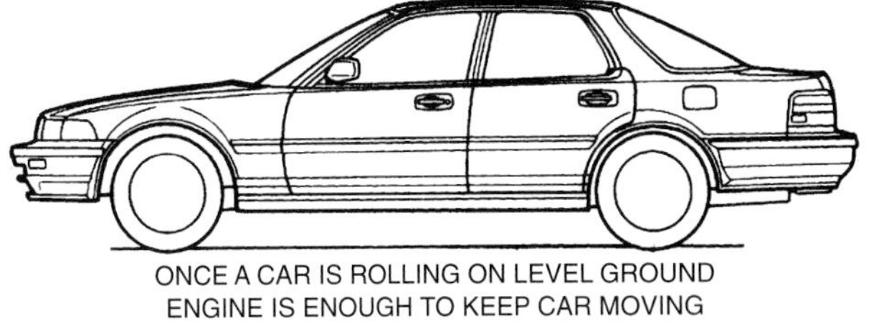

ONCE A CAR IS ROLLING ON LEVEL GROUND
ENGINE IS ENOUGH TO KEEP CAR MOVING

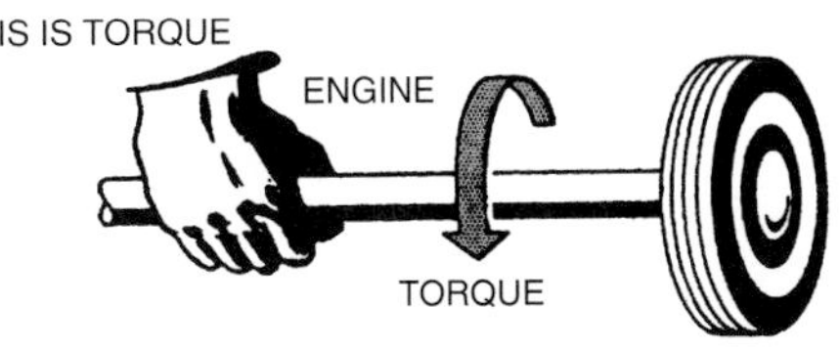

Figure 19-1. A minimal amount of torque is required to move a vehicle on level ground. Torque is the twisting force on an axle. (Chevrolet and Honda)

How Is Torque Multiplied?

If you (an engine) exerted all your strength (power) and still failed to lift a certain object (drive a vehicle), you could multiply your strength by using a ***lever and fulcrum***, **Figure 19-3**.

If you weigh 200 lbs. (90.72 kg) and get on the lever 5 ft. (1.52 m) from the fulcrum, a force of 1000 ft.-lbs. (1350 N•m) will be exerted. Then if a weight of 500 lbs. (226.8 kg) is placed on the other end of the lever 1 ft. (30.48 cm) from the fulcrum, it would exert a force of 500 ft.-lbs. (675 N•m). By using a simple torque converter or multiplier, you can now raise 500 lbs. (226.8 kg) with ease. How much weight could you raise by standing on the lever at point A in **Figure 19-4**?

To give another example, if you weigh 200 lbs. (90.72 kg) and want to lift 4000 lbs. (1800 kg) you would have to stand 20 ft. (6.09 m) from the fulcrum. In order to raise the weight a short distance, you must depress the end of the lever a distance 20 times as much as the distance the weight has been raised. See **Figure 19-5**.

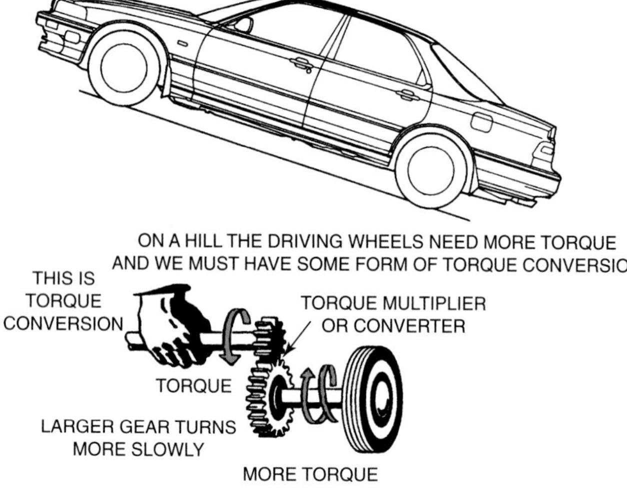

Figure 19-2. Additional torque is required when a vehicle is on a steep grade. The original torque from the engine is not sufficient, so it is necessary to increase torque by means of a torque converter or multiplier. Torque is multiplied through the use of transmission gears. (Chevrolet and Honda)

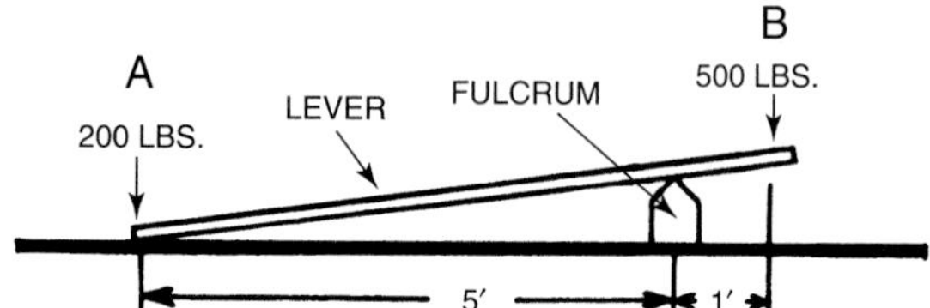

Figure 19-3. Lever and fulcrum. A weight of 200 lbs. at A will raise weight of 500 lbs. at B because the original weight is multiplied by using a lever and fulcrum.

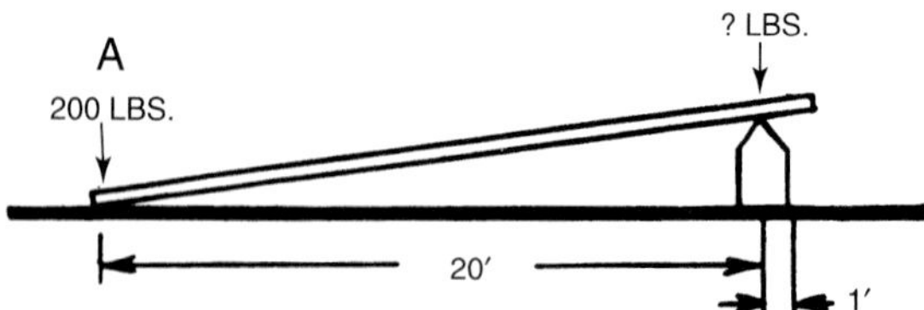

Figure 19-4. Problem: If you are standing on the lever at A, how much weight will you be able to lift?

You will note that as your (engine's) strength (torque) is increased, the speed with which the weight (vehicle) is moved becomes slower. This is true, as you would find it necessary to depress the lever a greater distance as you moved farther from the fulcrum. From a study of this simple lever and fulcrum, it is possible to make these deductions:

- If the speed of an engine remains constant and its torque is multiplied, the engine will lift more weight, but will take longer to do it.
- If the speed of an engine remains constant and its torque is reduced, it will lift less weight, but will lift it faster.

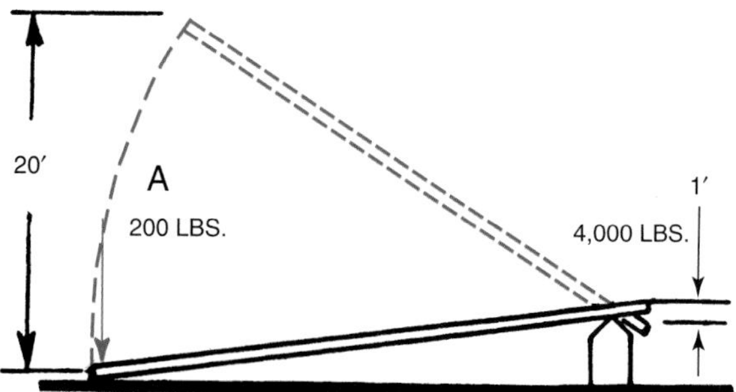

Figure 19-5. If the speed with which you depress the lever remains constant, the more you multiply your strength and the slower the weight will rise. To raise 4000 pounds, it is necessary to depress the level 20′ to raise the weight 1′.

Therefore, as engine torque multiplication increases, the vehicle road speed will decrease and as torque multiplication decreases, the vehicle road speed will increase, with the engine RPM remaining constant.

Torque Multiplication through Gears

The same principles used in the lever and fulcrum example can be applied to automotive transmission gears. Consider the fulcrum as the vehicle drive shaft, the amount of weight to be lifted as the torque required to turn the shaft, the engine torque as the person's weight, and the body of the gear as the lever. Then it is easy to see how the gear multiples the torque. **Figure 19-6** shows this principle.

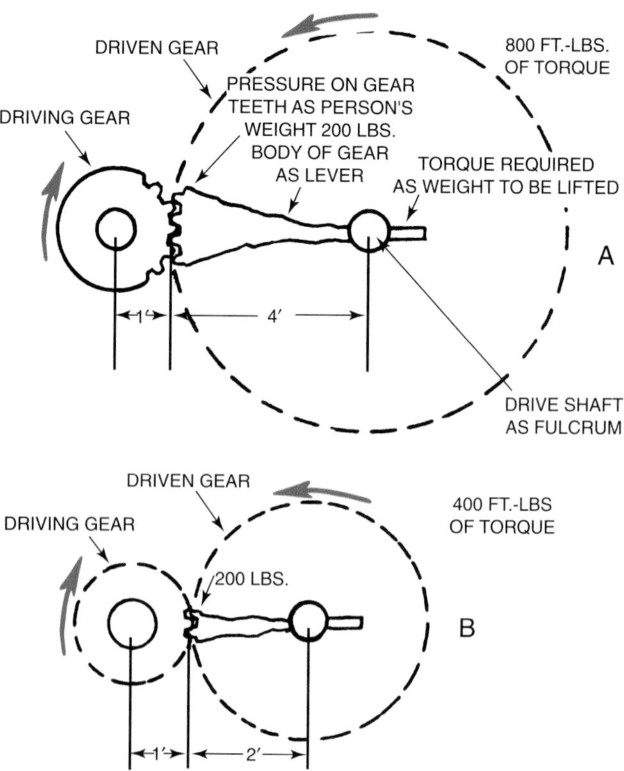

Figure 19-6. Principle of gear torque multiplication. The larger the driven gear is in relation to the driving gear, the greater the torque multiplication. As torque is increased, the driving gear must turn more times to turn the driven gear once.

You can see that as the size of the gear is increased, the same amount of pressure on a tooth will produce an increase in torque. In **Figure 19-6A**, the driving gear tooth is 1 ft. (30.48 cm) from the center of the shaft. If it were pushing down with a 200 lb. (90.72 kg) force, the driving tooth would be exerting a 200 ft.-lb. (270 N•m) push on the tooth of the driven gear.

As the driven tooth is 4 ft. (1.22 m) from the center of the drive shaft, it would exert an 800 ft.-lb. (1080 N•m) force on the drive shaft. In **Figure 19-6B**, the force of the driving gear would be the same, but as the driven tooth is now 2 ft. (60.96 cm) from the center, it would exert a force of only 400 ft.-lbs. (540 N•m) on the drive shaft.

Transmission Gear Ratios

The modern manual transmission can provide the driver with up to six forward ***gear ratios*** (difference in the number of teeth on each gear). The ***reduction gears*** provide gear ratios of approximately 3.5:1 (stated as 3.5 to 1) for the lowest gear, to about 1.5:1 for the highest. The ***direct drive gear*** has a 1:1 gear ratio. The ***overdrive gears*** have a gear ratio of about .7:1.

By selecting one of the ratios, it is possible to operate the vehicle under all normal conditions. In addition, torque is multiplied more through the differential gears. This will be discussed in Chapter 22. Ratios vary from vehicle to vehicle, depending on engine horsepower and vehicle weight. A ***reverse gear*** is also used. The reverse ratio is usually about 3:1.

Typical gear ratios for manual transmissions and transaxles are shown in **Figure 19-7**. Note that the five-speed transmission has an overdrive ratio. The overdrive ratio is incorporated into newer transmissions to increase fuel mileage and lower emissions.

Transmission Gear Construction

Transmission gears are made of high quality steel, carefully heat-treated to produce smooth, hard surface gear teeth with a softer, but very tough interior. They are drop-forged (machine hammered into shape) while red hot. The teeth and other critical areas are cut on precision machinery.

TYPICAL MANUAL TRANSMISSION/TRANSAXLE GEAR RATIOS		
THREE-SPEED	FOUR-SPEED	FIVE-SPEED
1ST GEAR, 3:1	1ST GEAR, 3.5:1	1ST GEAR, 3.2:1
2ND GEAR, 2:1	2RD GEAR, 2:1	2ND GEAR, 2:1
3RD GEAR, 1:1	3RD GEAR, 1.5:1	3RD GEAR, 1.4:1
REVERSE, 2.5:1	4TH GEAR, 1.00:1	4TH GEAR, 1:1 (DIRECT DRIVE)
	REVERSE, 3:1	5TH GEAR, 0.853:1 (OVERDRIVE)
		REVERSE, 3:1

Figure 19-7. Average transmission gear ratios for three-, four-, and five-speed transmissions. Actual transmission gear ratios will be different from these.

The teeth on transmission gears are cut into ***spur*** and ***helical*** patterns. The helical gear is superior in that it runs more quietly and is stronger because more tooth area is in contact. Helical gears must be mounted firmly, since there is a tendency for them to slide apart due to the spiral shape, **Figure 19-8**. ***Gear end play*** in the cluster gear and some other gears and shafts is controlled by the use of bronze and steel ***thrust washers***. These washers are installed on the end of the gear shafts between the moving gears and the stationary transmission case, **Figure 19-9**.

There must be some clearance between the gear teeth to allow for lubrication, expansion and possible size irregularity. This clearance is very small (a few thousandths of an inch). **Figure 19-10** illustrates backlash (clearance between face and flank areas) and also gives gear tooth nomenclature.

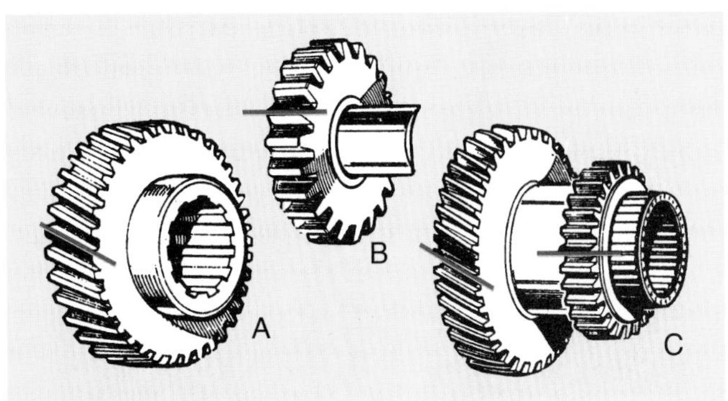

Figure 19-8. Gear tooth shapes. A—The helical tooth is cut at an angle to the centerline of the gear. B—The spur gear is cut straight across (parallel to gear centerline). C—The left side gear is helical gear, the right side is a spur gear. (Land-Rover)

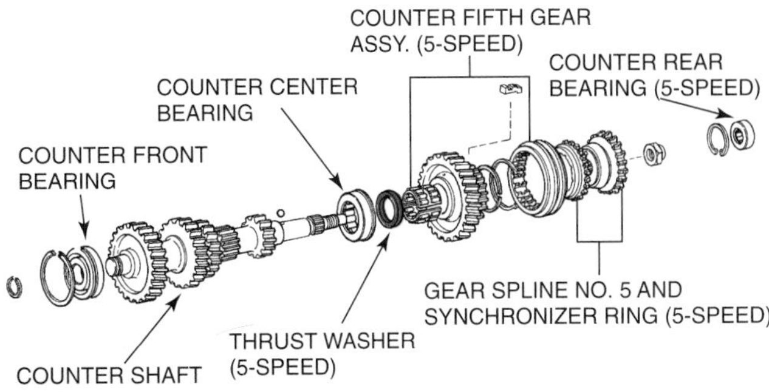

Figure 19-9. A thrust washer location on one Five-speed manual transmission. The locations and numbers used will vary. (Toyota Motor Corp.)

Transmission Cases

The ***transmission case*** holds the transmission gears, shafts, bearings, and washers. It bolts to the rear of the engine, or to the clutch housing. Many transmission cases and clutch housings are one-piece units. Most transmission cases are made of cast iron or aluminum and have a separate

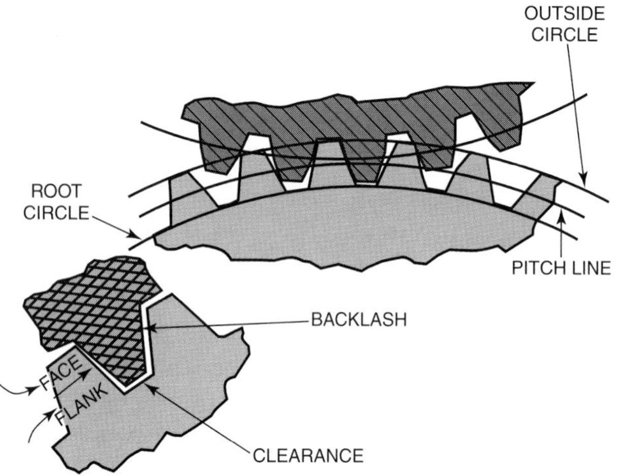

Figure 19-10. Elementary gear tooth nomenclature. Pitch line or pitch diameter is the true diameter used in calculating exact gear ratios. The outside circle is the full diameter of the gear. The root circle is the minimum diameter of the gear.

extension housing, which supports the output shaft. This housing may also contain the speedometer gear. The rear engine mount is usually attached to the extension housing.

Transmission Construction

The gears in all manual transmissions and transaxles are ***sliding gear*** types, which are moved in and out of engagement by the driver through ***shift linkage***. The four main classes of transmission/transaxle gears are reduction, direct drive, overdrive, and reverse, **Figure 19-11**. The following sections explain the workings of various manual transmissions and transaxles.

A typical transmission consists of a housing, four shafts, bearings, gears, synchronizers, and a shifting mechanism. **Figure 19-12** shows the four shafts and bearings in their relative positions in the housing. The shafts are held in place by pins pressed through the case and shaft or by the presence of external housings over the shaft openings.

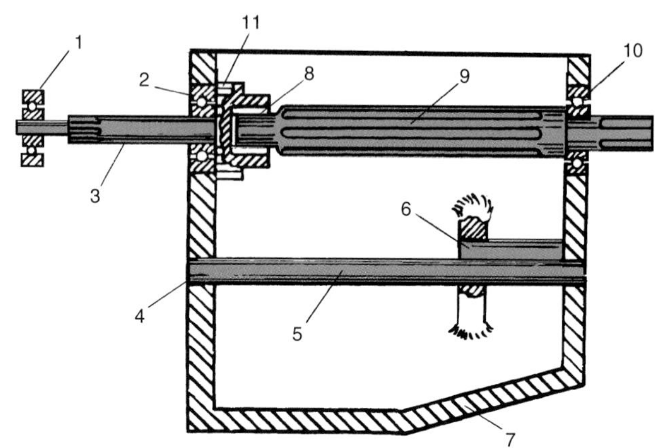

Figure 19-12. Housing shafts and bearings. The input and output shafts turn while the countershaft and reverse idler shaft remain fixed. 1—Clutch pilot bearing. 2—Input shaft bearing. 3—Input shaft. 4—Countershaft supported in housing. 5—Countershaft. 6—Reverse idler shaft supported in housing. 7—Housing. 8—Output shaft roller bearing. 9—Output shaft. 10—shaft ball bearing. 11—Input shaft gear teeth.

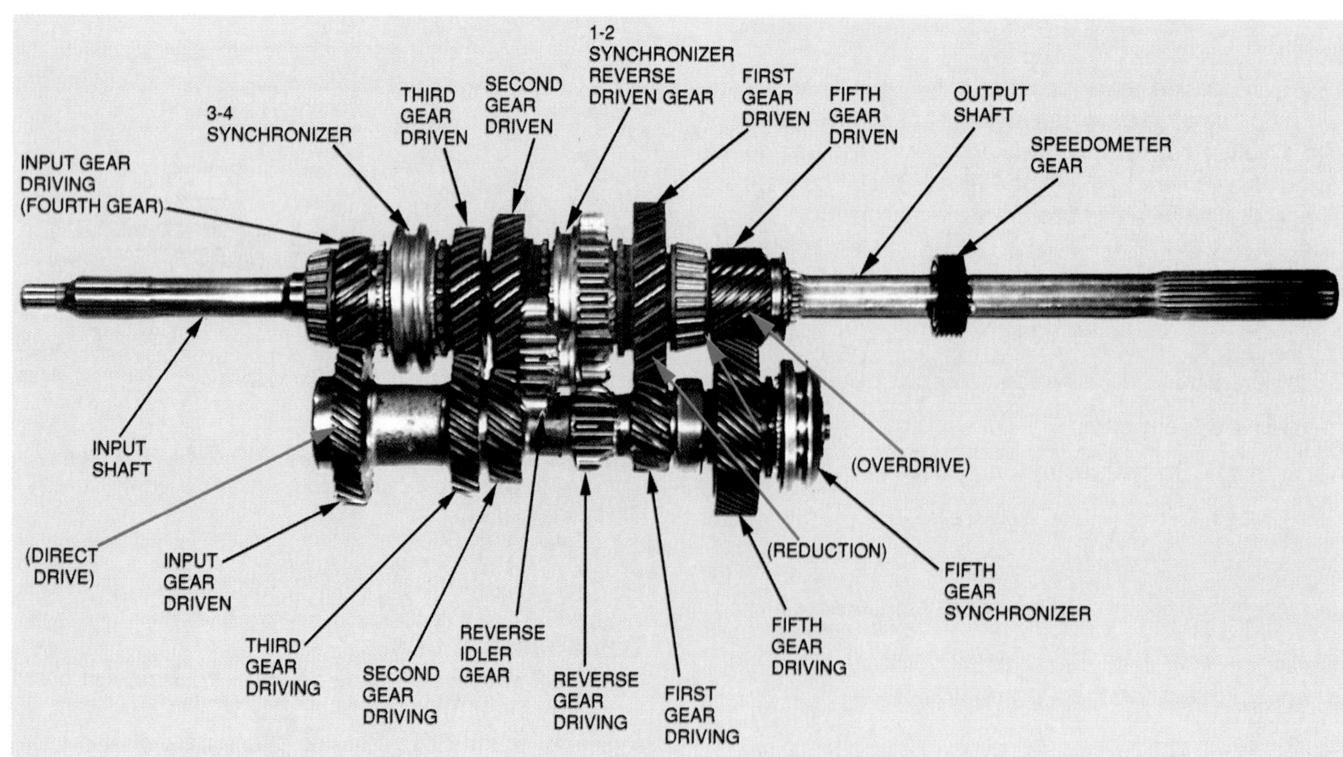

Figure 19-11. Reduction, direct drive, overdrive, and reverse gears as mounted on the shaft. (Ford Motor Co.)

Synchronizing Mechanism

Once the vehicle is in motion, the drive line will turn the output shaft continuously. As a result, the sliding gears will also be turning. When an attempt is made to mesh them with any of the cluster gears (which tend to stop when the clutch is depressed), the gear teeth will be subjected to damaging impact forces. The sound of ***gear clash*** when shifting results from the sliding gear teeth literally smashing against the cluster gear.

It is obvious that for one gear to mesh with another quietly and without damage, both gears must be rotating at nearly the same speed. Modern manual transmissions and transaxles have ***synchronized*** gears, with special internal clutches to prevent gear clash when shifting. The purpose of the synchronizer is to move ahead of the unit that is to be meshed, seize the other unit, and bring the rotational speed of both units together. Once both units are rotating at the same speed, they may be meshed.

Synchronizer Operation

Figure 19-13 is a simplified version, illustrating the operating principle of the synchronizing device. A is the input shaft and it is stopped with the clutch disengaged. In order to mesh the spinning splined hub B on the output shaft with the splined recess D on the input shaft, it will be necessary to start turning the input shaft. Notice that cone ring E is an extension of hub B. The ring is supported on pins that fit into holes in the hub. The ring is held out by springs G.

As sliding hub B is moved toward splined recess D, ring E will touch cone-shaped bottom H of recess. As the hub is moved closer to the recess, the springs G will force cone ring E tightly against the bottom of the cone recess. As the cone ring is turning, it will impart torque to the input shaft and start it turning. As the hub moves closer, spring pressure increases and cone ring will be jammed into cone recess with enough force to spin the input shaft at nearly the same speed of hub B. The hub may then be meshed into the recess with no grinding or shock.

The simplified synchronizer as shown in **Figure 19-13** would be for direct drive. By putting a cone ring (called a blocking ring or synchronizing cone) on both sides of the hub, **Figure 19-14**, it could be shifted the other way and engage the splined recess in the output shaft second gear. For this arrangement, the second gear would no longer be a sliding gear and would not be splined to the output shaft. The second gear would become a constant mesh gear and would be turned whenever the cluster gear is revolving. It would rotate on the output shaft either on a bushing or roller bearings.

When direct drive or overdrive is desired, the sliding clutch hub would be shifted into the input shaft recess. For second gear, the clutch hub would be shifted back until meshed with the second gear. When the hub engages the second gear, the output shaft would be locked to the second gear through the clutch hub. The output shaft would be driven by the cluster second speed gear driving the locked output shaft second gear. Both shifts would be quiet, as the synchronizer would equalize or synchronize their speeds before meshing as in **Figure 19-14**.

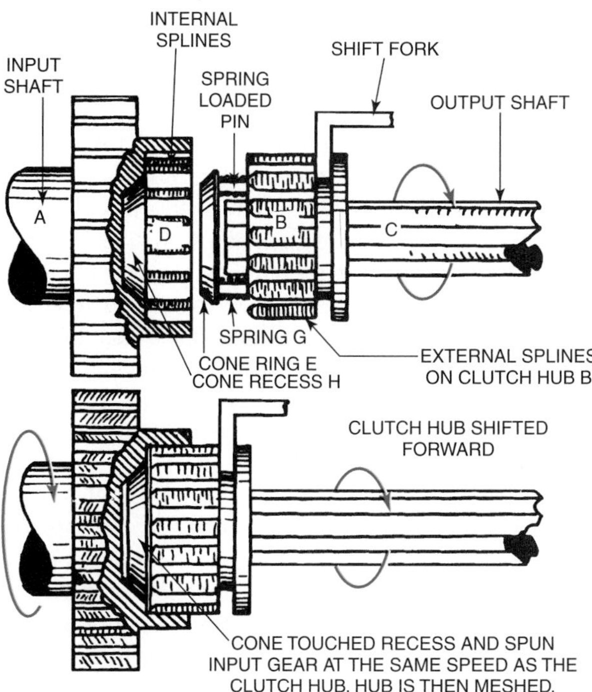

Figure 19-13. Simplified synchronizer. As clutch hub B is shifted toward splined recess D, cone ring E will strike cone recess H and spin the input shaft. Hub B is then meshed into splined recess D without grinding. Cone ring pins and springs are forced into the clutch hub.

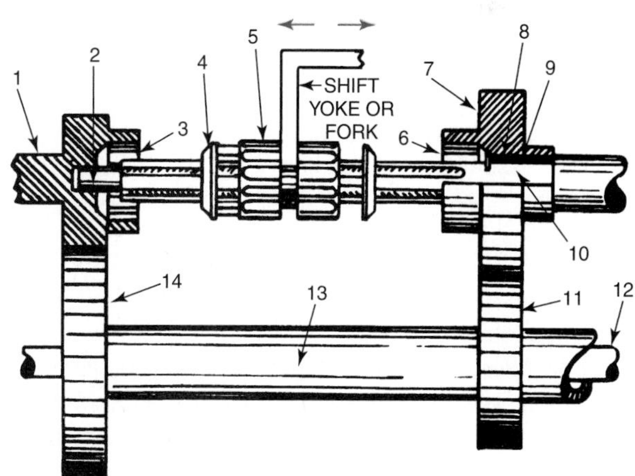

Figure 19-14. Constant mesh second gear with synchromesh in second gear and direct drive. Output shaft second gear 7, is free to rotate on output shaft 10. A roller bearing or bushing is provided at 9. A snap ring 8 keeps the second gear from moving back and forth. When synchromesh hub 5 is moved into the second gear splines 6, second gear is locked to output. Other parts include: 1—Input shaft. 2—Supported inner end of output shaft. 3—Input splines. 4—Cone synchronizer ring. 11—Cluster second gear. 12—Countershaft. 13—Cluster assembly. 14—Cluster drive.

Study the ***synchronizer assemblies*** (unit containing the synchronizers) illustrated in **Figure 19-15**. This popular type splines the hub to the shaft, but does not permit end movement of the hub. The synchronizer assembly is shifted forward or backward on the hub. This action forces (by means of inserts) the blocking rings to equalize gear speeds before the synchronizer assembly engages the other gear to lock the gear to the hub. Current practice is toward full synchronization for all forward gears and reverse. This permits first or reverse gear to be engaged while the vehicle is still in motion.

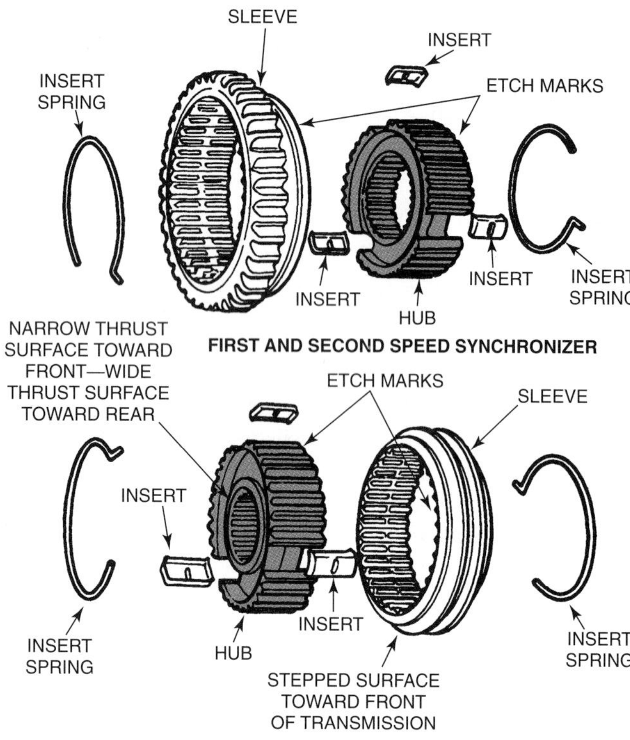

Figure 19-15. Two synchronizer units used for first, second, third, and overdrive gears. The hub is splined to the shaft. The synchronizer assembly is moved back and forth on hub to connect it with the selected gear. (Ford)

Various Types of Synchronizer Shapes

Although shapes vary, all synchronizer units in use today use a cone (blocking ring) that precedes the movement of either the clutch hub, or in many cases, the synchronizer assembly. The cone engages a tapered surface on the part to be engaged to synchronize the speed before meshing. As you study the transmissions and transaxles in this chapter, determine the location and operation of the synchronizers.

Transmission Operation

To show how the transmission works, the four shafts with gears are shown in **Figure 19-16**. By studying the illustration, you will see that the transmission is in neutral. Notice that the ***input shaft*** is turning, but the ***output shaft*** is standing still. Also note that the input shaft gear is always driving a shaft referred to as the ***counter gear*** or ***cluster gear***. The cluster gear revolves on roller bearings. The cluster gear is constantly in mesh with the ***reverse idler gear shaft***.

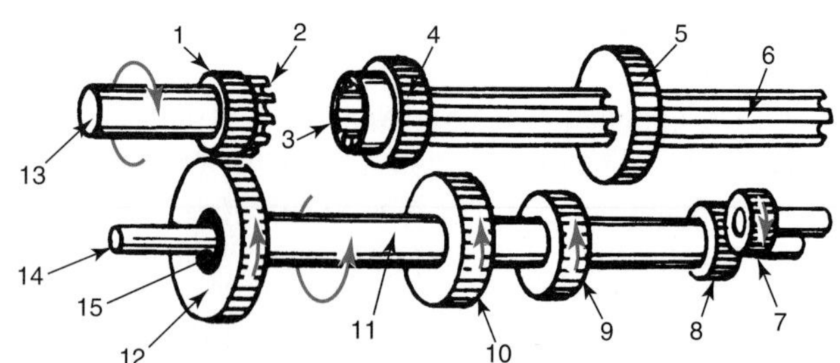

Figure 19-16. Transmission in neutral. Input shaft 13 is revolving. Input gear 1 is driving the entire cluster gear assembly. All cluster gears are connected to this cluster body. The reverse idler cluster gear 8 is driving reverse idler gear 7. Output shaft 6 is not moving as both sliding low and reverse gear 5 and sliding high and second gear 4 are in neutral positions. 2—Splined stub. 3—Splined recess on sliding gear 4. 6—Output shaft is splined to allow the gears to be shifted back and forth. 8—Cluster reverse idler gear. 9—Cluster low gear. 10—Cluster second gear. 11—Cluster body. 12—Large constant mesh cluster gear. 14—Countershaft. 15—Cluster bearing.

The cluster gear assembly has gears which are all cut on the same steel blank. The large gear at the front is the drive gear and the following gears are used in reduction gear ratios. On transmissions with overdrive, one or two of the cluster gears will be used to provide overdrive ratios. The last gear on the cluster is usually the reverse gear. When the input shaft is turning, the cluster gear also turns, but in an opposite or counter direction.

The reverse idler gear turns when the cluster gear is turning. The reverse idler gear has a pressed-in internal bushing and rotates on the stationary reverse idler shaft. There are two gears that are always in contact: the input shaft gear and cluster drive gear, and the cluster reverse gear and reverse idler gear. These are called ***constant mesh gears***.

The output shaft is splined and two gears are placed on it. Both gears are free to move back and forth on the splines. The smaller of the two is the second speed gear. Notice that it has a splined recess in its face. This splined recess is designed to slip over a splined stub protruding from the input shaft. The larger sliding gear is the low and reverse gear. It can be shifted into mesh with either the reverse idler gear or the low cluster gear. Study **Figure 19-16** and trace the flow of power (indicated by arrows).

The following sections explain the operation of a typical manual transmission with synchronizers. Note the presence of the counter gear and reverse idler gear under the input and output shafts. Although three-speed manual transmissions are increasingly rare, a three-speed transmission is used in this section to simplify the explanation of gear and synchronizer operation.

Neutral

Figure 19-17 shows the transmission in neutral. Notice that the input shaft drives the cluster gear, which in turn drives the constant mesh second speed gear. As the second gear

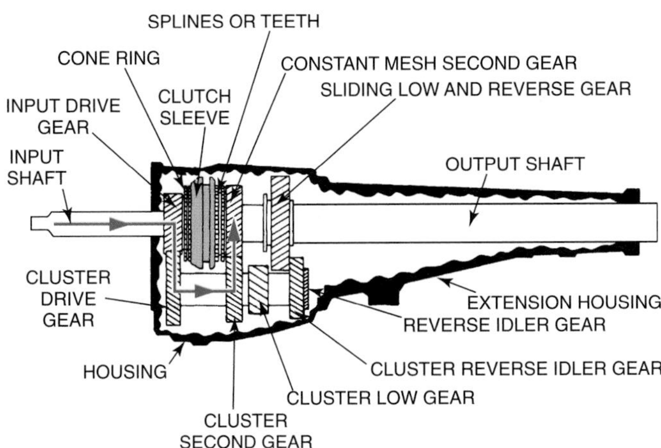

Figure 19-17. Manual transmission in neutral. Power flow is from the input gear to the cluster drive gear to the output constant mesh second gear. Since the synchromesh synchronizer assembly is in neutral, constant mesh second gear will revolve on the output shaft and no power will be transmitted beyond this point. (Pontiac)

turns on the output shaft, no power is transmitted. This transmission uses a strut-type synchronizer. The synchronizer assemblies are in a neutral position. The sliding low and reverse gear is also in a neutral position.

First Gear

When the transmission is shifted into first gear, the clutch is depressed, releasing the clutch disc. The input shaft and cluster gear will stop turning. The output sliding low gear is shifted forward and meshed with the cluster low gear. The output second gear is left out of mesh. When the clutch is engaged, the input shaft will turn the cluster gear. The cluster gear will revolve at a slower speed. The cluster low gear, being smaller than the output shaft low sliding gear, will turn the output shaft at a slower speed, **Figure 19-18.**

Figure 19-19 illustrates the transmission in first gear. The synchronizer assembly is still in the neutral position, but the sliding low and reverse gear has been shifted into mesh with the cluster low gear. The output shaft speed will be

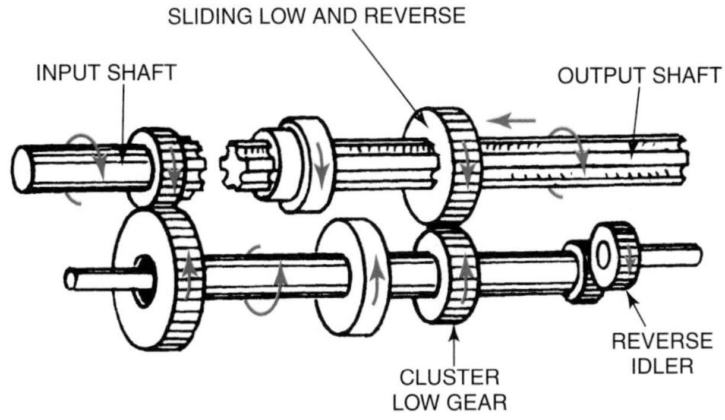

Figure 19-18. Transmission in first gear. The sliding low and reverse gear has been shifted forward and is meshed with the cluster low gear. Output shaft is now turning. All gears that are in mesh with another gear are shown with simulated teeth.

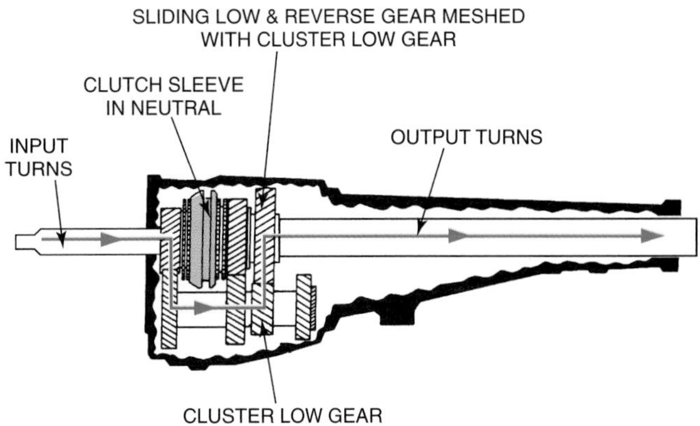

Figure 19-19. Manual transmission in first gear. Power flow is from the input gear to the cluster drive gear, from cluster low gear to output sliding low and reverse gear. As the sliding low and reverse gear is meshed, power is transmitted to output shaft. (Pontiac)

roughly one-third that of the input shaft, giving a gear reduction of 3:1. This reduction multiplies the engine torque, but gives less forward speed.

Second Gear

When ready for second gear, the clutch is depressed to break the flow of power from the engine. The low sliding gear is moved to a neutral position and the second sliding gear is meshed with the cluster second gear. The clutch is then released and the flow of power is resumed with less torque multiplication. See **Figure 19-20**.

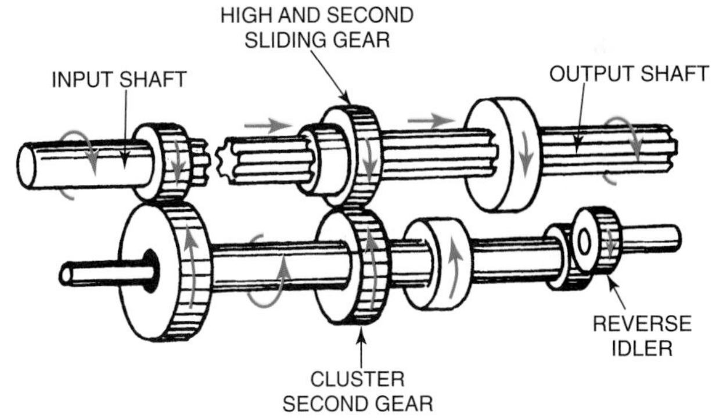

Figure 19-20. Transmission in second gear. The sliding low gear has returned to the neutral position. The sliding second gear is moved into mesh with the cluster second gear. Output shaft now turns faster with less torque multiplication.

Notice that the cluster second gear is larger in proportion to the sliding second gear. **Figure 19-21** shows the transmission in second gear. The synchronizer assembly has been shifted back until it engages the splines or small teeth on the second gear. The gear reduction is now around 2:1 and torque falls off, but speed picks up.

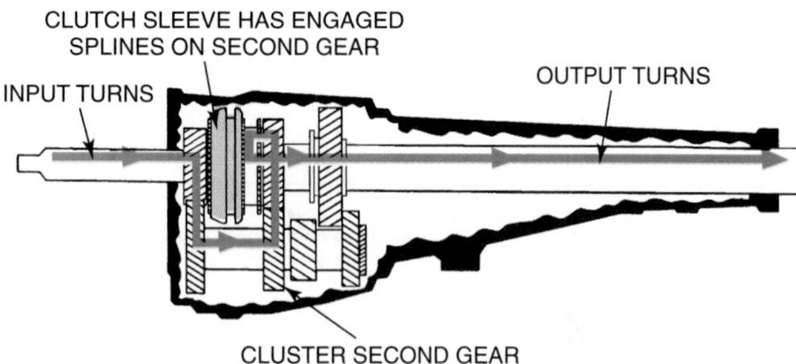

Figure 19-21. Manual transmission in second gear. Power flow is from the input gear to the cluster drive gear, to cluster second gear to constant mesh output second gear. As the clutch hub engages the teeth or splines on the second gear, it is locked to the output shaft and will transmit power. (Pontiac)

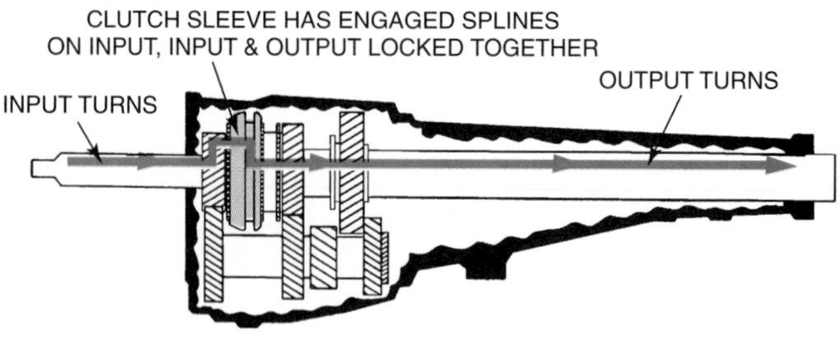

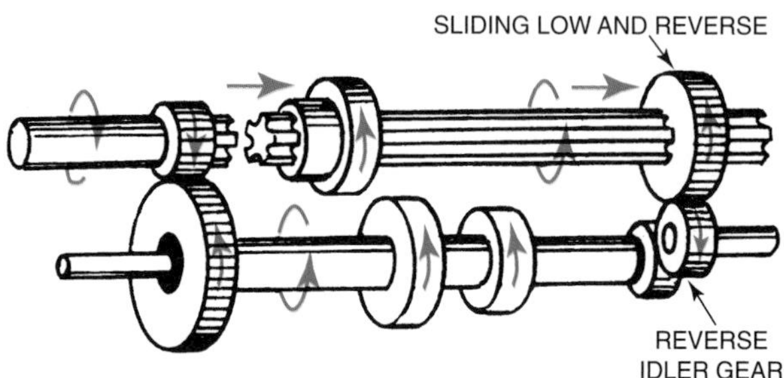

Figure 19-23. Manual transmission in direct drive. The clutch hub has now engaged splines on the end of the input shaft. This will lock the input and output shafts together and power is transmitting straight through with no torque multiplication. The cluster is driven but will not transmit power. (Pontiac)

Third Gear

When the vehicle has attained sufficient speed, the clutch is again depressed. The sliding second gear is moved forward until the splined recess has slipped over the splined stub on the input shaft. The low sliding gear remains unmeshed. When the clutch is released, the transmission is in direct drive with no torque multiplication. See **Figure 19-22.** The input and output shafts are locked together through the splined stub and recessed second gear.

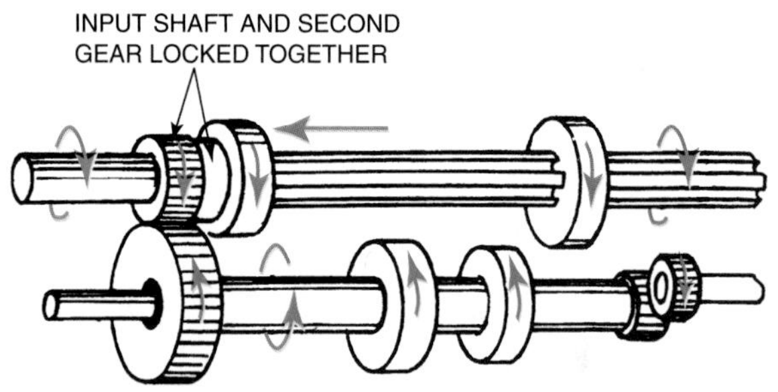

Figure 19-22. Transmission in direct drive. Sliding second gear has moved toward the input until the splined recess in the gear engages the splined stub on the input shaft. The sliding low gear is in a neutral position. The input and output shafts are now locked together and drive is direct.

Figure 19-23 shows the transmission in third gear. The synchronizer assembly has been shifted away from the second gear and has engaged the small teeth on the input shaft. As this locks the input and output shafts together, this gives direct drive through the transmission.

Reverse

In **Figure 19-24**, the sliding second gear is placed in the neutral position. The low reverse sliding gear is moved back until it engages the reverse idler gear. As the reverse idler is revolving in the same direction as the input shaft, it will impart a reverse rotation to the output shaft. This will cause the output shaft to drive the vehicle backward. **Figure 19-25** illustrates the transmission in reverse gear. The gear reduction or torque multiplication is comparable to that of the low gear, around 3:1.

Figure 19-24. Transmission in reverse gear. The sliding second gear has been returned to neutral. The low and reverse sliding gear is meshed with the reverse idler gear. The output shaft now rotates in a reverse direction. The gear ratio is about same as first gear.

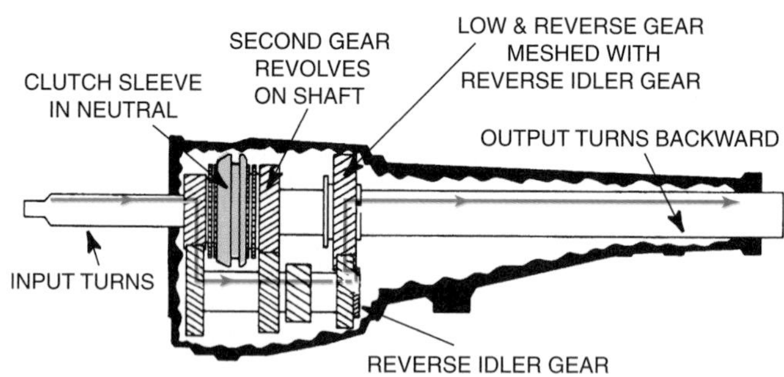

Figure 19-25. Manual transmission in reverse gear. The synchronizer assembly has now been returned to a neutral position. The sliding low and reverse gear has been meshed with reverse idler gear. The input gear drives the cluster. The cluster drives the reverse idler and reverse idler drives the output shaft in a reverse direction. (Pontiac)

Four-Speed Transmission

Figure 19-26 illustrates a typical four-speed transmission. All gears are synchronized. Gear teeth are helical cut with the exception of the reverse sliding and reverse idler gears. Notice the bearing locations and the method used to

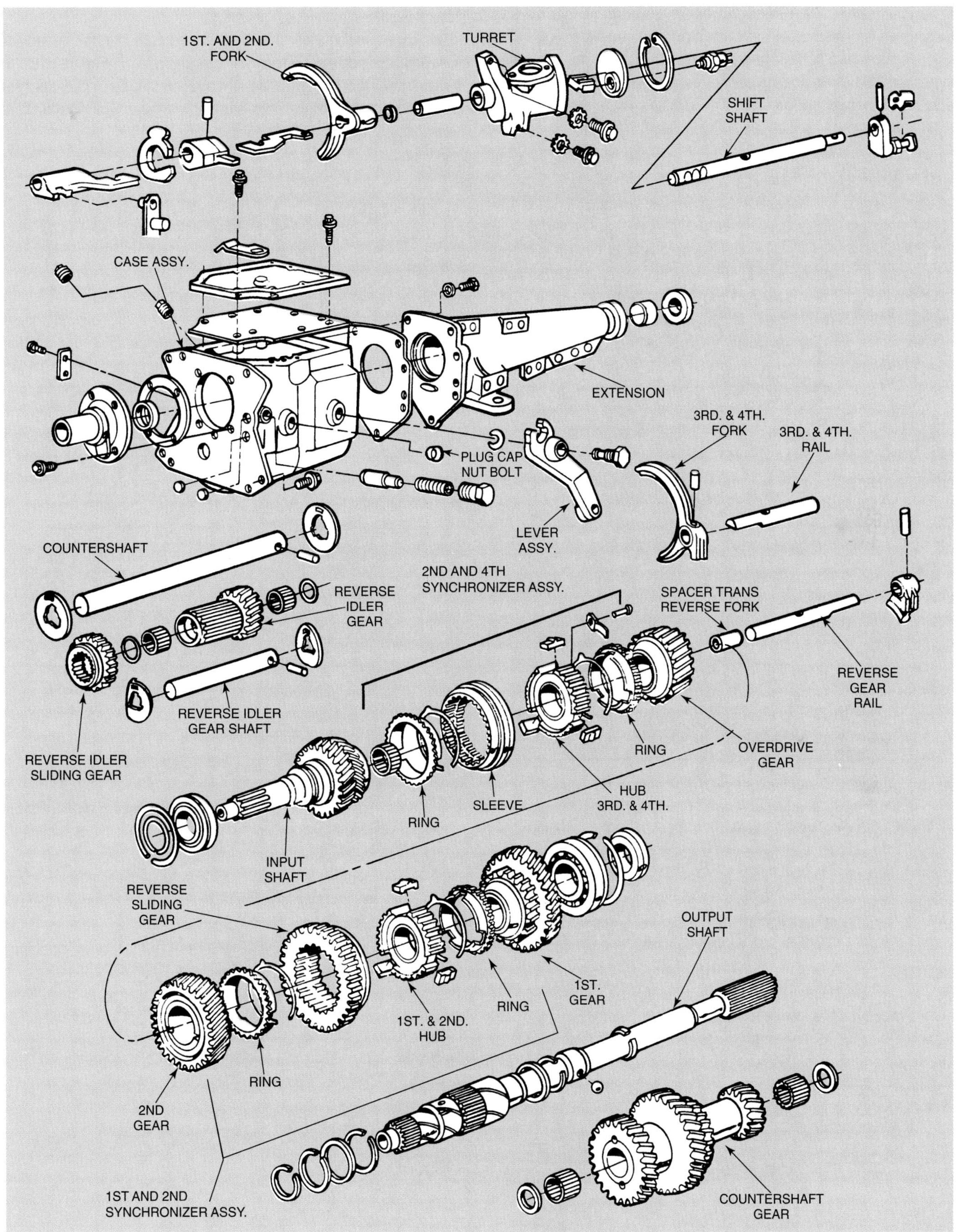

Figure 19-26. An exploded view of a four-speed transmission. Study the overall construction carefully. (Ford Motor Co.)

secure all shafts. The second gear is a constant mesh gear turning on a bushing on the output or main shaft.

Figure 19-27 illustrates the gear drive positions in a fully-synchronized, four-speed transmission. Study the gear drive positions for the various speeds.

Five-Speed Transmission

An exploded view of a five-speed, fully synchronized (all forward speeds) transmission is shown in **Figure 19-28**. Notice that the fifth gear is an overdrive gear. The counter gear is again used to produce a gear ratio. A typical overdrive gear ratio is about .7:1.

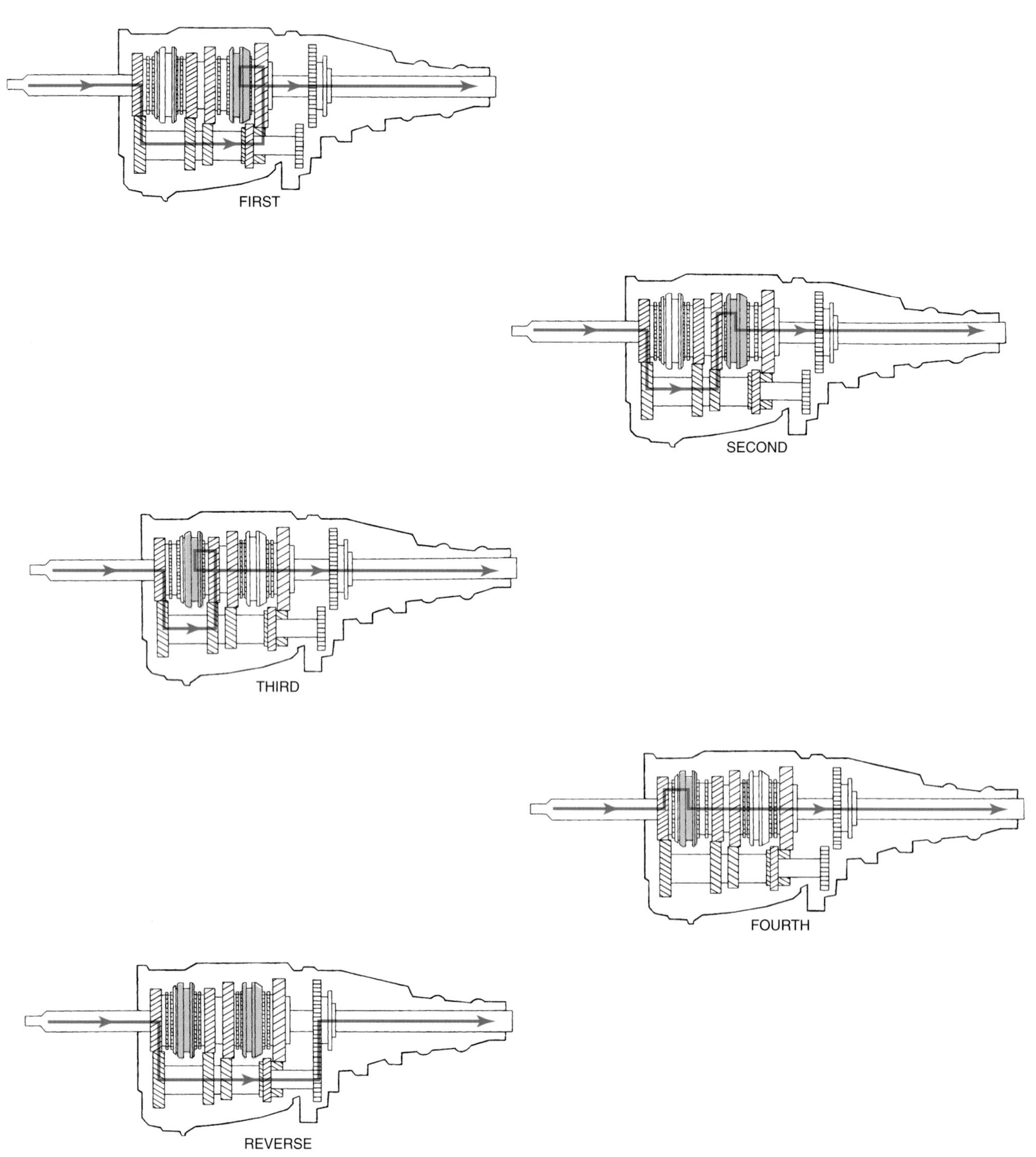

Figure 19-27. Gear drive positions in a fully synchronized four-speed transmission. (General Motors)

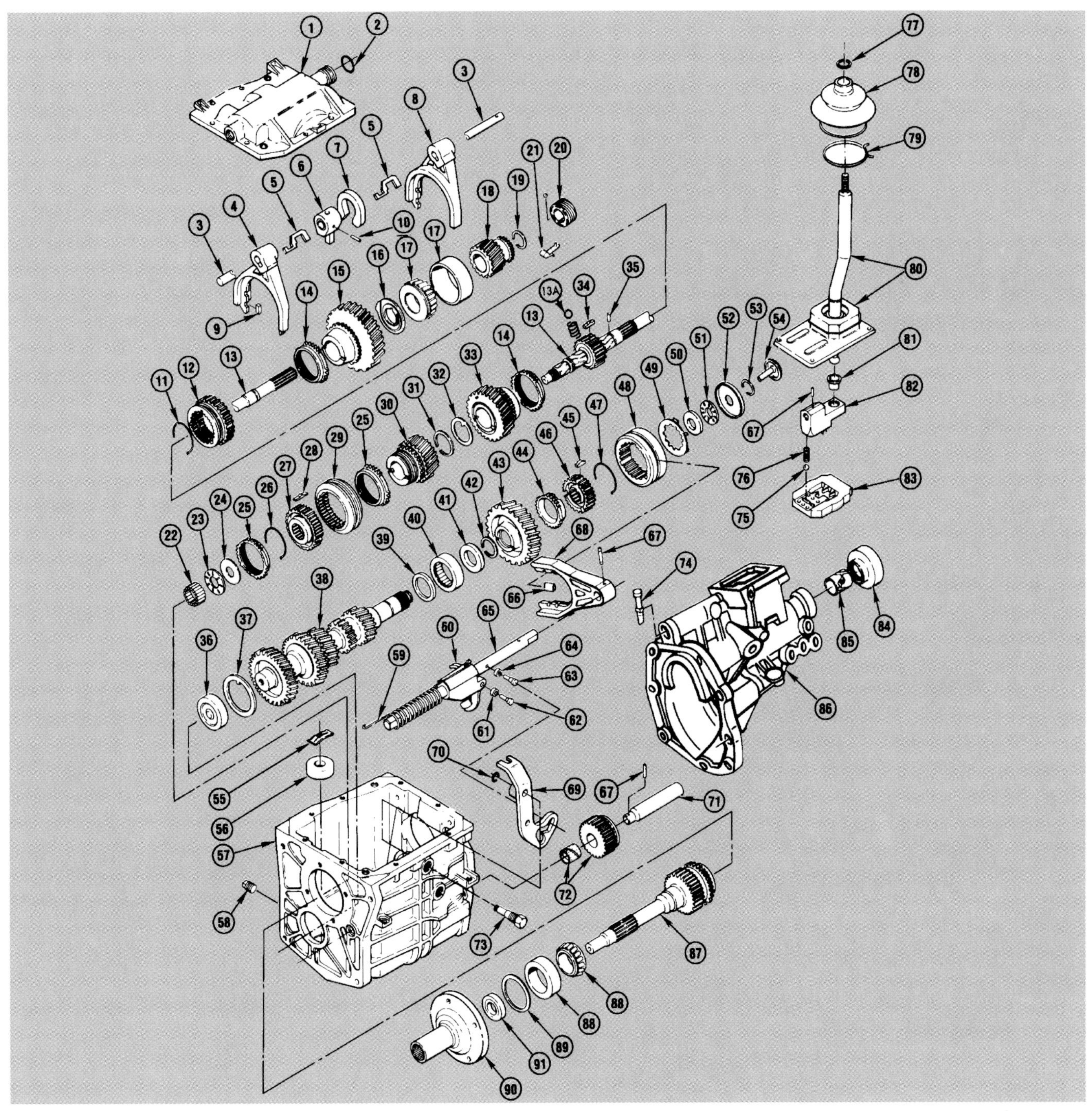

Figure 19-28. A five-speed transmission shown in an exploded view. 1—Transmission cover. 2—O-ring seal. 3—Shift shaft. 4—3rd and 4th shift fork. 5—Shift fork plate. 6—Control selector arm. 7—Gear selector interlock plate. 8—1st and 2nd shift fork. 9—Shift fork insert. 10—Roll pin. 11—Synchronizer spring. 12—Reverse sliding gear. 13—Output 1 and 2 synchronizer shaft. 14—1st and 2nd synchronizer blocking ring. 15—1st gear. 16—1st gear thrust washer. 17—Rear bearing. 18—5th speed driven gear. 19—Snap ring. 20—Speedometer drive gear. 21—Speedometer drive gear clip. 22—Main shaft bearing. 23—Main drive gear bearing race. 24—Thrust washer. 25—3rd and 4th synchronizer ring. 26—3rd and 4th synchronizer spring. 27—3rd and 4th synchronizer hub. 28—3rd and 4th synchronizer key. 29—3rd and 4th synchronizer sleeve. 30—3rd speed gear. 31—Snap ring. 32—2nd speed gear thrust washer. 33—2nd sped gear. 34—1st and 2nd synchronizer key. 35—1st speed gear thrust washer retainer pin. 36—Bearing. 37—Thrust washer. 38—Counter gear. 39—Bearing space. 40—Bearing. 41—Bearing spacer. 42—Snap ring. 43—5th speed drive gear. 44—5th synchronizer ring. 45—5th synchronizer key. 46—5th synchronizer hub. 47—5th synchronizer spring. 48—5th synchronizer sleeve. 49—5th synchronizer key retainer. 50—5th synchronizer thrust bearing front race. 51—5th synchronizer needle thrust bearing. 52—5th synchronizer thrust bearing rear race. 53—Snap ring. 54—Oiling funnel. 55—Magnet nut. 56—Magnet. 57—Transmission case. 58—Fill and drain plug. 59—Reverse lock spring. 60—Reverse shift fork. 61—Fork roller. 62—Reverse fork pin. 63—Shift rail pin. 64—Rail pin roller 65—5th and reverse shift rail. 66—Shift fork insert. 67—Roll pin. 68—5th shift fork. 69—5th and reverse relay lever. 70—Reverse relay lever retaining ring. 71—Reverse idler gear shaft. 72—Reverse idler gear. 73—5th speed shift lever pivot pin. 74—Exterior ventilator. 75—Steel ball. 76—Detent spring. 77—Boot retainer. 78—Boot. 79—Boot retainer. 80—Shift lever. 81—Shift lever damper sleeve. 82—Offset shift lever. 83—Detent and guide plate. 84—Rear oil seal. 85—Extension housing bushing. 86—Extension housing. 87—Main drive gear. 88—Front bearing. 89—Bearing adjust shim. 90—Rear drive gear bearing retainer. 91—Drive gear bearing oil seal.

Figure 19-29 illustrates the power flow and gears of a five-speed transmission. Study all the transmission parts and their construction.

Manual Transaxles

On front-wheel drive vehicles, the transmission and differential assemblies are installed in a single housing, referred to as a transaxle. Transaxles operate in the same way as rear drive transmissions and rear axle differentials, but place all parts together. The manual transaxle has the same kinds of gears, synchronizers, and shifting mechanisms as a manual transmission. The number of forward gear ratios varies from three to five. Transaxles are also used on vehicles with rear engines and rear-wheel drive.

The major differences between the transmission and transaxle are the shape of the case and the placement

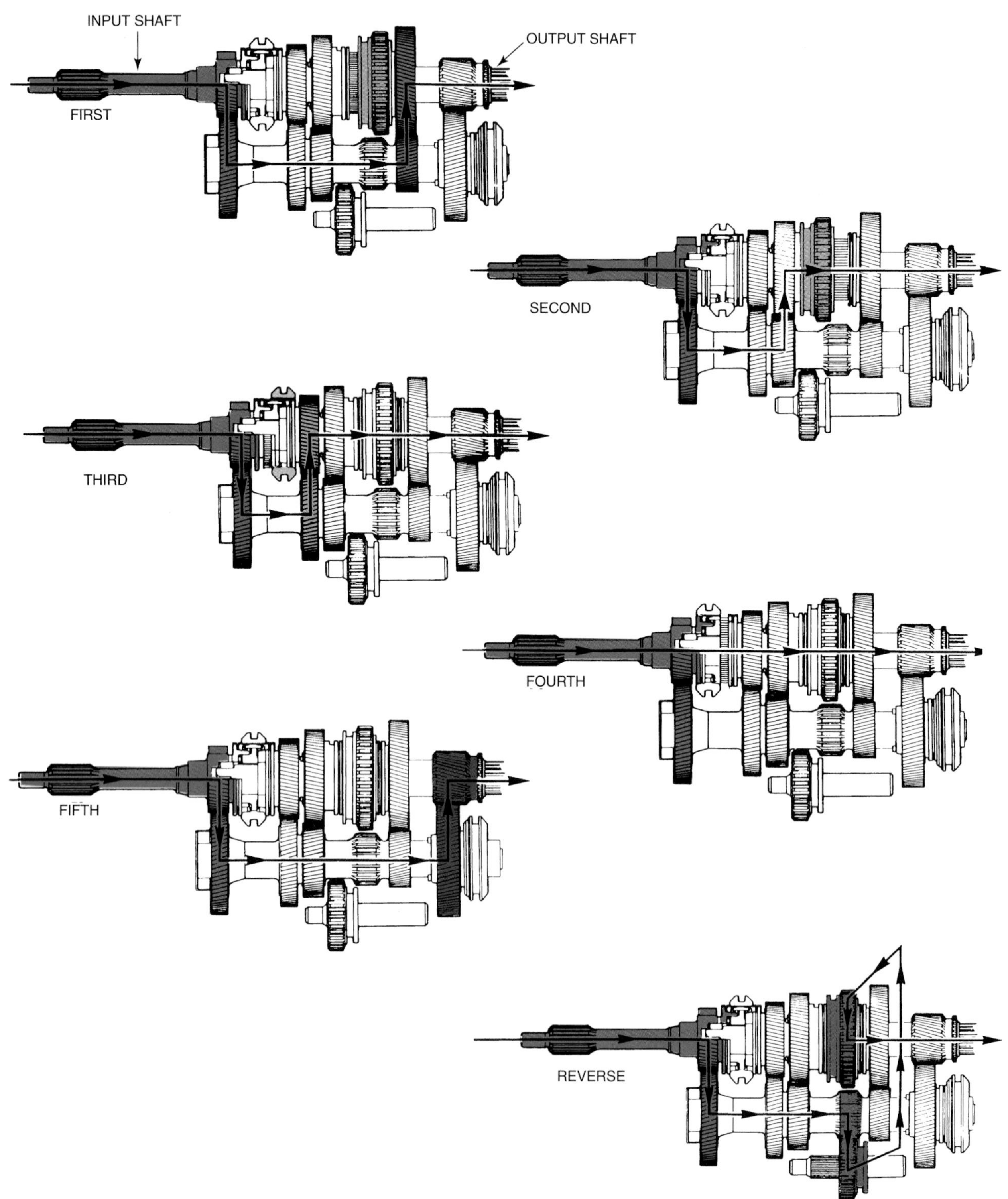

Figure 19-29. Power flow and gears in a typical five-speed manual transmission. (Borg-Warner)

of parts within the case. The transaxle may also contain differential gears and may use two gears or a drive chain to transfer power inside the transaxle. The obvious external difference is the presence of two axles connecting the transaxle to the front wheels. See **Figure 19-30**.

There are two basic variations of front-wheel drive transaxle construction. Which type is used depends on engine placement in the vehicle. Engines can be installed sideways in the vehicle, called a transverse engine, or facing forward, called a longitudinal engine. **Figure 19-30** shows a manual transaxle for use in a vehicle with a transverse engine. Power flow takes place in a straight line without any change in the angle once leaving the engine.

The power flow through a transaxle on a vehicle with longitudinal engine placement is shown in **Figure 19-31**. In this design, the output shaft drives the pinion and ring gear in the differential. This ring and pinion gear is a hypoid type, like the ones used on rear-wheel drive vehicles. The design of the ring and pinion cause the power to make a 90° turn.

Transaxle Cases

The ***transaxle case*** holds the internal parts of the transaxle. The case bolts to the rear of the engine, or to the clutch housing. Most transaxle cases are one- or two-piece units incorporating the clutch housing, **Figure 19-32**. Some of the engine mounts may be attached to parts of the transaxle.

Transaxle Gear and Synchronizer Operation

The following sections explain the operation of a typical four-speed manual transaxle with synchronizers. Note the similarities between the operation of the gears in the transaxle and the gears in a rear-wheel drive transmission. The major

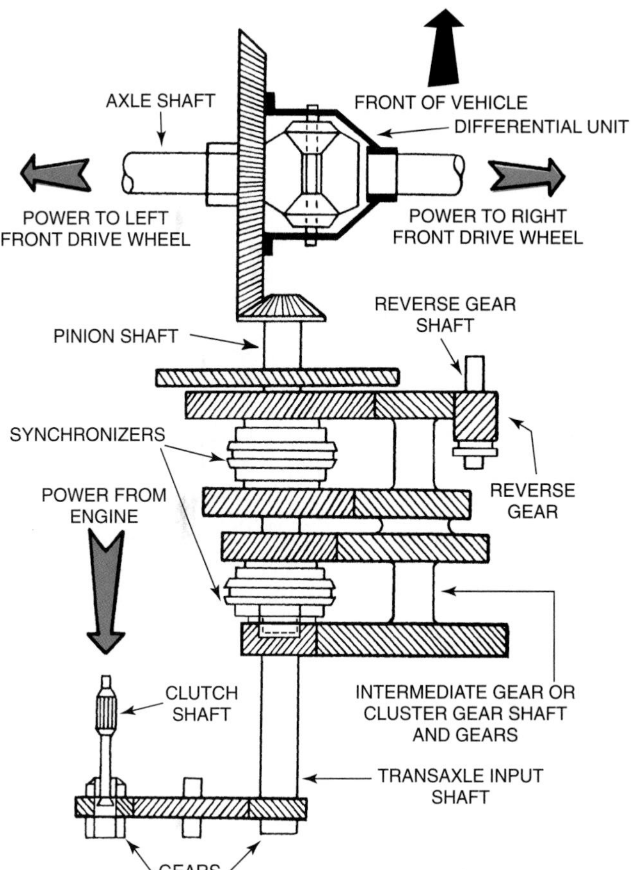

Figure 19-31. Longitudinal engine and transaxle power flow for a front-wheel drive car. Drive gears are needed to transmit engine power from the clutch shaft to the transaxle input shaft. (Saab)

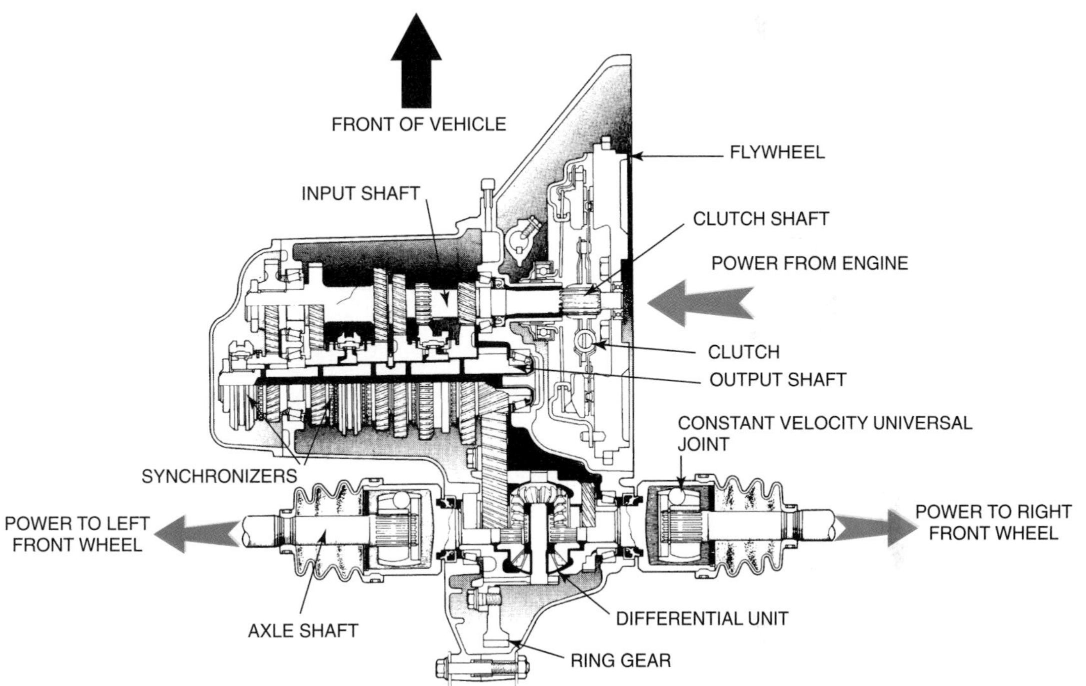

Figure 19-30. Cutaway view of a five-speed transaxle assembly. Note differential and drive axle assembly. The engine crankshaft and axle shafts are parallel, so the power flow does not have to take a 90° turn. (Mazda Motor Corporation)

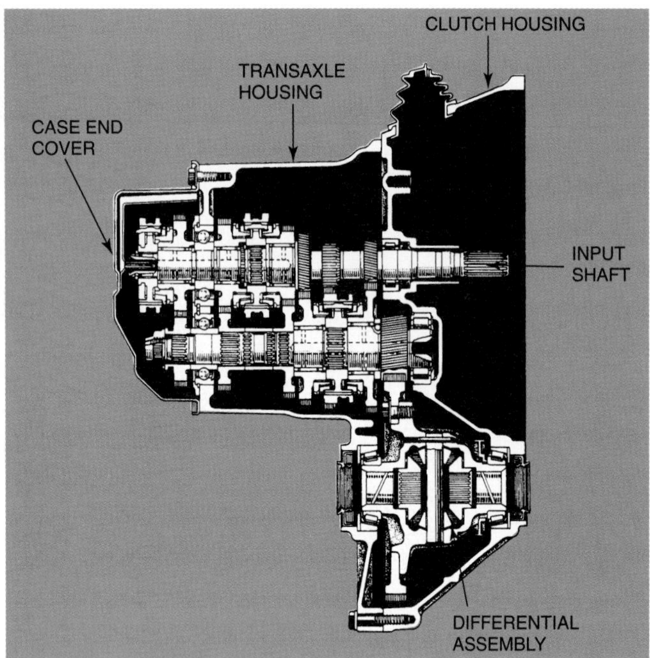

Figure 19-32. Cutaway view of a manual transaxle housing and clutch housing assembly. (Toyota Motor Corporation)

difference is that there is no cluster or counter gear. The input and output shafts are placed next to each other.

Neutral

In **Figure 19-33**, the transaxle is in neutral. Notice that the input shaft is driving the gears on the output shaft. Since none of the gears are connected to the output shaft, no power is being transmitted. The synchronizer assemblies are in a neutral position. The reverse idler gear is also in neutral.

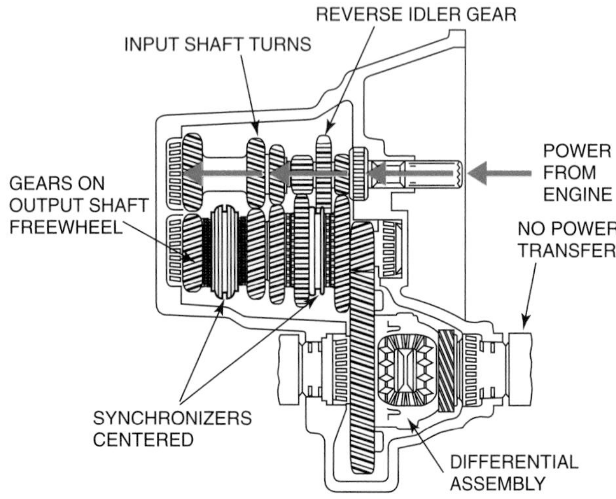

Figure 19-33. One particular four-speed transaxle that is mounted to a transverse engine. The power flow is in neutral. (Ford Motor Co.)

First Gear

Figure 19-34 shows the transaxle in first gear. The rear synchronizer assembly is still in the neutral position, but the front synchronizer has been shifted forward to engage the first gear and the output shaft. Power flows from the input shaft to the output shaft through the gears. Since the input gear is much smaller than the output gear, the first gear ratio is high, about 3:1.

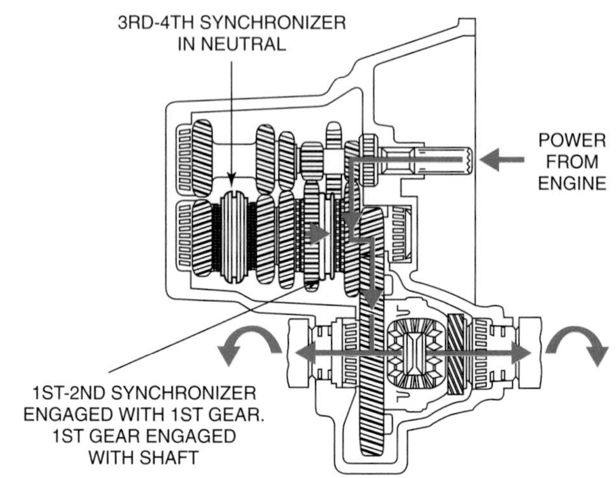

Figure 19-34. Transaxle illustrating power flow in *first* gear. Ratio is about 3:1. (Ford Motor Co.)

Second Gear

In **Figure 19-35** the transaxle is in second gear. The rear synchronizer assembly remains in the neutral position. The front synchronizer assembly has been shifted back to engage second gear and the output shaft. Power flows from the input shaft to the output shaft at the second gear ratio of 2.5:1.

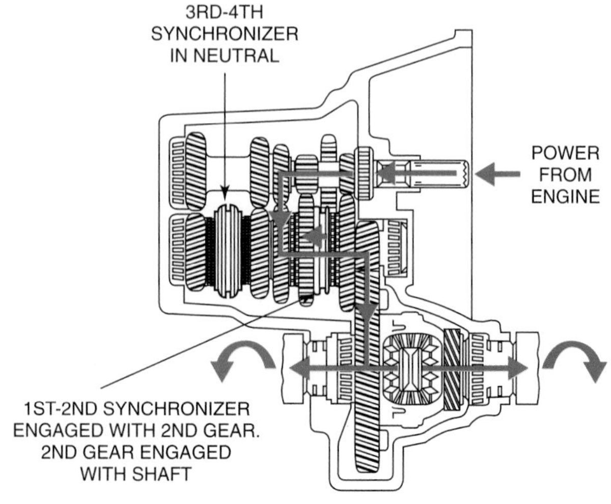

Figure 19-35. Same transaxle depicting power flow while in *second* gear. Ratio is about 2.5:1. (Ford Motor Co.)

Third Gear

Figure 19-36 shows the transaxle in third gear. The front synchronizer assembly has been returned to the neutral position. The rear synchronizer assembly has been shifted forward to engage the third gear and the output shaft. Power flows from the input shaft to the output shaft at the third gear ratio of 2:1.

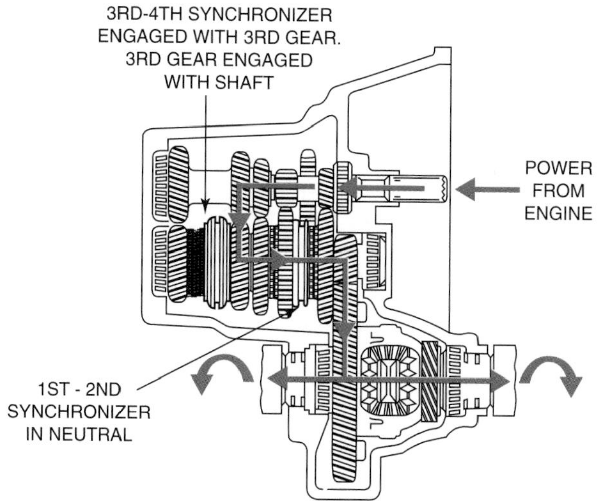

Figure 19-36. Power flow through transaxle in *third* gear. Ratio is about 2:1. (Ford Motor Co.)

Fourth Gear

Figure 19-37 shows the transaxle in fourth gear. The front synchronizer assembly remains in the neutral position. The rear synchronizer assembly has been shifted to the rear to engage the fourth gear and the output shaft. Power flows from the input shaft to the output shaft at the fourth gear ratio. Since the input and output gears are the same size, the ratio is 1:1.

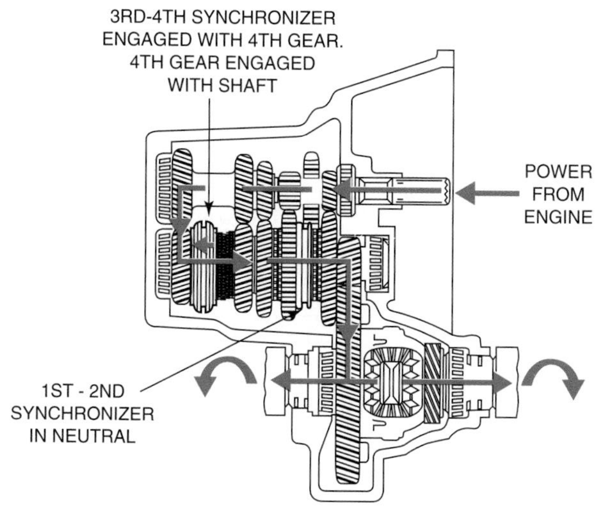

Figure 19-37. Power flow through transaxle in *fourth* gear. Ratio is 1:1. (Ford Motor Co.)

Reverse Gear

The transaxle in **Figure 19-38** is in reverse gear. Both synchronizer assemblies are in the neutral position. The reverse idler gear has been moved into mesh with both the input and output shafts. The input shaft drives the idler gear, which turns the output shafts in the opposite direction of engine rotation. This will cause the output shafts to drive the vehicle backward.

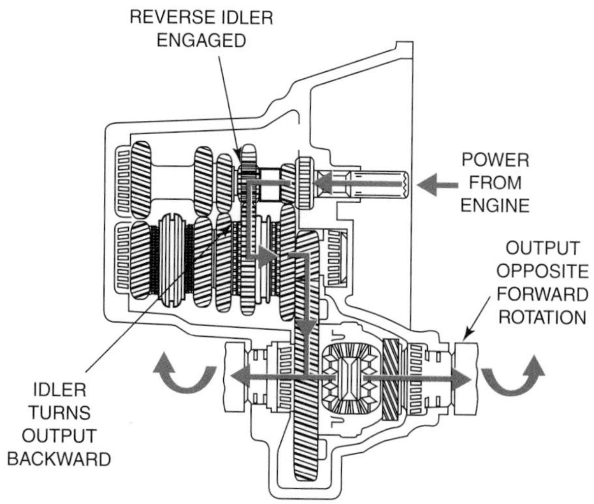

Figure 19-38. The transaxle in *reverse*. The reverse idler gear is now engaged and the output shaft and differential turn in a reverse direction. (Ford Motor Co.)

Five-Speed Transmission and Transaxle

An exploded transaxle is shown in **Figure 19-39**. Study the parts and learn their names.

Transmission and Transaxle Shift Mechanism

All manual transmissions and transaxles are shifted by means of ***shifter forks*** that ride in a groove cut into the synchronizer assembly and sliding gear. The forks are attached to a cam and shaft assembly. Spring-loaded steel balls pop into notches cut in the cam assembly to hold the shift mechanism into whatever gear is selected. The shafts pass through the housing or housing cover and are fastened to shift levers, **Figure 19-40**.

On older vehicles with three-speed transmissions, the shift levers can be operated via ***shift linkage*** attached to a steering wheel shift control. Newer vehicles use a console or floor mounted shift stick. Most late model four-, five-, and six-speed transmissions operate the shift forks by means of a floor stick shift that enters the transmission housing and engages sliding shift fork rails. Other floor shift designs use either linkage or cables. Some shift sticks are pivoted in a ball-like assembly. **Figures 19-41** and **19-42** illustrate different shift mechanism arrangements. Almost all shifters make some provision for adjustment.

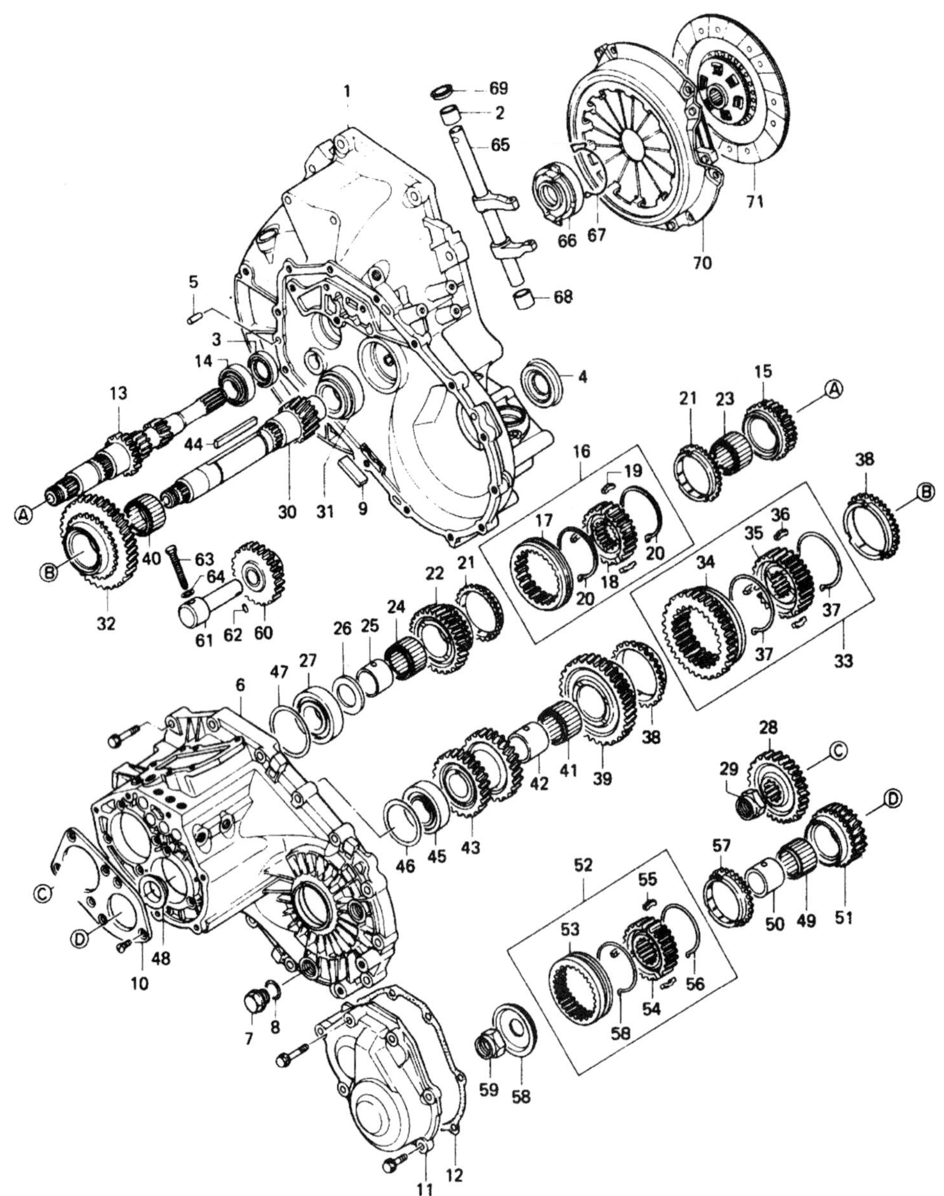

1. CLUTCH AND DIFF. HOUSING
2. CLUTCH SHAFT BUSHING
3. INPUT SHAFT OIL SEAL
4. DRIVE SHAFT OIL SEAL
5. STRAIGHT KNOCK PIN
6. TRANSAXLE CASE
7. DRAIN PLUG
8. GASKET
9. MAGNET
10. BEARING RETAINER
11. REAR COVER
12. GASKET
13. INPUT SHAFT
14. INPUT SHAFT FRONT BEARING
15. 3RD GEAR ASSEMBLY
16. 3RD/4TH SYNCHRONIZER ASSEMBLY
17. SYNCHRONIZER SLEEVE
18. CLUTCH HUB
19. INSERT
20. INSERT SPRING
21. 3RD/4TH BLOCKER RING
22. 4TH GEAR ASSEMBLY
23. 3RD NEEDLE BEARING
24. 4TH NEEDLE BEARING
25. 4TH COLLAR
26. 4TH GEAR THRUST WASHER
27. INPUT SHAFT REAR BEARING
28. 5TH GEAR
29. INPUT SHAFT END NUT
30. OUTPUT SHAFT
31. OUTPUT SHAFT FRONT BEARING
32. 1ST GEAR ASSEMBLY
33. 1ST/2ND SYNCHRONIZER ASSEMBLY
34. REVERSE GEAR
35. CLUTCH HUB
36. INSERT
37. INSERT SPRING
38. 1ST/2ND BLOCKER RING
39. 2ND GEAR ASSEMBLY
40. 1ST NEEDLE BEARING
41. 2ND NEEDLE BEARING
42. 2ND COLLAR
43. 3RD/4TH OUTPUT GEAR
44. KEY
45. OUTPUT SHAFT REAR BEARING
46. OUTPUT SHAFT BEARING SHIM
47. INPUT SHAFT BEARING SHIM
48. 5TH GEAR THRUST WASHER
49. 5TH NEEDLE BEARING
50. 5TH COLLAR
51. 5TH GEAR ASSEMBLY
52. 5TH SYNCHRONIZER ASSEMBLY
53. SYNCHRONIZER SLEEVE
54. CLUTCH HUB
55. INSERT
56. INSERT SPRING
57. 5TH BLOCKER RING
58. INSERT STOPPER PLATE
59. OUTPUT SHAFT END NUT
60. REVERSE IDLER GEAR ASSEMBLY
61. REVERSE IDLER SHAFT
62. STRAIGHT PIN
63. REVERSE IDLER SHAFT BOLT
64. GASKET
65. CLUTCH FORK SHAFT ASSEMBLY
66. CLUTCH RELEASE BEARING
67. RELEASE BEARING SPRING
68. CLUTCH SHAFT BUSHING
69. CLUTCH SHAFT SEAL
70. CLUTCH PRESSURE PLATE ASSEMBLY
71. CLUTCH DISC ASSEMBLY

Figure 19-39. Cutaway view of a five-speed manual transaxle assembly. Note the two-piece housing and shift linkage arrangement. (Oldsmobile)

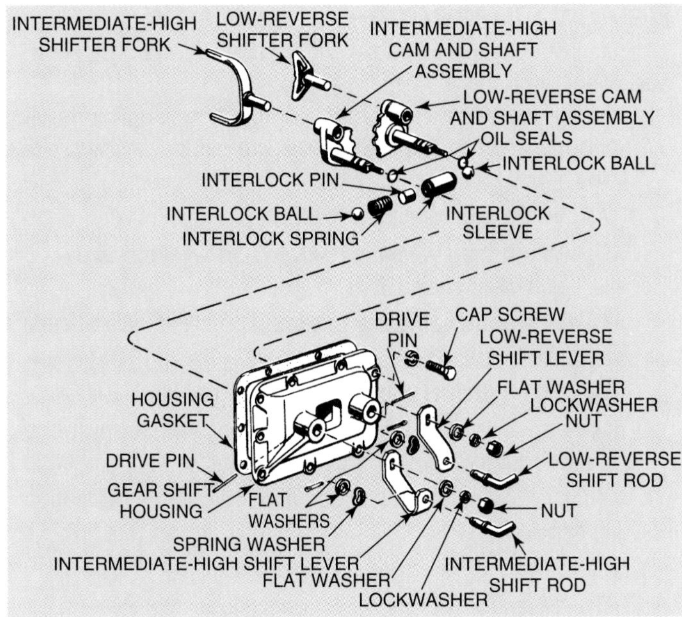

Figure 19-40. Gearshift housing assembly—exploded view. The shifter fork engages the low and reverse sliding gear and synchronizer assembly. The forks are actuated by moving shift levers. Each fork must be returned to neutral before another can be moved. This is accomplished by the interlock pin. The interlock balls hold cam into whatever position it has been placed. (Ford Motor Co.)

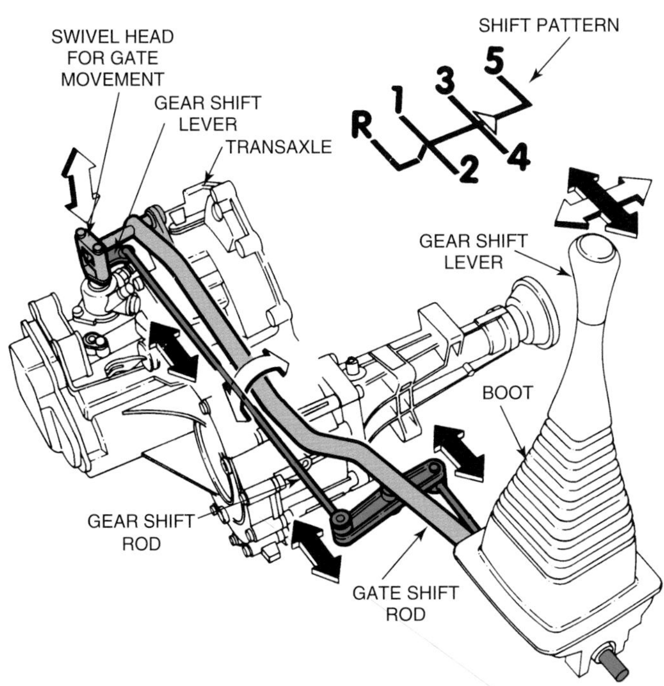

Figure 19-41. Overall view of a five-speed transaxle floor shift assembly. These layouts vary by vehicle and manufacturer. (Volkswagen)

Manual Transmission and Transaxle Lubrication

The transmission is lubricated by filling the housing partially full of lubricant, **Figure 19-42**. As the gears turn, they throw the oil throughout the case to lubricate all parts. This resembles the splash lubrication system used in engines. Manual transmissions can be lubricated by gear oil, usually 90 or 140 weight, or automatic transmission fluid. Leakage is controlled by oil seals at the input shaft and output shaft(s), and by gaskets on the shift covers and between the case and extension housing. Seals are installed on the manual shaft and the speedometer gear housing.

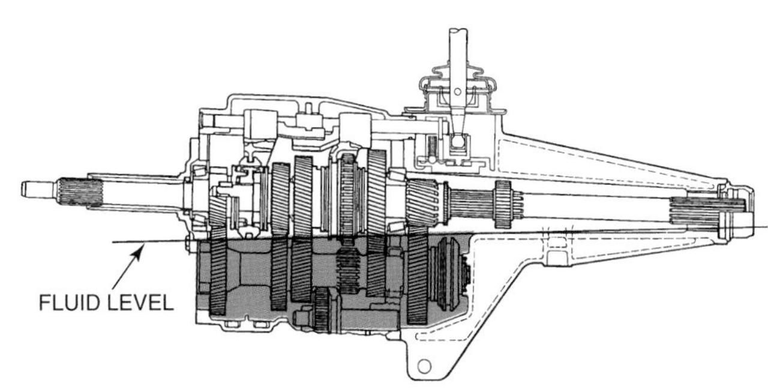

Figure 19-42. A five-speed transmission with a floor shift assembly. This transmission cutaway illustrates the correct transmission fluid level. Note how the stick (shift lever) is connected to the shift shaft . (General Motors)

Speed Shifting

When shifting the transmission, it is wise to hesitate in the neutral range for a brief moment. The synchromesh unit depends upon a limited amount of friction to synchronize the speeds. By hesitating a moment and then shifting smoothly into the desired gear, the synchromesh unit will have time to function.

So-called ***speed shifting*** will cause the synchronizer cones to wear very quickly from excessive friction. Speed shifting is also very damaging to the gears and for that matter, the entire drive train. Speed shifting may be justified when a vehicle is entered into competition (the transmission is also modified in most cases to allow it to withstand the abuse). At any other time, it is considered the mark of a poor driver, generally one who knows nothing of how a vehicle is built. Good drivers and good technicians (good technicians are almost always excellent drivers) have a sincere respect for the work and skill involved in good machinery.

Summary

The engine must have a transmission or transaxle to multiply engine torque to adapt to changing road and load conditions. Torque multiplication is accomplished through the use of gears. By arranging a movable set of various sized gears, it is possible to obtain a reasonable degree of torque multiplication.

Most transmissions have four or five forward gear ratios and one reverse. A few have six forward speeds. Transmissions are generally equipped with helical gears to assure longer life and quieter operation. One gear (usually low or reverse) may be of the sliding type, or all forward gears may be of constant mesh design.

A synchromesh unit is designed to provide smooth gear shifting without clashing or grinding. The transmission consists of the input and output shaft assemblies, the cluster gear, and the reverse idler assembly. All are housed in a strong steel or aluminum housing.

The shafts revolve on roller and ball bearings. The second speed gear and reverse idler gear are sometimes mounted on bronze bushings. The input and output shafts are secured by means of snap rings and bearing retainers. End play in the cluster gear is controlled by the use of bronze and steel thrust washers.

The transmission housing can be bolted to the clutch housing or cast as an integral part. The transmission housing often acts as a base for the rear engine mount. On front-wheel drive vehicles, the transmission and differential assemblies are installed in a single housing, referred to as a transaxle. Transaxles are also used on vehicles with rear engines and rear wheel drive.

Transaxles operate in the same way as rear drive transmissions and rear axle differentials, but place all parts together. The manual transaxle has the same kinds of gears, synchronizers, and shifting mechanisms as a manual transmission. The major differences between the transmission and transaxle are the shape of the case, and the placement of the parts within the case. The transaxle may use a drive chain or two gears to transfer power inside of the transaxle. The transaxle has two axles connecting the transaxle to the front wheels.

The gears are selected either by a console or floor mounted shift stick or steering wheel column shift control. The transmission is lubricated by filling the housing partially full of gear oil or automatic transmission fluid. As the gears turn, they throw the oil to lubricate all parts. Leakage is controlled by oil seals at both front and rear ends.

Know These Terms

Manual transmission
Manual transaxle
Lever and fulcrum
Gear ratio
Reduction gears
Direct drive gear
Overdrive gears
Reverse
Spur
Helical
Gear end play
Thrust washer
Backlash
Transmission case
Extension housing
Sliding gears
Shift linkage
Gear clash
Synchronized gears
Synchronizer assemblies
Input shaft
Output gear
Counter
Cluster gear
Reverse idler gear shaft
Constant mesh gears
Transaxle case
Shifter forks
Shift linkage
Speed shifting

Review Questions—Chapter 19

Do not write in this book. Write your answers on a separate sheet of paper.

1. When would an internal combustion engine have enough power to operate a vehicle without a transmission?
   (A) When the vehicle is started on a level road.
   (B) When the vehicle is operated on level roads at sufficient speed.
   (C) When the vehicle is operated on steep hills at sufficient speed.
   (D) When the vehicle is operated in reverse.
2. If the speed of an engine is constant and its torque is multiplied, the engine will be able to move _____ weight.
3. What is the relationship between torque multiplication and speed?
   (A) When torque goes up, speed goes up.
   (B) When torque goes up, speed goes down.
   (C) When torque goes down, speed goes down.
   (D) Changes in torque do not affect speed.
4. A gear ratio of 2.5:1 is a(n) _____ gear.
   (A) overdrive
   (B) direct drive
   (C) reduction
   (D) neutral
5. A gear ratio of .8:1 is a(n) _____ gear.
   (A) overdrive
   (B) direct drive
   (C) reduction
   (D) neutral
6. Name the four shafts used in the transmission.
7. Technician A says that synchronizing the gears prevents gear clash. Technician B says that synchronizing the gears reduces gear damage. Who is right?
   (A) A only.
   (B) B only.
   (C) Both A & B.
   (D) Neither A nor B.
8. The manual transmission is lubricated by _____ .
   (A) an oil pump
   (B) oil pressure from the engine
   (C) splash lubrication
   (D) Both A & C.
9. Two gears that are always engaged with each other are called __________ gears.
10. What problem do thrust washers prevent?
11. The parts that are located inside of the transmission or transaxle are called shift _____ .

12. All of the following statements about four- and five-speed transmissions are true, EXCEPT:
    (A) they increase fuel mileage.
    (B) they reduce acceleration.
    (C) they are more efficient than a three-speed unit.
    (D) they are more efficient than an automatic transmission.
13. Technician A says that the inner bearing on the input shaft is a bronze bushing. Technician B says that the reverse idler gear is always supported on roller bearings. Who is right?
    (A) A only.
    (B) B only.
    (C) Both A & B.
    (D) Neither A nor B.
14. Transmission housings are made of _____ or _____ .
15. Why is speed shifting considered a foolish practice by expert drivers?

A transmission is but one component in a vehicle's drive train. However, the entire drive train can be affected if the transmission is in poor operating condition. (Ford)

Cutaway of a four-wheel drive sport utility vehicle. Can you find any four-wheel drive components in this illustration? (General Motors)

20

Four-Wheel Drive

After studying this chapter, you will be able to:

- Explain the purpose of four-wheel drive.
- Explain the operation of four-wheel drive transfer cases.
- Identify the components of four-wheel drive transfer cases.

Four-wheel drive is becoming more popular as the number of off-road and sport utility vehicles increases. Four-wheel drive is also useful for driving on wet or ice covered roads. This chapter covers the design and operation of four-wheel drive units and the differences between full-time and part-time four-wheel drive systems. Front wheel locking hubs are also discussed.

Purposes of the Four-Wheel Drive

Four-wheel drive is useful for driving off-road, when traction to all four wheels is desirable. Until recently, most four-wheel drive vehicles were either trucks or off-road vehicles. Since powering all of the wheels adds to vehicle stability when driving on roads that are covered with ice, many passenger cars and vans are now equipped with four-wheel drive. Four-wheel drive systems are used with both manual and automatic transmissions.

Four-Wheel Drive Components

The main four-wheel drive components are the transfer case, drive shafts, front and rear differentials, and a drive selection system. Some also use locking devices at the front hubs to engage the front-wheel drive. The drive shafts and differentials are identical to those used on two-wheel drive vehicles. For more information on differentials and drive shafts, refer to Chapter 22.

Four-Wheel Drive Designs

There are two types of four-wheel drive, based on engine placement. In a vehicle with conventional engine placement, the rear-wheel drive components are the same as they would be with a two-wheel drive setup. A ***transfer case*** is installed between the transmission output shaft and the drive shafts.

The transfer case directs power to the front wheels when needed. Both drive shafts are attached to the transfer case. The rear drive shaft sends power to the rear differential assembly as in a two-wheel drive system. The front drive shaft sends power to the front differential assembly. This design is by far the most common type of front-wheel drive design.

In the front-wheel drive vehicle, the front wheels are driven in the same manner as they would be on a two-wheel drive vehicle. The transfer case is installed on the transaxle, or may be a part of it. On some designs, both axles have separate differentials. Other designs have the front differential as part of the transaxle assembly. The front wheels are driven at all times. In four-wheel drive, power travels through the rear drive shaft to the rear differential assembly. The two designs are shown in **Figure 20-1.**

Transfer Case

The purpose of the transfer case is to divide the power flow so that both the front and rear wheels are driven by engine power when necessary. Engine power enters the transfer case from the transmission output shaft. Power is transmitted to each drive shaft either through a set of ***gears*** or a ***chain and sprocket*** set. **Figure 20-2** shows the power flow in four-wheel drive for the two transfer case types.

Transfer cases can be either full- or part-time units. The full-time unit with a chain drive is the most common today. Full-time systems have a ***differential assembly*** or a viscous coupling to compensate for changes in front and rear axle speeds. Both systems are explained below.

Full-Time Transfer Case

Most automobiles, vans, and some light trucks use a ***full-time transfer case***. The full-time unit delivers power to the front and rear axle assemblies all of the time. The case is constructed from either aluminum or cast iron. Most cases include a high and low range. High range is about 1:1 and low range is 2:1. **Figure 20-3** shows an exploded view of a full-time transfer case. Note the viscous coupling. Study and learn the part names and locations.

Full-Time Transfer Case Operation

The full-time transfer case drives all four wheels whenever the vehicle is moving. These transfer cases use either a differential assembly or viscous coupling to compensate for differences in front and rear axle speeds during turning and driving over irregular surfaced terrain. By compensating for different driving conditions, the full-time transfer case will permit the vehicle to be continuously operated in four-wheel drive on hard and dry surfaced highways.

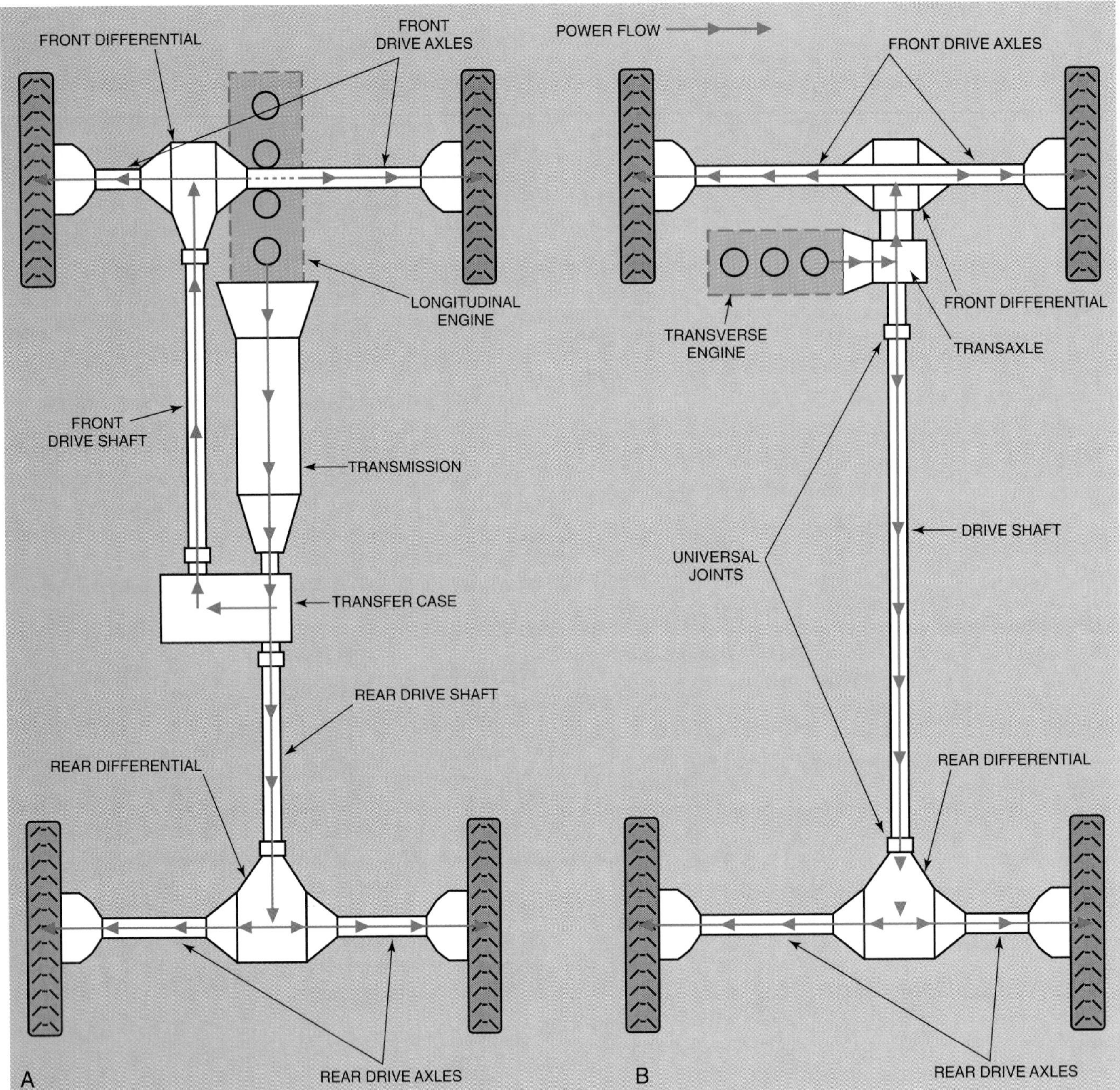

Figure 20-1. Two similar four-wheel drive systems and the power flow through each. A—A longitudinal engine and transmission with primary rear-wheel drive. B—Transverse engine position. No separate transfer case is used with this particular transaxle arrangement.

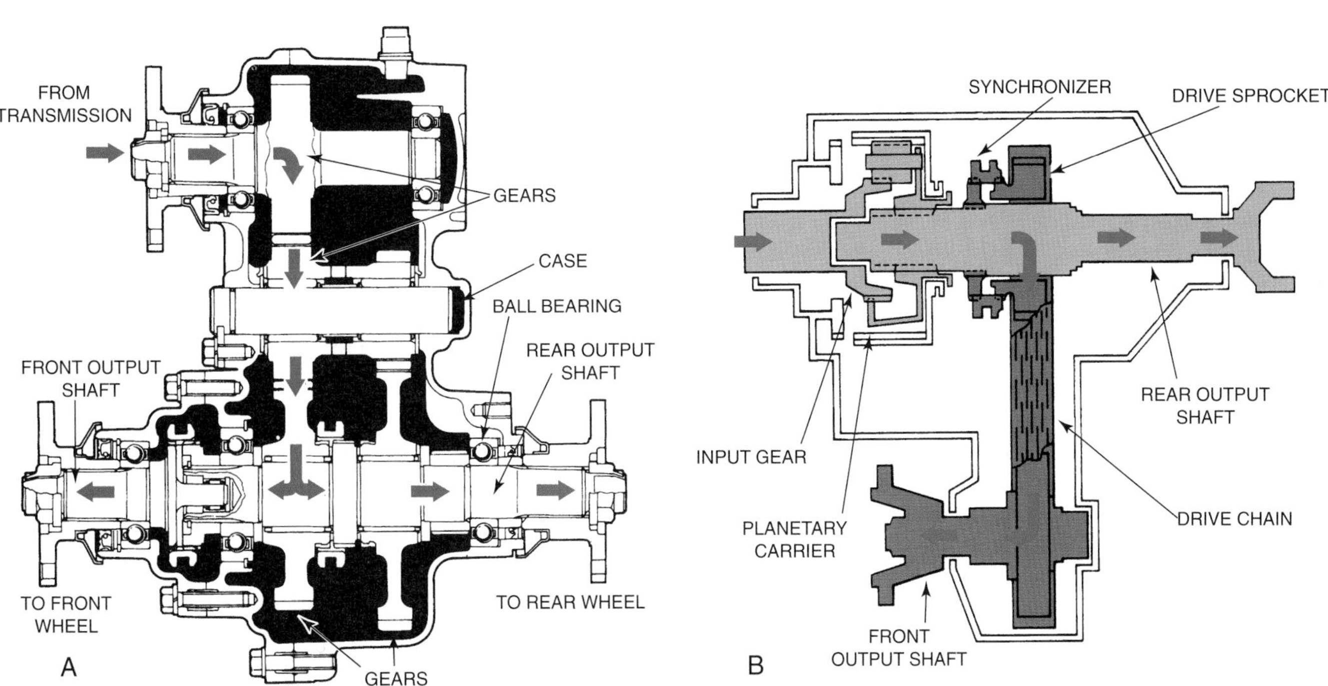

Figure 20-2. A—Power flow through a part-time gear set transfer case. Shown is the pattern for four-wheel drive high range. The front shift fork slides the sleeve of the front clutch onto the toothed clutch ring. This couples the rear output shaft to the front output shaft. The front and rear shafts now turn together—power flows to all wheels. B—Four-wheel drive, high-range power flow through a part-time, chain drive transfer case. (Suzuki and General Motors)

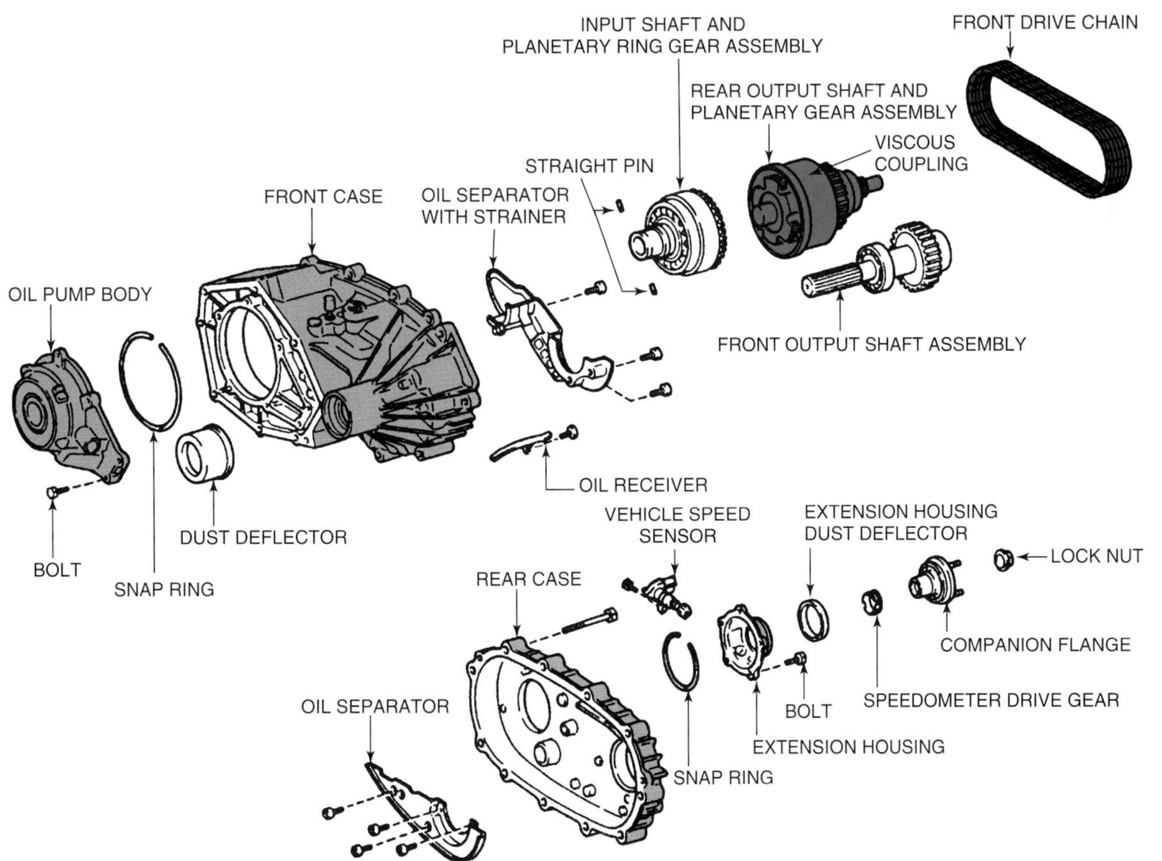

Figure 20-3. An exploded view of a viscous coupling full-time case. Note this unit uses a drive chain in place of a gear set. When the speed difference between the front and rear wheels becomes higher because of vehicle driving conditions, the speed difference is absorbed by the operation of the viscous coupling. (Toyota Motor Corporation)

Full-Time Transfer Case with Differential Unit

Note in **Figure 20-4** that the drive chain transmits power from the drive sprocket to the differential case sprocket. A differential unit inside the case applies driving torque to both front and rear transfer case output shafts. The differential allows the front and back shafts to rotate at different speeds while still applying power. This permits constant use of four-wheel drive.

This unit is equipped with a limited-slip differential that will provide some drive to either the front or rear axle if one wheel is spinning. For very severe traction situations, the unit has a mechanism to lock out (stop) the differential action. This provides direct drive to both front and rear drive shafts. An exploded view of another full-time transfer case is illustrated in **Figure 20-5**. Note the use of a viscous coupling. This transfer case is available with or without a low range reduction unit. **Figure 20-5** does not show the reduction unit.

Another type of full-time transfer case is shown in **Figure 20-6**. It uses a chain drive and a standard differential unit. Low and high range is accomplished through gearing. This unit may be locked out whenever excessive wheel spin is experienced. The lockout must be disengaged whenever the vehicle is driven on dry paved roads.

Full- and Part-Time Transfer Cases with Viscous Couplings

These transfer cases operate somewhat like the differential equipped unit above. The basic difference is the ***viscous coupling*** used in place of the mechanical limited-slip differential.

The viscous coupling consists of a series of plates with ridges or grooves. The plates do not touch. The coupling is filled with a thick silicone fluid that fills the spaces between the plates. The coupling is placed so that power to the rear wheels must pass through it. See **Figure 20-7**.

During normal operation (no wheel spin), the viscous coupling does not operate. When excessive wheel spin is present (causing pronounced speed variations between front and rear axles), the coupling transfers torque to the axle having the best traction.

The silicone fluid in the coupling is very thick, or viscous, and will not thin due to heat. As one axle overspeeds because of wheel slippage, the coupling rotational speed also increases. As the coupling speed becomes higher, the fixed clutch plates are forced to rotate in the silicone fluid as they increase in speed. As fluid is forced between the clutch plates and displaced (moved), it expands, creating friction and more resistance to higher input speed.

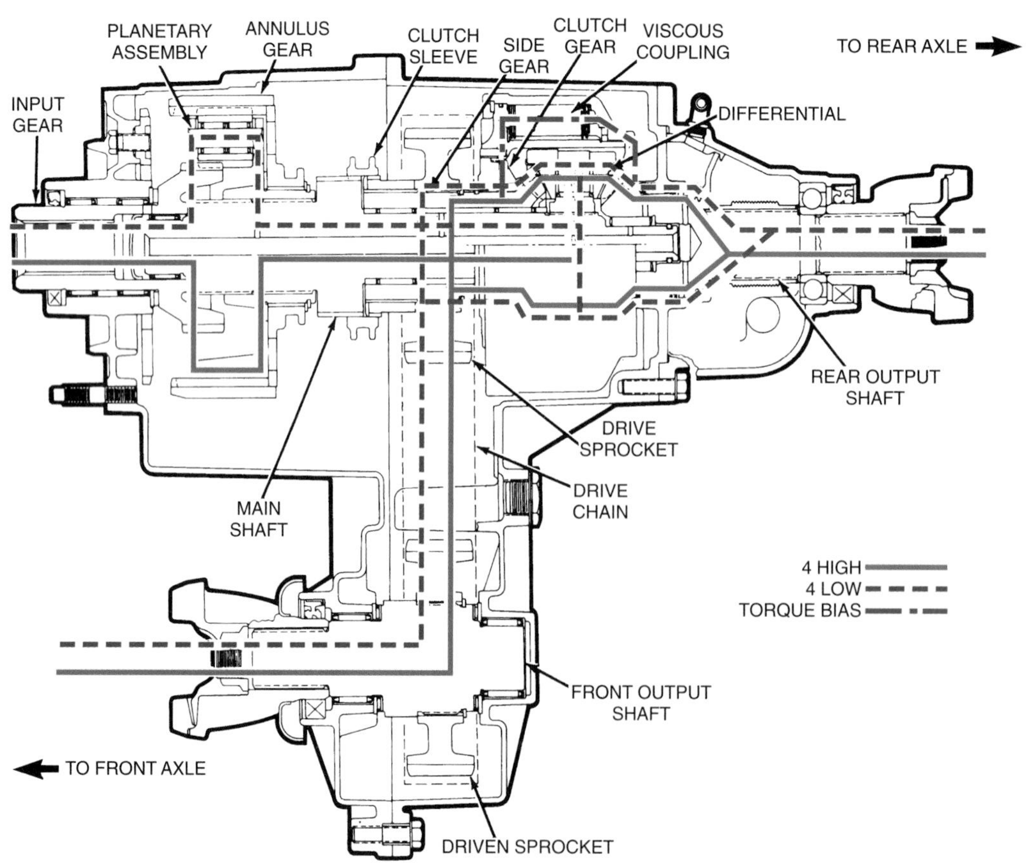

Figure 20-4. Schematic showing power flow through a full-time transfer case. Note that the power is supplied to the front and rear axles in both low and high range. (Jeep)

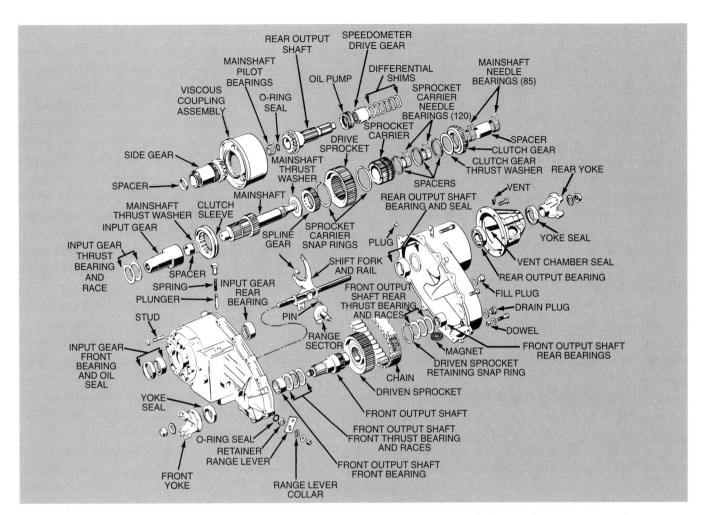

Figure 20-5. An exploded view of a full-time, select-drive transfer case. Note the magnet which is used to catch and hold any ferrous (iron) particles in the fluid. This helps to prevent premature wear on the bearings, chain, etc. (Chrysler)

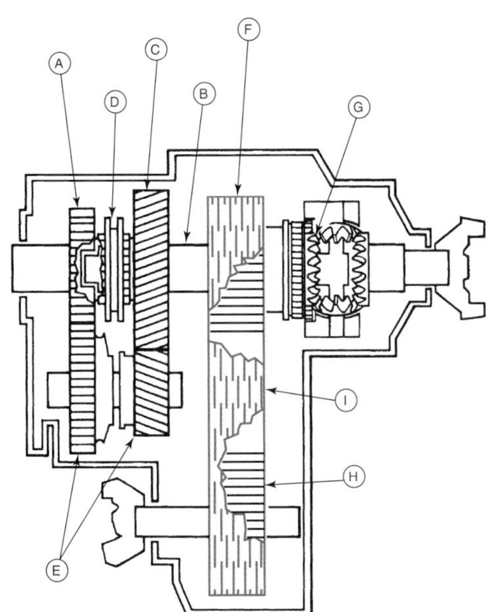

Figure 20-6. Cross-sectional view of a full-time transfer case. A—High-range input gear. B—Input shaft. C—Low-range input gear. D—Sliding clutch. E—Counter (or cluster) gears. F—Front axle drive sprocket. G—Interaxle differential front gear. H—Driven sprocket. I—Drive chain. (General Motors.)

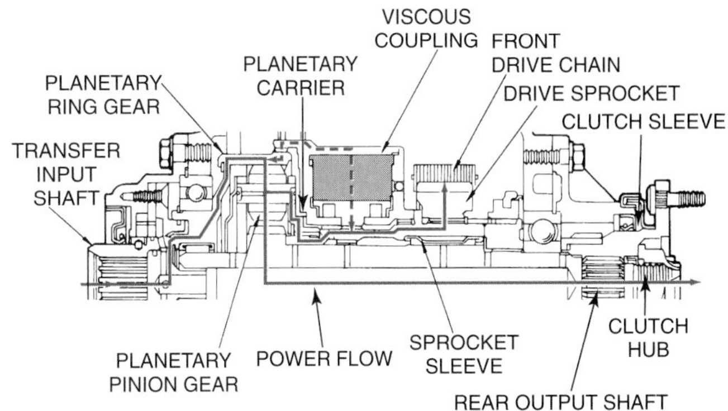

Figure 20-7. A cross-sectional view of the power flow through a full-time transfer case. Note the viscous coupling action. (Toyota Motor Corporation)

The coupling does not lock the axles together, but controls the amount of axle slippage by sending maximum torque to the slower moving axle. Note that some viscous coupling transfer cases also have an internal differential assembly. An exploded view of a viscous coupling transfer case is shown in **Figure 20-8**.

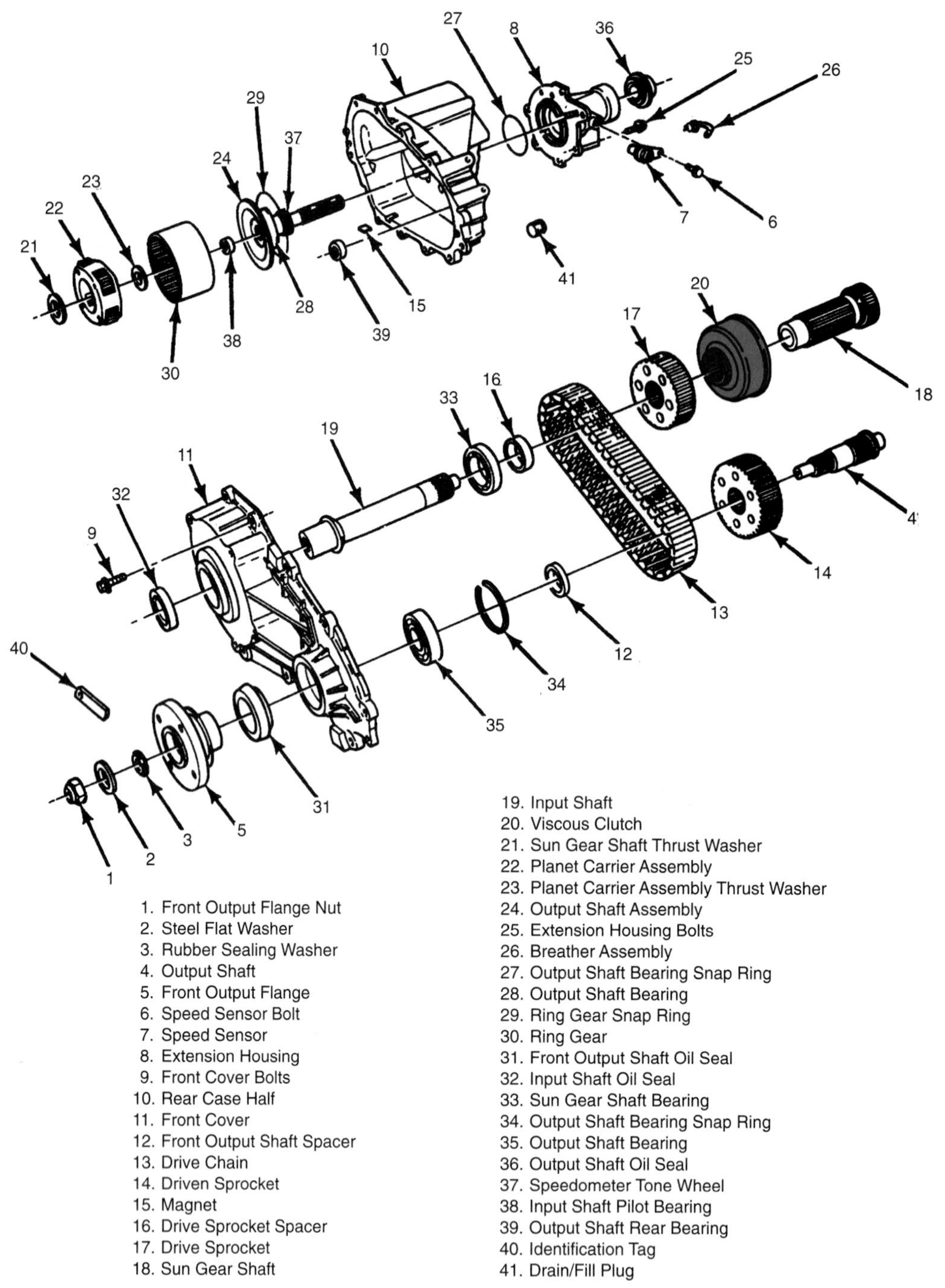

Figure 20-8. Exploded view of a chain drive transfer case which uses a viscous clutch assembly. (Chevrolet)

Part-Time Transfer Case

Many older four-wheel drive systems used a ***part-time transfer case***. The part-time transfer case is also used on some late model vehicles. The part-time unit will deliver power to the rear wheels only, or to the front and rear wheels, as determined by the driver. Part-time four-wheel drive transfer cases should not be operated on dry, hard surfaced roads.

The part-time transfer case housing is made of cast iron or aluminum. Power is transmitted through a series of gears and/or chains. Most incorporate a high and low range, which provides a gearing arrangement around 1:1 (high range) or 2.6:1 (low range). **Figure 20-9** shows an exploded view of one type of part-time transfer case. Study the parts and learn their names and locations.

Part-Time Transfer Case Selection Control

On all part-time four-wheel drive vehicles, a drive range selection control is included. The part-time transfer case is

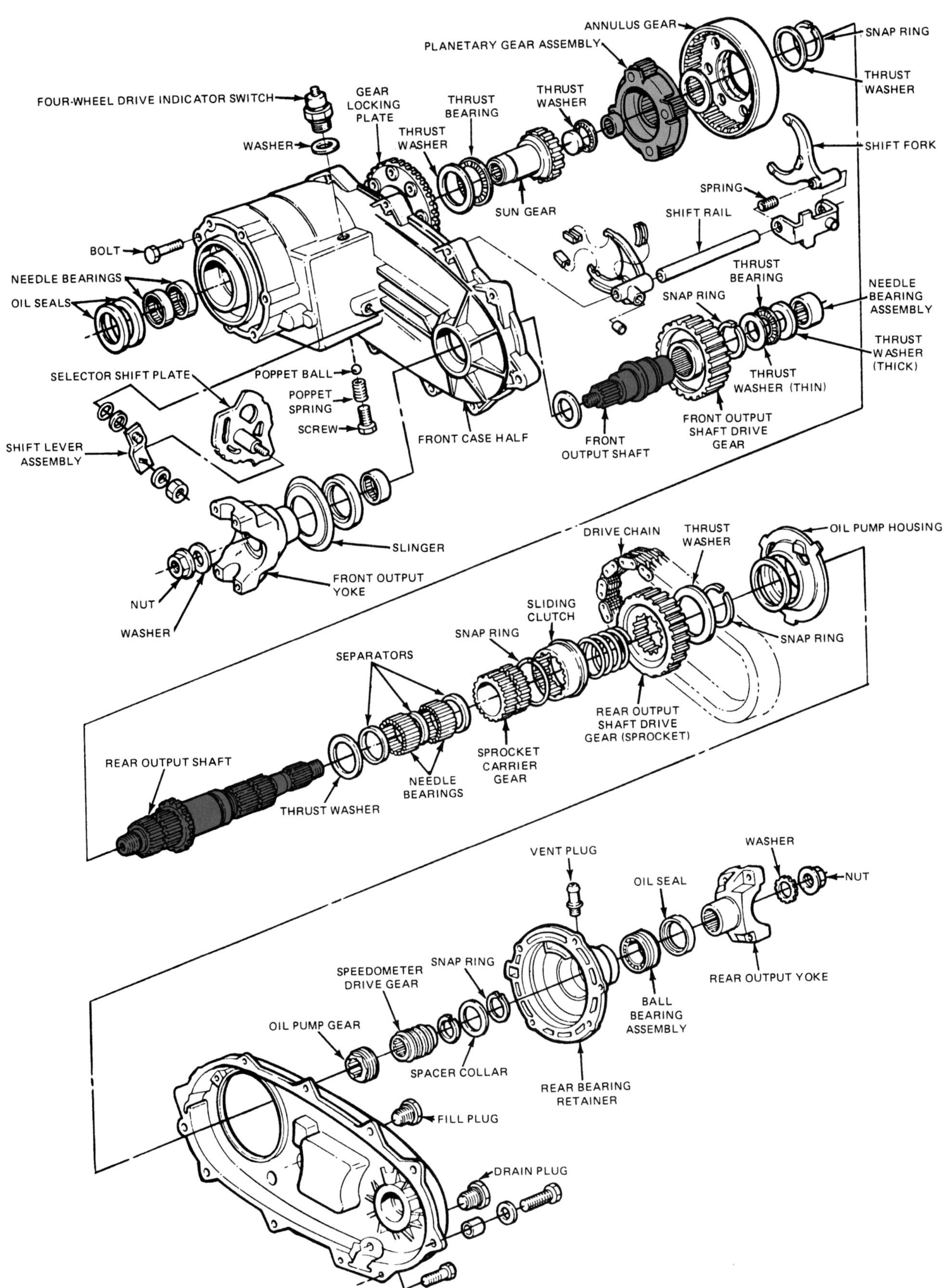

Figure 20-9. Exploded view of a transfer case which uses a two-piece input-output shaft. Note the drive chain and planetary gear assembly. (Ford Motor Company)

normally in two-wheel drive and is placed in four-wheel drive only when needed. The driver can select one of three different drive modes:

- Two-wheel drive high range.
- Four-wheel drive high range.
- Four-wheel drive low range.

A neutral range is also available for towing. These ranges are operated by either moving a lever on the vehicle center console, floor, by pressing a button, or turning a switch on the dash. Although it would be possible to operate in two-wheel drive low range, transfer case shift linkage does not allow selecting this range. A lever control for a part-time transfer case is shown in **Figure 20-10**.

Figure 20-10. This four-wheel drive truck uses a lever control to shift the part-time transfer case into its different ranges. While some vehicles use this type of lever control, electronic controlled push button or switch control is also used. (GMC)

Transfer Case Action—Low Range, High Range

Follow the power flow in **Figure 20-11**. Start with the transmission output shaft, which is splined to the transfer case main drive gear. The main drive gear is in constant mesh with the idler cluster gear. The idler turns freely in needle bearings on the idler shaft. Note that the idler cluster is made up of two gears, one large (high range) and one small (low range). These gears are in constant mesh with the output gears. Both output gears turn freely in bushings on the transfer case output shaft.

The ***sliding clutch*** is splined to the output shaft. In the neutral position, it is centered between, but not engaging the output gears. Low range (high engine RPM, low wheel RPM) is engaged when the sliding clutch is shifted into engagement with the low range output gear. High range (low engine speed RPM, high drive wheel RPM) is engaged when the sliding clutch is moved into engagement with the high range output gear. The output gear that is engaged with the clutch turns the sliding clutch. Since the clutch is splined to the output shaft, the shaft also turns.

Part-Time Four-Wheel Operation

The transfer case output shaft is made of two parts. The long section is driven by the sliding splined clutch. This long section of the output shaft, in turn, will drive the rear drive shaft. The shorter front section is connected to the long section only when the four-wheel drive sliding splined clutch (splined to and constantly turned by long section) is moved into engagement with the splined end of the short shaft.

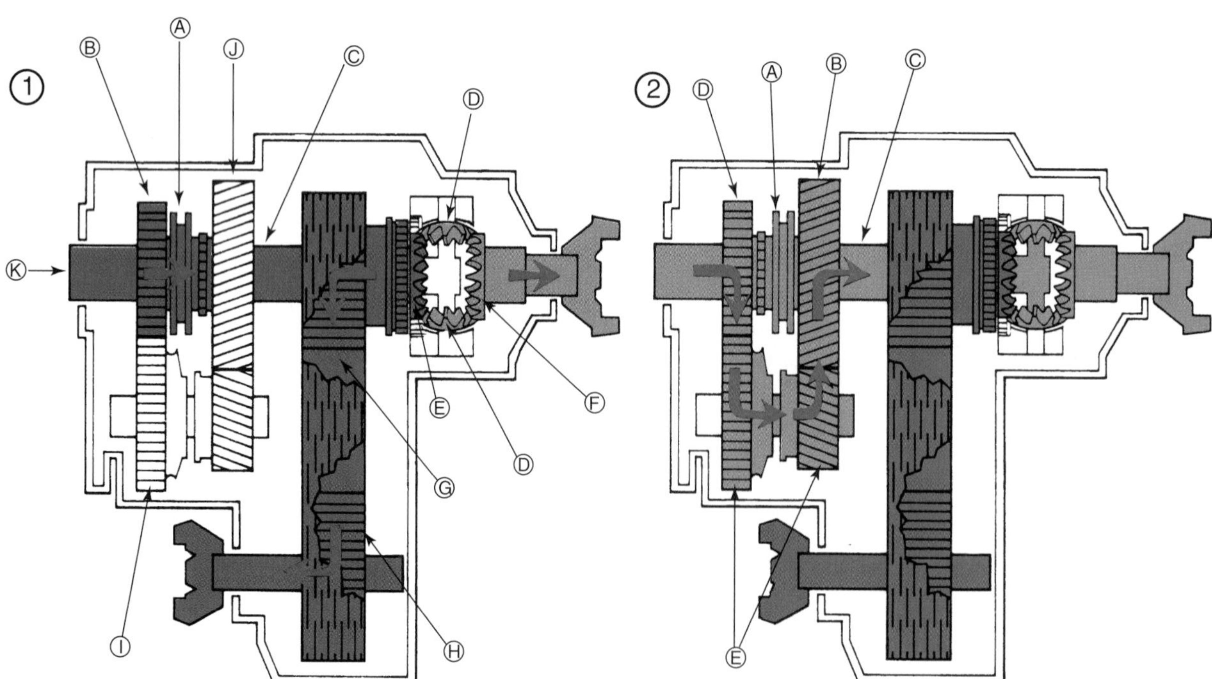

Figure 20-11. Power flow through a part-time transfer case in four-wheel, high range (1) and four-wheel low range (2). A—Sliding clutch. B—High range input gear. C—Input shaft. D—Pinion gears. E and F—Side gears. G—Drive sprocket. H—Driven sprocket. I—Counter gear. J—Low range input gear. K—Transmission output shaft. (Chevrolet)

When two-wheel drive is desired, the short output shaft is disconnected by moving the four-wheel sliding clutch out of engagement. The long shaft will then drive the rear drive shaft, but will not apply torque to the front. Four-wheel drive is engaged by moving the four-wheel clutch into engagement with the short shaft. Torque will then be applied to both front and rear drive shafts.

Vehicles equipped with this type of transfer case must never be operated in four-wheel drive on dry, hard-surfaced roads. The vehicle's front wheels will rotate slightly faster (follow a more curved path) than the rear. As a result, windup increases in the entire drive train until something breaks.

Windup is the development of internal stresses between the front and rear axle parts, caused by differences in speeds between the front and rear axles. Either the tires will break loose (slip) on the pavement, or a vehicle drive train part will be damaged. Since it is very difficult for the tires to break loose on hard, dry road surfaces, the vehicle will probably be damaged.

On another type of transfer case, **Figure 20-12**, the front drive shaft and ring gear assembly is driven at all times. A ***vacuum motor***, **Figure 20-13**, is used to disconnect the front wheels from the front axle. The vacuum motor operates a shift collar, similar to the collar used with a manual transmission synchronizer, installed in the front drive axle. Some transfer cases use a cable system to disconnect the front axle.

In **Figure 20-14**, an electrical actuator moves the shift fork and differential sleeve to engage the front axle. The actuator is filled with a gas charge which expands when heated. As the gas expands, it moves the shift fork to engage the front axle. When the electrical current to the actuator is turned off, the gas cools. The shift fork return spring moves the fork back to its resting position, disengaging the axle, **Figure 20-13**.

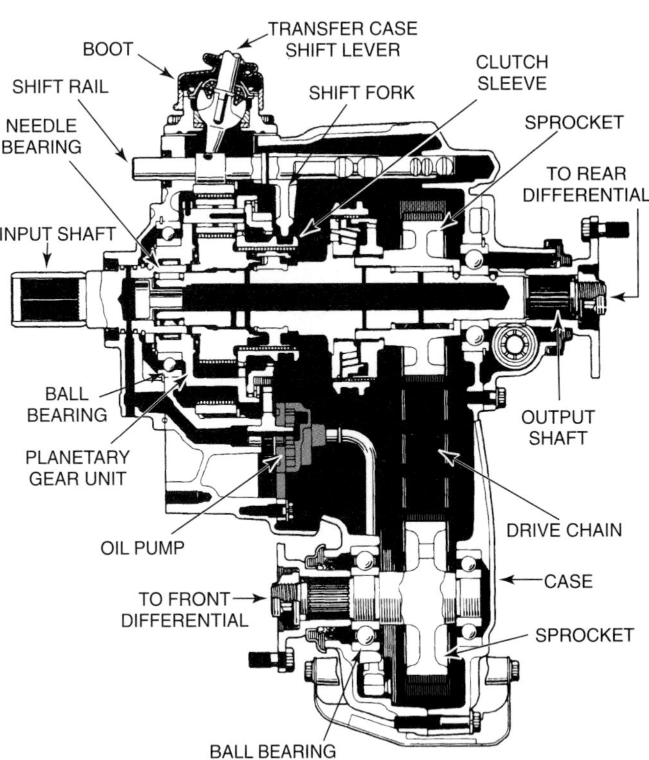

Figure 20-12. Cross-sectional view of a planetary gear-type transfer case. Gear ratios are: Two-wheel and four-wheel high range —1.00:1. Four-wheel low range—2.566:1. The case holds 1.1 quarts (1.0 liters) of SAE 75W-90 gear lubricant. Note the drive chain and the oil pump. (Toyota Motor Corporation)

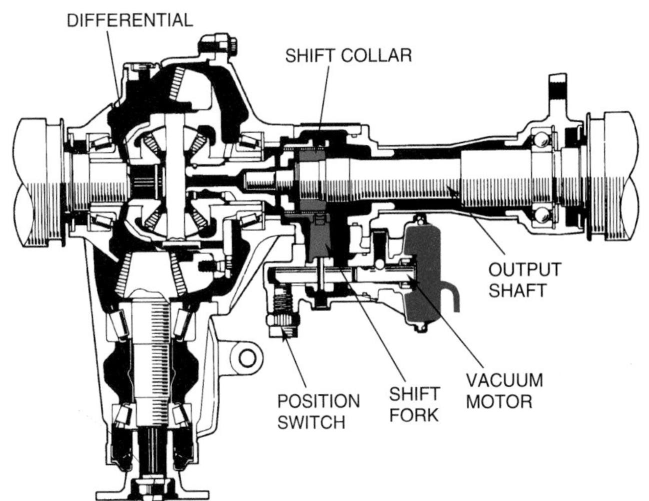

Figure 20-13. A cutaway view of the front axle and vacuum motor assembly. The vacuum motor operates the shift collar to engage or disengage the front axle. (Toyota Motor Corporation)

Transfer Case Lubrication

The transfer case is lubricated by filling the housing partially full of lubricant. The movement of the chain and gears throws lubricant throughout the case to lubricate all parts. Most have their own internal pumps. Transfer cases are lubricated by gear oil, usually 75W-90 weight or automatic transmission fluid. Transfer case leakage is controlled by oil seals at the front and rear output shafts, and by other seals, sealants, and gaskets where necessary. Viscous couplings, where used, are filled with silicone in a sealed assembly and are not serviceable.

Locking Hubs

Some four-wheel drive vehicles use front ***locking hubs*** to engage and disengage the front wheels and front drive axle. This type of arrangement is considered a part-time four-wheel drive, but is not controlled by internal parts in the transfer case. The locking hub is a device which locks the front wheels to the front axle at the wheel hub.

Some locking hubs are engaged and disengaged automatically by the use of one-way or overrunning clutches in the front drive axles. These one-way clutches will transfer power in one direction, but will freewheel (overrun) if power is applied in the other direction. These clutches lock up whenever the engine is driving the front wheels through the transfer case and unlock when the transfer case is returned to two-wheel drive operation. Other locking hubs can be automatically released by moving the transfer case to the two-wheel position and backing up the vehicle a short distance. A typical locking hub is shown in **Figure 20-15**.

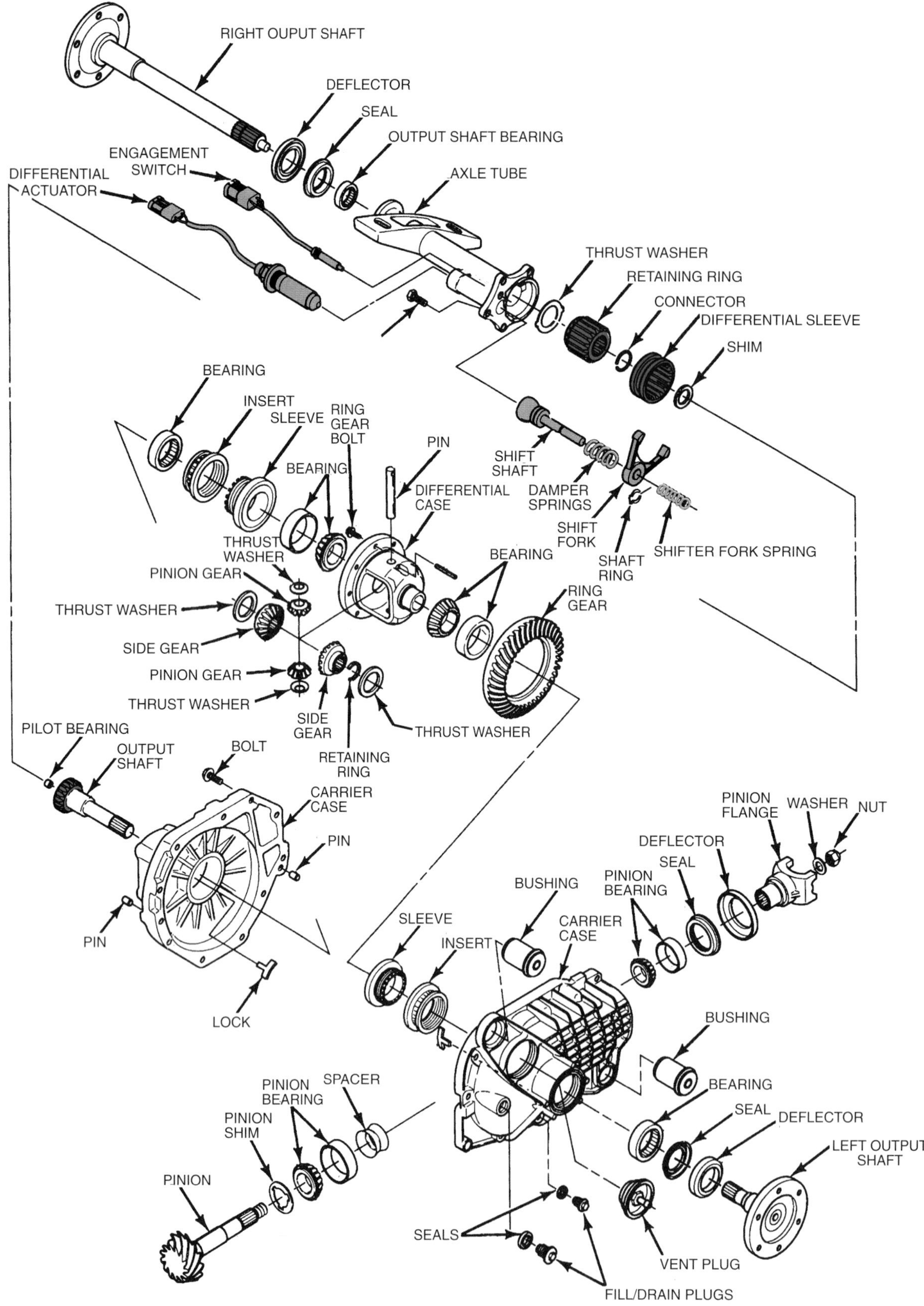

Figure 20-14. Exploded view of a four-wheel drive front axle assembly. The electric/gas differential actuator moves the shift fork, which slides the differential sleeve on and off the output shaft to engage and disengage the front axle. (General Motors)

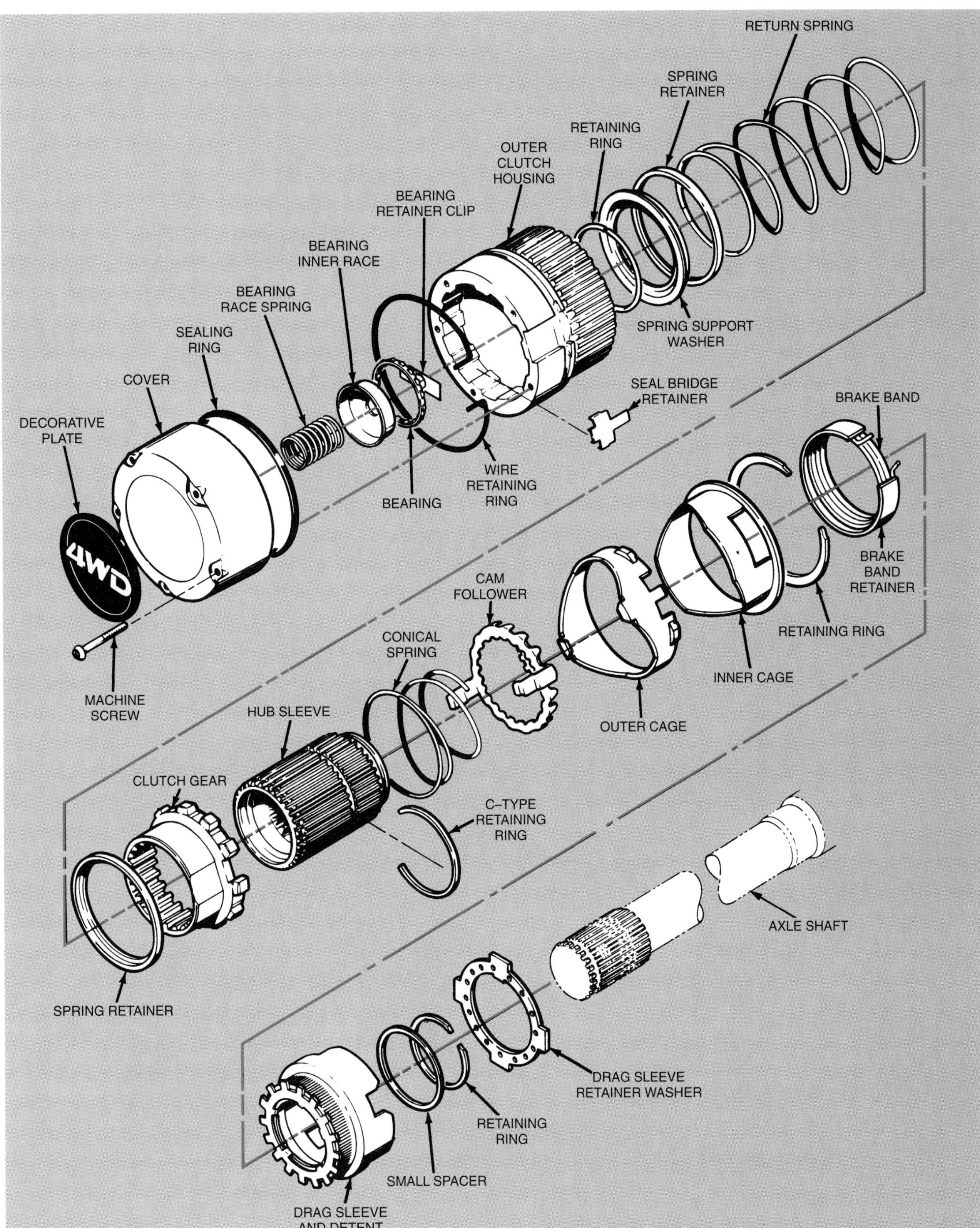

Figure 20-15. Exploded view of an automatic locking hub assembly. Study the various parts, their locations, and names. Manual locking hubs are also used. (General Motors)

Summary

Four-wheel drives are useful for driving off-road and on wet or icy roads. All four-wheel drive systems use a transfer case and most have two drive shafts. The transfer case provides a means of driving the front and rear axles at the same time. Transfer cases can be part-time or full-time designs. Transfer cases are self-contained units which attach to the rear of the transmission. Transfer case outputs are to the front and rear differential units. The front and rear differentials and drive shafts are identical to the designs used in two-wheel drive vehicles.

Transfer cases transmit power from the transmission output shaft through a set of gears or a chain and sprocket setup. In the full-time design, windup or internal stresses caused by differences in front and rear axle speeds are prevented by the use of a differential assembly or viscous clutch. The transfer case differential assembly operates in the same manner as a rear axle differential assembly. Some transfer case differentials are limited-slip units. The viscous clutch contains silicone fluid which allows differences in axle speed to be absorbed.

Part-time transfer cases use a sliding clutch to select drive positions. They are operated by moving a lever on the vehicle center console or floor or by pressing a button or turning a switch on the dash. They should never be operated in four-wheel drive on dry, paved roads. Some vehicles with full-time transfer cases make use of a vacuum motor to connect and disconnect the front axles from the front wheels. Other vehicles have locking hubs to engage and disengage the front drive axles.

Know These Terms

Four-wheel drive	Viscous coupling
Transfer case	Part-time transfer case
Gear	Sliding clutch
Chain and sprocket	Windup
Differential assembly	Vacuum motor
Full-time transfer case	Locking hubs

Review Questions—Chapter 20

Do not write in this book. Write your answers on a separate sheet of paper.

1. Technician A says that all four-wheel drive systems use a transfer case. Technician B says that all four-wheel drive systems have two drive shafts. Who is right?
   (A) A only.
   (B) B only.
   (C) Both A & B.
   (D) Neither A nor B.
2. The purpose of the transfer case is to divide the power flow so that both the front and rear wheels are driven by _____ .
3. If the vehicle has part-time four-wheel drive, what determines the drive position?
4. Some full-time transfer cases allow the vehicle to be driven on dry, hard surfaced roads constantly. They are equipped with a _____ .
   (A) differential assembly
   (B) viscous coupling
   (C) locking hub
   (D) Both A & B.
5. Transfer cases are generally attached to the _____ .
6. The most common type of internal transfer case drive system is the _____ drive.
   (A) chain
   (B) viscous coupling
   (C) gear
   (D) shaft
7. All of the following drive combinations can be selected on a part-time four-wheel drive system, EXCEPT:
   (A) two-wheel drive high range.
   (B) two-wheel drive low range.
   (C) four-wheel drive high range.
   (D) four-wheel drive low range.
8. Technician A says that a transfer case internal differential unit can be equipped with a limited-slip type differential. Technician B says that the viscous coupling is filled with heavy oil for added traction. Who is right?
   (A) A only.
   (B) B only.
   (C) Both A & B.
   (D) Neither A nor B.
9. One type of four-wheel drive system uses a vacuum motor mounted on the _____ .
   (A) transfer case
   (B) transmission
   (C) front axle
   (D) rear axle
10. All of the following statements about front locking hubs are true, EXCEPT:
   (A) the locking hubs are controlled by internal parts in the transfer case.
   (B) the locking hub locks the front wheels to the front axle at the wheel hub.
   (C) some locking hubs are operated by overrunning clutches.
   (D) some locking hubs can be automatically released by selecting the two-wheel position and backing up the vehicle.

21

Automatic Transmissions and Transaxles

After studying this chapter, you will be able to:

- Describe the parts of an automatic transmission.
- Describe the parts of an automatic transaxle.
- Explain how both automatic transmissions and transaxles function.
- Identify the major components of an automatic transmission or transaxle.
- Compare the different types of automatic transmissions and transaxles.

This chapter discusses the basic operating principles of automatic transmissions and transaxles. The automatic transmission and transaxle is in almost universal use and the technician must have a good basic understanding of its operation to successfully diagnose and repair other parts of the vehicle.

In most shops, transmission repair is a specialty area that pays very well and will always be in demand. This chapter will allow you to gain a basic knowledge of the inner workings of the automatic transmission and transaxle.

Transmissions and Transaxles

Like the manual transmission, ***automatic transmissions***, **Figure 21-1**, and ***automatic transaxles*** are designed to adapt the power of the engine to meet varying road and load conditions. In this case, however, the transmission or transaxle does this automatically. Instead of using set forward gear ratios, it can produce an infinite number of ratios between the engine and wheels. After the driver has selected the range, the transmission or transaxle shifts itself up or down depending on road speed, throttle position, and engine loading.

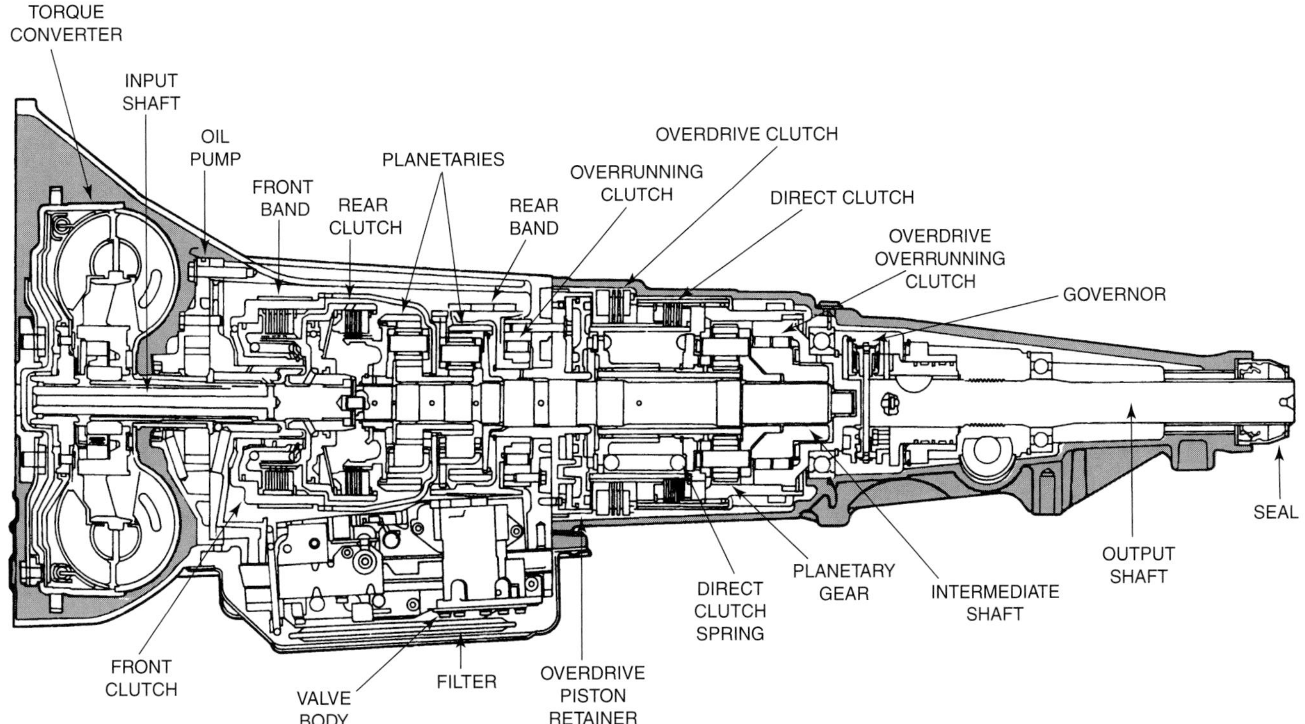

Figure 21-1. Cross-sectional view of a four-speed automatic overdrive transmission. Study the various parts carefully. (Chrysler)

Automatic Transmission and Transaxle Components

The automatic transmission and transaxle assembly use many similar parts. Every automatic transmission and transaxle contain the following parts:

- A torque converter to transmit power from the engine to the transmission.
- One or more planetary gearsets and shafts.
- A series of holding members, such as bands, multiple disc clutches, and overrunning clutches to control the planetary gearsets.
- Hydraulic servos and pistons to actuate the bands and clutches.
- One or more oil pumps to provide the necessary hydraulic pressure.
- Hydraulic control valves to control and direct hydraulic pressures.
- A means of cooling the fluid.
- Manual control system used by the driver to select the drive range.

Compare **Figures 21-1** and **21-2**. Note that while automatic transmissions and transaxles look completely different, the internal components, and their operation are remarkably similar. The major differences in the transaxle are:

- The transaxle has two output axle shafts.
- The transaxle valve body may be mounted on the side of the case, instead of the bottom, **Figure 21-2**.
- Some transaxle torque converters are separate from the rest of the unit. The torque converter drives the transmission via a drive chain.
- The transaxle usually contains an integral differential assembly, **Figure 21-2**.
- Some transaxles used on vehicles with longitudinal engines may have a separate differential bolted to the transmission section.

A unique transaxle design is the continuously variable transmission (CVT), **Figure 21-3**. This automatic transmission provides "stepless" shifting, which cannot be felt like those in a conventional automatic transmission. A multi-segmented metal drive belt with variable diameter pulleys is used to

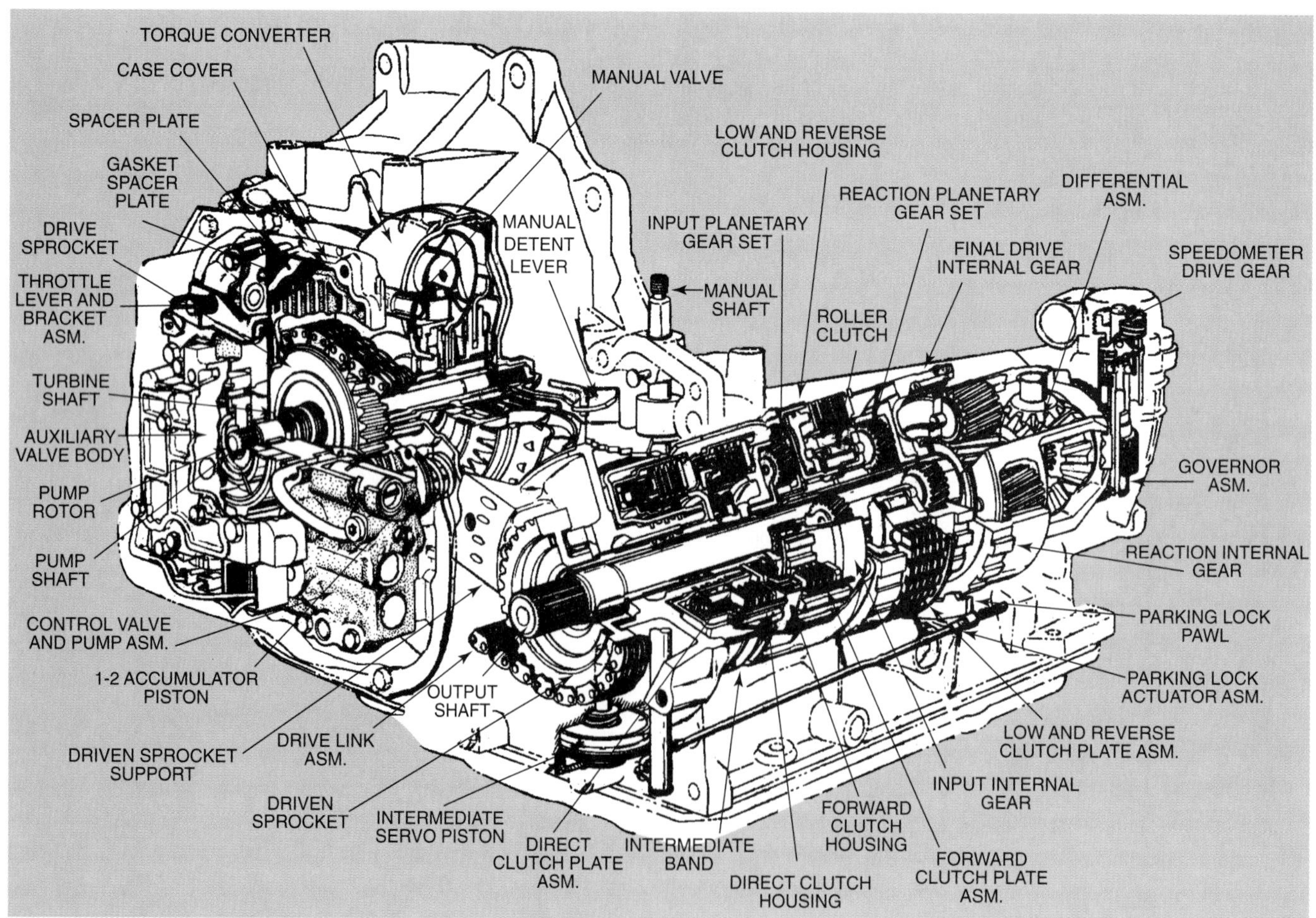

Figure 21-2. Cutaway view of a three-speed automatic transaxle. Study the part arrangements and their relationship to one another. (General Motors)

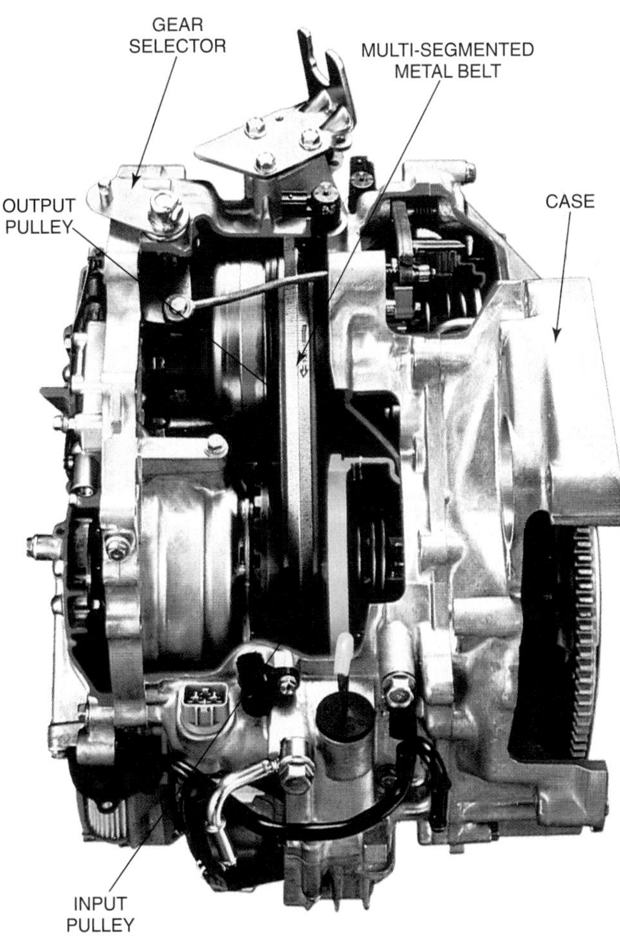

Figure 21-3. A cutaway view of a Continuously Variable Automatic Transmission (CVT). Note the use of the multi-segmented steel drive belt and pulleys. (Honda)

provide different gear ratios. The width of the pulleys are adjusted electronically to provide an infinite number of gear ratios.

Torque Converter

The automatic transmission and transaxle uses a fluid torque converter instead of a solid conventional clutch between the engine and transmission. All modern vehicles use ***torque converters*** as a way of transmitting power through the use of hydraulic fluid.

The advantage of the torque converter is that it allows the engine to continue running while the vehicle is stopped and in gear, eliminating the clutch pedal. Many older vehicles used a similar device called a ***fluid coupling***. However, the torque converter is more efficient than the fluid coupling and has been used on all vehicles with automatic transmissions manufactured in the last 30 years.

Building a Torque Converter

In the beginning of this text, you created an engine on paper. Here, we will construct a torque converter. To begin making a simple torque converter, pretend that you have a hollow steel doughnut. Start by cutting the hollow steel doughnut down the center, **Figure 21-4**. You now have two halves, each shaped like the unit shown in **Figure 21-5**. Vanes or fins are placed in each half, spaced at equal intervals, **Figure 21-6**.

Figure 21-4. Cutting through a hollow steel doughnut.

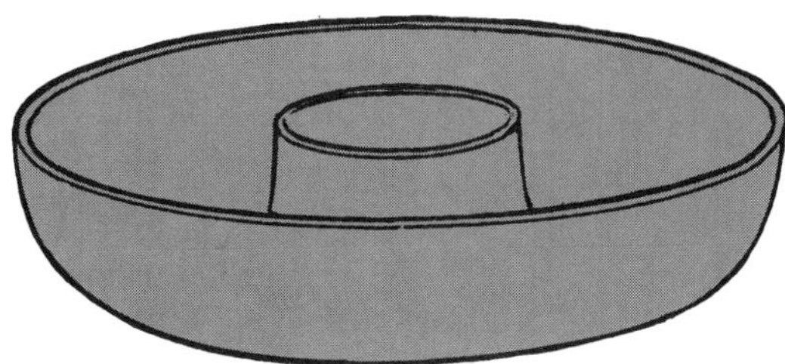

Figure 21-5. One half of the hollow steel doughnut.

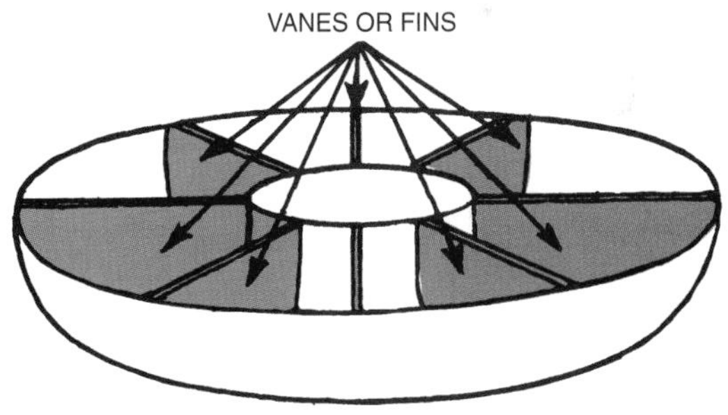

Figure 21-6. Vanes are placed in each half of the torque converter shell. Notice that the vanes are straight and equally spaced. In an actual torque converter, a greater number of vanes are used.

The two halves of the doughnut are fastened to shafts. One half is attached to the engine crankshaft through the flywheel. The half attached to the flywheel is called the ***impeller***, or pump. The other half is splined to the transmission input shaft. The housing that holds the impeller also extends around the half that is attached to the transmission input shaft. This half, called the ***turbine***, can turn inside of the housing. They are placed face-to-face with a slight clearance between them, **Figure 21-7**. The housing is filled with fluid, so there is no mechanical connection between the impeller and turbine.

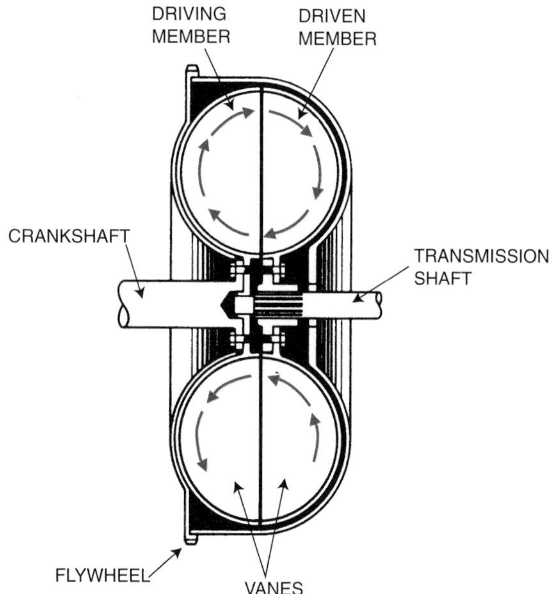

Figure 21-7. Torque converter members in place. One member is fastened to each shaft. Notice how the transmission shaft is supported by a bushing inside the end of the crankshaft.

Hydraulic Coupling

What we have constructed so far is a simple ***hydraulic coupling***. Although it is not very efficient, it is the basis for the torque converter. To understand how it works, imagine two fans facing each other closely. If one fan were running, the air blast from the driving fan would cause the other fan to spin. Air is the medium of power transfer. Since the two fans are not in an enclosed area, the power transfer would not be very efficient, **Figure 21-8**.

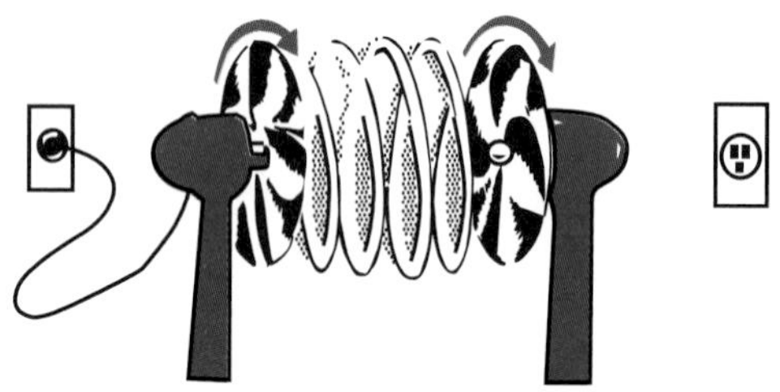

Figure 21-8. An example of a coupling drive using fans and air. Air from the driving fan will cause the other fan to spin. (Chevrolet)

The hydraulic coupling works on the same principle as the fan. However, fluid is used instead of air. When fluid is placed in a spinning horizontal impeller, the vanes carry the fluid around at high speed. This fluid motion is called ***rotary flow***. The centrifugal force of the rotating impeller will cause the fluid to fly outward and upward, **Figure 21-9**.

When the turbine is placed next to the impeller, the fluid thrown out by the impeller strikes the vanes in the turbine. See **Figure 21-10**. After striking the vanes, it travels up, over, and back down into the impeller. Since the impeller is continuing to turn, the fluid would again be thrown outward and upward against the vanes of the turbine. This circular motion of the fluid is termed ***vortex flow***.

Note this action in **Figure 21-10**. Engine power is transferred by the fluid from the impeller to the turbine. The impeller moves the fluid, which strikes the turbine and causes it to move. Therefore, power transfer is from impeller to fluid to turbine. To assist the fluid in maintaining a smooth vortex flow, a hollow ring is placed in each member. The fluid is guided around, reducing the tendency for the fluid to work against itself in the center area, **Figure 21-11**.

The hydraulic coupling can be used instead of a manual clutch. At idle, the movement of the fluid striking the turbine cannot overcome the vehicle brakes. There is no power transfer, even though the engine continues to run. When the brakes are released and the engine is accelerated, power

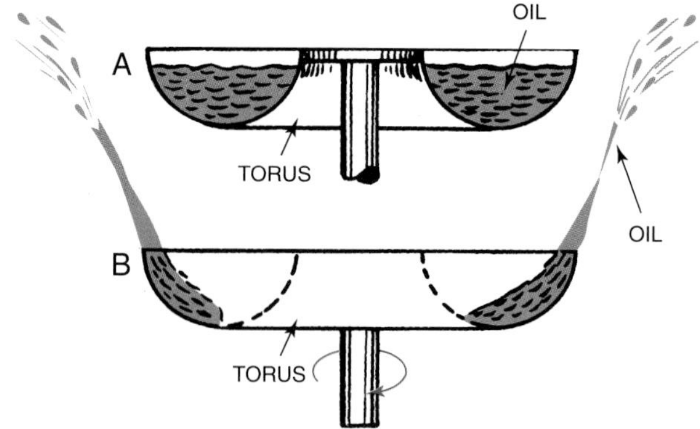

Figure 21-9. Oil action in a torque converter. A—Cutaway of torque converter at rest. Oil is level. B—As the torque converter is spinning, oil is thrown outward and is directed upward by the curved torque converter edge.

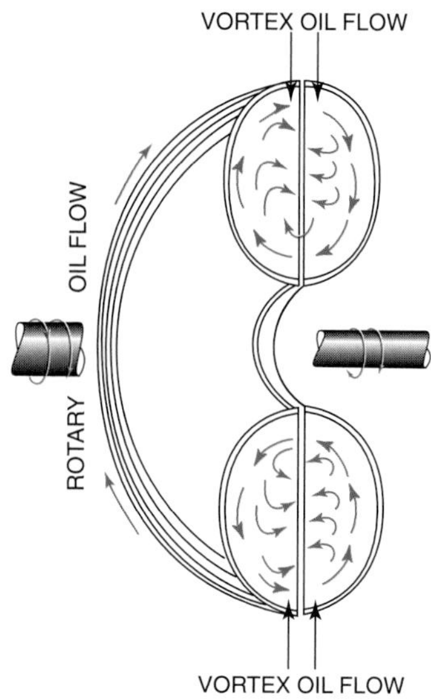

Figure 21-10. Rotary and vortex oil flow. When the impeller starts to spin, oil is thrown out and up into the turbine, around and back into the impeller. This is termed vortex action.

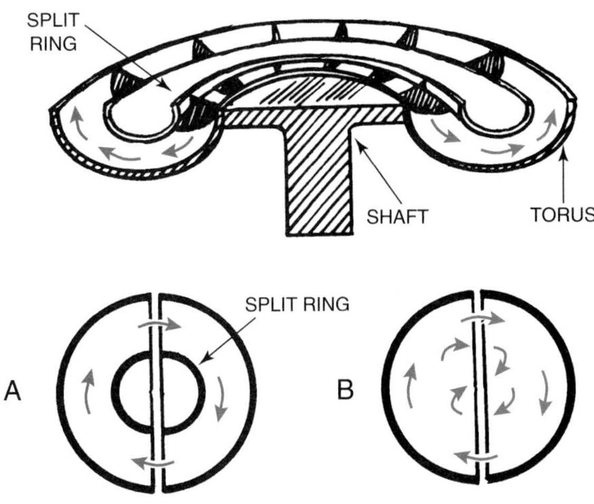

Figure 21-11. A split ring is used to smooth vortex flow. Top view illustrates split ring as it is installed in a torque converter. Vanes are cut away to receive ring. A—Note how the oil vortex flow is guided and smoothed by the split ring. B—This shows the turbulence present in the center of a torque converter when a split ring is not used.

flows smoothly from the impeller to the turbine and out through the transmission. This hydraulic coupling provides a smooth power transfer and reduces drive train wear. When coasting, the turbine tries to drive the impeller and allows the engine to act as a brake.

Hydraulic Coupling Efficiency

At normal vehicle speed, the coupling is very efficient. ***Slippage*** (the difference between the speeds of the impeller and turbine) is often less than 1%. However, at low engine speeds or during heavy acceleration, the fluid coupling is very inefficient. When the fluid strikes the turbine blades, it transfers its energy to the turbine. Therefore, when the fluid reenters the impeller, it is traveling at turbine speed.

At low vehicle speeds, turbine speed is much slower than the impeller. In fact, the fluid from the turbine is actually turning in the opposite direction of impeller rotation. The impeller must use engine power to reverse the fluid flow before returning it to the turbine. This wastes energy and makes torque multiplication impossible. This constant slippage in the coupling creates friction between the blades and fluid, which can overheat the fluid.

Stator

To make the impeller and turbine more efficient, we must add a third part called a ***stator*** to our torque converter. The stator is a small assembly made of curved blades, **Figure 21-12**. It is placed between the impeller and turbine. The job of the stator is to intercept the fluid thrown off by the turbine and redirect the fluid's path so it will enter the impeller in the same direction as impeller rotation. See **Figure 21-13**.

The impeller and turbine blades are curved to work more efficiently with the stator blades. As the impeller begins to spin, fluid is thrown outward into the curved vanes of the turbine. The fluid then circulates around through the turbine vanes. Instead of being discharged back into the impeller vanes, as in the simple hydraulic coupling, the fluid strikes the stator.

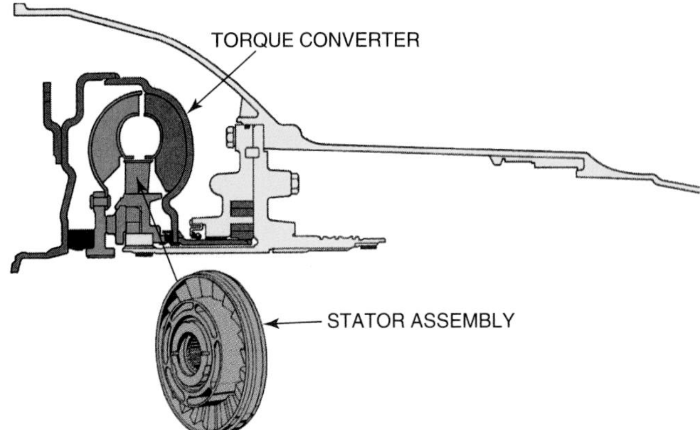

Figure 21-12. A stator assembly and its mounting location inside a torque converter. (General Motors)

The stator blades are curved to intercept the fluid leaving the turbine blades, **Figure 21-13**. The stator blades change the direction of flow fluid so that it enters the impeller in the same direction as impeller rotation. Instead of wasting engine power to reverse the fluid, the impeller is actually helped to turn. This reduces power loss and allows the engine torque to be multiplied. A typical torque multiplication figure is about 2:1.

The stator is mounted on a stationary shaft and can only spin in the direction of impeller rotation. However, at low vehicle speeds, the fluid coming from the turbine will attempt to move it in the opposite direction. To solve this problem, the stator will lock to the shaft at low vehicle speeds.

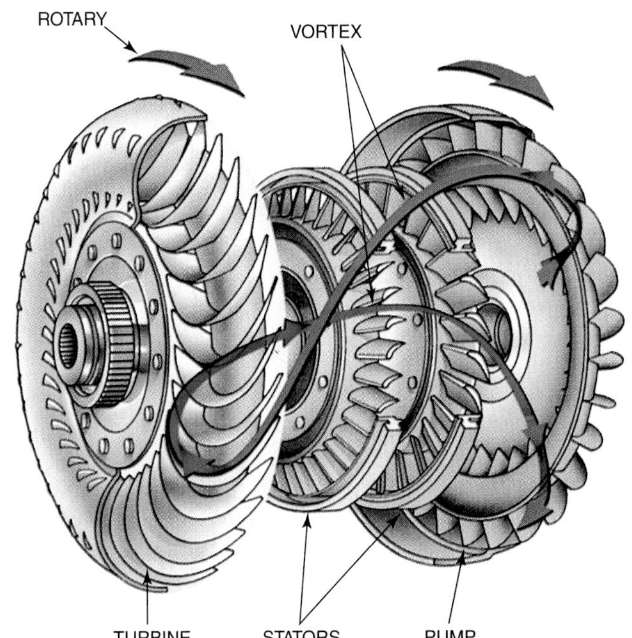

Figure 21-13. Rotary and vortex flow with stators. (Hydramatic Div. of General Motors)

When the vehicle reaches cruising speed, the turbine will be rotating at almost the same speed as the impeller and the stator will begin to impede the flow of fluid. However, at cruising speeds, fluid will strike the stator in the same direction as impeller rotation. This will cause it to unlock and the converter will function as a simple hydraulic coupling.

Stator locking and unlocking action is produced by using a one-way or overrunning clutch. Overrunning clutches can be of the sprag or roller type. These will be discussed in more detail under holding members in this chapter. Note the location of the stator one-way clutch in **Figure 21-14**.

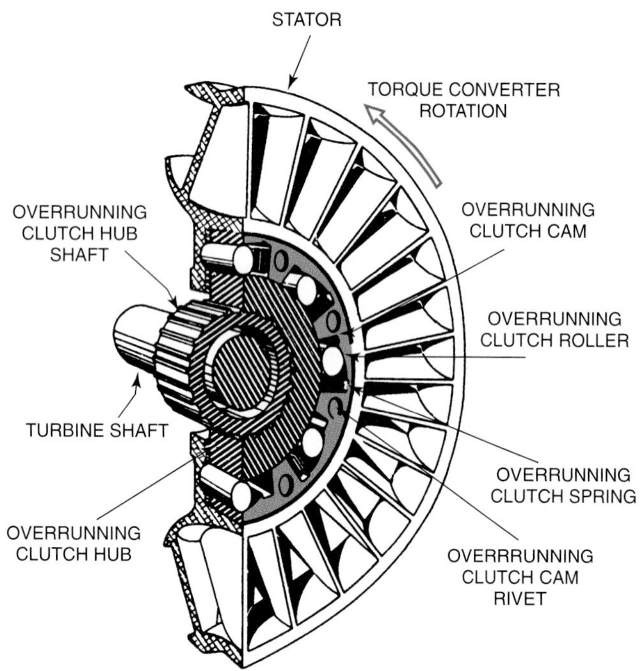

Figure 21-14. This stator unit uses a one-way roller clutch. Notice that the stator can turn in only one direction similar to an overrunning clutch.

Torque Converter Operation

The preceding sections covered the individual components of the torque converter. The following sections discuss the operation of the converter assembly. The following section describes the action of the converter when it is accelerating from rest and when it reaches cruising speed.

Accelerating from a Stop

With the transmission in drive range and engine idling (rest), there is very little transfer of torque from the impeller to the turbine when the vehicle is standing still. As the engine is accelerated from a stop, impeller speed increases rapidly. As impeller speed increases, fluid is thrown into the turbine with increasing force. Leaving the turbine, the fluid strikes the stator, **Figure 21-15**.

As the stator is forced backward by the fluid, the one-way clutch will lock it up. The fluid flow is then diverted before it reaches the impeller. As this vortex flow increases in speed, more and more torque is applied to the turbine. The maximum torque multiplication is delivered when the impeller has reached its highest velocity, and the turbine is standing still or at ***stall***. This condition is shown in **Figure 21-15**.

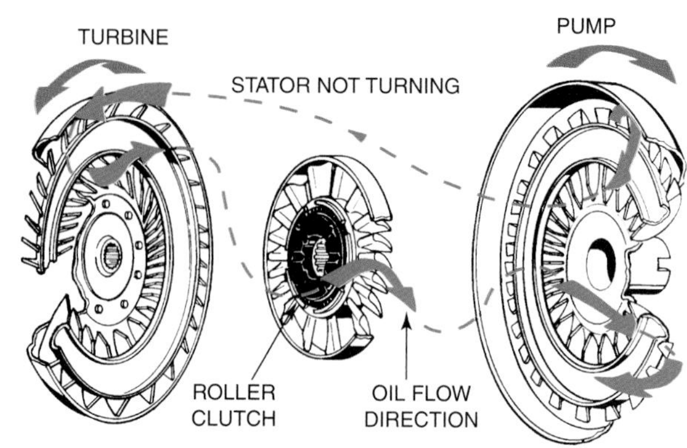

Figure 21-15. The turbine is at stall and the stator is not turning at this time. The roller clutch prevents the stator from turning in a counterclockwise direction. (General Motors)

Cruising Speed

As the turbine begins to turn, torque multiplication tapers off. When turbine speed increases to about 90% of impeller speed, the fluid leaving the trailing edges of the turbine will change its angle so that it begins to strike the rear face of the stator.

This change of angle is caused by the fact that even though the fluid is being thrown toward the stator, the turbine is moving by the stator faster than the fluid is moving toward the stator. As the turbine speed surpasses the fluid speed, the fluid strikes the back of the stator blades, causing the stator overrunning clutch to unlock and freewheel with the turbine. Since there is no torque multiplication, the converter performs much like a fluid coupling. This condition is known as the ***coupling point***, **Figure 21-16**.

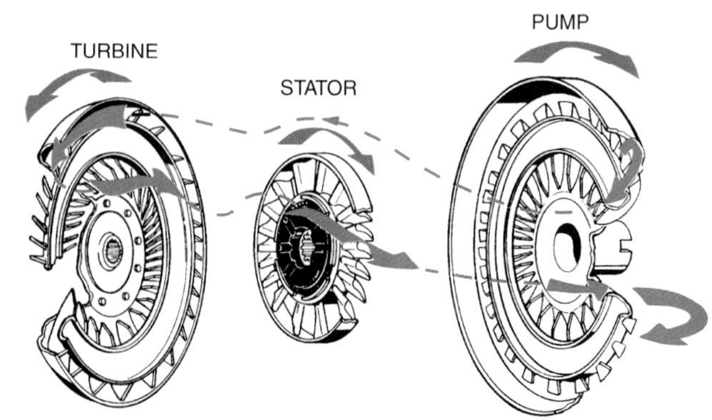

Figure 21-16. Torque converter "coupling" point. At this time, the converter is acting as a fluid coupling. The turbine and pump are turning at approximately the same speed. The stator has become neutral (inactive). (General Motors)

Torque Multiplication Curve

The ratio between ***torque multiplication*** with the turbine at stall and the coupling point is illustrated in **Figure 21-17**. Notice how torque multiplication drops off as turbine speed increases. The torque converter provides a variable drive ratio as opposed to the manual transmission's three or four fixed ratios. This provides a smooth flow of power that automatically adjusts to varying load conditions within the limit of the unit.

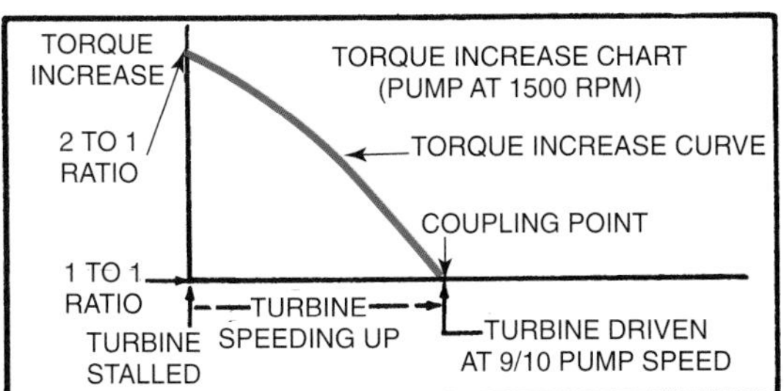

Figure 21-17. A torque curve chart. (Ford)

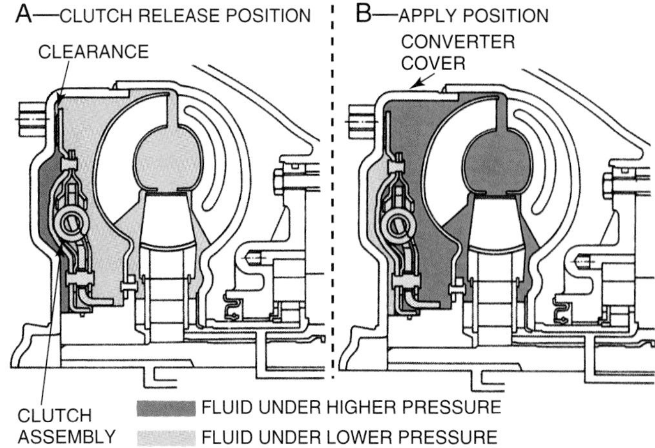

Figure 21-18 A—Lock-up clutch in the released position. Note the clearance between the clutch friction surface and converter housing. B—Clutch is in the applied position. A mechanical connection between the pump housing and turbine now exists. (LUK Automotive Systems)

Lockup Torque Converters

The normal converter allows some slippage, even at cruising speeds. This is due to the fact that the only connection between the pump and turbine is the transmission fluid. To prevent this slippage and improve fuel economy, most modern converters are equipped with a ***lockup clutch*** or ***torque converter clutch***. The entire assembly is called a ***lockup torque converter***.

Lockup Torque Converter Operation

Before lockup, the turbine and impeller are mechanically free of each other and drive is through the transmission fluid. There is no contact between the turbine and clutch friction surface. The lockup converter operates like a conventional torque converter. See **Figure 21-18A**.

When lockup conditions are present, fluid will move through a passage in the transmission input shaft. It will flow into the space between the turbine and clutch apply piston. The clutch apply piston will engage the clutch friction surface to lock the converter housing and turbine together. This action is shown in **Figure 21-18B**.

When the lockup clutch is activated, the engine and transmission input shaft are mechanically locked together. Slippage is reduced to zero and the vehicle gets better mileage. The elimination of slippage between the converter blades and the fluid also eliminates the major source of transmission fluid heating, extending fluid life.

Lockup Converter Control System

The lockup clutch is never applied until the vehicle is at a certain speed, or the transmission is in at least second gear. If the converter was locked at idle, it would stall the engine. To allow converter lockup, most computer control systems are designed so that the transmission must be in third or fourth gear, with vehicle speed above 35 to 40 mph (56-64 kmh) and without an excessive engine load (such as from steep hills or heavy acceleration). There are several lockup control systems, depending on the vehicle manufacturer. This will be discussed in more detail under hydraulic control systems.

Planetary Gearsets

Additional torque multiplication must be obtained for acceptable vehicle operation. Gears are needed to accomplish this. To accomplish this, the automatic transmission makes use of planetary gears.

A ***planetary gearset*** can be used to obtain reduction, overdrive, and direct drive gear ratios. The planetary gearset can also be used to reverse direction. The planetary gearset is compact and is in constant mesh. **Figure 21-19** illustrates a typical planetary gearset.

The planetary gearset is so named because of its resemblance to the solar system. The central gear is the ***sun gear***. Circling around the sun gear on their own shafts are the ***planet gears*** or pinions. The planet gears are held in a ***planet***

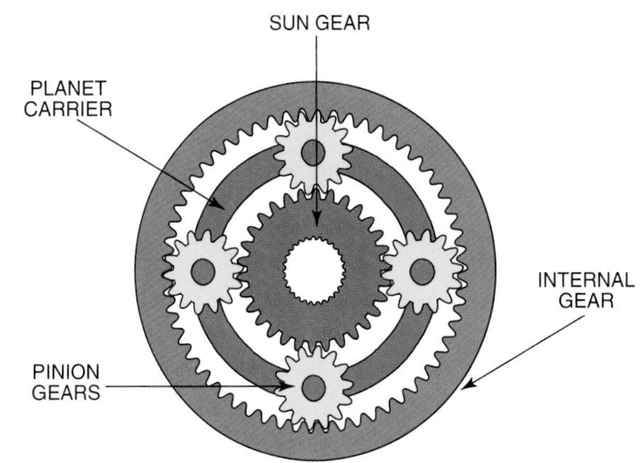

Figure 21-19. A typical planetary gearset. (General Motors)

carrier, but can rotate on their shafts. The outer gear is called the internal gear or ***ring gear***. All planetary gearsets use this basic arrangement.

Torque Increase

If the ring gear is held and power is applied to the sun gear, the planet gears are forced to turn and walk around the internal gear. This action causes the pinion carrier to revolve at a lower speed than the sun gear. Speed reduction and torque increase are large, **Figure 21-20**.

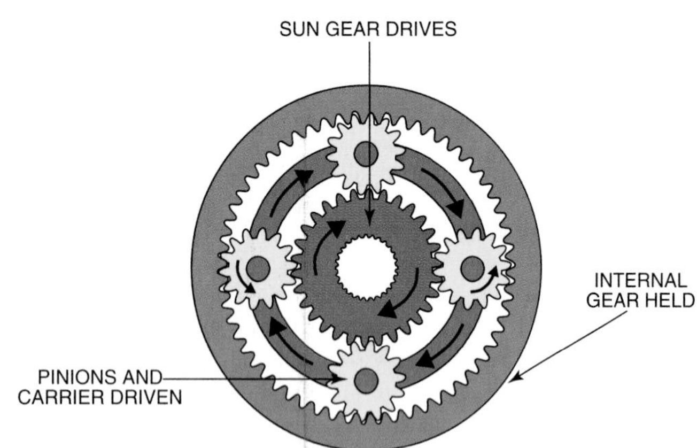

Figure 21-20. A planetary gearset in a large torque increase. (General Motors)

If the sun gear is held and the ring gear is driven, the planet gears will walk around the sun gear. This causes the pinion carrier to move slower than the internal gear. Torque will increase, however, the speed decrease is small. See **Figure 21-21**.

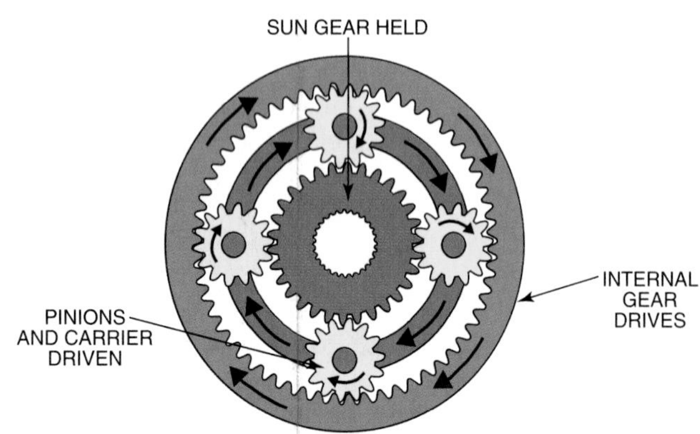

Figure 21-21. A planetary gearset in a small torque increase. (General Motors)

Direct Drive

When any two members of the planetary gearset are locked together, planetary action is stopped. Under these conditions, the gearset will revolve as a solid unit, providing direct (1:1) drive with no change in torque or speed, **Figure 21-22**.

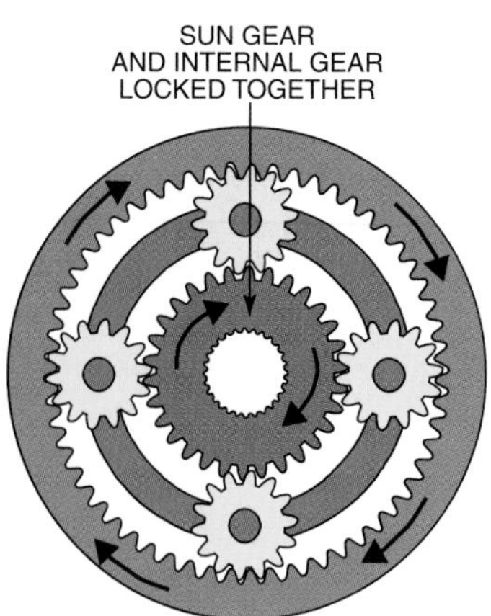

Figure 21-22. A planetary gearset in direct drive. (General Motors)

Overdrive

If the ring gear is held, and the planet carrier is driven, the planet gears must walk around inside the ring gear. This causes the sun gear to move faster than the planet carrier. Speed will increase and torque will decrease. See **Figure 21-23**.

Reverse

By holding the planet carrier and driving the sun gear, the planet gears are forced to rotate about their pins. This causes them to drive the internal gear in a reverse direction at a reduced speed, **Figure 21-24**.

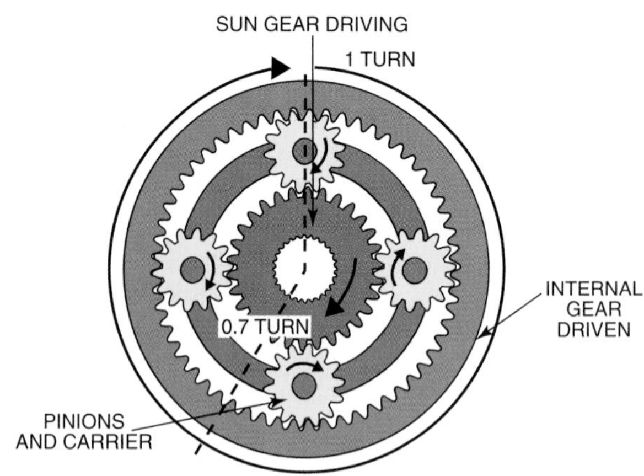

Figure 21-23. Relationship between rpm of the pinion carrier and internal gear. Every time transmission output shaft turns the pinion carrier .7 of a turn, the internal gear is forced to rotate a full turn. This causes the overdrive output shaft to revolve faster than the transmission output shaft. (General Motors)

Planetary Gearsets in Automatic Transmissions

The action of the three- and four-speed transmission gearsets in **Figures 21-25** and **21-26** illustrate general gearset operation.

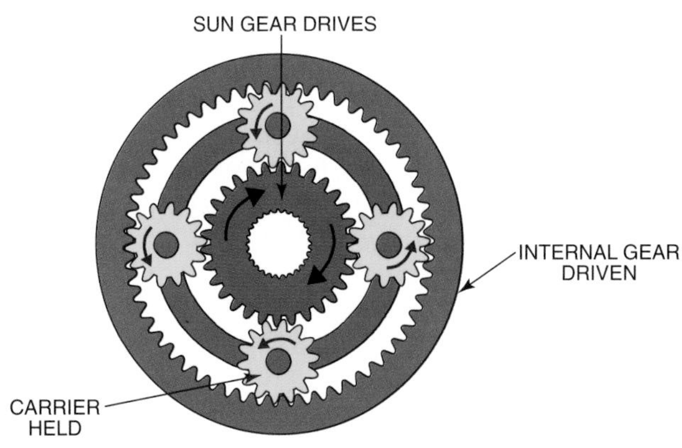

Figure 21-24. A planetary gearset in reverse. (General Motors)

Simpson Geartrain

The ***Simpson geartrain*** was used on three-speed transmission for many years. The Simpson geartrain is composed of two planetary gearsets. The sun gear of each set is connected together to form what is called a common sun gear. The action of the Simpson geartrain is shown in **Figure 21-25**. Note that the input shaft is turning, but none of the internal clutches have been applied (clutches and bands will be explained in the next section) when the transmission is in neutral. Power cannot flow through the planetary gears.

In first gear, the rear clutch is applied, which sends power through the front ring gear. The rotation of the ring gear rotates the front planet gears, driving the sun gear, but in the reverse direction. The sun gear drives the rear planetary gearset and the rear ring gear. The rear planet carrier is held by an overrunning clutch, similar in operation to the one used in the torque converter stator. The action of the rear planet gears reverses the rotation of the power flow. This reversal of power flow rotates the output shaft in the same direction as engine rotation. The combined action of the two planetary gearsets gives a reduction of about 2.5:1.

In second gear, a band is applied to stop the sun gear by stopping the rotation of the sun gear shell. With the sun gear stopped, power from the front ring gear turns the planetary gears and carrier around the stopped sun gear. Since the front planet carrier is splined to the output shaft, it is carried around in the same direction as engine rotation with a gear reduction of about 1.5:1. Note that the rear planet and ring gears are turning, but have no effect on power transfer.

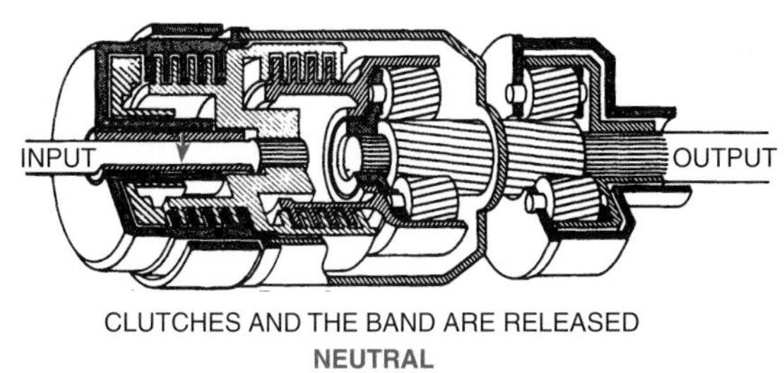

CLUTCHES AND THE BAND ARE RELEASED
NEUTRAL

THE INTERMEDIATE BAND IS APPLIED. THE REVERSE AND HIGH CLUTCH DRUM, THE INPUT SHELL, AND THE SUN GEAR ARE HELD STATIONARY.

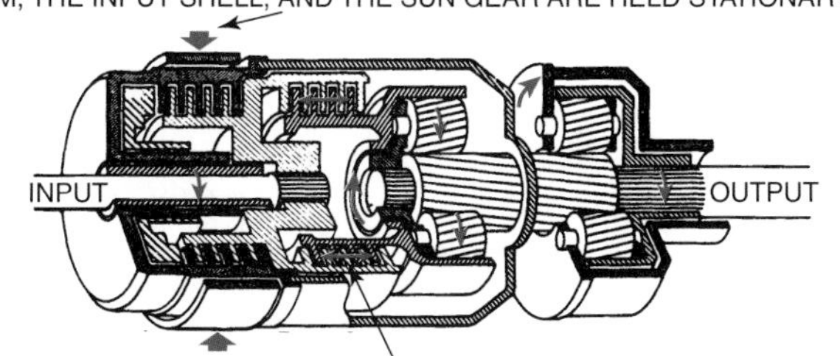

THE FORWARD CLUTCH IS APPLIED. THE FRONT PLANETARY UNIT RING GEAR IS LOCKED TO THE INPUT SHAFT.
SECOND GEAR

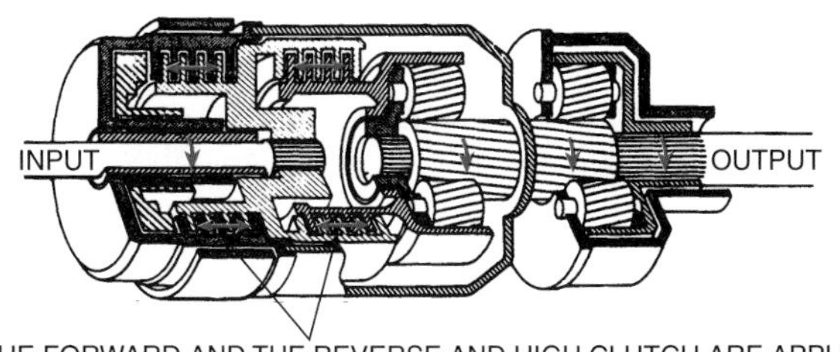

BOTH THE FORWARD AND THE REVERSE AND HIGH CLUTCH ARE APPLIED. ALL PLANETARY GEAR MEMBERS ARE LOCKED TO EACH OTHER AND ARE LOCKED TO THE OUTPUT SHAFT.
HIGH GEAR

THE FORWARD CLUTCH IS APPLIED. THE FRONT PLANETARY UNIT RING GEAR IS LOCKED TO THE INPUT SHAFT.

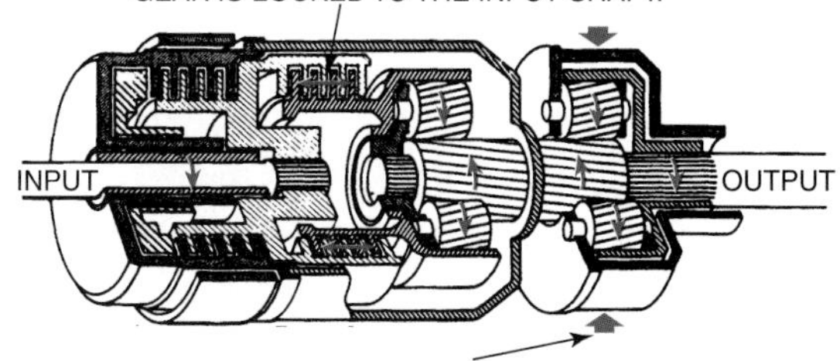

THE LOW AND REVERSE CLUTCH (LOW RANGE) OR THE ONE-WAY CLUTCH (D1 RANGE) IS HOLDING THE REVERSE UNIT PLANET CARRIER STATIONARY.
FIRST GEAR

GEAR RATIOS
FIRST 2.46:1
SECOND 1.46:1
HIGH 1.00:1
REVERSE 2.17:1

THE REVERSE AND HIGH CLUTCH IS APPLIED. THE INPUT SHAFT IS LOCKED TO THE REVERSE AND HIGH CLUTCH DRUM, THE INPUT SHELL, AND THE SUN GEAR.

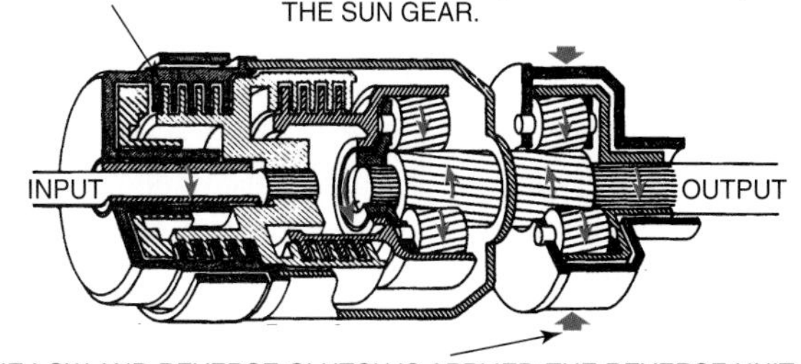

THE LOW AND REVERSE CLUTCH IS APPLIED. THE REVERSE UNIT PLANET CARRIER IS HELD STATIONARY.
REVERSE

Figure 21-25. Transmission action during the various speed ranges in a three-speed automatic transmission. (Ford)

In high gear, the band is released and the front clutch is applied. This causes the sun gear and front ring gear to be locked together, becoming a solid unit driving at a ratio of 1:1. Again, the rear planet gears and ring gears are turning, but have no effect on power transfer.

In reverse, the front clutch is applied, and the sun gear is turned by engine power. The rear planet carrier is held stationary by a band or clutch. The planet gears reverse the rotation of the ring gear as well as producing a reduction of about 2:1. In reverse, the front ring gear and planet gears turn, but have no effect on power transfer.

Compound Planetary Gearset

Figure 21-26 shows the start of a planetary gearset as used in a four-speed transmission. Notice that the primary sun gear and shaft are one piece and are splined to the converter turbine. The short planet gears are engaged with the sun gear. The long planet gears will engage with the reverse sun gear. The single ring gear engages the long planet gears. This type of planetary gear arrangement is known as a ***compound planetary gearset***, sometimes called a *Ravigneaux gearset.* The compound planetary gearset was used for many years in two- and three-speed transmissions.

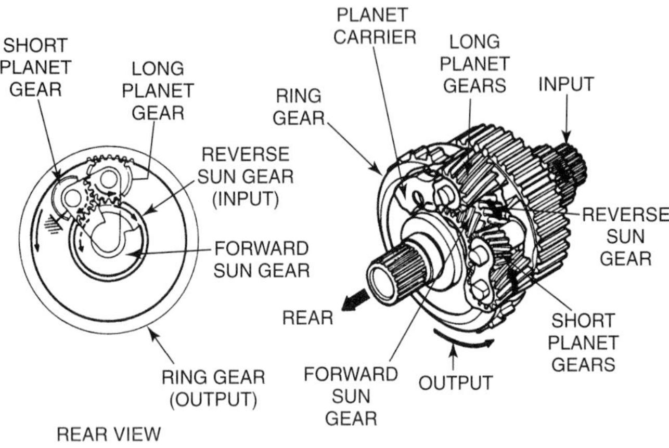

Figure 21-26. A compound planetary gearset from a four-speed automatic transmission. Power from the engine is transmitted to the input shaft by the torque converter clutch. All forward gears and reverse can be attained by this one gearset. (Subaru)

Holding Members

To operate the planetary gearsets, ***holding members*** are needed. The holding members are used to either direct engine power through one of the members of the planetary gearset, or to lock the member to the transmission case. Modern three- and four-speed automatics need at least two holding members applied in every gear. The different types of holding members are discussed in the following sections.

Bands

A ***band*** is a flexible steel device that encircles a rotating part, usually a clutch drum. One end of the band is secured to the case and may have an adjuster. The other end is fastened to a hydraulic piston or servo. The band face is covered with a friction lining. When the band is tightened, it will stop the part. A band always stops the rotation of a part by locking it to the transmission case. Bands can be either single wrap, **Figure 21-27**, or double wrap, **Figure 21-28**. The double wrap holds tighter, but the single wrap can be applied and released more smoothly and quickly.

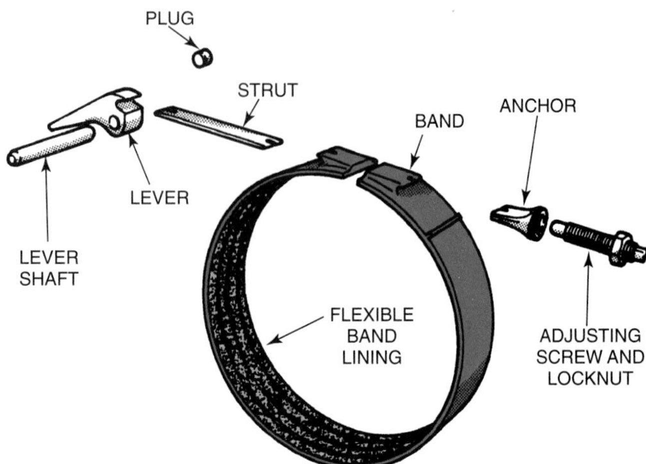

Figure 21-27. A "single-wrap" band with linkage parts. (Dodge)

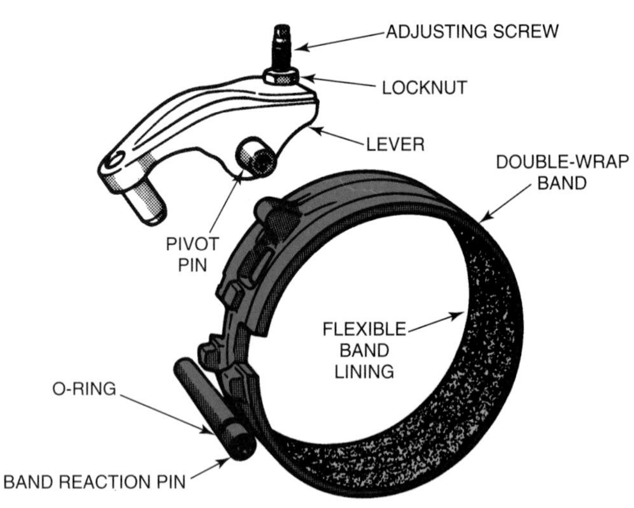

Figure 21-28. A "double-wrap" band and linkage parts. (Dodge)

Band Servos

A hydraulic piston called a ***servo*** is used to apply the band. It consists of a cylinder in which a piston is placed. An opening is provided at one end to admit fluid. The piston is held in the released position by spring pressure. Servos may be installed in the case or inside the transmission or transaxle itself.

When pressurized fluid enters the cylinder, the piston is forced forward, either actuating the band through direct contact or by actuating linkage. When the fluid pressure is reduced, the spring will return the piston and the band will release.

Some servos have openings on both sides of the piston. In this case, fluid can be used to assist in servo release. These servos are said to have an ***apply side*** and a ***release side***. Other servos have two or more internal pistons. **Figure 21-29** shows a cross-section of a servo. This unit acts directly on the band through a piston rod or stem.

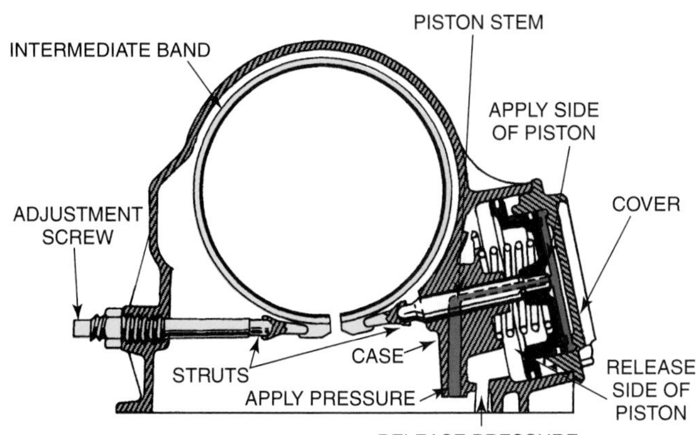

Figure 21-29. Cutaway of a direct acting servo. Oil pressure may be applied to either side of the piston, causing it to move in the desired direction. (Ford)

The servo in **Figure 21-29** is of the direct acting type. Note the band adjusting screw and how the piston may be applied and released by fluid pressure. **Figure 21-30** shows a servo operated through an actuating lever. This type is used to save space and provide more leverage on the band.

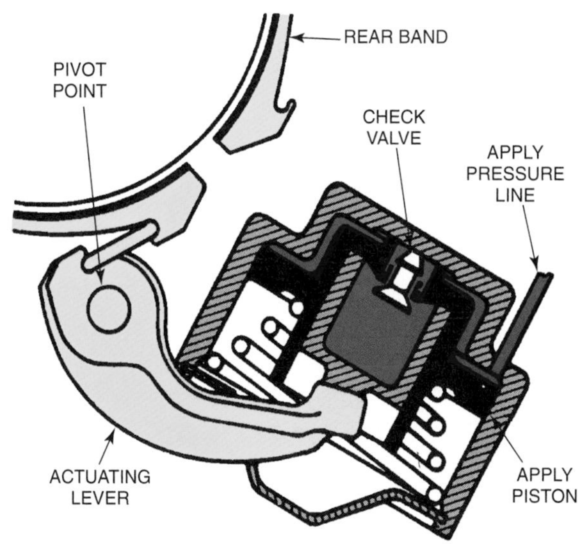

Figure 21-30. Cutaway of a lever-type band servo. Note that the servo movement is transmitted through a lever. The pivot point is positioned to cause the lever to increase the force of the apply piston.

Multiple Disc Clutch

Figure 21-31 shows a ***multiple disc clutch*** or ***clutch pack***. The ***clutch hub*** is splined to the primary sun gear shaft. A stack of ***clutch discs*** (the number varies) is arranged in a ***drum***. One set of discs is splined to the clutch hub; the other set is splined to the clutch drum. The internal splined and external splined discs alternate in the drum. The clutch pack is usually used to drive moving parts. However, a clutch pack is used in some transmissions to lock parts to the case in a manner similar to a band. See **Figure 21-32**.

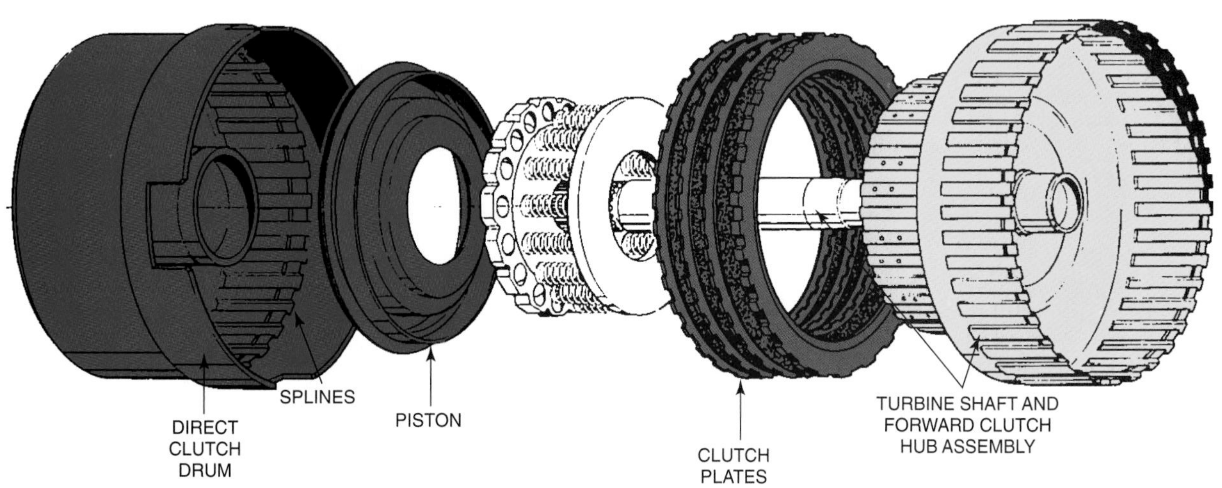

Figure 21-31. A multiple disc clutch (direct type) which is splined (connected with grooves or slots) to the clutch drum. (General Motors)

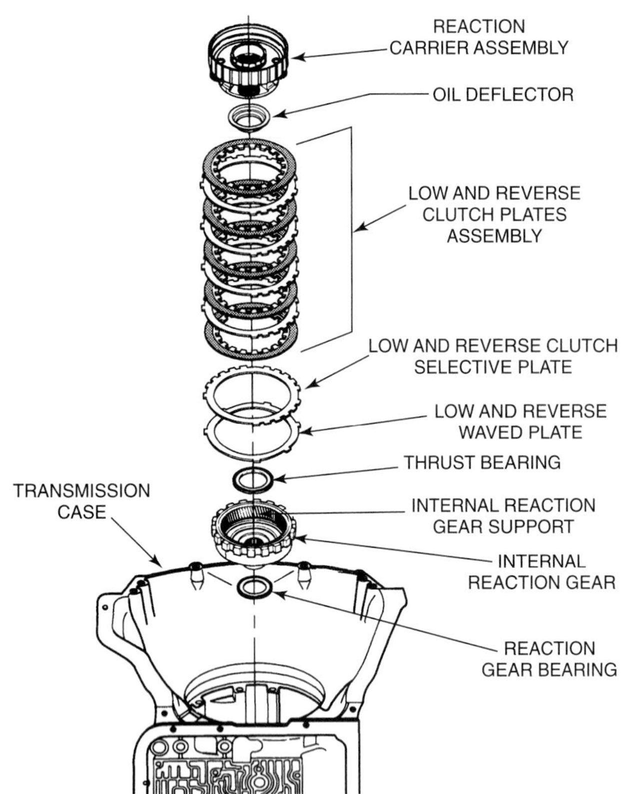

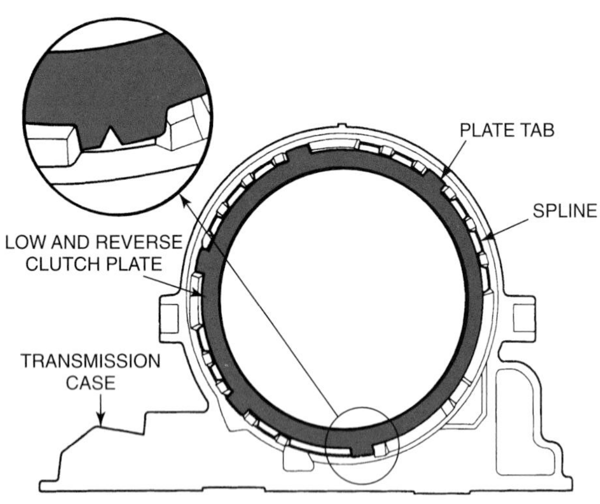

Figure 21-32. A clutch pack can lock certain parts to the transmission case when applied much like a band does. Note how the clutch plate tabs fit into the splines cast into the case. (Chevrolet)

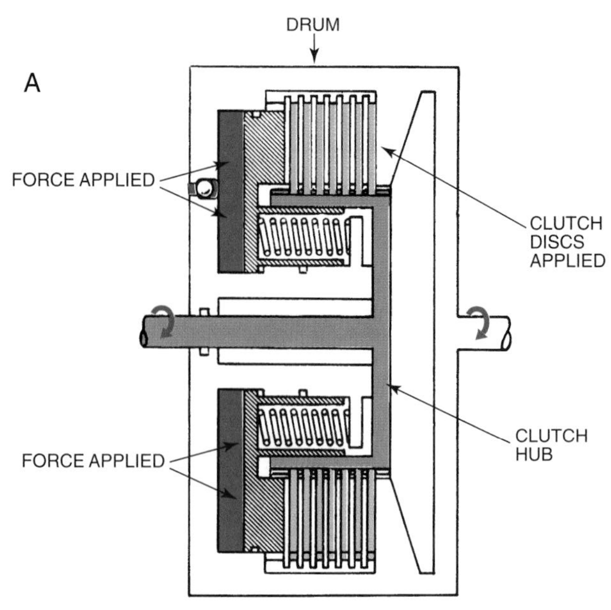

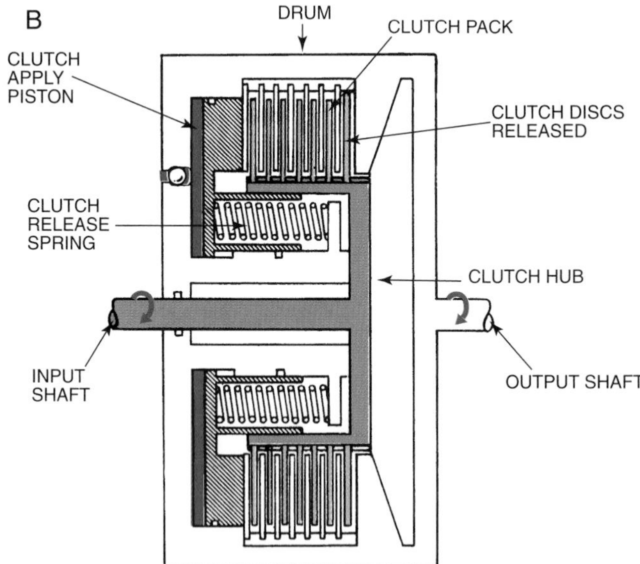

Figure 21-33. A—Multiple disc clutch schematic. Oil pressure is now acting against the clutch apply piston. The piston overcomes spring pressure and squeezes the clutch discs and plates together. As some discs are attached to the drum output shaft assembly and others to the hub (in an alternating arrangement), this squeezing action locks the hub and drum together. The hub now turns the output shaft. B—In this illustration, the hydraulic pressure is not acting on the clutch apply piston. With no oil pressure, the clutch release springs separate the clutch discs. The clutch hub turns but cannot drive the output shaft-drum assembly. (General Motors)

A ***clutch apply piston*** in the clutch drum squeezes the plates together when fluid pressure is applied to the piston. When the clutch piston is applied, the discs are pressed together to lock the clutch hub and drum together, **Figure 21-33A**. The clutch hub splined to the input shaft drum is part of the output shaft. With the clutch applied, both shafts are locked together and revolve as a solid unit. A heavy spring releases the discs when fluid pressure drops. See **Figure 21-33B**. Study the four-speed transmission in **Figure 21-34**. Note the number and layout of clutch packs.

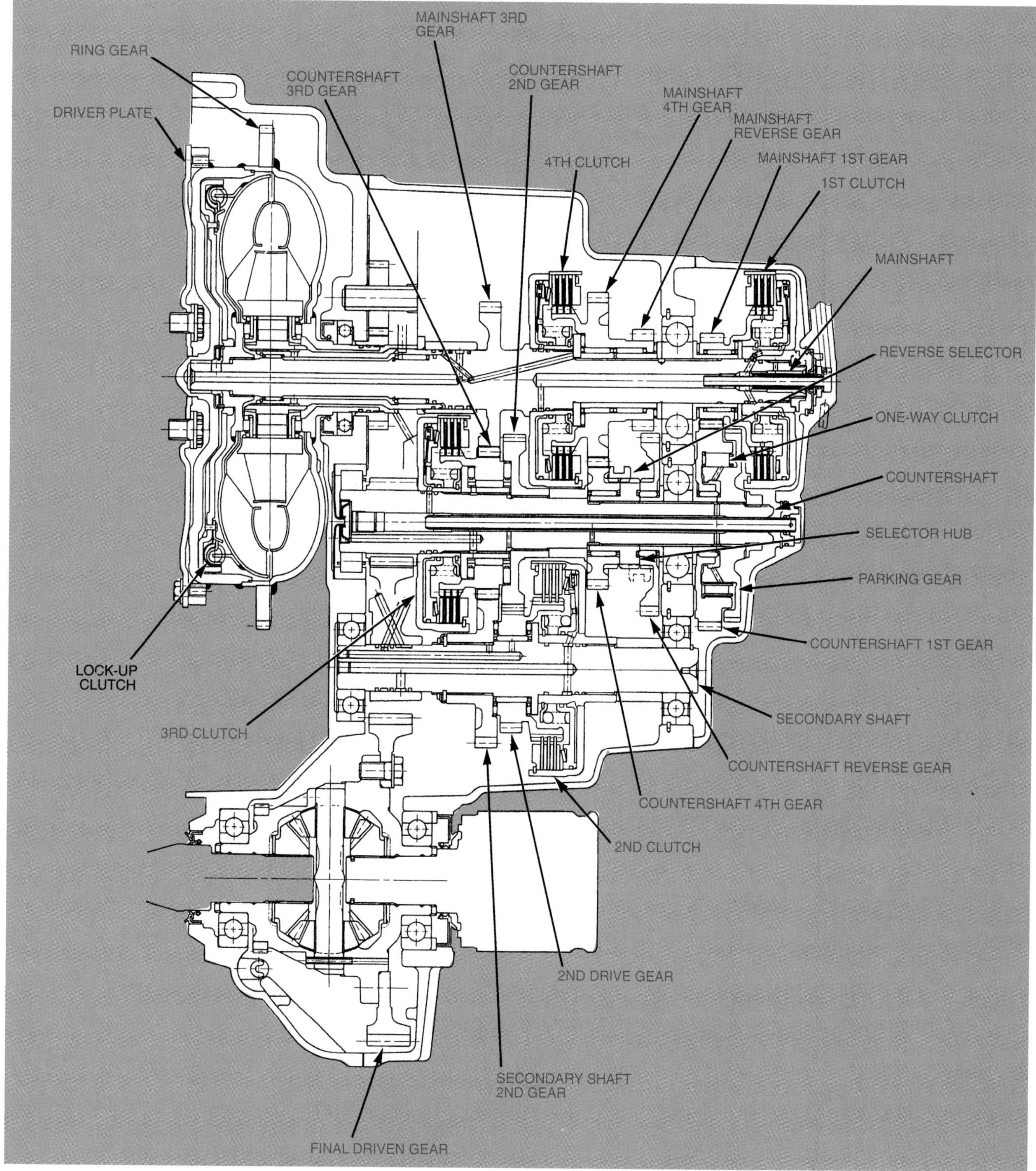

Figure 21-34. A cross-sectional view of a lock-up torque converter and clutch pack members mounted on a four-speed automatic transaxle. (Honda)

Overrunning Clutches

Overrunning clutches, sometimes called ***one-way clutches***, have the ability to lock when they are rotated in one direction and unlock when rotated in the other direction. The two types of overrunning clutches are the roller clutch and the sprag clutch. They are explained in the following sections.

Roller Clutch

The ***roller clutch***, **Figure 21-35**, consists of three parts: an inner hub with tapered cams on the outside surface, a series of hardened steel rollers (one per cam surface), and an outer ring or race. The rollers are held in position by a steel cage. Each roller has a return spring to keep it from jamming in the released position.

Note in **Figure 21-35A** that the outer race is attempting to turn counterclockwise. This forces the rollers up the cam surfaces until they are wedged tightly against the outer race. The outer race will then be held stationary. When the race attempts to turn in the clockwise direction, **Figure 21-35B**, the rollers disengage and return to their low positions on the cams. The outer race is now free to turn. Therefore, the roller clutch becomes a holding member if it is turned in the counterclockwise direction, but freewheels if turned in the clockwise direction.

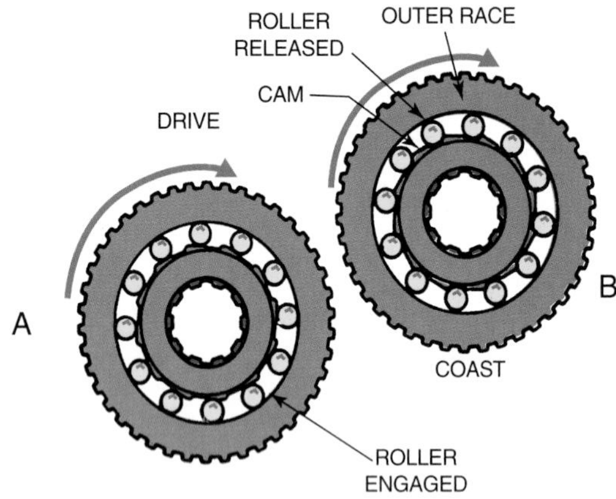

Figure 21-35. Roller clutch action. A—As the hub is being driven by the transmission output shaft, the rollers have traveled up the cams and seized the outer race thus driving it. B—The accelerator has been released and the inner hub has slowed down. This causes the clutch rollers to travel back down the cams and release the outer race, allowing it to freewheel or overrun. (Chevrolet)

Sprag Clutch

Sprag clutches are occasionally used, but they have mostly been replaced by the more durable roller clutch. Instead of using round rollers that walk up a tapered step to lock up the clutch, the sprag clutch uses a flat unit with curved edges. When the outer race is turned one way, the sprags tip and present their narrow diameter. This allows the outer race to turn.

When the race attempts to rotate in the opposite direction, the sprags tip the other way and present the long diameter. This causes them to jam between the two races, locking them together and preventing rotation. See **Figure 21-36**.

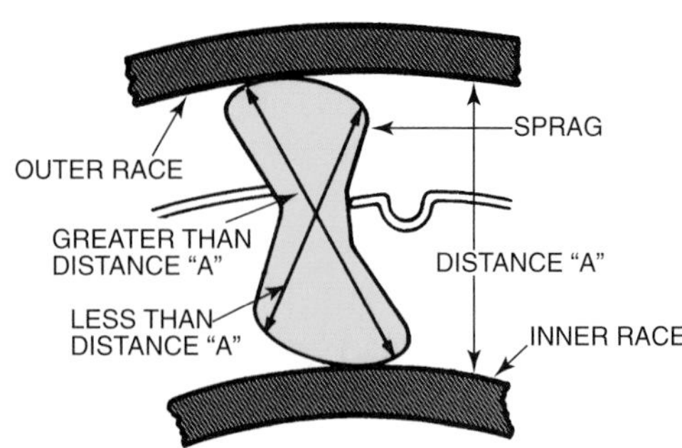

Figure 21-36. Cross-section view showing sprag clutch construction. (Cadillac)

Hydraulic Control System

The hydraulic control system dictates the operation of the other parts of the transmission. While an entire book could be dedicated to the study of the principles of fluid power or hydraulics, we will confine ourselves to saying that:

- Unlike air, liquids cannot be compressed.
- Since liquids cannot be compressed, they are an excellent way to transfer power.
- This hydraulic power can be used to operate holding members.
- All transmission operations are controlled with, or through, hydraulic power.

The following sections will explain how hydraulic pressure is developed and used in the automatic transmission and transaxle. The section begins with an outline of the major hydraulic system components common to all transmissions and transaxles.

Fluid Pump

Pressure is developed by a transmission ***fluid pump***, sometimes called an oil pump. The pump draws fluid from the transmission ***fluid pan*** and creates ***hydraulic pressure***, which is then directed to other parts of the transmission to fill the converter, operate the holding members, and control shifting. The pump is driven by the engine, usually through driving lugs on the rear of the converter. Therefore, whenever the engine is running, the pump is producing power to operate the rest of the hydraulic system.

Several types of fluid pumps have been used in automatic transmissions. One is the variable output, or ***vane pump***. Output is altered by moving the outer vane cylinder, **Figure 21-37**. Another is the ***rotor pump***, **Figure 21-38**. Another widely used pump is the ***internal-external gear pump***. Look at **Figure 21-39**.

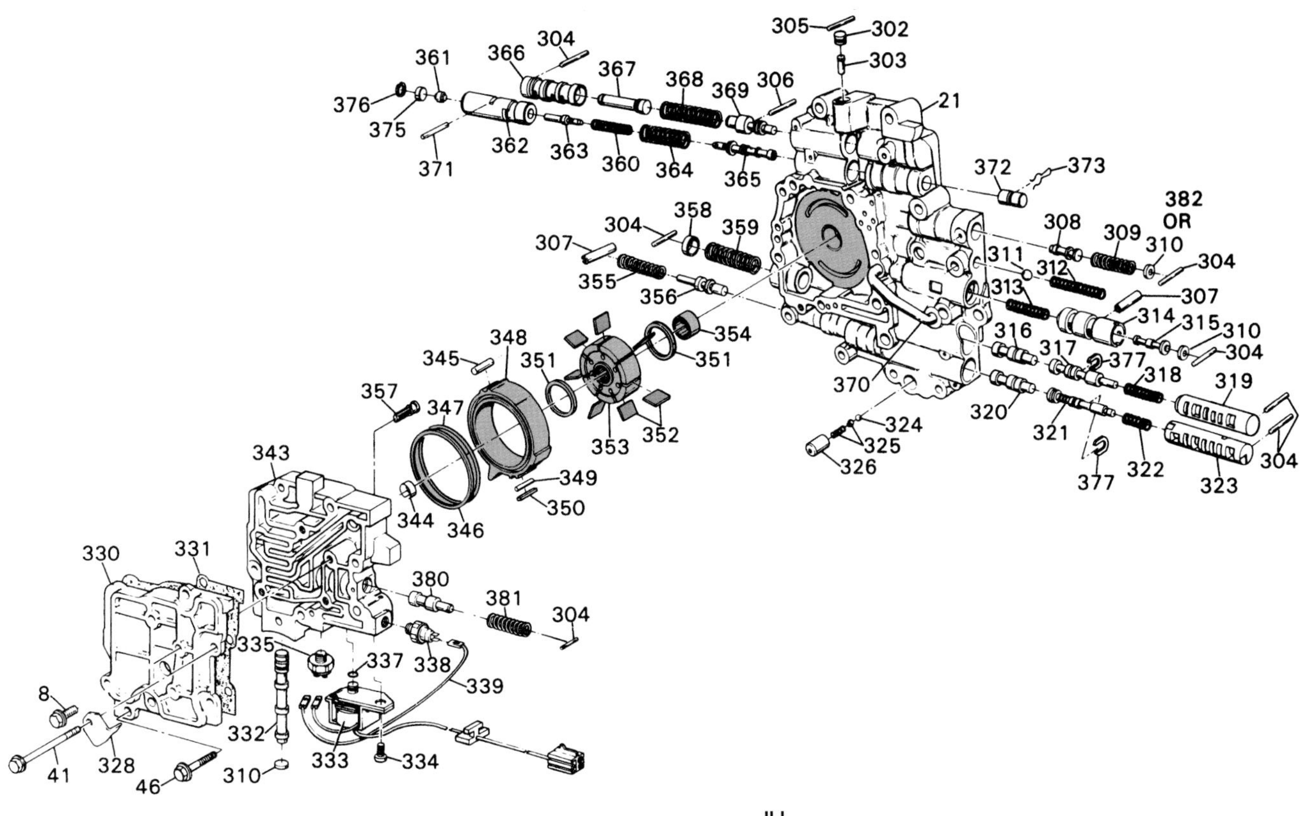

ILL. NO.	DESCRIPTION
8	Screw, Auxiliary Valve Body Cover/Auxiliary Valve Body
41	Screw, Auxiliary Valve Body to Case Cover
46	Screw, Aux. Valve Body Cover/Valve Body
21	Control Valve & Fluid Pump Assembly
302	Plug, Bore Line Boost Valve
303	Valve, Line Boost
304	Pin, Valve & Spring Retaining
305	Pin, Coiled Spring Line Boost Plug
306	Pin, Throttle Valve Bushing
307	Sleeve, Spring Retaining
308	Valve, Shift T.V.
309	Spring, Shift T.V.
310	Plug, Valve Bore
311	Ball, Pressure Relief
312	Spring, Pressure Relief
313	Spring, 1-2 Accumulator
314	Bushing, 1-2 Accumulator
315	Valve, 1-2 Accumulator
316	Valve, 2-3 Shift
317	Valve, 2-3 Throttle
318	Spring, 2-3 Throttle Valve
319	Bushing, 2-3 Throttle Valve
320	Valve, 1-2 Shift
321	Valve, 1-2 Throttle
322	Spring, 1-2 Throttle Valve
323	Bushing, 1-2 Throttle Valve
324	Ball, Lo Blow Off
325	Spring & Plug Assembly, Lo Blow Off
326	Plug, Lo Blow Off Valve
328	Retainer, Valve Body Pipe
330	Cover, Auxiliary Valve Body
331	Gasket, Auxiliary Valve Body Cover
332	Valve, Conv. Cl. Control
333	Solenoid Assembly
334	Screw, Solenoid
335	Switch, Pressure — 3rd Cl.

ILL. NO.	DESCRIPTION
337	Seal, O-Ring (Solenoid)
338	Switch, Gov. Pressure — 2nd Cl.
339	Harness, Solenoid Wire
343	Body, Auxiliary Valve
344	Sleeve, Auxiliary Valve Body
345	Pin, Slide Pivot
346	Ring, Fluid Seal (Slide to Cover)
347	Seal, O-Ring (Pump Slide)
348	Slide, Pump
349	Support, Pump Slide Seal
350	Seal, Pump Slide
351	Ring, Pump Vane
352	Vane Pump
353	Rotor, Fluid Pump
354	Bearing Assembly, Pump Shaft & Seal
355	Spring, 3-2
356	Valve, 3-2 Regulating Control
357	Filter, Auxiliary Valve Body
358	Plug, Pump Spring Retaining
359	Spring, Pump Priming
360	Spring, T.V. & Reverse Boost Valve
361	Valve, T.V. Boost
362	Bushing, T.V. & Reverse Boost Valve
363	Valve, Reverse Boost
364	Spring, Pressure Regulator
365	Valve, Pressure Regulator
366	Bushing, T.V. Plunger
367	Plunger, Throttle Valve
368	Spring, Throttle Valve
369	Valve, Throttle
370	Pipe, Valve Body
371	Pin, T.V. & Reverse Boost Bushing
372	Plug, Bore (Pressure Regulator)
373	Retainer, Bore Plug Pressure Regulator
375	Plug, Valve Bore Isolator
376	Ring, Valve Retaining
377	Retainer, Throttle Valve
380	Valve, Orifice Control
381	Spring, Orifice Control Valve
382	Plug, Valve Bore

Figure 21-37. Exploded view of a vane-type oil pump and the many related parts as used by one manufacturer. (Chevrolet)

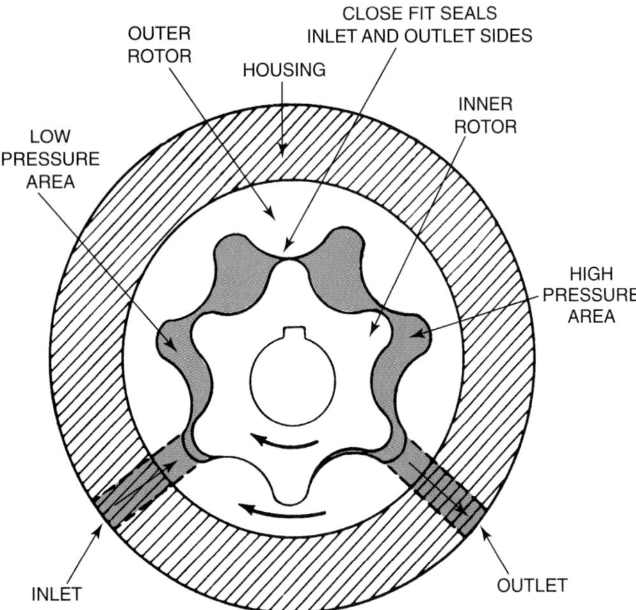

Figure 21-38. A rotor style oil pump. Note that the arrangement of gears resembles a gear oil pump.

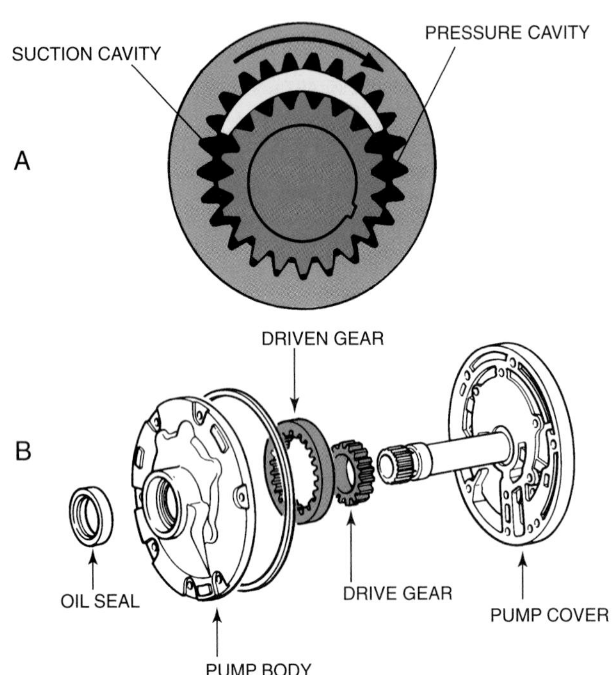

Figure 21-39. A—Note the action in a transmission internal-external gear oil pump. B—Exploded view of the gear-type oil pump. (Toyota)

Control Valves

The flow of hydraulic pressure is controlled by ***spool valves***. Spool valves are used to perform different functions, such as filling the torque converter and operating the transmission as needed. The major spool valves used in all automatic transmissions are discussed in the following sections.

Many of the spool valves and other components are installed in a ***valve body*** attached to the bottom of the transmission case. Some transmissions and transaxles have two or more valve bodies. Many transaxles have the valve body mounted on the side of the case to save space.

Main Pressure Regulator

The ***main pressure regulator*** monitors the pump output to control the system hydraulic pressure. When pressure exceeds the amount required, the valve opens and bypasses fluid back to the sump. This produces an overall system pressure, usually called ***line pressure***.

The main pressure regulator valve setting is controlled by spring tension on the valve. See **Figure 21-40**. To increase line pressure during heavy acceleration or in reverse, pressure from other parts of the hydraulic system is used to assist the pressure regulator spring. Some transmissions have a separate regulator valve to control torque converter pressure.

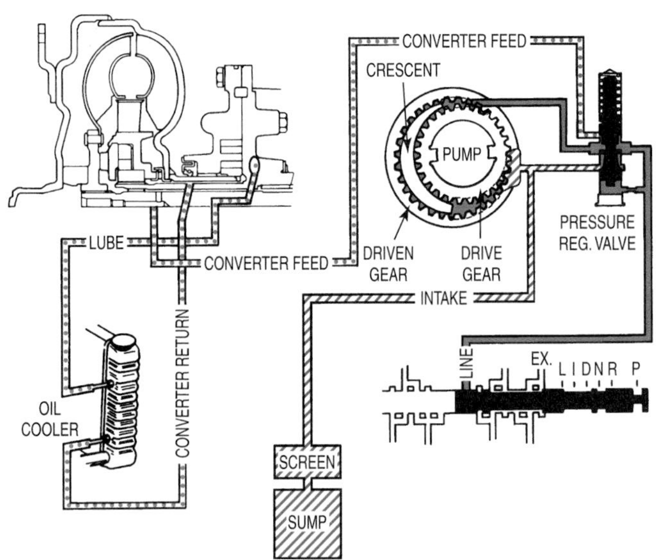

Figure 21-40. A gear oil pump, pressure regulator, and hydraulic circuit. These vary by manufacturer. (General Motors)

Manual Gear Selection

The transmission may be placed in any desired gear by moving a gearshift lever, located on the steering column or console. This moves the linkage connected to the manual valve in the transmission. The shift indicator shows the operator which gear has been selected. See **Figure 21-41**. A console floor shift mechanism is shown in **Figure 21-42**. Manual gear selection is often use to test automatic transmissions for proper operation.

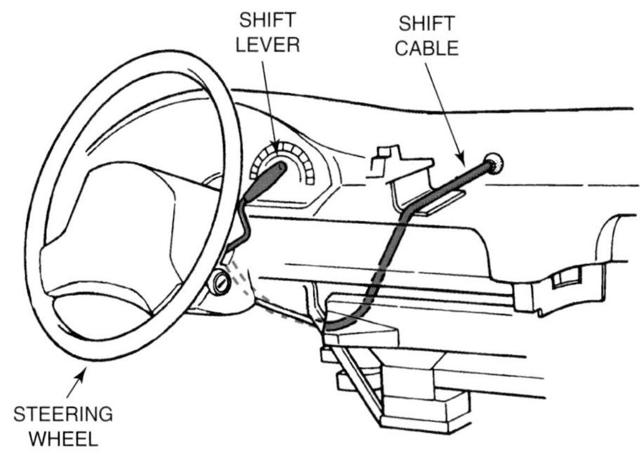

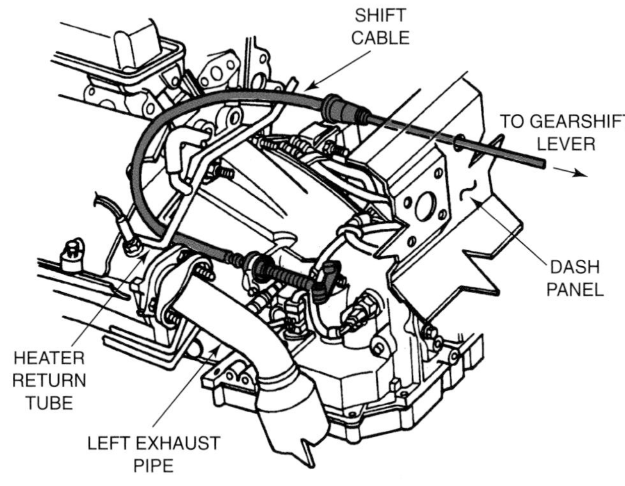

Figure 21-41. A steering column mounted gear shift cable arrangement and its path to the transmission. (Dodge)

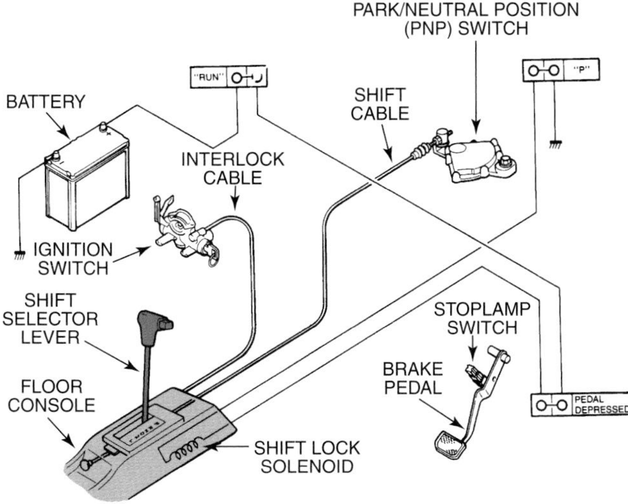

Figure 21-42. A console (floor) gearshift linkage arrangement. This system uses a brake transaxle shift interlock (BTSI) to prevent the transmission from being placed into gear without depressing the brake pedal. (Geo)

Simple Hydraulic System

The simple hydraulic circuit shown in **Figure 21-43** is but one part of the overall hydraulic system. Modern hydraulic systems are much more complex, but operate from the same basic principles shown here. Other valves are used to cushion shifts, provide passing gear, and control the lockup torque converter. The principles shown here apply to all automatic transmissions.

Manual Valve

The ***manual valve*** is operated by the driver to engage the transmission/transaxle in park, reverse, neutral, drive, second, or low, and other available gears, **Figure 21-43**. The manual valve linkage can be a series of rods and links or a cable.

The manual valve will move to the selected drive position. From the manual valve, the line pressure will go to the low band servo. The fluid pressure will apply the servo and tighten the band. When the vehicle is accelerated, torque will be imparted and the vehicle will move forward in the low gear. Note that the servo applies the band by overcoming a return spring on the band side of the servo.

Shift Valve

To enable the transmission to change gears, another line must be added that will run from the main line to a ***shift valve***. The shift valves redirect hydraulic pressure to the transmission or transaxle holding members to obtain different gears in drive range. The movement of one shift valve will cause one up or down shift between two different gears. The fluid will leave the shift valve and branch out. One line will go to the high range clutch unit, another to the low range servo. The shift valves are moved by the throttle and governor valves.

Throttle Valve

The ***throttle valve*** tries to move the shift valve to the downshift position. The throttle valve is controlled by linkage from the engine throttle plates or by a vacuum modulator operated by engine manifold vacuum. The throttle valve is connected to the foot throttle through linkage or a cable. The throttle valve linkage is often referred to as the throttle valve rod or cable, or just ***TV linkage***.

When the throttle is depressed, the throttle valve is moved, transmitting fluid pressure to the shift valve. The amount of pressure depends on the distance the throttle is moved. With either system, the more the throttle is depressed, the higher the throttle pressure.

Governor Valve

The ***governor valve*** tries to move the shift valve to the upshifted (higher gear) position. Some governor valves are mounted directly on the output shaft. Other governors are installed in the case and driven by the output shaft through gears. The higher the shaft speed, the higher the governor pressure.

As the vehicle accelerates, the governor unit will spin faster and faster. At a predetermined speed, the governor valve will open and fluid pressure will travel to the shift valve. See **Figure 21-44A**. When governor pressure reaches a

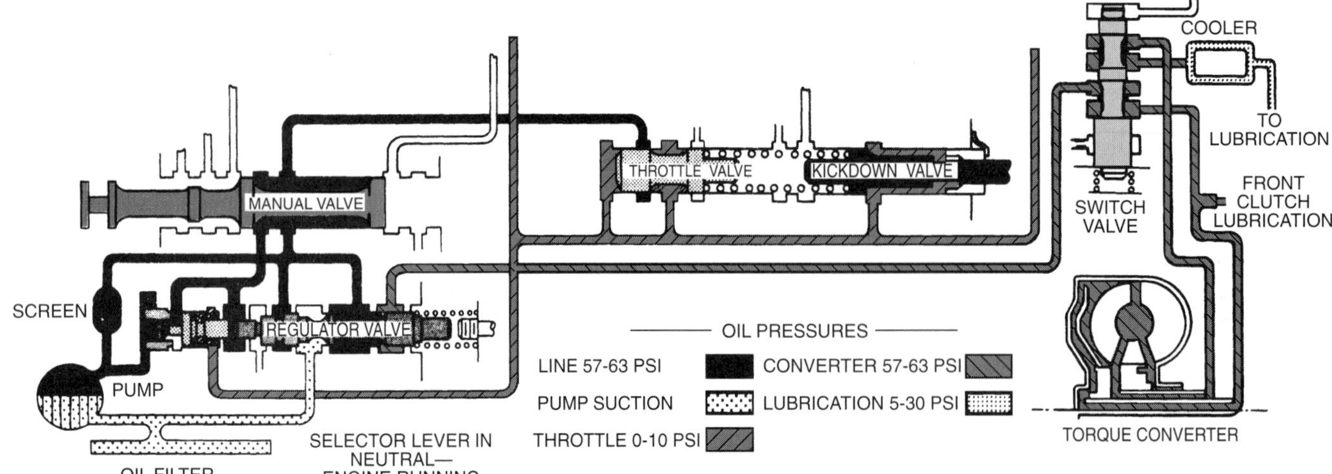

Figure 21-43. A manual control valve. This is one valve of the many which are used in the valve body of the modern automatic transmission. (Chrysler)

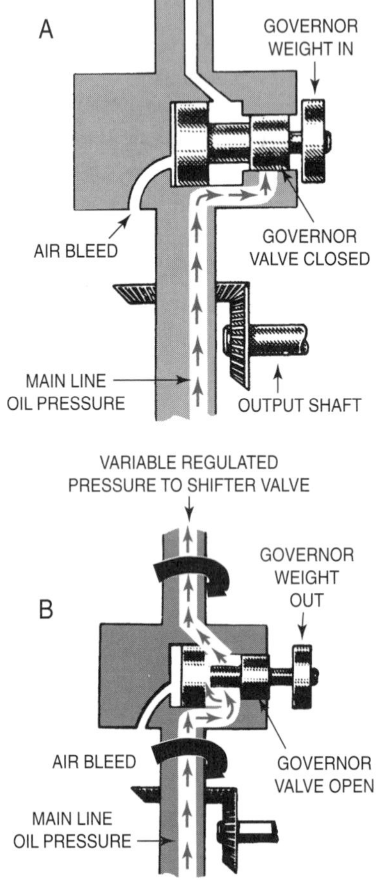

Figure 21-44. A—Schematic of governor valve in closed position. Note that the main line oil pressure cannot pass the governor valve. B—A schematic of a governor valve in the open position. Note that the main line oil pressure is now passing through the valve. (GMC)

certain level, the tension of the shift valve spring will be overcome and the shift valve will move to the left. **Figure 21-44B** shows the governor pressure building up against the shift valve. Note the different types of governors in **Figure 21-45**.

Detent Valve

The ***detent valve*** is used to achieve a forced downshift, sometimes called the detent or passing gear. The detent valve may be operated by mechanical linkage or by an electric solenoid. The detent linkage will be the same as the TV linkage, while the solenoid will be operated by a switch on the accelerator linkage.

The detent valve is used to force the transmission to downshift to the next lower gear. It supplies extra pressure to the shift valves to overcome governor pressure and move the shift valve to the downshifted position. The transmission detent will feel the same as a downshift when the vehicle is coasting to a stop.

Lockup Torque Converter Control Valve

The ***lockup torque converter control valve*** is used to ensure that the lockup converter clutch applies at cruising speeds and releases when the vehicle comes to a stop. The hydraulic control system may have a valve and an electric solenoid. The solenoid is computer-controlled, while the valve is operated by the throttle and governor pressures. On late model vehicles, the converter lockup clutch will apply without fluid pressure or control from the hydraulic system.

Transmission and Transaxle Shifting

Now let's put these circuits together. When the manual valve is opened, fluid pressure actuates the low band servo. The vehicle then moves forward in the low range. Fluid pressure also travels to the shift valve and the governor, **Figure 21-46**. When the vehicle reaches a predetermined road speed, the governor will have opened enough to allow fluid pressure to flow to the shift valve and force it open. When the shift valve opens, fluid pressure flows to the clutch piston and band servo.

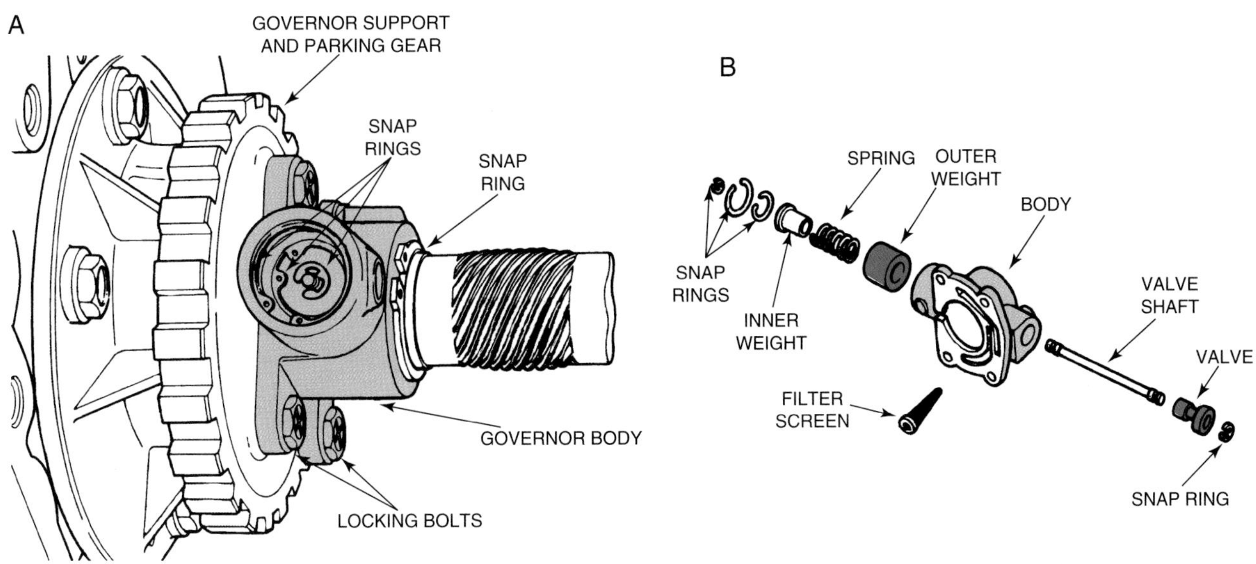

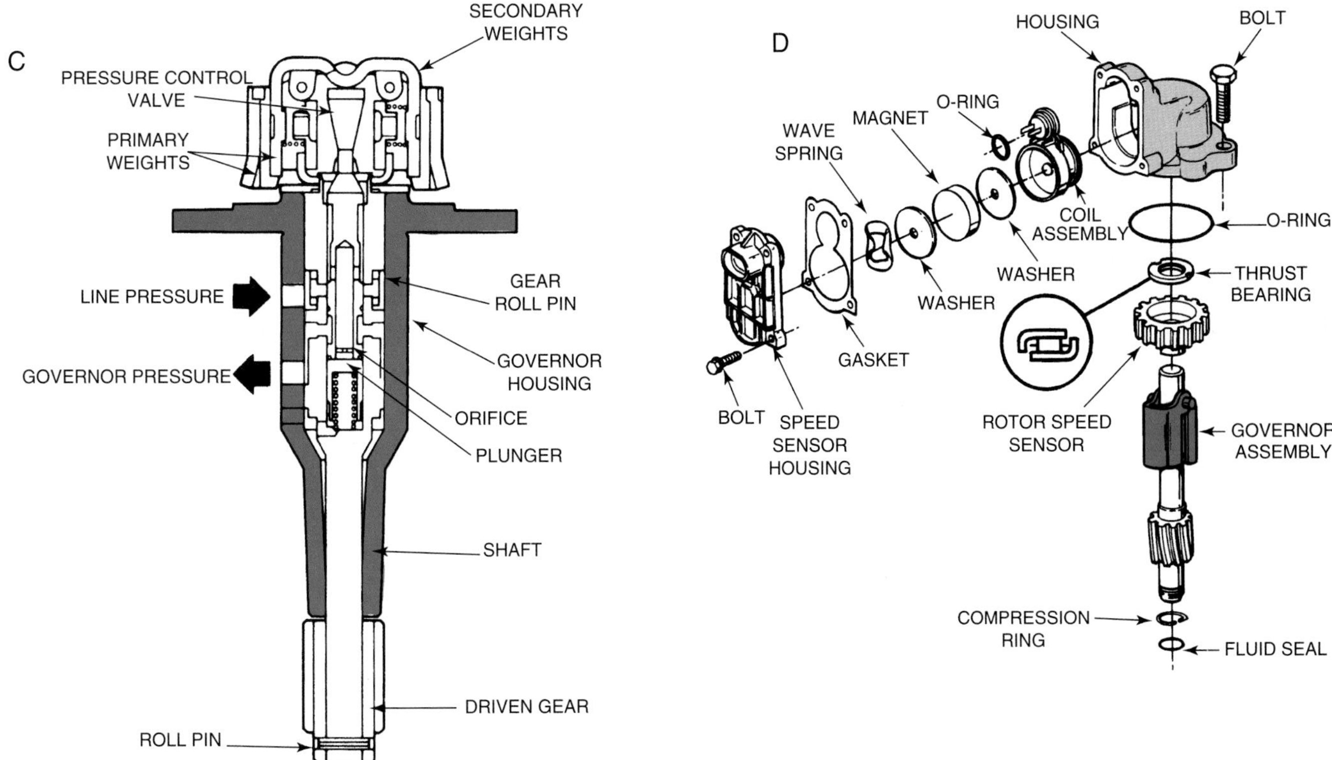

Figure 21-45. Various transmission governors used by different manufacturers. A—Governor assembly mounted on the transmission output shaft. B—Explolded view of an output shaft driven governor. This is the same unit shown in A. C—Cross-sectional view of a gear driven governor. D—Exploded view of a gear driven governor assembly with a speed sensor. (General Motors & Dodge)

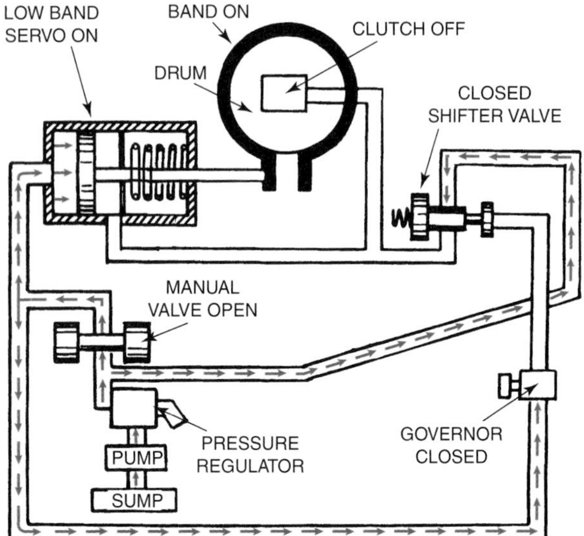

Figure 21-46. Schematic of transmission fluid flow in low range. The manual valve is open and pressure is applied to the low band servo, governor, and shifter valve. When the governor is closed, no pressure is applied to the open shifter valve. When the shifter valve closes, no pressure is applied to the low band servo or clutch.

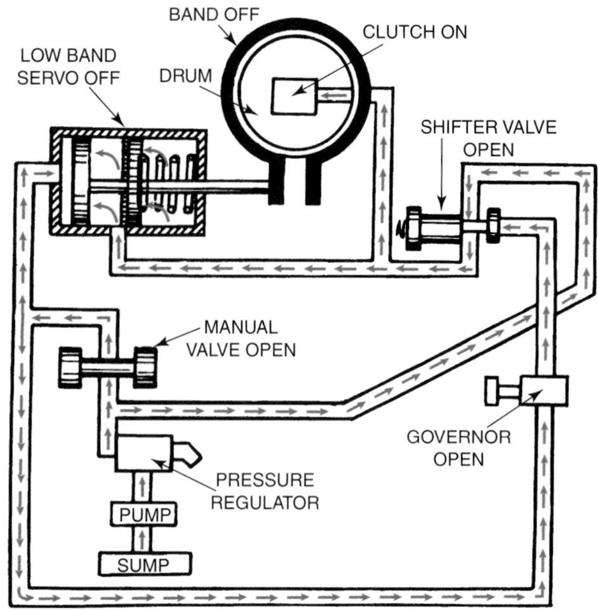

Figure 21-47. Schematic of transmission fluid flow in high range. Vehicle speed has opened the governor. Governor pressure opens the shifter valve, sending line pressure to both servo pistons and to the clutch. Servo releases and clutch applies. The transmission is now in high range.

As the pressure builds up in the clutch, a corresponding pressure builds up in the servo. The pressure from the shift valve is equal to the pressure on the other side of the servo. With the pressure equal on both sides of the servo piston, the return spring can release the band.

As the servo piston moves to the left, the pressure in the clutch unit will build up rapidly, eventually locking the clutch unit. The transmission is now in high range and will remain there until vehicle speed falls below the shift point. When this happens, governor pressure is lost, the clutch disengages and manual valve pressure reapplies the low band servo. All this is done automatically. **Figure 21-47** shows the circuit in high gear.

Governor Operation

As the vehicle moves forward in low gear, the output shaft spins a governor unit. Centrifugal force throws the governor weight outward, opening the governor valve. Fluid pressure channeled to the governor unit flows to the shift valve, **Figure 21-48**. When the vehicle is not moving, the governor weight is in and fluid pressure is stopped at the governor.

Varying the Shift Point

In the simple systems shown in **Figures 21-46** and **21-47**, the point at which the transmission shifts into high and downshifts into low are controlled by the governor. These points are always constant. In a real transmission, however, the shift points must vary according to load requirements. For rapid acceleration, the transmission should stay in the low range for a longer period. When attempting to pass at low speeds just above the downshift point, it is desirable to have the downshift occur when needed.

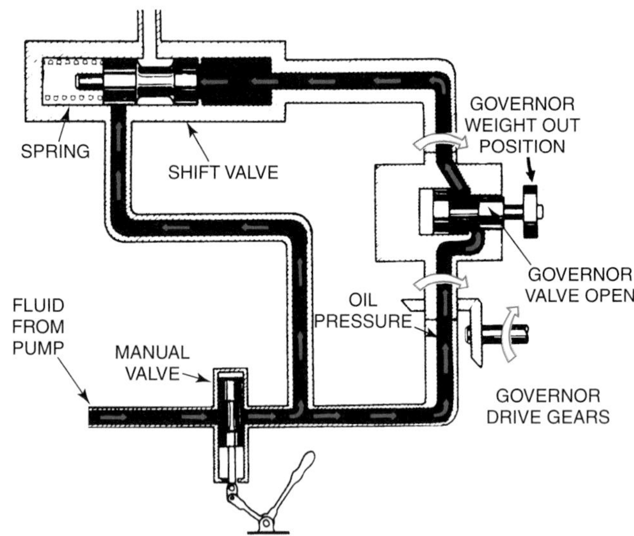

Figure 21-48 Governor pressure acting on shifter valve. Governor weight has been thrown out by centrifugal force. The governor valve is open and main line pressure is passing through the governor to the shifter valve.

Using Throttle Pressure to Vary the Shift Point

To delay the upshift, throttle pressure opposes governor pressure on the other side of the shift valve. Throttle pressure varies with throttle opening. The more the accelerator pedal is depressed, the more throttle pressure is sent to the shift valve.

A typical shift valve is shown in **Figure 21-49.** Note that governor pressure is available on one side of the shift valve. Governor pressure is attempting to move the valve to the upshifted position. On the other side of the valve, throttle pressure tries to keep the shift valve in the downshifted position. The interaction of governor pressure and throttle pressure causes the shift valve to change gears at a speed that is best suited to the load on the engine.

Throttle Valve

Pressure enters the valve at (A), **Figure 21-50**. It travels around the small spool section and exits through (B). It is bypassed to chamber (C) at the spool end (D). The pressure pushes on the shifter valve regulator plug and on the spool at (D). Spring pressure (E) is controlled by the TV cable. When the throttle opening is small, spring pressure (E) is light. Fluid pressure at (D) forces the spool valve to the right against the light spring pressure. As the spool valve moves to the right, it starts to block opening (A). As the size of opening (A) is reduced, the fluid pressure entering the valve is also reduced.

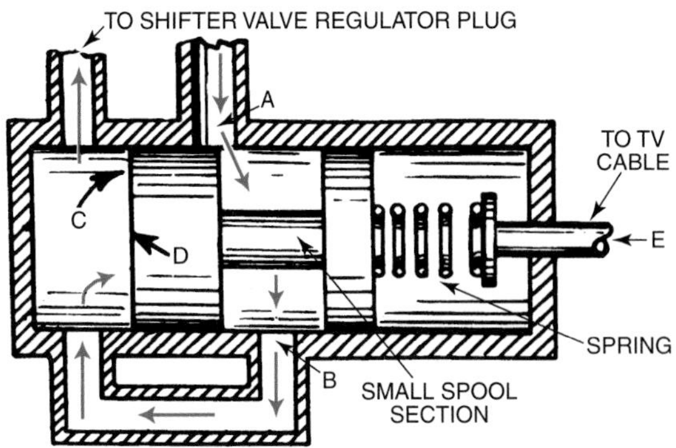

Figure 21-50. Throttle valve operation. Oil pressure balances spring pressure by moving the spool valve until opening "A" admits the proper pressure. A throttle valve (TV) rod or cable adjusts the spring pressure.

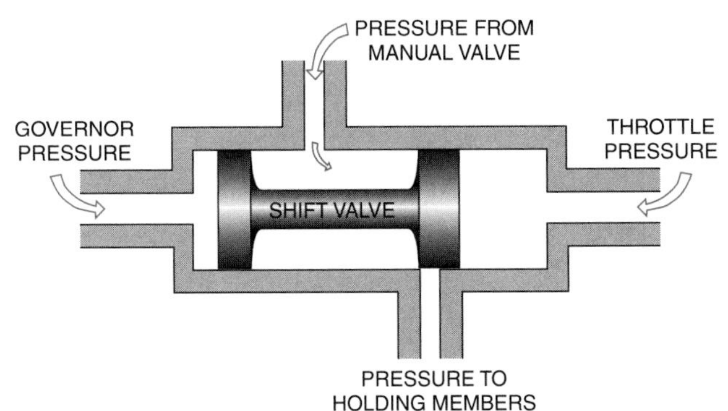

Figure 21-49. Typical shift valve arrangement. Throttle pressure opposes governor pressure, causing the valve to change gears at a speed best suited to engine load.

If the spring is pushing to the left with a force of 15 lbs (66.75 N), the spool will be forced to the right until entrance (A) is small enough to admit enough fluid to overcome spring pressure. The valve will then be balanced. If spring pressure is increased, it will force the spool to the left, increasing opening (A) and fluid pressure will build up in (C). This will force the spool to the right until opening (A) is again closed far enough to balance fluid pressure with spring pressure.

Vacuum Modulator

Many older vehicles use a ***vacuum modulator*** in place of the TV rod or cable. Some newer transmissions use a vacuum modulator as well as a TV cable. The modulator utilizes engine vacuum, a very reliable indicator of engine loading. The modulator provides an accurate means of controlling the throttle valve. The throttle valve is also called the modulator valve.

The vacuum modulator consists of a container separated into two areas by a flexible diaphragm. A spring forces the diaphragm rod to apply pressure to the throttle valve to increase TV pressure. The modulator is constantly altering throttle valve pressure in accordance with engine vacuum, **Figure 21-51A**. In operation, engine vacuum causes the diaphragm to move away (against spring pressure) from the throttle valve, lowering the pressure. As engine loading varies, vacuum fluctuates.

An altitude sensitive modulator is illustrated in **Figure 21-51B**. It incorporates an evacuated (air removed) bellows. The collapsing bellows tends to force the diaphragm toward the throttle valve. At sea level, atmospheric pressure is high,

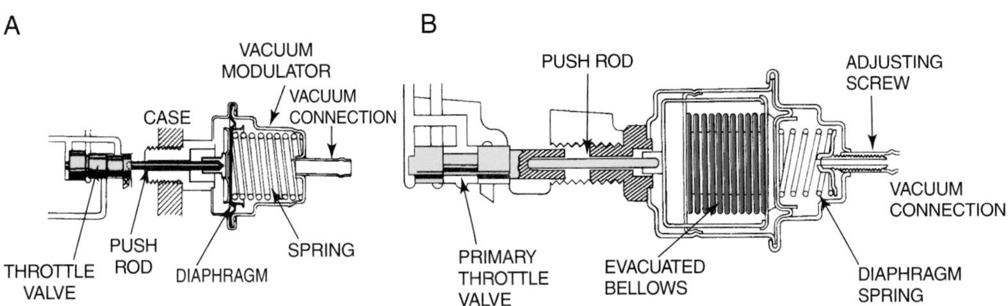

Figure 21-51. A—A vacuum modulator can control the throttle valve by varying pressure on the push rod. B—Altitude sensitive modulator. Note the evacuated bellows. (Ford)

adding to diaphragm pressure and raising TV pressure. As altitude increases, atmospheric pressure decreases, resulting in a decrease in TV pressure. The altitude compensating bellows is capable of providing a more uniform shift feel regardless of elevation.

Detent Valve

When the driver depresses the accelerator to the floor, the kickdown or detent valve is moved by either the linkage or solenoid and opens an additional passage to the shift valves. If governor pressure is low enough, the shift valve will be moved to the downshifted position. If the vehicle is moving too fast, the governor pressure will be too high to overcome and the transmission will stay in high. See **Figure 21-52**. To prevent band and clutch slippage, another passage from the detent valve opens to assist the main pressure regulator in raising line pressure.

Additional Shift Valves

Three- and four-speed automatic transmissions and transaxles require more shift valves. A two-speed unit requires one shift valve, a three-speed automatic needs two shift valves, and a four-speed unit needs three shift valves. All shift valves operate in the same manner. The higher gear shift valves have stronger springs and different sized valve areas. Therefore, the governor must be turning faster (higher road speed) to cause the valve to move to the upshifted position.

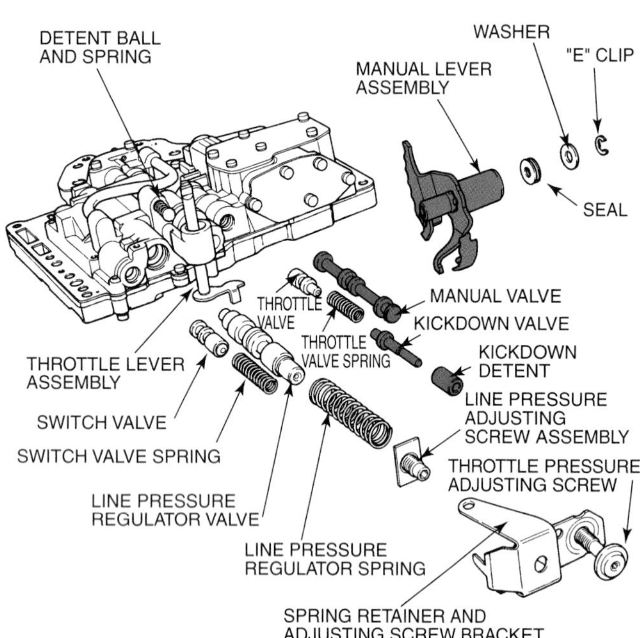

Figure 21-52. A manual detent downshift valve and related parts. (Geo & Dodge)

Lockup Converter Control Valves

The converter control valve can be thought of as another shift valve. Governor and throttle valve pressures operate on the valve to move it to the applied and released positions as needed. The major difference is that this valve usually receives its apply side pressure from one of the shift valves when it moves to the upshifted position. This ensures that the converter clutch will not be applied until the transmission/transaxle is out of first or second gear, depending on the design. See **Figure 21-53**.

On other designs, the converter clutch is operated by a solenoid, **Figure 21-54**. The solenoid may be controlled by an electric speed or gear relay or by an ECU. Some early systems use a vacuum switch that disengages the clutch when the engine vacuum drops. The transmission hydraulic system must still act on the lockup valve to make sure that the clutch is not applied at low speeds.

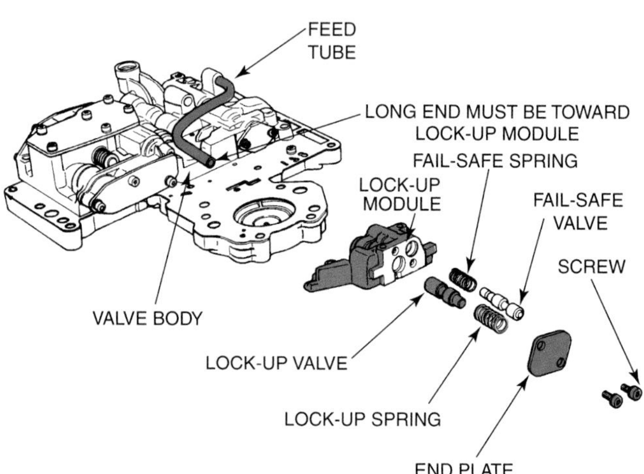

Figure 21-53. Exploded view of a lock-up torque converter control valve. Note the hydraulic fluid feed tube. (Chrysler Corp.)

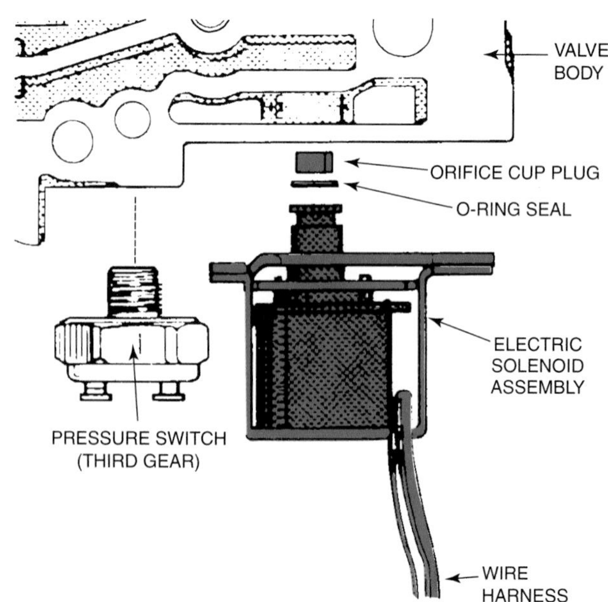

Figure 21-54. A torque converter clutch (TCC) electric solenoid valve. This solenoid is operated by the electronic control unit (ECU). By locking the converter clutch, slippage is reduced between the two, increasing fuel economy. (Chevrolet)

Computer-Controlled Shifting

On many vehicles produced during the last few years, an electronic control unit is used to shift the transmission or transaxle and apply the converter clutch. The transmission still has shift valves, however, they are operated by solenoids controlled by the ECU. There is no TV linkage or governor valve.

The transmission hydraulic system, including the shift pattern, converter lockup, and sometimes the line pressure is controlled by the ECU. A computer-controlled transmission schematic is shown in **Figure 21-55**. The ECU used to control transmission operation is often incorporated with the engine ECU and is sometimes referred to as a powertrain control module.

Accumulators

Almost every transmission or transaxle has at least one ***accumulator*** to cushion the servo or piston apply. **Figure 21-56** illustrates one type. The accumulator reduces the initial apply pressure by bleeding off some of the fluid to the servo or clutch. Fluid entering the servo or clutch piston is also sent to the accumulator piston. The accumulator spring contains a spring that is gradually compressed by the entering fluid. The accumulator allows a soft initial band or clutch application that is immediately followed by additional heavy pressure for firm engagement.

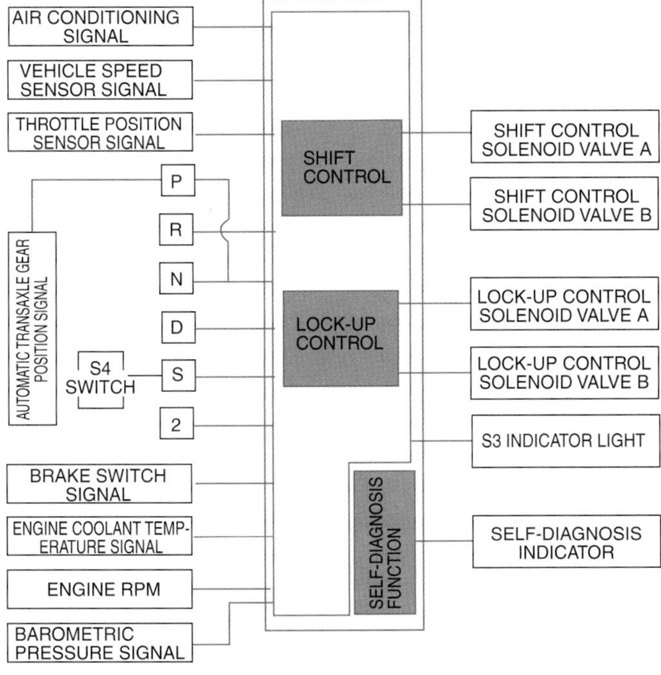

Figure 21-55. One particular computer-controlled, automatic transaxle operation schematic. (Honda Motor Co.)

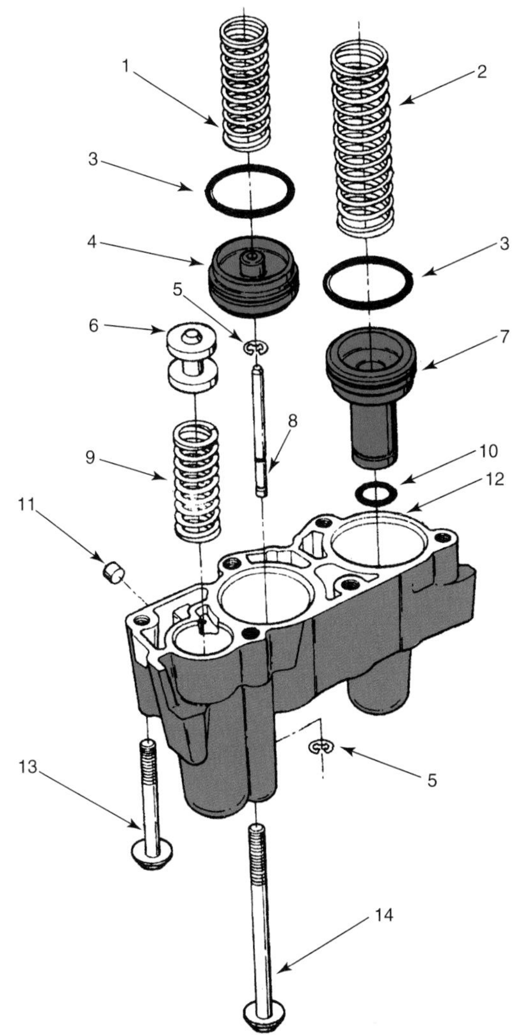

Figure 21-56. Exploded view of an accumulator housing and internal parts from a four-speed automatic transmission. The accumulators help reduce the shock (harshness) when the transmission shifts. 1—Piston spring for 4th clutch accumulator. 2—3rd clutch accumulator piston spring. 3—Square cut seal. 4—4th clutch accumulator piston. 5—Snap ring. 6—Torque signal compensator valve. 7—3rd clutch accumulator piston. 8—4th clutch accumulator piston pin. 9—Torque signal compensator spring. 10—Square cut seal. 11—Direct oil gallery plug. 12—Accumulator housing. 13—Accumulator housing valve body bolt. 14—Accumulator housing valve body bolt. (Chevrolet)

Complete Control Circuit

Figure 21-57 shows the complete hydraulic system as used in a four-speed automatic transmission. Note the extra shift valves. The various hydraulic system circuits must be compact and every possible means used to avoid external tubes or other awkward fluid transfer arrangements. Many of the shafts and other units are drilled to carry fluid.

Trace each valve and circuit and compare its operation to the simple system we looked at earlier. Most of the control valves are housed in the valve body. An exploded view of just some of the many components of a control valve body is shown in **Figure 21-58**.

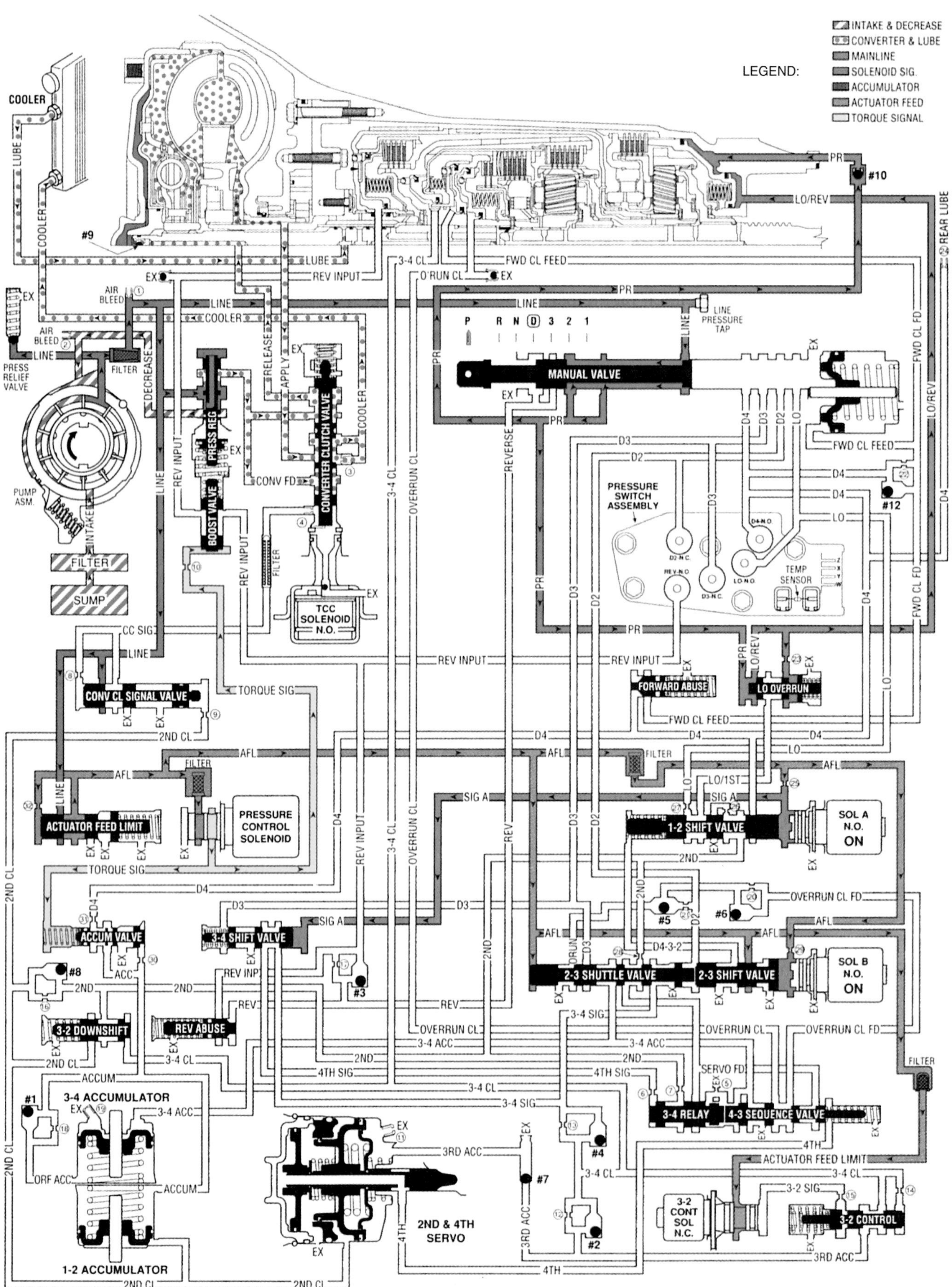

Figure 21-57. A complete hydraulic control circuit for a four-speed automatic transmission. This schematic shows transmission operation with the engine running and the transmission in park. Study all the circuits very carefully. (General Motors)

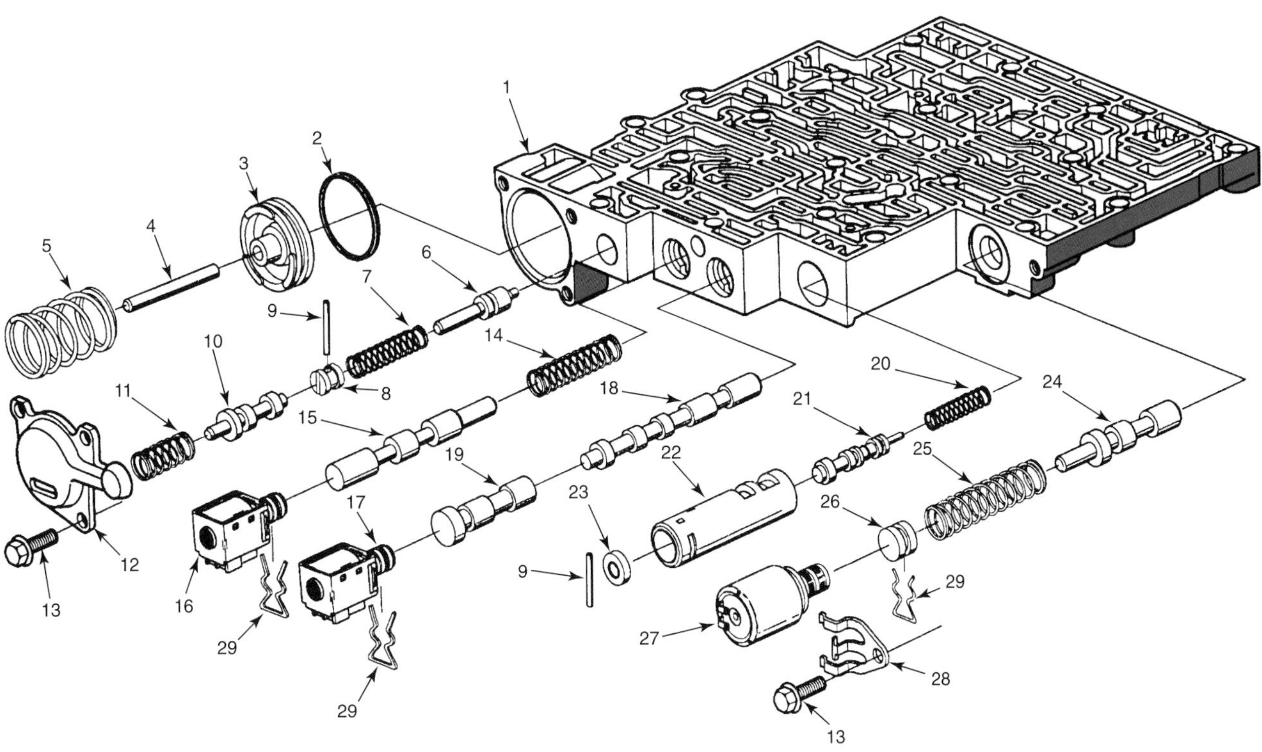

Figure 21-58. Exploded view of the accumulator end of an electronic control valve body for a four-speed automatic transmission. 1—Control body valve assembly. 2—Forward accumulator oil seal. 3—Forward accumulator piston. 4—Forward accumulator pin. 5—Forward spring. 6—Forward abuse valve. 7—Forward abuse valve spring. 8—Bore plug. 9—Coil spring pin. 10—Low overrun valve. 11—Low overrun valve spring. 12—Forward accumulator cover. 13—Bolt. 14—1—2 Shift valve spring. 15—1–2 Shift valve. 16—1–2 Shift solenoid. 17—2–3 Shift solenoid. 18—2–3 Shift valve. 19—2–3 Shuttle valve. 20—1–2 Accumulator valve spring. 21—1–2 Accumulator valve. 22—1–2 Accumulator valve sleeve. 23—Bore plug. 24—Actuator feed limit valve. 25—Actuator feed limit valve spring. 26—Bore plug. 27—Pressure control solenoid. 28—Pressure control solenoid retainer. 29—Retainer.

Transmission Case

The transmission case contains the internal shafts, gears, and holding members. It is also the mounting surface for the valve body, servos, pump, and linkage. Modern transmission cases are made of aluminum. Cases are vented to relieve internal pressure.

Transmission Lubrication

The transmission is lubricated by special fluid which circulates through the unit. ***Automatic transmission fluid*** or ***ATF***, is composed of mineral oil combined with additional additives that assist in lubrication and keep the oil from breaking down. Most modern automatic transmissions and transaxles use ***Dexron III/Mercon***. Dexron III replaces Dexron II, which had been in use for many years and is still available in some areas.

Some manufacturers, such as Honda, specify a different type of fluid. Type F fluid was used in Ford transmissions manufactured until the late 1970s. Do not use fluids marked with the letters AQ-ATF or Type A, Suffix A. These fluids are obsolete and are not suitable for modern vehicles. Dexron II should never be used in a transmission designed to use Dexron III/Mercon.

Fluid Filter

Another important part of the transmission is the ***filter***, which keeps dirt and metal out of the other parts. The transmission filter contains a paper filtering element inside of a metal housing. Filter shapes can vary and a few filters contain bypass valves. The filter is always placed in the pump intake, or suction, so that it can filter the transmission fluid before it enters the hydraulic system. It is always installed in the bottom of the transmission, where it can pick up fluid from the pan.

Fluid Cooling

Transmission fluid is cooled by circulating it, via tubes, to a external transmission cooler installed in the outlet side tank of the radiator. These coolers were discussed in more detail in Chapter 11. The fluid gives up its heat to the radiator coolant. The fluid always comes directly from the torque converter, which is the hottest part of the transmission. On downflow radiators, the cooler is located in the bottom tank. Extra coolers are sometimes installed in front of the vehicle radiator. These coolers remove heat by direct contact with the air. See **Figure 21-59**.

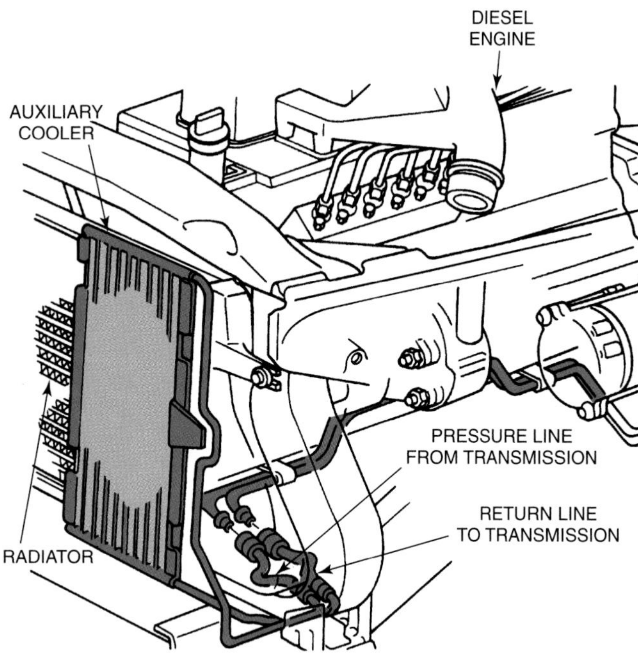

Figure 21-59. One type of automatic transmission fluid cooler which is mounted in front of the radiator. This particular cooler setup is for a transmission that is used with a diesel engine. (Dodge)

Summary

The automatic transmission eliminates the clutch pedal and manual gearshifting. After the driver has selected the range, the transmission shifts itself up or down depending on road speed, throttle position, and engine load. The operation of the automatic transmission and automatic transaxle is similar. The difference is in the placement of component parts and power flow.

The automatic transmission/transaxle assembly consists of a torque converter, planetary gearsets, holding members, oil pumps, and a hydraulic control system. The fluid must be cooled, and some actions of the transmission are controlled by the operator to select certain speed ranges. The torque converter can multiply the engine torque, thereby giving a wide range of power. The amount of multiplication is dependent on converter design and position between stall and coupling point.

The transmission or transaxle has one or more planetary gearsets to produce reduction, direct drive, overdrive, and reverse. Holding members, called bands, multiple disc clutches, or overrunning clutches, are used to control the planetary gearsets. The clutches and bands are actuated by hydraulic pressure produced by an engine-driven fluid pump. The fluid pressure is directed and controlled by a series of valves. Hydraulic pressure is distributed by fluid passageways in the valve body, case, and shafts.

The automatic transmission shifts in response to speed and load conditions. Speed ranges such as drive, low, and reverse can be selected by the driver through mechanical linkage connected to a control valve in transmission. Fluid is cooled by piping it to a cooler unit in one of the radiator tanks. Lubrication is provided by transmission fluid as it circulates through the unit.

Know These Terms

Automatic transmission
Automatic transaxle
Torque converter
Fluid coupling
Impeller
Turbine
Hydraulic coupling
Rotary flow
Vortex flow
Slippage
Stator
Stall
Coupling point
Torque multiplication
Lockup clutch
Torque converter clutch
Lockup torque converter
Planetary gearsets
Sun gear
Planet gears
Planet carrier
Ring gear
Simpson geartrain
Compound planetary gearset
Holding members
Band
Servo
Apply side
Release side
Multiple disc clutch
Clutch pack
Clutch hub
Clutch discs
Drum
Clutch apply piston
Overrunning clutch
One-way clutch
Roller clutch
Sprag clutch
Fluid pump
Fluid pan
Hydraulic pressure
Vane pump
Rotor pump
Internal-external gear pump
Spool valves
Valve body
Main pressure regulator
Line pressure
Manual valve
Shift valve
Throttle valve
TV linkage
Governor valve
Detent valve
Lockup torque converter control valve
Vacuum modulator
Accumulator
Automatic transmission fluid (ATF)
Dexron III/Mercon
Filter

Review Questions—Chapter 21

Do not write in this book. Write your answers on a separate sheet of paper.

1. Technician A says that the torque converter takes place of the conventional clutch used on standard transmissions. Technician B says that the torque converter is less efficient than the hydraulic coupling at high speeds. Who is right?
   (A) A only.
   (B) B only.
   (C) Both A & B.
   (D) Neither A nor B.

2. Which of the following torque converter parts is turned directly by the engine crankshaft?
   (A) Turbine.
   (B) Stator.
   (C) Impeller.
   (D) All of the above.
3. Which of the following parts is installed to turn a hydraulic coupling into a torque converter?
   (A) Turbine.
   (B) Stator.
   (C) Impeller.
   (D) All of the above.
4. When the torque converter lockup clutch is applied which of the following parts does not transmit power?
   (A) Transmission input shaft.
   (B) Turbine.
   (C) Impeller.
   (D) Converter housing.
5. The stator is unlocked when the turbine speed is _____.
   (A) zero
   (B) one-half of impeller speed
   (C) almost the same as impeller speed
   (D) Both A & B.
6. What mechanical device allows the stator to lock and unlock?
7. Stall speed is greatest when the turbine speed is _____.
   (A) zero
   (B) one-half of impeller speed
   (C) almost the same as impeller speed
   (D) Both A & B.
8. Converter coupling point is reached when the turbine speed is _____ .
   (A) zero
   (B) one-half of impeller speed
   (C) almost the same as impeller speed
   (D) Both A & B.
9. All of the following statements about planetary gearsets are true, EXCEPT:
   (A) the sun gear is always in the center of the gearset.
   (B) the planetary gears are always in mesh.
   (C) planetary gears take up more room than sliding gears.
   (D) a single planetary gearset can provide reduction and overdrive.
10. A planetary gearset can be used to _____ or _____ torque.
11. A planetary gearset can be used to _____ the direction of rotation.
12. If the sun gear and ring gear are locked together, what gear ratio is produced?
    (A) 1:1.
    (B) 1.5:1.
    (C) 2:1.
    (D) 3.5:1.
13. If the planet carrier is held stationary, and the sun gear drives, which part of the gearset is driven?
14. Technician A says that, when the planet carrier is held stationary, and the sun gear drives, the driven member will rotate in direct drive. Technician B says that a planetary gear member may be locked to the case by a band or a clutch. Who is right?
    (A) A only.
    (B) B only.
    (C) Both A & B.
    (D) Neither A nor B.
15. Bands are applied by _____ .
    (A) servos
    (B) accumulators
    (C) clutch apply pistons
    (D) overrunning clutches
16. A servo may have an _____ side and a _____ side.
17. The two commonly used types of one-way or overrunning clutches are the _____ and the _____ .
18. Technician A says that a multiple disc clutch may be applied by a servo. Technician B says that the multiple disc clutch plates have alternating inner and outer splines. Who is right?
    (A) A only.
    (B) B only.
    (C) Both A & B.
    (D) Neither A nor B.
19. The three types of pumps in common use in modern automatic transmissions and transaxles are the:
20. What does the shift valve control?
    (A) Band and clutch operation.
    (B) Gear changes.
    (C) Upshifts and downshifts.
    (D) All of the above.
21. Which valve controls overall system (line) pressure?
    (A) Manual valve.
    (B) Throttle valve.
    (C) Main pressure regulator.
    (D) Detent valve.
22. Which valve tries to move the shift valve to the upshifted position?
    (A) Throttle valve.
    (B) Governor valve.
    (C) Detent valve.
    (D) Manual valve.

23. What is the purpose of the accumulator?
    (A) Cushion shifts.
    (B) Cushion band and clutch apply.
    (C) Increase line pressure.
    (D) Both A & B.
24. Which valve tries to keep the shift valve in the downshifted position?
    (A) Throttle valve.
    (B) Main pressure regulator.
    (C) Governor valve.
    (D) Manual valve.
25. Technician A says that the turning output shaft operates the throttle valve. Technician B says that the governor valve can be operated by TV linkage or a modulator. Who is right?
    (A) A only.
    (B) B only.
    (C) Both A & B.
    (D) Neither A nor B.
26. State three things that make the automatic transaxle different from the automatic transmission.
27. Transmission fluid goes to the fluid cooler directly from the _____ .
    (A) torque converter
    (B) sump
    (C) servos
    (D) valve body
28. The add-on transmission cooler transfers fluid heat directly to the _____ .
    (A) coolant in the radiator
    (B) coolant in the engine
    (C) outside air
    (D) cooler fluid in the pan
29. None of the following transmission fluids should be used in the latest model automatic transmissions and transaxles, EXCEPT:
    (A) AQ-ATF.
    (B) DEXRON III/MERCON.
    (C) Type A Suffix A.
    (D) Type F.
30. The CVT transmission uses a metal drive belt with _____ to provide the different gear ratios.
    (A) a Simpson gear train
    (B) compound planetary gearsets
    (C) variable diameter pulleys
    (D) sliding gears

22

Axles and Drives

After studying this chapter, you will be able to:

- List the basic parts of a rear-wheel driveline.
- Explain the function of each rear-wheel drive shaft part.
- List the basic parts of a front-wheel driveline.
- Explain the function of each front-wheel drive axle part.
- Describe various types of front- and rear-wheel drivelines.
- Explain the construction of the solid rear-wheel drive axle assemblies.
- Explain the construction of the independent rear-wheel drive axle assemblies.
- Describe the front drive axles used on four-wheel drives.
- Explain the operation of the ring and pinion.
- Explain the operation of the differential unit.
- Explain the operation of limited-slip differentials.
- Compare rear-wheel and front-wheel drive differential assemblies.

This chapter will cover the differential assemblies and drive axles used on rear-wheel, front-wheel, and four-wheel drive vehicles. After studying this chapter, you will be familiar with the principles and component parts of rear-wheel drive shafts, CV axles, and the operation of the rear-wheel drive ring and pinion and differential assembly. Some of the information in this chapter is useful in understanding the operation of the four-wheel drive systems discussed in Chapter 20.

Rear-Wheel Drive Shafts

For the transmission output shaft to drive the differential pinion gear shaft, it is necessary to connect these two units by means of a ***drive shaft***, **Figure 22-1.** All rear-wheel drive vehicles have a drive shaft.

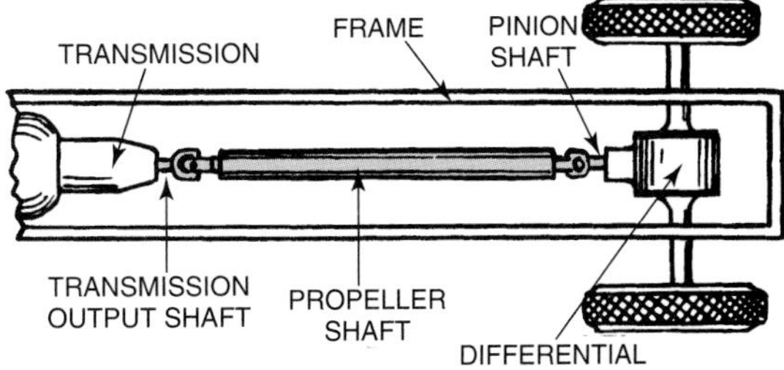

Figure 22-1. Top view showing drive shaft connecting the transmission to the differential.

Drive Shafts

The rear-wheel drive shaft is hollow to reduce weight, but has a diameter large enough to give the shaft great strength, **Figure 22-2.** Steel, aluminum, and graphite are used in drive shaft construction. Some have a rubber-mounted torsional damper.

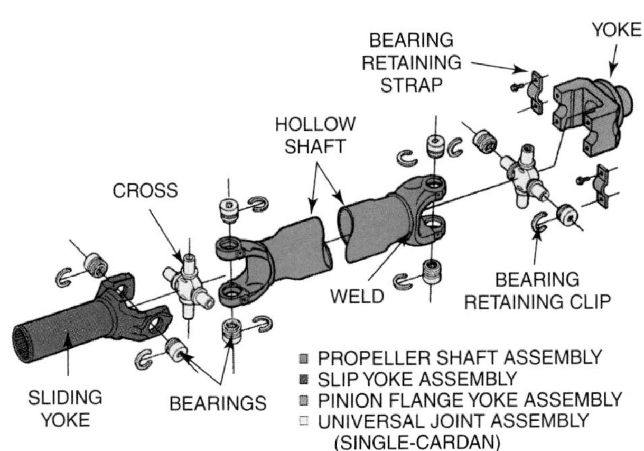

Figure 22-2. Typical drive shaft. Yokes are welded to a hollow shaft. Note the sliding yoke on one end. This allows the driveline to adapt to changes in length as the suspension moves up and down. (General Motors)

A yoke and splined stub (where used) are welded to the ends of a hollow shaft. The shaft must run true and be carefully balanced to avoid vibrations. The drive shaft is often turning at high speeds and can cause great damage if bent, imbalanced, or if there is wear in the flexible joints.

Drive Shaft Must be Flexible

When the rear wheels strike a bump or a hole, the rear axle housing moves up or down in relation to the frame. Since the transmission is mounted to the frame, it becomes obvious that to allow the rear axle to move up and down, the driveline or drive shaft must be able to flex, **Figure 22-3.**

The driveline angle changes as the vehicle travels over different grades, bumps in the road, and as the engine mounts wear over time. A solid driveline would begin to vibrate as the driveline angle changes, eventually resulting in a bent drive

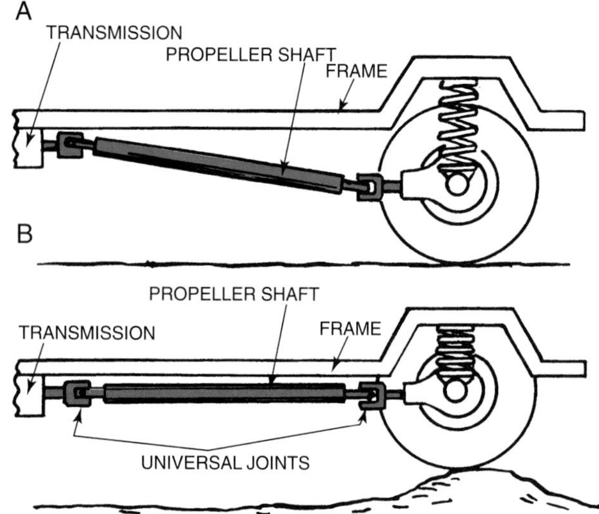

Figure 22-3. All drive shafts must flex. A—Normal operating angle of drive shaft. B—This shows how a bump forces the differential up. Notice how the drive shaft angle changes. Since the transmission is fixed to the frame and the differential has limited movement, the drive shaft must flex.

shaft or broken driveline component. To allow the drive shaft to move without breaking, a flexible joint is used. This is called a ***universal joint***, **Figure 22-4.** By using a universal joint, torque can be transmitted constantly while the rear axle is moving up and down, **Figure 22-5.**

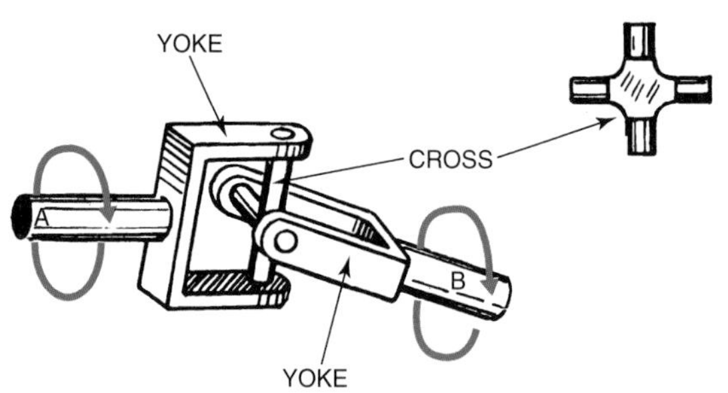

Figure 22-4. Simple universal joint. Shaft A is transmitting torque in a horizontal plane. Shaft B is transmitting the same torque, but at a different angle. The yokes swivel around the ends of the cross, imparting a degree of flexibility to the universal joint.

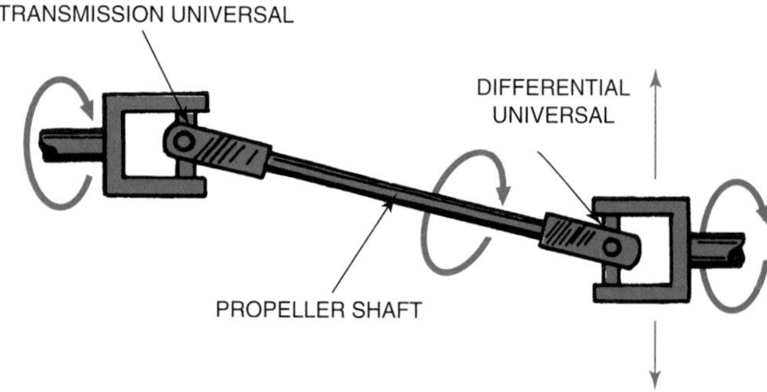

Figure 22-5. Universal joints in action. As the rear axle moves up and down, the universal joints allow the driveline angles to change without binding.

Universal Joint

The ***cross and roller*** universal joint, sometimes called a ***cardan universal joint***, consists of a ***center cross*** and two ***yokes***. The yokes are attached to the cross through needle bearing assemblies, usually called ***bearing caps***. The bearing caps are retained in the yoke by snap rings, injected plastic, U-bolts, or bolted plates. The bearing cap rollers surround the cross ends. These cross ends are called ***trunnions***. This allows the yokes to swivel on the cross with a minimum amount of friction.

Figure 22-6 shows an exploded view of a typical cross and roller universal joint. Notice that the trunnions are smoothly finished. A hollow in each trunnion leads to the center of the cross. A grease fitting is placed in the center. When lubricated, grease will flow out to each needle roller

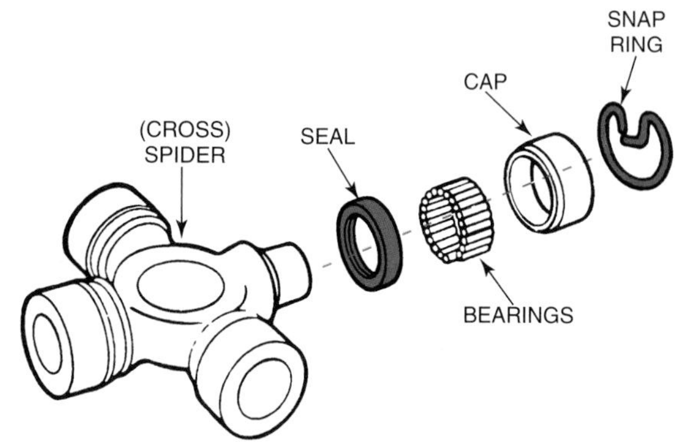

Figure 22-6. Universal joint cross with bushing and roller assemblies. (General Motors)

A cutaway view of a cross and roller universal joint is shown in **Figure 22-7.** Note that this particular setup employs injection molded plastic rings to secure the bearing cups or bushings in the yoke. Many original equipment universal joints do not have a grease fitting. If the joint needs lubrication, it must be disassembled and packed with grease.

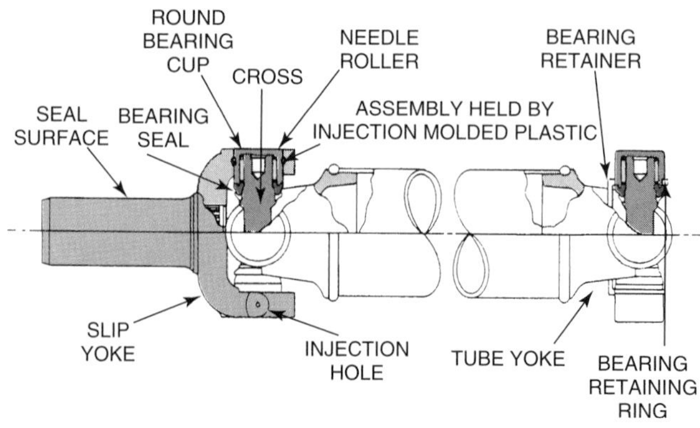

Figure 22-7. Cross and roller universal joint employing injection molded plastic "rings" to secure the bearing cup or bushing to the slip yoke. Plastic is forced into an injection hole and flows into a ring groove cut in the bushing and yoke. The plastic hardens and secures the bushing. (General Motors)

Slip Yoke

As the rear axle moves up and down, it swings on an arc that is different from that of the driveline. As a result, the distance between the transmission and rear axle will change to some extent, **Figures 22-7** and **22-8.**

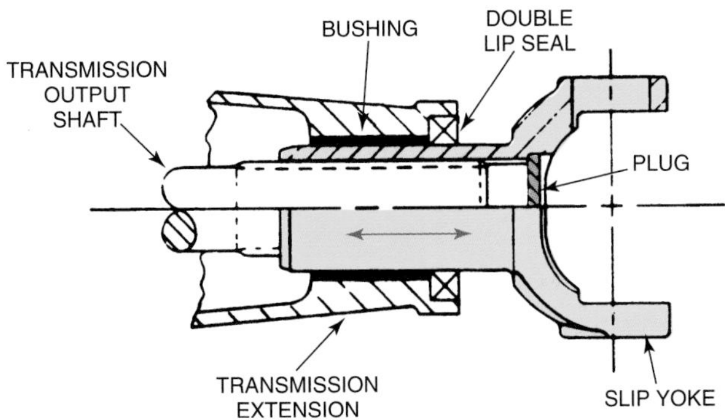

Figure 22-8. A slip yoke in position on the transmission output shaft, extending inside the transmission housing. The plug in the yoke helps to retain grease and prevents the entry of dirt into the splines. (General Motors)

To allow the driveline to adjust to these variations in length, a splined sleeve or ***slip yoke*** is needed. The end of the transmission stub shaft has a series of splines cut into it. The universal slip yoke sleeve has corresponding splines cut into it. The slip yoke sleeve is slipped over the splined transmission drive shaft stub and the universal joint is bolted to the rear axle output shaft. The shaft can move by the slip yoke sliding in the sleeve splines. Some two-piece shafts have a slip yoke on both shafts, **Figure 22-9.**

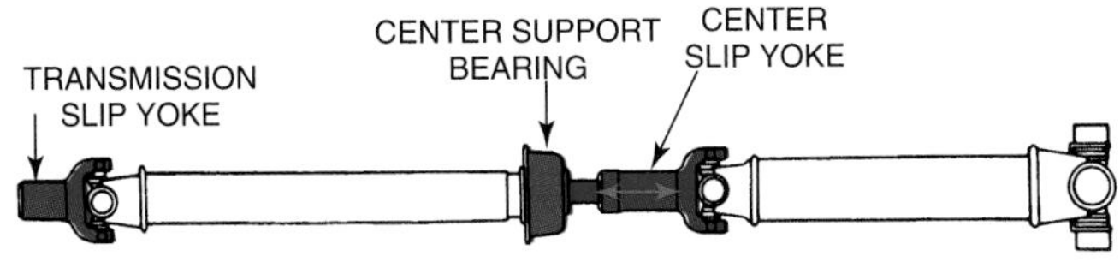

Figure 22-9. A two-piece drive shaft assembly which incorporates two slip yokes, one on each shaft. Note the center supporting bearing. (General Motors)

How the Wheels Drive the Vehicle

When the drive shaft turns the differential, the axles and wheels are driven forward. The driving force developed between the tires and the road is first transferred to the rear axle housing. From the axle housing, it is transmitted to the vehicle in one of two ways:

- Through leaf springs that are bolted to the housing and shackled to the frame.
- Through control or torque arms shackled (bolted, but free to swivel) to both frame and axle housing. This system is common on vehicles with rear coil springs.

These drive systems are referred to as a ***Hotchkiss drive***. See **Figure 22-10** for a side view of these drive systems. Some older vehicles used a drive system called a torque tube drive. This system transferred power through a torque tube that surrounded the drive shaft and which was bolted to the axle housing and pivoted to the transmission by means of a large ball socket.

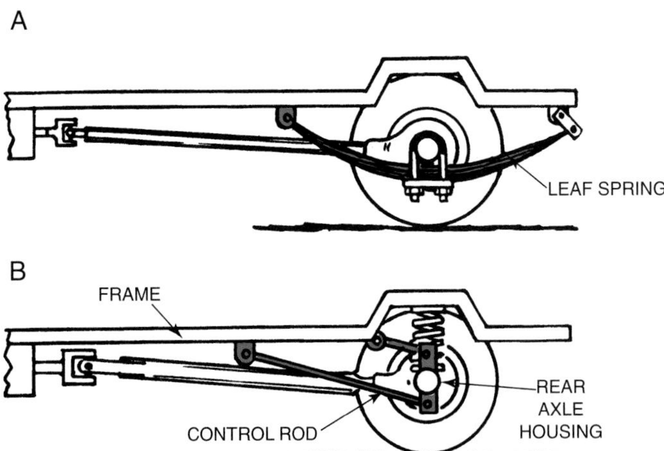

Figure 22-10. Methods used to transfer driving force from the rear axle housing to the vehicle frame. A—Force transmitted through leaf springs. B—Force transmitted through control rods.

Hotchkiss Drive

The Hotchkiss drive is in almost universal use today. It consists of an open drive shaft secured to the transmission output shaft and differential pinion gear shaft. Front and rear universal joints are used. If both joints are of the cross and roller type, a slip yoke must be provided.

The slip yoke allows the driveshaft length to change as the vehicle goes over road irregularities. The internal splines of the slip yoke engage the external splines of the transmission output shaft. As the length changes, the splines slide in relation to each other. The rear axle drive thrust is transmitted to the frame through springs or control arms.

The main task performed by the Hotchkiss drive is that of transmitting transmission output shaft torque to the differential. Wheel drive thrust is not transmitted through the driveline. A typical Hotchkiss drive is shown in **Figure 22-11.**

Two-Piece Hotchkiss

On some vans, trucks, and large cars, the drive shaft would be so long that vibration problems would almost always occur. To reduce excessive drive shaft length, a ***two-piece Hotchkiss drive*** is often used. This design uses three universal joints. With a two-piece drive shaft, it is necessary to use a ***center support bearing***. This support consists of a ball bearing held in a dustproof housing that is supported by the frame. Rubber is generally incorporated in the mount to reduce noise and vibration transfer to the frame. The rubber will also allow some additional movement of the shaft without binding.

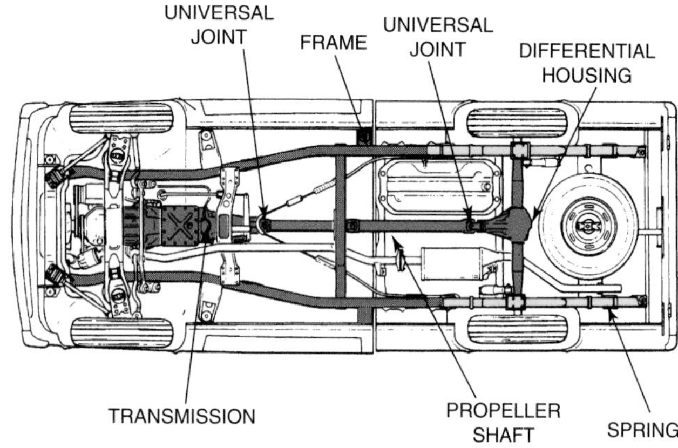

Figure 22-11. A typical Hotchkiss drive setup as used by one vehicle manufacturer. (Chrysler)

Figure 22-12 illustrates a typical center support bearing. Note that the front drive shaft rests in the rubber cushioned bearing. Many two-piece drive shafts install the slip yoke at the center support, as shown in **Figure 22-13.** This center support is also on rubber mounts. Both front and rear shafts in a two-piece Hotchkiss drive are illustrated in **Figure 22-13.**

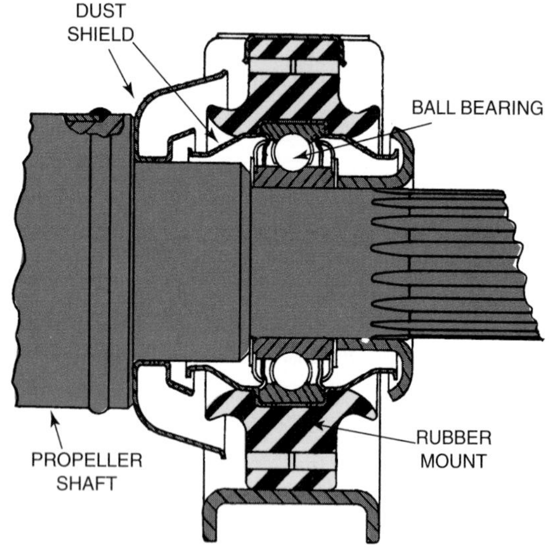

Figure 22-12. One type of two-piece drive shaft center support bearing. A sealed ball bearing, mounted in rubber, is used to support the two shafts where they meet. (GMC)

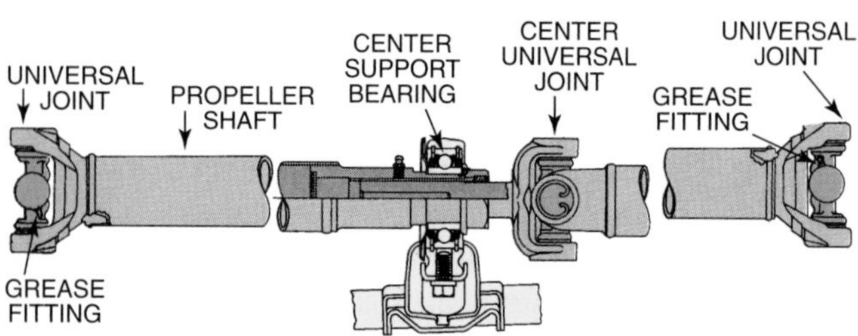

Figure 22-13. Two-piece Hotchkiss driveline—assembled. This setup uses three universal joints; two on one shaft, one on the other. A rubber cushioned center support bearing is used where the two shafts connect. (Chrysler)

Universal Joint Can Cause Fluctuating Shaft Speed

If a conventional universal joint is driving at an angle and the speed of the driving shaft is constant, the speed of the driven shaft will fluctuate. The driven shaft speed will rise and fall two times for every revolution. This is due to the fact that the angle of drive is not equally divided between the two shafts, **Figure 22-14.**

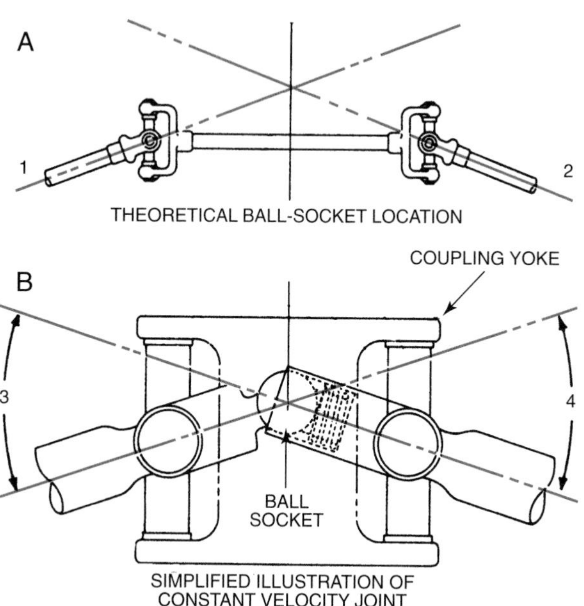

Figure 22-14. Schematic showing principle of constant velocity universal joint. A—The necessary location for the ball socket that will keep the drive angle evenly divided between shafts 1 and 2. B—Coupling yoke and ball socket. Note how both shafts form the same angle—3 and 4. (Chevrolet)

When using a universal joint on each end of a drive shaft, the shaft speed fluctuation can be greatly reduced if both universal joint yokes are lined up in the same plane. The chart in **Figure 22-15** illustrates drive shaft speed fluctuations.

With the joint yokes in the same plane, **Figure 22-16**, the torque of the drive shaft will be smoothly transmitted to the pinion shaft. The drive shaft speed will rise and fall from the front joint action, but the fluctuation will be canceled out by the action of the rear joint.

Rear-Wheel Drive Constant Velocity Universal Joint

A ***constant velocity universal joint*** uses two conventional universal joints in a single housing to produce an even transfer of torque without speed fluctuations. Constant velocity universal joints provide a smooth transfer of power.

The joint design makes constant speed transfer possible by automatically dividing the drive angle equally between two shafts. **Figure 22-17** shows a cutaway view of an assembled joint. **Figure 22-18** shows a one-piece drive shaft utilizing two constant speed universal joints.

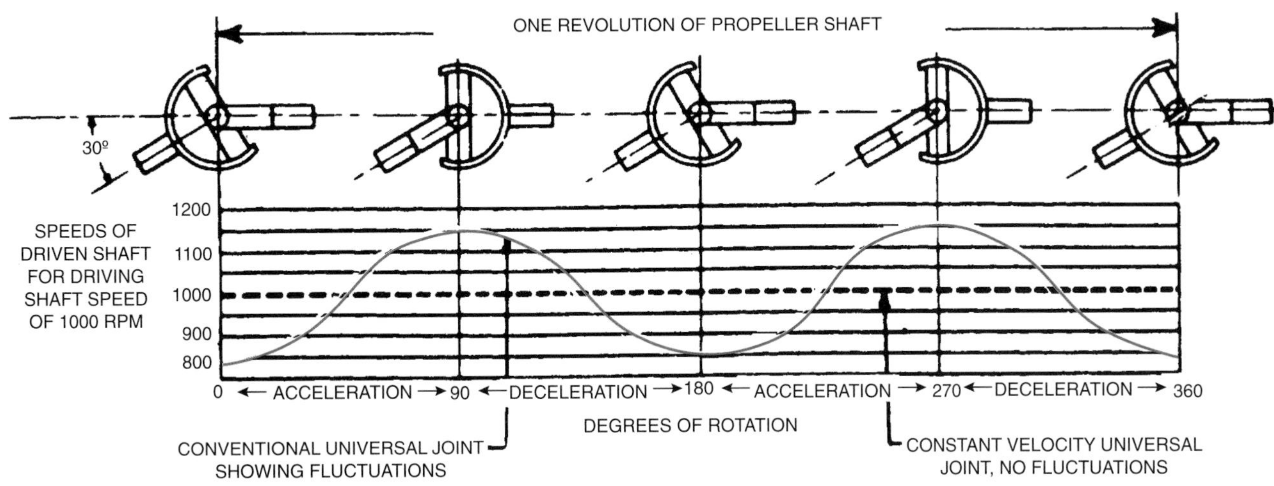

Figure 22-15. Drive shaft speed fluctuation chart. As the universal joint turns, the angle formed by the cross changes. It never divides the angle equally between two shafts, (except for one brief moment as cross angle shifts from one extreme to the other). This causes acceleration-deceleration fluctuation that transmits torque in a jerky fashion. Dotted line illustrates even speed when a constant velocity universal joint is used. Undulating line (waving) represents speed fluctuation with conventional universal joint.

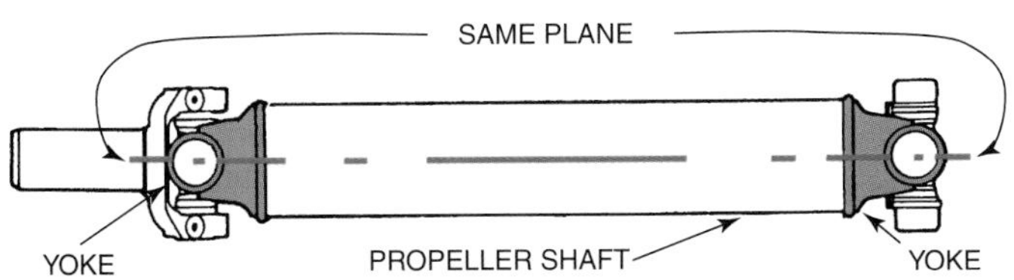

Figure 22-16. To prevent driveline fluctuations, universal joint yokes must be in the same plane. (General Motors)

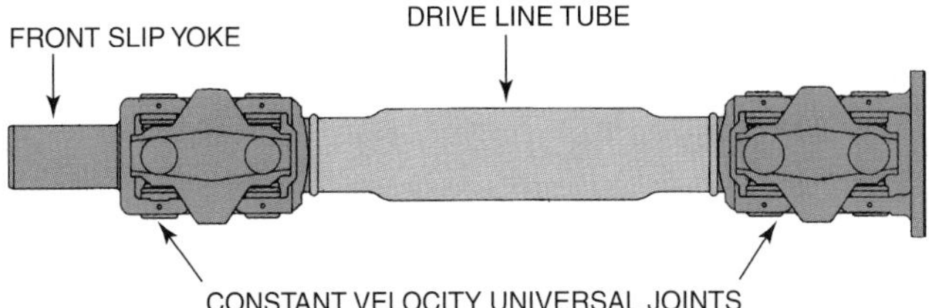

Figure 22-18. A one-piece drive shaft incorporating a constant velocity universal joint at each end. This produces a very smooth transfer of torque. (Cadillac)

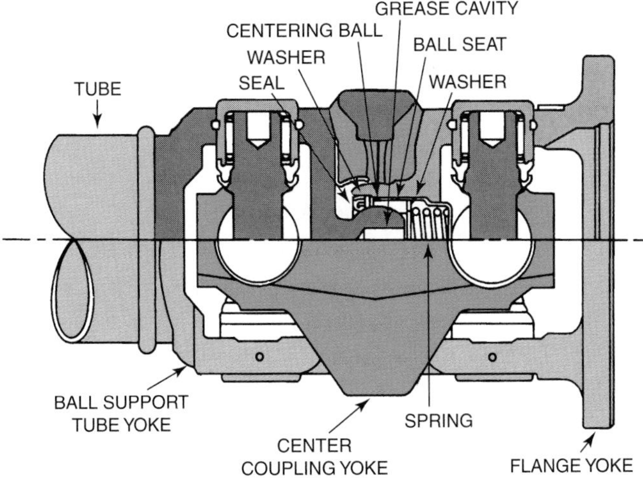

Figure 22-17. Cutaway of an assembled constant velocity joint. Note the centering ball and seat arrangement. This joint uses injected plastic retaining rings. (General Motors)

Constant Velocity (CV) Axles

One of the most obvious differences between vehicles with front-wheel drive and those with rear-wheel drive is the presence of two drive axles with flexible joints. These are called ***constant velocity axles***, or ***CV axles***.

CV axles allow the front wheels and front suspension to move up and down over bumps and to be turned while still delivering power. The CV axles can also change length to accommodate suspension movement without breaking. The CV axles must do all of this without transmitting excessive vibration into the passenger compartment.

CV Axle Shafts

The typical CV shaft is solid, although some CV shafts are hollow tubes similar to the rear-wheel drive shaft. CV axle shafts are always smaller in diameter than the rear drive shaft. Most CV axle shafts are one-piece units, as in **Figure 22-19.**

On high performance, luxury, and some four-wheel drive vehicles, the longest axle shaft may be a two-piece unit, as in **Figure 22-20.** This shaft has a center support bearing similar to the bearing used in a two-piece Hotchkiss drive shaft. The axles are held in place by nuts at the wheels, which tighten the axle to the front wheel bearing assembly, **Figure 22-21.**

Types of CV Joints

CV joints perform the same functions as the universal joint on rear-wheel drive vehicles. CV joints can drive the vehicle and compensate for steering actions while producing less vibration than a universal joint. **Figure 22-22** illustrates one particular front-wheel drive CV joint and shaft assembly.

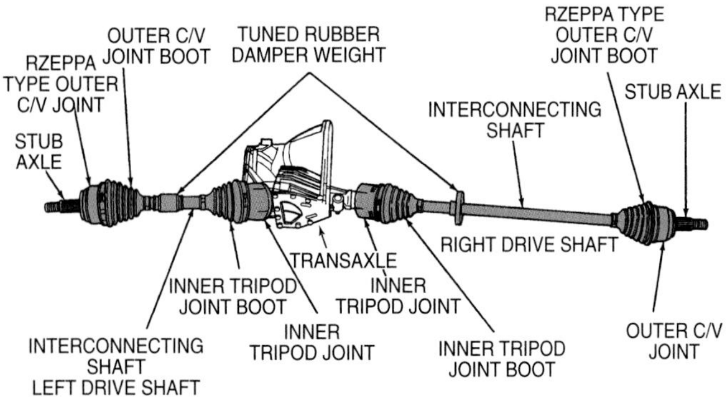

Figure 22-19. One type of front-wheel drive axle setup. This one uses two one-piece drive shafts of unequal length. (Chrysler)

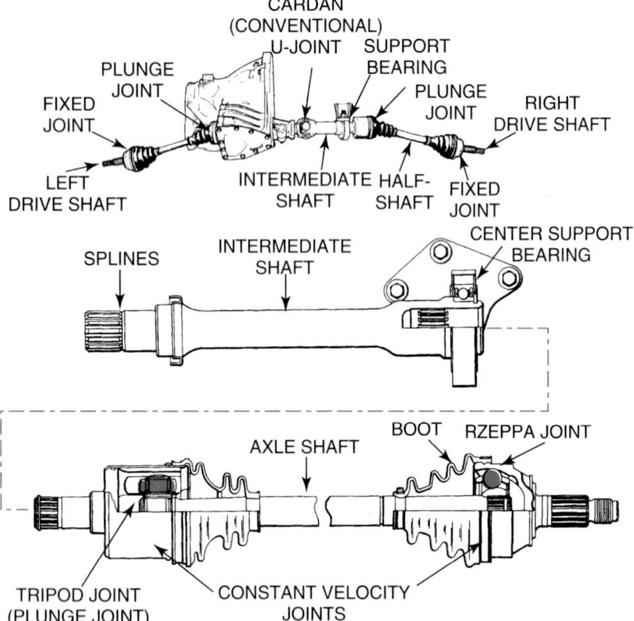

Figure 22-20. A two-piece axle shaft assembly which uses constant velocity (CV) joints. This CV axle also uses a center support bearing. (Sterling)

Front-wheel drive axles are equipped with inboard (inside) and outboard (outside) CV joints. **Figure 22-23** shows one such arrangement. CV joints are used on the front axles of some four-wheel drive vehicles and rear-wheel drive vehicles with independent rear suspensions.

There are two basic CV joint types: ***tripod*** and ***Rzeppa*** (also called ball and channel) type. Both of these are constant velocity designs. The Rzeppa joint is by far the most common. Some two-piece shafts may use a cross and roller joint in applications where driveline flexing is not a factor.

Tripod Joint

Figure 22-24 shows the tripod joint. It is composed of three trunnions inside of a three channel housing. Unlike the cross and roller universal joint, the tripod joint trunnions can slide back and forth in the channels as the joint rotates.

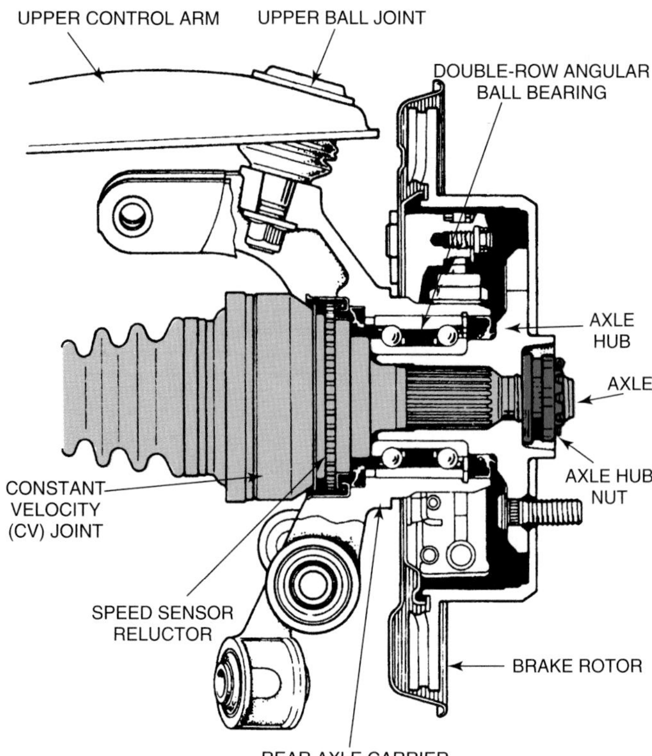

Figure 22-21. A cutaway view of a front hub assembly and drive axle. The axle is secured to the hub with a nut. The hub is pressed into a double-row angular ball bearing which requires no service or adjustment. (Lexus)

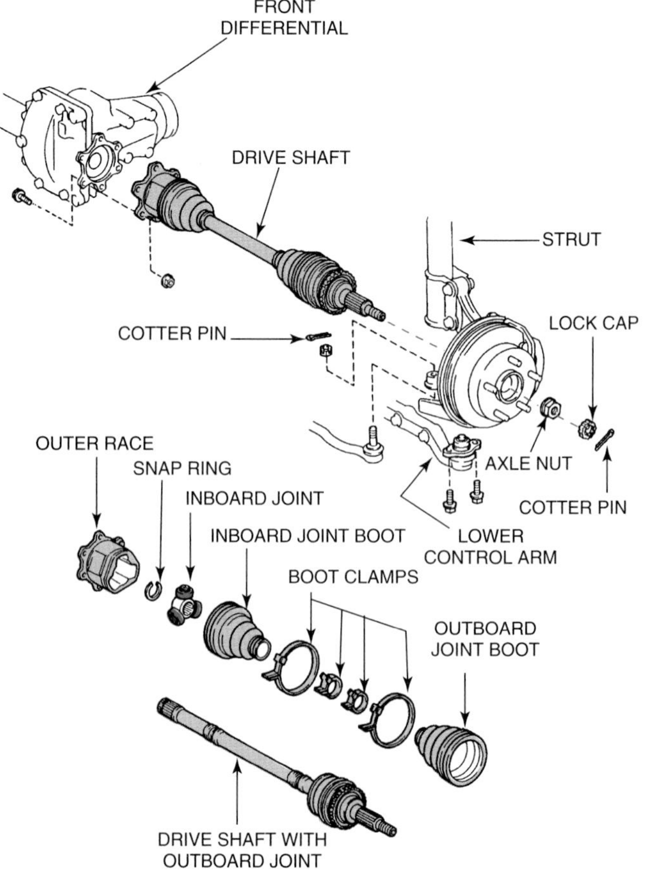

Figure 22-22. An exploded view of a four-wheel drive front axle and constant velocity joint. (Toyota)

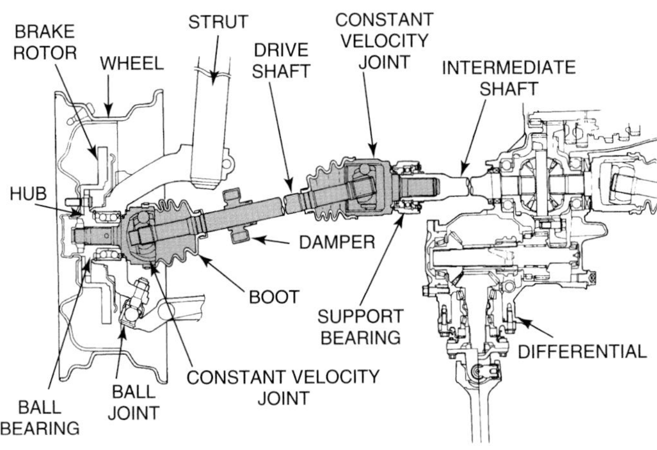

Figure 22-23. Cutaway view of inboard and outboard CV joints. The short CV axle can transmit engine power flow with minimal vibration. (Honda)

This maintains the same drive angle through the joint, allowing power to be transmitted without speed fluctuations. Inboard CV joints are usually tripod joints.

Rzeppa Joint

Figure 22-25 illustrates the Rzeppa CV joint, which was invented in 1926 by Alfred H. Rzeppa. It consists of a series of balls in a slotted cross and housing assembly. During operation, the balls slide back and forth in the slots as the joint rotates. Since the drive angle through the joint is not subject to fluctuations, power flows smoothly through the joint at any angle. Outboard CV joints are usually Rzeppa joints.

CV Joint Boots

High stress is constantly placed on both types of CV joints, so they must be continuously lubricated. Although the grease used in CV joints is heavy, it would be flung from the turning joint if a means of retaining it was not used. This is

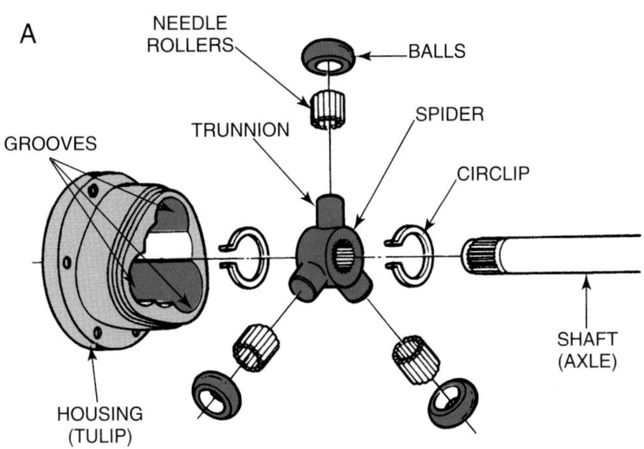

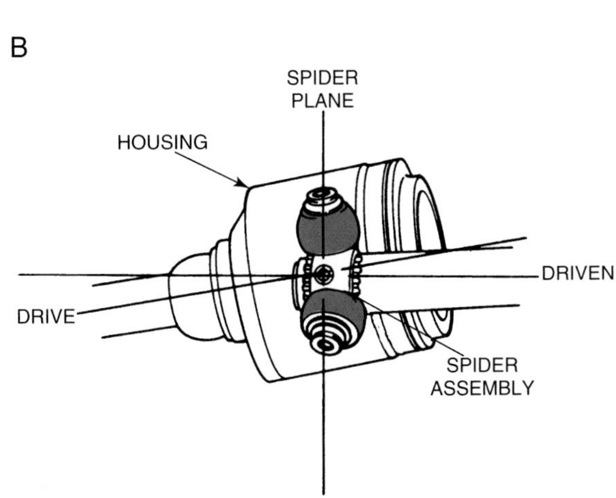

Figure 22-24. A—Exploded view of a tripot constant velocity joint. B—Tripot joint operation. The tripot joint uses three balls on needle bearings to transmit torque between the spider (splined to the axle shaft), and the housing. As the joint revolves, the three balls will change their positions to stay in the same plane with the spider. The balls can move in and out to allow for length changes as the suspension travels up and down. (AC Delco)

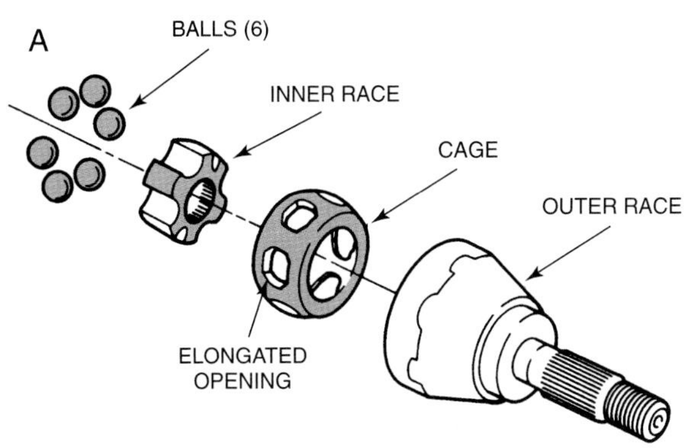

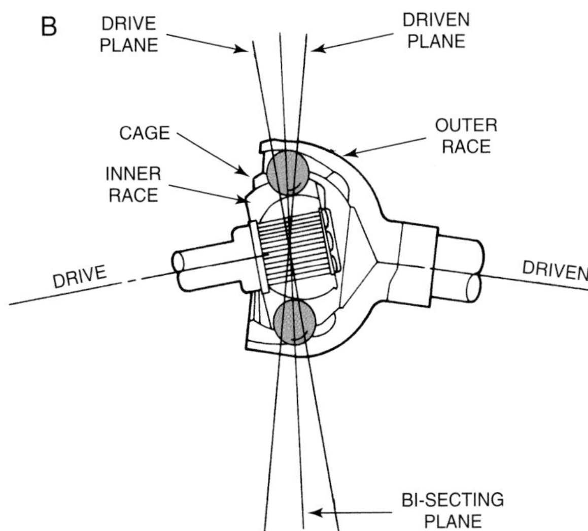

Figure 22-25. A—Exploded view of a Rzeppa (ball style) constant velocity joint. B—The Rzeppa joint as the balls operate in a bisecting plane between the angles of the axle shaft. As the joint revolves, the position of the balls will change so they can remain in this bi-secting plane. The position of the balls is controlled by the cage's elongated openings. (AC Delco)

accomplished by ***CV joint boots***, **Figure 22-26.** CV boots are bellows-shaped covers made of rubber or plastic. These boots completely seal the interior parts of the CV joint. Grease flung outward is caught by the boot and flows back into the joint when the vehicle speed decreases.

CV joints will usually last the life of the vehicle if the boots are not damaged. If the boot becomes damaged, the CV joint lubricant will be thrown out and dirt and water will enter, eventually resulting in joint failure. The boots should be inspected frequently and replaced if they are damaged.

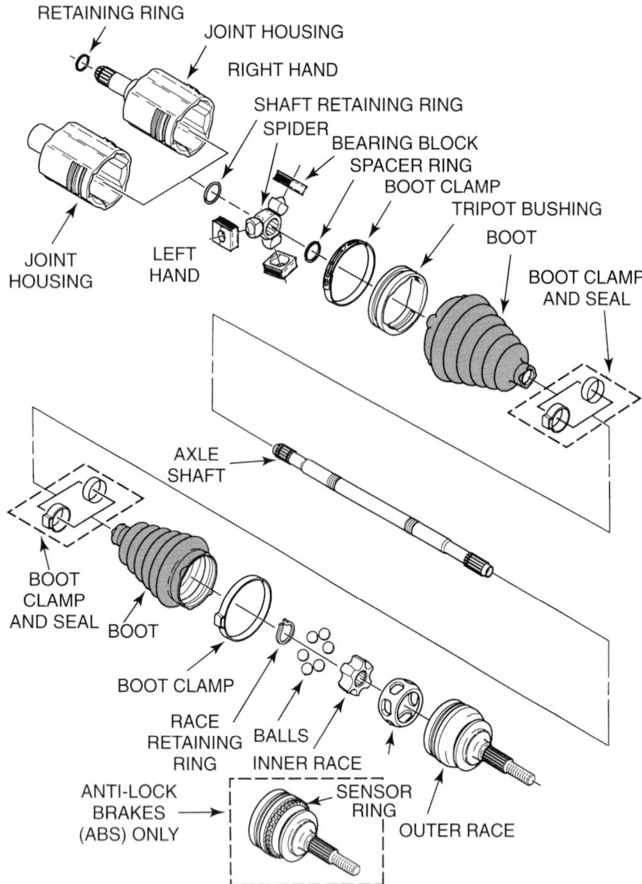

Figure 22-26. An exploded view of a front-wheel drive axle that incorporates a CV joint at each end. The boots protect the joint from dirt and water while providing a leak-proof grease retaining cover. (Cadillac)

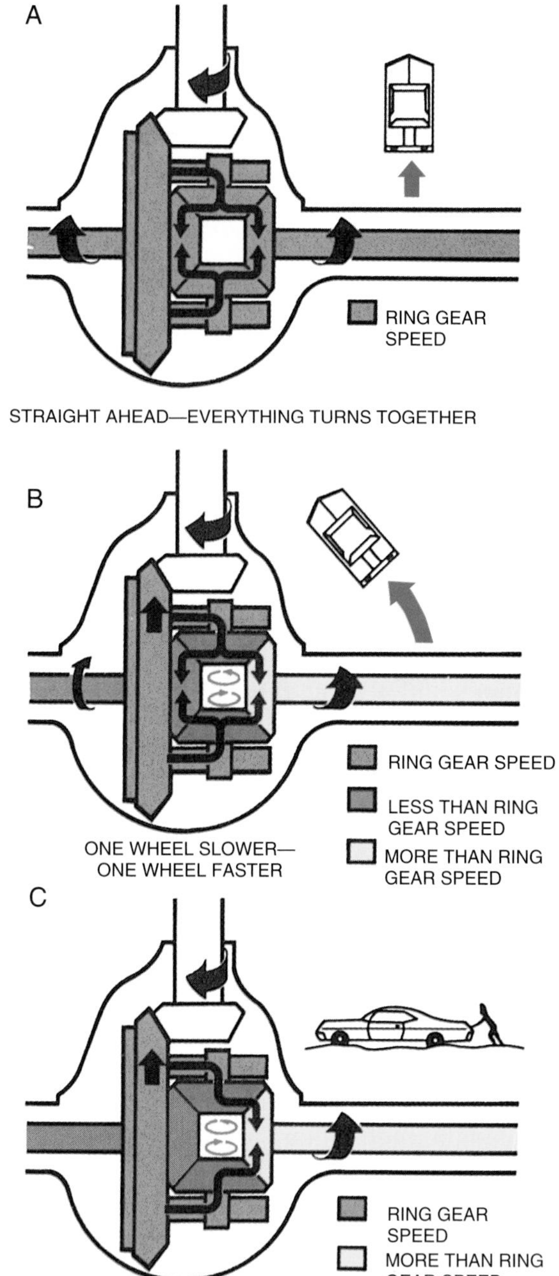

Figure 22-27. A—Differential and axle operation as a rear-wheel drive vehicle is traveling straight ahead. The differential case turns as the ring gear is driven by the pinion. The differential pinion shaft and gears rotate along with the differential case. The wheels and axles rotate at the same speed, so the side gears receive torque (force) from the differential pinions equally. B—As the vehicle turns left, the differential case turns as the ring gear is driven by the pinion. The differential pinion shaft and gears rotate along with the differential case. The "outer" wheel turns at a speed which is faster than the "inner" wheel, causing the side gears to rotate at different speeds. The side gears cause the differential pinion gears to rotate on the differential pinion shaft. C—Differential action when one wheel has very little, or no traction. The power flow is the same as when turning, but with only a single axle and side gear revolving. This is an "open" (not traction) type differential. (General Motors)

Rear-Wheel Drive Axle Assemblies

Rear-wheel drive axle assemblies have several important functions. They must hold the wheels on, keep them upright, and propel the vehicle forward or backward. They must drive the wheels in such a manner that one can turn faster than the other, while allowing both to receive torque, **Figure 22-27.**

Drive axle assemblies must distribute the driving force of the drive shaft to the wheels, and transmit it to the frame through the rear suspension. The axle assembly provides an anchor for springs, supports the weight of the vehicle, and forms the foundation for the rear brakes. Obviously, the axle assemblies must be well constructed.

Rear-Wheel Drive Axle Components

The rear axle assembly is used on all vehicle with rear-wheel drive. The basic axle assembly may be broken into several major sections: axle housing, axles, ring and pinion, and differential. These components form the basic rear-wheel drive axle unit.

Axle Housing

The ***axle housing*** is usually made of cast steel, aluminum, or stamped steel parts welded together. Two basic types of axle housings are used: the ***solid axle*** and the ***independent axle***. An independent rear axle is shown in **Figure 22-28.** This type of axle is sometimes referred to as a ***swing axle.*** The solid axle is shown in **Figure 22-29.**

The manner in how the axle retains the differential assembly determines many service procedures. The ***integral carrier***, **Figure 22-29**, is installed in the housing, and is serviced by removing an ***inspection cover*** (43). The ***removable carrier*** contains the ring and pinion and differential assembly and is removed from the housing for service, **Figure 22-30.**

Solid Axles

On solid axle housings, two steel axles are placed inside the housing. Their inner ends almost touch. The outer ends protrude out of the housing and form a base upon which the wheels and hubs are attached. The inner ends are splined and are supported by the differential assembly. The outer ends are supported in straight roller, tapered roller, or ball bearings, **Figure 22-31.**

Solid Axle Retention Methods

The axles are held in the housing by ***retainer plates*** mounted on the outside of the axle, under the brake components, as shown in **Figure 22-31.** On other rear axle assem-

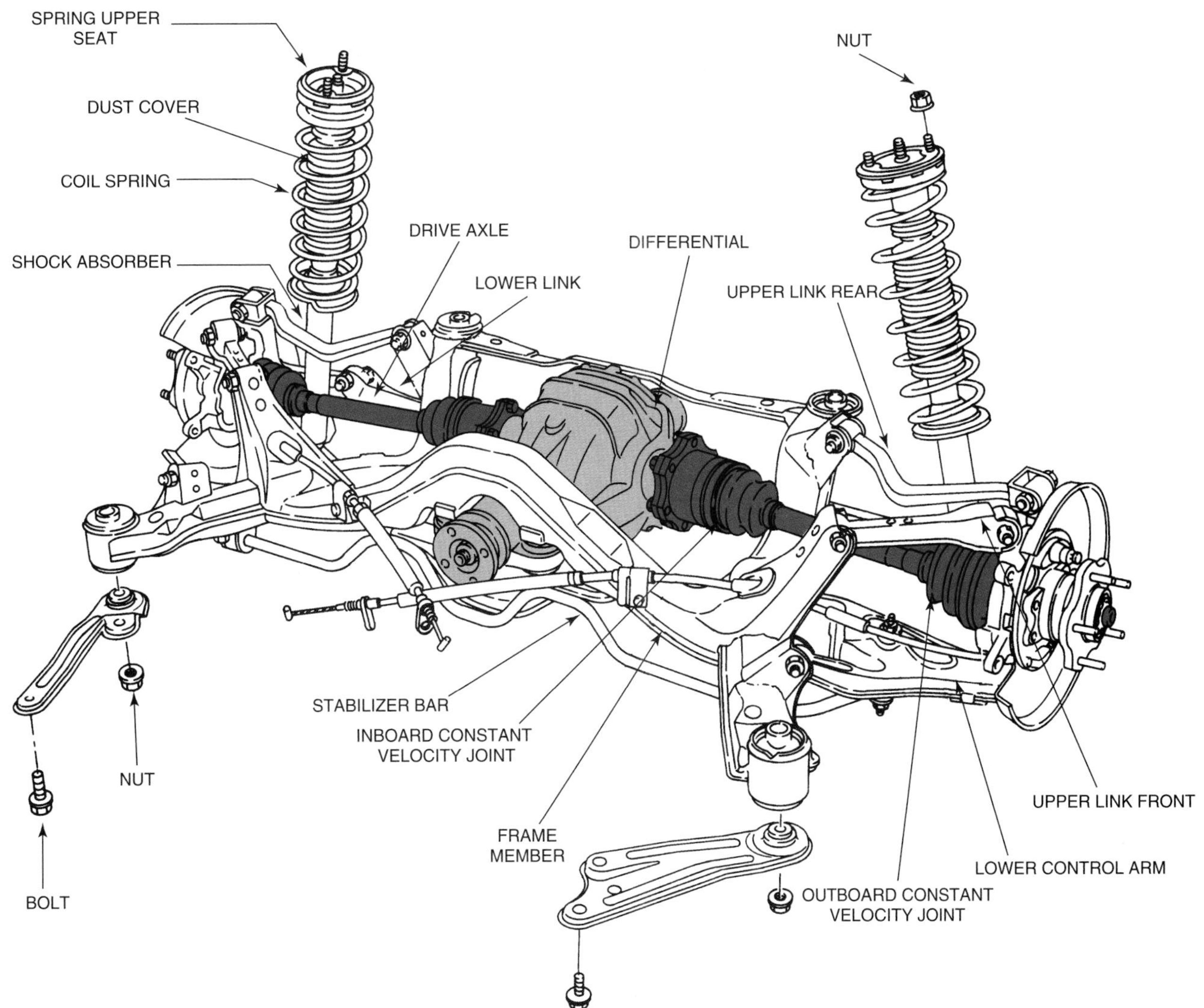

Figure 22-28. Overall view of one type of independent rear suspension and axles which use constant velocity (CV) joints on both the inboard and outboard ends of the axles. The differential assembly is fastened to the frame members, and does not move with the axles or suspension. (Nissan Motor Co.)

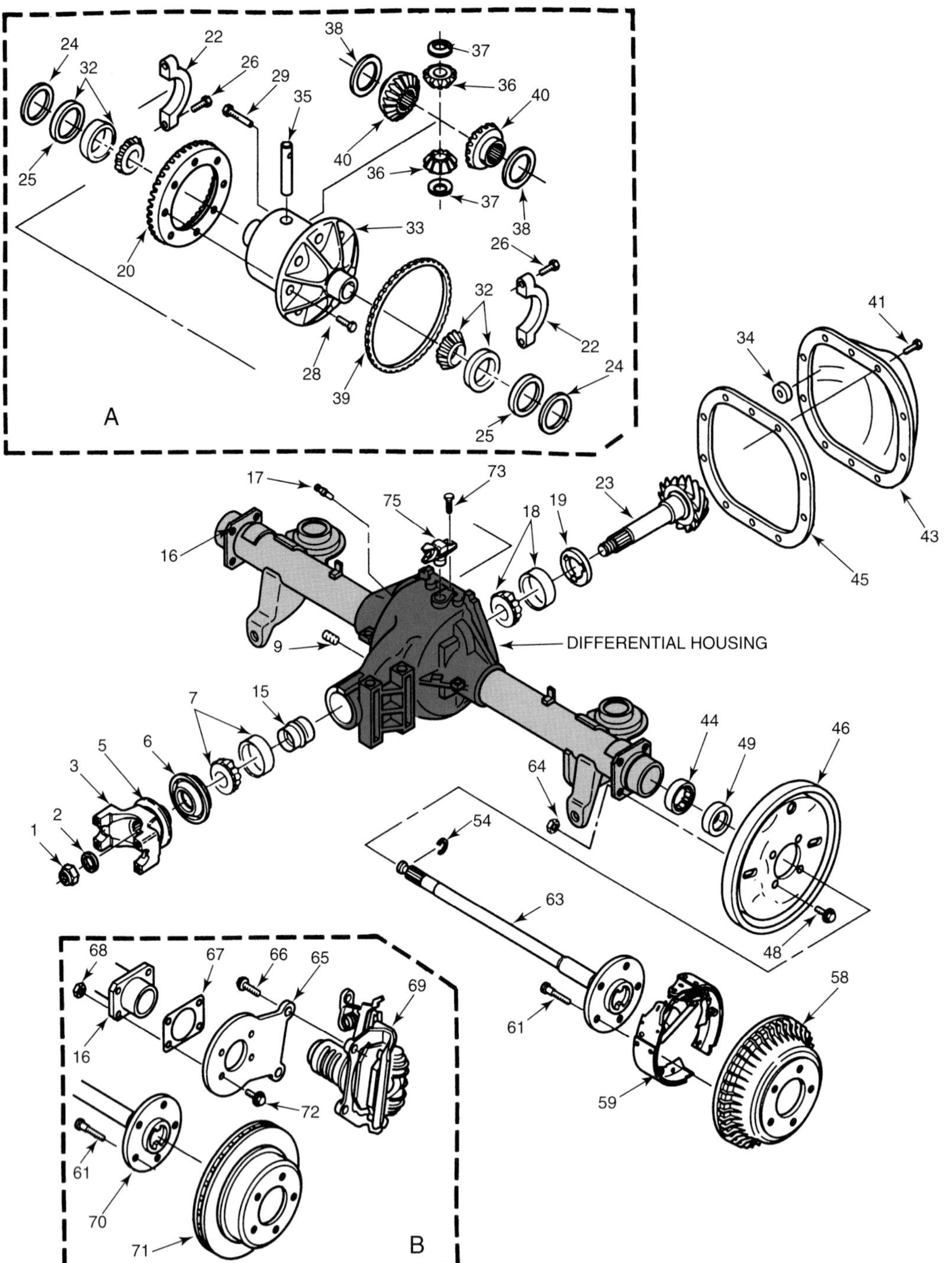

Figure 22-29. An exploded view of a semi-floating, hypoid type, rear axle assembly. This unit has an integral ring and pinion carrier (differential) housing. A—Exploded view of the differential assembly. B—Exploded view of the optional disc brake assembly. 1—Differential drive pinion gear nut. 2—Differential drive pinion gear washer. 3—Differential drive pinion gear yoke. 5—Differential drive pinion gear dirt deflector. 6—Differential drive pinion gear seal.7—Differential drive pinion gear outer bearing assembly. 9—Rear axle housing drain plug. 15—Differential drive pinion gear spacer. 16—Rear axle housing assembly. 17—Rear axle vent. 18—Differential drive pinion bearing assembly. 19—Shim differential drive pinion gear. 20—Gear differential ring. 22—Differential carrier bearing cap. 23—Differential drive pinion mounting gear. 24—Differential bearing shim. 25—Differential bearing spacer. 26—Differential bearing cap bolt/screw. 28—Differential ring gear bolt/screw. 29—Differential pinion gear shaft bolt/screw. 32—Differential bearing assembly. 33—Differential case. 34—Rear axle housing chip collecting magnet. 35—Differential pinion gear shaft. 36—Differential pinion gear. 37—Differential pinion gear thrust washer. 38—Differential side gear thrust washer. 39—Rear wheel speed sensor reluctor wheel. 40—Differential side gear. 41—Rear axle housing bolt/screw. 43—Rear axle housing cover. 44—Rear axle shaft bearing assembly. 45—Rear axle housing cover gasket. 46—Rear brake backing plate. 48—Rear axle housing bolt/screw. 49—Rear axle shaft bearing seal. 54—Rear axle shaft lock. 58—Rear brake drum. 59—Rear brake assembly. 61—Rear wheel bolt/screw. 63—Rear axle (drum brake assemblies) shaft. 64—Rear brake backing plate nut. 65—Rear brake caliper assembly plate. 66—Caliper assembly mounting bolt/screw. 67—Axle tube flange shim. 68—Rear brake caliper mounting plate nut. 69—Rear brake caliper assembly. 70—Rear axle (disc brake assemblies) shaft. 71—Rear brake rotor. 72—Rear brake backing plate bolt/screw. 73—Sensor plug mounting bolt/screw. 75—Wheel speed sensor assembly. (Pontiac)

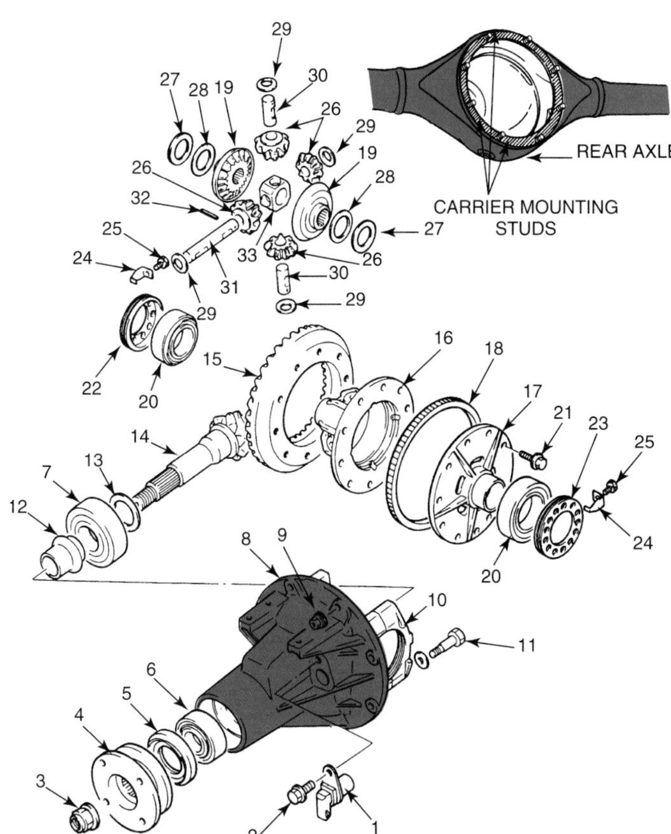

Figure 22-30. An exploded view of a removable ring and pinion carrier assembly. This unit can be removed from the rear axle housing. 1—Rear-wheel anti-lock brake speed sensor. 2—Rear-wheel anti-lock speed sensor bolt. 3—Pinion flange nut. 4—Pinion flange. 5—Pinion seal. 6—Outer pinion bearing. 7—Inner pinion bearing. 8—Differential carrier. 9—Differential carrier nuts. 10—Side bearing cap. 11—Side bearing cap bolt. 12—Collapsible spacer. 13— Pinion selective shim. 14—Pinion gear. 15—Ring gear. 16—Right differential case. 17—Left differential case. 18—Rear-wheel anti-lock (RWAL) excitor ring. 19—Side gear. 20—Side bearing. 21—Ring gear bolt. 22—Right side bearing adjuster. 23—Left side bearing adjuster. 24—Side bearing lock plate. 25—Side bearing lock plate bolt. 26—Differential pinion gear. 27—Side gear washer. 28—Side gear spring washer. 29—Pinion washer. 30—Pinion cross shaft—short. 31—Pinion cross shaft—long. 32—Cross shaft roll pin. 33—Cross shaft joint. (Geo)

blies, the axles are held in place by ***C-locks*** in the differential assembly, **Figure 22-32.**

Independent Axles

On independent rear suspensions, the axles resemble the front wheel CV axle shafts. These axles are usually equipped with CV joints at their inner and outer ends, **Figure 22-33.** The differential housing is bolted to the frame. The outer axle shafts are supported by bearing assemblies that are attached to the rear suspension.

The axles are held in place by nuts which tighten the outer CV axle shaft to the rear bearing assembly. The independent suspension allows the rear wheels to be individually sprung, improving vehicle handling. A variation of this axle

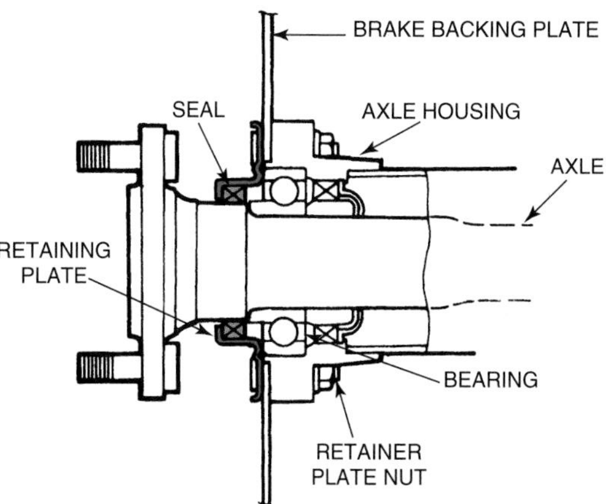

Figure 22-31. A cross-sectional view of a rear axle shaft, bearing, housing, seal, brake backing plate, and the bearing retainer. (Geo)

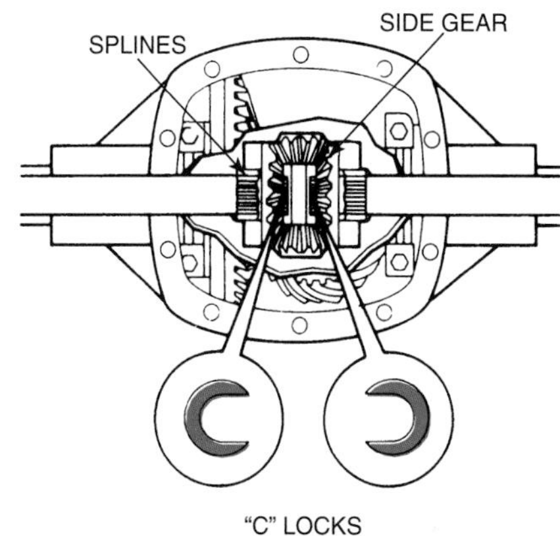

Figure 22-32. Axle shafts are retained in some differentials with "C" locks. The "C" lock is held in position by a groove cut in the side gear. The "C" lock cannot be removed until it is free of the side gear groove. (Chrysler)

setup is used on front- and four-wheel drive vehicles with the engine conventionally placed (engine facing forward).

Attaching Wheel Hubs

The wheel hub is welded to the axle. The axle end is formed into a flange upon which the wheel is bolted. The wheel lugs are pressed into drilled holes in the hub. The flange may have a hole to reach the backing plate nuts, **Figure 22-34.**

Types of Axles

Axles that turn with the wheels are called live axles. The two basic types include the ***full-floating axle*** and ***semi-floating axle***. The full-floating axle transmits engine power, but does not carry any of the vehicle weight. All weight is supported through the outer bearing assembly.

The semi-floating axle delivers power and supports the vehicle weight through the shaft bearing in the axle housing. Most cars utilize the semi-floating axle. Most heavy trucks

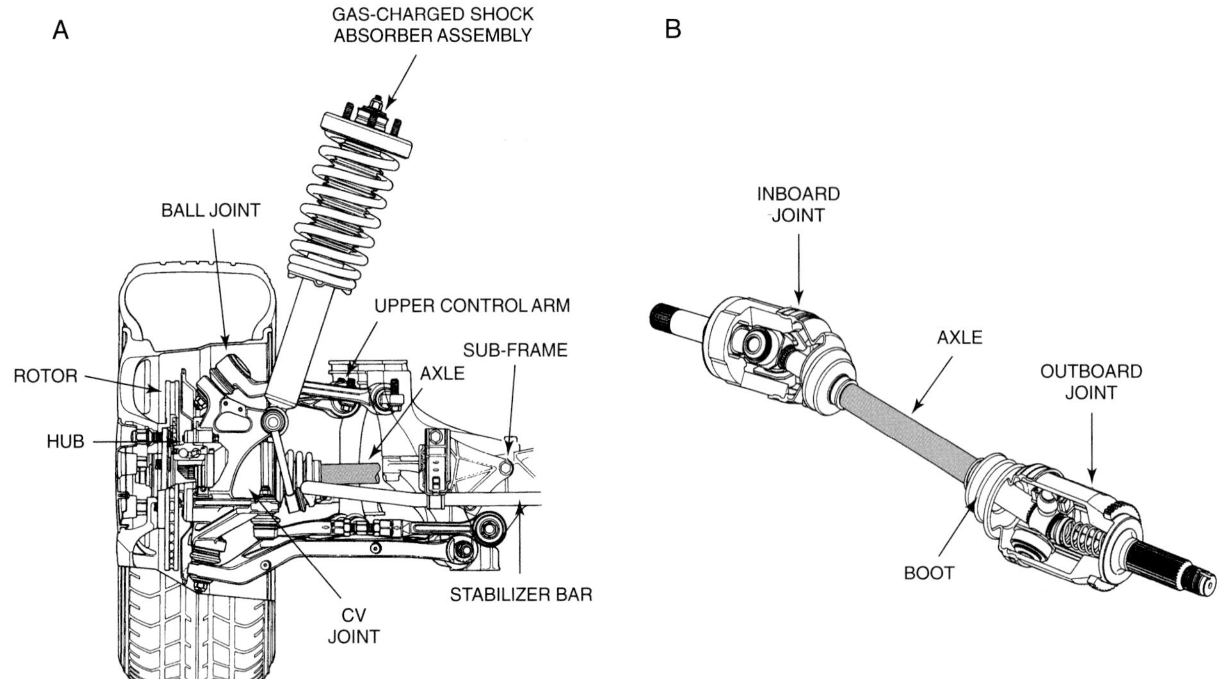

Figure 22-33. A—Cutaway view of one type of independent rear suspension system, left-hand side only shown. B—Axle shaft and cutaway views of the inboard and outboard constant velocity (CV) joints. (Honda)

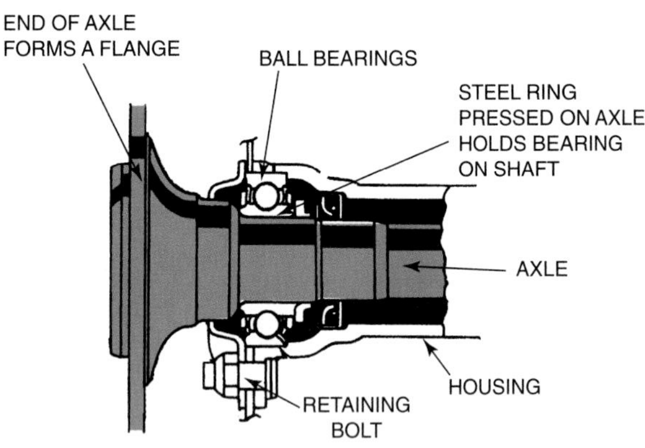

Figure 22-34. This flanged axle end is used to attach wheels and hubs. The wheel is bolted directly to the axle flange. (John Deere)

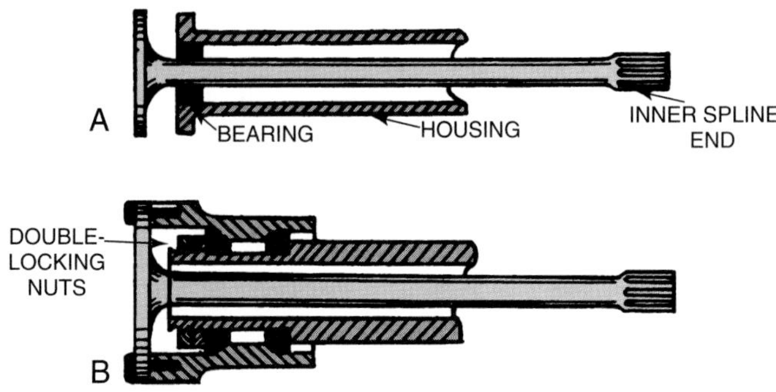

Figure 22-35. Axle types. A—The semi-floating axle drives, retains, and supports the wheel. The outer end of the axle is supported by a bearing. B—A full-floating axle drives the wheel. The housing-hub bearings support the load and retain the wheel. Note the double-locking nuts that draw the hub and tapered roller bearings together.

have full-floating axles. If the axle breaks, the wheel will not come off. The two types are illustrated in **Figure 22-35.**

Differential Construction

The rear wheels of a vehicle must turn at different speeds when rounding the slightest corner (outside wheel must roll farther). The rear wheels must also be able to continue driving the vehicle while turning at different speeds. Therefore, it is necessary to use a ***differential*** to drive the axles, so both receive power, yet they are free to turn at different speeds when needed.

A splined axle ***side gear*** is placed on the inner splined end of each axle. The axle side gear is supported by the ***differential case***. The side gear is free to turn in the case, **Figure 22-36.**

As shown in **Figure 22-37**, the differential case may be turned. It will revolve about the axle side gears. The differential pinion shaft will turn with the case, but the axle side gears will not be driven. By bolting a large ring gear to the differential case and connecting it to a pinion gear and shaft, it will be possible to turn the case. The drive shaft will be attached to the pinion gear shaft. The ring and pinion are discussed later in this chapter.

When the drive shaft turns the ring gear pinion, the pinion will turn the ring gear. The ring gear, in turn will revolve the differential case and ***pinion shaft***. The axle side gears will still not turn. By adding two ***differential pinion gears*** (the differential pinion shaft will pass through these gears) that mesh

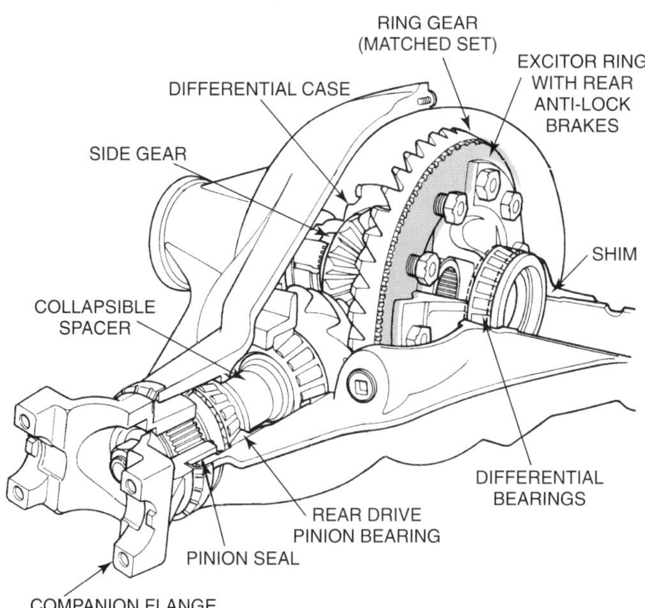

Figure 22-36. Cutaway view of a rear axle differential unit. The excitor ring is used with an anti-lock brake system. (Ford)

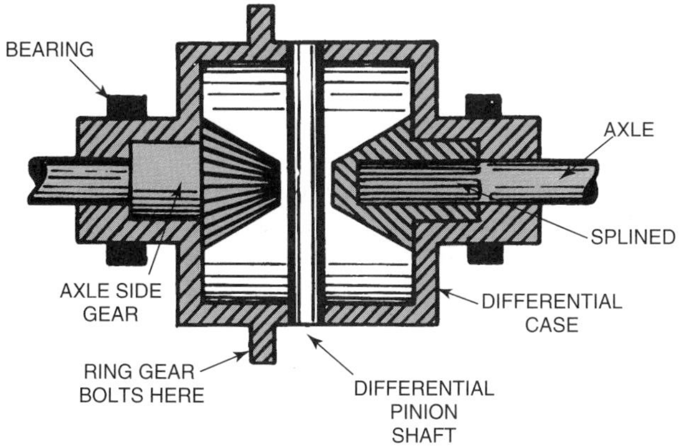

Figure 22-37. Axle side gears. The axle side gear is free to turn with the axle inside the differential case. The pinion shaft turns with the case. Both axle side gears are splined to the axle end. See right-hand axle gear.

with the side gears, the revolving case will turn the axle side gears with it.

Ring and Pinion

The ***ring*** and ***pinion gears*** are used to transfer the engine power from the drive shaft to the rear wheels. In addition to transferring power, the ring and pinion must cause the power to make a 90° turn between the drive shaft and wheels. The ring and pinion must also provide the correct axle ratio. The pinion is the smaller driving gear; the ring is always the larger driven gear.

Differential Operation

The drive shaft turns the ring gear pinion shaft. The pinion gear turns the ring gear which, in turn, revolves the differential case. When the case turns, the differential pinion shaft turns with it. As the differential pinions are mounted on this shaft, they are forced to move with the case. Being meshed with the axle side gears, the pinions will pull the axle side gears along with them.

When the vehicle is moving in a straight line, the ring gear is turning the case. The differential pinions and axle side gears are moving around with the case with no movement between the teeth of the pinions and axle side gears. The entire movement is like a solid unit.

When turning a corner, the case continues pulling the pinions around on the shaft. As the outer wheel must turn over a larger area, the outer axle side gear is now moving faster than the inner axle side gear. The spinning pinions not only pull on both axle side gears, but also begin to rotate on their shafts while walking in the axle side gears. This allows them

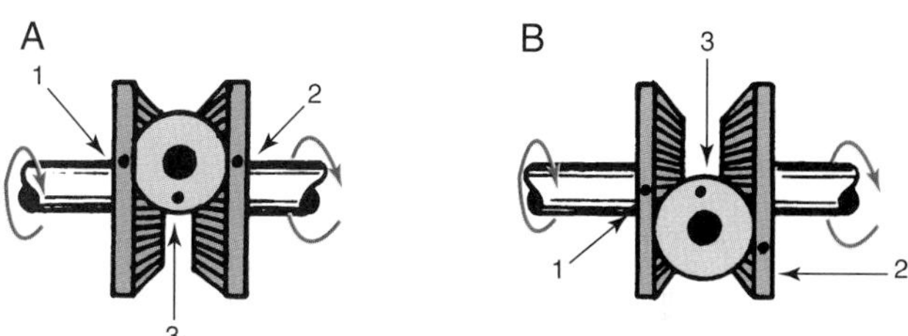

Figure 22-38. A— Differential side gear action. The pinion gear and shaft revolve as a unit when driving in a straight line. B—When turning, the axle side gears move forward but at different speeds. Here, pinion gear 3 is turning on its shaft. Notice that dot 2 has moved ahead of the pinion shaft. Dot 1, even though it is moving forward, is doing so at a slower speed than dot 2. It is moving slower than the pinion shaft. This would be the inside axle on a turn.

to pull on both axle side gears, while at the same time, compensating for differences in speed by rotating around their shafts. This action is shown in **Figure 22-38.**

In **Figure 22-38A,** the vehicle is moving in a straight line. The pinion is pulling both gears, but it is not turning. In **Figure 22-38B,** the right side axle gear is moving faster than the left axle gear. The pinion gear is still moving at the same speed. It is still pulling on both gears, but has now started to turn on the pinion shaft. This turning action, added to the forward rotational speed of the shaft, has caused the right-hand side gear to speed up and actually begin to pass the pinion shaft.

Study **Figures 22-39** and **22-40.** The reverse walking effect on the left-hand side gear has caused it to slow down. The differential action adjusts itself to any axle speed variation. If one wheel begins to slip, the axle on firm ground will stand still. The case continues spinning the pinions, but will walk around the stopped axle gear and drive the spinning axle. A limited-slip differential, which will be covered later in this chapter, is often used to overcome wheel slip.

Differential Carrier and Bearings

A heavy and rigid section is bolted to the housing. It contains the pinion gear, shaft, and bearings. This part is called the ***differential carrier***. Two large bearing holders are provided to support the spinning differential case. These are

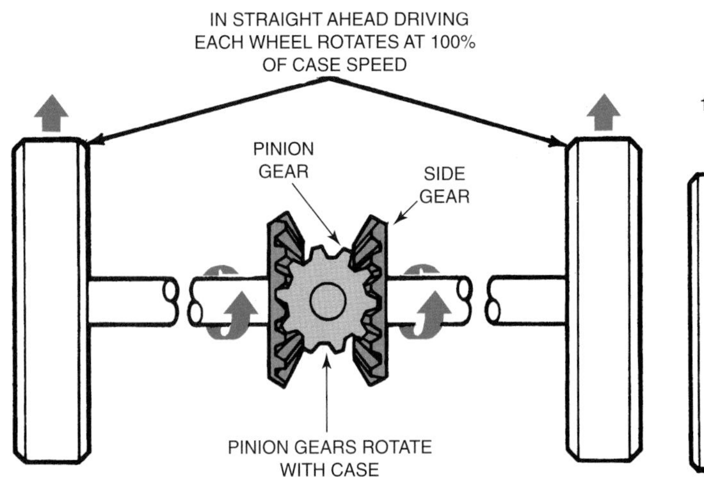

Figure 22-39. Differential action during straight ahead driving. Note that the pinion gears do not rotate and they drive both side gears at the same speed. (Chrysler)

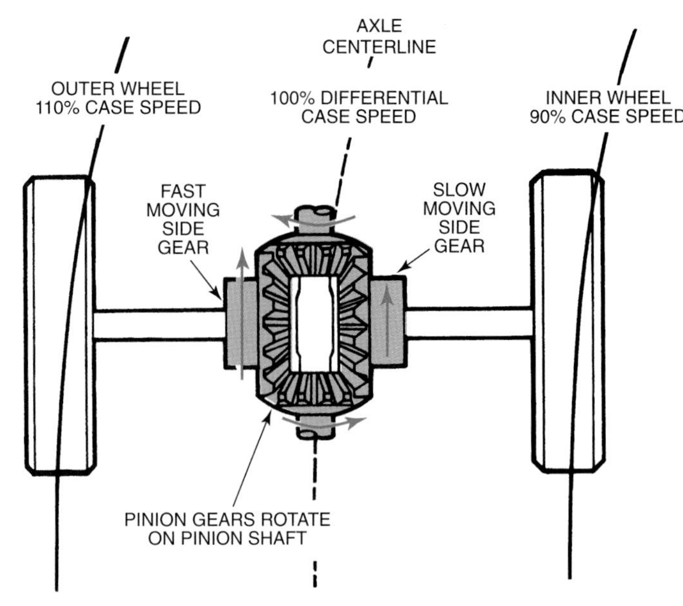

Figure 22-40. Differential action during a turn. The pinion gears rotate in opposite directions and "walk" around in the side gear. This allows the gears to turn (drive) the axles in the same direction, but at different speeds. (Chrysler)

called ***differential*** or ***carrier bearings***. In some applications, the carrier is made as a solid part of the axle housing. An assembled view of a differential is shown in **Figure 22-41.**

Limited-Slip Differential

To avoid the loss of driving force that occurs when one wheel begins to slip, ***limited-slip differentials*** contain internal components that automatically transfer the torque to the wheel that is not slipping. This enables the vehicle to continue its forward motion. This type of differential will provide better traction than the standard differential. It is particularly useful when roads are slippery. It is also valuable by producing faster acceleration in high performance vehicles. A high horsepower engine will often cause one wheel to spin during heavy acceleration if a standard differential is used.

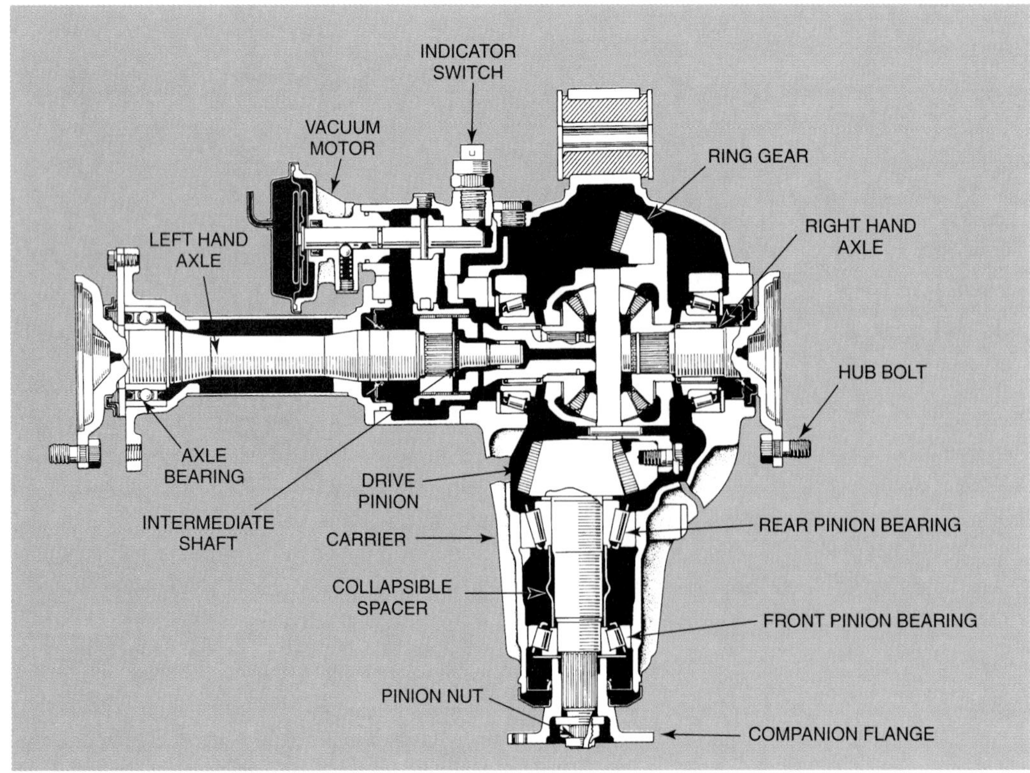

Figure 22-41. Cutaway view of an assembled front differential assembly that is used with a four-wheel drive vehicle. Note the vacuum motor which shifts the front axle into and out of engagement. (Toyota)

Most limited-slip differentials employ the principle of friction to provide some resistance to normal differential action. Some systems use worm and wheel gears. These are discussed in the following sections.

Clutch Plate Limited-Slip Differential

The ***clutch plate limited-slip differential***, sometimes called a Sure-Grip, Positraction, or Traction-Lok differential, is typical of friction limited-slip differentials. It resembles the standard differential, but with several important additions. The axle side gears are driven by four differential pinions. This requires two separate pinion shafts. The two shafts cross, but are free to move independently of each other. The shaft outer ends are not round, but have two flat surfaces that form a shallow V. These ramp-like surfaces engage similar ramps cut in the differential case, **Figure 22-42.**

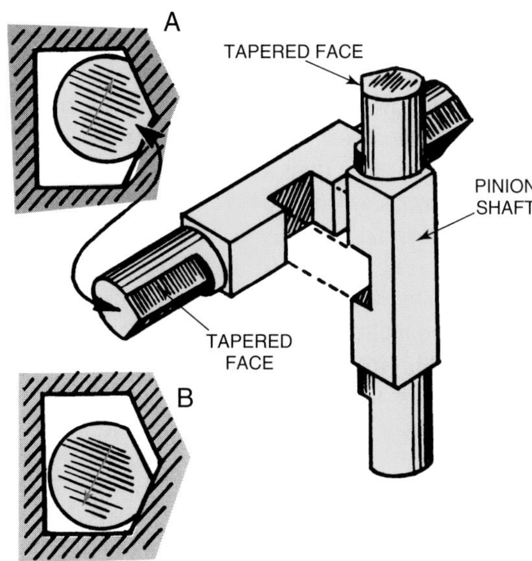

Figure 22-42. Pinion shafts and case ramps. Notice that the shafts fit together but are free to move outward. A—End view of a pinion shaft at rest in the bottom of case ramp. B—When the case pulls on the pinion shaft, the pinion end slides up the ramp. This forces the pinion shaft in an outward direction and moves the pinion gear outward, locking the clutch. Both shafts react but in opposite directions.

A series of four clutch discs are placed behind each axle side gear thrust member. Two of these discs are splined to the differential case and two are splined to the thrust member. The thrust member is splined to the axle. When the thrust members push outward, the clutch discs are forced together, locking the axle to the case, **Figure 22-43.**

Clutch Plate Operation

When the drive shaft drives the pinion gear, the torque thrust is transmitted to the ring gear. As the ring gear drives the differential case, the pinion shafts are forced to rotate with the case.

The differential pinions encounter resistance when they attempt to turn the axle side gears. **See Figure 22-44.** This resistance is transferred to the pinion shafts that are driving

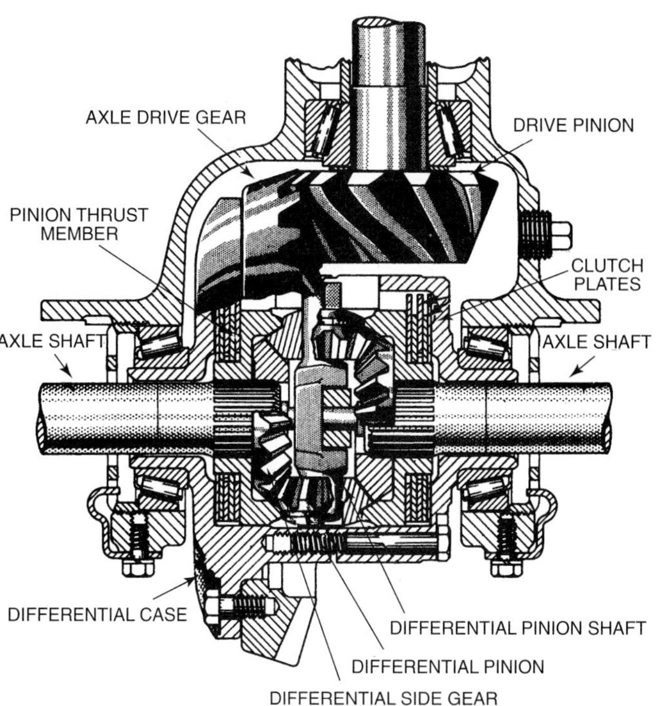

Figure 22-43. Cross-section of a Sure-Grip differential. This differential uses clutch plates by the pinion gears. Pinion thrust members are moved against the clutch plates by pinion gears.

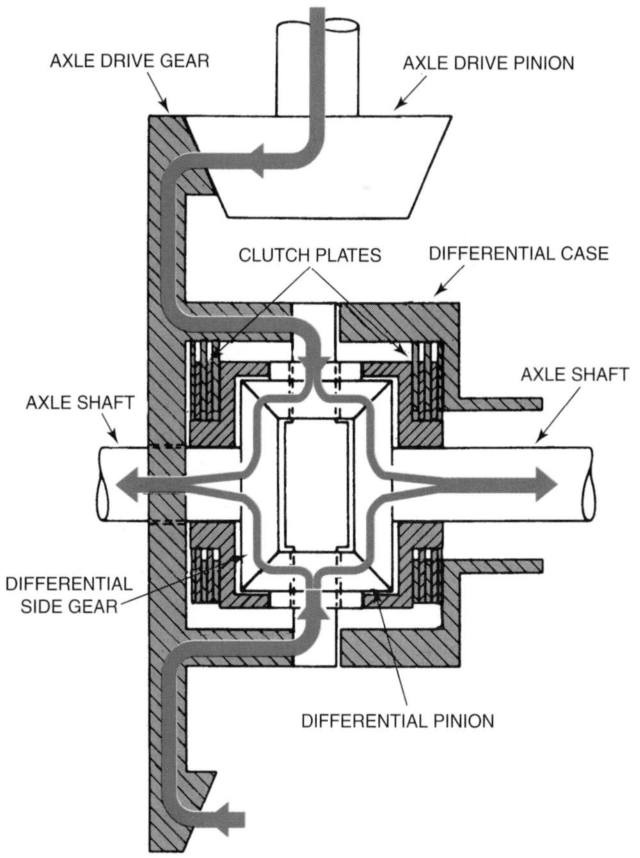

Figure 22-44. Sure-Grip differential operation when the vehicle is moving in a straight line. Both pinion shafts are in the outer positions, locking up the clutch plates. The differential pinions are not walking and both shafts are receiving equal torque. (Chrysler)

the pinions. As both ends of each pinion shaft are seated in tapered ramps, this forces the shafts to slide up the ramp surfaces. This sliding movement forces both shafts in an outward direction. As each shaft moves outward, it moves its pinions in the same direction. The pinions press against the side gears, forcing them to lock up the clutches. This is the action when the vehicle is traveling in a straight line.

When the vehicle turns a corner, the inner shaft slows down. When this happens, the pinion gears will start turning on their shafts. They will walk around the slower shaft and speed up the other shaft. This walking causes the outer shaft to rotate faster than the differential case, allowing the pinion shaft on the outer side to slide down its ramp. This releases the pressure on the outer clutches and lets the differential unit operate much like the standard model. **Figure 22-45** shows differential action when one axle is moving faster than the other. Note that the slower moving axle is receiving most of the torque since it remains clutched to the case.

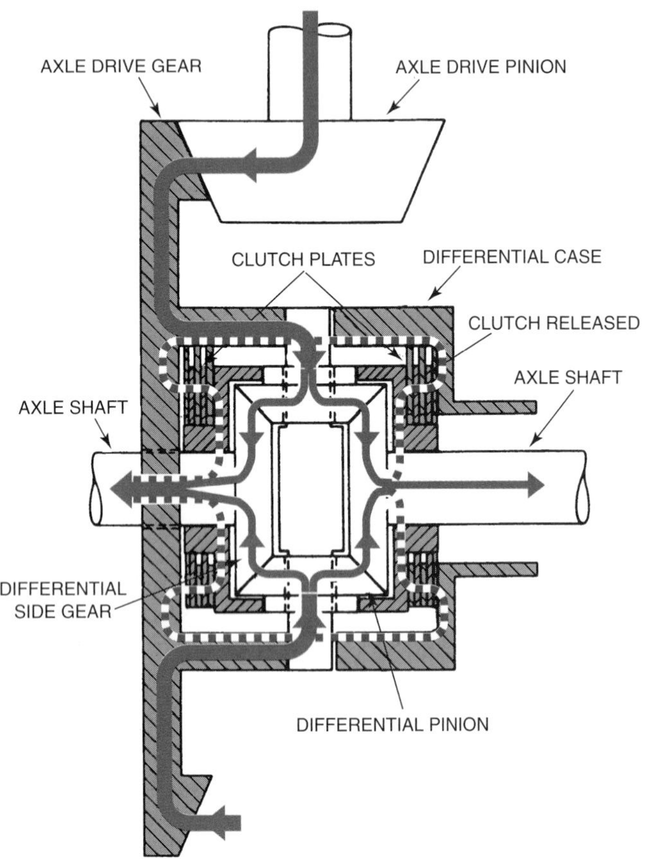

Figure 22-45. Sure-Grip differential operation when the vehicle is moving around a corner. The left axle has slowed down, which allows the pinions to walk and release pressure on the outer or right-hand clutch plate assembly. This causes the differential to transfer a greater amount of driving torque to the left-hand, or inner axle.

Cone Clutch Differential Operation

The ***cone clutch limited-slip differential*** operates in the same manner as the clutch plate differential, but uses cone clutches to provide the friction surface. The assembly shown in **Figure 22-46** uses cone clutches under coil spring pressure. The pressure of the coil springs forces the clutch cones into tight engagement with the case. This action tends to lock the axles to the case. In order for differential action to occur, the cones must be forced to slip. If one wheel slips, the other will still receive some driving force via the cone.

Torsen® Differential

The ***Torsen differential***, **Figure 22-47**, uses ***worm gears*** and ***wheels*** or ***spur gears*** to provide traction under all conditions. The operation of the Torsen differential depends on the fact that the worm gears can turn the wheels easily, but the wheels have a hard time turning the worm gears. This allows the Torsen differential to transmit power to the wheel with the most traction.

Hypoid Gearing

To facilitate lowering the drive shaft tunnel in the floor of the car and to allow lowering the body of the car, pinion gears enter and drive the ring gear somewhat below the centerline of the axles. This gearing setup is referred to as ***hypoid gearing***. A special hypoid lubricant is necessary to prevent premature wear due to the sliding and wiping action that takes place between the ring and pinion gear teeth, **Figure 22-48.**

Study the construction used in the pinion shaft assembly in **Figure 22-48.** Note that two tapered roller bearings are used to support the shaft. This type of bearing withstands both ***radial load*** (forces working at right angles to the shaft) and ***thrust load*** (lengthwise).

Ring and Pinion Adjustments

The ring and pinion adjustment in relation to each other are of critical importance to quiet operation and durability. The tooth contact position, as well as ***clearance*** and ***backlash*** (distance one gear will move back and forth without moving the other gear), is of critical importance. Ring and pinions are always matched and must be installed as a pair. Never replace one without the other. **Figure 22-49** illustrates tooth clearance, backlash, and other gear tooth nomenclature.

Correct and incorrect ring and pinion tooth contact patterns are shown in **Figure 22-50.** The correct contact pattern is very important for strength, wear, and quiet operation. After cleaning, these patterns are brought out by coating the teeth with ferric oxide or other suitable compounds. These compounds are included with many replacement ring and pinion sets. The gears are revolved in both directions and the contact pattern becomes visible. The drive side is the side that contacts when the pinion is driving the ring. The coast pattern is when the ring is driving the pinion.

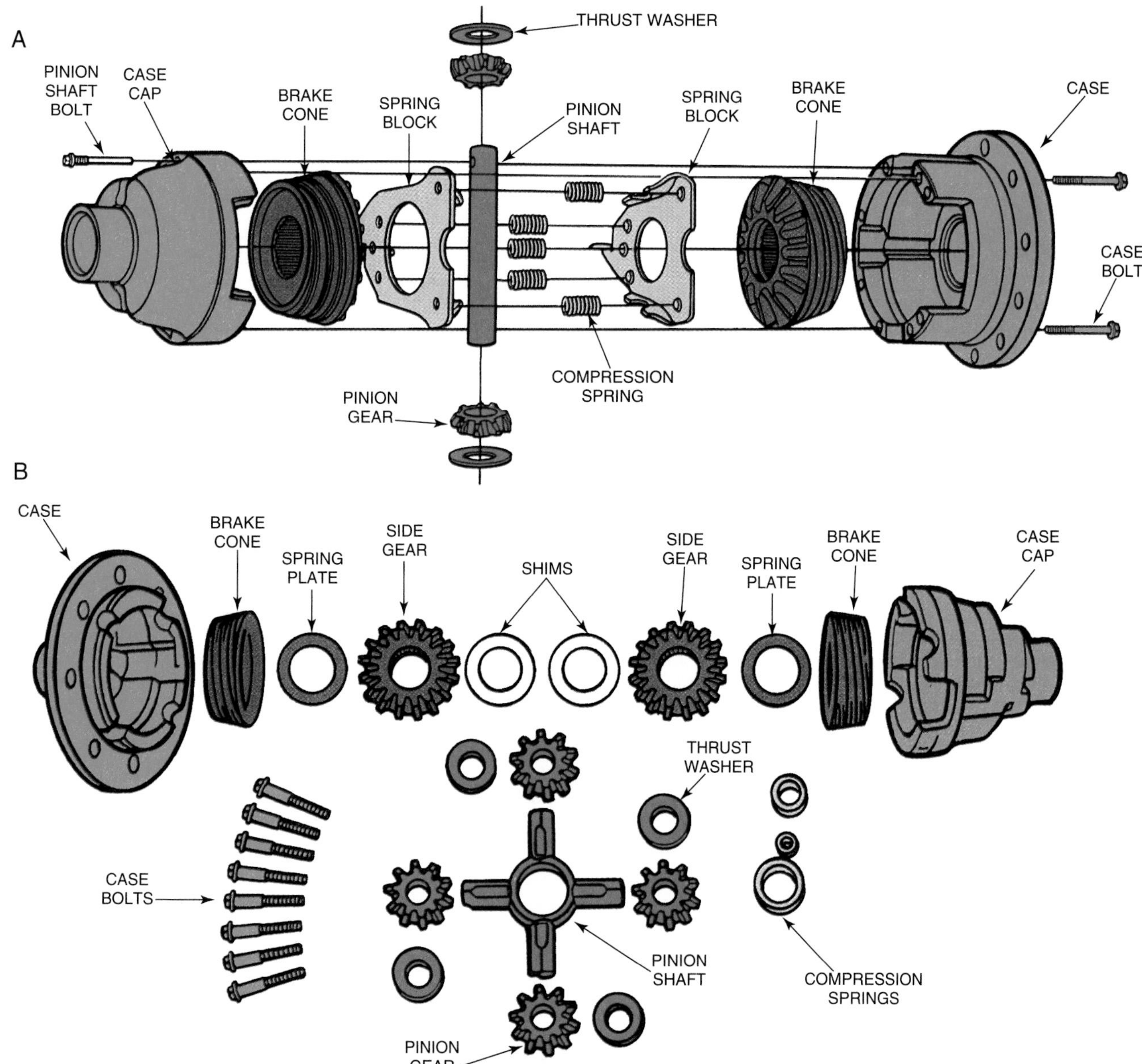

Figure 22-46. The cone clutch limited-slip differential operates in the same manner as the disc clutch types. Spring and side gear force push the cones into a machined surface on the differential case, driving both axles. A and B illustrate two different styles of cone clutch construction. (General Motors)

Gear Ratio

The engine speed must be reduced to provide sufficient power to move the vehicle. This reduction is accomplished by making the ring gear with several times the number of teeth as that of the pinion. The difference is expressed as a ***gear ratio***. The gear ratio of the ring and pinion is sometimes called the ***final drive ratio***. Gear ratios vary but commonly cover a range of around 2.56:1 to nearly 4.11:1. Higher gear ratios are used on large trucks.

When gear ratios are even, such as 3:1 or 4:1, the same teeth on the pinion would contact the same teeth on the ring every third or fourth revolution of the pinion. Any tooth damage or imperfection would then tend to be transferred to the constantly mating tooth, reducing service life. To avoid this, most ratios are uneven. For example, 2.56:1 and 3.27:1 are two commonly used gear ratios. The uneven ratio prevents frequent contact between the same teeth, providing a varying contact wear pattern.

Lower number final drive ratios will reduce engine speed and increase fuel economy, but will produce poorer acceleration. The correct ratio is determined by a careful consideration of the vehicle weight, the type of service it will have, the

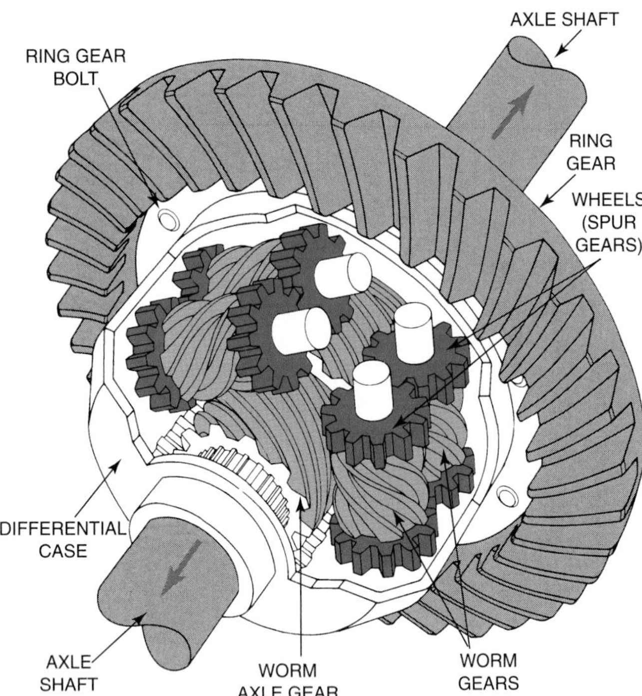

Figure 22-47. A cutaway view of a Torsen®/Gleason No-Spin gear and wheel positive traction differential unit. This differential will transmit power (torque) to the wheel with the best traction. (Torsen/Gleason Co.)

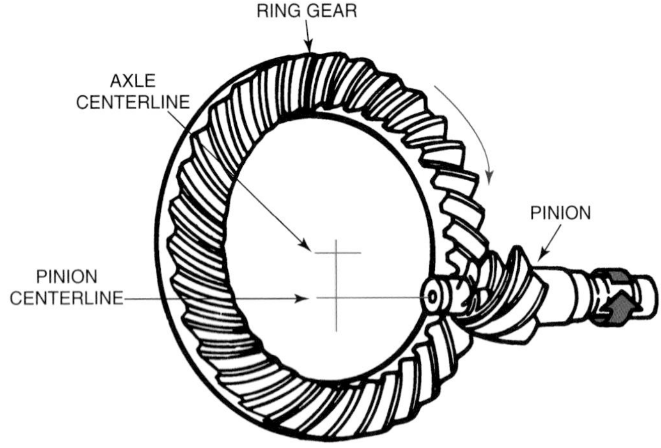

Figure 22-48. Cross-section of a typical hypoid gearset. Note how the pinion gear engages the ring gear below the axle centerline. (General Motors)

engine horsepower, and whether the transmission has an overdrive gear. See **Figure 22-51.**

Calculating Final Gear Ratio

To calculate the final gear ratio, remember that the pinion gear (drive gear) always has fewer teeth than the ring gear (driven gear). Count the number of teeth on the ring and pinion and divide the number of ring gear teeth by the number of pinion gear teeth.

If a pinion gear with 9 teeth is used with a ring gear with 27 teeth, for example, the drive ratio is 27 divided by 9, or 3:1.

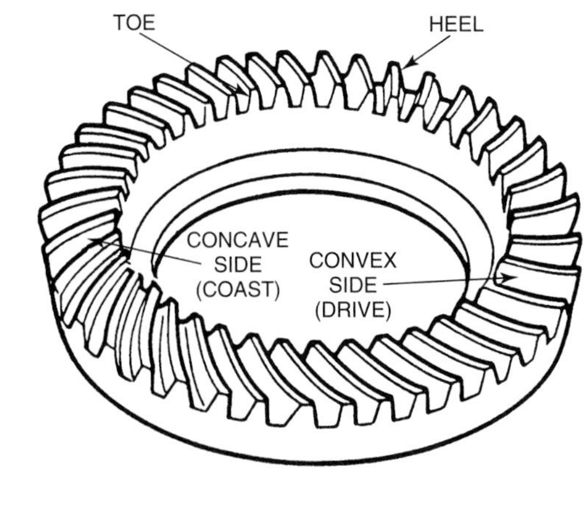

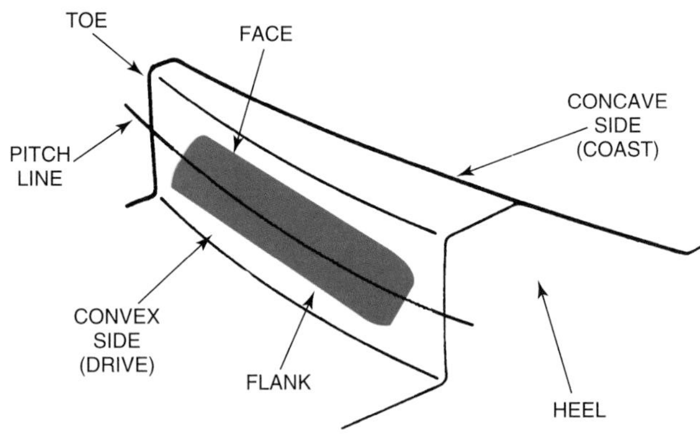

Figure 22-49. Gear tooth nomenclature. The toe is the small part of the gear that faces inward. (Chevrolet)

The final drive ratio is usually not a whole number. A more realistic situation would be a pinion gear with 11 teeth and a ring gear with 40 teeth. The final drive ratio would be 3.636363:1. For easier reading, such a number is rounded to two decimal places, or 3.64:1.

Common final gear ratios today range from 1.8:1 to 3.2:1. This permits better fuel economy. In older cars, factory gear ratios were usually lower (as much as 4.11:1) for greater pulling power and acceleration. Aftermarket high performance rear axles can have gear ratios as low as 8:1.

Axle Lubrication

A heavy viscosity oil, either petroleum-based or synthetic, is normally used with hypoid gears in rear axle housings. The viscosity rating for oil used in most rear axle assemblies is between SAE 80 and SAE 120. Special additives are required for limited-slip differentials.

In operation, oil is thrown over all the parts by the ring gear. Some installations make use of troughs to guide the returning lubricant to specific spots, such as the ring and pinion contact area and the pinion bearings. The lubricant is retained by gaskets or sealers and oil seals. A drain plug is provided at the bottom of some housings. A filler plug is located on the side of the housing or on the access cover.

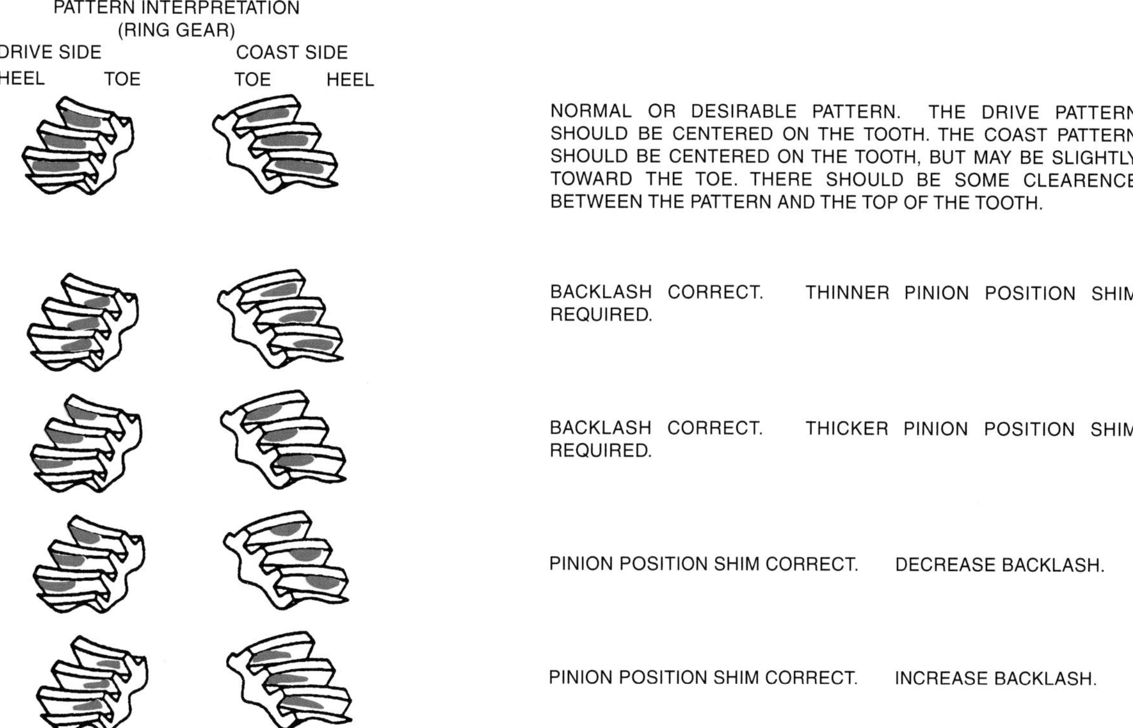

PATTERN INTERPRETATION

COAST SIDE

NORMAL OR DESIRABLE PATTERN. THE DRIVE PATTERN SHOULD BE CENTERED ON THE TOOTH. THE COAST PATTERN SHOULD BE CENTERED ON THE TOOTH, BUT MAY BE SLIGHTLY TOWARD THE TOE. THERE SHOULD BE SOME CLEARENCE BETWEEN THE PATTERN AND THE TOP OF THE TOOTH.
BACKLASH CORRECT. THINNER PINION POSITION SHIM REQUIRED.
BACKLASH CORRECT. THICKER PINION POSITION SHIM REQUIRED.
PINION POSITION SHIM CORRECT. DECREASE BACKLASH.
PINION POSITION SHIM CORRECT. INCREASE BACKLASH.

Figure 22-50. Ring and pinion gear tooth contact patterns. Pinion and backlash adjustments will be reversed on removable carrier differentials. (Chevrolet)

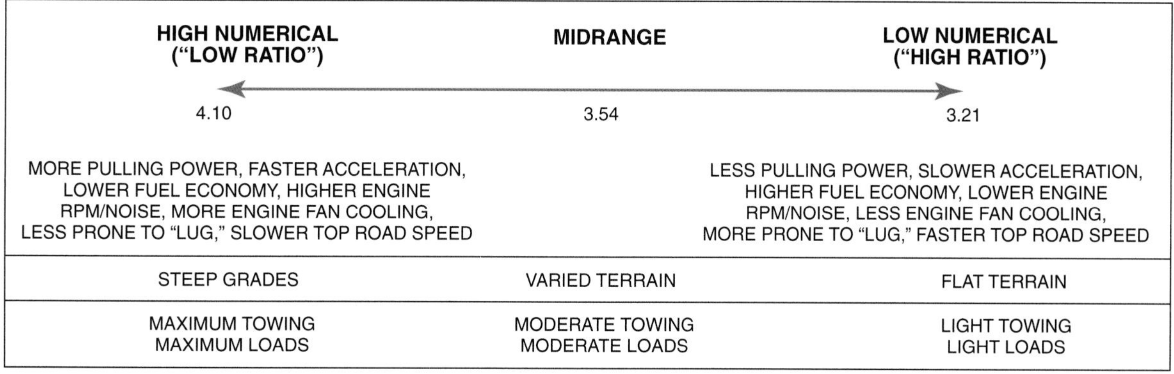

HIGH NUMERICAL
("LOW RATIO")
MIDRANGE
LOW NUMERICAL
("HIGH RATIO")
4.10
3.54
3.21
MORE PULLING POWER, FASTER ACCELERATION, LOWER FUEL ECONOMY, HIGHER ENGINE RPM/NOISE, MORE ENGINE FAN COOLING, LESS PRONE TO "LUG," SLOWER TOP ROAD SPEED
LESS PULLING POWER, SLOWER ACCELERATION, HIGHER FUEL ECONOMY, LOWER ENGINE RPM/NOISE, LESS ENGINE FAN COOLING, MORE PRONE TO "LUG," FASTER TOP ROAD SPEED
STEEP GRADES
VARIED TERRAIN
FLAT TERRAIN
MAXIMUM TOWING
MAXIMUM LOADS
MODERATE TOWING
MODERATE LOADS
LIGHT TOWING
LIGHT LOADS

Figure 22-51. An axle ratio versus power and fuel economy chart. As the axle ratios increase in number, power increases, but fuel economy decreases. As the number decreases, fuel economy increases, but there is less power. (Dodge)

Transaxle Differential Assembly

The front-wheel drive transaxle contains the differential assembly and some means of reducing engine speed. The differential assembly used with longitudinal engines is identical internally with the rear-wheel drive system. The differential assembly used with transverse engines is different in appearance, but accomplishes the same objectives.

In either type of transaxle, the operation of the differential assembly is the same as differentials used in rear-wheel drive axles. Front-wheel drive systems do not use limited-slip differentials, since the engine weight over the driving wheels usually provides sufficient traction in most conditions.

The transaxle final drive ratio may be accomplished by using a separate planetary gearset, such as the one shown in **Figure 22-52.** This planetary gearset is locked in one gear ratio at all times. Other transaxles provide the final drive ratio by varying the gear ratios inside the transaxle.

Transaxle Operation

Power (torque) is fed to the transaxle input shaft, via the clutch or torque converter. It then travels through the transaxle and is carried to the differential assembly by gears or a drive chain. The torque then passes through the differential gears, through the axles, onto the wheels, **Figure 22-53.** Transaxle parts and operation were discussed in Chapter 21.

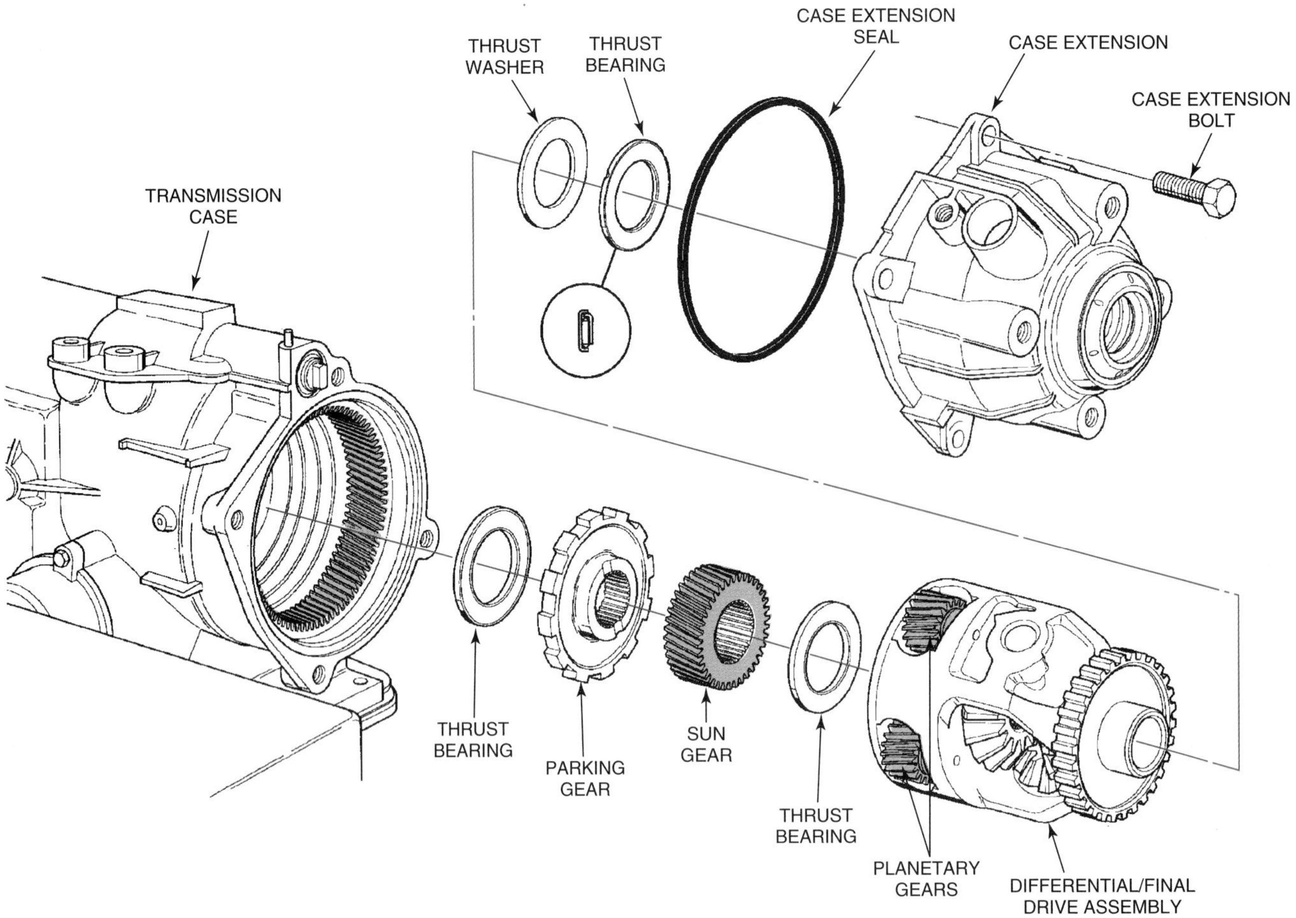

Figure 22-52. The final drive ratio in this automatic transaxle is accomplished by using a planetary gearset. (Cadillac)

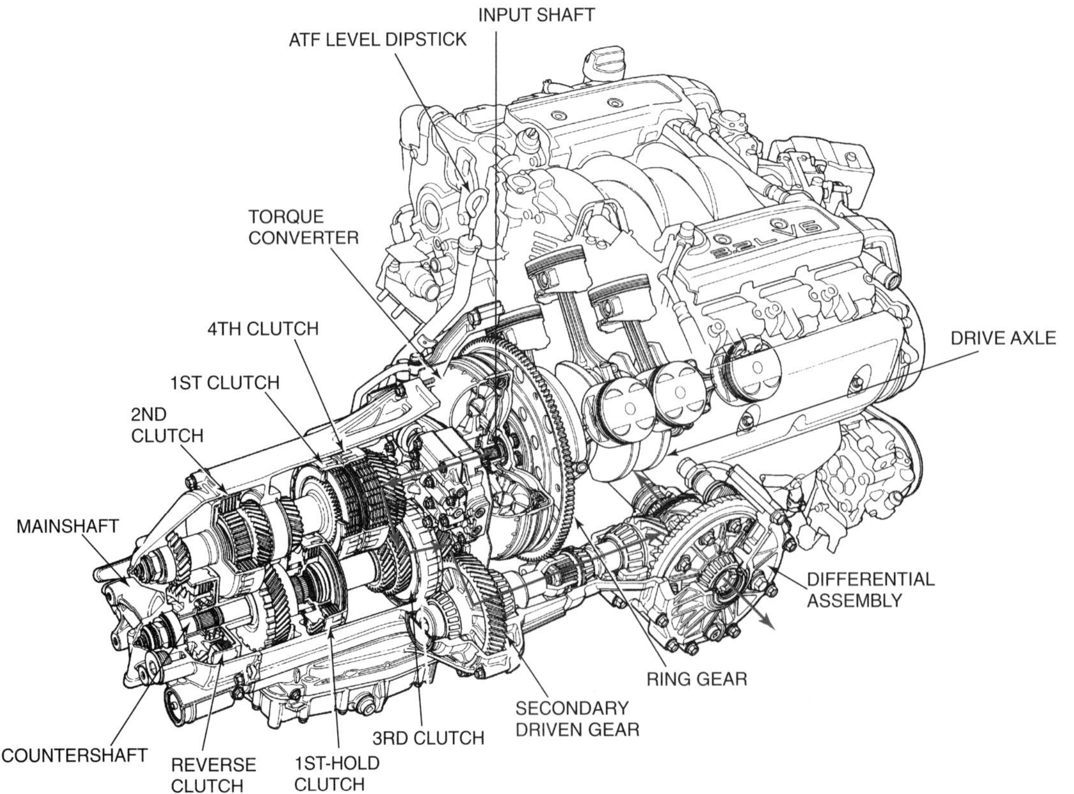

Figure 22-53. Power (torque) flow through one particular four-speed automatic transaxle in 4th gear. (Honda)

Four-Wheel Drive Front Drive Axles

Vehicles equipped with four-wheel drive use a front drive axle that is similar in construction to rear-wheel drive axles. Torque is supplied to the front axle by the transfer case, discussed in Chapter 20, and a front drive shaft. These axle assemblies can be either a one-piece housing setup, **Figure 22-54**, or an independent suspension type with CV axles that resemble the ones used on front-wheel drive vehicles. This type of front drive axle unit is illustrated in **Figure 22-55.**

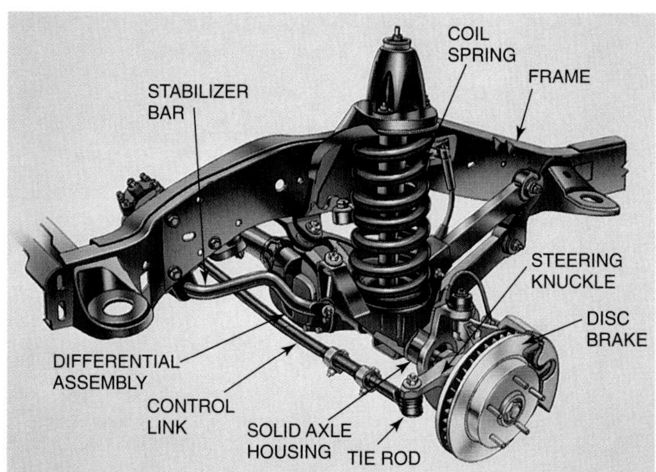

Figure 22-54. A four-wheel drive, solid one-piece front axle and differential assembly. This system uses a four link coil spring-type front suspension. An axle disconnect is used instead of manual or automatic hubs. (Dodge)

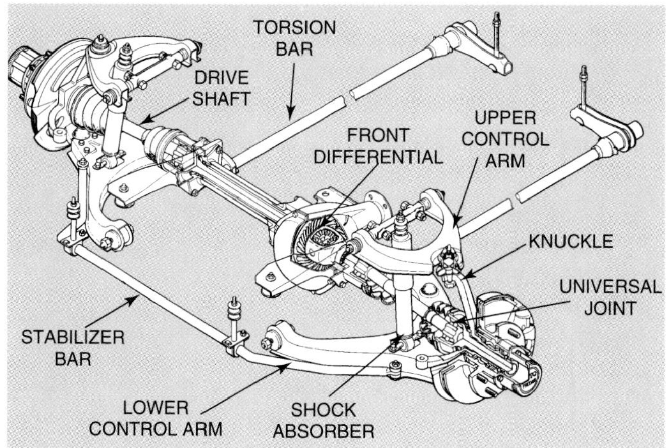

Figure 22-55. Cutaway view of a four-wheel drive, independent front suspension. Note the position of the torsion bars.

Front Axle Steering Provision

The differential assembly and other internal parts of the four-wheel drive front axle are identical to those used in the rear drive axle. However, the front axle assembly must allow the wheels to be turned as well as driven. To accomplish this, a yoke or strut and steering knuckle assembly is added to the solid front axle.

The wheel components are attached to the yoke, which is welded into place on the axle housing. To allow the drive axle to be turned with the steering knuckle, a universal joint is used. On the independent suspension front drive axle, the CV joints allow the outer axle to be turned with the wheels, **Figure 22-55.** Operation is similar to that of the front-wheel drive CV axle discussed earlier in this chapter.

Summary

Drivelines must operate at high speeds. Therefore, they must be accurately constructed and carefully balanced. The driveline must flex to accommodate movement in the rear suspension. Universal joints are used to allow this movement without damaging the drive shaft. The rear-wheel drive universal joint is called a cross and roller or Cardan joint. Some universal joints are lubricated upon construction and are sealed for life. Other types utilize a grease fitting for periodic lubrication.

To prevent speed fluctuation transfer to the pinion shaft, it is essential that both front and rear drive shaft yokes be in the same plane. As driveline length changes, it is necessary to use a slip joint to allow movement of the drive shaft. Constant velocity universal joints are used to provide smooth torque transfer without speed fluctuations.

The Hotchkiss drive is the most widely used driveline. When the rear axle housing drive thrust is carried by the springs or control arms, an open shaft Hotchkiss drive may be used. The Hotchkiss driveline uses two or more universal joints. The drive shaft may be of two-piece design. On some older cars, the torque tube system was employed.

CV axles are used on front-wheel drive vehicles and on some rear-wheel drive systems with independent rear suspensions. Most CV axle shafts are single solid shafts, although a few are hollow tubes. Some CV shafts are two-piece units with a center support bearing. The two basic CV joint types are the tripod and Rzeppa joints. Some two-piece shafts may use a cross and roller universal joint.

The rear axle assembly provides a base for springs, holds the vehicle up, provides a foundation for the brakes, holds the wheels in line, and drives the rear wheels while allowing them to turn at different speeds. The rear axle assembly is divided into several major sections: housing, axles, ring and pinion, and differential. Axles are of two types: semi-floating and full-floating. The differential unit contains axle side gears, pinions, a pinion shaft, and the case itself. The differential drives both axles while allowing their speeds to vary. Limited-slip differentials improve the performance of the differential under wet or icy conditions.

The differential unit is supported by the differential carrier and it revolves in tapered roller bearings. The ring gear is bolted to the differential case. The pinion gear, supported on a shaft in the carrier, drives the ring gear. The rear axle assembly is lubricated by a heavy hypoid gear oil. Limited-slip differentials require a special lubricant additive. Follow the manufacturer's specifications for proper lubricant type and viscosity.

Transaxle differentials use the same principles as the rear drive systems, but house all parts in the transaxle case. The differential assembly used with longitudinal engines is identical internally with the rear-wheel drive system. The differential assembly used with transverse engines is different in appearance, but works in the same manner.

On vehicles equipped with four-wheel drive, a front drive axle must be used. This enables the vehicle to go over rough and soft terrain such as rocks, mud, and sand. The four-wheel drive front axle must have a provision for steering. Four-wheel drive vehicles with solid front axles use universal joints, while independent front suspensions use CV joints.

Know These Terms

Drive shaft
Universal joint
Cross and roller
Cardan universal joint
Center cross
Yokes
Bearing caps
Trunnions
Slip yoke
Hotchkiss drive
Two-piece Hotchkiss drive
Center support bearing
Constant velocity universal joint
Constant velocity axle
CV axle
CV joint
Tripod joint
Rzeppa joint
CV joint boot
Axle housing
Solid axle
Independent axle
Swing axle
Integral carrier
Inspection cover
Removable carrier
Retainer plate
C-locks
Full-floating axle
Semi-floating axle
Differential
Side gear
Differential case
Pinion shaft
Differential pinion gears
Ring gear
Pinion gear
Differential carrier
Carrier bearings
Limited-slip differential
Clutch plate limited-slip differential
Cone clutch limited-slip differential
Torsen differential
Worm gears
Wheels
Spur gears
Hypoid gearing
Radial load
Thrust load
Clearance
Backlash
Gear ratio
Final drive ratio

Review Questions—Chapter 22

Do not write in this book. Write your answers on a separate sheet of paper.

1. The cross and roller universal joint allows the driveline _____ to change without damaging the drive shaft.
   (A) length
   (B) angle
   (C) shape
   (D) speed
2. Technician A says that the most common type of driveline arrangement is the torque tube drive. Technician B says that the torque tube drive uses two universal joints. Who is right?
   (A) A only.
   (B) B only.
   (C) Both A & B.
   (D) Neither A nor B.
3. A center support bearing is used on the __________ driveline.
4. When the drive thrust is carried by the rear springs or control arms, the _____ drive must be used.
5. When the cross and roller universal joint is used, it is necessary to incorporate a __________ in the drive system.
6. Technician A says that most CV axle shafts are solid. Technician B says that two-piece CV axle shafts use a center support bearing. Who is right?
   (A) A only.
   (B) B only.
   (C) Both A & B.
   (D) Neither A nor B.
7. All of the following statements about rear-wheel drivelines are true, EXCEPT:
   (A) if a universal joint is at an angle and the driving shaft speed is constant, the driven shaft speed will rise and fall twice during each revolution.
   (B) if the angle of drive is not equally divided between two shafts, the speed of the driven shaft will be constant.
   (C) with a universal joint on each end of a drive shaft, shaft speed fluctuation can be reduced if both universal joint yokes are in the same plane.
   (D) two universal joints in a single housing will produce an even transfer of torque without speed fluctuations.
8. Universal joint friction is reduced by using roller bearings between the bearing caps and _____.
9. If a universal joint does not have a grease fitting, how can it be lubricated?
10. Technician A says that an integral carrier has an inspection cover to gain access to the internal components. Technician B says that C-locks are used to hold the axles in place on an independent rear suspension. Who is right?
    (A) A only.
    (B) B only.
    (C) Both A & B.
    (D) Neither A nor B.
11. Name the two different types of axles.
12. Of the above axle types, which one is primarily used on automobiles and light trucks?

13. Of the above axle types, which one carries no vehicle weight?
14. When must the rear wheels of a vehicle turn at different speeds?
    (A) When rounding a corner.
    (B) When accelerating in a straight line.
    (C) When braking on wet or icy pavement.
    (D) Both A & C.
15. The differential ______ gears are splined to the axles.
16. All of the following statements about standard (non limited-slip) differential units are true, EXCEPT:
    (A) when the differential case turns, the differential pinion shaft turns with it.
    (B) the turning differential pinion gears will pull the axle side gears along with them.
    (C) during a turn, the differential pinion gears begin to rotate on the pinion shaft.
    (D) if one wheel begins to slip, the differential pinion gears will walk around the spinning side gear and drive the stopped side gear.
17. A clutch plate limited-slip differential will drive the _________ when one wheel begins to slip.
    (A) turning side gear
    (B) stationary side gear
    (C) turning pinion gear
    (D) differential carrier
18. What is the reason for using hypoid gears in the rear axle?
    (A) Quieter operation.
    (B) Longer gear life.
    (C) Using lighter lubricant.
    (D) Lowering the vehicle body.
19. If a pinion has 13 teeth, and the matching ring gear has 39 teeth, what is the gear ratio?
20. If a pinion has 11 teeth, and the matching ring gear has 40 teeth, what is the gear ratio?
21. If a pinion has 9 teeth, and the matching ring gear has 42 teeth, what is the gear ratio?
22. Technician A says that a good quality SAE 40 engine oil is satisfactory as a rear axle lubricant. Technician B says that limited-slip differentials must have a special type of gear oil. Who is right?
    (A) A only.
    (B) B only.
    (C) Both A & B.
    (D) Neither A nor B.
23. Technician A says that a transaxle differential is identical in operation to a rear-wheel drive differential. Technician B says that final drive ratio in a transaxle is sometimes obtained by using an extra planetary gearset. Who is right?
    (A) A only.
    (B) B only
    (C) Both A & B.
    (D) Neither A nor B.
24. What extra function must be provided for the front drive axle of a four-wheel drive vehicle?
25. Four-wheel drive vehicles can use both _____ or _____ front axle setups.

Rear disc brakes are becoming more common on new vehicles. (Chrysler)

23

Brakes

After studying this chapter, you will be able to:

- Identify the basic parts of the brake hydraulic system.
- Describe the principles used for brake hydraulic system operation.
- Identify the basic parts of the brake friction system.
- Describe the principles used for brake friction system operation.
- Explain the differences between drum and disc brakes.
- Describe the principles and components of the vacuum power brake system.
- Describe the principles and components of the hydraulic power brake system.
- Describe the principles and components of anti-lock brake systems.
- List the safety hazards and precautions involved in brake system repairs.

This chapter covers the components and operation of modern disc and drum brake systems. The brake system is subdivided into the hydraulic system (master cylinder, wheel cylinders, calipers, lines and hoses, and valves), and friction system (brake shoes and drums, disc brake rotors and pads). This chapter also covers drum and disc parking brakes, vacuum and hydraulic power brake assists, and anti-lock brake and traction control systems. After studying this chapter, you will understand the operating principles and components of modern brake systems.

Warning: Brake friction materials contain *asbestos*—a known carcinogen. Brake assemblies can produce small airborne particles of asbestos during cleaning, which are easily inhaled. Breathing these particles may cause emphysema or cancer. The following safety rules should be observed at all times:

- Never use compressed air to blow brake assemblies clean. Use a contained vacuum cleaning system or flush with cleaner or water.
- When some exposure might be unavoidable, wear an approved filter mask.

Hydraulic Basics

Early cars used complex linkages to operate the brakes. Modern vehicles use the principles of ***hydraulics*** to transfer power to the brakes. Hydraulics is the practical application of the principles of liquids in motion.

Liquids can include water, oil, transmission fluid, and for our purposes in this chapter, brake fluid. Brake fluid, confined in a ***hydraulic system***, is used to transmit both motion and pressure from the brake pedal to the wheels. A liquid under confinement can be used to:

- Transmit pressure
- Increase pressure
- Decrease pressure
- Transmit motion

Air Is Compressible, Liquids Are Not

Air confined under pressure will compress, thereby reducing its volume, **Figure 23-1.** When a liquid is confined and placed under pressure, it cannot be compressed. **Figure 23-2A** shows a cylinder filled with oil. A leakproof piston is placed on top of the oil. When a downward force is applied, as shown in **Figure 23-2B**, the force will not compress the oil.

Pascal's Law

A French scientist, Blaise Pascal, discovered that when pressure was exerted on a confined fluid, the pressure was transmitted undiminished throughout the fluid. **Figure 23-3** illustrates ***Pascal's law***. Notice that the original pressure (force) placed on the liquid is the same at all outlets.

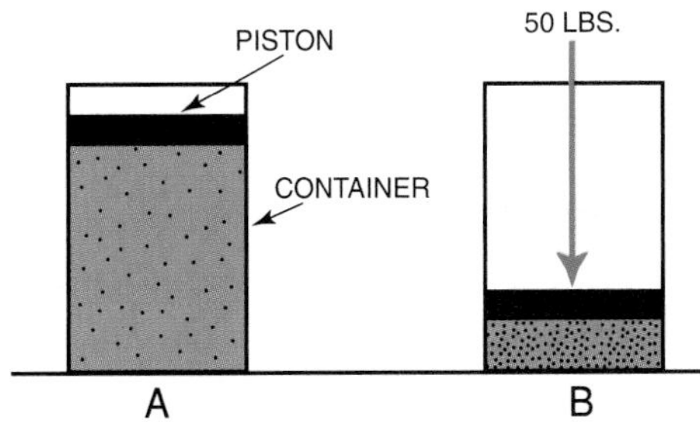

Figure 23-1. Air is compressible. A—There is no pressure on the piston. B—Pressure has forced the piston down, compressing the air trapped in the container.

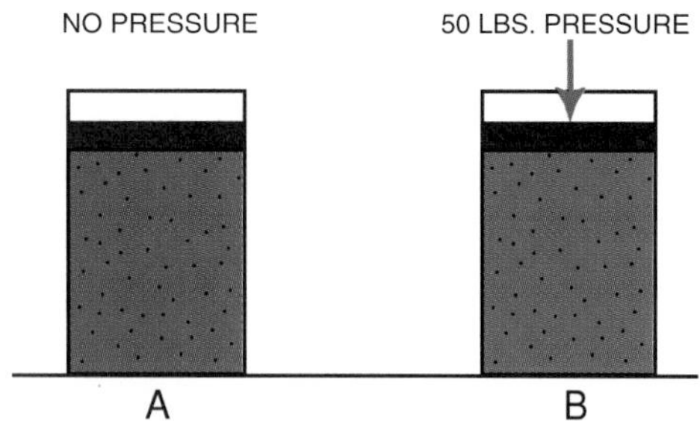

Figure 23-2. Liquids cannot be compressed. When pressure is applied to the piston in B, it does not compress the liquid.

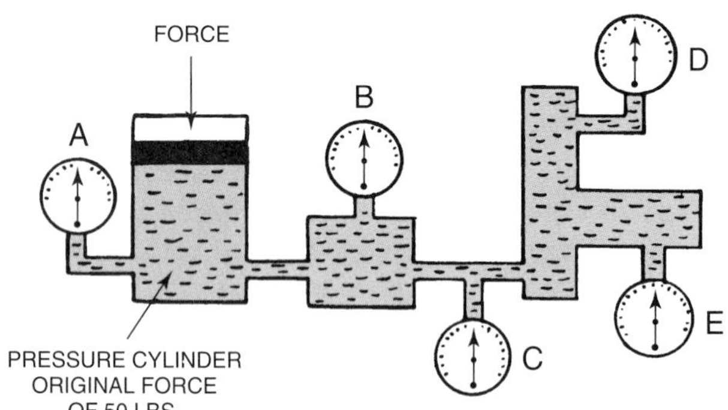

Figure 23-3. When pressure is exerted on a confined liquid, it is transmitted undiminished. Force on the piston has created a pressure of 50 psi (pounds per square inch), upon the liquid in the pressure cylinder. Notice that all gauges read the same throughout the system. Pressure is transmitted undiminished to all parts of the system. If gauge A reads 50 psi, gauges B, C, D, and E will also read 50 psi.

Liquids Can Transmit Motion and Force

In **Figure 23-4,** you will see that any movement of piston A will cause piston B to move an equal amount. This is a transmission of motion through a liquid. If a 200 lb. force is placed on piston A in **Figure 23-5,** piston B will support 200 lbs. Both pistons are the same size. In a hydraulic system, this force is usually expressed as ***hydraulic pressure***.

Liquids Can Increase Force

When a force is applied to piston A in **Figure 23-6,** it can be increased if it is transmitted to a larger piston B. If piston A has a surface area of 1 in^2, the 200 lb. force on piston A represents a pressure of 200 pounds per square inch (psi). According to Pascal's Law, this 200 psi force will be transmitted undiminished.

If piston B has a surface area of 20 in^2, piston A will exert a 200 lb. force on each square inch of piston B. This would produce a mechanical advantage (MA) of twenty, and the original 200 lb. force would be increased to 4000 lbs. The force may be further increased by either making piston A smaller or piston B larger.

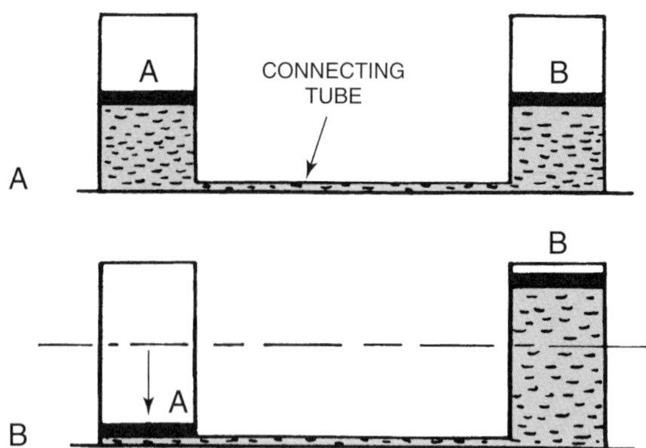

Figure 23-4. Liquids can transmit motion. A—Both pistons are at rest. B—Piston in left cylinder has been forced down. When one piston travels down, it forces liquid into the other cylinder. Since the other cylinder is the same size, the distance that the piston will be raised is equal to the distance that the other piston was lowered.

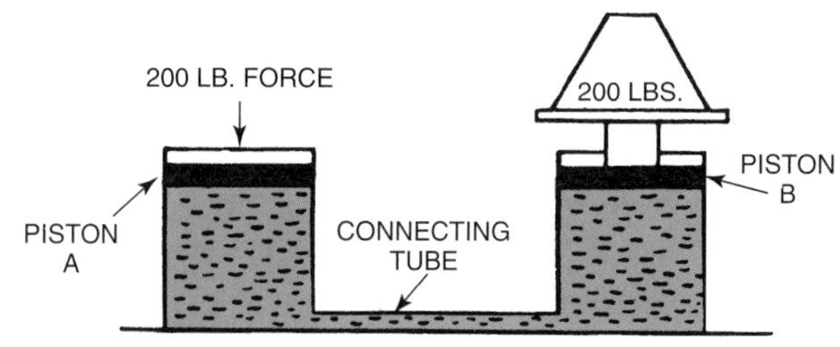

Figure 23-5. Using a liquid to transmit force. A 200 lb. force on piston A is transmitted, via liquid, to piston B.

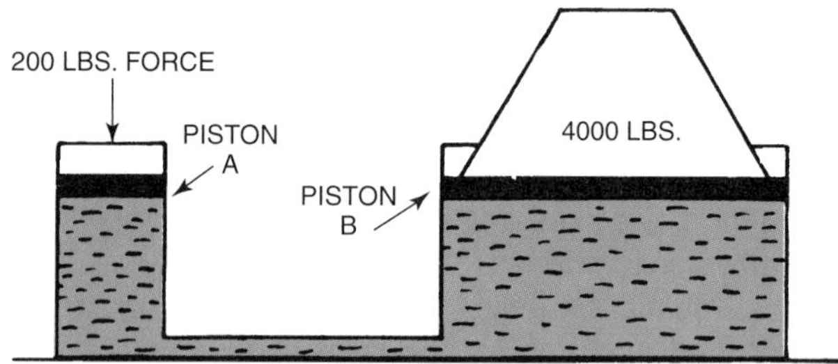

Figure 23-6. Liquids can be used to increase force. A 200 lb. force on piston A with an area of 1 in^2 is increased to 4000 lbs. on piston B with an area of 20 in^2.

Liquids Can Decrease Force

If a force is applied to piston B, the original force of 200 lbs. would have to be raised to 80,000 lbs. to produce a pressure of 4000 lbs. on piston A. By applying the force to the large piston, the force increase has been reversed and the original force, as applied to the smaller piston, will be diminished.

This force's ability to produce pressure can be increased or decreased by applying it to a larger or smaller area. Notice in **Figure 23-7** that pressures in the pump and lifting pistons are the same. However, the force applied to the pump piston has been increased many times on the head of the lift.

A Hydraulic Jack

Figure 23-7 shows how a fluid can be used to produce a powerful lifting force using the principles of hydraulics. When the jack handle raises piston A, piston A will form a vacuum. This will draw check valve 1 open and close check valve 2. When the handle is depressed with a force that exerts 200 lbs. pressure (or any force) on piston A, check valve 1 will close, check valve 2 will open and 200 psi will be transmitted to piston B. If piston B has a surface area one hundred times greater than the 1 in^2 area of A, piston B will raise a weight of 20,000 lbs.

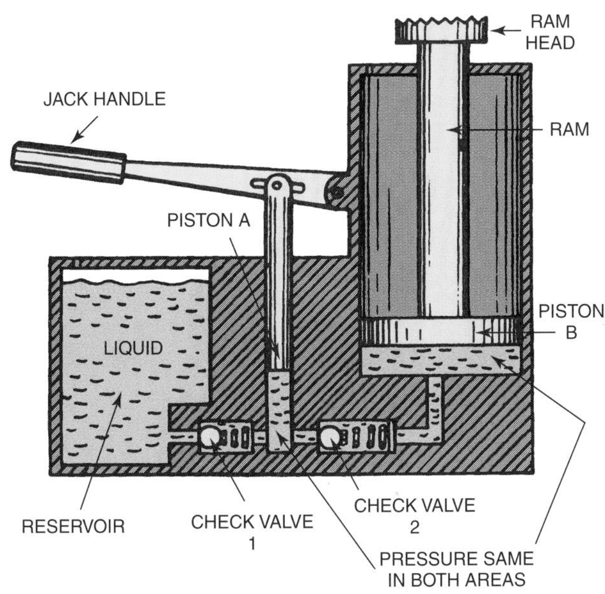

Figure 23-7. Simple hydraulic jack. If piston B has an area 100 times greater than piston A, the ram will lift 100 times more than the pressure generated by piston A.

Hydraulic Principles in Vehicle Brake Systems

When the driver depresses the brake pedal, the force is transmitted undiminished to each caliper or wheel cylinder. The caliper pistons or wheel cylinders transfer this force (increased or decreased, depending on piston area) to the friction linings.

When the master cylinder piston moves, the caliper pistons or wheel cylinders will move until the brake friction components are engaged. Further movement is impossible as any attempt to depress the master cylinder piston beyond this point will transmit additional pressure, not motion.

Brake Fluid

Brake fluid is used to transmit motion and pressure through the hydraulic system. Not any fluid can be used in a brake system. Some of the more important characteristics of quality brake fluid are:

- Maintains even viscosity throughout a wide temperature variation.
- Does not freeze at the coldest possible temperature that the vehicle may encounter.
- Boiling point is above the highest operating temperature of the brake system parts.
- Is hygroscopic (has the ability to absorb and retain moisture) to prevent internal freezing of parts.
- Acts as a lubricant for pistons, seals, and cups to reduce internal wear and friction.
- Must not corrode the brake system's metal parts.
- Must not swell or deteriorate the brake system's plastic and rubber parts.

Many brake fluids contain alcohol, glycerin, and other non-petroleum ingredients. Some heavy duty commercial and high performance vehicles use ***silicone brake fluid***. This fluid offers extended service life and top performance under a wide range of operating conditions.

Although most modern brake fluids are designed to be interchanged, do not mix different fluids without checking the manufacturer's service manual. Use high quality brake fluid only. It should meet or exceed current SAE recommendations and conform to the latest DOT (Department of Transportation) standards.

Under no circumstances put anything but brake fluid into the brake system. Any mineral or petroleum-based oils such as motor oil, transmission fluid, power steering fluid, kerosene, or gasoline in even the smallest amounts will swell and destroy the rubber cups and seals in the system.

Warning: Brake fluid is poison. Keep it away from skin and eyes. Do not allow brake fluid to splash on painted surfaces.

Master Cylinder

The ***master cylinder*** is the central unit in which hydraulic pressure is developed. Pressure from the driver's foot pressing on the brake pedal is transmitted to the master cylinder piston. As the piston is forced forward in the cylinder, it pushes brake fluid ahead of it. Since the system is airtight, the piston is acting upon a solid column of fluid.

When the friction members have moved far enough to fully apply, the fluid movement ceases and pressure rises in response to the amount of force on the master cylinder piston. **Figure 23-8** illustrates brake pedal linkage, master cylinder, hydraulic valves, lines, and brake assemblies. All of these components will be discussed later in this chapter.

Master Cylinder Construction

Most master cylinders are manufactured of cast iron or aluminum. The master cylinder has a drilled bracket as part of the casting for mounting. Modern master cylinders are installed on a vacuum or hydraulic power booster mounted to the vehicle's firewall, **Figure 23-9.** With the master cylinder in this location, it can be inspected and serviced easily and reduces the chances of contamination by water, oil, or dirt. It is operated by a suspended pedal.

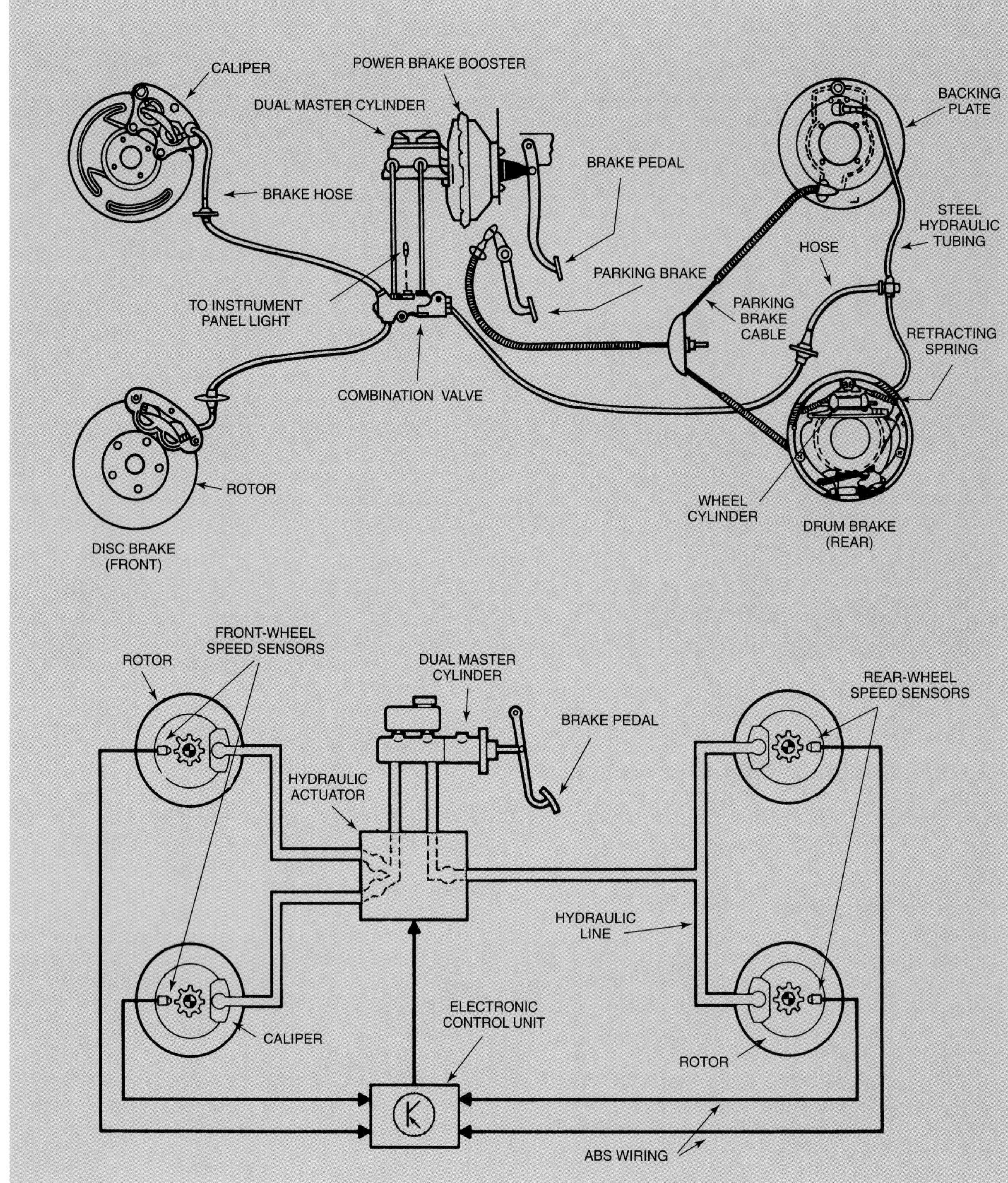

Figure 23-8. Schematics of hydraulic brake systems. A–Standard four-wheel brake system. B–Schematic of a four-wheel disc brake system that incorporates four-wheel anti-lock braking (ABS). (Audi/Bendix)

Brake Fluid Reservoir

The master cylinder has a ***reservoir*** for brake fluid. This will provide additional fluid to compensate for minute leaks or lining wear. The reservoir cover is vented (has an air hole in it) to allow fluid expansion and contraction without forming pressure or vacuum. A rubber diaphragm contacts the fluid and

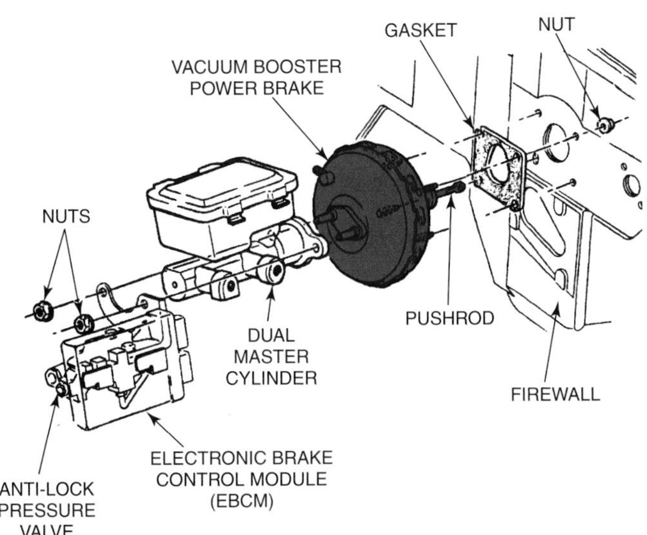

Figure 23-9. An exploded view of a power brake assembly. This assembly includes an anti-lock brake pressure valve and control module. (General Motors)

follows the level up and down. The diaphragm seals out dust and moisture.

Some reservoirs are made of clear plastic, or have clear windows so the fluid level can be checked without removing the cover. A typical reservoir cover is illustrated in **Figure 23-10.** A few master cylinders have a remote (separately located) reservoir, such as the one in **Figure 23-11.** A separate reservoir is used when the master cylinder is in a location that would be difficult to reach. Note that the reservoir in **Figure 23-11** has a ***low fluid warning switch***. This switch illuminates a dashboard light to warn the driver if the fluid level drops below the safe point.

Reservoir Fluid Level

The master cylinder reservoir should be filled to within 1/4″ to 1/2″ (6.35-12.7 mm) of the top of the reservoir,

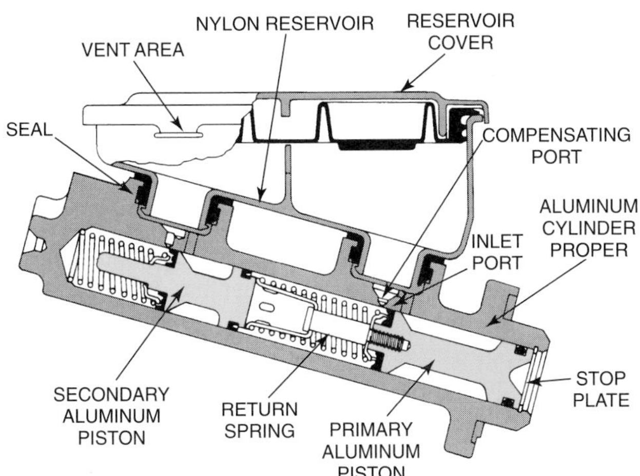

Figure 23-10. Typical dual master cylinder setup. This master cylinder uses a nylon reservoir and an aluminum cylinder body. Note how this setup is angled. (Delco)

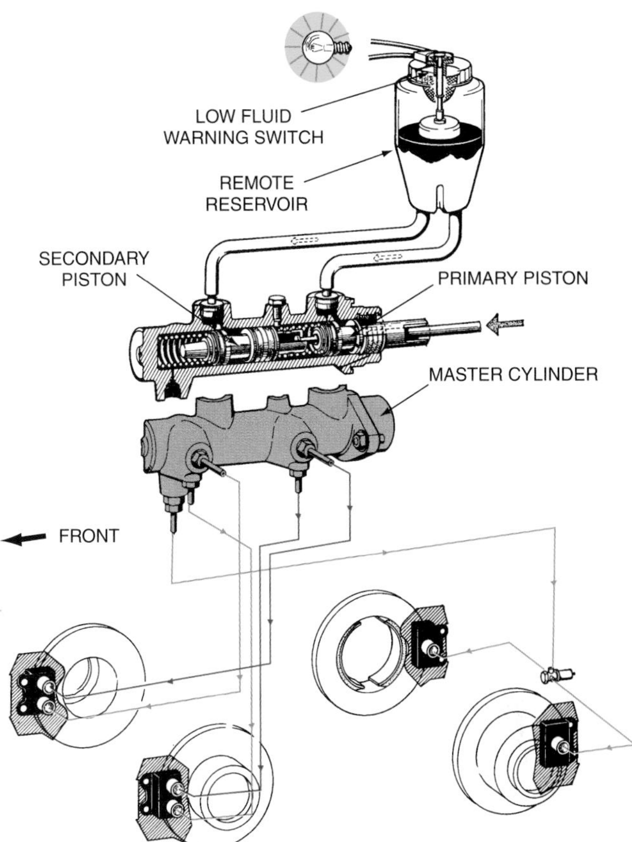

Figure 23-11. Split braking system employing dual (separate) front and single rear hydraulic lines. Some front braking is always possible unless both master cylinder pistons fail.

depending upon manufacturer's recommendations, **Figure 23-12.** Note that a separate reservoir is used for each piston intake port. Some master cylinders have one reservoir with an internal separator between each piston intake port.

Pressure Cylinder

A cylinder with a very smooth wall is provided in the master cylinder. This cylinder contains two close fitting aluminum ***pistons***. Some older vehicles use a single piston design. The cylinder is connected to each reservoir by two ports (holes), the compensating port and intake or breather port, **Figures 23-10** and **23-12.** Pressure is developed in this cylinder for use in the rest of the brake system. The ***piston return springs*** reposition the pistons when the brakes are released.

Piston Assembly

The inner face of each piston presses against a rubber ***primary cup***. This cup prevents fluid leakage past the piston. The outer piston end has a rubber ***secondary cup*** to prevent fluid from leaving the master cylinder. The inner piston head has several small ***bleeder ports*** that pass through the head to the base of the rubber primary cup.

Both piston assemblies rest in the cylinder. They are retained by a stop plate and/or a snap ring in the end of the cylinder. Pressure is applied to the pistons by means of a

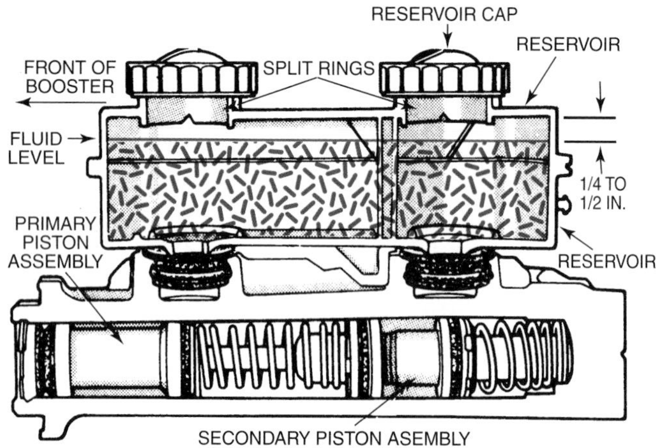

Figure 23-12. Master cylinder fluid level. Air space is important so that reservoir fluid can expand without leaking or creating pressure in the master cylinder. A flexible diaphragm may ride on top of the fluid. Note the double-piston construction. (Chrysler)

push rod that connects to the brake linkage. The push rod can be seen in **Figure 23-13.**

Dual Brake System

As discussed earlier, the master cylinder has two pistons. Master cylinders with two pistons provide two separate (often called "split" or "twin") braking systems. This is often called a ***dual brake system***. If one part of the system loses its hydraulic fluid from a leaking seal, ruptured hose, or cracked line, the other system will still provide some braking. Although braking performance is significantly reduced, the dual system provides an important margin of safety.

The dual brake system uses a ***dual master cylinder*** (also referred to as a tandem or double-piston master cylinder). An exploded view of a dual master cylinder is shown in **Figure 23-14.** The action of the cylinder when either the front or rear hydraulic system loses pressure is explained below.

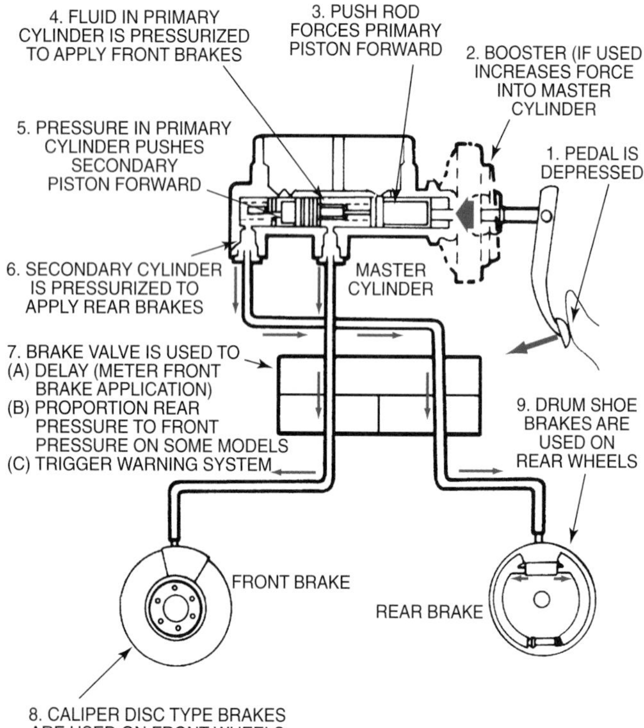

Figure 23-13. A dual master cylinder in the applied position. There is fluid pressure in the master cylinder. The fluid reservoir is not shown. (Ford)

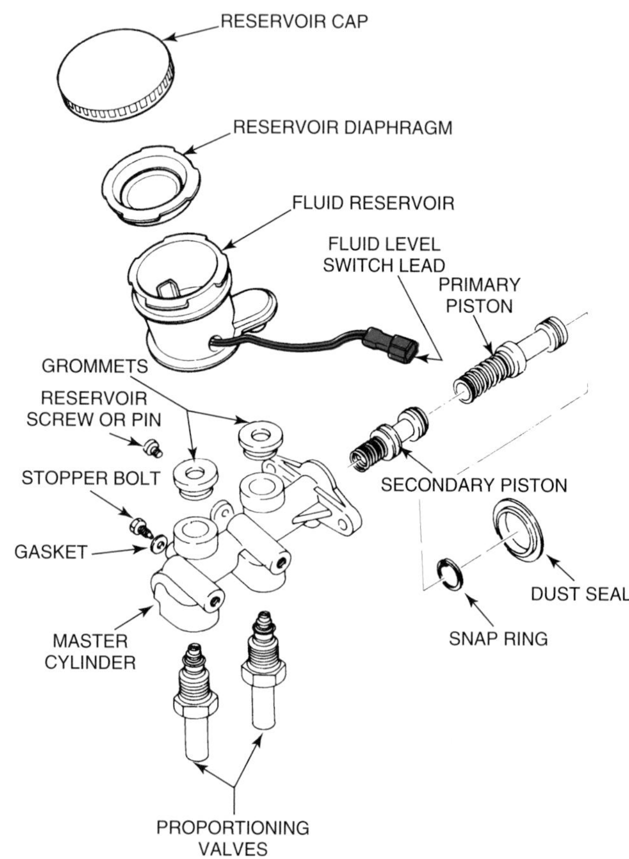

Figure 23-14. Exploded view of an aluminum master cylinder with nylon reservoir and screw on reservoir caps. Note the fluid level switch lead. (Chevrolet)

Systems may be arranged so that one master cylinder piston provides pressure for the front brakes, the other the rear. See **Figure 23-15A**. Another arrangement is the ***diagonal split brake system***, in which one master cylinder piston operates one front brake and one rear brake on opposite sides, **Figure 23-15B**.

Figure 23-16 shows the configuration and operation of a typical dual master cylinder under normal conditions. Note that movement of the push rod applies both the front and rear brakes.

Quick Take-Up Master Cylinder

The ***quick take-up master cylinder*** is used with the diagonal split brake system. Opposed front and rear brakes

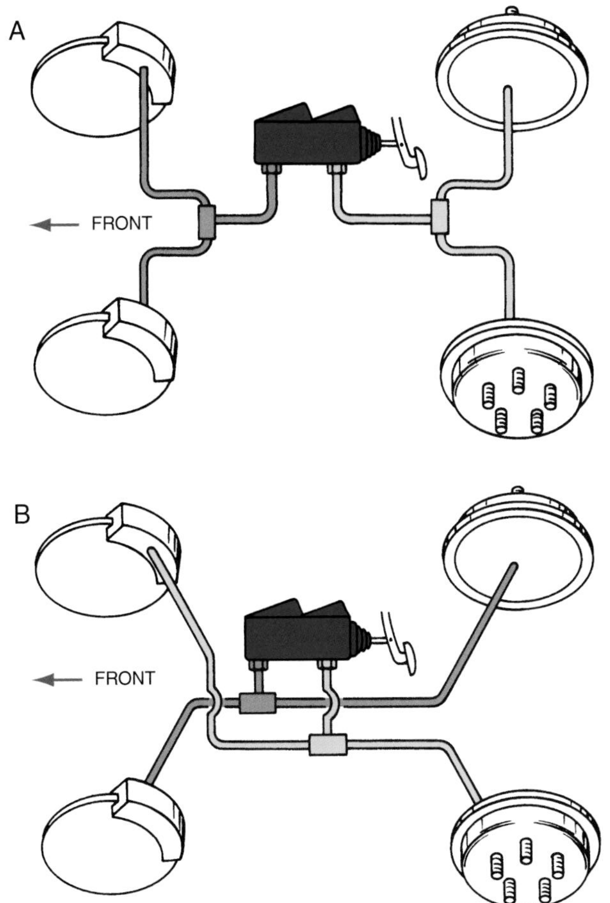

Figure 23-15. Two dual or "split" braking system designs. A—Typical front and rear setup is widely used. Note how each master cylinder piston serves only one set of brakes. B—Diagonal setup where each piston serves one front and one rear brake on opposite sides. Other designs such as right side-left side, dual front-dual rear, have been used. (EIS)

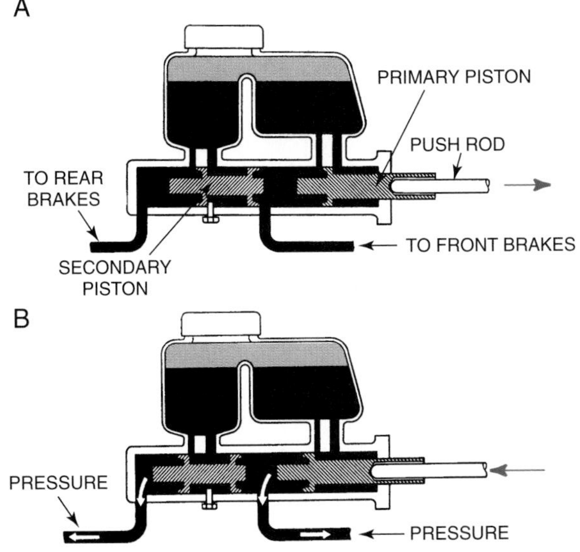

Figure 23-16. Dual master cylinder operation. A—Released position. B—Applied position. (Mercedes).

are both operated by the primary or secondary piston only. This master cylinder, **Figure 23-17**, operates in the same manner as the standard cylinder, with some exceptions. Proportioners (proportioning valves) and a warning light switch are attached directly to the cylinder. However, the most important feature is the ***quick take-up valve***.

Master Cylinder Operation

The following sections explain the operation of the master cylinder in the released position (brake pedal not being pressed by the driver) and applied positions (brake pedal being pressed by the driver). Also covered is master cylinder operation as the brake pedal is released.

Brake Pedal Released

Figure 23-18 shows the master cylinder in the released position. The primary piston is pushed back against the stop plate. There is static pressure in the brake line and wheel cylinder. The primary cups are free of the compensating ports and the breather port is open to the center section of the primary piston. No pressure is present within the master cylinder itself.

Brake Pedal Applied

Refer to **Figure 23-19.** When pedal pressure is applied to the primary piston, it moves forward and blocks off the compensating port, sealing the fluid in front of it. As the piston continues to move forward, it will transmit fluid pressure to the front calipers and to the base of the secondary piston. This causes the secondary piston to move forward, blocking off its compensating port and applying pressure to the rear calipers or wheel cylinders. Note that the pedal pressure has moved both pistons and has created pressure to both chambers.

Brake Release, Start

When the driver removes his or her foot from the brake pedal, push rod pressure is removed from the pistons, **Figure 23-20.** As the pistons start to move outward in the cylinder, they will move faster than the fluid can return through the lines. This creates a mild vacuum in the pressure chambers and fluid will flow through the bleeder holes in the head of the pistons. This fluid pressure will bend the lips of the cups away from the cylinder wall, and fluid will flow into the cylinder ahead of the pistons.

This action also allows the brakes to be pumped. ***Pumping*** is the repeated application of the pedal in quick movements. It is used when one full application fails to expand the friction members or when stopping on slick or icy roads. Pumping the brakes should not be necessary under normal driving conditions and usually indicates a hydraulic system leak or part failure, faulty brake adjustment or worn brake linings. The brakes should be serviced whenever pumping becomes necessary. The flow of fluid through the bleeder holes will also prevent the possible entry of air by keeping the cylinder filled at all times.

Brake Release, Finish

When pressure drops in the master cylinder, the friction linings begin to retract. As the linings retract, they force the

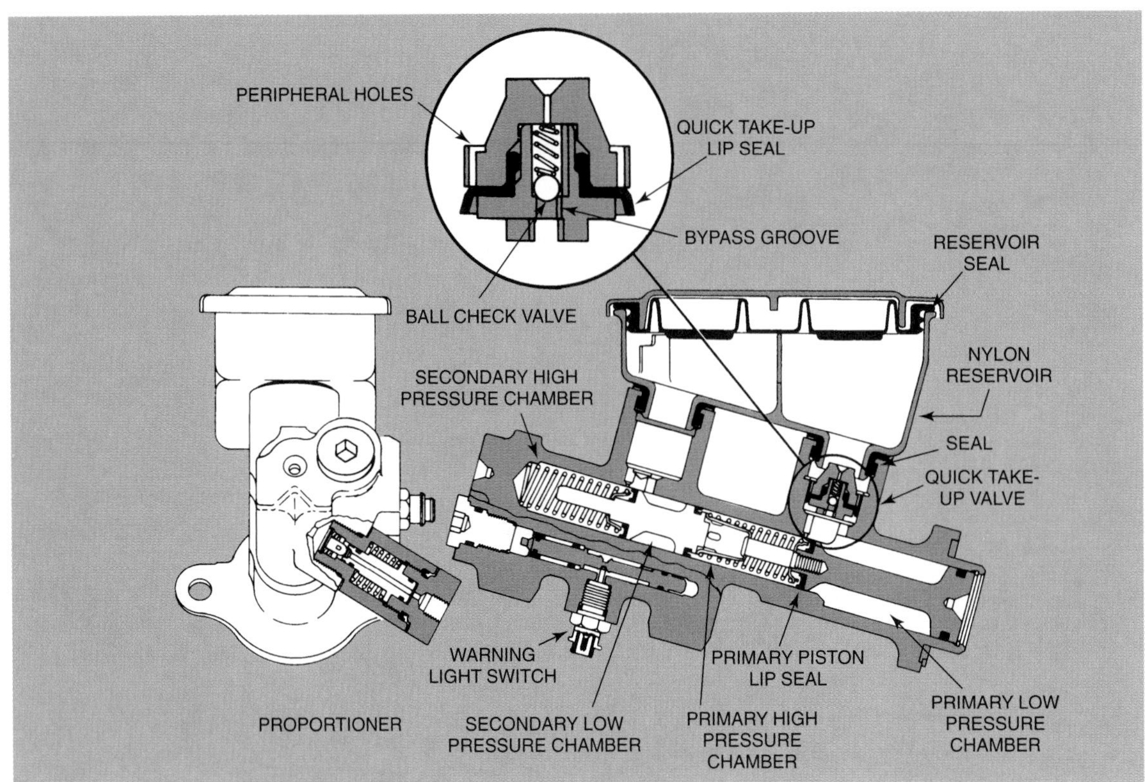

Figure 23-17. Cutaway view of a "quick take-up" master cylinder. Study the part names, construction and locations. (Delco)

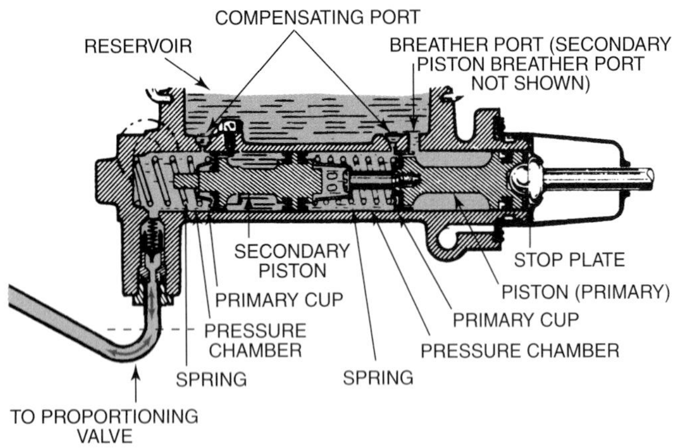

Figure 23-18. Dual master cylinder in the fully released position. The primary piston is against the stop plate and there is no pressure in the master cylinder. (Buick)

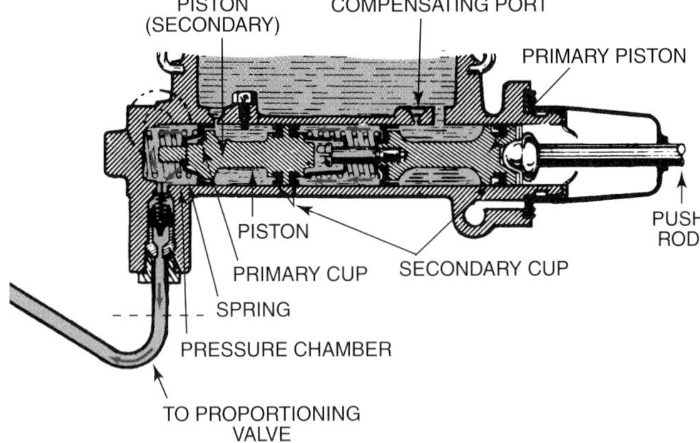

Figure 23-19. A dual master cylinder in the applied position. Pressure is being generated in the brake lines, forcing the caliper and wheel cylinder pistons out. (Buick)

brake fluid to flow back into the master cylinder. See **Figure 23-21.** As the pistons return to their fully released position against the stop plate, the primary cups uncover the compensating ports and any excess fluid will flow into the reservoir.

Brake System Hydraulic Control Valves

The brake hydraulic system of modern vehicles contains many valves to provide even braking and warn of problems. The most common valves are the ***proportioning valve***, the ***metering valve***, and the ***pressure differential switch***. These valves are sometimes installed in a single housing, called a ***combination valve***. These valves are discussed below.

Proportioning Valve

A proportioning valve is normally used in brake systems using disc brakes in the front and drum brakes in the rear. Under mild stops, braking effort is about equal front and rear. As pedal pressure is increased, the proportioning valve

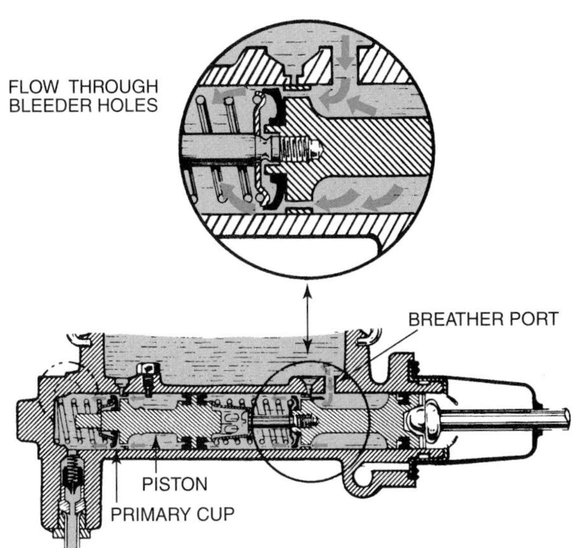

Figure 23-20. A master cylinder at the start of fast release. Note how the fluid flows through the bleeder holes in the head of the piston. This will force the rubber seal cup away from the cylinder wall. This will help to prevent a mild vacuum from retarding (holding back) the piston withdrawal. Cross-sectional view of the pistons in position in the master cylinder bore. (Buick)

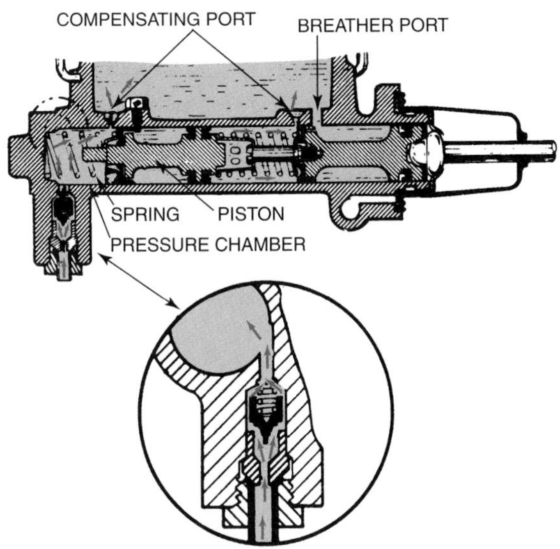

Figure 23-21. Master cylinder at the finish of brake release. Both pistons have uncovered their compensating ports. As the brake shoes are retracted, they squeeze the wheel cylinder pistons together and the fluid is forced back through the lines, into the master cylinder reservoir. (Buick)

controls (and finally limits) pressure to the rear wheels. This reduces the possibility of rear wheel lockup during heavy braking. The proportioning valve can be a separate unit or it can be incorporated into a combination valve. **Figure 23-22** shows two different proportioning valves.

Some vehicles utilize a diagonal split brake system with a ***dual proportioning valve***. The master cylinder is connected directly to the valve. From there, the system is divided diagonally. This arrangement is shown in **Figure 23-22.**

Height Sensing Proportioning Valve

The ***height sensing proportioning valve*** uses a variable pressure range feature, which increases the pressure to the rear brakes as the vehicle's weight (cargo) increases. This pressure will diminish as the vehicle's weight decreases. Most valves are located on the vehicle's chassis and is connected to the rear axle with a calibrated tension spring or a rod-type linkage, **Figure 23-23.**

Vehicle weight transfer during a stop will cause chassis height-to-axle distance to change. The spring or rod linkage will also change in length. This, in turn, adjusts the valve, limiting pressure to the rear brakes. Loading the vehicle (wood in the bed of a truck for example) will also actuate the valve.

Disc Brake Metering Valve

Vehicles with front disc and rear drum brakes require the use of a metering valve. See **Figure 23-24.** The metering valve closes off pressure to the front disc brakes until a specified pressure is developed in the hydraulic system. This allows pressure to force the back brake shoes to overcome retracting spring pressure and move into contact with the drum. Pressure beyond this opens the metering valve, sending fluid to both front and rear brakes.

Pressure Differential Switch

On systems with a dual master cylinder, failure of either the front or rear hydraulic system will allow the brake pedal to travel closer to the floorboard before applying the brakes. In addition, braking power is significantly reduced. Any hydraulic system failure must be corrected as soon as possible.

All dual brake systems use a pressure differential switch to warn the driver that one-half of the split brake system has failed, **Figure 23-24.** A small piston floats in a cylinder separating two pressure chambers. One side of each chamber is connected to one side of the master cylinder. The piston is centered by a spring on each end. An electrical switch is placed in the center of the piston. The switch will be grounded whenever the piston moves to one side. This completes an electrical circuit through the dashboard-mounted brake warning light.

The differential pressure switch in **Figure 23-24** is in the normal open or "light out" position. Each side has equal pressure and the piston remains centered. When one side of the system develops a leak, the pressure drops on that side of the valve. The piston is forced toward the low pressure side. It then touches the electrical plunger and provides the ground needed to light the warning light.

Combination Valve

Combination valves contain either two or three of the valves discussed above. They are called the two function valve and the three function valve. The two function valve combines the metering valve and the brake warning light switch in one unit. Some units may contain a proportioning valve instead of the metering valve.

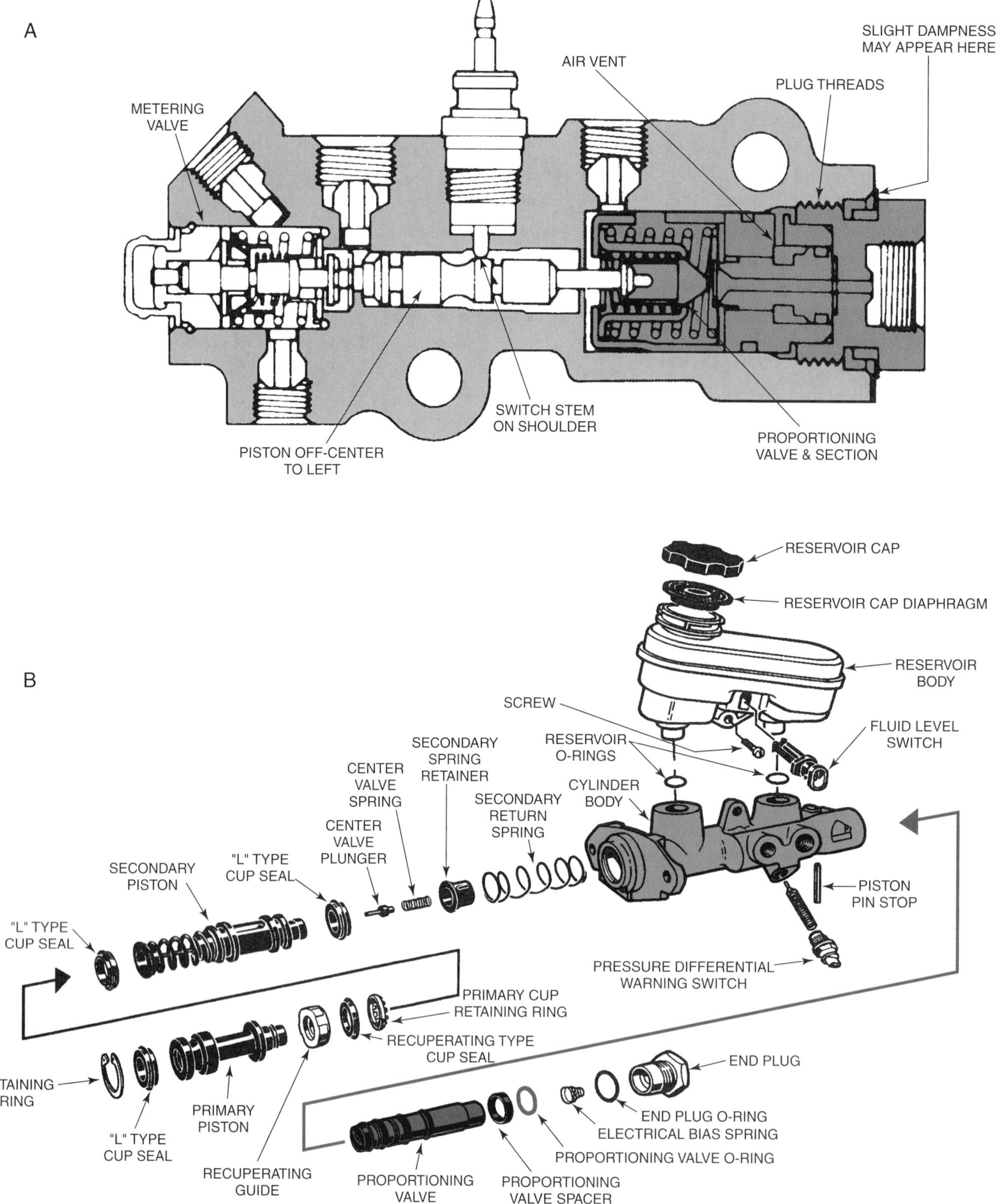

Figure 23-22. Proportioning valves. A–A proportioning valve inside a three-function combination valve assembly. B–An exploded view of a dual master cylinder that houses an internal proportioning valve in its cylinder body. Note the brake fluid level warning light switch. (Allied-Signal and General Motors)

The three function valve houses the metering valve, the proportioning valve, and the brake warning light switch. These valves cannot be adjusted or repaired. If they are defective, the entire valve must be replaced.

Brake Lines

The master cylinder is connected to the wheel cylinders by ***brake lines*** made of high quality, double-walled, steel tubing. The tubing is copper-plated and lead-coated to prevent

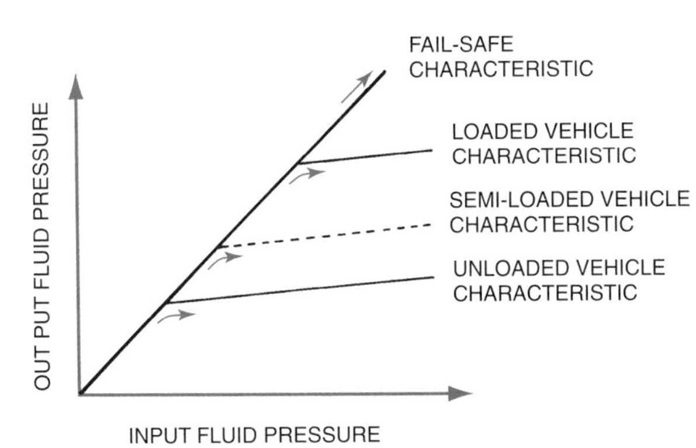

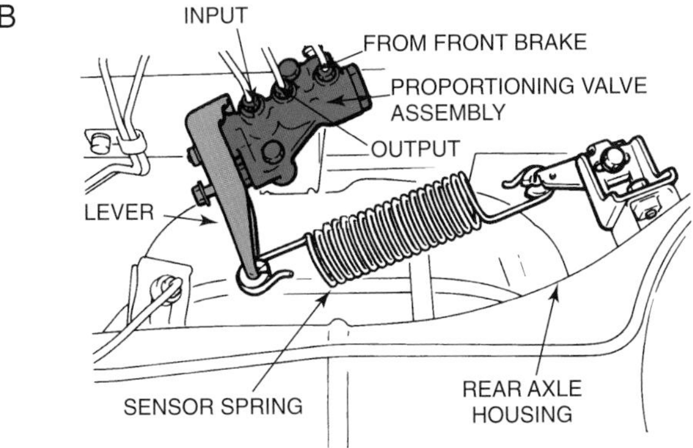

Figure 23-23. Load sensing proportioning valve assembly. A–Vehicle loaded (body height-to-ground) fluid output and input pressures. B–Proportioning valve and sensor springs shown in their mounting positions on the frame and rear axle housing. The sensor spring will pull the valve operating lever in farther, which increases brake fluid pressure to the rear brakes helping them overcome and stop the heavier than normal load. (Chrysler Corp.)

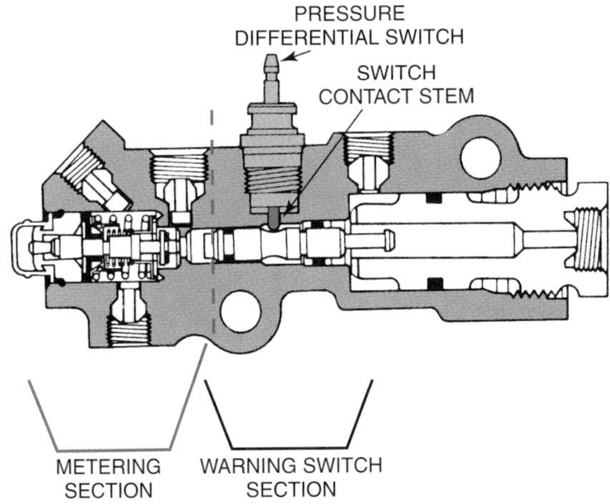

Figure 23-24. A cross-sectional view of a metering valve and warning light switch. The metering section limits hydraulic fluid pressure to the front brakes, until a certain predetermined rear brake apply pressure is obtained. The pressure is equalized when the brakes are not applied. The electrical switch turns on a dash mounted warning lamp if there is a malfunction (loss of fluid pressure in the front or rear system, etc.). (Allied-Signal)

rust and corrosion. Where brake tubing is connected, it uses a double lap flare or a special I.S.O. flare. **Figure 23-25** illustrates the procedure for brake tube flaring.

Brake lines must be replaced with high quality double-walled steel tubing only. Use of poor quality or incorrect materials such as copper tubing or standard steel tubing can result in a potentially fatal accident.

Brake Hoses

Since the wheels move up and down in relation to the body, flexible, high pressure ***brake hoses*** are used to carry fluid to each wheel cylinder or caliper piston. Using flexible hoses prevents line breakage as the wheels move up and down. **Figure 23-26** shows a flex hose leading from the brake line on its way to the rear wheel brake cylinder. Note that the line and hose, where connected, are securely held in a bracket.

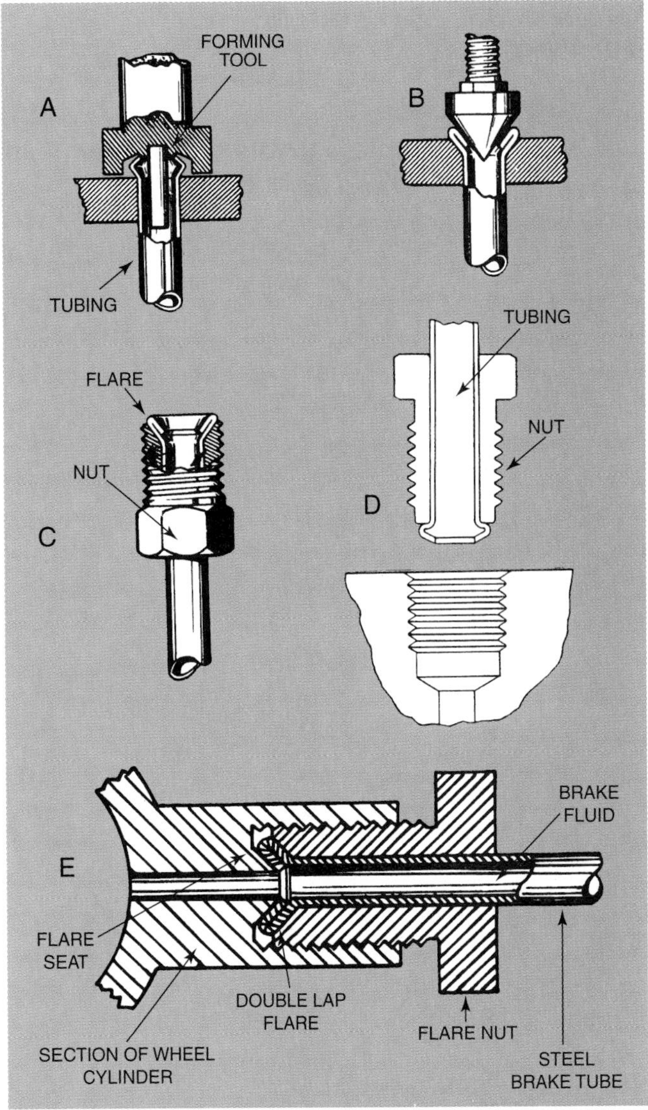

Figure 23-25. Brake tubing flares, A–Start of flare. B–Flare being finished. C–Completed flare with nut in place. D–I.S.O. flare. E–Double lap flare assembled in a wheel cylinder. (Bendix)

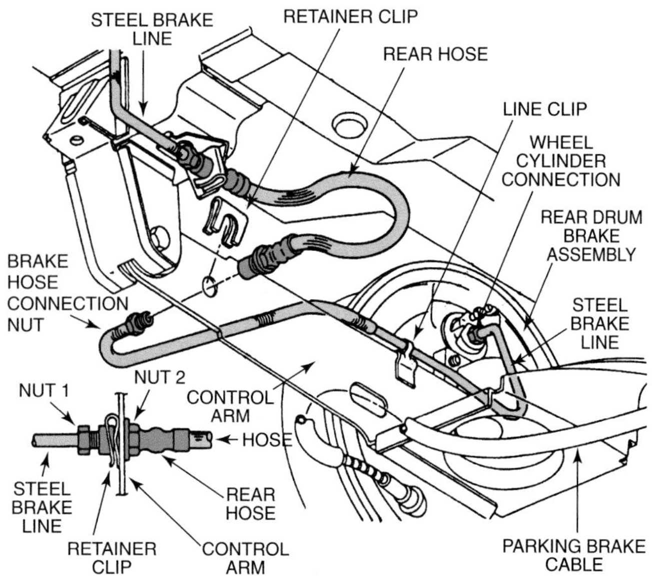

Figure 23-26. Flexible brake hose. To disconnect the brake line, hold nut 2 and loosen nut 1. Notice the use of retaining clips to keep line secured to mounting brackets. (Pontiac)

Figure 23-27 illustrates a typical hydraulic brake system. All brake lines and hoses must be secured by brackets or clips. At no point must they be free to rub or vibrate. This will cause wear and fatigue with resulting line or hose failure.

Disc Brake Calipers

On modern vehicles with disc brakes on the front wheels, and sometimes on the rear, a ***disc brake caliper*** having one or more hydraulic cylinders is used, **Figure 23-28.** The caliper is bolted securely to the spindle. ***Caliper pistons***, which can be constructed from cast iron, aluminum, or ceramic materials, are fitted to the caliper cylinders with the outer ends resting against the friction linings. A bleeder screw is connected to the cylinder. Rubber boots exclude the entry of dirt and moisture.

Caliper Operation

The caliper piston shown in **Figure 23-29** operates against a ***seal ring*** installed in a groove in the cylinder wall. On some calipers, the seal ring may be installed on the piston. When the brake is applied, the piston moves outward. In so doing, it stretches the seal to one side, **Figure 23-29A**.

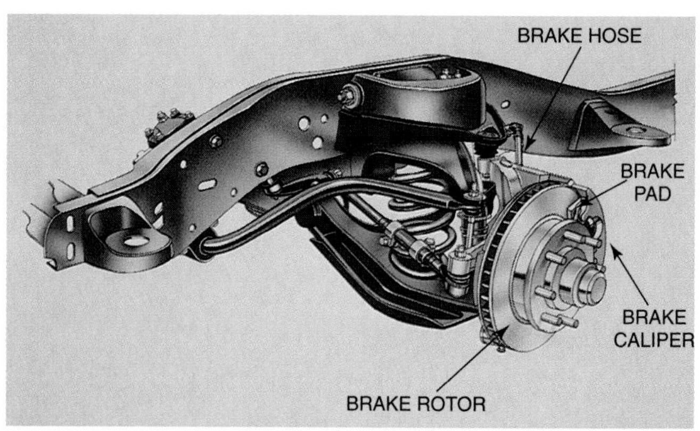

Figure 23-28. One particular front wheel disc brake assembly as used on a rear-wheel drive truck. Disc brakes have excellent cooling characteristics, making them very resistant to brake fade. (Dodge)

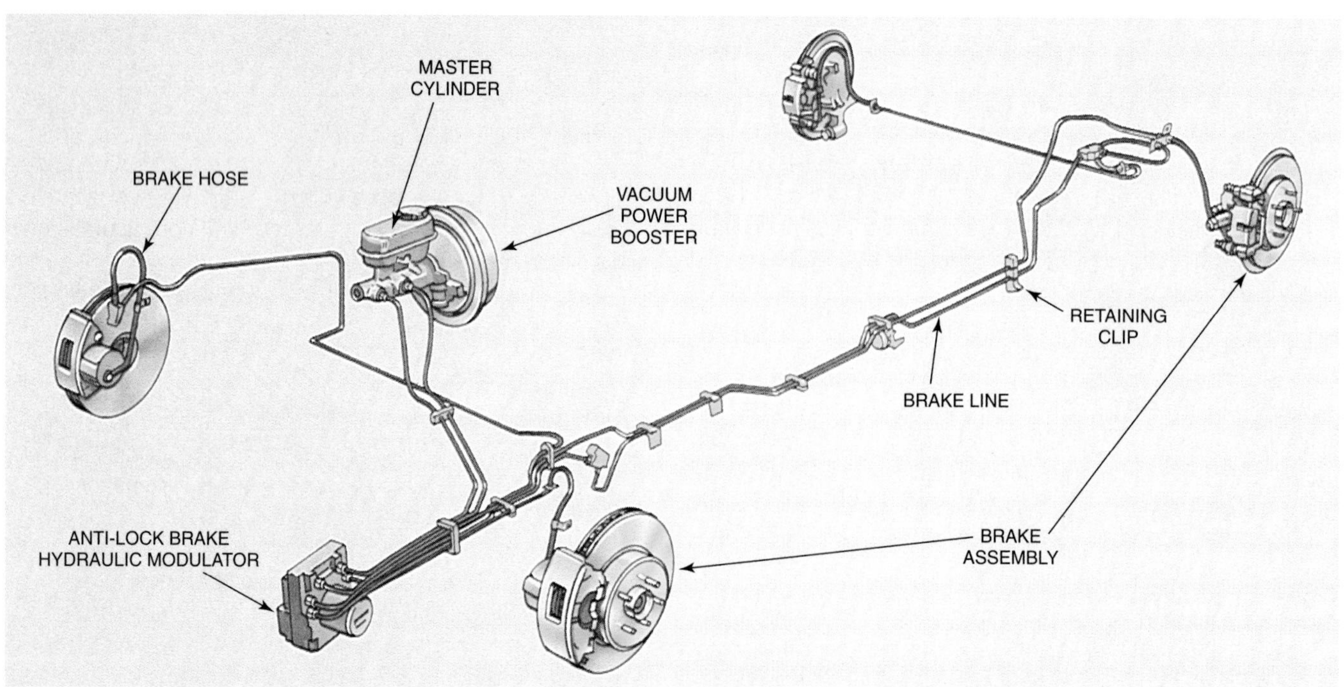

Figure 23-27. Overall view of an anti-lock brake hydraulic system. This system provides improved traction for stopping on slippery road surfaces. (Cadillac)

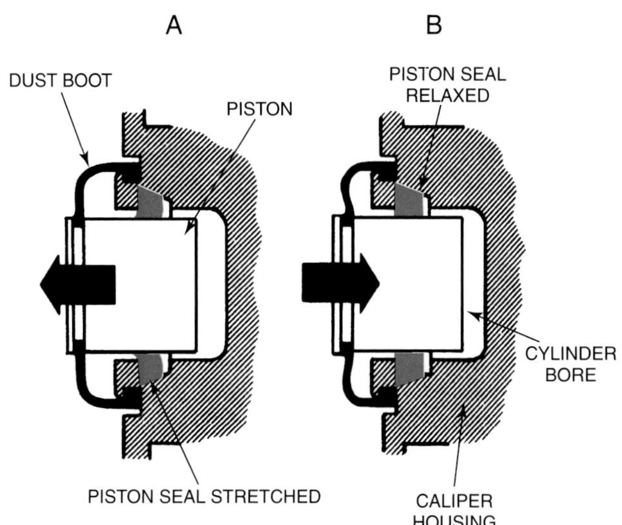

Figure 23-29. Piston seal pulls the piston back, providing pad-to-disc clearance. A—Brake piston applied. The piston moves outward and stretches the seal. B—Brake piston release. The seal rolls back to original shape, drawing the piston back with it. (Wagner)

When brake pressure is released, the seal returns to its normal position, **Figure 23-29B**. This seal roll action pulls the piston back around .005″ (0.13 mm), providing a small amount of lining-to-rotor clearance. As the linings wear, the piston moves out through the seal, automatically keeping the proper clearance.

Modern disc brake calipers are called ***floating calipers***. The caliper can slide, or "float," on its mounting surface. When the piston pushes the inside pad into contact with the rotor, the entire caliper slides back, causing the outside pad to contact the rotor. As long as the caliper can slide, both pads will be applied whenever the brake pedal is depressed.

Non-floating brake calipers usually have two or four pistons. Most pistons in multiple arrangements are of equal size on each side of the rotor. **Figure 23-30** is a comparison of sliding and non-sliding disc brake calipers. All modern vehicles have disc brakes on the front axle. Most vehicles have drum brakes on the rear, but some models are offered with disc brakes on all four wheels.

Other Caliper Parts

Some calipers slide on separate pins which also serve to attach the caliper to the spindle assembly. Many calipers have clips which keep the pads from rattling when the brakes are not applied. These clips are usually attached to the inner pad and piston. Dampening devices are sometimes placed between the pad and caliper piston or housing to reduce the possibility of brake squealing.

Wheel Cylinders

Wheel cylinders are used with all drum brakes. The wheel cylinder is used to transmit the master cylinder pressure to the friction linings and force them outward against the drum. One wheel cylinder (two, in some older systems) is used in each drum brake assembly.

Wheel Cylinder Construction

The wheel cylinder assembly is of rather simple construction. It consists of a cast iron housing, two aluminum ***wheel cylinder pistons*** (iron pistons are sometimes used), two rubber ***cups***, ***cup expanders***, a lightweight coil spring, two push rods, and two rubber ***dust boots***. These parts are shown in the cross section in **Figure 23-31.** On some systems, the cup expanders and push rods are not used. The cylinder is drilled to provide for a bleeder screw (covered later) and the brake line connection. The cylinder is usually bolted to the brake backing plate (covered later).

Single-piston cylinders were used on a few older vehicles in which two wheel cylinders per wheel were used. Since each cylinder operated only one of the shoes, it was necessary to direct the pressure in only one direction. Another design sometimes used was the ***stepped wheel cylinder***, in which one-half of the cylinder was one size, and the other half was another size.

Wheel Cylinder Operation

When the master cylinder forces fluid into the wheel cylinder, the two pistons move apart. Push rods, or links, connect each piston to a brake shoe. On some newer systems, the shoe is connected directly to the piston. As the pistons move outward, they force the shoes against the drum. Study the simplified illustration in **Figure 23-32.**

You will notice the cup design in the wheel cylinder is similar to the primary cup in the master cylinder. When the fluid exerts pressure against the cup, the flanged edges are pressed tightly against the cylinder. See **Figure 23-32.**

Hydraulic System Failure

A leak in the front wheel portion of the system would allow fluid to escape from the system and prevent pressure buildup. Note in **Figure 23-33A** that the primary piston moves in, forcing fluid from the burst portion of the system until the primary piston strikes the secondary piston. Additional pedal movement will cause the primary piston to physically move the secondary piston forward, transmitting normal brake pressure to the rear wheels.

If the rear brake system fails, the primary piston will force the secondary piston in until it strikes the end of the cylinder. The primary piston will then apply normal pressure to the front system as in **Figure 23-33B**. By providing separate systems for the front and rear brakes, the dual master cylinder will provide some braking force regardless of line failure. Although both front and rear systems could fail at the same time, such an event is highly unlikely.

Hydraulic System Must Be Free of Air

Since air is compressible, any air in a hydraulic system would render the system unfit for service. Air must be removed, so each caliper and wheel cylinder is equipped with a ***bleeder screw***. The bleeder screw is threaded into an opening leading into the center of the caliper or wheel cylin-

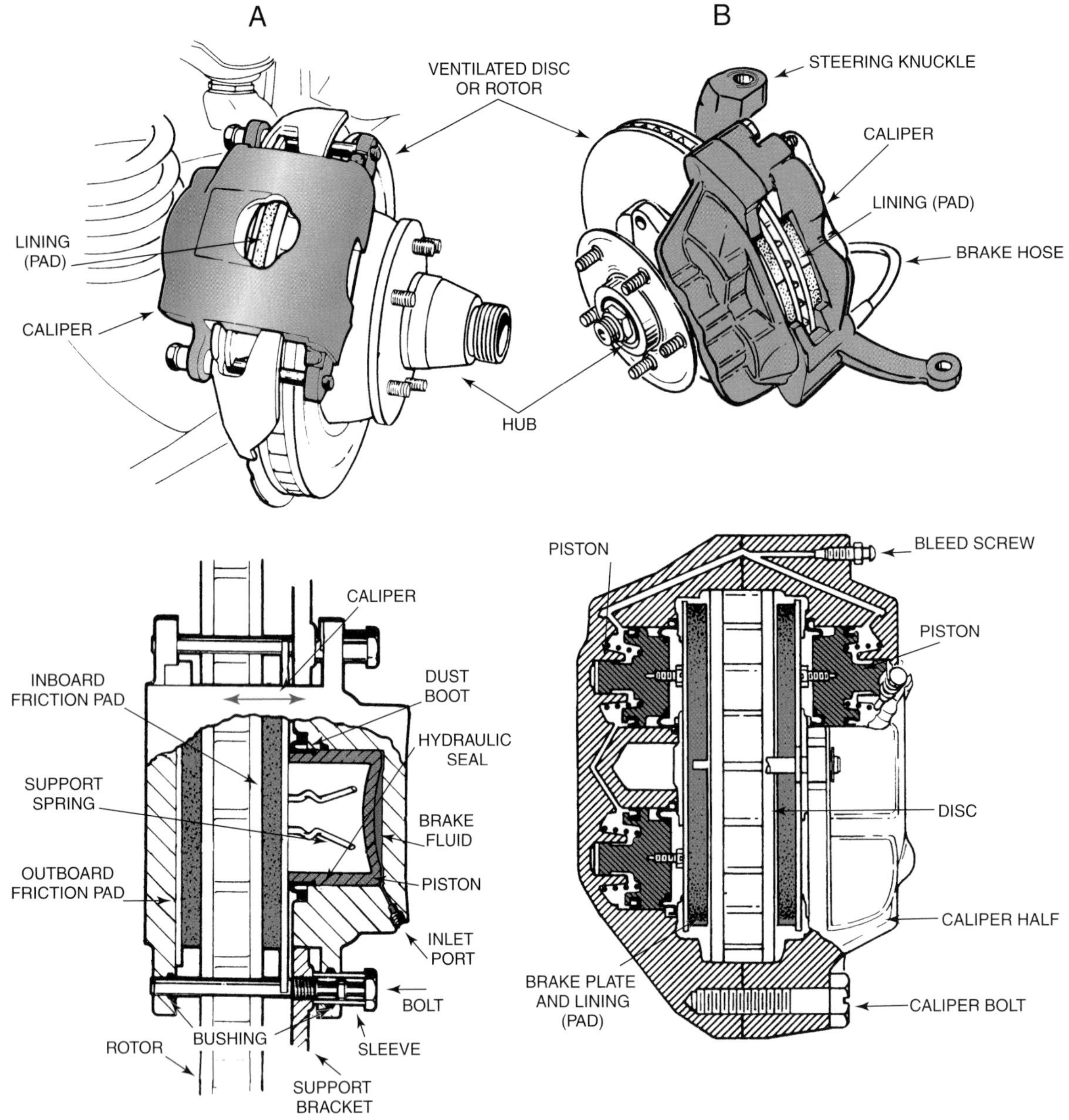

Figure 23-30. Two different disc brake assemblies. A–A sliding floating caliper. B–Non-sliding (fixed) caliper. The caliper in A uses one piston, while the caliper in B uses four. (FMC)

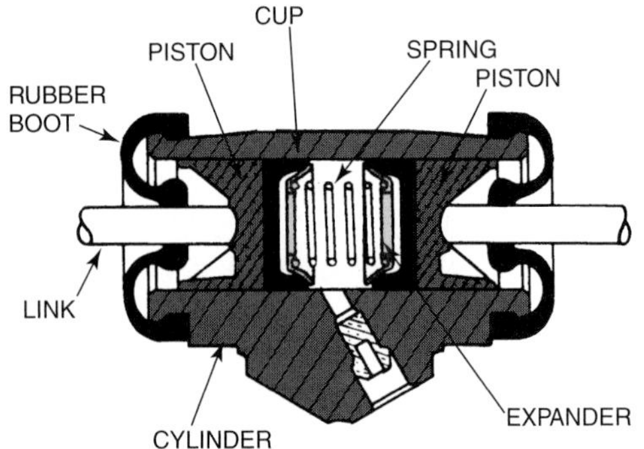

Figure 23-31. Cross section of a typical double-piston wheel cylinder. Stamped metal expander cups at each end of the spring tend to keep the lips of the rubber cups in constant contact with the cylinder wall. (Oldsmobile)

der. A bleeder screw installed in a wheel cylinder is shown in **Figure 23-34.**

The brake pedal is pumped several times and held to apply pressure on the master cylinder. The bleeder screw is loosened, allowing trapped air to escape. Then the bleeder screw is tightened before pressure is removed from the master cylinder. This process is repeated at each wheel until all of the air in the system is removed. When the screw is loosened, the tapered point uncovers the bleeder hole. This permits air to move up around the point, through the cross hole, and out the center passageway. When the screw is tightened, the point seals the opening.

One method of bleeding brakes is shown in **Figure 23-35.** One end of a rubber hose is connected to the bleeder screw. The other end is inserted in a partially filled jar of brake

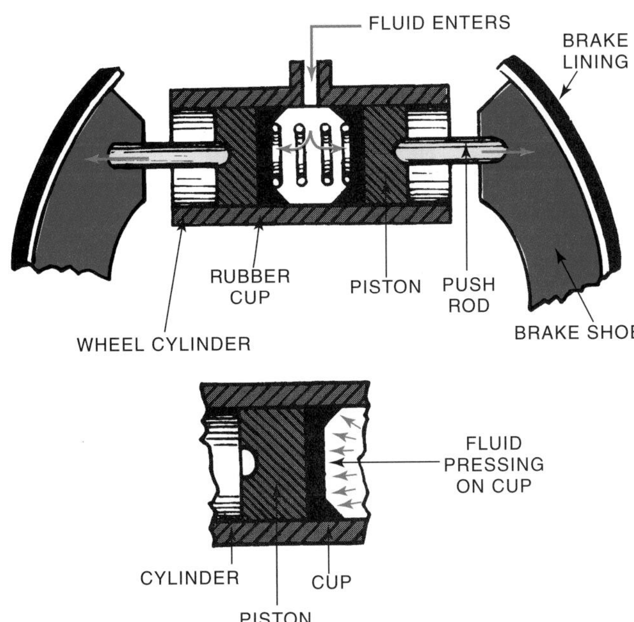

Figure 23-32. Simplified wheel cylinder action. Fluid under pressure enters the wheel cylinder and forces the pistons and push rods outward by pressing on the rubber cups.

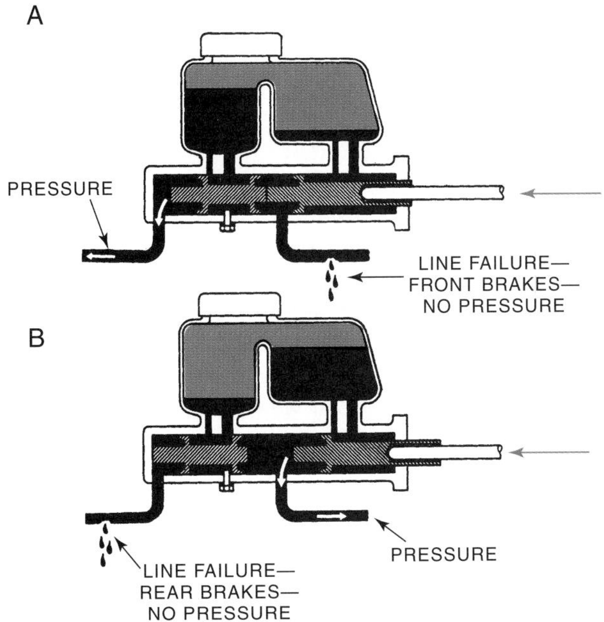

Figure 23-33. Hydraulic failures can be disastrous. With the dual hydraulic system, some braking force can be maintained. A—A line failure in the front brakes will allow fluid to escape, however, the rear brakes retains its supply of fluid and braking action is maintained at a reduced level. B—The rear brakes are affected, but the principles explained earlier still apply.

fluid. As the air leaves the end of the submerged hose, bubbles rise in the fluid. Air cannot be drawn back into the system.

Hydraulic System Vapor Lock

Fresh, quality brake fluid will remain liquid at any brake system temperature. Old, low quality brake fluid or fluid that is

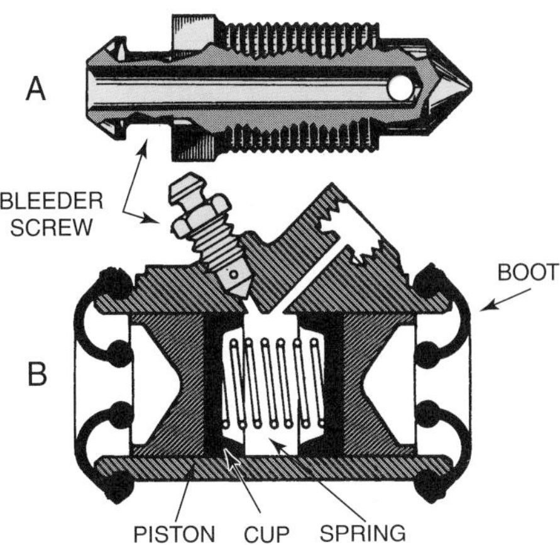

Figure 23-34. A—Cutaway of bleeder screw. B—Wheel cylinder showing bleeder screw.

contaminated can boil at normal braking temperatures. Boiling brake fluid becomes a gas, which can be compressed.

Since the brake hydraulic system depends on having noncompressible fluid, vaporized brake fluid will make it impossible to transmit the proper braking force from the brake pedal to the wheels. Many accidents have been caused by boiling brake fluid. After the accident, the fluid will cool off and recondense, making it impossible to locate as the problem. This is why high quality brake fluid should always be used.

Brake Friction Members

The purpose of the brake ***friction members*** is to physically stop the vehicle. These units, including disc brake pads and rotors and brake shoes and drums, change physical movement into heat by the use of friction. Every wheel contains either a disc or drum brake assembly.

The ***kinetic energy*** (energy in a moving object) in the moving vehicle is converted into heat by the brakes. A fast-moving vehicle has tremendous energy. To bring it to a stop will produce a great amount of heat. Most of the heat is given off by the friction members to the surrounding air. The components and function of each type of friction member is covered below.

Coefficient of Friction

As you learned in Chapter 3, friction is created whenever two objects are rubbed together. The ability of materials to slide across each other is called the ***coefficient of friction***. The coefficient of friction is calculated by dividing the force needed to push a load across a given surface.

For example, if it takes 5 pounds (2.25 kg) of force to move a 10 pound (4.5 kg) block across a smooth surface, the coefficient of friction between the two surfaces is 5 divided by 10 or .5. If it takes 20 pounds (9 kg) of pressure to move the same block across another surface, then the coefficient of friction is 20 divided by 10 or 2. This system is used to determine how various friction materials will act in combination with a brake rotor or drum to stop a vehicle.

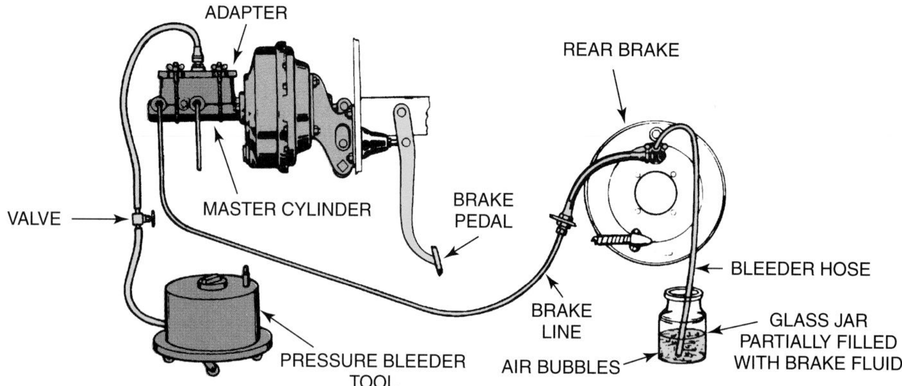

Figure 23-35. Bleeding hydraulic system with a pressurized bleeder tool. Note air bubbles appearing in the glass jar as fluid forces air from the lines. (Bendix)

Friction Linings

Brake linings are made from materials which will produce friction with the rotor or drum and withstand the high temperatures developed during braking. In the past, brake linings were made from ground asbestos fibers pressed into shape. Since asbestos has been shown to cause respiratory problems and cancer, most modern brake pads and shoes are made without asbestos.

Current linings are made from synthetic and steel fibers, and iron, ceramic, and metallic powders. Linings made from these materials are usually referred to as ***semi-metallic***. For heavy duty braking, special metallic linings, which are very resistant to brake fade, are used.

Brake Fade

Each time the brake friction members slow or stop vehicle movement, tremendous heat is created. If some method was not incorporated to dissipate this heat, the brake components would become so hot that they would cause other parts to catch fire. Therefore, brake friction materials are designed to melt at a certain temperature. When heat causes the friction material to melt, the rotor or drum slides over the melted friction material. No more friction is available to produce more heat. This is called ***brake fade***.

With no friction, braking is reduced, or even vanishes entirely. Even hard pedal pressure will not produce fast, even stops. If the brakes fade under normal driving, the cause must be found and corrected. Vehicles used in high speed and heavy duty applications have metallic linings that have great resistance to fade.

Disc Brake Assembly

The disc brake uses a ***rotor*** as its friction surface. The rotor is bolted to and revolves with the wheel hub. ***Disc brake pads*** are installed in the caliper which surrounds the rotor. The rotor is stopped when the brake pads are pressed against it by the caliper piston (discussed earlier). Most calipers are single piston types.

Older designs allowed the brake pads to drag very lightly against the rotor at all times. To reduce drag and improve fuel economy, modern disc brake systems operate with only a minimal pad to disc clearance (about .005″ or 0.13 mm). The disc brake pads are self-centering and press on each side of the rotor with equal pressure.

Disc brakes resist brake fade because the rotor's larger surface area is exposed to the open air. The rotor may be solid or have fins and/or drilled holes for cooling, **Figure 23-36.** Despite the fact that the rotor is exposed to water and dirt, the disc brakes work well during wet or dry operation without grabbing. Due to the relatively small friction area and the lack of servo or self-energizing action common in drum brakes, disc brakes require a higher operating hydraulic pressure.

Brake Pad Construction

Brake pads consist of friction materials ***bonded*** or ***riveted*** (glued) to a metal plate. Past friction materials included asbestos, which has been replaced with various high temperature compounds. These compounds are mixed with metallic compounds into a matrix to make a semi-metallic friction material. Some heavy duty high temperature brake applications use friction materials made up almost entirely of various metals. Sometimes, a metal wear sensor is riveted to the metal plate. When the pads are worn sufficiently, the wear sensor will contact the rotor, producing a high pitched squeal.

Braking Ratio

When a vehicle is stopped, there is a transfer of weight to the front of the vehicle due to its inertia. This forces the vehicle's weight against the front tires while it relieves weight off the rear tires somewhat.

To compensate for, and to take advantage of, this effect-the front brakes usually are designed to produce more stopping power than the rear. This is called the vehicle's ***braking ratio***. An average ratio would be: Front brakes—55% to 60%. Rear brakes—40% to 45% of the stopping force. This ratio will vary depending on the type and size of vehicle, vehicle loading, and whether the vehicle uses front- or rear-wheel drive. The hydraulic system may also be designed to modify the braking ratio for increased braking efficiency.

Backing Plate

A round, stamped sheet metal or steel plate is bolted to the front spindle, or to the end of the rear axle housing. This

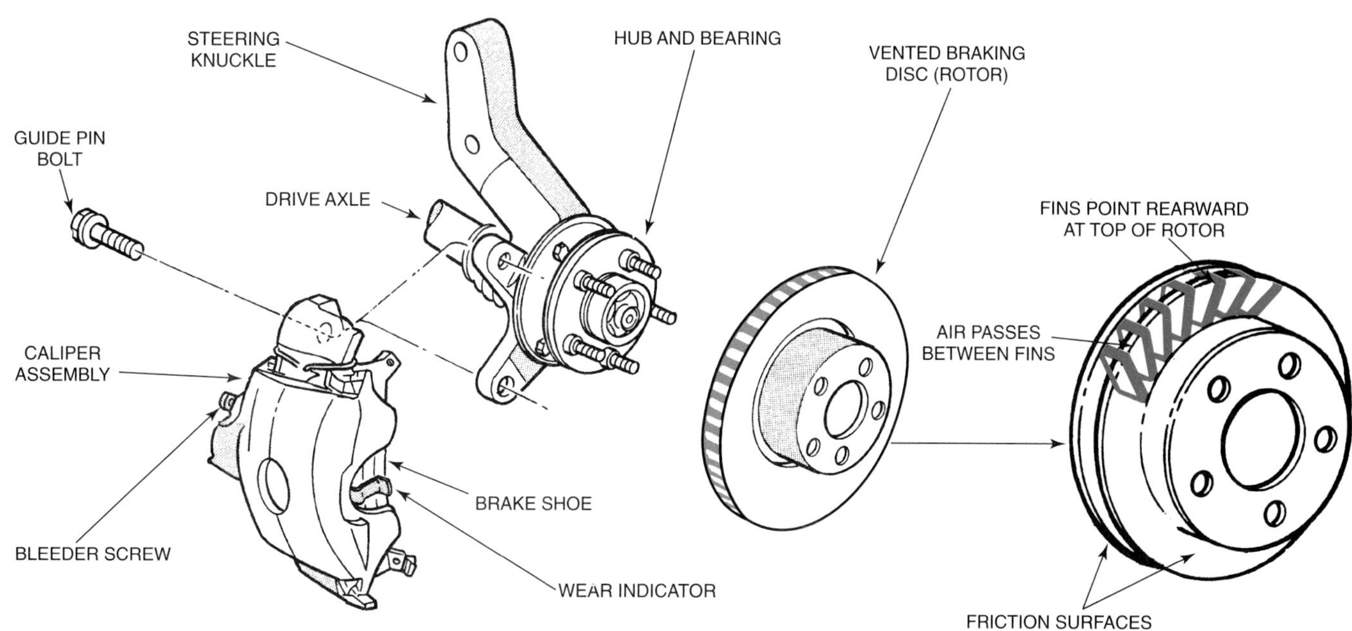

Figure 23-36. A disc brake assembly from a front-wheel drive vehicle which uses finned brake rotors. Air passing between the rotor fins while the vehicle is moving, helps to dissipate the heat. (Bendix/Chrysler)

backing plate, when used for disc brakes, serves as a shield for the caliper and rotor. When it is used for drum brakes, it serves as the foundation upon which the wheel cylinder and brake shoe assembly is fastened. The backing plate is rigid and cannot move in any direction, **Figure 23-37.**

Drum Brake Assembly

Each ***drum brake assembly*** consists of the brake backing plate, a set of brake shoes, shoe retainer clips and return springs, and a brake drum. These components are covered in the following sections.

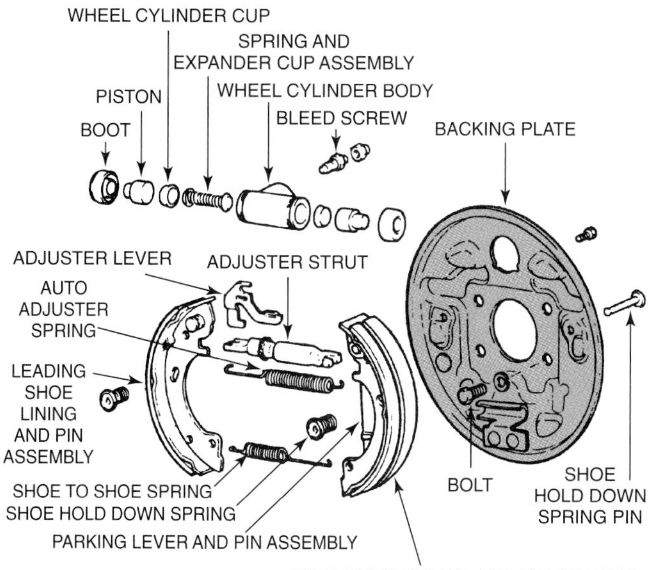

Figure 23-37. Typical brake backing plate. The backing plate forms a foundation for the brake shoes and wheel cylinder. Disc brakes use a similar plate to shield against debris. (Hyundai)

Brake Shoe Assembly

All drum brake assemblies use two ***brake shoes***. These shoes are made of stamped steel and have brake linings either riveted or bonded to the outer surface. Although of similar design, there are some minor differences in the shapes of the shoes. All utilize a T-shape cross-section, **Figure 23-38.** A web is used to give the shoe rigidity. When forced against the drum, it will exert braking pressure over the full lining width and length.

The brake shoes ends may be free-floating or they may have one end fastened to an anchor. The ***primary*** (front or leading) ***brake shoe*** faces toward the front of the vehicle. It often has a different size lining than the other shoe. The ***secondary*** (rear or trailing) ***brake shoe*** faces the rear of the vehicle. Some newer drum brake assemblies use brake shoes that can be mounted in either position.

Brake return or retracting springs are used to pull the shoes together when hydraulic pressure is released. Small

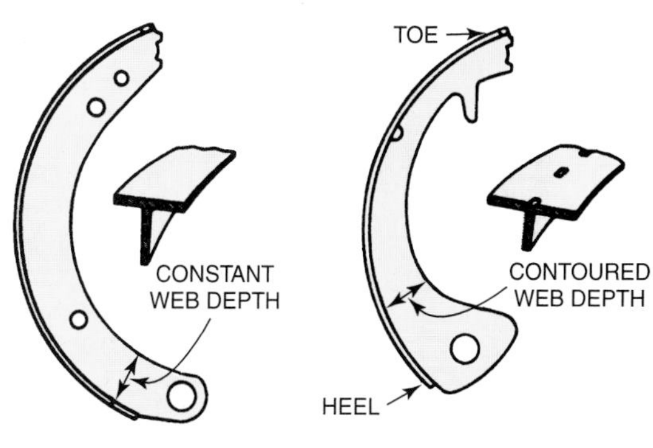

Figure 23-38. Two brake shoe designs. (Dodge)

spring clips of various designs are used to keep the shoes against the backing plate to ensure shoe alignment and prevent rattle. Study the arrangement of parts shown in the exploded view of a drum brake assembly in **Figure 23-39.**

Brake Drum

The ***brake drum*** fits between the wheel and hub and completely surrounds the brake shoe assembly. It comes very close to the backing plate so that water and dust cannot enter easily. The center section is constructed of stamped steel, with an outer cast iron braking rim. An aluminum outer rim is sometimes used to aid in cooling. When aluminum is used, a cast iron braking surface is fused to the aluminum rim.

The heavy casting enables the drum to absorb and dissipate heat from the braking process without distorting. Cooling fins are often cast into the rim to assist in heat dissipation, **Figure 23-40.** The braking area of the drum must be smooth, round, and parallel to the shoe surface. When the wheels turn, the drum revolves around the stationary brake shoes.

Drum Brake Operation

When hydraulic system operation creates pressure in the wheel cylinder, the wheel cylinder pistons and links move the brake shoes outward into contact with the revolving brake drum. Since the shoes cannot revolve, they will stop the drum and wheel.

Drum Brake Arrangement

There are many arrangements used in mounting drum brakes. Several popular ones are illustrated in this chapter. Most utilize ***servo action*** (one shoe helps to apply the other), as well as ***self-energizing*** (using frictional force to increase shoe-to-drum pressure).

Servo Action, Self-Energizing Brakes

Servo and self-energizing action is produced by hooking the heel of the primary shoe to the toe of the secondary shoe. When the wheel cylinder forces the top ends of the shoes against the revolving brake drum, it will try to carry the forward shoe around. As the primary shoe attempts to revolve, it will jam the secondary shoe against the single anchor pin. This stops both shoes and produces a binding effect that actually helps the shoes apply themselves. This servo and self-energizing action reduces the amount of needed pedal pressure.

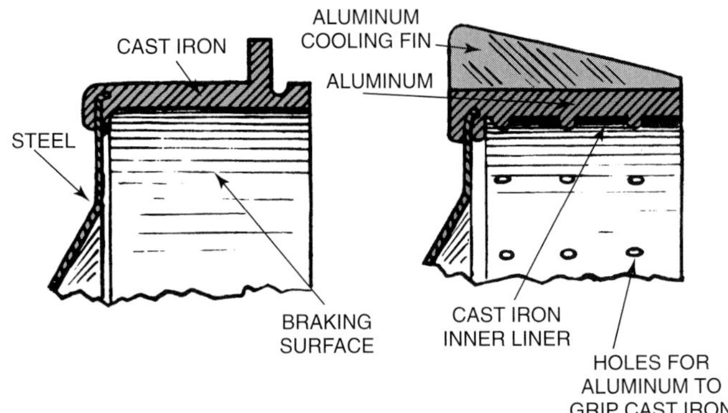

Figure 23-40. Cross-sections of cast iron and aluminum brake drums. Note the cast iron inner liner when aluminum is used to facilitate cooling.

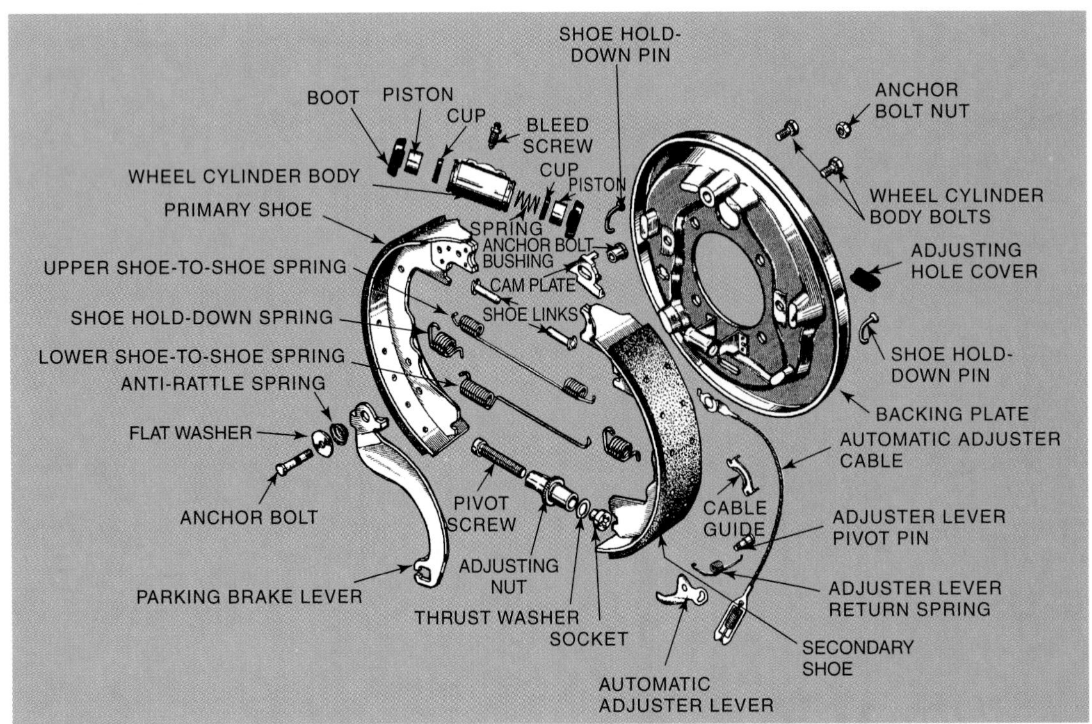

Figure 23-39. Exploded view of a backing plate and brake shoe assembly. (Dodge)

Note how the primary shoe in **Figure 23-41** attempts to rotate in the direction of the drum. Since the adjusting screw connects it to the toe of the secondary shoe, the heel of the secondary shoe is jammed against the anchor pin. The arrows illustrate the braking force direction. When the vehicle is traveling in reverse, the secondary shoe applies the primary shoe.

The primary shoe lining on self-energizing brakes generally is smaller and of a different composition. The secondary shoe does more of the braking, so less lining material is needed on the primary shoe. A servo action, self-energizing brake assembly is shown in **Figure 23-42.** Note that the primary shoe has less lining. The star wheel at the bottom is used to adjust the shoes to keep them close to the drum.

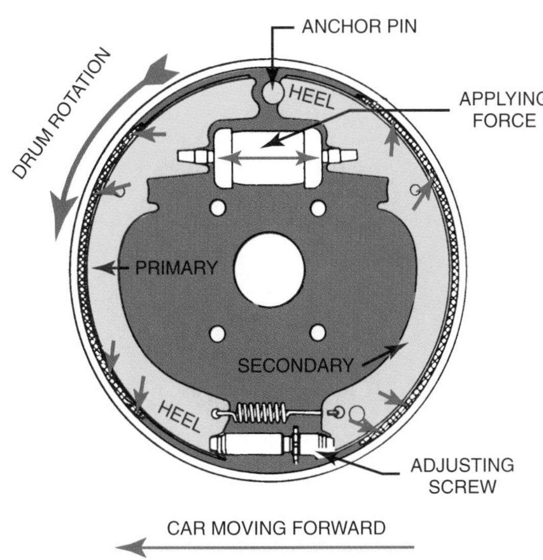

Figure 23-41. Servo action, self-energizing brake. When actuated, the primary shoe starts to move with the drum and in so doing, applies the secondary shoe. (Buick)

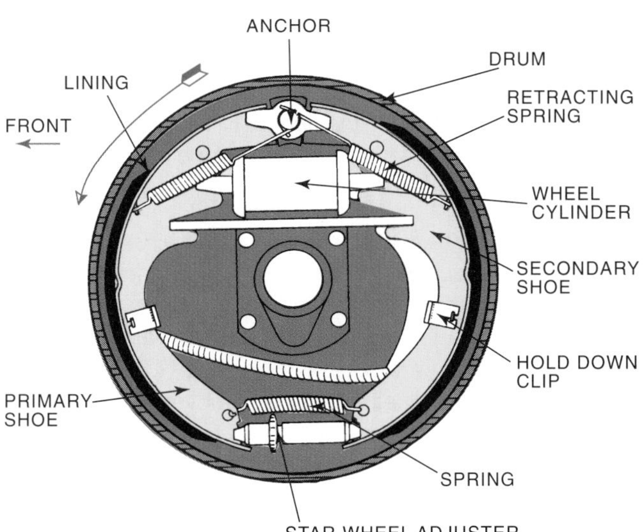

Figure 23-42. Servo action, self-energizing drum brake. The primary shoe (left) has a short lining when used in this setup (Wagner Electric Corp.).

Self-Adjusting Brakes

Modern drum brake systems are ***self-adjusting***. Each utilizes a series of springs, levers, and cables that automatically adjusts the drum-to-lining clearance as the linings wear down. Manual adjustment is provided by turning the star wheel with a special adjusting tool.

The adjustment systems are of three general types, which are identified below. Some systems operate only when braking in reverse, others in a forward direction, or braking in either direction.

Lever Adjuster

An ***adjustment lever*** is attached to one of the brake shoes in such a manner that it is free to pivot back and forth. One end engages the teeth in the star wheel, while the other is attached to a link, as in **Figure 23-43.** The link is attached either to the anchor (adjustment will take place when braking in one direction only) or to the other brake shoe (will function when braking in either direction).

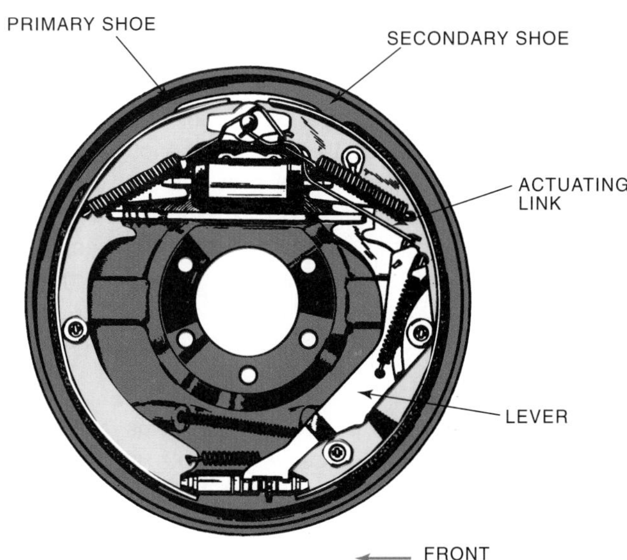

Figure 23-43. Lever type automatic brake adjuster. This unit functions only when braking in reverse.

A spring holds the adjustment lever in a downward position. Since the actuating link is attached to the anchor, it will operate only in one direction. On brake assemblies where the lever is pivoted to the secondary shoe, it will function only during reverse braking.

Cable Adjuster

Study the construction of the ***cable adjuster*** in **Figure 23-44.** This particular unit passes the actuating cable up over the anchor and attaches it to the secondary shoe, causing it to function when braking in either direction.

When the brakes are applied, the shoes move out until they contact the rotating drum. They will then move around with the drum until the secondary shoe heel engages the anchor. The small out and around movement tightens the

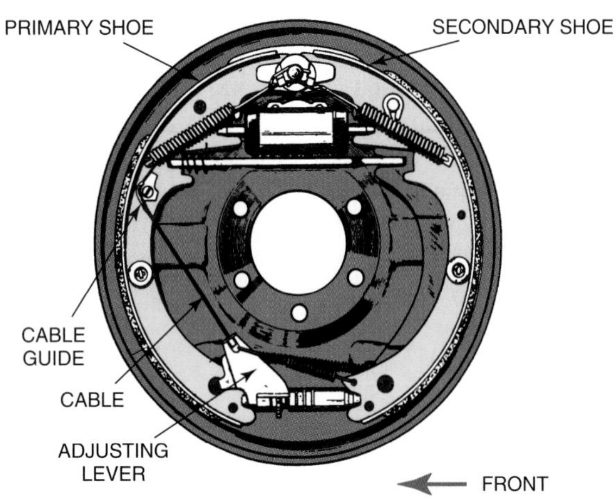

Figure 23-44. Cable type automatic brake adjuster. This specific drum brake setup works when braking in either direction.

actuating cable and pulls it upward on the adjustment lever a small amount. When the brakes are released, the shoes are retracted, causing the actuating cable to become slack and the return spring pulls the adjustment lever back down. This action is repeated during each brake application.

Since the amount of lever movement is very small, the cable will not normally draw the lever up far enough to engage a new tooth on the star wheel. However, when sufficient lining wear has taken place, the cable movement will be enough to draw the lever up to the next tooth on the star wheel. When the brakes are released, the return spring will draw the lever down, forcing it to rotate the star wheel one tooth. This adjustment action will take place during the entire life of the lining.

Link Adjuster

The ***link adjuster*** is very similar to the cable type. Instead of using a cable, it uses two separate links which perform the same task. The utilization of links instead of the cable is the basic difference, **Figure 23-45.**

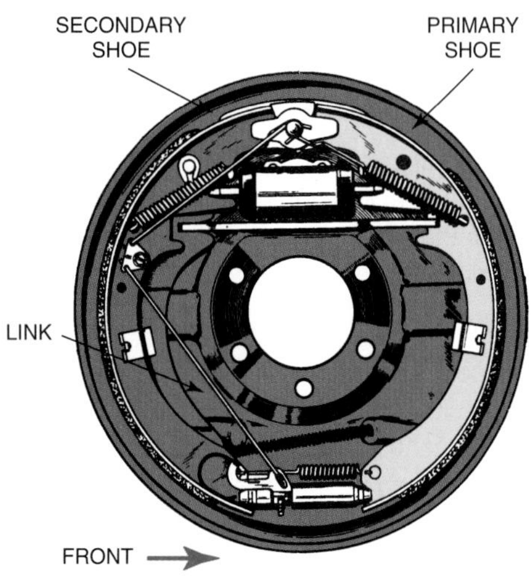

Figure 23-45. Link type automatic brake adjuster. This type of adjuster functions during reverse braking.

Parking Brakes

On most modern vehicles, the ***parking brakes***, sometimes called the ***emergency brakes***, is part of the rear brake assembly. This basic system is used with both drum and disc rear brakes. In most parking brakes, a foot pedal or hand-operated lever is connected through linkage and cables to both rear brake assemblies. When the emergency brake is applied, the cables pull a lever in each rear brake assembly.

Drum Parking Brake Operation

The top of the lever is bolted to one shoe and one end of a strut (steel bar) is notched into the lever below the top bolt. The other end of the strut is notched into the opposite shoe. The emergency brake cable is attached to the lower end of the lever. See **Figure 23-46.**

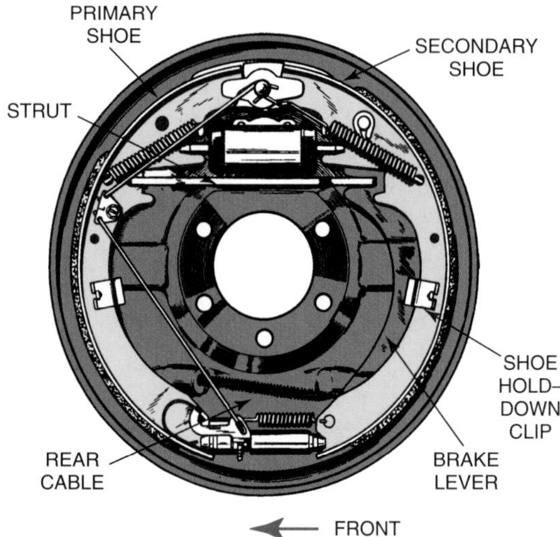

Figure 23-46. Rear-wheel parking brake assembly. Rear cable pulls on brake lever, thus causing strut to spread and apply the brake shoes. (Bendix)

When the cable pulls the bottom of the lever toward the left, the lever forces the strut to the left until it presses the shoe against the drum. At this point, continued movement of the lever causes it to pivot on the strut. This moves the lever and shoe to which it is bolted to the right until it engages the drum. Any further lever movement will now apply force to both shoes.

Caliper Parking Brake Operation

Vehicles equipped with rear disc brakes incorporate an integral thrust screw mechanism in the caliper piston. This caliper is similar to the front caliper in design and operation. The parking brake is actuated by a lever located on the inboard side of the caliper. The lever is moved by the parking brake cable.

As the lever is applied, it rotates the thrust screw which forces the piston assembly against the shoe and lining. Continued application moves the caliper, bringing the outboard shoe and lining into contact with the rotor, firmly holding it in position. An exploded view of this caliper setup is shown in **Figure 23-47.**

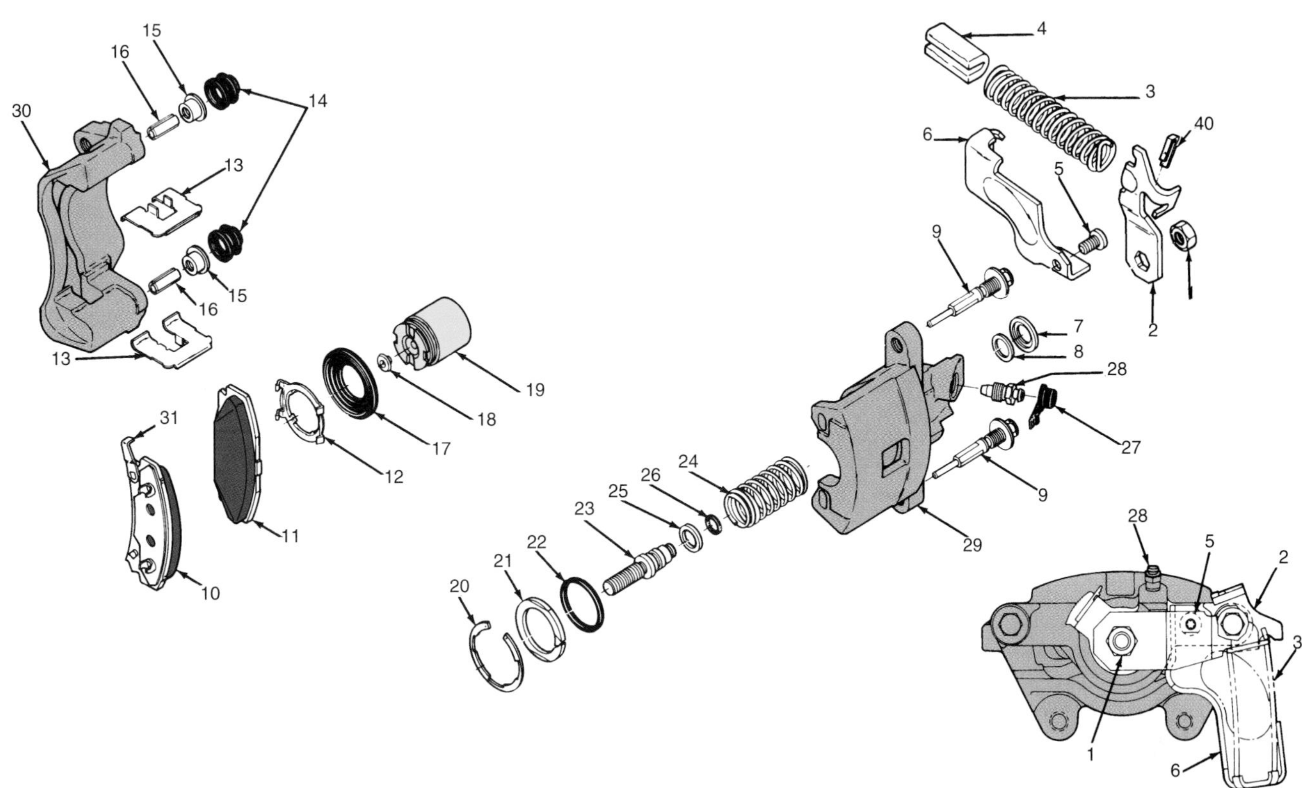

Figure 23-47. Exploded view of a rear disc caliper assembly. Note that this unit incorporates an integral parking brake setup. Study all part names and locations. 1—Nut. 2—Park brake level. 3—Return spring. 4—Damper. 5—Bolt. 6—Bracket. 7—Lever seal. 8—Anti-friction washer. 9—Mounting bolt. 10—Outboard shoe and lining. 11—Inboard shoe and lining. 12—Shoe retainer. 13—Insulator. 14—Bolt boot. 15—Support bushing. 16—Bushing. 17—Caliper piston boot. 18—Two-way check valve. 19—Piston assembly. 20—Retainer. 21—Piston locator. 22—Piston seal. 23—Actuator screw. 24—Balance spring and retainer. 25—Thrust washer. 26—Shaft seal. 27—Cap. 28—Bleeder valve. 29—Caliper housing. 30—Bracket. 31—Wear sensor. 32—Retaining clip.

Power Brake Systems

Most modern vehicles are equipped with ***power brakes*** as standard equipment. The power brake unit is an additional part that operates with the hydraulic system. Power brakes reduce the amount of pedal pressure necessary to stop the vehicle. Pedal travel (distance from release position to full brake application) can also be shortened. Power brakes can be vacuum or hydraulically assisted.

Vacuum-Assisted Power Booster

The ***vacuum power brake booster*** is a closed cylinder with a piston inside. One side of the piston is connected to the master cylinder piston. The other side is connected to the brake pedal. A ***vacuum control valve*** is placed between the brake and the piston.

Vacuum Power Booster Construction

A power brake booster operates from the vacuum produced in a gasoline engine. If the vehicle has a diesel engine, an engine- or electrically-driven vacuum pump is used, as diesel engines do not provide sufficient vacuum for efficient brake booster operation.

The vacuum control valve admits vacuum to one side of the piston, while the normal atmospheric pressure is allowed to exist on the other. This valve can also allow atmospheric pressure to reach both sides of the piston, **Figure 23-48.**

Vacuum and Atmospheric Suspended Booster

The ***vacuum suspended booster*** has a vacuum on both sides of the diaphragm or piston when the booster is in the released position. When the booster is applied, atmospheric pressure is admitted to one side to cause the necessary movement of the piston. See the schematic in **Figure 23-49A**. An ***atmospheric suspended booster*** has atmospheric pressure on each side of the piston in the released position. To cause booster action, vacuum is admitted to one side, **Figure 23-49B**. Most modern brake systems use the vacuum suspended booster.

Vacuum Power Booster Operation

When the driver presses the brake pedal down, the vacuum control valve closes off the atmospheric pressure to the brake cylinder side of the piston. Further movement of the brake pedal opens a vacuum inlet passage to this same side. As there is atmospheric pressure on one side of the piston and a partial vacuum on the other, the piston will be forced toward the vacuum side. Since the piston is connected to the master cylinder piston, it will apply pressure to the brake system. The vacuum power brake has four stages of operation:

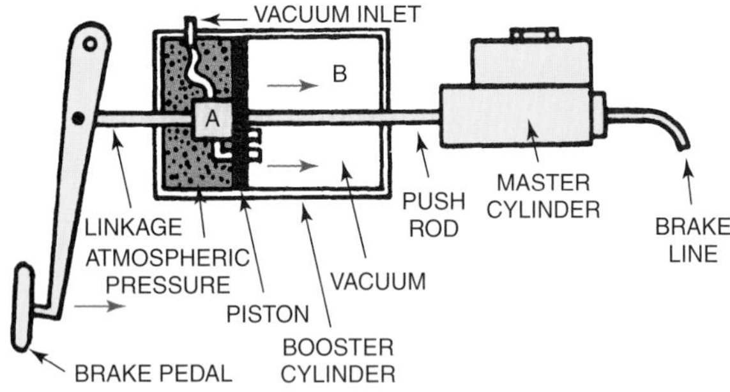

Figure 23-48. Schematic of a simplified power brake booster. Engine vacuum is admitted through the vacuum inlet. Control valve A can apply either vacuum or atmospheric pressure to the B section of the booster cylinder. When the brake pedal is being applied, the pedal linkage causes control valve A to cut off atmospheric pressure and to apply vacuum to the B side. Atmospheric pressure is forcing the piston and push rod to the right, which builds pressure in the master cylinder.

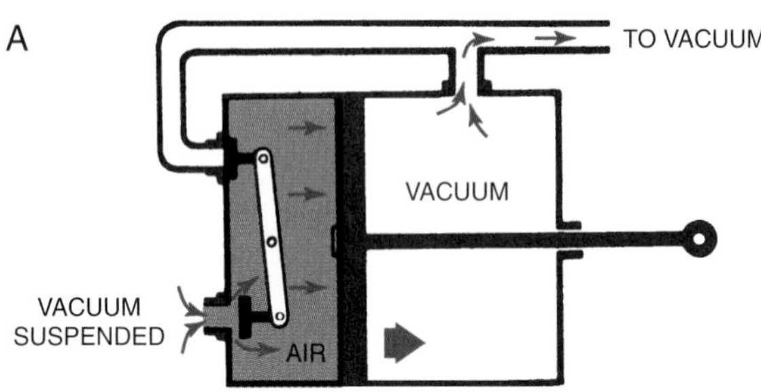

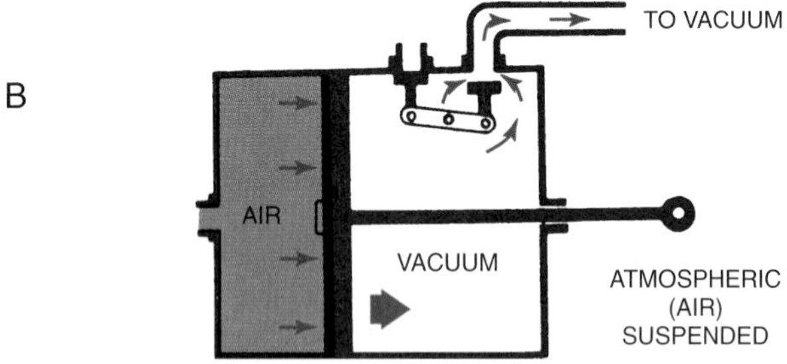

Figure 23-49. Power boosters. A–Vacuum suspended. B–Atmospheric (air) suspended type. Study the differences. (Wagner)

- Released position (brake pedal not being depressed).
- Applied position (brake pedal being depressed).
- Holding position (brake pedal with driver applying constant pressure).
- Releasing position (brake pedal being released).

These stages of operation are discussed in the following sections. Refer to the sectional view in **Figure 23-50**, and the operational cross-sections in **Figure 23-51.**

Released Position

With the engine running and the brake pedal fully released, the valve operating rod and plunger are moved to

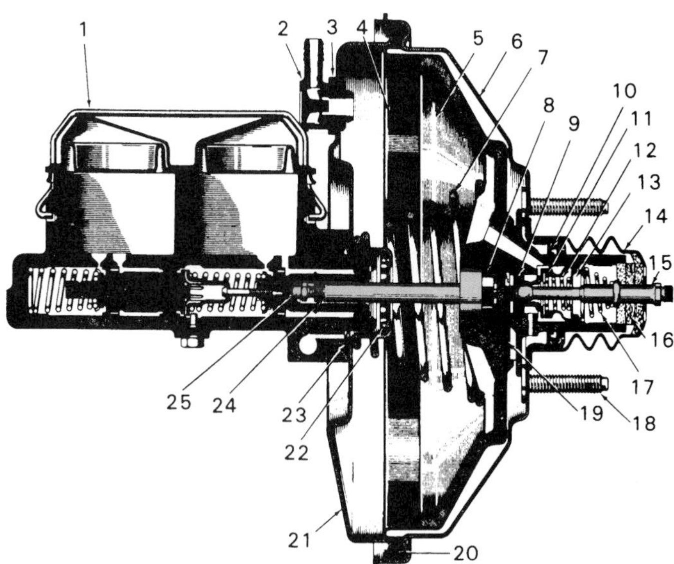

Figure 23-50. Cutaway of a typical vacuum power booster. 1—Master cylinder. 2—Vacuum check valve. 3—Grommet. 4—Diaphragm. 5—Diaphragm plate. 6—Rear shell. 7—Diaphragm spring. 8—Reaction plate. 9—Valve plunger. 10—Bearing seal. 11—Poppet valve. 12—Poppet valve spring. 13—Poppet retainer. 14—Dust boot. 15—Valve operating rod. 16—Filter and silencers. 17—Valve rod and plunger return spring. 18—Mounting stud. 19—Air valve lock plate. 20—Diaphragm. 21—Front shell. 22—Push rod seal. 23—Cylinder-to-shell seal. 24—Hydraulic push rod. 25—Adjusting screw. (GMC)

the right by the return spring, **Figure 23-51A**. This presses the right end of the valve plunger against the face of the poppet valve, closing off the atmospheric port. The vacuum port is open, which allows vacuum to form on both sides of the diaphragm plate. The left side of the diaphragm is subjected to constant vacuum, regardless of plunger position. With a vacuum on both sides, the diaphragm return spring moves the diaphragm plate to the right, releasing all pressure on the master cylinder.

Applied Position

When the driver depresses the brake pedal, the valve operating rod is moved to the left, **Figure 23-51B**. This causes the valve plunger to move to the left. As the plunger moves, it will compress the return spring and move the poppet valve into contact with the vacuum port seat in the valve housing. This closes the vacuum port leading to the right side of the diaphragm.

Further application of the brake pedal will cause the valve rod to force the valve plunger away from the poppet, opening the atmospheric port. As the port is opened, atmospheric air will rush into the right side of the diaphragm (control vacuum area) and force the diaphragm plate to the left. This applies pressure to the master cylinder push rod.

Providing Brake Feel

Note in **Figure 23-51A** how the rubber reaction disc is compressed when the brakes are applied. As the master

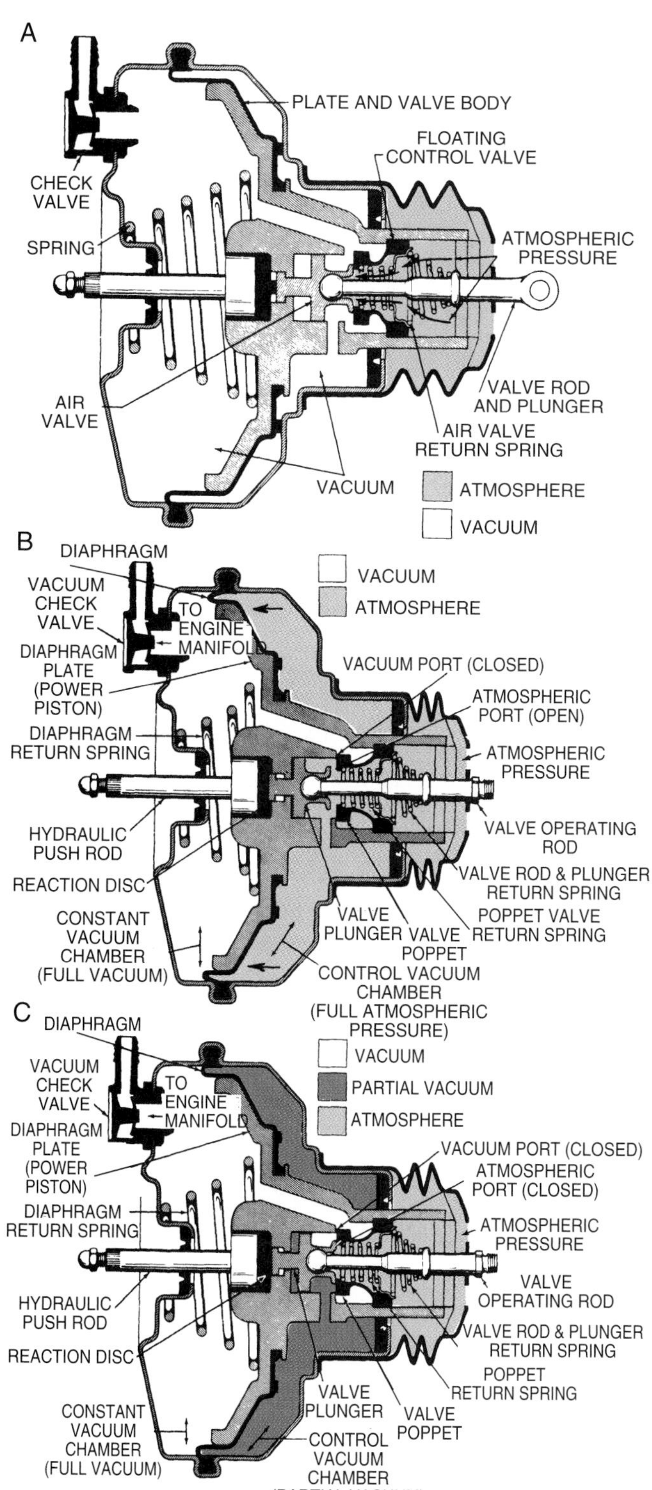

Figure 23-51. Typical vacuum power booster action. A–Released position. Vacuum is on both sides of the diaphragm. The spring moves the diaphragm to the right. Pressure is not applied to the master cylinder. B–Applied position. The poppet valve closes the vacuum port and the valve plunger moves to the left, opening the port. Atmospheric pressure forces the diaphragm plate toward the vacuum side. C–Holding position. Both vacuum and atmospheric ports are closed. This will retain the desired amount of braking. (GMC)

cylinder piston builds up hydraulic pressure in the brake system, a reaction pressure is transmitted back through the master cylinder push rod. The compressed rubber reaction disc transmits this pressure to both the diaphragm plate and the valve plunger. The pressure on the plunger valve tends to force it to the right (without moving the diaphragm plate), causing the valve to close off the atmospheric port.

The movement of the valve plunger is transmitted to the valve operating rod and on to the brake pedal. This provides the driver with ***brake feel*** to determine the amount of braking effort being exerted. This reaction force is directly proportional to pressure created within the master cylinder. Some power boosters use a reaction plate, levers, and diaphragms instead of the rubber disc. Note that the reaction disc in **Figure 23-51A** is not compressed and that a portion engages the valve plunger. Compare this with the compressed reaction disc shown in **Figure 23-51B**.

Vacuum Runout Point

The ***vacuum runout point*** occurs when vacuum has built up the maximum pressure possible on the power piston or diaphragm plate. If more pressure than full vacuum can produce is desired, additional pressure must be exerted by the driver on the pedal. This will hold the vacuum port wide open and transmit all of the driver's effort directly to the master cylinder piston. Wheel lockup (tires will stop revolving and slide) or anti-lock cycling will usually occur before the vacuum runout point is reached.

Holding Position

As long as the driver increases pedal pressure, the valve plunger will move to the left. When the driver exerts the desired amount of pressure, plunger movement will cease. The reaction force, transmitted to the reaction disc, will move the valve plunger (without moving the diaphragm plate) a slight amount to the right. This shuts off the atmospheric port. The vacuum port also remains shut.

When this balanced condition occurs, the right side of the diaphragm plate will be subjected to a specific and unchanging amount of atmospheric pressure. This maintains a constant pressure in the brake system. If more pedal force is applied, the atmospheric valve is reopened, causing the diaphragm plate to exert additional pressure on the master cylinder. It will maintain this pressure until pedal and reaction forces are again balanced.

Figure 23-51C shows a power booster in the holding position. Both vacuum and atmospheric ports are closed. Note how the reaction disc has moved the valve plunger a little to the right (in relation to diaphragm plate), as compared with the applied position in **Figure 23-51B**. This action closes the atmospheric port.

Released Position

If the pedal is released, the atmospheric port closes and the vacuum port opens, exposing both sides of the diaphragm plate to vacuum. As the force is removed from the plate, the diaphragm return spring will push the plate and valve assembly to the right of the booster cylinder.

Other Vacuum Booster Types

There are many variations in booster design, however, the basic operating principles are much the same. The paragraphs below briefly discuss some of the vacuum power booster designs used in the past, and on some vehicles today.

Tandem Booster

The ***tandem booster***, shown in **Figure 23-52**, uses two diaphragm plates to increase booster-to-master cylinder pressure. This has the effect of doubling the force of the vacuum in the same size booster. Tandem boosters are often found on larger cars, vans, and light trucks.

Vacuum Reservoir

As long as the engine is running, vacuum is available for the booster. When the engine stops, vacuum is gone. To provide several brake applications without engine vacuum, the booster is designed to maintain a ***vacuum reservoir***. Even though engine vacuum may be lost, enough vacuum will be retained in the booster body to provide one or more power-assisted stops.

Most modern vacuum booster systems have a check valve in the booster body itself. Carefully study the check valve arrangement in **Figure 23-52.** A separate reservoir, with a vacuum check valve, may be found on some cars and a few large trucks. One type is shown in **Figure 23-53.**

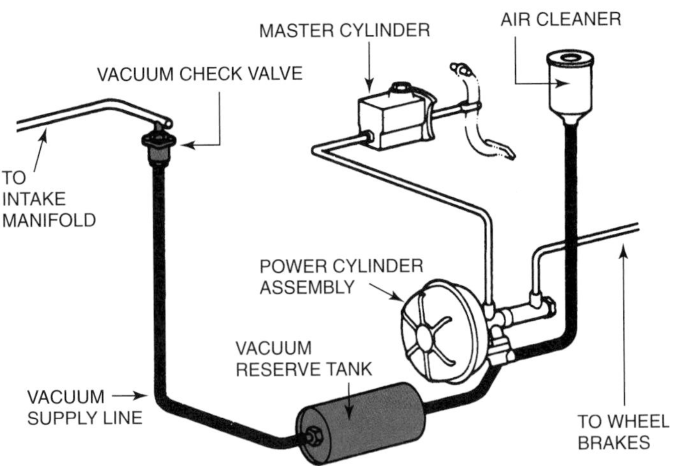

Figure 23-53. Vacuum reservoir tank. Reservoir allows one or two brake applications after engine vacuum is gone. (Wagner Div. of Cooper Industries)

Booster Failure

In the event the booster or the vacuum supply fails, the brakes may still be applied by foot pressure alone. It will require somewhat more pressure than when the booster is working, but the system will still function. Stopping distance will be greatly increased. Engine performance may also be affected if the failure is due to the loss of vacuum.

Hydraulic Pressure-Operated Power Booster

The ***hydraulic power booster*** (also called a Hydro-Boost) is used on some vehicles, usually those with diesel engines. The hydraulic booster is operated by hydraulic pressure developed by the power steering pump. Power steering

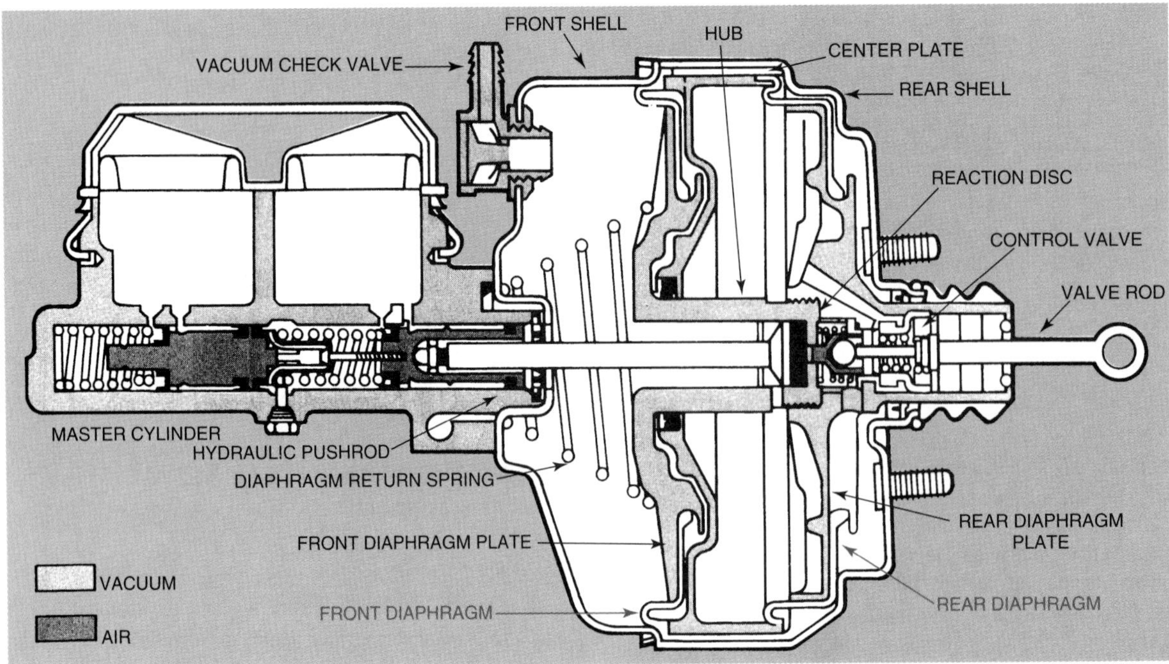

Figure 23-52. A cross-sectional view of a dual (tandem) vacuum power booster unit with attached master cylinder. Study the overall construction carefully. (Bendix)

pump pressure is delivered to the booster by high pressure hoses. Some heavy duty trucks employ an electro-hydraulic emergency pump as a backup source if the power steering pump becomes inoperative.

The hydraulic power booster system consists of a hydraulic pressure source and the hydraulic booster. **Figure 23-54** shows a typical pressure supply from the power steering pump. As with the vacuum booster, the hydraulic booster is coupled directly to the master cylinder. Movement of the brake pedal operates the booster valve, causing the booster to exert pressure on the master cylinder primary piston. The paragraphs below discuss the operation of the hydraulic booster in the released and applied positions.

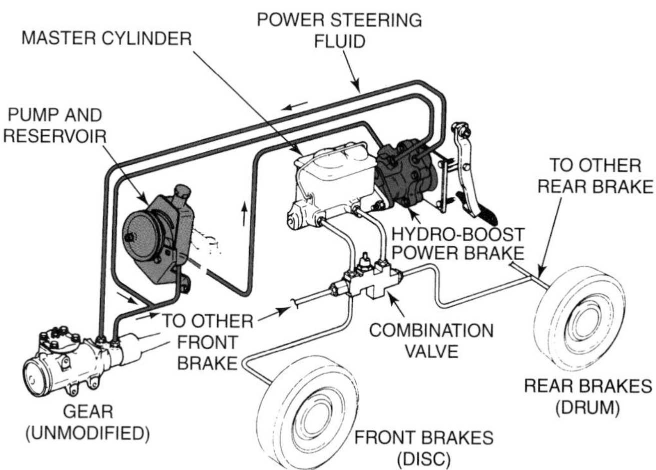

Figure 23-54. Hydraulic booster system. The pressurized fluid from the power steering pump is directed to the steering gear and then to the master cylinder. The brake and power steering fluid are separate from each other.

Hydraulic Power Booster Operation

In the fully released position, **Figure 23-55A**, power steering fluid flows through the booster power section on to the steering gearbox. No pressure is built up by the booster.

As the brake pedal is depressed, **Figure 23-55B**, the input rod and piston are forced forward. This action causes the lever to move the spool valve forward (towards the master cylinder), admitting more fluid into the space behind the power piston. As pressure builds, the power piston is forced forward, actuating the master cylinder.

The amount of pressure exerted by the booster (up to its maximum output) is proportional to the pressure exerted on the brake pedal. Brake feel is provided by fluid pressure generated in the space between the inboard end of the input rod and the power piston. This fluid pressure attempts to force the input rod back against pedal pressure. This gives the driver the needed pedal feel so that correct foot pressure may be exerted. Excess hydraulic fluid pressure is first vented to the accumulator; any remaining pressure goes back to the power steering pump reservoir.

Hydraulic Accumulator

All hydraulic ***accumulators*** are either spring-loaded or contain gas under pressure. The accumulator is filled with fluid by the power steering pump, and is pressurized whenever the brakes are applied. If the engine stalls or a problem occurs, such as a power steering pump failure or leak, or the drive belt breaks, the accumulator will contain enough pressurized fluid to provide between one and three power-assisted brake applications.

Even after the accumulator charge has been depleted, the brakes can still be operated by using pedal pressure only. As with all boosters that become inoperative, pedal resistance and stopping distance will be noticeably increased. **Figure 23-56** shows an exploded view of a hydraulic power booster unit. Note the accumulator.

STOP **Warning: Accumulators contain strong springs or pressurized gas and can fly apart with lethal force. Special training is recommended before servicing these units. Do not apply heat to an accumulator. For proper disposal, follow manufacturer's instructions.**

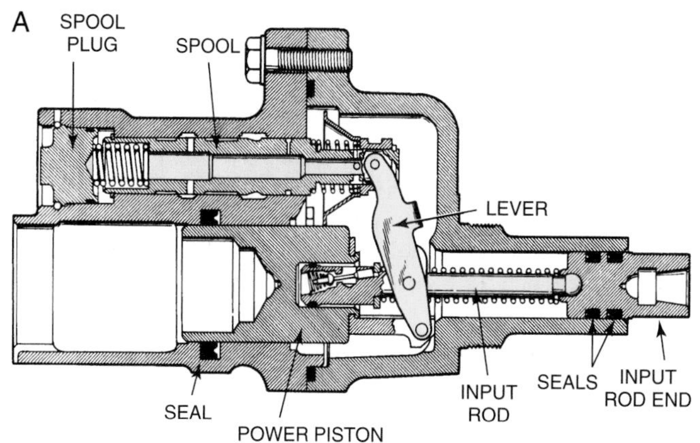

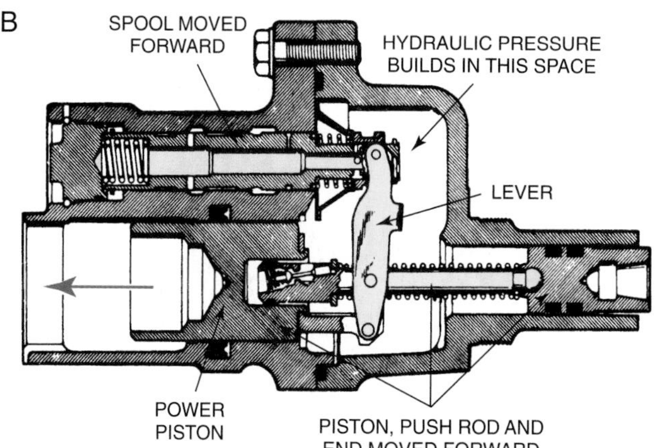

Figure 23-55. A—Cross-sectional view of one Hydro-Boost design. This booster is in the fully released position. B—Hydraulic booster in applied position. The lever has moved the spool forward, raising the fluid pressure behind the power piston. The piston moves forward to exert pressure on the master cylinder piston. (Bendix)

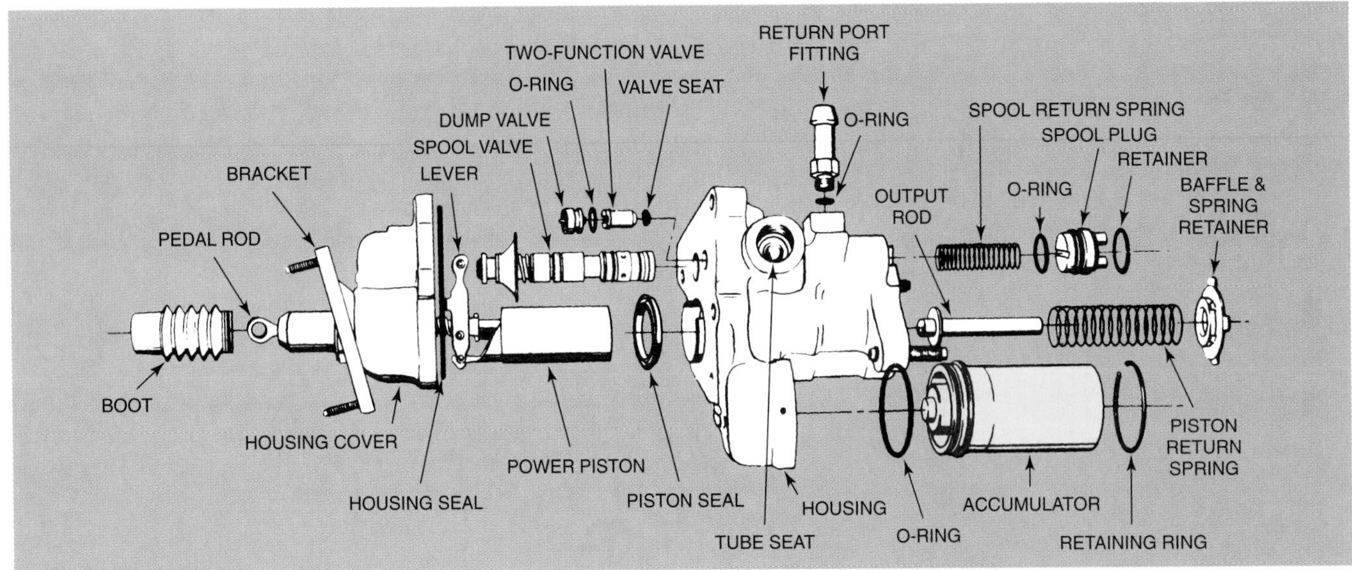

Figure 23-56. Exploded view of a hydraulic power booster. Note the accumulator that retains pressure for one or more brake applications in the event pump pressure is lost. (Chevrolet)

Anti-Lock Brake Systems (ABS)

Many vehicles are now being equipped with ***anti-lock brake systems***, usually called ***ABS***. Most ABS systems used on automobiles are similar to the one shown in **Figure 23-57** and control all four wheels. The ABS systems used on some light trucks operate only the rear wheels (rear-wheel anti-lock or RWAL), **Figure 23-58.** These systems, whether two- or four-wheel, use electronic and hydraulic components to help prevent wheel lockup during hard braking.

Anti-lock brakes allow the driver to maintain directional control while providing maximum braking efficiency. The ABS system does this by ***pulsing*** (applying and releasing) the brakes much more quickly than the driver could. This keeps any of the wheels from locking and causing a skid.

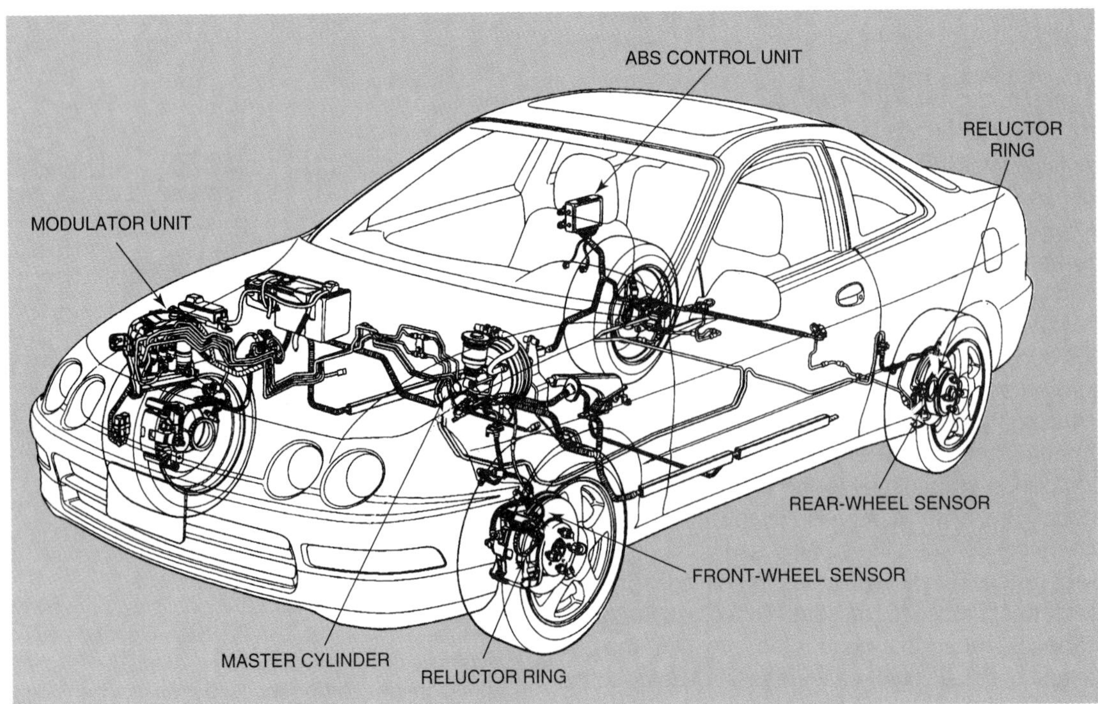

Figure 23-57. Overall view of a vehicle which is equipped with a four-wheel anti-lock braking system. (Honda)

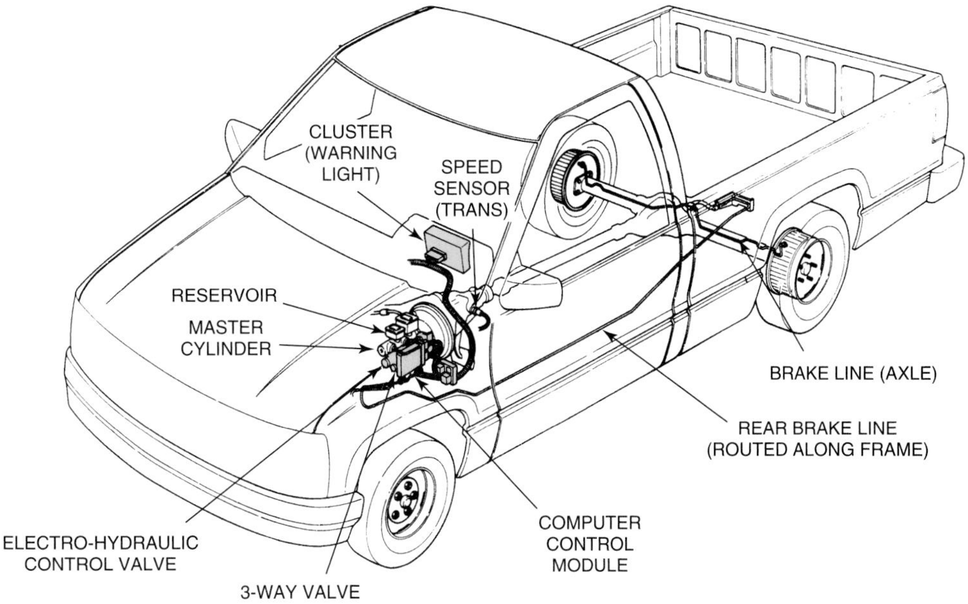

Figure 23-58. This pickup truck is equipped with a rear-wheel anti-lock (RWAL) braking system. A transmission mounted vehicle speed sensor is used in place of sensors being located at each rear wheel. (Chevrolet)

Anti-Lock Brake System Components

Anti-lock brake systems use an electronic control system to modify the operation of the brake hydraulic system. The electronic and hydraulic components work together to prevent wheel lockup during periods of hard braking. Most anti-lock brake systems, no matter who manufacturers them, contain several common components. These components, shown in **Figure 23-59**, include ***wheel speed sensors***, a ***control module***, and a ***hydraulic modulator***. Some vehicles also use G-force sensors and a pedal travel switch. Some anti-lock brake systems use an accumulator that is pressurized with brake fluid automatically by an electric pump. Most modern anti-lock brake systems have self-diagnostic capabilities.

Note that the brake friction components and most of the hydraulic components, such as wheel cylinders, caliper pistons, master cylinder, hydraulic lines, and power brake system components, are the same as those used on vehicles without ABS. When discussing ABS systems, the other friction and hydraulic components are referred to as the ***foundation brakes***, or the ***base brakes***.

Anti-Lock Brake Operation

Under normal braking conditions, the anti-lock portion of the brake system does not operate. The sensors continuously monitor wheel rotation and send signals to the anti-lock controller. When the brake pedal is pressed, fluid flows from the master cylinder, through the hydraulic modulator, and into the wheel cylinder or caliper, **Figure 23-60.** The wheel speed sensors monitor wheel speed to determine whether any of the wheels are ceasing to turn (locking).

If a wheel begins to lock up, the speed sensor reading alerts the ABS electronic control unit. When the ABS controller senses that a wheel is nearing lockup, it signals the appropriate solenoid valve in the hydraulic modulator to block the fluid passage between the master cylinder and the locking wheel assembly. When this occurs, pressure is trapped between the brake and the hydraulic modulator. Master cylinder fluid pressure cannot flow through the solenoid valve and the brake pressure at the affected wheel is held constant, **Figure 23-60.**

If the control module detects further lockup on the affected wheel, it will command the hydraulic modulator to decrease brake pressure to the appropriate wheel, **Figure 23-61.** To accomplish this, the solenoid valve in the modulator moves to cut off fluid pressure from the master cylinder and allows brake fluid at the caliper to flow back into the accumulator, master cylinder reservoir, or pump intake. When this occurs, pressure at the wheel is decreased.

With the piston type hydraulic modulator, the solenoid valves isolate the circuit from the master cylinder, and the pistons are moved to reduce pressure. **Figure 23-62** is a cutaway of one particular unit used to control fluid pressure to the front and rear brakes.

After reducing pressure, the wheel will accelerate. When the wheel accelerates to a preset level, the control module will signal the hydraulic modulator to increase pressure. This causes wheel deceleration and the cycle begins again.

When all the wheels are rotating normally, the solenoid valves in the modulator return to their original position and the foundation braking system takes over. At the same time, the modulator pump delivers any excess fluid back into the accumulator or master cylinder. Most anti-lock brake systems can repeat this cycle up to 15 times a second.

Traction Control Systems

To reduce wheel spin when accelerating on slippery surfaces, some vehicles are equipped with ***traction control systems***, also called acceleration slip regulation. These systems are able to reduce engine power and operate the brake

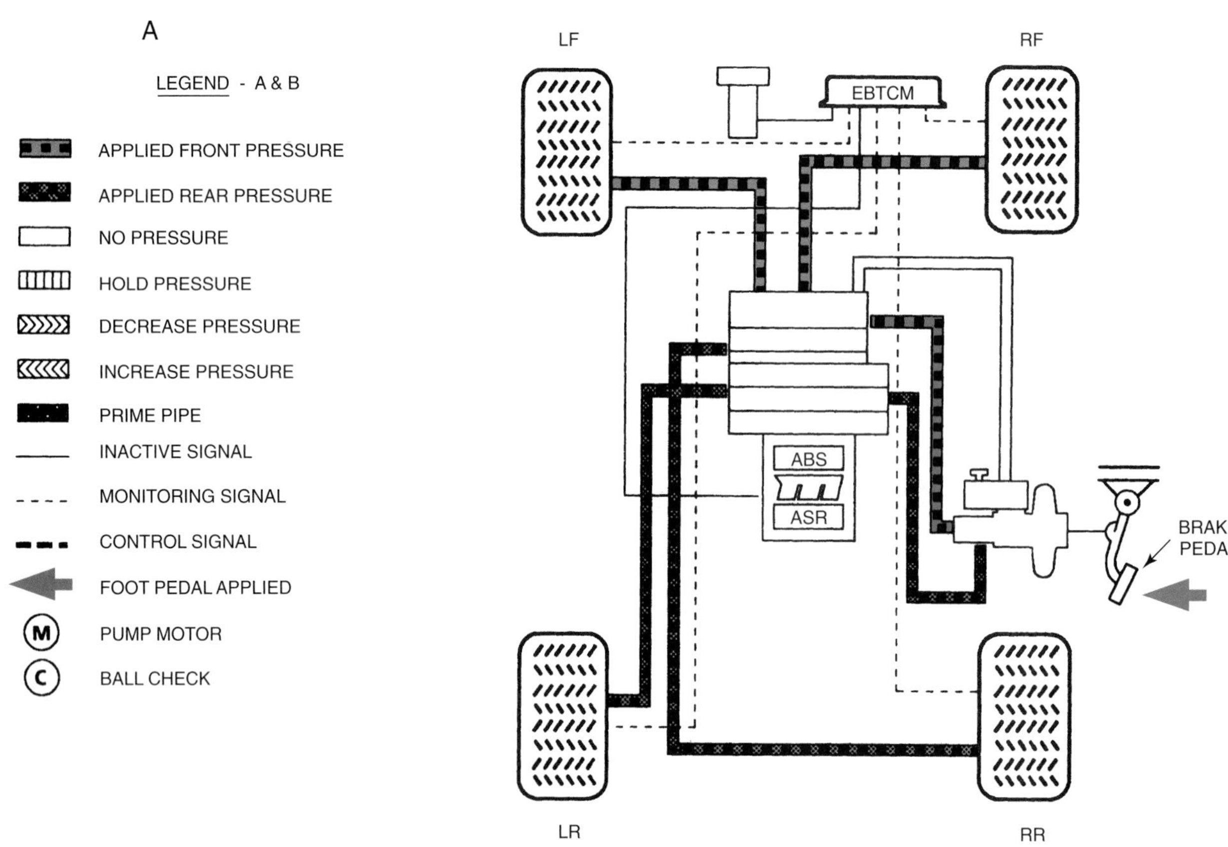

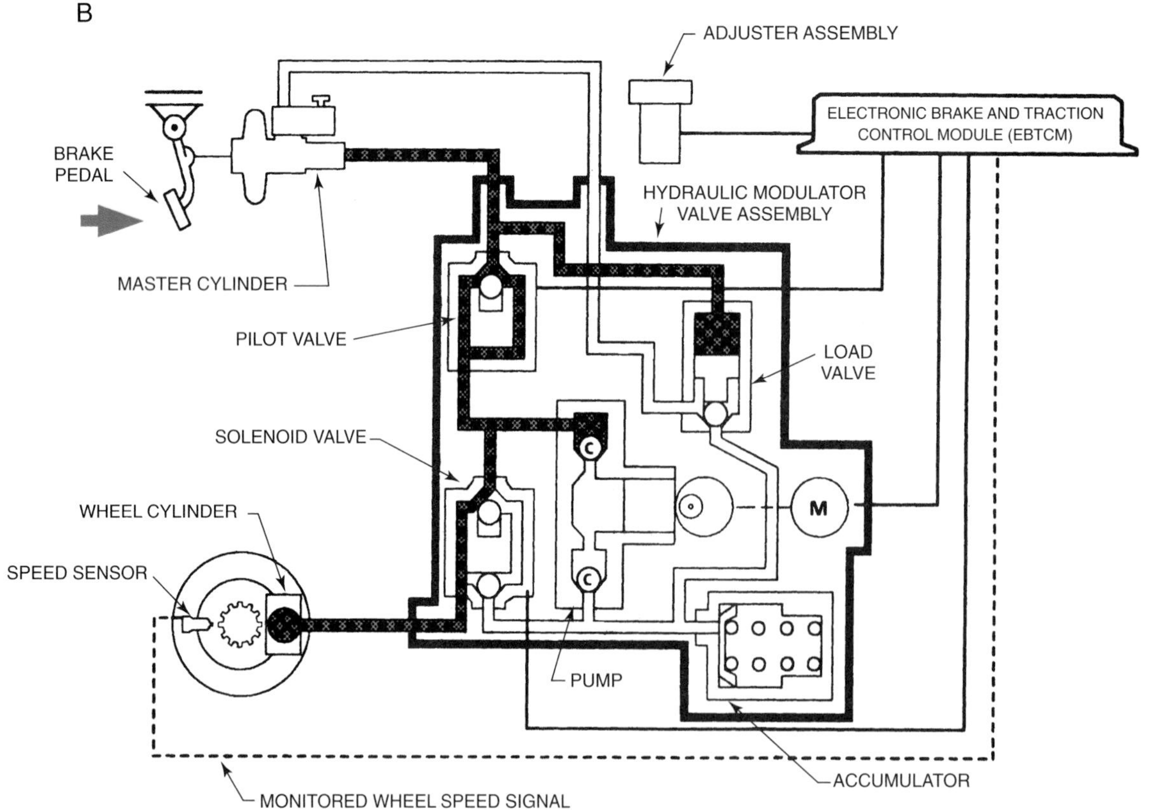

Figure 23-59. Schematics of four wheel anti-lock braking systems during normal (non-anti-lock mode) braking. A–Illustrates normal brake pressure traveling to the wheels. B–Fluid flow through the various valves of the system. Note that this system also includes an electronic traction control module. (Chevrolet)

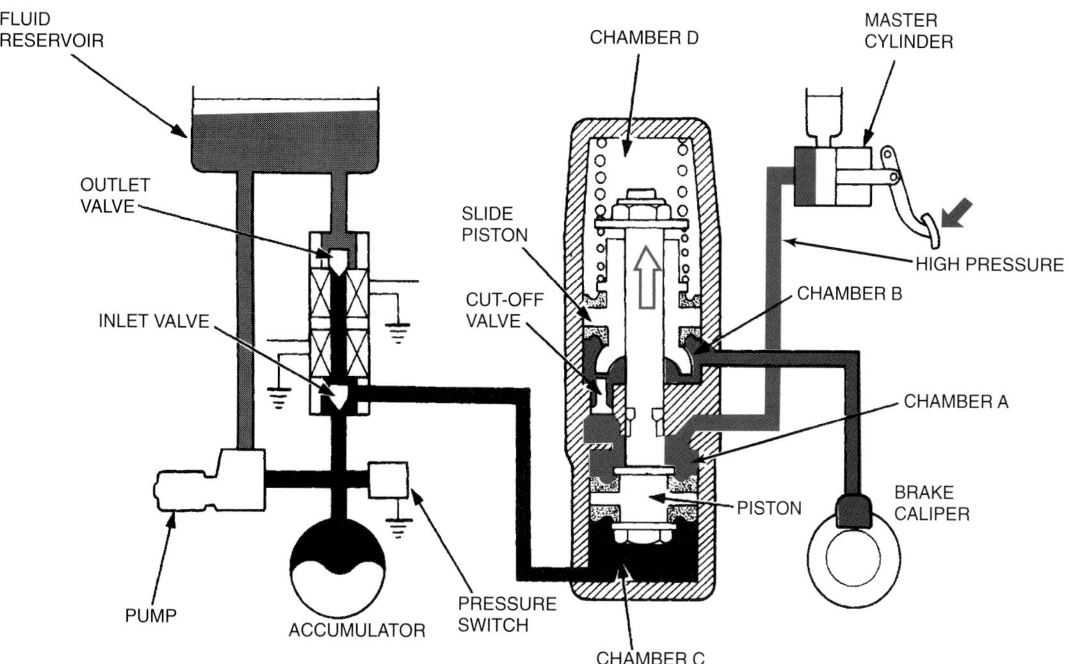

Figure 23-60. Schematic of an anti-lock brake system. When the brake pedal force is excessively heavy, wheel lockup can occur. When lockup is detected, the ABS control unit operates the solenoid valve, the outlet valve is closed, and the inlet valve is opened. High fluid pressure is sent into chamber C, and the piston is pushed up, causing the slide piston to travel upward and the cut-off valve to close. As the cut-off valve begins to close, fluid flow from the master cylinder to the brake caliper is interrupted. At the same time, the volume of chamber B increases and the fluid pressure in the brake caliper decreases. When both inlet and outlet valves are closed, caliper pressure is constant. When wheel lockup ceases, normal fluid pressure is resumed to the brake caliper. The outlet valve is open, the inlet valve is closed, and the solenoid valve is turned off. (Honda)

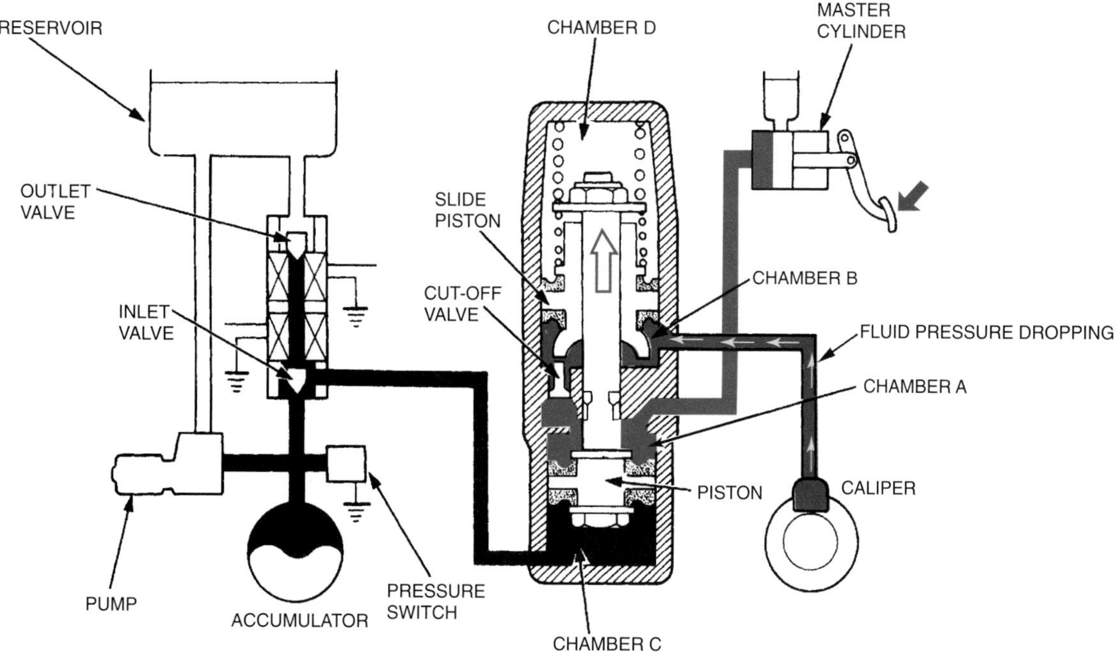

Figure 23-61. This schematic illustrates the brake fluid flow away from the brake caliper, decreasing braking ability of that wheel. This is part of normal ABS cycling. (Honda)

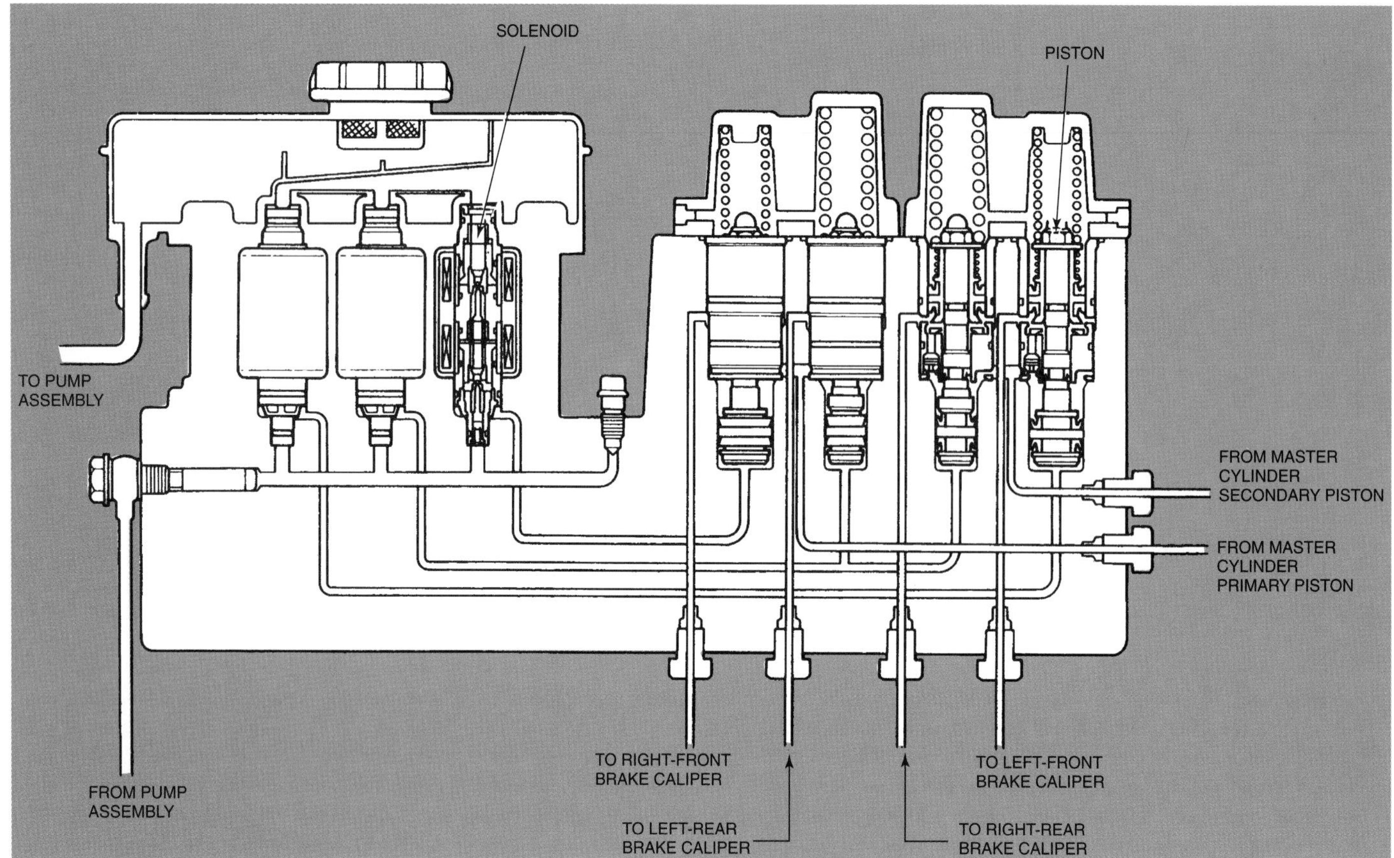

Figure 23-62. A cross-sectional view of a piston type modulator assembly which controls the fluid pressure to both the front and rear wheel brakes. (Honda)

system to increase vehicle acceleration and stability on wet, icy, or uneven road surfaces. Traction control systems also provide higher levels of cornering performance.

The traction control system can apply the brakes on the drive axles. On a two-wheel drive system, it will control only the driving wheels. On a full-time four-wheel drive system, it can apply any one of the four brakes. If the system detects one drive wheel spinning at a faster rate than the others, it will apply the appropriate amount of braking force to slow the wheel to the correct speed.

If the system determines that one or more drive wheels are spinning excessively, it can close the throttle or briefly retard ignition timing to prevent further spinning. Most vehicles with traction control also have an anti-lock brake system and are usually controlled by a single electronic control unit, **Figure 23-63.**

Brake Lights

The vehicle brake lights are operated either by a hydraulically operated switch placed somewhere in the brake line, or by a mechanically operated switch actuated by the brake pedal. The mechanical switch is the most common switch on newer vehicles. Brake lights use dual filament bulbs, one filament for the tail lights, and a brighter filament for the brake lights. Brake lights and switches are covered in more detail in Chapter 17.

Summary

Brake lining dust contains minute particles of asbestos, which can cause cancer. When cleaning brake assemblies, use care to avoid breathing the dust. Have proper ventilation. Have a dust disposal system and when needed, wear a suitable respirator.

Liquids, under confinement, can be used to transmit motion and to increase or decrease pressure. Air is compressible, liquids are not. Pascal's law states that when pressure is exerted on a confined liquid, it is transmitted undiminished throughout the liquid.

The brake system can be divided into two principal parts, the hydraulic system and the friction members. When the brake pedal is depressed, the master cylinder piston compresses brake fluid in the master cylinder. This causes fluid to move in the brake lines. This fluid movement will expand the wheel cylinders and calipers, causing them to apply the brake pads and shoes to the brake rotors and drums. Both movement and pressure are used in the hydraulic system.

Almost all vehicles use a double-piston or dual master cylinder, which provides a brake fluid source to the front and rear brakes. The front brakes generally provide somewhat more braking force than the rear brakes. This is to compensate for the transfer of weight to the front during stops. Hydraulic control valves are used to provide maximum braking power under all conditions. These valves are the metering

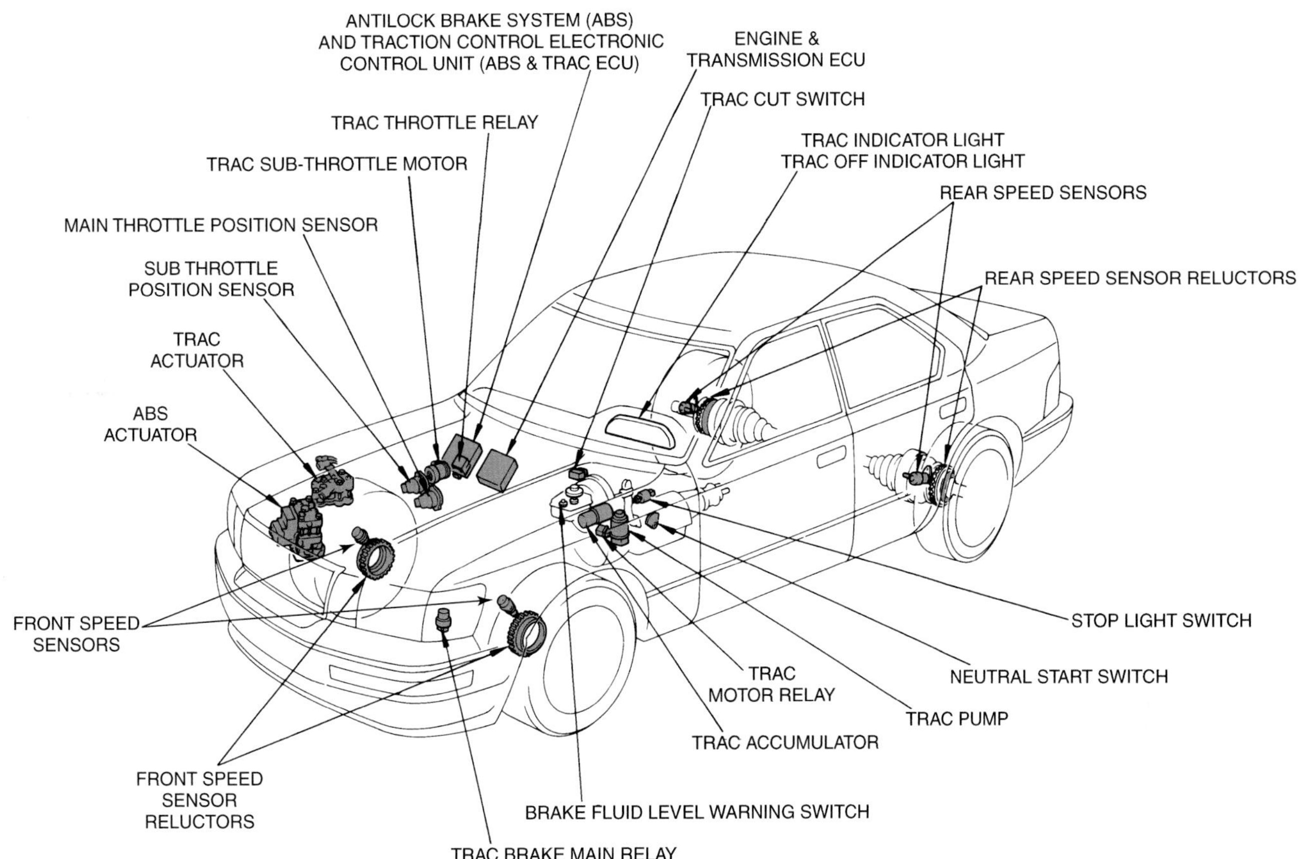

Figure 23-63. One particular traction control system and components. This traction control system manages the engine torque and braking of the drive wheels. Tire slippage (spinning) during takeoff on slippery road surfaces is carefully controlled. The system eliminates the need for the driver to control the accelerator pedal when starting off, turning, or accelerating on slick roads. The traction control and anti-lock brake electronic control units are combined into one control unit on this system. (Lexus)

valve, proportioning valve, and pressure differential switch. Two or more of these valves can be combined into a combination valve.

All air must be expelled from a hydraulic system. The removal of this air is termed bleeding. Never use any fluid other than high quality brake fluid in the system. Anything else will quickly ruin the rubber cups and seals, making the brakes inoperative.

Brake lines are made of double-walled steel tubing. Flexible hoses are used in both the front and in the rear. There are two types of wheel cylinders, single and double-piston.

The disc brake provides high resistance to both brake fade and grabbing. A caliper attached to the spindle or axle tightens and presses the brake pads against the rotor to stop it from turning. Disc brakes stay cooler, but require higher pedal pressure.

The drum brake assembly consists of a backing plate, brake shoes and springs, and a brake drum. The brake shoes are attached to the backing plate. The shoes are arranged so they may be expanded by the wheel cylinders. As the shoes expand, they will stop the brake drum and wheel. There is one primary shoe and one secondary shoe per wheel.

Some shoes are arranged so that servo action assists in applying the brakes. Brake shoe servo action is when one shoe helps apply the other. Self-energizing shoes are arranged so that shoe-to-drum friction helps apply the shoe. Some shoes are self-adjusting; others may be adjusted periodically to compensate for shoe lining wear.

There are two types of emergency or parking brakes in general use. One operates the regular service brake shoes on the rear axle, while the other operates rear brake pads on the rear axle. An older parking brake system uses a drive line brake.

Power brakes are used on almost all vehicles today. The power brake is essentially the regular brake system with a vacuum or hydraulic power booster added to reduce the braking effort by the driver. A vacuum reservoir is provided so that one or two brake applications can be made after the engine stops. Some systems used with diesel engines provide an engine- or electrically-driven pump to produce sufficient vacuum for brake application.

The diaphragm type vacuum power booster is in common use today. They may be either vacuum or atmospheric suspended. All work in essentially the same manner.

Atmospheric pressure can be applied to one side of the piston, upon application of the brake pedal. With atmospheric pressure on one side and vacuum on the other, the piston is forced to move in the direction of the vacuum. As it moves, it applies force either to the master cylinder piston or to the brake pedal itself.

A valve is used to control the admission of atmospheric pressure and the application of the vacuum. The booster works in four stages: released, being applied, holding, releasing. The power booster is designed to provide a braking "feel" for the driver, which is essential in controlling the amount of brake pressure. The vacuum runout point for a booster is when full vacuum is applied and the booster has exerted all possible pressure on the system.

Hydraulic power boosters are also used, especially on vehicles with diesel engines. They employ pressure from the power steering pump. The hydraulic booster system has an accumulator to provide reserve braking power if the hydraulic system fails.

Anti-lock brake systems use electronic and hydraulic components to help prevent wheel lockup during hard braking. The anti-lock brake system is a combination of electronic and hydraulic components which pulse the brakes quickly to prevent wheel lockup. Many anti-lock brake systems also provide for traction control to help with acceleration on slippery roads.

Know These Terms

Asbestos
Hydraulics
Hydraulic system
Pascal's Law
Hydraulic pressure
Brake fluid
Silicone brake fluid
Master cylinder
Reservoir
Low fluid warning switch
Pistons
Piston return springs
Primary cup
Secondary cup
Bleeder ports
Push rod
Dual brake system
Dual master cylinder
Diagonal split brake system
Quick take-up master cylinder
Quick take-up valve
Pumping
Proportioning valve
Metering valve
Pressure differential switch
Combination valve
Dual proportioning valve
Height sensing proportioning valve
Brake lines
Brake hoses
Disc brake caliper
Caliper pistons
Seal ring
Floating calipers
Wheel cylinder
Wheel cylinder pistons
Cups
Cup expanders
Dust boots
Single-piston cylinders
Stepped wheel cylinder
Bleeder screw
Friction members
Kinetic energy
Coefficient of friction
Semi-metallic
Brake fade
Rotor
Disc brake pads
Bonded
Riveted
Braking ratio
Backing plate
Drum brake assembly
Brake shoes
Primary brake shoe
Secondary brake shoe
Brake drum
Servo action
Self-energizing brakes
Self-adjusting brakes
Adjustment lever
Cable adjuster
Link adjuster
Parking brakes
Emergency brakes
Power brakes
Vacuum power brake booster
Vacuum control valve
Vacuum suspended booster
Atmospheric suspended booster
Brake feel
Vacuum runout point
Tandem booster
Vacuum reservoir
Hydraulic power booster
Accumulator
Anti-lock brake system (ABS)
Pulsing
Wheel speed sensors
Control module
Hydraulic modulator
Foundation brakes
Base brake
Traction control system

Review Questions—Chapter 23

Do not write in this text. Write your answers on a separate sheet of paper.

1. Technician A says that air is not compressible. Technician B says that liquids can transmit both pressure and motion. Who is right?
   (A) A only.
   (B) B only.
   (C) Both A & B.
   (D) Neither A nor B.
2. According to Pascal's law, the original pressure placed on a confined liquid is _________ throughout the liquid.
   (A) transferred with increased pressure
   (B) transferred with decreased pressure
   (C) transferred undiminished
   (D) transferred increase or decrease, depending on the shape of the container.
3. If a 100 lb. (46 kg) force is applied to a piston with 1 in^2 (6.5 mm^2)of area, how much pressure would a piston with an area of 10 in^2 (65 mm^2) exert, if both of these pistons rested with an airtight seal on a column of oil? Assume that both cylinders or columns are connected via a pipe.
4. How many pistons does the modern master cylinder have? _______
5. A dual master cylinder _____.
   (A) provides greater braking power
   (B) applies pressure to separate front and rear brake hydraulic systems
   (C) works the clutch also
   (D) has two cylinders, one on top of the other
6. The master cylinder reservoir must be filled to within about _____ of the reservoir cover or covers.

7. Technician A says that the reservoir cover must be vented. Technician B says that the reservoir cover must always be removed to check the fluid level. Who is right?
   (A) A only.
   (B) B only.
   (C) Both A & B.
   (D) Neither A nor B.
8. The reservoir provides extra fluid to _____.
   (A) replace losses due to seepage
   (B) provide extra fluid to compensate for lining wear
   (C) absorb heat from the master cylinder
   (D) Both A & B.
9. Brake lines must be made of _____.
   (A) double-walled steel
   (B) flared copper
   (C) double-walled brass
   (D) Any of the above.
10. Brake lines, where connections are made, must have a _____ flare.
11. Name three parts of a wheel cylinder.
12. Technician A says that a small amount of air in the brake system is permissible. Technician B says that automatic transmission fluid is a satisfactory substitute for brake fluid. Who is right?
   (A) A only.
   (B) B only.
   (C) Both A & B.
   (D) Neither A nor B.
13. Which of the following valves controls pressure to the rear wheels?
   (A) Proportioning valve.
   (B) Master cylinder check valve.
   (C) Pressure differential switch.
   (D) Metering valve.
14. Which of the following valves warns the driver that one side of the split hydraulic system has failed?
   (A) Proportioning valve.
   (B) Master cylinder check valve.
   (C) Pressure differential switch.
   (D) Metering valve.
15. Wheel cylinders are fastened to the _____.
   (A) brake drum
   (B) backing plate
   (C) spindle
   (D) All of the above.
16. How many single wheel cylinders would be used on a single brake drum assembly?
17. The secondary brake shoe faces _____.
   (A) the back of the vehicle
   (B) the front of the vehicle
   (C) to the front on the left side, to the rear on the right side
   (D) in either direction, depending on the manufacturer
18. The brake shoe configuration (design) commonly used today is the _____.
   (A) duo-servo, fixed anchor, self-adjusting
   (B) non-servo, single adjustable anchor
   (C) self-energizing, duo-anchor, self-centering
   (D) single servo
19. Technician A says that front and rear brake shoe return springs push the wheel cylinder pistons inward when hydraulic pressure is removed. Technician B says that the brakes produce a lot of heat and the brake drums and rotors should be thick to absorb it. Who is right?
   (A) A only.
   (B) B only.
   (C) Both A & B.
   (D) Neither A nor B.
20. All of the following statements are true, EXCEPT:
   (A) the backing plate holds the wheel cylinder and brake shoes in position.
   (B) the backing plate keeps water out of the shoe and drum assembly.
   (C) the caliper is the mounting surface for the rotor.
   (D) the caliper holds the pads in position.
21. The _________ valve closes off pressure to the front disc brakes until the hydraulic system pressure can overcome rear brake shoe retracting spring pressure.
22. Explain servo action.
23. What is meant by self-energizing?
24. Brake shoe self adjusters are used on _____ vehicles.
   (A) very old
   (B) off-road
   (C) disc brake
   (D) all modern
25. Why do disc brakes require higher hydraulic pressures as compared to drum brakes?
26. A type of parking brake that is not used much any more is the _____ brake.
   (A) rear drum
   (B) rear disc
   (C) transmission
   (D) None of the above.
27. The vacuum power brake requires a vacuum source other than the intake manifold on ________ engines.

28. The vacuum power brake vacuum control valve is operated by _____.
    (A) manifold vacuum
    (B) engine speed
    (C) master cylinder hydraulic pressure
    (D) movement of the brake pedal
29. What does vacuum suspended mean when referring to the power booster?
30. What type of power booster is in popular use today?
    (A) Bellows.
    (B) Diaphragm.
    (C) Piston.
    (D) All of the above.
31. A _______________ prevents loss of vacuum when the engine stops running.
32. The hydraulic brake booster utilizes pressure generated by the _____ system.
33. The hydraulic brake booster uses a _____ to provide reserve pressure in case the regular hydraulic system fails.
    (A) motor driven pump
    (B) vacuum pump
    (C) accumulator
    (D) vacuum back-up booster
34. Name the three main components of an anti-lock braking system.
35. Most anti-lock brake systems have ____________ capabilities.

The modern engine compartment is tightly packed, however fluid fill and check points are clearly marked. Can you find all of the fluid fill points for this engine? (Ford Motor Co.)

24

Suspension Systems

After studying this chapter, you will be able to:

- Identify the purpose of the suspension system.
- Explain the difference between sprung and unsprung weight.
- Name the different types of vehicle frames.
- Describe the function and operation of chassis springs.
- Describe the function of control arms and struts.
- Describe the function and operation of ball joints.
- Describe the function and operation of shock absorbers.
- Explain the differences between conventional and MacPherson strut suspensions.
- Explain the operation of front suspension components.
- Explain the operation of rear suspension components.
- Explain the operation of automatic level controls.

This chapter will cover the principles and components of the suspension system. There are many variations in the design of modern suspension systems. In addition, there are many variations between front and rear suspensions. However, they all operate on basic principles and have the same basic purpose—reducing road shock transferred to the passenger compartment and improving handling. After studying this chapter, you will be familiar with the purposes and designs of modern suspension systems.

The Need for a Suspension System

If a vehicle's axles were bolted directly to its frame or body, every rough spot in the road would transmit a jarring force throughout the vehicle. Riding would be uncomfortable, and handling at freeway speeds would be impossible. The fact that the modern vehicle rides and handles well is a direct result of a ***suspension system***.

Even though the tires and wheels must follow the road contour, the body should be influenced as little as possible. The purpose of any suspension system is to allow the body of the vehicle to travel forward with a minimum amount of up-and-down movement. The suspension should also permit the vehicle to make turns without excessive body roll or tire skidding. See **Figure 24-1.**

Sprung Weight

Sprung weight refers to the weight of all vehicle parts that are supported by the suspension system. The frame, engine, drive train, body, and all parts mounted on the body make up the sprung weight.

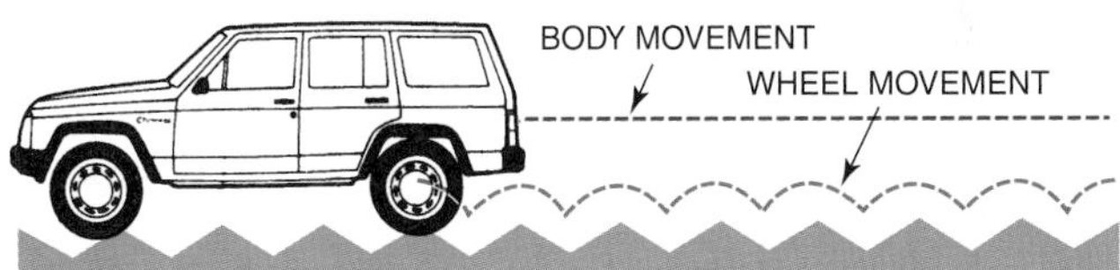

Figure 24-1. The vehicle body must be insulated against road shock. The blue dotted line shows how the vehicle body moves forward smoothly despite the rapid up-and-down motion of the suspension and wheels.

Unsprung Weight

Unsprung weight refers to all parts of the vehicle that are not supported by the suspension system, such as the wheels, tires, steering knuckles, and rear axle. The ideal situation is to have as little unsprung weight as possible in relation to the sprung weight. The unsprung weight will then have a minimal effect on the riding characteristics of the vehicle.

Suspension System Components

The following sections cover the various parts of the modern suspension system and discusses the relationship between them.

Vehicle Frames

A vehicle's frame or body must form a rigid structural foundation and provide solid anchorage points for the suspension system. There are two types of vehicle construction in common use today: ***body-over-frame construction***, **Figure 24-2,** which uses a separate steel frame to which the body is bolted at various points and ***unibody construction,*** **Figure 24-3,** in which the body sections serve as structural members. Unibody construction is the most common, but

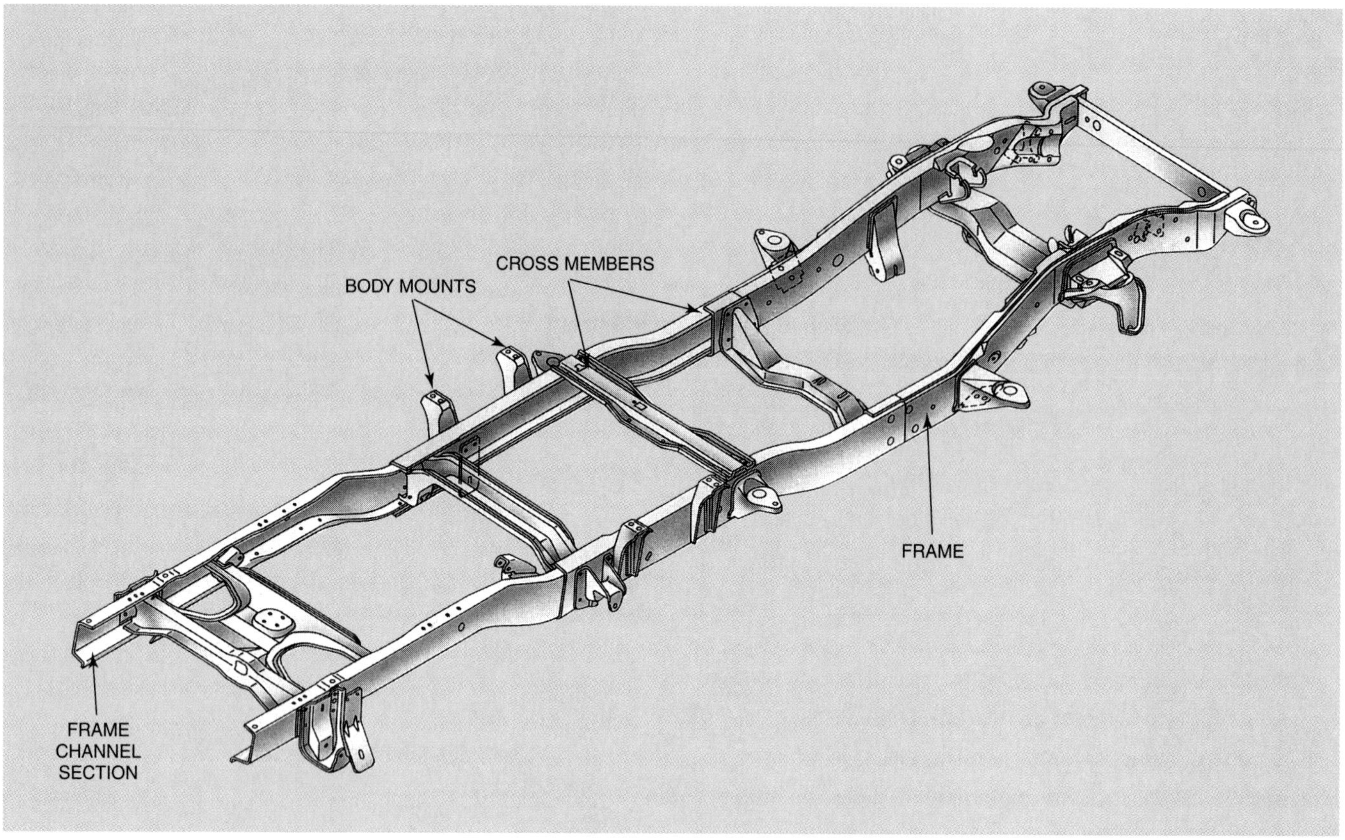

CROSS MEMBERS
BODY MOUNTS
FRAME
FRAME CHANNEL SECTION

Figure 24-2. A steel pickup truck channel type frame which is welded and riveted together. They may be coated with paint and/or wax to resist corrosion. (Dodge)

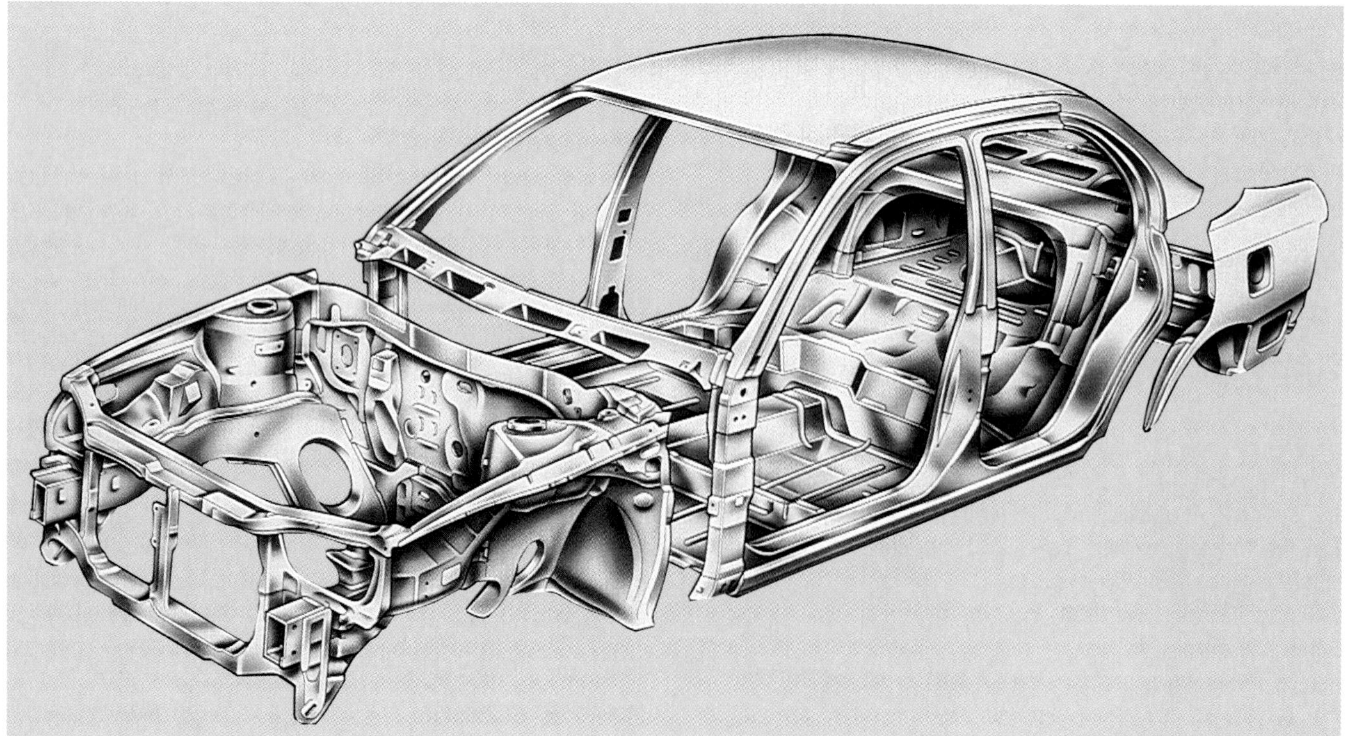

Figure 24-3. Unibody construction. By using the body sections as strength members, great rigidity is obtained. This type of body construction generally does not use a frame. All the sections are welded together. (Chevrolet)

body-over-frame construction is still used on pickup trucks and large cars.

Springs

The ***springs*** are the most obvious part of the suspension system. Every vehicle has a spring of some kind between the frame or body and the axles. There are three types of springs in general use today: leaf, coil, and torsion bar. Two different types of springs can be used on one vehicle. Air springs were once used in place of the other types of springs, but are now obsolete. Many modern vehicles have air-operated suspensions, but they are used to supplement the springs.

Leaf Springs

Leaf springs are usually used on the rear axles of large cars and pickup trucks. The most common leaf spring configuration consists of a number of flat steel leaves of varying lengths that are bolted together to form a single unit. This design is called the semi-elliptical leaf spring.

The spring is fastened to the axle by U-bolts, **Figure 24-4.** Notice the use of rubber isolators to eliminate spring vibration transfer. The ends of the spring are shackled to the frame. The ends of the top leaf, or main leaf, are curled around to form the spring eyes. See **Figure 24-5.** On some springs, the ends of the second leaf are curled around the eye to provide additional strength. The spring center bolt passes through the spring, keeping the leaves properly spaced in a lengthwise position. The top of the center bolt fits into a depression in the mounting pad, pinning the spring firmly to the axle mounting pad when the U-bolts are tightened.

When the spring is bent downward, the leaves slide against each other. The same action takes place on the rebound (springs snapping back to their original shape). To prevent squeaks and to ease spring action, neoprene or nylon pads are often used between the ends of the leaves. The springs then flex these small rubber pads. Several rebound clips are placed around the spring leaves at various points. These clips prevent the leaves from separating when the spring rebounds upward. Note the rebound clips and the leaf rubber pads shown in **Figure 24-5.**

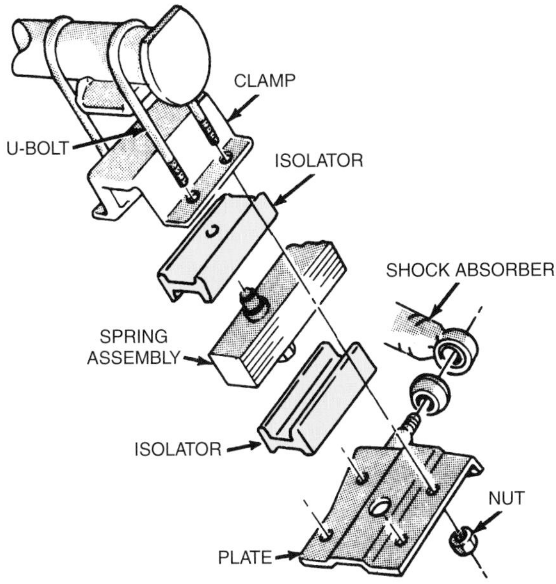

Figure 24-4. Spring to axle housing mounting. Note the use of rubber isolator blocks. (Dodge)

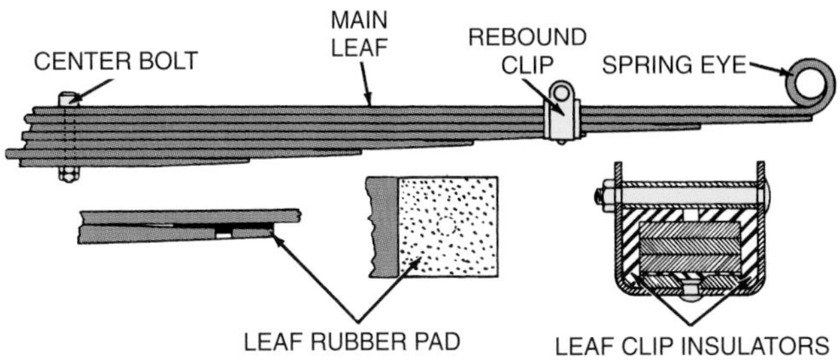

Figure 24-5. Partial view of a semi-elliptical leaf spring. The center bolt keeps the leaves in line and properly spaced.

Spring Shackles

The leaf spring is held in place by two ***spring shackles.*** One end of the spring is mounted on a pivot shackle that will not allow fore-and-aft movement. The other end of the spring is mounted on a hinge-type shackle that allows the spring to shorten and lengthen as it flexes up and down. See **Figure 24-6.**

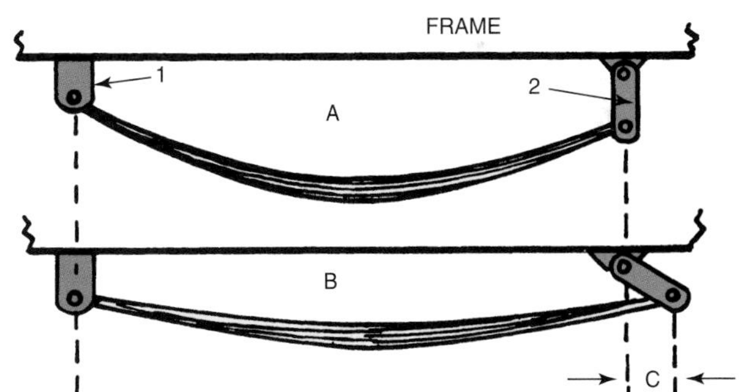

Figure 24-6. Flexing causes a variation in spring length. 1—Stationary spring hanger. 2—Hinge-type shackle. In A, normal load may produce this spring shape. In B, the spring has been compressed. End 1 has merely pivoted, but end 2 has moved out as the spring lengthens. The difference in length is shown at C.

Modern spring shackles use rubber bushings between the spring eye and the spring eye bolt. See **Figure 24-7.** The spring and bolt do not touch. The pivoting motion is absorbed by the flexible rubber bushing. The bushings require no lubrication. A complete semi-elliptical spring is shown in **Figure 24-8.** The left end pivots in a spring hanger. The right end is supported on a hinge shackle. Rubber bushings are used in both ends.

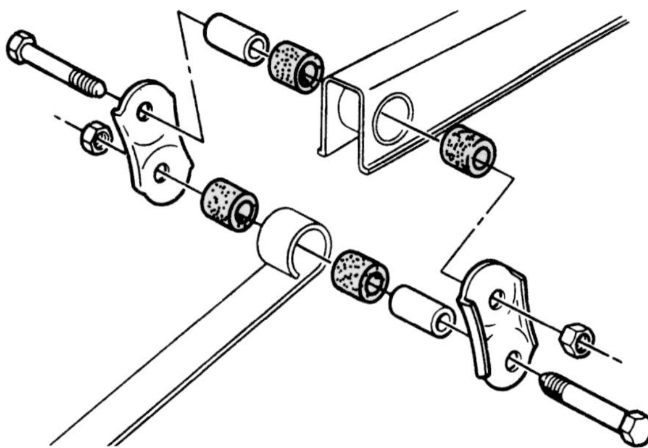

Figure 24-7. Rubber-bushed spring shackle. Bushings are in two parts. A metal sleeve fits inside the bushings and a shackle bolt passes through this sleeve. When the shackle bolts are tightened, the rubber bushings are somewhat compressed, causing them to expand and grip both the spring eye and the bushing sleeve tightly. All movement is taken within the rubber bushings.

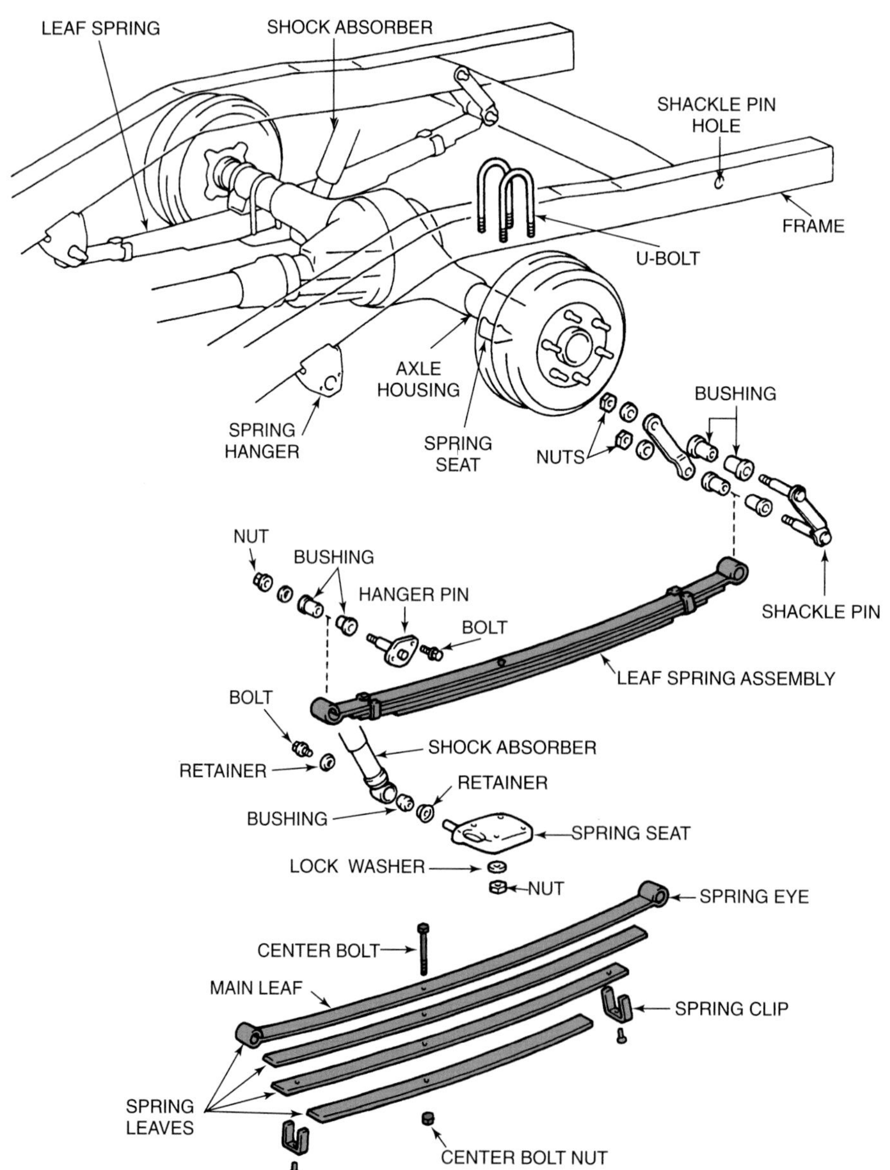

Figure 24-8. A semi-elliptical leaf spring rear suspension. The center bolt holds the leaves in line and properly spaced. (Toyota)

Single Leaf Spring

A single-leaf, semi-elliptical spring is shown in **Figure 24-9.** The spring is long and wide, is made of steel or fiberglass, and is relatively thin. The entire spring consists of one leaf.

Figure 24-9. Single leaf spring. Long, wide, and quite thin. Entire spring is made up of one leaf.

Coil Springs

Coil springs are used in many modern vehicles as part of conventional or MacPherson strut suspensions. They can be installed at the front or rear axle. Coil springs are made of special spring steel rods that are heated and wound in the shape of a spiral coil. Coil springs are carefully tempered to achieve the proper tension.

One end of the coil spring is secured to the frame or body, and the other is fastened to the axle or another suspension device. Pads are used on the ends to prevent vibration transfer and squeaks. See **Figure 24-10.**

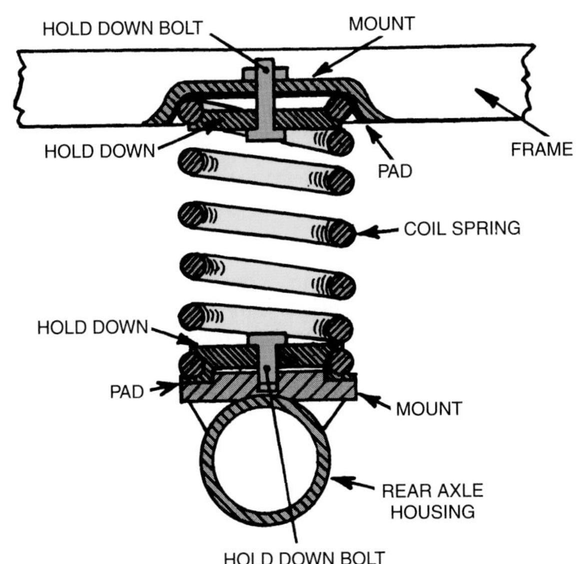

Figure 24-10. Coil spring. Each end is bolted to mount. Rubber or fabric pads are used at ends.

Torsion Bar

The ***torsion bar*** is a long spring-steel bar. One end is mounted to the vehicle's frame or body. The other end is free to turn. A lever arm, which is at right angles to the bar, is secured to the free end of the bar. This lever arm is fastened to the suspension device. When the wheel strikes a bump and the lever arm is pushed upward, it will twist the bar. The bar will have a resistance to twisting that has been carefully calibrated to the load forces it is designed to bear. **Figure 24-11** shows a typical torsion bar. **Figure 24-12** illustrates the operation of a simplified front torsion bar suspension.

Shock Absorbers

When the vehicle is traveling forward on a level surface and the wheels strike a bump, the spring is rapidly compressed (coil springs) or twisted (leaf springs and torsion bars). The spring will attempt to return to its normal loaded length. In so doing, it will rebound, causing the body of the vehicle to be lifted. Since the spring has stored energy, it will rebound past its normal length. The upward movement of the vehicle also assists in rebounding past the spring's normal length.

The weight of the vehicle then pushes the spring down after the spring rebounds. The weight of the vehicle will push the spring down, but since the vehicle is traveling downward, the energy built up by the descending body will push the spring below its normal loaded height. This causes the spring to rebound again. This process, called ***spring oscillation***, gradually diminishes until the vehicle is finally still. **Figure 24-13.** Spring oscillation can affect handling and ride quality and must be controlled.

Shock Absorber Control Action

To control spring oscillation, a dampening device called a ***shock absorber*** is used. Many types of shock absorbers have been used in the past. Modern vehicles use the ***telescoping shock absorber***. See **Figure 24-14.** One end of the shock is mounted to the frame, the other end is attached to the axle. Rubber bushings are used to fasten the shock. When the spring attempts to compress or rebound, its action is hindered by the shock absorber.

As the body attempts to rise and fall in relation to the axle, the shock must telescope out and in. The shock's resistance to the telescopic movement hinders this up-and-down motion. Instead of a long series of uncontrolled oscillations, spring action is smoothed out, and the vehicle will quickly return to its normal loaded level.

Shock Absorber Construction

The telescoping shock absorber consists of an inner cylinder, an outer cylinder, a piston and piston rod, and, in some cases, an outer dust and rock shield. A series of valves in the piston and at the bottom of the inner cylinder control the movement of the hydraulic fluid within the shock.

Shock Absorber Operation

Figure 24-15 illustrates a typical telescoping shock absorber in three positions. The left-hand illustration shows the shock at rest with the vehicle at normal loaded height. Notice that the inner cylinder (pressure tube) is completely

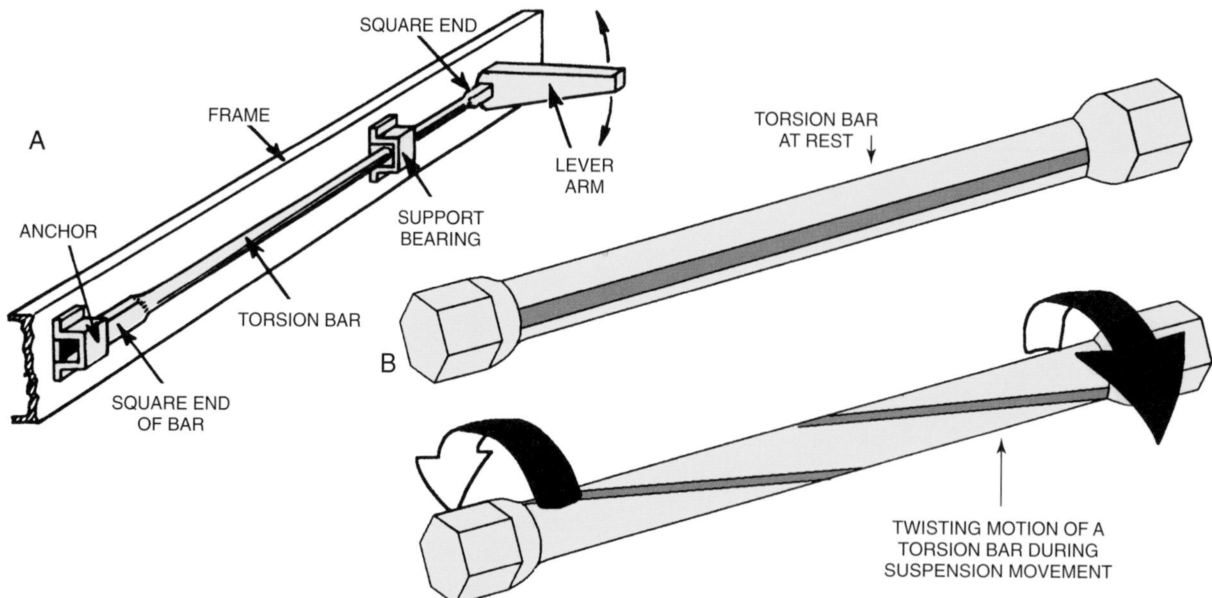

Figure 24-11. Operation of a longitudinal (parallel with vehicle) torsion bar. A—A simple torsion bar mounted to a vehicle frame. B—Torsion bar at rest and bar movement (twist) from the vehicle suspension jounce and/or rebound. The elasticity of the spring steel tries to return the torsion bar to its rest position. (FMC Corp.)

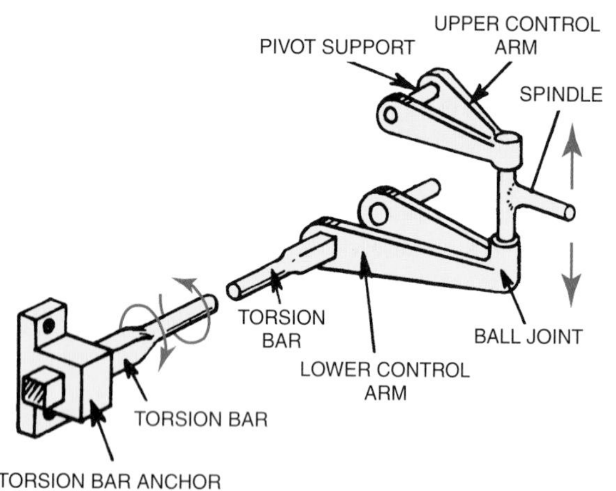

Figure 24-12. Simplified torsion bar front suspension. When the wheel spindle moves up or down, the torsion bar must be twisted. The torsion bar anchor is bolted to frame. Both lower control arm support and upper control arm pivot support are fastened to the frame by using support brackets.

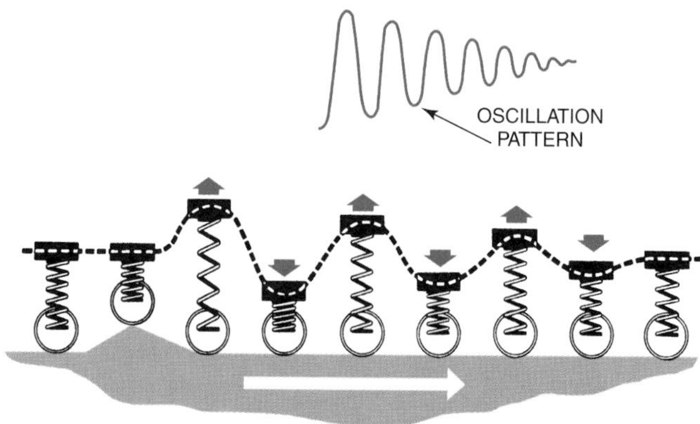

Figure 24-13. Spring oscillation. When the wheel strikes a bump, the spring is compressed. Even though the wheel returns to the level roadway, the spring oscillates. Oscillation gradually diminishes and finally stops. (Ford)

filled with fluid, while the outer cylinder (reservoir tube) is only partially filled. The piston is near the midpoint of its travel in order to allow up-and-down movement.

The middle illustration, **Figure 24-15,** shows the action of the shock during spring compression. Note that the piston and piston rod have moved down in the inner cylinder, forcing the fluid to pass through the piston valve into the upper section of the cylinder. Since the piston rod occupies space formerly filled with fluid, some fluid must be displaced through the base valve into the outer cylinder. The valves are calibrated to cause a certain resistance to the passage of fluid. Although the fluid passes through the valves quite easily when the shock is compressed slowly, any increase in compression speed causes a rapid buildup of resistance. The shock thus cushions the violent force of the spring compression.

The right-hand illustration, **Figure 24-15,** shows shock action during spring rebound. The piston is forced upward in the inner cylinder and the fluid trapped above the piston

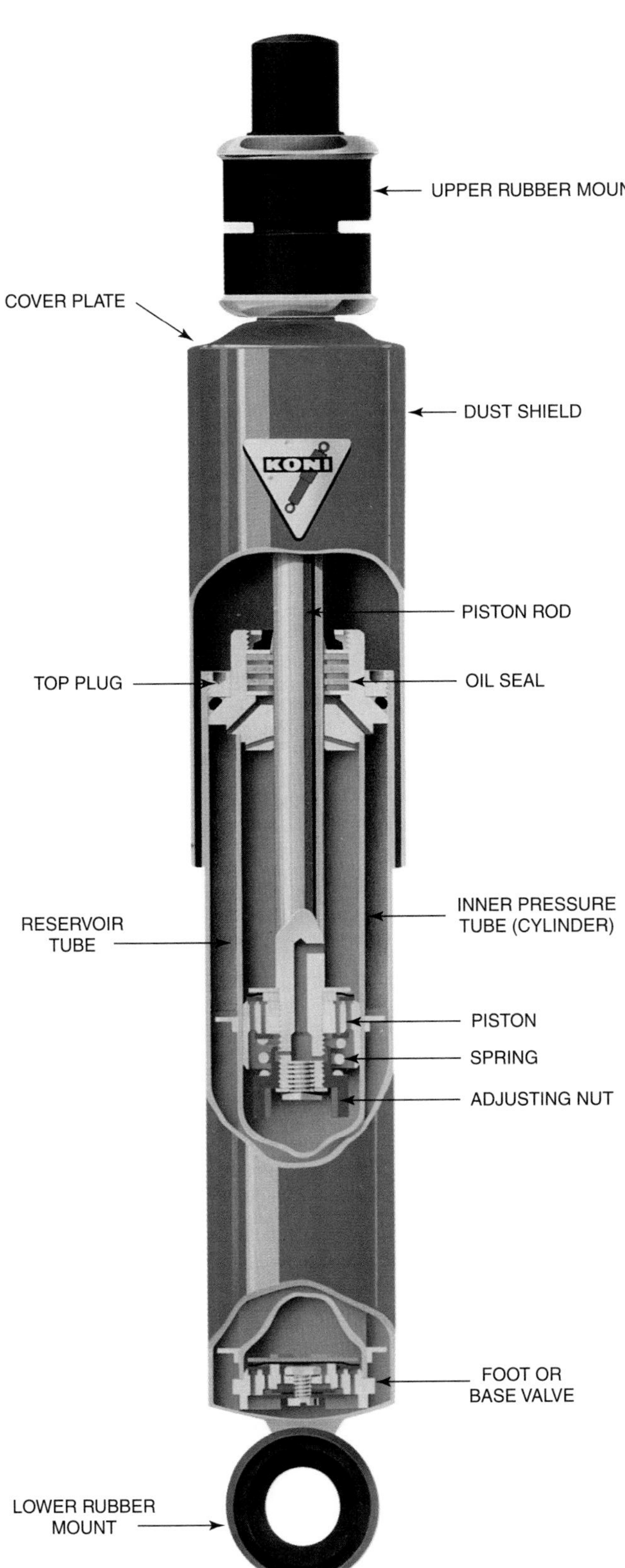

Figure 24-14. A cutaway of a typical shock absorber showing construction. (Koni)

moves down through the piston via a different valve. To compensate for the reduced amount of piston rod now in the pressure tube, additional fluid is pulled in from the outer cylinder through the base valve. Some shock valves are calibrated to provide more resistance on rebound than on compression.

Figure 24-16 shows a typical mounting setup for a telescoping shock absorber. The top end is fastened by rubber bushings to the frame. The bottom is bolted to the lower arm of the front suspension system. The up-and-down movement of the suspension arm will operate the shock. Some shock absorbers can be set to provide a soft, normal or a firm action, others provide this feature automatically. Many modern shock absorbers are gas-filled to improve rebound control without stiffening the ride. One is shown on a rear suspension system in **Figure 24-17.**

Air Shock Absorbers

Some suspension systems incorporate two adjustable ***air shock absorbers*** that are attached to the rear suspension and connected to an air valve with flexible tubing.

Air operated shock absorbers have hydraulic dampening systems which operate in the same manner as those on conventional shocks. In addition, they contain a sealed air chamber, which is acted on by pressure from a height control sensor. Varying the pressure to the air chamber causes the air shock to increase or decrease its length or operating range.

Air pressure is delivered to the air shocks through plastic tubing. The tubing connects the shocks to an air valve. Air pressure for raising the shocks is generally obtained from an outside source, such as a service station compressor, and is admitted through the air valve. To deplete the shocks of unwanted air (lower vehicle curb height), the air valve core is depressed, allowing air to escape.

Control Arms and Struts

All vehicles have either ***control arms*** or ***struts*** to keep the wheel assembly in the proper position. The control arms and struts allow the wheel to move up and down while preventing it from moving in any other direction. The wheel will tend to move in undesirable directions whenever the vehicle is accelerated, braked, or turned. Vehicle suspensions may have control arms only or a combination of control arms and struts.

Figure 24-18 shows a typical control arm used on many front-wheel drive vehicles. The front and rear attachment points pivot on rubber bushings. The attachment points absorb the tendency of the control arm to move forward and rearward as the wheel moves. The control arm design keeps the wheel from moving inward or outward.

By designing the upper and lower control arms carefully, it is possible to have a suspension that allows the wheel to move up and down while causing it to remain in the straight up and down position. The upper and lower control arms

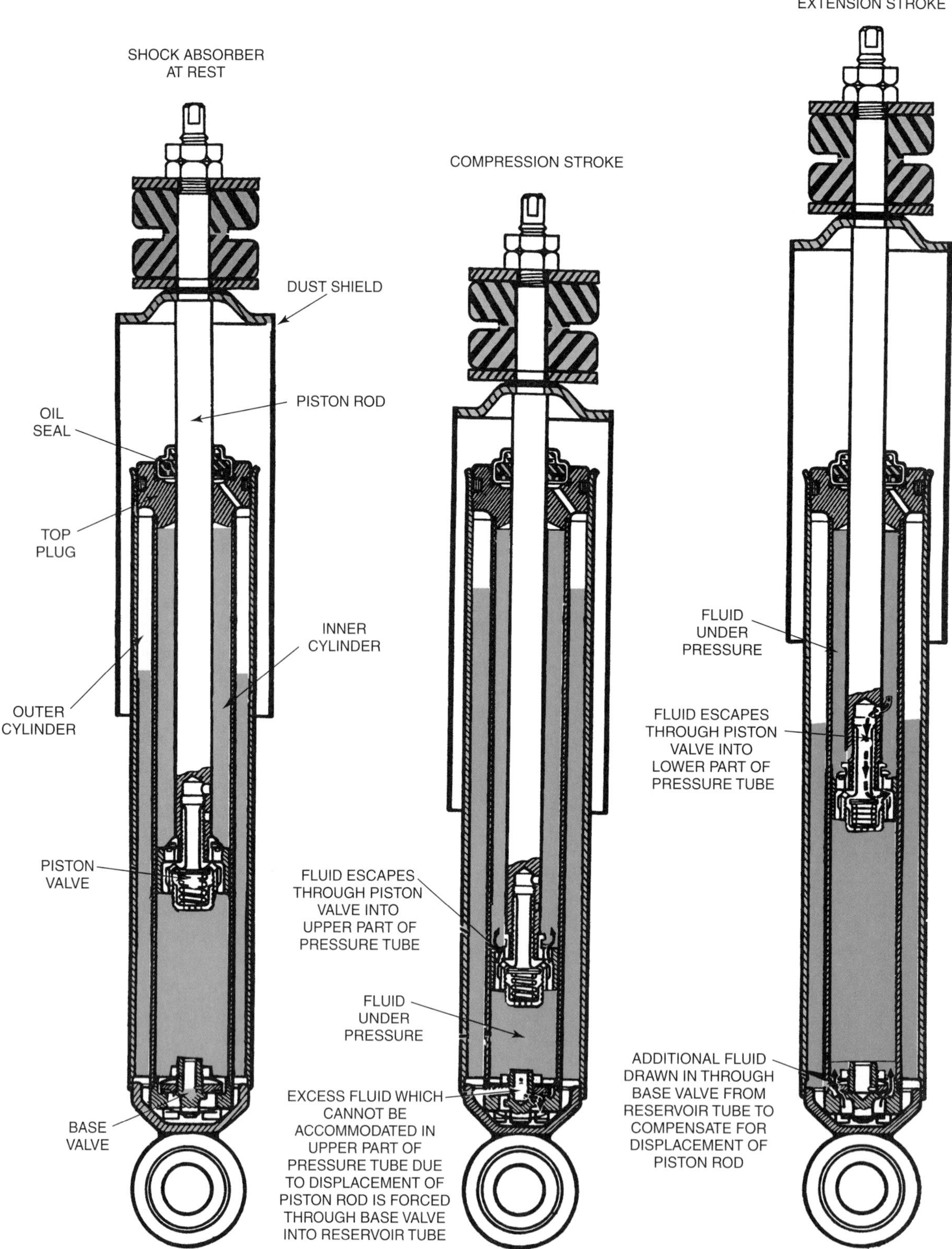
SHOCK ABSORBER AT REST
COMPRESSION STROKE
EXTENSION STROKE
DUST SHIELD
PISTON ROD
OIL SEAL
TOP PLUG
INNER CYLINDER
OUTER CYLINDER
PISTON VALVE
BASE VALVE
FLUID ESCAPES THROUGH PISTON VALVE INTO UPPER PART OF PRESSURE TUBE
FLUID UNDER PRESSURE
EXCESS FLUID WHICH CANNOT BE ACCOMMODATED IN UPPER PART OF PRESSURE TUBE DUE TO DISPLACEMENT OF PISTON ROD IS FORCED THROUGH BASE VALVE INTO RESERVOIR TUBE
FLUID UNDER PRESSURE
FLUID ESCAPES THROUGH PISTON VALVE INTO LOWER PART OF PRESSURE TUBE
ADDITIONAL FLUID DRAWN IN THROUGH BASE VALVE FROM RESERVOIR TUBE TO COMPENSATE FOR DISPLACEMENT OF PISTON ROD

Figure 24-15. Shock absorber action—at rest, compression, extension. Study each illustration and determine shock action shown in each.

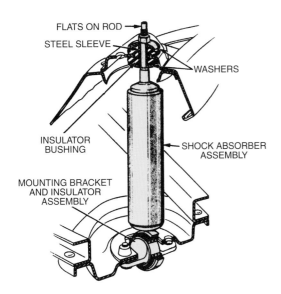

Figure 24-16. Shock absorber mounting. (Dodge)

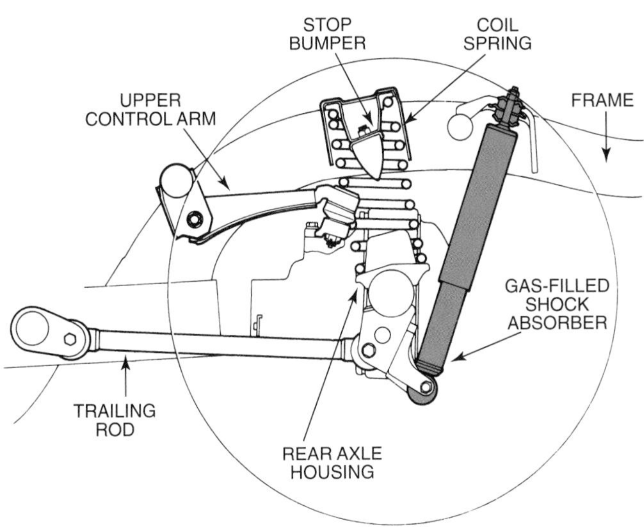

Figure 24-17. A gas-filled shock absorber mounted to a coil spring rear suspension system. (Geo)

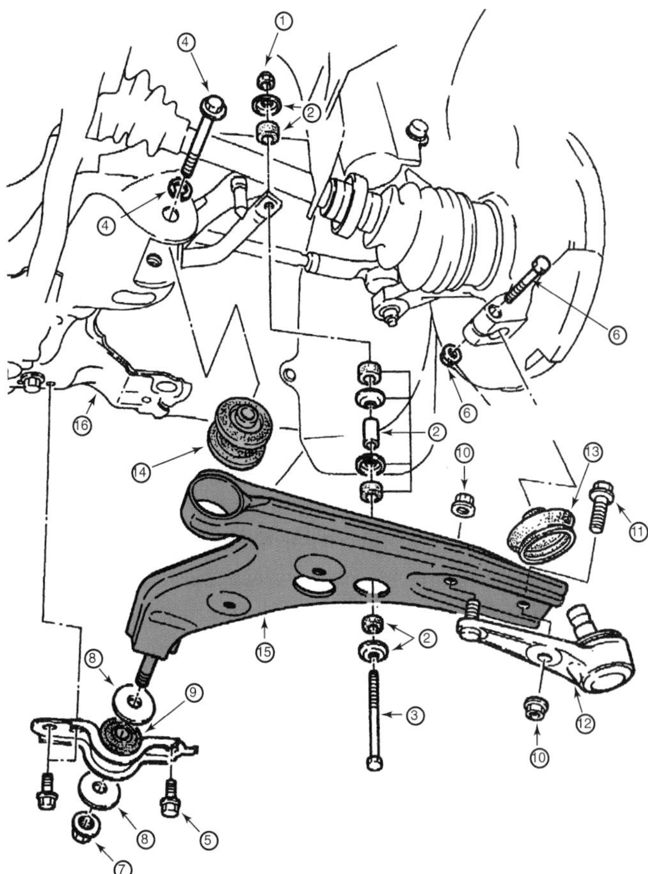

Figure 24-18. An exploded view of a front-wheel drive suspension assembly showing a typical control arm (lower), mounting bushings and locations. 1—Stabilizer nut. 2—Link assembly. 3—Stabilizer bolt. 4—Washer. 5—Bolt. 6—Bolt. 7—Nut. 8—Washer. 9—Lower arm rubber bushing. 10—Nut. 11—Bolt. 12—Lower arm ball joint. 13—Ball joint dust boot. 14—Lower arm rubber bushing. 15—Lower control arm. 16—Frame/body. (Kia Motors)

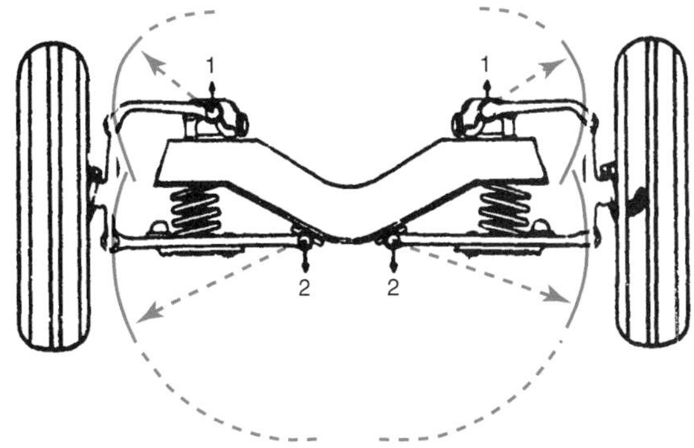

Figure 24-19. Upper and lower arm pivot arcs of individual front wheel suspension system.

move through different arcs, keeping the outer pivot points in alignment. See **Figure 24-19.** This improves handling over bumps.

Figure 24-20 shows a typical control arm and strut assembly. All contact points with the frame are mounted through rubber bushings. The strut rod (also called the tension strut or the strut bar) keeps the control arm from moving forward and rearward. The control arm itself keeps the wheel from moving inward or outward.

Ball Joints

All modern vehicles use ***ball joints***, **Figure 24-21.** The function of the ball joint is to carry the vehicle load to the wheels while allowing relative movement between the wheel assembly and the body. The ball joint that carries the majority of the vehicle weight is called the ***load-carrying ball joint***. The other joint is called the ***follower joint.*** Depending on the suspension arrangement, ball joints can be placed under either ***tension loading*** (forces attempt to pull joint apart) or

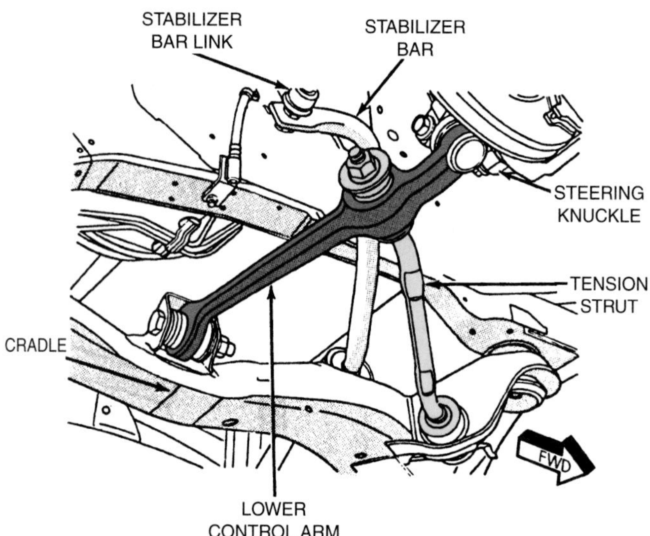

Figure 24-20. A control arm and strut type front suspension system. (Chrysler Corp.)

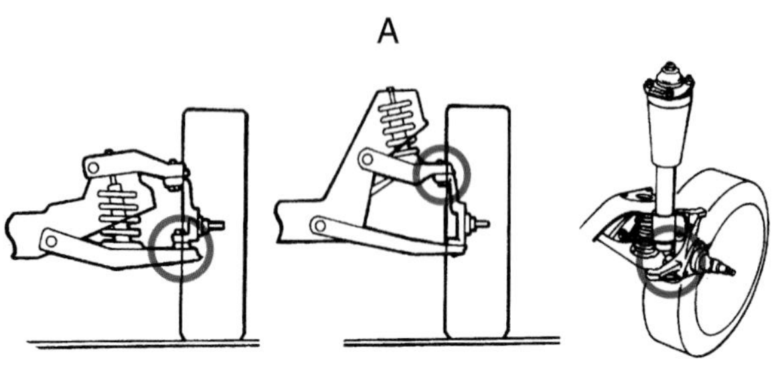

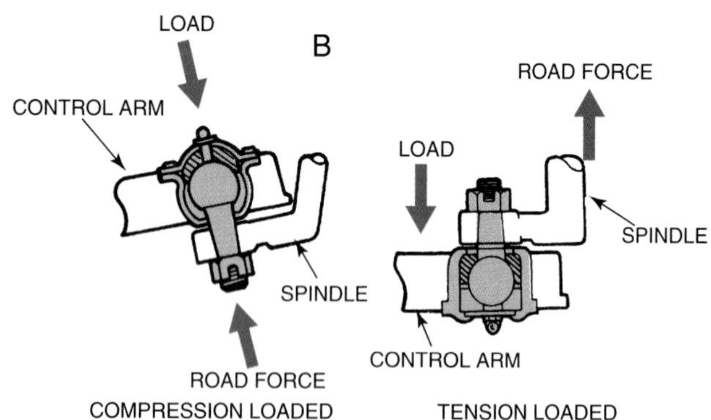

Figure 24-21. Load carrying ball joints. A—The location of the spring will determine which joint will be the load carrier. B—The different styles of load carrying ball joints—compression loaded and tension loaded. Study the construction, road and load force. (Moog)

compression loading (forces attempt to compress joint). **Figure 24-22** shows one that is ***preloaded*** by a coil spring. Preloading will keep the joint bearing surfaces in constant contact. One arrangement (coil spring placed between the upper control arm and body) places the main joint at the top. In this setup, the main joint is compression loaded. The follower joint is tension loaded.

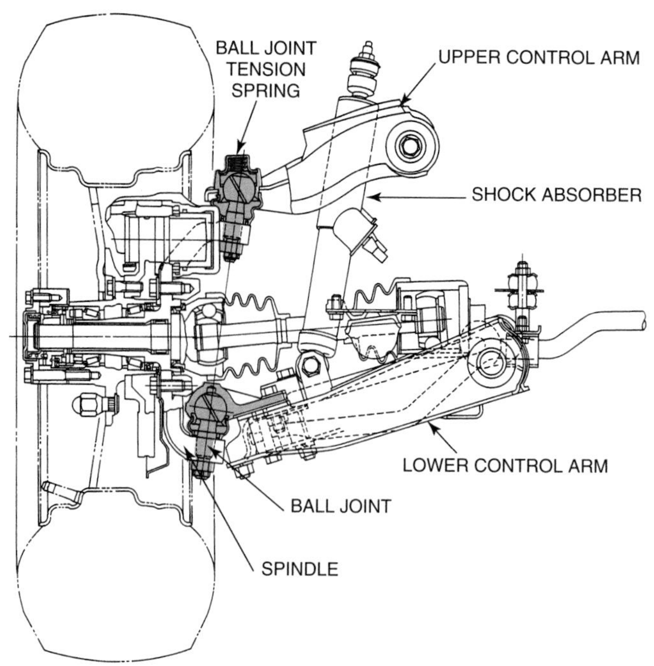

Figure 24-22. Cross-sectional view of a four-wheel drive front suspension system which uses a tension spring loaded upper ball joint. Study the layout. (Isuzu)

Figure 24-23 illustrates one basic arrangement in which both joints are tension loaded. Note that the lower joint is the load-carrying joint. Another arrangement tension loads the follower joint and compression loads the load-carrying joint.

Stabilizer Bars

To overcome the tendency of a vehicle's body to lean on corners, a ***stabilizer bar*** is incorporated into the suspension system. Made of spring steel, the stabilizer bar passes across the frame and is secured to the frame in two places. The outer ends of the stabilizer bar are fastened to the lower control arms (when used in front) through short linkage arms. Stabilizer bars can be used on both the front and rear suspensions.

Figure 24-23 illustrates a stabilizer bar and its location on the front cradle. The two bushing brackets are used to fasten the bar to the frame. A short linkage (spacer) connects the stabilizer bar to the control arms. **Figure 24-24** shows one side of a stabilizer bar. Notice how it is connected directly to the lower control arm, eliminating additional linkage.

Stabilizer Bar Action

As long as the body dips up and down in a level manner, the stabilizer bar does not affect its action. When the body attempts to tip to one side, one end of the bar bends down and the other end bends up. This imparts a twisting force to the bar. The bar resists this twisting force. In so doing, it helps to keep the vehicle level. See **Figure 24-25.**

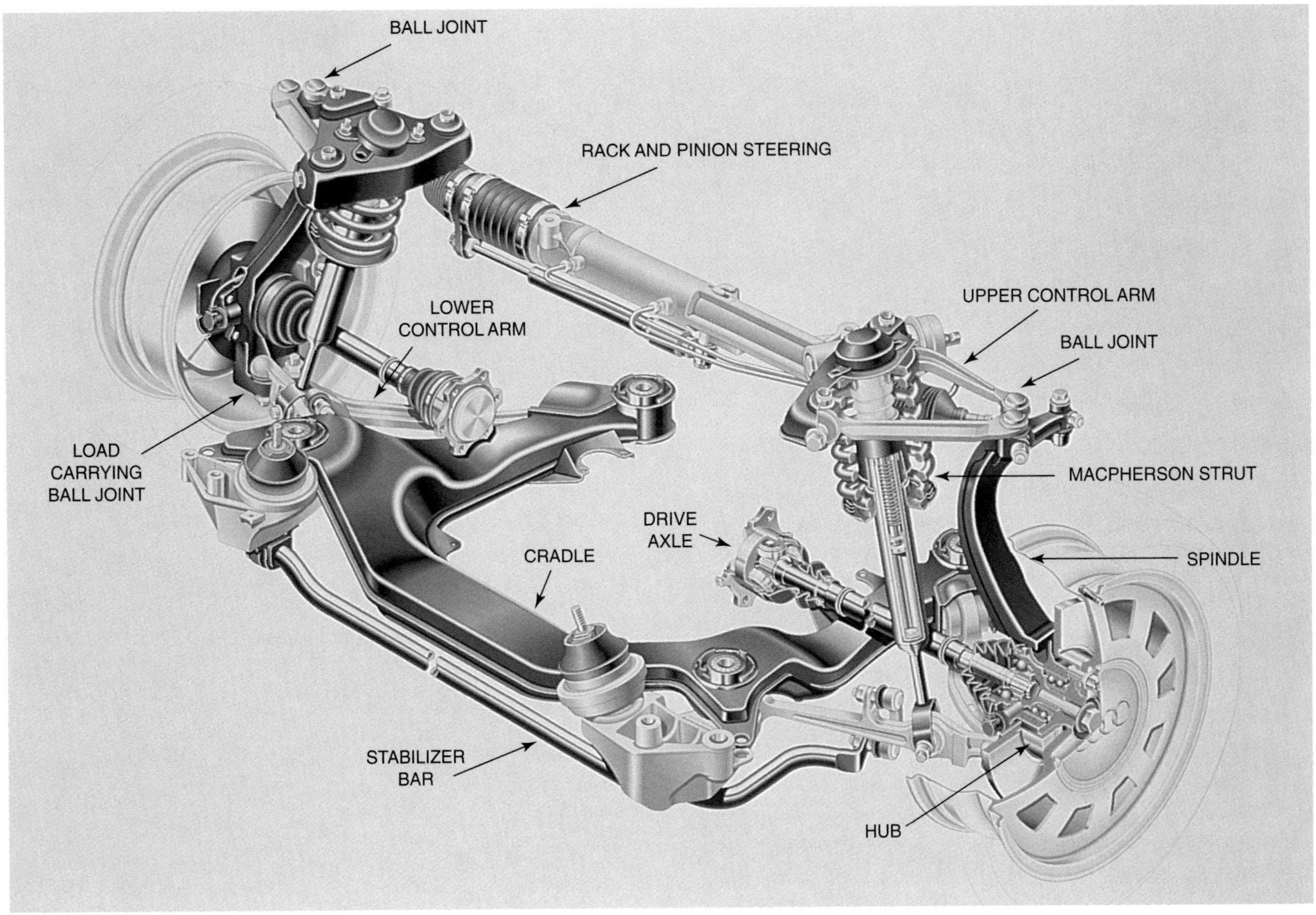

Figure 24-23. Cutaway view of an independent suspension as used on a front-wheel drive vehicle. Study the entire suspension layout. (Audi)

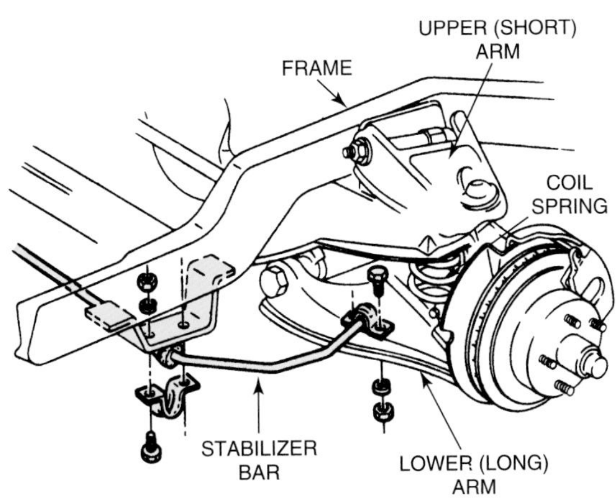

Figure 24-24. Stabilizer bar connectors. Other end of bar (not shown) is mounted in the same manner. (Chevrolet)

Front Suspension Systems

Almost all modern front suspension systems are independent. With an independent suspension, each front wheel is free to move up and down with a minimum effect on the other wheel. In an independent suspension system, there is also far less twisting motion imposed on the frame than in a system with a solid axle. Nevertheless, a few off-road, four-wheel drive vehicles and large trucks continue to use a solid axle front suspension. The two major types of independent front suspension are the ***conventional front suspension*** and the ***MacPherson strut front suspension***.

Conventional Independent Front Suspension

In the conventional front suspension system, one or two control arms are used at each wheel. In most systems, the coil springs are mounted between the vehicle's frame and the lower control arm. See **Figure 24-26.** In older systems, coil springs are mounted between the upper control arm and vehicle body, **Figure 24-27.**

A torsion bar front suspension system is shown in **Figure 24-28.** When the lower arm moves upward, it twists the torsion bar. A transverse-mounted torsion bar arrangement is illustrated in **Figure 24-29.**

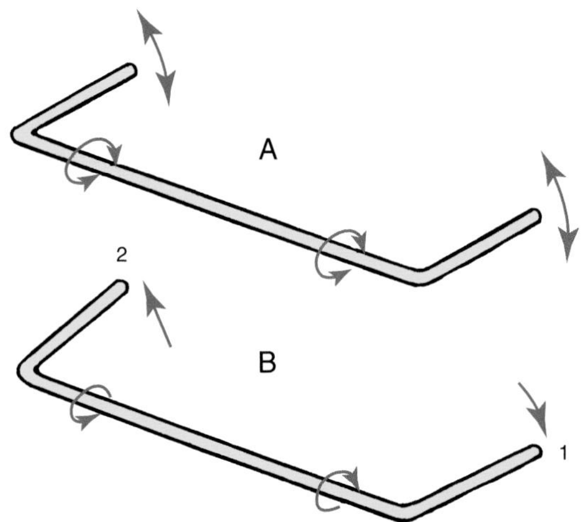

Figure 24-25. Stabilizer bar action. In A, the frame is moving up and down, but is not tipping. The bar offers no torsional resistance. In B, the frame is tipping to the right, pressing down on 1 and lifting up on 2. The bar must be twisted to allow this action. Resistance to twisting will minimize rolling or tipping motion of frame

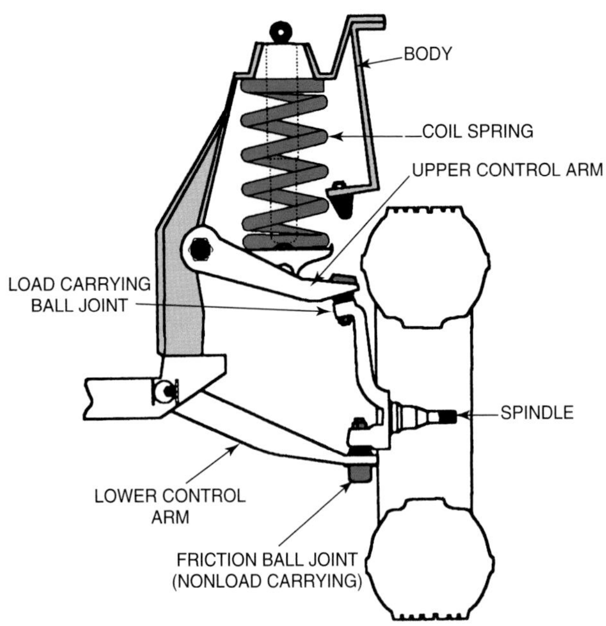

Figure 24-27. Coil spring mounted above the upper control arm. Note the narrow lower control arm. (FMC)

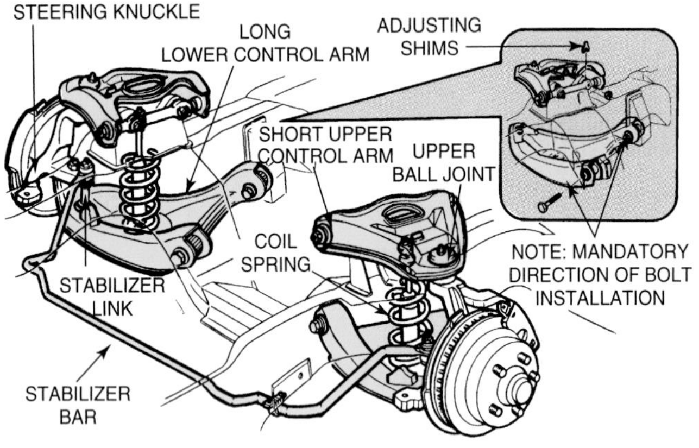

Figure 24-26. Independent front wheel suspension, using long and short control arms and coil springs. Note stabilizer bar. (General Motors)

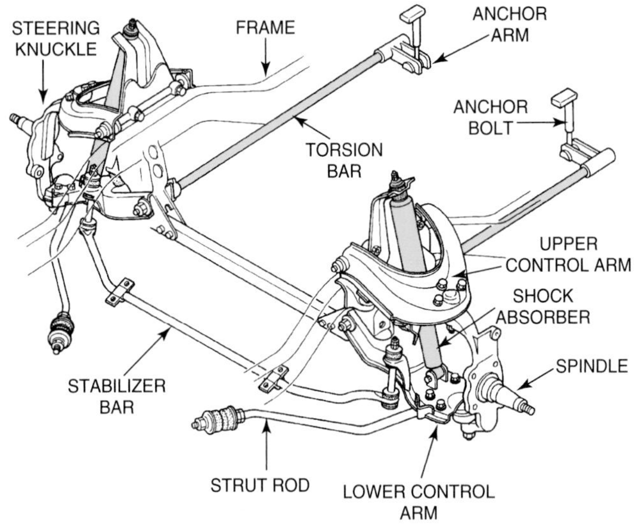

Figure 24-28. A torsion bar, independent front suspension system as used on a rear-wheel drive truck. (Mazda)

Coil Spring Front Suspension

Figure 24-30 shows a typical independent front suspension that uses rubber bushing control arm pivots. The top of the coil spring rests in a cup-like spot against the frame. The bottom of the coil spring is supported by a pad on the lower control arm. The top of each shock absorber is fastened to the frame; the bottom is attached to the lower control arm.

When the wheel strikes a bump, it is driven upward. This causes arms to pivot upward, compressing the spring and shock. Rubber bumpers limit control arm travel and soften the blow when the limit is reached. For steering, the front wheel steering knuckle pivots on ball joints.

A current front suspension system (one side) is shown in **Figure 24-31.** The spindle, or steering knuckle, pivots on ball joints. Rubber bushings are used for the control arms. In **Figure 24-32,** the coil spring is mounted on top of the upper control arm, with the upper end supported in a tall housing that is an integral part of the body and frame.

Torsion Bar Front Suspension

A torsion bar is located on each side of the frame in the front of the vehicle. The lower control arm is attached to the

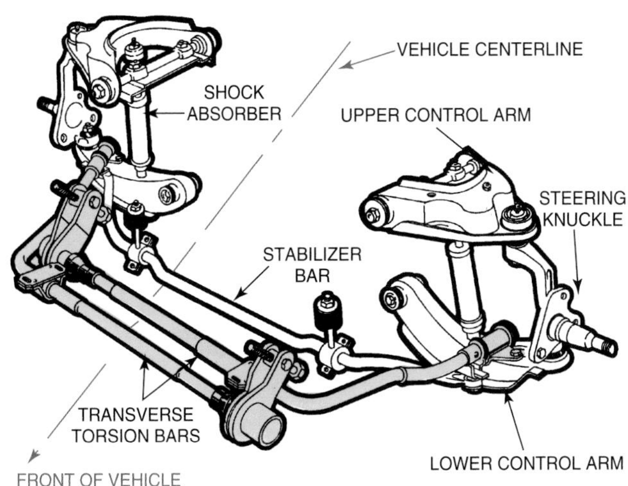

Figure 24-29. A transverse torsion bar front suspension assembly. The torsion bar runs across the width of the chassis. This setup also eliminates the need for strut rods. (Moog)

free end of the torsion bar. When the wheel is driven upward, the lower control arm moves upward, twisting the long spring-steel bar. **Figure 24-33** shows a view of a complete torsion bar suspension. **Figure 24-34** illustrates a similar setup for hinging the lower control arm. Notice that the arm pivots on a stub shaft. The torsion bar fits into the end and does not support the arm. It merely produces spring resistance to up-and-down movement.

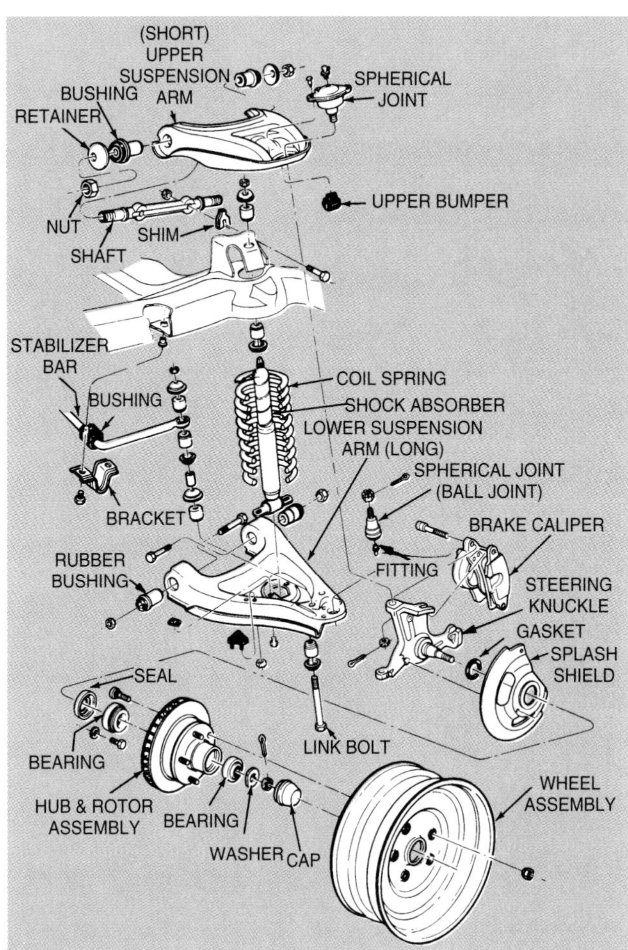

Figure 24-30. Exploded view of long and short arm suspension setup. (Chrysler)

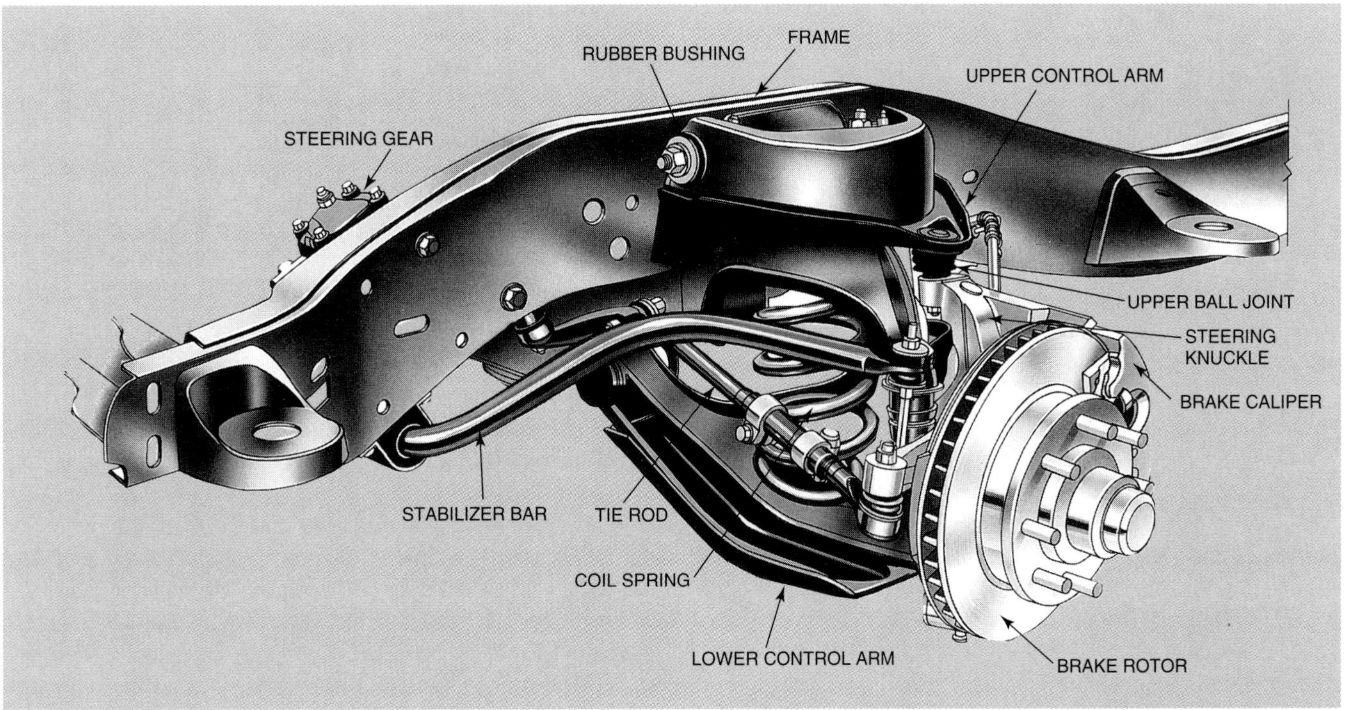

Figure 24-31. One side of a front suspension system that is used on a two-wheel drive truck. Study the construction. (Dodge)

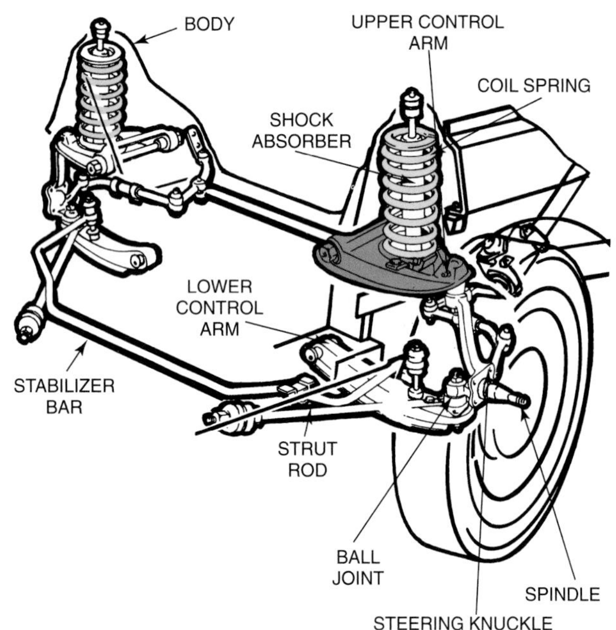

Figure 24-32. A coil spring mounted on top of the upper control arm and attached to the body. (Moog)

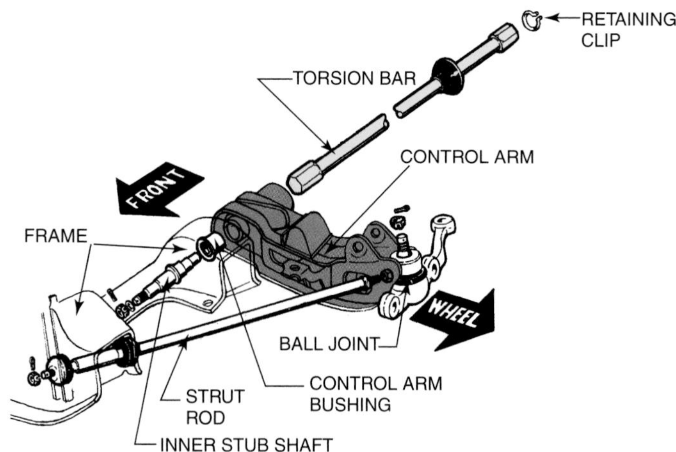

Figure 24-34. A torsion bar with lower control arm. The control arm is supported on the inner stub shaft, not on the torsion bar itself. (Moog)

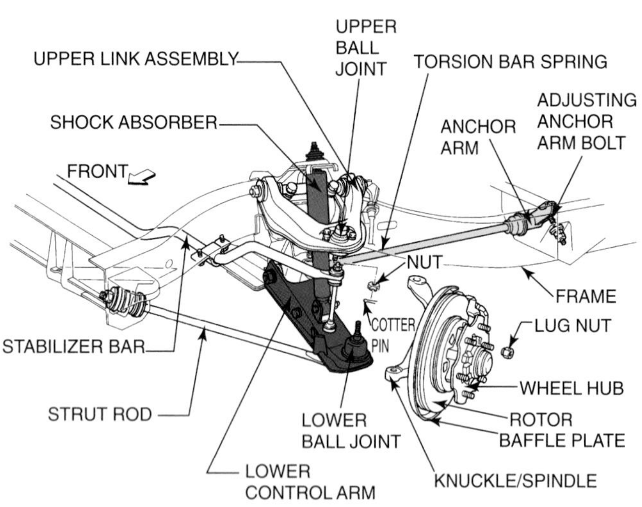

Figure 24-33. A torsion bar front suspension for a rear-wheel drive vehicle. The torsion bar is fastened to the lower control arm in this setup. Some four-wheel drive vehicles, with torsion bars, have them connected to the upper control arm. This provides increased ground clearance. (Nissan)

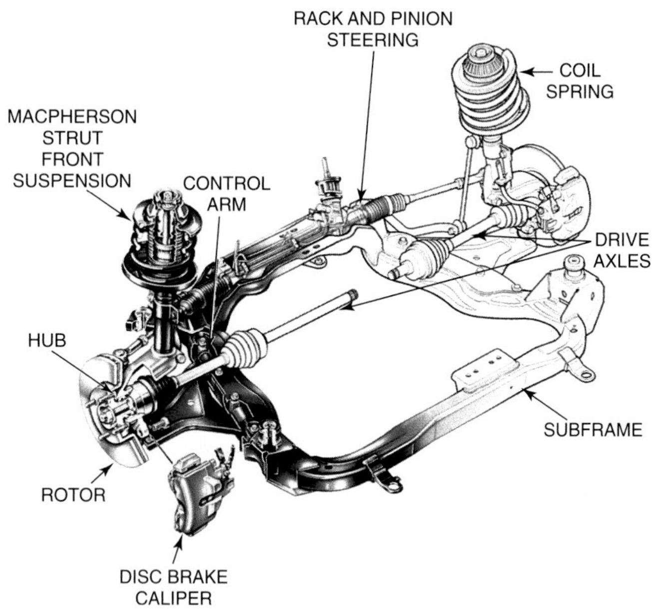

Figure 24-35. Cutaway view of a MacPherson strut front suspension assembly that is mounted to the vehicle subframe. This setup is used with a front-wheel drive car. (Ford Motor Company)

MacPherson Strut Front Suspension

Most modern vehicles, especially those with front-wheel drive, use the MacPherson strut front suspension systems, **Figure 24-35.** Note that the MacPherson strut contains a coil spring, which is mounted on top of the heavy strut-and-pedestal assembly. Rubber pads at the top and bottom of the coil spring cushion spring application and reduce squeaks, **Figure 24-36.** The entire MacPherson strut assembly is attached to the steering knuckle at the lower part of the pedestal, **Figure 24-37.** The bottom of the MacPherson strut assembly is attached to the single control arm through a ball joint.

The entire strut assembly turns when the wheel is turned. A bearing or thrust plate at the top of the strut assembly allows relative movement between the assembly and the vehicle body. The ball joint allows the strut assembly to turn in relation to the control arm. The strut contains a ***damper***, which operates in the same manner as a conventional shock absorber. Most damper assemblies have a protective cover that keeps dirt and water away from the damper piston rod.

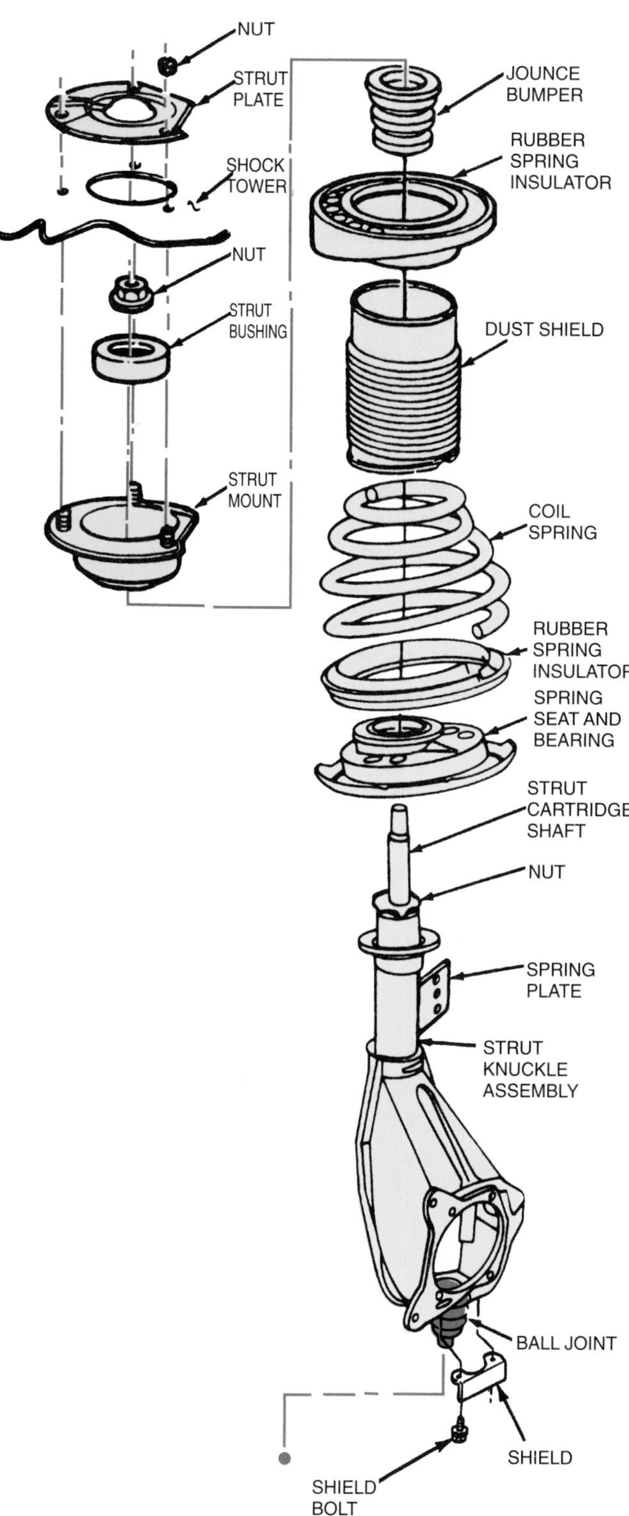

Figure 24-36. Exploded view of a MacPherson strut assembly. (Pontiac)

The advantage of the MacPherson strut is its compact design, which allows more room for service on small car bodies. In addition to the common MacPherson strut coil spring arrangement discussed above, a slightly different design places the coil spring between the frame or body and the lower control arm. While this design is not common, it does qualify as a MacPherson strut suspension.

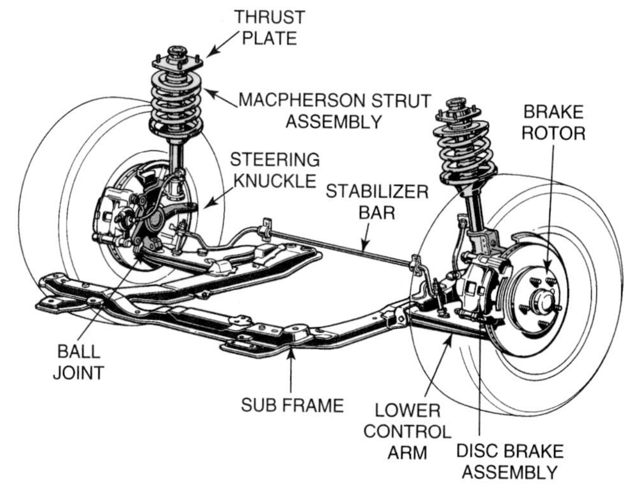

Figure 24-37. A MacPherson strut attached to the steering knuckle. Note the stabilizer bar. (Ford Motor Company)

Solid Axle Front Suspension

The use of the solid axle front suspension is generally confined to trucks and off-road vehicles. This system uses a solid steel dead axle (does not turn with wheels) with a leaf spring at each side. Pivot arrangements between the axle and the wheel spindles allow the wheels to swivel on each end, **Figure 24-38.** A solid axle, four-wheel drive setup is shown in **Figure 24-39.**

Any up or down movement of either front wheel causes a vertical tipping effect of both wheels because they share a common axle. This effect upsets the steering geometry and

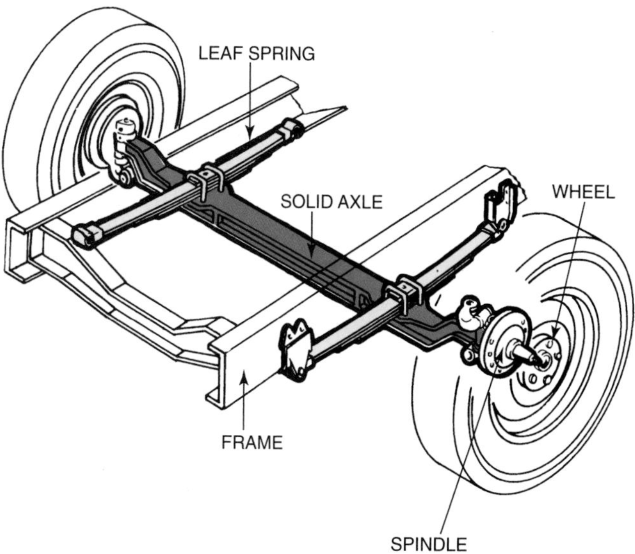

Figure 24-38. A solid front axle, leaf spring, suspension assembly. (Moog)

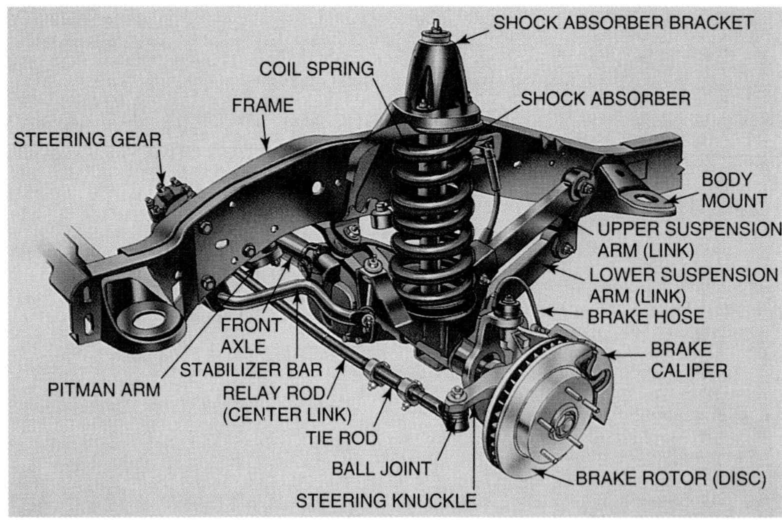

Figure 24-39. A solid front axle, link and coil spring front suspension as used with one particular four-wheel drive truck. (Dodge)

imposes a twisting motion to the frame. See **Figure 24-40.** One exception to this is the Twin-I-beam front axle used in some light trucks. In this setup, two beams are used. Each beam pivots independently, **Figure 24-41.** Coil springs are placed between the I-beam axles and the vehicle's frame.

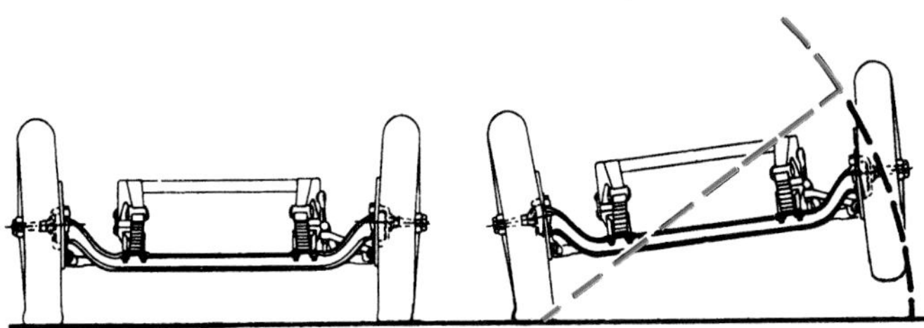

Figure 24-40. Tipping effect of solid front axle.

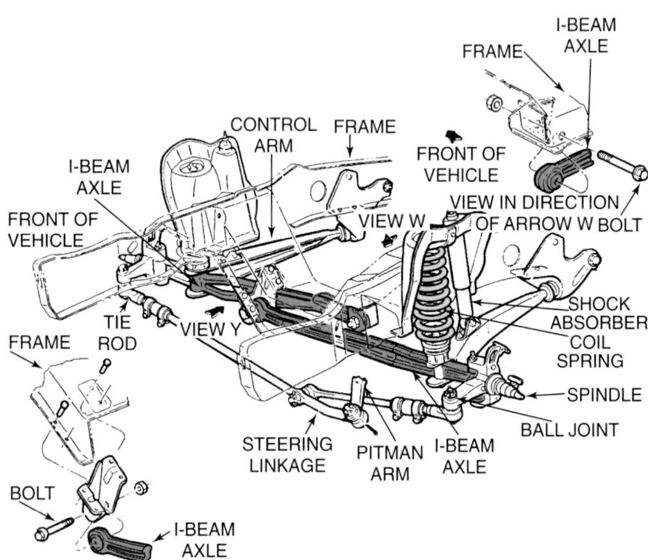

Figure 24-41. Ford Twin I-Beam front suspension as used on light duty (1/2, 3/4, and 1 ton) trucks. A coil spring rides between each individual I-beam and the frame. Each axle is free to pivot up and down without affecting the other. (Ford)

Rear Suspension Systems

Rear suspensions on vehicles with a solid rear axle housing generally utilize coil springs or leaf springs. When the vehicle has an independent rear suspension system, coil springs, MacPherson struts, a single transverse leaf spring, or even torsion bars can be used.

Solid Rear Axle Suspensions

Figure 24-42 illustrates a typical suspension system utilizing coil springs. The rear axle housing is mounted on

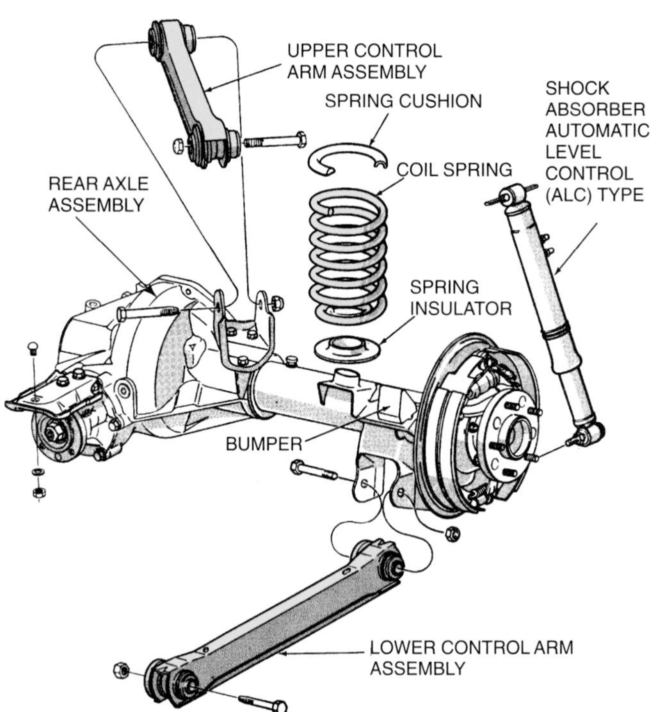

Figure 24-42. View of a rear suspension system that uses coil springs, with upper and lower control arms. This setup is also termed a four-link suspension. (Cadillac)

springs and is attached to a set of upper and lower control arms (also called control links or control struts) to provide proper rear axle housing alignment. The control arm pivot points are insulated by rubber bushings. One end of the arm is connected to the housing; the other end is connected to the frame. The arm arrangement allows the rear axle housing to move up and down but prevents excessive front-to-back and side-to-side movement.

The shock absorbers in **Figure 24-43** are straddle mounted (shocks are closer together at the top than the bottom). In addition to controlling oscillation, straddle mounting assists in controlling side movement between the frame and the axle housing and helps resist tipping on turns. Another type of solid rear suspension is shown in **Figure 24-44.** This type is used with the solid rear axles found on many front-wheel drive vehicles.

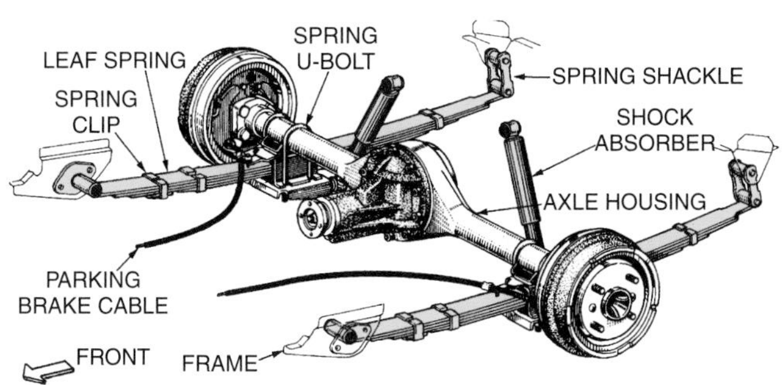

Figure 24-43. Leaf spring rear suspension system. Springs control axle housing rotation, fore-and-aft and side-to-side movement. Strattle mounted (angled) shock absorbers also help control side-to-side movement. (Toyota)

Independent Rear Suspensions

Some rear end designs have differential housing attached to the vehicle's frame and drive the rear wheels through open (or enclosed) axles, using universal joints to allow flexibility. These assemblies were discussed in more detail in Chapter 22. The independent rear suspension system in **Figure 24-45** uses a MacPherson strut assembly for each rear wheel. These strut assemblies are mounted to the spindles. Suspension movement is controlled with lateral links and trailing arms. A stabilizer bar is also used.

Coil springs and MacPherson struts are also widely used. There are other variations in the use of control arms. **Figure 24-46** illustrates a semi-independent rear suspension that incorporates a twisting cross beam axle setup. The axle connects to the vehicle's underbody and holds the wheels in correct alignment.

Suspension Lubrication

Chassis grease is compounded to enter fittings readily, lubricate well, stay flexible at low temperatures, adhere to the moving parts, and resist water. Where chassis grease is needed for the suspension system, ***grease fittings*** are provided. Grease fittings are often used on the ball joints. Some vehicles have fittings on the control arm bushings. Some suspension parts have ***plugs*** instead of fittings. If plugs are installed, the technician must replace them with grease fittings. After greasing is completed, the plugs can be replaced or the fittings can be left in place for the next lubrication.

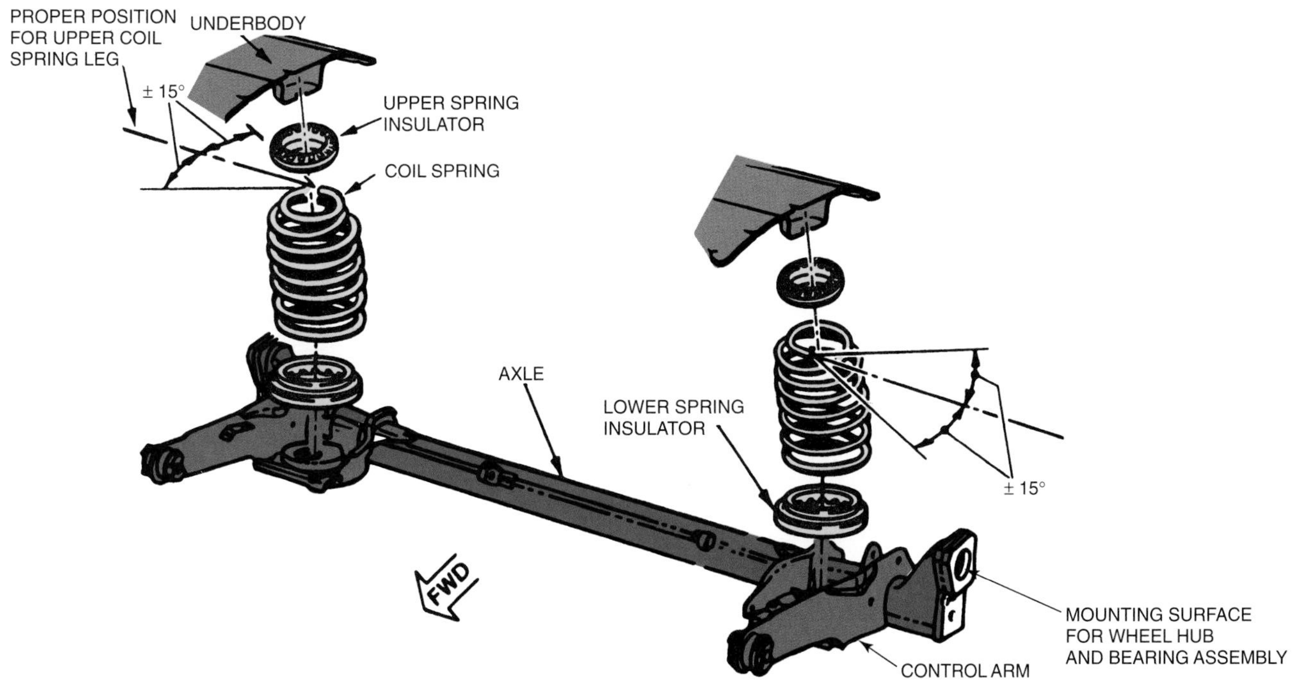

Figure 24-44. A solid rear axle assembly as used with one particular front-wheel drive car. This axle is used on standard, as well as with electronic level control (ELC) suspensions. (Pontiac)

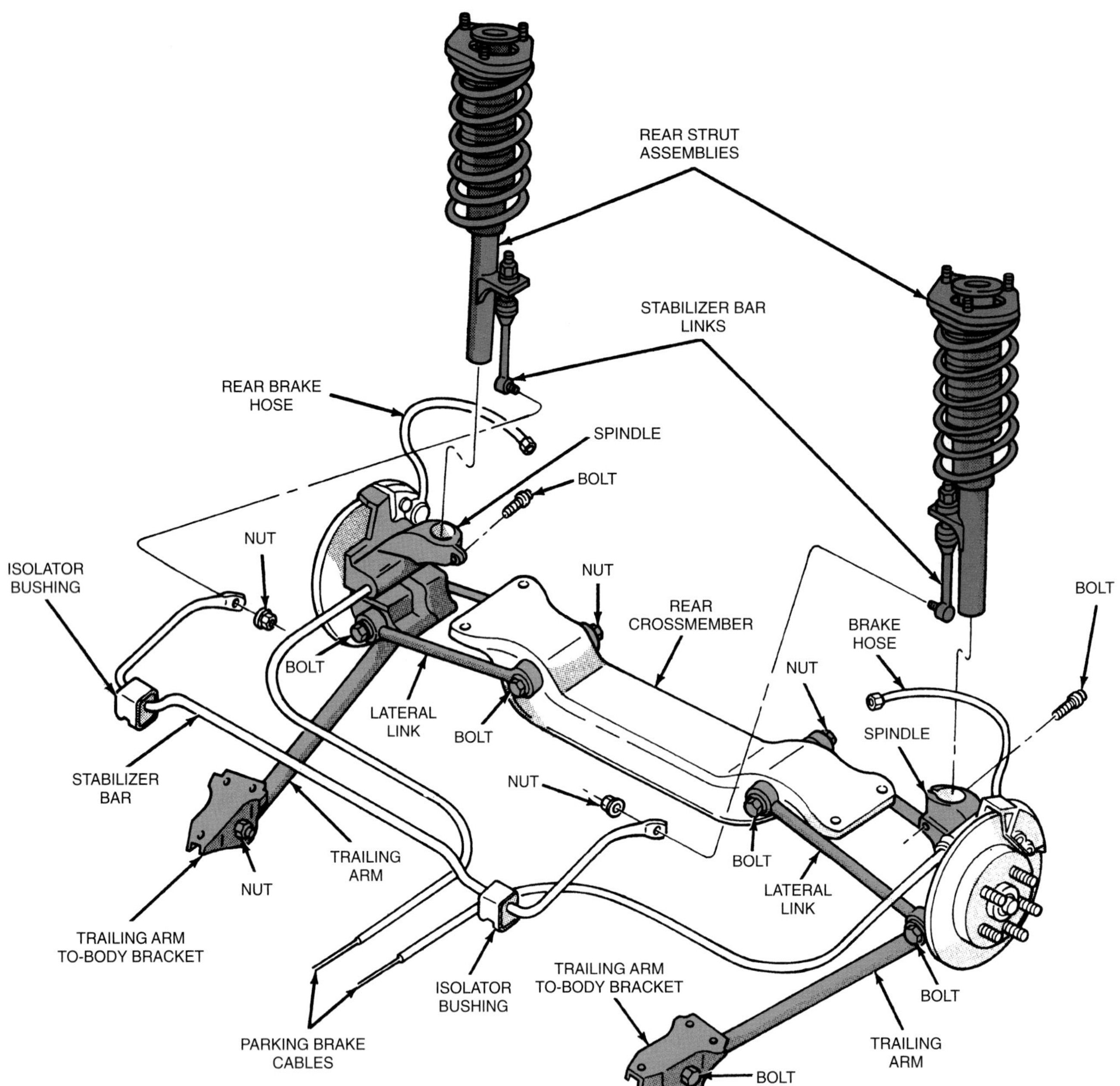

Figure 24-45. A fully independent rear suspension which incorporates a MacPherson strut unit for each wheel. Note the use of rear disc brakes. (Chrysler)

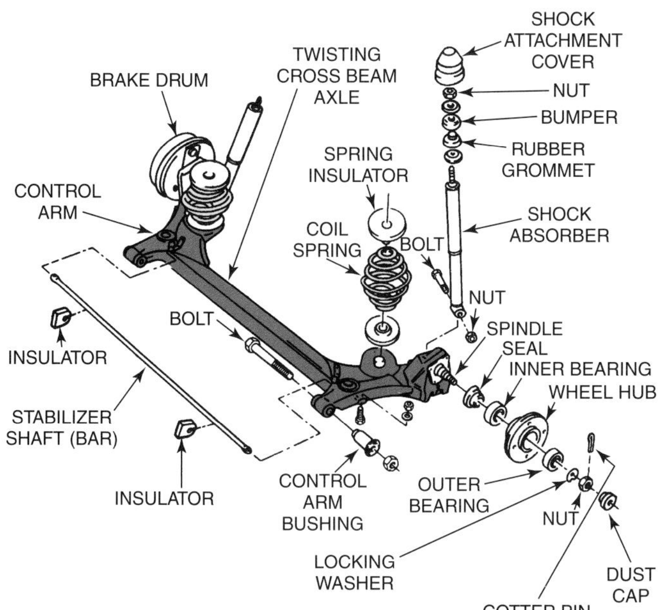

Figure 24-46. A semi-independent rear suspension which uses a twisting cross beam axle, coil springs, shock absorbers, and a stabilizer shaft (bar). (Pontiac)

Many newer cars have joints that utilize Teflon® plastic in their construction. This, coupled with improved lubricants and seals, has extended greasing intervals to as many as 30,000 miles (48 270 km). Some pivot points are permanently lubricated.

Automatic Level Control

The ***automatic level control system*** is designed to maintain a nearly constant rear curb height (distance from the frame to the ground), regardless of load changes over the rear axle.

A basic level control system will generally include an air compressor (vacuum or electrically operated air pump), an air tank (reservoir), a pressure regulator, a height control valve, special shock absorbers, connecting wire, tubing, hose, and an air dryer. See **Figure 24-47.** A hydraulic level control system using an oil pump is shown in **Figure 24-48.**

Automatic Level Control Operation

The basic control method of the automatic level control system is to develop air pressure and direct this pressure to air-operated shock absorbers. The amount of air pressure in the shock absorbers is then modified in response to load changes in the vehicle. Most of the level control systems on late model vehicles use air-hydraulic MacPherson struts instead of air shocks.

Air Compressor

Air pressure is developed by an vacuum- or electric-operated ***air compressor,*** **Figure 24-49.** The vacuum air compressor is operated by intake manifold vacuum, delivered by a rubber hose to the unit. The electric compressor's motor is powered by battery voltage through a switch in a height sensor or from a control module. As seen in **Figure 24-50,** the arrangement of check valves causes a diaphragm to flex and

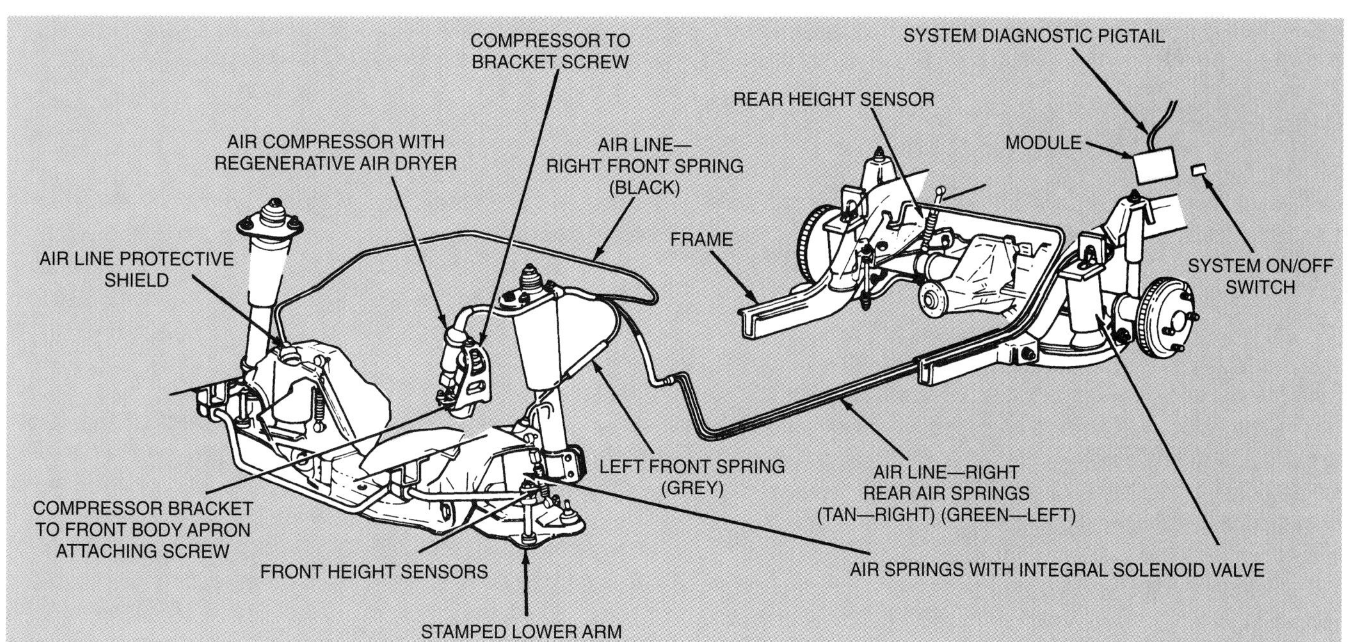

Figure 24-47. Schematic view of an air spring suspension system. Necessary air pressure to operate the air springs is provided by an electric compressor. Compressor is triggered to start or stop by electrical signals from a control module. The control module is activated by electric signals from three height sensors connected between the suspension and body. Any distance fluctuations between suspension and body (weight added or removed from the vehicle) will cause the system to add or deplete air in the springs to maintain a predetermined ride height. (Ford)

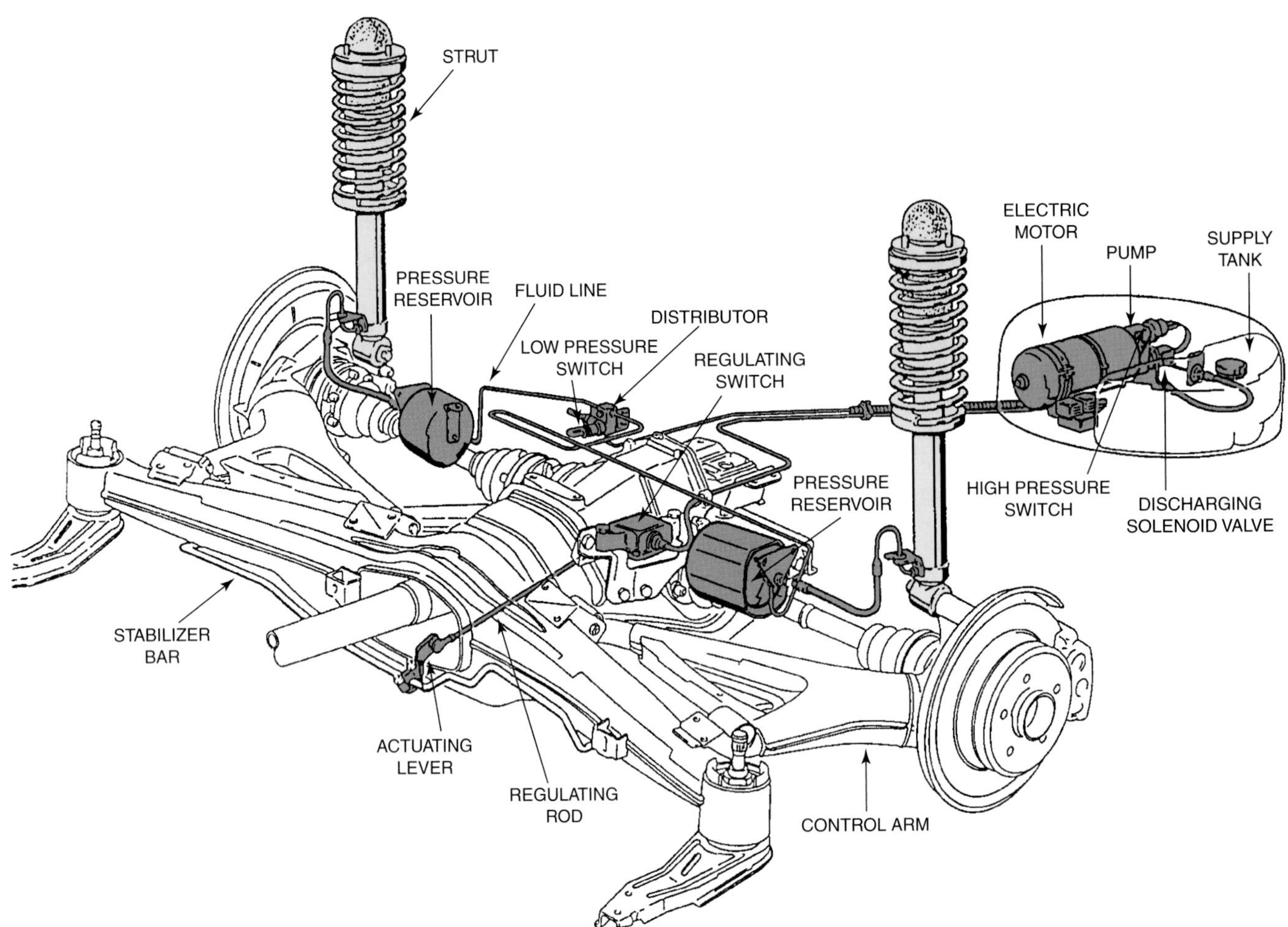

Figure 24-48. A hydraulic level control system used on an independent rear suspension. The electrically driven hydraulic pump supplies fluid under pressure to the struts, when turned on by the regulating switch. As the suspension is forced downward from a load in the vehicle, the actuating lever and regulating rod are moved at the same time. When the suspension travel reaches a predetermined point, the regulating rod will trigger the regulating switch. The hydraulic pump will now run until the vehicle is elevated to its normal ride height, at which time the pump will be shut off as the regulating rod turns the regulating switch off. (BMW)

operate a piston, which develops air pressure. Electric compressor output is controlled by an electric pressure switch. Compressor operation can also be controlled by a switch in the passenger compartment.

This air pressure enters a reservoir, which holds a reserve of pressure to operate the system. When the pressure in the reservoir is equal to the pressure generated by the pump, the compressor stops operating. When air is used by the system, the decrease in pressure will cause the pump to begin operating again. Air pressure from the reservoir to the system is limited by the pressure regulator valve to protect the pump and shock absorbers, **Figure 24-51.**

Some systems use an air dryer/reservoir combination, **Figure 24-52.** The air dryer/reservoir contains a chemical which absorbs moisture from the air before it can cause corrosion in other parts of the system. The moisture is returned to the atmosphere when air is released from the system. It also contains a valve that is designed to maintain a predetermined air pressure in the system when the compressor is turned off.

Height Control Valve

The amount of air pressure delivered to the shock absorbers is controlled through a ***height control sensor*** (sometimes called a level control valve or sensor). The height control sensor is attached between the vehicle frame and a moveable suspension part, usually the rear axle. It is designed to sense a distance change between the frame and suspension through an overtravel lever, or actuating arm, attached to the rear axle housing. As the distance changes,

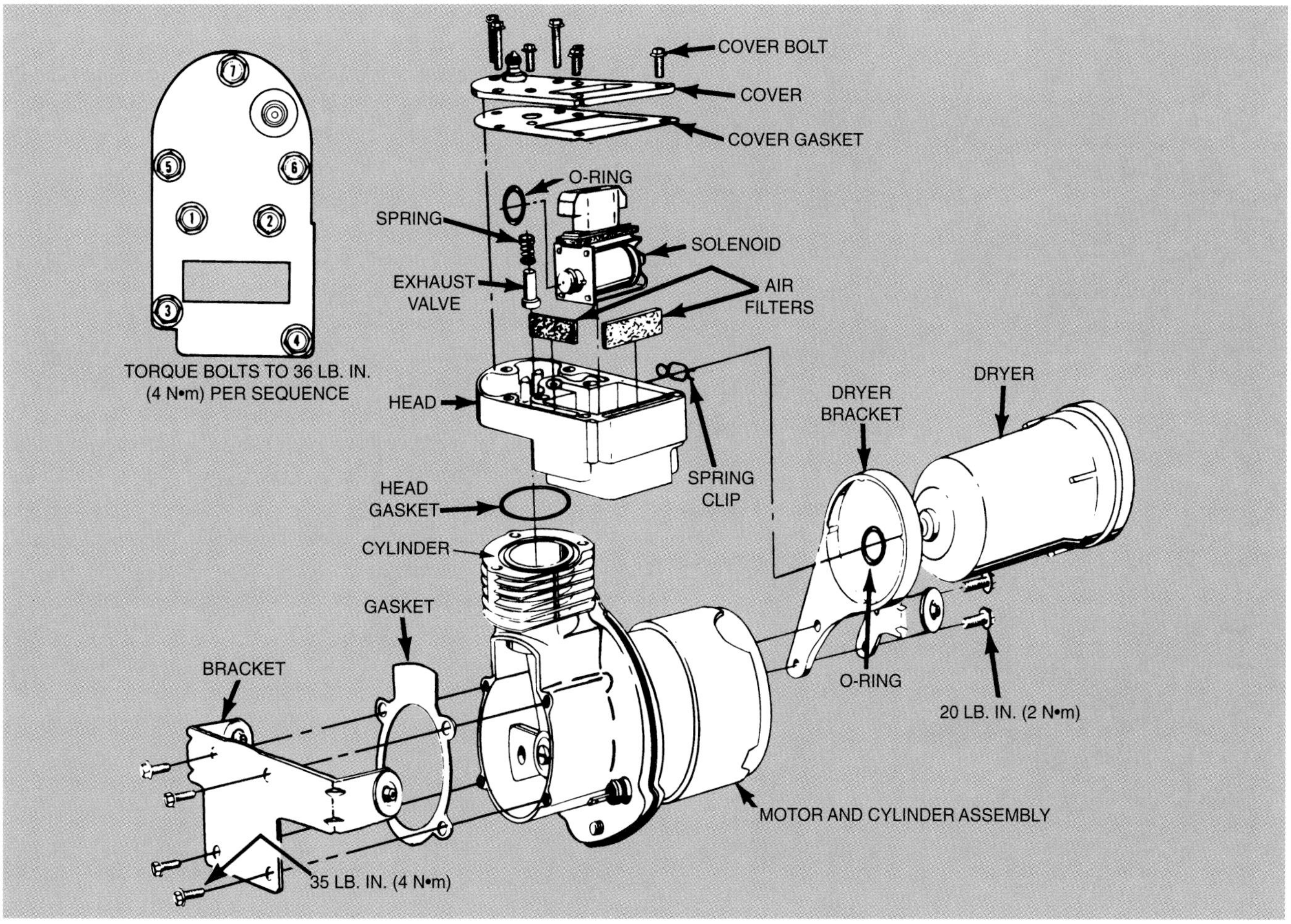

Figure 24-49. Exploded view of a single piston air compressor used to furnish air for an electronic level control (ELC) system. (Pontiac)

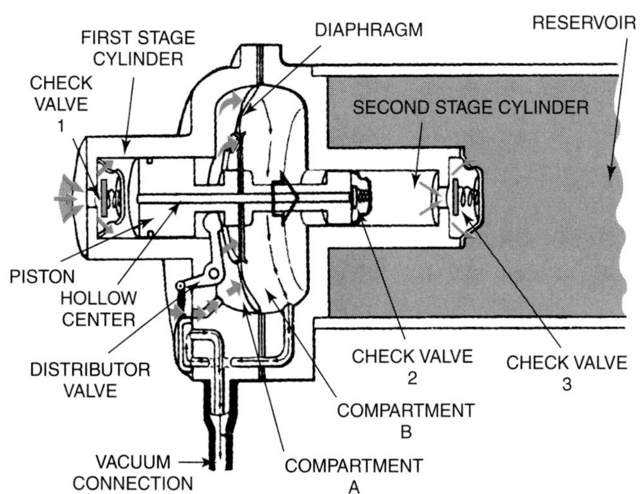

Figure 24-50. Vacuum-operated air compressor and action. As the diaphragm forces the piston toward the reservoir, fresh air is drawn in through check valve 1. Air is compressed in the second stage cylinder and forced through check valve 3 into the reservoir. Reverse stroke forces air from the first stage cylinder through a hollow cylinder into the second stage cylinder. (Pontiac)

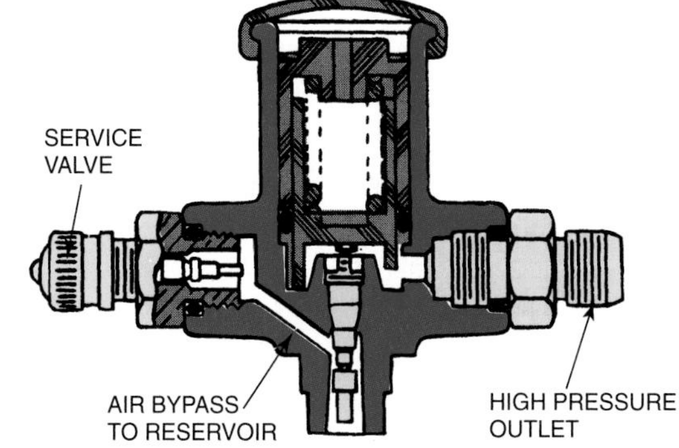

Figure 24-51. Pressure regulator valve.

the overtravel lever will cause the air compressor to admit or exhaust air. **Figure 24-53** illustrates a typical mechanical control unit with linkage.

In **Figure 24-54**, the overtravel lever is compensating for weight added to the vehicle by opening the height control

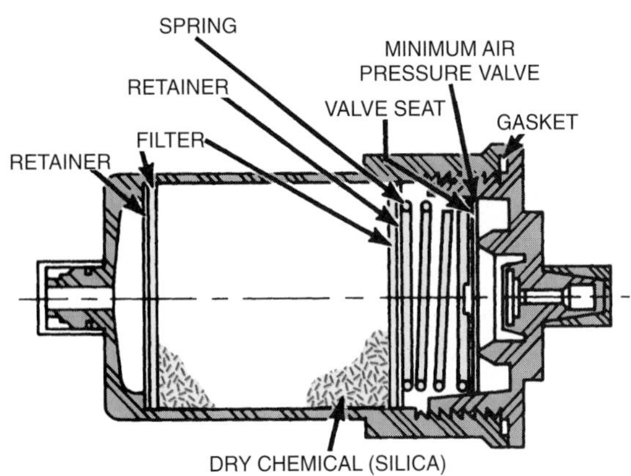

Figure 24-52. Level control system dryer unit. Note the use of silica as a drying agent. (Oldsmobile)

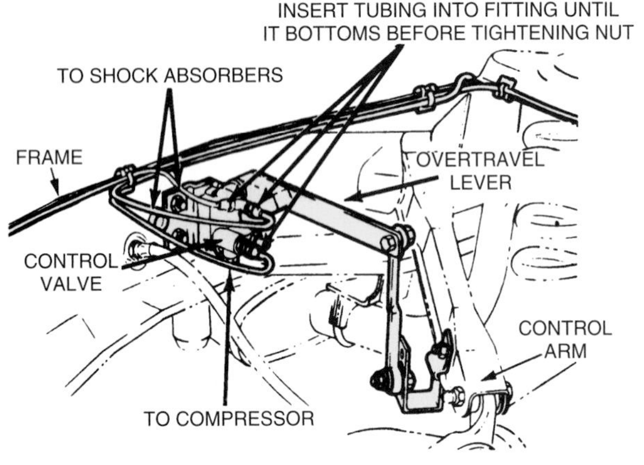

Figure 24-53. Automatic level control valve is actuated by over-travel level. (Pontiac)

valve to admit more air to the shocks. When the air shocks have raised the vehicle to the distance that the vehicle was at before the load was added, the lever will cause the valve to close. To prevent bumps from actuating the control valve, a time delay mechanism is built into the sensor.

From the control unit, air passes to the shock absorbers or struts. These operate in the same manner as the air shock absorbers discussed earlier in this chapter. The only difference is that varying air pressures are controlled by the sensor instead of an outside air pressure source. A typical air shock is illustrated in **Figure 24-55.**

Computerized Ride Control Systems

Computerized ride control systems, also called ***active suspensions***, are computer-controlled electronic suspension systems that can adapt to specific driving conditions. See **Figure 24-56.**

Typical ride control systems contain steering sensors, which monitor the direction and speed of steering wheel rotation; brake sensors, which monitor brake system applications; and an acceleration signal sensor, which monitors the rate of acceleration. A control module receives signals from these sensors and uses them to determine output signals to the actuators. The actuators control the flow of hydraulic fluid to adjustable shock absorbers. Many systems have a mode select switch, which allows the driver to adjust the ride control system.

Under normal driving conditions, a driver can pick the ride best suited for handling and comfort. For example, if a vehicle is being driven on a winding road, the driver can pick a firm setting. In this mode, the ride control module will signal actuators to close valves on the adjustable shocks. This will produce a stiff ride and the vehicle will corner well. If the

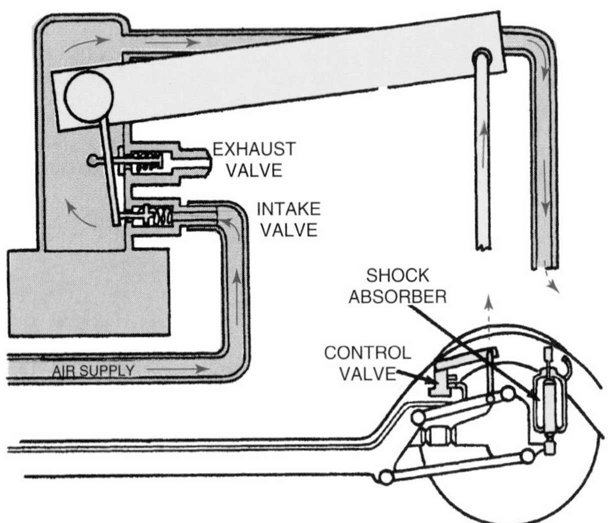

Figure 24-54. Control sensor action when a load is added to the vehicle. Vehicle curb height is decreased, causing the over-travel lever to move upward to open the intake valve and pass compressed air to the shock absorbers. (Buick)

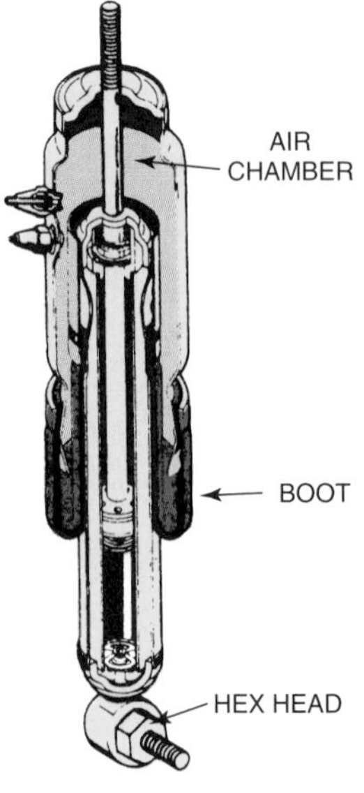

Figure 24-55. Special air shock absorber used with an automatic level control system. Note rubber boot. (Buick)

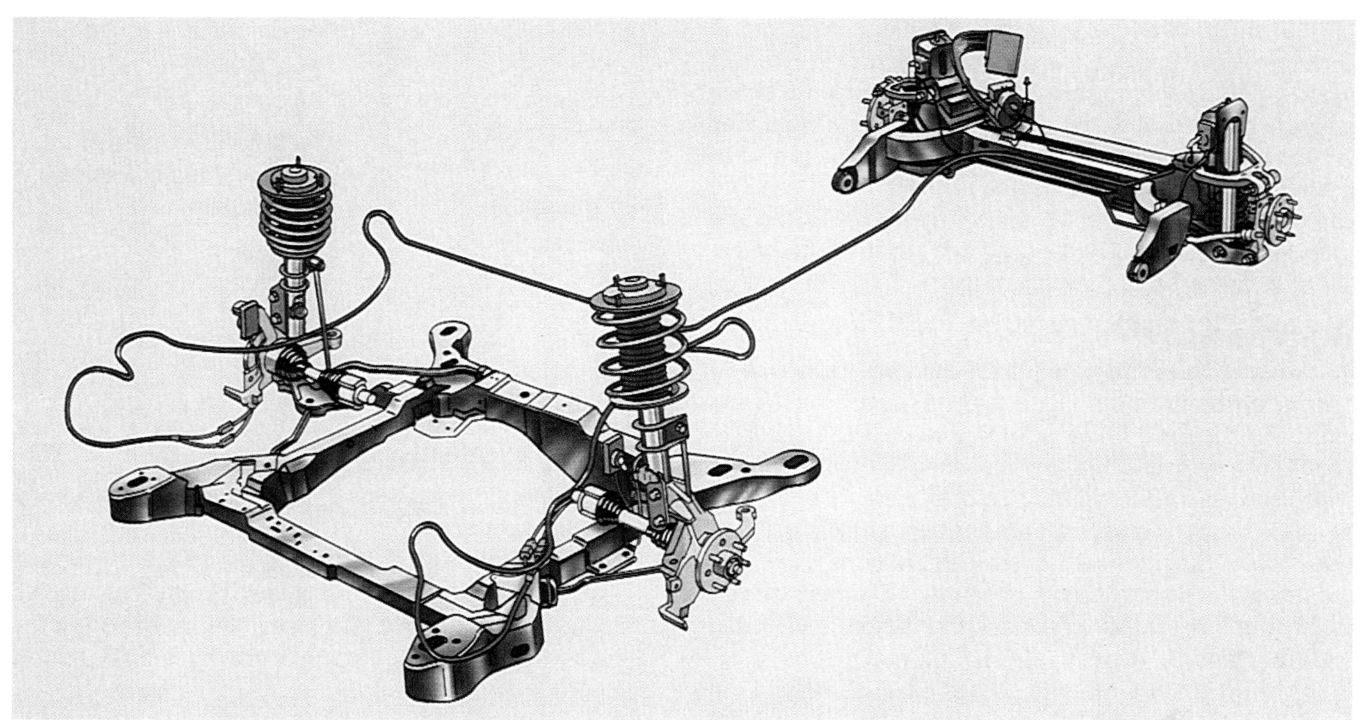

Figure 24-56. A computerized, Continuously Variable Road Sensing Suspension (CV-RSS) system. These suspensions adjust automatically to the ever changing road conditions—smooth, rough, corners, etc. (Cadillac)

vehicle is being driven on a highway, however, the driver can switch to a soft mode. In this mode, the valves in the adjustable shocks are opened and a soft, comfortable ride is produced.

Under certain driving conditions, the control module can override the driver's selection. During periods of hard braking, for example, a brake sensor signals the control module to close the valves on the adjustable front shocks or struts. This stiffens front shocks and minimizes front-end diving. Under heavy acceleration, the acceleration sensor signals the control module to stiffen the adjustable rear shocks or strut, preventing rear-end squatting. A road-sensing suspension system is shown in **Figure 24-57.** System design and operation vary from manufacturer to manufacturer.

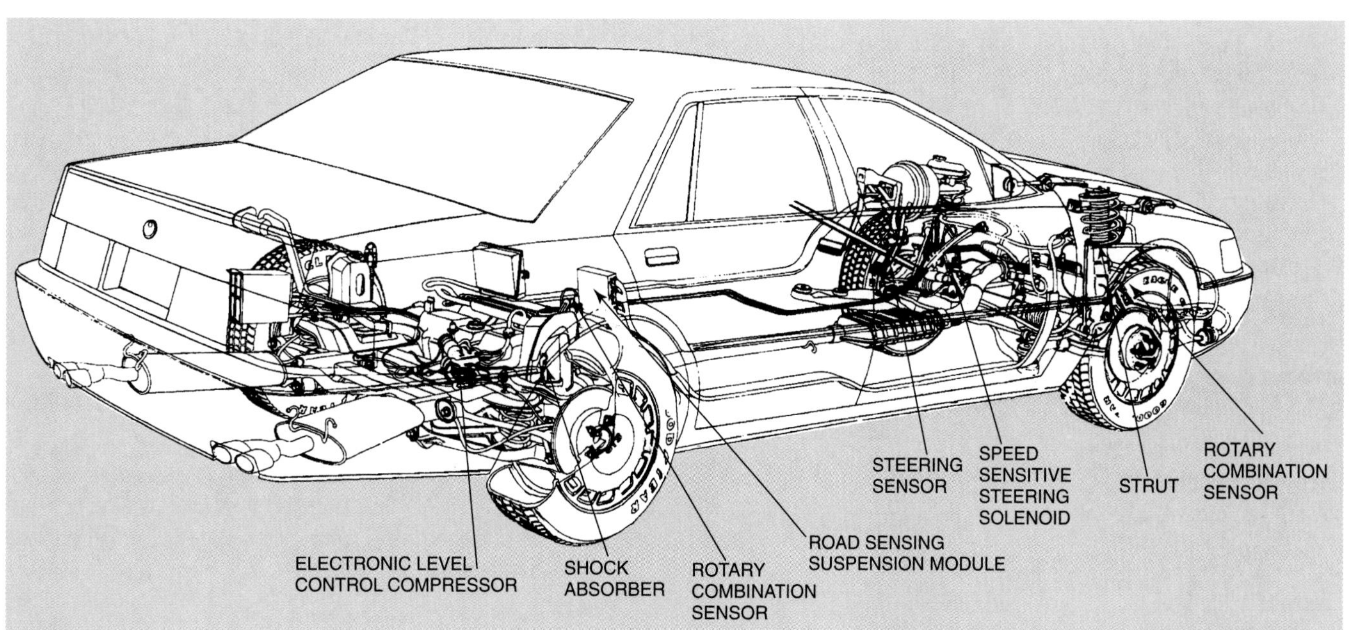

Figure 24-57. A schematic of a road sensing suspension (RSS) system, which controls shock absorber and strut damping forces, in relation to changing road and driving conditions. This system is able to make these changes in 15-20 milliseconds (thousandths of a second). Speed sensitive steering on this vehicle is also controlled by this system. (Cadillac)

When the driver steers the vehicle into a sharp turn, a steering sensor sends an appropriate signal to the control module. The module signals the actuators to increase shock pressure on the outside of the vehicle. This action decreases vehicle leaning in turns. Other systems also provide programmed ride control. The degree of stiffness and handling can be programmed into the system by the driver. Modern systems are usually computer-controlled, with a control pad or switch in the passenger compartment.

Summary

Vehicle bodies must be suspended on some type of spring devices to isolate them from the irregularities of the road. The body and frame unit must be rigid and strong to provide secure anchorage for the suspension system and to provide alignment and the securing of all parts.

Some cars use separate frames, while others use unibody construction. Where the separate frame is used, it is constructed of steel channels or square tubing and welded together. It has various crossmembers to provide mounting and bracing points.

There are three types of springs used to suspend the car: leaf springs, coil springs, and torsion bars. The suspension system depends on these springs, as well as control arms and strut rods, to control suspension operation. Stabilizer bars help prevent tipping or rolling on corners. All modern vehicles use ball joints to provide for movement between the spindle and the control arm. Hydraulic shock absorbers are used to control spring oscillation. Most vehicles have independent wheel suspensions on the front. This is accomplished by mounting the wheel assembly on the ends of pivoting control arms. Any of the three springs can be used in this system.

The rear axle can be independent or solid. Common springs used in the rear axle assembly are the coil spring, leaf spring, or torsion bar. Various combinations of control arms or stabilizer bars are used, depending on the type of drive. Many modern rear suspensions make use of MacPherson strut assemblies.

Some air shocks are manually pumped up by the driver. Ride control systems are usually air powered. They may be controlled by a leveling valve or by an on-board computer.

Know These Terms

Suspension system
Sprung weight
Unsprung weight
Body-over-frame construction
Unibody construction
Spring
Leaf spring
Spring shackles
Coil spring
Torsion bar
Spring oscillation
Shock absorber
Telescoping shock absorber
Air shock absorber
Control arm
Strut
Ball joint
Load-carrying ball joint
Follower joint
Tension loading
Compression loading
Preloaded
Stabilizer bar
Conventional front suspension
MacPherson strut front suspension
Damper
Rear suspension
Grease fitting
Plugs
Automatic level control system
Air compressor
Height control sensor
Computerized ride control system
Active suspension

Review Questions—Chapter 24

Do not write in this book. Write your answers on a separate sheet of paper.

1. Technician A says that the wheels, tires, and rear axle are part of the vehicle's unsprung weight. Technician B says the parts of the vehicle that are supported by the suspension system are the unsprung weight. Who is right?
   (A) A only.
   (B) B only.
   (C) Both A & B.
   (D) Neither A nor B.
2. In _____ construction, the body sections serve as structural members.
   (A) stub
   (B) unibody
   (C) conventional
   (D) None of the above.
3. The hinge shackle used with a leaf spring allows it to change _____ as it moves up and down.
4. All the following statements about coil springs are true, EXCEPT:
   (A) coil springs can be part of a conventional suspension.
   (B) coil springs are not used as part of a MacPherson strut suspension.
   (C) coil springs can be installed at the front axle.
   (D) coil springs can be installed at the rear axle.
5. A torsion bar absorbs shocks by _____ .
   (A) compressing
   (B) expanding
   (C) twisting
   (D) bending
6. Technician A says that both ends of the torsion bar are anchored to the frame. Technician B says that air springs are obsolete. Who is right?
   (A) A only.
   (B) B only.
   (C) Both A & B.
   (D) Neither A nor B.

7. The coil spring can be mounted in all of the following ways, EXCEPT:
   (A) between the frame and the rear axle.
   (B) between the frame and the lower control arm.
   (C) between the body and the upper control arm.
   (D) between the upper and lower control arms.
8. If the shock absorber did not control spring oscillation, what would happen to the oscillations?
9. How does the shock absorber control spring oscillation?
10. Technician A says that some ball joints are tension loaded. Technician B says that some ball joints are compression loaded. Who is right?
   (A) A only.
   (B) B only.
   (C) Both A & B.
   (D) Neither A nor B.
11. When the solid front axle is used, it is generally used on _____ .
   (A) older trucks
   (B) large cars
   (C) four-wheel drive vehicles
   (D) Both A & C.
12. What is the greatest advantage of the independent front wheel suspension system?
13. What does a stabilizer bar prevent?
14. Name the two types of loading that ball joints can be placed under.
15. Which of the following statements about a load-carrying ball joint is true?
   (A) It may be tension or compression loaded.
   (B) It carries the majority of the vehicle weight.
   (C) It does not swivel as well as move up and down.
   (D) It is the only ball joint on a MacPherson strut assembly.
16. Name the major internal parts of the telescoping shock absorber.
17. Technician A says that some of the newer suspension joints can forego greasing for periods up to 30,000 miles (48 270 km). Technician B says that rubber bushed shackles should be oiled every 5000 miles (8000 km). Who is right?
   (A) A only.
   (B) B only.
   (C) Both A & B.
   (D) Neither A nor B.
18. What kind of springs can be used on rear axle assemblies?
   (A) Leaf.
   (B) Coil.
   (C) Torsion bar.
   (D) All of the above.
19. All of the following statements about the automatic level control height control valve are true, EXCEPT:
   (A) the valve is attached to the vehicle frame.
   (B) the valve is connected to a suspension arm or rear axle housing.
   (C) the valve senses any distance changes between the frame and drivetrain.
   (D) the valve is operated by an overtravel lever.
20. An active suspension system can be controlled by __________.
   (A) a computer
   (B) the driver through a control pad or switch
   (C) the engine control computer
   (D) Both A & B.

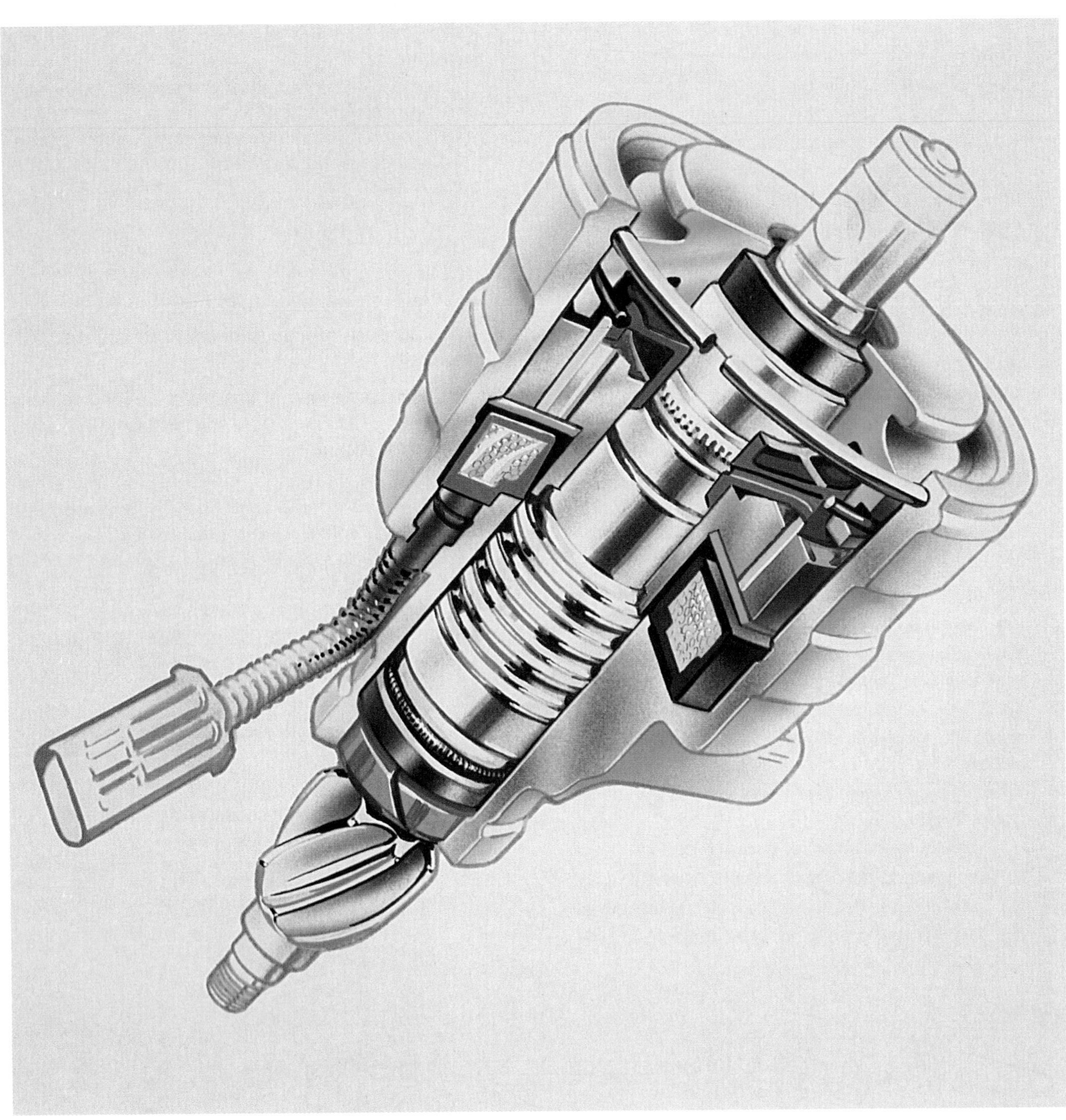

Cutaway of a power rack-and-pinion spool valve assembly that uses an electromagnet to vary the steering effort required in response to vehicle speed and road conditions. (General Motors)

25

Steering Systems

After studying this chapter, you will be able to:

- Identify the major parts of a parallelogram steering system.
- Identify the major parts of a rack-and-pinion steering system.
- Compare the parallelogram and rack-and-pinion steering systems.
- Explain the operation of a parallelogram power steering gear.
- Explain the operation of a rack-and-pinion power steering gear.
- Describe the different types of power steering pumps.

This chapter is concerned with the principles and components of the steering system. Modern steering systems are either parallelogram systems or rack-and-pinion systems. Both are available in manual and power-assisted versions. Studying this chapter will prepare you for Chapter 27, which discusses wheel alignment.

Steering System

The steering system is designed to allow the driver to move the front wheels to the right or left with a minimum of effort and without excessive movement of the steering wheel. Although the driver can move the wheels easily, road shocks are not transmitted to the driver. This absence of road shock transfer is referred to as the nonreversible feature of steering systems.

Many types of steering systems have been used in the past, **Figure 25-1**. This chapter will cover the steering systems currently used. Although there are some variations, all steering systems have some principles and components in

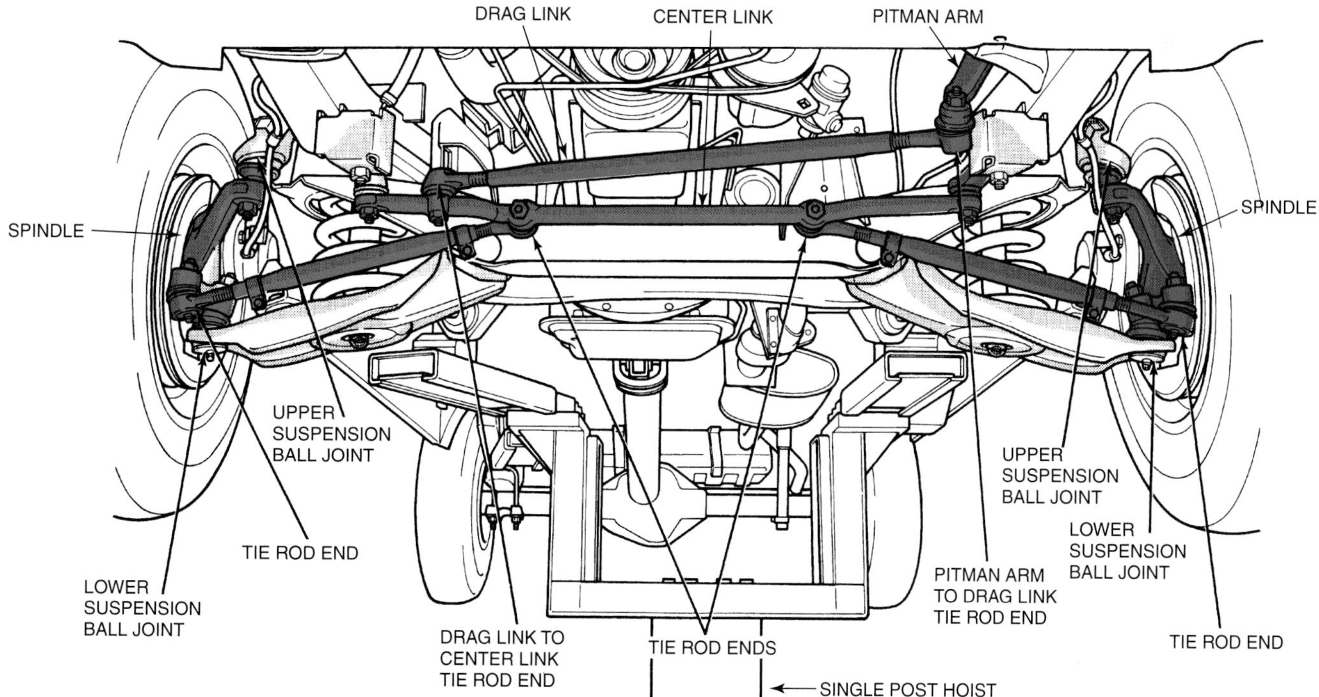

Figure 25-1. One particular conventional steering linkage arrangement as used with a rear-wheel drive pickup truck. The truck has been raised on a single-post lift to expose the steering linkage. (Dodge)

common. The basic steering system can be divided into three main assemblies:

- The spindle and steering arm assemblies.
- The linkage assembly connecting the steering arms and steering gear.
- The steering wheel, steering shaft, and steering gear assembly.

These assemblies are used in all steering systems. The principles and components of each assembly are discussed in the following sections.

Spindle and Steering Arm Assembly

When the front wheels are turned, the ***spindle assembly*** units (sometimes called the steering knuckles) swivel on the ball joints of a conventional suspension system or on the ball joint and upper bearing assembly of a MacPherson strut assembly. The ball joints and other suspension system components were discussed in Chapter 24. **Figure 25-2** illustrates a MacPherson strut suspension.

Spindle

The spindle is attached to the wheel rim and tire, as well as the brake drum or rotor. These parts rotate on the spindle. The spindle is also the mounting surface for the stationary brake components, such as the caliper and pads or the backing plate, wheel cylinder, and shoes. A typical spindle assembly is shown in **Figure 25-3.**

Steering Arms

The ***steering arms*** are either bolted to the spindle or forged as an integral part of the spindle. The outer ends of the steering arms have a tapered hole to which the tie rod end ball sockets are attached. When the steering arms are moved to the right or left, they force the spindle to pivot on the ball joints or on the ball joint and strut top bearing, depending on system design. See **Figure 25-4.**

Steering Linkage

Linkage is necessary to connect the steering gear pitman arm to the steering arms. The two linkages most commonly used today are the ***parallelogram linkage*** and the ***rack-and-pinion linkage***. The parallelogram linkage was the most common arrangement for many years and is still used on large cars and pickup trucks. The rack-and-pinion has been used on most small and mid-size vehicles for the past 15 years. The two types of linkage are discussed in the following sections.

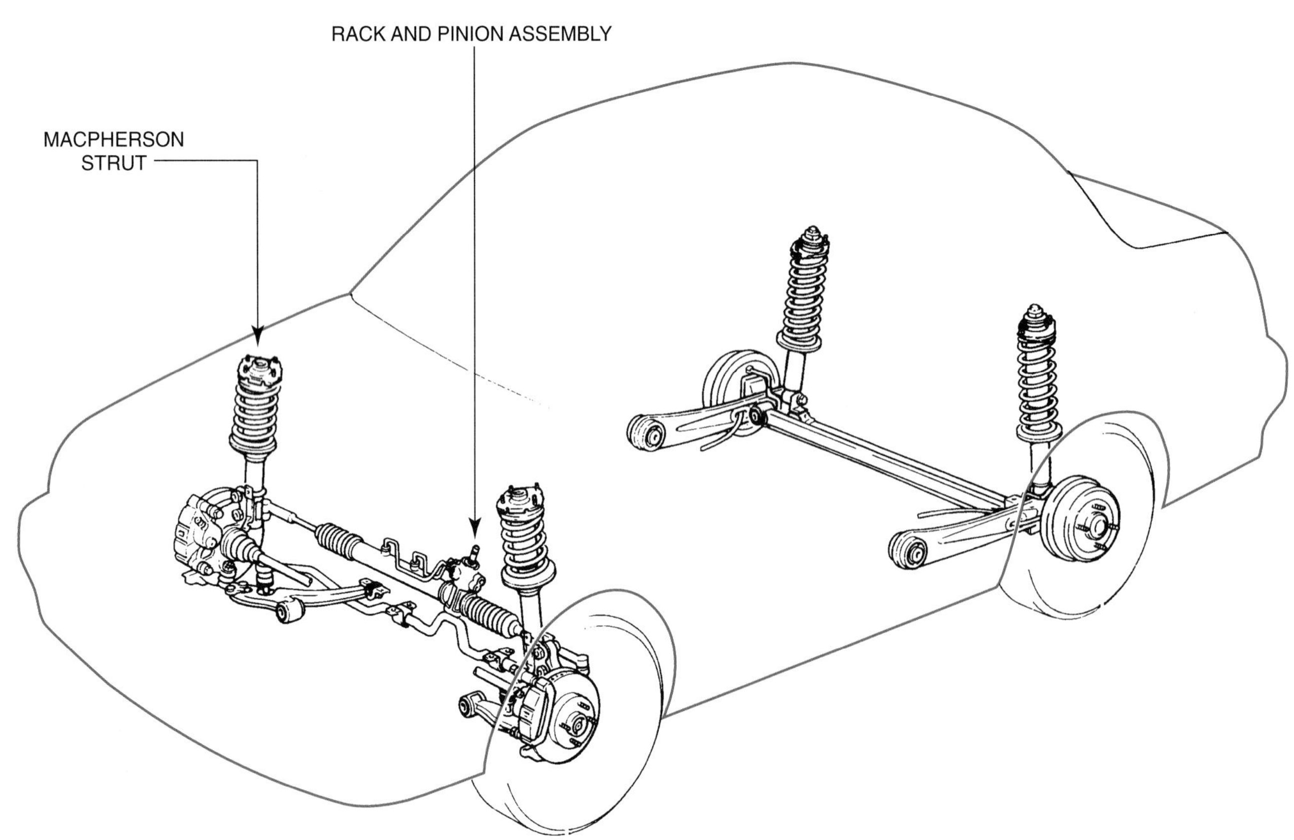

Figure 25-2. An overall view of a MacPherson strut steering setup. (Toyota Motor Corporation)

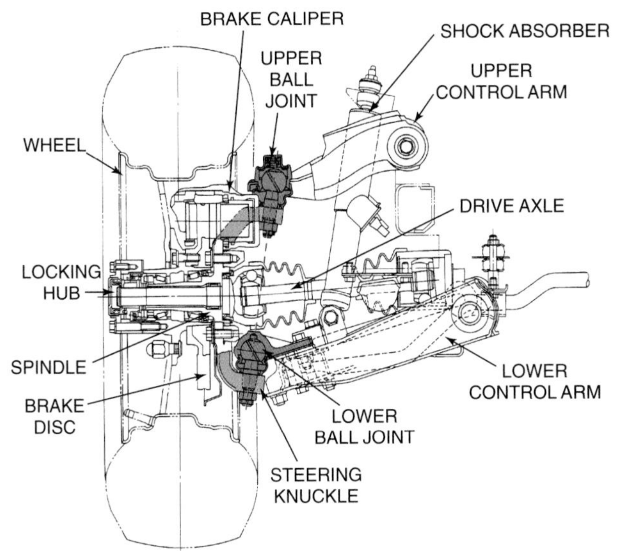

Figure 25-3. A ball joint steering knuckle assembly as used with a four-wheel drive vehicle. No kingpin is used. The steering knuckle swivels on the upper and lower ball joints. Note the centerline drawn through the ball joints. (Isuzu)

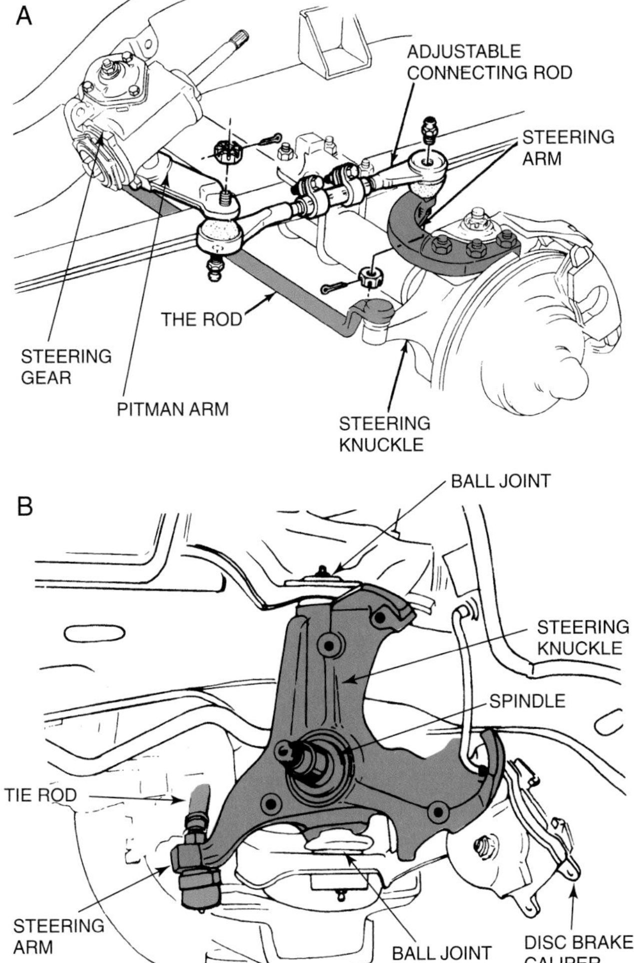

Figure 25-4. A —A steering arm, which is bolted to the spindle on a four-wheel drive front axle. B—This steering arm is forged as an integral part of the steering knuckle and spindle. Note the location of the tie rod connection. (General Motors)

Parallelogram Linkage

When viewed from the top, the parallelogram linkage is arranged in the shape of a parallelogram, **Figure 25-5.** Movement at the connecting points of the linkage pieces is handled by ***ball-and-socket connections***, such as those shown in **Figures 25-6** and **25-7.** The parts of the parallelogram linkage include the pitman arm, the center link, the idler arm, and the tie rods.

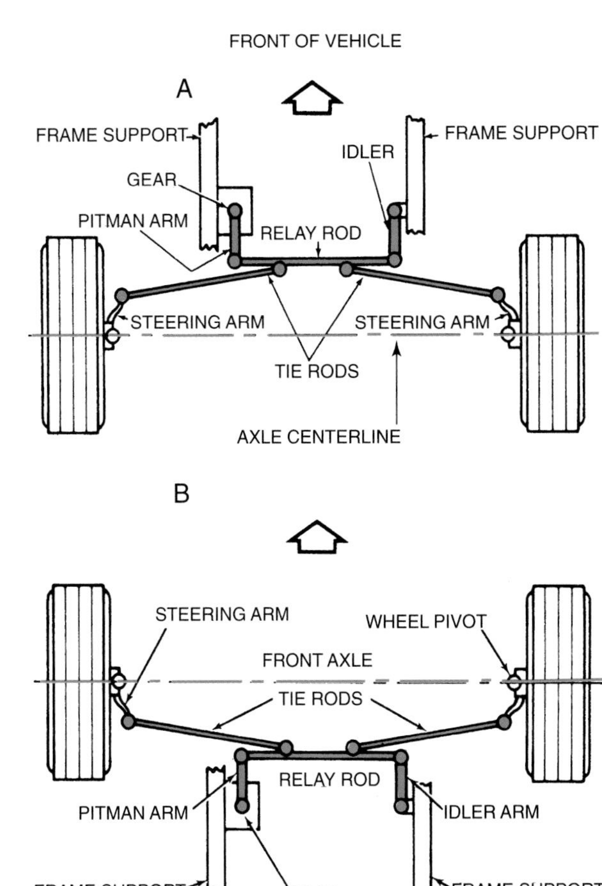

Figure 25-5. A—A parallelogram steering linkage located in front of the axle centerline. B—This parallelogram linkage is placed behind the axle centerline. These different linkage setups will help to maintain the suspension geometry as the steering and suspension move, keeping all parts in proper relationship to one another. (General Motors)

Pitman Arm

The ***pitman arm*** (also called the steering gear arm) is attached to the output shaft of the steering gear. The pitman arm is usually splined to the output shaft and is held in place by a large nut. See **Figure 25-8.** Movement of the steering gear causes the pitman arm to swing from side to side. The pitman arm may be equipped with a ball joint connection where it is attached to the center link, or it may have a hole to accept a ball stud.

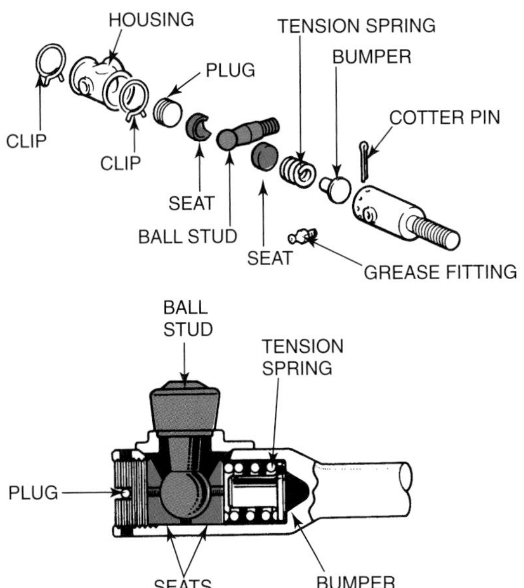

Figure 25-6. Two styles of ball-and-socket connections. If the tension spring on the side of the ball stud seat breaks on either type, it cannot collapse far enough to permit the socket seats to spread apart and drop the ball stud. Seat spreading is controlled by the steel block or bumper inside the spring. (Toyota)

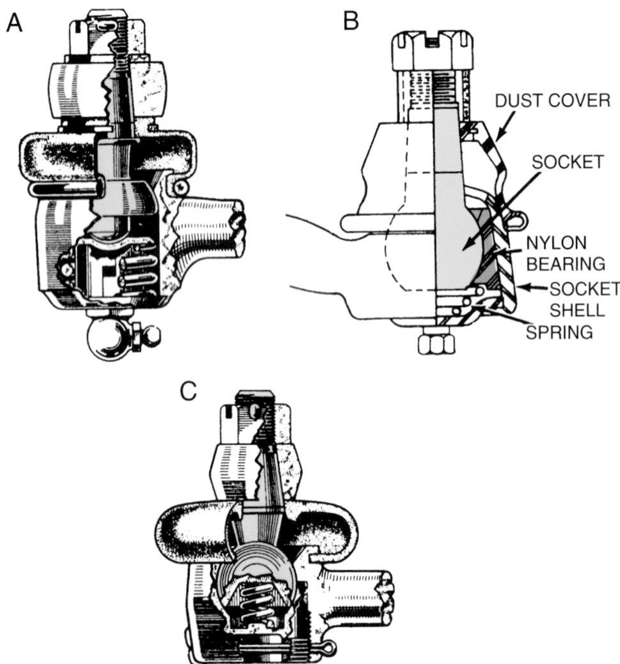

Figure 25-7. Different types of ball socket tie rod ends. B uses a nylon bearing, while A and C employ a metal-to-metal bearing surface.

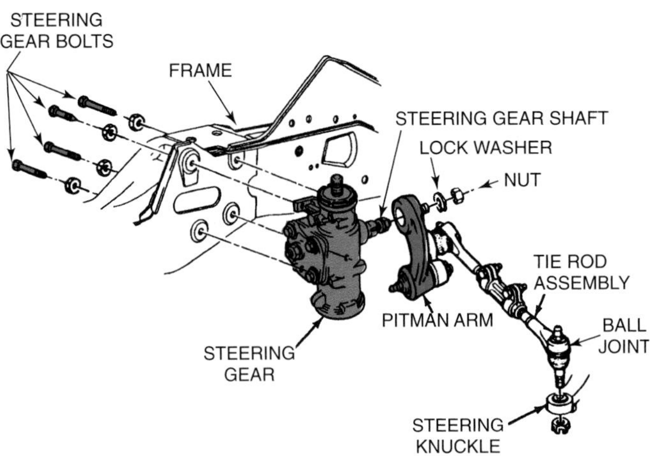

Figure 25-8. A pitman arm which is removed from the steering gear. The pitman arm is fastened to the steering gear shaft with a nut and lock washer. (Chevrolet)

Center Link

The ***center link*** (also called a relay rod, or drag link) is a long bar that transfers movement from the pitman arm to the other parts of the steering linkage. The pitman arm is attached to one end of the center link by means of a ball joint connection, **Figure 25-9.**

Idler Arm

The end of the center link opposite the pitman arm is attached to an ***idler arm*** with a ball joint connection. The idler arm pivots on metal bushings or bearings that are placed on a solid shaft that is bolted to the frame. Some idler arms pivot on rubber bushings. The idler arm is arranged so it is parallel to the pitman arm. The purpose of the idler arm is to keep the center link in alignment as it is moved by the pitman arm. Two idler arms used in one particular steering setup are shown in **Figure 25-9.**

Tie Rods

Tie rods, **Figure 25-10,** are used to connect the steering arms to the center link. The parallelogram steering system has two inner tie rods and two outer tie rods. One end of each tie rod has a ball joint socket, which is used to connect the rod to the steering arms or center link. The ball sockets are commonly called the tie rod ends. The other end of each rod is threaded.

A threaded ***tie rod sleeve*** is installed between the tie rods to form a tie rod assembly. The threads on one side of the assembly are left-hand threads, while the threads on the other side are right-hand threads. Therefore, turning the sleeve will change the length of the assembly. Tie rod assembly length can be changed to adjust toe. This is discussed in more detail in Chapter 26. One type of adjustable tie rod end is shown in **Figure 25-11.** A typical tie rod adjustment procedure is shown in **Figure 25-12.**

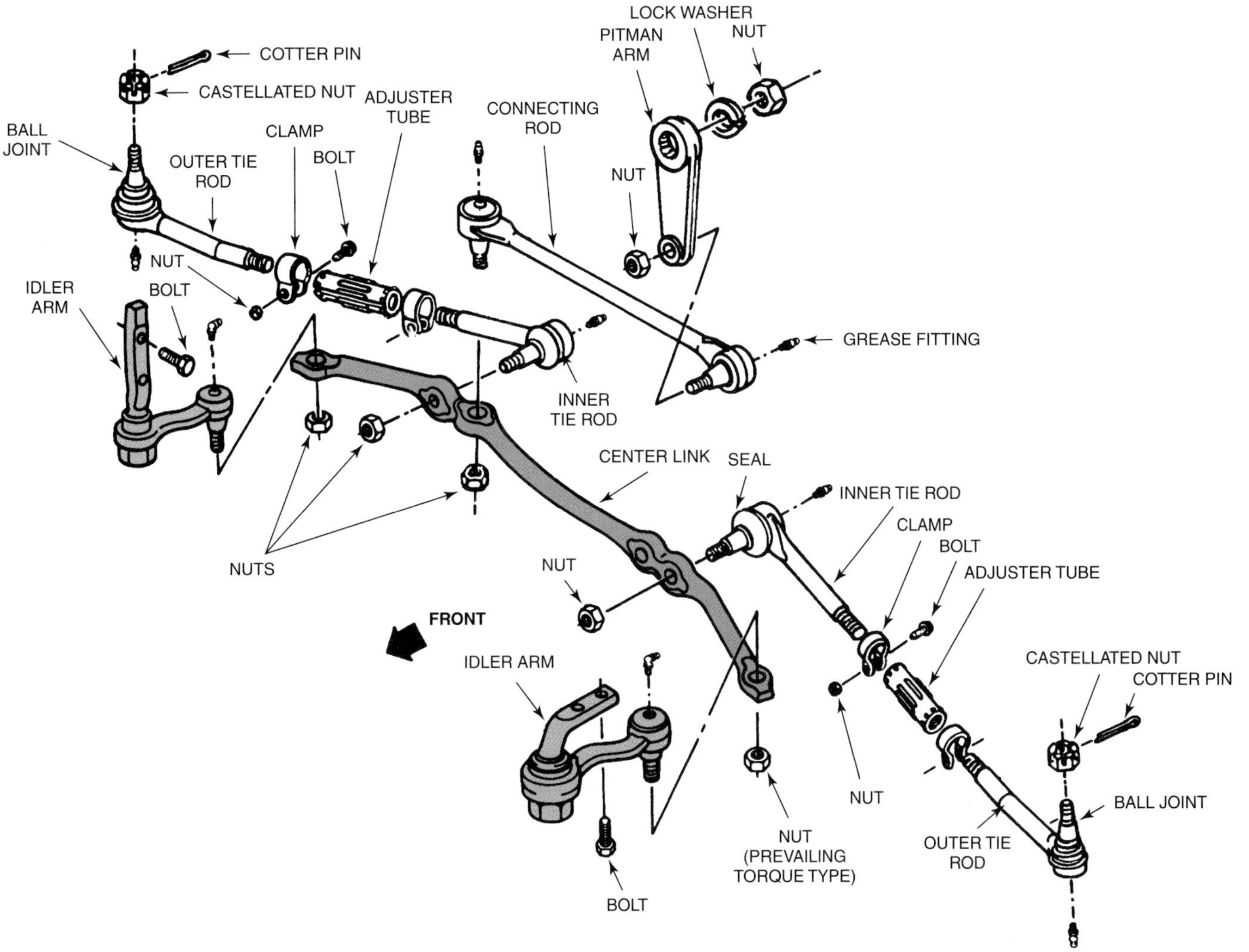

Figure 25-9. A center link with other related steering linkage parts. (Chevrolet)

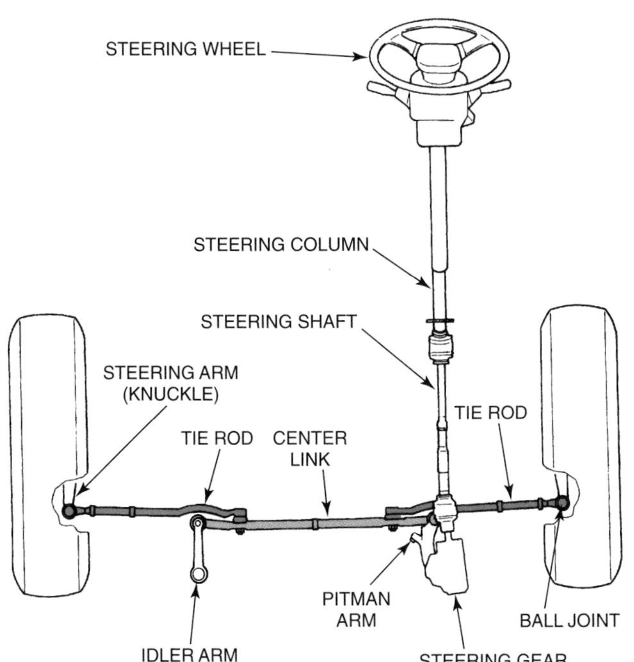

Figure 25-10. Tie rods connecting steering arms (knuckles) to the center link (relay rod). (Geo)

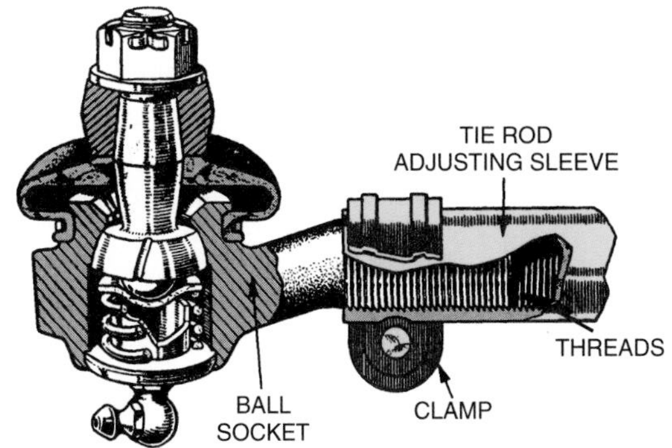

Figure 25-11. Adjustable tie rod end. The ball socket can be screwed in or out of tie rod. When proper adjustment is achieved, it is retained by tightening the clamp on the tie rod. This will bind the tie rod threads against those on the ball socket end.

Tie rod assemblies are usually equal length and about the same length as the lower control arms. As the frame (to which the center link is fastened via the pitman arm and idler arm) moves up and down in relation to the spindle, there will

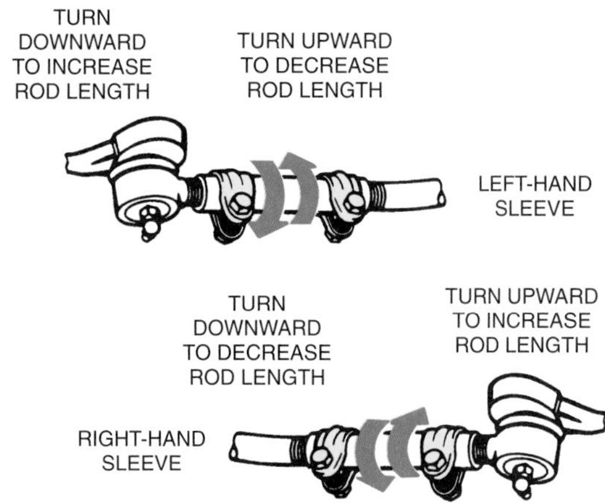

Figure 25-12. Tie rod adjustment. This type uses a double-end sleeve with two clamps. (Mercury)

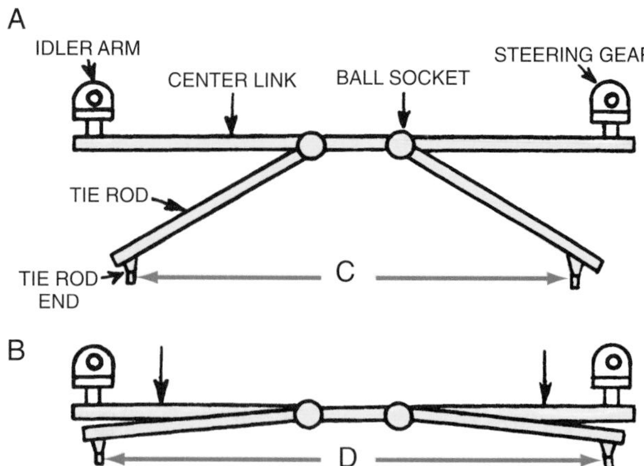

Figure 25-13. The center link and tie rods should be nearly parallel. A—Setup with the tie rods and center link not parallel (front view). Notice distance C between tie rod ends. B—When the body of the vehicle dips downward, the center link is carried down with the frame. The tie rod ends are fastened to the steering arms and cannot move down. As the center link travels down, the angle between the link and the rods narrows. The tie rods must spread out (distance D), forcing the steering arms out and toeing the wheels in. When a vehicle is at normal curb height, the link and rods should be almost parallel, as in B. In this position, the link can move up and down with a minimal effect on toe-in and toe-out.

be minimum disturbance in toe-in and toe-out, since the control arm and tie rod assemblies will be moving through almost identical arcs. This similarity of arcs keeps the tie rods and spindle in the same relative position when driving over rough roads.

Parallelogram Component Arrangement

The center link and the tie rods should be as close to parallel as possible (when viewed from front). This will help eliminate toe-in and toe-out changes as the frame moves up and down, **Figure 25-13.** A complete parallelogram linkage system is shown in **Figure 25-14.**

Another type of linkage arrangement used on trucks with a solid front axle is shown in **Figure 25-15.** Notice that the tie rod is one piece and is connected at each end to the steering arms. A drag link, working fore and aft, connects the pitman arm (sometimes referred to as the steering sector shaft arm) to an extension on the spindle. When the drag link swivels this spindle, the action is transferred to the steering arms by the tie rod.

Rack-and-Pinion Linkage

The rack-and-pinion steering system contains fewer linkage components than the parallelogram system. There is no pitman arm, idler arm, or center link. The steering column shaft is connected directly to the pinion, or sector shaft. The pinion operates the rack section, which operates the tie rods and, therefore, the steering knuckles. See **Figure 25-16.**

The only rack-and-pinion linkage parts are inner and outer tie rods. The outer tie rod ends resemble those discussed earlier in this chapter. The inner tie rod ends are a simple ball socket arrangement that is covered with a bellows or boot to keep out water and dirt. The inner tie rod can turn

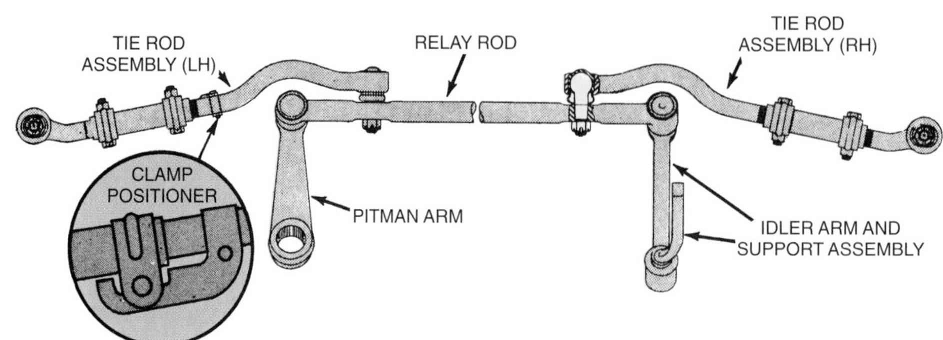

Figure 25-14. Top view of a parallelogram linkage. Each tie rod can be adjusted without changing toe-in or toe-out of the opposite wheel. This allows a proper setting and leaves the steering wheel in the straight ahead position. (Oldsmobile)

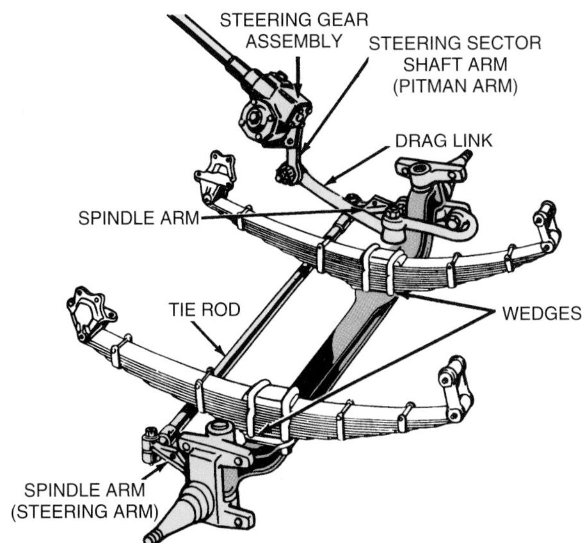

Figure 25-15. Steering linkage. This drawing shows a typical linkage setup on a solid axle, light truck application. The steering arm (spindle arm) on the driver's side acts as a bell crank. (Ford)

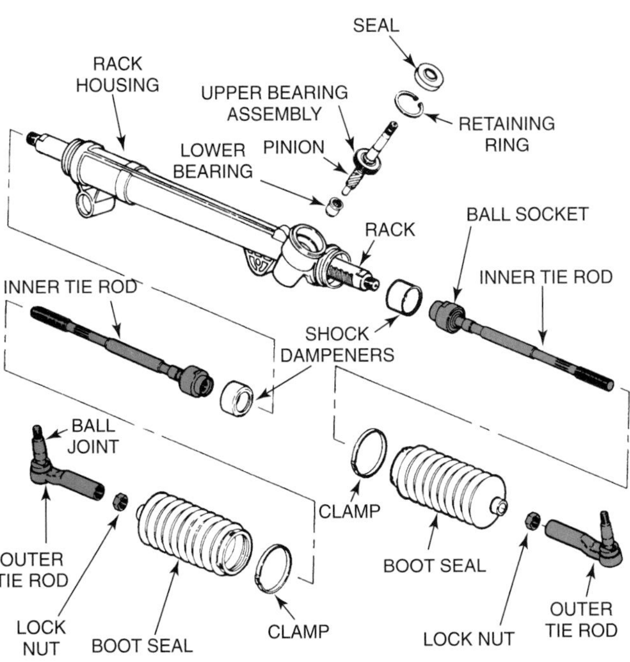

Figure 25-16. Exploded view of a rack-and-pinion steering gearbox assembly. Both ends of the rack are attached to the steering arms (not shown) by means of the tie rods. (General Motors)

in its socket, and toe adjustments are made by loosening a locknut and turning the inner tie rod in the outer tie rod. A few rack-and-pinion units have adjustment sleeves that are similar to those used in parallelogram linkage systems.

Steering Wheel, Steering Shaft, and Steering Gear

The ***steering wheel***, ***steering shaft***, and ***steering gear*** are important parts of the steering system. Steering effort begins at the steering wheel and shaft. The steering gear transforms the turning motion of the shaft to the linear movement of the linkage.

Steering Wheel

Besides the steering wheel's obvious job of allowing the driver to input steering motion, it contributes to steering ease by its size. The larger the steering wheel, the less effort required to make turns. This is important on vehicles without power steering.

Steering Shaft

The steering shaft connects the steering wheel and the steering gear. In the past, it consisted of an upper shaft made of hardened steel, a flexible coupling, and a lower, or stub, shaft splined to match the input shaft of the steering gear. The modern steering shaft has a two-piece upper shaft, which is designed to collapse if the driver is thrown against the steering wheel during an accident. The steering shaft often has a flexible joint to allow steering effort to be transmitted through at an angle. A modern steering shaft is shown in **Figure 25-17.**

Steering Shaft Coupling

The ***steering coupling*** absorbs road vibrations that get past the other steering system linkage. It consists of a rubber disc and is connected between the upper and lower steering shafts. The disc is firm enough to accurately transmit steering effort between the shafts. Vibrations and road shocks transmitted from the lower shaft twist the rubber instead of traveling to the upper shaft. A typical steering shaft coupling assembly is shown in **Figure 25-18A.**

Steering Shaft Universal Joint

Many steering shafts have one or more ***universal joints,*** which allow the shaft to be located at various convenient angles without moving the steering gear. Universal joints also relieve the mounting stresses and deflecting (bending) forces sometimes encountered in a solid shaft. In addition, the universal joint dampens steering shock and vibration transfer to the steering wheel. See **Figure 25-18B**.

Steering Gear

The steering gear is designed to multiply the driver's turning torque so the front wheels may be turned easily. When the parallelogram linkage is used, the torque developed by the driver is multiplied through gears and is then transmitted to the wheel spindle assemblies through the linkage. On the rack-and-pinion steering system, the steering column shaft is

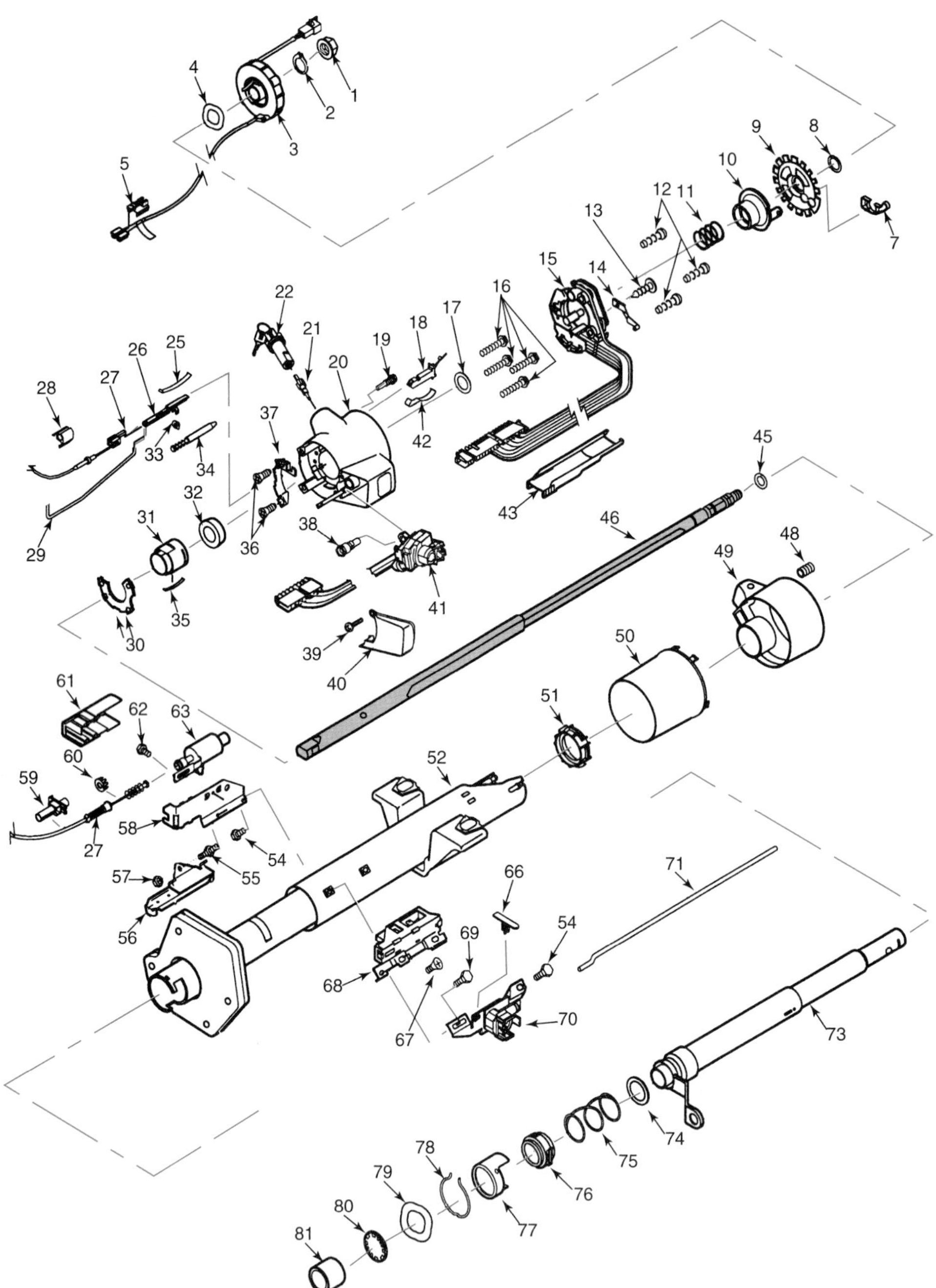

Figure 25-17. An exploded view of a standard (non-tilt) steering column assembly used with an airbag. Columns vary in design and construction. 1—Locking hex nut. 2—Retaining ring. 3—Airbag coil assembly. 4—Wave washer. 5—Connector shroud. 7—Shaft lock spacer. 8—Retaining ring. 9—Shaft lock. 10—Turn signal cancel cam assembly. 11—Upper bearing spring. 12—Binding HD cross recess screw. 13—RD wash HD screw. 14—Switch actuator arm. 15—Turn signal switch assembly. 16—HD Screw. 17—Thrust washer. 18—Buzzer switch assembly. 19—Lock retaining screw. 20—Steering column housing assembly. 21—Sector. 22—Steering column lock cylinder set. 25—Rack preload spring. 26—Switch actuator rack. 27—Solenoid cable assembly. 28—Connector clip. 29—Switch actuator rod. 30—Upper bearing retainer. 31—Bearing retaining bushing. 32—Bearing assembly. 33—Spring thrust washer. 34—Spring and bolt assembly. 35—Horn circuit contact. 36—Flathead cross recess screw. 37—Shift lever gate. 38—Switch actuator pivot pin. 39—Binding HD cross recess screw. 40—Housing cover. 41—Pivot and pulse switch assembly. 42—Buzzer switch retaining clip. 43—Wiring protector. 45—Retaining ring. 46—Steering shaft assembly. 48—Shift lever spring. 49—Gearshift lever bowl assembly. 50—Gearshift bowl shroud. 51—Bowl lower bearing. 52—Steering column jacket assembly. 54—Wash HD screw. 55—Dimmer and ignition switch mounting stud. 56—Connector body bracket assembly. 57—Hexagon nut. 58—Solenoid bracket. 59—Adjuster housing assembly. 60—Ball joint cap. 61—Solenoid protector. 62—Hex washer head screw. 63—Interlock solenoid assembly. 66—Cable retainer. 67—Flat HD screw. 68—Ignition switch assembly. 69—Hex washer HD tapping screw. 70—Dimmer switch assembly. 71—Dimmer switch actuator rod. 73—Shift tube assembly. 74—Spring thrust washer. 75—Shift tube return spring. 76—Adapter and bearing assembly. 77—Bearing retainer. 78—Lower bearing adapter clip. 79—Bearing preload washer. 80—Lower spring retainer. 81—Shaft sleeve.

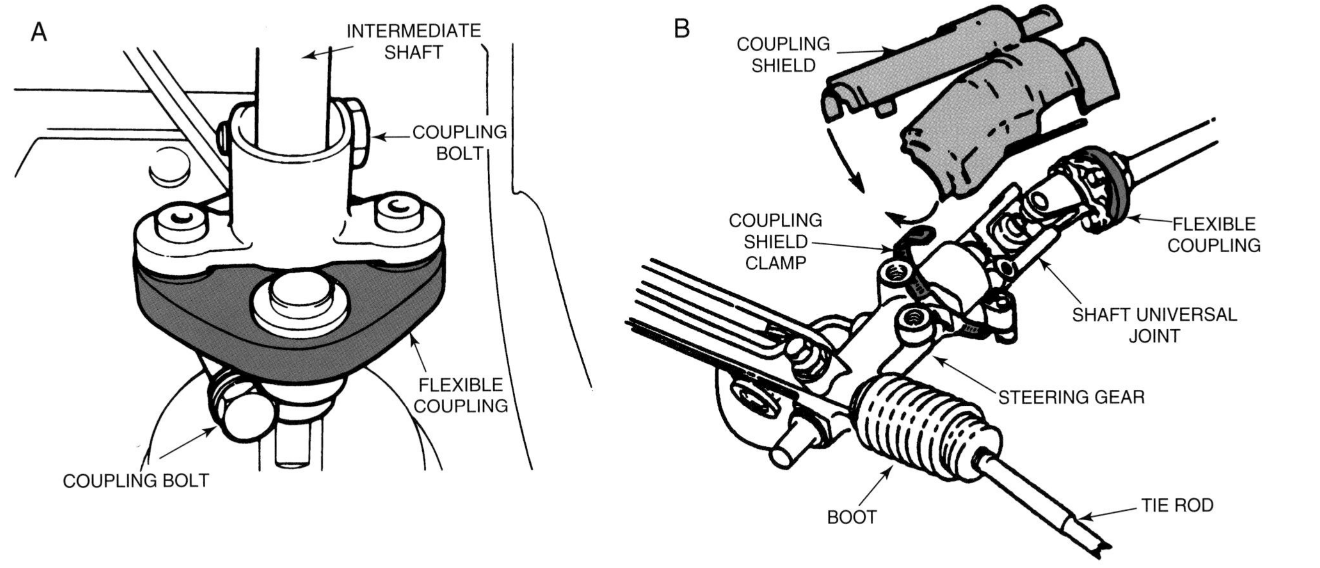

Figure 25-18. Two different types of steering shaft flexible coupling designs. Compare A and B. Note the protective coupling shield used in B. This shield protects the joint from rocks, mud, etc. (General Motors)

connected directly to the pinion shaft. Turning the pinion moves the rack section, which moves the linkage. Late-model vehicles use either manual steering gears or power steering gears.

Manual Parallelogram Linkage Steering Gear

The simplest form of the manual steering gear used with parallelogram steering systems is the ***conventional steering gear,*** **Figure 25-19.** It consists of a steel shaft with a ***worm***

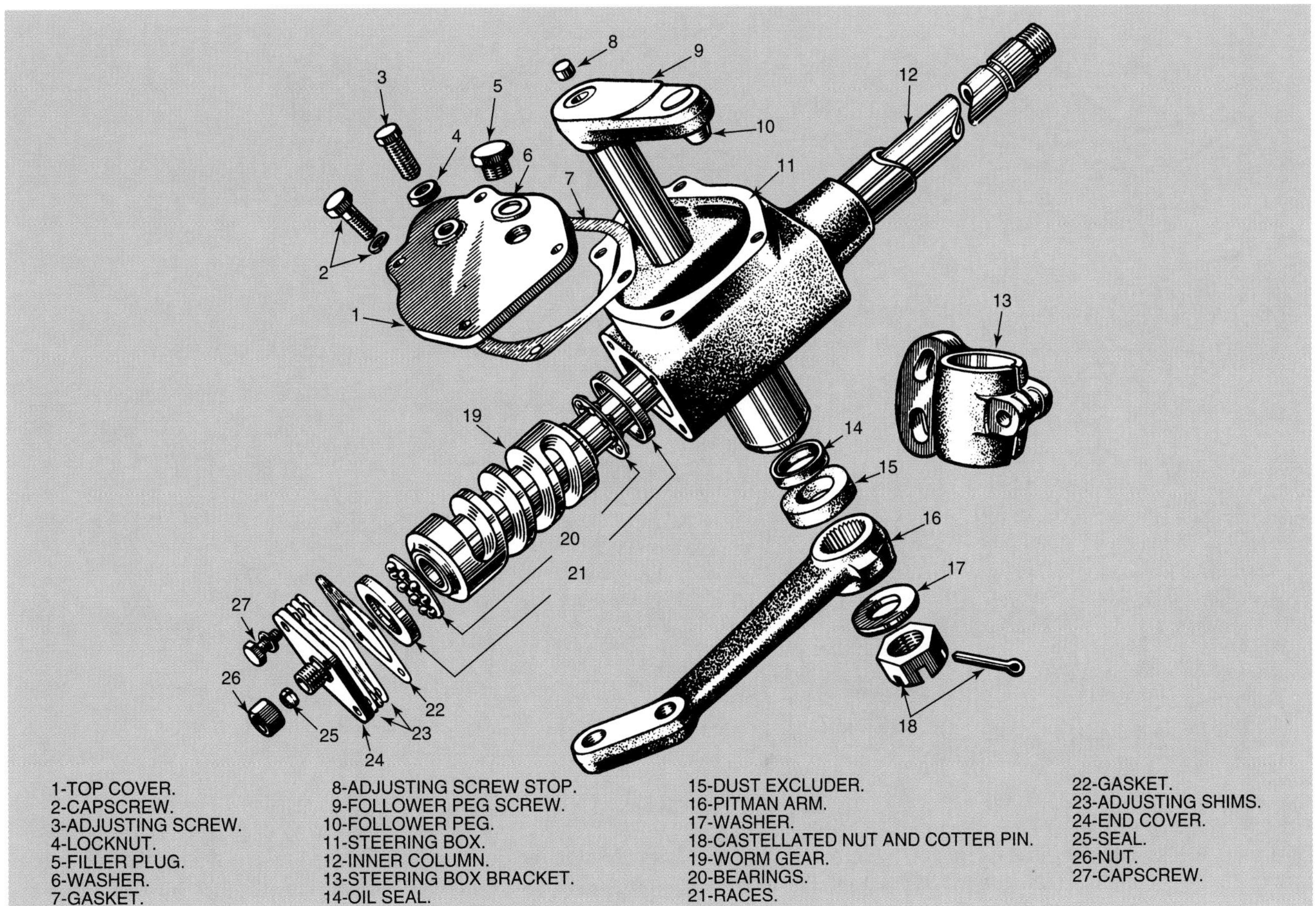

Figure 25-19. Exploded view of a worm and tapered pin (follower peg) type steering box.

gear on one end. The other end is attached to the steering shaft. A ***pitman arm shaft*** passes at right angles to the worm gear. The inner end of the pitman shaft (also called cross shaft, roller shaft, sector shaft, or gear shaft) engages the worm gear either through gear teeth, a round steel peg, a roller gear, or a notched arm that passes around a peg affixed to a ball nut that rides on the worm.

In **Figure 25-19,** a tapered peg, (10), rides in the worm grooves. When the worm is revolved, it causes the peg to follow along the worm grooves, imparting a turning motion to the pitman shaft assembly. Splines in the pitman arm attach it securely to the pitman shaft.

The worm gear end of the steering shaft is supported by ball or roller bearings located on each side of the worm. These bearings remove end and side play in the worm. Bearing adjustment is provided by shims. The pitman arm shaft operates in bushings or needle bearings. There is an adjustment screw to control worm and pitman shaft clearance.

A cast housing surrounds and holds all parts in alignment. The housing is partially filled with gear oil for lubrication. Seals prevent the escape of lubricant and keep out dust and dirt. The housing is bolted to the frame or some other rigid area of the vehicle. The steering shaft passes up through a steering column to the steering wheel.

Recirculating Ball Steering Gear

Figure 25-20 is a cross-section of the ***recirculating ball steering gear.*** Instead of a tapered peg riding in the worm grooves, this steering gear uses a ball nut that rides up and down the worm. Ball bearings ride half in the worm and half in the nut. When the worm is turned, the ball bearings impart an axial (lengthwise of shaft) thrust to the ball nut. Because the

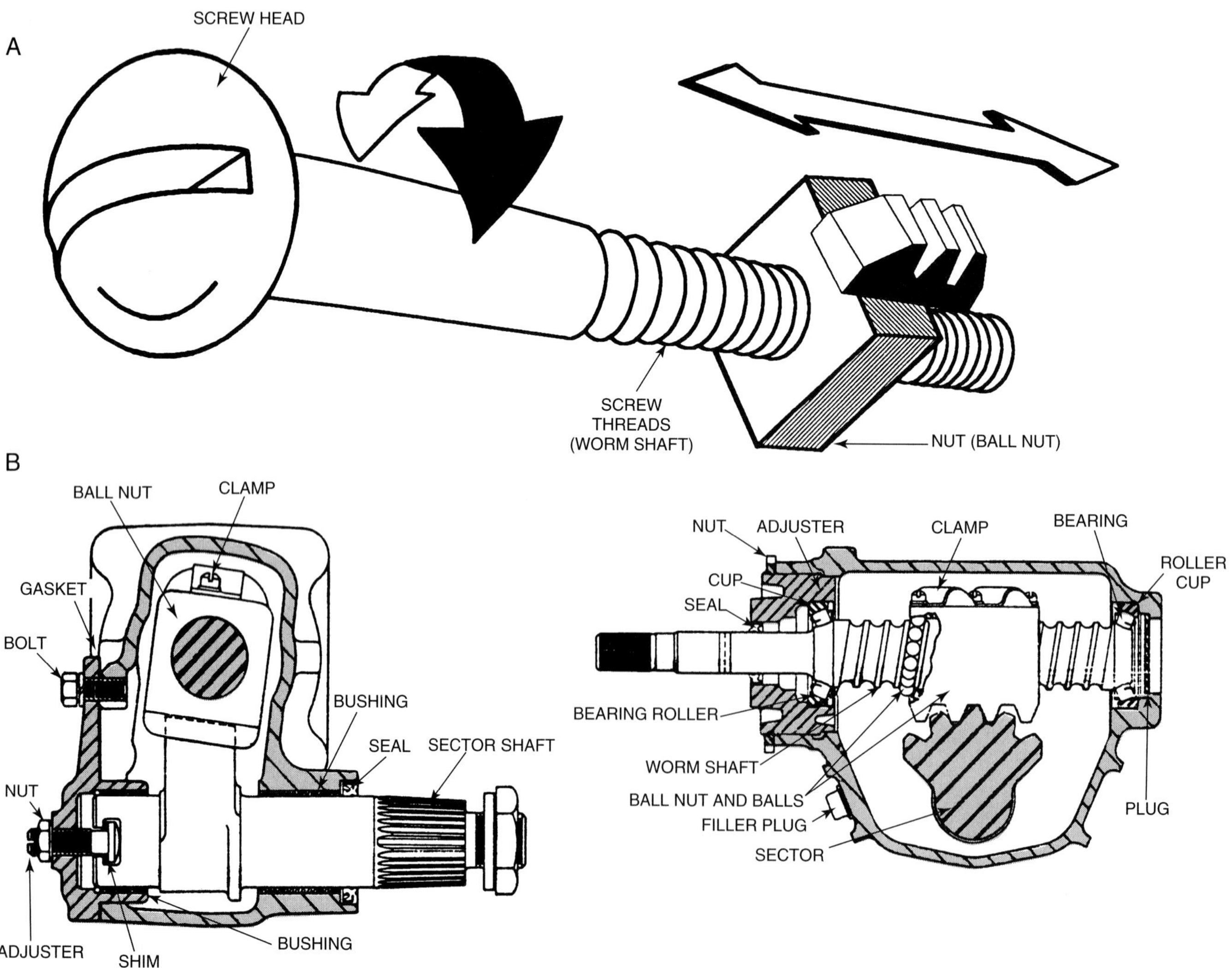

Figure 25-20. Recirculating ball worm and nut steering gear. A—A simple steering gear action produced from a screw and nut. By turning the screw, the nut (ball nut) will travel up or down the threads of the screw (worm shaft). This turns the steering wheel rotary (around-and-around) motion into a lateral (side-to-side) movement of the steering linkage. B—Cross-sections of the recirculating ball worm and nut steering gear. Study part names, locations, and functions. (Dodge and FMC Corp.)

nut is prevented from turning, it will travel up or down, depending on the worm direction.

As the ball bearings roll, they reach the end of the groove in the nut and are forced through a return tube. Eventually, they re-enter the nut groove at the opposite end. This process accounts for the name recirculating ball steering gear.

One side of the nut has gear teeth cut into it. The pitman arm sector gear teeth engage the nut teeth. When the nut travels up or down, the pitman shaft is forced to rotate. The recirculating ball worm and nut steering gear reduces friction to a very low level, making it a very popular design.

Worm-and-Roller Steering Gear

A ***worm-and-roller steering gear*** is shown in **Figure 25-21.** The pitman shaft engages the worm by means of a roller with gear teeth cut around it. The roller is mounted on antifriction bearings. Because the roller is free to turn, the friction developed in a wiping gear action is bypassed. When the worm is rotated, the roller teeth follow the worm and impart a rotary motion to the pitman shaft.

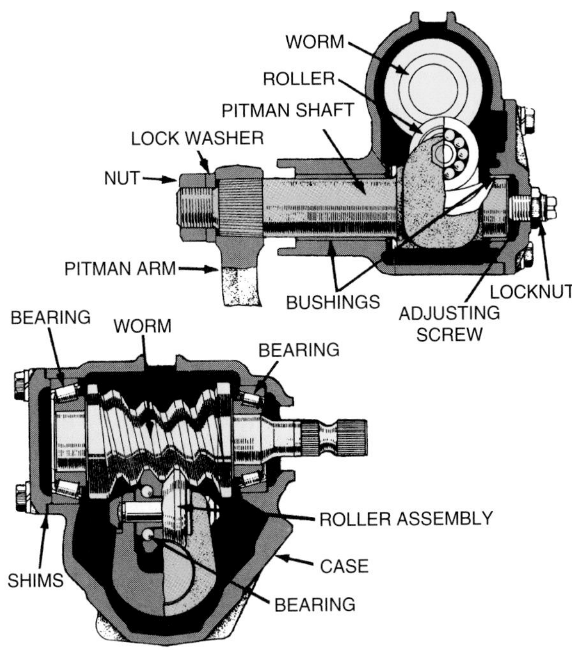

Figure 25-21. Cross-sections of a worm and roller manual steering gear. (Toyota)

Manual Rack-and-Pinion Steering Gear

A manual rack-and-pinion steering gear is shown in **Figures 25-22** and **25-23.** A steering shaft from the steering wheel fastens to the ***pinion shaft,*** usually through a universal joint. The teeth on the pinion are meshed with the teeth on the ***rack.*** When the pinion is turned, it forces the rack to move, either to the right or left. The ends of the rack are connected to the steering arms by flexible joints and short tie rods. All gear rack movement is transferred directly to the steering arms.

Steering Gear Reduction

The amount of ***gear reduction*** in steering gears depends on the vehicle type and weight, the steering characteristics (fast or slow) desired, and the ultimate use of the vehicle. Steering gear ratios will vary from 15:1 to 24:1. Steering wheel movement will average from 4 to 6 turns to move the wheels from full left to full right for standard manual steering gears. The more turns needed to move from full left to full right, the easier the steering. The modern steering gear ratio is a compromise between steering ease and handling.

Power Steering

Power steering is designed to reduce the effort needed to turn the steering wheel by utilizing hydraulic pressure to bolster (strengthen) the normal torque developed by the steering gear. Power steering systems should ease steering wheel manipulation and, at the same time, offer enough resistance that the driver can retain some road feel. Power steering is used with both conventional and rack-and-pinion systems. Both types of power steering systems are discussed in the following sections.

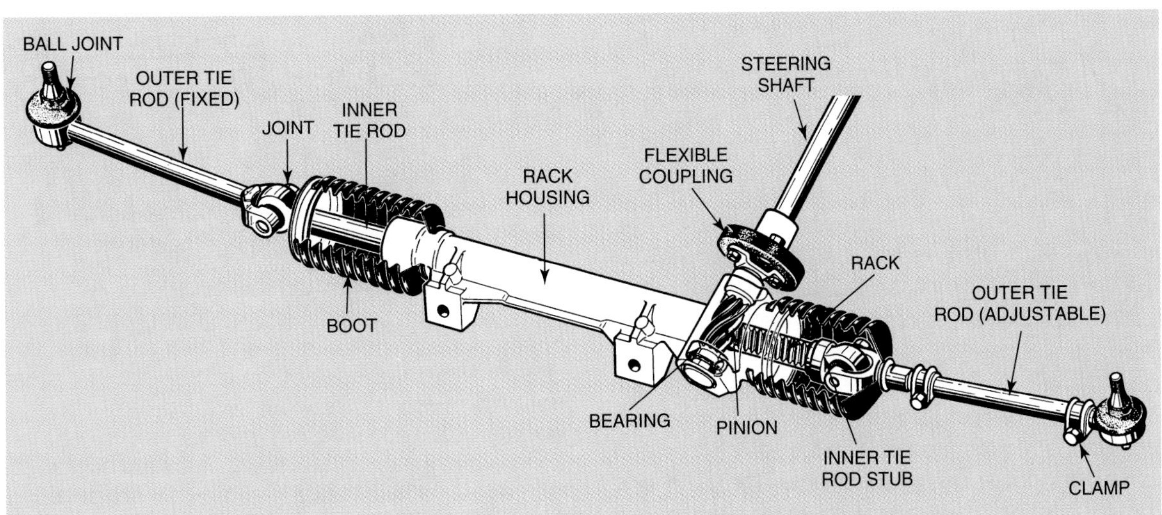

Figure 25-22. Cutaway view of a rack-and-pinion assembly. Note how the rack-and-pinion gears mesh.

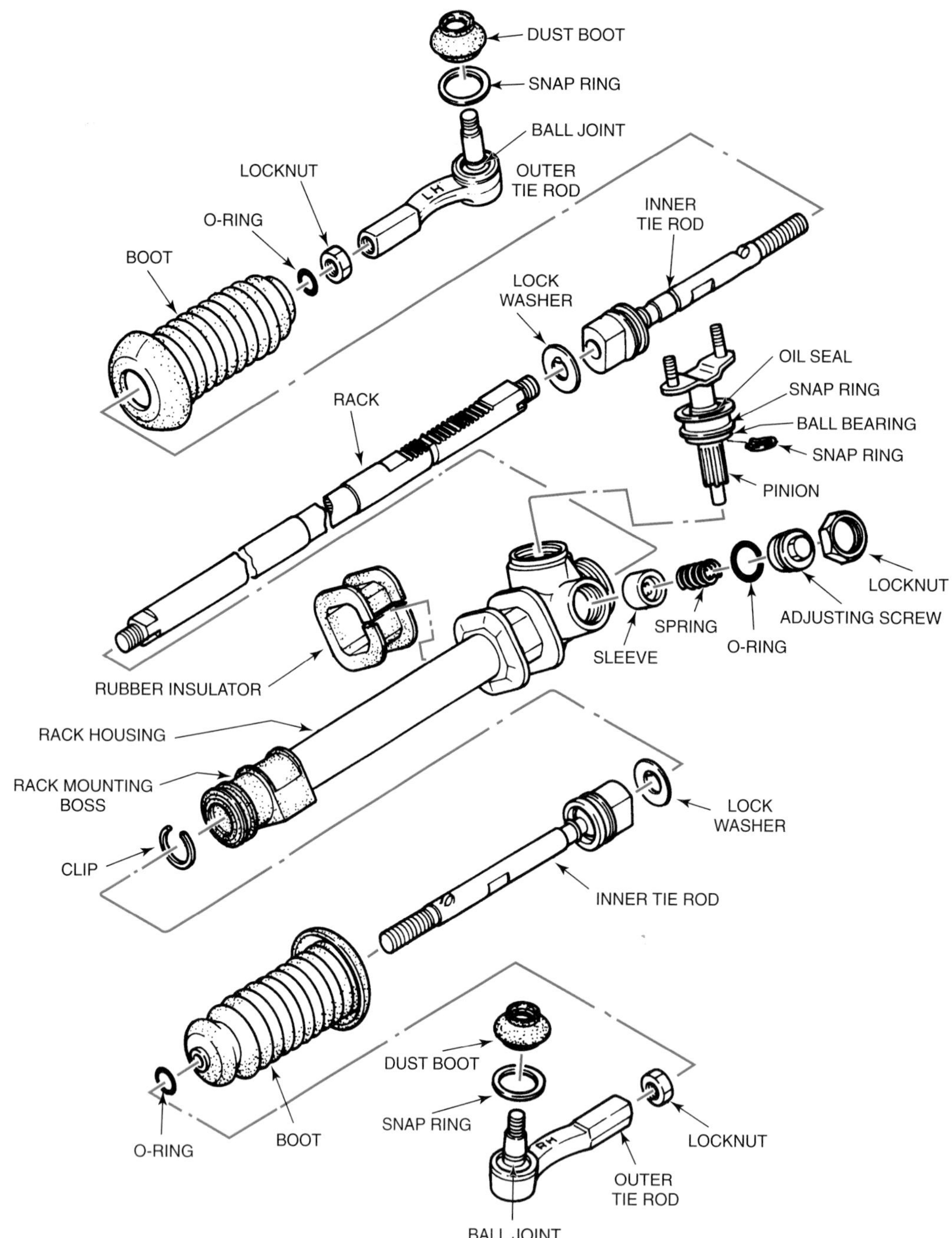

Figure 25-23. An exploded view of a manual rack-and-pinion steering assembly. Study the overall construction and part layout carefully. (Subaru)

Road Feel

Road feel is the feeling imparted to the steering wheel by the wheels of a vehicle in motion. This feeling is very important to the driver in sensing and predetermining vehicle steering response.

A good driver can tell when the front wheels are approaching the walk (slipping sideways) point and if more or less turning effort is required to stabilize the vehicle body on turns or when driving in windy conditions. Complete and effortless power steering would rob the driver of valuable road feel. To maintain road feel, power steering systems require some wheel effort. Wheel effort of approximately 1 to 9 lbs. pull on the steering wheel rim is acceptable.

Power Steering Components

All modern power steering systems are integral systems. These systems use control and use the hydraulic pressure directly within the steering gear housing. The steering gear housings used in integral systems are known as ***self-contained power steering gears***. Pressure is provided by a hydraulic pump, usually belt driven by the engine, to produce the necessary hydraulic pressure. A few are driven by an electric motor. A typical integral power steering system is shown in **Figure 25-24.**

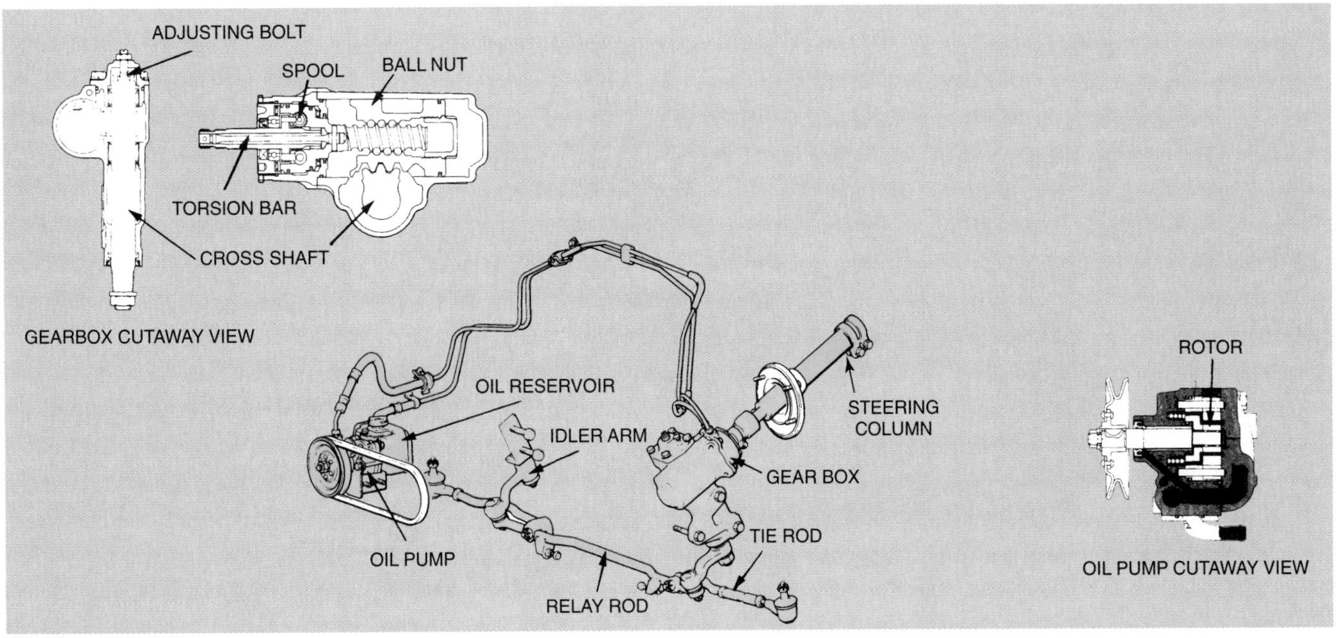

Figure 25-24. Integral power steering system. Note the gearbox and oil pump cutaways. (Chrysler)

In all self-contained power steering gears, a ***control valve*** is actuated by the driver's steering effort. This valve admits oil, under pressure, to one side of a ***hydraulic piston*** or the other. The pressure the oil creates against the piston is transferred to the pitman shaft, assisting the driver in turning the front wheels.

Self-Contained Power Steering Gear Components

The self-contained steering gear contains the control valve mechanism, the power piston, and the gears. Pressure developed by the unit is applied to the pitman shaft. All modern conventional power steering gears are ***inline*** units. **Figure 25-25** shows a basic inline unit. Notice that all the basic elements of the conventional manual steering gear are used. A pitman shaft (steering gear cross shaft) engages a recirculating ball nut that rides on a typical worm gear or shaft. The basic differences include the utilization of the ball nut as a hydraulic power piston and the addition of a ***reaction unit***, a ***pivot lever***, and a valve body containing a ***spool valve.***

Unit Operation

In operation, oil under pressure from the pump enters the valve body through the inlet passage. When no effort (above 1 to 9 lb.) is being applied to the steering wheel, the reaction unit is centered and the pivot lever is straight up and down. This keeps the spool valve in a neutral position.

Two round washer-like reaction springs hold the center thrust bearing race in a neutral position, **Figure 25-25.** The center thrust bearing is attached to the worm shaft and secured between two thrust bearings. These bearings are, in turn, affixed to the worm shaft by a shoulder on the left end and a locknut on the right end. The bearings revolve with the worm shaft, but the center thrust bearing remains stationary. The center thrust bearing can move back and forth with the worm shaft, but it cannot move around with it.

The lower end of the pivot lever rests in a cutout in the center thrust bearing. When the center thrust bearing is in the neutral position, the pivot lever is straight up and holds the spool valve in the neutral position, **Figure 25-26A.** With the spool valve in the neutral position, oil is fed to both sides of the power piston and to both reaction rings.

Oil also works its way around the balls between the power piston and the worm shaft. This oil collects in the hollow formed between the left end of the worm shaft and the left end of the power piston. Pressure created on the worm shaft end is counterbalanced by pressure created on the worm shaft balancing ring. In the neutral position, oil pressure is equalized on all sides and no hydraulic pressure is applied to the pitman shaft by the power piston.

Left Turn

When the steering wheel is turned to the left, the worm shaft attempts to impart motion to the power piston. Because the power piston has teeth that engage the pitman arm teeth, the resistance to turning offered by the wheels holds the power piston stationary.

Because the power piston does not want to move, the worm shaft (which has some lateral play) will start to screw out of the power piston. As it screws out just a few thousandths of an inch, it moves the center thrust bearing race with it. The center thrust bearing race will then move the bottom of the pivot lever to the right. This causes the top of the pivot lever to move the spool valve to the left.

When the spool valve moves to the left turn position, **Figure 25-26B,** it opens the left turn oil passageway and oil

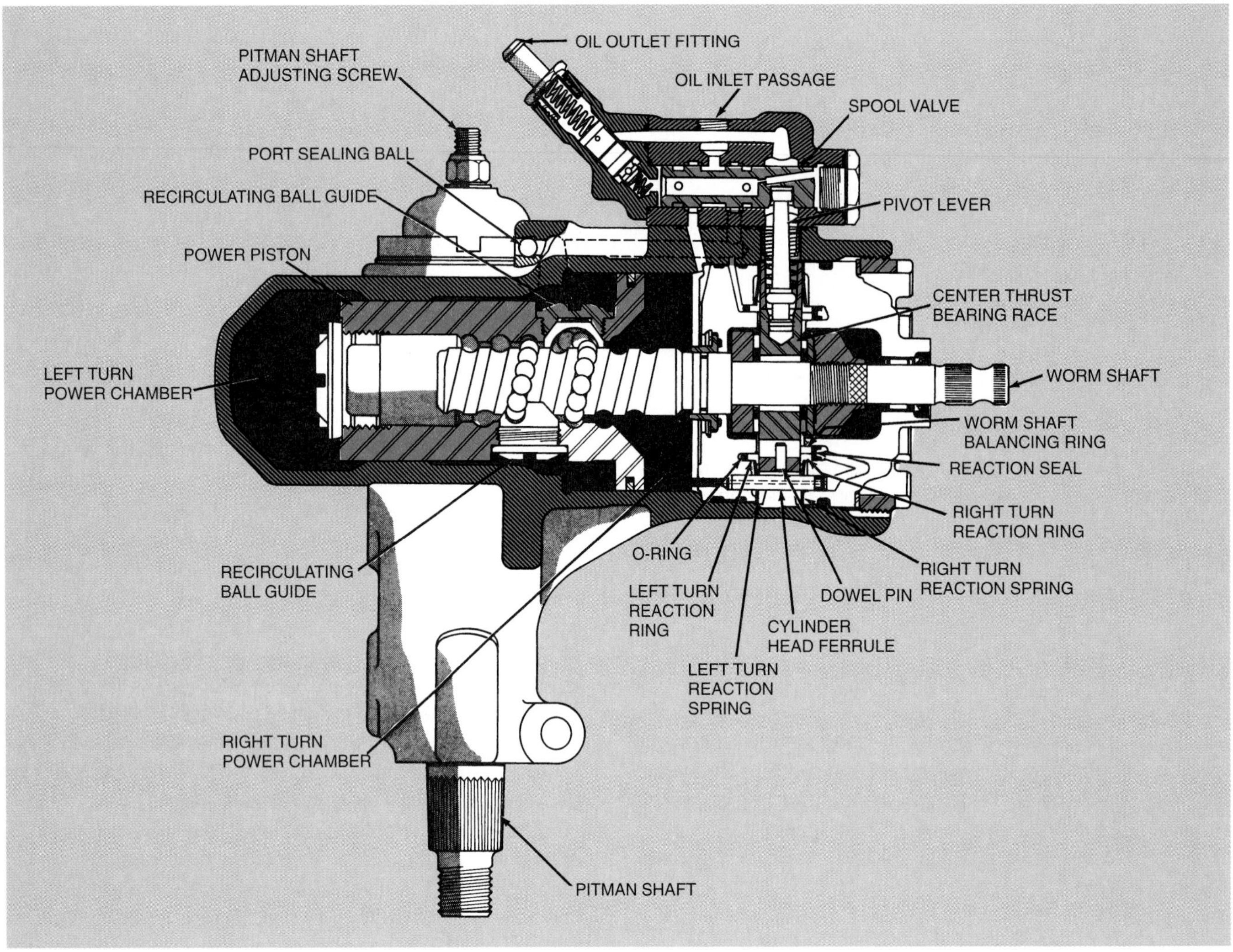

Figure 25-25. Cross-section of a power steering unit. This is a self-contained, inline unit.

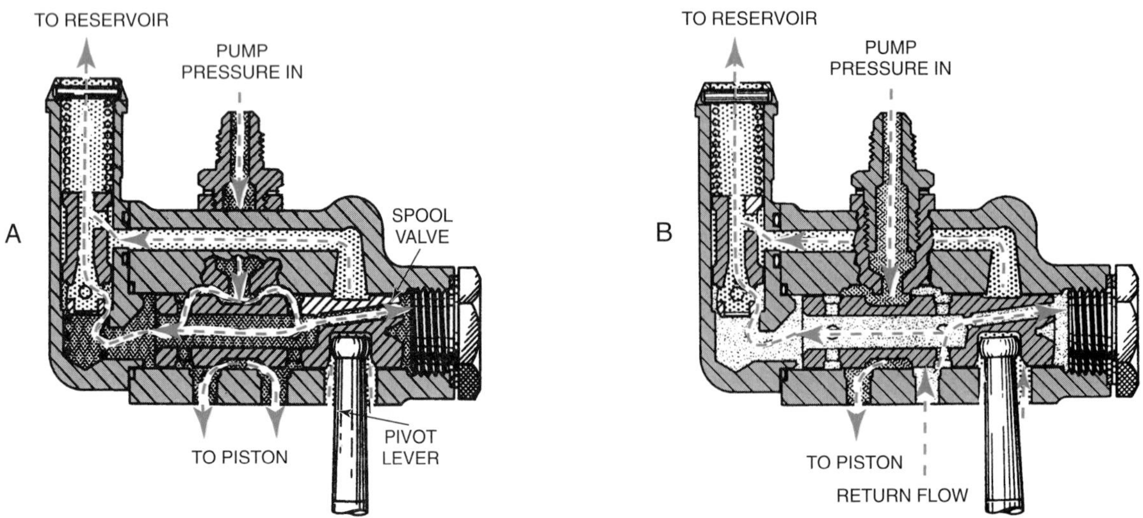

Figure 25-26. A—A spool valve in the neutral position. Note that the spool valve is centrally located and that pressurized oil passes to the power piston through both piston outlets. B—Spool valve in the left turn position. The valve has moved to the left. The right piston outlet is cut off from the source of pressurized oil. Return flow will now pass up through the right-side piston outlet. (Plymouth)

travels to the left turn power chamber. The right turn passageway can no longer receive pressurized oil, but it has opened to allow the oil to be returned from the right turn power chamber.

As soon as the spool valve is moved to the left, oil flows to the left turn power chamber and exerts a force on the power piston. Since the right turn power chamber oil is no longer pressurized, the power piston will be forced to the left. This squeezes the right turn power chamber oil out through the oil return passageways and back to the pump reservoir. Look at **Figure 25-27.**

As the oil pressurizes the left turn power chamber, it also flows to the right side reaction ring. This creates pressure that attempts to return the center thrust bearing to the neutral position. The round reaction springs also try to force the center thrust bearing back to the neutral position. These two forces on the center thrust bearing fight against the driver's turning force on the wheel, producing the essential road feel.

When the driver's turning force no longer exceeds the reaction ring pressure, the center thrust bearing will be forced back to the neutral position. This pivots the spool valve to neutral and equalizes the pressure on each side of the power piston.

Right Turn

A right turn produces the same action as a left turn, but the spool valve moves to the right and oil pressure is applied to the right turn power chamber. The worm shaft is connected to the steering shaft by a flexible connector. This connector allows the necessary worm shaft end movement. Maximum pump pressure for this unit is around 950 psi (6550 kPa). Oil leakage is prevented by neoprene O-rings that ride in grooves. Study **Figures 25-25** through **25-27.** Trace the oil flow patterns. Study the various parts and their actions.

Inline Unit with Torsion Bar and Rotating Spool Valve

A popular variation of the inline, self-contained power steering unit is shown in **Figure 25-28.** It employs a spool valve, but instead of sliding back and forth, the spool rotates a small amount inside the valve body. This type of power steering unit is referred to as an open-center, rotary-type, three-way valve or simply a ***rotary spool valve***. The rotary spool valve is a very close fit in the valve body.

The steering shaft is connected to a coupling, which, in turn, is splined to the lower stub shaft. The stub shaft passes through and is pinned to the spool valve. The outer end of the stub shaft is pinned to a torsion bar (tempered steel bar designed to twist a certain amount when a turning force is applied). The inner end of the stub shaft floats (is free) on the bar.

The inner end of the torsion bar is pinned to the valve body cap. The valve body cap, in turn, is pinned to the worm shaft. A pressure line and return line from the steering pump are attached to the gear valve housing.

Oil pressure in the pressure cylinder will run approximately 125 psi (861.87 kPa) for normal driving, 400 psi (2758 kPa) for cornering, and up to 1200 psi (8274 kPa) for parking.

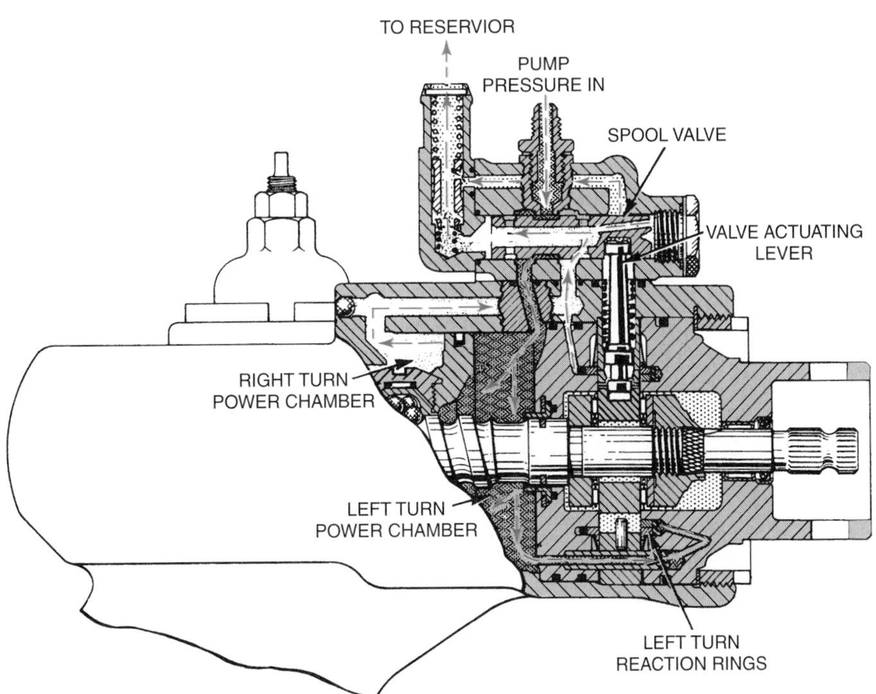

Figure 25-27. Oil flow during a left turn. Study the oil flow for this turn position. A right turn is merely a reversal of this when the spool valve is moved to the right by the valve-actuating lever.

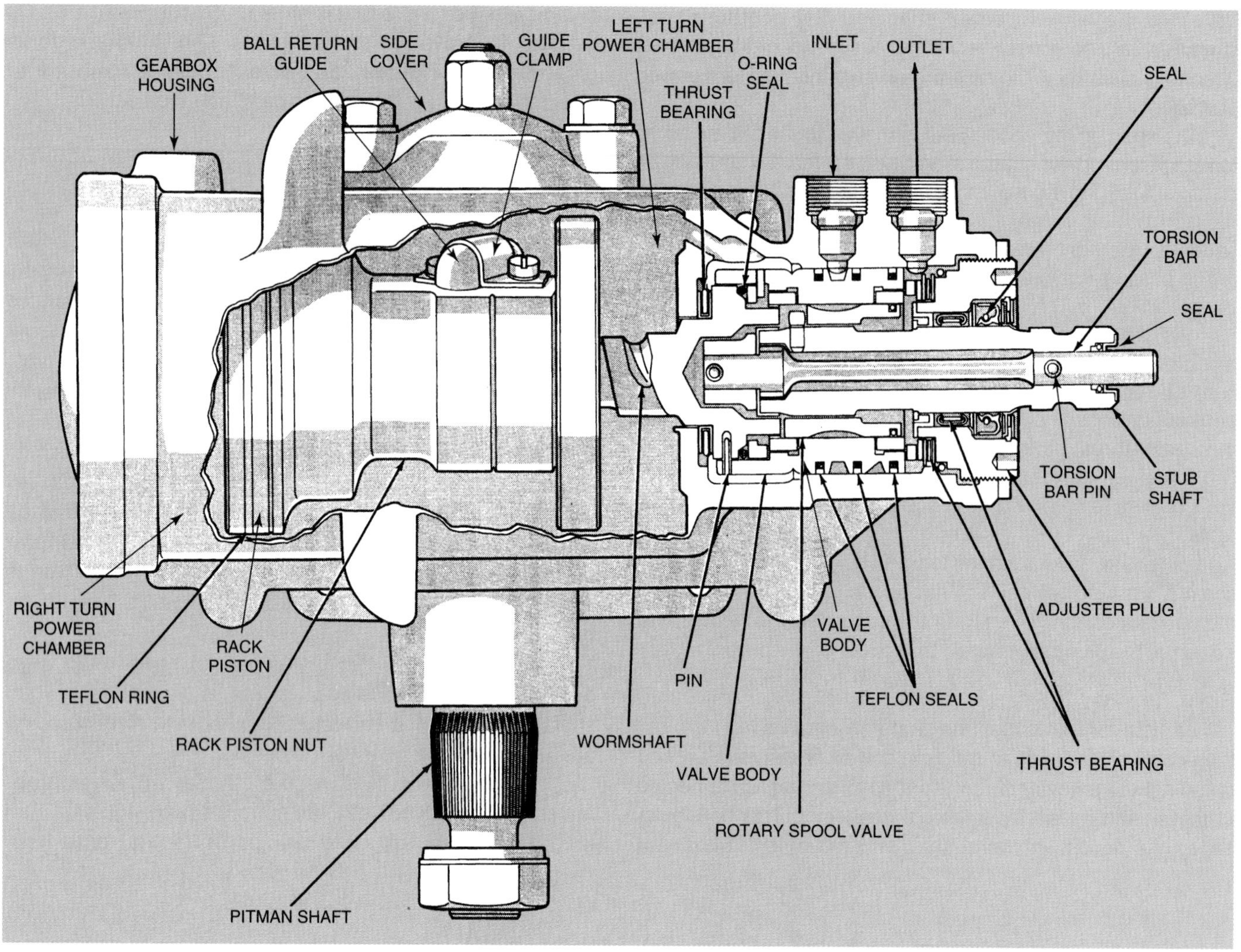

Figure 25-28. This power steering gear employs a rotary valve to control oil flow to the power rack-piston. (Dodge)

Rotary Valve Action, Straight Ahead

When the vehicle is traveling straight ahead and no appreciable turning effort is applied to the steering wheel, the torsion bar holds the spool valve in the neutral position. In this position, the spool valve is aligned to cause oil entering the valve body from the pressure port to pass through the slots in the spool valve. From the spool valve, oil flows to the oil return port. No pressure is applied to either side of the power rack piston. Oil flow is shown in **Figure 25-29.**

Rotary Valve Action, Right Turn

When the steering wheel is moved to the right, a turning force is applied to the stub shaft and through the pin to the torsion bar. Remember that the inner end of the torsion bar is pinned to the valve body cap. (Valve body cap is pinned to the valve body, which is pinned to the worm shaft.) The resistance of the front wheels to movement causes the turning effort to twist the torsion bar. As the bar twists, the stub shaft revolves enough to cause the spool valve to rotate.

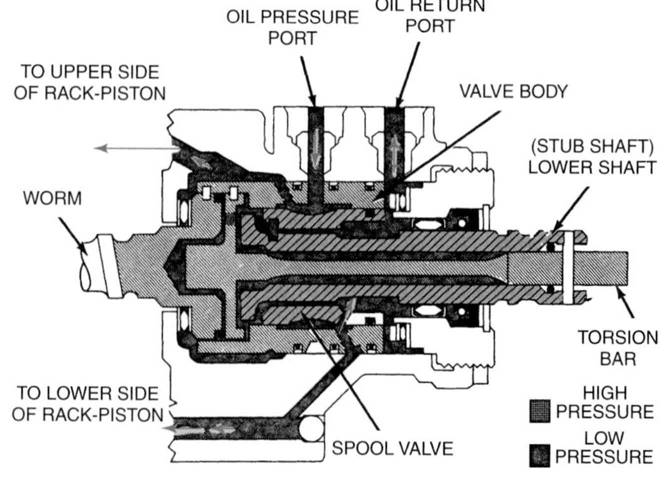

Figure 25-29. Oil flows through a rotary valve in the neutral position. Study valve construction. (Oldsmobile)

As the spool valve rotates to the right, the right turn grooves are closed off from the return grooves and opened to the pressure grooves. The left turn spool grooves are opened to the return grooves and closed to the pressure grooves. This action permits oil flow to the right turn power chamber. The oil pressure forces the power rack-piston upward, applying additional turning force to the pitman shaft sector teeth.

On this particular unit, the initial power assist will start with a 1 lb. pull on the rim of the steering wheel. This turns the valve about 0.3°. Full power assist will result with about 3 1/2 lb. pull on the steering wheel, rotating the spool about 4°. See **Figure 25-30.**

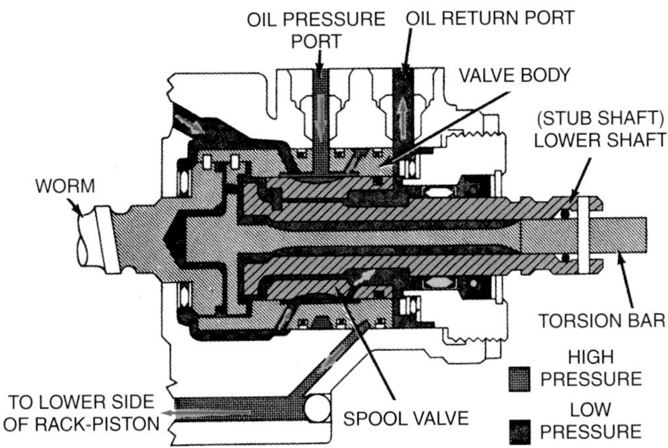

Figure 25-30. Oil flow through the rotary valve in the right turn position.

Rotary Valve Action, Left Turn

Moving the steering wheel to the left twists the stub shaft to the left, causing the spool valve to rotate to the left. This closes off the left turn grooves from the return grooves and opens them to the pressure grooves. The right turn slots are closed to pressure and opened to return. Oil flows to the left turn power chamber and forces the power rack-piston downward, **Figure 25-31.**

What Happens if the Torsion Bar Breaks?

The stub shaft is connected to the worm shaft by two tangs, (some movement is allowed to permit torsion bar action). In the unlikely event the torsion bar should break, the tangs will still provide manual turning force. An exploded view of a rotary valve power steering gear is shown in **Figure 25-32.**

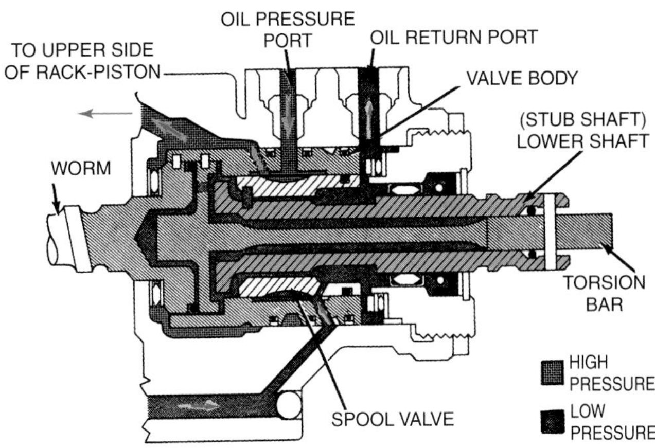

Figure 25-31. Oil flow through the rotary valve in the left turn position.

Rack-and-Pinion Power Steering

The power rack-and-pinion power steering system also uses a rotary control valve that directs the hydraulic fluid from the pump to either side of the rack piston. An overall view of this setup is shown in **Figure 25-33.** Steering wheel motion is transferred to the pinion. From there, it is sent through the pinion teeth, which are in mesh with the rack teeth. See **Figure 25-34.**

The integral rack piston, which is connected to the rack, changes hydraulic pressure to a linear force (back and forth movement in a straight line). This, in turn, moves the rack in a right or left direction. The force is transmitted by the inner and outer tie rods to the steering knuckles, which, in turn, move the wheels. **Figure 25-35** shows the rack-and-pinion power assist in the straight ahead, left turn, and right turn positions.

Oil Flow Regulation

Hydraulic oil sent to the right or left side of the rack piston is regulated by the spool valve located in the valve housing, **Figure 25-36.** The oil is transmitted via tubing and hose from the pump. Spool valve construction varies somewhat, but all valves perform in basically the same manner.

Power Steering Pumps

The ***power steering pump*** provides the hydraulic pressure needed to operate the power steering system. The pump can be mounted as a separate unit and driven by a belt, or it can be built as an integral part of the front engine plate and driven directly by the crankshaft. Most pumps are belt driven.

Oil returning from the gearbox enters a reservoir. From the reservoir, the oil flows into the pump intake. The power steering pump can be a ***vane***, ***rotor***, ***roller-vane***, ***slipper-vane***, or ***gear-type pump.*** These pumps operate in the same manner as the equivalent pumps in the engine oiling system

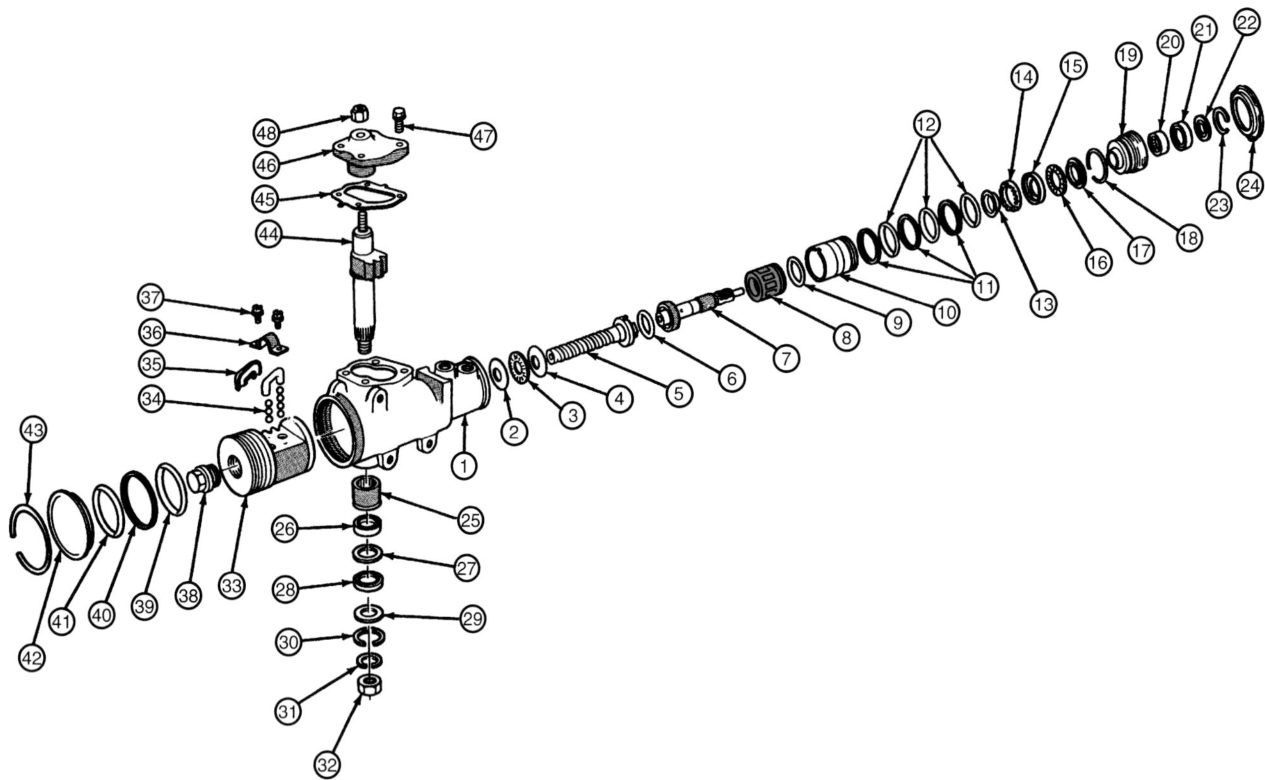

Figure 25-32. Exploded view of a power steering gear utilizing a rotary spool valve. 1—Steering gear housing. 2—Thrust bearing (worm) race. 3—Roller thrust (worm) bearing assembly. 4—Thrust bearing (worm) race. 5—Steering worm. 6—O-ring (stub shaft) seal. 7—Stub shaft. 8—Valve spool. 9—O-ring (spool) seal. 10—Valve body. 11—Valve body ring. 12—O-ring (valve body) seal. 13—Bearing (adjuster) retainer. 14—Thrust bearing spacer. 15—Upper thrust bearing (small) race. 16—Upper thrust bearing. 17—Upper thrust bearing (large) race. 18—O-ring seal. 19—Adjuster plug. 20—Needle bearing. 21—Stub shaft seal. 22—Stub shaft dust seal. 23—Retaining ring. 24—Adjuster plug lock nut. 25—Needle (pitman shaft) bearing assembly. 26—Pitman shaft (single lip) seal. 27—Seal back-up (pitman shaft) washer. 28—Pitman shaft (double lip) seal. 29—Pitman shaft lock washer. 30—Retaining (pitman shaft seal) ring. 31—Pitman shaft lock washer. 32—Pitman shaft nut. 33—Rack piston nut. 34—Ball. 35—Ball return guide. 36—Ball return guide clamp. 37—Lock washer & screw assembly. 38—Rack piston plug. 39—O-ring seal (housing end plug). 40—Rack piston ring. 41—O-ring (housing end plug) seal. 42—Housing end plug. 43—Retaining (housing end plug) ring. 44—Pitman shaft gear assembly 45—Gasket seal assembly. 46—Housing side cover assembly. 47—Hex head (side cover) bolt. 48—Lash adjuster nut. (Chrysler)

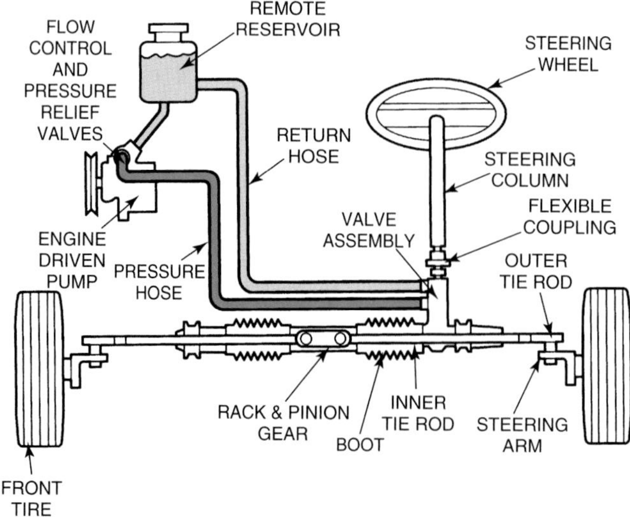

Figure 25-33. Overall view of a power rack-and-pinion steering assembly. Note the remote fluid reservoir, and low and high pressure fluid hoses. (General Motors)

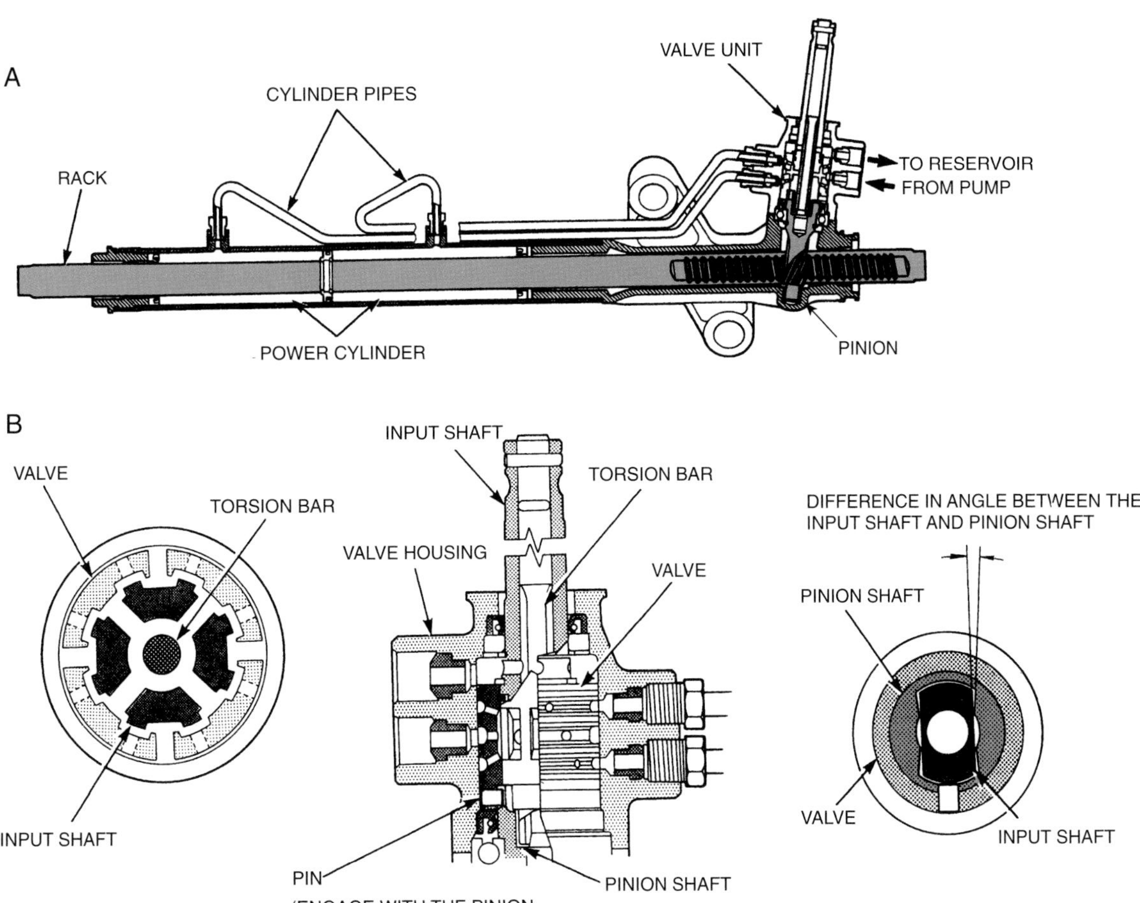

Figure 25-34. Cutaway views of one particular type of power rack-and-pinion steering assembly. A—Cutaway view of the steering gearbox and power rack unit. B—Cutaway showing the valve unit. This valve controls the steering fluid pressure. This system is able to function as a manual rack-and-pinion unit if fluid operating pressure is lost due to a faulty pump, broken or loose line, etc. Study carefully. (Honda)

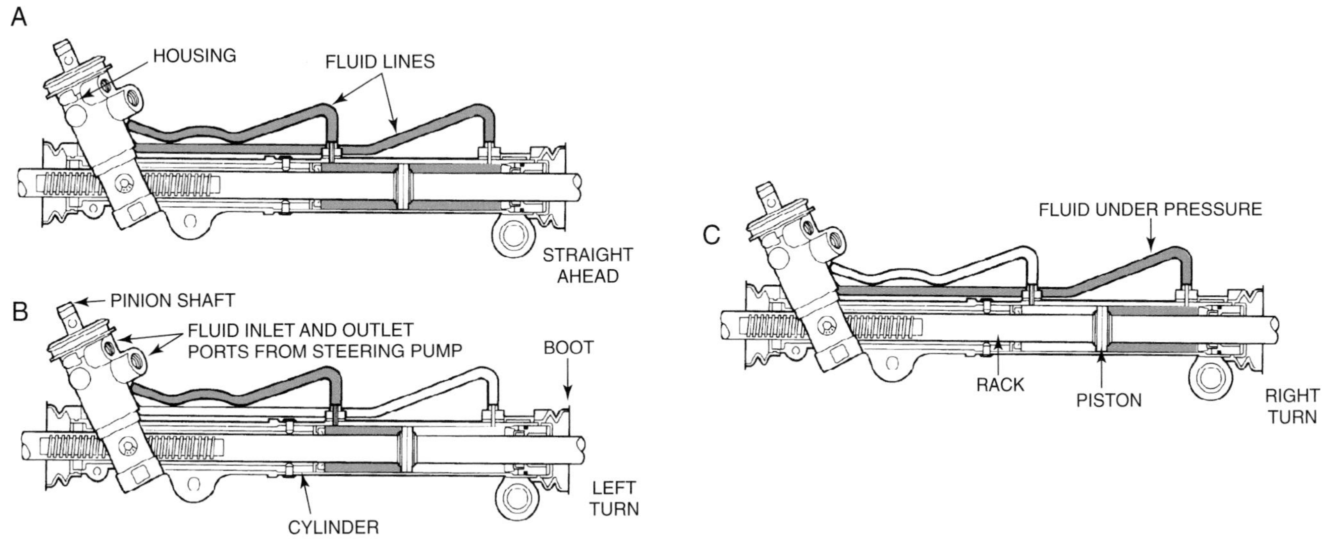

Figure 25-35. Turning positions of the power rack-and-pinion assembly, which is mounted in front of the axle centerline. This is termed "front steer." A—Straight ahead (no turn). B—Left turn position. C—Right turn position. (Saginaw Division of General Motors)

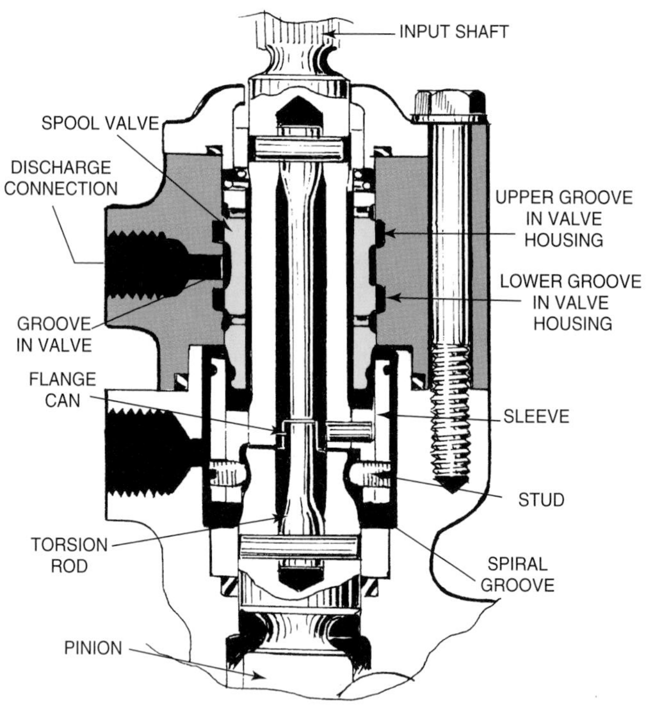

Figure 25-36. Cutaway view of a spool valve and housing used on one type of power rack-and-pinion steering system. (Volvo)

and in automatic transmissions and transaxles. The maximum pump pressure is about 1000 psi (6895 kPa) on newer vehicles and over 2000 psi (13 790 kPa) on older vehicles. Pressure is kept within normal specifications by a pressure regulator valve. The oil (usually special power steering fluid or, on older vehicles, automatic transmission fluid) provides the lubrication for the pump and steering mechanism. **Figure 25-37** is an exploded view of a vane-type power steering pump.

Figure 25-38 shows vane operation in another vane-type pump. Notice that some fluid, which is under pressure, is admitted behind the vane ends to force them outward to contact the pump. An exploded view of a roller-vane pump is shown in **Figure 25-39.** An exploded view of a gear pump is shown in **Figure 25-40.**

Power Steering Hoses and Tubing

The power steering gear and pump are connected by high-pressure hoses and steel tubing. These hoses are able to withstand very high pressure and vibration between the engine-mounted pump and the frame-mounted steering gear. **Figure 25-41** shows the routing of typical power steering hoses.

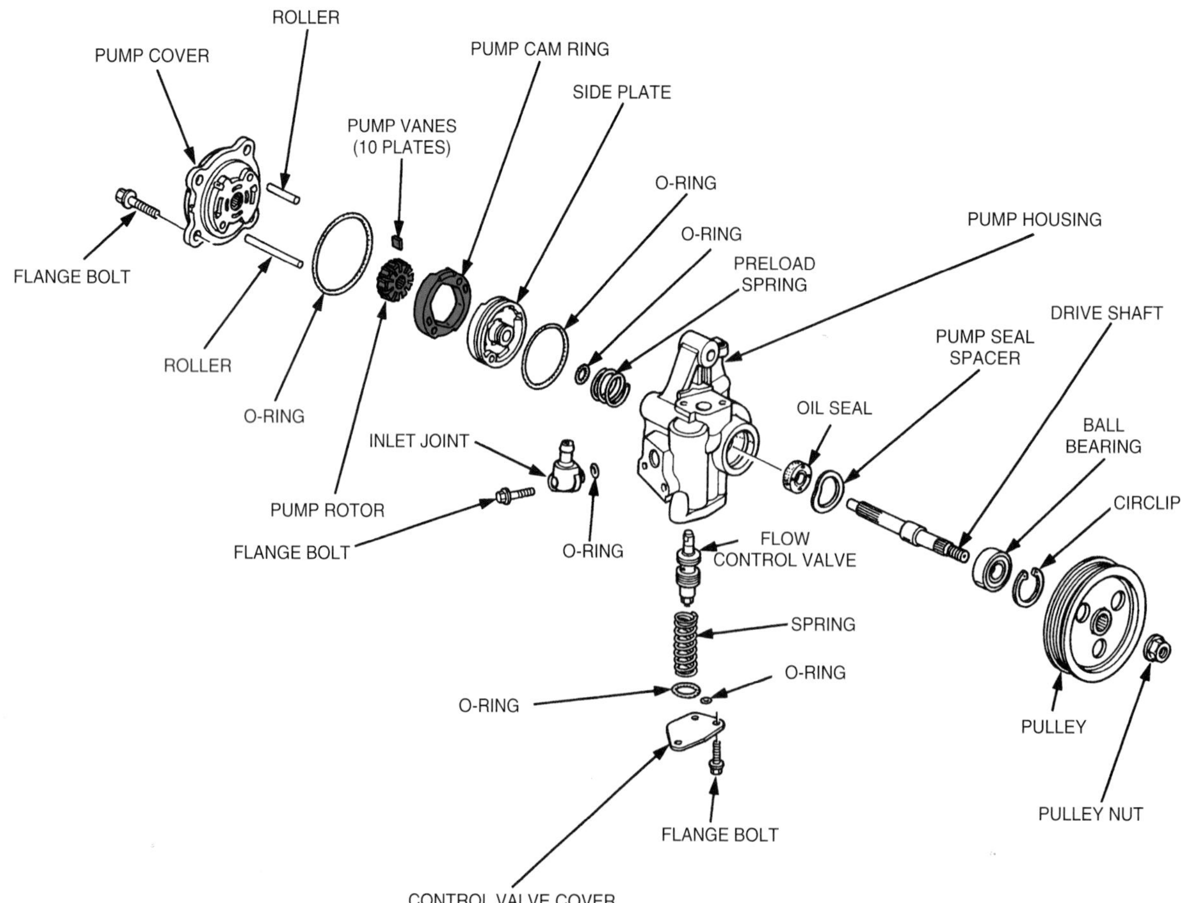

Figure 25-37. Exploded view of a vane type power steering pump. This pump is supplied with power steering fluid from a remote mounted reservoir (not shown). Study the construction. Note that 10 vanes are used. (Honda)

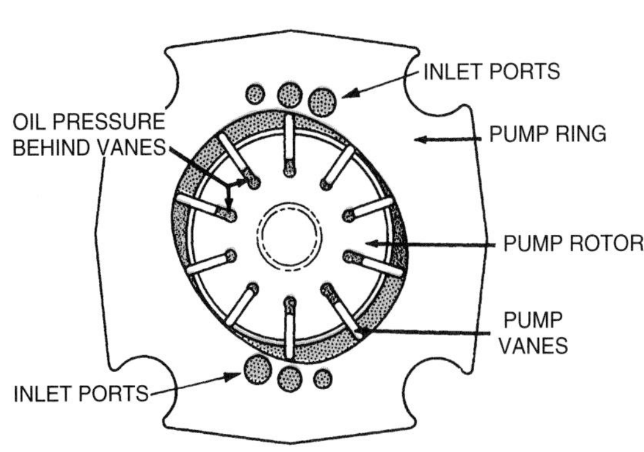

Figure 25-38. Vane-type power steering pump. (GMC)

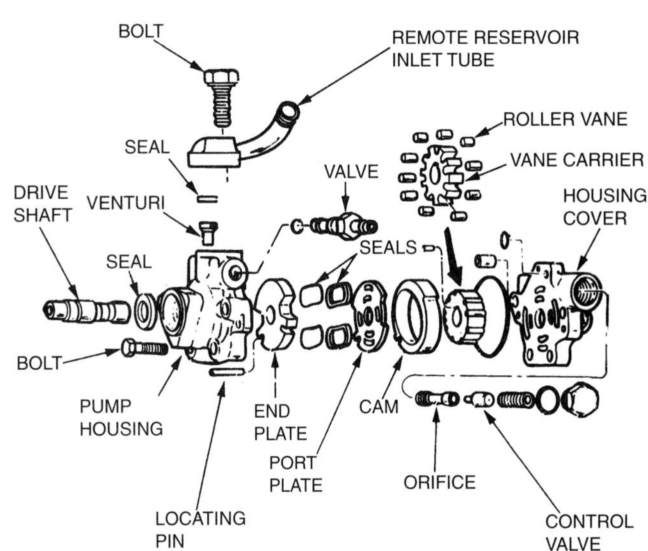

Figure 25-39. Exploded view of a roller-vane power steering pump. This unit uses a remote reservoir (not shown)

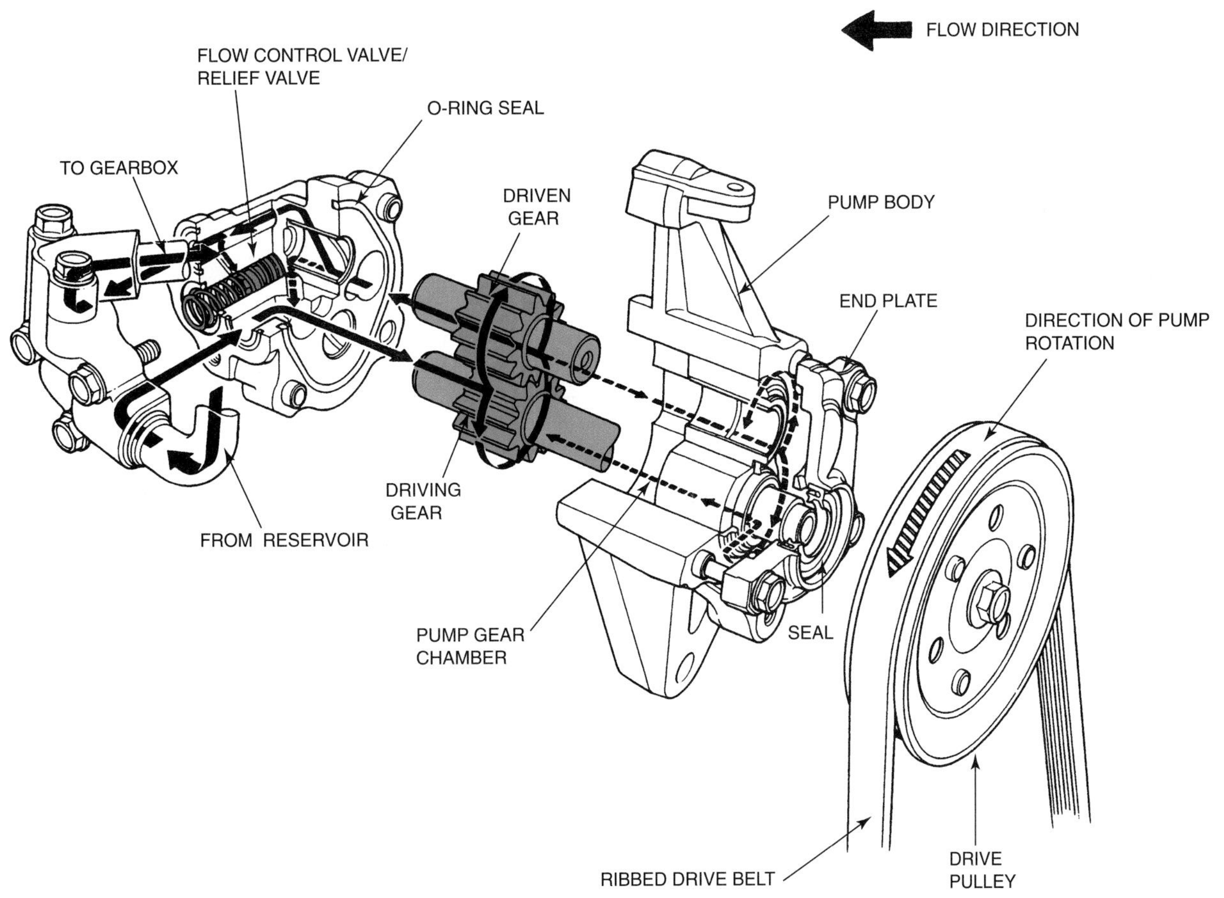

Figure 25-40. Exploded view of a gear type power steering pump. Note the flow control-relief valve which maintains a pump output fluid pressure of 1135 psi-1280 psi (8000 kPa - 9000 kPa). This pump is made from aluminum to help reduce weight and aid in cooling. Follow the flow direction arrows. (Honda)

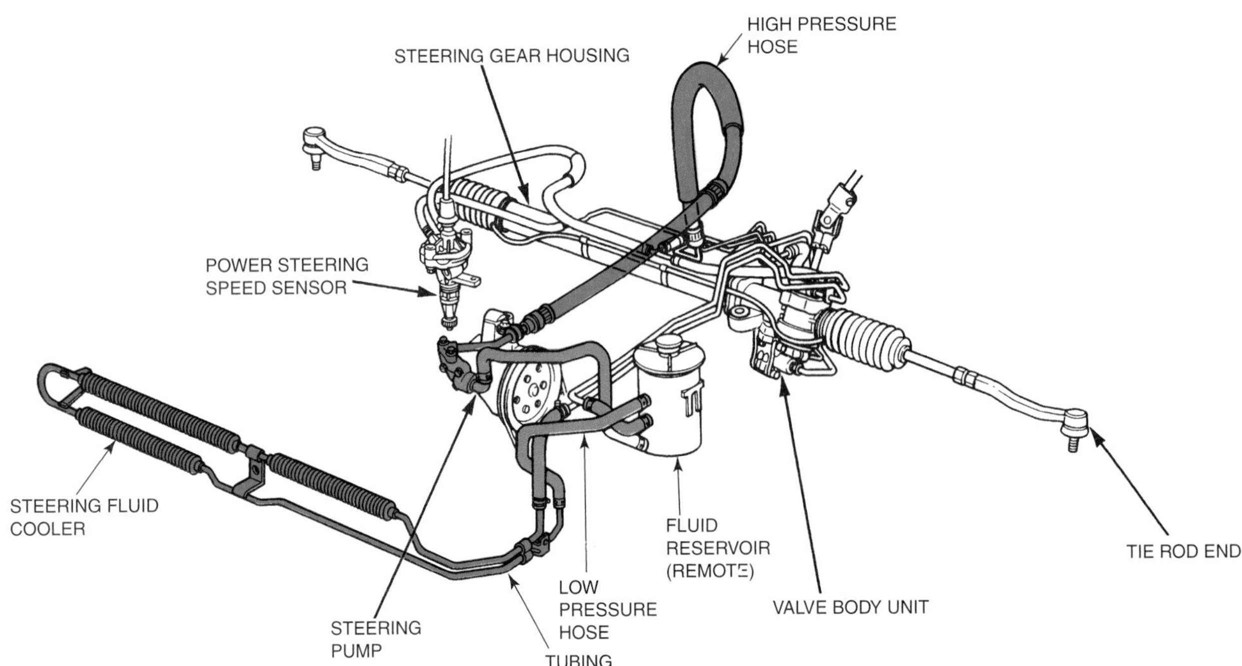

Figure 25-41. Overall view of a power rack-and-pinion steering system showing hose and tubing routing. Note the power steering fluid cooler. This is an oil-to-air type unit. (Honda)

Power Steering Systems Can Be Operated Manually

In the event of pump pressure failure, power steering systems can be operated manually with somewhat more effort than required with a conventional manual steering gear. This is because there is always a mechanical connection through the gears, even when no hydraulic pressure is available. Extra effort is required to move the hydraulic fluid in the gearbox, as well as turn the wheels.

Power Steering Fluids

The fluid level in the power steering pump should be checked frequently. Check the level when the fluid is at its normal operating temperature. Most pumps have a dipstick for checking the fluid level, **Figure 25-42.** Fluid should be between the add and full marks. If the fluid level is low, inspect the entire system for leaks. If fluid must be added, always use the power steering fluid intended for the system. Most modern vehicles are designed to use special power steering fluid, not automatic transmission fluid. Always check manufacturer's specifications before adding fluid.

Electronic Steering Control

Electronic steering control systems are sometimes used on vehicles equipped with power steering. These systems reduce power steering assist at cruising speeds to increase control and road feel, and increase assist at low speeds to facilitate parking. If the electronic system fails, the power steering system operates normally.

Electronic steering systems are operated by a control module that relies on inputs from the vehicle speed sensor and a steering wheel rotation sensor. The control module uses information on steering wheel movement and vehicle speed to determine the amount of power steering assist provided by a flow control solenoid. The flow control solenoid is mounted in the high-pressure side of the power steering system, usually in the pump output passage. In some systems, a magnetic rack-and-pinion spool valve assembly is used instead of a flow control solenoid to vary steering assist.

In the near future, some vehicles will be equipped with fully electronic steering. With fully electronic steering, there will be no mechanical connection between the steering wheel and the steering gear. Sensors in the steering wheel will operate motors that turn the wheels.

Summary

Steering linkage is provided to transfer the force from the pitman arm to the steering arms. The two most popular linkage arrangements are the parallelogram arrangement and the rack-and-pinion arrangement. The various linkage parts are connected by ball-and-socket connections, which allow swiveling movement without end play.

The steering wheel and steering shaft deliver the turning effort to the steering gear. Vehicles can use either a manual steering gear or a power steering unit. Some popular types of gearing used in both manual and power systems include the recirculating ball, worm-and-sector, worm-and-taper pin, and worm-and-roller steering gears. Most new cars use rack-and-pinion steering.

Both conventional and rack-and-pinion steering gears can be power assisted. Both types of power steering systems are actuated by oil pressure. The pressurized oil is admitted to either side of the power cylinder as controlled by a valving setup.

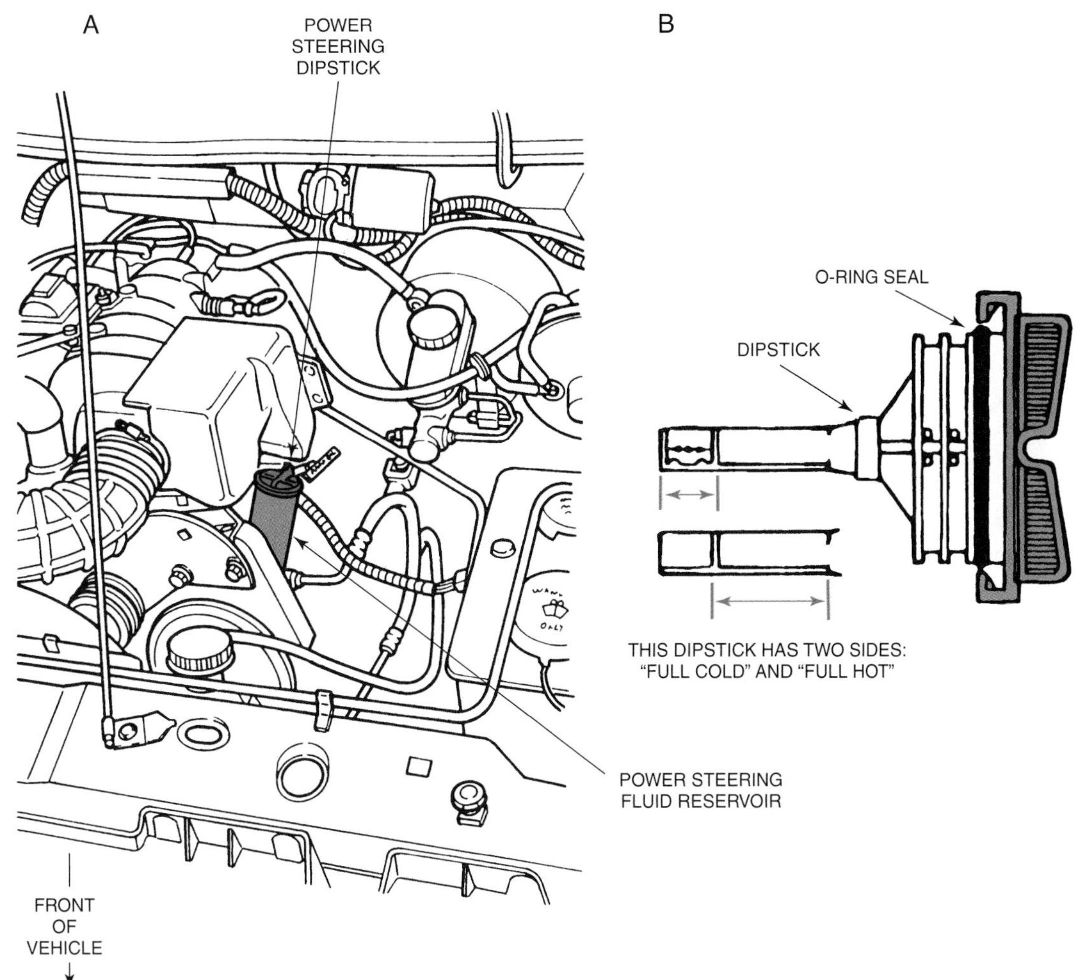

Figure 25-42. Power steering fluid reservoir. A—Power steering pump reservoir and dipstick location on one engine. B—Dipstick showing the full cold side and the full hot side fluid levels. (Ford Motor Co.)

Road feel is maintained by the use of a torsion bar, balanced oil pressure, and return springs. The power steering system uses oil pressure to assist the driver only when the pull on the steering wheel exceeds 1 to 9 lbs. When the turning effort on the steering wheel falls below this amount, oil pressure is equalized on both sides of the power piston.

Oil pressure is provided by a high-pressure oil pump. The system can use the vane, rotor, roller-vane, or slipper-vane pump unit. The pump is connected to the power unit by high-pressure hoses. The correct fluid must always be used in the power steering system.

Know These Terms

Spindle assembly
Steering arm
Parallelogram linkage
Rack-and-pinion linkage
Ball and socket connections
Pitman arm
Center link
Idler arm
Tie rod
Tie rod sleeve
Steering wheel
Steering shaft
Steering gear
Steering coupling
Universal joint
Conventional steering gear
Worm gear
Pitman arm shaft
Recirculating ball steering gear
Worm and roller steering gear
Pinion shaft
Rack
Gear reduction
Power steering
Road feel
Self-contained power steering gear
Control valve
Hydraulic piston
Inline
Reaction unit
Pivot lever
Spool valve
Rotary spool valve
Power steering pump
Vane pump
Rotor pump
Roller vane pump
Slipper vane pump
Gear pump
Electronic steering control systems

Review Questions—Chapter 25

Do not write in this book. Write your answers on a separate sheet of paper.

1. The front wheels swivel on one or two _____ .
   (A) bearings
   (B) ball joints
   (C) rubber bushings
   (D) Both A & B.
2. Name four major components of the parallelogram steering linkage.
3. The steering arms are attached to the _____ .
   (A) tie rod ends
   (B) spindles
   (C) center link
   (D) Both A & B.
4. Technician A says that a ball-and-socket connection is used to attach the pitman arm to the center link. Technician B says that a ball-and-socket connection is used to attach the idler arm to the frame. Who is right?
   (A) A only.
   (B) B only.
   (C) Both A & B.
   (D) Neither A nor B.
5. Why does the modern steering shaft have a two-piece upper shaft?
6. The side of the idler arm opposite the center link is bolted to the _____________.
   (A) frame
   (B) steering gear
   (C) spindle
   (D) None of the above.
7. The modern steering gear ratio is a compromise between _____ and _____ .
8. A self-contained power steering gear contains all the following hydraulic units, EXCEPT:
   (A) control valve.
   (B) hydraulic piston.
   (C) pump.
   (D) worm gear.
9. What part serves as the hydraulic power piston on a parallelogram steering gear?
10. What part serves as the hydraulic power piston on a rack-and-pinion steering gear?
11. Name three types of power steering pumps.
12. Power steering pressures on older vehicles could be as high as _____ pounds or _____ kPa.
13. Technician A says that automatic transmission fluid can be used to top off any power steering system. Technician B says that the power steering fluid level should only be checked when there is evidence of a leak. Who is right?
    (A) A only.
    (B) B only.
    (C) Both A & B.
    (D) Neither A nor B.
14. Why is a flexible coupling installed between the upper and lower steering shaft?
15. Technician A says that rack-and-pinion steering makes use of a worm and roller assembly to assist rapid rotation. Technician B says that recirculating balls are used in some steering gears to reduce friction. Who is right?
    (A) A only.
    (B) B only.
    (C) Both A & B.
    (D) Neither A nor B.

26

Wheels and Tires

After studying this chapter, you will be able to:

- Describe basic wheel rim design and construction.
- Describe various types of wheel hubs and bearings.
- Explain the methods of modern tire construction.
- Explain different tire construction and size designations.
- Identify tire and wheel size designations.
- Select appropriate tire inflation and rotation procedures.

This chapter will cover wheel rims, wheel hubs and bearings, and tires. While these items are common to all vehicles, enough variations exist to make a thorough understanding of these parts necessary. Upon completion of this chapter, you will be able to identify the major types of wheel rims, identify the differences in wheel hubs and bearings used in both front- and rear-wheel drive vehicles, and how to use the most current tire rating system.

Wheels

In the past, almost all factory supplied ***wheel rims*** (the assembly on which the tire is mounted) were made of ***stamped steel***, either painted to match the vehicle's color or simply painted black. In the past, the only way to obtain any other type of wheel was from an aftermarket supplier. Today, ***custom wheels*** made of aluminum, magnesium, graphite, and other materials are available not only from the aftermarket, but as original equipment from vehicle manufacturers. These two major types of wheels are discussed in the following two sections.

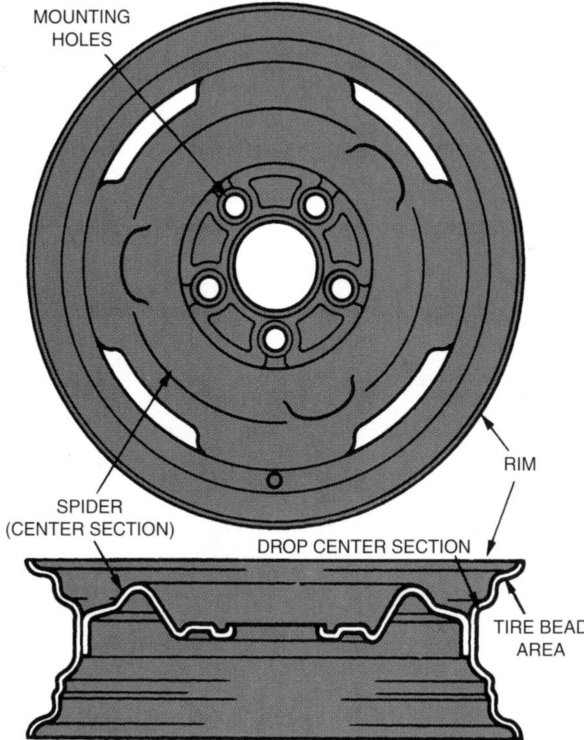

Figure 26-1. Typical drop center wheel. Center section (spider) can be riveted or welded to the rim. Note the drop center section to permit tire removal. Lightweight cast or forged aluminum wheels are finding increased use. (Chrysler)

Steel Wheels

Most stamped steel wheels are ***drop center*** types, as shown in **Figure 26-1**. It is made in two sections. The outer part forms the rim. The center section (sometimes called the spider) is spot welded to the rim. The center portion has four, five, or sometimes six holes used in mounting the wheel to the vehicle.

The center section of the rim is recessed or "dropped" to allow the tire to be removed easily. When one bead of the tire is placed in the dropped area, it is then possible to pull the opposite bead off the rim. Some wheels have ***safety ridges*** near the lips of the rim. In the event of a blowout, the safety ridges tend to keep the tire from moving into the dropped center, then coming off the wheel, **Figure 26-2**. The wheel must be manufactured to relatively close tolerances so that no vibration results when it is mounted, **Figure 26-3**.

Custom Wheels

Custom wheels made of aluminum, aluminum alloy, composite (graphite or plastic), or chromed steel are available as original equipment from many vehicle and aftermarket manufacturers. Some wheels are solid aluminum, while others have a steel core covered with aluminum. Alloy wheels are usually a mixture of aluminum and magnesium. Wire wheels are also available. **Figure 26-4** shows one particular original equipment custom wheel.

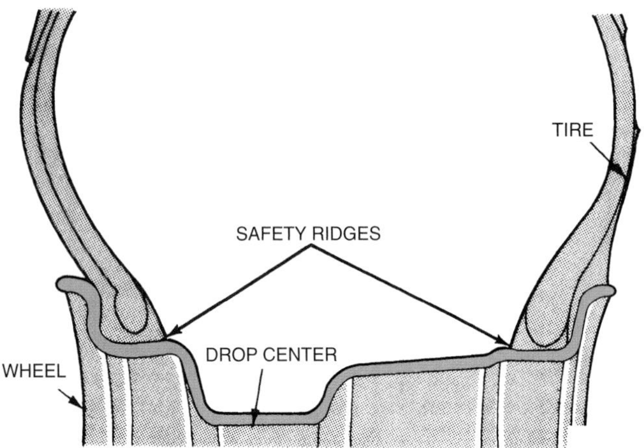

Figure 26-2. The safety rim has a ridge on each side of the wheel which retains the tire in case of a flat or blowout (ruptured tire). (Chrysler)

Figure 26-4. With improvements in materials and manufacturing, many vehicle manufacturers are offering aluminum custom wheels as standard equipment. One type of aluminum wheel used on a luxury sedan is shown here. (Oldsmobile)

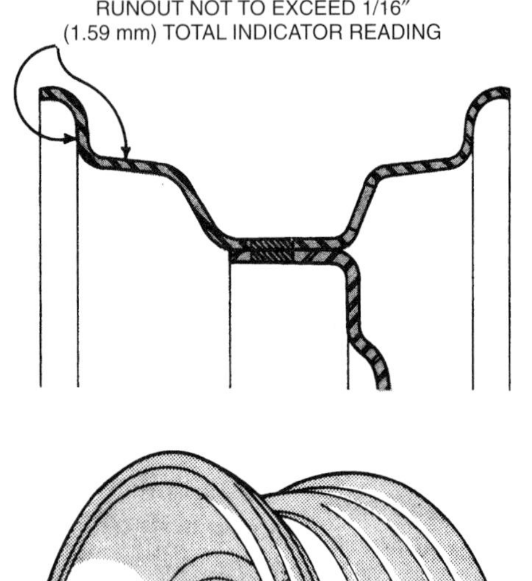

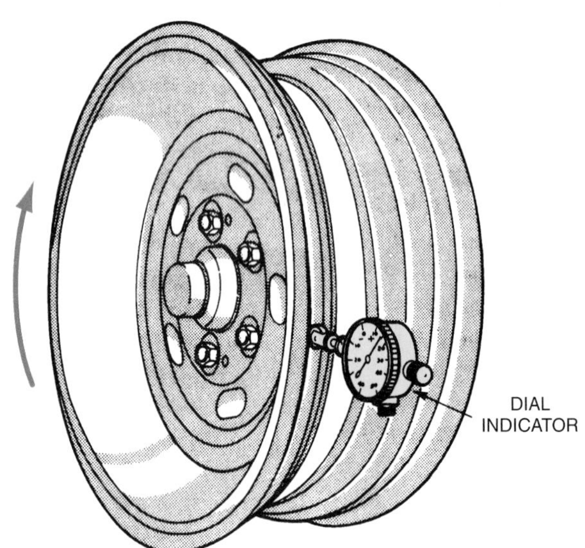

Figure 26-3. All wheels must be true. In the upper illustration, left-handed arrow indicates the surface to be checked to determine side movement; right-hand arrow indicates area checked for out-of-roundness. The wheel is spun and a dial indicator is used to read any movement in .001″ (0.025 mm) increments. Runout should not exceed 1/16 or .0625″ (1.59 mm). (Chrysler)

Wheel Rim Size

Rim size is determined by three measurements: rim width, rim diameter, and flange height. Rim width and diameter are measured in inches. A letter (J, K, and so on) is used to indicate a definite flange height. For example, a K rim flange is 0.77″ (19.5 mm) high. See **Figure 26-5**. These measurements apply to rims made of any material.

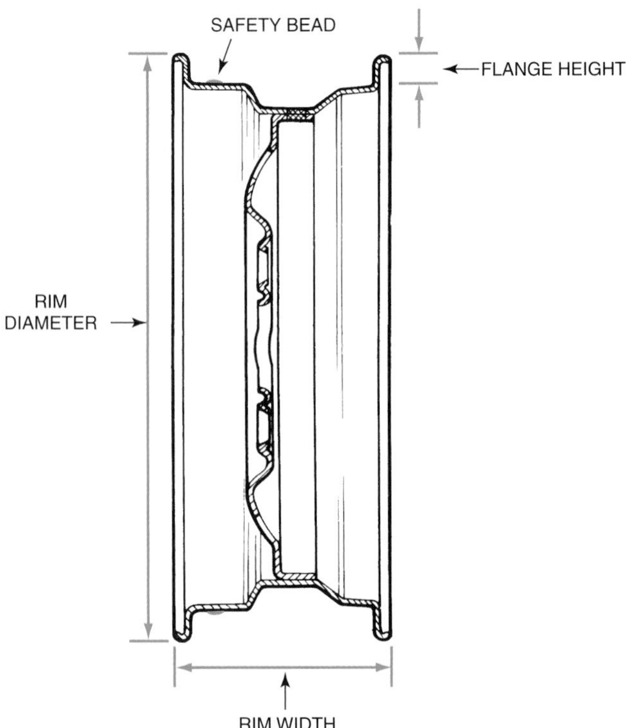

Figure 26-5. Measuring points used to determine rim size. These numbers are sometimes cast on the inner side of the rim. (Jaguar)

Wheel Lug Bolts and Nuts

The wheel is bolted to the hub either by ***lug bolts*** that pass through the wheel and thread into the hub, or by threaded studs, called ***wheel lugs***, that protrude from the hub. In the case of threaded studs, special tapered nuts called ***lug nuts*** are required. The lug nut taper matches the taper of the wheel rim mounting holes.

Most vehicles use right-hand thread wheel lugs on both sides. A few vehicles use left-hand threads (nut or bolt turns counterclockwise to tighten) on the driver's side, and right-hand threads (turn clockwise to tighten) on the passenger's side.

Torquing Wheel Nuts

When the wheel rims are made of aluminum, or a composite of aluminum and steel, proper ***wheel nut torque*** is critical. Always torque to the manufacturers' specifications. Although correct torque is not as important with steel rims, most modern vehicles have front disc brakes with thin rotors that can be deformed by overtorquing. A complete wheel, tire, and hub assembly is shown in **Figure 26-6**.

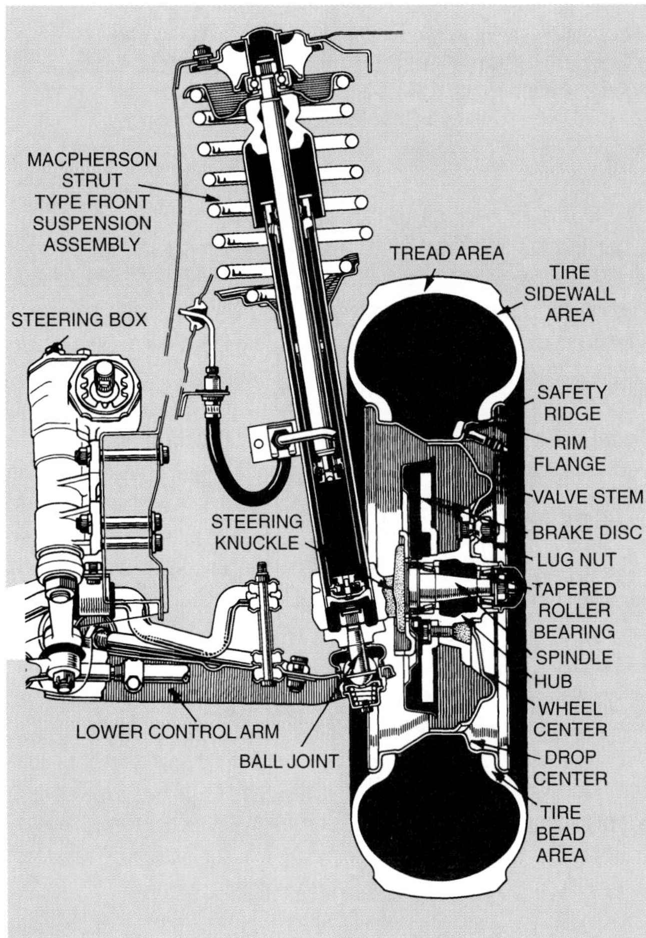

Figure 26-6. Front wheel and tire assembly. Safety ridges are used to prevent the tire from leaving the wheel when flat. (Toyota)

Do not use an impact wrench on any aluminum or alloy wheel, unless it is equipped with a ***torque stick***. Torque sticks are flexible extensions which limit the output of the impact wrench to a preset value. When a certain torque value is reached, the stick will flex instead of transmitting further torque. A torque wrench should be used to tighten the rim if torque sticks are not available.

Wheel Hubs and Bearings

The ***wheel hubs*** form the mounting surface for the wheel and tire. Front-wheel hubs revolve about the steering knuckle spindle on either ball or tapered roller bearings. Rear-wheel hubs are usually part of the axle assembly on solid axle rear-wheel drive vehicles. Front-wheel drive vehicles, and vehicles with independent rear suspensions have separate hubs containing the wheel bearings.

Wheel Bearings

The drive and non-drive axles of a vehicle are supported by bearings. Two main types of bearings are used on modern vehicles:

- ***Tapered roller bearings***, **Figure 26-7**.
- ***Straight roller*** and ***ball bearings***, **Figure 26-8**.

Tapered roller bearings can be cleaned, regreased, and adjusted. The straight roller and ball bearings used on most modern front-wheel drive vehicles cannot be adjusted and cleaned, and are replaced as a unit. Wheel bearings use a quality, high temperature grease. **Figure 26-9** shows the location of this grease in a typical tapered roller bearing assembly. Notice that the hollow hub is not completely filled. The bearings, however, are thoroughly ***packed***.

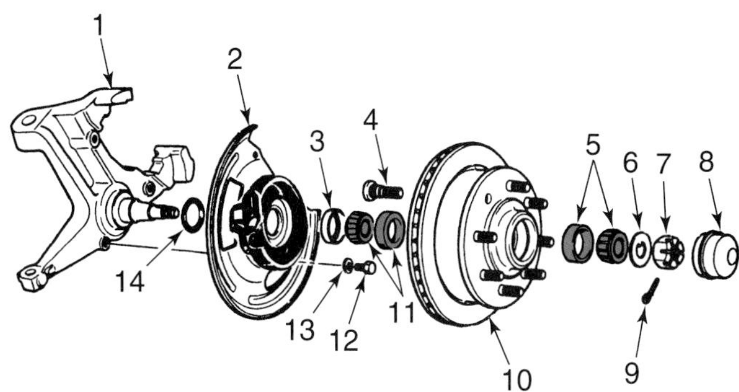

Figure 26-7. Exploded view of a front hub assembly which uses tapered roller bearings. 1—Steering knuckle. 2—Splash shield. 3—Grease seal. 4—Hub bolt. 5—Outer tapered roller bearing assembly. 6—Washer. 7—Castellated nut. 8—Dust cap. 9—Cotter pin. 10—Brake rotor assembly. 11—Inner tapered roller bearing assembly. 12—Bolt. 13—Lock washer. 14—Gasket. (Chevrolet)

Bearing Retaining Devices

Some method of holding the bearing on the wheels must be used. On front and rear wheels with tapered roller bearings, a ***castellated nut*** (slotted) is used on the end of the

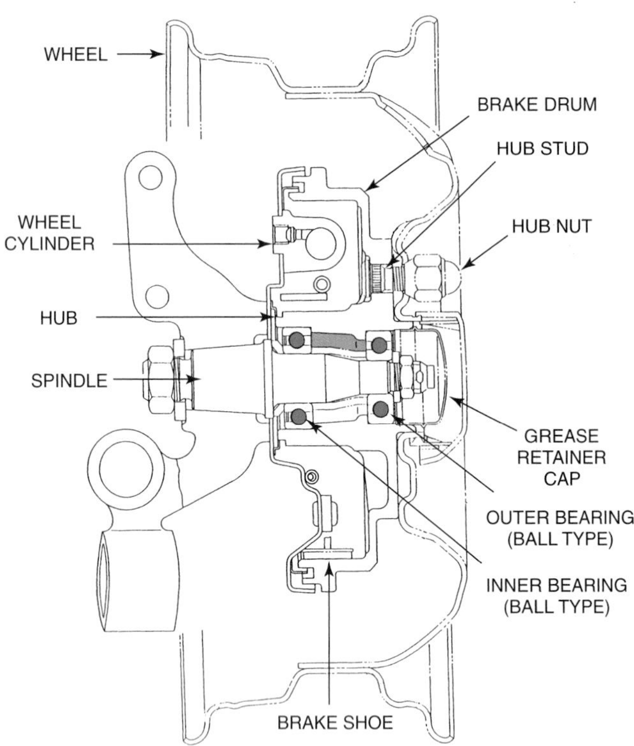

Figure 26-8. Cross-sectional view of a rear hub assembly which uses inner and outer ball bearings. (Daihatsu)

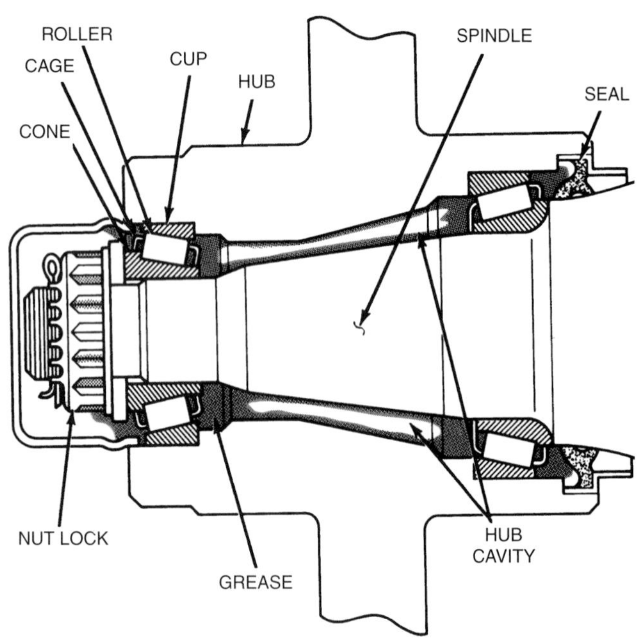

Figure 26-9. Front hub showing placement of wheel bearing grease. (Dodge)

spindle to remove excessive play between the bearings and hub. A ***safety washer*** is used between the nut and the bearing. It has a metal tab that rides in a groove in the threaded area of the spindle.

When the hub and bearing spin, the rotating motion is not transferred to the nut, since the washer tab prevents it from turning. A new ***cotter pin*** is installed through the holes in the nut and spindle to keep the nut from turning. **Figure 26-10** shows the spindle nut, cotter pin and safety washer.

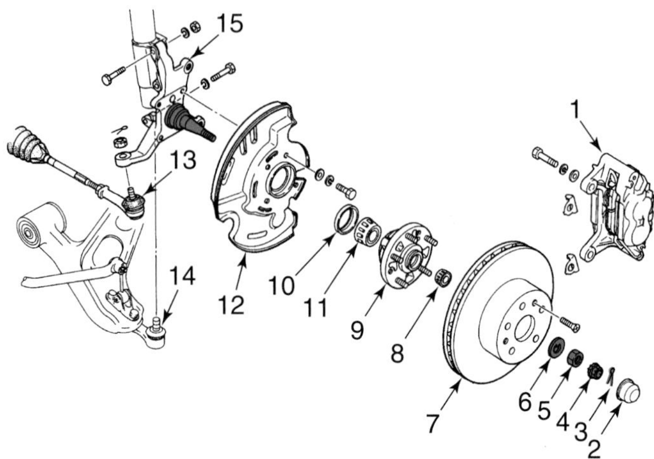

Figure 26-10. Exploded view of a front suspension and spindle assembly showing the spindle nut (5), cotter pin (3), safety washer (6), and fine adjustment nut set cover (4). 1—Brake caliper. 2—Hub cap. 3—Cotter pin. 4—Fine adjustment nut cover. 5—Spindle nut. 6—Safety washer. 7—Brake disc. 8—Outer bearing. 9—Wheel hub. 10—Oil seal. 11—Inner bearing 12—Dust cover. 13—Tie-rod end. 14—Lower control arm ball joint. 15—Knuckle spindle. (Mazda Motor Corp.)

Some adjusting nuts are staked into position by denting a thin lip on the nut itself. The lip is forced into a special groove in the spindle thread area. This is called a ***staked nut***. The method allows for near zero play bearing adjustment. A new nut must be used each time bearing service is performed. **Figure 26-11** illustrates a staked nut.

A hub and bearing arrangement for one type of front-wheel drive is shown in **Figure 26-12**. The hub is splined to the outer drive shaft and is held in place by a locknut. The shaft (and hub) revolve in a special double-row ball bearing. Most bearings used on front-wheel drive axles are pressed onto the hub. The rear bearings of many front-wheel drive vehicles use a similar design.

Tires

Tires serve two main functions. They provide a cushioning action that softens the jolts caused by road irregularities. In this respect, they are actually serving as part of the suspension system. They also provide proper ***traction***. Traction enables the vehicle to drive itself forward, provides a means of steering, and allows reasonably fast stopping. Therefore, quality tires are essential to safe vehicle operation.

Tire Construction

Tire treads and sidewalls are constructed of various types of natural and synthetic rubber (neoprene). The tire cords, which support the tread, can be made of rayon, nylon,

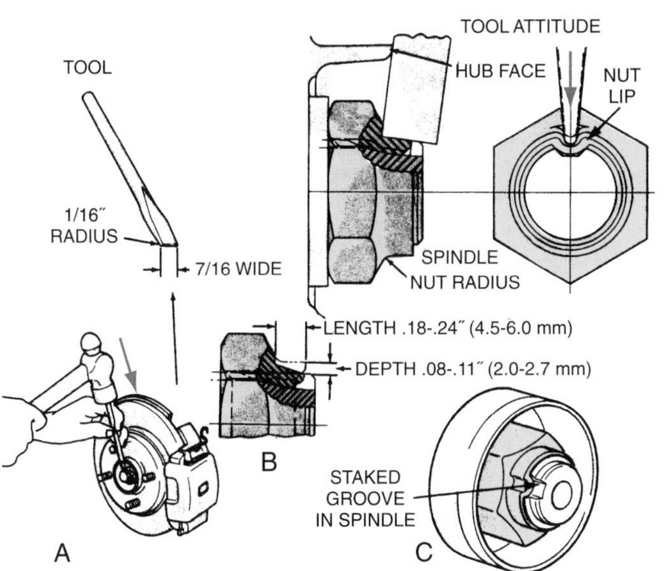

Figure 26-11. Staking a hub nut. A—A chisel is struck with a hammer, causing the nut lip to be bent into the spindle groove. B—Length and depth of proper stake. C—Finished staked nut. (Plymouth)

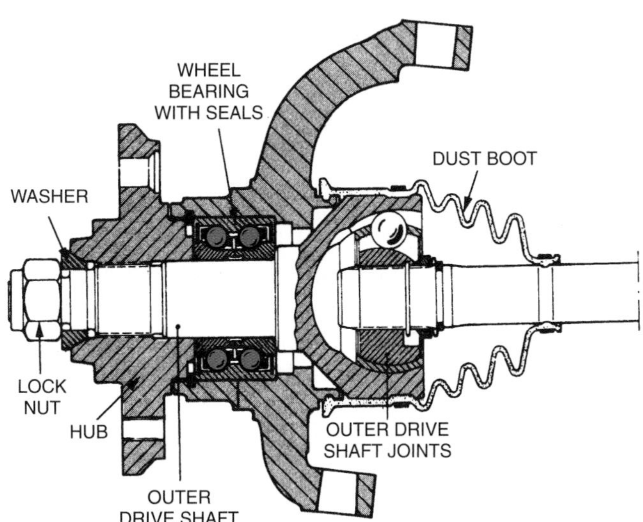

Figure 26-12. Cross-sectional view of a front-wheel drive front hub showing part arrangement. Study carefully. (Saab)

polyester, aramid, Kevlar, and fiberglass. Steel wire is often used in the belt section of tires. **Figure 26-13** shows typical tire construction.

The fabric casing uses layers of interwoven cord material to provide the tire with shape and strength. The cord material is wrapped around steel wires located at the point where the tire attaches to the rim. This contact point is called the ***bead***. The beads provide strength to prevent the tire from opening up and leaving the wheel. Filler strips reinforce the bead section. The bead also provides the air seal between the rim and tire.

Many older tires were constructed with two or four layers, or ***plies*** of cord material, at an angle to the tire centerline.

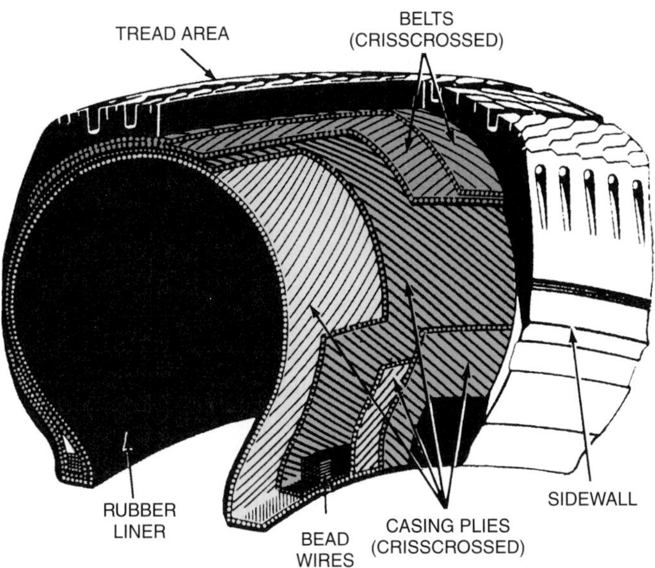

Figure 26-13. Typical bias belted tire construction. This specific tire uses four sidewall plies (bias arrangement) and two belts beneath the tread. (BF Goodrich)

Look at **Figure 26-14A**. This is termed ***bias construction***. Another method of arranging the plies is illustrated in **Figure 26-14B**. This is called ***bias-belted tire construction***. Note the difference in cord angles between A and B.

The third type of construction is ***radial tire construction***, in which the body cords cross the tire centerline almost at right angles to the belts, **Figure 26-14C**. Radial tires may also be belted. Today, radial tires are used for almost all applications, including retrofitting older vehicles that originally came with bias ply tires.

Tire Rating Information

Tire ratings are determined by a system which uses a series of letters and numbers molded on the side of the tire, or listed in the vehicle owners manual. The rating system identifies the tire size, rim size, and type of construction, as well as the maximum speed and load handling capabilities. A typical modern tire might have the series of letters and numbers shown in **Figure 26-15** on its sidewall.

The letter **P** indicates the tire is designed for a passenger vehicle, such as a car or light truck. Other possible tire designations are **T** for temporary, such as a space saver tire, **LT** for light truck, and **C** for commercial vehicles, such as large trucks. This letter may be absent on some tires.

The number **205** represents the tire's section width (width of the tread) in millimeters, measured at its widest point. This is the actual tire size. Sizes range from 145 to 315, with most sizes from about 185 to 235. The higher the number, the larger the tire.

The number **55** is the aspect ratio (relationship of a tire's cross-sectional height to its width). This means that this tire's height is 55 percent of its width, **Figure 26-15**. Other common aspect ratios are 50, 60, 65, 70, 75, and 80. The higher the aspect ratio, the taller the tire. Low aspect ratios provide more traction, while high aspect ratios give better mileage.

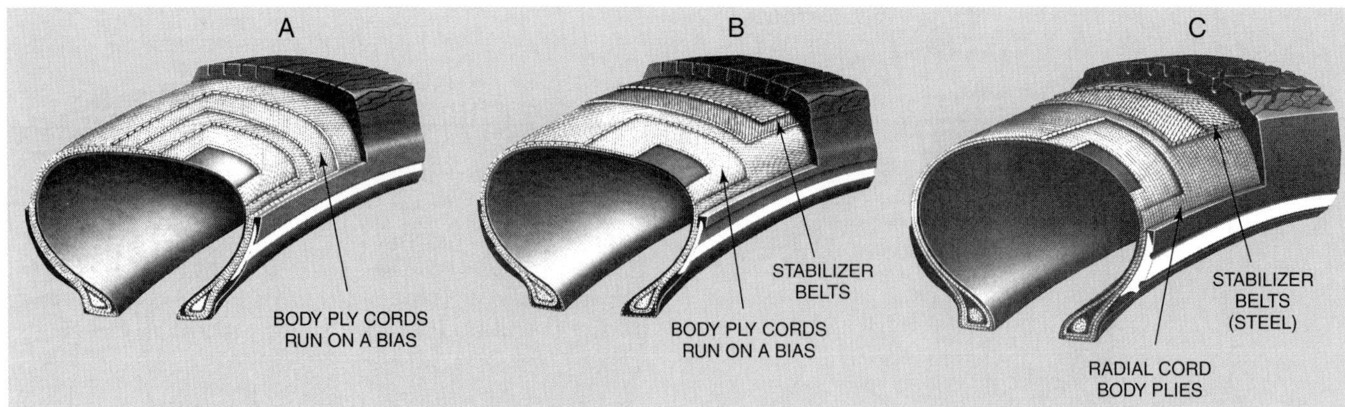

Figure 26-14. Typical tire construction. A—Bias ply (sidewall plies crisscrossed, no belts). B—Bias-belted (sidewall plies crisscrossed, belts beneath tread area). C—Radial construction. (BF Goodrich)

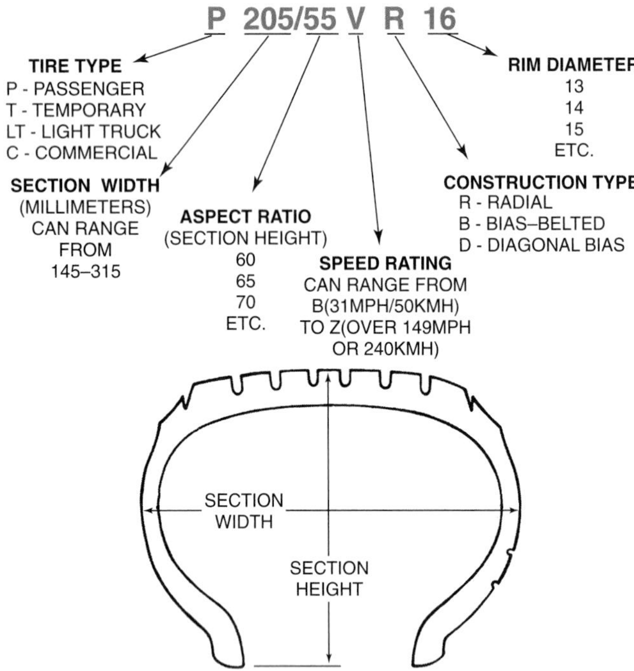

Figure 26-15. A typical metric tire size breakdown. Size (P 205/55VR16 for example) will be molded into the tire sidewall area. (Chevrolet)

The letter **V** is the ***speed rating***. The speed rating gives the maximum speed at which the tire can be operated. This ranges from B (31 mph or 50 kmh) to Z (over 149 mph or 240 kmh). The letter **R** indicates radial ply construction. A bias ply tire will be marked **B**. Almost all passenger car and light truck tires made today are radial tires.

The number **16** is the rim diameter in inches. Modern wheel rims range from 12 inches to 17 inches. The most common rim sizes are 13, 14, and 15 inches. Other rim dimensions are discussed in the section on wheel rims.

Tire Quality Grading

All passenger and light truck vehicle tires currently produced are graded by the US Department of Transportation's (DOT) ***uniform tire quality grading system***. The quality grading system is applicable to the three areas shown below.

The ***temperature resistance rating*** uses three letter grades: A, B, and C. While A offers the greatest resistance to heat generation, C provides the least. All tires must meet the C rating. Tire ***traction rating*** is also graded A, B, and C. A offers the best traction (wet roads), while C offers the minimum amount.

Tire ***tread wear rating*** is graded by using a set of numbers ranging from 100 to about 500. A tire that has a tread wear grade of 150 should supply approximately 50% more mileage than a tire with a tread wear rating of 100.

Other Tire Rating Numbers and Letters

Some tires have a number, such as 80, which indicates the tire's load index. The load index is the load carrying capacity of a tire. This particular number means that the tire would have a load carrying capacity of 992 pounds (450 kg). Therefore four of these tires could be installed on a vehicle which weighs, fully loaded, no more than 3968 pounds (1800 kg). Many tires have the speed rating letter after the load index number.

Some tires have the letters **M & S** on their sidewalls. This designates that it is an all-season tire. The letters M & S stand for mud and snow. Tires with this designation can be used on wet, muddy, or icy roads and will give a reasonable amount of traction in these conditions.

Special Service Tires

Many safety and high speed tires are available. High speed tires generally use steel wires which resist expansion from centrifugal force in the outer ply and tread areas. Steel, which helps to control distortion, also resists punctures better than fabric belts.

Most vehicles are now equipped with a compact, or ***space saver spare tire***. It is designed to take up less cargo space and is lighter than a normal tire. The compact spare tire is for temporary use only. Use it to replace a flat until it is repaired or replaced. Do not rotate a compact spare in with the other tires. When replacing this tire, mount only on a special wheel designed for its use. Do not use hub caps or wheel covers on the rim as possible tire damage could result. Keep the space saver tire inflated to recommended pressure. A space saver spare tire is shown in **Figure 26-16**.

Determining the Correct Tire for a Vehicle

When replacing original equipment tires, the technician should take into consideration vehicle weight, type of suspension system, and horsepower. A tire that is too small will be overloaded. A tire that is too large will add to the vehicle's unsprung weight, upsetting the suspension system balance.

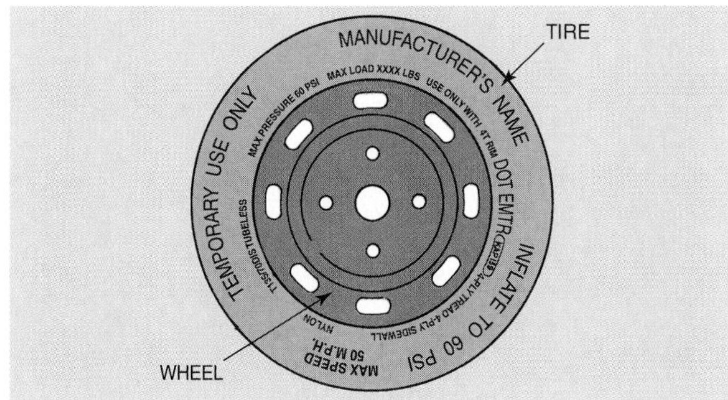

Figure 26-16. A compact spare tire is designed to be light, small, and most importantly, used only as a temporary tire.

A tire load information label can be found on most vehicles. It is generally located on the driver's door, on the door pillar, or inside the glove box. The load information label contains specific information on maximum vehicle load, tire size (including spare tire), and inflation pressures. Speed rated tires should be replaced with tires that have an equal or higher speed rating. Older vehicles with bias or bias-belted tires can be equipped with radial tires by the use of interchange charts, such as the one in **Figure 26-17**.

Tire Valves

The ***tire valve***, or ***valve stem***, **Figure 26-18**, is the means of adding air to the tire and wheel assembly. The tire valve is a rubber and metal assembly with a small ***valve core*** threaded into the center. The valve core contains the actual valve mechanism. Depressing the valve allows air to enter or leave the tire. Tire valves are always installed in the wheel rim.

Tire Pressure is Important

Vehicle manufacturers determine the proper tire and ***inflation pressure*** after extensive research. Tires must be inflated to recommended pressures. Proper inflation pressure is important for proper traction, riding comfort, fuel mileage, and tire wear. Note the effects of over, under, and proper inflation in **Figure 26-19**. Note how the full tread width of the correctly inflated tire (center illustration in **Figure 26-19**) contacts the surface of the road.

Overinflation will make the tire ride hard and more susceptible to casing (ply) breakage. Note how the overinflated tire bulges in the center, pulling the edges of the tire away from the road. This produces rapid wear in the center and will not allow all of the tread to contact the road for best traction.

Underinflation will cause tire heating, produce mushy steering, and will not properly present the full tread area for traction. An underinflated tire will cause the sidewalls to flex excessively and will quickly generate damaging heat. In addition, the center of the tread will buckle up away from the road and cause rapid wear on the tire's outer edges.

Figure 26-20 shows an inflation pressure conversion chart. To provide proper steering, ride, wear, and dependability, keep the tire pressures within the manufacturer's recommended range. Inflate tires to recommended pressure when cold (preferably after the vehicle has been sitting overnight). A tire may be considered cold after standing out of direct sunlight for three or four hours.

When the vehicle is driven, the tire temperature rises. The rise will depend upon speed, load, road smoothness, and prevailing temperature. A cold tire at 24 psi (165 kPa) will build up pressure to about 29 psi (200 kPa) after four miles (6.4 km) at speeds over 40 mph (64.4 kmh).

When checking hot tires, the pressure will exceed cold specifications. Never let air out of a hot tire to get the specified cold inflation pressure. For heavy loads or sustained high speed driving, some tire manufacturers recommend increasing cold pressure 4 psi (25 kPa). Do not increase tire pressure above this.

Tire Rotation

Tire life can be extended by ***tire rotation*** at regular intervals. The front tires are placed at the rear before any front wheel misalignment can cause serious tire imbalance or tire wear. Tire rotation should be done about every 5000 miles (8000 km). **Figure 26-21** shows some popular methods of rotation, using both conventional and radial tires. Some tire manufacturers recommend that radial tires be rotated so they always turn in the same direction, as in the right hand illustration in **Figure 26-21**. The compact space saver spare tire should never be rotated.

Tire Runout and Balance

Tires must run reasonably true to prevent vibrations. **Figure 26-22** shows one set of ***tire runout*** measurements. Notice that tire runout is not as critical as wheel runout. As the tire is flexible, it can absorb some of its own inaccuracy. Wheels should also be checked for radial and lateral runout, **Figure 26-23**.

To operate smoothly, it is essential that a tire and wheel be properly ***balanced***. Tire imbalance will cause poor ride, excessive tire wear, and steering and suspension component wear. The two major effects of imbalance are ***wheel tramp*** (tire and wheel hopping up and down), and ***wheel shimmy*** (shaking from side to side). For maximum tire wear and ride quality, all tires must be in static and dynamic balance.

Static Balance

When a tire is in proper ***static balance***, the weight mass will be evenly distributed around the axis of rotation. If the wheel is raised from the floor and spun several times and it

CONSTRUCTION TYPES

DIAGONAL	DIAGONAL AND BELTED BIAS				RADIAL				
	78 Series	70 Series	60 Series	50 Series	Metric	78 Series	70 Series	60 Series	50 Series
6.15/155-13									
					155R13				
6.00-13					165R13				
	A78-13	A70-13	A60-13			AR78-13	AR70-13	AR60-13	
6.50-13	B78-13	B70-13	B60-13	B50-13	175R13	BR78-13	BR70-13	BR60-13	BR50-13
	C78-13	C70-13	C60-13	C50-13		CR78-13	CR70-13		CR50-13
7.00-13					185R13				
	D78-13	D70-13	D60-13	D50-13		DR78-13	DR70-13		
					195R13	ER78-13	195/70R-13	ER60-13	
					155R14				
	A78-14					AR78-14		AR60-14	
6.45-14	B78-14		B60-14		165R14	BR78-14			
6.95-14	C78-14	C70-14	C60-14		175R14	CR78-14	CR70-14		
	D78-14	D70-14	D60-14			DR78-14	DR70-14		
7.35-14	E78-14	E70-14	E60-14		185R14	ER78-14	ER70-14	ER60-14	
7.75-14	F78-14	F70-14	F60-14	F50-14	195R14	FR78-14	FR70-14	FR60-14	
8.25-14	G78-14	G70-14	G60-14	G50-14	205R14	GR78-14	GR70-14	GR60-14	GR50-14
8.55-14	H78-14	H70-14	H60-14	H50-14	215R14	HR78-14	HR70-14	HR60-14	
8.85-14	J78-14	J70-14	J60-14		225R14	JR78-14	JR70-14	JR60-14	JR50-14
		L70-14	L60-14				LR70-14	LR60-14	
				M50-14					
				N50-14					
	A78-15	A70-15				AR78-15			
	B78-15		B60-15	B50-15	165R15	BR78-15	BR70-15		
6.85-15	C78-15	C70-15	C60-15		175R15	CR78-15	CR70-15		
	D78-15	D70-15				DR78-15	DR70-15		
7.35-15	E78-15	E70-15	E60-15	E50-15	185R15	ER78-15	ER70-15	ER60-15	
7.75-15	F78-15	F70-15	F60-15		195R15	FR78-15	FR70-15	FR60-15	
8.25-15	G78-15	G70-15	G60-15	G50-15	205R15	GR78-15	GR70-15	GR60-15	GR50-15
8.55-15	H78-15	H70-15	H60-15	H50-15	215R15	HR78-15	HR70-15	HR60-15	HR50-15
8.85-15	J78-15	J70-15	J60-15		225R15	JR78-15	JR70-15	JR60-15	JR50-15
9.00-15		K70-15				KR78-15	KR70-15		
9.15-15	L78-15	L70-15	L60-15	L50-15	235R15	LR78-15	LR70-15	LR60-15	LR50-15
	M78-15					MR78-15	MR70-15		
8.90-15									
	N78-15			N50-15		NR78-15			
7.00-15									
6.00-16									

Figure 26-17. A tire size and construction "interchangeability" chart. There are a number of different charts used, listing almost any interchange that may be needed. (Rubber Manufacturers Assoc.)

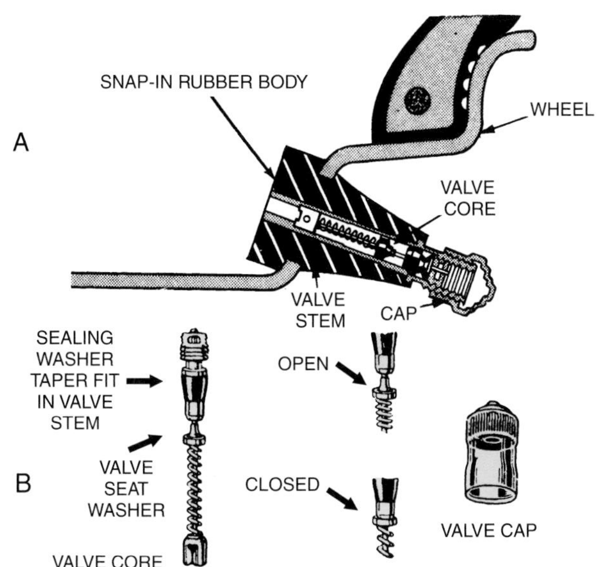

Figure 26-18. A—In typical valve stem construction, the rubber body or stem houses the core. B—Valve core action. Note that the valve cap acts as second seal against air leakage, as well as preventing entry of dust. (Cadillac)

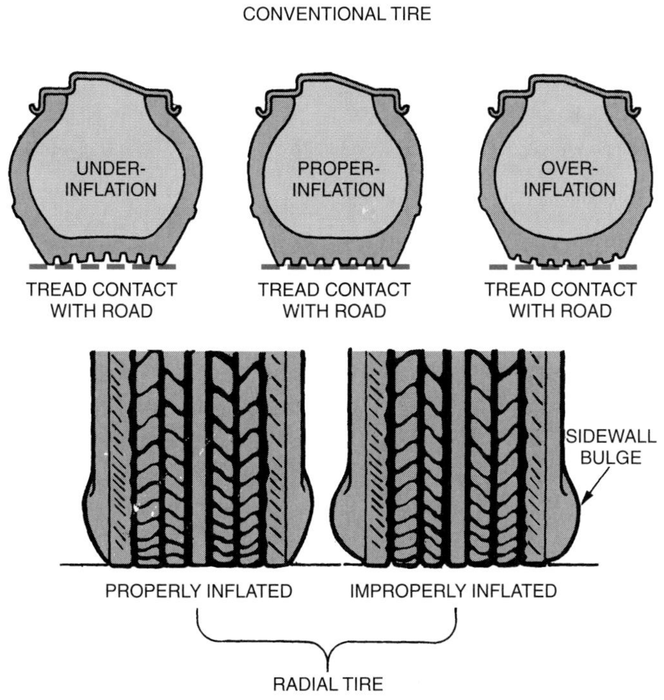

Figure 26-19. Inflation effect on tire and tread contact. Proper road contact can only be obtained with correct inflation. Notice the slight difference in sidewall bulge between the properly and improperly inflated radial tire. (Chevrolet and Ford)

always stops in a different place, it is statically balanced, **Figure 26-24**.

If the wheel revolves until one side is down, it is obvious that this side is heavier than the opposite side. The tire and wheel is statically imbalanced. Static imbalance is corrected

PSI	kPa	PSI	kPa
20	140	31	215
21	145	32	220
22	155	33	230
23	160	34	235
24	165	35	240
25	170	36	250
26	180	40	275
27	185	45	310
28	190	50	345
29	200	55	380
30	205	60	415
Conversion: 6.9 kPa = 1 psi			

Figure 26-20. Typical tire pressure conversion chart: psi = pounds per square inch and kPa = kilopascals (metric pressure measurement). You should also see the "Useful Conversion Tables" in the back of this text for exact conversion figures. (Chevrolet)

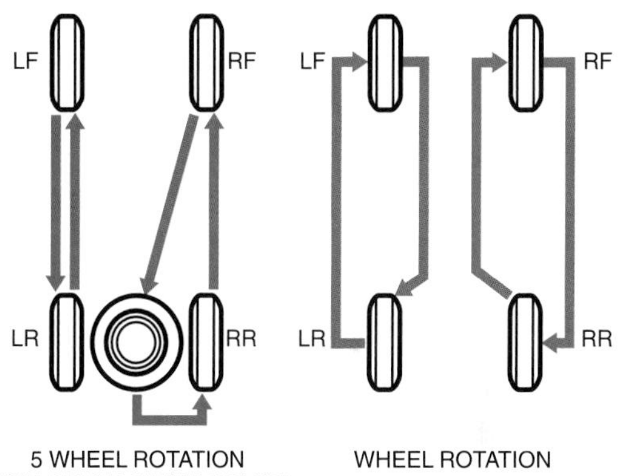

Figure 26-21. Tire rotation recommendations as used by one manufacturer. Note that radial tires are only moved from front-to-back or back-to-front on the same side. Cross-switching (which reverses tire rotation) should not be done on radials. (Chevrolet)

by clipping lead weights to the rim edge opposite the heavy side.

Dynamic Balance

Dynamic balance is achieved when the weight mass is in the same plane as the centerline of the wheel. Note in **Figure 26-25** that the centerline of the weight mass is not in the same plane as the centerline of the wheel. When the wheel spins, centrifugal force will attempt to bring the weight mass centerline into the wheel centerline plane.

Weight mass 2 will attempt to move to the left. Weight mass 3 will attempt to move to the right. As the wheel revolves, the weight masses will reverse their positions, 2 will move to the right, 3 to the left. Each half turn of the wheel will

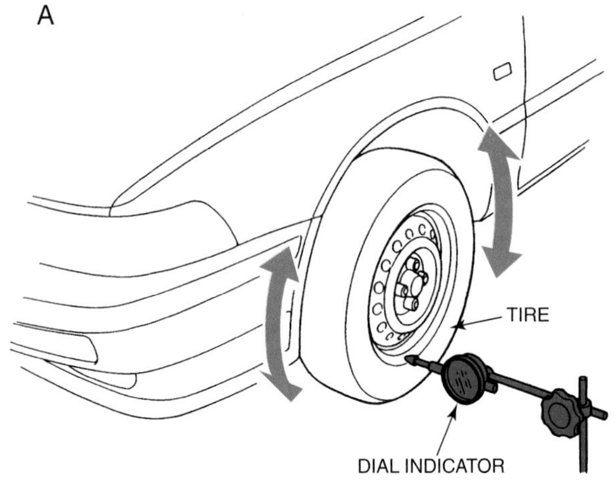

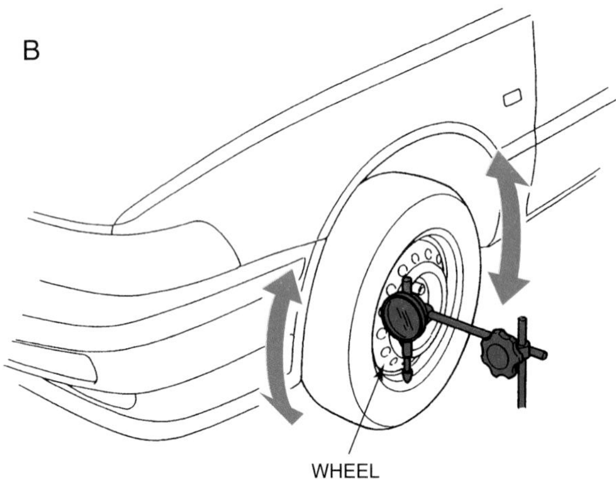

Figure 26-22. The wheel and tire runout must be within specific limits as set by the manufacturer. These limits will vary by manufacturer, and will differ depending on the material used for wheel construction—steel, aluminum, etc. Note the use of a dial indicator to obtain precise readings. (Honda)

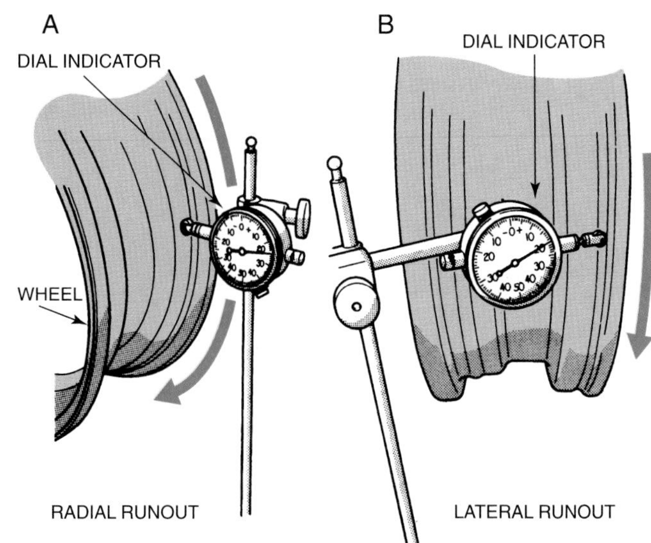

Figure 26-23. A—Checking a wheel for radial runout with a dial indicator. B—Checking the same wheel for excessive lateral runout. Wheels should be checked before the tire is mounted. (Chrysler)

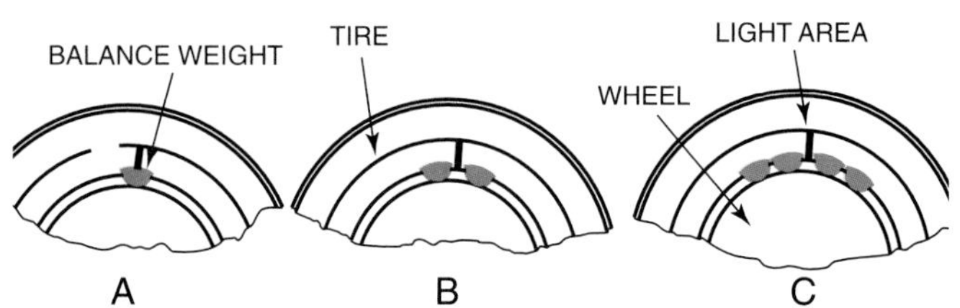

Figure 26-24. A—Static balance weight. B and C—If more than one weight is necessary, it is essential that they be evenly distributed on each side of the light area (indicated by small vertical line). (Plymouth)

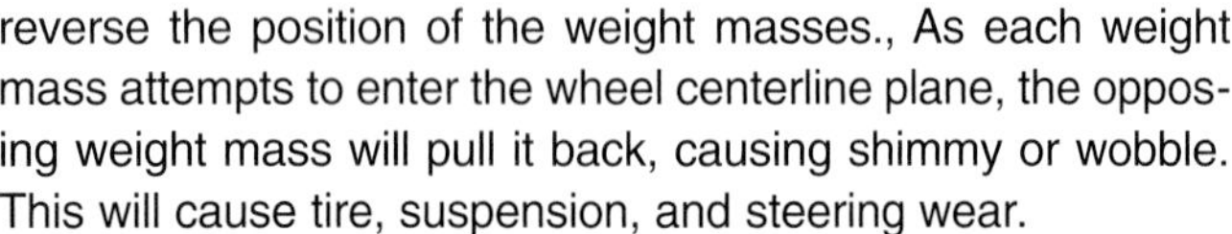

reverse the position of the weight masses., As each weight mass attempts to enter the wheel centerline plane, the opposing weight mass will pull it back, causing shimmy or wobble. This will cause tire, suspension, and steering wear.

Dynamic imbalance is corrected by adding lead wheelweights to both sides of the rim, until the weight mass centerline is in the same plane as the wheel centerline, **Figure 26-26**. When this happens, the assembly is now dynamically balanced.

High Speeds and Overloading Produce Heat and Excessive Wear

The relationship between vehicle speed, overloading, and tire temperature is shown in **Figure 26-27**. Note how the heat goes up faster when a given tire is subjected to a 20% overload as opposed to the normal load. When tires operate at 250°-275°F (121°-135°C), they lose a great deal of strength and wear resistance.

High speed driving also affects tire tread wear. Higher speeds result in higher tire temperatures and increased tread

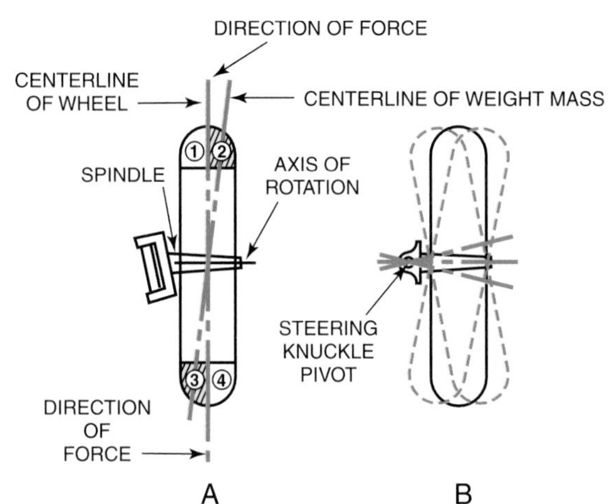

Figure 26-25. This tire is dynamically imbalanced. A is a front view and B is a top view. The weights on each side try to pull the tire into the centerline. As the tire approaches its centerline, the opposing weight pulls it away, creating wheel shimmy. (Plymouth)

wear. At 80 miles per hour, (128 kph) for example, tire temperatures build to the critical level, even under normal load. See **Figure 26-27**.

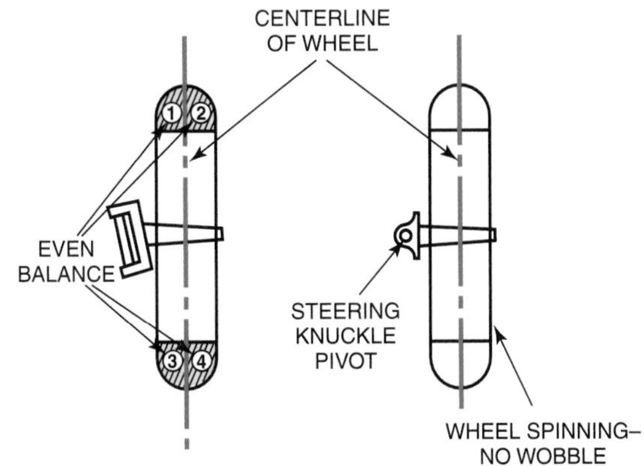

Figure 26-26. Dynamic balance. Note the absence of wheel wobble when dynamic balance is achieved.

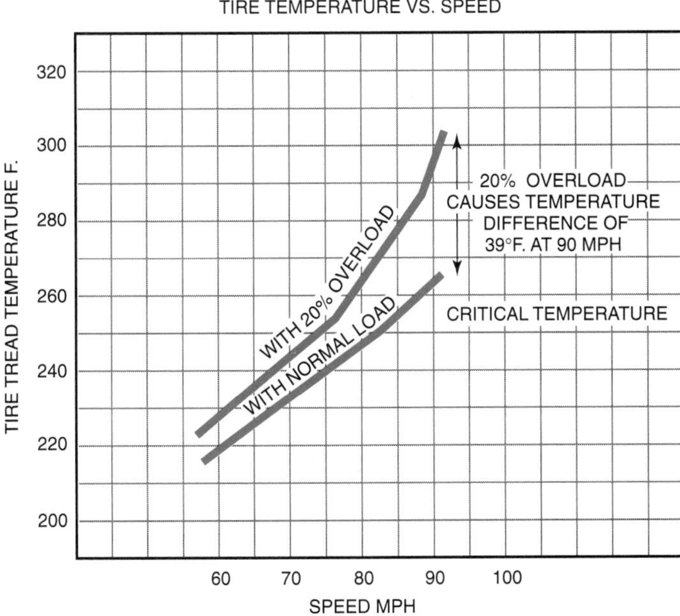

Figure 26-27. Tire speed, load, and temperature graph. At high speeds, a small speed increase causes a large increase in the temperature. Note how overload raises temperature and reduces tire life. (Rubber Manufacturers Assoc.)

Summary

Wheel rims can be made of stamped steel, or custom wheels made of various materials. The wheel rims are fastened to the wheel hub using four to six lug nuts or bolts.

The wheel hubs rotate on tapered roller, straight roller, or ball bearings, and are lubricated by wheel bearing grease. Most front wheels use tapered roller bearings. Rear wheels are bolted to integral or detachable hubs.

Tires are an important part of the suspension system. They provide a cushioning effect to absorb some road shocks and supply the traction necessary for driving and braking. Rubber, fabric, and steel wires are used in tire construction. Special safety type tires are available. A wide variety of tread designs are manufactured. Most tires are the radial type.

Newly manufactured tires are graded using the Uniform Tire Quality Grading System. Tires and wheels must run true, and be balanced statically and dynamically. Lead wheel weights are used in balancing a tire and wheel assembly. Proper tire inflation is vital to good steering, riding quality, and tire life.

High speeds can ruin a tire. Excessive speed promotes heating resulting in a loss of tire strength and wear quality. Safe vehicle operation is also drastically reduced. Tires should be rotated every 5000 miles (8 046 km). Always double check when mounting wheels on a vehicle to make certain they are bolted on securely.

Know These Terms

Wheel rim
Stamped steel
Custom wheel
Drop center
Safety ridges
Lug bolts
Wheel lugs
Lug nuts
Wheel nut torque
Torque stick
Wheel hub
Tapered roller bearing
Straight roller bearing
Ball bearing
Packed
Castellated nut
Safety washer
Cotter pin
Staked nut
Traction
Bead
Plies
Bias tire construction
Bias-belted tire construction
Radial tire construction
Tire rating
Speed rating
Uniform tire quality grading system
Temperature resistance rating
Traction rating
Tread wear rating
Space saver spare tire
Tire valve
Valve stem
Valve core
Inflation pressure
Overinflation
Underinflation
Tire rotation
Tire runout
Tire balance
Wheel tramp
Wheel shimmy
Static balance
Dynamic balance

Review Questions—Chapter 26

Do not write in this book. Write your answers on a separate sheet of paper.

1. Custom wheel rims can be made from _____ .
   (A) aluminum
   (B) magnesium
   (C) steel and aluminum
   (D) All of the above.

2. All of the following statements about wheel bearings are true, EXCEPT:
   (A) tapered roller bearings can be cleaned, regreased, and adjusted.
   (B) most newer vehicles use greasable ball bearings.
   (C) straight roller bearings used on front-wheel drive vehicles are replaced as a unit.
   (D) wheel bearings must be packed with quality, high temperature grease.
3. Air is sealed in the tire by the rim and tire ________ .
4. Which of the following parts should be replaced when tapered roller bearings are greased?
   (A) Safety washer.
   (B) Slotted nut.
   (C) Cotter pin.
   (D) Both A & C.
5. What is the advantage of the drop center steel wheel?
6. What are the two main functions of vehicle tires?
7. Technician A says that air is injected into the tire through a valve stem and core assembly. Technician B says that the valve core contains the actual valve mechanism. Who is right?
   (A) A only.
   (B) B only.
   (C) Both A & B.
   (D) Neither A nor B.
8. If an older vehicle was originally equipped with bias ply tires, what kind of replacement tires are usually available?
   (A) Bias tires only.
   (B) Bias or belted tires only.
   (C) Bias, belted, or radial tires.
   (D) Radial tires.
9. Technician A says that a tire in dynamic balance will also be in static balance. Technician B says that static balancing a tire will correct shimmy. Who is right?
   (A) A only.
   (B) B only.
   (C) Both A & B.
   (D) Neither A nor B.
10. Extreme wear on the outside on both sides of the tire tread indicates _____ .
    (A) overinflation
    (B) underinflation
    (C) incorrect toe
    (D) a bent wheel rim
11. The most commonly used tire construction type is the _____ .
    (A) radial
    (B) bias
    (C) bias belted
    (D) biased radial
12. List the three different Uniform Tire Quality Grading System categories.
13. Draw a diagram showing how radial tires should be rotated.
14. When rotating tires, the compact (space saver) spare must be used _____ .
    (A) in the front
    (B) in the back
    (C) for emergencies only
    (D) Either A or B.
15. How does high speed affect tire life?
16. Tires should be rotated every _____ miles/kilometers.
    (A) 5000/8000
    (B) 7500/12 000
    (C) 10,000/16 000
    (D) 15,000/32 000
17. Proper tire inflation pressure is important in four ways. Name them.
18. All of the following statements about tire inflation are true, EXCEPT:
    (A) a tire is considered cold after standing out of direct sunlight for four hours.
    (B) when the vehicle is driven, the temperature of its tires rises.
    (C) always let air out of hot tires to get specified cold pressure.
    (D) for heavy loads or sustained high speed driving, tire pressure can be increased by 4 psi (25 kPa).
19. Draw a cross section of a typical tire and name the various parts.
20. What part of the tire tread will wear out when a tire is underinflated?

27

Wheel Alignment

After studying this chapter, you will be able to:

- Explain the purpose of wheel alignment.
- Identify the major wheel alignment angles.
- Identify the most common related wheel alignment angles.
- Identify vehicle alignment adjustment devices.
- Identify suspension, steering, and tire factors which could affect alignment.

In Chapters 24 through 26, you learned about the design and components of the vehicle suspension system, steering system, and wheels and tires. This chapter discusses the principles of aligning all of these parts to provide optimum handling and tire wear. Although this process involves almost all of the steering and suspension components, it is commonly known as wheel alignment. After completion of this chapter, you will be able to explain wheel alignment angles and be familiar with the devices used to measure these angles.

Purpose of Wheel Alignment

It is not enough to merely place the front wheels on hubs, stand them up straight and devise a method of turning them from left to right. The vehicle could be driven, but it would steer poorly. At higher speeds, it would become dangerous to handle and tire life would be short. For good steering, handling, and smooth operation, the wheels must be precisely aligned.

Wheel Alignment Angles

Wheel alignment is the process of measuring and correcting the various angles formed by the front and rear wheels, spindles, and steering arms. Correct alignment is vital. Improper alignment can cause hard steering, pulling to one side, wandering, noise, and rapid tire wear.

Types of Wheel Alignment

Rear-wheel drive vehicles require a ***two-wheel alignment***, or front-wheel alignment only. The rear wheels are attached to a solid rear axle assembly which cannot be adjusted, and generally stays in alignment throughout the life of the vehicle.

Today, most front-wheel drive vehicles have provisions for adjusting the rear wheels. In addition, many rear-wheel drive vehicles are equipped with independent rear suspensions, which also must be adjusted. Also, modern solid rear axles and suspension systems are lightweight and can become misaligned. Therefore, the ***four-wheel alignment***, in which the front and rear wheel alignment angles are checked and adjusted, is commonly performed.

Adjustable front wheel settings on most modern vehicles are caster, camber, and toe. Nonadjustable settings are steering axis inclination and toe-out on turns. Rear wheel settings that can be made on many modern vehicles are camber and toe. Modern practice is to check both front and rear wheel alignment. Note that the various alignment angles are all related. A change in one can alter the others.

Alignment Measurement Values

All alignment values, except for toe, are measured in degrees. Degrees is the unit of measurement for angles. One degree is 1/360 of a circle. Toe is measured in fractions of an inch or millimeters.

Caster

Everyone is familiar with the common caster used on some furniture. You may have wondered why the little wheel always swivels around and trails almost obediently behind a chair or dresser leg. The reason is simple.

Referring to **Figure 27-1**, you will see that the centerline about which the caster wheel swivels strikes the floor ahead of it. If the caster attempts to move to either side, the point of resistance between the caster and the floor will be to the side of the swivel centerline.

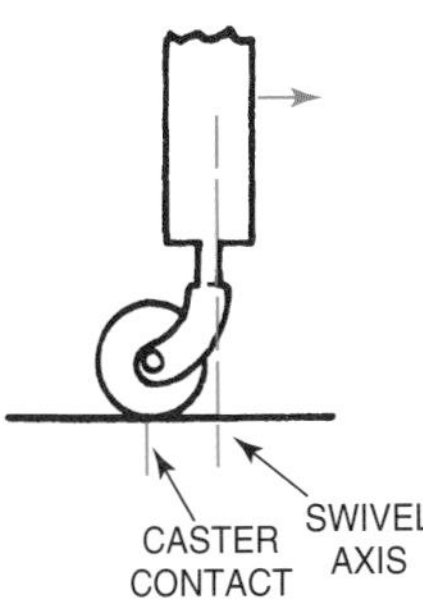

Figure 27-1. A chair leg caster. The centerline of the swivel axis contacts the floor ahead of the caster wheel contact point.

Since the caster is encountering some resistance to forward motion, it will not move until the moving weight mass actually drags it forward. However, it cannot be dragged forward until all the slack is taken up. In this case, it will not move straight forward until it is directly in back of the swivel centerline. **Figure 27-2** illustrates how resistance to forward motion forces a trailing object (caster wheel) to line up behind the moving swivel line.

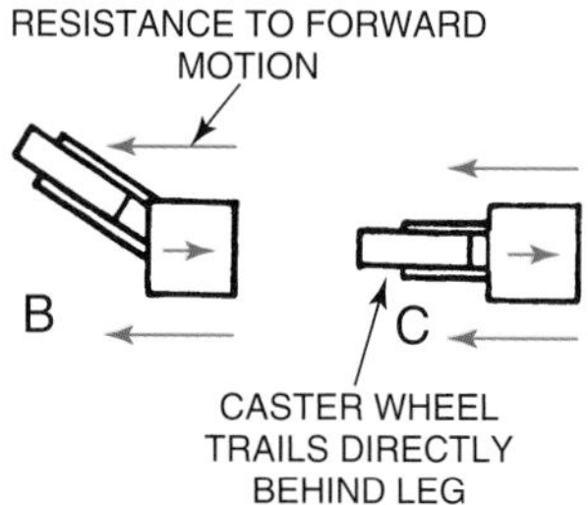

Figure 27-2. To demonstrate the caster effect, as the table leg B starts to move to the right, the caster wheel, which is off to one side encounters a tooling resistance. This resistance forces the wheel to line up and trail directly behind the table leg in C.

The application of the caster principle to vehicles involves the placement of the ***steering axis***. The steering axis is formed by the upper and lower ball joints, strut mount, etc., **Figure 27-3**. This creates ***caster***, which can be either positive or negative.

Positive Caster

By tilting the top of the steering knuckle toward the rear of the car, the centerline of the axis, or steering point, will be in front of the point at which the tire encounters the road. As this places the point of resistance to forward motion behind the steering centerline, the wheel is forced to track behind, automatically lining up with the moving car. This is referred to as *positive caster*, **Figure 27-4**.

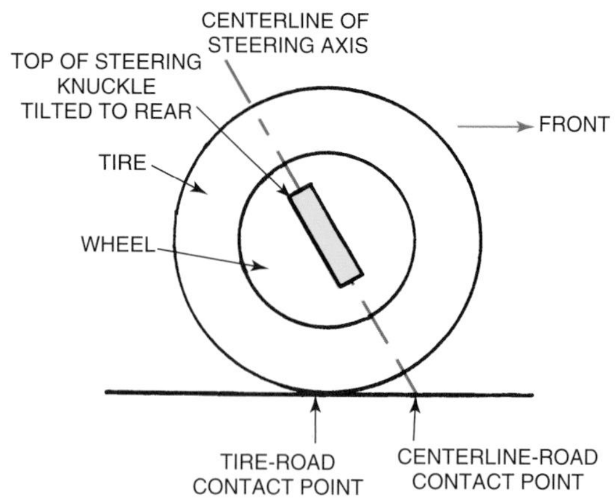

Figure 27-4. Positive caster. Steering axis (swivel axis) strikes the road in front of the tire-to-road contact point. The wheel will trail behind the steering axis centerline-to-road contact point. (Tilt shown is exaggerated.)

Positive caster tends to force the wheels to travel in a straight ahead position. It also assists in recovery (wheels turning back to straight ahead position) after making a turn. On late model cars, there is often little or no positive caster. Positive caster makes it more difficult to turn the wheels from the straight ahead position than when no caster angle is present.

Another aspect of positive caster is a mild tipping effect when cornering. When making a right turn, the right wheel will cause the steering knuckle to raise slightly, while the left wheel will allow the knuckle to lower, creating the tipping effect. If the left side of the vehicle was allowed to rise during a right turn, it would have an adverse effect on the vehicle's cornering ability. When the wheels are turned to the left, the left side raises and the right side drops.

Negative Caster

To ease turning, many cars employ a *negative caster* angle, which angles the top of the steering knuckle to front of

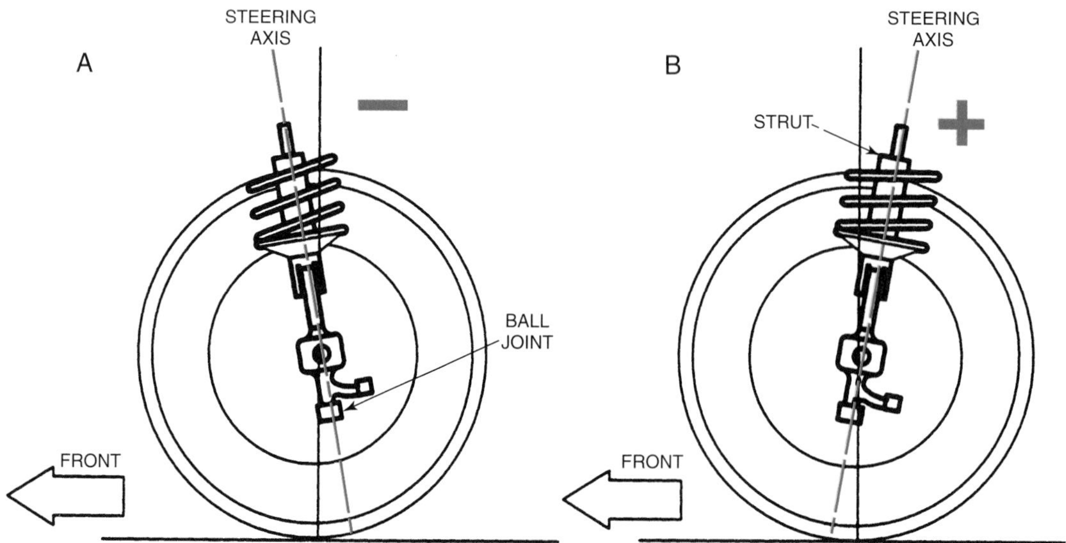

Figure 27-3. The steering axis is formed by the ball joint and strut position. A—Negative caster. B—Postive caster. (Hunter Engineering Company.)

car. This will ease steering while also causing the mild tipping effect needed when cornering. Proper tracking is still provided by the included angle (to be discussed later), **Figure 27-5**.

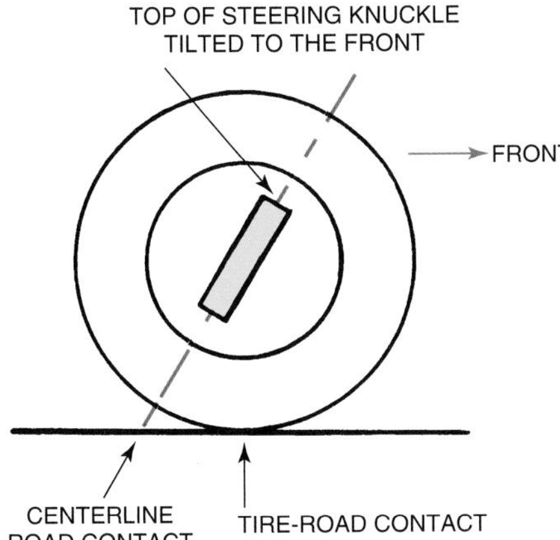

Figure 27-5. Negative caster. Steering axis centerline strikes the road behind the tire-to-road contact point. (Tilt shown is exaggerated.)

Caster has little effect on tire wear, but can cause handling problems, especially if there is a great difference (more than 1°) in caster between the two front wheels. A vehicle will always tend to drift toward the side with the least positive caster. This is called ***pulling***. Pulling occurs when the vehicle moves from a straight line to one side of the road or the other, with no pressure applied to the steering wheel. If one wheel is more negative than the other (even if both are negative), the vehicle will pull toward that wheel.

Camber

Camber is the tilting of the wheel centerline, viewed from the front of the vehicle, away from a true vertical line. Camber angles are usually small, usually no more than 1° positive or negative from zero. An incorrect camber setting will cause pulling and tire wear.

Positive and Negative Camber

Positive camber is provided by angling the top of the wheel away from the vehicle. When viewing the car from the front, the tops of the wheels are farther apart than the bottom. In short, the centerline of the wheel is no longer in a vertical plane (straight up and down). Look at **Figure 27-6**. If the tops of the wheels are closer together than the bottoms, the camber is negative.

Positive camber places the wheel centerline closer to the centerline of the steering axis where both intersect the road. This assists in reducing road shock to the steering system and the tendency of the front wheels to spread apart at the front (toe-out). Many vehicles use a small amount of positive camber (less than 1°). Some front-wheel and off-road vehicles have the camber set at zero or slightly negative.

If the wheel has excessive positive camber, it will make the vehicle pull to one side. If the wheel has too much

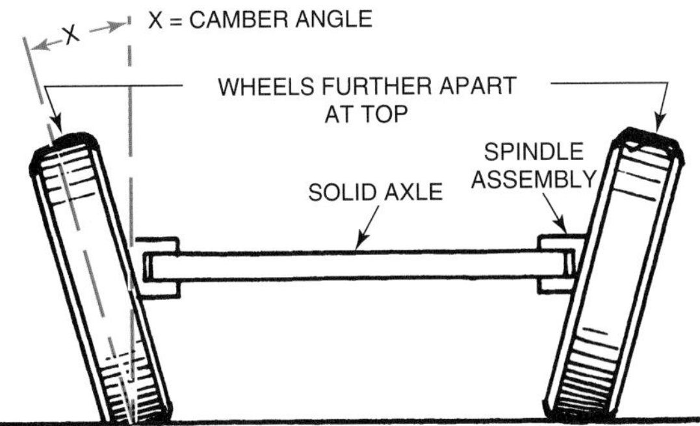

Figure 27-6. Positive camber. When viewing the car from the front, the tops of the wheels are farther apart than the bottoms. (Camber angle shown is exaggerated.)

negative camber, it will make the vehicle pull to the other side. See **Figure 27-7**.

Steering Axis Inclination

The steering knuckle ball joints are closer together at the top than the bottom. The steering axis (view from the front of the car) places the centerline of the steering ball joints closer to the centerline of the wheel, **Figure 27-8**. This angle is known as ***steering axis inclination***, or ***SAI***.

When the wheel centerline is to the outside of the center line of the steering axis (where they intersect the road), the wheels tend to toe-out. This is caused by the road-tire resistance pushing back on the spindle, causing it to swivel backward on the ball joints or toe-out, **Figure 27-9**. When the centerline of the wheel intersects the road at a point inside of the steering axis centerline intersection, the wheels tend to toe-in (tires are closer together in the front than in the back), **Figure 27-10**.

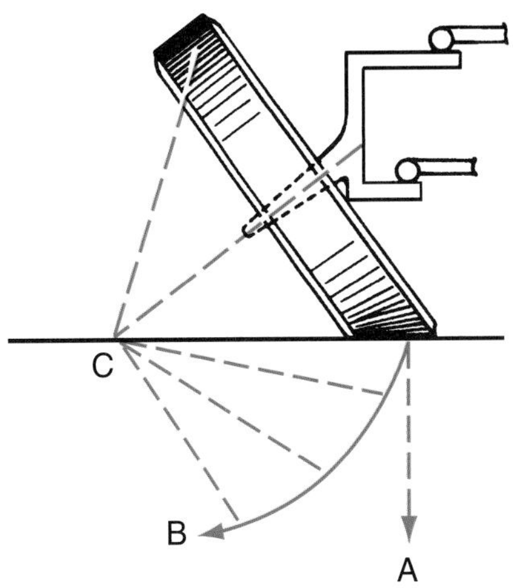

Figure 27-7. Camber toe-out tendency. A is the path that the tire should take when pointed straight ahead. B is the actual path taken by the tire. When the wheel is cambered, the surface of the tire will engage the road at an angle. The tire will adapt itself to the road, producing a cone shape with an apex at C. The tire will roll on path B as though it were a cone. (The angle is exaggerated.)

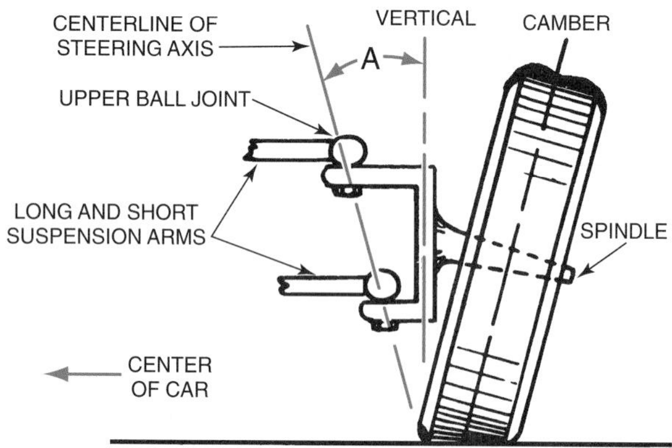

Figure 27-8. Steering axis inclination, front view, driver's side wheel and spindle assembly. Note how the top ball joint is closer to the center of the vehicle than the bottom ball joint. This tips the steering axis centerline as shown. Steering axis inclination angle shown at A. (The angles are exaggerated.)

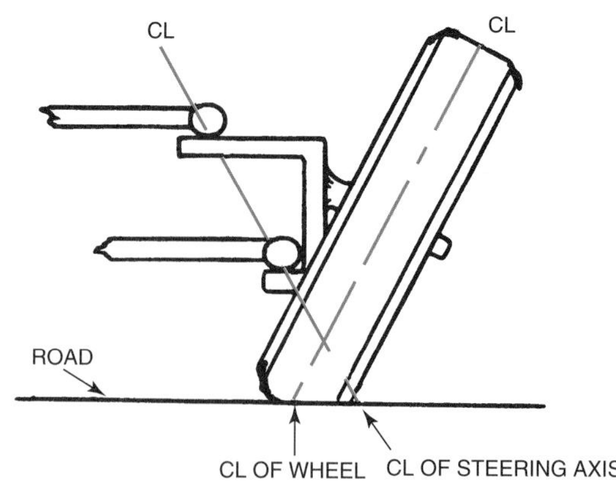

Figure 27-10. Toe-in effect. You will note that the CL of the wheel now intersects the road inside of the point where the steering axis CL strikes the road. Resistance to forward motion will now cause the wheel to toe-in when the car moves forward. (Angles are exaggerated.)

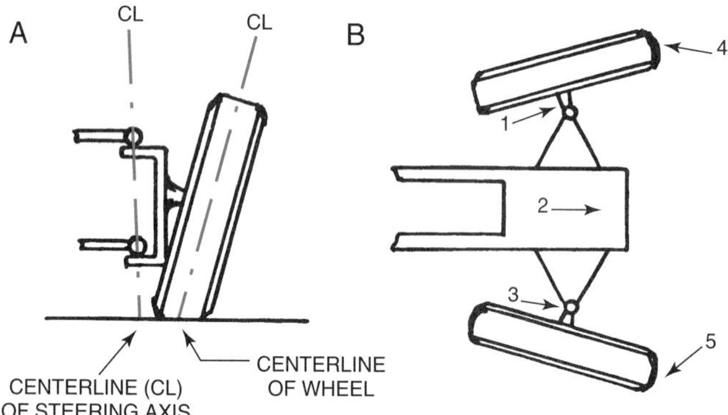

Figure 27-9. Toe-out effect. A is front view, showing centerline (CL) of wheel striking the road outside of the point where CL (centerline) of the steering axis strikes the road. B is a top view showing how the wheels are forced to toe-out when the vehicle moves forward. Arrows 1, 2, and 3 represent the direction of force pushing the car forward. Arrows 4 and 5 represent the force of resistance to forward motion. As the resistance forces are outside of the steering axis-to-road intersection, the wheels toe-out. (Angles are exaggerated.)

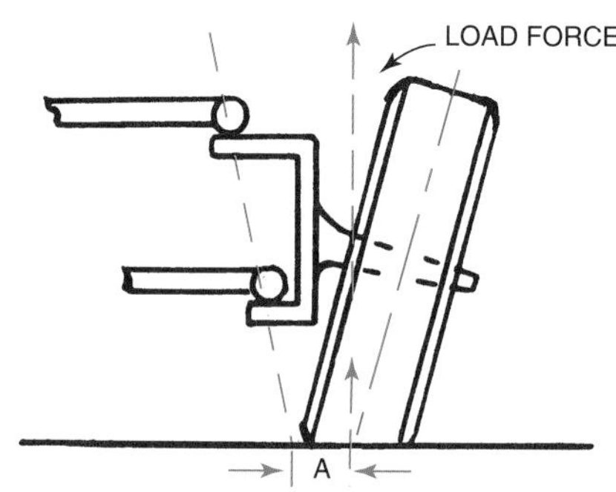

Figure 27-11. Suspension shock transfer. Load force, transmitted in a vertical plane will now strike the heavy inner end of the spindle. A shows that the distance between the wheel centerline and steering axis centerline is reduced by camber and steering axis inclination. Resistance to forward motion is now closer to the steering axis, reducing shock transfer to the tie rod ends and steering system. (Angles are exaggerated.)

Steering axis inclination helps to bring the centerlines of the wheel and the steering axis closer together where they intersect the road surface. This also causes suspension shocks to be transmitted to and absorbed by the heavy inner spindle and knuckle assembly, **Figure 27-11**.

Steering axis inclination also helps the wheels follow the straight ahead position. This tracking effect is caused by the fact that whenever the wheels are turned from the straight ahead position, the car is actually lifted up. Remember that the wheel spindle, when turned, is moved about the steering axis.

As the axis is tilted, the spindle will move in a downward direction. Since a downward movement of the spindle is impossible (it is supported by the wheel assembly), the steering knuckle is forced upward. This will raise the car a small amount. The effect is similar on both sides. The weight of the car will force the spindles to swivel back, returning the wheels to the straight ahead position, **Figure 27-12**.

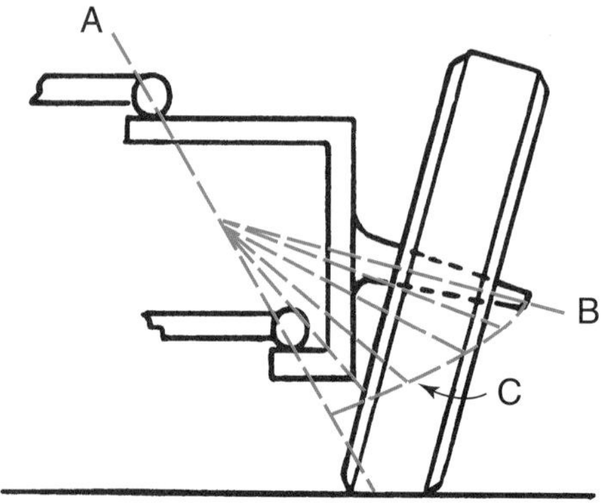

Figure 27-12. Lifting effect of steering axis inclination. A–Steering axis CL. B–Path of spindle end as the wheel is turned. Note that as the spindle revolves, it travels downward. Since it cannot come closer to the road (wheel holds it up), the steering axis will be lifted. This action affects both wheels regardless of the turning direction.

Included Angle

Included angle is the total of the SAI and the camber on any one wheel. Included angle can be determined by adding the SAI and camber angles together. The resulting figure is the included angle. The combination of SAI and camber to create included angle is shown in **Figure 27-13**.

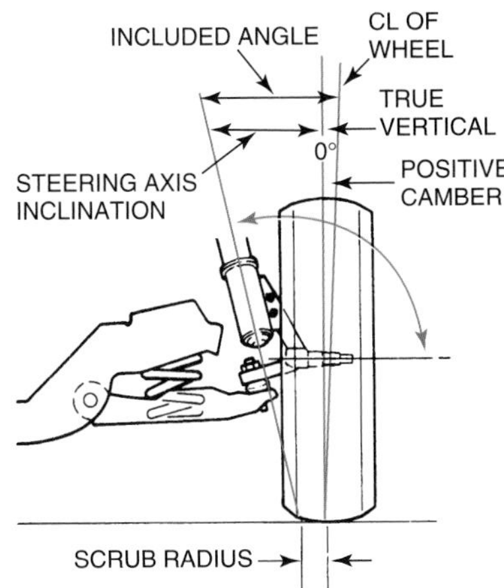

Figure 27-13. The steering included angle is determined by combining both the steering axis inclination and camber angles together. (Chevrolet)

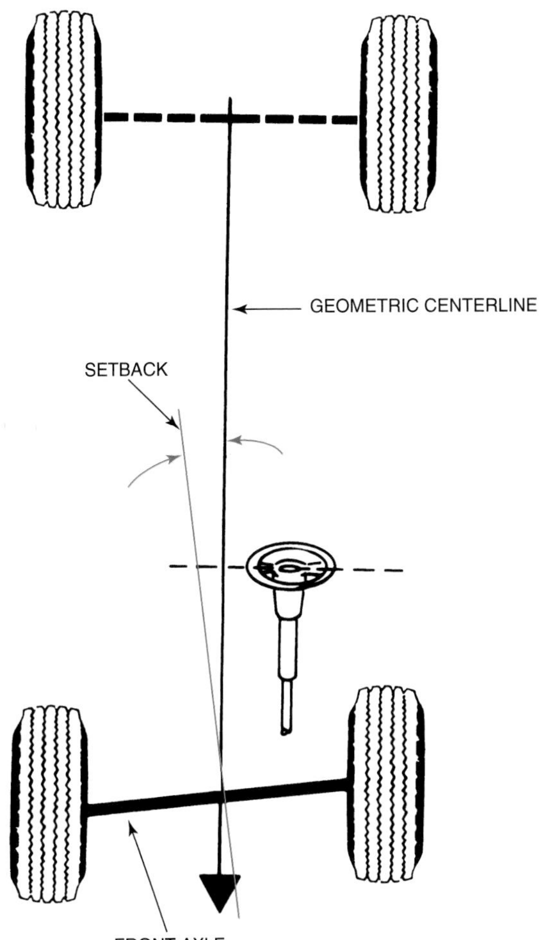

Figure 27-14. Front wheel setback. This is the angle formed between the geometric centerline (CL) and a line that is perpendicular to the front axle. Setback can be positive—right front wheel is setback farther than the left, or negative—left front wheel is setback farther than the right. Extreme setback will cause the vehicle to have very poor handling. (Perfect Equipment Corporation)

Setback

Setback is a condition in which one front wheel spindle is positioned behind the spindle on the opposite side. This condition can also be present in rear wheel assemblies. Excessive setback is usually caused by collision damage. One indication of excessive setback is a caster reading that varies by more than 1° from one side of the vehicle to the other. Many types of alignment equipment cannot measure setback.

Severe setback can sometimes be detected by measuring the distance between the rear of each tire and the wheel opening on both sides and comparing. Setback can sometimes cause pulling toward the wheel that is the farthest back, **Figure 27-14**.

Toe

Toe is the relative positions of the front and rear of a tire in relation to the other tire on the axle. Look at **Figure 27-15**. Note that distance at the back of the tires is greater than distance in the front. Rear-wheel drive vehicles are toed in to compensate for the natural tendency of the road-to-tire friction to force the wheels apart.

On some front-wheel drive vehicles, the front tires are toed out. This is done to offset the force created by the drive axles, which tend to pull the tires inward during operation. The toe setting compensates for this and allows the front tires to run parallel to one another while rolling straight down the road.

Toe compensates for the wheel movement tendencies plus any wear or play in the steering linkage. Proper toe will allow the tires to move forward without a scrubbing, scraping action between the tire and road. Excessive toe conditions will cause a rapid tire wear condition called ***feathering*** as shown in **Figure 27-16**. Toe can also be adjusted on most vehicles to set the steering wheel in the centered position when driving straight ahead.

Toe-Out on Turns

When a car turns a corner, the inner wheel must turn in a shorter radius (smaller circle) than the outer wheel. To allow the inner wheel to cut more sharply, both wheels must be able to ***toe-out on turns*** automatically. This essential action is accomplished by bending both steering arms so that they angle slightly toward the center of the vehicle, **Figure 27-17**.

When the wheels are turned, the steering arm on the inside turns more sharply. This is due to the angle of the turn. **Figure 27-18** illustrates the difference in wheel angles while turning. Note that when driving straight ahead, the wheels are set to the normal toe position.

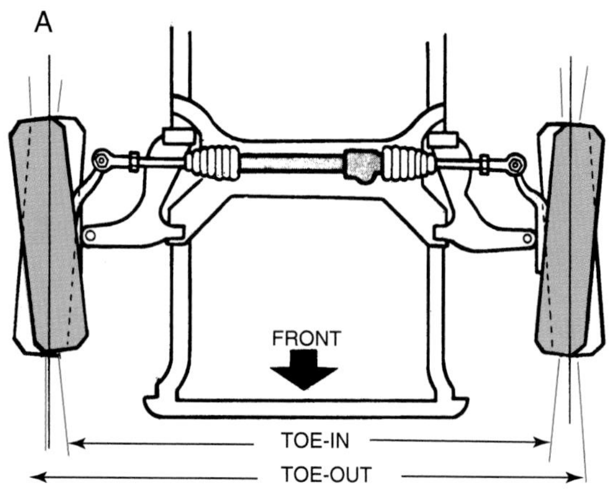

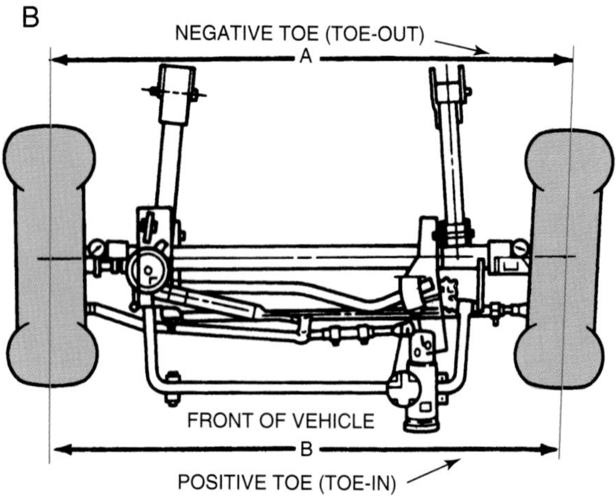

Figure 27-15. A—Toe-in and toe-out on a front-wheel drive vehicle. B—Toe-out (negative toe) and toe-in (positive toe) on a rear-wheel drive, four-wheel drive vehicle. Note the toe-in on the right-hand tire as opposed to the neutral straight ahead position on the left-hand tire (Chrysler Corp.)

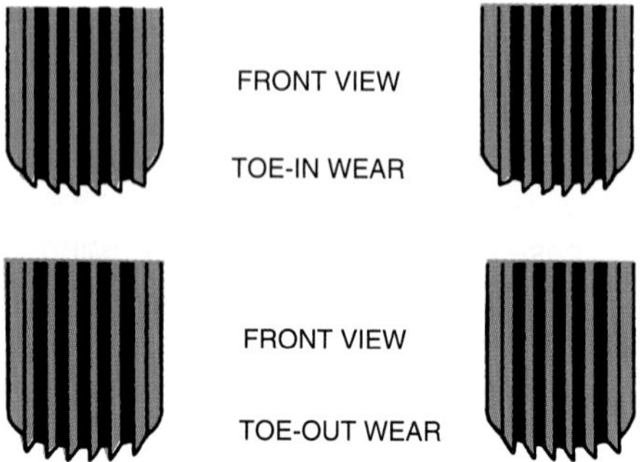

Figure 27-16. Toe-in and toe-out tire wear. If you ran your hand across one of these tires, you would be able to feel the feathered wear pattern. (Cadillac)

Slip Angles

When a vehicle rounds a corner at high speed, the actual path taken by the tires is somewhat different than if ***inertial force*** was not present. Inertial force is the tendency of a moving object (such as a vehicle) to continue moving in the same direction.

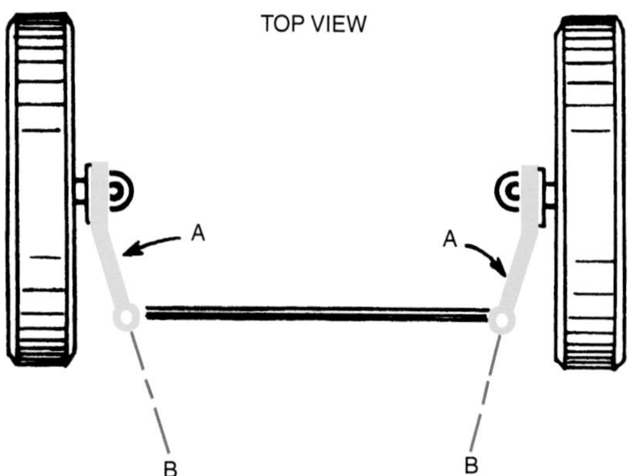

Figure 27-17. Steering arms, A, on most vehicles are angled inward. If the centerlines B were continued, they would meet somewhere near the differential.

As the wheels turn around a corner, a cornering force is created that tries to move the car in the exact direction the wheels are aimed. This cornering force is counteracted by inertial force, which tries to keep the vehicle body moving in the same direction as it was before the turn. The actual path the wheels take is determined by a balance between cornering and inertial forces.

The difference in the actual path and the path with no inertial force is the ***slip angle***. The higher the vehicle speed, the greater the slip angle. The greater the slip angle, the more rapid the tire wear. **Figure 27-19** depicts typical slip angles. Slip angle cannot be changed by alignment adjustments, and can be reduced only by careful cornering.

Steering Geometry

The term ***steering geometry*** is often used to describe the various angles formed by the suspension alignment setup. In other words, caster, camber, toe, steering axis inclination, and other angles are sometimes collectively referred to as steering geometry.

Alignment Adjustments

Various means are provided for adjusting the different front and rear wheel alignment angles. If these angles vary from their initial factory settings, the adjustment procedures must be followed exactly to reset the angles. The following sections discuss some of the more common ways in which alignment can be adjusted. Always consult the proper service manual for exact adjustment procedures.

In some cases, the experienced technician will deviate from the factory settings to increase tire life and vehicle handling. The experienced technician can also compensate for misalignment of a non-adjustable angle by setting an adjustable angle slightly outside of factory specifications.

Thrust Angle

Aligning all four wheels makes it possible to set the ***thrust angle*** to ensure perfect wheel tracking. The thrust angle is an imaginary line at a right angle (90°) to the rear axle. Ideally, the thrust angle should be parallel with the vehicle's centerline. The vehicle should follow this line with no

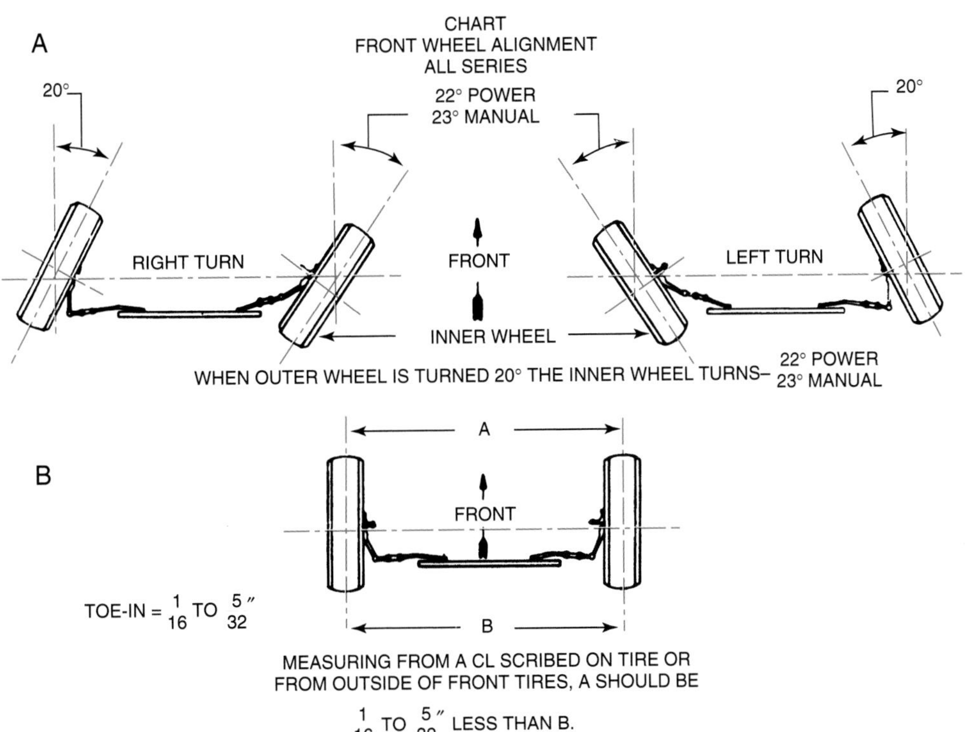

Figure 27-18. A—Toe-out on turns. Note that the inner wheel on a curve always turns a few degrees more sharply than the outer wheel. B—Toe-in is set with the wheels in the straight ahead position. (Buick)

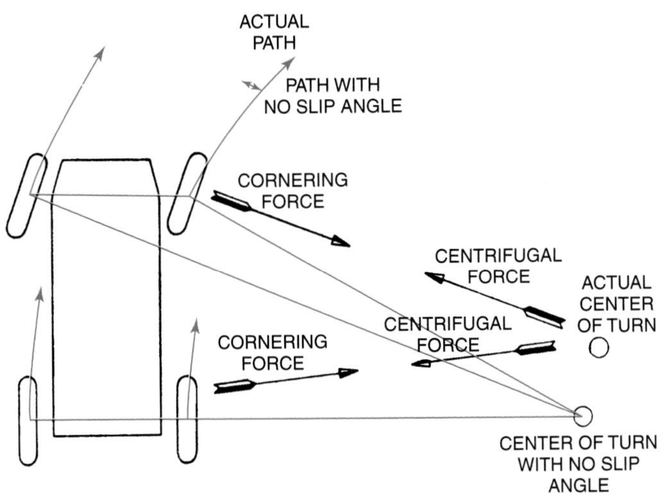

Figure 27-19. Cornering slip angles. Actual center of the turn is different from the theoretical turning center with no slip angle.

deviation when driving straight ahead. This is known as ***wheel tracking***. If the rear wheels are not tracking correctly, the vehicle will not travel in a straight line unless the front wheels are turned to compensate for the misalignment. This misalignment is often referred to as ***dog tracking***.

Tracking is set by aligning the vehicle's thrust angle with its geometric centerline to obtain perfect tracking. However, factory tolerances, accident damage, and normal wear make perfect tracking rare. Perfect thrust angle alignment is also more difficult on vehicles with front-wheel drive, four-wheel drive, and four-wheel steering. The thrust angle should be aligned as close as possible to the vehicle's geometric centerline. This will reduce tire wear, increase fuel economy, and improve handling.

Front Wheel Adjustments

Figure 27-20 shows how caster is adjusted on some vehicles by moving the lower strut rod in or out. The rod has a threaded section and locknuts to make adjustment easier. **Figure 27-21** illustrates how caster is adjusted on some vehicles with MacPherson strut suspensions by loosening the nuts holding the top of the strut tower and sliding the tower forward or backward. The strut tower mounting holes may be cut or filed out to allow enough movement.

Figure 27-22 shows one common way of changing the camber and caster by loosening the nuts holding the top of the strut tower and sliding the tower in or out. In this design, the rivets holding the top of the strut tower to the body are drilled out or chiseled off and the tower moved to the desired position. Another method uses an egg-shaped washer, sometimes called an ***eccentric,*** or eccentric cam, attached to the lower control arm, **Figure 27-23**.

In **Figure 27-24**, the camber is adjusted by moving an eccentric located on the top or bottom bolt holding the strut assembly to the spindle. On other designs, camber is adjusted by loosening the strut bolts and pushing or pulling the wheel into position. The strut bolts are then retightened. The strut rod mounting slots may require filing or cutting to allow enough movement to reach the desired camber angle.

On many vehicles, adjusting caster will affect camber and vice versa. For this reason, caster and camber are

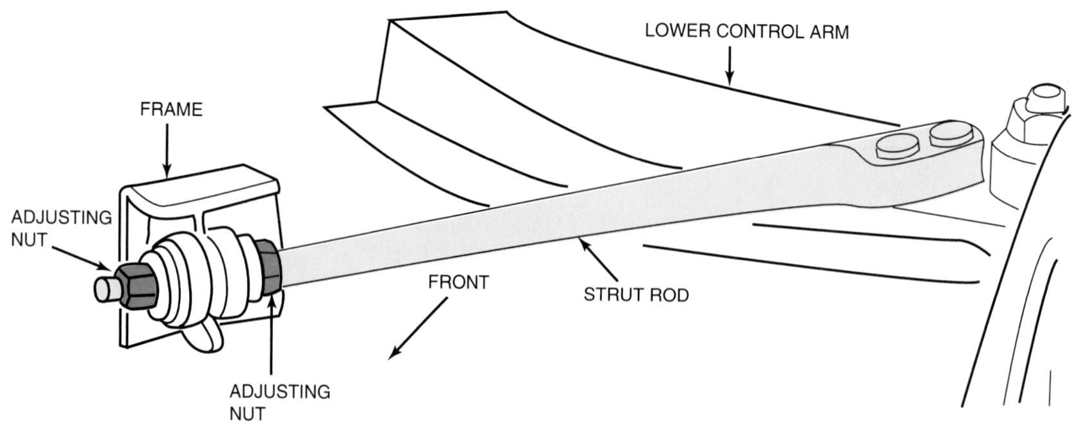

Figure 27-20. Front caster can be adjusted here by moving the lower strut rod in or out by turning adjusting nuts. (Hunter Engineering Company)

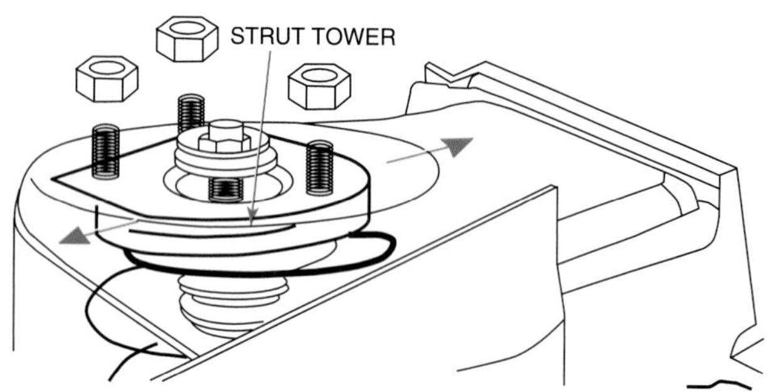

Figure 27-21. Adjusting front caster by moving the strut tower. (Hunter Engineering Company)

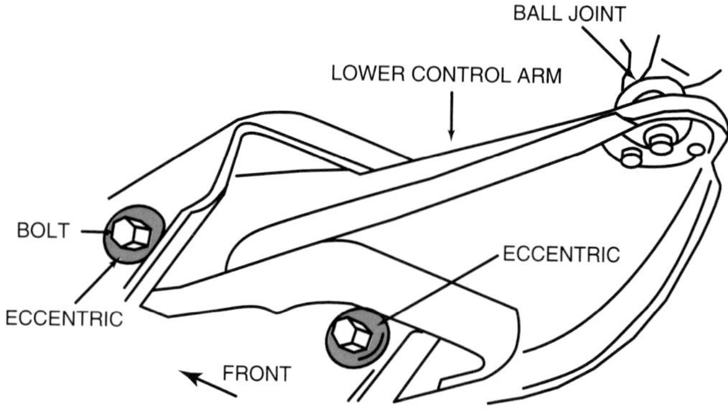

Figure 27-23. Camber and/or caster can be adjusted by rotating the eccentrics on this lower control arm. (Hunter Engineering Company)

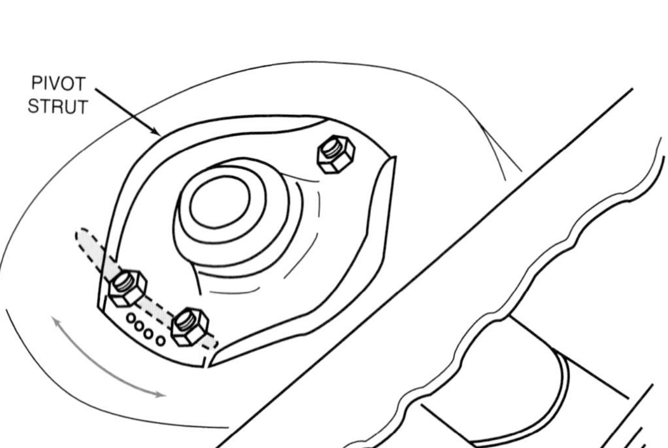

Figure 27-22. Sliding strut tower in or out to change camber. Once set, locknuts are tightened to specifications. (Hunter Engineering Company)

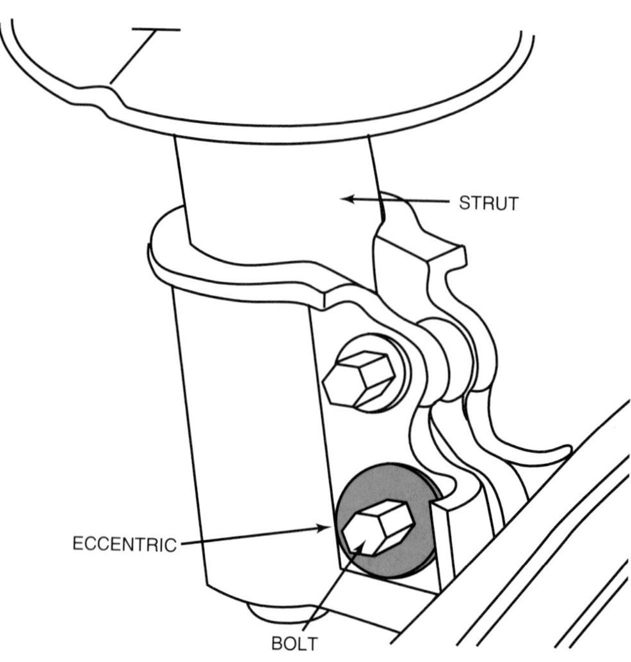

Figure 27-24. Adjusting front camber by loosening and turning the eccentric. (Hunter Engineering Company)

usually adjusted and checked together. There are a number of methods used to provide for caster and camber adjustment.

On many older light trucks and large rear-wheel drive domestic cars, ***shims*** are used at both sides of the upper control arms, **Figure 27-25**. Another method uses slotted holes in the vehicle frame or control arm, **Figure 27-26**, to allow the

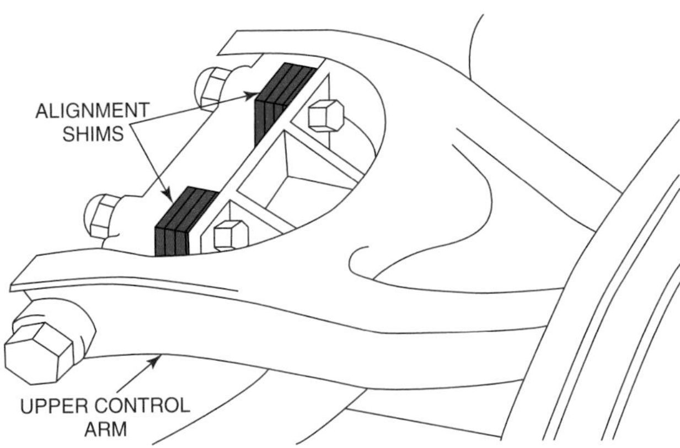

Figure 27-25. Adjustment shims used on both sides of the upper control arm to correctly set camber/caster. (Hunter Engineering Company)

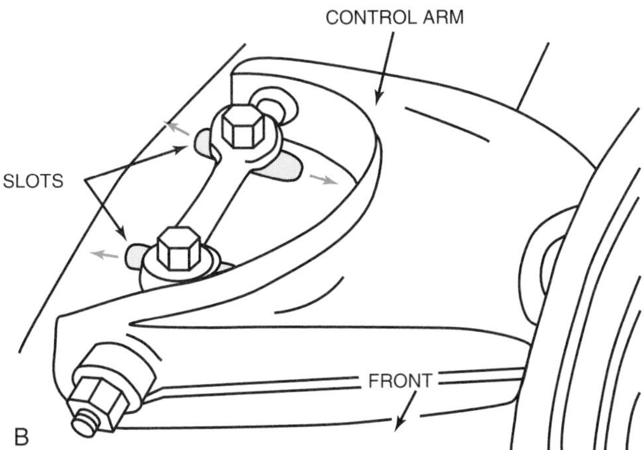

Figure 27-26. Adjusting camber/caster by moving the control arm in or out in the alignment slots. Usually, a special tool is needed to help control the amount of adjustment. (Hunter Engineering Company)

entire control arm to be moved. Eccentric cams or washers on the upper control arm bushings may be used to make adjustments, **Figure 27-27**. Or the strut tower nuts are removed, and the strut assembly lowered and rotated to obtain the correct reading, **Figure 27-28**.

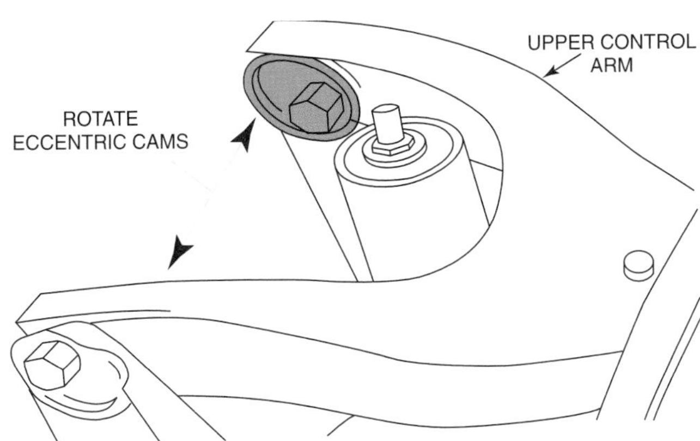

Figure 27-27. Camber/caster adjustment by rotating the eccentric cams on an upper control arm. (Hunter Engineering Company)

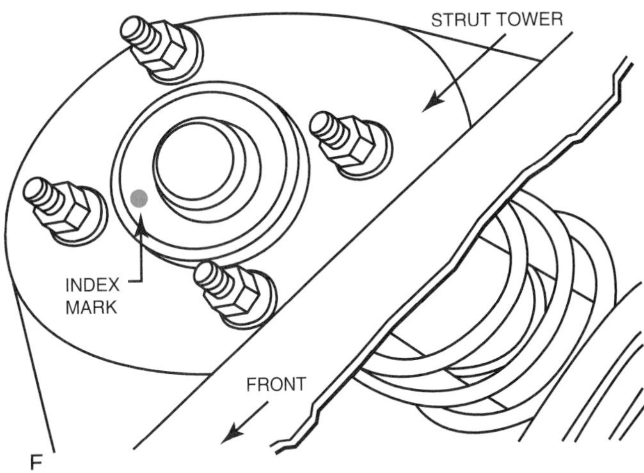

Figure 27-28. Camber/caster adjustment by removing the strut mounting nuts, lowering the strut to clear holes, and then rotate to properly align the correct index marks. Bolts are then passed back through the holes and nuts are installed and torqued to specifications. (Hunter Engineering Company)

Certain control arms are designed with the ball joint off center, as in **Figure 27-29**. Adjusting one end will have more effect on the camber, while adjusting the other end will have more effect on the caster.

Front toe is set by adjusting the tie rods. The design of the threaded tie rod adjusting sleeves was covered in Chapter 25. **Figure 27-30** shows the typical adjusting methods on parallelogram linkage. **Figure 27-31** shows the typical method of adjusting the tie rod assemblies used with rack and pinion steering systems.

Rear Wheel Adjustment

Since the rear wheels on most vehicles do not affect steering, the most common effects of improper settings are tire wear and noise. However, if the rear toe is off, it may be difficult to center the steering wheel for straight ahead driving. Note that there is no adjustment for caster on the rear wheels, since this is an alignment angle that affects steering, and is not needed on the rear.

Figure 27-32 shows one method of adjusting the camber and toe using eccentric cams or washers. One way of adjusting the toe by using threaded rods is shown in **Figure 27-33**. Other designs provide for loosening the rear strut bolts, pushing or pulling the wheel into position, and retightening the bolts, **Figure 27-34**.

Like rear camber, rear toe can be set by either eccentric cams or threaded rods. On a few designs, **Figure 27-35**, toe is adjusted by loosening a lock bolt and pushing or pulling the suspension part into position. The lock bolt is then retightened. **Figure 27-36** illustrates a method of setting toe by lengthening or shortening the control arm.

Some vehicles with ***four-wheel steering*** (the rear wheels turn with the front wheels) must have the rear toe set by a different method than that used by vehicles with two wheel steering. To make this adjustment, the technician must have a special tool to lock the rear steering gearbox.

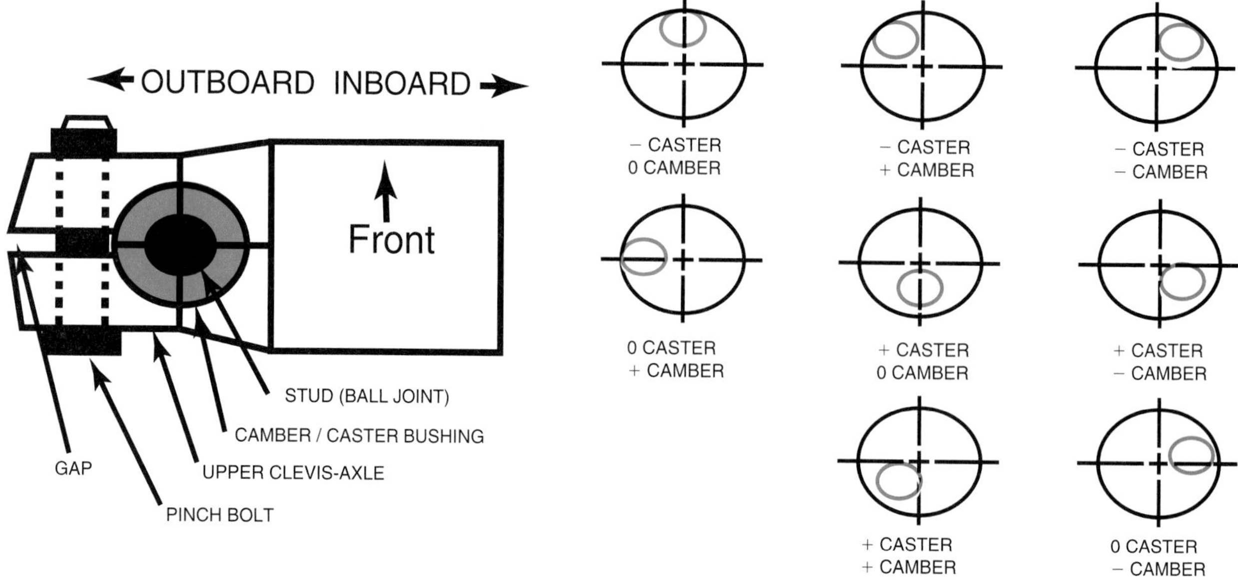

Figure 27-29. A control arm which is designed with the ball joint off center. Note the different positions available by rotating the ball joint. (Hunter Engineering Company)

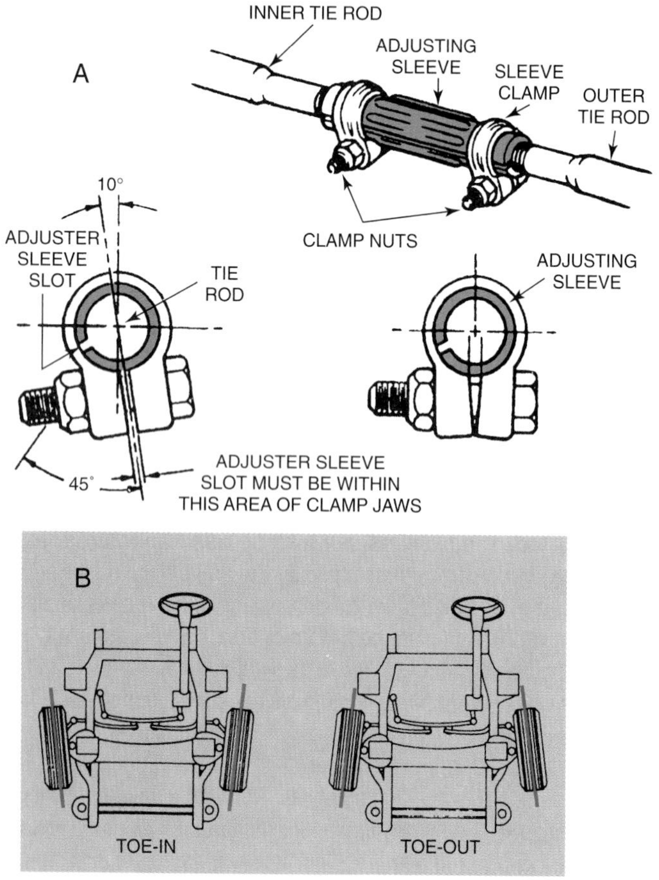

Figure 27-30. Parallelogram steering linkage toe adjustment. A—Tie rod sleeve clamps are loosened and the sleeve (tube) is turned to either lengthen or shorten the tie rod to obtain the correct toe setting. B—Excessive toe-in and excessive toe-out. (General Motors/Monroe)

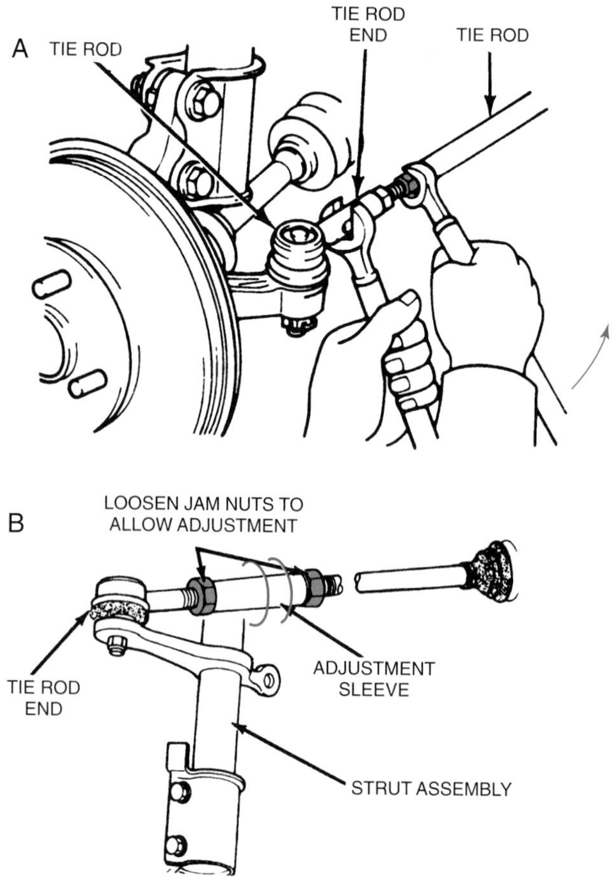

Figure 27-31. A—Adjusting the tie rod in or out to correctly set toe. B—Struts are turned in or out by loosening and rotating adjustments eccentrics. (Hunter Engineering Company)

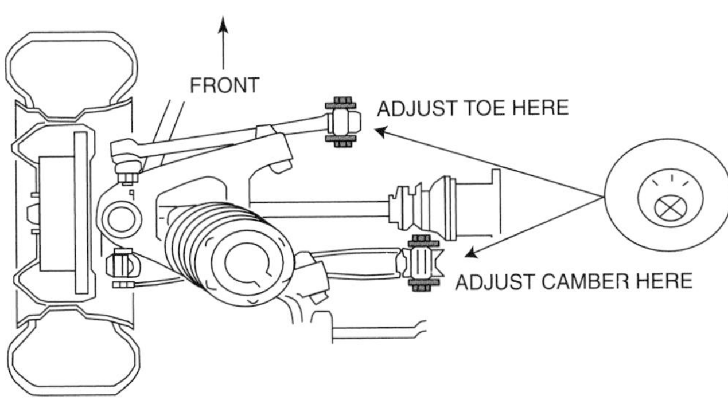

Figure 27-32. Setting rear wheel camber/toe by loosening and rotating adjustment eccentrics. (Hunter Engineering Company)

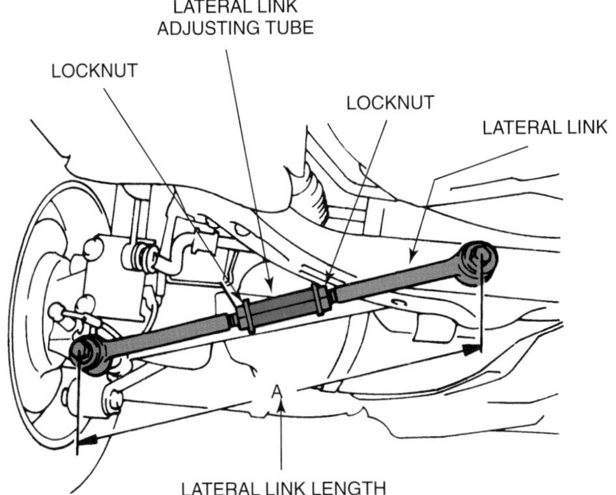

Figure 27-33. Rear toe is adjusted by turning the lateral link adjusting tube (sleeve). Once the toe is correct, the locknuts are tightened. On this particular vehicle, one complete rotation of the adjusting tube will change the toe setting about 1.2°. (Geo)

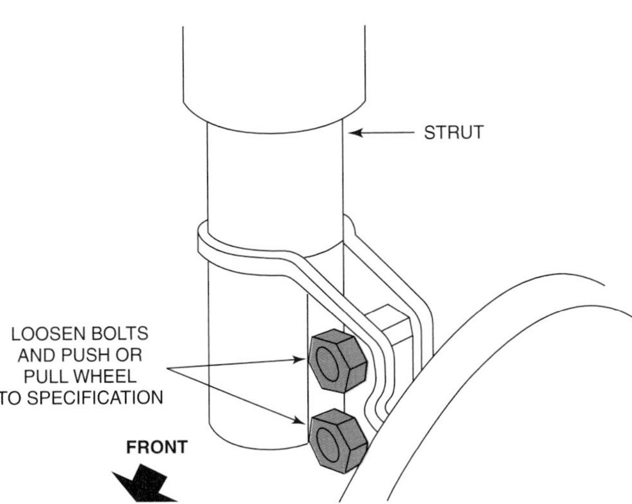

Figure 27-34. Setting rear camber by loosening the bolts and pulling or pushing the wheel in or out until the correct specifications are obtained. (Hunter Engineering Company)

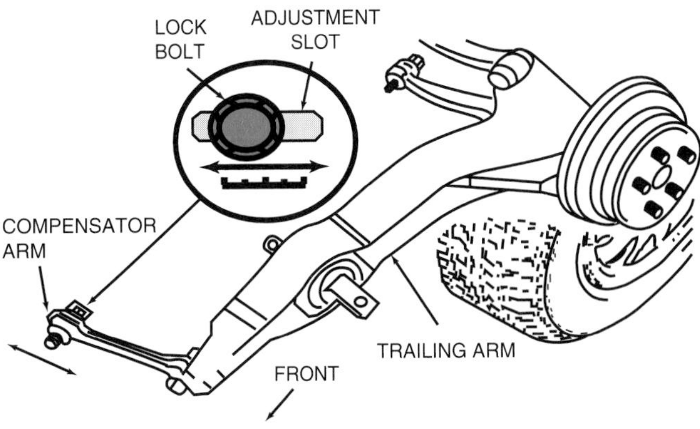

Figure 27-35. Setting the rear toe by moving the compensator arm in or out of its slot. Once set, the bolt is tightened to specifications. (Hunter Engineering Company)

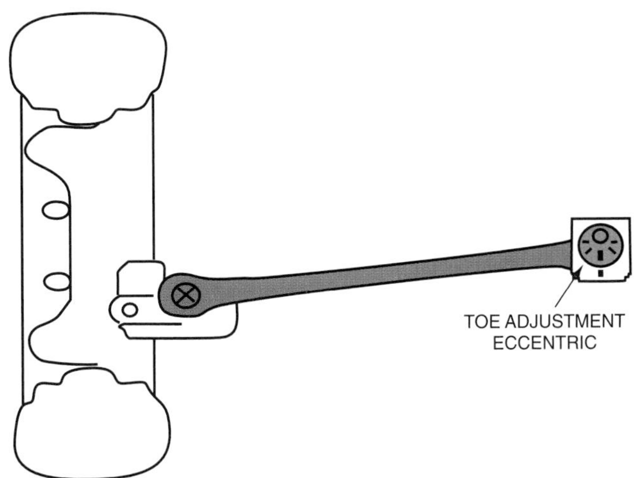

Figure 27-36. This rear suspension uses an eccentric cam to lengthen or shorten the control arm to set rear toe. (Hunter Engineering Company)

Using Shims to Set Rear Alignment

On many late model vehicles, there are no factory installed adjusting devices for setting rear wheel camber and toe. A method of adjusting alignment using ***rear wheel shims*** has been devised. Shims can also be used on the rear axles of some vehicles in which the factory adjustment is used up. Rear wheel shims are round metal or plastic discs that are thicker on one side than the other, as seen in **Figure 27-37**.

The amount of needed caster and toe change is determined by recording the actual readings and looking up the factory specifications. The type of shim and its exact placement is then calculated.

The rear brake drum or rotor must be removed, and the bearing housing and brake backing plate are separated from the rear axle. The shim is then placed between the rear bearing/backing plate and the axle, as in **Figure 27-38**. This shim will tilt the wheel in the correct direction to give the proper camber and toe specifications. All rear axle parts are reinstalled in their original positions and the alignment is rechecked.

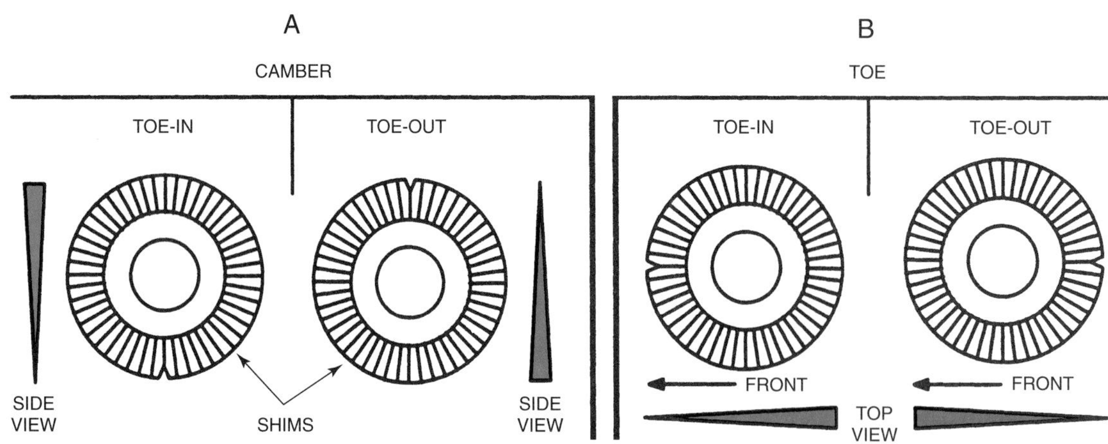

Figure 27-37. The placement of the tapered shim will determine the amount of change obtained for camber, toe, or a combination of both. A—Typical shim placement for adjusting rear camber. B—Typical shim placement for adjusting rear toe. (Hunter Engineering Company)

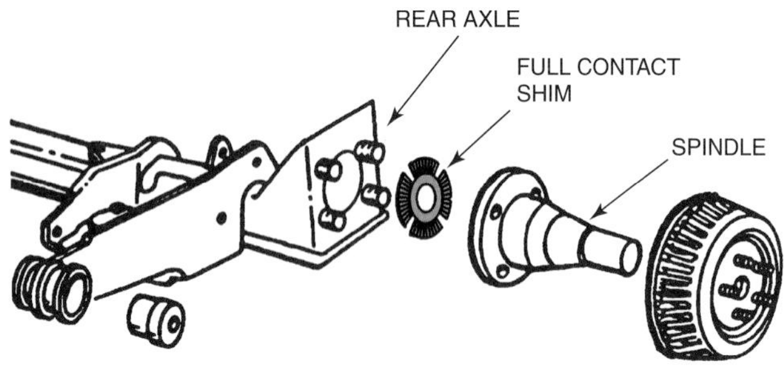

Figure 27-38. Full contact shim placement for rear suspension alignment. The tapered shim will tilt the spindle into the correct position. (Hunter Engineering Company)

Tire Wear Patterns and Alignment

Tires are a reliable indicator of alignment problems. Misadjusted camber and toe have the largest effect on tire wear. However, excessive cornering speeds, lack of tire rotation, or tire spin can also cause abnormal wear. **Figure 27-39** shows the various types of tire wear, and the alignment problems which can cause them.

Alignment Equipment

Wheel alignment equipment varies greatly, from simple mechanical gauges to complex electronic devices. To do any kind of alignments, the technician must have equipment capable of checking all of the alignment angles listed above.

Simple camber and caster checking devices, held by magnets to the wheel hub, will allow fairly accurate checking, as seen in **Figure 27-40**. Toe can be checked by using a trammel bar, **Figure 27-41**, or even a tape measure. However, to do an accurate four-wheel alignment on a modern vehicle, more elaborate equipment is needed.

The basic alignment rack will be equipped with ***turning plates*** to allow the front wheels to be turned for measuring caster, and to allow the vehicle suspension to settle into its normal riding position after being raised. The ideal alignment rack will have a pit deep enough or can be raised high enough to permit easy access to the underside of the vehicle. The best alignment racks are attached to a hydraulic lift, so they may be lowered to drive the vehicle on and off and raised to gain access to the vehicle's underside for extended periods.

The alignment machine itself should be able to accurately measure all of the alignment angles. Modern alignment machines must be capable of measuring the rear-wheel alignment as well as front-wheel alignment. This necessitates the use of rear wheel sensing devices. The wheel mounted sensing devices used on modern alignment machines are usually referred to as ***alignment heads***.

Many alignment machines use wheel-mounted light beam generators to measure the alignment angles, **Figure 27-42**. The latest alignment machines use wheel mounted electronic sensors and a self-contained computer to provide readouts on a screen, **Figure 27-43**. There are many manufacturers of alignment equipment. Obtain and study their equipment manuals to familiarize yourself with the range of available equipment.

When checking and adjusting wheel alignment, use quality equipment that is in good condition. Wheel alignment is a precision operation; both equipment and techniques must be perfect.

Check for Worn Parts Before Alignment

Many steering and suspension parts can wear or bend, causing the alignment to change. Before performing an alignment, always check for worn, loose, or bent parts.

Examples of parts that commonly wear and affect alignment are ball joints, control arm and strut rod bushings, tie rod ends, and idler arms. Parts that can become bent are control arms, strut rods, and parts of the steering linkage. Springs can sag and throw the alignment out of specifications. The fasteners holding steering and suspension parts together can become loose, especially on off-road and other vehicles subjected to heavy shock loads and vibration. Remember that an alignment performed with loose, worn, or otherwise defective parts is a waste of time.

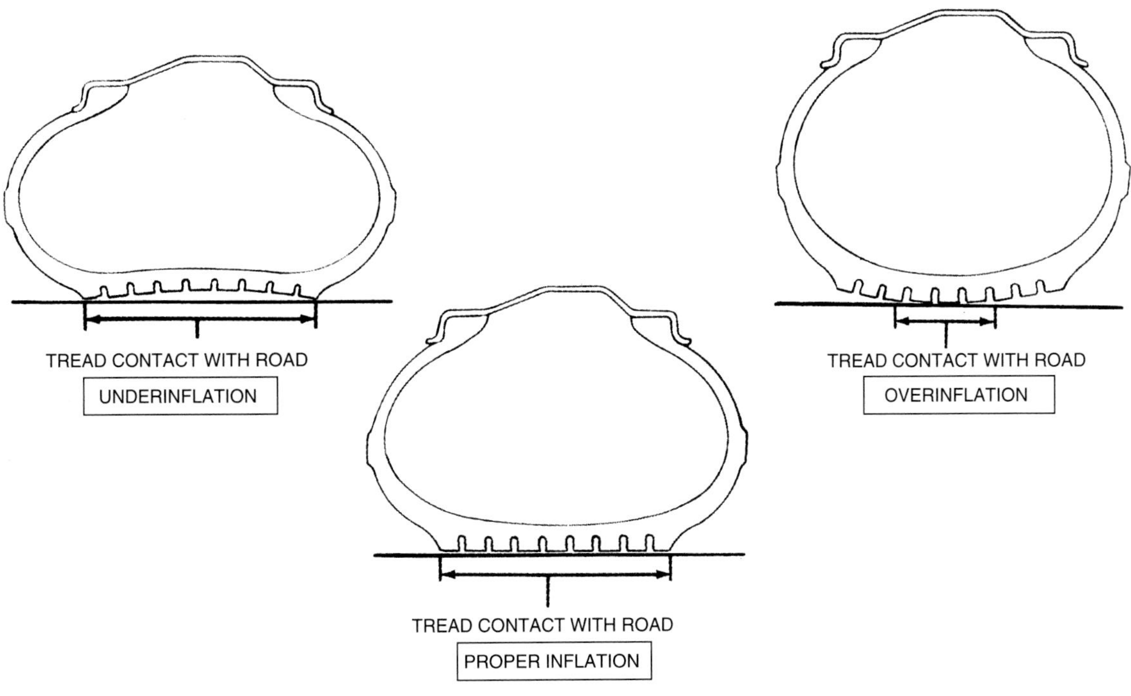

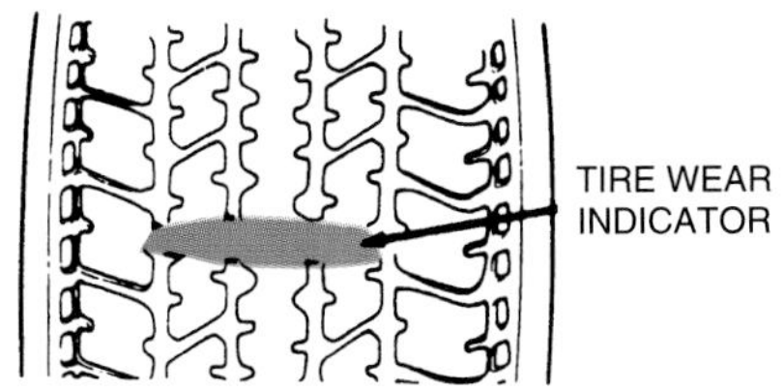

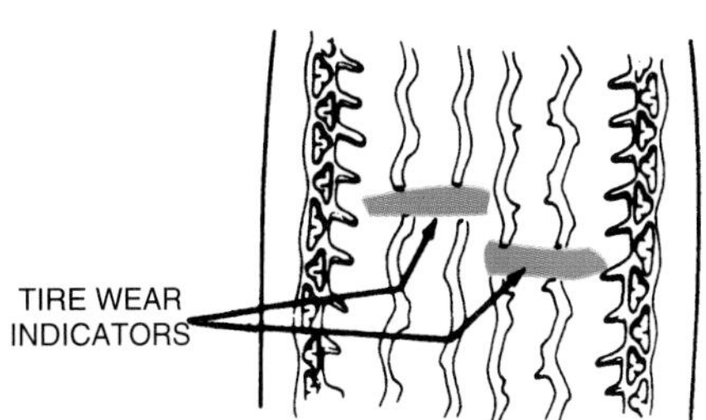

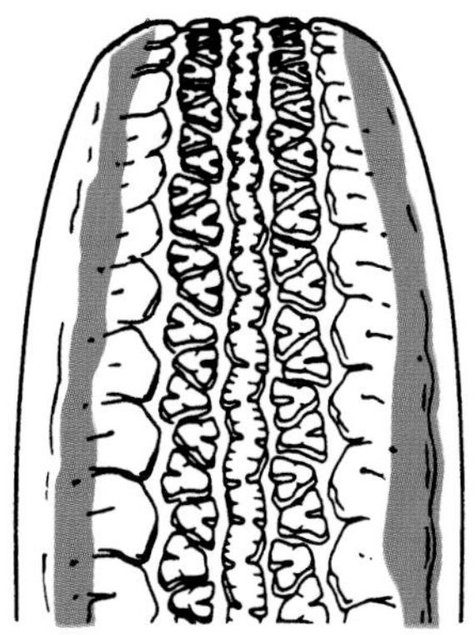

HARD CORNERING
UNDERINFLATION
LACK OF ROTATION

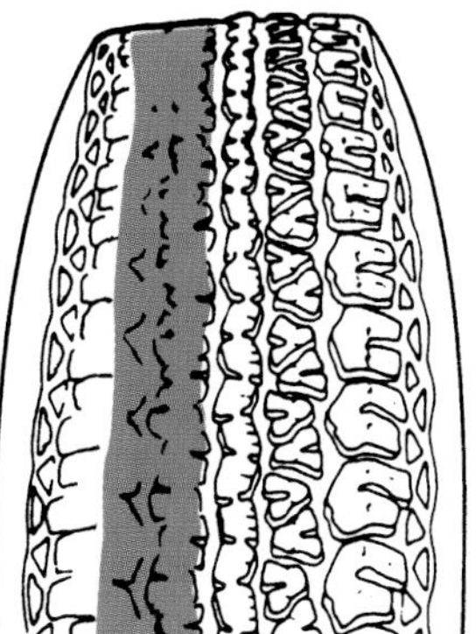

EXCESSIVE TOE ON
THE NON-DRIVE AXLE
LACK OF ROTATION

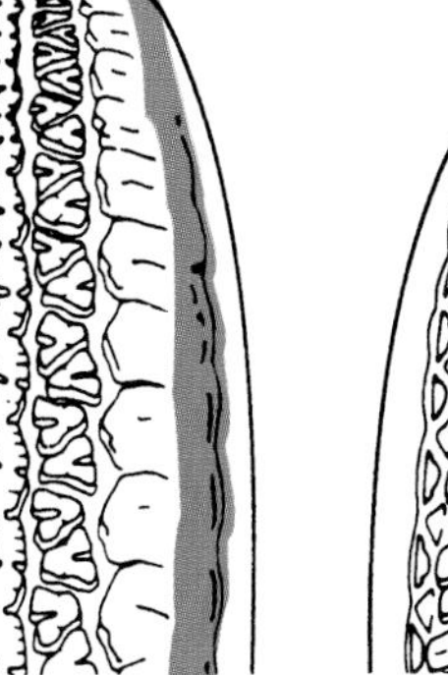

HEAVY ACCELERATION
ON DRIVE AXLE
EXCESSIVE TOE ON
DRIVE AXLE
LACK OF ROTATION
OVERINFLATION

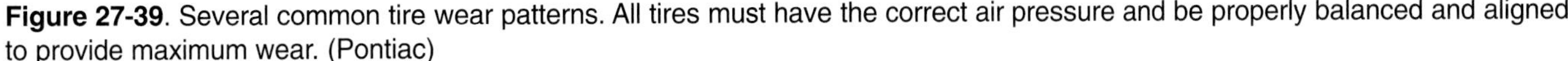

Figure 27-39. Several common tire wear patterns. All tires must have the correct air pressure and be properly balanced and aligned to provide maximum wear. (Pontiac)

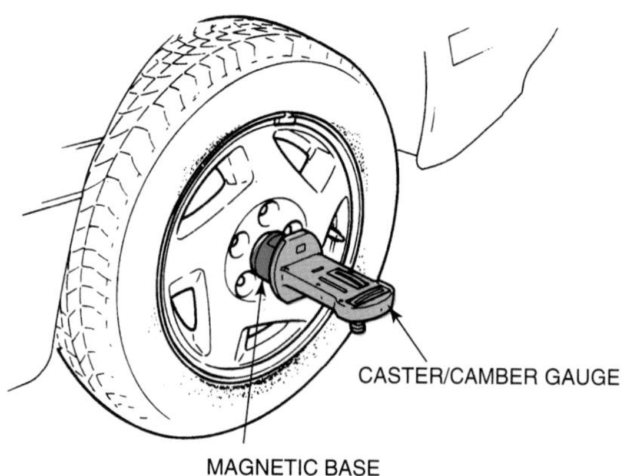

Figure 27-40. A caster/camber gauge which is held to the wheel hub by a magnetic base. Some screw onto the spindle threads on front-wheel drive vehicles. (Mazda)

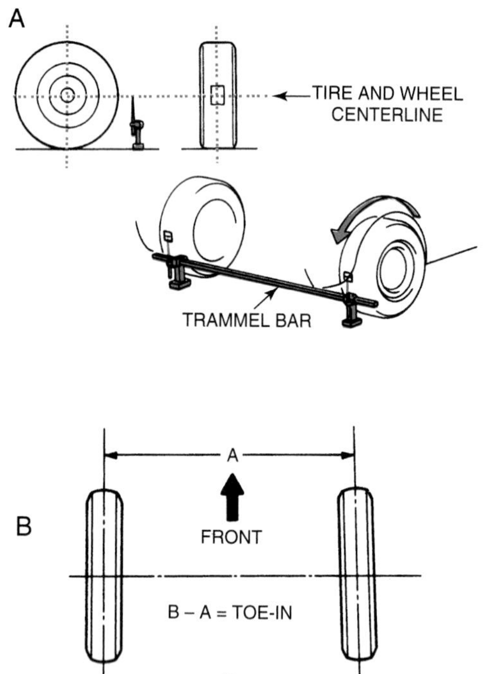

Figure 27-41. A—Toe setting being checked by using a trammel bar. Note how the trammel bar is aligned with the tire and wheel centerline. B—This provides the most accurate toe reading. While not as accurate as electronic alignment equipment, it is still widely used. (Toyota)

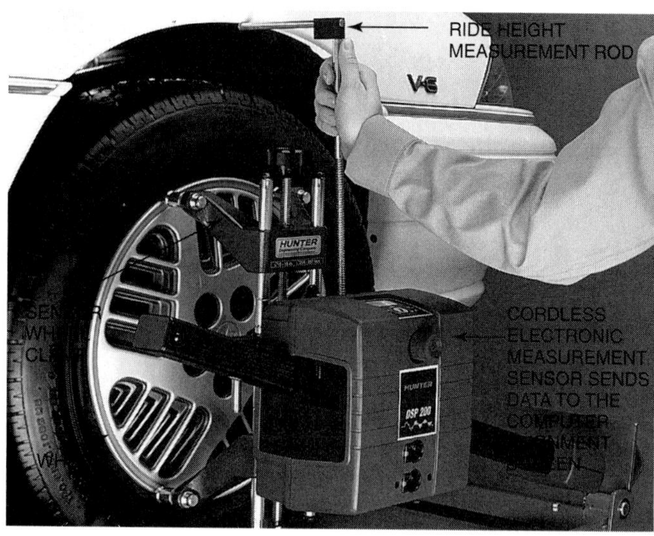

Figure 27-42. An ASE (Automatic Service Excellence) certified alignment technician installing a wheel-mounted light beam generator for obtaining wheel alignment specs. (Hunter Engineering Company)

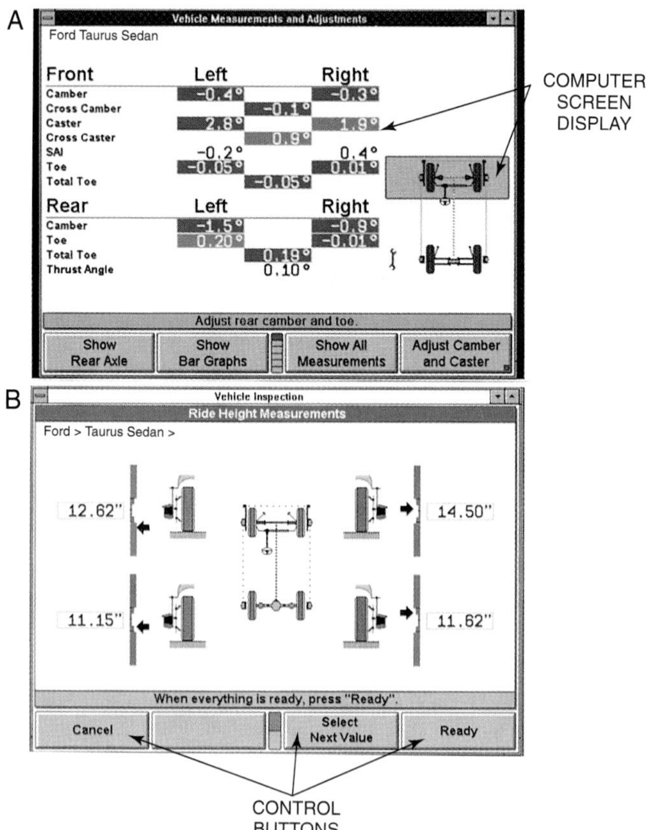

Figure 27-43. A computer wheel alignment display screen. A—Displays left and right specifications for both front and rear wheel alignment. B—Displays ride height measurements. (Hunter Engineering Company)

Other Factors That Can Affect Alignment

Many vehicles brought in for alignment are for a pulling complaint. The alignment is adjusted, but the vehicle continues to pull. In many cases, the problem is ***radial tire pull.*** Radial tire pull is caused by a defect in a radial tire's construction, and can only be corrected by replacing the tire. The tire on the pulling side is usually the defective tire. A quick check for radial tire pull is to switch the front tires. If the vehicle stops pulling or pulls in the other direction, the tire is the source of the problem.

Some vehicles may come in with a complaint of pulling to the right, however, when the vehicle is driven on a relatively flat road, there is no pull. This is due to ***road crown,*** **Figure 27-44.** Most roads are angled or crowned to allow water

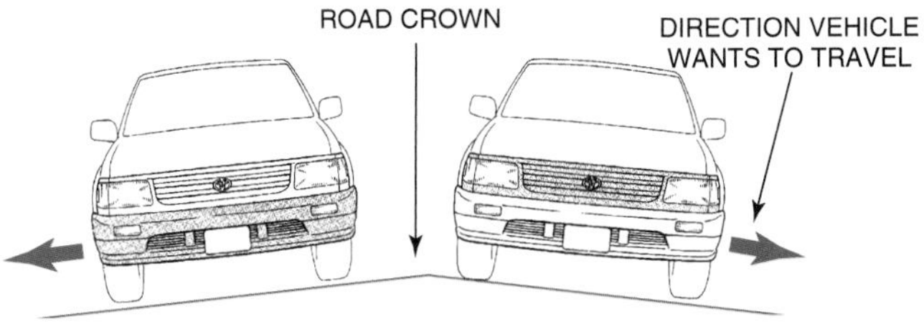

Figure 27-44. Many road surfaces are crowned (higher in the center than the edges), to aid in shedding rainwater. Caster or a minimal amount of camber can be used to offset this crown effect. (Toyota)

runoff. The amount of road crown can vary from area to area. An experienced alignment technician will set a vehicle's alignment to compensate for the amount of road crown that is most prevalent in that area. In most situations, however, no additional compensation for road crown is necessary.

Summary

To ensure easy and safe handling of a car, the front wheels must be properly aligned. Several angles must be considered. The front and back wheels on many newer vehicles must be checked and adjusted if necessary. Alignment is the result of many angles used to increase handling and tire life.

Caster is achieved by tilting the steering axis forward or backward. When the top of the axis is tipped toward the rear of the vehicle, the caster is positive. When the top is tipped toward the front of the vehicle, the caster is negative.

Camber is the tilting of the wheel centerline, viewed from the front of the vehicle, away from a true vertical line. Positive camber spaces the tops farther apart than the bottoms, while negative camber spaces the tops closer together.

Steering axis inclination is produced by tipping the top of the steering axis toward the center of the car. Toe-in is when the wheels are closer together at the front than the rear. Toe-out is when the wheels are closer together at the back than the front.

The angle formed between the steering axis and the centerline of the spindle is called the included angle. Toe-out on turns is accomplished by angling the steering arms in toward the center of the car. When cornering, the actual path taken by the car and the path upon which the wheels are aimed are different. This difference is expressed in terms of slip angle.

Vehicle manufacturers have devised many methods to change caster and camber. Typical methods are eccentric washers or cams, shims, threaded rods, and slotted holes.

Modern alignment equipment ranges from simple turntables to elaborate electronic devices. Check the tires for wear patterns indicating problems such as worn, loose, or bent parts before performing an alignment. Also check for radial tire pull, a common source of problems.

Know These Terms

Wheel alignment
Two-wheel alignment
Four-wheel alignment
Steering axis
Caster
Pulling
Camber
Steering axis inclination (SAI)
Included angle
Setback
Toe
Feathering
Toe-out on turns
Inertial force
Slip angle
Steering geometry
Thrust angle
Wheel tracking
Dog tracking
Eccentric
Shim
Four-wheel steering
Rear wheel shim
Turning plates
Heads
Radial tire pull
Road crown

Review Questions—Chapter 27

Do not write in this text. Write your answers on a separate sheet of paper.

1. What does proper wheel alignment ensure?
2. All of the following statements about caster are true, EXCEPT:
   (A) positive caster helps the wheels to turn back to the straight ahead position after a turn.
   (B) the steering axis is formed by the upper and lower ball joints
   (C) positive caster makes it easier to turn the wheels from the straight ahead position.
   (D) positive caster tends to force the wheels to travel in a straight ahead position.
3. Excessive negative camber on a vehicle's left front wheel can cause which of the following?
   (A) Pulling to the left side.
   (B) Pulling to the right side.
   (C) Left tire wear.
   (D) Both B & C.
4. Technician A says that included angle is the total of the steering axis inclination and camber on any one wheel. Technician B says that included angle is the total of the steering axis inclination and camber on both front wheels. Who is right?
   (A) A only.
   (B) B only.
   (C) Both A & B.
   (D) Neither A nor B.
5. Which of the following angles, if incorrect, can cause pulling?
   (A) Camber.
   (B) Caster.
   (C) Setback.
   (D) All of the above.

6. All of the following statements about vehicle toe are correct, EXCEPT:
   (A) on modern vehicles, toe can be set in or out.
   (B) on many modern vehicles, the toe can be set on the rear wheels.
   (C) toe can be adjusted to straighten the steering wheel.
   (D) toe has little effect on tire wear.
7. Proper steering axis inclination will allow the vehicle's _____ to help return the front wheels to the straight ahead position.
   (A) speed
   (B) weight
   (C) turning radius
   (D) Both A & B.
8. What is the usual cause of excessive setback?
9. The major factor in the size of the tire slip angle is the _____ .
   (A) camber
   (B) caster
   (C) toe
   (D) vehicle speed
10. Why does the experienced technician sometimes make an adjustment that is outside factory specifications?
11. Adjustable strut rods are often used to adjust front wheel _____ .
   (A) caster
   (B) camber
   (C) setback
   (D) included angle
12. On some vehicles, the front caster and camber can be adjusted at the same time by moving the ____________.
13. Three methods are used to move control arms to adjust caster and camber. What are they?
14. Technician A says that caster is not adjustable on the rear wheels of most vehicles. Technician B says that a special tool is needed to adjust toe on a four wheel steering vehicle. Who is right?
   (A) A only.
   (B) B only.
   (C) Both A & B.
   (D) Neither A nor B.
15. In an ideal situation, the thrust angle will exactly match the ____________ of the vehicle.
16. Dog tracking means that ____________.
   (A) the vehicle caster is too positive
   (B) the vehicle caster is too negative
   (C) the vehicle caster is neutral
   (D) the vehicle caster does not affect dog tracking
17. All of the following statements about rear wheel shims are true, EXCEPT:
   (A) there are no factory installed rear wheel adjusters on many late model vehicles.
   (B) rear wheel shims are round metal or plastic discs that are thicker on one side.
   (C) rear wheel shims cannot be used on vehicles with a factory adjustment.
   (D) rear wheel shims can be used to set rear wheel camber and toe.
18. Alignment equipment turn plates are used to measure _____ .
   (A) caster
   (B) camber
   (C) SAI
   (D) Both A & B.
19. Name four parts that can wear out and affect alignment.
20. Radial tire pull can be compensated for by ________.
   (A) increasing caster
   (B) increasing camber
   (C) replacing the tire
   (D) aligning the thrust angle with the vehicle centerline

28

Air Conditioning and Heating

After studying this chapter, you will be able to:

- Identify the three methods of heat transfer.
- Explain the principles of latent and sensible heat.
- Identify modern refrigerants.
- Explain the relationship between pressure and temperature in modern refrigerants.
- Identify the basic refrigeration cycle.
- Identify the basic parts of an air conditioning system.
- Describe the operation of each air conditioning system component.
- Explain the basic operation of the heater system.
- Compare differences in air conditioning control system design.
- State basic refrigerant safety rules.

This chapter covers the automotive air conditioning system, including the refrigeration system, heater system, airflow system, and electrical systems. Also included are the basic principles of heat transfer, latent heat, and pressure-temperature relationships. Vital refrigerant safety precautions are also covered. Studying this chapter will enable you to recognize the basic operation of an automotive air conditioner and how it relates to outside temperature and humidity and to the operation of other vehicle systems.

Purposes of Automotive Air Conditioning

Automotive air conditioning makes driving far more comfortable in hot and humid weather. It is a method whereby air entering the vehicle is cooled, cleaned, and dehumidified.

Principles of Refrigeration

The automotive air conditioning system makes use of the principles of refrigeration. The ***refrigeration system*** is a specific part of the air conditioner, separate from other systems which will be discussed later.

What Is Cooling?

The term ***cooling*** may be defined as a process of removing heat from an object. Actually, there is no such thing as cold. When we speak of something being cold, we are really saying that the object is not as hot as something else.

Cold is a reference to a certain portion of the temperature range. The normal use of the word cold is generally determined by a comparison to the human sense of comfort. Cold and hot are actually different degrees of ***heat.*** Heat can be defined as a form of energy that can increase, decrease, or maintain a specific temperature.

Effects of Heat

Some of the more commonly observed effects of heat would be the expansion that takes place when a substance is heated or raised to a higher temperature. The added heat creates a more violent agitation of the molecules that make up the substance, increasing the length of their travel with a resulting increase in physical size.

Water, when heated sufficiently, vaporizes. When heat is lowered enough, water becomes a solid. Wood will eventually ignite if enough heat is added. Water freezing and wood burning are actual changes produced by heat. In many ways, the operation of the refrigeration system parallels the action of some of the Earth's natural cooling processes, as shown in **Figure 28-1.**

Transference of Heat

Heat may be transferred from one object to another in three ways:

- Conduction.
- Convection.
- Radiation.

These three methods of heat transfer are discussed in the following sections.

Conduction

Conduction is accomplished by actual physical contact between two bodies of different temperature. The agitated action of the molecules of the hotter body are transferred to the cooler body. If a copper wire is held in a flame, the heat of the flame is rapidly conducted through the wire. It travels from molecule to molecule. Some substances, such as metals, transfer heat much more rapidly than others. See **Figure 28-2A**.

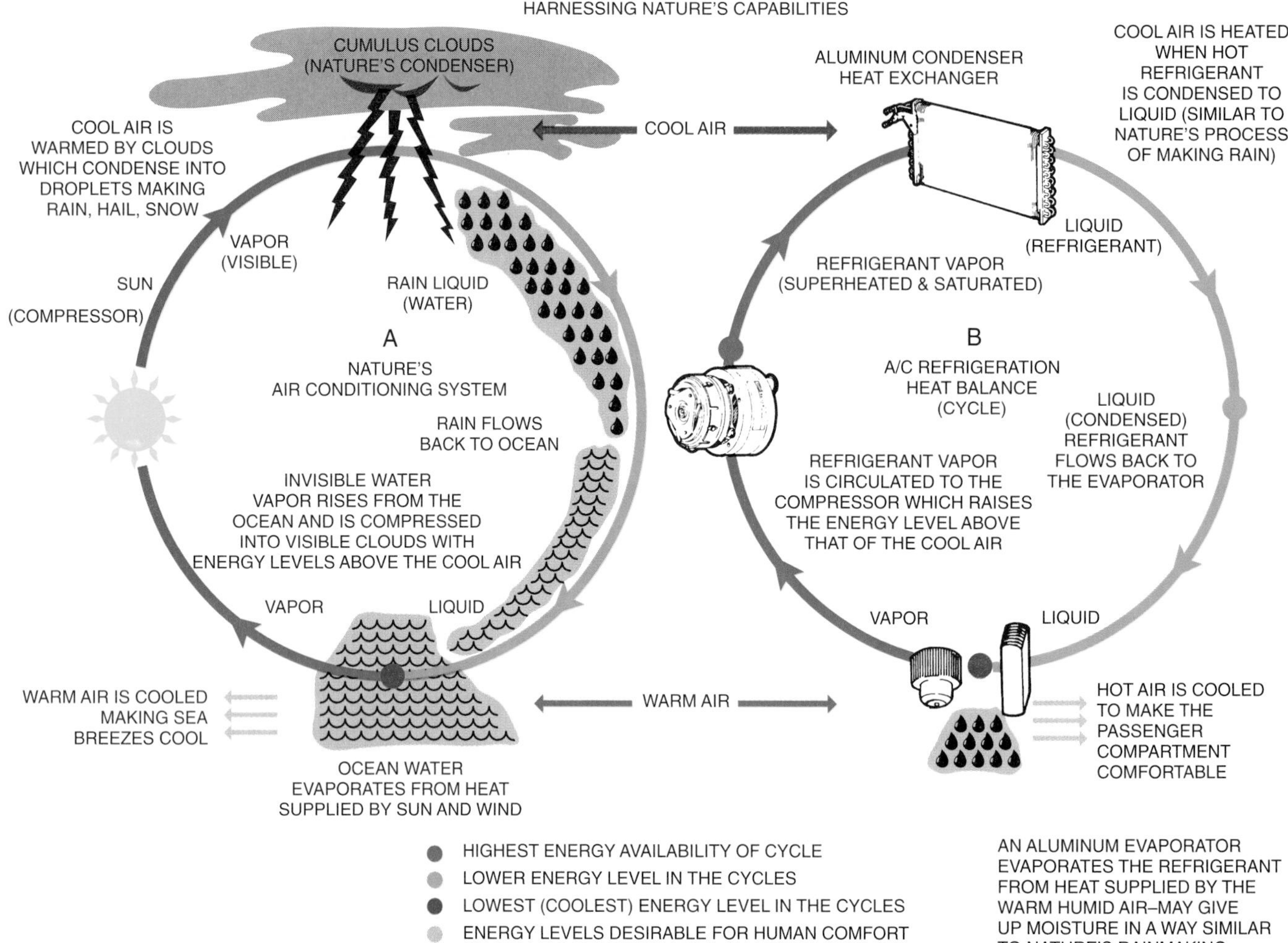

Figure 28-1. The refrigeration principles employed by an automotive air conditioning system B, are very similar to those found in nature's own air conditioning system, A. Study both cycles. (General Motors)

Convection

Heat can be transferred from one object to another by the surrounding air. This is one form of ***convection.*** As the heated air from the hotter body rises, it comes in contact with the cooler body, raising its temperature. This is shown in **Figure 28-2B**. In a room, the air circulation will make a full circle. Even though hot air rises, circulation as it cools will finally bring it to the floor level and heat all objects in the room.

Radiation

Even in the absence of air or physical contact, heat can be transferred through space by ***infrared rays,*** which are not visible to the human eye. Energy is emitted from molecules and atoms because of internal changes. To get an idea of how heat is transmitted by ***radiation,*** assume a body containing heat gives off infrared rays. These rays travel through space and have the ability to cause molecules of other substances to increase their speed. Even though the rays may not contain heat, they impart heat into the other object. An example of this is how the sun heats the Earth through millions of miles of empty space, **Figure 28-2C**.

Heat Transfer is From Hot to Cold

By studying convection, conduction, and radiation, it can be seen that when two bodies are placed in reasonable proximity to each other, the hotter body will give up heat to the colder body. Eventually the hot body will cool and the cool body will heat until they reach an equal temperature level.

States of Matter

All matter exists in one of three ***states: solid, liquid,*** or ***gas.*** The state of a substance is dependent on the speed of the molecules which make it up. Changing between states requires the addition or subtraction of heat.

Latent Heat

If heat is imparted to ice, the ice will start to melt. As it is melting, it will not grow warmer even if a great amount of heat is applied to it. 32° ice will become 32° water. This heat that disappears into the ice with no temperature rise is called hidden or ***latent heat.*** It is there, but it cannot be detected. All of the heat is used to cause a ***change of state,*** from solid to liquid in this case.

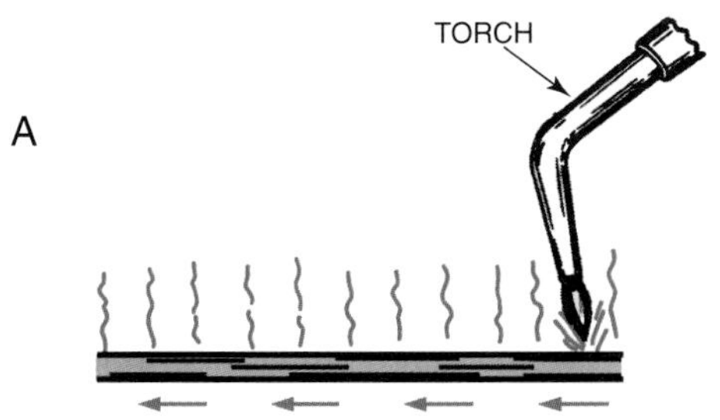

B
TRANSFER OF HEAT BY CONVECTION AND RADIATION. AIR ABOVE HOT COALS IS HEATED, RISES, AND TRANSFERS HEAT TO BIRD ON SPIT. A GREAT DEAL OF HEAT IS ALSO TRANSFERRED BY RADIATION.

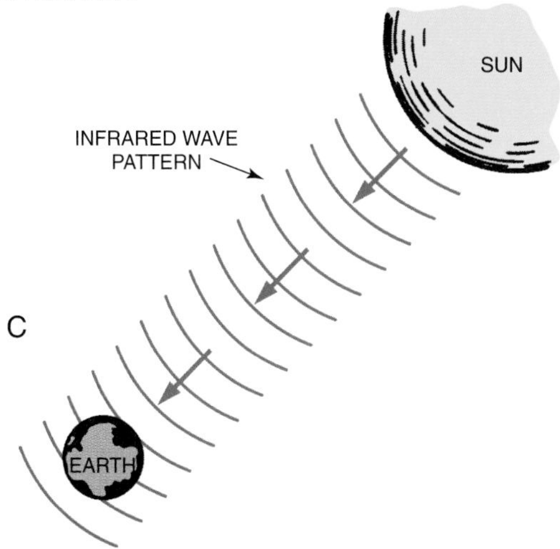

Figure 28-2. Study three methods of heat transfer. A—Conduction. B—Convection. C—Radiation.

Any addition of heat that can be seen as a rise in temperature is called ***sensible heat*** (heat which can be sensed). Water will reach the boiling point at 212°F or 100°C (at sea level). Heat applied at this point, even a great amount, will not cause a further increase in temperature. Once again heat is absorbed, but is not apparent.

The amount of heat being absorbed to ***vaporize*** (change from liquid to gas) the water is great, but if one were to measure the temperature of the vapor, it too would register 212°F (100°C). The heat then, is "hidden" in the water vapor.

When this vapor comes in contact with cooler air, any latent heat will transfer to the cooler air and the vapor will ***condense*** (change from gas to liquid). This latent heat of vaporization (heat transfer involved in changing a liquid to a vapor and vapor back to a liquid) is the basic principle of the refrigeration process used in air conditioning.

Pressure-Temperature Relationship

At sea level, water changes to a vapor (boils) at 212°F (100°C). If water at 20,000 feet (6096 m) in the air is heated, it will boil at a lower temperature. The atmospheric pressure on the water at 20,000 feet (6096 m) is less, so the boiling point will be lower.

When the pressure is raised above normal atmospheric pressure, the boiling point will also rise. The condensation point is likewise affected. The fact that pressure changes can affect the vaporization and condensation of a substance is called the ***pressure-temperature relationship.*** The refrigeration process makes use of this vaporization and condensation effect.

Types of Refrigerant

Refrigerant is the liquid used to provide the cooling effect in the refrigeration system. Two types of refrigerant are used in modern vehicle refrigeration systems: R-134a and R-12. ***R-134a,*** sometimes called HFC-134a, was introduced in 1992 and has been used in all original-equipment air conditioners since 1994. Older vehicles use ***R-12***, or ***Freon***. Before the introduction of R-134a, R-12 was the only refrigerant used in vehicle air conditioners.

R-12 has been phased out because it damages the earth's ozone layer. The ozone layer is located in the upper atmosphere and protects the earth from ultraviolet radiation. The Environmental Protection Agency (EPA) banned the manufacture of R-12 at the end of 1995.

R-134a and R-12 are not interchangeable. An air conditioner using R-12 must be recharged with R-12, and an air conditioner using R-134a must be recharged with R-134a. Some EPA-approved substitute refrigerants are available. However, EPA approval means only that the refrigerant is less damaging to the ozone layer than R-12, not that it works well in an air conditioner. Most substitute refrigerants have drawbacks. A few have been found to damage refrigeration systems.

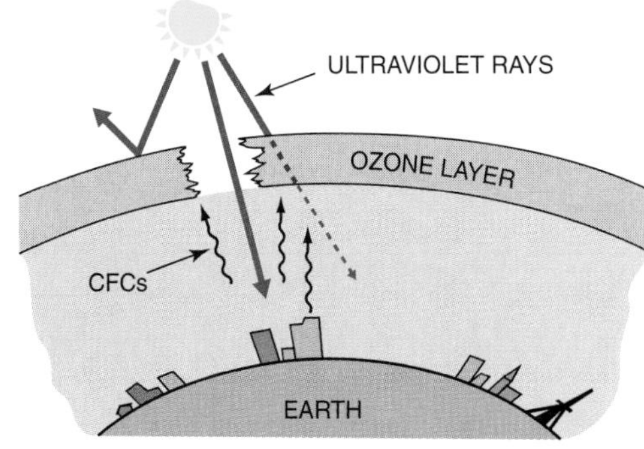

Figure 28-3. A hole in the ozone layer, which helps to protect the Earth from the sun's ultraviolet rays. It is thought that the chlorofluorocarbons (CFC's) from automobile air conditioner refrigerant (R-12), is part of the cause. (Toyota Motor Corp.)

There is a distinct pressure-temperature relationship for refrigerants. As the pressure is increased, the boiling point rises. A pressure-temperature chart for R-12 is shown in **Figure 28-4** and R-134a in **Figure 28-5.** When pressure is lowered, the refrigerant boils. When pressure is increased, the vaporized refrigerant returns to a liquid state.

Refrigerant is used over and over in the system and will maintain its efficiency indefinitely unless contaminated with dirt, water, or air. It is transparent and colorless in both the vapor and liquid state. Refrigerant is nonpoisonous, except when in direct contact with an open flame. Unless combined with moisture, it is noncorrosive. It is heavier than air and will become a vapor when released into the atmosphere.

How the Refrigeration System Works

Figure 28-6 illustrates the four basic parts making up the typical refrigeration system. Study this schematic and note the direction of refrigerant flow as well as the sequence of

REFRIGERANT —12 PRESSURE — TEMPERATURE RELATIONSHIP

The table indicates the pressure of Refrigerant—12 at various temperatures. For instance, a drum of Refrigerant at a temperature of 80°F (26.6°C) will have a pressure of 84.1 PSI (579.9 kPa). If it is heated to 125°F (51.6°C), the pressure will increase to 167.5 PSI (1154.9 kPa). It also can be used conversely to determine the temperature at which Refrigerant—12 boils under various pressures. For example, at a pressure of 30.1 PSI (207.5 kPa), Refrigerant—12 boils at 32°F (0°C).

(°F)	(°C)	(PSIG)	(kPa)	(°F)	(°C)	(PSIG)	(kPa)
-21.7	-29.8C	O(ATMOSPHERIC PRESSURE)	O(kPa)	55	12.7C	52.0	358.5
				60	15.5C	57.7	397.8
-20	-28.8C	2.4	16.5	65	18.3C	63.7	439.2
-10	-23.3C	4.5	31.0	70	21.1C	70.1	482.7
-5	-20.5C	6.8	46.9	75	23.8C	76.9	530.2
0	-17.7C	9.2	63.4	80	26.6C	84.1	579.9
5	-15.0C	11.8	81.4	85	29.4C	91.7	632.3
10	-12.2C	14.7	101.4	90	32.2C	99.6	686.7
15	-9.4C	17.7	122.0	95	35.0C	108.1	745.3
20	-6.6C	21.1	145.5	100	37.7C	116.9	806.0
25	-3.8C	24.6	169.6	105	40.5C	126.2	870.2
30	-1.1C	28.5	196.5	110	43.3C	136.0	937.7
32	0C	30.1	207.5	115	46.1C	146.5	1010.1
35	1.6C	32.6	224.8	120	48.8C	157.1	1083.2
40	4.4C	37.0	255.1	125	51.6C	167.5	1154.9
45	7.2C	41.7	287.5	130	54.4C	179.0	1234.2
50	10.0C	46.7	322.0	140	60.0C	204.5	1410.0

Figure 28-4. Chart illustrating relationship between pressure and temperature for Refrigerant-12. (Oldsmobile)

PSIG	TEMP (°F) R-134A	PSIG	TEMP (°F) R-134A	PSIG	TEMP (°F) R-134A	PSIG	TEMP (°F) R-134A
0	−14.7	25	29.3	70	69.6	200	130.1
1	−12.1	26	30.5	75	72.9	210	133.5
2	−9.6	27	31.7	80	76.1	220	136.7
3	−7.2	28	32.9	85	79.2	230	139.8
4	−4.9	29	34.0	90	82.2	240	142.9
5	−2.7	30	35.1	95	85.0	250	145.9
6	−0.6	32	37.4	100	87.8	260	148.8
7	1.4	34	39.5	105	90.5	270	151.6
8	3.4	36	41.6	110	93.1	280	154.3
9	5.3	38	43.6	115	95.6	290	157.0
10	7.1	40	45.6	120	98.0	300	159.6
11	8.9	42	47.4	125	100.4	310	162.2
12	10.6	44	49.2	130	102.7	320	164.7
13	12.3	46	51.0	135	104.9	330	167.2
14	13.9	48	52.8	140	107.1	340	169.6
15	15.4	50	54.5	145	109.3	350	171.9
16	17.0	52	56.4	150	111.4	360	174.2
17	18.5	54	57.8	155	113.3	370	176.5
18	19.9	56	59.3	160	115.4	380	178.7
19	21.4	58	60.8	165	117.4	390	180.7
20	22.8	60	62.4	170	119.3	400	183.1
21	24.1	62	63.9	175	121.2		
22	25.5	64	65.4	180	123.0		
23	26.8	66	66.8	185	124.8		
24	28.0	68	68.2	190	126.6		

Figure 28-5. A pressure-temperature relationship chart for refrigerant R-134a.

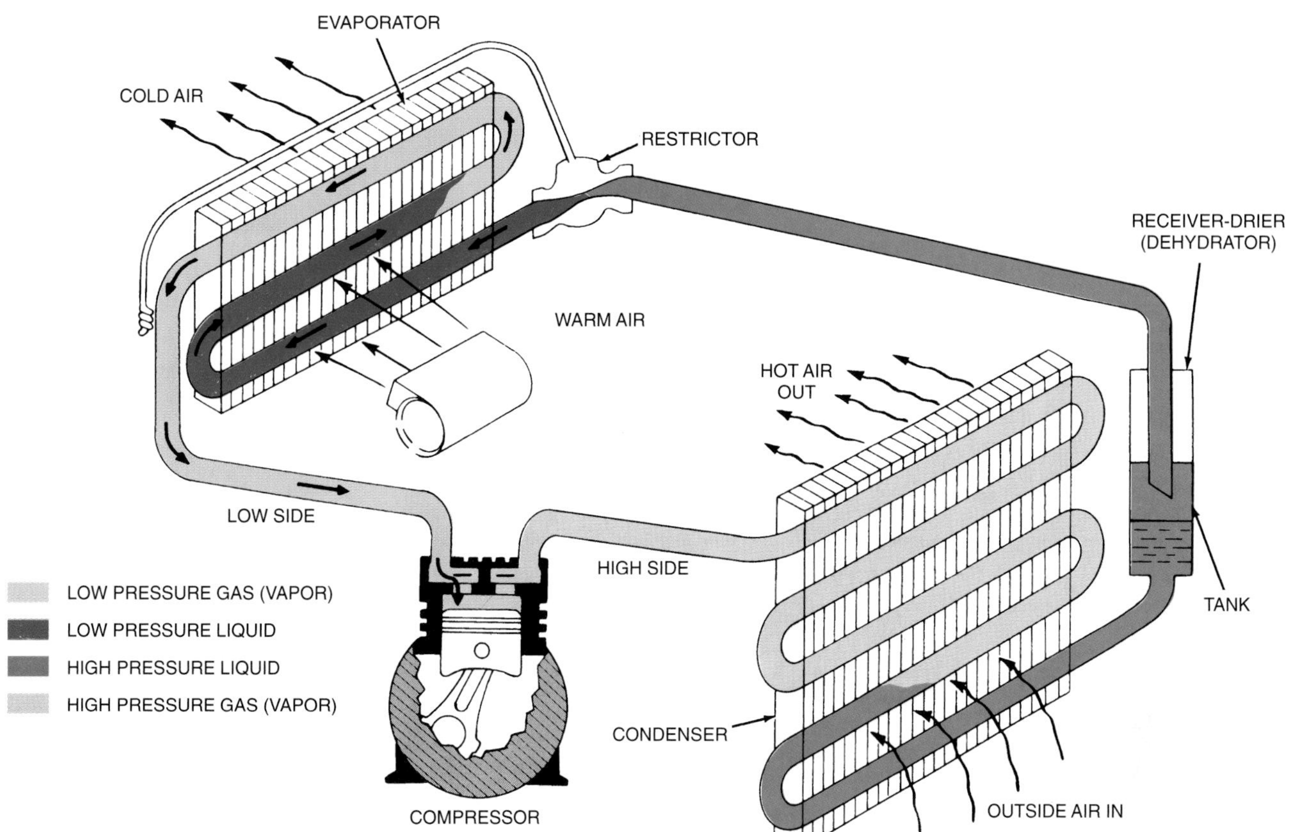

Figure 28-6. Schematic of a typical air conditioning system showing refrigerant flow. Study each phase. (Deere & Co.)

operations involved. Also note the color code introduced with this illustration for high and low pressure liquid and gas (vapor).

Figure 28-7 shows the refrigerant state (low or high pressure, gas, or liquid) in various parts of the system during the refrigeration cycle. With the system operating, high pressure liquid refrigerant collects on the high pressure side of the system. The refrigerant moves through a restriction (top) into the part called the evaporator (left side). The restriction causes the liquid refrigerant to enter the evaporator at low pressure. The lowered pressure lowers the temperature at which it will boil or vaporize.

In the evaporator, the refrigerant is warmed by air forced through the coils by a fan. As more heat from the air is absorbed, the refrigerant begins to boil, turning into a vapor. By the time it reaches the evaporator outlet, the refrigerant is completely vaporized. The cooled air then enters the passenger compartment.

From the evaporator, refrigerant vapor is drawn into the compressor (bottom), which raises the pressure of the vapor and, at the same time, causes a rise in temperature. The vapor is pumped to the part called the condenser (right side) under high pressure, sometimes as much as 400 psi (2400 kPa). The pressure rise raises the boiling point of the refrigerant to a temperature higher than the outside air.

In the condenser, the hot, high pressure refrigerant vapor gives up its heat to the airstream moving over the condenser

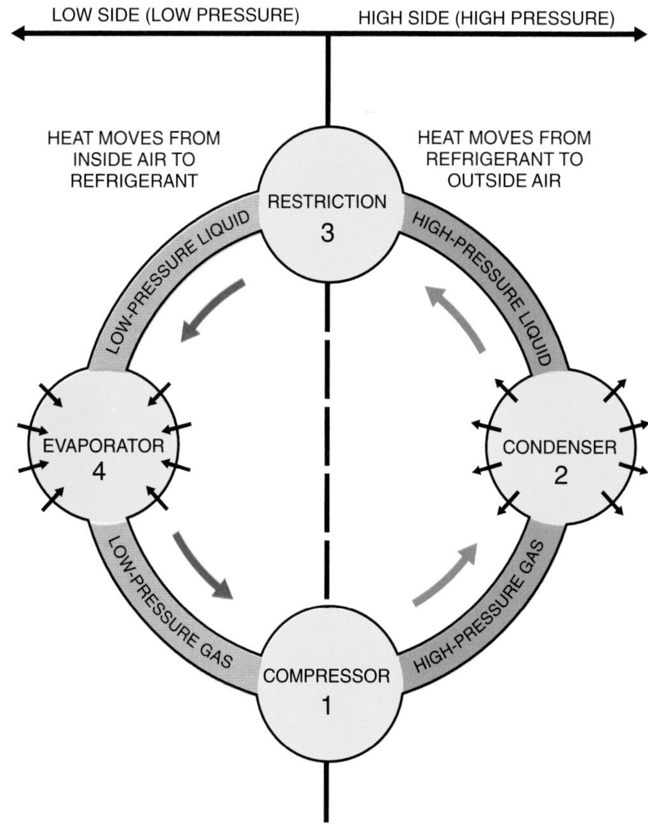

Figure 28-7. Schematic illustrating the four basic parts employed in the refrigeration cycle. Note refrigerant flow direction. (Deere & Co.)

fins. Air is forced through the condenser by vehicle movement or by the radiator cooling fan when the vehicle is stationary. The change in temperature causes the refrigerant to condense or return to its liquid state. Still under pressure from the compressor, the liquid refrigerant flows to the restrictor for another cycle through the system.

Types of Refrigeration Systems

There are two major types of refrigeration systems used in automotive air conditioning, depending on the type of restrictor used between the condenser and evaporator. One system controls the entrance of liquid refrigerant into the evaporator depending on the temperature at the evaporator outlet. In this system, the compressor runs whenever the air conditioner is on.

On the other type of system, the restrictor opening is not variable. The refrigerant flow is controlled by turning the compressor on and off as needed. On some systems, the pumping capacity of the compressor is varied as needed.

Major Components of the Air Conditioning System

All air conditioner refrigeration systems contain the following parts:

- Compressor.
- Condenser.
- Restrictor.
- Evaporator.
- Blower motor.
- Lines and hoses.

In addition, some air conditioner refrigeration systems contain the following components:

- Receiver-dehydrator.
- Accumulator.
- Evaporator pressure control.
- Control switches.
- Sight glass.
- Mufflers.

This chapter will cover each of these devices in turn.

Compressor

The refrigerant leaves the evaporator as a low pressure vapor and travels to the ***compressor*** through the flexible low pressure vapor line. Look at **Figure 28-8.** The compressor draws in the vapor and forces it through the high pressure vapor line to the condenser. The compressor is used to both move and pressurize the refrigerant.

Basic Compressor Designs

Most automotive refrigerant compressors are piston types. Internally, they resemble small engines. Many older air conditioning systems use two-cylinder compressors. These compressors are either inline designs, or V-type compressors.

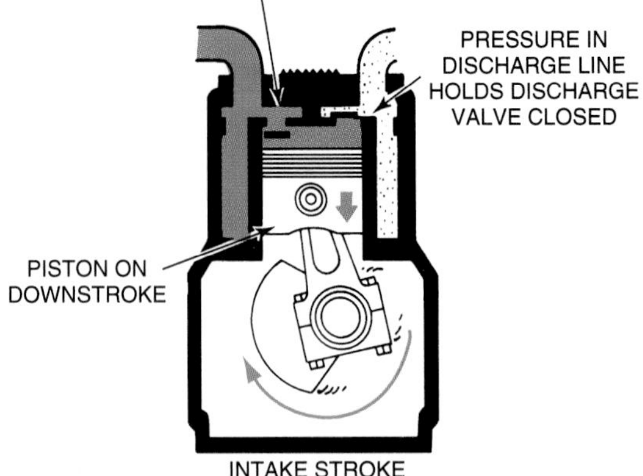

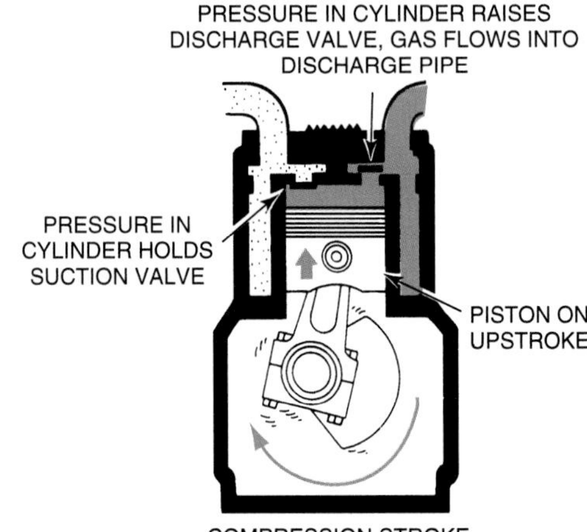

Figure 28-8. Pumping action of a typical reciprocating piston compressor. (Deere & Co.)

The pistons in these compressors are attached to a rotating crankshaft in the same manner as an engine. The cylinders are at right angles to the rotating crankshaft.

Other air conditioning systems employ radial compressors, such as the one in **Figure 28-9.** The ***radial*** type also uses a type of crankshaft to operate the pistons. Note that these pistons are also at right angles to the crankshaft.

Another common type of compressor is the ***axial*** type, **Figure 28-10.** In the axial type the pistons are parallel to the rotating input shaft. The pistons are forced to move back and forth in an axial direction by the action of a rotating swash or wobble plate. On this six-cylinder compressor, there are three double end pistons operating in six separate cylinders. In addition to the piston compressors, a few units are rotary types. These compressors use rotary vanes to compress the refrigerant.

Variable Displacement Compressors

Some compressors are designed with internal valves that allow their pumping capacity to be varied. This is used to

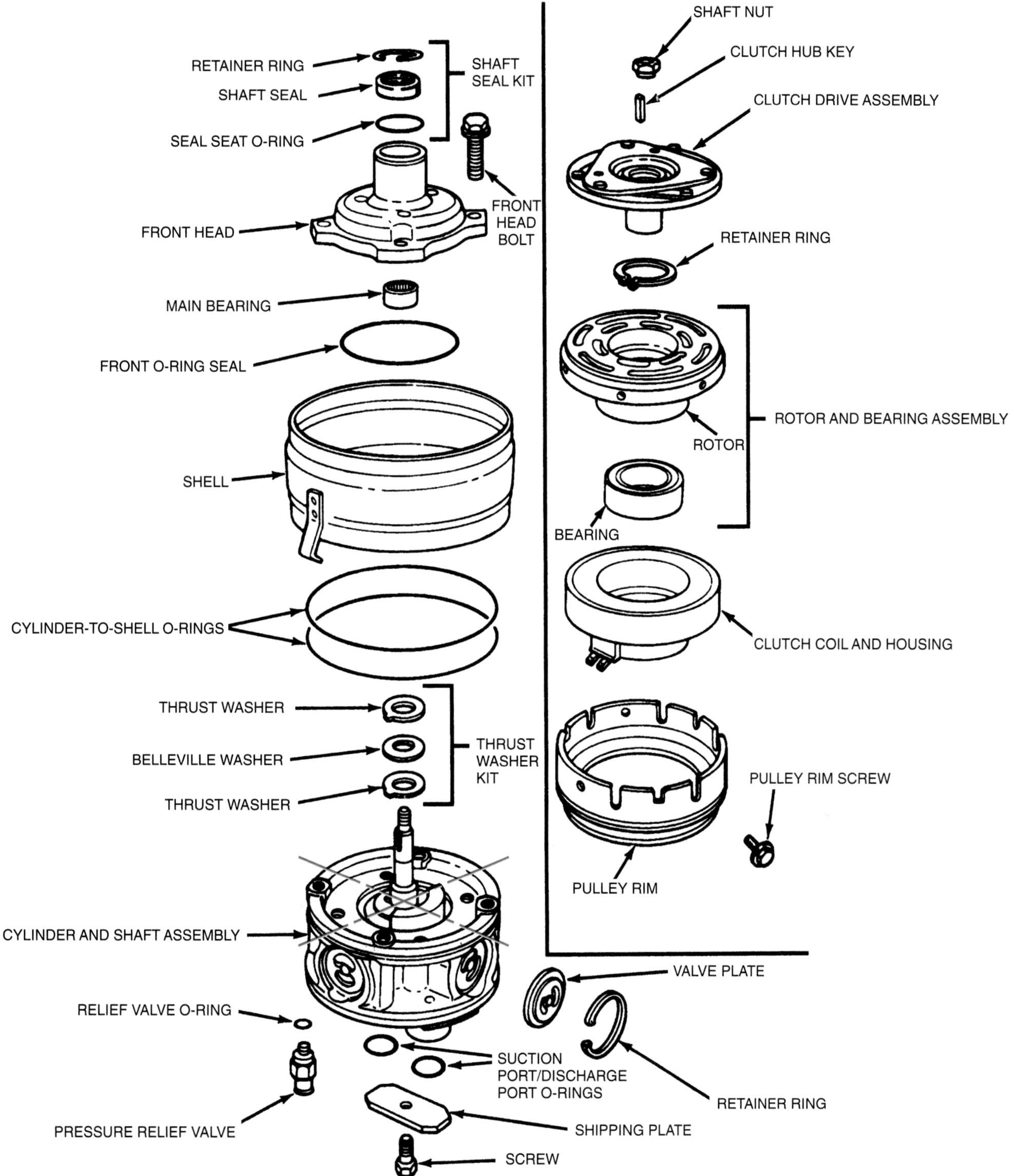

Figure 28-9. Exploded view of a four-cylinder radial compressor. (AC-Delco)

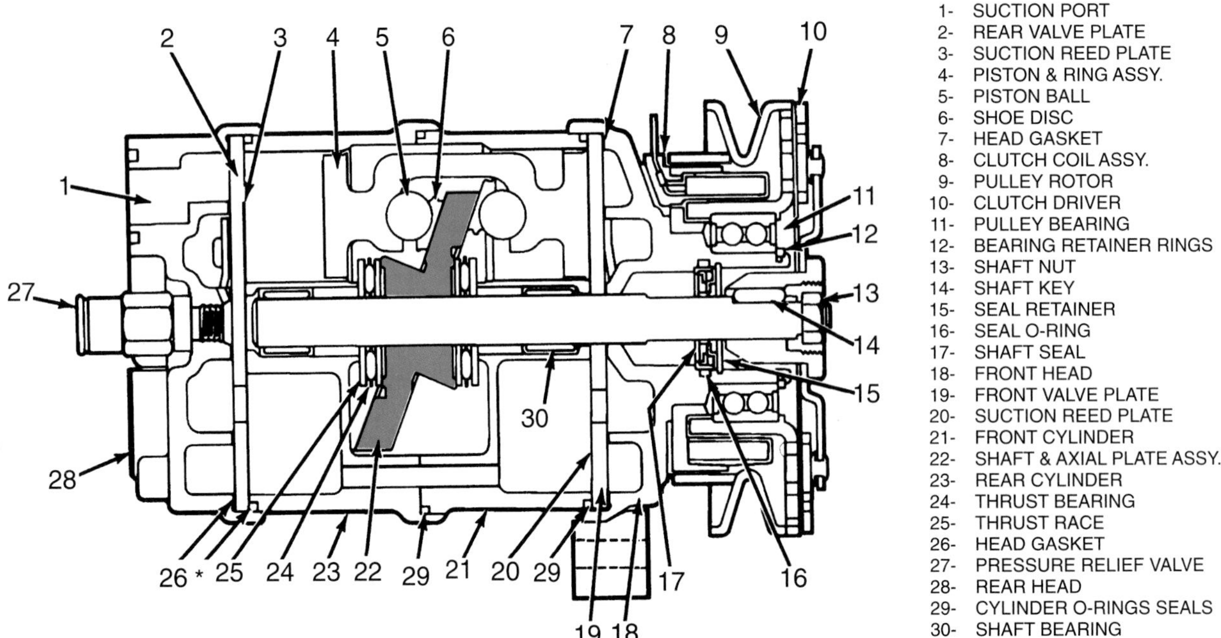

Figure 28-10. A cross-sectional view of a six-cylinder, fixed displacement, axial type air conditioning compressor. As the axial (swash) plate turns, it moves the double-ended pistons back and forth. Study carefully. (AC-Delco)

control the system temperature to prevent evaporator icing (condition where the water on the surface of the evaporator freezes, which can block air movement). This type is called a ***variable displacement compressor.***

A five-cylinder, variable displacement compressor is shown in **Figure 28-11.** An internal valve in the compressor intake reduces the capacity of the compressor whenever the evaporator pressure becomes too low. This reduces the load on the engine without the shock of the on and off clutch application. This compressor can meet all air conditioning needs without cycling on and off.

Pressure Relief Valve

The compressor can be equipped with a ***pressure relief valve.*** Under unusual conditions, the compressor pressure may exceed safe limits. If this happens, the relief valve will pop off (open) and reduce pressure to a safe level. If the relief valve is forced open, the condition which is responsible should be corrected.

Compressor Magnetic Clutch

To place the air conditioning system into operation, the compressor's ***magnetic clutch*** is energized. The clutch coil draws the clutch hub inward, locking the revolving pulley to the compressor shaft. Note in **Figure 28-12** that the magnetic coil does not revolve. Its magnetic pull is transmitted through the pulley to the armature. The armature plate and hub assembly are fastened to the compressor drive shaft. The magnetic clutch pulley, when not driving the compressor, spins on ball bearings.

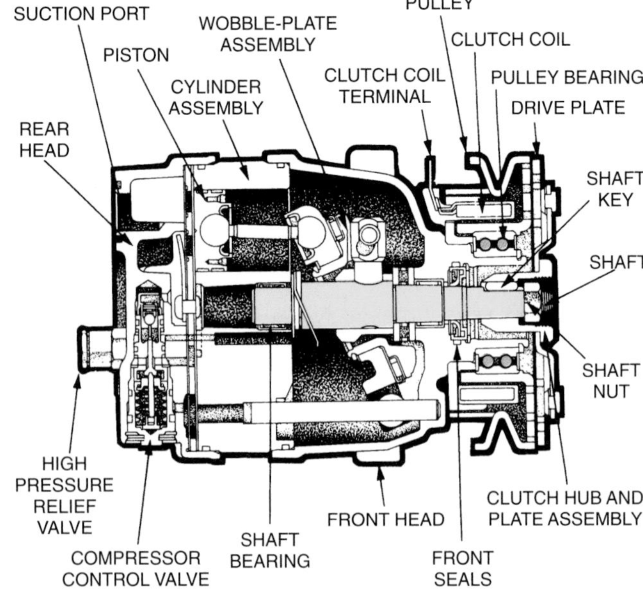

Figure 28-11. A cutaway view of a variable displacement, five-cylinder compressor. The wobble (swash) plate angle is increased or decreased to meet the demands of the system. When the demands are high, angle of wobble plate increases (longer piston stroke). When demands are low, piston stroke is shorter. This eliminates the clutch cycling on and off.

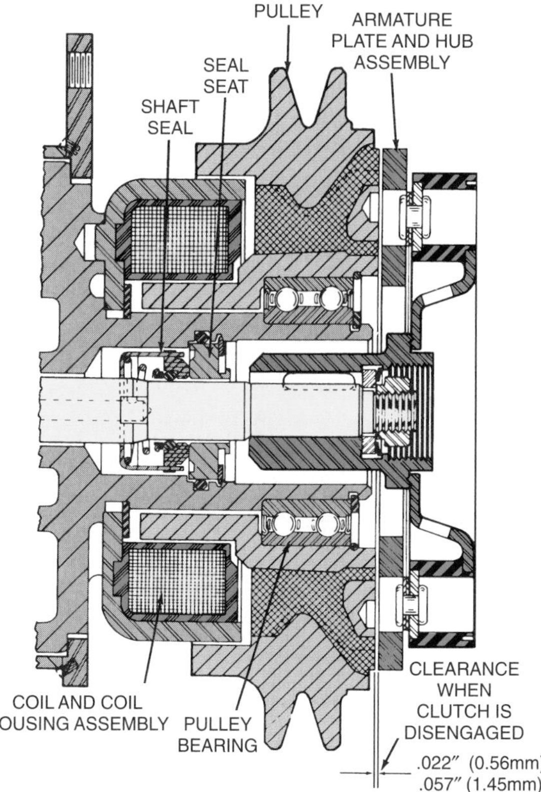

Figure 28-12. A magnetic clutch. The pulley revolves whenever the engine is running. When electrical current is passed through the electromagnetic coil, magnetic pull draws the armature plate up against the pulley, driving the compressor. (Pontiac)

On variable displacement and other systems, the compressor clutch is applied whenever the air conditioner is on. On cycling clutch systems, the compressor is turned on and off to control evaporator pressure. Various designs are used for magnetic clutches. One type is shown on the compressor in **Figure 28-13.**

Condenser

The heated, high pressure vapor from the compressor is forced into the ***condenser,*** where heat is removed by air passing over the cooling fins. As the vapor travels through the condenser coils, it gives up enough heat to the passing airstream to cause the vapor to return to the liquid state. As the liquid leaves the condenser, it is stored in the receiver-dehydrator. The condenser is mounted in front of the radiator, so it will be exposed to the stream of cooling air. Air flow is provided by the engine fan, an auxiliary electric fan, or by the air forced through the condenser when the vehicle is moving.

Condensers (and evaporators) are often made of aluminum. During system operation, condenser pressures and temperatures are both high. Condensers are mounted in front of the radiator, as shown in **Figure 28-14.** Follow the path of the high pressure vapor in **Figure 28-15** from the compressor as it travels through the condenser. Note how it changes into a liquid.

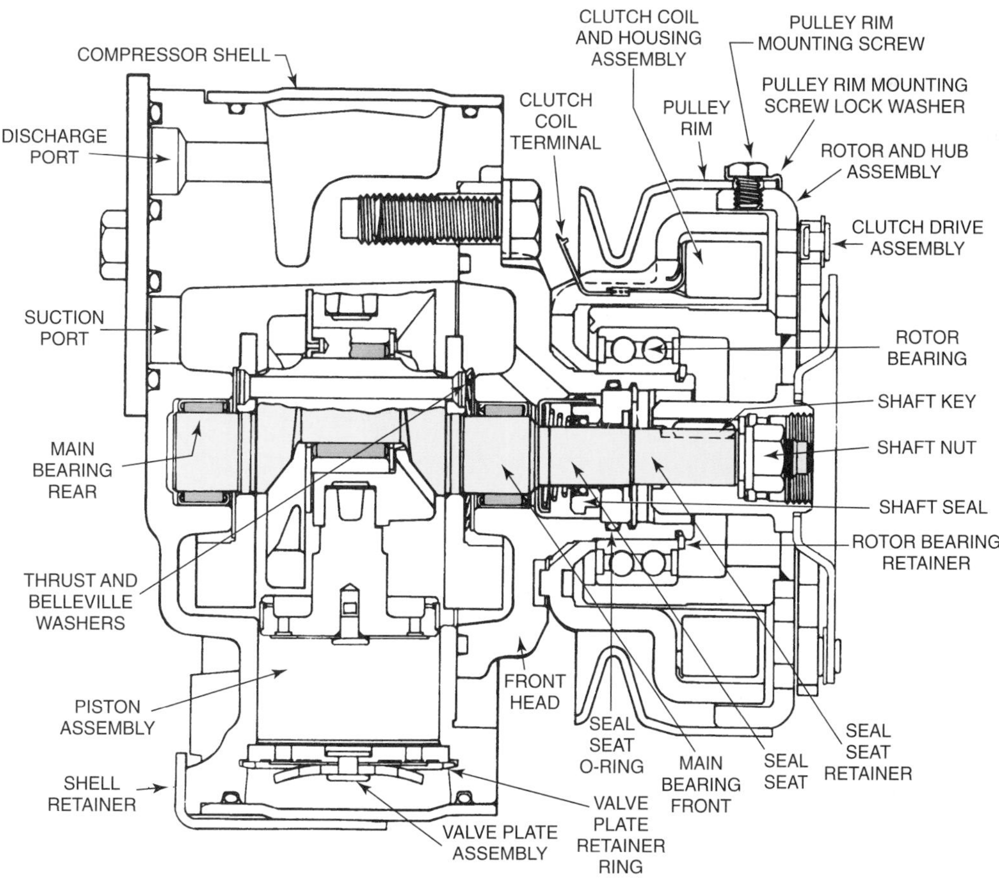

Figure 28-13. A typical four-cylinder, radial type compressor. Study the construction. (Buick)

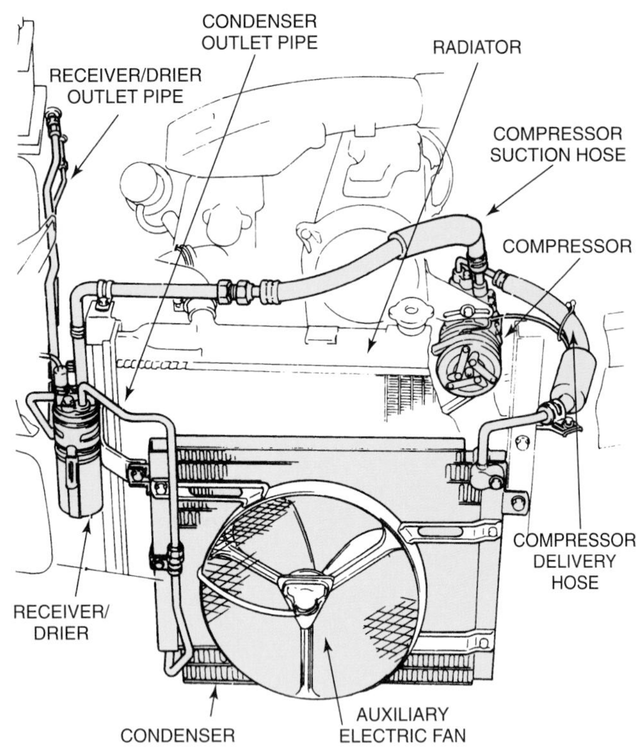

Figure 28-14. Typical air conditioning condenser mounting position in front of the radiator. (General Motors)

Refrigerant Flow Restrictor

Refrigerant reaches the ***flow restrictor*** as a high pressure liquid. The restrictor causes the refrigerant to enter the evaporator in small amounts, reducing its pressure. The flow restrictor can be an ***expansion valve*** or a ***fixed orifice tube,*** depending on the system design.

Expansion Valve

The expansion valve admits a metered amount of refrigerant into the evaporator. The refrigerant moves into the evaporator as a relatively low temperature, low pressure liquid. See **Figure 28-16.**

The temperature sensitive thermal bulb is attached to the evaporator outlet. A capillary tube (length of tubing of small diameter which acts as a throttle on refrigerant) connects the bulb to the expansion valve. As the temperature of the evaporator outlet rises, the expansion valve is opened, admitting a greater amount of refrigerant. When the temperature drops, the valve begins to close. The action of the valve is also affected by the internal spring and by evaporator pressure.

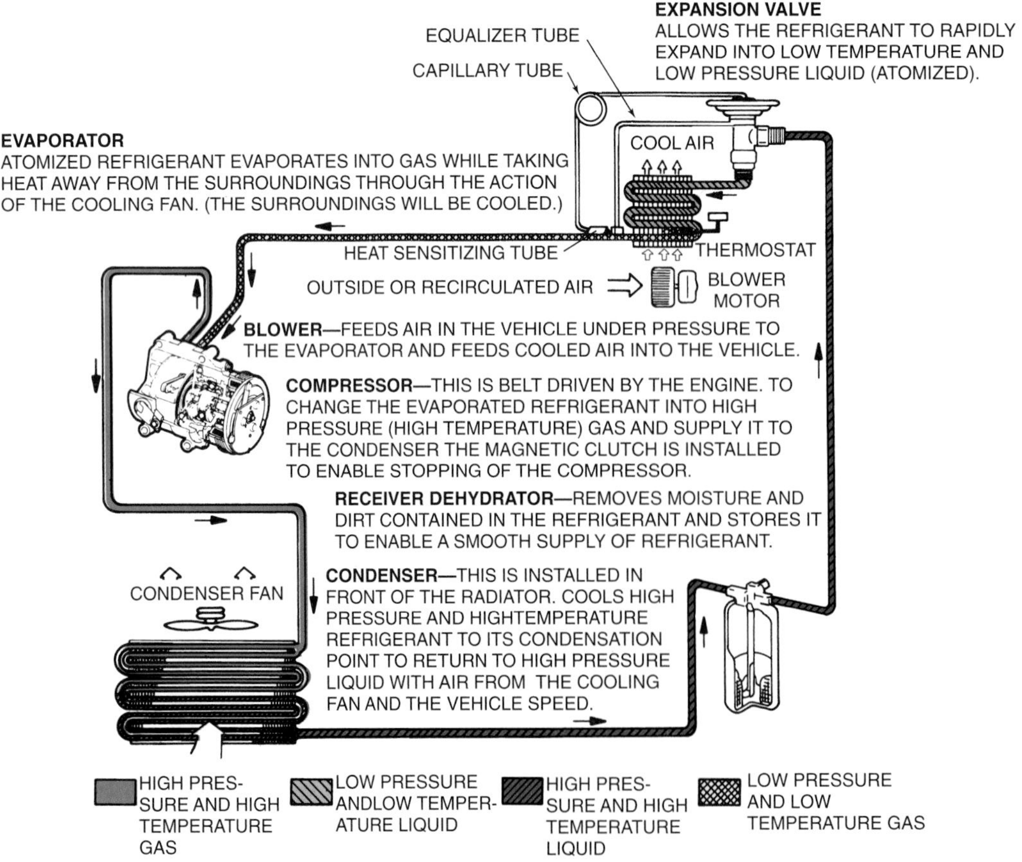

Figure 28-15. Air conditioning system refrigerant flow schematic. Follow refrigerant flow from the receiver-dehydrator, expansion valve, evaporator, compressor, condenser, and back to the receiver-dehydrator. (Hyundai)

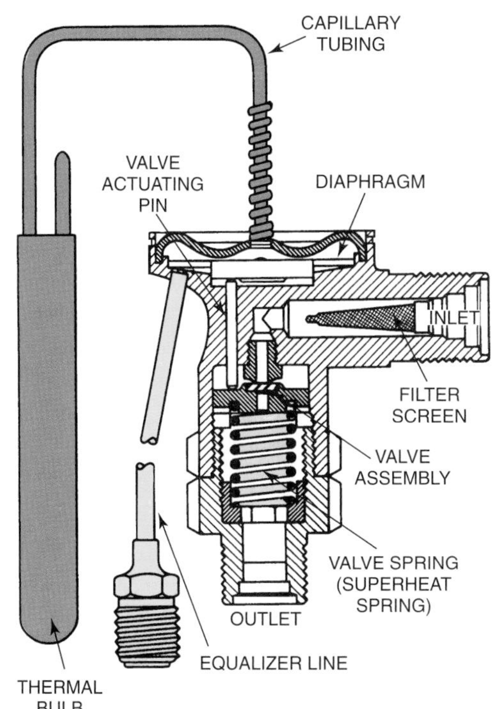

Figure 28-16. A typical expansion valve. This valve is controlled by three forces: 1—Valve spring pressure; 2—Equalizer line pressure; 3—Thermal bulb pressure. (GMC)

Fixed Orifice

Some systems utilize a fixed orifice as an expansion valve. The plastic or metal tube houses a fine screen and a small fixed orifice through which a metered flow of liquid refrigerant can pass, **Figure 28-17.** The orifice is placed in a line that travels from the condenser outlet to the evaporator inlet. The fixed orifice diameter restricts the flow of refrigerant, providing a low pressure liquid flow to the evaporator.

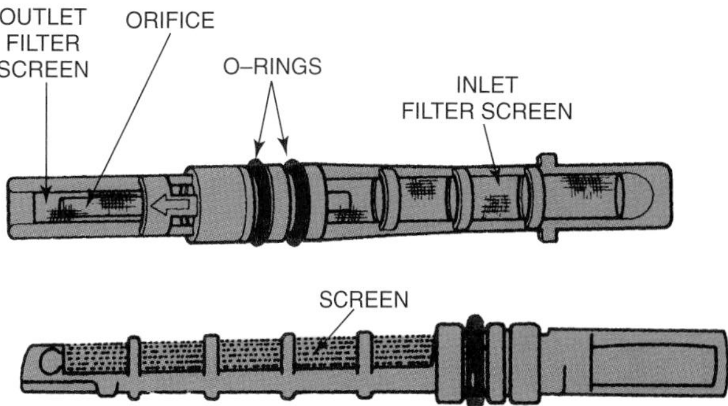

Figure 28-17. Air conditioning system utilizing a fixed orifice as an expansion valve. These valves vary somewhat in design and construction. (Sun Electric Corporation)

A typical system using an orifice tube is shown in **Figure 28-18.** To control the flow of refrigerant through the evaporator, a cycling clutch compressor is used. The system shown in **Figure 28-18** uses a thermostatic cycling switch to turn the compressor clutch on or off.

Evaporator

The ***evaporator*** is a heat exchanger which transfers heat from the incoming air to the refrigerant. As the low pressure liquid refrigerant enters the evaporator from the expansion valve, it begins to vaporize (boil). This vaporizing action absorbs heat from the tubes and cooling fins. Air passing over the cold tubes and fins gives up heat to the core and a stream of cooled air enters the vehicle, **Figure 28-19.**

When the air strikes the cold fins, some of the moisture in the air condenses and drains off the core. This ***dehumidifies*** (removes moisture from) the air. Particles of dust and

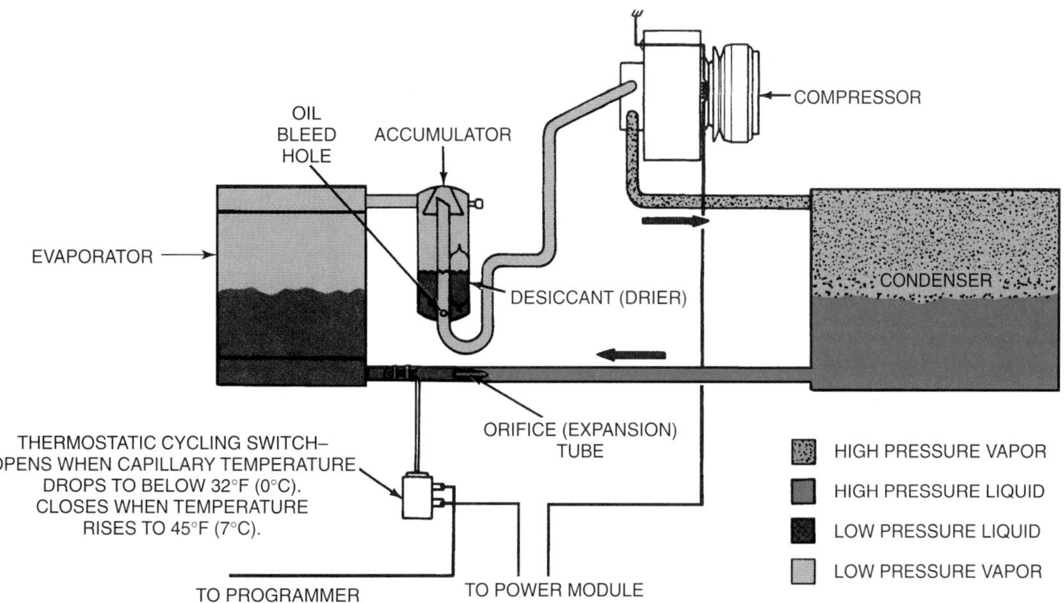

Figure 28-18. A schematic showing the refrigerant cycle for a CCOT (cycling clutch orifice tube) system. Note the cycling switch. (Cadillac)

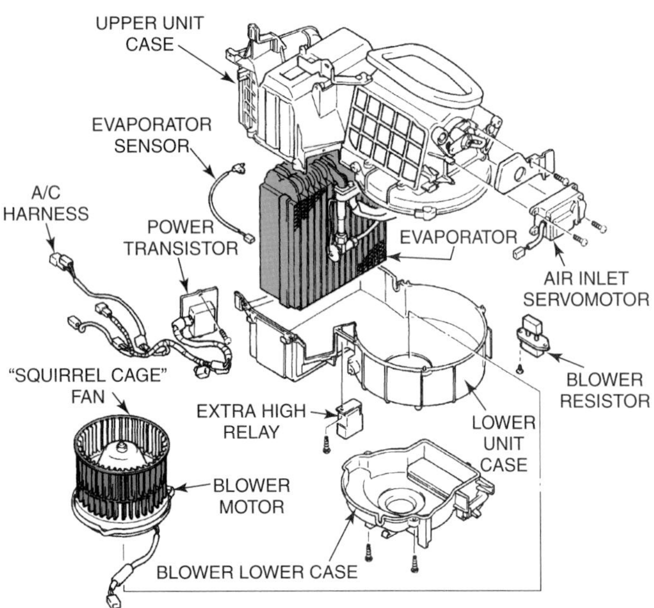

Figure 28-19. A "squirrel cage" fan, blower motor, housing, evaporator and related parts for one system. These evaporator units are produced from aluminum. Inlet and outlet pipes are welded on. They are chemically coated to help retard corrosion. (Toyota Motor Corporation)

pollen tend to stick to the wet fins and drain off with the water, removing a significant amount of pollutants from the air entering the vehicle. The water seen dripping from a parked vehicle on a hot day is the water which has condensed on the evaporator and drained out of the evaporator case.

Some evaporators have a line placed at the bottom of the housing and connected to the suction side of the compressor. This line returns any refrigerant oil which may have settled to the bottom of the evaporator to the compressor.

Blower Motor

Air is forced over the evaporator by a fan called a ***blower.*** The blower used in most vehicles is called a ***squirrel cage,*** **Figure 28-19.** The squirrel cage blower is compact and delivers a large volume of air for its size. The squirrel cage blower is driven by a ***blower motor.*** The blower motor is a high speed dc motor coupled directly to the squirrel cage.

Blower motor speed is usually controlled by a dashboard switch and a series of ***blower resistors.*** The blower resistors, **Figure 28-20A,** reduce the electrical flow through the motor, reducing its speed. In the high speed position, full battery current flows through the motor. The resistors are often selected through a ***blower relay,*** **Figures 28-20A** and **28-20B.** A schematic of a typical blower electrical system is shown in **Figure 28-21.** Blower motors are sometimes computer-controlled. These computer control systems can control the blower motor by varying the current to the motor, eliminating the blower resistors.

Lines and Hoses

To connect the parts of the refrigeration system shown above, ***refrigeration lines*** and ***hoses*** are used. Metal lines made of aluminum or steel are often used to connect parts that are solidly mounted, such as the condenser and evaporator. Connections between parts that have relative movement, such as the engine-mounted compressor and the body-mounted condenser, use flexible hoses.

All fittings, including lines and hoses, use one or more O-rings. The fittings alone do not provide a sufficient seal to prevent air from entering the system. Fittings on older systems use one O-ring. Newer vehicles can use up to three O-rings on any one fitting. Line fittings to the compressor normally use only one O-ring.

Muffler

A small ***muffler*** is often placed between the compressor and the condenser to reduce pumping noise. The muffler is sometimes part of the hose itself. This type of hose is usually referred to as a muffler hose. A muffler may also be used to reduce line vibrations. Mufflers are installed with the outlet side down so that refrigeration oil will not be trapped.

Receiver-Dehydrator

The ***receiver-dehydrator*** is used only on air conditioning systems with expansion valves. It is located between the condenser and restrictor. It is a reservoir for liquid refrigerant after it has passed through the condenser, **Figure 28-22.** The receiver-dehydrator filters and removes moisture through means of a ***desiccant*** (drying agent). It is often the location of the sight glass, when one is used.

The receiver may contain a fusible safety plug, which releases pressure when the refrigerant temperature exceeds a specified point (about 212°F or 100°C.). When mounted on the low side, a pressure cycling switch can also be installed.

Sight Glass

A ***sight glass*** may be installed in the receiver-dehydrator or in a high pressure liquid line. The sight glass is used to inspect the refrigerant for the presence of bubbles or foam. One type of sight glass construction is shown in **Figure 28-23.** The sight glass allows the technician to determine the state of charge by checking for the presence of bubbles. Under most operating conditions, bubbles in the refrigerant means that the system charge is low. Some systems, especially those without an expansion valve, do not have a sight glass.

Accumulator

Accumulators are used only on vehicles with fixed orifice air conditioning systems. The accumulator is installed between the evaporator and compressor suction line and acts

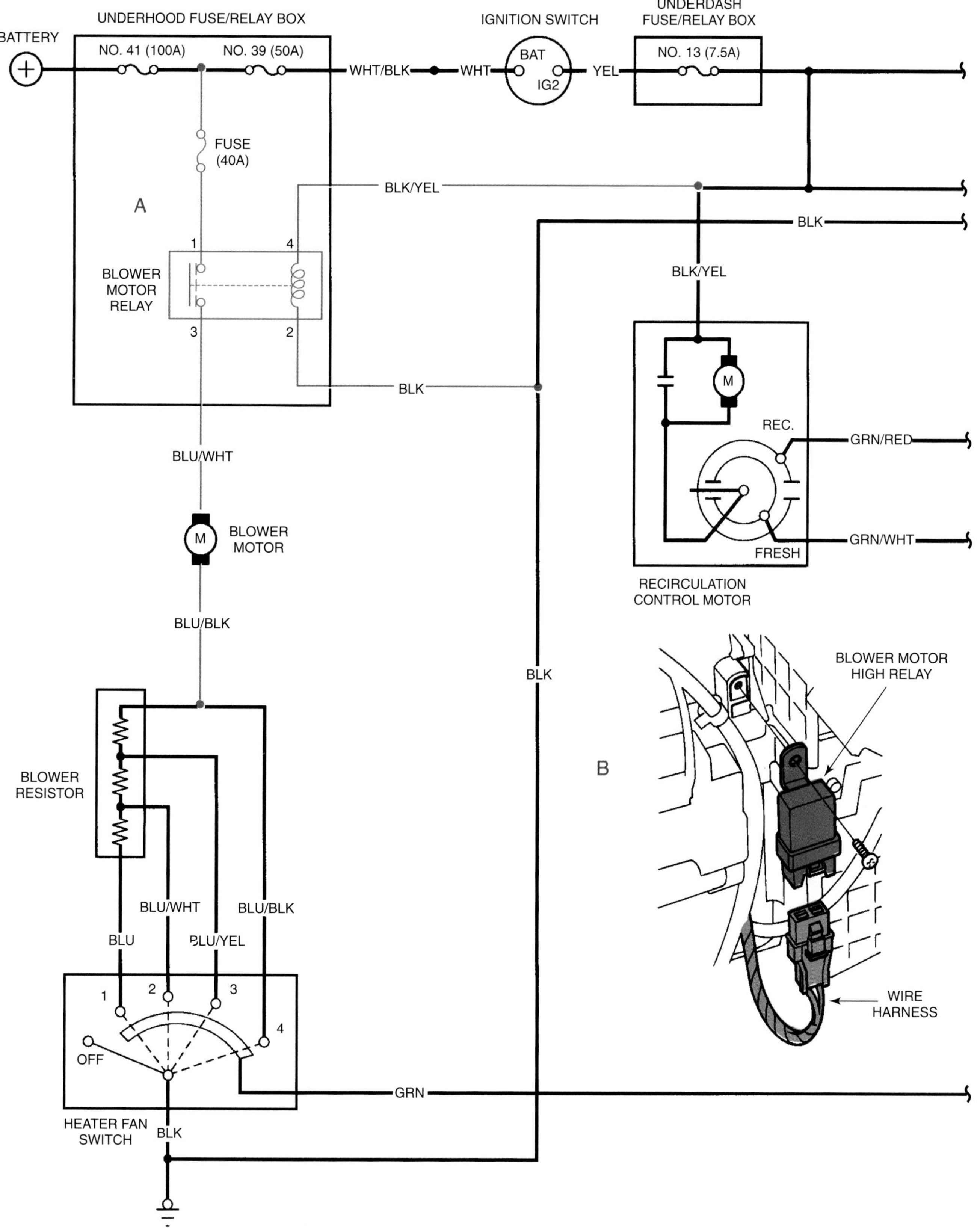

Figure 28-20. A—An electrical schematic showing a blower motor relay in the open (off) position, and related wiring. These resistors, number of resistors, and locations will vary from one manufacturer to another. B—A blower motor relay and wiring harness. (Honda Motor Co.)

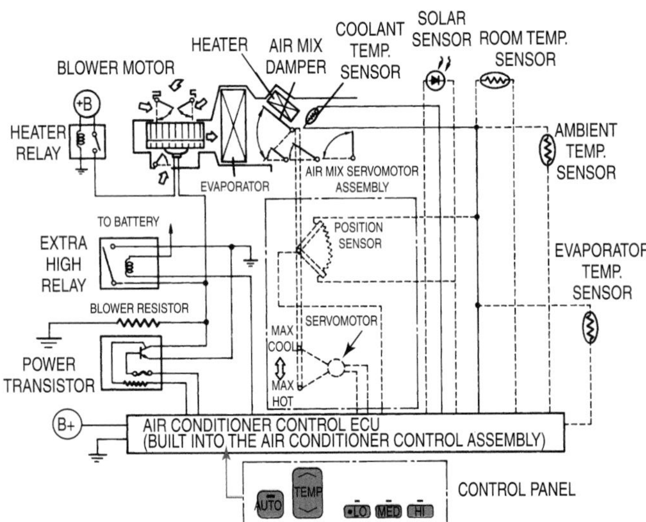

Figure 28-21. Electrical schematic for one type of blower motor system. Selection is made on the control panel, then the air conditioner control unit (ECU) electrically triggers the correct motors, relays, etc. (Toyota Motor Corporation)

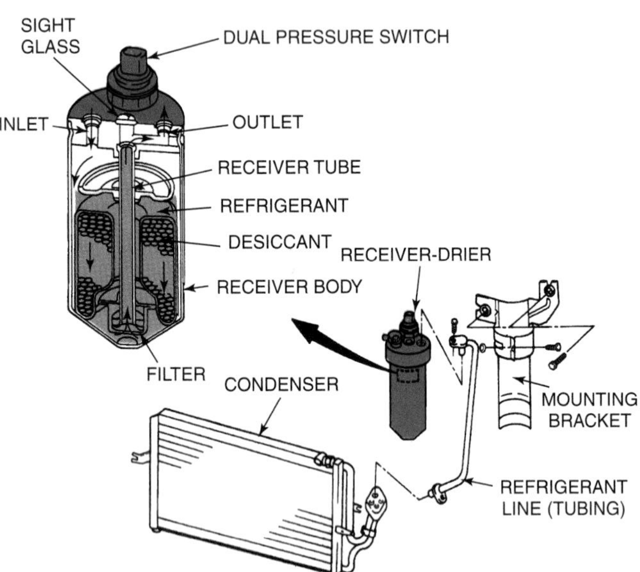

Figure 28-22. Cutaway view of a receiver-dehydrator. Follow refrigerant flow arrows. (Hyundai)

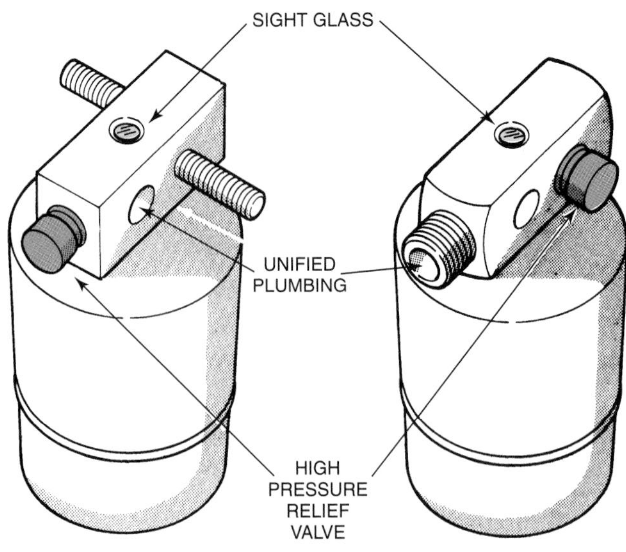

Figure 28-23. Two possible locations for a high pressure relief valve. This valve protects against excessive system pressure.

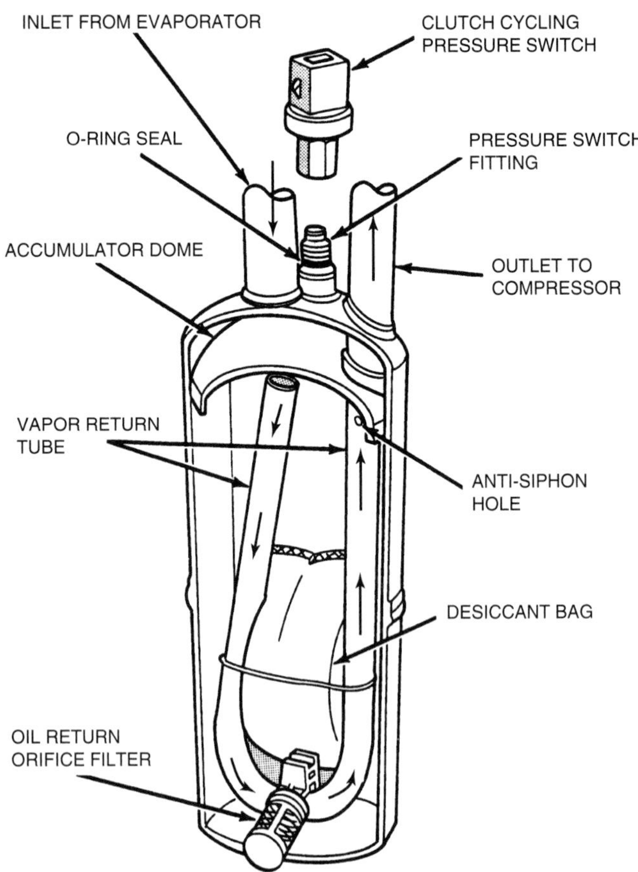

Figure 28-24. Cutaway of an accumulator assembly that uses a screen filter, desiccant, and pressure switch. (Dodge)

as a reservoir for pressurized refrigerant vapor. The design of the accumulator causes any liquid refrigerant to fall to the bottom of the accumulator body, where it evaporates before it is drawn into the compressor. This prevents damage to the compressor as it cycles on and off. The accumulator also contains the system desiccant for moisture removal.

An accumulator containing a filter screen, liquid bleed hole, desiccant, and pressure switch fitting is shown in **Figure 28-24.** Some accumulators used on older vehicles may also incorporate pressure control valve, expansion valve, and a service fitting. See **Figure 28-25.**

Evaporator Pressure Controls

Evaporator temperature will vary from 33° to 60°F (1° to 16°C). If the temperature drops to freezing (32°F or 0°C), moisture that is condensed on the evaporator core will freeze

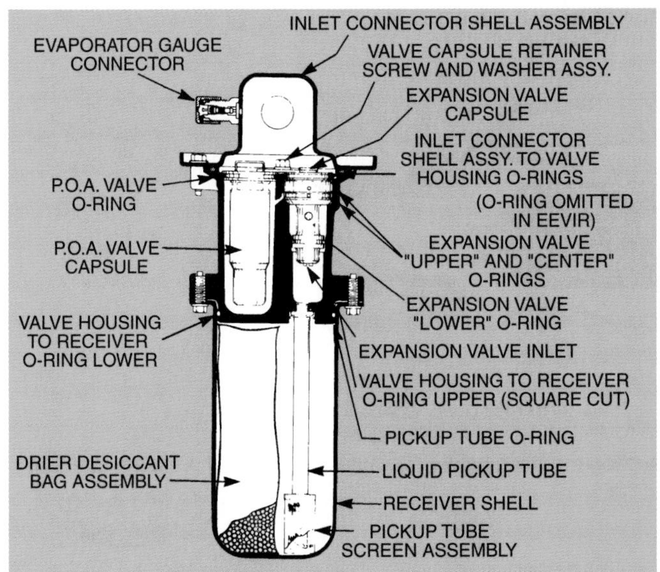

Figure 28-25. Valves in Receiver combines POA valve, expansion valve, and desiccant. (Harrison Radiator)

and block off the airflow. Pressure in the evaporator must be controlled to prevent freezing. If pressure can be held between 29 and 30 psig (200 and 207 kPa), freezing will be eliminated. There are several types of ***evaporator pressure controls.***

Hot Gas Bypass Valve

A ***hot gas bypass valve*** was used on some older vehicles. The valve was placed in the outlet side of the evaporator and metered high pressure, hot refrigerant from the compressor into the evaporator. This maintained the evaporator pressure at a point high enough to prevent icing. A hot gas bypass valve is illustrated in **Figure 28-26.**

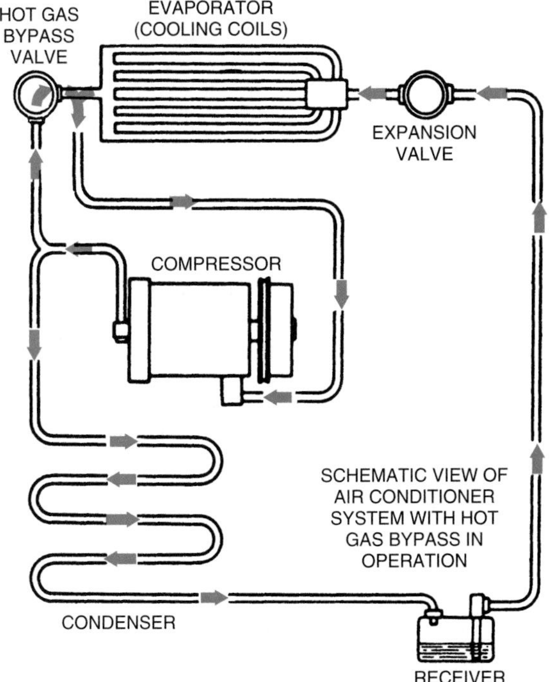

Figure 28-26. Hot gas bypass valve. Note how the bypass valve feeds hot, high pressure refrigerant back into the evaporator outlet. (Buick)

Suction Throttling Valve

The ***suction throttling valve*** was also used on older vehicles. It was placed at the evaporator outlet. When evaporator pressure rises, the valve releases pressure into the low pressure vapor line to the compressor. If the pressure drops down, the valve restricts the flow, raising evaporator pressure.

Pilot Operated Absolute

The ***pilot operated absolute valve,*** or POA valve was used on a few vehicles until recently. A POA valve is shown in **Figure 28-27.** The bellows shown in **Figure 28-27** is under a vacuum. The bellows operates the pilot needle valve. Opening or closing of the needle valve alters the relative pressure between the compressor and evaporator sides of the piston, causing the piston to open and close the cylinder ports and maintain a constant evaporator pressure.

Compressor Cycling Switch

The method of preventing evaporator icing on most newer vehicles is a ***thermostatic compressor switch*** that stops driving the compressor when evaporator core temperature drops to a specific point. It does this by deenergizing the compressor magnetic clutch directly or by use of a body control computer.

When the temperature rises to a set level, the switch energizes the magnetic clutch and the compressor is again driven. This on and off action of the compressor prevents evaporator core freezing. Refer to **Figure 28-28** for a compressor cycling switch. Similar systems use a pressure switch at the compressor inlet to cut out the compressor when low side pressure drops too far. They may also be located on the accumulator.

Compressor Control Switches

The air conditioning system may have several electrical switches. These switches control various functions in the system or prevent system damage. The most common class of switches control the air conditioning compressor.

Some compressors are equipped with a ***low pressure switch*** to prevent compressor damage in the event the system charge is lost. If system pressure drops below 25 psi (172 kPa), the switch contacts open, de-energizing the magnetic clutch. This provides compressor protection in the event the system refrigerant charge is lost. It also prevents system operation when ambient (surrounding) air temperature is below freezing. Some low pressure switches are monitored by the vehicle's engine or body ECU.

The compressor magnetic clutch can also be de-energized by a ***full throttle cut-out switch,*** sometimes called a ***wide open throttle switch.*** This switch eliminates compressor drag when the vehicle is being accelerated at full throttle. It is attached to the throttle linkage. Another air conditioner related device attached to the throttle is the anti-dieseling switch, which momentarily energizes the compressor clutch when the ignition is switched off. This loads the engine and reduces the chance of dieseling.

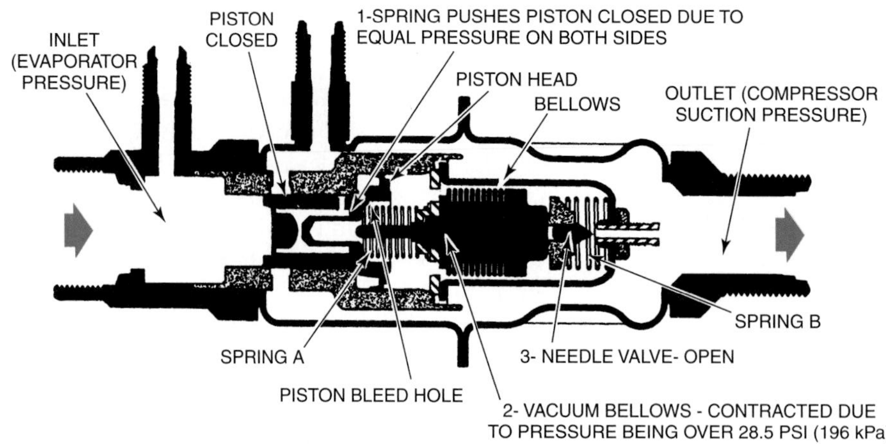

Figure 28-27. Schematic of a typical Pilot Operated Absolute (POA) suction throttling valve. The system is off and pressure is equal on both sides of the piston, so spring A has forced the piston closed. (Pontiac)

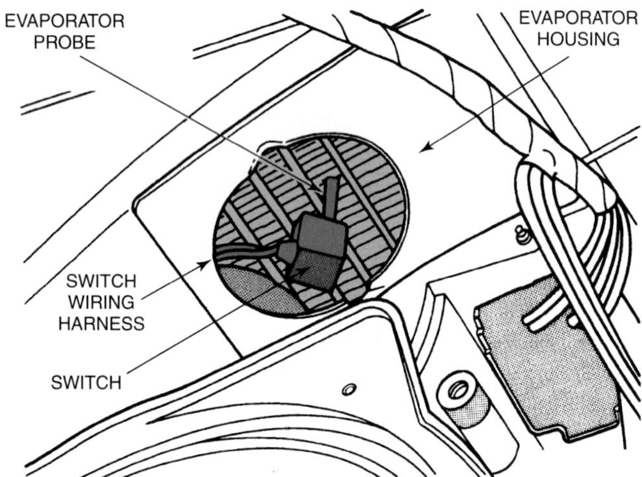

Figure 28-28. Air conditioning compressor cycling switch and its probe, as mounted on the evaporator core. (Chrysler)

Below about 35°F (2°C) ambient (surrounding) air temperatures are too low for the air conditioner to be effective. The compressor can be damaged by operation at low temperatures. To eliminate unneeded compressor operation when the ambient temperature is too low, the ***ambient air temperature switch*** shuts off the compressor when the temperature drops below a certain setting. Several switches are shown in **Figure 28-29.**

Other Control Switches

Other switches control non-air conditioner related components when the air conditioning system is used. ***Idle speed-up solenoids*** are designed to improve engine operation by raising idle when the air conditioner is turned on. Other solenoids close the throttle plates when the ignition is turned off. These are called ***anti-dieseling solenoids,*** which prevent dieseling when the engine is hot due to air conditioner operation. Refer to Chapters 9, 10, and 14 for information on engine speed up devices.

Some air conditioning systems employ an extra fan in front of the condenser. This will provide additional air flow when idling in hot weather. Other vehicles use an electric fan for condenser and radiator cooling. These fans are normally turned on automatically whenever the air conditioning system is operating.

Service Valves

All automotive air conditioning systems provide discharge and suction ***service valves.*** These valves permit the system charge to be recovered (emptied), as well as charged (filled) and checked with pressure gauges.

The service valves used in R-134a systems are known as ***quick-disconnect valves***. See **Figure 28-30A.** Pressing the hose connector onto the fitting and turning the handle depresses the valve. R-134a fittings are different sizes on the high and low pressure sides of the system. This prevents the technician from accidentally connecting the wrong hose to the valve.

Most R-12 systems use a spring-loaded valve called a ***Schrader valve,*** **Figure 28-30B.** The hose connection is threaded over the valve, causing the pin to be depressed. Schrader valves on late-model vehicles have different size threads on the high- and low-pressure sides to ensure that the hoses are connected to the proper valve.

A few older R-12 systems use the type of service valve shown in **Figure 28-31.** These valves were usually located on the compressor. Once the hoses are connected, the valves can be turned to evacuate or charge the system.

Refrigerant Oil

A specific amount of system oil, usually called ***refrigerant oil,*** is placed in the system to provide lubrication for the compressor. Since refrigerant has great affinity (attraction) for the oil, a certain amount circulates through the system.

R-12 systems use a ***mineral oil*** based lubricant in most applications. In R-134a systems, an ***ester*** or ***polyalkylene glycol*** (PAG) oil is used. Refrigeration oil is highly refined and must be absolutely free of moisture. No other type of oil can be substituted.

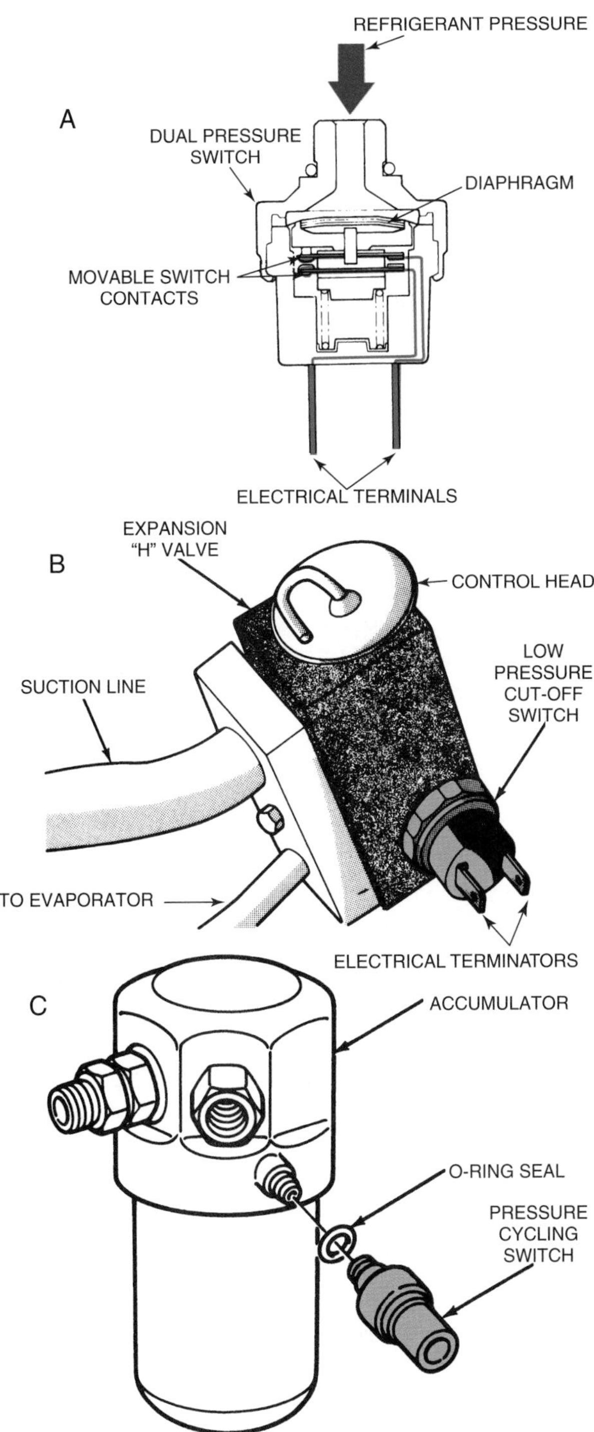

Figure 28-29. Several air conditioning pressure cycling switches. A—Dual pressure switch cutaway. This switch will turn the compressor on or off under conditions of abnormally low refrigerant pressure 30 psi (207 kPa) or high pressure 384 psi (2641 kPa). Reasons for low or high pressures may be insufficient refrigerant, low refrigerant temperatures, high refrigerant temperatures, and overcharging with refrigerant during service. B—Low pressure cutoff switch. This switch is located on the expansion "H" valve and is connected to the compressor clutch. It will turn off the electrical power to the compressor clutch if the refrigerant pressure falls below the switch control point. The switch is factory calibrated and sealed. C—Pressure cycling switch. This spring-loaded diaphragm switch controls the electrical power to the compressor clutch and is often mounted on the accumulator. (AC-Delco, Dodge, Geo)

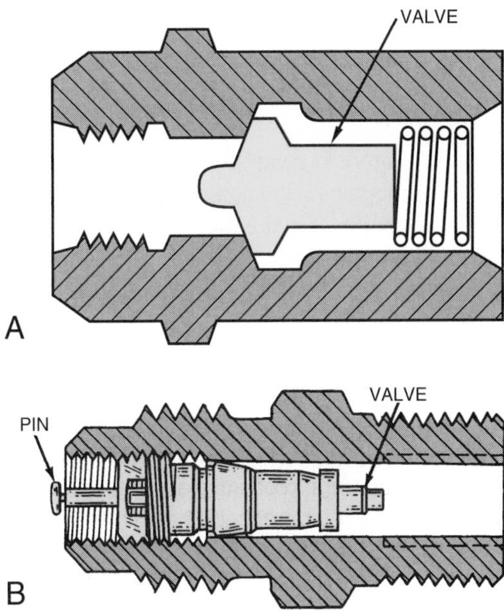

Figure 28-30. Typical air conditioning service valves. A—Quick-disconnect valve. B—Schrader valve.

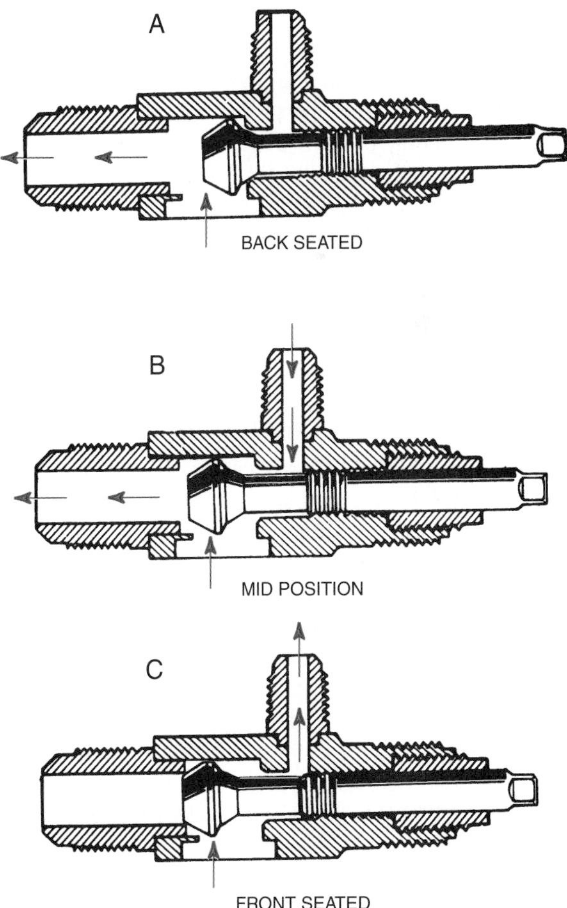

Figure 28-31. One type of refrigerant service valve. A—Normal operating position, "back seated." B—Charging position. C—Evacuating position, "front seated." The arrows indicate the directions of refrigerant flow. (General Motors)

Refrigerant Handling Precautions

Refrigerant must be handled with extreme care. The following safety precautions must always be observed whenever refrigerant is handled.

- Keep refrigerant in a well-ventilated area and wear protective goggles. Refrigerant is heavier than air. It will displace the air in the room and can cause suffocation. If the refrigerant comes into contact with an open flame, it will form poisonous ***phosgene gas.***
- Keep liquid refrigerant away from the skin and eyes as it will freeze on contact. The result is similar to frostbite. If refrigerant enters the eyes, do not panic. Splash large amounts of cold water into the eyes as a means of raising the temperature. Do not rub the eyes. Apply several drops of clean mineral oil. The oil will absorb the refrigerant and help flush it from the eyes. The same procedure may be followed if the skin has been frozen. Seek a physician at once—even if the pain seems to have subsided.
- Always use an approved ***refrigerant recovery system*** to discharge the refrigeration system. Discharging refrigerant into the atmosphere is a violation of Federal and state law.

Retrofitting

When it was discovered that R-12 was contributing to the depletion of the Earth's ozone layer, an international treaty was signed which banned the production of R-12 by the mid 1990s. As the supply of R-12 diminishes, it will become necessary to adapt older air conditioning systems to use R-134a. This procedure is called ***retrofitting.***

As long as R-12 is available, it may be used to recharge systems that use this refrigerant. If the refrigeration system requires a major repair, the option to retrofit to R-134a may be given to the customer. Once the supply of R-12 begins to diminish, the price will increase to the point that retrofitting becomes an economical alternative. Once the supply of R-12 is gone, retrofitting will become a necessity to repair these systems. Each manufacturer publishes guidelines for retrofitting their vehicles.

Heater Systems

Vehicle heaters may be used separately or in combination with an air conditioner. All modern heaters depend on hot engine coolant flowing through a copper, aluminum, or aluminum and plastic ***heater core,*** **Figure 28-32,** to provide heat to the passenger compartment. Coolant is piped from the engine to the heater core through ***heater hoses.*** Heater hoses are smaller diameter versions of the radiator hoses. The squirrel cage blower and blower motor provide the air flow through the heater core.

Many heater systems are equipped with a coolant ***shut-off valve,*** **Figure 28-33,** which stops the flow of engine coolant through the heater core when it is not needed. The shut-off valve is usually installed in the heater hoses. They are generally manually or vacuum operated.

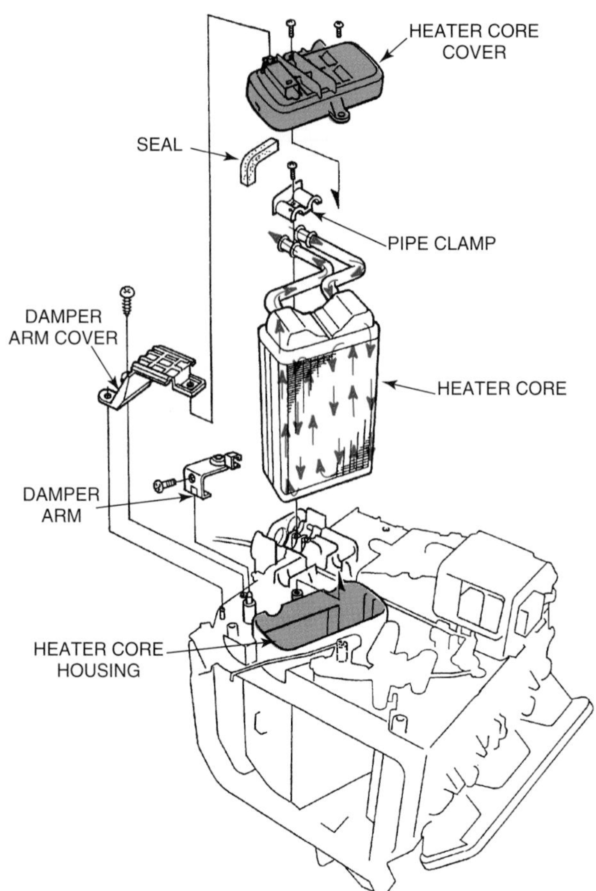

Figure 28-32. A heater core illustrating coolant flow and heater core location. Some cores have the coolant flow in an up and down direction, while others travel from side to side. (Honda Motor Co.)

Air Distribution and Control

Modern air conditioning systems incorporate the heater core in the same case as the evaporator. By means of a control door, often called a ***blend door,*** the air can be directed through the heater core, the evaporator, or both. Control of the door may be by manually-operated dash controls or controlled automatically.

The cooled air is passed into the passenger compartment through ***ductwork*** and ***air control doors.*** The ductwork and doors are arranged to allow the air to be recirculated, or to draw in fresh air. Some systems pass all air, either from outside or recirculated, through the evaporator core first. If heating is necessary, heated air is blended in to produce the required outlet air temperature. This method has the advantage of dehumidifying the air, which prevents window fogging during damp weather conditions.

The air is exhausted through floor ducts, dashboard vents, or defroster outlets. The air control doors are usually controlled by ***vacuum diaphragms.*** Most systems are arranged so that at least three air settings are possible:

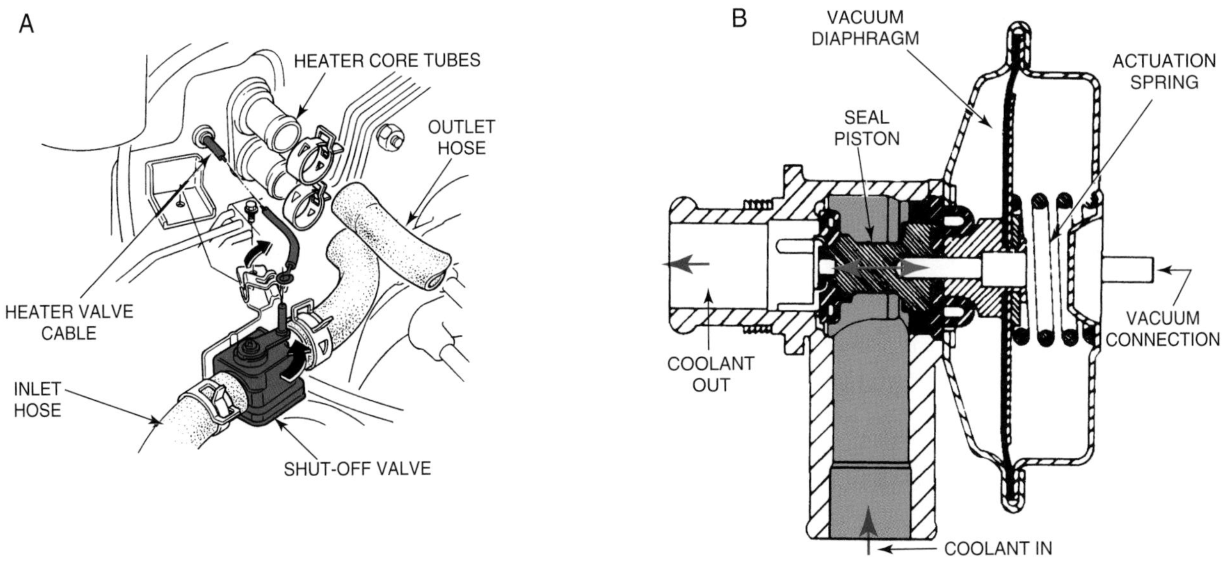

Figure 28-33. A—A manually controlled heater hose coolant shut-off valve. B—Cutaway of a heater hose shut-off valve. This valve is vacuum controlled. The valve is shown in the closed position. (AC-Delco and Honda Motor Co.)

- Air within the vehicle can be recirculated.
- All air entering the vehicle can be drawn from outside.
- Some outside air can be mixed with inside air that is recirculating.

In addition, most systems make provisions for defrosters, blower operation through the vents when the air conditioner is not used, heater operation, and bi-level operation (air blows through vents, heater outlets, and defroster outlets). A few systems can separately control the air temperature that comes out of the driver and passenger vents.

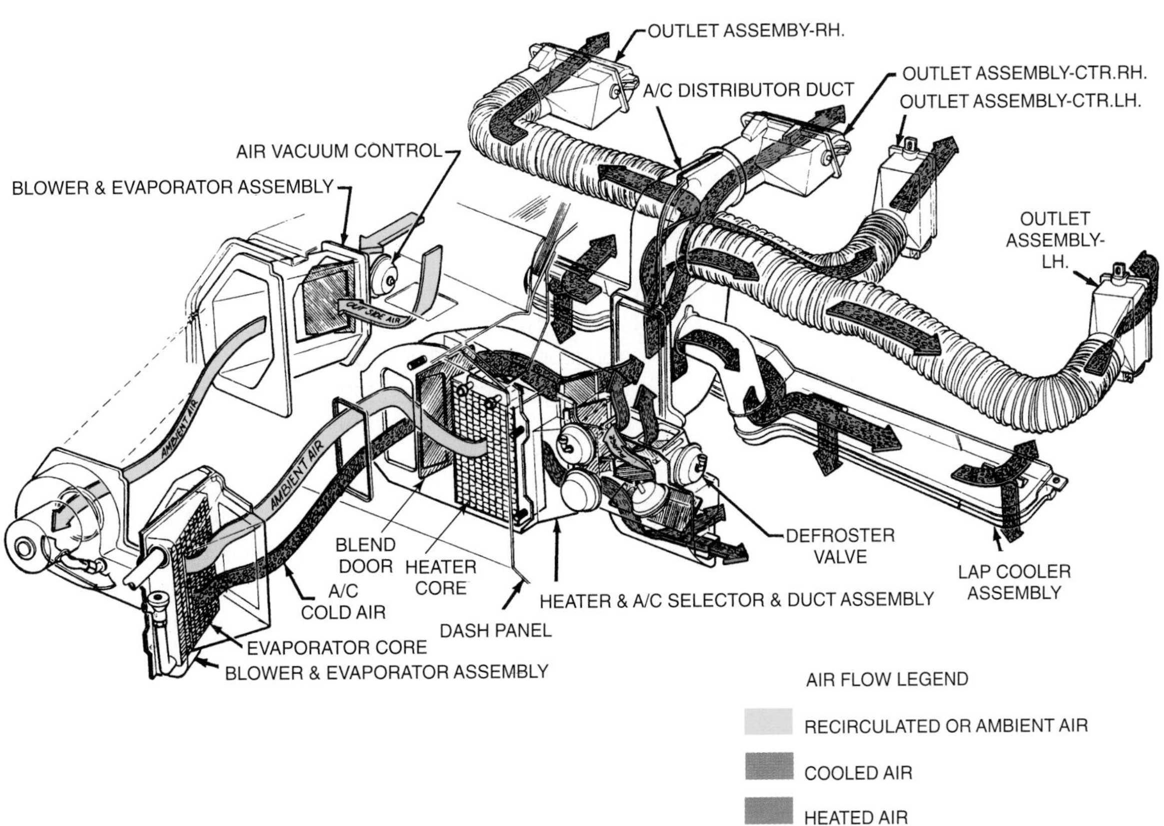

Figure 28-34. Schematic showing air conditioning system airflow. Air amount, direction, and blending are controlled by ducts, doors, flaps, louvers, speed control switches, and vacuum controls.

A schematic illustrating a typical system air flow pattern is shown in **Figure 28-34.** Note that all air is first passed through the evaporator core. Remember that air flow varies widely, depending upon the complexity of each manufacturers' individual system.

Heater-Air Conditioning Control Systems

Modern heating and air conditioning systems have elaborate control systems using electrical switches and relays, electronic sensors and controls, vacuum diaphragms, reser-

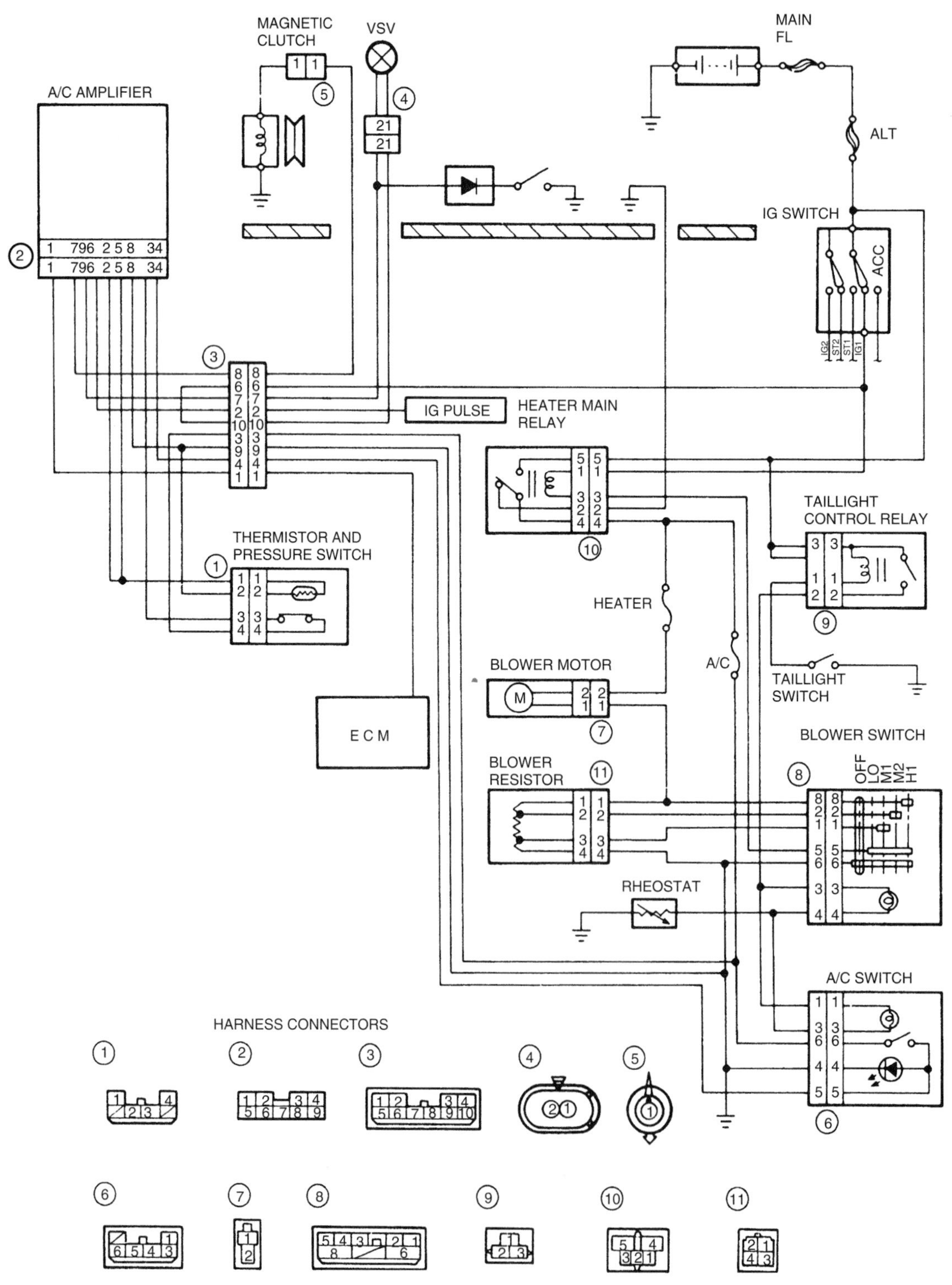

Figure 28-35. Typical wiring diagram for one specific air conditioning system. Such diagrams are very important in diagnostic and service work. (Toyota)

voirs, hoses, and mechanical linkage. All of these controls work together to operate the heating and air conditioning system. Modern auto air conditioning systems, especially in the computer-controlled systems employing ***thermistors,*** have relatively complex wiring and vacuum control systems. **Figure 28-35** shows a wiring schematic for a cycling clutch air conditioning system.

Wiring and Vacuum Systems

The air mixture or blending abilities of an air flow control system, such as illustrated in **Figure 28-36,** permits a discharge of temperature-regulated air into the vehicle. In hot weather, this blending system will permit the evaporator to operate constantly near the freezing point, permitting maximum efficiency.

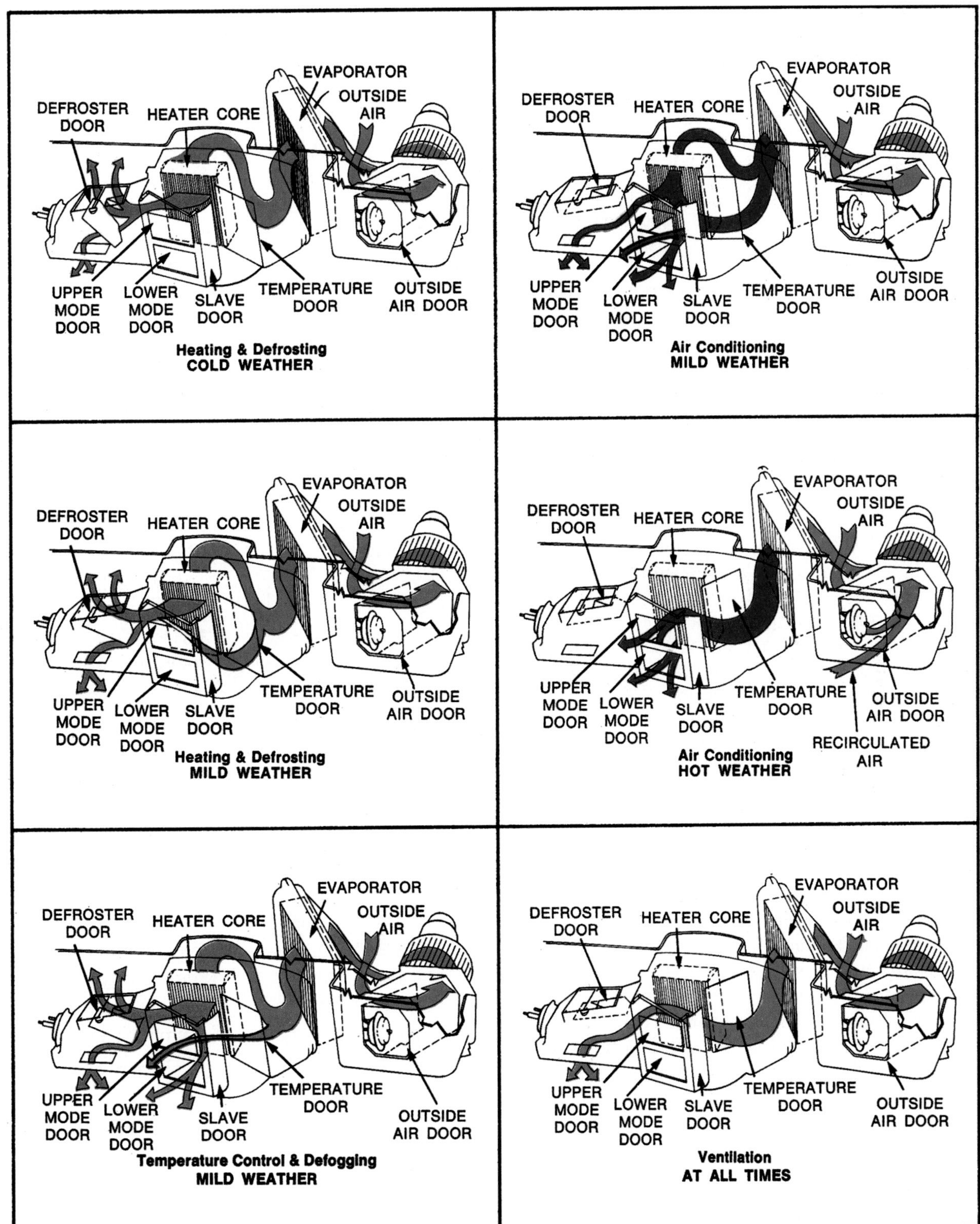

Figure 28-36. One particular air mixing or blending system illustrating air flow and mixture during various settings and outside weather conditions. (Harrison Radiator Division)

In colder weather, the controls can be adjusted to produce warm air from the same ducts. When the weather is cold and humid, the system can be adjusted to use the evaporator to remove humidity from the incoming air, and the heater core to warm it. This air is then directed to the windshield to reduce fogging.

Summary

Air conditioning is the process of cooling, cleaning, and dehumidifying air before it enters the passenger compartment. Cooling is a process of removing heat from an object. Technically speaking, there is no such thing as cold, only degrees of heat. Heat is a form of energy that can affect a temperature change in any object.

Heat produces various effects. Expansion, contraction, vaporization, burning, and freezing result from the removal or addition of heat. Heat transfer occurs through convection, conduction, or radiation. Heat always transfers from a hot object to a colder object.

Latent heat is the hidden heat absorbed during a change of state. Sensible heat is heat that can be felt as a change in temperature. When water is changed from a liquid to a vapor, a great deal of heat can be applied without a change in temperature. This latent heat is hidden in the vapor.

Refrigerant is used in refrigeration systems. It changes state at pressures which make it suitable for use in a refrigeration system. It is colorless, tasteless, and heavier than air in either the liquid or vapor state. It must be handled with great care.

With the system operating, high pressure liquid refrigerant moves to the flow restrictor which meters a certain amount of low pressure liquid refrigerant into the evaporator. As the low pressure liquid refrigerant moves through the evaporator, it begins to boil. Heat is absorbed from the passing air. When refrigerant reaches the evaporator outlet, it is completely vaporized. The vapor contains latent heat. From the evaporator, the refrigerant vapor is drawn into the compressor. The compressor raises the pressure and temperature of the vapor. The vapor still retains all the latent heat of vaporization that it picked up on its way through the evaporator.

When the hot, high pressure refrigerant vapor enters the condenser, it begins to give up heat to the airstream moving over the condenser fins. In the condenser, it loses sufficient heat to return to the liquid state. Still under pressure from the compressor, the liquid refrigerant travels from the condenser to a flow control device to start another cycle.

This continual cycling of liquid refrigerant to the evaporator, where it vaporizes and is drawn out, compressed and condensed, produces a steady cooling effect on the air passing through the evaporator core. In this way, heat is taken from air entering the vehicle and is discharged into the atmosphere at the condenser.

The automotive refrigeration system consists of a compressor, condenser, flow restrictor, evaporator, blower motor, and lines and hoses. Other parts which may be included in the refrigeration system are the receiver-dehydrator, accumulator, evaporator pressure control, control switches, sight glass, and mufflers.

Various switches are installed on the refrigeration system to protect the compressor and improve refrigeration system action. Other switches protect the vehicle engine. Switches can also be used to shut off the compressor during heavy acceleration and during times when ambient air temperature is below certain limits.

The heater system uses the heat from the engine to warm the passenger compartment. Many heater cores are equipped with shut-off valves to reduce unwanted heat. Refrigerant is dangerous. Observe all safety precautions when handling refrigerant, or when working on or near the system.

Know These Terms

Refrigeration system
Cooling
Cold
Heat
Conduction
Convection
Infrared rays
Radiation
State
Solid
Liquid
Gas
Latent heat
Change of state
Sensible heat
Vaporize
Condense
Pressure-temperature relationship
Refrigerant
R-134a
R-12
Freon
Compressor
Radial
Axial
Variable displacement compressor
Pressure relief valve
Magnetic clutch
Condenser
Flow restrictor
Expansion valve
Fixed orifice tube
Evaporator
Dehumidify
Blower
Squirrel cage
Blower motor
Blower resistors
Blower relay
Refrigeration line
Refrigeration hose
Muffler
Receiver-dehydrator
Desiccant
Sight glass
Accumulator
Evaporator pressure control
Hot gas bypass valve
Suction throttling valve
Pilot operated absolute valve
Thermostatic compressor switch
Low pressure switch
Full throttle cut-out switch
Wide open throttle switch
Ambient air temperature switch
Idle speed-up solenoid
Anti-dieseling solenoid
Service valves
Schrader valve
Refrigerant oil
Mineral oil
Ester oil
Polyalkylene glycol oil (PAG)
Phosgene gas
Refrigerant recovery system
Retrofitting
Heater core
Heater hoses
Shut-off valve
Blend door
Ductwork
Air control doors
Vacuum diaphragm
Thermistor

Review Questions—Chapter 28

Do not write in this book. Write your answers on a separate sheet of paper.

1. A good air conditioning system will _____, _____ and _____ the air entering the vehicle.
2. Cold and hot are actually different degrees of _____ .
   (A) state
   (B) heat
   (C) moisture content
   (D) Both A & B.
3. All of the following statements about heat transfer are true, EXCEPT:
   (A) some substances transfer heat much more rapidly than others.
   (B) heat can be transferred from one object to another by the surrounding air.
   (C) heat transferred by convection travels at the speed of light.
   (D) the sun heats the earth because of radiation.
4. All matter exists in one of three states. Name them.
5. Explain the difference between latent heat and sensible heat.
6. If all of the latent heat was taken from steam at 212°F (100°C), what would it become?
   (A) Steam at 211°F (99.45°C).
   (B) Water at 213°F (100.56°C).
   (C) Water at 212°F (100°C).
   (D) Water at 211°F (99.45°C).
7. Inside of the condenser of an operating refrigeration system, what happens to the refrigerant?
   (A) It gives off heat to the outside air.
   (B) It absorbs heat from the outside air.
   (C) It changes from a vapor to a liquid.
   (D) Both A & C.
8. Technician A says that heat is always transferred from a hot object to a colder one. Technician B says that conduction is heat transfer through the surrounding air. Who is right?
   (A) A only.
   (B) B only.
   (C) Both A & B.
   (D) Neither A nor B.
9. Name the most common types of refrigerant.
10. Of the two refrigerants above, which one is becoming obsolete?
11. What is the purpose of the receiver in the system?
   (A) Store liquid refrigerant.
   (B) Remove moisture from the refrigerant.
   (C) Cool the refrigerant.
   (D) Both A & B.
12. What does the presence of bubbles in the sight glass mean?
   (A) Refrigerant level is too low.
   (B) Refrigerant level is too high.
   (C) The wrong oil is being used.
   (D) Either A or C.
13. The evaporator removes _____ from the air.
   (A) heat
   (B) moisture
   (C) dust
   (D) All of the above.
14. The refrigerant entering the compressor is a _____ .
   (A) high pressure vapor
   (B) low pressure vapor
   (C) high pressure liquid
   (D) low pressure liquid
15. The refrigerant leaving the condenser is a _____ .
   (A) high pressure vapor
   (B) low pressure vapor
   (C) high pressure liquid
   (D) low pressure liquid
16. The refrigerant entering the evaporator is a _____ .
   (A) high pressure vapor
   (B) low pressure vapor
   (C) high pressure liquid
   (D) low pressure liquid
17. Technician A says that the compressor magnetic clutch cycles on and off on some air conditioner refrigeration systems. Technician B says that the compressor pumping capacity can be varied on some air conditioner refrigeration systems. Who is right?
   (A) A only.
   (B) B only.
   (C) Both A & B.
   (D) Neither A nor B.
18. Name the valves which have been used in the past to prevent evaporator coil icing.
19. What are the discharge and suction service valves used for?
20. Technician A says that accumulators are used only on vehicles with fixed orifice refrigeration systems. Technician B says that the system desiccant can be located in the accumulator or evaporator. Who is right?
   (A) A only.
   (B) B only.
   (C) Both A & B.
   (D) Neither A nor B.

21. The compressor can be automatically disengaged (magnetic clutch de-energized) under all of the following conditions, EXCEPT:
    (A) wide open throttle operation.
    (B) engine overheating.
    (C) loss of refrigerant charge.
    (D) low ambient air temperature.
22. What is the purpose of the blower resistors?
23. The accumulator is located _____ .
    (A) at the inlet to the evaporator
    (B) at the outlet of the evaporator
    (C) at the inlet to the condenser
    (D) at the outlet of the compressor
24. The fixed orifice type expansion tube reduces refrigerant pressure before it enters the _____ .
    (A) compressor
    (B) condenser
    (C) evaporator
    (D) None of the above.
25. The air control doors are usually operated by _____ .
    (A) duct air flow
    (B) vacuum diaphragms
    (C) electric solenoids
    (D) electric motors

29

ASE Certification

After studying this chapter, you will be able to:

- Explain why technician certification is necessary.
- Explain the process of registering for ASE tests.
- Explain how to take the ASE tests.
- Identify typical ASE test questions.
- Explain what is done with ASE test results.

This chapter will explain the purposes of the National Institute for Automotive Service Excellence (ASE). Also discussed are the advantages of ASE certification. This chapter also explains how to apply for and take the ASE tests. After finishing this chapter, you will be familiar with the purposes of ASE, the general layout of ASE tests, how to take the ASE tests, and the purpose and distribution of ASE test results.

Reasons for ASE Tests

The concept of setting standards of excellence for skilled jobs is not new. Ancient societies had associations of workers who set standards and enforced rules of conduct. Many modern labor unions are descended from early associations of skilled workers. Certification processes for aerospace workers and electronics technicians have existed since the beginnings of these industries.

Due to the fragmented, decentralized nature of the automotive repair industry, standards were difficult to establish and enforce. Anyone could claim to be an automotive technician, no matter how unqualified. A large segment of the public came to regard technicians as unintelligent, dishonest, or both.

The ***National Institute for Automotive Service Excellence***, now called ***ASE***, was established in 1975 to provide a ***certification*** process for automobile technicians. ASE is a non-profit corporation formed to encourage and promote high standards of automotive repair service. ASE does this by providing a series of written tests on various subjects in the Automotive Repair, Truck Repair, Body/Paint, Engine Machinist, and Parts Specialist areas. New tests in Advanced Engine Performance and Compressed Natural Gas Vehicles are now being offered.

These tests are called ***standardized tests***, which means that the same test in a particular subject is given to each participant. Any person passing one or more of these tests and meeting certain experience requirements, is certified in the subject covered by that test. If a technician can pass all of the tests in the automotive, heavy truck, or body/paint areas, he or she is certified as a ***master technician*** in that area.

The ASE certification test program is designed to identify and reward skilled and knowledgeable technicians. Periodically, technicians must recertify, which provides an incentive for updating skills and for keeping up with current technology. ASE certification allows potential employers and the consumer to identify good technicians, while helping the technician advance his or her career.

The ASE test and certification program is not mandatory on a national level. However, some states are considering legislation to make certification a requirement, and many repair shops now hire only ASE certified technicians. Current ASE certification is necessary to attend most manufacturer sponsored or state vehicle inspector training classes. Over 500,000 persons are now ASE certified in one or more areas.

Other ASE activities include encouraging the development of effective automotive service training programs, conducting research on the best instruction methods, and publicizing the advantages of technician certification.

The ASE certification program has brought many advantages to the automotive industry, including increased respect and trust of automotive technicians, at least of those who are ASE certified. This has resulted in better pay and working conditions for technicians and increased standing in the community. Because of ASE, automotive technicians are taking their place next to other skilled artisans.

Applying for the ASE Tests

Anyone may apply for and take an ASE test. To become certified, however, the applicant must have two years of work experience as an automobile, truck, or auto body technician. In some cases, training programs or courses, an apprenticeship program, or time spent performing similar work may be substituted for all or part of the work experience. One exception to these guidelines is the Advanced Engine Performance test, which is administered only to technicians currently certified in Auto Engine Performance.

ASE tests are given in May and November. Tests are usually held during a two week period at night during the work week. The tests are given by a separate organization called ACT. ACT is a non-profit organization experienced in administering standardized tests. The tests are given at designated test centers in over 300 locations. If necessary, special test centers can be organized in remote locations. However, there must be enough potential applicants for the establishment of a special test center to be practical.

To apply for the ASE tests, begin by obtaining an ***application form*** like the one shown in **Figure 29-1**. Contact ASE at the following address to obtain the most current application form:

National Institute for Automotive Service Excellence
13505 Dulles Technology Drive
Herndon, VA 22071

ASE will send the proper form, which is enclosed in a ***registration booklet*** explaining how to complete the form. The form should be filled out carefully, recording all needed information. You may apply to take as many tests as you wish, one test, a few tests, or all of the tests being given. Proof of work experience, or qualified substitutes, should also be included with the form.

Determine the closest test center and record its number in the appropriate space. Most test centers are located at local colleges, high schools, or vocational schools. A fee is charged to register for the test series and for each test to be taken. The current fee structure is in the registration booklet and all fees must be included with the application. In some cases, your employer may pay the registration and test fees before the test or reimburse you based on the number of tests passed.

To be accepted for either the May or November ASE tests, the application and payment must arrive at ASE headquarters at least one month before the test date. Since all applications are handled on a first come-first serve basis, it should be sent in as early as possible.

After sending the application and fees, you should receive an ***admission ticket*** to the test center by mail within two weeks. Contact ACT using the phone number given in the registration booklet if the admission ticket has an error or has not arrived and it is less than two weeks until the first test date.

If your first choice for a test center is filled when ACT receives the application, you will be directed to report to the nearest center that has an opening. You should contact ACT immediately if it is not possible to go to the alternate test center that was assigned.

ASE

Registration Form—ASE Tests

Important: Read the Booklet first. Then print neatly when you fill out this form.

1. a. Social Security Number: ___-__-____

1. b. Telephone Number: (During the day) ___ ___-____
Area Code Number

1. c. Previous Tests: Have you ever registered for any ASE certification tests before? ❑ Yes ❑ No

2. Last Name: **First** **Middle Initial**

3. Mailing Address:
Number, Street, and Apt. Number
City
State Zip Code

4. Date of Birth: Month Day Year

5. Sex: Blacken one Circle ❍ Male ❍ Female

6. Race or Ethnic Group: Blacken one circle. (For research purposes. It will not be reported to anyone and you need not answer if you prefer.)

1 ❍ American Indian 2 ❍ Black 3 ❍ Caucasian/White 4 ❍ Hispanic/Mexican
5 ❍ Oriental/Asian 6 ❍ Puerto Rican 7 ❍ Other (Specify)

7. Education: Blacken only one circle for the highest grade or year you completed.

Grade School and High School (including Vocational): ❍ 7 ❍ 8 ❍ 9 ❍ 10 ❍ 11 ❍ 12
After High School: Trade School or Technical (Vocational) School: ❍ 1 ❍ 2 ❍ 3 ❍ 4
College: ❍ 1 ❍ 2 ❍ 3 ❍ 4 ❍ More

8. Employer: See pp. 18-19 for the code which describes your employer.

9. Test Center: (See pp 20-21.) Center Number City State

10. Tests: Blacken the circles (regular) or squares (recertification) for the test(s) you plan to take. (Do not register for a recertification test unless you passed that regular test before.) **Note:** Do not register for more than four regular tests on any single test day.

Regular	Recertification	Regular	Recertification
❍ A2 Auto: Automatic Trans/Transaxle	❑ E	❍ T3 Med/Hvy Truck: Drive Train	❑ L
❍ A3 Auto: Manual Drive Train and Axles	❑ F	❍ T4 Med/Hvy Truck: Brakes	❑ M
❍ A6 Auto: Electrical/Electronic Systems	❑ G	❍ T5 Med/Hvy Truck: Suspension/Steering	❑ N
❍ A7 Auto: Heating & Air Conditioning	❑ H	**November 9**	
❍ M1 Machinist: Cylinder Head Specialist		❍ T1 Med/Hvy Truck: Gasoline Engines	❑ I
❍ M2 Machinist: Cylinder Block Specialist		❍ T2 Med/Hvy Truck: Diesel Engines	❑ J
❍ M3 Machinist: Assembly Specialist		❍ T6 M/Hvy Truck: Elec./Electronic Systems	❑ K
		❍ F1 Alt. Fuels: Lt. Vehicle CNG	
❍ A1 Auto: Engine Repair	❑ A	❍ B2 Body: Painting & Refinish	❑ P
❍ A8 Auto: Engine Performance	❑ B	❍ B3 Body: Nonstructural Analysis	
❍ A4 Auto: Suspension & Steering	❑ C	❍ B4 Body: Structural Analysis	
❍ A5 Auto: Brakes	❑ D	❍ B5 Body: Mechanical & Electrical Components	
❍ L1 Adv. Level: Auto Adv. Engine Perf. Spec.			

11. Fees:

Number of **Regular** tests **(except L1)** marked above_____ x $15 = $_______
If you are taking the L1 Advanced Level Test, add $40 here + $_______
If you marked **any** of the **Recertification** tests, add $20 here + $_______
Registration Fee + $ 20
TOTAL FEE = $_______

12. Fee Paid By: 1 ❍ Employer 2 ❍ Technician

13. Experience: Blacken one circle. (To substitute training or other appropriate experience, see page 3.)

1 ❍ I certify that I have two years or more of full-time experience (or equivalent) as an automobile, medium/heavy truck, or collision repair/refinish technician or as an engine machinist. (Fill in 14 below.)
2 ❍ I don't yet have the required experience. (Skip 14 below.)

14. Job History: If you claimed two years of experience in 13 above, list **all** your employers for the two years. List your current employer first. Then list your previous employer. Your experience will **not** be accepted without job details listed for every employer. If self-employed, list the name and address of your shop. **Important:** If you've had more than two employers during the two years, list the other employer(s) on a separate sheet. Attach the sheet to this form.

1 ______________ From ______ To ______
Current Employer Month Year Month Year
Number & Street
City State Zip Code Telephone Number

Job Details: In this space, give details about the kinds of repair work you have done during the two-year qualifying period listed above. (If you sent in the details earlier, you can leave this space blank.)

2 ______________ From ______ To ______
Previous Employer Month Year Month Year
Number & Street

Job Details: In this space, give details about the kinds of repair work you have done during the two-year qualifying period listed above. (If you sent in the details earlier, you can leave this space blank.)

15. Authorization: **By signing and sending in this form, I accept the terms set forth in the *Registration Booklet* about the tests and the reporting of results.**

Signature of Registrant Date

Figure 29-1. A typical registration form for the National Institute for Automotive Service Excellence or ASE. (ASE)

Taking the ASE Test

Arrive at the test center early; no one will be permitted to enter once the test starts. Bring your admission ticket and a drivers' license or other photographic identification to the test center. In addition to these items, bring some extra Number 2 pencils. Although pencils will be made available at the test center, the extra pencils will save you time if one breaks.

Listen to and follow all instructions given by the test administrators. During the actual test, carefully read all test questions before making a decision as to the proper answer. The ASE tests are designed to measure three things:

- Basic information on how automotive systems and components work.
- Diagnosis and testing of systems and components.
- Repairing automotive systems and components.

Each ASE test will contain between 40 and 80 test questions, depending on the subject. All test questions are ***multiple-choice***, with four possible answers. These types of multiple choice questions are similar to the multiple-choice questions used in this textbook.

Test Question Examples

The following section discusses the types of questions that you may encounter. Studying these sample test questions will help you with the real ASE test questions. Study the sample test questions that follow. Try to answer them and then study the explanations following them.

One-Part Question

1. The cooling system part that allows the engine to warm up quickly is the _____ .
   (A) water pump
   (B) coolant recovery reservoir
   (C) thermostat
   (D) radiator

Notice that the question calls for the best answer out of all of the possibilities. The thermostat is the part that remains closed to prevent coolant flow through the radiator until the engine warms up. Therefore, "C" is correct.

Two-Part Question

1. Technician A says that the ignition coil changes high voltage into low voltage. Technician B says that the distributor pickup coil produces a low voltage signal. Who is right?
   (A) A only.
   (B) B only.
   (C) Both A and B.
   (D) Neither A nor B.

This question asks you to read two statements and decide if they are true. Both statements can be true, both can be false, or only one of them can be false. In this case, the statement of technician A is wrong, since the ignition coil produces high voltage from low voltage. The statement of technician B is correct, since the pickup coil does produce a low voltage signal. Therefore, the correct answer is "B."

Negative Questions

Some questions are called ***negative questions***. These questions ask you to identify the *wrong* answer from several choices. They will usually have the word "except" in capital letters in the question. An example of a negative question is given below.

1. All of the following valves can be installed as part of the air injection (smog) pump, EXCEPT:
   (A) diverter valve.
   (B) air switching valve.
   (C) check valve.
   (D) EGR valve.

Since the EGR valve is not part of the air injection system, while the other valves are, the correct answer is "D."

A variation of the negative question will use the word "least," such as the one below.

1. Which of the following rear-wheel drive shafts is the LEAST likely to be found on a late model vehicle?
   (A) Hotchkiss.
   (B) Hotchkiss with constant-velocity universal joint.
   (C) Torque tube.
   (D) Two-piece Hotchkiss.

In this case, the drive shaft that is least likely to be found on a late model vehicle is the torque tube drive. Therefore, the correct answer is "C."

Incomplete Sentence Questions

Some test questions are ***incomplete sentences***, with one of the four possible answers correctly completing the sentence. An example of an incomplete sentence question is given below.

1. The coolant temperature sensor is used to measure _____ temperature.
   (A) exhaust gas
   (B) engine
   (C) incoming air
   (D) ambient (outside) air

Once again the question calls for the best answer. The coolant temperature sensor measures the temperature in the engine by monitoring coolant temperature, so "B" is correct.

After completing all the questions in a particular test, each answer should be checked one time. This ensures that you missed nothing that would change an answer and that no careless errors were made on the answer sheet. In most cases, rechecking your answers more than once is unnecessary; many times this results in changing correct answers to incorrect ones. The time allowed for each test session is usually about four hours. However, you may leave after completing your last test and handing in all test material.

Test Results

ACT requires six to eight weeks to receive the tests from the various centers, process and grade them, and mail the results. You will receive a confidential report of your performance on the tests, either "pass" or "more preparation needed." An official diagnostic report will follow in about two to three weeks. The diagnostic report will show your actual score on the tests.

The test questions are subdivided into general areas to help determine any areas in which you need to study. For instance, the engine performance test questions will be divided into such subsections as ignition, fuel and exhaust, emission control, engine mechanical and electrical, and computerized engine control.

Included with the report is a certificate of certification for all of the tests that you have passed, an ASE shoulder patch, and a pocket card listing all of the areas that you are certified in. See **Figure 29-2**.

All ASE test results are confidential information and ASE will only provide them to the person who took the test. This is true even if your employer has paid the test fees. The only test information that ASE will release is to confirm to an employer that you are certified in a particular skill area. Test results will be mailed to your home address. If you wish your employer to know exactly how you performed on the tests, you must provide him or her with a copy of your test results.

Figure 29-2. An ASE certified master auto technician. The ASE patch shows dedication and knowledge. Technicians who earn them should wear them with pride. (Hunter Engineering)

If you fail a certification test, you can retake it again as many times as you would like. However, you (or your employer) must pay all of the applicable registration and test fees again. You should study all available information in the areas where you did poorly. A copy of the ASE Preparation Guide may be helpful to sharpen your skills in these areas. The ASE Preparation Guide is free and can be obtained by filling out the coupon at the back of the registration booklet.

Recertification Tests

Once you have passed the certification test in any area, you must take a ***recertification test*** every five years to keep your certification. This ensures that your certification remains current, and is proof that you have kept up with current technology. The process of applying to take the recertification tests is similar to that for the original certification tests. Use the same form and enclose the proper recertification test fees. If the person's certification has lapsed, they must take the regular certification test.

Summary

The automotive industry was one of the few major industries that did not have testing and certification programs. This lack of professionalism in the automobile industry led to poor or unneeded repairs as well as decreased status and pay for automobile technicians. The National Institute for Automotive Service Excellence, or ASE, was formed in 1975 to overcome these problems. ASE tests and certifies automotive technicians in major areas of automotive repair. This has increased the skill level of technicians, resulting in better service, and increased benefits for technicians.

ASE tests are given in the Spring and Fall. Anyone can register to take the tests by filling out the proper registration form and paying the test fees. The applicant must also select the test center that he or she would like to go to. To be considered for certification, the applicant must have two years of hands-on experience as an automotive technician. About three weeks after applying for the test, the applicant will receive a test entry ticket which he or she must bring to the test center.

The actual test questions will test your knowledge of general system operation, diagnosing problems, and repair techniques. All of the questions are multiple choice questions. The questions must be read carefully. The entire test should be reviewed one time only to catch careless mistakes.

Test results will arrive within six to eight weeks after the test session. Results are confidential, and will be sent only to the home address of the person who took the test. If a test was passed, and the experience requirement has been met, the technician will be certified for five years. Anyone who fails a test can take it again in the next session. Tests can be taken as many times as necessary. Recertification tests can be taken at the end of the five year certification period.

Know These Terms

National Institute for Automotive Service Excellence
ASE
Certification
Standardized tests
Master Technician
Application form
Registration booklet
Admission ticket
Multiple choice
Negative question
Incomplete sentences
Recertification test

Review Questions—Chapter 29

Do not write in this book. Write your answers on a separate sheet of paper.

1. Technician A says that ASE encourages high standards of automotive service and repair by providing classes in automotive service. Technician B says that all ASE tests are multiple choice written tests. Who is right?
   (A) A only.
   (B) B only.
   (C) Both A & B.
   (D) Neither A nor B.
2. An ASE Master Automobile Technician has passed _____ tests in the automotive test series.
   (A) at least two
   (B) at least four
   (C) all eight
   (D) one
3. ASE tests are given _____ each year.
   (A) once
   (B) twice
   (C) four times
   (D) twelve times
4. The advantages that ASE certification has brought to automotive technicians include all of the following, EXCEPT:
   (A) increased respect.
   (B) better working conditions.
   (C) lower pay.
   (D) increased standing in the community.
5. Technician A says that a promise to work in the automotive field for two years can be substituted for the minimum experience requirement. Technician B says that some training programs or courses can be substituted for the experience requirement. Who is right?
   (A) A only.
   (B) B only.
   (C) Both A & B.
   (D) Neither A nor B.
6. ASE tests are held _____ .
   (A) during normal working hours
   (B) at night during the work week
   (C) on national holidays
   (D) any time
7. ASE tests are designed to measure your knowledge of what three things?
8. ASE test questions resemble the ones in _____ .
   (A) college level courses
   (B) essay-type tests
   (C) verbal examinations
   (D) this textbook
9. A technician can retake any certification test _____ times.
   (A) two times
   (B) four times
   (C) five times
   (D) any number of times
10. ASE provides test results to _____ .
    (A) the technician who took the test
    (B) whoever paid for the test
    (C) the technician's employer
    (D) Both A & C.

In order for the modern technician to be successful, a good education is vital. Even after graduation, technicians must attend training courses in order to keep pace with the many changes in automotive systems. (Ford Motor Co.)

30

Career Opportunities

After studying this chapter, you will be able to:

- Describe the various types of careers available in automotive service.
- Compare the different types of specialized automotive technicians.
- Identify sources of training in automotive service.
- Explain why it is important to finish school before seeking a full-time job as a technician.

This chapter will discuss the various jobs available in the automotive field one. Upon completion of this chapter you will know what steps to take to become an automotive technician.

A Vast Field

The very essence of economic life and growth depends in a great part upon the continued improvement and advancement of the automobile. The automotive field has become so large that today it employs about one out of every seven workers. The possibilities are almost unlimited if you are willing to apply yourself, get the proper training, and keep up with the changes in the field of automotive technology.

Types of Automotive Service Careers

This chapter will concern itself with careers directly related to automotive maintenance and repair. The typical automotive service career can involve working on automobiles and light trucks, heavy trucks, tractor-trailers, construction equipment, farm equipment, and small engines.

Light Repair Technician

As the term implies, the ***light repair technician*** performs minor repair and service tasks, such as exhaust system service, cooling system service, etc. The light repair technician may check out new vehicles before delivery to make certain that all systems are functioning and in proper adjustment.

The light repair technician is often responsible for performing periodic services and checkups that are normally made after a certain mileage has been reached. The technician would then discuss the vehicle's performance with the owner and check up on any complaints regarding its operation.

The experience gained while performing this job, even though it generally involves minor repair and adjustment, is invaluable. It will form the foundation for more complicated repair work or for specialty work, such as driveability, alignment, engine overhaul, or brake work.

Heavy Repair Technician

The ***heavy repair technician*** services, dismantles, checks, repairs, and reassembles engines, transmissions, transaxles, and differentials. See **Figure 30-1**. This job requires a great deal of study, practice, and experience.

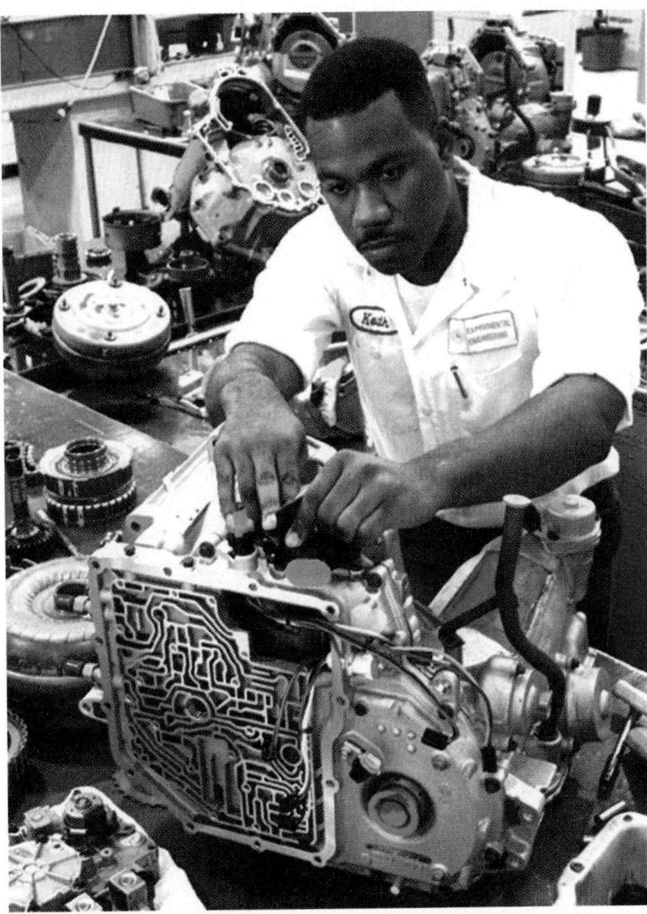

Figure 30-1. This heavy repair technician performs major repair work. Here, the technician is rebuilding a front-wheel drive transaxle. (Ford Motor Co.)

At first, the apprentice technician will work under a qualified and experienced heavy repair technician. Success in this area will largely depend upon your aptitude (natural ability), interest, and ambition. Having mastered the work involved, you will be a valuable asset in any shop and will have placed yourself in a good position for advancement.

Areas of Specialization

Modern automobiles and light trucks are constantly being improved. With the addition of electronic engine and drive train controls, anti-lock brakes, air bags, electronic steering, ride and traction controls, and other new devices and features, it is becoming more and more difficult for a technician to master the service and repair of the entire vehicle.

If the volume of work permits, full-time work in one area will help the technician become highly proficient. Additionally, ***specialization*** in one or more areas will provide the customer with faster and more efficient service. The technician handling the job will have advanced and concentrated training and experience in a chosen specialty. Areas of specialization commonly include driveability, brakes, transmissions/transaxles, alignment, electrical/electronics, air conditioning, and auto body.

Driveability Technician

The ***driveability technician*** diagnoses and isolates engine problems and services the ignition system, fuel system, and many other engine and vehicle systems. A thorough understanding of engine and drivetrain fundamentals, electricity and electronics, and electronic engine controls is a must for the driveability technician. Alternator and starter diagnosis and repair may also be required. Driveability work requires a high level of technical competency. It is a good paying job that provides an excellent chance for further advancement, **Figure 30-2**.

Brake Technician

The ***brake technician*** does such work as disc and drum turning; shoe adjustment; caliper, master cylinder, and wheel cylinder repair; bleeding; and line replacement. Special training in brake system diagnosis, power brake units, anti-lock brakes, and traction control systems is required.

Transmission/Transaxle Technician

The ***transmission/transaxle technician*** must be able to work on both standard and automatic units. This technician will also service electronic transmission/transaxle control systems. Necessary skills include testing, diagnosis, disassembly, repair, and reassembly of transmissions and transaxles.

The transmission/transaxle technician must have a thorough background in the fundamentals of engine and transmission/transaxle operation, as well as special training in this field. This is a highly specialized area, with excellent pay and opportunities for advancement.

Alignment Technician

The ***alignment technician*** performs wheel balancing and wheel alignment. See **Figure 30-3**. This technician also works on the steering gearbox, steering linkage, spindles, springs, ball joints, shocks, struts, and control arms. It is the alignment technician's responsibility to see that the vehicle steers and handles easily and safely, and that the tires run smoothly and wear properly.

To be a successful alignment technician, you must have a thorough understanding of rack-and-pinion and parallelogram steering, MacPherson struts, power steering systems, steering geometry, modern alignment equipment and adjusters, and electronic and pneumatic ride controls.

Electrical/Electronic Technician

The modern vehicle makes wide use of electrical units. The ***electrical/electronic technician*** will handle work on

Figure 30-2. This driveability technician is reinstalling the intake plenum on a fuel injected engine. Other duties include electrical, computer, and emission related repairs. (Black and Decker)

Figure 30-3. An alignment technician makes sure that the steering geometry angles on a vehicle are correct for maximum handling and tire wear. (Hunter)

radios, tape and CD systems, electric seats, window operating devices, instruments, alternators, regulators, and starters. Knowledge of wiring and lighting is also required.

The job of electrical/electronic technician requires an extensive knowledge of electricity and electronics, as well as the repair and adjustment procedures for all electrical and electronic units.

Air Conditioning Technician

The ***air conditioning technician*** checks, adjusts, and repairs the vehicle air conditioning system. There are many jobs for technicians with special training in this area.

Success in this area requires a knowledge of the various types of refrigerants used and methods of recovering and reusing them. The air conditioning technician must also know about heating systems, control systems, and ductwork.

Body and Fender Technician

Body and fender work, although separate from mechanical repair, is an important area of automotive service. The ***body and fender technician*** repairs damage to the vehicle's metal structure, the upholstery, and the inside trim. The body and fender technician installs glass and repairs locks, handles, and various trim pieces.

This job requires proficiency in cutting, welding, metal bumping, filling, priming, and painting. Painting is becoming a specialty within the field of body and fender work.

Supervisory Positions

Upon entering a particular field, most workers look forward to advancement in pay and position. Advancement depends upon your knowledge and skill, as well as your ability to cope with the problems connected with the position.

Shop Supervisor

The ***shop supervisor*** is in charge of the technicians in the repair department. The supervisor is held directly responsible for the work turned out by the technicians. This calls for a highly competent technician who is familiar with and able to perform all jobs that enter the shop.

In larger repair shops, the supervisor's time will be spent directing the technicians: checking their work, giving suggestions, and in general, seeing that the shop runs smoothly and efficiently. In a small shop, the supervisor may spend part of the day doing repair work, with the remainder of the time devoted to supervision.

Service Manager

The ***service manager*** holds a very responsible position. This title is held by the person in charge of the overall service operation. The service manager must see that customers get prompt, efficient, and fair service. The service manager's primary responsibility is to see that customers are pleased and that the technicians and others in the service department are satisfied and doing good work.

The service manager's job usually entails handling employee training programs. This requires close cooperation with factory representatives to see that the latest and best service techniques are employed. The service manager must have insight into the job itself. This calls for leadership ability, a good personality, training, knowledge, and the ability to work with others.

Small Shop Technician

In small shops, the technician must perform several functions. The ***small shop technician*** will often be called upon to meet customers, diagnose problems, and prepare cost estimates and bills. A thorough knowledge of all work done in the shop is required.

Overall Job Picture

Other jobs found in the modern automotive repair facility, in addition to those discussed, are shown on the dealer organization chart in **Figure 30-4**.

Working Conditions

The automotive service center of today is a far cry from the crowded, dark, cold, and poorly ventilated garage of yesterday. The modern shop is well lighted, roomy, and well ventilated. In general, modern automotive service facilities are pleasing places in which to work.

Many modern shops feature lunchrooms, showers, and lockers for their employees. The addition of heavy power equipment has eliminated much of the heavy labor involved and has provided easier and more comfortable access to the various parts of the vehicle.

Most technicians work inside and most modern shops are heated in the winter. However, when a job requires roadside repair, the technician will be faced with varying weather conditions.

Wages

It is difficult to say how much money you will make as an automotive technician. Much depends on the type of work performed, the geographical location of the job, the employer, the prevailing business conditions, and the job supply and demand.

Pay can be based on an hourly wage or it can be based on commission. Under the wage system, the technician is paid by the number of hours he or she is actually at the shop, no matter what the work load. Under the commission, or flat rate system, the amount the technician is paid depends on the work that comes in and how fast he or she can complete it. Many automotive technicians receive a certain percentage of the customer labor charge. Because the rates for various jobs are fixed, the skilled technician will be able to complete more jobs in one day than the unskilled technician and, as a result, earn more money.

Most technicians earn a good living. Many shops pay by a combination of an hourly wage and commission, and most shops offer a package of benefits, such as paid insurance, vacation, and a pension or profit sharing setup. Wages and commissions are competitive between shops in the same area. The earnings of most automotive technicians are in line with the earnings of individuals employed in other technical jobs.

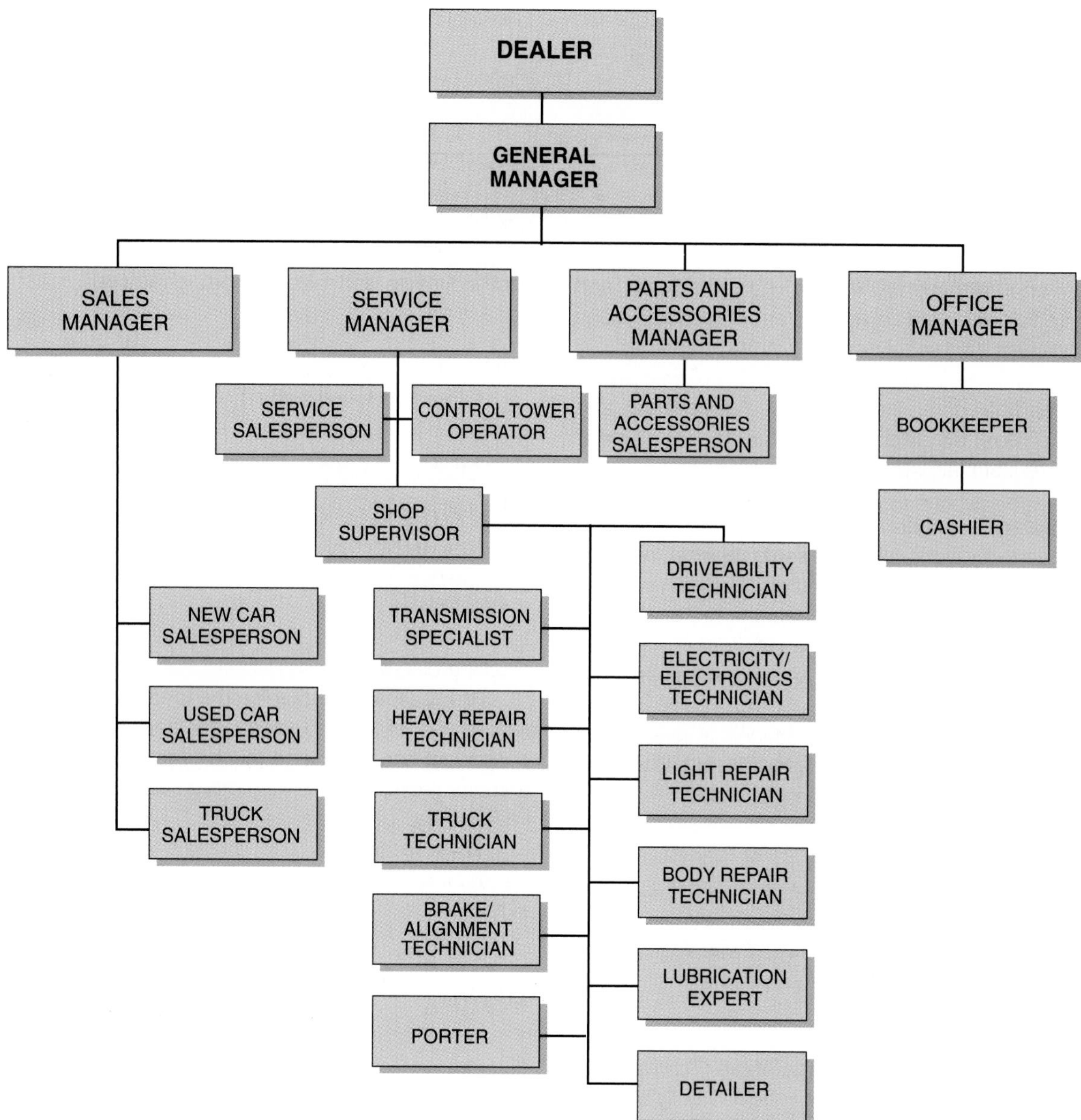

Figure 30-4. This chart shows how a modern automobile dealership is organized.

Working Hours

Working hours per week usually vary from 40 to 48 hours. In some instances, technicians work beyond these hours. If the shop is unionized, overtime salary is generally paid for all hours over those set by the union. Holidays, vacations with pay, and sick time are becoming increasingly available to technicians.

Availability of Employment

The increasing use of vehicles and the addition of more complex systems on vehicles has created a lucrative field for qualified technicians. If you really learn the trade and are a hard working, conscientious person, you should be able to find employment.

About a third of all technicians work in new and used vehicle dealer service departments. Another third work in repair shops that specialize in a single aspect of automotive repair, such as body and fender work, driveability, brake, front end, radiator, and other areas. Many technicians work in service stations that perform light repairs or in small, privately owned shops that perform a variety of repairs. Other technicians work for large fleet owners, such as utility companies, rental companies, dairies, produce companies, and federal, state, county, and city organizations.

If you are sincerely interested in all types of vehicles, enjoy working on and studying about them, and are in relatively good health, the chances are good that you will like this work and do well in it.

How Does a Person Become an Automotive Technician?

A person develops into a good technician by study, observation, instruction, and experience. The traditional method of starting in the automotive field by pumping gas at a local service station has all but disappeared. However, many other opportunities are available and will be discussed below.

Several sound reasons for finishing high school, as well as recommended courses, will be discussed first. Remember that many high schools offer fine training programs in automotive subjects, either on site or at separate vocational training centers. Stay in school and take advantage of them.

Stay in High School

Above all else, graduate from high school. High school dropouts are considered poor risks by employers and for good reasons. The vast majority of students who drop out of school do so for one of several reasons. Some of the reasons why students drop out include:

- They lack the intelligence to compete in school.
- They are just plain lazy.
- They are bored and feel that schoolwork does nothing for them.
- They want to start earning money so that they may feel grown up and independent.
- They cannot get along with others and rebel against authority.

The actual reasons for dropping out may be one or more of those mentioned. Almost all automotive training programs require a high school diploma for admittance. Also, most corporate and manufacturer training courses and programs require a high school diploma.

The Dropout in the Job Market

When you look for a job, you will find that your prospective employer also has high school graduates applying for the job. If high school graduates are available, why should an employer take a chance on a dropout who did not have what it takes to finish high school?

We know that there are some fine, intelligent, hardworking students who drop out of school, but most employers cannot be expected to track down these people. Actual practice has shown that, on the average, it is much safer to hire a high school graduate than a dropout.

National figures on salaries and employment show that high school dropouts are unemployed a great deal of the time. When they do find work, it is usually at a low-paying, low skill job. The world today is moving forward at an ever-increasing speed, and there will be little chance for the unskilled and uneducated dropout. Stay in school.

What to Take While in School

Contrary to what many young people in school think, courses in language arts, math, science, and social studies are extremely valuable to anyone planning to enter one of the trades.

Language Arts

The ability to express yourself, both orally and in writing, is important, since you will have many opportunities to meet the public and to make written reports. The impression you make through your speaking and writing will have a definite bearing on your worth to any employer, as well as on your chances for advancement. Work in your language courses—it will pay off.

Math

You often hear the statement, "What good does math do? I'm going to be an automotive technician." It does a great deal of good. Technicians make many precision measurements and fill out customer bills and cost estimates. A good grasp of mathematics is needed to fully understand automotive theory. Do not forget that there are many times outside the shop when you will need reasonable skill in math. Business math, algebra, geometry, and trigonometry will all be useful tools for your mental toolbox.

Science

Chemistry and physics will be invaluable aids in understanding the construction and operation of the automobile. The vehicle uses the principles of chemistry and physics in hundreds of ways. Also, a good foundation in basic science principles will allow you to better understand the world around you.

Social Studies

A thorough knowledge of our world, its people, its problems, and its aims is essential if you hope to make an intelligent contribution to society in general. This knowledge will set you apart as a knowledgeable person and will allow you to discuss countless things in an intelligent fashion. Your contributions to your work, your country, and your chances of advancement will be greater with study in this area.

Other Important Courses

You should take such elective courses as mechanical drawing, reading, art, electricity and electronics, machine shop, welding, automotive theory, and automotive shop. With the increasing use of computers in automotive systems, as well as in diagnostic equipment, you should try to familiarize yourself with computers as much as possible. Not only will this help you better understand and be able to diagnose computer control systems, it will provide you with an additional job skill.

Use Your Counselor and Instructors

To make certain you are on the right track and that you are getting the classes you need, see your school guidance counselor regularly. Most high school guidance offices will provide you with valuable assistance in determining your assets and liabilities, your intended career path, and the high school courses necessary.

The school guidance counselor can also provide you with information on furthering your training after graduation from high school. Also, talk with your shop instructors, as they often have local contacts and can be of great help in explaining what you will need to know to get and keep your first job.

Studies after Graduation

If you desire additional training after graduation, many educational outlets are available. Vocational schools and colleges, the armed forces, and apprentice programs offer excellent opportunities to acquire the training necessary to gain employment in this field. The following section contains a brief description of each.

Vocational Schools and Colleges

Many ***vocational schools*** (sometimes called technical or trade schools), ***technical colleges,*** and ***junior colleges*** offer excellent courses aimed at teaching automotive technology. Some of these schools are supported by state and local governments, while others are private institutions. Many corporate sponsored apprentice programs are located in these schools. Before selecting any school or college, it is wise to make a careful check of the school itself. You should determine several things:

- Is the school a reputable institution?
- Does it offer the courses you need?
- Does it have quality instructors and plenty of late-model equipment?
- Does it have a job placement agency?
- How far along the road to complete training will it take you?
- What are the costs?
- Does the school offer aid in securing a part-time job if you must work to help pay expenses?

To help you check on these items, contact the school itself, the Better Business Bureau in the city in which the school is located, your high school guidance office, the labor union to which the technicians in the area of the school belong (if applicable), and the Vocational Department of the State Department of Education. A thorough investigation of the school of your choice will be well worth your time.

Armed Forces

If you have decided to join one of the ***armed forces***, it may be possible for you to receive training in the field of automotive technology. The armed forces have top-quality schools. If you have the opportunity to receive the training they offer and the experience of enlistment, you will be well on your way to becoming a good technician. For information regarding training in the armed forces, see your local recruiting agent and your high school guidance counselor.

Apprentice Programs

Many technicians learn the trade by enrolling in a manufacturer or corporate sponsored ***apprentice program***. The apprentice program is designed to teach a specific trade through on-the-job training and study. The apprentice is exposed to all facets of the trade. The program generally includes some off-the-job training in automotive systems, as well as physics, math, and other related subjects.

As apprentices gradually develop skills, they will be exposed to the more technical aspects of the job. In a good training program, the apprentices will learn the trade thoroughly. Upon completion of the apprenticeship, the graduate will be a highly skilled and valued employee. Many graduates are often retained by the shops that they worked at during their apprenticeship.

How to Enter an Apprentice Program

If you are interested in becoming an apprentice, contact employers in the trade you wish to enter. One may have an apprenticeship program and be willing to hire you. The local labor union representing the trade of your choice may also be able to help. Many cities have local joint apprenticeship committees. You may want to contact your local state employment service office.

Most apprentice programs give credit for previous experience and training. How much credit is given will depend on the individual employer or the apprenticeship committee involved.

Qualifications Essential for Apprenticeship

You should have a sincere interest in the trade and be willing to study, observe, take instruction, and work. You should have average or above average ability to work with your hands. Learning a trade will equip you with valuable skills and place you in demand.

Additional Training Information

If you are interested in becoming an automotive technician and think you would like to learn the trade, ask your high school guidance counselors for all the information they have available. They will be glad to help.

A good collection of pamphlets, brochures, training manuals, and other materials will form the basis for a very useful library. Remember that today, as always, learning and earning go hand in hand. The following associations may also provide interesting and informative materials

American Petroleum Institute (API)
1220 L Street NW
Washington, DC 20005

Automotive Safety Foundation
1776 Massachusetts Avenue NW
Washington, DC 20036

Battery Council International (BCI)
401 N. Michigan Avenue
Chicago, IL 60611

Equipment And Tool Institute (ETI)
1806 Johns Drive
Glenview, IL 60025-1657

Mechanics Education Association (MEA)
1805 Springfield Avenue
Maplewood, NJ 07040-2910

Motor And Equipment Manufacturers Association
(MEMA)
P.O. Box 13966
Research Triangle Park, NC 27709-3966

Motor Vehicle Manufacturers Association Of The U.S., Inc.
300 New Center Building
Detroit, MI 48202

National Automobile Dealers Association (NADA)
8400 Westpark Drive
McLean, VA 22102

Professional Mechanics Association (PMA)
P.O. Box 840
Stewartstown, PA 17363-0840

Rubber Manufacturers Association (RMA)
1400 K Street NW, Suite 900
Washington, DC 20005

The following vehicle manufacturers, as well as others, may be able to provide you with brochures, booklets, and other materials.

BMW of North America, Inc.
300 Chestnut Ridge Road
Woodcliff Lake, NJ 07675

Chrysler Corporation
College Graduates:
College Recruitment Services
Personnel Department
Chrysler Corporation
Chrysler Center
12000 Chrysler Drive
Highland Park, MI 48288-1919

High School /Vocational School Graduates:
Manager, Personnel Department
Chrysler Corporation
Chrysler Center
12000 Chrysler Drive
Highland Park, MI 48288-1919

Ford Motor Company
Placement and College Recruitment
Ford Motor Company
The American Road
Dearborn, MI 48121

General Motors Corporation
Salaried Placement Personnel
General Motors Corporation
General Motors Building
3044 W. Grand Blvd.
Detroit, MI 48202

Honda Motor Co., Inc.
1919 Torrance Blvd.
Torrance, CA 90501-2746

Mazda Motor Of America, Inc.
7755 Irvine Center Drive
Irvine, CA 92630

Mercedes-Benz Of North America, Inc.
P.O. Box 350
One Mercedes Drive
Montvale, NJ 07645-0350

Mitsubishi Motor Sales Of America, Inc.
6400 Katella Avenue
Cypress, CA 90630

Nissan Motor Manufacturing Co., USA
P.O. Box 191
18501 S. Figueroa Street
Carson, CA 90248-0191

Porsche Cars North America, Inc.
100 W. Liberty Street
Reno, NV 89501

Saab Cars USA, Inc.
4405-A Saab Drive
P.O. Box 9000
Norcross, GA 30091

Subaru Of America, Inc.
P.O. Box 6000
Subaru Plaza
Cherry Hill, NJ 08034-6000

Toyota Motor Sales USA, Inc.
19001 S. Western Avenue
Torrance, CA 90509

Volkswagen Of America, Inc.
3800 Hamlin Road
Auburn Hills, MI 48236

Volvo Cars of North America, Inc.
P.O. Box 913
One Volvo Drive
Rockleigh, NJ 07647

Plan Your Career Early

Do not "float" through high school wondering what you want to do. Make every effort to set your sights on a goal: college, technical school, armed forces, apprenticeship, or other career decision.

Guidance counselors will be able to help identify your talents and can give helpful information and guidance in choosing an appropriate career. Also discuss your ideas and plans with your automotive instructor, who is vitally interested in this field and is in good position to answer your questions.

Start your high school career by taking subjects that will allow you to select a college, commercial, technical, or vocational course in your junior or senior year. By doing this, you can always change your plans. You will then have a background of studies that will enable you to go any one of several ways.

Keep Up-to-Date

There is little room in this fast moving age for unskilled labor. Unless you learn a trade, you may find yourself out of work most of the time and when you do get employment you may be poorly paid and perhaps doing work that you do not enjoy.

Remember the words ***think***, ***plan***, ***study***, ***work***, ***learn***, ***practice***, and ***question***. These are good words and are significant to anyone entering a trade. If you decide to be an automotive technician, decide to be the best. If you do, these words will become very familiar to you.

Summary

Career opportunities in the automotive field are very diverse. However, to enter this field, you must be willing to learn and apply yourself. Each area of automotive technology requires a good education not only in the systems, but in basic subjects such as math and science.

There are numerous programs that a prospective technician can take to learn their chosen trade. Many programs are located at trade or technical schools. Some are operated by corporate or manufacturer sponsorship. The armed forces also provides another route for receiving technical training. Always think and plan ahead, this will ensure that you have a bright future. Whether you choose to pursue a career as an automotive technician or not, stay in school.

Know These Terms

Light repair technician
Heavy repair technician
Specialization
Driveability technician
Brake technician
Transmission/transaxle technician
Alignment technician
Electrical/electronic technician
Air conditioning technician
Body and fender technician
Shop supervisor
Service manager
Small shop technician
Vocational school
Technical college
Junior college
Armed forces
Apprentice programs
Think
Plan
Study
Work
Learn
Practice
Question

Review Questions—Chapter 30

Do not write in this book. Write your answers on a separate sheet of paper.

1. Which of the following shop employees overhauls engines, transmissions, and other major vehicle components?
   (A) Shop manager.
   (B) Shop foreman.
   (C) Light repair technician.
   (D) Heavy repair technician.
2. The alignment technician or front end technician must know a great deal about _____ .
   (A) ball joint replacement
   (B) steering linkage
   (C) strut rods and control arms
   (D) All of the above.
3. Technician A says that all technicians are paid a combination of salary and commission. Technician B says that all technicians are paid by commission. Who is right?
   (A) A only.
   (B) B only.
   (C) Both A & B.
   (D) Neither A nor B.
4. In a large shop, the shop supervisor may spend part of the day doing all of the following, EXCEPT:
   (A) repairs to vehicles.
   (B) directing the technicians.
   (C) doing paperwork.
   (D) seeing that the shop runs efficiently.
5. All of the following statements about modern repair facilities are true, EXCEPT:
   (A) hydraulic lifts and power tools have removed much of the heavy labor.
   (B) most automotive technicians work inside.
   (C) the service facility of today is crowded, dark, cold, and poorly lighted.
   (D) many modern repair facilities have lunchrooms and showers.
6. Why would an employer prefer to hire a high school graduate?
   (A) Better chance to get a good worker.
   (B) More educated workers.
   (C) Have to pay more.
   (D) Both A & B.
7. In what two areas will a knowledge of mathematics come in handy for the automobile technician?
8. Name three places where you could receive additional training in automotive repair after graduating from high school.
9. Which of the above is designed to teach a specific trade through a combination of on-the-job training and study? _____
10. Most high school guidance offices will provide you with valuable assistance in determining _____ .
   (A) your assets and liabilities
   (B) your intended career
   (C) the high school courses necessary to reach your goals
   (D) All of the above.

Conversion Tables

SOME COMMON ABBREVIATIONS/SYMBOLS			
U.S. CUSTOMARY		METRIC	
UNIT	ABBREVIATION	UNIT	ABBREVIATION
inch	in.	kilometer	km
feet	ft.	hectometer	hm
yard	yd.	dekameter	dkm
mile	mi.	meter	m
grain	gr.	decimeter	dm
ounce	oz.	centimeter	cm
pound	lb.	millimeter	mm
teaspoon	tsp.	cubic centimeter	cm^3
tablespoon	tbsp.	kilogram	kg
fluid ounce	fl. oz.	hectogram	hg
cup	c.	dekagram	dkg
pint	pt.	gram	g
quart	qt.	decigram	dg
gallon	gal.	centigram	cg
cubic inch	$in.^3$	milligram	mg
cubic foot	$ft.^3$	kiloliter	kl
cubic yard	$yd.^3$	hectoliter	hl
square inch	$in.^2$	dekaliter	dal
square foot	$ft.^2$	liter	L
square yard	$yd.^2$	centiliter	cl
square mile	$mi.^2$	milliliter	ml
Fahrenheit	F	square kilometer	km^2
barrel	bbl.	hectare	ha
fluid dram	fl. dr.	are	a
board foot	bd. ft.	centare	ca
rod	rd.	tonne	t
dram	dr.	Celsius	C
bushel	bu.		

MEASURING SYSTEMS

U.S. CUSTOMARY	METRIC
LENGTH	
12 inches = 1 foot	1 kilometer = 1000 meters
36 inches = 1 yard	1 hectometer = 100 meters
3 feet = 1 yard	1 dekameter = 10 meters
5,280 = 1 mile	1 meter = 1 meter
16.5 feet = 1 rod	1 decimeter = 0.1 meter
320 rods = 1 mile	1 centimeter = 0.01 meter
6 feet = 1 fathom	1 millimeter = 0.001 meter
WEIGHT	
27.34 grains = 1 dram	1 tonne =1,000,000 grams
438 grains = 1 ounce	1 kilogram = 1000 grams
16 drams = 1 ounce	1 hectogram = 100 grams
16 ounces = 1 pound	1 dekagram = 10 grams
2000 pounds = 1 short ton	1 gram = 1 gram
2240 pounds = 1 long ton	1 decigram = 0.1 gram
25 pounds = 1 quarter	1 centigram = 0.01 gram
4 quarters = 1 cwt	1 milligram = 0.001 gram
VOLUME	
8 ounces = 1 cup	1 hectoliter = 00 liters
16 ounces = 1 pint	1 dekaliter = 10 liters
32 ounces = 1 quart	1 liter = 1 liter
2 cups = 1 pint	1 deciliter = 0.1 liter
2 pints = 1 quart	1 centiliter = 0.01 liter
4 quarts = 1 gallon	1 milliliter = 0.001 liter
8 pints = 1 gallon	1000 milliliter = 1 liter
AREA	
144 sq. inches = 1 sq. foot	100 sq. millimeters = 1 sq. centimeter
9 sq. feet = 1 sq. yard	100 sq. centimeters = 1 sq. decimeter
43,560 sq. ft. = 160 sq. rods	100 sq. decimeters = 1 sq. meter
160 sq. rods = 1 acre	10,000 sq. meters = 1 hectare
640 acres = 1 sq. mile	

TEMPERATURE

FAHRENHEIT		CELSIUS
32° F	Water freezes	0° C
68° F	Reasonable room temperature	20° C
98.6° F	Normal body temperature	37° C
173° F	Alcohol boils	78.34° C
212° F	Water boils	100° C

USEFUL CONVERSIONS

WHEN YOU KNOW:  MULTIPLY BY:  TO FIND:

WHEN YOU KNOW:	MULTIPLY BY:	TO FIND:
	TORQUE	
Pound - inch Pound - foot	0.11298 1.3558	newton-meter (N•m) newton-meters
	LIGHT	
Foot candles	1.0764	lumen/meters^2 (lm/m^2)
	FUEL PERFORMANCE	
Miles/gallon	0.4251	kilometers/liter (km/L)
	SPEED	
Miles/hour	1.6093	kilometers/hr (km/h)
	FORCE	
kilogram ounce pound	9.807 0.278 4.448	newtons (n) newtons newtons
	POWER	
Horsepower	0.746	kilowatts (kw)
	PRESSURE OF STRESS	
Inches of water pounds/sq. in.	0.2491 6.895	kilopascals (kPa) kilopascals
	ENERGY OR WORK	
Btu Foot - pound Kilowatt-hour	1055.0 1.3558 3600000.0	joules (J) joules joules

CONVERSION TABLE
METRIC TO U.S. CONVENTIONAL

WHEN YOU KNOW	MULTIPLY BY: * = Exact		TO FIND
	VERY ACCURATE	APPROXIMATE	
LENGTH			
millimeters	0.0393701	0.04	inches
centimeters	0.3937008	0.4	inches
meters	3.280840	3.3	feet
meters	1.093613	1.1	yards
kilometers	0.621371	0.6	miles
WEIGHT			
grains	0.00228571	0.0023	ounces
grams	0.03527396	0.035	ounces
kilograms	2.204623	2.2	pounds
tonnes	1.1023113	1.1	short tons
VOLUME			
milliliters	0.20001	0.2	teaspoons
milliliters	0.06667	0.067	tablespoons
milliliters	0.03381402	0.03	fluid ounces
liters	61.02374	61.024	cubic inches
liters	2.113376	2.1	pints
liters	1.056688	1.06	quarts
liters	0.26417205	0.26	gallons
liters	0.03531467	0.035	cubic feet
cubic meters	61023.74	61023.7	cubic inches
cubic meters	35.31467	35.0	cubic feet
cubic meters	1.3079506	1.3	cubic yards
cubic meters	264.17205	264.0	gallons
AREA			
square centimeters	0.1550003	0.16	square inches
square centimeters	0.00107639	0.001	square feet
square meters	10.76391	10.8	square feet
square meters	1.195990	1.2	square yards
square kilometers	0.3861019	0.4	square miles
hectares	2.471054	2.5	acres
TEMPERATURE			
Celsius	*9/5 (then add 32)		Fahrenheit

CONVERSION TABLE
U. S. CONVENTIONAL TO METRIC

WHEN YOU KNOW	MULTIPLY BY: * = Exact		TO FIND
	VERY ACCURATE	APPROXIMATE	
LENGTH			
inches	*25.4		millimeters
inches	*2.54		centimeters
feet	*0.3048		meters
feet	*30.48		centimeters
yards	*0.9144	0.9	meters
miles	*1.609344	1.6	kilometers
WEIGHT			
grains	15.43236	15.4	grams
ounces	*28.349523125	28.0	grams
ounces	*0.028349523125	.028	kilograms
pounds	*0.45359237	0.45	kilograms
short ton	*0.90718474	0.9	tonnes
VOLUME			
teaspoons	*4.97512	5.0	milliliters
tablespoons	*14.92537	15.0	milliliters
fluid ounces	29.57353	30.0	milliliters
cups	*0.236588240	0.24	liters
pints	*0.473176473	0.47	liters
quarts	*0.946352946	0.95	liters
gallons	*3.785411784	3.8	liters
cubic inches	*0.016387064	0.02	liters
cubic feet	*0.028316846592	0.03	cubic meters
cubic yards	*0.764554857984	0.76	cubic meters
AREA			
square inches	*6.4516	6.5	square centimeters
square feet	*0.09290304	0.09	square meters
square yards	*0.83612736	0.8	square meters
square miles	*2.589989	2.6	square kilometers
acres	*0.40468564224	0.4	hectares
TEMPERATURE			
Fahrenheit	* 5/9 (after subtracting 32)		Celsius

Metric Tables

DIMENSIONAL AND TEMPERATURE CONVERSION CHART

INCHES	DECIMALS	MILLI-METERS	INCHES TO MILLIMETERS		MILLIMETERS TO INCHES		FAHRENHEIT & CELSIUS			
			in.	m.m.	m.m.	in.	°F	°C	°C	°F
1/64	.015625	.3969	.0001	.00254	0.001	.000039	-20	-28.9	-30	-22
1/32	.03125	.7937	.0002	.00508	0.002	.000079	-15	-26.1	-28	-18.4
3/64	.046875	1.1906	.0003	.00762	0.003	.000118	-10	-23.3	-26	-14.8
1/16	.0625	1.5875	.0004	.01016	0.004	.000157	-5	-20.6	-24	-11.2
5/64	.078125	1.9844	.0005	.01270	0.005	.000197	0	-17.8	-22	-7.6
3/32	.09375	2.3812	.0006	.01524	0.006	.000236	1	-17.2	-20	-4
7/64	.109375	2.7781	.0007	.01778	0.007	.000276	2	-16.7	-18	-0.4
1/8	.125	3.1750	.0008	.02032	0.008	.000315	3	-16.1	-16	3.2
9/64	.140625	3.5719	.0009	.02286	0.009	.000354	4	-15.6	-14	6.8
5/32	.15625	3.9687	.001	.0254	0.01	.00039	5	-15.0	-12	10.4
11/64	.171875	4.3656	.002	.0508	0.02	.00079	10	-12.2	-10	14
3/16	.1875	4.7625	.003	.0762	0.03	.00118	15	-9.4	-8	17.6
13/64	.203125	5.1594	.004	.1016	0.04	.00157	20	-6.7	-6	21.2
7/32	.21875	5.5562	.005	.1270	0.05	.00197	25	-3.9	-4	24.8
15/64	.234375	5.9531	.006	.1524	0.06	.00236	30	-1.1	-2	28.4
1/4	.25	6.3500	.007	.1778	0.07	.00276	35	1.7	0	32
17/64	.265625	6.7469	.008	.2032	0.08	.00315	40	4.4	2	35.6
9/32	.28125	7.1437	.009	.2286	0.09	.00354	45	7.2	4	39.2
19/64	.296875	7.5406	.01	.254	0.1	.00394	50	10.0	6	42.8
5/16	.3125	7.9375	.02	.508	0.2	.00787	55	12.8	8	46.4
21/64	.328125	8.3344	.03	.762	0.3	.01181	60	15.6	10	50
11/32	.34375	8.7312	.04	1.016	0.4	.01575	65	18.3	12	53.6
23/64	.359375	9.1281	.05	1.270	0.5	.01969	70	21.1	14	57.2
3/8	.375	9.5250	.06	1.524	0.6	.02362	75	23.9	16	60.8
25/64	.390625	9.9219	.07	1.778	0.7	.02756	80	26.7	18	64.4
13/32	.40625	10.3187	.08	2.032	0.8	.03150	85	29.4	20	68
27/64	3421875	10.7156	.09	2.286	0.9	.03543	90	32.2	22	71.6
7/16	.4375	11.1125	.1	2.54	1	.03937	95	35.0	24	75.2
29/64	.453125	11.5094	.2	5.08	2	.07874	100	.37.8	26	78.8
15/32	.46875	11.9062	.3	7.62	3	.11811	105	40.6	28	82.4
31/64	.484375	12.3031	.4	10.16	4	.15748	110	43.3	30	86
1/2	.5	12.7000	.5	12.70	5	.19685	115	46.1	32	89.6
33/64	.515625	13.0969	.6	15.24	6	.23622	120	48.9	34	93.2
17/32	.53125	13.4937	.7	17.78	7	.27559	125	51.7	36	96.8
35/64	.546875	13.8906	.8	20.32	8	.31496	130	54.4	38	100.4
9/16	.5625	14.2875	.9	22.86	9	.35433	135	57.2	40	104
37/64	.578125	14.6844	1	25.4	10	.39370	140	60.0	42	107.6
19/32	.59375	15.0812	2	50.8	11	.43307	145	62.8	44	112.2
39/64	.609375	15.4781	3	76.2	12	.47244	150	65.6	46	114.8
5/8	.625	15.8750	4	101.6	13	.51181	155	68.3	48	118.4
41/64	.640625	16.2719	5	127.0	14	.55118	160	71.1	50	122
21/32	.65625	16.6687	6	152.4	15	.59055	165	73.9	52	125.6
43/64	.671875	17.0656	7	177.8	16	.62992	170	76.7	54	129.2
11/16	.6875	17.4625	8	203.2	17	.66929	175	79.4	56	132.8
45/64	.703125	17.8594	9	228.6	18	.70866	180	82.2	58	136.4
23/32	.71875	18.2562	10	254.0	19	.74803	185	85.0	60	140
47/64	.734375	18.6531	11	279.4	20	.78740	190	87.8	62	143.6
3/4	.75	19.0500	12	304.8	21	.82677	195	90.6	64	147.2
49/64	.765625	19.4469	13	330.2	22	.86614	200	93.3	66	150.8
25/32	.78125	19.8437	14	355.6	23	.90551	205	96.1	68	154.4
51/64	.796875	20.2406	15	381.0	24	.94488	210	98.9	70	158
13/16	.8125	20.6375	16	406.4	25	.98425	215	100.0	75	167
53/64	.828125	21.0344	17	431.8	26	1.02362	212	101.7	80	176
27/32	.84375	21.4312	18	457.2	27	1.06299	220	104.4	85	185
55/64	.859375	21.8281	19	482.6	28	1.10236	225	107.2	90	194
7/8	.875	22.2250	20	508.0	29	1.14173	230	110.0	95	203
57/64	.890625	22.6219	21	533.4	30	1.18110	235	112.8	100	212
29/32	.90625	23.0187	22	558.8	31	1.22047	240	115.6	105	221
59/64	.921875	23.4156	23	584.2	32	1.25984	245	118.3	110	230
15/16	.9375	23.8125	24	609.6	33	1.29921	250	121.1	115	239
61/64	.953125	24.2094	25	635.0	34	1.33858	255	123.9	120	248
31/32	.96875	24.6062	26	660.4	35	1.37795	260	126.6	125	257
63/64	.984375	25.0031	27	690.6	36	1.41732	265	129.4	130	266

CAPACITY CONVERSION U.S. GALLONS TO LITERS

Gallons	0 Liters	1 Liters	2 Liters	3 Liters	4 Liters	5 Liters
0	00.0000	3.7853	7.5707	11.3560	15.1413	18.9267
10	37.8533	41.6387	45.4240	49.2098	52.9947	56.7800
20	75.7066	79.4920	83.2773	87.0626	90.8480	94.6333
30	113.5600	117.3453	121.1306	124.9160	128.7013	132.4866
40	151.4133	155.1986	158.9840	162.7693	166.5546	170.3400

MILLIMETER/INCH CONVERSION CHART

mm	in	mm	in	mm	in	mm	in	mm	in	mm	in	mm	in	mm	in	mm	in
mm	in	15	= .5905	30	=1.1811	45	=1.7716	60	=2.3622	75	=2.9527	90	=3.5433	105	=4.1338	120	=4.7244
.25	=.0098	15.25	= .6004	30.25	=1.1909	45.25	=1.7815	60.25	=2.3720	75.25	=2.9626	90.25	=3.5531	105.25	=4.1437	120.25	=4.7342
.50	=.0197	15.50	= .6102	30.50	=1.2008	45.50	=1.7913	60.50	=2.3819	75.50	=2.9724	80.50	=3.5630	105.50	=4.1535	120.50	=4.7441
.75	=.0295	15.75	= .6201	30.75	=1.2106	45.75	=1.8012	60.75	=2.3917	75.75	=2.9823	90.75	=3.5728	105.75	=4.1634	120.75	=4.7539
1	=.0394	16	= .6299	31	=1.2205	46	=1.8110	61	=2.4016	76	=2.9921	91	=3.5827	106	=4.1732	121	=4.7638
1.25	=.0492	16.25	= .6398	31.25	=1.2302	46.25	=1.8209	61.25	=2.4114	76.25	=3.0020	91.25	=3.5925	106.25	=4.1831	121.25	=4.7736
1.50	=.0591	16.50	= .6496	31.50	=1.2402	46.50	=1.8307	61.50	=2.4213	76.50	=3.0118	91.50	=3.6024	106.50	=4.1929	121.50	=4.7885
1.75	=.0689	16.75	= .6594	31.75	=1.2500	46.75	=1.8405	61.75	=2.4311	76.75	=3.0216	91.75	=3.6122	106.75	=4.2027	127.75	=4.7933
2	=.0787	17	= .6693	32	=1.2598	47	=1.8504	62	=2.4409	77	=3.0315	92	=3.6220	107	=4.2126	122	=4.8031
2.25	=.0886	17.25	= .6791	32.25	=1.2697	47.25	=1.8602	62.25	=2.4508	77.25	=3.0413	92.25	=3.6319	107.25	=4.2224	122.25	=4.8130
2.50	=.0984	17.50	= .6890	32.50	=1.2795	47.50	=1.8701	62.50	=2.4606	77.50	=3.0512	92.50	=3.6417	107.50	=4.2323	122.50	=4.8228
2.75	=.1083	17.75	= .6988	32.75	=1.2894	47.75	=1.8799	62.75	=2.4705	77.75	=3.0610	92.75	=3.6516	107.75	=4.2421	122.75	=4.8327
3	=.1181	18	= .7087	33	=1.2992	48	=1.8898	63	=2.4803	78	=3.0709	93	=3.6614	108	=4.2520	123	=4.8425
3.25	=.1280	18.25	= .7185	33.25	=1.3091	48.25	=1.8996	63.25	=2.4901	78.25	=3.0807	93.25	=3.6713	108.25	=4.2618	123.25	=4.8524
3.50	=.1378	18.50	= .7283	33.50	=1.3189	48.50	=1.9094	63.50	=2.5000	78.50	=3.0905	93.50	=3.6811	108.50	=4.2716	123.50	=4.8622
3.75	=.1476	18.75	= .7382	33.75	=1.3287	48.75	=1.9193	63.75	=2.5098	78.75	=3.1004	93.75	=3.6909	108.75	=4.2815	123.75	=4.8720
4	=.1575	19	= .7480	34	=1.3386	49	=1.9291	64	=2.5197	79	=3.1102	94	=3.7008	109	=4.2913	124	=4.8819
4.25	=.1673	19.25	= .7579	34.25	=1.3484	49.25	=1.9390	64.25	=2.5295	79.25	=3.1201	94.25	=3.7106	109.25	=4.3012	124.25	=4.8917
4.50	=.1772	19.50	= .7677	34.50	=1.3583	49.50	=1.9488	64.50	=2.5394	79.50	=3.1299	94.50	=3.7205	109.50	=4.3110	124.50	=4.9016
4.75	=.1870	19.75	= .7776	34.75	=1.3681	49.75	=1.9587	64.75	=2.5492	79.75	=3.1398	94.75	=3.7303	109.75	=4.3209	124.75	=4.9114
5	=.1968	20	= .7874	35	=1.3779	50	=1.9685	65	=2.5590	80	=3.1496	95	=3.7401	110	=4.3307	125	=4.9212
5.25	=.2067	20.25	= .7972	35.25	=1.3878	50.25	=1.9783	65.25	=2.5689	80.25	=3.1594	95.25	=3.7500	110.25	=4.3405	125.25	=4.9311
5.50	=.2165	20.50	= .8071	35.50	=1.3976	50.50	=1.9882	65.50	=2.5787	80.50	=3.1693	95.50	3.7598	110.50	=4.3504	125.50	=4.9409
5.75	=.2264	20.75	= .8169	35.75	=1.4075	50.75	=1.9980	65.75	=2.5886	80.75	=3.1791	95.75	=3.7697	110.75	=4.3602	125.75	=4.9508
6	=.2362	21	= .8268	36	=1.4173	51	=2.0079	66	=2.5984	81	=3.1890	96	=3.7795	111	=4.3701	126	=4.9606
6.25	=.2461	21.25	= .8366	36.25	=1.4272	51.25	=2.0177	66.25	=2.6083	81.25	=3.1988	96.25	=3.7894	111.25	=4.3799	126.25	=4.9705
6.50	=.2559	21.50	= .8465	36.50	=1.4370	51.50	=2.0276	66.50	=2.6181	81.50	=3.2087	96.50	=3.7992	111.50	=4.3898	126.50	=4.9803
6.75	=.2657	21.75	= .8563	36.75	=1.4468	51.75	=2.0374	66.75	=2.6279	81.75	=3.2185	96.75	=3.8090	111.75	=4.3996	126.75	=4.9901
7	=.2756	22	= .8661	37	=1.4567	52	=2.0472	67	=2.6378	82	=3.2283	97	=3.8189	112	=4.4094	127	=5.000
7.25	=.2854	22.25	= .8760	37.25	=1.4665	52.25	=2.0571	67.25	=2.6476	82.25	=3.2382	97.25	=3.8287	112.25	=4.4193		
7.50	=.2953	22.50	= .8858	37.50	=1.4764	52.50	=2.0669	67.50	=2.6575	82.50	=3.2480	97.50	=3.8386	112.50	=4.4291		
7.75	=.3051	22.75	= .8957	37.75	=1.4862	52.75	=2.0768	67.75	=2.6673	82.75	=3.2579	97.75	3.8484	112.75	=4.4390		
8	=.3150	23	= .9055	38	=1.4961	53	=2.0866	68	=2.6772	83	=3.2677	98	=3.8583	113	=4.4488		
8.25	=.3248	23.25	= .9153	38.25	=1.5059	53.25	=2.0965	68.25	=2.6870	83025	=3.2776	98.25	=3.8681	113.25	=4.4587		
8.50	=.3346	23.50	= .9252	38.50	=1.5157	53.50	=2.1063	68.50	=2.6968	83.50	=3.2874	98.50	=3.8779	113.50	=4.4685		
8.75	=.3445	23.75	= .9350	38.75	=1.5256	53.75	=2.1161	68.75	=2.7067	83.75	=3.2972	98.75	=3.8878	113.75	=4.4783		
9	=.3543	24	= .9449	39	=1.5354	54	=2.1260	69	=2.7165	84	=3.3071	99	=3.8976	114	=4.4882		
9.25	=.3642	24.25	= .9547	39.25	=1.5453	54.25	=2.1358	69.25	=2.7264	84.25	=3.3169	99.25	=3.9075	114.25	=4.4980		
9.50	=.3740	24.50	= .9646	39.50	=1.5551	54.50	=2.1457	69.50	=2.7362	84.50	=3.3268	99.50	=3.9173	114.50	=4.5079		
9.75	=.3839	24.75	= .9744	39.75	=1.5650	54.75	=2.1555	69.75	=2.7461	84.72	=3.3366	99.75	=3.9272	114.75	=4.5177		
10	=.3937	25	= .9842	40	=1.5748	55	=2.1653	70	=2.7559	85	=3.3464	100	=3.9370	115	=4.5275		
10.25	=.4035	25.25	= .9941	40.25	=1.5846	55.25	=2.1752	70.25	=2.7657	85.25	=3.3563	100.25	=3.9468	115.25	=4.5374		
10.50	=.4134	25.50	=1.0039	40.50	=1.5945	55.50	=2.1850	70.50	=2.7756	85.50	=3.3661	100.50	=3.9567	115.50	=4.5472		
10.75	=.4232	25.75	=1.0138	40.75	=1.6043	55.75	=2.1949	70.75	=2.7854	85.75	=3.3760	100.75	=3.9665	115.75	=4.5571		
11	=.4331	26	=1.0236	41	=1.6142	56	=2.2047	71	=2.7953	86	=3.3858	101	=3.9764	116	=4.5669		
11.25	=.4429	26.25	=1.0335	41.25	=1.6240	56.25	=2.2146	71.25	=2.8051	86.25	=3.3957	101.25	=3.9862	116.25	=4.5768		
11.50	=.4528	26.50	=1.0433	41.50	=1.6339	56.50	=2.2244	71.50	=2.8150	86.50	=3.4055	101.50	3.9661	116.50	=4.5866		
11.75	=.4626	26.75	=1.0531	41.75	=1.6437	56.75	=2.2342	71.75	=2.8248	86.75	=3.4153	101.75	=4.0059	116.75	=4.5964		
12	=.4724	27	=1.0630	42	=1.6535	57	=2.2441	72	=2.8346	87	=3.4252	102	=4.0157	117	=4.6063		
12.25	=.4823	27.25	=1.0728	42.25	=1.6634	57.25	=2.2539	72.25	=2.8445	87.25	=3.4350	102.25	=4.0256	117.25	=4.6161		
12.50	=.4921	27.50	=1.0827	42.50	=1.6732	57.50	=2.2638	72.50	=2.8543	87.50	=3.4449	102.50	=4.0354	117.50	=4.6260		
12.75	=.5020	27.75	=1.0925	42.75	=1.6831	57.75	=2.2736	72.75	=2.8642	87.75	=3.4547	102.75	=4.0453	117.75	=4.6358		
13	=.5118	28	=1.1024	43	=1.6929	58	=2.2835	73	=2.8740	88	=3.4646	103	=4.0551	118	=4.6457		
13.25	=.5217	28.25	=1.1122	43.25	=1.7028	58.25	=2.2933	73.25	=2.8839	88.25	=3.4744	103.25	4.0650	118.25	=4.6555		
13.50	=.5315	28.50	=1.1220	43.50	=1.7126	58.50	=2.3031	73.50	=2.8937	88.50	=3.4842	103.50	=4.0748	118.50	=4.6653		
13.75	=.5413	28.75	=1.1319	43.75	=1.7224	58.75	=2.3130	73.75	=2.9035	88.75	=3.4941	103.75	=4.0846	118.75	=4.6752		
14	=.5512	29	=1.1417	44	=1.7323	59	=2.3228	74	=2.9134	89	=3.5039	104	=4.0945	119	=4.6850		
14.25	=.5610	29.25	=1.1516	44.25	=1.7421	59.25	=2.3327	74.25	=2.9232	89.25	=3.5138	104.25	=4.1043	119.25	=4.6949		
14.50	=.5709	29.50	=1.1614	44.50	=1.7520	59.50	=2.3425	74.50	=2.9331	89.50	=3.5236	104.50	=4.1142	119.50	=4.7047		
14.75	=.5807	29.75	=1.1713	44.75	=1.7618	59.75	=2.3524	74.75	=2.9429	89.75	=3.5335	104.75	=4.1240	119.75	=4.7146		

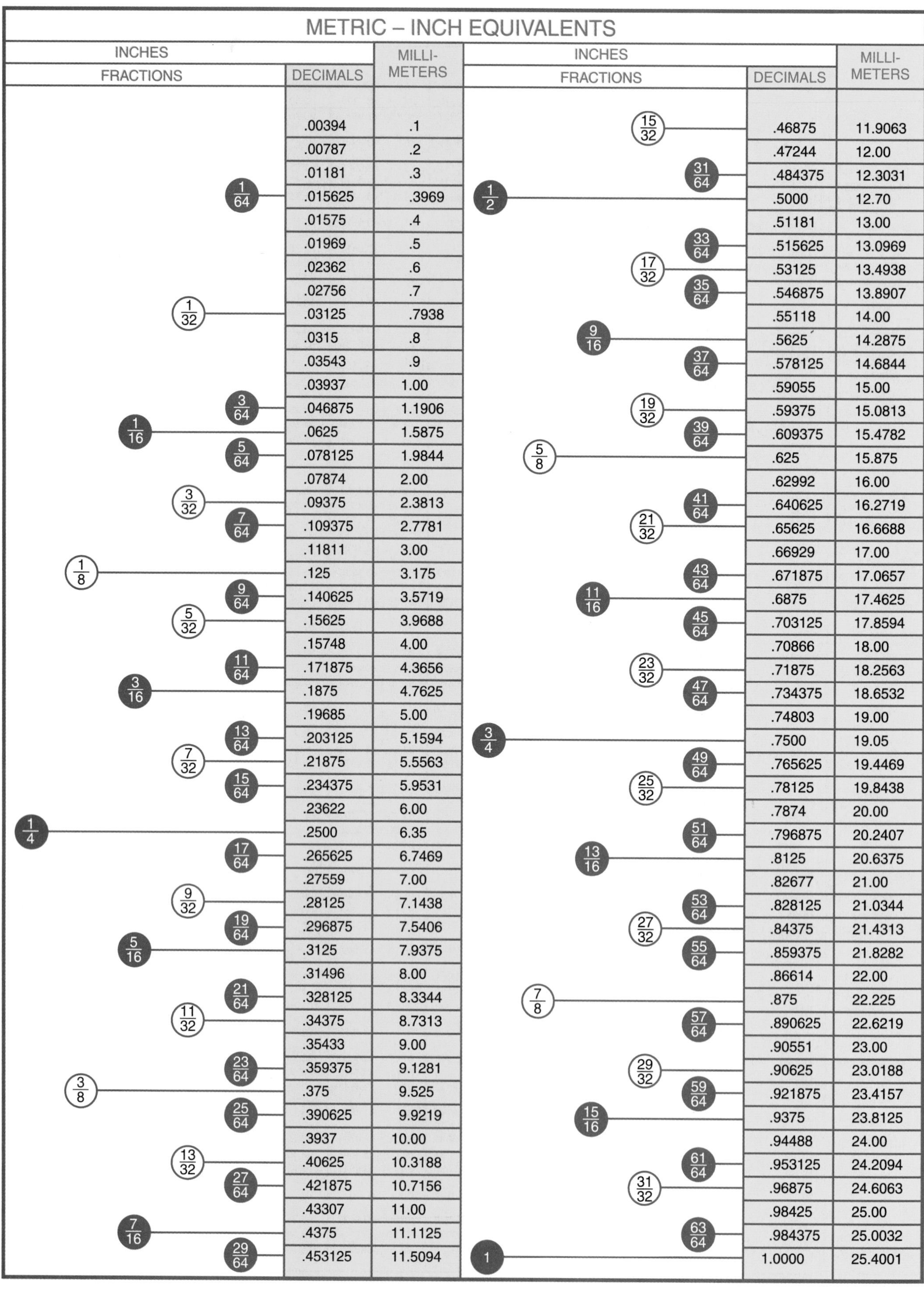

METRIC – INCH EQUIVALENTS

INCHES		MILLI-METERS	INCHES		MILLI-METERS
FRACTIONS	DECIMALS		FRACTIONS	DECIMALS	
	.00394	.1	15/32	.46875	11.9063
	.00787	.2		.47244	12.00
	.01181	.3	31/64	.484375	12.3031
1/64	.015625	.3969	1/2	.5000	12.70
	.01575	.4		.51181	13.00
	.01969	.5	33/64	.515625	13.0969
	.02362	.6	17/32	.53125	13.4938
	.02756	.7	35/64	.546875	13.8907
1/32	.03125	.7938		.55118	14.00
	.0315	.8	9/16	.5625	14.2875
	.03543	.9	37/64	.578125	14.6844
	.03937	1.00		.59055	15.00
3/64	.046875	1.1906	19/32	.59375	15.0813
1/16	.0625	1.5875	39/64	.609375	15.4782
5/64	.078125	1.9844	5/8	.625	15.875
	.07874	2.00		.62992	16.00
3/32	.09375	2.3813	41/64	.640625	16.2719
7/64	.109375	2.7781	21/32	.65625	16.6688
	.11811	3.00		.66929	17.00
1/8	.125	3.175	43/64	.671875	17.0657
9/64	.140625	3.5719	11/16	.6875	17.4625
5/32	.15625	3.9688	45/64	.703125	17.8594
	.15748	4.00		.70866	18.00
11/64	.171875	4.3656	23/32	.71875	18.2563
3/16	.1875	4.7625	47/64	.734375	18.6532
	.19685	5.00		.74803	19.00
13/64	.203125	5.1594	3/4	.7500	19.05
7/32	.21875	5.5563	49/64	.765625	19.4469
15/64	.234375	5.9531	25/32	.78125	19.8438
	.23622	6.00		.7874	20.00
1/4	.2500	6.35	51/64	.796875	20.2407
17/64	.265625	6.7469	13/16	.8125	20.6375
	.27559	7.00		.82677	21.00
9/32	.28125	7.1438	53/64	.828125	21.0344
19/64	.296875	7.5406	27/32	.84375	21.4313
5/16	.3125	7.9375	55/64	.859375	21.8282
	.31496	8.00		.86614	22.00
21/64	.328125	8.3344	7/8	.875	22.225
11/32	.34375	8.7313	57/64	.890625	22.6219
	.35433	9.00		.90551	23.00
23/64	.359375	9.1281	29/32	.90625	23.0188
3/8	.375	9.525	59/64	.921875	23.4157
25/64	.390625	9.9219	15/16	.9375	23.8125
	.3937	10.00		.94488	24.00
13/32	.40625	10.3188	61/64	.953125	24.2094
27/64	.421875	10.7156	31/32	.96875	24.6063
	.43307	11.00		.98425	25.00
7/16	.4375	11.1125	63/64	.984375	25.0032
29/64	.453125	11.5094	1	1.0000	25.4001

Dictionary of Automotive Terms

A

Abbreviations: Letters or letter combinations that stand for words. Used extensively in the automotive industry.

ABDC: After Bottom Dead Center.

ABS: Anti-lock Braking System.

ABS warning light: Amber indicator lamp mounted in the instrument cluster. Illuminates when there is a problem with the anti-lock brake system.

Accelerator: A driver-operated pedal used to control engine speed.

Accelerator pump: A small pump located in the carburetor, usually made of plastic, rubber, or Viton. Designed to supply additional fuel as needed for vehicle acceleration.

Accident: An unplanned event that results in damage and/or injury.

Accumulator (Air conditioning): Receiver-dehydrator combination used to remove moisture and store additional refrigerant.

Accumulator (Brake): Gas-charged cylinder or chamber used to store brake fluid. Normally used in anti-lock brake systems.

Accumulator piston: Assembly used to cushion the application of a band or clutch in an automatic transmission.

Acetylene: Gas commonly used in Oxyacetylene welding or cutting operations.

Ackerman principle: Design which has the steering knuckles mounted so that the steering arms can be turned inward. This principle produces toe-out on turns.

ACT: American College Testing program.

Active suspension: A method of ride control that uses struts that are controlled electronically by the driver. A switch is used to adjust the vehicle's ride firmness to meet various driving conditions.

Actuator: Any output device controlled by a computer.

ADS: Association Of Diesel Specialists.

Add-on transmission cooler: Heat exchanger installed in front of the radiator to provide extra cooling for an automatic transmission.

Additives: Any material added to automotive oils to improve its performance under certain conditions, such as to decrease an engine oil's tendency to break down when heated.

Advance (Ignition timing): Ignition timing setting so that a spark occurs earlier or more degrees before Top Dead Center.

AEA: Automotive Electronic Association.

AERA: American Engine Rebuilders Association.

Air bag: A protective device that inflates in a frontal collision. Designed to act as a cushion between the front seat occupants and the vehicle's interior.

Air bleed: An orifice or small passageway in a carburetor designed to allow a specific amount of air to enter a moving column of gasoline.

Air cleaner: A device for filtering, cleaning, and removing dust, dirt, and foreign debris from the air being drawn into an engine, air compressor, etc.

Air-cooled: An engine or other object that is cooled by passing a stream of air over its surface.

Air conditioning: Process in which the air coming into the passenger compartment is cleaned, cooled, and dehumidified.

Air conditioning technician: A person who is trained in the proper diagnosis and repair of automotive air conditioning systems.

Air control door: One or more doors used to control the direction of airflow in the air conditioning air distribution system.

Air dam: A metal or plastic shroud placed beneath the front of a vehicle. Designed to channel air flow over the radiator and condenser.

Air filter: A pleated paper element installed inside the air cleaner. Used to filter dirt, dust, and foreign debris from the air stream entering the engine.

Airflow meter: Measures rate at which air enters the engine.

Airflow sensor: Sensor that monitors the amount of air entering an engine's throttle body.

Air-fuel ratio: Ratio by weight between the air and fuel that makes up the engine's fuel charge.

Air gap: The space between the electrodes of a spark plug, starting motor and alternator armatures, etc.

Air horn: The air inlet on the carburetor to which the air cleaner is normally attached.

Air injection system: An emissions control system used to lower levels of carbon monoxide and hydrocarbons. Air is injected in the exhaust manifold near the exhaust valves. Also called an Air Injection Reaction system or AIR.

Air pollution: Any release of harmful substances into the atmosphere by engine operation or other causes.

Air spring: A container and plunger separated by air under pressure. Used on some suspension systems in place of coil springs.

ALDL: Assembly Line Diagnostic Link. Also referred to as a diagnostic connector.

Align: To bring various parts of a unit into their correct positions with respect to each other or to a predetermined location.

Alignment heads: One or more mechanical or electronic sensors mounted on a vehicle's rims. Used to measure steering geometry angles. Can be part of a computerized system.

Alignment technician: A person who specializes in suspension and steering repairs, as well as vehicle alignment.

Allen wrench: A hexagonal wrench, which is usually îlî shaped, designed to fit into a hexagonal hole.

Alloy: A mixture of two or more metals. An example is solder, which is a mixture of lead and tin.

Alpha numeric: Older system used to designate tire size. Example: HR 78-15. The letter H=load/size relationship. R=radial. 78=height-to-width ratio. 15=rim diameter in inches.

Alternating current (ac): An electrical current that moves in one direction and then the other.

Alternator: A generator in which alternating current is first generated, then changed into direct (dc) current.

Altitude compensator: Any sensor or device that automatically compensates for changes in altitude.

Aluminum: A metal known for its lightweight and good heat dissipation characteristics. It is often alloyed with other metals.

Ambient temperature: The temperature of the air surrounding an object.

Ambient air temperature switch: Temperature sensitive switch that is designed to prevent air conditioning compressor operation if the outside ambient air temperature is below a certain point.

Ammeter: Instrument used to measure the flow of electric current in a circuit in amperes. Normally connected in series in a circuit.

Ampere: The unit of measurement for the flow of electric current.

Anaerobic sealer: A chemical sealer that cures in the absence of air.

Analog: A signal that continually changes in strength.

Aneroid: A device, such as a barometer bellows that does not contain or use a liquid.

Annealing: A process used to soften metals. Metal is heated and then cooled gradually over a period of time.

Anode: The positive pole in an electrical circuit.

Anti-dieseling solenoid: Electric solenoid mounted next to a carburetor's throttle plate lever. Used to set the engine's hot idle rate and designed to close the throttle plates when the engine is shut off.

Anti-lock brake system (ABS): A computer-controlled system that is part of the base brake system. The system "cycles" the brakes on and off to prevent wheel lockup and skidding.

Anti-rattle clip: One or more metal components designed to keep brake pads from vibrating and rattling.

Antifreeze: A liquid, usually ethylene glycol, that is added to water in a vehicle's cooling system. This mixture serves as the engine's coolant.

Antifriction bearing: A bearing that uses balls or rollers between a journal and bearing surface to decrease friction.

Apprentice programs: Any educational program designed to teach a trade through a combination of on-the-job training and classroom study

Apprentice Technician: A beginner who is learning under the direction of one or more experienced auto technicians.

Arcing: Electricity leaping the gap between two electrodes.

Armature: The revolving part in an alternator or motor. The moving part in a relay or horn.

ASE: Abbreviation for National Institute For Automotive Service Excellence.

ASIA: Automotive Service Industry Association.

ASME: American Society Of Mechanical Engineers.

Asbestos: A mineral material that has great heat resistant characteristics. Once widely used in brake and clutch linings. Asbestos is a known cancer causing agent.

Aspect ratio: The ratio between tire section height and section width.

Asphyxiation: Death caused by lack of oxygen.

ATA: American Trucking Association.

ATDC: After Top Dead Center.

Atmospheric pressure: The pressure exerted by the earth's atmosphere on all objects. Measured with reference to the pressure at sea level, which is around 14.7 psi (101 kPa).

Atom: A tiny particle of matter made up of electrons, protons, and neutrons. The electrons orbit around the center or nucleus, made up of protons and neutrons. Combinations of atoms make up molecules.

Atomizes: The process of breaking a liquid, such as gasoline, into tiny drops and mixing with air.

Automatic level control system: A hydraulic or pneumatic system that adjusts the vehicle's body height in response to changes in vehicle load.

Automatic transmission: A transmission that uses fluid pressure to shift itself. Shift points and range is determined by road speed, engine loading, altitude, and throttle position. Can also be computer-controlled.

Automatic transmission fluid (ATF): Mineral oil with special additives to make it compatible with clutch and band materials and to prevent foaming and breakdown.

Auxiliary oil cooler: A system designed to remove excess heat from engine oil by either passing it through a heat exchanger exposed to the air or by channeling engine coolant through an oil filter adapter mounted on the engine.

AWG: American Wire Gage, which is a standard measure of wire size. The smaller the wire, the larger the number.

Axle: Shaft or shafts used to transmit driving force to the wheels.

Axle housing: A metal enclosure for the drive axles and differential. It is partially filled with differential fluid and other additives as needed.

Axle ratio: The relationship or ratio between the number of times the propeller shaft must revolve to turn the axle drive shafts once.

B

Backfire: A condition where unburned fuel in the intake or exhaust manifold ignites.

Backlash: The clearance or "play" between two parts, such as the teeth of two gears.

Backing plate: A metal plate that holds the wheel cylinder, shoes, and other parts inside a drum brake.

Back pressure: The exhaust system's resistance to the flow of exhaust gases.

Backup lights: White lights in the rear of the vehicle. Lights turn on whenever the transmission is placed in reverse.

Baffle: An obstruction used to check of deflect the flow of gases, liquids, sounds, etc.

Ball bearing (Antifriction): A bearing consisting of an inner and outer hardened steel race separated by a series of hardened steel balls.

Ball joint: A flexible joint which utilizes a ball and socket type of construction. Used primarily in steering linkage and steering knuckle pivot support.

Ballast resistor: A specially constructed wire whose properties tend to increase or decrease voltage in direct proportion to the heat of the wire. Used in older point and electronic ignition systems.

Band: Holding member in an automatic transmission. Used to stop clutch rotation during certain transmission gear ranges.

Band and clutch application chart: Diagnostic chart used to identify transmission member operating conditions in each gear range.

Barometric pressure sensor: A sensor that measures atmospheric pressure.

Base brakes: The hydraulic and friction members of the automotive brake system. Does not include any boosters or ABS components.

Battery: A device containing one or more cells that produce electricity through electrochemical action.

Battery cables: Heavy wires used to connect the battery to the vehicle's electrical system.

Battery capacity: The amount of current a battery is capable of delivering.

Battery charging: The process of restoring a battery's charge by passing current through it in a reverse direction (positive to negative).

Battery plate: Battery components made of lead peroxide in sponge form and porous lead.

Battery rating: Standardized measurement of a battery's ability to deliver an acceptable level of energy under specified conditions. Standards established by the Battery Council International (BCI).

Battery temperature sensor: Sensor installed in some batteries to control overcharging.

Battery voltage: Determined by the number of cells in the battery. Each cell is capable of producing 2.1v. Batteries used in modern vehicles produce approximately 12.6v.

BBDC: Before Bottom Dead Center.

BCI: Battery Council International.

BDC: Bottom Dead Center.

Bead: The steel reinforced edge of a tire that engages the wheel rim.

Bearing: A part on which a shaft, journal, or pivot turns.

Bell housing: Metal cover installed around the flywheel and clutch assembly.

Bendix drive: A self-engaging starter drive gear. It is mounted on a screw shaft attached to the starter motor armature. Also is used as a general term to describe the overrunning clutch on other starters.

Bezel: Crimped edge of metal used to secure a transparent instrument cover.

Bias tire: A type of tire construction in which the plies crisscross from bead to bead. No additional belts are used under the tread area.

Bias-belted tire: Tire construction method using crisscrossed plies with additional belts under the tread area.

Bleed: The process of removing air or fluid from a closed system, such as the brakes.

Blend door: Door installed in the air conditioning blower case. Controls temperature of the air going to the passenger compartment.
Block: The part of the engine containing the crankshaft, pistons, and cylinders. It is made of cast iron or aluminum and is the foundation for the entire engine.
Block heater: A heating device installed in place of one of the block's freeze plugs. Used to heat an engine during cold weather.
Blowby: Any leakage or loss of compression past the piston rings. Often accompanied by oil from the crankcase passing through into the combustion chamber.
Blower motor: An electric motor used to force air to move through the evaporator in an air conditioning system.
Blueprinting: Process of assembling an engine to *exact* specifications.
Body computer module (BCM): On-board computer responsible for controlling such functions as interior climate, radio, instrument cluster readings, and on some vehicles, cellular telephones. May also interact with the engine's electronic control unit.
Body technician: A person who specializes in body panel and frame repair.
Boost pressure: A measure of engine power provided by a supercharger or turbocharger.
Bore: The diameter of a cylinder. Sometimes used to refer to the cylinder itself.
Boring bar: A machine that uses a bar equipped with cutting bits to cut engine cylinders to specific size or to a new diameter.
Boss: A rib or enlarged area designed to strengthen a portion or area of an object.
Bottled gas: Petroleum gas compressed into liquefied form and contained in strong metal cylinders.
Bourdon tube: A circular hollow piece of metal used in some instruments. Pressure on the hollow end causes it to attempt to straighten. The free end moves the needle on the gauge face.
Bowl vent: A passageway in a carburetor that prevents pressure buildup in the fuel bowl.
Brake anchor: Steel stud upon which one end of the brake shoes is attached to or firmly rests against. Anchor is normally cast as part of the backing plate.
Brake booster: Brake component operated by engine vacuum or power steering fluid to decrease the braking effort needed by the driver.
Brake drum: A cast iron or aluminum housing upon which the brake shoes contact to provide braking action.
Brake fade: A condition in which repeated severe brake application causes the braking surfaces to lose their frictional ability, resulting in impaired braking efficiency.
Brake feel: The discernible relationship between the amount of brake pressure and the actual force being exerted by the driver.
Brake fluid: A special fluid compound used in hydraulic brake systems. It must meet exacting specifications, such as resistance to heating, freezing, and thickening.
Brake horsepower (BHP): The measurement of useable horsepower delivered at the crankshaft. Usually computed by placing the engine on a chassis dynamometer.
Brake hoses: Flexible hose which connects solid brake lines to wheel brake assemblies.
Brake lines: Double walled steel tubing used to carry brake fluid from the master cylinder to the wheel brake assemblies.
Brake shoes: Metal carrier to which friction linings are attached. Used with drum brake assemblies.
Brake technician: A person who is specially trained to diagnose and repair brake system problems
Brake System Warning Light: Red light mounted in the instrument cluster. Indicates a hydraulic system malfunction in the brake system or application of the parking brake. Different from the amber light used to indicate ABS system problems.
Braking ratio: A comparison of the front wheel to rear wheel braking effort. On most vehicles, the ratio is 55%-60% for the front brakes and 45%-50% for the rear brakes.
Braze: To join two pieces of metal together by heating the edges to be joined and then using brass or bronze on the area as a solder.
Break-in: The process of wearing in two or more parts that have been replaced or reconditioned.
Breaker arm: The movable part of a pair of distributor contact points.
Breaker points: A pair of movable points that are opened and closed to make and break the ignition primary circuit.
Broach: To finish a metal surface by forcing a multiple edge cutting tool across its surface.
Bronze: An alloy consisting of tin and copper. Normally used in bushing and bearing material.
Brown and Sharpe (B & S) Gage: A standard measure of wire gage, similar to AWG standard.
Brushes: A bar of carbon, copper, or other conductive material that rides on the slip rings of a motor or alternator.
British thermal unit (BTU): Measurement of the amount of heat required to raise the temperature of one pound of water by 1°F.
BTDC: Before Top Dead Center.
Burnish: To bring the surface of a metal to a high shine by rubbing it with a hard, smooth object.
Bus: Pathway for data inside a computer. Can also be used to refer to a circuit used to connect two on-board computers.
Bushing: A smooth, removable liner used as a bearing for a shaft, piston pin, etc.
Butane: Petroleum gas that is liquefied under pressure. Sometimes referred to as liquefied petroleum gas and often combined with propane.
Bypass: To move around or detour from the normal route taken by a flowing substance, such as electricity, air, or fluid.
Bypass valve: Valve used to permit coolant flow when the thermostat is closed.

C

CAFE: Corporate Average Fuel Economy.
Calibrate: An adjustment made to produce a desired effect.
Caliper: A disc brake component which forms the cylinder and contains the piston and brake pads. It provides braking by producing a clamping action on a rotating disc.
Calipers: An adjustable measuring tool placed around or within an object and adjusted until it just contacts. The tool is then withdrawn and the distance between the contact points is measured.
Calorie: Metric unit of measurement for heat. The amount of heat needed to raise the temperature of one gram of water by 1°C.
Cam angle: The number of degrees of rotation of the distributor shaft when the contact points are closed.
Cam ground piston: Piston ground slightly egg-shaped. When heated, it becomes round.
Cam lobe: The offset portion of a shaft that, as it turns, will impart motion to another part.
Camber: The outward or inward tilt of a wheel from its centerline.
Camshaft: A shaft with offset lobes used to operate valves.
Camshaft gear: A gear that is used to drive the camshaft.
Camshaft position sensor: A sensor used by a computer control system to reference camshaft position in order to adjust spark timing, fuel, etc. It is similar to a hall-effect switch.
Capacitance: The property of a capacitor or condenser that permits it to receive and retain an electrical charge.
Carbon (C): Hard or soft black deposits found in combustion chamber, spark plugs, valves, etc. An excellent conductor of electricity.
Carbon dioxide (CO_2): A dry chemical mixture that is an excellent fire retardant and is used to make dry ice. Also is produced as an end product of catalyzed exhaust emissions.
Carbon monoxide (CO): A deadly colorless, odorless gas that is formed when fuel is not burned completely.
Carbon tracking: Lines of carbon found inside a distributor or on an ignition coil. Usually an indication of arcing.
Carburetor: A device used for many years to mix gasoline with air in correct proportions and deliver it to the engine.
Carburetor circuits: A series of passageways and units in a carburetor designed to perform a specific function.
Carburetor icing: The formation of ice on the throttle plate or valve during cold, humid weather conditions.

Carcinogen: Any substance, such as asbestos and carbon tetrachloride, that can cause cancer.

Cardan universal joint: A type of universal joint that has yokes at right angles to each other.

Case-harden: A process of heating a piece of steel to harden its surface while the inside remains relatively soft.

Castellated nut: A nut that has a series of slots cut into it, into which a cotter pin or safety wire may be passed to secure the nut. Top of nut resembles a castle battlement.

Caster: The backward or forward tilt of a wheel away from its centerline.

Catalytic converter: An emissions control device which uses the elements platinum, palladium, and rhodium as catalysts to reduce harmful exhaust emissions.

Cathode: The negative pole of an electrical circuit.

Cell: The individual compartments in a battery which contain positive and negative plates suspended in electrolyte.

Celsius: Metric unit of measurement for temperature. Under standard atmospheric conditions, water freezes at 0°C and boils at 100°C.

Center link: Also called a "drag link", it is used to connect the pitman arm to the tie rods in a parallelogram steering system.

Center of gravity: The point on an object on which it could be balanced.

Center support bearing: A bearing used to mount a two-piece Hotchkiss driveshaft.

Central processing unit (CPU): A microprocessor inside an electronic control unit responsible for controlling the ECU's operations.

Centrifugal advance mechanism: A series of weights used to advance engine timing as engine speed increases.

Centrifugal force: A force which tends to keep a rotating object away from the center of rotation.

Cetane number: A measurement of the performance characteristics of diesel fuel in cold weather.

CFM: Cubic Feet per Minute. Used as a measurement for the amount of air entering an engine's intake.

Chamfer: To bevel an edge on an object at the edge of a hole.

Change of state: Any condition where a substance changes from a solid to liquid or from liquid to vapor.

Charcoal canister: An emissions control device which contains activated charcoal granules. Used to trap and store fuel vapor when the engine is not running.

Charge: A given or specified weight of refrigerant in an air conditioning system. Also the electrical rate of flow that passes through the battery to restore it to full power.

Charging station: A cart containing a manifold gauge, refrigerant tank, and pump. Used to recover and recharge air conditioning systems.

Chase: Process of using a tap to repair slightly damaged threads.

Chassis: The framework of a vehicle without its body panels.

Chassis dynamometer: A machine used to measure engine power at the vehicle's wheels.

Chatter: A condition where a disc vibrates severely as it accelerates.

Check engine light: An amber light used to indicate that a problem exists in a vehicle's engine computer control system. Also see Malfunction Indicator Light.

Check valve: A valve that permits flow in only one direction.

Chilled iron: Cast iron possessing a hard outer surface.

Choke plate: A plate in the carburetor that reduces the volume of air admitted to the engine.

Choke stove: A compartment in or on the exhaust manifold used to divert hot air to the automatic choke device.

Chromium: A substance added to steel to help it resist oxidation and corrosion.

Circuit: A source, resistance, and path to carry a flow from the source to the resistance and back to the source. An example is an electrical circuit. The wires (path) allow electricity from the battery (source) to flow to a headlamp (resistance) and back to the battery.

Circuit breaker: Protective device that will open and close when current draw becomes excessive. Unlike a fuse, it does not blow out.

Clearance: A given space between two parts such as a piston and cylinder.

Clockwise: Rotation to the right as with clock hands.

Closed cooling system: A type of cooling system which uses an overflow bottle.

Closed loop: Engine ECU operation mode used when an engine has reached its optimum operating temperature and conditions. When in closed loop, most ECU's rely on the oxygen sensor to determine if engine operating conditions need to be changed.

Cluster gear: A cluster of gears that are all cut on one long gear blank. The cluster gears ride in the bottom of a manual transmission, providing a connection between the transmission input and output shaft.

Clutch: A friction device used to connect and disconnect the flow of power from one unit to another.

Clutch disc: The part of the clutch assembly that is splined to the transmission input shaft and presses against the flywheel face.

Clutch fork: A lever which forces the throw-out bearing into the clutch pressure plate.

Clutch housing: Cast iron or aluminum housing that surrounds the flywheel and clutch mechanism.

Clutch linkage: A cable or other mechanism which transfers movement from the clutch pedal to the throw-out fork.

Clutch pilot bearing: A small bronze bushing or ball bearing placed in the end of the crankshaft or in the center of the flywheel. Used to support the outboard end of the transmission input shaft.

Clutch start switch: Switch that prevents the engine from starting unless the clutch pedal is depressed.

Coefficient of friction: The amount of friction developed when two objects are moved across each other. Coefficient of friction is calculated by dividing the force needed to push a load across a given surface.

Coil: Electrical transformer designed to increase voltage.

Coil spring: A section of spring steel wound in a spiral pattern.

Coil spring pressure plate: A type of pressure plate used with diesel engines. Contains coil springs to soften the clutch's impact.

Cold cranking amps: Measurement of cranking amperes that a battery can deliver over a period of 30 seconds at 0°F (-18°C)

Cold patching: Repair method used to seal leaks in plastic fuel tanks.

Cold solder: An electrical joint that has had solder applied to it before it reached proper soldering temperature. Characteristics include clumping of solder on the joint.

Cold start valve: A temperature controlled valve or fuel injector which sprays extra fuel for cold starts.

Combination valve: A brake system hydraulic valve which includes a pressure differential valve, metering valve, and proportioning valve.

Combustion: The process of burning.

Combustion chamber: The area above a piston with the piston at TDC. Measured in cubic centimeters.

Commutator: A series of copper bars connected to an armature winding. The bars are insulated from each other and from the armature.

Compensating port: A small hole placed in a master cylinder to permit fluid to return to the reservoir.

Composite headlamp: Reflector and lens assembly used with halogen bulbs. Designed to be used in a specific vehicle.

Compound: A combination of two or more ingredients mixed together.

Compression: Opposite of tension. Reduction in volume, such as compressing a gas. Also applying pressure to a spring to reduce its length.

Compression gauge: A gauge used to test compression in engine cylinders.

Compression ignition: Principle used in diesel engines to ignite an air-fuel mixture by using heat of compression.

Compression ratio: The relationship between cylinder volume when the piston is at TDC and cylinder volume when the piston is at BDC.

Compression rings: The top set of piston rings designed to seal between the piston and cylinder to prevent escape of gases from the combustion chamber.

Compression stroke: The portion of the piston's movement devoted to compressing the fuel mixture in the engine's cylinder.

Compressor: An engine-driven device used to compress refrigerant and causes it to flow in the air conditioning system.

Compressor wheel: A fan like wheel that forces air into the intake manifold.

Condense: Process of turning vapor back to liquid.

Condenser: A device for turning hot refrigerant from vapor to liquid. Also, a capacitor installed between the breaker points and coil to absorb surges of electricity.

Conduction: The transfer of heat from one object to another by being in direct contact.

Conductor: Any material that can form a path for electrical current.

Connecting rod: The connecting link between the crankshaft and the pistons.

Connecting rod bearing: Inserts which fit into the connecting rods and rides on the crankshaft journals.

Connecting rod cap: The removable lower part of the connecting rod.

Constant mesh gears: Gears that are always in mesh with each other.

Constant velocity axle: An axle that utilizes two constant velocity joints to effect torque transfer to the driving wheels.

Constant velocity universal joint: A universal joint designed to effect a smooth transfer of torque from a driven shaft to the driving shaft without any fluctuations in speed.

Contact pattern: The area on a ring gear tooth that matches with the pinion gear tooth.

Continuous fuel injection: A type of fuel injection in which the fuel injectors are constantly open. The amount of fuel to the cylinders is controlled by the ECU, which monitors an airflow sensor.

Contraction: Reduction in the size of an object when it cools.

Control arm: A movable arm that is used as part of a suspension system.

Control module: An electronic device used in ignition systems to switch the primary current on and off.

Control rack: A toothed rod inside a mechanical injector pump which rotates the pump plunger to control the quantity of injected fuel.

Control valve: A valve used to control the pressure of fluid in a circuit.

Convection: A transfer of heat from one object to another by heating the air surrounding the object.

Coolant: Ethylene glycol or other liquid used in a cooling system.

Coolant nozzles: Holes placed in the cooling passages of an engine block to provide coolant to hard to reach areas.

Coolant recovery system: A plastic bottle and hose that is used with a closed cooling system to recover and provide additional coolant when needed.

Coolant temperature sensor (CTS): A sensor that monitors engine coolant temperature. Used by the engine control ECU to determine when to enter closed loop operation.

Cooling fins: A series of thin metal strips placed between cooling passages to help dissipate heat.

Cooling system: A system that allows air or liquid to circulate to maintain a constant temperature.

Core hole plugs: See *Freeze plugs.*

Corrosion inhibitors: Any chemical added to a substance that prevents the formation of oxidation or corrosion.

Cotter pin: A soft metal pin that is used to secure components in place.

Counter shaft: An intermediate transmission shaft that transfers motion from one shaft to another.

Counterbalance shaft: A shaft that is geared with the crankshaft and camshaft. Designed to cancel out vibrations from these two shafts.

Counterclockwise rotation: Rotation to the left, opposite to that of clock hands.

Countersink: To make a counterbore so that the head of a screw may set flush or below the surface.

Cradle: Segmented frame on a front-wheel drive vehicle that supports the engine and front suspension.

Crankcase: The part of the engine that surrounds the crankshaft. Different from the pan, which simply covers the crankcase.

Crankcase dilution: Condition where fuel passes the piston rings and enters the crankcase, diluting the engine oil.

Crankshaft: The main shaft which supports the connecting rods and turns piston reciprocation into motion.

Crankshaft gear: A gear that is pressed on the crankshaft. Used to drive a chain or belt which drives the camshaft gear.

Crankshaft position sensor: A sensor similar to a hall-effect switch that is used to monitor crankshaft position and speed.

Crankshaft throw: The offset part of the crankshaft where the connecting rods fasten.

Cross and roller: A type of universal joint that uses a center cross (spider) mounted in needle bearings.

Crossmember: Part of the vehicle frame of unibody that runs crosswise in relation to the vehicle's length.

CRT: Cathode Ray Tube.

Cu. In.: Cubic Inches.

Curb weight: The weight of an unmanned vehicle with fuel, oil, coolant, and all standard equipment installed.

Current: The movement of free electrons in a conductor.

CVT: Continuously Variable Transmission.

Cycle: A reoccurring period in which a series of actions take place in a definite order.

Cylinder: A hole or holes that have a set depth and contain pistons.

Cylinder head: Metal section bolted on top of the engine block and forms part of the combustion chamber.

D

Damper: A unit or device used to reduce or eliminate vibration, oscillation, of a moving part, fluid, etc.

Dashboard: Part of vehicle interior containing the instrument cluster, air conditioning controls, switches, etc.

Dashpot: A cylinder with a piston or diaphragm with a small vent hole used to retard the movement of some other part.

Dead axle: An axle that does not rotate, but merely forms a base on which to attach wheels.

Dead center (DC): Piston at the extreme top or bottom of its stroke.

Deceleration: The process of slowing down in rotational speed, forward speed, etc.

Deflection rate: The measurement of the amount of force required to compress a leaf spring one inch.

Deglazer: An abrasive tool used to remove the glaze from cylinder walls before new piston rings are installed.

Degree: 1/360 of a circle.

Degree wheel: Wheel-like tool attached to the crankshaft. Used to time valves to a high degree of accuracy.

Density: The relative mass of an object in a given volume.

Deploy: Term used to describe the activation of an air bag system.

Desiccant: A material used to absorb and remove excess moisture.

Detergent oil: Engine oil that has extra chemical detergents added to remove engine deposits and hold them in suspension.

Detonation: Condition where the fuel charge fires or burns too rapidly. Audible through the combustion chamber walls as a knocking noise.

Dexron III/Mercon: Automatic transmission fluid that replaces Dexron II, which was in use for many years.

Diagnosis: The process of analyzing symptoms, test results, etc., to determine the cause of a problem.

Diagnostic control module: Electronic computer that controls the air bag system. Also referred to as an Diagnostic Energy Reserve Module (DERM).

Diagnostic trouble codes: A code that is displayed that can be used to determine where a malfunction is located in a computer-controlled system.

Diagonally split brake system: A brake system that has the master cylinder pistons actuating diagonally opposed brake assemblies. Example: One piston will actuate the left front and right rear brakes.

Dial gauge: A precision micrometer which indicates exact readings via needle movements on a dial.

Diaphragm: A flexible partition used to separate two different compartments.

Die: A tool for cutting threads.

Diesel engine: An engine that uses diesel oil for fuel. A diesel engine injects fuel oil directly into the cylinders. The compression is so great that the air itself is hot enough to ignite the diesel fuel without a spark.

Diesel injection pump: A mechanical pump that develops high pressure to force fuel out of the injectors and into the combustion chambers.

Dieseling: Condition in which the engine continues to run after the ignition key is turned off. Also referred to as run-on.

Differential: A gear system that drives both axles at the same time, but allows them to rotate at different speeds when turning a corner.

Differential case: A steel unit to which the ring gear is attached. The case drives the pinion gears and forms an inner bearing surface for the axle pinion gears.

Differential yoke: Component that connects the rear universal to the pinion shaft.

Digital EGR valve: EGR valve that can be operated by the vehicle's engine ECU.

Digital signal: An electronic signal that uses on and off pulses.

Dimmer switch: A foot- or hand-operated switch that operates the high and low beam headlights.

Diode: A semiconductor device that allows current flow in one direction but resists it in the other.

Dipstick: A metal or plastic rod used to determine the quantity of oil or fluid in a system.

Direct current (dc): Electric current that flows steadily in one direction only.

Direct drive: Gear condition in an automatic transmission where the crankshaft and driveshaft turn at the same speed.

Direct fuel injection: Fuel injection system where fuel is sprayed directly into the combustion chamber. Also see Multiport Fuel Injection.

Direct ignition system: A type of computer-controlled ignition system similar to the distributorless system, but does not use spark plug wires.

Disc brake assembly: A brake assembly that uses a hydraulic caliper to actuate brake pads against a metal rotor. Used for both front and rear brakes.

Discharge: Process of drawing electric current from a battery.

Discharging (air conditioning): Procedure for draining a vehicle's air conditioning system of its refrigerant charge.

Displacement: The total volume displaced by the cylinders when moving from BDC to TDC.

Distillation: The process of heating a liquid to the point that its components vaporize and then capturing each separate component as it condenses in a tower or column.

Distribution tubes: Tubes used in the engine cooling system to circulate and direct flow of coolant to hard to reach areas.

Distributor: A unit designed to open and close the ignition primary circuit, either mechanically or electronically. Also used to distribute secondary voltage to the proper cylinder at the correct time.

Distributor cap: A plastic insulated cap containing one central terminal and a series of evenly spaced outer terminals in a circular pattern around the center terminal. Used to distribute secondary voltage to the spark plugs.

Distributor injection pump: A pump that uses one or two cylinders to inject diesel fuel into an engine.

Distributor rotor: Part designed to transfer secondary current to the distributor cap outer terminals.

Distributorless ignition system: A type of computer-controlled ignition system that eliminates the distributor by using a sensor mounted on the crankshaft and/or camshaft to provide timing.

Diverter valve: A valve used in the air injection system to divert air away from the injection nozzles during periods of deceleration to prevent backfiring.

Documentation: Repair orders or other means used to record work performed on a vehicle.

Dog tracking: Condition where the rear wheels of a vehicle are not aligned with the front wheels.

DOHC: Abbreviation for a Dual Overhead Camshaft Engine.

Double lap flare: A flare which, when made, utilizes two wall thicknesses. Used primarily in solid steel brake lines.

Dowel pin: A steel pin that is passed through two matching holes to maintain the parts in proper alignment.

Draw: The amount of current required to operate an electrical device.

Draw-filing: Filing by passing the file at right angles to the work.

Driveability technician: A person who is trained in the diagnosis and repair of engine performance and computer control system problems.

Drive-fit: A fit between two parts that is so tight that they must be driven together.

Driveline: The drive shaft, universal joints, etc., that connect the transmission to the rear differential.

Drive shaft: One or more shafts connecting the transmission output shaft to the differential pinion shaft.

Drive shaft angle: The angle at which the drive shaft meets the rear differential or front wheels.

Drive shaft imbalance: Rapid oscillation of the drive shaft due to an unbalanced drive shaft or excessive shaft runout.

Drive train: All of the collective parts (engine, transmission, drive shaft, differential, drive axles, etc.) that generate power and transmit it to the wheels.

Driveability: The process of diagnosing, troubleshooting, isolating, and repairing a problem on a vehicle. Generalized term normally used when describing a repair of an engine performance problem.

Drop center: A type of rim in which the center section is lower than the two outer edges. This allows the bead of the tire to be pushed into the low area while the other side is pulled over and off the flange.

Drum brake assembly: A brake system that uses a wheel cylinder to force two brake shoes against a rotating drum. Used primarily as rear brakes, but have been used in the front in older vehicles.

Dry charged battery: A battery that is charged, but lacks electrolyte. It is filled with electrolyte when it is placed into service.

Dry friction: Any resistance to movement between two unlubricated surfaces.

Dual exhaust system: An exhaust system that uses separate pipes, mufflers, and catalytic converters for each bank of cylinders.

Dust cap: Metal cap installed on the end of an axle or spindle to keep dirt out and grease in.

Dwell: The amount of time the distributor points remain closed between openings. Also see *Cam angle.*

Dye penetrant: A testing material that can be sprayed or painted onto iron and aluminum parts to locate cracks.

Dynamic balance: Condition when the centerline of a revolving object is in the same plane as the object itself.

Dynamic imbalance: Unbalanced condition when the centerline of a revolving object is not in the same plane as the object itself.

Dynamometer: A machine used to measure engine horsepower output. An engine dynamometer measures horsepower at the crankshaft and the chassis dynamometer measures horsepower at the wheels.

E

ECC: Electronic Climate Control.

Eccentric: A circle within a circle that has a different shape and center.

ECM: Electronic or Engine Control Module.

Economizer valve: Fuel flow device that controls the flow of fuel in a carburetor under certain conditions.

EEC: Evaporative Emissions Control.

EFE: Early Fuel Evaporative System.

ELC: Electronic Level Control.

Electrical technician: A person who is trained in the diagnosis and repair of automotive electrical systems.

Electrode: The insulated center and rod that is attached to the center of a spark plug. Also refers to a welding rod.

Electrolyte: A solution of sulfuric acid and water in a battery.

Electromagnet: A magnet that is produced by placing a coil of wire around a steel or iron bar. When current flows through the wire, the bar becomes magnetized.

Electromagnetic interference (EMI): Electronic noise that is created when two or more wires carrying a strong voltage or signal are allowed to cross. Can also be created by radio waves.

Electromotive force (EMF): Voltage.

Electron: A negatively charged particle that makes up part of the atom.

Electron theory: The accepted theory of electronics that states that electricity flows from positive to negative.

Electronic control unit (ECU): General term used for any computer that controls a vehicle system.

Electronically erasable programmable read only memory (EEPROM): A type of microprocessor whose programming can be changed by special electronic equipment that "burns in" the new programming.

Element: One complete set of positive and negative plates with separators.

Elliot axle: Solid front axle in which the ends straddle the steering knuckles.

Emissions: Any release of harmful materials into the environment.

Emissions information label: Label normally located in the engine compartment that gives timing, idle speed, and vacuum hose routing information.

Emissions tune-up: See *Maintenance tune-up.*

End play: Amount of lengthwise movement between two parts.

Energy: Any capacity for doing work.

Engine: A device that converts energy into useful mechanical motion.

Engine displacement: The volume of space in which the piston moves in a full stroke multiplied by the number of cylinders in the engine.

Engine dynamometer: A device that tests engine output at the crankshaft.

Engine mounts: Pads made of metal, rubber, and plastic. Designed to hold engine to the frame. May also be liquid-filled.

Engine speed sensor: A sensor that sends information on engine speed and piston location to the ECU.

EPA: Environmental Protection Agency.

Equalizer pipe: A short pipe placed between the exhaust pipes on a dual exhaust system to equalize exhaust back pressure. May be used as location for oxygen sensor.

Erasable programmable read only memory (EPROM): A type of microprocessor whose programming can be altered only by erasing it with special equipment and reprogramming.

ESC: Electronic Spark Control.

EST: Electronic Spark Timing.

Ester oil: A type of refrigerant oil used in some R-134a air conditioning systems.

Ethyl alcohol: Grain alcohol or ethanol, a gasoline additive.

Ethylene glycol: A liquid chemical used in engine coolant.

Evacuation (air conditioning): Process of removing oxygen from an air conditioning system by pumping air out of the system, creating a vacuum.

Evaporation: The process of a liquid turning into vapor.

Evaporation control system: An emissions control system designed to prevent fuel vapor from escaping into the atmosphere.

Evaporator: A heat exchanger installed in the air conditioner blower case. Absorbs heat and humidity from the incoming air and transfers it to the circulating refrigerant.

Exhaust back pressure: Reverse pressure exerted in the exhaust system.

Exhaust gas analyzer: Electronic calibrated device used to measure the amount of pollutants in exhaust emissions.

Exhaust gas recirculation (EGR) valve: A valve that allows a controlled amount of exhaust gas into the intake manifold during certain period of engine operation. Used to lower exhaust emissions.

Exhaust manifold: Connecting pipes between the cylinder head exhaust ports and the exhaust pipes.

Exhaust pipe: Pipe used to connect the exhaust manifold to the catalytic converter and muffler.

Exhaust stroke: The portion of the piston's travel that is devoted to expelling the burned fuel charge.

Exhaust system: Components which carry exhaust emissions to the rear of the vehicle. These components include the muffler, exhaust pipes, catalytic converter, and exhaust manifold.

Expansion tank: A plastic tank used to recover excess coolant as part of the coolant recovery system.

Expansion valve: A temperature sensitive device used to regulate the flow of refrigerant into the evaporator. Also called a thermostatic expansion valve (TXV).

Extension housing: A housing bolted to the transmission case. Contains the transmission output shaft and oil seal.

Extreme pressure lubricants (EP): A lubricant compounded to withstand heavy loads such as that imposed between gear teeth.

F

F-head: An engine that has one valve in the block and the other in the cylinder head.

Fahrenheit: Unit of measurement for temperature of which the boiling point of water is 212°F and the freezing point is 32°F.

Fan: A mechanically or electrically operated device designed to create a moving stream of air, generally for cooling purposes.

Fan clutch: A temperature controlled device mounted to an engine-driven fan. It allows the fan to freewheel when the engine is cold or when the vehicle is at highway speed.

Fast idle cam: A carburetor cam that increases idle speed when the choke is closed.

Feathering: Tire wear pattern that is raised on one side. Appears and feels like the edges of feathers on a bird.

Feeler gauge: A thin strip of metal used for measuring clearance between two parts.

Ferrous metal: Metal containing iron or steel.

Fiber optic: A path for electricity or data transmission in which light acts as the carrier.

Field: Area surrounded by a magnetic force.

Filter: A device designed to remove foreign substances from air, oil, or fuel.

Final drive ratio: Overall gear reduction at the front or rear wheels.

Firewall: Metal bulkhead between the engine and passenger compartment.

Firing order: The order in which engine cylinders must be fired.

Fit: Contact area between two machined surfaces.

Flange: A projecting edge designed to keep a part in place.

Flare: A flange applied to tubing in order to provide a seal and to keep a fitting in place.

Flash point: The point at which an oil will flash or burn.

Flashover: An arc or jump of ignition current across an open space.

Float: A hollow device made of plastic or metal that is lighter than gasoline. Used to control the inlet needle valve to produce the proper fuel level in the bowl.

Float bowl: Part of the carburetor housing that acts as a reservoir for gasoline and contains the float and other carburetor parts.

Float circuit: The portion of the carburetor devoted to maintaining a constant fuel level.

Float level: Height of fuel in the fuel bowl or the specific float setting.

Flooding: Condition where the fuel mixture is overly rich or an excessive amount has reached the cylinders.

Flux: Lines of magnetic force moving through a magnetic field. Also, material used to join two pieces of metal being soldered or brazed.

Flywheel: A large heavy wheel that forms the base for the starter ring gear and provides a mounting surface for the torque converter or clutch assembly.

Foot-pound (ft.-lb.): The measurement of the amount of work involved in lifting one pound one foot.

Force: The amount of push or pull force exerted upon a body.

Foundation brakes: See *Base brakes.*

Four-gas analyzer: Exhaust gas analyzer that measures the amount of hydrocarbons, carbon monoxide, carbon dioxide, and oxygen in exhaust emissions.

Four-stroke cycle engine: An engine requiring two complete cycles of the crankshaft to fire each piston once.

Four-wheel alignment: Process in which all the wheels on a vehicle are aligned with each other.

Four-wheel drive: A vehicle in which the front wheels as well as the rear may be driven.

Four-wheel steering: System used to provide limited steering for the rear wheels. Operates in relation to the front wheels.

Frame: The steel structure that supports or is part of the vehicle's body.

Frame rails: Structural sections of the car frame.

Free electrons: Electrons in outer orbits around the nucleus of an atom. Can be moved out of their orbits relatively easily.

Freeze plugs: A stamped metal disc installed in the holes where the core was removed from the casting. Allows for expansion if coolant should freeze, preventing a cracked casting. Also called core plugs.

Freezing point: The temperature at which a liquid turns into a solid.

Frequency: The rate of change in direction, oscillation, cycles, etc., in a given time span.

Friction: Any resistance to movement between two objects placed in contact with each other.

Friction bearing: A bearing made of babbitt or bronze with a smooth surface.

Front end technician: See *Alignment technician.*

Fuel: Any substance that will burn and release heat.

Fuel bowl: Storage area in carburetor for extra fuel.

Fuel distributor: A device which meters fuel to the injectors at the correct flow rate for engine conditions.

Fuel injection: A system that sprays fuel directly into or just ahead of the cylinders.

Fuel injector: Valve controlled by an electronic solenoid or spring pressure.

Fuel lines: The portion of the fuel system that carries fuel from the tank to the filter and on to the carburetor or fuel injectors. Can be made of metal, rubber, or plastic.

Fuel pulsation: Fuel pressure variations due to pump action.

Fuel pump: A mechanically or electrically driven vacuum device used to draw fuel from the tank and force it into the fuel system.

Fuel rail: A metal tube that connects the injectors to the main fuel lines.

Fuel tank: A large tank made of steel or plastic used to store the vehicle's supply of fuel.

Fulcrum: A support on which a lever pivots in raising an object.

Full-floating axle: A rear drive axle that does not hold the wheel or support any weight, but merely drives the wheel.

Full-flow filter system: An oil filter that filters *all* of the oil passing through the engine

Full-time transfer case: A four-wheel drive unit that drives all of the wheels all of the time.

Full pressure system: A type of oiling system that draws oil from the sump and forces it through passages in the engine.

Full throttle cut-out switch: A switch or sensor that is monitored by the ECU. Allows the ECU to disengage the air conditioning compressor when a wide open throttle condition is detected.

Fuse: Protective device that interrupts current flow if an overload condition is present in the circuit.

Fusible link: A special calibrated wire installed in a circuit. Will allow an overload condition for short periods. A constant overload will melt the wire and break the circuit.

G

Galvanize: To coat a metal with a molten alloy mixture of lead and tin. Used to prevent corrosion.

Galvanometer: A device used for location, direction, and amount of an electric current.

Gas: A nonsolid material that can be compressed. When heated, it will expand and condense when cooled.

Gas line freeze: Condition caused by water in fuel turning to ice in cold weather and blocking the fuel lines.

Gasahol: Automotive engine fuel which contains 2-20% ethanol alcohol.

Gasket: A material placed between two parts to prevent leakage.

Gasoline: A hydrocarbon fuel used in internal combustion engines.

Gassing: Bubbles that rise to the top of battery electrolyte during charging. The bubbles are caused by hydrogen gas produced as a by-product of the charging process.

Gear: A circular object, either flat or cone shaped, upon which a series of teeth are cut.

Gear backlash: The small amount of clearance between gear teeth.

Gear clash: Noise that is heard when gears fail to mesh properly.

Gear oil: A high viscosity oil used in manual transmissions, transfer cases, and some differentials.

Gear pump: A type of pump that uses meshing gears to provide pressure and fluid movement.

Gear ratio: The relationship between the number of turns made by a driving gear to complete one full turn of the driven gear. If the driving gear makes four revolutions in order to turn the driven gear once, the gear ratio would be four to one (4:1).

Gear reduction: Setup in which a small gear is used to drive a larger gear. Produces an increase in torque.

Gearbox ratio: The number of turns between the steering wheel and the sector gear.

Generator: An electromagnetic device consisting of an armature, field windings, and other parts that produces electricity when turned. Also see *Alternator.*

Glass pack muffler: A type of muffler that uses a straight through design with no baffles. It uses fiberglass packing around a perforated pipe to deaden exhaust noise.

Glaze: A highly smooth, glassy finish on a cylinder wall. Produced over a long period of time by piston ring friction.

Glaze breaker: An abrasive tool used to remove glaze from a cylinder wall prior to the installation of new piston rings.

Glow plugs: A heating element used to help diesel engines start initially or in cold weather.

Governor: A device designed to control and regulate speed.

GPM: Gallons Per Minute.

Grade markings: Lines placed on the heads of some bolts to indicate tensile strength.

Gram (G): A metric unit of measurement for weight or mass equal to 0.03527 oz.

Grid: Lead screen or plate to which the battery plate active material is affixed.

Grind: To remove metal from an object by means of an abrasive wheel.

Groove: The space between two adjacent tire tread ribs.

Gross axle weight rating (GAWR): The total load carrying capacity of a given axle. It can be expressed as a rating at the springs or at the ground.

Gross horsepower: The brake horsepower of an engine with optimum fuel and ignition settings and without allowing for power absorbed by the engine-driven accessories.

Gross torque: The maximum torque produced when measured at the engine's crankshaft. Does not allow for torque consumed by the engine-driven accessories.

Gross vehicle weight rating (GVWR): The total weight of the vehicle including passengers and load. It is used as an indicator of how much weight can be safely loaded in a certain vehicle.

Ground: The terminal of the battery connected to the vehicle's frame.

Growler: A device used to test electric motor armatures.

Gum: Any oxidized portions of petroleum products that accumulate in the engine and fuel system.

Gut: To strip the interior of a car or internals of a part, such as a muffler.

H

Hall effect switch: A type of distributor pickup used in many ignition systems.

Halogen bulb: Quartz light bulb in which a tungsten filament is surrounded by a halogen gas such as iodine, bromine, etc.

Harmonic balancer: A device that is mounted on one end of a crankshaft. Used to reduce torsional vibration.

Harmonic vibration: A high frequency vibration caused by the crankshaft.

Hazard flashers: Circuit used to flash the turn signal lamps as a warning to other drivers.

Hazardous waste: Any chemical or material that has one or more characteristics that makes it hazardous to health, life, and/or the environment.

Header: A steel pipe that connects the exhaust manifold to the catalytic converter.

Headlights: Main driving lights on the front of the vehicle.

Headlight delay switch: An electronic switch that is used to control the on-off operation of the headlights after the ignition is turned off.

Heat exchanger: A device, such as a radiator, used to cool or heat by transferring heat from it to the air.

Heat range: Rating given for the operating temperature of spark plugs.

Heat riser: An area surrounding the intake manifold through which exhaust heat can rise to warm the fuel mixture during warmup.

Heat-shrink tubing: Plastic tube used to insulate electrical solder joints.

Heat shield: A sheet of metal or fiberglass used to shelter heat sensitive components, such as wiring, from excessive engine and exhaust heat.

Heat stove: A sheet metal housing placed around an exhaust manifold. An intake pipe to the carburetor air cleaner provides hot air when needed.

Heat treatment: Application of controlled heating and cooling to a metal that is timed in order to produce certain properties.

Heater core: A radiator-like unit in the blower case in which coolant circulates. Used to heat the vehicle interior.

Heavy repair technician: A person who specializes in the repair of engines and/or transmissions/transaxles.

Heel: The outside or larger part of a gear tooth.

Height sensing proportioning valve: A brake valve that can change the braking ratio in relation to vehicle load.

Height sensor: An electronic switch used to measure changes in vehicle ride height. Normally used with electronic ride control systems.

Helical gear: A gear that has teeth cut at an angle to its centerline.

Helicoil: An insert used to repair damaged threads.

Hemispherical combustion chamber: A dome shaped combustion chamber that provides increased engine aspiration while suffering less heat loss than other designs. Often referred to as a "hemi."

Herringbone gear: A pair of helical gears that operate together and are placed so that they form a "V" shape.

High side: Section of an air conditioning system in which the refrigerant is under high pressure.

Hone: Process of removing metal with a fine abrasive stone. Used to achieve close tolerances.

Hood pins: Metal pins designed to keep hood closed.

Horizontal-opposed engine: An engine that is placed perpendicular to the vehicle chassis.

Horn: A circuit that provides the driver with an audible warning signal.

Horsepower (hp): Measurement of an engine's ability to perform work. One horsepower is defined as the ability to move 33,000 pounds one foot in one minute.

Hose: A flexible rubber or neoprene tube for carrying water, oil, and other fluids.

Hose clamp: A device used to secure hoses to their fittings.

Hot idle compensator: A solenoid that increases idle speed when the engine is hot to prevent stalling.

Hot tank: A heated chemical cleaning tank that is used to clean parts that are immersed into it. Should not be used with aluminum parts.

Hotchkiss drive: A drive axle in which the driving force is transmitted to the frame through rear springs or link arms that connect the rear housing to the frame.

Hub: Mounting plate for the wheels on the end of an axle, spindle, or bearing.

Hydraulics: The science of fluids in motion.

Hydraulic modulator: An anti-lock brake system component that contains valves that are electronically controlled. Used to increase, decrease, or maintain brake system pressure to control wheel lockup (skidding). Also called a hydraulic actuator.

Hydraulic power booster: A brake booster that uses power steering fluid pressure to provide assistance in brake actuation.

Hydraulic system: An arrangement of tubing and pistons used to transmit force through fluid from one location to another.

Hydraulic valve lifters: A valve lifter that uses hydraulic pressure from the engine's oil system to keep it in constant contact with both the camshaft and pushrod/valve stem.

Hydrocarbons (HC): A combination of hydrogen and carbon atoms. An unwanted exhaust pollutant resulting from unburned fuel.

Hydrometer: A float device used to measure the specific gravity of battery electrolyte to determine its state of charge.

Hygroscopic: The ability to absorb moisture from the air.

Hypoid gearing: A system of gearing in which the pinion gear meshes with the ring gear below the centerline of the ring gear.

I

I-head: An engine that houses both valves in the head.

ID: Inside Diameter.

Idle air control valve (IAC): A computer-controlled valve used to supply air to a fuel injected engine at idle. Also regulates idle speed.

Idle mixture adjustment screw: One or more needle valves used to adjust the air-fuel mixture on a carburetor.

Idle speed: The crankshaft rotational speed in an engine with a closed throttle plate.

Idle speed-up solenoid: A device that increases engine idle speed when an engine-driven accessory, such as the air conditioning compressor, is used.

Idler arm: A steering system component that supports one end of the center link.

Ignition: Lighting or igniting a fuel charge by means of a spark or compression.

Ignition coil: An electrical component used to increase battery voltage to fire spark plugs.

Ignition module: An electronic component that controls ignition spark sequence and fires the coil(s) when needed.

Ignition switch: A key operated switch mounted on the steering column for connecting and disconnecting power to the ignition and electrical system.

Ignition timing: The relationship between the exact time a plug is fired and the position of the piston in terms of degrees of crankshaft rotation.

Impact sensor: An open switch that is designed to close when an impact of sufficient force is encountered. Used in air bag systems to detect vehicle impact.

Impeller: A wheel-like device that has fins cast into it. It is mounted on a water pump shaft to turn for pumping coolant.

Inclined engine: An engine that is set at an angle. Allows engine to be installed in a smaller space.

Included angle: The angle formed by the steering knuckle and the center of the wheel if center lines were drawn through it. Combines both camber and steering axis inclination angles.

Incoming air temperature sensor (IAT): A thermistor installed in the air cleaner or air intake tube. Sensor used by the ECU to monitor the temperature of the air entering the engine.

Independent suspension: A suspension system that allows each wheel to move up and down without influence from the other wheel on that axle.

Indicated horsepower (IHP): The measure of power produced by burning fuel within the cylinders.

Indirect fuel injection: Fuel injection system that sprays fuel into the intake manifold.

Induction: The imparting of electricity to an object by magnetic fields.

Inertia: Force which tends to keep stationary objects from moving and keeps moving objects in motion.

Inertia switch: A switch that is designed to operate only if a sudden movement occurs, such as a collision.

Inflation pressure: The amount of air pressure a tire can handle safely.

Information center: Display that shows vehicle condition to the driver.

Infrared rays: Rays of light that are invisible to the human eye. Used with some dyes to trace fluid leaks.

Inhibitor: A material added to another material to control or prevent some unwanted action, such as corrosion, foaming, etc.

Injector: A valve that is controlled by a solenoid or spring pressure to inject fuel into the engine.

Inline: Any group of components in a straight row.

Inline engine: An engine in which all of its cylinders are in a straight row.

Input sensors: Any sensor that provides information to an ECU.

Input shaft: A shaft that delivers power to a mechanism, such as the transmission input shaft delivering power to the transmission gears.

Insulation: Any material that resists the flow of electrons, heat, or noise.

Intake air heaters: A wire grid that is placed as a spacer between the throttle plate and the intake manifold. Electricity is used to heat the air-fuel charge as it enters the intake manifold.

Intake manifold: Series of connecting tubes or housing between the throttle plate and the openings to the intake valves.

Intake stroke: The portion of the piston's movement that is devoted to drawing the air-fuel mixture into the combustion chamber.
Intake valve: Valve through which air and fuel is admitted to the cylinder.
Integral: A device that is formed as part of another unit.
Integrated circuit (IC): A single chip of semiconductor material which contains various electrical components in miniaturized form.
Intermediate gear: Any transmission gear between low and high.
Intermittent: An event that occurs at different intervals.
Intermittent codes: Computer diagnostic code that does not return immediately after it has been cleared.
Internal combustion engine: An engine that burns fuel within itself as a means of developing power.
Internal gear: A gear with teeth cut on its inward facing surface.
Ion: An electrically charged atom or molecule produced by an electrical field, high temperature, etc.
Ionize: To convert partially or completely into ions.

J

Jet: An orifice used to control the flow of gasoline in various parts of a carburetor.
Joule: Metric unit of measurement for energy or work equal to a force of one Newton applied through a distance of one meter. One joule is equivalent to 0.737324 ft-lbs.
Jounce bumper: A rubber block that keeps suspension parts from contacting the frame when the vehicle encounters a large bump or hole.
Journal: The part of a shaft or axle that actually contacts a bearing.
Jump start: Method of starting a vehicle with a weak or dead battery through the use of jumper cables.
Jumper cables: A pair of electrical cables used to start a car with a weak or dead battery.
Jumper wire: A wire used to make a temporary electrical connection.

K

Keep alive memory (KAM): A type of computer memory that stores changes in sensor values. Provides information to the computer's CPU so it can properly adjust the signals to the various output devices. Sometimes referred to as adaptive strategy.
Key: A small metal piece that fits into a groove partially cut into two parts to allow them to turn together.
Keyway: A slot cut into a shaft, pulley, or hub that permits the insertion of a key.
Kickdown valve: Spool valve inside an automatic transmission's valve body that causes the transmission to shift into a lower gear during fast acceleration.
Kilometer (km): A metric unit of measurement for distance equivalent to 5/8 of a mile.
Kinetic energy: Any energy associated with motion.
Kingpin: A hardened steel shaft around which the steering knuckle pivots. Normally used on large trucks.
Knock: Engine noise caused by detonation, preignition, or a worn mechanical part.
Knock sensor: Engine sensor that detects preignition, detonation, and knocking.
Knurl: The process of roughing a piece of metal by pressing a series of cross-hatched lines into the finished surface, which raises the area between the lines.

L

L-head engine: An engine design that incorporated both valves on one side of the engine cylinder.
Lacquer: A solution of solids, such as paint, that evaporate with great rapidity.
Laminate: To build up or construct something out of a number of thin sheets.
Land: A portion of metal separating the grooves that rings ride against.
Lapping: The process of fitting two surfaces by rubbing them together with an abrasive between them.
Latent heat: The amount of heat beyond a substance's boiling or melting point required to change a solid to liquid or liquid to vapor.
Lateral runout: Side-to-side movement of a wheel, tire, or rotor.
Lathe: Machine on which a solid piece of material is spun and shaped by a fixed cutting tool.
Lead burning: Process of connecting two pieces of metal by melting the edges together.
Leaf spring: Suspension springs made of steel leaves bound together. A varying number of steel leaves are used depending on its intended use. Some leaf springs are solid units made of fiberglass.
Leak detector: Electronic tool used to find refrigerant leaks.
Lean mixture: A fuel mixture with an excessive amount of air in relation to fuel.
LED code: A diagnostic trouble code indicated by a pattern of flashes by a light emitting diode.
Lever: A rigid bar or shaft that pivots on a fixed fulcrum. It is used to increase force or to transmit a change in motion.
Lift: Maximum distance a valve head is raised off its seat.
Light emitting diode (LED): A special function diode that lights when forward biased.
Light repair technician: A person who specializes in preparing cars for delivery and performing minor repairs and maintenance. Normally these are apprentices who are working under the supervision of one or more experienced technicians.
Limited-slip differential: A differential unit designed to provide superior traction by transferring the driving torque to the wheel with the best traction. Extremely useful in wet or icy conditions.
Liner: A thin section, such as a cylinder liner, placed between two parts.
Linkage: Any movable series of rods, levers, and cables used to transmit motion from one unit to another.
Liquid: A substance that is neither a solid nor a liquid. Can assume the shape of the vessel in which it is placed without a change in volume.
Liquid withdrawal: A system which draws LPG gas from the bottom of the tank to ensure delivery of liquefied gas.
Liter: Metric unit of measurement for liquid equal to 2.11 pints. Also used as a measure of volume equal to 61.027 cubic inches.
Live axle: An axle in which power travels from the differential to the wheels.
Load range: Tire designation which uses a letter system (A,B,C, etc.) to indicate specific tire load and inflation limits.
Lock washer: A type of washer used to prevent accompanying nut from working loose.
Locking hub: Component used in four-wheel drive vehicles to transfer power from the driving axles to the wheels.
Lockup torque converter: A torque converter with an internal clutch that locks the turbine to the impeller when the automatic transmission is in direct drive or overdrive.
Longitudinal: Parallel to the length of the vehicle.
Louver: Slots cut into the hood or body, usually for ventilation.
Low brake pedal: Condition where the brake pedal approaches too close to the floorboard before actuating the brakes.
Low fluid warning switch: A sensor used in various systems to notify the driver of low or no fluid in a system, such as washer fluid or engine coolant.
Low pressure switch: One that prevents air conditioning compressor clutch engagement if the system pressure falls below a specified point. May also be included as an input sensor to the engine's electronic control unit.
Low side: Low pressure side of an air conditioning system.
Lug bolts: Threaded bolts or studs that are press-fit into an axle, hub, or brake rotor and accept lug nuts to mount wheels.
Lug nuts: Large steel nuts used to hold a wheel to the axle hub.

M

MacPherson strut suspension: A suspension system in which the wheel assemblies are attached to a long telescopic, shock-like strut. It permits the wheels to pivot while acting as a shock absorber.
Magnetic clutch: An electromagnetic clutch that engages and disengages the air conditioning pulley.
Magnetic field: Area encompassed by magnetic lines of force surrounding a magnet.

Magnetic timing meter: A tachometer that uses magnetism as a triggering device.

Magnetism: Invisible lines of force that attract ferrous metals.

Main bearing: Series of bearings that support the crankshaft in the engine.

Main body: The central portion of the carburetor that forms the fuel bowl and air horn.

Main cap: Metal pieces that bolt to the block. Used to support the crankshaft.

Main discharge tube: Carburetor fuel passage from the bowl to the air horn.

Maintenance-free battery: A sealed battery that requires no additional water or electrolyte during its useful life.

Maintenance tune-up: Tune-up that includes replacing the spark plugs and air, fuel, and emissions filters.

Malfunction indicator light (MIL): Amber-colored light in the instrument cluster used to indicate that a problem exists in a vehicle's computer control system. Also called a Check Engine or Service Engine Soon light. Generalized term used for any instrument cluster light used to indicate a problem in a system.

Manifold: A pipe or series of pipes connecting a series of ports to a common opening.

Manifold absolute pressure sensor (MAP): Computer sensor used to measure the barometric pressure in relation to intake vacuum. Sometimes called a barometric pressure sensor.

Manifold air temperature sensor (MAT): Computer sensor used to measure the temperature of the air coming into the intake manifold. Also called an Intake Air Temperature (IAT) sensor.

Manifold heat control valve: A valve placed in the exhaust manifold or exhaust pipe. Used to route hot air from the exhaust manifolds to the base of the carburetor to aid in warmup.

Manifold vacuum: Vacuum that exists in the intake manifold under the carburetor or throttle body throttle plate.

Manual transmission/transaxle: A driver operated (shifted) transmission or transaxle.

Mass airflow sensor (MAF): Computer sensor used to measure the amount of air entering the intake manifold. Also called an Airflow Meter.

Master cylinder: The part of the hydraulic brake system in which system pressure is generated.

Master Technician: An experienced technician. Usually used to refer to someone who has passed all of the ASE certification tests in a skill area.

Material Safety Data Sheet (MSDS): Information on a chemical or material that must be provided by the material's manufacturer. Lists potential health risks and proper handling procedures.

Matter: Any substance that makes up anything that occupies space, has weight, and is perceptible to the senses.

Mechanical efficiency: An engine's rating as to how much potential horsepower is lost through friction within its moving parts.

Mechanical fuel injection: A system that utilizes a mechanically driven fuel pump to force fuel into the engine.

Mechanical fuel pump: An engine mounted fuel pump that is driven by an eccentric device.

Mechanical lifter: Solid valve lifter used in some engines. Requires that valve train be adjusted periodically.

Melting point: Temperature at which a solid becomes a liquid.

Meter: Metric unit of measurement of length. One meter is equal to 39.37 inches.

Metering rod: Movable rod that is used to vary the opening through the carburetor main jet.

Metering valve: A valve that limits hydraulic pressure to the front brakes until a predetermined line pressure is reached.

Methanol: Methyl alcohol or wood alcohol.

Metric system: A decimal system of measure that is based on 10. Used by most of the world and in some professions worldwide.

Micrometer: A precision measuring tool that will give readings to within thousandths or ten thousandths of an inch.

Microprocessor: A small silicon chip that contains elements in a computer. Often referred to as an integrated circuit or "IC."

Millimeter (mm): Metric unit of measurement equivalent to .039370 of an inch. One inch is equivalent to 25.4 mm.

Milling: Process of cutting metal with a multitooth rotating cutting wheel.

Mineral oil: A lightweight oil used as a refrigerant oil in R-12 air conditioning systems and as a base oil for power steering and automatic transmission fluid.

Misfiring: Failure of one or more cylinders to fire.

Mixture control (MC) solenoid: An electronic solenoid located in the carburetor. Used to move the metering rods in and out.

Modulator: Pressure control or regulating device used in automatic transmissions.

Molecule: The smallest portion that matter may be divided into and still retain all of its properties.

Monoblock: A block in which the cylinders are cast as a unit.

Motor: A device which converts electrical or fluid energy to mechanical energy.

MPH: Miles Per Hour.

Muffler: A chambered unit attached to a pipe or hose to deaden noise.

Multi-viscosity oil: An engine oil that can exhibit different viscosity characteristics when heated or cooled.

Multimeter: An electrical test meter that can be used to test for voltage, current, or resistance.

Multiple disc clutch: A clutch assembly that contains several clutch discs in its construction.

Multiplexing: A method of using one communications path to carry two or more signals simultaneously.

Multiport fuel injection: Fuel injection system in which there is one injector per cylinder. Also called Multi-Point, Tuned Port, or Sequential Fuel Injection.

MVMA: Motor Vehicle Manufacturers Association.

N

NADA: National Automobile Dealers Association.

Natural gas: General term for any petroleum based gas, such as propane and liquefied petroleum gas.

NC threads: National Coarse thread sizes.

Needle bearing: An antifriction roller bearing that uses many small diameter rollers in relation to their length.

Needle valve: A valve with a long, thin tapered point that operates in a small hole or jet. The hole size is changed by moving the needle in and out.

Negative terminal: A terminal from which current flows on its path to the positive terminal. Usually designated by a minus sign (-).

Net horsepower: Maximum horsepower at the flywheel with all engine-driven accessories in use.

Net torque: Maximum torque at the flywheel with all engine-driven accessories in use.

Neutral safety switch: A switch that prevents starter engagement if the transmission is in gear.

Neutron: A particle of an atom that has a neutral charge. Forms the central core of an atom along with protons.

Newton meter (N•m): Metric unit of measurement for torque equivalent to .7376 foot pounds.

NF thread: National Fine thread sizes.

NHTSA: National Highway Traffic Safety Administration.

Nonferrous metal: A metal that contains no iron or very little iron.

North pole: One of the poles of a magnet from which lines of force originate.

Nozzle: An opening through which fuel mixture is directed into the air stream flowing through a carburetor.

NSC: National Safety Council.

Nut lock: A slotted nut which fits over a standard nut on a wheel spindle. Normally used on nondriving axles.

O

OBD II: On-Board Diagnostics Generation Two. Protocol adapted by vehicle manufacturers for standardization of diagnostic trouble codes and automotive terminology.

Octane: Rating indicating a fuel's tendency to resist detonation. Does not have a bearing on the fuel's quality.
OD: Outside Diameter.
Odometer: A device used to measure and register the number of miles a vehicle has traveled.
OEM: Original Equipment Manufacturer.
Ohm: Unit of measurement for resistance to the flow of electric current in a given unit or circuit.
Ohm's Law: (E = I x R) Formula for computing unknown voltage, resistance, or current in a circuit by using two known factors to find the unknown value.
Ohmmeter: An electrical instrument used to measure the amount of resistance in a given unit or circuit.
Oil classification CA, CB, CC, CD, CE: Classification for oil designed for use in diesel engines.
Oil classification SA, SB, SC, SD, SE, SF, SG, SH: Classification for oil designed for use in automotive gasoline engines.
Oil cooler: An air or liquid cooled device used to remove excess heat from the engine or transmission oil.
Oil gallery: Cast or drilled passageway or pipe in an engine. Used to carry oil from one part of the engine to another.
Oil pickup: Tube and screen that connects to the oil pump and extends to the bottom of the oil pan. Used by oil pump to pickup oil.
Oil pressure gauge: A dash mounted instrument that provides oil pressure readings in pounds per square inch or in kilopascals.
Oil pump: A device that forces oil under pressure to the oil galleries for distribution throughout the engine.
Oil ring: The bottom piston ring which scrapes oil off the engine cylinder walls.
Oil seal: A device used to prevent oil leakage past certain areas, such as around a rotating shaft.
Oil slinger: An impeller or other device attached to a revolving shaft that will throw any oil passing it outward so that it can return to its point of origin.
Open circuit: A circuit that is broken or disconnected.
Open loop: Computer operation mode used when an engine has not reached its optimum operating temperature. The ECU operates the engine using a basic set of fixed variables provided by the ECU's PROM.
Optical sensor: A light sensitive device used in some distributors to determine when to close the ignition circuit. Its operation is similar to a Hall-Effect switch.
Orifice: A small hole or restricted opening used to control the flow of gasoline, oil, refrigerant, etc.
Orifice tube: A tube with a calibrated opening used to restrict the amount of liquid refrigerant. Used in place of an expansion valve in some systems.
Oscillate: Any back and forth swinging action like that of a pendulum.
Oscilloscope: Test device used to observe voltage by displaying line patterns in relation to time.
Otto cycle: The four-stroke cycle consisting of intake, compression, power, and exhaust.
Out-of-round: Condition where a cylinder or other round object, such as a tire, has greater wear at one diameter than another.
Output device: A computer-controlled device used to change an engine setting, such as fuel and spark timing.
Output shaft: A shaft that delivers power from within a mechanism, such as a transmission.
Overdrive: An arrangement of transmission gears that results in the driven shaft turning more revolutions than the driving shaft.
Overflow tank: A plastic container used to hold extra coolant in a closed cooling system.
Overhead camshaft (OHC): A camshaft mounted above the cylinder head. Usually driven by a timing chain, belt, or a combination of the two.
Overhead valves (OHV): An engine design where the valves are located in the cylinder head.
Overrunning clutch: A clutch mechanism that will drive in one direction and slip in the other.
Oversteer: The tendency of a car to turn more sharply in a corner than the driver intends.
Oxides of nitrogen (NO_x): An undesirable compound of nitrogen and oxygen in exhaust gases. Usually produced when combustion chamber temperatures are excessively high.
Oxydize: To combine an element with oxygen or a catalyst which converts it to its oxide form. This action can form rust in ferrous metals.
Oxidizing catalyst: A chemical or other substance which causes oxygen to combine with the metal to produce rust and scale.
Oxygen sensor (O_2S): An exhaust sensor used to measure the amount of oxygen in the exhaust gases produced by the engine. The signal sent to the engine ECU is used to determine fuel mixture, spark timing, etc.

P

Packing: The process of filling a bearing with grease.
Pad: Disc brake friction lining.
Pan: A thin metal cover bolted to the bottom or side of an engine or transmission/transaxle to contain oil.
Pancake Engine: An engine that is laid horizontally to reduce overall height.
Parallel circuit: An electrical circuit that has two or more resistance units wired so that current can flow through them at the same time.
Parallelogram steering linkage: A steering system utilizing two short tie rods connected to steering arms and to a long center link. The link is supported on one end of an idler arm and the other end is attached to a pitman arm. The arrangement forms a parallelogram shape.
Parasitic load: Normal electrical load from the ECU, radio, and other electrical components placed on a vehicle's battery when the engine is not operating.
Parking brakes: A hand- or foot-operated brake which prevents vehicle movement while parked by actuating the rear brakes.
Parking lights: Small lights on the front and rear of the vehicle. Used to make vehicle more visible during nighttime driving.
Part-time transfer case: A four-wheel drive transfer case that permits either four-wheel or two-wheel drive.
Pascal's Law: A principle of fluids that states that when pressure is applied to a confined fluid, it is transferred undiminished throughout the fluid.
Pawl: A stud or bar adapted to engage with teeth cut on another part, such as on a transmission gear.
Peen: To flatten out by pounding with the round end of a hammer.
Penetrating oil: Special oil used to free corroded parts so that they can be removed.
Permanent magnet: A magnet capable of retaining its magnetic properties over a very long period of time.
Petcock: A valve placed in a tank or line for draining purposes.
Petroleum: Raw oil taken out of the ground to be refined into gasoline, kerosene, etc.
Phillips head screw: A screw head having a fairly deep cross slot.
Phosgene gas: A toxic, colorless, poisonous gas produced when refrigerant is burned.
Phosphor-bronze: A bearing material composed of lead, tin, and copper.
Photo sensitive diode: A semiconductor device that allows current to flow when exposed to light.
Pickup coil: An electronic distributor component that sends electrical pulses to the ignition module when the distributor shaft rotates.
Pilot bearing: The bushing that supports the end of the transmission input shaft.
Pilot shaft: A "dummy" shaft used temporarily to align parts for assembly.
Pinging: A metallic rattling sound produced by the engine during acceleration. Sound associated with detonation or preignition.
Pinion carrier: Part of the rear axle that contains and supports the pinion gear shaft.
Pinion gear: A small gear that is either driven by, or driving, a larger gear.
Piston: Cylindrical plug that is closed on one end and attached to a connecting rod at the other. When the fuel charge is fired, the force of the explosion will transfer from the closed end, through the piston to the connecting rod and crankshaft.

Piston boss: Built-up area around the piston pin hole.
Piston collapse: A reduction in the diameter of the piston skirt caused by heat and impact stress.
Piston displacement: The amount of air displaced by the piston when it is moved through the full length of its stroke.
Piston expansion: An increase in piston diameter due to normal heating.
Piston head: The portion of the piston above the top ring.
Piston lands: The portion of the piston between the ring grooves.
Piston pin: A steel pin that is passed through the piston. Used as a base for attaching the connecting rod. Also called a wrist pin.
Piston rings: A split ring installed in a groove in the piston. Seals the compression chamber from the crankcase.
Piston ring end gap: Distance left between the ends of the rings when installed in the cylinder.
Piston ring grooves: Series of slots in the piston in which the rings are fitted.
Piston skirt: The portion of the piston below the rings.
Piston skirt expander: Spring device placed inside a piston skirt to increase its diameter.
Pitman arm: A short lever arm splined to the steering gear cross shaft.
Pivot lever: A pin or shaft on which a part rests or turns.
Planet carrier: Part of a planetary gearset upon which the planetary gears are affixed.
Planetary gears: The gears in a planetary gearset that mesh with the ring and sun gears. Referred to as planetary gears because they orbit or move around the center, or sun gear.
Plastigage: A measuring tool that is compressed between two tightly fitting surfaces, such as bearings, to measure clearance.
Plates: Thin sections of lead peroxide or porous lead used to make up a battery's positive and negative cells.
Platinum (Pt): A precious metal used in the construction of catalytic converters. It is sometimes used as an electrical conductor.
Play: Any movement between two parts.
Plies: Layers of rubber impregnated with fabric or steel that make up the carcass of the tire.
Plug gap: The distance between the center and side electrodes of a spark plug.
Plug heat range: A numeric valve assigned to a spark plug to indicate how hot a spark it is capable of producing.
POA: Pilot Operated Absolute Valve.
Polarity: The positive or negative terminals of a battery. Also the north and south poles of a magnet.
Pole Piece: Component of motor that keeps the armature rotating.
Pole shoes: Metal pieces around which field coil winding are placed.
Polyalkylene glycol (PAG) oil: Oil used in air conditioning systems that have R-134a as a refrigerant.
Polymer: A chemical chain of simple molecules added to a substance, such as oil, to increase its performance characteristics.
Poppet valve: A valve used to open and close a circular port or hole.
Port fuel injection: A type of fuel injection system that uses one injector per cylinder. Also called Multiport injection.
Positive crankcase ventilation (PCV) system: A system that uses a valve to clear the engine of blowby gases.
Positive: Terminal to which electrons flow. Usually indicated by a plus (+) sign.
Potential horsepower: A measurement of the maximum amount of horsepower available.
Potentiometer: A variable resistor that can be used to adjust voltage in a circuit.
Power booster: A vacuum or hydraulic operated device used to increase brake pedal force on the master cylinder during stops.
Power brakes: A brake system that has a vacuum or hydraulic powered booster as part of the overall system.
Power steering: A steering system that utilizes hydraulic pressure to increase the driver's turning effort.
Power steering pump: A belt-driven pump that provides hydraulic pressure to assist in the driver's turning efforts.
Power stroke: The downward movement of the piston that occurs after the air-fuel charge has been ignited.
Powertrain: Group of components such as the transmission, axles, or joints, used to provide driving force to the wheels.
Power valve: Device that allows more fuel to enter the carburetor when more power is needed.
PPM: Parts Per Million.
Practical efficiency: The amount of horsepower delivered to the driving wheels.
Preheating: The application of heat in preparation for some further treatment, such as welding.
Preignition: Condition where the fuel charge is ignited before it is intended.
Preloading: Process of adjusting a bearing so that it has a mild pressure placed upon it.
Press-fit: Condition of fit between two parts that requires pressure to force the two parts together.
Pressure: Any force per unit of area placed on a surface.
Pressure bleeding: A method of bleeding a system by using additional pressure from an external source.
Pressure cap: A cap that is designed to hold a preset amount of pressure.
Pressure differential switch: A hydraulic switch in the brake system that operates the brake warning light in the dashboard.
Pressure plate: A spring-loaded assembly that keeps pressure on the clutch disc against the flywheel.
Pressure plate cover: A metal cover that bolts onto the pressure plate to hold its components together.
Pressure regulator: A device in an automatic transmission that limits maximum line pressure.
Pressure relief valve: A valve designed to open at a specific pressure to prevent excessive pressure on a closed system.
Pressure sensor: A sensor that is used to detect excessive high or low pressures in a system.
Pressure-splash system: A system that uses an oil pump to supply oil to the camshaft and crankshaft bearings and the movement of the crankshaft to splash oil onto the cylinder walls and other nearby parts.
Pressure-temperature relationship: The relationship between ambient temperature and pressure of refrigerant in an air conditioning system.
Primary brake shoe: Brake shoe installed facing the front of the vehicle.
Primary circuit: A low voltage circuit that is part of the ignition system.
Primary winding: Low voltage winding in a coil. It is made of heavy wire.
Primary wiring: Small insulated wires which serve the low voltage needs of the ignition and vehicle systems.
Printed circuit: An electrical circuit that is made by conductive strips printed on a board or panel.
Programmable read only memory (PROM): A semiconductor chip that contains instructions that are permanently encoded into the chip. Instructions contain base operating information for how components should operate under various conditions.
Prony brake: A device that utilizes friction to measure the horsepower output of the engine.
Propane: Hydrocarbon-based gas that is mixed with butane and is sometimes used as an engine fuel. Also known as LPG or CNG.
Propeller shaft: See *Drive shaft.*
Proportioning valve: A valve in the brake lines that equalizes system pressure between the front and rear brakes.
Proton: A positively charged particle that is part of the atom.
PSI: Pounds per Square Inch.
Pulsation damper: A metal or plastic sleeve used to smooth out fuel pump pulsations or surges to the rest of the fuel system.
Pulse air injection system: An emission control system which feeds air into the exhaust system by using exhaust pulses.
Pulse fuel injection: Fuel system in which injectors are only open for a short period and remain closed the rest of the time. The amount of fuel delivered is controlled by how long the injector is open.
Pulse width: The length of time a fuel injector is held open by the engine control computer.
Pump: A device that is designed to move coolant, oil, fuel, etc., from one area to another.
Purge: The process of removing air or impurities from a system. Also see Bleeding.

Push rod: A rod that connects the valve lifter to the rocker arm.

Q

Quadrant: A gear position indicator often using a shift lever actuated pointer. Can be marked PRNDD21 (four-speed).

Quenching: Process of dipping a heated object into water, oil, or other substance to quickly reduce its temperature.

Quick charge test: Method of determining if a battery's plates are sulfated.

R

R-12 (CFC-12): Refrigerant used in older air conditioning systems. Gradually being replaced by R-134a in newer vehicles. Also called dichlorodifluoromethane (CCl_2F_2).

R-134a (HFC-134a): Refrigerant used in the air conditioning systems of most vehicles manufactured after 1992. Replaced R-12 due to environmental concerns. Also called tetrafluroethane (CH_3CH_2F).

Rack: A flat toothed bar inside a rack-and-pinion steering gear that is moved left or right by the pinion gear.

Rack-and-pinion: A steering gear that utilizes a pinion gear on the end of the steering shaft. The pinion gear engages a long rack, which is connected to the steering arms via tie rods.

Radial: A line at right angles to a shaft, cylinder, etc., centerline.

Radial compressor: An air conditioning compressor that uses reciprocating pistons set at right angles to the drive shaft and spaced around the shaft in a radial fashion.

Radial runout: Difference in rotation caused by uneven diameter.

Radial tire construction: A tire that has its plies set parallel and at right angles to the centerline of the tire.

Radial tire pull: A condition where a defect in a radial tire causes a vehicle to pull left or right.

Radiation: Any transfer of heat from one object to another when the hotter object radiates invisible waves of heat that strike surrounding objects, causing them to vibrate and heat.

Radiator: A heat exchanger used to remove heat from engine coolant. It is comprised of a series of finned passageways. As coolant moves through the passageways, heat is given off to the fins, which is then dissipated by passing air through the fins.

Radiator core: A series of finned passages made of copper or aluminum in a radiator through which coolant passes and gives off its excess heat.

Radius: Distance in a straight line from the center of a circle.

Radius rods: Rods attached to the axle and pivoted on the frame. Used to keep an axle at a right angle to the frame.

Random access memory (RAM): A portion of computer memory that serves as temporary storage for data. This data is lost if power to the computer is lost. It is used to store sensor information and any diagnostic trouble codes.

Ram air: Air that is forced through a condenser or radiator or into the engine by vehicle movement.

Rated horsepower: Indication of horsepower that can be safely placed upon an engine for a prolonged period of time.

Ratio: A fixed relationship between things in number, quantity, or degree. An example is an air-fuel mixture of one part of fuel for fifteen parts of air or 15:1.

Read only memory (ROM): A type of computer memory that cannot be changed and is not lost if power is cut off. Contains the general information to operate a computer.

Ream: To enlarge or smooth a hole by using a round cutting tool with fluted edges.

Rear wheel shim: A beveled metal disc placed between the hub bearing and rear axle on some front-wheel drive vehicles to adjust rear camber and/or toe.

Receiver-dehydrator: An air conditioning system component that is used to dry and store refrigerant.

Reciprocating motion: Any back and forth motion such as the action of pistons in an engine.

Recirculating ball worm and nut: A steering gear that utilizes a series of ball bearings that feed through and around grooves in a worm and nut.

Rectified: A term used to describe alternating current (ac) that is changed to direct current (dc).

Reduction gear: A gear that increases torque by reducing the speed of a driven shaft in relation to the driving shaft.

Reed switch: An electronic switch that consists of two metal strips or reeds. The reeds are influenced by a magnetic field, which causes them to open or close, depending on the application.

Reference voltage: A known voltage (can vary from 0.5-5V) that is sent to a sensor by a computer. The changes in resistance in the sensor will change the voltage, which is read by the computer as a change in temperature, airflow, etc.

Refrigerant: A liquid used in refrigeration systems to remove heat from the evaporator and carry it to the condenser. Automotive systems use R-12 and R-134a.

Refrigerant hose: Any neoprene hoses that are used to connect the fixed components of a refrigeration system to components that are not fixed, such as the compressor to accumulator connection.

Refrigerant line: Any metal line that is used to connect two components of a refrigeration system that are fixed, such as the evaporator to condenser line.

Refrigerant oil: A special oil (mineral, ester, or polyalkylene glycol) which lubricates the air conditioning compressor.

Refrigerant recovery unit: Electronic station that combines a refrigerant storage tank, vacuum pump, gauges, and service valves. Used to recover, recycle, and recharge refrigerant in automotive air conditioning systems.

Relative volatility: The point at which a substance or one of its components turns to vapor or "flashes".

Relay: A magnetically operated switch used to make or break current flow in a circuit.

Relieve: The process of removing metal from the valve seat area and cylinder to improve the flow of fuel into the cylinder.

Reluctor: A component in an electronic ignition system distributor. It is affixed to the distributor shaft and triggers a magnetic pickup, which triggers the control module to fire the coil.

Remote keyless entry: Electronic system that is added to some vehicles to enable the vehicle's owner to lock and unlock the doors and open the trunk using a key fob transmitter.

Removable carrier: A rear axle assembly in which the differential housing, ring, and side gears can be removed as a single unit.

Reserve capacity: The amount of time a battery can produce an acceptable current when not charged by the alternator.

Reservoir: A tank or bottle used to hold a reserve of fluid, such as coolant or washer fluid.

Resistor: A device placed in a circuit to lower voltage and current flow.

Resistance: The measure of opposition to electrical flow in a circuit.

Resonator: A small muffler-like device that is placed in an exhaust system to further reduce exhaust noise.

Retard: To set the ignition timing so that the spark occurs later or less degrees before TDC.

Retread: A used tire that has new rubber bonded to it.

Retrofitting: Process of converting an air conditioning system that uses R-12 to handle R-134a refrigerant.

Return spring: A spring positioned to close a valve or return a brake shoe back to its resting position.

Reverse bias: A condition or arrangement where a diode acts as an insulator.

Reverse elliot axle: A solid bar front axle on which the steering knuckles span or straddle the axle ends.

Reverse flow muffler: A muffler that has its internal pipes arranged so that exhaust gases flow in a reverse direction inside the muffler before exiting to the tailpipe.

Reverse flush: Method of cleaning by flushing a cleansing agent through a system in the reverse direction of normal fluid flow.

Reverse idler shaft: A shaft in a transmission that holds the reverse gear.

Rheostat: A variable resistor used to control current flow.

Ribbed belt: A V-type drive belt that has small ridges added along its length.

Rich mixture: An air-fuel mixture that has an excessive amount of fuel in relation to air.

Ridge reamer: A device used to remove the metal ridge that forms at the top of a cylinder due to wear prior to piston removal.

Rim: The metal wheel upon which a tire is mounted.

Ring expander: A spring device used to hold a ring snugly against the cylinder walls.

Ring gap: Distance between the end of the piston rings when installed in the cylinder.

Ring gear: Large gear attached to the differential carrier or to the outer gear in a planetary gear setup.

Rivet: A metal pin used to hold two objects together. One end has a head while the other end must be set or peened over.

RMA: Rubber Manufacturers Association.

Road crown: Angle or slope of the roadway to allow water to run off.

Road feel: The feeling imparted to the steering wheel by the wheels of the vehicle in motion. This "feeling" is important in sensing and predetermining steering response.

Rocker arm: A lever arm used to direct downward motion on a valve stem.

Rocker shaft: A shaft upon which rocker arms are mounted in some engines.

Rod cap: The lower removable half of a connecting rod.

Roller bearing: A bearing which contains hardened roller ball bearings between two races.

Roller clutch: A clutch that utilizes a series of rollers placed in ramps. The clutch will provide driving force in one direction, but will slip in the other.

Roller lifter: A valve lifter that has a roller bearing which rides on the cam lobe to reduce friction.

Roller vane pump: A pump which uses spring-loaded vanes that are shaped like rollers to provide the pumping action.

Rollover valve: A valve in the fuel tank or delivery lines that prevents the escape of raw fuel in the event of a vehicle rollover.

Room temperature vulcanizing (RTV): A type of sealant that cures at room temperature.

Rotary engine: An internal combustion engine that uses one or more triangular rotors instead of pistons to accomplish the intake, compression, power, and exhaust cycles. Also referred to as a Wankel engine.

Rotary motion: A continuous motion in a circular direction, such as performed by a crankshaft.

Rotary pump: A pump that uses a star-shaped rotor.

Rotary valve: A valve that exposes ported holes to allow the entrance and exit of gases.

Rotor (Brake): A flat metal disc that serves as the friction surface for the front brake assemblies.

Rotor (Ignition): A rotating contact that routes secondary current from the coil to the individual spark plug wires.

Rotor housing: A circular metal housing with evenly spaced slots cut into it to provide a housing for rotor vanes.

Rotor pump: A type of pump that uses a central rotor with spring-loaded vanes.

RPM: Revolutions Per Minute.

Running-fit: A fit in which sufficient clearance has been provided to enable parts to turn freely and to receive lubrication.

Run-on: See *Dieseling.*

Runout: The side-to-side distortion or play of a rotating part.

Rzeppa joint: A type of constant velocity joint that uses a ball-and-cage assembly to provide joint motion.

S

SAE: Society of Automotive Engineers.

SAE thread: Commonly referred to as a standard thread. The standard counts the number of threads per inch.

Safety rim: A wheel designed with two safety ridges that prevent the tire from dropping into the center of the wheel in the event of a blowout.

Safety valve: A valve designed to open when internal pressures within a container exceed a predetermined level.

Safety washer: A flat washer installed on a nondriving wheel bearing to keep the bearing from contacting the spindle nut.

Saybolt test: Test used to determine the viscosity of a fluid.

Scale: Accumulation of corrosion and mineral deposits within a cooling system.

Scanner: An electronic tool, usually hand held, that is used to read and interpret diagnostic codes and engine sensor information. Commonly referred to as a scan tool.

Scatter shield: A metal shield that surrounds a manual transmission bell or clutch housing. It is bolted or welded to the frame and is designed to protect the driver and spectator from flying debris in the event of a clutch explosion.

Scavenging: Refers to the cleaning or blowing out of exhaust gas.

Schrader valve: A valve that is similar to the spring-loaded valve used in tire stems. Normally used in R-12 air conditioning system service fittings.

Score: A scratch or groove on a finished surface.

SCR: Silicon Controlled Rectifier.

Screw extractor: A device used to remove broken bolts, screws, etc.

SCS: Speed Control Switch. (Speed sensitive spark advance control.)

Scuffing: A roughening of the cylinder wall caused when there is no oil film separating the moving parts and metal-to-metal contact is made.

Seal: A formed device made of plastic, rope, neoprene, or Viton. Used to prevent oil leakage around a moving part, such as a shaft.

Sealant: A liquid or paste material applied to a surface along with, or in place of, a gasket to prevent oil leaks.

Sealed beam headlight: A headlight lens that has its lens, bulb, and reflector fused together into a single unit.

Sealed bearing: A bearing that has been sealed at the factory and cannot be serviced during its useful life.

Seat: A surface upon which another part rests or seats. An example would be a valve face resting on its valve seat.

Secondary: The portion of some two- and all four-barrel carburetors that operates under heavy engine load.

Secondary brake shoe: The brake shoe whose friction surface faces the rear of the vehicle.

Secondary circuit: The high voltage portion of the ignition system.

Secondary wires: The high voltage wires from the coil to the distributor and from the distributor to the spark plugs.

Secondary winding: The high voltage winding in a coil. It is made of very fine wire.

Section height: The overall height of a tire from the bottom of the tread to the top of the bead.

Section width: The overall width of the tire measured at the exterior sidewall's widest points.

Sector shaft: The output gear in a steering gearbox.

Sediment: An accumulation of matter or foreign debris, which settles to the bottom of a liquid.

Seize: Condition where two or more parts have forced the lubricant out from between them due to excessive heat and/or friction. When this happens, the parts stick together or freeze.

Self-adjusting brakes: A brake assembly that, through normal application, will maintain the friction members in close adjustment.

Self-diagnostics: The ability of a computer to not only check the operation of all of its sensors and output devices, but to check its own internal circuitry and indicate any problems via diagnostic trouble codes.

Self-energizing brakes: A drum brake assembly that, when applied, develops a wedging action that actually assists or boosts the braking force developed by the wheel cylinders.

Self-induction: The creation of voltage in a circuit by varying current in the circuit.

Self test: A diagnostic test that a computer runs on itself and its associated systems (input sensors and output devices) to ensure that there are no faults.

SEMA: Specialty Equipment Manufacturers Association.

Semi-floating axle: A type of axle used in most modern vehicles. The outer end turns the wheels and supports the weight of the vehicle. The splined inner end "floats" in the differential gear.

Semi-metallic: A friction material that has metal particles added to an organic compound to increase its useful life.

Semiconductor: A substance, such as silicon, that acts as a conductor or insulator, depending on its operating condition and application.

Semielliptical leaf spring: A spring commonly used on truck rear axles. It consists of one main leaf and a number of progressively shorter leaf springs.

Sensible heat: Any additional heat that can be seen in a rise of temperature.

Sensor: A device that monitors a condition and reports on that particular condition to a computer.

Separator: Plastic, rubber, or other insulating material placed between a battery's plates.

Series circuit: A circuit with only one path for current to flow.

Series-parallel circuit: A circuit in which a series and parallel circuits are combined.

Serpentine belt: A single belt used to drive all of the engine-driven accessories.

Service engine soon (SES) light: See *Malfunction Indicator Light.*

Service valves: Points on an air conditioning system where a manifold gauge set can be attached and refrigerant recovered or replaced.

Servo: An automatic transmission hydraulic piston assembly used to apply bands. Also a vacuum or fluid operated device used to push or pull another part.

Servo action: A drum brake assembly constructed so that the primary shoe bears against the secondary shoe. When the brakes are applied, the primary shoe attempts to move in the direction of the rotating drum, which applies force to the secondary shoe.

Setback: A measurable condition in which one wheel spindle is positioned behind the spindle on the opposite side.

Shackle: A link for connecting a leaf spring to the frame.

Shift forks: Devices in a manual transmission that straddle and move the synchronizers and gears back and forth on their shafts.

Shift lever: The driver-operated handle used to shift the transmission through its various gears.

Shift linkage: A cable or mechanical linkage used to shift a transmission into its various gears.

Shift point: The points, at either engine rpm or vehicle speed, when a transmission should be shifted to the next gear.

Shim: A thin piece of brass, steel, or plastic inserted between two parts to adjust the distance between them.

Shimmy: A condition where a wheel shakes from side-to-side.

Shock absorber: An oil- or gas-filled device used to control spring oscillation in suspension systems.

Short circuit: An accidental grounding of an electrical circuit or electrical device.

Shrink fit: A fit so tight that one part must be cooled or heated to fit on another part.

Shroud: A plastic or metal enclosure around a fan to guide and facilitate air flow.

Shunt: An alternate or bypass portion of a circuit.

Shunt winding: A wire coil that forms an alternate path through which electrical current can flow.

Sidedraft: A type of carburetor that takes in air in a horizontal plane.

Sidewall: The portion of a tire between the bead and tread.

Sight glass: A clear glass window in a receiver-dehydrator that can be used to check the refrigerant level.

Silicon (Si): An element that is neither a good conductor or insulator. By doping silicon with different elements, its characteristics can be changed. It is used to make transistors, integrated circuits, and other semiconductor devices.

Single barrel: A carburetor that has only one barrel or throttle opening to the intake.

Single exhaust system: An exhaust system that has only one pipe leading from the exhaust manifold to the catalytic converter, muffler, and out to the tailpipe.

Skid plate: A stout metal plate attached to the underside of a vehicle to protect the engine and transmission oil pans, drive shaft, etc., from damage from "grounding out" on rocks, curbs, and road surface.

Slant engine: An inline engine in which the cylinders have been tilted at an angle from the vertical plane.

Slave cylinder: A hydraulic cylinder used to produce movement of the cutch fork.

Sleeve: A replaceable pipe-like section that is pressed or pushed into a block.

Sliding gear: A transmission gear that is splined to the shaft. This gear can be moved back and forth for shifting purposes.

Slip angle: The difference in the actual path taken by a vehicle and the path it would have taken if it had followed the direction the wheels were pointed.

Slip rings: Metal rings mounted on an alternator drive shaft in which brushes make continuous contact.

Slip yoke: Driveline component that allows back and forth drive shaft movement in response to road conditions.

Sludge: Black, mushy deposits found in the interior of the engine. Caused by oxidized petroleum products mixed with dirt and other contaminants.

Smog: Generalized term used to describe air pollution caused by chemical fumes and smoke.

Snap ring: A split ring snapped in a groove to hold a bearing, thrust washer, gear, etc. in place.

Sodium valve: An engine valve that has metallic sodium added to its stem to speed up the transfer of heat from the valve head to the stem and from the guide and block.

SOHC: Abbreviation for Single Overhead Camshaft engine.

Solenoid: An electrically operated magnetic device used to operate some unit. An iron core is placed inside a coil. When electricity is applied to the coil, the iron core moves in the coil and, as a result, it will exert some force on anything it is connected to.

Solid axle: A single beam which runs between both wheels. Can be used on either the front or rear of the vehicle.

Solid axle suspension: A type of suspension that incorporates a solid axle as an integral part of its construction.

Solid state: Any electrical device that has no moving parts, such as a transistor, diode, or resistor.

Solvent: A liquid used to dissolve or thin another material.

South pole: One of the poles of a magnet from which lines of force originate.

Spacer: A piece of metal or other material placed between two parts to provide clearance, or thrust force, for a fastener.

Space saver spare tire: A spare tire that is smaller than a standard vehicle tire. Used for emergency purposes only.

Spark: A bridging or jumping of an air gap between two electrodes by electrical current.

Spark advance: To cause a spark plug to fire earlier by altering the ignition timing by advancing the distributor or by firing the coil earlier.

Spark gap: Air gap between the center and side spark plug electrodes.

Spark ignition: A system that uses electricity to create a spark to ignite the engine fuel.

Spark knock: See *Preignition* and *Detonation.*

Spark plug: A device which contains two electrodes across which electricity jumps to produce a spark.

Spark plug wires: A series of heavy gage insulated wires used to carry high voltage from the distributor and/or coil(s) to the spark plugs.

Specific gravity: A relative weight of a given volume of a specific material as compared to an equal volume of water.

Speedometer: Instrument used to determine vehicle speed in miles or kilometers per hour.

Speed control: Method of maintaining a set speed as determined by the driver. Usually referred to as cruise control.

Speed density: Method of determining the amount of air going into the intake manifold by monitoring sensor inputs and calculating the amount of airflow based on the sensor readings.

Speed rating: A letter designation indicating the maximum safe speed that a tire is designed to handle.

Spider gears: Small gears mounted on a shaft pinned to the differential case. They mesh with, and drive, the axle end gears.

Spindle: A machined shaft on which bearing races rest. Normally an integral part of a steering knuckle.

Spiral bevel gears: A ring and pinion gear setup in which the teeth are tapered and cut on a spiral so that they are at an angle to the centerline of the pinion shaft.
Splash system: An oiling system which supplies oil to moving parts by attaching dippers to the bottom of the connecting rods. These dippers can either dip into shallow trays or into the sump itself. The spinning dippers splash oil over the inside of the engine.
Splines: Metal grooves cut into two mating parts.
Split hydraulic system: Brake system that is setup so that there are two separate hydraulic circuits. This provides some braking action if one section fails.
Spool valve: A hydraulic control valve shaped like a thread spool. Used primarily in automatic transmission valve bodies.
Sprag clutch: A clutch that will allow rotation in one direction, but not in the other. Commonly referred to as an overrunning clutch.
Spring: A suspension component that supports the vehicle chassis and compensates for uneven surfaces. Can use leaf, coil, or torsion bar construction. Other springs are used to close valves and provide tension in other components.
Spring booster: A device used to increase the load capacity of standard springs.
Spring loaded: A device or other component held in place or under tension by one or more springs.
Spring oscillation: A process when a vehicle's spring is rapidly compressed or twisted and rebounds past its normal length and height.
Spring pressure: A component that is held in place by a spring.
Spring steel: Heat-treated steel that has the ability to withstand a great amount of deflection and return to its original shape.
Sprung weight: The weight of all the parts of the vehicle that is supported by the springs and suspension system.
Spur gear: A gear which has its teeth cut parallel to the shaft.
Spurt hole: A small hole drilled through the connecting rod that lines up with the oil hole in the crankshaft journal. When the holes line up, oil spurts out to lubricate the cylinder walls.
Sq.ft.: Square Foot
Sq.In.: Square Inch.
Squirrel cage: A type of circular fan blade attached to the blower motor in an air distribution system.
Squish area: The area between the piston head at TDC and the cylinder head, which makes up the combustion chamber.
Stabilizer bar: A transverse mounted spring steel bar that controls and minimizes body lean or tipping in corners.
Staked nut: Type of nut whose edges can be bent downward into a slot to secure it on a shaft.
Stall: To stop rotation or operation.
Stall speed: The highest possible speed of torque converter impeller rotation without turbine rotation.
Stall test: A method used by technicians to test for automatic transmission slippage.
Stamped steel: A sheet metal part that is formed by pressing between metal dies.
Starter: An electric motor which uses a geardrive to crank (start) the engine.
Starter pinion gear: A small gear on the end of the starter shaft that engages and turns the flywheel ring gear.
Static balance: Condition where a tire's weight mass is evenly distributed around the axis of rotation. If the wheel is raised from the floor and spun several times, it should always stop in a different place.
Static imbalance: Condition where a tire's weight mass is not evenly distributed. This condition is characterized by up and down vibration as the tire's weight mass tries to bring the tire back into balance.
Static electricity: A charge of electricity generated by friction between two objects.
Stator: A small hub in the torque converter that improves oil flow. Also, the stationary wire field in an alternator.
Steel pack muffler: A straight through muffler which utilizes metal shavings surrounding a perforated pipe.
Steering arm: An arm bolted or forged to the steering knuckle. Used to transmit force from the tie rod to the knuckle.
Steering axis inclination (SAI): The angle formed by the ball joints, steering knuckle, kingpin, etc.
Steering column: Assembly containing the steering shaft, steering wheel, turn signal mechanism, and related wiring.
Steering gear: The assembly containing the gears, valves, and other components used to multiply turning force.
Steering geometry: Term sometimes used to describe the various angles formed by the components making up the vehicle suspension, such as caster, camber, toe, thrust angle, SAI, etc.
Steering knuckle: The inner portion of the spindle, which is affixed to a kingpin, ball joints, or a ball joint and strut.
Steering linkage: The various arms, rods, joints, etc., that connect the steering gear to the wheels.
Steering shaft: A two-piece shaft that transfers turning motion from the steering wheel to the steering gearbox.
Stoichiometric: A perfect or chemically correct air-fuel mixture.
Stoplight: Red warning lights attached to the rear of a vehicle. Used to indicate that the vehicle is slowing down or stopping.
Straight roller bearing: A bearing which uses straight, non-tapered rollers in its construction.
Stroke: The distance the piston moves between TDC and BDC.
Strut: A suspension component containing a shock damper cartridge and coil spring. It is used in many vehicles as a replacement for the shock absorber, front upper control arm, and some rear axle control arms.
Strut damper cartridge: A replaceable shock absorber-like component which is installed inside a strut housing.
Stud: A fastener that has threaded rods at both ends.
STV: Suction Throttling Valve.
Sulfated: Condition where the lead in a battery's plates deteriorate and combine with the sulfur from the battery electrolyte to form a sulfate which coats the plates.
Sump: The part of an oil pan that contains the oil.
Sun gear: The center gear in a planetary gear assembly around which the other gears revolve.
Supercharger: A belt-driven compressor which pumps air into an engine's intake manifold. Superchargers are dependent on engine speed and are most efficient at high engine speeds.
Supplemental restraint systems: See *Air bag.*
Suspension system: The series of components that allow the vehicle to move up and down, turn, and compensate for variations in the road surface.
Sway bar: See *Stabilizer bar.*
Sweating: Process of joining two pieces of metal together by placing solder between them and, while clamping them tightly, applying sufficient heat to melt the solder.
Switch: A device to make or break the flow of current through a circuit.
Synchronizer assembly: An assembly that permits the meshing of gears without grinding.

T

T-head: An engine that has the intake valve on one side of the cylinder and the exhaust valve on the other side.
Tachometer: Device used to measure engine speed in RPM.
Tailpipe: The exhaust pipe running from the muffler to the rear of the car.
Tandem booster: A brake booster with two internal diaphragms that increase vacuum boost pressure.
Tank gauge unit: A variable resistor float device placed inside the fuel tank to monitor fuel level.
Tap: To cut threads in a hole with a threaded, tapered tool. Also to repair badly damaged threads.
Taper: Wear condition in which a cylinder is worn more at the top than at the bottom.
Tapered roller bearing: A bearing that utilizes a series of tapered steel rollers that operate between an outer and inner race.
Tappet: Screw used to adjust valve clearance between the valve stem and the rocker arm.
TCS: Transmission Controlled Spark.
TDC: Top Dead Center.

Technical service bulletins (TSB): Information published by vehicle manufacturers in response to vehicle conditions, problems, etc., that may not be diagnosed by normal methods.

Temperature gauge: A dash mounted gauge that is used to indicate engine temperature.

Tension: Any pulling or stretching stress placed on an object.

Terminals: The connecting points in an electrical circuit.

Test light: A device that will show the presence of current by lighting a small light.

Thermal efficiency: The percentage of heat developed in a burning fuel charge that is used to develop power.

Thermistor: A device that changes its resistance in relation to heat.

Thermostat: A temperature sensitive device used in cooling systems to control coolant flow in relation to temperature.

Thermostatic air cleaner: An emission control device used to control the temperature of the air entering the engine.

Thermostatic compressor switch: A switch that prevents air conditioning compressor engagement if the ambient air temperature is below a certain point.

Thermostatic expansion valve (TVX): See *Expansion valve.*

Throttle body fuel injection (TBI): Fuel injection system that uses one or more fuel injectors mounted above or in the throttle body itself.

Throttle body: Throttle plate assembly that contains sensors and vacuum connectors. Used in place of a carburetor throttle plate on fuel injected vehicles.

Throttle plate: Movable valve inside a throttle body or at the base of a carburetor. Opens and closes to admit air to the engine.

Throttle position sensor (TPS): Input sensor to the engine control ECU. Used to monitor throttle position.

Throttle return dashpot: A carburetor solenoid that slows throttle closing to prevent stalling.

Throttle valve: A valve controlled by linkage from the engine throttle plates or by a vacuum modulator operated by engine manifold vacuum. The throttle valve linkage is often referred to as the throttle valve rod or cable, or just TV linkage. It controls the amount of fluid pressure to the shift valve depending on the distance the throttle is moved.

Throw: Offset portion of the crankshaft designed to accept a connecting rod.

Throw-out bearing: Bearing that is used to minimize pressure between the clutch surface and the throw-out fork.

Thrust angle: Imaginary lines of force that cross lengthwise through a vehicle's tires.

Thrust bearing: A bearing designed to resist side pressure.

Thrust load: A pushing or shoving force exerted against one body by another.

Thrust washer: A bronze or hardened steel washer placed between two moving parts to prevent longitudinal movement.

Tie rod: One or more rods used to connect steering arms together.

Timing belt: A flexible toothed belt used to rotate the camshaft.

Timing chain: Drive chain used to operate the camshaft by turning off the crankshaft.

Timing gear cover: Metal cover placed over the timing gears.

Timing light: A stroboscopic unit. It is connected to the secondary circuit to produce flashes of light in unison with the firing of a specific spark plug.

Timing marks: A series of calibrating marks usually located on the harmonic balancer. Used to set engine timing.

Timing sprocket: Chain or belt sprockets on the crankshaft and camshaft.

Tire: A rubber covered carcass made of steel and fiber cords.

Tire rating: A series of numbers and letters that designate how a particular tire will perform under different conditions.

Toe: The narrow part of a gear tooth. Also the angle at which opposing wheels are converging.

Toe-out on turns: The angle the front wheels assume when turning in relation to each other.

Tolerance: The amount of variation permitted from an exact size or measurement.

Torque: Any turning or twisting force.

Torque converter: A fluid coupling used to transfer engine torque to the transmission input shaft. Can also multiply engine torque.

Torque multiplication: Increasing engine torque through the use of a torque converter or gears.

Torque stick: Calibrated tool used with an impact wrench to remove wheel lug nuts.

Torque wrench: A calibrated wrench designed to indicate the amount of torque applied to a fastener.

Torsen differential: A limited-slip differential that uses worm gears and wheels or spur gears to provide traction under all conditions. The differential operation depends on the fact that the worm gears can turn the wheels easily, but the wheels have a hard time turning the worm gears.

Torsion bar: A long bar made of spring steel attached in such a way that one end is anchored while the other is free to twist.

Tracking: The distance between the front or rear wheels.

Traction: The frictional force generated between the tires and the road.

Traction control system (TCS): A computer-controlled system that reduces idle speed and selectively applies the brakes to reduce excessive wheel spin.

Tramp: An up and down or hopping motion of the front wheels.

Transducer: A vacuum regulator that is controlled electronically.

Transfer case: A transmission driven gearbox that provides driving force to both front and rear propeller shafts on a four-wheel drive vehicle.

Transistor: A semiconductor that is used as a switching device.

Transaxle: A transmission and differential combined into one unit.

Transmission: A device that uses gearing and torque conversion to change the ratio between the engine RPM and driving wheel RPM.

Transmission control solenoid: A computer-controlled solenoid used to control transmission shift patterns.

Transmission oil cooler: A heat exchanger located inside the radiator or mounted in front of the condenser to cool automatic transmission fluid.

Transmission/transaxle technician: A person who is trained in the diagnosis and repair of transmissions and transaxles. A highly specialized area, this person usually performs transmission repairs exclusively.

Tripod CV-joint: A constant velocity joint used on the inner part of a CV axle. It is triangular shaped and has three sets of needle bearings on a spider.

Troubleshooting chart: Diagnostic flow chart that provides step-by-step procedures to test automotive systems.

Trunnions: One or two bearings placed opposite to each other to permit a swivel or tilting action of some part.

Tune-up: The process of checking and adjusting engine timing and replacing spark plugs, filters, etc.

Turbine wheel: A wheel that has fixed vanes so that a moving column of liquid or air will impart a turning motion to the wheel.

Turbocharger: A turbine device that utilizes exhaust pressure to increase the air pressure going into the cylinders.

Turbulence: A violent irregular movement or agitation of a fluid or gas.

Turning radius: Diameter of a circle transcribed by the outer front wheel when making a turn.

Two-stroke cycle engine: An engine that requires one complete revolution of the crankshaft to fire each piston once.

Two-wheel alignment: A suspension alignment in which only the front wheel angles are checked and adjusted.

U

UAW: United Auto Workers.

UIC: Universal Integrated Circuit.

Undercoating: A soft material sprayed on the underside of a vehicle to deaden noise and to resist corrosion.

Understeer: The tendency of a vehicle, when turning a corner, to turn less sharply than the driver intends.

Unibody: A vehicle design in which the frame and body is one unit.

Uniform Tire Quality Grading System: A quality grading system that uses letters and numbers to grade a tire's temperature resistance, traction, and tread wear.

Universal joint: A flexible joint that permits changes in driving angles between a driving and driven shaft.

Unleaded gas: Automotive fuel that contains no tetraethyl lead.

Unsprung weight: The weight of all of the vehicle's parts not supported by the springs, such as wheels and tires.
Upset: The widening of the diameter of a metal part, usually by pounding.
Upshift: A shift into a higher transmission gear.

V

V-Belt: A V-shaped belt that is used to turn engine-driven accessories such as the alternator, water pump, and air conditioning compressor.
V-Engine: An engine in which the cylinders are arranged in two separate banks set at a 60° or 90° angle to each other.
Vacuum: A pressure in an enclosed area that is lower than atmospheric.
Vacuum advance: A mechanism installed inside a distributor that can advance and retard ignition timing in response to engine vacuum.
Vacuum booster: A brake system booster that uses engine vacuum to provide braking assist.
Vacuum gauge: A test gauge used to determine the degree of vacuum existing in a chamber.
Vacuum modulated EGR: An exhaust gas recirculation valve that will open and close in response to engine vacuum.
Vacuum motor: A device which utilizes a vacuum operated diaphragm which causes movement of some other unit.
Vacuum pump: A motor that creates extra vacuum to operate vehicle accessories.
Vacuum reservoir: A tank that is used to store vacuum for situations when the engine does not provide enough vacuum for vacuum-operated vehicle accessories.
Vacuum runout point: The point in a vacuum brake booster when it is providing the maximum boost pressure possible.
Vacuum switch: A switch that opens or closes when vacuum is applied.
Valve: A metal device for opening and closing an aperture or port.
Valve clearance: The space between the end of the valve stem and its actuating mechanism, (rocker arm or camshaft).
Valve core: A threaded valve that screws into a tire's valve stem.
Valve duration: The length of time in degrees of camshaft movement that the valve remains open.
Valve face: The outer edge of the lower part of the valve. Mates with the valve seat.
Valve float: Condition in which valves are forced back open before they have had a chance to seat. Usually occurs at extremely high engine speeds.
Valve guide: A hole machined into the head to support the valves as they ride up and down. May be machined oversized to accept a removable guide.
Valve head: The part of the valve below the stem. The valve face is machined on this part.
Valve keeper (or **key**): Device which snaps into a groove in the valve stem. Used to retain valve and spring assembly in the head.
Valve lash: Valve tappet clearance or total clearance in the valve train with the cam follower on the camshaft's base circle.
Valve lift: The distance that the valve moves from fully closed to fully open.
Valve lifter: A solid or hydraulic plunger that is moved by the camshaft to open the valves.
Valve margin: The width of the edge of the valve between the top of the valve and edge of the face.
Valve overlap: Period in degrees of camshaft rotation in which both the intake and exhaust valves are partially open.
Valve ports: The openings through the head from the intake and exhaust manifolds to the combustion chamber.
Valve seal: A seal that is placed over the valve stem to prevent oil leakage between the stem and the guide.
Valve seat: The area on which the face of the valves rest when closed.
Valve spring: A coil spring used to keep valves closed.
Valve spring seat: Cup shaped washer in which the valve spring sits.
Valve spring shim: A precision machined washer used to adjust valve spring tension.
Valve stem: The portion of the valve that is inside the head. The stem rides in the valve guide.
Valve timing: The relation of the position of the camshaft to crankshaft position so that the valves will open and close at the proper time.
Valve train: The various parts that make up the valve and operating mechanism.
Vane pump: A type of pump that uses spring loaded vanes that throw off, or is moved by, liquid or air.
Vapor lock: Boiling or vaporization of fuel in the fuel lines due to overheating. Can cause hard starting, stalling, or failure to start condition.
Vaporize: A rapid change in state from a liquid to a gas.
Variable displacement compressor: An air conditioning compressor that is designed with internal valves that allow its pumping capacity to be varied. Used to control the system temperature to prevent evaporator icing (condition where the water on the surface of the evaporator freezes, which can block air movement).
Vehicle identification number (VIN): Individual series of letters and numbers assigned to a vehicle by the manufacturer at the factory.
Vehicle speed sensor (VSS): Sensor placed in the transmission/transaxle or the rear axle assembly. Used by the engine's ECU to monitor vehicle speed.
Venturi: A tube that is tapered so that it forms a small area. Used primarily in the air horn of a carburetor.
Venturi vacuum: A vacuum that is created as the air entering the venturi suddenly speeds up, which causes it to be "stretched." The more the air speeds up, the stronger the vacuum.
Vibration damper: A round metal or rubber weight attached to a crankshaft or other part to minimize vibration.
Viscosimeter: A device used to determine the viscosity of a given liquid. The length of time that it takes a heated liquid to flow through a set orifice determines its viscosity.
Viscosity: A measure of a fluid's ability to flow or its thickness.
Viscosity index: A rating given to oils and other fluids to indicate their resistance to changes in viscosity when heated.
Viscous coupling: A fluid-filled clutch used in some full-time four-wheel drive transfer cases. During periods of wheel slippage, it transfers torque to the axle that has the best traction.
Volatility: The tendency of a fluid to evaporate in relation to temperature.
Volt: Unit of measurement of electrical pressure or force that will move a current of one ampere through a resistance of one ohm.
Voltage drop: A lowering of circuit voltage due to excessive lengths of wire, undersize wire, or through a resistance.
Voltage regulator: A mechanical or electrical device used to control alternator output.
Voltmeter: Instrument used to measure voltage in a given circuit.
Volume: Unit of measurement of space in cubic inches or cubic centimeters.
Volumetric efficiency: A comparison between the actual and ideal efficiency of an internal combustion engine. The comparison is based on the actual volume of fuel mixture drawn in and what would be drawn in if the cylinder were completely filled on each stroke.
Vortex: A mass of whirling liquid or gas.
Vulcanization: Process of heating rubber to alter its characteristics.

W

Wandering: Condition in which the front wheels of a vehicle tend to steer in one direction and then the other.
Wankel engine: A rotary internal combustion engine.
Waste gate: A valve that vents excess exhaust gas to limit the amount of boost delivered by a turbocharger.
Waste spark: A spark produced by a distributorless or direct ignition system in a cylinder during its exhaust stroke.
Water detector: A sensor installed in a diesel fuel system which warns the driver of water contamination of the fuel.
Water jacket: The area around the engine cylinders that is left hollow so that coolant may be admitted.
Water pump: The coolant pump; any pump used to circulate coolant through an engine.
Watt: Unit of measurement of electrical power. It is obtained by multiplying volts by amperes.

Wear bar: A solid rubber bar that appears across a tire when it has reached its safety limit.

Wedge: An engine in which the combustion chamber shape forms a wedge.

Wedge chamber: A combustion chamber which utilizes a wedge shape. Its design is efficient and lends itself to mass production.

Wet compression test: Compression test made by placing a small amount of engine oil in a cylinder that has a low reading. Used to determine if a low cylinder reading is caused by worn rings.

Wet friction: The resistance to movement between two lubricated surfaces.

Wet sleeve: A thick metal barrel inserted into an engine cylinder. Is constantly in contact with engine coolant.

Wheel alignment: Refers to checking and adjusting the various steering angles of both the front and rear wheels.

Wheel balancer: A device used to check wheel and tire static and dynamic balance.

Wheelbase: The distance between the center of the front tires and the center of the rear tires.

Wheel bearing: Ball or roller bearing assemblies that reduce friction and support the wheels and axles as they rotate.

Wheel cover: A metal or plastic cover that fits over the center section of a steel wheel.

Wheel cylinder: A hydraulic cylinder used with drum brake systems to actuate the brake shoes.

Wheel hop: A hopping action of the front or rear wheels during heavy acceleration.

Wheel rim locks: Locking lug nuts or bolts used to deter theft of custom wheels.

Wheel speed sensor: Magnetic sensor used in an anti-lock brake system to measure wheel speed.

Wheel shimmy: Lateral (side-to-side) vibration of a tire and wheel assembly.

Wheel tracking: Ability of the rear wheels to follow directly behind the front wheels.

Wheel tramp: Hopping (up and down) vibration of a tire and wheel assembly. Different from wheel hop, as it is not torque related.

Wheel weight: A small lead weight used to balance a tire and wheel assembly. Can be clipped or taped onto the wheel.

Windup: A buildup of internal stresses between the front and rear axle parts, caused by differences in speeds between the front and rear axles.

Wide open throttle switch: A switch that signals the engine ECU to shut off the air conditioning compressor clutch during wide open throttle acceleration.

Wire harness: A group of primary wires encased in a paper or plastic sleeve. Used to ease installation and to prevent wire damage.

Wiring diagram: A detailed drawing showing the location of electronic components and devices that are connected together in a circuit.

Work: A force applied to a body, causing it to move. Measured in foot-pounds, watts, or joules.

Worm gear: A coarse, spiral gear cut into a shaft. Used to engage with or drive another, gear or portion of a gear.

Worm and roller: Type of steering gear that utilizes a worm gear on a rotating shaft. A roller on one end of the cross shaft engages the worm.

Worm and sector: A type of steering gear that utilizes a worm gear engaging a portion of a gear on a cross shaft.

WOT: Wide Open Throttle.

Wrist pin: See *Piston pin.*

Z

Zener diode: A silicon diode that serves as a rectifier. It will allow current to flow in one direction only until the applied voltage reaches a certain level. Once it reaches this point, the diode allows current to flow in the opposite direction.

Acknowledgements

The production of a book of this nature would not be possible without the cooperation of the automotive industry. In preparing the mauscript for ***Auto Fundamentals***, the industry has been most cooperative. The authors acknowledge the cooperation of these companies with great appreciation:

Accurate Products, Inc.; AC Spark Plug Div. of General Motors Corp.; AE Clevite; Aeroquip Corp.; Aimco; Air Lift Co.; Air Reduction; Al-Beck Forbes, Inc.; Albertson and Co.; Alemite Div. of Stewart-Warner; Alfa Romeo Cars, Allen Test Products; All-Lock Co., Inc.; Alondra, Inc.; Aluminum Co. of America; A.L.C. Co.; Amco Mfg. Corp.; American Brake Shoe Co; American Bosch Arma Corp.; American Hammered Automotive Replacement Div.; American Iron and Steel Institute; American manufacturers Assn.; American Optical Co.; American Petroleum Institute; American-Standard; American Standards Assn., Inc.; Amerimac, Inc.; Ammco Tools, Inc.; Anti-Friction Bearing Manufacturers Assn., Inc.; AP Parts Corp.; Appleton Electric Co.; Armstrong Patents Co., Ltd.; Armstrong Tire & Rubber; Armstrong Tool Co.; Arnolt Corp.; Aro Corp.; Audi; Ausco Co.; Automotive Electric Assn.; Automotive Products, Inc.; Automotive Service Industry Assn.; Baldwin, J. A., Mfg. Co.; Band-it Company; Barbee Co., Inc.; Battery Council International; Beach Precision Parts Co.; Bear Mfg. Co.; Beckman Instruments, Inc.; Belden Mfg. Co.; Bendix Automotive Service Div. of Bendix Corp.; Benwil Industries; Bethlehem Steel Co.; BF Goodrich; Big Four Industries, Inc.; Binks Mfg. Co.; Black and Decker Mfg. Co.; Blackhawk Mfg. Co.; B & M Automotive Products; Bonney Forge and Tool Works; Borg & Beck; Bilstein Corp. of America; Borg Warner Corp.; Robert Bosch Corp.; Bowes Mfg., Inc.: Branick Mfg. Co., Inc.; Breeze Corp., Inc.; Bremen Bearing Co.; British Motor Corp. — Hambro, Inc.; Brown and Sharpe, Indus. Prod. Div.; Cadillac Div. of General Motors Corp.; Buick Div. of General Motors Corp.; Bundy Tubing; Burke Co.; Carter Div. of ACF Industries, Inc.; Cedar Rapids Eng. Co.; Central Tool Company; Champion Pneumatic Machinery Co.: Champion Spark Plug Co.; Chevrolet Div. of General Motors Corp.; Chicago Rawhide Mfg. Co.; Chief Industries, Inc.; Chrysler Corp.; Citroen Cars Corp.; Clayton Associates, Inc.; Clayton Manufacturing Co.; Cleveland Graphite; Cole-Hersee Co.; Colt Industries; Continental Air Tools; Continental Motors Corp.; Cooper Tire and Rubber Co.; Corbin Co.; Cornell, William Co.; Corning, Cox Instrument; CPI Engineering Services, Inc.; CRC Chemicals; Cummins Engine Co., Inc.; Dana Corp.; Deere and Co.; Delco-Remy Div. of General Motors Corp.; Detroit Diesel Allison; Deutz Corp.; DeVilbiss Co.; Dodge Div. of Chrysler Corp.; Dole Valve Co.; Dover Corp.; Dow Corning Corp.; Dual Drive, Inc.; Duff-Norton; Dunlop Tire Company; Dura-Bond Engine Parts Co.; Duralcan USA; Durke - Atwood Co.; Easco Tools; Eaton Corp.; Echlin Mfg. Co.; Edelmann, E., and Co.; E. I. duPont de Nemours and Co.; EIS Automotive Corp.; Electrodyne; Environmental Systems Products, Inc.; ESB Brands, Inc.; Ethyl Corp.; Eutectic Welding Alloys Corp.; Everco Industries, Inc.; Exxon Company USA; Fafnir Bearing Co.; FAG Bearing, Ltd.; Federal-Mogul Corp.; Fel-Pro, Inc.; Ferrari Cars; Fiat; Firestone Tire and Rubber Co.; Fiske Brothers Refining Co.; Fletch/Air Inc.; Fluke Corporation; FMC Corp; Ford Motor Co.; Fox Valley Instrument; Fram Corp.; Gates Rubber Co.; Gatke Corp.; General Electric; General Instrument Corp.; General Tire & Rubber Co.; Geo Division of General Motors; G. H. Meiser & Co.; GKN Automotive, Inc.; Glassinger & Company; Globe Hoist Co.; GMC Truck and Coach Div. of General Motors Corp.; Goodall Mfg. Co.; Goodrich Co.; Goodyear Tire and Rubber Co.; Gould Inc.; Graco, Inc.; Gray Co., Inc.; Graymills Corp.; Grey-Rock Div. of Raybestos-Manhattan, Inc.; Guaranteed Parts Co.; Guide Lamp Div. of General Motors; Gulf Oil Corp.; Gunite Foundries Div. of Kelsey-Hayes Co.; Gunk Chemical Div. of Radiator Specialty Co.; Halibrand Eng. Corp.; Hamilton Test Systems; Harrison Radiator Div. of General Motors; Hastings Mfg. Co.; Hayden, Inc.; H.E. Dreyer, Inc.; Hein-Werner Corp.; Heli-Coil Products; Helm, Inc.; Hickok Automotive Group; H. K. Porter, Inc.; Holley Carburetor Div. of Colt Ind.; Homestead Industries, Inc.; Honda; Hub City Iron Co.; Huck Mfg. Co.; Hunter Eng. Co.; Hydramatic Div. of General Motors; Ideal Corp.; Hyundai; Ignition Manufacturers Inst.; Imperial Eastman Corp.; Ingersoll-Rand; Inland Mfg. Co.; International Harvester Co.; International Mfg. Co.; Iskenderian Racing Cams; Isuzu; ITT Automotive; Jaguar Cars, Ltd.; Jeep; Johnson Bronze Co.; Johns-Manville; Kal-Equip. Co.; K-D Mfg. Co.; Kelly-Springfield Tire Co.; Kelsey-Hayes Co.; Kem Manufacturing, Inc.; Kent Moore Org.; Kester Solder Co.; Kia Motors; Kleer-Flo Co.; K. O. Lee Co.; Koni America, Inc.; Kwik-Way Afg. Co.; Land-Rover; Lear Siegler, Inc.; Leece-Neville Co.; Lenroc Co., Libby-Owens-Ford Co.; Lincoln Electric Co.; Lincoln Eng. Co.; Lincoln-Mercury Div. of Ford Motor Co.; Lisle Corporation; Littlefuse, Inc.; Loctite Corporation; Lufkin Rule Co.; Mack Trucks, Inc.; MacMillan Petroleum Corp.; Magnaflux Corp.; Mansfield Tire & Rubber Co.; Maremont Corp.; Marquette Corp.; Martin Senour Paints; Marvell-Schebler Products Div. of Borg-Warner Corp.; Maserati; Master Pneumatic-Detroit Inc.; Mazda; McCartney Manufacturing Co., Inc.; McCord Corp.; McCreary Tire & Rubber Corp.; Meco, Inc.; Mercedes-Benz; Merit Industries, Inc.; Meyer Hydraulics; Micro Test; Midland-Ross Corp.; Mobil Oil Corp.; Monitor Manufacturing; Monroe Auto Equipment Co.; Moog Industries, Inc.; Morton-Norwich Products, Inc.; Motorcraft; Motorola Automotive Products, Inc.; Motor Wheel Corp.; Murray Corp.; Muskegon Piston Ring Co.; Mustang Dynamometers; NAPA-Belden; National Board of Fire Underwriters; National Engines Co.; Nice Ball Bearing Co.; Nicholson File Co.; Nissan, Nugier, F. A., Co.; Oakite Products, Inc.; Oldsmobile Div. of General Motors Corp.; Omega Mfg. Co.; Owatonna Tool Co.; Packard Electric; P and G Mfg. Co.; Parker Fluid Connectors; Paxton Products; Pennsylvania Refining Co.; Perfect Circle Corp.; Perfect Equipment Corporation; Permatex Co., Inc.; Phillips Temco, Inc.; Pontiac Div. of General Motrs Corp.; Porsche; Porter, H. K., Inc.; Prestolite Co.; Pro-Cut International; Proto Tool Co.; P. T. Brake Lining Co., Inc.; Purolator Products, Inc.; Pyroil Co.; Quaker State Corp.; Questor; Raybestos Div. of Raybestos-Manhattan, Inc.; Realmarket Associates; Rexnord; Rinck-Mcllwaine, Inc.; Robertshaw Controls Co.; Robinair Division of SPX Corporation; Rochester Div. of General Motors; Rockford Clutch Div. of Bog-Warner Corp.; Rockwell International; Rodac Corp.; Rootes Motors, Inc.; Rottler Boring Bar Co.; RTI Technologies Inc.; Rubber manufacturers Assn.; Ruger Equipment Co.; Saab-Scania of America, Inc.; Saginaw Steering Gear; Salisbury Corp.; Saturn Div. of General Motors Corp; Schrader Div. of Scovill Mfg. Co., Inc.; Sealed Power Corp.; Semperit of America, Inc.; Shell Oil Co.; Sherwin-Williams Co.; Shim-A-Line Inc.; Simpson Electric Company; Sioux Tools Inc.; SKF Industries, Inc.; Skil; Slep Electronics; Snap-on Tools Corp.; Society of Automotive Eng., Inc.; Solex Ltd.; Sornberger Equip. Sales; South Bend Lathe, Inc.; Spicer; Standard Motor Products; Standard Oil Co. of Calif.; Standard-Thomson Corp.; Stant Mfg. Co., Inc.; Star Machine and Tool Co.; Starrett, L. S., Co.; Stemco Mfg. Co.; Stewart-Warner; Storm-Vulcan, Inc.; Straza Industries; Sturtevant, P. A., Co.; Subaru; Sun Electric Corp.; Sunnen Products Co.; Takato Total Safety Systems; Testing Systems, Inc.; Texaco, Inc. The Aluminum Association; Thexton Mfg. Co., Inc.; Thompson Products Replacement Div. of Thompson-Ramo-Woolridge, Inc.; Thor Power Tool Co.; 3-M Company; Timken Roller Bearing Co.; Torrington; Tomco Coupler; Toyota; Traction Master Co.; Trucut (Frank Wood and Co.); TRW, Inc.; Ultra-Violet Products, Inc.; Union Carbide Corp.; Uniroyal, Inc.; United Parts Div. of Echlin Mfg. Co.; UOP, Inc.; United Tool Processes Corp.; U.S. Chemicals; U.S. Cleaner Corp.; Utica-herbrand Div. of Kelsey-Hayes Co.; Vaco Products Co.; Valvoline Oil Co.; Van Norman Machine Co.; Vellumoid Co.; Vetronix Vehicle Systems; Victor Co.; Volkswagen of America, Inc.; Volvo of America Corp.; Voss Inc.; Wagner Electric Corp.; Walbro, Walker Mfg. Co.; Warner Gear-Warner Motive; Weatherhead Co.; Weaver Mfg. Div. of Dura Corp.; Werther International; Wessels Co.; Westberg Mfg. Co.; Wheelabrator-Frye Inc.; Whitaker Cable Corp.; White Engine Co.; Williams, J.H., and Co.; Wilton Corp.; Wix Corp.; Woodhill Permatex; World Bestos Div. of the Firestone Tire and Rubber Co.; Wright-Austin Co.; Wudel Mfg. Co.; Young Radiator Co.

Portions of the materials contained in this text have been reprinted with the permission of General Motors Corporation, Service Technology Group.

DOHC 16V
STUDIO COLLINS

Index

D

E

F

G

T

U

V

W